D1002908

A Treasury of Traditional Wisdom

A Treasury of Traditional Wisdom

Presented by

Whitall N. Perry

QUINTA ESSENTIA

First published 1971

This paperback edition first
published 1991 by Quinta Essentia,
5 Green Street, Cambridge CB2 3JU, UK.

A catalogue record of this book is available
from the British Library

ISBN 1870196 08 2 *paper*

Printed in the United Kingdom

FOREWORD

by Marco Pallis

One of the paradoxical aspects of the crisis of doubt, defeatism and disillusion through which the Western world is passing at the present time is the existence, among the educated public in all countries, of a never-ending demand for information concerning religious teachings and practices in all times and climes. Hardly a day passes without some freshly translated text, complete with editorial comments of greater or lesser competence, being added to the list of works in print. Metaphysical treatises of the profoundest kind, records of mystical experience in every conceivable form, instruction on yogic methods formerly only divulged to initiates, illustrations of the treasures of sacred art ranging from sumptuously produced volumes down to picture post-cards of no small quality—all this is continually going to swell the spate of religious documentation. To which must be added all that the anthropologists have been able to find out about tribal civilizations in Africa, Polynesia and the Americas, their spiritual lore and accompanying rites and customs. What for the Nineteenth Century amounted to sheer superstition, childish when not abominable and in any case beneath the notice of sophisticated men, is now, on closer examination, found to provide food for envy more readily than for contempt.

On the part of Christians, whose own faith has been partially or largely eroded, this wish to know how the various non-Christian traditions have severally regarded the function of man upon this earth and its eventual outcome is not an unhealthy sign under present circumstances, even though it may often result in further diluting an existing intellectual lukewarmness or else lead to an addiction for one of the many forms of religious charlatanism that characterizes times of general decadence. It is against such a background that a work like this great Anthology of the Spiritual Life can be seen to have an immediate usefulness inasmuch as it can help to canalize the attention of the serious seeker in directions likely, if intelligently and persistently followed, to be fruitful of positive results.

In this same connection it is noteworthy that this century of ours, for all its vast and various discontents and because of them, has been productive of some of the most clear-sighted and heart-searching commentaries that the West has ever known, among which those of René Guénon and, more especially, Frithjof Schuon provide pre-eminent examples: it is indeed a long time since anything of this intellectual order has been heard in Europe and the fact that such a thing could happen now of all times is yet another paradox worthy of careful attention. Small wonder that the author of our Anthology, in his preambles to the various sections into which this immense work is subdivided, has quoted repeatedly and with telling effect from the above-mentioned writers.

One other contemporary writer has been freely quoted by the present anthologist in the course of various explanatory notes: this is Ananda K. Coomaraswamy, who first became known through his masterly studies of Art and later through studies of the Indian doctrines of rare lucidity. Coomaraswamy made no secret of his respect for the two writers mentioned above: speaking as an Oriental (he was really of mixed parentage) he declared that, with one addition, he knew no other writers who were fully qualified in our time to build an intellectual two-ways bridge between the Indian and European wisdoms. Like them, Coomaraswamy stands for the 'Philosophia Perennis', not by way of any personal eclecticism but as one for whom traditionally diverse forms, still as valid as ever, have become transparent to the point of ceasing to oppose any impenetrable barrier to metaphysical insight. Having corresponded, myself, with A. K. Coomaraswamy during all the last part of his life, I can say with certainty that had he lived to see the present Anthology published he would have welcomed it and done his utmost to get it known and used both in America, where he made his home, and in India where he felt he belonged by ancestry and temperament.

The selecting and assembling of the diverse material from which the present collection has been formed was the work of some fifteen years. The idea of offering the world a collection of texts, not merely relating to a given religion, but representing the concurrent testimony of all the principal spiritual currents of mankind was not altogether new when our author started on his enormous task. Already in the year 1939 there had appeared Robert Ballou's *Bible of the World,* an excellent compilation in its way confined, however, to strictly 'scriptural' texts. Lin Yutang's *The Wisdom of China and India* covered much ground within Asian limits. To these must be added Aldous Huxley's *Perennial Philosophy,* a work which, despite its wealth of splendid extracts from sacred literature disposed under given headings as in the present book, remains too eclectic and personal to match the claim implied in its title; its author's preferences, not to say prejudices, become in effect the criteria of validity, both positively and by what they exclude. It was partly from a wish to show what a work bearing such a title might, and indeed must be that Whitall Perry was eventually led into starting the book we now have before us. But his strongest inspiration came to him through a saying of Coomaraswamy contained in the final note of that author's *Hinduism and Buddhism* (New York, 1943): 'The time is coming when a Summa of the Philosophia Perennis will have to be written, impartially based on all orthodox sources whatever.' When he made this remark Coomaraswamy little knew that the hope he was expressing would begin to be fulfilled within less than ten years of his uttering it. To Whitall Perry these words came as a vocational summons, one which once received allowed of no refusal.

As the author himself has explained in his Introduction, a scheme imposed itself from the outset patterned upon basic modalities of knowledge and being and arranged in the sequence of a spiritual quest. Carrying out such a scheme called for a very special kind of skill which, in my opinion, is exemplified by the present work in a most satisfying degree. This skill might be described as 'the art of spiritual juxtaposition', which means far more than just choosing out and classifying texts relating to a particular religious theme or themes; it means that these texts are so placed in relation to one another as to throw into relief each other's truth—and their common truth—as an evocation from their very contrast.

The author's disposition of his material can fittingly be described as an arabesque of mutually revealing quotations. It is a common property of designs bearing this name that the geometrical scheme, whereby their various elements consisting either of other geometrical or else plant forms are held together and unified, is itself so disposed as to allow foreground and background features to exchange functions freely when the design is viewed in a different light or from a different angle. As anyone will agree who has visited the mosques of Cairo or any other place where this art is displayed in all its perfection, the completeness of the transformation taking place with a shift of viewpoint can be breathtaking for the onlooker. Yet this apparent displacement of the axes of the design entails not the slightest feeling of alienation; the arabesque's centre of gravity seems to be nowhere fixed, yet everywhere recognizable. No wonder that a particular generic name was coined to describe this mode of clothing surfaces in forms and colours, in recognition of the fact that this has been one of the most original, as well as characteristic expressions of the Arab genius—as also of the Islamic genius with which it is inseparably linked. Every true arabesque is like an ideogramme, at once simple and complex, of the Unitary Faith.

So it is with the arrangement of texts here before us. Imagine a reader intent on developing his insight into a certain spiritual theme, let it be Creation or the Destiny of Man or what else you will: on turning up the appropriate section he will find before him a certain number of quotations of mixed origin which, at first sight, might be supposed to have been picked out and placed in haphazard fashion. Let us say that his eye first lights upon a Taoist sentence of baffling conciseness, placed between two other passages taken respectively from the sayings of a Red Indian and a Tibetan Buddhist sage; and again, as his eye is allowed to range further he may come upon a Christian comment on the same theme and a Hebraic one no less arresting: by dint of giving attention to each of the different traditional expressions here present it will begin to dawn upon that reader's mind that each of the texts he has been concerned with has in fact functioned as focal point in relation to its neighbours and vice versa.

In the imaginary case we have pictured above, the persuasive force of the Taoist message will have been thrown into sharpest relief by the more 'personalist' presentation of the same truth in Christian or Jewish mode and also, but in a different way, by the 'naturist' vision of the Red Indian master; while in the impersonal language of the Buddhist example a closer kinship of form, as well as content, will be apparent. A similar experience, *mutatis mutandis,* will result when one of the other texts is treated as central: contrast or even conflict of form and concordance of content will, between them, throw up an identity of truth. The great art of the compiler has been to favour such an experience at every turn; this it is which I previously described as the art of 'spiritual juxtaposition' and this is also why I have compared the result to an arabesque: in a scheme of this kind there are no parts, strictly speaking; every element functions at once as focal point or as concurrent and confirmatory evidence according to circumstances. The net impression left on the mind is the unanimity of witness.

However, there is yet another way in which an over-all sense of unity is maintained and transmitted throughout this book: not only do the texts associated under a given heading create a synthesis of mutual corroboration across, and thanks to, the very extent of their formal and stylistic differences, but also the main headings themselves are similarly interlocked thanks to an ingenious method of cross-references that will allow whatever aspect of spirituality happens to be studied at a given time to be continually kept in touch with other aspects, in such a way as to lead inexorably to awareness that there is but a single theme being propounded by each and all the 1,044 pages of this book and that so it would be were these pages to be multiplied tenfold: 'there is no theme but THE THEME' would be a fitting motto for this whole work; the primary formula of Islam provides a perfect model here.

Moreover, this referring back and forth between the various traditional wisdoms is itself an exercise in cultivating 'metaphysical plasticity' of mind, in which readiness to recognize the presence of unity, wheresoever it chooses to reveal itself, is blended with an accurate discernment in regard to forms and more especially to those qualitative factors on which the intrinsic distinctiveness of each form rests. What one always needs to remember is that traditional forms, including those bearing the now unpopular name of dogmas, are keys to unlock the gate of Unitive Truth; but they are also (since a key can close, as well as open, a gate) possible obstacles to its profoundest knowledge: all depends on how the forms are envisaged and used at different stages of the Way. It is by his active exercise of the faculty of metaphysical insight conferred on him that man is enabled to situate, and eventually to transcend, distinctions of religious form, but without blurring or betrayal of that which provides its efficient cause for each form as such. Living his own form faithfully and intensely a man will live all 'other' forms by implication. This is yet another lesson to be learned with the help of the present book, by those who find the way to reading it 'in depth', through the 'heart's eye'—the faculty of intuitive intelligence previously mentioned—and not merely 'on the surface', through the brain's eye which reasons and analyses, but never actually sees.

It is hardly necessary to point out, in one's appreciation of the vast amount of ground covered by the author, that even a document of spiritual science as wide-reaching as this one could never hope to be exhaustive, nor does he himself put forward any claims of this kind; even for the most informed and discriminating of editors there will always be a problem of selection that will inevitably leave a number of gaps unfilled in what is actually offered to his prospective readers. The most anyone could hope to accomplish when undertaking such a work is to arrive at a reasonably representative choice out of the immense total of man's spiritual heritage as preserved in writings.

One further question presents itself before this foreword is brought to a close: to what kind of readers can this Anthology chiefly be of service?

To anyone who has had time to examine the text in detail and to notice the author's own passing comments it will be plainly evident which kind of reader he primarily had in mind, namely the man of serious spiritual intent, often distressed because of the religious disintegration of the times, but also, for this reason, doubly determined to seek out the truth at all costs—the man who is intent on truth because he is convinced that truth alone is what counts in its own right or is able to save, in the long run, either oneself or others.

Next to the man or woman who is 'spiritually committed' in the very complete sense described above must be placed, as far as the usefulness of this book is concerned, the numerous and various persons whose aspirations, genuine as far as they go, do not as yet imply more than an average degree of perceptiveness, though often (did they but know it) they would in fact be capable of much more than that. For them, a work like the present one provides an undreamed-of opportunity to find out what the scope of a religious life can be when lived intensely so that, emboldened by this discovery, they may be led on further to discover in themselves possibilities of which they had been hardly conscious hitherto.

Every kind and degree of religious aspiration should surely find here something answering to its present needs.

We must now pass on to consider the interests of a rapidly growing class of readers who will find in the present work a fertile source of reference and inspiration: one is thinking of those who, from any motive and angle, concern themselves with 'Comparative Religion'. If the historically or academically minded student will be able to find here much that is conducive to his own branch of research, this will apply far more to those for whom the subject of spiritual 'oecumenism' has become important—those, that is to say, who envisage the possibility of some kind of fruitful dialogue across what formerly amounted to quasi-impassable religious frontiers. This is a characteristic phenomenon of our times, due firstly to the fact that old positions of conventional prejudice can no longer be maintained in the face of the vast amount of information about other people's religions now become generally available, and secondly due to the fact that all religions are equally under attack by a militant atheism such as, in former times, was a thing hardly conceivable. This common danger has made many former antagonists feel themselves allies, with the result that they all wish, and indeed need, to have an effective forum of inter-communication, such as can only be created on a basis of honest (which includes factually accurate) consideration of each others' point of view. To all advocates of religious understanding at an oecumenical level this Anthology comes as a ready-to-hand manual wherewith to equip oneself for meeting with others like-minded drawn from all parts of the world; it is the nearest thing to a whole religious library compressed between the covers of a single book that has appeared hitherto—unless one regards as such the work of the Greek Stobaeus in the fifth century A.D. who started out with similarly encyclopaedic aims, though the range of his subject-matter was necessarily limited to the sum of knowledge available in the Mediterranean regions at that date.

Just as Greek had become a practically universal language for the post-classical Western world, so also English has become a universal medium of communication for our present world. It therefore seems most appropriate to these times that an author having English for his mother tongue should have elected to carry to fulfilment a project first conceived in the mind of the prescient compiler of some sixteen centuries ago.

CONTENTS

CONTENTS

INTRODUCTION

For by one Spirit are we all baptized into one body, whether we be Jews or Gentiles, whether we be bond or free: and have been all made to drink into one Spirit.

I. Corinthians, XII. 13

And We shall take from every nation a witness.

Qur'ân, XXVIII. 75

The reader is invited to enter upon a spiritual journey. In this book he will encounter the heritage he shares in common with all humanity, in what is essentially timeless and enduring and pertinent to his final ends. Out of this myriad mosaic of material emerges a pattern of the human personality in the cosmos that is unerringly consistent, clear, and struck through with a resonance infallible in its ever renewed reverberations of the one same Reality.

These texts with the aid of the introductory passages likewise serve as an illustrative commentary to the exposition of contemporary authorities on doctrines relating to the Primordial and Universal Tradition (Philosophia Perennis, Lex Aeterna, Hagia Sophia, Dîn al-Ḥaqq, Dhamma, Tao, Sanâtana Dharma). They are not intended as a *proof* of One Spirit behind many forms; and in fact the Truth cannot be proven, for as Ananda K. Coomaraswamy says, 'a first cause, being itself uncaused, is not *prob*-able but axiomatic' (*Time and Eternity*, p. 42), i.e. self-evident, to the Self that alone can comprehend the Self.[1] Frithjof Schuon expresses the same idea from a different viewpoint: 'One can assuredly prove every truth;[2] but every proof does not enter into all minds. Nothing is more arbitrary than to reject the classic proofs of God, for each one is valid in relation to a certain need of causality. This need of causality increases, not in proportion to knowledge, but in proportion to ignorance. For the sage, each star, each flower, metaphysically proves the Infinite' (*Perspectives spirituelles et Faits humains*, p. 10).[3] But they are intended as an intellectual aid to faith, that ghostly compass pointing the way to the Divine Arcana or Mysterium Magnum, where, once the 'Flight of the Alone to the Alone'[4] has been accomplished, the Self will know its own Identity as clearly and infallibly as a river 'knows' its element upon merging with the sea.

Reality 775
773
Knowl. 749
Realiz. 890
Sp. Drown. 71

[1] It is a question here of intellectual intuition and not discursive reasoning: cf. René Guénon: *Le Symbolisme de la Croix*, p. 138, note 3.

[2] Cf. Hermes: 'Concerning God, no question is insoluble' (*Lib.* XI. ii. 16 b).

[3] 'As long as a man argues about God, he has not realized Him' (Sri Ramakrishna, *Gospel*, p. 735).

[4] Plotinus: *Enneads*, VI. ix. 11.

Neither are these citations offered as an apology for spiritual laxity, a universal tolerance (which 'implies indifference, and becomes intolerable'—Coomaraswamy: *Am I My Brother's Keeper?*, p. 38), or fraternity of mankind united under One Religion. A supraformal synthesis cannot be recast in terms of a formal syncretism; and on the contingent plane of existence each traditional form has its own inexorable laws and logic and integral cohesion adapted to the time, place and people involved, to the exclusion of other forms, being in its own form and at its level an all-inclusive whole.[1] 'Gnosis[2] must therefore discard all these doctrines (with metaphysical pretensions) and base itseif alone on the Orthodox Tradition contained in the Sacred Books of all peoples, a Tradition which in reality is everywhere the same, in spite of the diverse forms it assumes in order to adapt itself to each race and each period. But here again one must have great care to distinguish this true Tradition from all the erroneous interpretations and fantastic commentaries made on it in our time by a flood of schools more or less occultist in character, which unfortunately have wished too often to speak on matters where they had no knowledge' (René Guénon: *La Gnose,* Dec., 1909).[3]

Wr. Side 459 *Reality* 790

Wr. Side 442

Hence no text has been included which gives proof of traditional irregularity. On the other hand, the Spirit, transcending form, has the power to break through it, for 'the wind bloweth where it listeth'; and occasional citations—above all from the Christian sphere—are entered when strikingly appropriate to the context, even though the traditional affiliations regarding ritual transmission (never doctrinal orthodoxy) might be hard to trace.[4] In general, the

[1] Because identical, through analogical transposition, with the Supreme Center.

[2] This word is normally used in this book in its etymological sense, without reference to any particular sect.

[3] Those who follow the accepted revelations of Tradition can have little sympathy for the thinly-disguised intellectual pretension of certain groups that would put themselves, like God, at the Center of all traditions, without even knowing what it is to participate fully—as regards doctrine, method and ritual—in the form of any one tradition. The function of *Chakravartî* belongs to the Pantocrator alone—a position contested only by Satan or the Antichrist.

[4] The neo-paganism of the Renaissance was not confined to Protestant countries. Nor were men like Boehme, John Smith, Sterry, Law or Angelus Silesius the champions of Protestant aberrations any more than Boethius, Epictetus or Plotinus were the defenders of pagan heresies. On the contrary, the former like their illustrious predecessors were among the intellectual and spiritually-illumined witnesses to metaphysical truths of a universal order still operative in the West, a gnosis not limited to time or place or form, but part of a profound esoteric current that was bound to exteriorize in extranormal modalities when the normal channels would no longer carry it.

The decline of intellectuality from natural causes alone could hardly explain the rejection of Clement of Alexandria from the Martyrology in the 17th Century, as just one example of the strange hostility to gnosis which characterizes large segments of the Western world particularly since the fall of the Middle Ages. It would be interesting to know the response of Pope Gregory XIV to the proposal of the 16th Century Ferraran scholar, Patrizzi, for a restoration of true philosophical learning to reunite the scattered Church, based primarily on Egyptian and Greek esoterism, including the *Theologia* of Aristotle, evidently well known to the Arabs and also to the Scholastics (a Latin version appeared at Rome in 1519 under the patronage of Pope Leo X), and presumably a direct link between essential Platonic doctrine and the Neoplatonic school (cf. Scott: *Hermetica,* I. 36 ff.). One is reminded of the 'Hesychast controversy', which according to the Catholic Encyclopedia Dictionary (N.Y. 1941), 'may be considered an issue between Greek Platonist philosophy and Latin rationalist Aristoteleanism' (Patrizzi held that Aristotle's *Theologia* represented a return to the true Platonic faith after years of relative dissension during which his other works were produced). What is essential to retain in all this is that 'a church or society—the Hindu would make no distinction—that does not provide a way of escape from its own regimen, and will not let its people go, is defeating its own ultimate purpose' (Coomaraswamy: *Hinduism and Buddhism,* p. 29).

essential characteristic of tradition is an unchanging reality, universal constant or norm (the Qur'ânic *dînu 'l-qayyim*), polarizing in the World Axis where the contingent and principial domains conjoin, and whose traces will always be discernible in doctrine, symbol and rite to certain objective criteria, however diverse or even contradictory in appearance may seem the outward forms.[1]

Introds. *P. State* *Peace*

The three main divisions and six parts of the book are arranged in accordance with fundamental spiritual perspectives, and the chapters in each part carry further developments of different possible approaches to the Truth. It is not a question of systematizing that which by definition transcends and eludes formal limitations, but rather of presenting the subject in a way that will facilitate the reader's orientation.[2] Various themes will necessarily overlap, since in the end they all originate from and lead back to the same Source; nonetheless there is nothing arbitrary in the divisions, for they follow basic modalities of knowledge and being.[3] Also, the point of departure is 'subjective' or microcosmic, taking man in relation to spiritual verities, rather than 'objective' or macrocosmic with the universe as the starting point, so that cosmological doctrine will only enter in occasionally, where relevant to the context. The logical sequence[4] is thus in terms of a progression or journey towards recovery of the Primordial Self, 'lost' or obscured through the process of individuation implied in existence, but which in essence 'we' have never ceased to be. *Duo sunt in homine:* 'This distinction of an immortal spirit from the mortal soul . . . is in fact the fundamental doctrine of the Philosophia Perennis wherever we find it' (Coomaraswamy: *Hinduism and Buddhism,* p. 57).

Reality 803

Creation 42 *Metanoia* 480

An admirable argument for the material in the following chapters is to be found in the opening passages of the preface to another work—a seventeenth century English translation[5] of Boehme's *Signatura Rerum*: 'This book is a true mystical mirror of the highest wisdom. The best treasure that a man can attain unto in this world is true knowledge; even *the knowledge of himself*: For *man* is the great mystery of God, the microcosm, or the complete abridgment of the whole universe: He is the *mirandum Dei opus,* God's masterpiece, a living emblem and hieroglyphic of eternity and time; and therefore to know whence he is, and what his temporal and eternal being and well-being are, must needs be that ONE necessary thing, to which all our chief study should aim, and in comparison of which all the wealth of this world is but dross, and a loss to us.

Knowl. 761 *Realiz.* 859 *Holiness* 924

Reality 775 *Center* 847

'Hence Solomon, the wisest of the kings of Israel, says: "Happy is the man that findeth

[1] By extension a traditional society is one primarily oriented to ultimate, spiritual ends, in contradistinction to a secular society frankly established in view of material benefits. The first possibility may include the substantial amplitude of the second, but the inverse could never be realized, as the lesser cannot contain the greater.

[2] In fact, the impossibility of systematizing is abundantly evidenced in the complementary logic of the numerous cross-references between chapters, which are at best approximations, aiding as spiritual 'flashes' or insights.

It must also be emphasized that the criterion determining the arrangement of texts is that of relevancy and logical sequence, without reference to hierarchy of authority or degree of inspiration.

[3] This ternary conception paired into six attributes is explicitly formulated by Schuon in the last chapter of *L'Oeil du Coeur* and the last chapter of *Les Stations de la Sagesse.* It may be pointed out that the Hindu Divinity, *Bhagavân,* is envisaged as comprising six attributes, namely, Infinite Detachment, Strength, Treasure, Splendour, Knowledge, and Glory.

[4] Similar, *mutatis mutandis,* to the scheme employed in the *Divine Comedy.*

[5] Presumably by John Sparrow (cf. Hobhouse: *Law,* pp. 267, 414). The *Everyman* text from which these passages are taken gives no reference, except that it follows the text of Law's edition.

wisdom, and the man that getteth understanding; for the merchandise of it is better than the merchandise of silver, and the gain thereof than fine gold; she is more precious than rubies, and all things that can be desired are not to be compared unto her".

Void 724 Inv. 1031

'This is that wisdom which dwells in nothing, and yet possesses all things, and the humble resigned soul is its playfellow; this is the divine alloquy, the inspiration of the Almighty, the breath of God, the holy unction, which sanctifies the soul to be the temple of the Holy Ghost, which instructs it aright in all things, and searches τὰ βάθη τοῦ θεοῦ,[1] the depths of God.

Inv. 1017

'This is the precious pearl, whose beauty is more glorious, and whose virtue more sovereign than the sun: It is a never-failing comfort in all afflictions, a balsam for all sores, a panacea for all diseases, a sure antidote against all poison, and death itself; it is that joyful and assured companion and guide, which never forsakes a man, but convoys him through this valley of misery and death into the blessed paradise of perfect bliss.

Death 208 Knowl. 743 Ecstasy 641

'If you ask, What is the way to attain to this wisdom? Behold! . . . "If any man will come after me, let him deny himself". . . . "Unless you be born again, you cannot see the kingdom of heaven". . . . "If any man seemeth to be wise in this world, let him become a fool that he may be wise".

Wr. Side 448 P. State 563 Suff. 126

'Herein lies that simple childlike way to the highest wisdom, which no sharp reason or worldly learning can reach unto; nay, it is foolishness to reason, and therefore so few go the way to find it: The proud sophisters and wiselings of this world have always trampled it under foot with scorn and contempt, and have called it enthusiasm, madness, melancholy, whimsy, fancy, etc., but wisdom is justified of her children.

Introd. Orthod.

'Indeed, every one is not fit for or capable of the knowledge of the eternal and temporal nature in its mysterious operation, neither is the proud covetous world worthy to receive a clear manifestation of it; and therefore the only wise God (who giveth wisdom to every one that asketh it aright of him) has locked up the jewel in his blessed treasury, which none can open but those that have the key; which is this, viz., "Ask, and it shall be given you; seek, and ye shall find; knock, and it shall be opened unto you:" The Father will give the Spirit to them that ask him for it.'[2]

[1] *I. Cor.* II. 10.

[2] One might object that this book renders esoteric material too readily accessible in a single volume, and that these texts in their peculiar juxtaposition tend to unveil and place certain truths at the disposal of everyone apart from all question of intellectual or spiritual qualification. Hence, if it can serve the spiritual interests of some, the inverse is equally implicit in the nature of the thing. The apology or excuse for presenting this material is the same as that put forward by others (e.g. Schuon: *L'Oeil,* p. 191; Guénon: *Le Roi du Monde,* ch. XII; Suzuki: *Mysticism: Christian and Buddhist,* pp. 148–149; Joseph Epes Brown: *The Sacred Pipe,* p. xii) writing today from a traditional viewpoint: that the intellectual chaos of the modern world permits of drastic responses which formerly might have induced something of the same anarchy they are now employed to combat, if not overcome. Specious parodies of these truths are appearing everywhere, and can only be answered by proper formulations of the true doctrines. This material will serve as a mirror to the inner complexion of each reader, who alone must be responsible for the use to which he turns it—at least insofar as he departs from the original intention of the authors cited, which intention, however inadequately realized, it is the sole aim here to disclose. As Shabistari says of his Secret Rose Garden:

'You will see tradition, earthly and mystical truths,
All arranged clearly in knowledge of detail.
Do not seek with cold eyes to find blemishes,
Or the roses will turn to thorns as you gaze.'

PRELUDE

CREATION

Tao begets One; one begets two; two begets three; three begets all things. All things are backed by *yin* and faced by *yang*, and harmonized by the immaterial Breath (*ch'i*).

Tao Te Ching, XLII

God became man, that man might become God.

St Athanasius, St Augustine, St Cyril of Alexandria, Meister Eckhart, Jacob Boehme, and others.[1]

From the metaphysical perspective, 'creation (or manifestation) is rigorously implied in the infinity of Principle. . . . The world cannot not exist, since it is a possible and therefore necessary aspect of the absolute necessity of Being' (Schuon: *De l'Unité transcendante des Religions*, p. 66).

Introd.
Sin
M.M.
978

From the spiritual perspective, 'what we call the world-process and a creation is . . . a game (*krîḍâ, lîlâ, παιδιά, dolce gioco*) that the Spirit plays with itself, and as sunlight "plays" upon whatever it illuminates and quickens, although unaffected by its apparent contacts.' The sensible world is 'the consequence of the Spirit's awareness of "the diversified world-picture painted by itself on the vast canvas of itself" (Śankarâcârya).[2] It is not by means of this All that he knows himself, but by his knowledge of himself that he becomes this All' (Coomaraswamy: *Hinduism and Buddhism*, pp. 14–15, commenting *Bṛihad-Âraṇyaka Upanishad*, I. iv. 10).

From the cosmological perspective, creation is a progressive exteriorization of that which is principially interior, an alternation between the essential pole (*purusha, yang*) and the substantial pole (*prakriti, yin*) of a single Supreme Principle (Self, Âtmâ), which Itself as 'Motionless Mover' (Aristotle) is not involved in Its productions: 'It is Universal Being which, relatively to the manifestation of which It is the Principle, polarizes Itself into "Essence" and "Substance", without Its intrinsic unity being in any way affected thereby' (Guénon: *L'Homme et son Devenir selon le Vêdânta*, 3rd ed., p. 48). 'The Dragon-Father remains a Pleroma, no more diminished by what he exhales than he is increased by what is repossest' (Coomaraswamy: *Hinduism and Buddhism*, p. 7). Manifestation is by way of progressive individuation, limitation, descent, or 'fall': 'To go from essence towards substance is to go from the center towards the circumference, from the interior towards the

Reality
773

[1] Cf. for example, Eckhart in Evans, I. 387, and Boehme, *Sig. Rerum*, X. 53.

[2] 'I must myself be Sun, and paint with my Rays the colourless Sea of all the Godhead' (Angelus Silesius: *Cherub*. I. 115). 'Thou art the brush that painted the hues of all the world' (Farîdu 'd-dîn Shakrgunj; Hughes: *Dictionary of Islam*, London, ed. 1935, p. 618).

exterior ... from unity towards multiplicity' (Guénon: *Le Règne de la Quantité et les Signes des Temps*, p. 160). 'He has exteriorized everything inasmuch as He is the Interior, and He has withdrawn the existence of everything inasmuch as He is the Exterior' (Ibn Aṭâ'illâh: *Ḥikam*, no. 152).

It can thus be seen that 'evolution' according to traditional science implies a deterioration: the world commences with a Golden Age, rather than terminating in a Millenium (which belongs to another world, another age, another cycle); and in fact the modern Theory of Evolution denies God the power to create something perfect.[1] Compare for example *Genesis*, I. 27 and 31: 'God created man in his own image', 'And God saw every thing that he had made, and, behold, it was very good', and supporting references cited below, with modern teachings on man's ancestry. 'To believe, as do certain "neo-yogis", that "evolution" will produce a superman "who will differ from man as much as man differs from animals, or animals from vegetables", is simply not to know what man is' (Schuon: *Perspectives spirituelles*, p. 150).[2] 'The virtue of each thing, whether body or soul, instrument or creature, does not reach a high pitch of perfection by chance, but as the result of the order and truth and art which are imparted to them' (Plato: *Gorgias*, 506 D). This bestowal is the legacy of *immutable* archetypes.

Wr. Side 464

Beauty 674

[1] Cf. Philo: *On the Creation of the World*, Why Heaven Was Created after Earth.

'As to the primeval condition of our race,' writes William F. Warren, first president of Boston University, in that profound and remarkable work of erudition, *Paradise Found, the Cradle of the Human Race at the North Pole, a Study of the Prehistoric World* (Boston, 1885), 'a truly scientific mind will wish to base its conception not on the air-hung speculations of mere theorists, but on an immovable foundation of fact, attested and confirmed by the widest, oldest, and most incontestable of all concurrences of divine and human testimony. According hereto, as in its beginning light was light, and water water, and the Spirit spirit, so in his beginning *Man was Man*. It says that the first men could not have been *men* without a human consciousness, and that they could not have had a human consciousness without rationality and freedom' (p. 418). Contrast a statement like this with that of an evolutionary scientist: 'However inconceivable it may be to us moderns, there truly was a time when people were *unable to make any conceivable distinction between a plant and a man'*—Dr Wilhelm Mannhardt, Berlin, 1876 (p. 409). For Warren this irresistibly raises the 'perplexing and ... fundamental problem as to the possibility and credibility of primitive procreation itself' (p. 410).

In another place and vein Warren deplores the 'unfortunate and ominous fact that the average biologist of the present day sees nothing worthy of his professional attention back of the present (19th) century. The intellectual history of the human race has not the slightest interest for him or value for his work. Ages on ages of human observation and thought and speculation touching the problems of life are to him as if they had never been. If he acquaints himself with them in the slightest degree, it is usually only for the sake of amusing his hearers with what he considers the grotesque and absurd ideas of former times, and impressing them with the contrast which latter-day "science" presents. For all that his race has done until just before his own immediate teachers began, he has little more than pity and contempt.

'Now, in any department of human learning, such an attitude of mind is certainly to be deplored. Its effects are detrimental in every aspect. In proportion as it prevails among any class of intellectual workers, in just that proportion does that class become isolated from the one collective and historic intellectual life of humanity' (p. 315).

Returning to the question of evolution, Warren reminds us that 'what the zoologist calls the "lowest" forms of organization are rather the *highest*, if by "highest" we mean those forms which are most inclusive, *lebenskräftig*, and susceptible of evolutionary differentiation' (p. 295).

[2] Emerson knew better than Aurobindo when he proclaimed that 'man is a god in ruins' (Warren, op. cit., p. 417).

One becomes identified with the object of one's knowledge. For those who believe the world emanates from God, there is a way back to God; for those who believe the world emanates from chaos, there is likewise a way that corresponds with this possibility.[1]

Introd.
Judgment

[1] 'Then shall the dust return to the earth as it was: and the spirit shall return unto God who gave it' (*Ecclesiastes,* XII. 7). 'In whom, when I go forth, shall I be going forth?' (*Praśna Upanishad,* VI. 3: tr. Coomaraswamy). Cf. also Cudworth: 'The Atomical way, which is to resolve the corporeal phenomena, not into forms, qualities, and species, but into figures, motions, and fancies' (*The True Intellectual System of the Universe,* 1st American Edition, 1837, I. 69).

The Process of Manifestation

There was then neither being nor non-being. . . .
Without breath breathed by its own power That One.

Rig-Veda, x. 129. 1, 2

M.M. 994; Infra 38

At the beginning of the beginning,[1] even nothing did not exist. Then came the period of the Nameless. When ONE came into existence, there was ONE, but it was formless. When things got that by which they came into existence, it was called their virtue. . . . By cultivating this nature, we are carried back to virtue; and if this is perfected, we become as all things were in the beginning. We become unconditioned.

Chuang-tse (ch. XII)

Introd. *Heterod.*

He was. Taaroa was his name, he stood in the void: no earth, no sky, no men. Taaroa calls to the four corners of the universe. Nothing replies. Alone existing, he changes himself into universe. Taaroa is the light, he is the seed, he is the base, he is the incorruptible. The universe is only the shell of Taaroa. It is he who puts it in motion and brings forth its harmony.

Polynesian Tradition

Beauty 670

Creation is only the projection into form of that which already exists.

Srimad Bhagavatam, III. 2

Hermes . . . says that when God changed His form, the universe was suddenly revealed and put forth in the Light of Actuality—this world being nothing but a visible Image of a Hidden God. This is what the Ancients meant when they said that Pallas leapt forth in divine perfection from the forehead of Jupiter, with the aid of Vulcan (or Divine Light).

The All-Wise Doorkeeper

Beauty 663

This world, with all its stars, elements, and creatures, is come out of the invisible world; it has not the smallest thing or the smallest quality of anything but what is come forth from thence.

William Law

[1] ' "In the beginning", or rather "at the summit", means "in the first cause" : just as in our still told myths, "once upon a time" does not mean "once" alone but "once for all" ' (Coomaraswamy: *Hinduism and Buddhism*, p. 6). '*In principio* signifies, in the beginning of all things. It also means the end of all things, since the first beginning is because of the last end' (Eckhart: Evans, I. 224). 'Manifestation always proceeds from a state of non-manifestation, although it is a question of an anteriority which is purely principial, and not temporal. God is too sublime for there to be any temporal relation between Himself and His creation. . . . His primacy is essentially His ulteriority. His anteriority is His posteriority' (Jili: *Al-Insân al-Kâmil*, tr. Titus Burckhardt, p. 58).

He acted Himself into being.

Plotinus

There is no origination of anything that is mortal, nor yet any end in baneful death: but only mixture and separation of what is mixed, but men call this 'origination'.

M.M. 978

Empedocles

The Greeks do not rightly use the terms 'coming into being' and 'perishing'. For nothing comes into being nor yet does anything perish, but there is mixture and separation of things that are. So they would do right in calling the coming into being 'mixture', and the perishing 'separation'.[1]

Introd. *Holiness*

Anaxagoras

'I spoke a word, and the word glorified Me: then of the glorifying of the word created I a light and a darkness. Of the light I created the spirits of such as believe, and of the darkness created I the spirits of such as disbelieve. Then I mingled the light with the darkness, and made it to be a stone-jewel: the jewelness was of the light, and the stoniness was of the darkness.'

Inv. 1003
Wr. Side 459

Niffari

The matter of this world is the materiality of the Kingdom of Heaven, brought down into a created state of grossness, death, and imprisonment, by occasion of the sin of those angels who first inhabited the place or extent of this material world. Now these heavenly properties which were brought into this created compaction lie in a continual desire to return to their first state of glory; and this is the groaning of the whole creation to be delivered from vanity which the Apostle speaks of.

Infra 38
Charity 608

William Law

The *Creation* of the World was a *Vail* cast upon the Face of God, with a figure of the Godhead wrought upon this Vail, and God Himself *seen* through it by a *dim transparency*; as the Sun in a morning, or Mist, is seen by a refracted Light through the thick *medium* of earthly *Vapours*.

Illusion 85

Peter Sterry

Observe the three parts into which God has divided the first substance. Of the first and pure part He created the Cherubin, Seraphin, Archangels, and all the other angels. Out of the second, which was not so pure, He created the heavens and all that belongs to them; of the third, impure part, the elements and their properties.

John A. Mehung

As a thing begins, so it ends. Out of one arise two, and out of two one—as of God the Father there was begotten God the Son, and from the two proceeded God the Holy Ghost. Thus was the world made, and so also shall it end. Consider carefully these few points, and

Reality 775

[1] The same process that Hermetic terminology describes as 'coagulation' and 'solution'.

you will find, firstly, the Father, then the Father and the Son, lastly, the Holy Spirit. You will find the four elements, the four luminaries, the two celestials, the two centrics. In a word, there is nothing, has been, and shall be nothing in the world which is otherwise than it appears in this symbol, and a volume might be filled with its mysteries.

Symb. 306

Michael Sendivogius

From point comes a line, then a circle;
When the circuit of this circle is complete,
Then the last is joined to the first.

Humility 199

Shabistarî

Tawhîd is the return of man to his origin, so that he will be as he was before he came into being.

Junayd

The knowledge of the ancients reached the highest point—the time before anything existed. This is the highest point. It is exhaustive. There is no adding to it.

Chuang-tse (ch. XXIII)

In old times the perfect man of Tao was subtle, penetrating and so profound that he can hardly be understood. . . .

Going back to the origin is called peace;
It means reversion to destiny.
Reversion to destiny is called eternity.
He who knows eternity is called enlightened. . . .
Being supreme he can attain Tao.
He who attains Tao is everlasting.
Though his body may decay he never perishes.[1]

Peace 705

P. State 563

Tao Te Ching, XV, XVI

Inexhaustibility of the Creator

He is from eternity to eternity, and to him nothing may be added,
Nor can he be diminished, and he hath no need of any counsellor.

Ecclesiasticus, XLII. 21–22

[1] 'When a society is perishing, the wholesome advice to give to those who would restore it is to recall it to the principles from which it sprang. . . . To fall away from its primal constitution implies disease, to go back to it, recovery' (Pope Leo XIII, *Rerum Novarum*—cited by Harold Robbins in *The Sun of Justice*, London, 1938, p. 26).

These participations neither augment nor diminish him, for they are communicated by him, not detached from him. *Void* 724

Chuang-tse (ch. II)

As milk is spontaneously changed into curds and water into ice, so *Brahma* modifies Itself in diverse ways, without the aid of instruments or external means of any kind whatever. Thus the spider spins its web out of its own substance, subtle beings take diverse forms, and the lotus grows from marsh to marsh without organs of locomotion. *Reality* 773

Brahma-Sûtra, II. i

In use it can never be exhausted.

Tao Te Ching, VI

My bounty is as boundless as the sea,
My love as deep; the more I give to thee,
The more I have, for both are infinite. *Renun.* 158

Shakespeare (*Romeo and Juliet*, II. ii. 133)

Shall I choose for a protecting friend other than Allâh, the Originator of the heavens and the earth, Who feedeth and is never fed?

Qur'ân, VI. 14

The Namu-amida-butsu inexhaustible,
However much one recites it, it is inexhaustible. *Inv.* 1031

Saichi

Do not think of the water failing; for this water is without end.

Dîvâni Shamsi Tabrîz, XII

The atoms of the universe may be counted, but not so my manifestations; for eternally I create innumerable worlds.

Srimad Bhagavatam, XI. x

The fire which serves as pyre for the Phoenix, and cradle where he resumes a new life . . . draws its origins from the highest mountain on earth. . . . This fire is the source of all light which illumines this vast universe: it imparts heat and life to all beings . . ., a Flame that is never consumed. *Center* 831 833

Michael Maier

The yon is fulness; fulness, this.
From fulness, fulness doth proceed.
Withdrawing fulness's fulness off,
E'en fulness then itself remains.

Bṛihad-Âraṇyaka Upanishad, V. 1

Introd.
Reality

And here you may behold the sure ground of the absolute impossibility of the annihilation of the soul. Its essences never began to be, and therefore can never cease to be; they had an eternal reality before they were in or became a distinct soul, and therefore they must have the same eternal reality in it.

William Law

To be poured into without becoming full, and pour out without becoming empty, without knowing how this is brought about,—this is the art of 'Concealing the Light'.

Chuang-tse (ch. II)

Charity
602

Gods boundlesse mercy is (to sinfull man)
Like to the ever-wealthy Ocean:
Which though it sends forth thousand streams, 'tis ne're
Known, or els seen to be the emptier:
And though it takes all in, 'tis yet no more
Full, and fild-full, then when full-fild before.

Robert Herrick

Infra
48

As God willed no longer to remain in Himself, alone, therefore created He the soul and gave Himself in great love to her alone. Whereof art thou made, O Soul, that thou soarest so high over all creatures and whilst mingling in the Holy Trinity, yet remainest complete in thyself?

Mechthild of Magdeburg

Who is Nature and Who is all that is manifested from her? We did not see her diminished by that which was manifested from her, or increased by the not-being of aught manifested that was other than she.

Ibn 'Arabî

The Principle is an infinity which nothing can augment or diminish.

Chuang-tse (ch. XXII)

Reality
803

There is no existence for the unreal and the real can never be non-existent. The Seers of Truth know the nature and final ends of both.

Know That to be indestructible by which all this is pervaded. No one is ever able to destroy that Immutable.

These bodies are perishable; but the Dweller in these bodies is eternal, indestructible and impenetrable. Therefore fight, O descendant of Bharata!

He who considers this (Self) as a slayer or he who thinks that this (Self) is slain, neither of these knows the Truth. For It does not slay, nor is It slain.

This (Self) is never born, nor does It die, nor after once having been, does It go into non-being. This (Self) is unborn, eternal, changeless, ancient. It is never destroyed even when the body is destroyed.

Bhagavad-Gîtâ, II. 16–20

A Fountain ever equally unexhaust, a Sea unbounded. . . .

Peter Sterry

In its essence it must have an abundance of blood, like the Pelican, which wounds its own breast, and, without any diminution of its strength, nourishes and rears up many young ones with its blood. This Tincture is the Rose of our Masters, of purple hue, called also the red blood of the Dragon.

Basil Valentine

Of another Liquor wise men tell,
Which is fresher than Water of the Well;
Fresher Liquor there is none in tast,
Yet it woll never consume ne waste;
Though it be occupied evermore,
It will never be lesse in store;
Which *Democrit* named for his intent,
Lux umbra carens, Water most Orient.

Thomas Norton

P. State 563

Unfathomable as the sea, wondrously ending only to begin again, informing all creation without being exhausted, the *Tao* of the perfect man is spontaneous in its operation.

Chuang-tse (ch. XXII)

Holiness 924

Simultaneity of the Divine Act

Do not fondly imagine that God, when he created the heavens and the earth and all creatures, made one thing one day and another the next. Moses describes it thus it is true, nevertheless he knew better: he did so merely on account of those who are incapable of understanding or conceiving otherwise. All God did was: he willed and they were.

Eckhart

Center 838

Verily His command, when He intendeth a thing, is that He saith unto it: Be! and it is.

Qur'ân, XXXVI. 82

The force with which God works is his will; and his very being consists in willing the existence of all things. . . . For God wills things to be, and, in that way, these things also have existence.

Hermes

Introd. *Waters*

That Lord having dwelt in that egg (*Brahmânda*) for a year, spontaneously, by his own meditation, split that egg in two.

And with those two shares he formed the heaven and the earth, in the middle the sky and the eight regions, and the perpetual place of waters.

Mânava-dharma-śâstra, I. 12, 13

For with God, to will is to accomplish, inasmuch as, when he wills, the doing is completed in the self-same moment as the willing.

Hermes

There is an end, then, of the notion that the universe came into being 'in six days'.

Philo

For nor before nor after was the process of God's outflowing over these waters.

Dante (*Paradiso*, XXIX. 20)

Beauty 670

God's thoughts are immediately translated into real existences. . . . God's conceptions are at once objective essences.

Michael Sendivogius

How great is the power of God! His mere will is creation; for God alone created, since He alone is truly God. By a bare wish His work is done, and the world's existence follows upon a single act of His will.

Clement of Alexandria

When God set forth to create the universe, his Thought encompassed all the worlds at once. . . . All were created in a single instant.

Zohar

Inv. 1003

He commanded, and they were created.

Psalm CXLVIII. 5

Without effort he sets in motion all things by mind and thought.

Xenophanes

Center 841

Since in God being and thinking are identical, the notion which God, in thinking Himself, generates as an exact image of Himself, is identical with God Himself.[1]

Marsilio Ficino

[1] Here Ficino is referring to the creation of the Archetypes.

Creation as Play or Sport

Brahma's creative activity is not undertaken by way of any need on his part, but simply by way of sport, in the common sense of the word.

Brahma Sûtra, II. i. 32, 33

Reality 773

God has created the world in play.

Sri Ramakrishna

What is willing in the Godhead? It is the Father watching the play of his own nature. What is this play? It is his eternal Son. There has always been this play going on in the Father-nature. Play and audience are the same. . . . As it is written in the Book of Wisdom, 'Prior to creatures, in the eternal now, I have played before the Father in his eternal stillness.' The Son has eternally been playing before the Father as the Father has before his Son. The playing of the twain is the Holy Ghost in whom they both disport themselves and he disports himself in both. Sport and players are the same. Their nature proceeding in itself. 'God is a fountain flowing into itself,' as St Dionysius says.

Eckhart

Center 838

Supra 28

Thou didst contrive this 'I' and 'we' in order that Thou mightst play the game of worship with Thyself,

That all 'I's' and 'thou's' should become one soul and at last should be submerged in the Beloved.[1]

Rûmî

Infra 48

Sp. Drown. 713

There's no such sport as sport by sport o'erthrown,
To make theirs ours and ours none but our own.

Shakespeare (*Love's Labour's Lost*, v. ii. 153)

Sin 62

O Lord of all, hail unto thee!
The Soul of all, causing all acts,
Enjoying all, all life art thou!
Lord of all pleasure and delight!

Maitri Upanishad, v. 1

As the spider weaves its thread out of its own mouth, plays with it, and then withdraws it again into itself, so the eternal, unchangeable Lord, who is formless and attributeless, who is absolute knowledge, and absolute bliss, evolves the whole universe out of himself, plays with it, and again withdraws it into himself.

Srimad Bhagavatam, XI. iii

Supra 26

[1] The same in Japanese Zen, *yukezammai*=samadhi in which all activity is play (*lîlâ*), or 'the samadhi of play': cf. R. Kita and K. Nagaya, *How Altruism Is Cultivated in Zen*, p. 135.

God's delight is in the communication of Himself, His own happiness to everything according to its capacity. He does everything that is good, righteous, and lovely for its own sake, because it is good, righteous, and lovely.

William Law

It will be given to us to behold the Bodhisatta's infinite Buddha-*liḷhâ* and to hear his word.

Jâtaka, I. 54

Illusion 85

People see his pleasure-ground;
Him no one sees at all.

Bṛihad-Âranyaka Upanishad, IV. iii. 14

The laughter of the Gods must be defined to be their exuberant energy in the universe, and the cause of the gladness of all mundane natures.

Proclus

Manvantaras are countless, as are also the creations and destructions; the Supreme Being does this again and again as if in sport.

Mânava-dharma-śâstra, I. 80

Beauty 679

This world, you see, is like a drum; there is a Being who plays all kinds of tunes on it.

Ananda Moyî

'From whence dost thou come?' the holy Râbi'a was asked.
'From the other world.'
'And where dost thou go?'
'To the other world.'
'What art thou doing in this world?'
'I am making a game of it.'

'Aṭṭâr

Pilg. 387

Go seek, O mind, go seek Vrindâvan[1] in your heart,
Where with His loving devotees
Sri Krishna sports eternally.

Hindu Song

Yogis sport (*rama*) in the eternal God, whose self is composed of *sat* (Being), *chit* (Knowledge), and *ananda* (Bliss). Hence the term *Râm* means the Supreme God.

Chaitanya

Metanoia 484

Man is made to be the plaything of God, and this, truly considered, is the best of him; wherefore also every man and woman should walk seriously, and pass life in the noblest of pastimes, and be of another mind from what they are at present. . . . And what is the right

[1] A town associated with Krishna's childhood.

way of living? . . . We ought to live sacrificing, and singing, and dancing, and then a man will be able to propitiate the Gods.[1]

Plato (*Laws*, 803)

It is God Himself who is sporting in the form of man.

Sri Ramakrishna

Illusion 94

I trow that whoso had grace to do and feel as I say, he should feel good gamesome play with him (God), as the father doth with the child, kissing and clasping, that well were him so!

The Cloud of Unknowing, XLVI

The soul is taken by God to a secret place where it must not ask nor pray for anyone, for God alone will play with it in a game of which the body knows nothing, any more than the peasant at the plough or the knight in the tourney; not even His loving mother Mary; she can do nothing here. Thus God and the soul soar further to a blissful place of which I neither can nor will say much. It is too great and I dare not speak of it for I am a sinful creature.

Mechthild of Magdeburg

M.M. 998

O peaceful and pleasant War! where the supream Love stands on both sides, where, as in a mysterious Love-sport, or a Divine *Love-play, it fights with it self.*

Peter Sterry

Center 835

Just as time is a play before God, so was the outward life of the (first) man a play before the inward and holy life which was the true image of God.

Boehme

When Thou, my Expectation, art not near,
Each moment is an age of grief and fear;
But while I may behold and hear Thee, all
My days are glad, and Life's a Festival!

Abu 'l-Ḥusayn al-Nûrî

Contem. 536

All the world's a stage,
And all the men and women merely players.

Shakespeare (*As You Like It*, II. vii. 139)

Remember that you are an actor in a play, and the Playwright chooses the manner of it. . . . Your business is to act the character that is given you and act it well; the choice of the cast is Another's.

Epictetus

Action 338

[1] This description still characterizes a people like the Balinese.

Center 816

The heart of man is the seat of this Lila, which can be reproduced at all times, in the heart of every true Bhakta.

Nilakantha, on the Krishna Lila

Metanoia 488

The dance of Nature proceeds at one side, that of Gnosis on the other.

Tiru-Aruḷ-Payan, IX. 3

I am a dancer and you are the manager of the theatre; I dance as you make me. My tongue is merely a harp, and you (Krishna) are the musician who plays on it. I utter whatever you think in your mind.

Râmânanda Ray

Contem. 528

The cause of human happiness and misery is a false representation of the understanding. This world is a stage stretched out by the mind, its chief actor, and *Atman* sits silent as a spectator of the scene.

Yoga-Vasishtha

Heaven is our heritage,
Earth but a players stage.

Thomas Nashe

Beauty 688

I bow to thee (Chaitanya, as Krishna incarnate). Who can fathom the depths of thy heart? As the juggler makes the wooden puppet dance, while it knows not what it plays or what it sings, so, too, does the man whom you inspire, dance without knowing why he is dancing or through whom.

Sanâtan

Action 340

He Himself is the Player and Witness of the play.

Swami Ramdas

Illusion 85
Symb. 317

For in and out, above, below,
'Tis nothing but a Magic Shadow-show,
Play'd in a Box whose Candle is the Sun,
Round which we Phantom Figures come and go.

Omar Khayyâm

Illusion 99

So this world has no substantial reality,
But exists as a shadowy pageant or a play.

Shabistarî

My play here is finished. My kingdom is established.

Srimad Bhagavatam, XI. i

Creation by Pairs

In the beginning this world was just Being, one only, without a second. It bethought itself: 'Would that I were many! Let me procreate myself!'

Chândogya Upanishad, VI. ii. 2, 3

Infra 48

He was, indeed, as large as a woman and a man closely embraced. He caused that self to fall into two pieces. Therefrom arose a husband and a wife. . . . He copulated with her. Therefrom human beings were produced.

And she then bethought herself: 'How now does he copulate with me after he has produced me just from himself? Come, let me hide myself.' She became a cow. He became a bull. With her he did indeed copulate. Then cattle were born. She became a mare, he a stallion. She became a female ass, he a male ass; with her he copulated, of a truth. Thence were born solid-hoofed animals. She became a she-goat, he a he-goat; she a ewe, he a ram. With her he did verily copulate. Therefrom were born goats and sheep. Thus, indeed, he created all, whatever pairs there are, even down to the ants.

Bṛihad-Âraṇyaka Upanishad, I. iv. 3, 4

Then she became a duck,
And he became a rose-kaimed drake.

The Twa Magicians

. . . Jove
Put on all shapes to get a Love:
As now a *Satyr*, then a *Swan;*
A *Bull* but then; and now a man.

Robert Herrick

The gods themselves,
Humbling their deities to love, have taken
The shapes of beasts upon them: Jupiter
Became a bull, and bellow'd; the green Neptune
A ram, and bleated; and the fire-rob'd god,
Golden Apollo, a poor humble swain.

Shakespeare (*Winter's Tale*, IV. iii. 25)

The Being of all beings is but one only Being, but in its generation it separates itself into two principles, viz. into light and darkness, into joy and sorrow, into evil and good, into love and anger, into fire and light, and out of these two eternal beginnings (or principles) into the third beginning, viz. into the creation, to its own love-play and melody, according to the property of both eternal desires.

Boehme

Reality 775

Supra 33

All things are double, one against another, and he hath made nothing defective.

Center 847

He hath established the good things of every one. And who shall be filled with beholding his glory?

Ecclesiasticus, XLII. 25, 26

Glory be to Him Who created all the sexual pairs, of that which the earth groweth, and of themselves, and of that which they know not!

Qur'ân, XXXVI. 36

The World Begins in Perfection

P. State 583

In the beginning, before there was any division of subject and object, there was one existence, Brahman alone, One without a second. That time is called the Krita yuga, or the golden age, when people skilled in knowledge and discrimination realized that one existence. . . . Men had but one caste, known as Hamsa. All were equally endowed with knowledge, all were born knowers of Truth; and since this was so the age was called Krita, which is to say, 'Attained'.

Srimad Bhagavatam, XI. xvii & xi

And the Lord God planted a garden eastward in Eden; and there he put the man whom he had formed.

Genesis, II. 8

He created the heavens and the earth with truth, and He shaped you and made good your shapes.

Qur'ân, LXIV. 3

Infra 42

God created Adam in His own form.

Muhammad

Intellect (νοῦς) the Father of all, He who is Life and Light, gave birth to Man, a Being like to Himself. And He took delight in Man, as being His own offspring. . . . With good reason then did God take delight in Man; for it was God's own form that God took delight in. And God delivered over to Man all things that had been made.

Hermes

Human birth . . . reflects my image.

Srimad Bhagavatam, XI. xix

When God created the soul he fell back upon himself and made her after his own likeness.

Eckhart

Man, is not he Creation's last appeal,
The light of Wisdom's eye? Behold the wheel
Of universal life as 'twere a ring,
But Man the superscription and the seal.

Omar Khayyâm

Holiness 924

All the Creatures stood together in Man, as in the Head, in a Divine Harmony of their Essences and Operations, of each with itself, of each with other, of all with God. From Man as the Head, this Harmony was propagated and maintain'd thro' the Creatures, subsisting apart by themselves. One Divine Life, mov'd, shin'd, sounded in and thro' all, as an unexpressible Love, Beauty, Musick, made up out of all, compleat in all, beginning and terminating in Man, as the Head of all.

Peter Sterry

Rev. 967

Verily We created man in the fairest rectitude.

Qur'ân, XCV. 4

Golden was that first age, which, with no one to compel, without a law, of its own will, kept faith and did the right.

Ovid

In those days . . . all times were pleasant. . . . People had no need of sacraments for the purification of their bodies, and their youth was permanent.

Mârkaṇḍeya Purâṇa, XLIX

Introd. *Suff.*

The world is the fairest of creations.

Plato (*Timaeus*, 29 A)

The highest Good, who himself alone doth please, made man good and for goodness, and gave this place to him as an earnest of eternal peace.

Dante (*Purgatorio*, XXVIII. 91)

This image made to the image of God in the first shaping was wonderly fair and bright, full of burning love and ghostly light.

Walter Hilton

Adam was both man and woman and yet neither one nor the other but a virgin, full of chastity and modesty and purity, such was the image of God; he had in himself the two principles of fire and light, and in their conjunction was found his love of himself, his virginal principle, which was the beautiful garden of pleasure planted with roses in which he loved himself; it is what we shall be at the resurrection of the dead, as Christ teaches us (*Matt.*, XXII. 30), saying that we shall no longer marry nor be given in marriage, but shall

P. State 579

be as the angels of God . . . and yet not only pure spirits like the angels, but clothed in a glorious body in which reposes the spiritual and angelic body. . . .

If God had created him in the terrestrial life, perishable, miserable, naked, sick, bestial and painful, He would not have introduced him into Paradise. If He had desired that there be pregnancy and bestial reproduction, He would from the beginning have created a man and a woman, and the two sexes would have been separated at the 'Verbum Fiat' following the two principles, male and female, as was the case with other terrestrial creatures.

Boehme

Reality 790

When God made man the innermost heart of the Godhead was put into man.

Eckhart

Christ was the first man.

Eckhart

Had not Christ been in our first father as a birth of life in him, Adam had been created a mere child of wrath in the same impurity of nature, in the same enmity with God, and in the same want of an atoning Saviour as we are at this day.

William Law

P. State 583

And so in the days when natural instincts prevailed, men moved quietly and gazed steadily. At that time, there were no roads over mountains, nor boats, nor bridges over water. All things were produced, each for its own proper sphere. Birds and beasts multiplied; trees and shrubs grew up. The former might be led by the hand; you could climb up and peep into the raven's nest. For then man dwelt with birds and beasts, and all creation was one. There were no distinctions of good and bad men. Being all equally without knowledge, their virtue could not go astray. Being all equally without evil desires, they were in a state of natural integrity, the perfection of human existence.

Chuang-tse (ch. IX)

The reason why the life of man was, as tradition says, spontaneous, is as follows: In those days God himself was their shepherd, and ruled over them, just as man, who is by comparison a divine being, still rules over the lower animals. Under him there were no forms of government or separate possession of women and children; for all men rose again from the earth, having no memory of the past. And although they had nothing of this sort, the earth gave them fruits in abundance, which grew on trees and shrubs unbidden, and were not planted by the hand of man. And they dwelt naked, and mostly in the open air, for the temperature of their seasons was mild; and they had no beds, but lay on soft couches of grass, which grew plentifully out of the earth. Such was the life of man in the days of Cronos.

Plato (*Statesman,* 272 A)

Before he ate of the tree of knowledge, Adam was all spirit and wore angelic clothing like Enoch and Elias. This is why he was worthy to eat the fruits of paradise, which are the fruits of the soul.

Ezra

Men would have gone naked upon the earth, for the Celestial interpenetrated the Exterior and was its garment:[1] and he (Adam) moved in great beauty, joy and pleasure, with a childlike heart. He would have drunken and eaten magically, not in the body as now. . . . He had no sleep in him, night was to him as the day: for he saw with glorious eyes by means of his own light; the interior man, interior eye, saw across the exterior; just as in the next world we shall have no need of the sun, for we shall see with divine vision, by the light of our own nature.

Center 816 833

Boehme

The (cosmic) forces do not work upward from below, but downward from above.

Hermes

God dwells in the nothing-at-all that was prior to nothing, in the hidden Godhead of pure gnosis whereof no man durst speak.

M.M. 994

Eckhart

Man was very goodly to look on, bearing the likeness of his Father. . . . And . . . downward-tending Nature . . . , seing the beauty of the form of God, smiled with insatiate love of Man, showing the reflection of that most beautiful form in the water, and its shadow on the earth. And he, seeing this form, a form like to his own, in earth and water, loved it, and willed to dwell there. And the deed followed close on the design; and he took up his abode in matter devoid of reason. And Nature, when she had got him with whom she was in love, wrapped him in her clasp.

Introd. *Waters*

Illusion 85

Hermes

For in the past, we were made of mind, we fed on rapture, self-luminous, we traversed the air in abiding loveliness; long long the period we so remained. For us sooner or later, after a long long while the savoury earth had arisen over the waters. Colour it had, and odour and taste. We set to work to make the earth into lumps, and feast on it. As we did so our self-luminance vanished away. When it was gone, moon and sun became manifest, star-shapes and constellations, night and day, the months and half-months, the seasons and the years.

P. State 583 *Flight* 949

Sin 66

Aggana Suttanta

Ye are inferior in stature in comparison with your predecessors; and so, also, (will be) your posterity than yourselves: even as creation is already grown old, and is already past the strength of youth.

IV Ezra, v. 55

[1] By this interpretation, the 'coats of skins' in which God clothed fallen man would correspond with the exteriorization of the individual and corporeal state—as is moreover indicated by Boehme himself in a later passage from the same work (XXI. 16). Cf. also the citation of Scotus Erigena in the chapter 'Holiness'.

Have they not travelled in the land and seen the nature of the consequences for those who were before them? They were stronger than these in power, and they dug the earth and built upon it more than these have built.

Qur'ân, XXX. 9

That which has a precedency is more honorable than that which is consequent in time. As for instance . . . , the east is more honorable than the west; the morning than the evening; the beginning than the end; and generation than corruption.

Pythagoras

Wr. Side 464

For the sake of those among people of the future who will appreciate, I am writing this—knowing that all those of later times will be of inferior calibre.

Ko Hung

Heterod. 420

That nature, which contemns its origin,
Cannot be border'd certain in itself.

Shakespeare (*King Lear*, IV. ii. 32)

Man's Primordial Birthright

Supra 26

In the beginning, verily, this world was non-existent. Therefrom, verily, Being (*sat*) was produced. That made itself a Self (*Âtman*). Therefore it is called the well-done. What that well-done is—that, verily, is the essence (*rasa*) of existence. For truly, on obtaining that essence, one becomes blissful.

Taittirîya Upanishad, II. 7

Realiz. 887

If you possess true knowledge (gnosis), O Soul, you will understand that you are akin to your Creator.

Hermes

Contem. 536

God . . . so copied forth himself into the whole life and energy of man's soul, as that the lovely characters of Divinity may be most easily seen and read of all men within themselves. . . . The *impresse* of souls is . . . nothing but God himself, who could not write his own name so as that it might be read, but only in rational natures.

John Smith the Platonist

Holiness 924

The being of man . . . is the noblest being of all made things.

The Epistle of Privy Counsel, III

... Adam, of whom Christ received the flesh. Christ therefore is in Adam and Adam in Christ.

St Augustine

You are a principal work, a fragment of God Himself, you have in yourself a part of Him. Why then are you ignorant of your high birth? ... You bear God about with you, poor wretch, and know it not. Do you think I speak of some external god of silver or gold? No, you bear Him about within you and are unaware that you are defiling Him with unclean thoughts and foul actions. If an image of God were present, you would not dare to do any of the things you do; yet when God Himself is present within you and sees and hears all things, you are not ashamed of thinking and acting thus: O slow to understand your nature, and estranged from God!

Faith 505

Epictetus

Realize thy Simple Self,
Embrace thy Original Nature.

Tao Te Ching, XIX

The trees of the New mountain were once beautiful. Being situated, however, in the borders of a large state, they were hewn down with axes and bills;—and could they retain their beauty? Still through the activity of the vegetative life day and night, and the nourishing influence of the rain and dew, they were not without buds and sprouts springing forth, but then came the cattle and goats and browsed upon them. To these things is owing the bare and stript appearance of the mountain, which when people see, they think it was never finely wooded. But is this the nature of the mountain?

And so also of what properly belongs to man—shall it be said that the mind of any man was without benevolence and righteousness? The way in which a man loses his proper goodness of mind is like the way in which the trees are denuded by axes and bills. Hewn down day after day, can it—the mind—retain its beauty? But there is a development of its life day and night, and in the calm air of the morning, just between night and day, the mind feels in a degree those desires and aversions which are proper to humanity, but the feeling is not strong, and it is fettered and destroyed by what takes place during the day. This fettering taking place again and again, the restorative influence of the night is not sufficient to preserve the proper goodness of the mind; and when this proves insufficient for that purpose, the nature becomes not much different from that of the irrational animals, which when people see, they think that it never had those powers which I assert. But does this condition represent the feelings proper to humanity?

Realiz. 873

Mencius

When we remain as we have been created, we are in a state of virtue. ... If we had to seek for virtue outside of ourselves, that would assuredly be difficult; but as it is within us, it suffices to avoid bad thoughts and to keep our souls turned towards the Lord.

Philokalia

Mind is consciousness which has put on limitations. You are originally unlimited and perfect. Later you take on limitations and become the mind.

Sri Ramana Maharshi

Someone said to Jesus (Peace be upon him!), 'Who trained you?' He said, 'No one trained me. I saw the ignorance of the ignorant man to be a blemish, so I avoided it.'

Christ in Islâm

There is nothing in the world easier than goodness and nothing more difficult than depravity. By 'goodness' is meant tranquillity of mind, undisturbed by cupidity. By 'depravity' is meant a grasping spirit with many cravings. He who is satisfied with the simple needs of his nature, refusing the superfluous delights of the world, will not be tempted by any seductions. He who follows the law of his nature, will preserve his soul, without any inward conflicts. Hence the statement 'It is easy to be good.' Clambering up city walls, scaling dangerous heights, thieving the official keys, forging and stealing official money, rebelling and murdering, lying and bearing false witness, are acts contrary to human nature. Hence the saying, 'It is difficult to be bad.'

Renun. 156

Huai Nan Tzû

Contem. 547

The soul is by nature made for heaven and God is her lawful heritage.

Eckhart

This glory and honor wherewith man is crowned ought to affect every person that is grateful, with celestial joy: and so much the rather because it is every man's proper and sole inheritance.

Thomas Traherne (Gloss on *Psalm* VIII)

As sure as man is called to this unity, purity, and perfection of love, so sure is it that it was at first his natural heavenly state and still has its seed or remains within him, as his only power and possibility of rising up to it again.

William Law

The plain truth is that of necessity God is bound to cherish us just as though his Godhood were at stake, as in fact it is. God can no more do without us than we can without him, nay, even if we turned from God it would be impossible for God to turn his back on us.

Love 618

Eckhart

There is nothing that is supernatural, however mysterious, in the whole system of our redemption; every part of it has its ground in the workings and powers of nature and all our redemption is only nature set right, or made to be that which it ought to be.

William Law

Art cannot change or overstep the natural order of the universe.

Henry Madathanas

Nothing can be got out of a thing which is not in it. Therefore every species, every genus, every natural order, is naturally developed within its own limits, bearing fruit after its own kind, and not within some other essentially different order: everything in which seed is sown must correspond to its own seed.[1]

Richard the Englishman

By burning anything to ashes you may gain its salt. If in this dissolution the sulphur and mercury be kept apart, and restored to its salt, you may once more obtain that form which was destroyed by the process of combustion. This assertion the wise of this world denounce as the greatest folly, and count as a rebellion, saying that such a transformation would amount to a new creation, and that God has denied such creative power to sinful man. But the folly is all on their side. For they do not understand that our Artist does not claim to create anything, but only to evolve new things from the seed made ready to his hand by the Creator.

Basil Valentine

Introd. *Grace Orthod.* 280

Know also that animals only multiply after their kind, and within their own species. Hence our Stone can only be prepared out of its own seed, from which it was taken in the beginning.

Basil Valentine

Thus we never find a vein of lead, for instance, which does not contain a few permanent grains, at least, of gold and silver . . . which are imparted to it by Nature for the purpose of multiplication and development, as I myself have experienced, and am able to testify.

Nicholas Flamel

If divine truth spring up only from the root of true goodness; how shall we ever endeavour to be good, before we know what it is to be so? or how shall we convince the gainsaying world of truth, unless we could also inspire virtue into it?

To both which we shall make this reply; that there are some radical principles of knowledge that are so deeply sunk into the souls of men, as that the impression cannot easily be obliterated, though it may be so much darkened.

John Smith the Platonist

A reproach is sometimes levelled at our Art, as though it claimed the power of creating gold; every attentive reader . . . will know that it only arrogates to itself the power of developing, through the removal of all defects and superfluities, the golden nature, which the baser metals possess in common with that highly-digested metallic substance. . . . I have shewn that the transmutation of metals is not a chimerical dream, but a sober possibility of Nature, who is perfectly capable of accomplishing it without the aid of magic;

Reality 775

[1] 'Evolution cannot exist for a moment without demonstrable transformations. But botanical phenomena provide us with no transformations—not even one' (Arthur P. Kelly, Botanist, Head of the Landenburg Laboratory).

'Evolution is a kind of dogma which the priests no longer believe, but which they maintain for their people' (Paul Lemoine, Geologist, Director of the National Museum at Paris).

and that this possibility of metallic transmutation is founded upon the fact that all metals derive their origin from the same source as gold, and have only been hindered from attaining the same degree of maturity by certain impurities, which our Magistery is able to remove.

Philalethes

Realiz. 873

The search 'Who am I?' . . . ends in the annihilation of the illusory 'I' and the Self which remains over will be as clear as a gooseberry in the palm of one's hand. . . . You are sure to realize the Self for it is your natural state.

Sri Ramana Maharshi

Infra 47

I have breathed into him (man) of My Spirit.

Qur'ân, xv. 29

M.M. 978

The mind from the beginning is of a pure nature, but since there is the finite aspect of it which is sullied by finite views, there is the sullied aspect of it. Although there is this defilement, yet the original pure nature is eternally unchanged. This mystery the Enlightened One alone understands.

Asvaghosha

Just as crystal, which is clear, becomes coloured from the colour of another object, so likewise the jewel of the mind becomes coloured with the colour of mental conceits. Like a jewel the mind is naturally free from the colour of these mental conceits; it is pure from the beginning, unproduced, immaculate and without any self-nature.

Cittaviśuddhiprakaraṇa

All *Metallick* Bodies are compounded of *Argentvive* and *Sulphur*, pure or impure, by accident, and not innate in their first Nature; therefore, by convenient *Preparation*, 'tis possible to take away such Impurity.

Geber

Transmutation is a great mystery, which is by no means—as fools suppose—contrary to the course of Nature, or the law of God. Without this Philosopher's Stone, the imperfect metals can be transmuted neither into gold nor silver.

Paracelsus

Supra 28 42 *Center* 847

The soul is a miraculous abyss of infinite abysses, an undrainable ocean, an inexhausted fountain of endless oceans, when it will exert itself to fill and fathom them. For if it were otherwise man is a creature of such noble principles and severe expectations, that could he perceive the least defect to be in the Deity, it would infinitely displease him.

Thomas Traherne

O Man! The Divine Image was thy *proper Habitation*. This Image in thy Person was a *Sun* of Beauty shining in every part of it; and a Shield of Power defending it on every side. In this Image all Blessed things were united to make a Paradise for thee, which thou didst carry about in thine own Person, as thy proper Form, thine inseparable Habitation, like to God, who carryeth his own Heaven into every place with him.

Peter Sterry

Let no one then think lightly of the Word, lest he be despising himself unawares.

Clement of Alexandria

Realiz. 859

So set thy course for religion as a man by nature upright—the nature of Allâh, in which He hath created man.

Qur'ân, XXX. 30

Spiritual Paternity

Call no man your father upon the earth: for one is your Father, which is in heaven.

St Matthew, XXIII. 9

It is that prescient-spiritual-Self (*prajñâtman*) that grasps and erects the flesh.

Kaushîtaki Upanishad, III. 3

It is the spirit that quickeneth; the flesh profiteth nothing.

St John, VI. 63

From the Sun, as the universal father, proceeds the quickening principle in nature.

Ohiyesa

Symb. 317

Man is begotten by man and by the sun as well.

Aristotle (*Physics*, II. 2. 194 b)

When the time comes for the embryo to receive the spirit, at that time the sun becomes its helper.

Rûmî

The sun is not only the author of visibility in all visible things, but of generation and nourishment and growth, though he himself is not generation.

Plato (*Republic*, 509 B)

When the father thus emits him as seed into the womb, it is really the Sun that emits him as seed into the womb.

Jaiminîya Upaniṣad Brâhmaṇa, III. x. 4

Light is the progenitive power.

Taittirîya Saṁhitâ, VII. 1.1.1

Allâh created you from dust, then from a little fluid, then He made you pairs (male and female). No female beareth or bringeth forth save with His knowledge.

Qur'ân, XXXV. 11

Are not the parents, as it were, concomitant causes only, while (the divine) Nature is the highest, elder and true cause of the begetting of children?

Philo

Reality 773

Now God, though He is absolutely immaterial, can alone by His own power produce matter by creation; and so He alone can produce a form in matter, without the aid of any preceding material form. . . . Therefore, as no pre-existing body had been formed, through whose power another body of the same species could be generated, the first human body was of necessity made immediately by God.

St Thomas Aquinas

I will pour my spirit upon thy seed.

Isaiah, XLIV. 3

The power of the soul, which is in the semen through the Spirit enclosed therein, fashions the body.

St Thomas Aquinas

On the conjunction of three things there is descent into the womb. As to this, there must be coitus of the parents, it must be the mother's season, and the gandharva must be present. For so long there is conception.

Majjhima-nikâya, I. 265

The power of generation belongs to God.

St Thomas Aquinas

Self-Revelation

Knowl. 761

To show forth his wisdom has He made all things.

Hermes

Divine Wisdom created the world in order that all things in His Knowledge should be revealed.

Rûmî

'I was a hidden treasure; I wished to be known; therefore I created the world.'

Muhammad (*ḥadîth qudsî*)

Contem. 547

I was among his hidden treasures.
From Nothing he called me forth.

Nahmanides

Introd. *Void*

The goodness of God breaking forth into a desire to communicate good was the cause and the beginning of the creation.

William Law

Love 618

For God has not brought forth the creation, that he should be thereby perfect, but for his own manifestation, viz. for the great joy and glory; not that this joy first began with the creation, no, for it was from eternity in the great mystery, yet only as a spiritual melody and sport in itself. The creation is the same sport out of himself, which he melodizes: and it is even as a great harmony of manifold instruments which are all tuned into one harmony.

Boehme

Beauty 679 Supra 33 *Beauty* 689

God created the soul according to his own most perfect nature that she might be the bride of his only-begotten Son. He knowing this full well decided to go forth out of the private chamber of his eternal Fatherhood where he has slept for aye and be proclaimed abroad while inwardly abiding in the first beginning of his primitive light-nature. . . . He proceeded out of the supreme in order to go in again accompanied by his bride and show her the hidden mystery of his secret Godhead, where he is at peace with himself and with all creatures.

Eckhart

Center 841

God (*al-ḥaqq*=the Truth) wished to see the essences of His most perfect Names, which number could never exhaust,—or if thou preferest, thou canst equally say: God wished to see His own essence—in a global object, which being endowed with existence, recapitulates the whole divine order,—and in this way to manifest His mystery to Himself.

Ibn 'Arabî

Man alone contains within himself as many species as exist on earth.

Boehme

Holiness 924 Supra 42

All the scriptures have been written, God has made the world and all angelic nature in order that God might be born in the soul and the soul be born into God. All cereal nature means wheat, all metal nature means gold, all generation means man. As one philosopher observes, 'We can find no animal without some likeness to man.'

Eckhart

Charity 608

'Only for my service have I manifested thee. If I reveal the secret of this, it is for my intercourse; and if I approach thee, it is for my companionship. I have not manifested thee to continue in that which veils thee from Me, nor have I built thee and fashioned thee to advance and recede in that which divides thee from my intercourse.'

Niffarî

Judgment 250

Center 828

The Image of God is the most perfect creature. Since there cannot be two Gods the utmost endeavour of Almighty Power is the Image of God. It is no blasphemy to say that God cannot make a God: the greatest thing that He can make is His Image: a most perfect creature, to enjoy the most perfect treasures, in the most perfect manner. A creature endued with the most divine and perfect powers: . . . able to see all eternity with all its objects, and as a mirror to contain all that it seeth: able to love all it contains, and as a Sun to shine upon its loves.

Thomas Traherne

Metanoia 480

The Great Self, having two natures, proceeds with intent to experience both the true and the false.

Maitri Upanishad, VII. xi. 8

While God is indeed one, His highest and chiefest powers are two, even goodness and sovereignty. Through His goodness He begat all that is, through His sovereignty He rules what He has begotten.

Philo

God cannot know himself without me.

Eckhart

Man is the link between God and Nature. . . . As God has descended into man, so man must ascend to God.

Jîlî

He hath brought me forth His son in the image of His eternal fatherhood, that I should also be a father and bring forth Him.

Eckhart

Jesus said: If the flesh has come into existence because of the spirit, it is a marvel; but if the spirit has come into existence because of the body, it is a marvel of marvels. But I marvel at how this great wealth has made its home in this poverty.

The Gospel according to Thomas, Log. 29

BOOK ONE: JUSTICE–FEAR–ACTION

PART I: *Sacrifice – Death*

SEPARATION–SIN

Why dost thou call me good? None is good but God alone.

St Luke, XVIII. 19 (Douay)

There is no sin greater than that of existence itself.

Muhammad[1]

This then, which is one and simple by nature, man's wickedness divideth, and while he endeavoureth to obtain part of that which hath no parts, he neither getteth a part, which is none, nor the whole, which he seeketh not after.

Boethius (*Consolat. Philosoph.*, III. ix)

Taoism regards the actual dichotomy between man and his primordial nature in terms of a *disequilibrium*. Vedanta starts from the perspective of *illusion*, while Buddhism speaks of the same thing in terms of *ignorance*. Judeo-Christianity teaches that man is in a state of *fall*, whereas Islam describes it from the viewpoint of *rebellion*.

'If we say that we have not sinned, we make Him a liar, and His word is not in us' (*I. John*, I. 10). 'Manifestation by definition implies imperfection, as the Infinite by definition implies manifestation; this ternary "Infinite, manifestation, imperfection", constitutes the explanatory formula for all that can seem "problematic" to the human mind in the vicissitudes of existence' (Schuon: *De l'Unité transcendante*, p. 66).[2] 'Deem not strange the occurrence of afflictions as long as thou art in this *perishable abode*, for verily it has begotten nothing except what merits its appellation—and inevitable is this designation' (Ibn 'Aṭâ'illâh: *Ḥikam*, no. 34). Likewise Boethius: 'Thou hast yielded thyself to fortune's sway; thou must be content with the conditions of thy mistress' (*Consolat. Philosoph.*, II. i). No individual as such in time and space is free from the conditions thereof. 'The man who has found reality, as well as the man who is still in the coils of the phenomenal, is like one travelling over a flooded road' (Hônen, p. 610); bearing in mind, however, 'that all things work together for good to them that love God, to them who are the called according to his purpose' (*Romans*, VIII. 28). 'Therefore if thou suffer persecution, wretchedness, and other dis-eases, thou hast that which accords to the place in the which thou dwellest' (Richard

Introd. *Creation*

Illusion 85

Suff. 118

[1] *Études Traditionnelles*, Paris, 1950, p. 109.

[2] 'The question: "Why does evil exist?" really comes down to the question of knowing why there is existence; the serpent is found in Paradise because Paradise exists. Paradise without the serpent would be God' (Schuon: *Études Traditionnelles*, 1956, p. 203). For a detailed metaphysical exposition of the subject, cf. Guénon: 'Le Démiurge' (*Études Traditionnelles*, 1951, p. 145 ff.).

Rolle: *The Fire of Love,* I. viii). 'It is not the world then that deceives men,' says Hermes ('De Castigatione Animae'; *Hermetica,* IV, p. 289); 'but men deceive themselves, and so bring themselves to ruin. They think their happiness consists in the goods which this world gives, and think that these goods will last for ever, forgetting that life in this world is an alternation of good and bad.'

'It must needs be that offences come; but woe to that man by whom the offence cometh!' (*Matt.,* XVIII. 7). Once the doctrine of 'original sin' is admitted,—that man is in a state of disequilibrium, fall, illusion, ignorance, or insubordination, then one is in a position to recognize the reality of sin or evil, and the necessity of responding adequately to a situation which does in fact exist. 'Ignorance is much more than mere lack of information on this or that subject. It includes every kind of sin against the Light, not only false beliefs, but unawareness, loose thinking, woolly-mindedness, obscurantism, and above all, indifference to knowledge, neglect of the duty of trying to be truthful and intelligent; a life organized in such a manner as to produce constant distractions, dishonest stifling of doubts, doubt as to the necessity of seeking knowledge at all, neglect of opportunities of listening to those who have a doctrine to teach, all these things fall within the scope of Ignorance' (Marco Pallis: *Peaks and Lamas,* 3rd ed., London, 1942, p. 157). Nineteenth century moralism derives from the sense of sin which still lingers 'biologically' when intellectually the comprehension has been lost; in the twentieth century it is the sense of sin itself (and *a fortiori* the sense of virtue) which practically disappears, along with the ability to envisage an ineluctable cause and effect sequence between man's actions and the state of his environment macrocosmically, and his own future states microcosmically.

Introd. *Beauty*

'Each transgression must be considered as expressing on the part of the agent the lack of a positive quality, such as wisdom, strength, or purity. Now if each positive quality relates to a divine aspect, the absence of such a quality must equally relate, if not to another divine aspect from which it proceeds though very indirectly, then at least to a cosmic center either luciferian or satanic in nature—a center which is the direct source of the negative quality, and which illusorily opposes itself to the divine aspect that it denies. Vice lives by the regular and somewhat rhythmic communication with the obscure center which determines its nature, and which, like an invisible vampire, attracts, clasps and engulfs the being in a state of transgression and disequilibrium. If it were not thus, a simple infraction would remain but an isolated case; but every infraction is by definition a precedent and establishes contact with a tenebrous center[1]—and this again throws light on the necessity for rites of purification, which have precisely the effect of disrupting such contacts and of re-establishing communication with the divine aspect, of which the transgression—like its cosmic center—has been the negation' (Schuon: *L'Oeil du Coeur,* pp. 114–115).

Introd. *Heterod.*

Passages are cited below giving credence to the possibility that sin might be turned to spiritual profit, but the fundamental key to this delicate point is the question of *volition* ('there is nothing unclean of itself: but to him that esteemeth any thing to be unclean, to him it is unclean'; *Romans,* XIV. 14), since all sin resides in the will. Evil as such is always

Infra 76

[1] Cf. Hermes: 'The reckless vehemence of irrational impulse is indivisible' (*Hermetica,* I. 247): and he explains that this adhesive unity in sins is the reason 'that they all depart together', when they do depart.

evil: 'Shall evil be recompensed for good?' (*Jeremiah,* XVIII. 20);[1] 'Woe unto them that call evil good, and good evil; that put darkness for light, and light for darkness; that put bitter for sweet, and sweet for bitter!' (*Isaiah,* V. 20). 'Can there indeed be a Law of the Buddha which encourages men to evil, and hinders them from pursuing the good? The thing is unthinkable' (Hônen, p. 350). When sin really is in question, the criterion will center upon the *contrition* engendered or knowledge gained, and its power to effect a *lasting* cure and true 'metanoia'; for other cases of apparent sin, where the intention is pure and the act conformable with God's will, the 'sin' is in the appearance only, as will be born out by the positive fruits stemming from the action.

[1] 'Sanctification through sin' was the heresy of Russia's *Klystis* sect, a doctrine made notorious by Rasputin. 'What shall we say then? Shall we continue in sin, that grace may abound? God forbid. How shall we, that are dead to sin, live any longer therein?' (*Romans,* VI. 1, 2). The antinomianism of a Sabbatai Zevi belongs to the same order of ideas. G. G. Scholem calls the doctrines of Jacob Frank 'a veritable *religious myth of nihilism*' (*Jewish Mysticism,* p. 316).

God Alone Is Good

God alone is good; all other things are incapable of containing such a thing as the Good.

Hermes

God alone is good and perfect.

Ananda Moyî

The seed of all mankind was in the loins of fallen Adam.

William Law

Man is sin. . . . The body-thought brings out the idea of sin. The birth of thought is itself sin.

Sri Ramana Maharshi

Humility 191

All our righteousnesses are as filthy rags.

Isaiah, LXIV. 6

Go, do not commit foulness, for even our fair deeds appear foul in the sight of our beauteous Loved One.

Rûmî

Judgment 242

If I justify myself, mine own mouth shall condemn me: if I say, I am perfect, it shall also prove me perverse.

Job, IX. 20

The very fact of being born is sufficient to provoke Thy justice and chastisement.

Calderón

Grace 552

No man can work well, and love God, and be chaste, except God give it to him.

Richard Rolle

The fact that men after having lived evilly should have renounced the worldly life and become saints worthy of the veneration of the world showeth the virtue of the Holy *Dharma*.

Gampopa

The Good cannot be in things that come into being, but only in that which is without beginning.

Yet as participation in all (the ideal archetypes of things) is distributed in the world of matter, so also participation in the Good. And in this way the Kosmos too is good.

Hermes

Beauty 670

Whatever of good befalleth thee, it is from Allâh, and whatever of ill befalleth thee, it is from thyself.

Qur'ân, IV. 79

Introd. *Conform.*

God's Mercy

Even if the most wicked worships Me with undivided devotion, he should be regarded as good, for he is rightly resolved.

Very soon he becomes a righteous soul and attains to eternal peace. Know thou, O son of Kunti, that my devotee never perishes.

Bhagavad-Gîtâ, IX. 30, 31

Conform. 170

Say: O My slaves who have been wasteful with their lives! Do not despair of the mercy of Allâh; verily, Allâh forgiveth the entirety of sins. Lo! He ever is the Forgiving, the Merciful.

Qur'ân, XXXIX. 53

Faith 514

All the wrong and the reproach that is brought on God by sin he gladly bears and has borne many a year in order that mortals might arrive at a lively understanding of his love and the more to rouse their love and gratitude and fan the flame of their devotion—the normal and proper reaction from their sin. For this reason God is willing to bear the brunt of sins and often winks at them, mostly sending them to people for whom he has provided some high destiny.[1]

Eckhart

Love 618

The times of this ignorance God winked at; but now commandeth all men every where to repent.

Acts, XVII. 30

God pardons men's sins out of an eternal design of destroying them; and whenever the sentence of death is taken off from a sinner, it is at the same time denounced against his sins.

John Smith the Platonist

[1] 'God sometimes permits weaknesses in order to be able then to raise up, by means of the contrast between these accidental infirmities and the essential being, virtues all the more profound' (Schuon: *Perspectives spirituelles,* p. 240).

For the merciful, the repentance of the offender is a sufficient advocate.

Hermes

Supra 56
Infra 61

By Him Who holds my soul, if ye did not sin, verily would Allâh do away with you, and bring forth a people who sin and who ask forgiveness of Allâh, and them would He forgive.

Muhammad

Beauty 689

All the evils of contrariety and disorder in fallen nature are only as so many materials in the hands of infinite love and wisdom, all made to work in their different ways as far as is possible to one and the same end, *viz.*, to turn temporal evil into eternal good.

William Law

Those saints, which God loves best,
The Devill tempts not least.

Robert Herrick

Infra 62

If the fierceness were not, there could be no mobility.

Boehme

Judgment 266

Strength and fire in the divine nature are nothing else but the strength and flame of love, and never can be anything else; but in the creature strength and fire may be separated from love, and then they are become an evil, they are wrath and darkness and all mischief.

William Law

Suff. 120

My strength is made perfect in weakness.

II. Corinthians, XII. 9

Metanoia 495

The more grievous a man's sins seem to him, the readier God is to forgive them, to enter the soul and drive them out; for everyone is most diligent in getting rid of what is most disagreeable to him. The more in number a man's sins are, the greater they are, the more immeasurably glad God is to forgive them. The more they irk him, the quicker he is about it. The sooner divine repentance reaches up to God, the sooner the sins are swallowed up in the abyss of God—as quickly as I can shut my eyes. They are annihilated as if they had never happened, if only the repentance be whole.[1]

Eckhart

Humility 199

He who is truly penitent and really sorry shall receive pardon without doubt or delay. The prayer that is made by a contrite and humble heart is quickly granted, a heart contrite by fear and humbled by sorrow.

Richard of Saint-Victor

[1] 'The best man can make his confession and attain purification in a single moment' (Coates and Ishizuka: *Hônen*, p. 410. note 2. commenting on Zendô's four-fold rule for practising the Nembutsu).

'To whosoever is sorry for sin I forgive it and whoso yet laments his sin, to him do I give My Grace. Whoso grieves for sin so that he would rather die than sin again, shall after this life be condemned to no further punishment even should he commit daily sins.'

Mechthild of Magdeburg

However sinful a person may be, if he would stop wailing inconsolably, 'Alas! I am a sinner, how shall I attain Salvation?' and casting away even the thought that he is a sinner, if he would zealously carry on meditation on the Self, he would most assuredly get reformed.

Sri Ramana Maharshi

When God loves a man, sin shall not hurt him.

Muhammad

Metanoia 493

Warwick: My gracious lord, you look beyond him quite:
The prince but studies his companions
Like a strange tongue, wherein, to gain the language,
'Tis needful that the most immodest word
Be look'd upon, and learn'd; which once attain'd,
Your highness knows, comes to no further use
But to be known and hated. So, like gross terms,
The prince will in the perfectness of time
Cast off his followers; and their memory
Shall as a pattern or a measure live,
By which his Grace must mete the lives of others,
Turning past evils to advantages.
King Henry: 'Tis seldom when the bee doth leave her comb
In the dead carrion.

Shakespeare (*Henry IV, Pt. 2,* IV. iv. 68)

Pilg. 366

Infra 62

Holiness 921

Supra Introd.

Also God shewed that sin shall be no shame to man, but worship. For right as to every sin is answering a pain by truth, right so for every sin, to the same soul is given a bliss by love: right as diverse sins are punished with diverse pains according as they be grievous, right so shall they be rewarded with diverse joys in Heaven according as they have been painful and sorrowful to the soul in earth. For the soul that shall come to Heaven is precious to God, and the place so worshipful that the goodness of God suffereth never that soul to sin that shall come there without that the which sin shall be rewarded; and it is made known without end, and blissfully restored by overpassing worship.

For in this Sight mine understanding was lifted up into Heaven, and then God brought merrily to my mind David, and others in the Old Law without number; and in the New Law He brought to my mind first Mary Magdalene, Peter and Paul . . . and others also without number.

Julian of Norwich

On no account let anyone suppose that he is far from God because of his infirmities or faults or for any other reason. If at any time thy great shortcomings make an outcast of thee and thou canst not take thyself as being nigh to God, take it then at any rate that he is nigh to thee, for it is most mischievous to set God at a distance. Man goes far away or near but God never goes far off; he is always standing close at hand, and even if he cannot stay within he goes no further than the door.

Faith 505 *Center* 841

Eckhart

Do not be troubled about whether your heart is good or bad, or your sin light or grievous. Only determine in your heart that you will be born into the Pure Land, and so repeat the '*Namu Amida Butsu*' with your lips, and let the conviction accompany the sound of your voice, that you will of a certainty be born into the Pure Land. Then, according to your determined faith, that *karma* will certainly be produced which will result in your birth into that land.... It all depends upon your own mental act, which makes everything easy. If in your mind you waver, you will end in wavering. But if you make up your mind to it, that settles your destiny to be born into the Pure Land.

P. State 583 *Inv.* 1009 *Action* 329

Hônen

'The sign of My forgiveness in the affliction is, that I make it a means to a knowledge.'

Suff. 128

Niffarî

Mayhap He opens to thee the door of obedience and does not open to thee the door of acceptance; and mayhap He ordains for thee sin, and it is an occasion for realization.

Ibn 'Aṭâ'illâh

So great is the Goodness of God that he has put nothing in this life, in whatever state one finds oneself, which is an obstacle to salvation.

Holiness 935

St Catherine of Siena

Though you be as a Tree twice dead, & pull'd up by the rootes, Hee can make you to live, & to grow, being Himselfe a Roote, and the Power of an Endless Life in you.... If you fall as a Man, ly not in yo[r] filth as a beast: Defend not evill, give not yo[r]self up to a rage in it, to an enmity against the Good, to a despair of ever being better, like a Devill.

Wr. Side 448

Peter Sterry

If thy dispersion be from one end of heaven to the other, the Lord will gather thee thence.

Faith 514

Deuteronomy, XXX. 4

O Lord, if Thou holdest me responsible for my sins
I shall cling to Thee for Thy Grace.
I with my sin, am an insignificant atom.
Thy Grace is resplendent as the Sun.

Humility 191

Ansârî

God's Justice

Forgiveness is only incumbent on Allâh toward those who do evil in ignorance, and then turn quickly in repentance. These are they toward whom Allâh turneth.

Qur'ân, IV. 17

Metanoia 493

Envy not the glory and riches of a sinner: for thou knowest not what his ruin shall be.

Ecclesiasticus, IX. 16

Judgment 250
Action 329

'But,' one says, 'I see the noble and good perishing of hunger and cold.'
Well, and do you not see those who are not noble and good perishing of luxury and ostentation and vulgarity?

Epictetus

Suff. 120

Say not: I have sinned, and what harm hath befallen me? for the Most High is a patient rewarder.[1]

Ecclesiasticus, V. 4

Let not their conduct grieve thee, who run easily to disbelief, for lo! they injure Allâh not at all. It is Allâh's Will to assign them no portion in the Hereafter, and theirs will be an awful doom. . . .

And let not those who disbelieve imagine that the rein We give them bodeth good unto their souls. We only give them rein that they may grow in sinfulness. And theirs will be a shameful doom.

Qur'ân, III. 176, 178

Neither is it possible to conceal fire in a garment, nor a base deviation from rectitude in time.

Pythagoric Saying

The place where Dharma and Adharma are done is one. The place where their fruits are experienced, Heaven or Hell, is elsewhere.

Sri Chandrasekhara Bhâratî Swâmigal

Action 335

Forsake the outwardness of sin and the inwardness thereof. Lo! those who garner sin will be awarded that which they have earned.

Qur'ân, VI. 120

Disorder also is subject to the Master, but he has not yet imposed order upon it.

Hermes

Introds. *Pilg. & Realiz.*

Reverence not thy neighbour in his fall.

Ecclesiasticus, IV. 27

[1] Both this presumption on the one hand and despair (as in the citations in the preceding section) on the other are classed as 'sins against the Holy Ghost'.

The Extraction of Poison with Poison

Creation 33
P. State 579

Nature sports with Nature; and Nature contains Nature; and Nature knows how to surmount Nature.

Hermetic Formula

Similia similibus curantur ('Like is cured by like').

Paracelsus

M.M. 978

Just as water that has entered the ear may be removed by water and just as a thorn may be removed by a thorn, so those who know how, remove passion by means of passion itself. Just as a washerman removes the grime from a garment by means of grime, so the wise man renders himself free of impurity by means of impurity itself.

Cittaviśuddhiprakaraṇa

Then of the venome handled thus a medicine I did make;
Which venome kills and saveth such as venome chance to take.
Glory be to him the graunter of such secret wayes,
Dominion, and Honour, both with Worship, and with Prayse.

George Ripley

Shiva, the Terrible God, has taught in the doctrine of the left hand that spiritual progress is only possible with the aid of the very things which are the cause of man's fall.

Kulârnava Tantra

O! mickle is the powerful grace that lies
In herbs, plants, stones, and their true qualities:
For nought so vile that on the earth doth live
But to the earth some special good doth give,
Nor aught so good but strain'd from that fair use
Revolts from true birth, stumbling on abuse:
Virtue itself turns vice, being misapplied,
And vice sometime's by action dignified.

Creation 42

Shakespeare (*Romeo and Juliet*, II. iii. 15)

Obscuring passions, being the means of reminding one of Divine Wisdom (which giveth deliverance from them), are not to be avoided (if rightly used to . . . reach disillusionment).

Gampopa

There is no evill that we do commit,
But hath th' extraction of some good from it:
As when we sin: God, the great *Chymist*, thence
Drawes out th' *Elixar* of true penitence.

Metanoia 484

Robert Herrick

There cannot be self-restraint in the absence of desire: when there is no adversary, what avails thy courage?

Hark, do not castrate thyself, do not become a monk: chastity depends on the existence of lust.

Beauty 676

Rûmî

He thought within himself that this world was far better than Paradise had men eyes to see its glory, and their advantages. For the very miseries and sins and offences that are in it are the materials of his joy and triumph and glory. So that he is to learn a diviner art that will now be happy, and that is like a royal chemist to reign among poisons, to turn scorpions into fishes, weeds into flowers, bruises into ornaments, poisons into cordials. And he that cannot learn this art, of extracting good out of evil, is to be accounted nothing. Heretofore, to enjoy beauties, and be grateful for benefits was all the art that was required to felicity, but now a man must, like a God, bring Light out of Darkness, and order out of confusion. Which we are taught to do by His wisdom, that ruleth in the midst of storms and tempests.

P. State 583

Wr. Side 464

Thomas Traherne

That man is more praiseworthy who, having an evil nature, controls and governs himself in opposition to the impulse of that nature, than he who, having a good disposition, leads a good life, and should he go astray, returns to the right; as it is more praiseworthy to manage a bad horse than one that is not vicious.

Pilg. 379

Dante (*Il Convito,* III. viii. 9)

Ambrosia can be extracted even from poison; elegant speech, even from a child; good conduct, even from an enemy; gold, even from impurity.

Mânava-dharma-śâstra, II. 239

'Lest I should be exalted,' he says (*II. Cor.* xii, 7), 'there was given me a sting of my flesh, an angel of Satan.' O poison, which is not cured save by poison! 'There was given me a sting of my flesh, an angel of Satan, to buffet me!' The head was beaten lest the head should be exalted. O antidote, which is made, as it were, from a serpent and therefore is called *theriaca*! For that serpent persuaded to pride. 'Eat, and you shall be as gods' (*Gen.* iii. 5). This is the persuasion of pride; whereby the devil fell, thereby he cast down. Justly therefore the serpent's poison is healed by the serpent.

Heterod. 428

St Augustine

Are the evils in the Universe necessary because it is of later origin than the Higher Sphere?

Perhaps rather because without evil the All would be incomplete. For most or even all forms of evil serve the Universe—much as the poisonous snake has its use—though in most cases their function is unknown. Vice itself has many useful sides: it brings about much that is beautiful, in artistic creations for example, and it stirs us to thoughtful living, not allowing us to drowse in security.

Supra *Introd.*

Plotinus

Neighbour: Why has God created wicked people?

Creation 33 Master: That is His will, His play. In His mâyâ there exists avidyâ as well as vidyâ. Darkness is needed too. It reveals all the more the glory of light. There is no doubt that anger, lust, and greed are evils. Why, then, has God created them? In order to create saints. . . . Again, see how His whole play of creation is perpetuated through lust.

Sri Ramakrishna

Thou art the first knave that e'er made a duke.

Shakespeare (*Measure for Measure*, v. i. 357)

The Mote in the Eye

Illusion 85 As long as the soul has not thrown off all her veils, however thin, she is unable to see God. Any medium, but a hair's-breadth, in betwixt the body and the soul stops actual union.

Eckhart

The truth is that you cannot attain God if you have even a trace of desire. Subtle is the *Realiz.* 890 way of dharma. If you are trying to thread a needle, you will not succeed if the thread has even a slight fibre sticking out.

Sri Ramakrishna

Although the sin which he had compassed was but a hair, yet that hair had grown in his eyes.

Adam was the eye of the Eternal Light: a hair in the eye is a great mountain.[1]

Rûmî

Renun. 160 That man who, having heard and touched, seen and eaten and smelled, neither rejoices nor ever is sad, he is to be known (as a man) who has conquered his senses.

Infra 77 But among all the senses, if one sense fails, by that his wisdom fails, as water (runs out) by one hole from a leather bag.

Mânava-dharma-śâstra, II. 98, 99

Supra Introd. He who conquers one passion, conquers many; and he who conquers many, conquers one.

Akaranga Sutra, I. i. 4

[1] These three passages help show how the contemplative can with perfect objectivity and justice regard himself as 'the worst of sinners'.

'No' is not necessarily 'No', nor is 'Yes' 'Yes';
But when you miss even a tenth of an inch, the difference widens up to one thousand miles;
When it is 'Yes', a young Naga girl in an instant attains Buddhahood,
When it is 'No', the most learned Zensho while alive falls into hell.

Yoka Daishi

Introd. *Judgment*

I am not happy till I return to God discarding every means of sin and all its brood together with all creatures.

Eckhart

Renun. 146

While believing that even the man who is so sinful that he has committed the ten evil deeds and the five deadly sins may be born into the Pure Land, as far as you are concerned, be not guilty even of the smallest sins.

Hônen

And Noah said: My Lord! Leave not one of the disbelievers in the land.
If Thou shouldst leave them, they will mislead Thy slaves and will beget none save lewd ingrates.

Qur'ân, LXXI. 26, 27

Judgment 242

There is but one way to worship God; it is to be devoid of evil.

Hermes

Snares of the Soul

This (elemental soul) (*bhûtâtman*), verily, is overcome by Nature's (*prakṛti*) qualities (*guṇa*).

Now, because of being overcome, he goes on to confusedness; because of confusedness, he sees not the blessed Lord (*prabhu*), the causer of action, who stands within oneself (*âtma-stha*). Borne along and defiled by the stream of qualities, unsteady, wavering, bewildered, full of desire, distracted, this one goes on to the state of self-conceit (*abhimânatva*). In thinking 'This is I' and 'That is mine', he binds himself with his self, as does a bird with a snare.

Maitri Upanishad, III. 2

Metanoia 480 *Illusion* 89 *Death* 220

As long as you were in the world of simple (incomposite) things, (O Soul), you had plenty of things to see and know; for you could see all the worlds displayed before you, pure and bright and shining; and down below, at the bottom of all, was situated the world

Creation 38

of things that come to be and cease to be,—a dark and murky world, having as little brightness in comparison with them as has a black stone that is seen amidst transparent water.

Well, you thought fit to enter this (dark) world, that you might find out by experience what sort of place it is; and having decided to do so, you forsook the high level of the simple things, and came down to the low level of the mixed things. And so, eagerly seeking the things you longed for, you departed (from your home above), and came into the world of things that come to be. In thus forsaking the world of simple things, and seeking after and desiring the world of composite (or concrete) things, you were like a bird that makes for the snare, to take the berry that is placed there, and finds its own life taken by the snare; or like a fish that wants to eat the fisherman's bait, and is itself eaten by the fisherman.

Infra 66

Hermes

At Kâmârpukur I have seen the mongoose living in its hole up in the wall. It feels snug there. Sometimes people tie a brick to its tail; then the pull of the brick makes it come out of its hole. Every time the mongoose tries to be comfortable inside the hole, it has to come out because of the pull of the brick. Such is the effect of brooding on worldly objects that it makes the yogi stray from the path of yoga.

Renun. 146

Sri Ramakrishna

A little Bird ty'd by the Leg with a String, often flutters and strives to raise itself: but still it is pull'd down to the Earth again: Thus a Soul fixt in a *Self-Principle,* may make attempts to Pray and Offer at the Bosom of God; but still it is snatch'd down by that String of *Self,* which ties it to the Ground.

Peter Sterry

Thou wert a favourite falcon, kept in captivity by an old woman:
When thou heard'st the falcon-drum thou didst fly away into the Void.

Illusion 108

Dîvâni Shamsi Tabrîz, XLVIII

Anatomy of the Fallen State

God created man incorruptible, and to the image of his own likeness he made him. But by the envy of the devil, death came into the world.

Wr. Side 441

Wisdom, II. 23, 24

Through his default he dwelt here but little; through his default into lamentation and anxiety changed honest laughter and sweet play.

Dante (*Purgatorio,* XXVIII. 94–96)

In the beginning of creation, the core of the Shekhinah was in the lower regions. And because the Shekhinah was below, heaven and earth were one and in perfect harmony. The wellsprings and the channels through which everything in the higher regions flows into the lower were still active, complete and unhindered, and thus God filled everything from above to below. But when Adam came and sinned, the order of things was turned into disorder, and the heavenly channels were broken.

Creation 38
Introd. *Flight*

Joseph Gikatila

For though God created man at the beginning in His own image, and made him more glorious and perfect than other creatures, and breathed into him a living and immortal soul, yet by the fall the image of God was defaced, and man was changed into the very reverse of what God had intended that he should be.

The Sophic Hydrolith

The soul . . . by reason of lust had become tne principal accomplice in her own captivity.

Plato (*Phaedo*, 82 E)

The origin and cause of thoughts lies in the splitting up, by man's transgression, of his single and simple memory, which has thus lost the memory of God and, becoming multiple instead of simple, and varied instead of single, has fallen a prey to its own forces.

Knowl. 775
P. State 573

St Gregory of Sinai

This consciousness (*citta*) is luminous, but it is defiled by adventitious defilements.

Death 220

Anguttara-nikâya, I. 10

Thus saith Saint John in the gospel. *Lux in tenebris lucet, et tenebrae eam non comprehenderunt.* That is, the light of grace shineth in murkness, that is to men's hearts that are murk through sin; but the murknesses take it not.[1]

Grace passim

Walter Hilton

The first Light which shined in my Infancy in its primitive and innocent clarity was totally eclipsed: insomuch that I was fain to learn all again. If you ask me how it was eclipsed? Truly by the customs and manners of men, which like contrary winds blew it out: by an innumerable company of other objects, rude, vulgar, and worthless things, that like so many loads of earth and dung did overwhelm and bury it: by the impetuous torrent of wrong desires in all others whom I saw or knew that carried me away and alienated me from it: by a whole sea of other matters and concernments that covered and drowned it: finally by the evil influence of a bad education that did not foster and cherish it. All men's thoughts and words were about other matters. They all prized new things which I did not dream of. I was a stranger and unacquainted with them; I was little and reverenced their authority; I was weak, and easily guided by their example: ambitious also, and desirous to approve myself unto them. And finding no one syllable in any man's mouth of those things, by degrees they vanished, my thoughts (as indeed what is more fleeting than a thought?)

Knowl. 734

[1] *St John*, I. 5.

were blotted out; and at last all the celestial, great, and stable treasures to which I was born, as wholly forgotten, as if they had never been.

Creation 42

Thomas Traherne

Suppose that gold belonging to a man on his travels
Had fallen into a place full of stinking dirt.
As it is indestructible by nature, it would stay there
For many hundreds of years.

A deity, with a pure heavenly eye, would see it there,
And say to people:
'When I have cleansed this gold, the most precious substance of all,
I will bring it back to its precious state.'

Ratnagotravibhâga, I. 108, 109

A Soul becomes ugly—by something foisted upon it, by sinking itself into the alien, by a fall, a descent into body, into Matter. The dishonour of the Soul is in its ceasing to be clean and apart. Gold is degraded when it is mixed with earthy particles; if these be worked out, the gold is left and is beautiful, isolated from all that is foreign, gold with gold alone. And so the Soul; let it be but cleared of the desires that come by its too intimate converse with the body, emancipated from all the passions, purged of all that embodiment has thrust upon it, withdrawn, a solitary, to itself again—in that moment the ugliness that came only from the alien is stripped away.[1]

Suff. 124

Plotinus

Creation 42

The eternal, blissful, and natural state has been smothered by this life of ignorance.

Sri Ramana Maharshi

Thy mind, thy reason, and the love of thy soul are so mickle set in beholding and in love of earthly things, that of ghostly things thou feelest right little. Thou feelest no reforming in thy self, but thou art so belapped with this black image of sin for aught that thou mayest do, that upon what side thou turnest thee thou feelest thyself defouled and spotted with fleshly stirrings of this foul image.

Walter Hilton

Such harmony is in immortal souls;
But, whilst this muddy vesture of decay
Doth grossly close it in, we cannot hear it.

Shakespeare (*Merchant of Venice*, v. i. 63)

Whether thou sleepest or wakest God goes on with his work. That we have no sense of it is because our tongue is furred with the slime of creatures and possesses not the salt of divine affection.

Eckhart

[1] 'Sin is to the spirit what dirt is to the physical body' (K. V. Rangaswami Aiyangar: *Some Aspects of the Hindu View of Life According to Dharmaśâstra*, Baroda, 1952, p. 93).

There was a time when . . . we beheld the beatific vision and were initiated into a mystery which may be truly called most blessed, celebrated by us in our state of innocence, before we had any experience of evils to come, when we were admitted to the sight of apparitions innocent and simple and calm and happy, which we beheld shining in pure light, pure ourselves and not yet enshrined in that living tomb which we carry about, now that we are imprisoned in the body, like an oyster in his shell.

P. State 583

Plato (*Phaedrus,* 250 C)

The ancient theologists and priests testify that the soul is conjoined to the body through a certain punishment, and that it is buried in this body as in a sepulchre.

Judgment 250

Philolaus

Bright and luminous as you are, O Soul, by your own nature, you went to the world of darkness, and engaged in combat with it; and the world of darkness obscured your light, and encompassed you with darkness, and blinded you, and made you lose sight of all that you had seen, and forget all that you had known; and in the end, you were captured and held prisoner.

Hermes

In the supreame, & inward part of itselfe it (the Soul) conteineth all fformes of things in their Originall, Eternall, Glorious Truths, & substances. In it's lower, & more outward part, which is still itselfe, & within itselfe it bringeth forth itselfe sportingly into a shadowie ffigure of itselfe & in this shadowie ffigure into innumerable shadowes, & ffigures of those glorious fformes in it's superior part. This shadowie ffigure is that, which wee call this world, & the body. The Soule often looking upon this, like Narcissus upon his owne fface in the ffountaine, forgets it to be itselfe, forgets that itselfe is the fface, the shadow, & the ffountaine, so it falls into a fond Love of itselfe in it's owne shadowie ffigure of itselfe. So it languisheth, & dys becoming only a Shadow of itselfe, in which itselfe with all it's superior, and true Glories ly buried.

Creation 33
Symb. 306
Illusion 109

Peter Sterry

The soul, once turned towards matter, was enamoured of it, and burning with desire to enjoy corporeal pleasures, was thenceforward unwilling to detach itself from matter.

Al-Kâtibi

While keeping my physical frame I lost sight of my real self. Gazing at muddy water, I lost sight of the clear abyss.

Void 721

Chuang-tse (ch. XX)

Weeping I said: 'Present things with their false pleasure turned away my steps soon as your face was hidden.'

Dante (*Purgatorio,* XXXI. 34)

As fire appears to have small or great volume, and beginning and end, through being falsely identified with the burning wood, so does the Atman appear to take on the attributes of the body by dwelling within it.

Srimad Bhagavatam, XI. iv

All sins are contained in this one category, that one turns away from things divine and truly enduring, and turns towards those which are mutable and uncertain. And although the latter are rightly placed each in its order, and work out that beauty proper to them it is nevertheless the mark of a perverted and ungoverned mind to be in subjection to them as things to be pursued, when by the divine order and law it is set above them as things to be directed.

Beauty 689 Introd. *Conform.*

St Augustine

Our own soul, created wise and thoughtful in the image of God, having refused to know God, has become bestial, senseless and almost insane through delighting in material things. For habit is wont to alter nature and change its action in accordance with the direction of the will.

Renun. 152a

St Gregory of Sinai

Now I would have you know that the Godly-minded man employs his soul-powers in his outward man no more than his five senses really need it; and his interior man only has recourse to the five senses so far as it is guide and keeper to these five senses and can stop them being put to bestial uses as they so often are by those who live according to the baser appetites, as do the mindless beasts, and who deserve the name of beast rather than that of man. What surplus energy she has beyond what she expends on her five senses the soul bestows upon her inner man, and supposing he has toward some right high endeavour she will call in all the powers she has loaned to the five senses and then the man is said to be senseless and rapt away, his object being either some unintelligible form or some formless intelligible. Remember, God requires every spiritual man to love him with all the powers of his soul. 'Thou shalt love the Lord thy God with all thy heart,' he says. Some squander all their soul-powers on their outward man. Namely, those whose thoughts and feelings hinge on temporal goods, all unwitting of an inner man. And even as the virtuous man will now and then deprive his outward self of all the powers of the soul what time he is embarking on some high adventure, so bestial man will rob his inner self of all its soul-powers to expend them on his outer man.

Infra 77 *Ecstasy* 636 *Faith* 505

Eckhart

It is certainly true that all composite substances are liable to decomposition: that this decomposition, when it takes place in the animal world, is called death; and that the human body is a substance compounded of the four elements. But it is also true that the elements of Paradise, where man was created, are not subject to this law, seeing that they are most pure and incorruptible heavenly essences; and if man had remained in this pure and celestial region, his body would have been incapable of natural decay. Adam, however, in an evil day for our race, disobeyed his Creator, and straightway was driven forth to the beasts, into the world of corruptible elements which God had created for the beasts only. From that day forward his food was derived from perishable substances, and death began to work in his members. The pure elements of his creation were gradually mingled and infected with the corruptible elements of the outer world, and thus his body became more and more gross, and liable, through its grossness, to natural decay and death. The process of degeneration was, of course, slow in the case of Adam and his first descendants; but, as

Creation 38

time went on, the seed out of which men were generated became more and more infected with perishable elements. The continued use of corruptible food rendered their bodies more and more gross—and human life was soon shortened to a very brief span indeed.

Michael Sendivogius

Lowest and unhappiest must be judged those who have closed their eyes to the rays of the highest good shining everywhere so that they cannot see in that very light, outside of which nothing good is seen, how great an evil it is always to be without that thing without which any visible thing is evil.

Marsilio Ficino

Center 833 *Symb.* 306

From the time that the first man opened his ears to the voice of the Enemy, he became deaf thereby, and all we after him, so that we cannot hear or understand the sweet voice of the Eternal Word. Yet we know that the Eternal Word is still . . . unutterably nigh to us inwardly, in the very principle of our being. . . . And it is ever speaking in man; but he hears it not by reason of the sore deafness that has come upon him. . . . The cause whereof is that the Enemy has whispered in his ear, and he has listened to the voice, and hence has he grown deaf and dumb. What is this most hurtful whispering of the Enemy? It is every disorderly image or suggestion that starts up in thy mind, whether belonging to thy creature likings and wishes, or the world and the things thereof; whether it be thy wealth, reputation, friends or relations, or thy own flesh, or whatever it be that lays hold of thy fancy, making thee to like or do somewhat. . . . Not only worldly but also religious men are liable to this deafness, if they make the creature their idol and aim, and their hearts are possessed therewith. The Devil has marked this, and suggests to them the imaginations to which he finds them inclined. With some their ears are stopped up with their own inventions, and the daily routine of habit with which they go through certain outward acts.

Tauler

Center 841 *Creation* 42

Contem. 528

It is indeed nothing else that makes men question the immortality of their souls, so much as their own base and earthly loves, which first makes them wish their souls were not immortal, and then to think they are not.

John Smith the Platonist

Illusion 89 *Heterod.* 420

Hast thou seen him who maketh his desire his god, and Allâh sendeth him astray purposely, and sealeth up his hearing and his heart, and setteth on his sight a covering?

Qur'ân, XLV. 23

Introd. *Conform.*

When (the faculty of the soul) is overwhelmed by too much commerce with the flesh and occupied with the sensible soul of the body, (it) is not worthy to command the divine substances.

Cornelius Agrippa

Infra 77

To overcome sorrow and win happiness men wander in vain, for they have not sanctified their thought, the mysterious essence of holiness. Then I must keep my thought well

Introd. *Holiness*

governed and well guarded; what need is there of any vows save the vow to guard the thought? . . .

Center 816
Holy War 407
Knowl. 755
Judgment 261
Orthod. 288
Renun. 152b

The thief Heedlessness, waiting to escape the eye of remembrance, robs men of the righteousness they have gathered, and they come to an evil lot. The Passions, a band of robbers, seek a lodging, and when they have found it they rob us and destroy our good estate of life. Then let remembrance never withdraw from the portal of the spirit; and if it depart, let it be brought back by remembering the anguish of hell. Remembrance grows easily in happy obedient souls from the reverence raised by their teachers' lore and from dwelling with their masters. . . . The thought thus must be kept ever under watch; I must always be as if without carnal sense, like a thing of wood.

Śânti-deva

Supra 65

He who looks toward the creation fails,
And he who returns to the Truth prevails.

Hujwîrî

Depart out of the earthly matter that encompasses thee: escape, man, from the foul prison-house thy body, with all thy might and main, and from the pleasures and lusts that act as its jailers; every terror that can vex and hurt them, leave none of them unused; menace the enemy with them all united and combined. Depart also out of sense-perception thy kin. For at present thou hast made a loan of thyself to each sense, and art become the property of others, a portion of the goods of those who have borrowed thee, and hast thrown away the good thing that was thine own.

Supra 64

Philo

Conform. 166
Introd. *Heterod.*

Sin is nothing better than a brat of darkness and deformity; it hath no other extraction or pedigree than may be derived from those unclean spirits that are nestled in hell. All men in reality converse either with God or with the devil, and walk in the confines either of heaven or of hell: they have their fellowship either with the Father and the Son, as St John speaks;[1] or else with the apostate and evil angels.

John Smith the Platonist

Behold My heart surrounded with the thorns which ungrateful men place therein at every moment by their blasphemies and ingratitude.

Apparition of the Virgin Mary to Lucy dos Santos

As a crop of grain overgrown by weeds sinks under disease, and thrives not, so a scion of the Buddha, if overcome by sin, cannot grow in grace.

Śânti-deva

But the wicked are like the troubled sea, when it cannot rest, whose waters cast up mire and dirt.

Isaiah, LVII. 20

[1] *I. John*, I. 3.

The misery of your fall ariseth naturally from the greatness of your sin. For to sin against infinite love, is to make oneself infinitely deformed: to be infinitely deformed, is to be infinitely odious in His eyes whose love of beauty is the hatred of deformity.

Thomas Traherne

Some learned people say that it is human to sin. In all temptations of my sinful body, all feelings of my heart, all understanding of my senses and all nobility of my soul, I could find none other but that it is devilish that man sins.

Mechthild of Magdeburg

Creation 42

Cast away your existence entirely,
For it is nought but weeds and refuse.

Shabistarî

Death 206

Those filthy lusts and corruptions which men foment and entertain in their minds, are the noisome vapours that ascend out of the bottomless pit; they are the thick mists and fogs of hellish darkness arising in their souls, as a preface and introduction of hell and death within. Where we find uncleanness, intemperance, covetousness, or any such impure or unhallowed behaviour, we may say, Here Satan's throne is.

Wr. Side 448

This sinful and corrupt nature being the true issue of hell itself, is continually dragging down men's souls thither. All sin and wickedness in man's spirit hath the central force and energy of hell in it, and is perpetually pressing down towards it, as towards its own place. There needs no fatal necessity or astral impulses to tumble wicked men down forcibly into hell: no, for sin itself, hastened by the mighty weight of its own nature, carries them down thither with the most swift and headlong motion.

Supra Introd.

John Smith the Platonist

O men, whither are you being swept away? You are drunken; you have drunk up the strong drink of ignorance; it has overpowered you, and now you are even vomiting it forth. Stand firm; turn sober; look upward with the eyes of the heart,—if you cannot all, yet those at least who can.

Center 816

This evil of ignorance floods all the land; its current sweeps along the soul which is penned up in the body, and prevents it from coming to anchor in the havens of salvation. Suffer not yourselves then to be borne along down stream by the strong current, but avail yourselves of a backflow, those of you who are able to reach the haven, and cast anchor there, and seek a guide to lead you to the door of the House of Knowledge. There you will find the bright light which is pure from darkness; there none is drunken, but all are sober, and they look up and see with the heart Him whose will it is that with the heart alone He should be seen. For He cannot be known by hearing, nor made known by speech; nor can He be seen with bodily eyes, but with mind and heart alone.

Pilg. 385
Orthod. 288
Knowl. 761
Center 833
M.M. 987

But first you must tear off this garment which you wear,—this cloak of darkness, this web of ignorance, this prop of evil, this bond of corruption,—this living death, this conscious corpse, this tomb you carry about with you,—this robber in the house, this enemy who hates the things you seek after, and grudges you the things which you desire. Such is the garment in which you have clothed yourself; and it grips you to itself and holds you down, that you may not look upward and behold the beauty of the Truth, and the Good that abides above, and hate the evil of this thing, discovering its ill designs against you. For it makes senseless what men deem to be their organs of sense, stuffing them up with the gross mass of matter, and cramming them with loathly pleasures, so that you may neither hear of the things you ought to hear of, nor see the things you ought to see.

Hermes

Wr. Side passim

Thou art violently carried away from grace: there is a devil haunts thee in the likeness of a fat old man; a tun of man is thy companion. Why dost thou converse with that trunk of humours, that bolting-hutch of beastliness, that swoln parcel of dropsies, that huge bombard of sack, that stuffed cloak-bag of guts, that roasted Manningtree ox with the pudding in his belly, that reverend vice, that grey iniquity, that father ruffian, that vanity in years? Wherein is he good but to taste sack and drink it? wherein neat and cleanly but to carve a capon and eat it? wherein cunning but in craft? wherein crafty but in villany? wherein villanous but in all things? wherein worthy but in nothing?

Shakespeare (*Henry IV, Pt. I,* II. iv. 497)

Judgment 261

Souls smell in Hades.

Heraclitus

Now when the spark of the love of God, or the divine light, was accordingly manifested in the soul, it presently saw itself with its will and works to be in hell, in the wrath of God, and found that it was a misshapen ugly monster in the divine presence and the kingdom of heaven; at which it was so affrighted, that it fell into the greatest anguish possible, for the judgement of God was manifested in it.

Boehme

Thou shalt loathe and be weary with all that thing that worketh in thy mind and in thy will, unless it be only God. For otherwise surely, whatsoever it be, it is betwixt thee and thy God. And no wonder if thou loathe and hate to think on thyself, when thou shalt always feel sin a foul stinking lump, thou knowest never what, betwixt thee and thy God: the which lump is none other thing than thyself. For thou shalt think it oned and congealed with the substance of thy being: yea, as it were without separation.

Humility 191
Supra 56

And therefore break down all knowing and feeling of all manner of creatures; but most busily of thyself. For on the knowing and the feeling of thyself hangeth the knowing and the feeling of all other creatures.

Death 220

The Cloud of Unknowing, XLIII

Pride

Dear brethren, rather than you should say or think yourselves to be different from or better than other men, I would that you should return to the world.

St Augustine

Humility 190

The Messiah (Peace be upon him!) said, 'O company of the disciples, how many lamps has the wind put out, and how many worshippers has self-conceit spoiled!'

Christ in Islâm

For they loved the praise of men more than the praise of God.

St John, XII. 43

Question: Which is the main pitfall a spiritual aspirant should be guarded against. when he is walking on the path of God-realization?
Answer: That is. spiritual pride due to a conceited consciousness of his progress on the path.

Swami Ramdas

Heterod. 423

The worst man is the one who sees himself as the best.

'Alî

Evil can have no beginning, but from pride; nor any end, but from humility.

William Law

The beginning of the pride of man is to fall off from God:
Because his heart is departed from him that made him: for pride is the beginning of all sin: he that holdeth it, shall be filled with maledictions, and it shall ruin him in the end.

Ecclesiasticus, x. 14, 15

Would you see the deepest root and iron strength of pride and self-adoration, you must enter into the dark chamber of man's fiery soul, where the light of God (which alone gives humility and meek submission to all created spirits) being extinguished by the death which Adam died, Satan, or which is the same thing, self-exaltation, became the strong man that kept possession of the house till a stronger than he should come upon him. . . . The fancied riches of parts, the glitter of genius, the flights of imagination, the glory of learning, and the self-conceited strength of natural reason: these are the strongholds of fallen nature, the master-builders of pride's temple in the heart of man, and which, as so many priests, keep up the daily worship of idol-self.

William Law

Knowl. 734

When the man in the soul, the intellect, is dead, unchecked evil prevails.

Eckhart

Metanoia 480
Introd. *Rev.*

His (leviathan's) scales are his pride, shut up together as with a close seal. . . .
Upon earth there is not his like, who is made without fear.
He beholdeth all high things: he is a king over all the children of pride.

Job, XLI. 15, 33, 34

Other vices fasten onto evil, to the end that it be done; pride alone fastens onto good, to the end that it perish.

St Augustine

Wr. Side passim

All the other vices flee God; pride alone rises up against Him.

Boethius

It is in respect to sinners that God designates Himself (in the *Torah*) as 'He who dwells in the midst of their impurities.' But as our sages (of the *Talmud*) teach, God says concerning the person of pride: 'I and he cannot live together in the world.'

Israel Baal Shem

God alone originates of Himself, He who can say: I AM. What is based on itself apart from Him, is pride. Herein consists the fall of Lucifer. For he wished to stand on that which has no foundation, and thus it is that he fell into the void.

St Hildegard

I repeat it then, beware of giving place to pride. Yes, indeed beware!

Hônen

Sin in Volition

Not the tasting of the tree was in itself the cause of so great exile, but only the transgressing of the mark.

Dante (*Paradiso,* XXVI. 115)

The will turned away from the unchangeable and universal good, and turned to its own, or some outward or inferior good, sins. It turns to its own, when it would be in its own power;[1] to an outward, when it strives to know what belongs to others, or not to itself; to an inferior, when it loves the pleasures of the body; and thus man, becoming proud, curious, fleshly, passes over into another life, which, in comparison of the former, is death.

Supra 66

St Augustine

[1] 'Mortal sin in metaphysics is the conviction or assertion of independent self-subsistence, as in Satan's case' (Coomaraswamy: *On the Pertinence of Philosophy,* Contemporary Indian Philosophy, London, 1936).

For all sin is fulfilled in three ways, viz., by suggestion, by delight, and by consent. Suggestion is occasioned by the Devil, delight is from the flesh, and consent from the mind. For the serpent suggested the first offence, and Eve, as flesh, was delighted with it, but Adam consented, as the spirit, or mind.

St Gregory the Great

Beautiful and *Blessed* is this Will, when its own Bridegroom shines upon it, and attracts it by beams, by sweet glimpses of it self, spread through all things, more *clear,* or more *obscure,* but *true.* Then do her *beauty* and *blessedness* both change into *deformity* and *misery,* when by hellish enchantments she is deluded, and drawn into the embraces of a wrinkled *Witch,* or a *Spirit,* from the *darkness* below, *transformed into the likeness of her heavenly Love.*

Peter Sterry

Wr. Side 459

Nothing hath separated us from God but our own will, or rather our own will is our separation from God. . . . The fall of man brought forth the kingdom of this world; sin in all shapes is nothing else but the will of man driving on in a state of self-motion and self-government, following the workings of a nature broken off from its dependency upon, and union with, the divine will. All the evil and misery in the creation arises only and solely from this one cause.

William Law

Conform. 166

To him that knoweth to do good, and doeth it not, to him it is sin.

James, IV. 17

Knowl. 745

As he knew his obligation to love God in all things, and as he endeavored so to do, he had no need of a director to advise him, but . . . he needed much a confessor to absolve him.

Brother Lawrence

We want not so much means of knowing what we ought to do, as wills to do that which we may know.

John Smith the Platonist

Closing the Doors of the Senses

Light up thy lantern and see in this image five windows by the which sin cometh into thy soul, as the prophet saith: *Mors ingreditur per fenestras nostras.* Death cometh in by our windows.[1] These windows are our five wits, by the which thy soul goeth out from himself and seeketh his delight and his feeding in earthly things, against his own kind; as by the eye for to see curious and fair things, by the ear for to hear wondrous and new tidings, and so of the other wits. By the unreasonable using of these wits unto vanity wilfully, thy soul is mickle letted from the ghostly wits within; therefore thee behoveth stop the windows and shut them, but only when need asketh for to open them. And that were little mastery if thou mightest once see thy soul by clear understanding, what it is, and how fair it is in his own kind, were not that it is overlaid with a black mantle of this foul image.

Supra 66 *Realiz.* 859

Walter Hilton

Now the orifices are doors and windows of the spirit: breath and will are the messengers and signals of the five viscera, i.e., the animal life. When the eyes and ears are under the allurements of colour and sound then the passions (five viscera) are moved and not in a state of rest. When this is the case flesh and blood sweep onward in their sensuousness unceasingly: then, in turn, the spirit gallops forth wildly into the outward world of sense and does not guard itself within its self-contained domain.[2] When this condition is reached, the coming of distress and joy, though mountainously great, is not understood. On the contrary, were the ear and eye clear and pure without the allurement of desires: the breath and will simple and unalloyed, happy and contented and few in its appetites: the animal life reposeful, not being wasted and scattered: the spirit self-possessed and centred within the bodily frame and not scattered without: when these conditions exist, it would follow that the past ages could be known and the coming events of the future could be seen, and even more than this could be done. Hence it is said a large exploration only gives a little knowledge. That is to say the spirit should not be exercised in the extraneous. Look within. The five colours confuse the eye, causing it to lose its clarity. The five tones derange the ear, leading to loss of true perception. The five flavours disorganize the palate, causing it to injure the taste, likes and dislikes confuse the heart, causing it to lose its proper course in action. These four, . . . though they are the means by which life is carried on, yet, involve all men in their toils.

Renun. 156 *Peace* 700 *P. State* 583 *Knowl.* 734 *Action* 335

Huai Nan Tzû

Close the door of your cell to the body, the door of your lips to conversation, and the inner door of the soul to evil spirits.

St John of the Ladder

The enemy is then more easily overcome, if he be not suffered to enter the door of our hearts, but be resisted at the very gate, on his first knocking.

Renun. 146

The Imitation of Christ, I. xiii

If you would perform our task rightly, take the spiritual water, in which the spirit was from the beginning, and preserve it in a closely shut chamber.

Contem. 542

Basil Valentine

[1] *Jer.*, IX. 21.

[2] This recalls Plato's doctrine of the charioteer and steeds in *Phaedrus*.

Learn how to be entirely unreceptive to sensations arising from external forms, thereby purging your bodies of receptivity to externals.

Huang Po

I myself am keeping a rest-house, whatever is within, I do not allow it to go out and whatever is without, I do not allow it to come in. If anyone comes in or goes out, he does not concern me, for I am contemplating my own heart, not mere clay.

Râbi'a of Baṣra

As for the monk who has not fully controlled his speech, mind, and intellect—his vows, austerities, and charity leak out like water from an unbaked jar.

Srimad Bhagavatam, XI. X

'Close that gate of thy heart by which otherness enters, for thy heart is My temple. Stand watchful over the closing, and remain in it, until thou meetest.'

Niffarî

Contem. 536

Look not round about thee in the ways of the city, nor wander up and down in the streets thereof.

Ecclesiasticus, IX. 7

Let a wise man, like a driver of horses, exert diligence in restraint of his senses straying among seductive sensual objects.[1]

Mânava-dharma-śâstra, II. 88

Flee from sins as from the face of a serpent, for if thou comest near them, they will take hold of thee.

The teeth thereof are the teeth of a lion, killing the souls of men.

Ecclesiasticus, XII. 2, 3

Holy War 407

Kungtutse asked Mencius, 'We are all human beings. Why is it that some are great men and some are small men?' Mencius replied, 'Those who attend to their greater selves become great men, and those who attend to their smaller selves become small men.' 'But we are all human beings. Why is it that some people attend to their greater selves and some attend to their smaller selves?' Mencius replied, 'When our senses of sight and hearing are distracted by the things outside, without the participation of thought, then the material things act upon the material senses and lead them astray. That is the explanation. The function of the mind is thinking: when you think, you keep your mind, and when you don't think, you lose your mind. This is what heaven has given to us. One who cultivates his higher self will find that his lower self follows in accord. That is how a man becomes a great man.'

Mencius

Metanoia 480

Conform. 166

Good men spiritualize their bodies: bad men incarnate their souls.

Benjamin Whichcote

Metanoia 488

[1] Cf. note 2, p. 78.

Supra 66

If you desire true pleasures and unceasing joys, you must put off your unclean garment, cast off the burden of your body, and guard against things repugnant to your (incorporeal) substance; and having done so, turn to the world of true pleasures and unceasing joys, clothe yourself in garments congruous with your true being, and surround yourself with forms appropriate to your own substance, forms everlasting and unchanging,—those forms, of which copies and semblances were seen by you while you were yet in the world of things that come to be and cease to be.

Creation 42 *Realiz.* 873

Hermes

Center 816

Things divine cannot be obtained by those whose intellectual eye is directed to body; but those only can arrive at the possession of them who stript of their garments hasten to the summit.

Proclus, citing the Chaldean Oracle

Close the eye that sees falsely and open the intellectual eye.

Dîvâni Shamsi Tabrîz, XLV

Peace of heart is disturbed by passions; so if you do not allow passions to approach the heart, it will always remain at peace. In the unseen warfare, the warrior stands fully armed at the gates of the heart and repulses all those who attempt to enter and disturb it. While the heart is at peace, victory over the attackers is not difficult. Peace of heart is both the aim of spiritual warfare, and the most powerful means to achieve victory in it. So, when passionate turmoil steals into the heart, do not jump to attack the passion in an effort to overcome it, but descend speedily into your heart and strive to restore quiet there. As soon as the heart is quietened, the struggle is over.

Holy War 407 *Renun.* 143 *Contem.* 528 *Peace* 698

Unseen Warfare, II. xiv

Beauty 679

As musick is conveyed sweetest, and furthest upon a river in ye Night: so is ye Musick of ye heavenly voice carried most clearly, pleasantly to ye understanding, when all ye outward senses ly wrapt up in darkness, and ye depth of night.

Peter Sterry

The Recognition of Sin

Q. What is the most harmful sin?
A. The sin thou dost not know to be a sin.[1]

Ahmad b. 'Âsim al-Antâkî

[1] 'A hesitating sinner, whose conscience is alive, is far preferable to one who is misnamed a sincere, but mistaken, doer' (Marco Pallis: *Peaks and Lamas*, 3rd edit., p. 158).

Monks, two dhamma-teachings of the wayfarer arahant, a rightly awakened one, take place one after the other. What two? 'Look at evil as evil' is the first dhamma-teaching. 'Seeing evil as evil, be disgusted therewith, be cleansed of it, be freed of it' is the second dhamma-teaching. These two dhamma-teachings of the wayfarer take place one after the other.

Realiz. 859

Itivuttaka

I know you all, and will awhile uphold
The unyok'd humour of your idleness:
Yet herein will I imitate the sun,
Who doth permit the base contagious clouds
To smother up his beauty from the world,
That when he please again to be himself,
Being wanted, he may be more wonder'd at,
By breaking through the foul and ugly mists
Of vapours that did seem to strangle him.

Supra 62

Shakespeare (*Henry IV, Pt. I*, I. ii. 219)

The greatest calamity that befalls the heedless is that they are ignorant of their own faults; for anyone who is ignorant here shall also be ignorant hereafter: '*Those who are blind in this world shall be blind in the next world*' (*Qur'ân*, XVII. 72).

Hujwiri

When God wishes well unto His servant He causes him to see the faults of his soul.

Muhammad

Realiz. 859

The first way to avoid an Evil, is to know it, and to know the Cause and Occasion of it.

Benjamin Whichcote

For a thousand offences which a man truly acknowledges and confesses himself to be guilty of, are not so perilous and so mischievous to a man as a single offence which thou wilt not recognize nor allow thyself to be convinced of.

Tauler

I would sooner have the man who sins a thousand mortal sins and knows it, than him who sins but one in ignorance: that man is lost.

Eckhart

And if they disobey thee, say: Lo! I am innocent of what they do.

Qur'ân, XXVI. 216

The Dispelling of Sin

Wr. Side passim

Their adversity among themselves is very great. Ye think of them as a whole whereas their hearts are divers.

Qur'ân, LIX. 14

Introd. *Illusion* *Contem.* 532

Dispersion (vikshepa) is very powerful. But spiritual virtue (sattva) is more powerful still. Increase it, and you will easily master the fluctuations of your mind.

Swami Sivananda

Renun. 146

Wr. Side 474

Let us endeavour to get our minds enlightened with divine truth, clear and practical truth, let us earnestly endeavour after a true participation of the divine nature: and then shall we find hell and death to flee away before us. Let us not impute the fruits of our own sluggishness to the power of the evil spirit without, or to God's neglecting of us: say not, Who shall stand against those mighty giants? . . . Open thy windows, thou sluggard, and let in the beams of divine light, that are there waiting upon thee till thou awake out of thy slothfulness; then shalt thou find the shadows of the night dispelled and scattered, and the warm beams of light and love infolding thee. . . . We need not go and beat the air to drive away those evil spirits from about us, as Herodotus reports the Caunians once to have beaten out the strange gods from amongst them: but let us turn within ourselves, and beat down that pride and passion, those holds of Satan there, which are therefore strong, because we oppose them weakly. Sin is nothing else but a degeneration from true goodness, conceived by a dark and cloudy understanding, and brought forth by a corrupt will: it hath no consistency in itself, or foundation of its own to support it. . . . Let us withdraw our will and affections from it, and it will soon fall into nothing.

John Smith the Platonist

ILLUSION

Come, sit thee down upon this flowery bed,
While I thy amiable cheeks do coy,
And stick musk-roses in thy sleek smooth head,
And kiss thy fair large ears, my gentle joy.

Shakespeare (*A Midsummer-Night's Dream,* IV. i. 1)

The *Mâyâ* of Hindu terminology is the 'art' or creative 'magic'[1] whereby the world of phenomena is manifested; *Mâyâ* becomes 'illusion' only when appearance is no longer understood as such, but mistaken for Reality (cf. Guénon: 'Mâyâ', *Études Traditionnelles,* 1947, and Schuon: 'Sur les traces de Mâyâ', id., 1961). The goddess Kâlî is often represented naked to show that she is free of her own *Mâyâ.*[2]

God is simple Essence; *Mâyâ* is His power of polarizing into dualities; desire originates in these pairs by their *want* of wholeness, which want is exacerbated by the same *Mâyâ* into increasing attraction for the now-unknown, hence exotic, counterpart—following a descending scale of manifestation. In the degree of their remotion from Principle these successive dualities engender desires which become increasingly urgent and finally uncouth; while inversely, desires purify and sublimate in the ascending direction towards Principle, as dualities are surmounted, universalized, and ultimately extinguished in their one unifying Essence.

Creation 37

'Individuality is motivated by and perpetuated by wanting; and the cause of all wanting is "ignorance" (*avidyâ*),—for we "ignore" that the objects of our desire can never be possessed in any real sense of the word, ignore that even when we have got what we want, we still "want" to keep it and are still "in want". The ignorance meant is of things as they really are, and the consequent attribution of substantiality to what is merely phenomenal; the seeing of Self in what is not-Self' (Coomaraswamy: *Hinduism and Buddhism,* p. 62).

Introd. *Renun.*

'In the iconography of Śiva, the demon on whom he tramples is called "the person of amnesia" (*apasmâra puruṣa*)' (Coomaraswamy: 'Recollection, Indian and Platonic', Supplement to the Journal of the American Oriental Society, Vol. 64, no. 2, p. 13).

Holy War 407
Knowl. 755

In Islamic metaphysical cosmology, following 'Abd al-Karîm al-Jîlî, 'the word *al-wahm*

[1] The Assyrian *Maga* would suggest that both of the above words derive from a common root, especially since *mâyâ* as 'art'=*sophia*.—Cf. also Guénon: 'Hermès' (*Le Voile d'Isis,* 1932), on assimilations between *Mâyâ-Dêvî* (mother of the Buddha), *Maïa* (mother of Hermes), and *Maria,* whose month is *May,* all of which really alludes to the substantial, feminine or '*shaktic*' pole of manifestation.

[2] Cf. Sir John Woodroffe: 'The Indian Magna Mater', *Indian Art and Letters,* vol. II, no. 2, 1926, p. 85.

designates the conjectural faculty, the active imagination or the power of illusion, which represents the most redoubtable cosmic power that man has received on loan, for it manifests the demiurgic tendency attracted to every possibility that is as yet unexhausted' (Titus Burckhardt: *De l'Homme universel,* pp. 20, 21).

Truth, however, remains more invincible than illusion, physically, psychically and spiritually. Illusion is not a second reality, but simply Reality veiled or distorted. That a desire can be quenched proves the existence of Reality, that it can remanifest proves the existence of illusion, that it can only be resolved finally in the Truth proves that all its phenomenal reality comes from the Truth.

Introd.
Sin

'Life is the traversing of a cosmic and collective dream by an individual dream, a consciousness, an ego. Death extracts the particular dream from the general dream and plucks out the roots which the former has sunk into the latter. The universe is a dream woven of dreams; the Self alone is awake.

Holiness 934

'The objective homogeneity of the world proves, not its absolute reality, but the collective character of the illusion, or of such an illusion, or such a world' (Schuon: *Perspectives spirituelles,* p. 228).[1]

[1] This must not serve as a pretext, however, for confusing different levels of reality. 'The Intellect's function is just the inverse: that is to say, insofar as it unifies "from within", it discerns "from without"; metaphysical synthesis is not a physical levelling' (Schuon: 'A propos de la doctrine de l'illusion', in *Sentiers de Gnose,* p. 69).

Mâyâ and Magic

The supreme Spirit, unlimited by time and space, of His own will and by the power of His omnipotence, takes upon Himself the limited forms of time and space. Know that the world, although appearing as substantial, has nothing substantial in it: it is a void, being merely an appearance created by the images and vagaries of the mind. Know the world to be an enchanted scene, presented by the magic of *maya*.

Yoga-Vasishtha

Magic is the formative power in the eternal wisdom . . . a mother in all three worlds, and makes each thing after the model of that thing's will. It is not the understanding, but a creatrix according to the understanding, and lends itself to good or to evil; . . . of use to the children for God's kingdom, and to the sorcerers for the devil's kingdom.

Boehme

Wr. Side 451

Magic makes a straw a mountain by artifice; again, it weaves a mountain like a straw.
It makes ugly things beautiful by means of sleight; it makes beautiful things ugly by means of opinion.
The work of magic is this, that it breathes and at every breath transforms realities.
At one time it shows a man in the guise of an ass, at another time it makes an ass look like a man and a notable.
Such a magician is within you and latent: truly, there is a concealed magic in temptation (exerted by the fleshly soul);
But in the world in which are these magic arts, there are magicians who defeat sorcery.
In the plain where this fresh poison grew, there has also grown the antidote, O son.
The antidote says to you, 'Seek from me a shield, for I am nearer than the poison to thee.
Her (the fleshly soul's) words are magic and thy ruin; my words are lawful magic and the counter-charm to her magic.'

Rûmî

Infra 94

Sin 62

Supra Introd.

Infra 108

God has created the world in play, as it were. This is called Mahâmâyâ, the Great Illusion. Therefore one must take refuge in the Divine Mother, the Cosmic Power Itself. It is She who has bound us with the shackles of illusion. The realization of God is possible only when those shackles are severed. . . .

In northwest India the bride holds a knife in her hand at the time of marriage; in Bengal, a nut-cutter. The meaning is that the bridegroom, with the help of the bride, who is the embodiment of the Divine Power, will sever the bondage of illusion.

Sri Ramakrishna

Creation 33

It is the power of appearance that leads us astray.

Plato (*Protagoras*, 356 D)

Nothing impels thee like illusion (*al-wahm*).

Ibn 'Aṭâ'illâh

Every moment Thou art delivering us, and again we are going to a snare, O Thou who art without want!

Rûmî

Action 329

Contem. 528

Judgment 250

By reason of the habit-energy stored up by false imagination since beginningless time, this world (*vishaya*) is subject to change and destruction from moment to moment; it is like a river, a seed, a lamp, wind, a cloud: (while the Vijnana[1] itself is) like a monkey who is always restless, like a fly who is ever in search of unclean things and defiled places, like a fire which is never satisfied. Again, it is like a water-drawing wheel or a machine, it (i.e. the Vijnana) goes on rolling the wheel of transmigration, carrying varieties of bodies and forms, resuscitating the dead like the demon Vetala, causing the wooden figures to move about as a magician moves them. Mahamati, a thorough understanding concerning these phenomena is called comprehending the egolessness of persons.[2]

Lankavatara Sutra, XXIV

M.M. 978

Ananda, if in this world disciples practiced meditation assiduously, though they attained all the nine stages of calmness in Dhyana, yet do not accomplish the attainment of Arhats free from the intoxicants arising from worldly contaminations and attachments, it is wholly due to their grasping this deceiving conception of discriminative thinking that is based on unrealities and mistaking the delusion as being a reality.

Surangama Sutra

Mysteries of the kingdom of God, you are less inexpressible than the mysteries of the kingdom of men.

Louis-Claude de Saint-Martin

Death 220
Sin 66

In the marvellous *lîlâ* of the Lord, it is evident that the consciousness of ego is what has made man fall from his supreme plane of benediction and peace.

Swami Ramdas

Supra Introd.

The concupiscence of matter . . . incessantly attracts a new form.

John A. Mehung

There are Gods and men who delight in becoming. When they are taught the Law for the cessation of becoming, their mind does not respond.

Buddhaghosa

[1] Rendered by Suzuki as 'consciousness' or 'thought'.
[2] 'Egolessness' here meaning 'unreality', devoid of a true Self or Center.

The divine Plato knew that there are three kinds of Sirens; the *celestial*, which is under the government of Jupiter; *that which produces generation*, and is under the government of Neptune; and *that which is cathartic*, and is under the government of Pluto. It is common to all these to incline all things through an harmonic motion to their ruling Gods. Hence, when the soul is in the heavens, the Sirens are desirous of uniting it to the divine life which flourishes there. But it is proper that souls living in generation should sail beyond them,[1] like the Homeric Ulysses, that they may not be allured by generation, of which the sea is an image. And when souls are in Hades, the Sirens are desirous of uniting them through intellectual conceptions to Pluto. So that Plato knew that in the kingdom of Hades there are Gods, daemons, and souls, who dance as it were round Pluto, allured by the Sirens that dwell there.

Pilg. 385

Proclus

This whole world the illusion-maker (*mâyin*) projects out of this (Brahma).
And in it by illusion (*mâyâ*) the other (the individual soul) is confined.
Now, one should know that nature (*prakṛiti*) is illusion (*mâyâ*),
And that the Mighty Lord (*maheśvara*) is the illusion-maker (*mâyin*).
This whole world is pervaded
With beings that are parts of Him.

Śvetâśvatara Upanishad, IV. 9, 10

You have heard much of this world,
Yet what have you seen of this world?
What is its form and substance?
What is Simurgh, and what is Mount Kaf?
What is Hades and what is Heaven and Hell?
What is that unseen world
A day of which equals a year of this?

Introd. *Flight*

Shabistarî

That which is neither in the beginning, nor in the end, but only in the middle, exists only in appearance. It is a mere name and form.

Srimad Bhagavatam, XI. XX

A thing appears in the world and then goes to destruction.
If it has no true existence, how may it appear again?
If it is free from both manifestation and destruction, what then arises?
Stay! Your master has spoken.

Reality 803

Saraha

Things that are not immutable, are not at all.

Infra 99

St Augustine

The Shaikh[2] (may God be well pleased with him) says in the *Faṣṣ i Shu'aibî*, that the universe consists of accidents all pertaining to a single substance, which is the Reality underlying all existences. This universe is changed and renewed unceasingly at every

Reality 775

[1] I.e., the Sirens under the dominion of Neptune.
[2] Muḥyi-d-dîn Ibn 'Arabi.

moment and at every breath. Every instant one universe is annihilated and another resembling it takes its place, though the majority of men do not perceive this, as God most glorious has said: 'But they are in doubt regarding the new creation' (*Qur'ân,* L. 15). . . . They have not grasped the fact that the universe, together with all its parts, is nothing but a number of accidents, ever changing and being renewed at every breath, and linked together in a single substance, and at each instant disappearing and being replaced by a similar set. In consequence of this rapid succession, the spectator is deceived into the belief that the universe is a permanent existence.

Jâmî

Now they say that the world is unreal. Of what degree of unreality is it? Is it like that of a son of a barren mother or a flower in the sky, mere words without any reference to facts? Whereas the world is a fact and not a mere word. The answer is that it is a superimposition on the one Reality, like the appearance of a snake on a coiled rope seen in dim light.

Sri Ramana Maharshi

P. State 579
Holiness 924

I acknowledge and confess that thou (Mother Nature) art the Mother and Empress of the great world, made for the little world of man's mind. Thou movest the bodies above, and transmutest the elements below. At the bidding of thy Lord thou dost accomplish both small things and great, and renewest, by ceaseless decay and generation, the face of the earth and of the heavens. . . . All that exists and is endued with being flows forth from thee by virtue of the power that God has given to thee. All matter is ruled by thee, and the elements are under thy governance. From them thou takest the first substance, and from the heavens thou dost obtain the form. That substance is formless and void until it is modified and individualized by thee. First thou givest it a substantial, and then an individual form.

John A. Mehung

The truth is that God alone is real and all else unreal. Men, universe, house, children—all these are like the magic of the magician. The magician strikes his wand and says: 'Come delusion! Come confusion!' Then he says to the audience, 'Open the lid of the pot; see the birds fly into the sky.' But the magician alone is real and his magic unreal. The unreal exists for a second and then vanishes.

Sri Ramakrishna

Death 220
Contem. 523

All visible things are Mâyâ. Mâyâ will vanish through the effect of knowledge (jnâna). One must strive to get rid of Mâyâ, which devastates the mind; the destruction of the mind (manas) means the annihilation of Mâyâ. Meditation is the only way in which to dominate Mâyâ.

Swami Sivananda

Verily this divine Mâyâ of mine, composed of Gunas,[1] is difficult to surmount; those who take refuge in Me alone, they cross over this Mâyâ.

Bhagavad-Gîtâ, VII. 14

[1] The three tendencies in *Prakriti* (the substantial pole of existence) pervading all manifestation: *Sattva,* ascending, luminous: *Rajas,* expansive, fiery: *Tamas,* descending, obscure.

The Hold of Passion

The expense of spirit in a waste of shame
Is lust in action; and till action, lust
Is perjur'd, murderous, bloody, full of blame,
Savage, extreme, rude, cruel, not to trust;
Enjoy'd no sooner but despised straight;
Past reason hunted; and no sooner had,
Past reason hated, as a swallow'd bait,
On purpose laid to make the taker mad:
Mad in pursuit, and in possession so;
Had, having, and in quest to have, extreme;
A bliss in proof,—and prov'd, a very woe;
Before, a joy propos'd; behind, a dream.
All this the world well knows; yet none knows well
To shun the heaven that leads men to this hell.

Shakespeare (*Sonnet* CXXIX)

Knowl. 745

It is not to be feared for thee lest the Way confuse thee. but verily it is to be feared for thee lest passion overpower thee.

Ibn 'Aṭâ'illâh

When the eye beholds the love of this world, the sight of the heart is extinguished.

Ibrâhim b. Adham

Center 816

I know full well what ills I mean to do
But passion overpowers what counsel bids me.[1]

Euripides (*Medea*, 1078)

This is the great trial or strife of human life, whether a man will live to the lusts of the beast, the guile of the serpent, the pride and wrath of the fiery dragon, or give himself up to the meekness, the patience, the sweetness, the simplicity, the humility, of the Lamb of God.

William Law

Conform. 166

The sword of holy sorrow pierces my soul because those who appear to be spiritual are so inconstant.

Mechthild of Magdeburg

Maya is a very huge saw. Lust, anger, greed, delusion, pride, jealousy, hatred, egoism, etc., are the teeth of this huge saw. All worldy-minded persons are caught up in the teeth of

[1] Cf. Casanova: 'The greatest part of my life was spent in trying to make myself ill. and when I had succeeded. in trying to recover my health. (Now) age. that cruel and unavoidable disease. compels me to be in good health in spite of myself'(*Memories*, cited in *Time*, October 19. 1959. p. 76).

this saw and are crushed. Those who are endowed with purity, humility, love, dispassion, devotion and enquiry are not hurt. They escape through divine grace. They pass smoothly below the saw and reach the other side of immortality.

Beauty 676

Swami Sivananda

Renun. 136

The world is so constructed, that if you wish to enjoy its pleasures, you must also endure its pains. Whether you like it or not, you cannot have one without the other.

Swâmi Brahmânanda

M.M. 978

Whom have you seen in the whole world
Who ever once acquired pleasure without pain?
Who, in attaining all his desires,
Has remained at his height of perfection?

Shabistarî

Sin 66

We forgot that the sensual objects were pleasant and cool only like the shade under the hissing hood of an angry serpent and we sought them as capable of giving us happiness. We only enlisted ourselves on the side of Duryodhana who saw and heard the Lord in person and yet did not profit thereby.

Sri Chandrasekhara Bhâratî Swâmigal

Renun. 146

Noble desires do not easily come to the soul. Know this: those who have them, have them by a special grace of God. In this world of *Mâyâ,* men receive innumerable blows and suffer unimaginable misfortunes; however, they do not want to change their orientation. It is strange, they go back each time to the same situation, only to receive blow after blow. And if someone gives them good advice, they become offended.

Swâmi Brahmânanda

As a dog returneth to his vomit, so a fool returneth to his folly.

Proverbs, XXVI. 11

A Ch'i individual stole some money at a crowded bazaar. He was walking away with it when the police asked him why it was that he stole the money in the market. The thief replied that the sight of the money filled his mind to the exclusion of the policeman. So his desires made him forgetful of the nature of his act.

Huai Nan Tzû

The craving for worldly things, which is chronic in man, is like the patient's craving for water. There is no end to this craving. The typhoid patient says, 'I shall drink a whole pitcher of water.' The situation is very difficult. There is so much confusion in the world.

Sri Ramakrishna

The eye of the covetous man is insatiable in his portion of iniquity: he will not be satisfied till he consume his own soul, drying it up.

Ecclesiasticus, XIV. 9

The fly runs towards the fire or lamp thinking that it is a flower and gets burnt up. Even so the passionate man runs towards a false beautiful form thinking that he can obtain real happiness and gets burnt up in the fire of lust.

Infra 109

Swami Sivananda

The fish looks eagerly at the red fly
With which the fisherman will take him:
But it does not see the hook—
So it is with the poison of the world
Its danger is not realized.

Mechthild of Magdeburg

The worldly man's yearning for God is momentary. It lasts as long as a drop of water on a red-hot frying-pan.

Sri Ramakrishna

Arjuna said:

But, O Descendant of Vrishni (Krishna), impelled by what power does a man commit sin even against his wish, constrained, as it were, by force?

The Blessed Lord said:

It is desire, it is anger, born of Rajo-Guna (quality of passion): of unappeasable craving and of great sin; know this as the foe in this world.

As fire is enveloped by smoke, as a mirror by dust, as an embryo by the womb, so is this (Self) covered by that.

Sin 66

Bhagavad-Gitâ, III. 36–38

Embellished for them is the evil of their deeds when they pursue their lusts.

Qur'ân, XLVII. 14

Some passions are of the body, others are passions of the soul; some are passions of lust, others passions of the excitable part, yet others are passions of thought.[1] Of the latter some are passions of the mind and others of reasoning. All of them combine with one another in various ways and affect one another.

Knowl. 734

St Gregory of Sinai

O son of Kunti, dangerous are the senses, they even carry away forcibly the mind of a discriminative man who is striving for perfection.

The man of steady wisdom, having subdued them all (senses), becomes fixed in Me, the Supreme. His wisdom is well-established whose senses are under control.

Peace 700

Thinking of sense-objects, man becomes attached thereto. From attachment arises longing and from longing anger is born.

Action 329

[1] This is important. People who appear dispassionate are often riddled with this passion, which predominates in the thinking of these times.

From anger arises delusion; from delusion, loss of memory is caused. From loss of memory, the discriminative faculty is ruined and from the ruin of discrimination, he perishes.

Bhagavad-Gîtâ, II. 60-63

Renun. 136

To be overcome by pleasure is ignorance in the highest degree.

Plato (*Protagoras,* 357 E)

Thus, Ânanda, on name and form depends consciousness;
On consciousness depend name and form;
On name and form depends contact;
On contact depends sensation;
On sensation depends desire;
On desire depends attachment;
On attachment depends existence;
On existence depends birth;
On birth depend old age and death, sorrow, lamentation, misery, grief, and despair.
Thus does this entire aggregation of misery arise.

Dîgha-Nikâya

Knowl. 755

The causes of passions are sinful deeds; of thoughts—passions; of imagination—thoughts; the cause of opinions is memory (become multiform); of memory—forgetfulness (of what is true and needful); the mother of forgetfulness is ignorance; of ignorance—laziness; laziness is born of lustful desire; desire is born of movement in a false direction; movement in a false direction comes from committing an action; such action is the fruit of a foolish tendency to evil and of adherence to the senses and sensory things.

St Gregory of Sinai

By attributing worth to tangible objects, man becomes attracted to them; attraction to them brings desire for them; desire leads to competition and dispute amongst men. These rouse violent anger, and the result is delusion. Delusion completely overcomes man's sense of right and wrong.

Srimad Bhagavatam, XI. xiv

Every man is tempted, when he is drawn away of his own lust, and enticed. Then when lust hath conceived, it bringeth forth sin: and sin, when it is finished, bringeth forth death.

James, I. 14, 15

For of a froward will, was a lust made; and a lust served, became custom; and custom not resisted, became necessity.

St Augustine

Introd. *Sin* *M.M.* 978

He who avoids one passion, avoids them all severally; and he who avoids them severally, avoids one. . . .

He who knows wrath, knows pride; he who knows pride, knows deceit; he who knows deceit, knows greed; he who knows greed, knows love; he who knows love, knows hate; he who knows hate, knows delusion; he who knows delusion, knows conception; he who knows conception, knows birth; he who knows birth, knows death; he who knows death, knows hell; he who knows hell, knows animal existence; he who knows animal existence, knows pain.

Therefore, a wise man should avoid wrath, pride, deceit, greed, love, hate, delusion, conception, birth, death, hell, animal existence, and pain.

Akaranga Sutra, I. i. 4

Judgment 250

Assume a virtue, if you have it not.
That monster, custom, who all sense doth eat,
Of habits devil, is angel yet in this,
That to the use of actions fair and good
He likewise gives a frock or livery,
That aptly is put on. Refrain to-night;
And that shall lend a kind of easiness
To the next abstinence: the next more easy;
For use almost can change the stamp of nature,
And master ev'n the devil or throw him out
With wondrous potency.

Shakespeare (*Hamlet*, III. iv. 160)

Sin 62

Renun. 152a

Every desire is insatiable, and therefore is always in want.

Sextus the Pythagorean

Supra Introd.

Whosoever will remember his lusts shall understand that the end of pleasure is sadness. Which if it be able to cause happiness, there is no reason why beasts should not be thought blessed, whose whole intention is bent to supply their corporal wants.

Boethius

Under the sway of strong impulse the man who is devoid of self-control wilfully commits deeds that he knows to be fraught with future misery. But the man of discrimination, even though moved by desires, at once becomes conscious of the evil that is in them, and does not yield to their influence but remains unattached.

Srimad Bhagavatam, XI. vii

If the old desires, the old demands try to make their whispers heard, destroy them with the cudgel of discernment (viveka) and with the sword of renunciation (vairâgya).

Swami Sivananda

Introd. *Renun.*

Of all the acts of devotion by which God's favour is sought none has greater value than resistance to passion, because it is easier for a man to destroy a mountain with his nails than to resist passion.

Junayd

Holy War 396

Knowl. 745

Who is the man of wisdom, that may rightly boast of wisdom? 'Tis he who, recognising the sinful thing, keepeth himself aloof from it.

Shekel Hakodesh, 129

If a man has acquired faith, takes delight in contemplation of me, is indifferent to work, and yet, though knowing their vanity, fails to give up all desires—let him with complete devotion continue to worship me with a cheerful heart. Though he may find it necessary to satisfy his desires, which he is unable for the time to give up, let him all the while ponder on the emptiness of such gratification and know it to be fraught with evil consequences.

Srimad Bhagavatam, XI. xiii

The Cosmic Dream

Supra 85

The objects of sense in the world ever changing—
These we adhere to as things of reality;
But in the ocean of birth and death, they drown us.
How long shall we wander in this path of dreams?
This world to us indeed seems permanent and fixed,
Yet after all, what is it but a road of dreams
To which life after life we must perforce return?

Seami Motokiyo

People sleep, and when they die they wake.

Muhammad

The whole of existence is imagination within imagination, while true Being is God alone.

Ibn 'Arabî

Sin 62

Vishnu incarnated Himself as a sow in order to kill the demon Hiranyâksha. After killing the demon, the sow remained quite happy with her young ones. Forgetting her real nature, she was suckling them very contentedly. The gods in heaven could not persuade Vishnu to relinquish His sow's body and return to the celestial regions. He was absorbed in the happiness of His beast form. After consulting among themselves, the gods sent Śiva to the sow. Śiva asked the sow, 'Why have you forgotten yourself?' Vishnu replied through the sow's body, 'Why, I am quite happy here.' Thereupon with a stroke of his trident Śiva destroyed the sow's body, and Vishnu went back to heaven.[1]

Sri Ramakrishna

[1] There appears a sly poke here by the Śaivas at the Vaishnavas, whose perspective (all reservations made) can in fact where conditions conspire lean to a certain terrestrial affluence of a nuance something less than spiritual.

My Oberon! what visions have I seen!
Methought I was enamour'd of an ass.
. . .

Come, my lord; and in our flight
Tell me how it came this night
That I sleeping here was found
With these mortals on the ground.
. . . .

. . . Think no more of this night's accidents
But as the fierce vexation of a dream.

Shakespeare (*Midsummer-Night's Dream,* IV. i)

Behind my work was ambition, behind my love was personality, behind my purity was fear, behind my guidance the thirst for power! Now they are vanishing and I drift. I come, Mother, I come, in Thy warm bosom—floating wheresoever Thou takest me—in the voiceless, in the strange, in the wonderland. I come, a spectator, no more an actor!

Swami Vivekananda

If there were another Vivekananda, he would have understood what Vivekananda has done! And yet,—how many Vivekanandas shall be born in time! !

Swami Vivekananda

A life devoted to the interests and enjoyments of this world, spent and wasted in the slavery of earthly desires, may be truly called a dream, as having all the shortness, vanity, and delusion of a dream; only with this great difference, that when a dream is over nothing is lost but fictions and fancies; but when the dream of life is ended only by death, all that eternity is lost, for which we were brought into being.

William Law

Creation 48

He who is forgetful of the Self, mistaking the physical body for it, and goes though innumerable births, is just like one who wanders all over the world in a dream; and then realizing the Self would only be like the waking up from the dream-wanderings.

Sri Ramana Maharshi

Judgment 250

When once the Soule has lost her way,
O then, how restlesse do's she stray!
And having not her God for light,
How do's she erre in endlesse night!

Robert Herrick

The veil between God and his servant is neither earth nor heaven, nor the Throne nor the Footstool: thy selfhood and illusions are the veil, and when thou removest these thou hast attained unto God.

Abû Sa'îd ibn Abi 'l-Khayr

Death 220

Mind is only a cloud that hides the sun of Truth. Man is, in fact, God playing the fool. When He chooses, He liberates Himself.

Grace 552

Swami Ramdas

You are asleep, and your vision is a dream,
All you are seeing is a mirage.
When you wake up on the morn of the last day
You will know all this to be Fancy's illusion;
When you have ceased to see double,
Earth and Heaven will become transformed;
When the real sun unveils his face to you,
The moon, the stars, and Venus will disappear;
If a ray shines on the hard rock
Like wool of many colours, it drops to pieces.

M.M. 978
Center 833

Shabistari

Even though apparently awake, one is still asleep if one sees multiplicity. Wake up from this dream of ignorance and see the one Self. The Self alone is real. . . .

This world to-day is, to-morrow is not—empty as a dream, shifting like a circle of fire. There is but one consciousness—pure, transcendental—though it appears as multiple in form.

Reality 803
775

Srimad Bhagavatam, XI. vii

The phenomena of life may be likened unto a dream, a phantasm, a bubble, a shadow, the glistening dew, or lightning flash, and thus they ought to be contemplated.

Infra 99

Prajñâ-Pâramitâ (*Diamond Sutra*)

All things are to be regarded as forms seen in a vision and a dream, empty of substance, un-born and without self-nature, . . . (existing) only by reason of a complicated network of causation which owes its rise to discrimination and attachment and which eventuates in the rise of the mind-system and its belongings and evolvements.

Action 329

Lankavatara Sutra, VII

All life is a dream,
And dreams themselves—
O mockery—
Are nothing but a dream.

Calderón

Just as dreams are unreal in comparison with the things seen in waking life, even so the things seen in waking life in this world are unreal in comparison with the thought-world, which alone is truly real.

Hermes

Those who dream of the banquet, wake to lamentation and sorrow. Those who dream of lamentation and sorrow wake to join the hunt. While they dream, they do not know that they dream. Some will even interpret the very dream they are dreaming; and only when they awake do they know it was a dream. By and by comes the Great Awakening, and then we find out that this life is really a great dream. Fools think they are awake now, and flatter themselves they know if they are really princes or peasants. Confucius and you are both dreams; and I who say you are dreams,—I am but a dream myself.

Chuang-tse (ch. II)

In reality, the entire terrestrial existence of the Prophet (Muhammad) passed thus, as a dream in a dream.[1]

Ibn 'Arabî

Once upon a time, I, Chuang-tse, dreamt I was a butterfly, fluttering hither and thither, to all intents and purposes a butterfly. I was conscious only of following my fancies as a butterfly, and was unconscious of my individuality as a man. Suddenly, I awaked, and there I lay, myself again. Now I do not know whether I was then a man dreaming I was a butterfly, or whether I am now a butterfly dreaming I am a man. Between a man and a butterfly there is necessarily a distinction. The transition is called the transformation of material things.

Chuang-tse (ch. II)

Death 223

How can you determine whether at this moment we are sleeping, and all our thoughts are a dream; or whether we are awake, and talking to one another in the waking state?

Plato (*Theaetetus*, 158 B)

If you enter the life of dreams, O Soul, do not place your happiness in it, nor in the spectacle it exhibits to you, and do not suppose that it is real; else, when you wake, you will be a laughing-stock.

The world of things that come to be and cease to be is a world of dreams. He who is asleep and dreaming (in the literal sense) in this world is in reality dreaming doubly; and when he wakes (in the literal sense), he is like a man who has been awakened from an 'incidental' sleep, but has given himself up again to his 'natural' sleep.

Hermes

When the heart weeps for what it has lost, the spirit laughs for what it has found.

Sufic Aphorism

Creation 33

As a dream when one awaketh; so, O Lord, when thou awakest, thou shalt despise their image.[2]

Psalm LXXIII. 20

[1] This, of course, is taking 'illusion' in its most exalted sense, 'an individual or inspired dream, involved in a macrocosmic dream' (Titus Burckhardt, in a note to his translation of the 'Hikmatun Nûriyah': *Études Traditionnelles,* 1951, p. 20).

[2] Cf. Heinrich Zimmer: *Myths and Symbols in Indian Art and Civilization,* N.Y., 1946, p. 194: 'The world-process is the materialization of Vishnu's dream.'

The attitude expressed in the psalm above recalls the figure of Kâlî thrusting out her tongue with shame upon discovering that she has been dancing on the inert (sleeping) form of Shiva: or again, Vishnu in the quality of Rudra come to swallow up the universe.

When God shall awaken himself upon the World, then shall it be known, that he alone is the Eye, the Light, the Life of the World. . . . When this Eternal Spirit, in whose sleep, like painted forms in a dream, they (the things of this world) vainly flutter about, awakeneth himself, he at once dissolveth them into their *own nothingness* and *despiseth them,* as having never been anything.

Peter Sterry

The more you meditate the more you will enjoy an intense inner spiritual life. . . . Sense-objects will no longer attract you. The world will appear to you as a long dream.

Swami Sivananda

Do not wander from contingency to contingency, for that is to be like the ass of the grinding mill—he trudges along, and the place he starts for is the one from which he has started:—instead, proceed from creation to the Creator.

Infra 109

Ibn 'Aṭâ'illâh

As the Pythagoreans and Platonists believe, during the whole time the sublime Soul lives in this base body, our mind, as though it were ill, is thrown into a continual disquiet—here and there, up and down—and is always asleep and delirious; and the individual movements, actions, passions of men are nothing but vertigos of the sick, dreams of the sleeping, deliriums of the insane, so that Euripides rightly called this life the dream of a shadow. But while all are deceived, usually those are less deceived who at some time, as happens occasionally during sleep, become suspicious and say to themselves: 'Perhaps those things are not true which now appear to us; perhaps we are now dreaming.' Whoever among the dreamers is so affected is in comparison to the others, as Tiresias is said by Homer to be, among the shadows. He alone is wise, the poet says; but all the others fly about like shadows, or rather fly about as shadows.

Contem. 528

Marsilio Ficino

Lo! the guilty are in error and madness.

Qur'ân, LIV. 47

Can a man ever understand God's ways? I too think of God sometimes as good and sometimes as bad. He has kept us deluded by His great illusion. Sometimes He wakes us up and sometimes He keeps us unconscious. One moment the ignorance disappears, and the next moment it covers our mind. If you throw a brick-bat into a pond covered with moss, you get a glimpse of the water. But a few moments later the moss comes dancing back and covers the water.

M.M. 978

One is aware of pleasure and pain, birth and death, disease and grief, as long as one is identified with the body. All these belong to the body alone, and not to the Soul. . . . Attaining Self-Knowledge, one looks on pleasure and pain, birth and death, as a dream.

Renun. 136

Sri Ramakrishna

The senses and thoughts are like weeds on the clear water—covering the surface of the water.

The hand of the intellect sweeps those weeds aside; then the water is revealed to the intellect. . . .

Unless God loose the hand of the intellect, the weeds on our water are increased by sensual desire. . . .

When piety has chained the hands of desire, God looses the hands of the intellect.

Rûmî

Realiz. 873

Knowl. 761

Men are like addicts of drink,
Unaware of their own state.
The wise are wide-awake,
Having cast out the spell.

Ansârî

Holiness 934

Would that you too, my son, had passed forth out of yourself, so that you might have seen, not as men see dreaming figures in their sleep, but as one who is awake.

Hermes

In a dream-state are all actions, however righteous they may seem;
Transcend deeds, and seek ye knowledge of the Real, O Tingri folk.

Phadampa Sangay

Action 346

If the mind contents itself with sense objects, the concept of the reality of the universe can only increase. If on the contrary, it thinks without cease of Atma, the world will appear to it as a dream.

Swami Sivananda

Reflect: is not the dreamer, sleeping or waking, one who likens dissimilar things, who puts the copy in the place of the real object?

I should certainly say that such an one was dreaming.

But take the case of the other, who recognises the existence of absolute beauty and is able to distinguish the Idea from the objects which participate in the Idea, neither putting the objects in the place of the Idea nor the Idea in the place of the objects—is he a dreamer, or is he awake?

Plato (*Republic*, 476 C)

Beauty 670

The Phantom Flux of Life

Know that the life of the world is but play[1] and diversion and pageantry, and boasting among you, and rivalry in wealth and children; as the likeness of vegetation after rain, the

[1] No longer *lîlâ* or divine sport, but merely the vain frivolity of 'those that worship ignorance', and for the victims of this parody, 'great is the destruction' (*Bṛihad-Âraṇyaka Upanishad*, IV. iv. 10. 14).

growth whereof dumbfounds the sowers;[1] then it drieth up and thou seest it yellowing, then it will be scattered. And in the Hereafter is grievous punishment, and forgiveness from Allâh and His good pleasure, whereas the life of the world is nought but the stuff of illusion.

Qur'ân, LVII. 20

Thus they who are possessed by desire suffer much and enjoy little, as the ox that drags a cart gets but a morsel of grass. For the sake of this morsel of enjoyment, which falls easily to the beast's lot, man, blinded by his destiny, wastes this brief fortune, that is so hard to win.

Judgment 250

Śanti-deva

Action 346 *Inv.* passim

The cares of this world, and the deceitfulness of riches, and the lusts of other things entering in, choke the word, and it becometh unfruitful.

St Mark, IV. 19

We see a clear proof of it in Adam and Eve, when the wrath awoke into the quality which was in them and took command: thus it appears in most men where the outward part of the soul rules over the entire body and where the bestial man seeks only the pleasure of this world, that is to say the outward honours, the power and beauty, the food and drink which make an animal of him; he is therefore proud to be a beast, as proud as if he were a god, although he is only a wretched and perishable beast in which the true man is shut up without life.

Sin 66 *Holiness* 924

Boehme

This world is a playground, and death is the night.

Rûmi

The number of the days of men at the most are a hundred years: as a drop of water of the sea are they esteemed: and as a pebble of the sand, so are a few years compared to eternity.

Ecclesiasticus, XVIII. 8

Supra 94

What is life? It is a flash of a firefly in the night. It is a breath of a buffalo in the winter time. It is as the little shadow that runs across the grass and loses itself in the sunset.

Chief Isapwo Muksika Crowfoot

You cannot have real love of God unless you know that the world is impermanent, only of two days' existence, while its Creator alone is real and eternal.

Reality 803

Sri Ramakrishna

There is little difference between what one calls a long life and a short one. After all, it is but a moment in the indefinity of time.

Chuang-tse (ch. XXII)

[1] *Kuffâr*—a special use of the word 'disbelievers'.

We vanish from the world as locusts,
Our life is as a breath.

IV Ezra, IV. 24

Death 223

The great glories of this world pass away in the twinkling of an eye.

T'u Lung

As for man, his days are as grass: as a flower of the field, so he flourisheth.
For the wind passeth over it, and it is gone; and the place thereof shall know it no more.

Psalm CIII. 15, 16

. . . Like the baseless fabric of this vision,
The cloud-capp'd towers, the gorgeous palaces,
The solemn temples, the great globe itself,
Yea, all which it inherit, shall dissolve
And, like this insubstantial pageant faded,
Leave not a rack behind. We are such stuff
As dreams are made on, and our little life
Is rounded with a sleep.

Shakespeare (*The Tempest*, IV. i. 151)

Wr. Side 464

Everything that is thereon passes away:
And there remains but the Face of thy Lord, Possessor of Majesty and Glory.

Qur'ân, LV. 26, 27

That which is dissoluble is destructible; only that which is indissoluble is everlasting. . . .
Nothing that is corporeal is real; only that which is incorporeal is devoid of illusion.

Hermes

Faith 505

The wind of mutability comes with its ruthless blast, dissolving all created things.

Hônen

For all that is not of God shall perish.

The Imitation of Christ, III. xxxii

Oh ye who so long feeding on the Husk
Forgo the Fruit, and doating on the Dusk
Of the false Dawn, are blinded to the True:
That in the Maidân of this world pursue
The Golden Ball which, driven to the Goal,
Wins the World's Game but loses your own Soul:
Or like to Children after Bubbles run
That still elude your Fingers; or, if won,
Burst in Derision at your Touch; all thin
Glitter without, and empty Wind within.

Beauty 663

So as a prosperous Worldling on the Bed
Of Death—'Behold, I am as one,' he said,
'Who all my Life long have been measuring Wind,
And, dying, now leave even that behind'—
This World's a Nest in which the Cockatrice
Is warm'd and hatcht of Vanity and Vice:
A false Bazâr whose Wares are all a lie,
Or never worth the Price at which you buy:
A many-headed Monster that, supplied
The faster, faster is unsatisfied;
So as one, hearing a rich Fool one day
To God for yet one other Blessing pray,
Bid him no longer bounteous Heaven tire
For Life to feed, but Death to quench, the Fire.
And what are all the Vanities and Wiles
In which the false World decks herself and smiles
To draw Men down into her harlot Lap?
Lusts of the Flesh that Soul and Body sap,
And, melting Soul down into carnal Lust,
Ev'n that for which 'tis sacrificed disgust:
Or Lust of worldly Glory—hollow more
Than the Drum beaten at the Sultan's Door,
And fluctuating with the Breath of Man
As the Vain Banner flapping in the Van.
And Lust of Gold—perhaps of Lusts the worst;
The mis-created Idol most accurst
That between Man and Him who made him stands:
The Felon that with suicidal hands
He sweats to dig and rescue from his Grave,
And sets at large to make Himself its Slave.[1]

Supra 89

Peace 700

Infra 108

For lo, to what worse than oblivion gone
Are some the cozening World most doated on.
Pharaoh tried *Glory*: and his Chariots drown'd:
Kârûn with all his Gold went underground:
Down toppled Nembroth with his airy Stair:
Schedâd among his Roses lived—but *where*?

And as the World upon her victims feeds
So She herself goes down the Way she leads.
For all her false allurements are the Threads

[1] The ancient Egyptians had the wisdom to bury their gold in the form of funerary offerings intended for the next world: in this way the gold was consecrated spiritually and served at the same time to anchor or fix the 'barakah' of the civilization substantially, and thus help assure its stability and perpetuity.—The Pharaoh mentioned later in the poem refers to a period of decline.

The Spider from her Entrail spins, and spreads
For Home and hunting-ground: And by and bye
Darts at due Signal on the tangled Fly,
Seizes, dis-wings, and drains the Life, and leaves
The swinging Carcase, and forthwith re-weaves
Her Web: each Victim adding to the store
Of poison'd Entrail to entangle more.
And so She bloats in Glory: till one Day
The Master of the House, passing that way,
Perceives, and with one flourish of his Broom
Of Web and Fly and Spider clears the Room.

'Aṭṭâr — *Wr. Side* 464

On Mount Yoshino each returning year,
How beautiful the cherries blossom gay.
Split the tree open wide and then draw near,
Tell me where is the flower now, I pray.

Japanese Ode

Irâm indeed is gone with all its Rose
And Jamshŷd's Sev'n-ring'd Cup where no one knows:
But still the Vine her ancient Ruby yields,
And still a Garden by the Water blows.

Omar Khayyâm

The past time has departed, so that it no longer is: and the future is not in existence, in that it has not yet arrived. And even the present is not . . . , in that it does not abide. For seeing that the present does not stand fast, and does not abide even for an instant, how can it be said to be 'present', when it cannot stand fast for one moment?

Hermes

The past is already past,
The future is not yet here,
The present never abides;
Things are constantly changing, with nothing on which to depend;
So many names and words confusingly self-created—
What is the use of wasting your life thus idly all day?

Ryôkwan

Thou hast nor youth nor age,
But, as it were, an after-dinner's sleep,
Dreaming on both.

Shakespeare (*Measure for Measure*, III. i. 32)

Tako tsubo ni	The octopuses in the jars:[1]
Hakanaki yume wo	Transient dreams,
Natsu no tsuki	The summer moon.

Zen *Haiku*

Souls that are entangled, involved in worldliness, never come to their senses. They lie in the net but are not even conscious that they are entangled. If you speak of God before them, they at once leave the place. They say: 'Why God now? We shall think of Him in the hour of death.' But when they lie on their death-beds, they say to their wives or children: 'Why have you put so many wicks in the lamp? Use only one wick. Otherwise too much oil will be burnt.'

Death 215

Sri Ramakrishna

Time is an unwholesome physician, for it deceives the patient daily with the expectation of the future, and before expelling the old pains, it adds new ones to the old and accumulates daily so many evils that through the fallacious hope of life it leads to death. We must live today; he who lives tomorrow never lives. If you want to live today, live for God, in whom yesterday and tomorrow are naught but today.

Center 838

Marsilio Ficino

Say not 'to-morrow' or 'the day after to-morrow'; for those that perished, perished because they abode always in their hopes, until the truth came upon them suddenly in their heedlessness, and wilful as they were they were carried to their dark, narrow graves, abandoned by all their kith and kin.

Judgment 239

Ibrâhîm b. Adham

To-morrow, and to-morrow, and to-morrow,
Creeps in this petty pace from day to day,
To the last syllable of recorded time;
And all our yesterdays have lighted fools
The way to dusty death.

Shakespeare (*Macbeth*, v. v. 19)

Oh! how the hours hasten to change into days, the days into months, the months into years, and those into life's annihilation!

'Alî

Enquire of the man who hath gotten himself wealth, if he hath also assured himself of the years of his life.

Shekel Hakodesh, 86

Boast not thyself of to-morrow; for thou knowest not what a day may bring forth.

Proverbs, XXVII. 1

[1] Fishermen trap octopuses in jars that have been sunk into the sea.

Man knoweth not what the morrow will be,
The events of the morrow are in the hand of God.

Egyptian Tradition

Who can say with certainty that one will live to see the morrow?

Lodan-Gawai-Roltso, verse 204

When thou passest the evening, do not expect the morning, and when thou passest the morning, do not expect the evening, and take from thy health for thy sickness, and from thy life for thy death.

Ibn Omar

Go to now, ye that say, Today or to morrow we will go into such a city, and continue there a year, and buy and sell, and get gain: Whereas ye know not what shall be on the morrow. For what is your life? It is even a vapour, that appeareth for a little time, and then vanisheth away. For that ye ought to say, If the Lord will, we shall live, and do this, or that.

James, IV. 13–15

And never say of anything: Lo! I shall do that tomorrow,
Except if Allâh will.

Qur'ân, XVIII. 23, 24

If one does not seek after the great *Tao* in order to release himself from the wrong mode of life, he will finally lose his chance of possessing the talent—and is not to be regarded as a wise man. A hundred years of living is but a transient moment, the length of which resembles only a spark struck from a stone. The fate of life is like a bubble floating on water. Those who know nothing but income, emolument, fame, and rank will soon see their faces turning pale and their bodies degenerating. I suppose that the gain of money is capable of filling the valleys; however this non-permanent possession cannot purchase the things which do not come back.

Judgment 250

Faith 501

Chang Po-tuan

It is told how Jesus, son of Mary, met an old man on a mountain, who lived in the open air without shelter against heat and cold. 'Why dost thou not build a house?' he asked him.

'O Spirit of God', replied the old man, 'prophets before thee predicted that I would only live for 700 years; therefore it is not worth my trouble to settle down.'

Ṣafûrî

Man in the world lodging for a single life-time
Passes suddenly like dust borne on the wind.
Then let us hurry out with high steps
And be the first to reach the highways and fords:
Rather than stay at home wretched and poor
For long years plunged in sordid grief.

Pilg. 365
Holy War 403

Old Chinese Poem

What is our life? a play of passion,
Our mirth the musicke of division,
Our mothers wombes the tyring houses be,
Where we are drest for this short Comedy,
Heaven the Judicious sharpe spectator is,
That sits and markes still who doth act amisse,
Our graves that hide us from the searching Sun,
Are like drawne curtaynes when the play is done,
Thus march we playing to our latest rest,
Onely we dye in earnest, that's no Jest.

Creation 33

Sir Walter Ralegh

Men think much of their own advancement and of many other worldly things; but there is no improvement in this decaying world, which is as a tempting dish, sweet-coated, yet full of deadly gall within.

It is as an extinguished lamp, whose flame is lost, fled we know not where. It is as intangible as a mist; try to lay hold of it, and it proves to be nothing!

Yoga-Vasishtha

The life of this world is but comfort of illusion.

Qur'ân, III. 185

Vanity of vanities; all is vanity.

Ecclesiastes, I. 2

After spending the best part of my life in the study of the liberal arts and sciences, and in the company of wise men and judicious scholars, I was compelled, as the result of my observation of mankind, to arrive at the melancholy conclusion that the hearts of most persons are set either on ambitious and vainglorious projects, on sensual pleasures, or on the accumulation of wealth by all and any means; and that few care either for God or for virtue. At first I did not quite know whether to become a disciple of the laughing or of the weeping philosopher, or whether to join in the exclamation of the wise Prince of Israel: 'All things are vanity.' But at length the Bible and experience taught me to take refuge in the study of the hidden secrets of Nature. . . .

Knowl. 734 *Suff.* 120 Introd. *Charity*

Michael Maier

Jesus, son of Mary (Peace be upon him!), said, 'By God, the world has not settled in the heart of a worshipper without three of its things sticking to it: labour whose distress does not cease, poverty which does not catch up on its wealth, and hope which does not attain its goal.'

Christ in Islâm

The world is a dwelling surrounded by scourges, and heaped with perfidy. Its state endures not, and all who come to it perish.

The world is a dwelling degrading to its owner, where the lawful is mixed with the unlawful, good with evil, sweet with bitter.

Look upon the world with the eye of the cloistered ascetic; not as one loving it blindly.

'Alî

Renun. 152b

Know this; the object you crave for is perishable and transient in itself. How then can lasting peace be derived from it?

Swami Ramdas

Peace 694

Our senses through ignorance of Reality, falsely tell us that what appears to be, is.

Plutarch

Thou shalt not let thy senses make a playground of thy mind.

Tibetan Precept

Contem. 528

The Messiah (Peace be upon him!) said, 'The world is a field belonging to Iblîs (Satan), and its people are his ploughmen.'

Christ in Islâm

Action 346

The enjoyments which are born through contact (with sense-objects) are ever generators of misery; (they are) with beginning and end. O son of Kunti, the wise do not seek pleasure in them.

Bhagavad-Gîtâ, v. 22

Nothing over which the firmament of heaven revolves maintains an unchanged existence, but every hour assumes some new form. Every moment a fresh picture is presented to the view; and one appearance is scarce complete ere another supervenes, obliterating all traces of the first, as wave follows wave upon the shore. No wise man would seek to build his house upon the waves, or hope to find a foundation for it there. . . .

Introd. *Reality*

The wisest of mankind are those who have renounced all worldly desires, and chosen the calm and peaceful lot of a recluse's life. Behind every pleasure lurk twenty pains; far better is it then to forego one fleeting joy and spare oneself a lifetime of regret.

'Azîz ibn Muhammad al-Nasafî

. . . Because each pleasure and pain is a sort of nail which nails and rivets the soul to the body, until she becomes like the body, and believes that to be true which the body affirms to be true; and from agreeing with the body and having the same delights she is obliged to have the same habits and haunts, and is not likely ever to be pure at her departure to the world below, but is always infected by the body; and so she sinks into another body and there germinates and grows, and has therefore no part in the communion of the divine and pure and simple.

Plato (*Phaedo*, 83 D, E)

Introd. *Sin*

Supra 89

Judgment 250

P. State 573

All worldly pursuits have but the one unavoidable and inevitable end, which is sorrow: acquisitions end in dispersion; buildings, in destruction; meetings, in separation; births, in death. Knowing this, one should, from the very first, renounce acquisition and heaping-up, and building, and meeting; and, faithful to the commands of an eminent *guru*, set about realizing the Truth.

Orthod. 288

Milarepa

The world is both seeking and sought. He who seeks the next world, this world seeks him until his provision in it is complete; and he who seeks the present world, the next world seeks him until death comes and seizes him by the neck.

Christ in Islâm

Infra 109

If thou desirest a glory that never vanishes, then seek not to embrace a glory that vanishes.

Ibn ʿAṭâ'illâh

Men of wisdom exhort us: 'Do not cry for the perishable.' Weep not for the dead. Stand firmly fixed in the consciousness of your immortal Existence and see as an unperturbed witness, the passing vicissitudes of life, just as you witness the clouds that pass before your gaze. Let us watch the divine play on the world stage. People appear on it and disappear from it. The world is a passing show. We are the sun of Truth and the world a flitting panorama like the clouds. Whatever has name and form must change and vanish. It is only that immortal Truth, Spirit or God, that nameless, formless, birthless and deathless Reality—with which we are one—that never changes and ever exists.

Creation 33

Realiz. 887

Swami Ramdas

He who knows that this body is like froth, and has learnt that it is as unsubstantial as a mirage, will break the flower-pointed arrow of Mâra, and never see the king of death.

Death 232

Dhammapada, IV. 46

The Snare of Woman

A man forgets God if he is entangled in the world of mâyâ, through a woman. It is the Mother of the Universe who has assumed the form of mâyâ, the form of woman. One who knows this rightly does not feel like leading the life of mâyâ in the world. But he who truly realizes that all women are manifestations of the Divine Mother may lead a spiritual life in the world. Without realizing God one cannot truly know what a woman is. . . . While practising sâdhanâ[1] a man should regard a woman as a raging forest fire or a black cobra. But in the state of perfection, after the realization of God, she appears as the Blissful Mother. Then you will look on her as a form of the Divine Mother.

Love 625

Sri Ramakrishna

[1] Spiritual discipline.

Give not the power of thy soul to a woman, lest she enter upon thy strength, and thou be confounded.

Ecclesiasticus, IX. 2

For the lips of a strange woman drop as an honeycomb, and her mouth is smoother than oil:

But her end is bitter as wormwood, sharp as a two-edged sword.
Her feet go down to death; her steps take hold on hell.

Proverbs, v. 3–5

He who falls into the snares of women is like a bird that falls into the hands of a little silly child; the child plays with it merrily and is glad, but meanwhile the bird endures the pains of death, and undergoes all manner of tortures.

Hermes

Fie upon thee! how is it that a woman, the vanity of the world,
Has been able to hold thee so long captive in her web?[1]

Angelus Silesius

Renun. 152b

'Woman and gold' alone is mâyâ. If mâyâ is once recognized, it feels ashamed of itself and takes to flight. A man put on a tiger skin and tried to frighten another man. But the latter said: 'Ah! I have recognized you! You are our Harê.' At that the man in the skin went away smiling to frighten someone else.

Sri Ramakrishna

Sin 82

What now, young men, do you think? Which were the better for you, to go tracking the woman, or to go tracking the Self?

Vinaya Piṭaka, I. 23

Symb. 306

The Consummation of Desire

The soul that finds not some solid and self-sufficient good to centre itself upon, is a boisterous and restless thing: and being without God, it wanders up and down the world, destitute, afflicted, tormented with vehement hunger and thirst after some satisfying good: and, as any one shall bring it tidings, 'Lo here, or Lo there is good,' it presently goes out towards it, and with a swift and speedy flight hastens after it. The sense of an inward indigency doth stimulate and enforce it to seek its contentment without itself, and so it

[1] Kâlî's 'only clothing is a girdle made of dead men's hands' (W. J. Wilkins: *Hindu Mythology*, Calcutta, 2nd edn., n.d., p. 309).

wanders up and down from one creature to another; and thus becomes distracted by a multiplicity of objects. And while it cannot find some one and only object upon which, as being perfectly adequate to its capacities, it may wholly bestow itself; while it is tossed with restless and vehement motions of desire and love through a world of painted beauties, false glozing excellencies; courting all, but matching nowhere; violently hurried everywhere, but finding nowhere *objectum par amori*; while it converseth only with these pinching particularities here below, and is not yet acquainted with the universal goodness; it is certainly far from true rest and satisfaction, from a fixed, composed temper of spirit.

Contem. 547

Peace 694

John Smith the Platonist

Holy War 407

If thou give to thy soul her desires, she will make thee a joy to thy enemies.

Ecclesiasticus, XVIII. 31

Introd. *Sin*

He that perversely desires a good of nature, though he attain it, is evil himself in the enjoyment of this good, and wretched, being deprived of a better.

St Augustine

Desire never rests by enjoyment of lusts, like as fire surely increases the more by butter (offered in it).

Mânava-dharma-śâstra, II. 94

Think that you suffer a great punishment when you obtain the object of corporeal desire; for the attainment of such objects never satisfies desire.

Sextus the Pythagorean

We may live for thousands of years and may obtain whatever we desire of the world, but we shall never be happy so long as our hunger for earthly things does not perish.

Swami Ramdas

Introd. *Conform.*

Bondage springs from worship of Devas and Devis, commission of sinful and virtuous actions, enjoyment of their consequent fruits, and attachment to sense objects.

Devikalottara-Jnanachara-Vichara-Patalam

Supra 89

Eager to escape sorrow, men rush into sorrow; from desire of happiness they blindly slay their own happiness, enemies to themselves.

Śânti-deva

The cloyed will,—
That satiate yet unsatisfied desire, that tub
Both fill'd and running,—ravening first the lamb,
Longs after for the garbage.

Shakespeare (*Cymbeline*, I. vi. 47)

The mourning for the dead is seven days: but for a fool and an ungodly man all the days of their life.

Ecclesiasticus, XXII. 13

Satan was ever man's deserter (in the hour of need).

Qur'ân, XXV. 29

Hell and destruction are never full; so the eyes of man are never satisfied.

Proverbs, XXVII. 20

Holy War 396

Could you from all the world all wealth procure,
More would remain, whose lack would leave you poor![1]

St Gregory Nazianzen

Renun. 156

The whole earth cannot satisfy the lust of the flesh; who can do its will? To him who longs for the impossible come guilt and bafflement of desire; but he who is utterly without desire has a happiness that ages not.

Śânti-deva

If a man understands the design of God by which all things are ordained, he will despise all material things, and his vices will be healed; but when folly and ignorance continue, all the vices grow in strength, and lacerate the soul with incurable sores; and infected and corrupted by the poison, the soul breaks out in tumours, so to speak, save in the case of those whose souls are cured by the sovereign remedy of knowledge and intelligence.

Hermes

Supra 89
Knowl. 761

He who knows God is disinterested in the gifts of God, and he who is negligent of God is insatiable for the gifts of God.

Shaykh Aḥmad al-ʿAlawî

Conform. 170

A vagabond mind running hither and thither among the varying and false delights of the world is tired out, not satisfied, by its vain exertion; while, starved, it counts as little whatever it gormandizes upon compared with what remains to be devoured, and ever it craves the things removed from it not less anxiously than it joyfully has and holds those that are at hand. For who is there who can gain the whole world? And although a man can never be certain when with anguish he may lose even the little which he has gained with toil, he is certain, nevertheless, that some time or other he will lose it. Thus a perverted will strains eagerly after a direct short-cut to the *best* and hurries on to that whereby it may be filled. Yea, in truth, by such tortuous routes as these does vanity amuse itself, does iniquity deceive itself. So if you would attain to the fulfilment of what you wish for, that is, if you would lay hold upon that which, once grasped, leaves no more to be desired—what is the necessity of putting the rest to the test? You run along bypaths and you will die long before you attain the object of your desires along this circuitous route.

St Bernard

Death 215
Knowl. 734

Creatures are all striving after their primitive pure nature, after their supreme perfection.

Eckhart

P. State 573

[1] 'An incessant "progress", never ending in contentment, means the condemnation of all men to a state of irremediable poverty' (Coomaraswamy: *Am I My Brother's Keeper?*, pp. 2-3).

The desire for perfection . . . is that desire which always makes every pleasure appear incomplete, for there is no joy or pleasure so great in this life that it can quench the thirst in our Soul.

Dante (*Il Convito*, III. vi. 3)

We always find a restless appetite within ourselves which craves for some supreme and chief good, and will not be satisfied with any thing less than infinity itself.

John Smith the Platonist

Creation 28
42, 48

Hearken: God is infinite and without end, but the soul's desire is an abyss which cannot be filled except by a Good which is infinite; and the more ardently the soul longeth after God, the more she wills to long after Him; for God is a Good without drawback, and a well of living water without bottom, and the soul is made in the image of God, and therefore it is created to know and love God.

Tauler

Charity
608

Every natural thing in its own way longs for the Divine and desires to share in the Divine Life as far as it can.

Aristotle

Void
721

This world is finite, and truly that other is infinite: image and form are a barrier to that Reality.

Rûmî

These same are real desires (*satya kâma*) with a covering of what is false. Although they are real, there is a covering that is false.

Chândogya Upanishad, VIII. iii. 1

Beauty
663

For all things the primal good is the cause of their being good. Therefore it is the cause of their being desired. But if for its sake all things are to be desired, consequently itself is to be desired more than all things and by all things.

Marsilio Ficino

These people are like . . . a frog living in a well, who has never seen the outside world. He knows only his well; so he will not believe that there is such a thing as the world. Likewise, people talk so much about the world because they have not known the joy of God.[1]

Sri Ramakrishna

Love
614

In every veil you see, the Divine Beauty is concealed, making every heart a slave to him. In love to Him the heart finds its life; in desire for Him, the soul finds its happiness. The heart which loves a fair one here, though it knows it not, is really His lover.

Jâmî

[1] This recalls exactly Plato's image of the prisoners in the cave (*Republic*, 514 f.).

The objects of earthly loves are mortal, hurtful and loves of shadows that change and pass, for these are not what we really love, not the good that we are really in search of. But there is the true object of our love, where we can be with it, grasp it and really possess, where no covering of flesh excludes.

Plotinus

Knowl. 745

All that exists tends towards perfection, and thus is the Philosopher's Stone prepared.[1]

Abraham Lambspring

Charity 608

Whether you like it or not, whether you know it or not, secretly all nature seeks God and works toward him.

Eckhart

Conform. 185

Perfection is of two kinds, inchoative or complete, partial or entire. Complete perfection (the complete digestion of all crudities and elimination of all impurities) is the ultimate aim of Nature; and she has reached it in our gold, which with its brilliancy lights up the whole earth.

Philalethes

P. State 583
Center 833

Perhaps you have noticed that even in the very lightest breeze you can hear the voice of the cottonwood tree; this we understand is its prayer to the Great Spirit, for not only men, but all things and all beings pray to Him continually in differing ways.

Black Elk

Charity 608

For we know that the whole creation groaneth and travaileth in pain together until now.

Romans, VIII. 22

All creatures desire to speak God in their works; they all of them speak him as well as they can but they cannot really pronounce him. Willy-nilly, in weal or in woe they are all trying to utter God who yet remains unspoken.

Eckhart

M.M. 994

The aspiration of a traveller lists not to pause upon a thing unveiled to it but what the appeals of the Truth call out to him: 'That which thou seekest lies yet before thee!'; and the adornments of creatures do not become conspicuous but what their true natures call out to thee: 'Verily are we but a seduction, wherefore do not fall into disbelief!'

Ibn 'Aṭâ'illâh

In the first days of my youth I tried to find it in the creatures, as I saw others do; but the more I sought, the less I found it, and the nearer I went to it, the further off it was. For of every image that appeared to me, before I had fully tested it, or abandoned myself to peace in it, an inner voice said to me: 'This is not what thou seekest.'

Henry Suso

[1] This doctrine of essential perfection latent in each existing form is not to be confused with the evolutionist's dream of substantial perfectibility progressing through mutations in form.

This world, rightly regarded, is a place for learning truth in. The visible forms of things which it presents to our senses are fleeting and perishable: but they are semblances or shadows of forms that are not apprehensible by sense, forms that are real and everlasting. . . . The deceptive and fleeting pleasures of the sense-world suggest to us that we should turn from them to the true and unceasing pleasures of the thought-world; the frail, transitory, and perishable forms of the sense-world bid us turn from them to the stable and constant forms of the thought-world.

Beauty 670

Hermes

Everything (transitory) is for the experience of the Self which is eternal and free from all adjuncts.

Śrî Śankarâchârya

Creation 42

By nature the soul loves nothing but good. The fact is, every man, if he is really sensible, will find he loves nothing but perfect good, and that is why God has not given perfect good to any creature. If the soul found perfect good in creature she would unite with it. . . . All creatures are crying aloud to man, 'Thou lookest for truth and goodness which we are not: seek God, he is both truth and goodness.' Man is always searching for holiness and happiness. Satisfaction, perfection, does not belong to creature. One leads on to another, food to drink and drink to clothes, etc. In all these things we are allowed to taste the joys of the perfect good in God. St Augustine says, 'Are ye seeking, then look in God.' (God) is the end of creature, or the soul could never have full satisfaction in God, or peace. In God the soul will have all perfection at once; if any of it were outside God she would want to have that as well: she would forget the most for the least and pay the penalty. St Augustine says, 'It is arrant nonsense that the soul is without him who is everywhere, that she is without him without whom she cannot be, that she loves not him apart from whom she cannot love at all.'

Center 847

Reality 803

Eckhart

Peace 694

Our Soul may never have rest in things that are beneath itself.

Julian of Norwich

(God) only is to be sought; for apart from Him all things are nothing. Possessing only Him, we possess all things, for He is all. Knowing Him we know all, for He is the Truth of all things. It is His will that the vast admirable contrivance of the universe should lead us to admiration of him. . . . But to anyone enquiring in learned ignorance what they are, or how, or to what purpose, all things must reply: 'We can answer nothing. . . . If thou wouldst know anything concerning us, seek it in our reason and cause, seek it not in us. There, while thou seekest but one, shalt thou find all things.'

Void 724

Reality 775

Nicholas of Cusa

Center 841

Supra 89

O wonder, all is wonder! how one flees Him from Whom there can be no separation, to pursue that which can have no duration.—'For verily it is not the eyes that grow blind, but it is the hearts within the breasts that grow blind'.[1]

Ibn 'Aṭâ'illâh

[1] *Qur'ân*, XXII, 46.

For what passes for the most truly existent is most truly non-existent, while this unseen First is the Principle of being and Sovereign over reality. . . . All that exists desires and aspires towards the Supreme by a necessity of nature, as if all had received the oracle that without It it is impossible to be.

Metanoia 488
Faith 505

Plotinus

All men's thoughts, which are turmoiled with manifold cares, take indeed divers courses, but yet endeavour to attain the same end of happiness, which is that good which, being once obtained, nothing can be further desired. Which is the chiefest of all goods, and containeth in itself whatsoever is good, and if it wanted anything it could not be the chiefest, because there would something remain besides it which might be wished for. Wherefore, it is manifest that blessedness is an estate replenished with all that is good. This, as we said, all men endeavour to obtain by divers ways. For there is naturally ingrafted in men's minds an earnest desire of that which is truly good: but deceitful error withdraweth it to that which falsely seemeth such. . . . Wherefore, since all things are desired in respect of goodness, they are not so much wished for as goodness itself. But we granted that to be blessedness for which other things are desired, wherefore in like manner only blessedness is sought after: by which it plainly appeareth, that goodness and blessedness have one and the self-same substance. . . . But we have showed that God and true blessedness are one and the self-same thing. . . . We may then securely conclude that the substance of God consisteth in nothing else but in goodness.

Center 847
Wr. Side 474

Boethius

I take refuge in the unity of Thy Quality against every quality.

Judgment 239

Niffari

Love that One Who, when you shall cease to be, will not Himself cease to be, that you may become one who will never cease to be.

Death 206

Abû Sa'îd ibn Abi 'l-Khayr

SACRIFICE AND SUFFERING

Every soul must taste of death, and We try you with evil and with good, for ordeal. And unto Us will ye be returned.

Qur'ân, XXI. 35

It can be seen in the chapters on Creation and Separation how the Infinite ineluctably connotes limitation, imperfection, or evil as one consequence deriving from its All-Possibility. Suffering in turn is the sense of privation following upon separation, which like a hook, has the power to pull one round, to manifest one's limitation, misery and destitution—hitherto perhaps hidden by vain illusions and pleasures. In suffering is concealed the mercy and compassion of the Divine Physician who best knows the malady and the cure. That is why some say there is nothing higher than suffering, when offered up to God.

Introd. *Realiz.*

'The Sacrifice (*yajña*) undertaken here below is a ritual mimesis of what was done by the Gods in the beginning, and in the same way both a sin and an expiation. . . . It is not only *our* passible nature that is involved, but also *his*. In this compatible nature he sympathizes with our miseries and our delights and is subjected to the consequences of things done as much as "we" are. He does not choose his wombs, but enters into births that may be haughty or naughty (*sadasat*) and in which his mortal nature is the fructuary (*bhoktṛ*) equally of good and evil, truth and falsity. That "he is the only seer, hearer, thinker, knower and fructuary" (*Aitareya Âraṇyaka* and *Bṛhadâraṇyaka Upaniṣad*) in us, and that "whoever sees, it is by *his* ray that he sees" (*Jaiminîya Upaniṣad Brâhmaṇa*), who looks forth in all beings, is the same as to say that "the Lord is the only transmigrator" (Śaṅkarâcârya), and it follows inevitably that by the very act with which he endows us with consciousness "he fetters himself like a bird in the net" (*Maitri Upaniṣad*), and is subject to the evil, Death,—or *seems* to be thus fettered and subjected' (Coomaraswamy: *Hinduism and Buddhism*, pp. 10, 16).

Sin 65

René Guénon examines the 'alchemy' of suffering (often associated with asceticism), employing the word '*ascèse*' as a Western equivalent for the Sanskrit *tapas:* 'The first meaning of *tapas* is really "heat"; . . . this heat is plainly that of an interior fire which must burn what the Kabbalists called the "rinds", that is, it must destroy everything in the being which makes itself an obstacle to spiritual realization. . . . If *tapas* often takes on the meaning of painful or grievous effort, it is not that a special value or importance is attributed to suffering as such, . . . but that in the nature of things, detachment from contingencies is forcibly always painful for the individual, whose very existence likewise belongs to the contingent order. . . . Basically one could say that all true ascesis is essentially a "sacrifice"; and . . . in all traditions, sacrifice, in whatever form it presents itself, properly constitutes the ritual act par excellence, the act in which are summarized as it were all the others. . . . Ascesis in its most profound and complete signification is nothing other than the sacrifice of the "self" accomplished to realize the consciousness of the "Self" ' ('Ascèse et ascètisme', *Études Traditionnelles,* 1947, pp. 272–274).

Introd. *Action*

Death 220

Frithjof Schuon studies the more spiritual nuances of the question:

'God tests the faith of man, which man must prove.

'Integral, sincere faith always implies a renunciation, a poverty, a privation, since the world—or the ego—is not God.

'God tests by removing, man proves by renouncing. . . .

'Man is the author of his misfortune insofar as it is felt as a suffering; the world is the author of it insofar as his misfortune endeavours to keep him in cosmic illusion; and God is the Author of it insofar as it comes to man as a sanction, but also as a purification, therefore as a trial. . . . It is evident that suffering in itself has no character of trial; the impious man may suffer, but in his case there is nothing spiritual to try' (*Perspectives spirituelles et Faits humains,* pp. 173–175).

The light of God shines through the interstices of a noble rigour; however, suffering, which like its prototype death, is inescapably linked with the individual states of existence, is rather to be accepted than cultivated, through risk of engendering intellectual suicide if carried to passional extremes, and of thwarting spiritual growth. Sacrifice as method is established by tradition or revelation, which fixes the formal limits; the rest is contingent and relative.[1]

Introd. *Conform.*

[1] Cf. *Love's Labour's Lost,* I. i. 72:

'All delights are vain; but that most vain
Which, with pain purchas'd doth inherit pain.'

—In the Golden Age, the only thing forbidden was the Apple.

Suffering Inseparable From the Fallen State

O Maghavan, verily, this body is mortal. It has been appropriated by Death. (But) it is the standing-ground of that deathless, unembodied Self (Âtman). Verily, he who dwells in the body has been appropriated by pleasure and pain. Verily, there is no freedom from pleasure and pain for one while he dwells in the body. Verily, while one is unembodied, pleasure and pain do not touch him.

Chândogya Upanishad, VIII. xii. 1

Introd. *Sin*

The world is afflicted with death and decay, therefore the wise do not grieve, knowing the terms of the world.

Buddhaghosa

Thy will is entered into self-hood; and all that does vex, plague, and annoy thee, is only thy self-hood; thou makest thyself thy own enemy, and bringest thyself into self-destruction or death.

***Renun.* 146**

Now if thou wilt get again out of death, then thou must wholly forsake thy own self-desire, which has introduced itself into strange essence, and become in self-hood, and the self-desire, as a nothing, so that thou dost no longer will or desire to thyself, but wholly and fully introduce thy desire again with the resignation into the eternal, viz. into God's will, that the same will may be thy will and desire.

Conform. 166

Without this there is nothing but misery and death, a continual dying and perishing.

Boehme

Judgment 266

Infra 126a

We find in the prophet Isaiah that the fire with which each one is punished is described as his own; for he says, 'Walk ye in the light of your own fire, and in the flame which ye have kindled' (*Is.* L. 11). By these words it seems to be indicated that every sinner kindles for himself the flame of his own fire and is not plunged into some fire already kindled by another. Of this fire the fuel and food are our sins.

Origen

Life is a poison that one absorbs, if one knows it not for a poison.

'Alî

Judgment 250

The whole human race is so miserable and above all so blind that it is not conscious of its own miseries.

Comenius

Miseries, though belonging to the world of dreams, are of a certainty painful, and do not vanish until we cease our dreaming. Nor does this dream of life come to an end for him whose thoughts are engrossed in transitory, sensuous things.

Illusion 94

Srimad Bhagavatam, XI. xv

If we want to be without any sort of pain, we must ascertain first the cause of pain. A little consideration will make it clear to us that the root cause of all sorrow is the mistaken identification of the Self with the body. . . . Ignorance alone is the prime cause of all misery.

Sri Chandrasekhara Bhâratî Swâmigal

So long as you consider yourself the body you see the world as external. The imperfection appears to you. God is perfection. His work is also perfection. But you see it as imperfection because of your wrong identification.

Sri Ramana Maharshi

Beauty 689

Though one thinketh joys and sorrows come of causes opposite,
Yet within oneself are found their roots and causes, Tingri folk.

Phadampa Sangay

Realiz. 859

Self is the root, the tree, and the branches of all the evils of our fallen state.

William Law

Death 220

The great ancients have laid down:

'This body of the Brâhmaṇa is not given to him for the sake of sensuous enjoyment. It is meant for undergoing great suffering in the world and for leading to limitless happiness after death.'

It is therefore the duty of the Brâhmaṇa to submit his body to suffering by making it perform the several Karmas prescribed for him by the Śâstras. . . . Just as the pleasure of tasting sweet sugar is possible only if the canes are submitted to the crushing, so if we want lasting happiness the body must be put to penance. If we hesitate to injure the sugarcane, there can be got no sugar. The way therefore to attain happiness is by the performance of our ancient Dharmas. We must not neglect them.

Sri Chandrasekhara Bhâratî Swâmigal

Orthod. 300

A man who would enter Paradise must go through fire and water, whether he be Peter, to whom the keys of heaven were entrusted, or Paul, a chosen vessel of God, or John, to whom all the secrets of God were revealed.

St Augustine

Pilg. 366

The fact is that when Heaven proposes to impose a great responsibility on a man, it is sure first to discipline his purposes by suffering, and his bones and sinews by bodily toil, to starve his limbs and flesh, to empty his very self, confounding all his undertakings.

Mencius

Those to whom Love draws nigh are the most severely proven.

'Aṭṭâr

Infra 128

In none of Christ's sufferings did his Godhead come to the help of his manhood.

Eckhart

Wr. Side 459

The Devil and his servants have at all times done to the followers of Christ what they did to Christ Himself.

The Sophic Hydrolith

Those inward blessings of the spiritual life ... are so many spurs, motives and incitements to live wholly unto God; yet they may, instead of that, fill us with self-satisfaction and self-esteem and prompt us to despise others that want them, as in a poor, mean and reprobate state; who yet may be higher advanced and stand in a nearer degree of union with God, by humility, faith, resignation and pure love, in their inward poverty and emptiness than we who live high upon spiritual satisfactions and can talk of nothing but our feasts of fat things.

All that I would here say of these inward delights and enjoyments is only this—they are not holiness, they are not piety, they are not perfection, but they are God's gracious allurements and calls to seek after holiness and spiritual perfection. They are not to be sought for their own sakes, ... but to be received as cordials that suppose us to be sick, faint, and languishing, and ought rather to convince us that we are as yet but babes than that we are really men of God.

Pilg. 379

William Law

Sin 61

It is better, if the will of God be so, that ye suffer for well doing, than for evil doing.

I. Peter, III. 17

The Divine Allopathy

I say that next to God there is no nobler thing than suffering.... I hold, if anything were nobler than suffering, God would have saved mankind therewith, for we might well accuse him of being unfriendly to his Son if he knew of something superior to suffering.... Further, I maintain, no man apart from God has ever been so holy or so good as to deserve the least nobility such as the smallest suffering would give.... I tell you, right suffering is the mother of all virtues, for right suffering so subdues the heart, it cannot rise to pride but perforce is lowly.

Humility 191

Eckhart

For as the sufferings of Christ abound in us, so our consolation also aboundeth by Christ. And whether we be afflicted, it is for your consolation and salvation, which is effectual in the enduring of the same sufferings which we also suffer: or whether we be comforted, it is for your consolation and salvation.

II. Corinthians, I. 5, 6

One must know that misfortune, being the means of leading one to the Doctrine, is also a *guru*.

Gampopa

Infra 128

Sweet are the uses of adversity,
Which like the toad, ugly and venomous,
Wears yet a precious jewel in his head.

Shakespeare (*As You Like It,* II. i. 12)

The soul that is without suffering does not feel the need of knowing the ultimate cause of the universe. Sickness, grief, hardships, etc., are all indispensable elements in the spiritual ascent.

Ananda Moyî

There is no work of mortification equal to long-suffering.

Śânti-deva

We must not attribute our losses, misfortunes, sufferings and humiliations to the evil spirit or to man; but to their true author, God.

St Alphonsus Liguori

Reality 803

And he that taketh not his cross, and followeth after me, is not worthy of me.

St Matthew, x. 38

Harkee, all rational souls! The swiftest steed to bear you to your goal is suffering; none shall ever taste eternal bliss but those who stand with Christ in depths of bitterness. Nothing is more gall-bitter than suffering, nothing so honey-sweet as to have suffered. The most sure foundation for this perfection is humility, for he whose nature here creeps in deepest depths shall soar in spirit to highest height of Deity; for joy brings sorrow and sorrow brings joy.

Eckhart

Humility 191
Metanoia 488

What life is so much to be dreaded as a life of worldly ease and prosperity? What a misery, nay, what a curse is there in everything that gratifies and nourishes our self-love, self-esteem, and self-seeking! On the other hand, what happiness is there in all inward and outward troubles and vexations when they force us to feel and know the hell that is hidden within us and the vanity of everything without us, when they turn all our self-love into self-abhorrence and force us to call upon God to save us from ourselves, to give us a new life, new light, and new spirit in Christ Jesus.

Pilg. 366
Illusion 99

'O happy famine,' might the poor Prodigal have well said, 'which, by reducing me to the necessity of asking to eat husks with swine, brought me to myself and caused my return to my first happiness in my father's house.' . . .

O happy famine, which leaves you not so much as the husk of one human comfort to feed upon! For this is the time and place for all that good and life and salvation to happen to you which happened to the Prodigal Son.

William Law

The wise, indeed, make lamentations at first; the foolish beat their heads at the last.

Rûmî

When I had emptied to the dregs the cup of human suffering, I was led to consider the wretchedness of this world, and the fearful consequences of our first parents' disobedience. Then I saw that there was no hope of repentance for mankind, that they were getting worse day by day, and that for their impenitence God's everlasting punishment was hanging over them; and I made haste to withdraw myself from the evil world, to bid farewell to it, and to devote myself to the service of God.

Creation 38

Basil Valentine

'Do not think that the afflictions of the world leave Me indifferent. I love souls and I want to save them. To attain My end I use hardship, but it is through pure mercy. In times of abundance, souls forget Me and are lost, whereas in distress they turn to Me and are saved.'

Sister Consolata

Introd. *Reality*
Center 835

Know that the word 'Unicity' (*al-wâḥidiyah*) denotes the following revelation of the Essence: the Essence appears as Quality and the Quality as Essence, so that under this aspect, each Divine Quality appears as the essential determination (*al-ʿayn*) of each of the others. Thus for example, the Avenger (*al-muntaqim*) here is God Himself, and God is the Avenger; on the other hand, the Avenger here is the Beneficent (*al-munʿim*) Himself. Similarly, Unicity manifests Itself in grace (*an-niʿmah*) and manifests essentially in vengeance. Grace, which is an aspect of mercy (*ar-raḥmah*), thus appears as the essence even of vengeance, which itself is an aspect of chastisement; and on the other hand vengeance, which is nothing other than chastisement, appears as an aspect of grace which is identified with mercy. All this is by virtue of the appearance of the Essence in the Qualities and in their effects.

Wr. Side 474

Jili

How often has it happened that what we consider a punishment and chastisement of God, was a special work of grace, an act of His infinite mercy!

St Alphonsus Liguori

It may be generally much better for us, while we are so apt to magnify and court any mundane beauty and glory, as we are, that providence should disorder and deface these things, that we might all be weaned from the love of them, than that their lovely looks should so bewitch and enchant our souls as to draw them off from better things. And I dare say, that a sober mind that shall contemplate the state and temper of men's minds, and the confused frame of this outward world, will rather admire the infinite wisdom of a gracious providence in permitting and ordering that ataxy which is in it, than he would were it to be beheld in a more comely frame and order.

Beauty 663
Illusion 109

John Smith the Platonist

If a man doomed to death be released with one hand cut off, is it not well for him? and if one through human tribulations escapes hell, is it not also well for him? If one cannot bear the small suffering of the moment, then why does he not put away the wrath that will bring upon him the agonies of hell? . . . I may be cleft, pierced, burnt, split open many and many a time for countless millions of aeons, and never win the Enlightenment. But this pain that wins me the Enlightenment is of brief term; it is like the pain of cutting out a buried arrow to heal its smart. All physicians restore health by painful courses; then to undo much suffering let us bear a little.

Judgment 261

Śânti-deva

Water, water I desire,
Here's a house of flesh on fire:
Ope' the fountains and the springs,
And come all to Buckittings:
What ye cannot quench, pull down;
Spoile a house, to save a towne:
Better tis that one shu'd fall,
Then by one, to hazard all.

Holiness 924

Robert Herrick

In the case of the irrational animals, intellect (νοῦς) co-operates with the special form of instinct which belongs to each several kind of beast; but in men, intellect works against the natural instincts. Every soul, as soon as it has been embodied, is depraved by pain and pleasure; for pain and pleasure belong to a composite body, and seethe like juices in it, and the soul steps into them and is plunged in them. Those souls then of which intellect takes command are illuminated by its light, and it counteracts their prepossessions; for as a good physician inflicts pain on the body, burning or cutting it, when disease has taken possession of it, even so intellect inflicts pain on the soul, ridding it of pleasure, from which spring all the soul's diseases. And godlessness is a great disease of the soul; for the beliefs of the godless bring in their train all kinds of evils, and nothing that is good. Clearly then, intellect, inasmuch as it counteracts this disease, confers good on the soul, just as the physician confers health on the body. But those human souls which have not got intellect to guide them are in the same case as the souls of the irrational animals. For intellect co-operates with them, and gives free course to their desires; and such souls are swept along by the rush of appetite to the gratification of their desires, and strive towards irrational ends; and like the irrational animals, they cease not from irrational anger and irrational desire, and are insatiable in their craving for evils.

Introd. *P. State* Supra 118
Sin 66
Illusion 89

Hermes

Did God know anyone willing to suffer the sum of human suffering, he would give it him to bear that his worth might be so much the greater in eternity.

Humility 199

Eckhart

Whoever blames God, despises his mercy.

Boehme

To say as some have said, If God is all love towards fallen man, how can He threaten or chastise sinners? is no better than saying, If God is all goodness in Himself and towards

Wr. Side 474

man, how can He do that in and to man which is for his good? . . . It is love alone in the holy Deity that will allow no peace to the wicked, nor ever cease its judgments till every sinner is forced to confess that it is good for him that he has been in trouble, and thankfully own that not the wrath but the love of God has plucked out that right eye, cut off that right hand, which he ought to have done but would not do for himself and his own salvation.

William Law

Beauty 689

We may rest assured that nothing whatever happens on earth without God's permission. What a source of consolation to know that even the sufferings and adversities which God sends us are for our very best, and have in view our eternal salvation. Ah, how great will be our shame when we stand before the judgment-seat of God and see clearly the loving intention of Divine Providence in sending us those trials which we tried to evade, thus battling against our own salvation!

St Alphonsus Liguori

Center 841

The divine understanding beholding all things most clearly, must needs beget the greatest freedom that may be; which freedom, as it is bred in it, so it never moves without the compass of it. And though the divine will be not determined always to this or that particular, yet it is never bereft of eternal light and truth to act by: and therefore, though we cannot see a reason for all God's actions, yet we may know they were neither done against it, nor without it.

John Smith the Platonist

It is a short lesson, that thou ever praise God, and with a true, not false heart say, 'I will bless the Lord at all times: his praise shall be always in my mouth.' It is a short lesson: it is, namely, that thou know that He giveth of His mercy, when He giveth; that He taketh away of His mercy, when He taketh away; and that thou must not believe that thou art abandoned by the mercy of Him who either comforteth thee by giving, lest thou fail, or punisheth thee when thou art uplifted, lest thou perish. Praise Him therefore, whether in His gifts or in His scourges. The praise of the scourges is the medicine for the wound.

St Augustine

Conform. 180

Bear that which is necessary, as it is necessary.

Sextus the Pythagorean

The Alchemy of Suffering

And the light of Israel shall be for a fire, and his Holy One for a flame: and it shall burn and devour his thorns and his briers in one day.

Isaiah, x. 17

But who may abide the day of his coming? and who shall stand when he appeareth? for he is like a refiner's fire, and like fullers' soap:

And he shall sit as a refiner and purifier of silver: and he shall purify the sons of Levi, and purge them as gold and silver, that they may offer unto the Lord an offering in righteousness.

Malachi, III. 2, 3

Christ sees it needful to put His children into the furnace, but He is seated beside it: His eye is steadily fixed on the work of refining and purifying.

St Alphonsus Liguori

The purpose of this discipline and this rough treatment is that the furnace may extract the dross from the silver.

The testing of good and bad is in order that the gold may boil and bring the scum to the top.

Pilg. 366

Rûmî

In alchemy, nothing bears fruit without having first been mortified.

Light cannot shine through matter, if the matter has not become subtle enough to let the rays pass through.

Death 208

Henry Madathanas

We have the Principle of Life in us, for the most part in Sensual Pleasures; as a piece of Gold in the Dirt; as the Sun in a Cloud; as the Brain or Fancy in a Mist or Fumes. Wipe the Dirt off the Gold, scatter the Cloud from before the Sun, the Mist on the Fancy; chase vain Delights out of the Soul: All these will shine in their proper Beauties.

Sin 66

Peter Sterry

Let not poverty and misfortune distress you; for as gold is tried in the fire, the believer is exposed to trials.

'Alî

No substance can be rendered perfect without a long suffering.

Great is the error of those who imagine that the philosophers' stone can be hardened without first having been dissolved; their time and their work are lost.

Infra 132
Pilg. 378

Henry Madathanas

The old nature is destroyed, dissolved, decomposed, and, in a longer or shorter period of time, transmuted into something else. Such a man is so well digested and melted in the fire of affliction that he despairs of his own strength and looks for help and comfort to the mercy of God alone. In this furnace of the Cross, a man, like earthly gold, attains to the true black Raven's Head, *i.e.,* loses all beauty and reputation in the eyes of the world; and that not only during forty days and nights, or forty years, but often during his whole life, which is thus often more full of sorrow and suffering than of comfort and joy.

Humility 191
197

The Sophic Hydrolith

Supra 120 M.M. 978

He gives thee *basṭ* (spiritual 'expansion') so as not to keep thee in *qabḍ* ('contraction'), and He gives thee *qabḍ* so as not to leave thee in *basṭ;* and He leads thee away from them both in order that thou mayest belong to none other than Him.

Ibn 'Aṭâ'illâh

No Soul Tasked Beyond Its Strength

Orthod. 296

Allâh tasketh not a soul beyond its scope.

Qur'ân, II. 286

God is faithful, who will not suffer you to be tempted above that ye are able.

I. Corinthians, X. 13

God sends us nothing that is too hard or too painful to bear. He proportions all to our strength and abilities. Our trials are suited to our needs as the glove to the hand of the wearer. All things will contribute to our sanctification if we but co-operate with the designs of Divine Providence.

St Alphonsus Liguori

Sin 62

Blessed is the man that endureth temptation.

James, I. 12

Contemptus Mundi

Renun. 136

A Brahman should ever shrink from honour as from poison, and should always be desirous of disrespect as if of ambrosia.

Mânava-dharma-śâstra, II. 162

Always flee from whatever you deem profitable to your (lower) self: drink poison and spill the water of life.

Death 206

Revile any one that praises you: lend (both) interest and capital to the destitute.

Let safety go, and dwell in the place of fear: leave reputation behind and be disgraced and notorious.

I have tried far-thinking (provident) intellect; henceforth I will make myself mad.

Rûmî

Ecstasy 641

Unless I prepare myself with cheerful willingness to be despised and forsaken of all creatures, and to be esteemed altogether nothing, I cannot obtain inward peace and stability, nor be spiritually enlightened, nor be fully united unto thee.

The Imitation of Christ, III. xli

For you must know I have found more of God in the least despisery than ever I did in the sweetness of creatures.

Eckhart

All that one calls honour, name, or reputation must be considered by the spiritual aspirant as filth or poison, whereas he must accept dishonour and scorn in the way that one wears a golden necklace. That is how he can reach the goal with certitude.

Swami Sivananda

Metanoia 488

If the world hate you, ye know that it hated me before it hated you.

St John, xv. 18

Blessed are they which are persecuted for righteousness' sake: for theirs is the kingdom of heaven.

St Matthew, v. 10

You are not yet blessed, if the multitude does not laugh at you.

Seneca

Holiness 910

This I have known, that the more men have raved against me with words of backbiting, so mickle the more I have grown in ghostly profit.

Richard Rolle

When a Christian hath begun to think of spiritual progress, he beginneth to suffer from the tongues of adversaries. Whoever hath not yet suffered from these, hath not yet made progress; and whoever suffereth them not, doth not even endeavour to progress.

St Augustine

Empty honour is deaf and blind before God,
Undeserved contempt sanctifies all God's children.

Mechthild of Magdeburg

One who praises you for qualities you lack, will next be found blaming you for faults not yours.

'Alî

I know now as I have known before that I must learn to drink many cups of gall to the dregs for alas! the devil has many friends among spiritual people, who are so full of poison that they not only drink it themselves but must needs hand it in bitterness to the children of God.

Mechthild of Magdeburg

No might nor greatness in mortality
Can censure 'scape: back-wounding calumny
The whitest virtue strikes. What king so strong
Can tie the gall up in the slanderous tongue?

Shakespeare (*Measure for Measure*, III. ii. 200)

Renun. 139

A prophet is not without honour, but in his own country, and among his own kin, and in his own house.

St Mark, VI. 4

Philosophers have no honour in their cities.

Plato (*Republic*, 489 B)

The seeds of Vajrabantul do not fall to the bottom of the tree. They are carried by the wind far off and take root there. So the spirit of a prophet manifests itself at a distance and he is appreciated there.

Sri Ramakrishna

As he (Pythagoras) easily saw the difficulty of complying with the laws of his country, and at the same time remaining at home and philosophizing, and considered that all philosophers before him had passed their life in foreign countries, he . . . went into Italy, conceiving that place to be his proper country, in which men well disposed towards learning were to be found in the greatest abundance.

Iamblichus

Your kinsmen are often farther from you than strangers.

'Alî

Knowledge Through Chastisement

Despise not the chastening of the Lord; neither be weary of his correction.

Proverbs, III. 11

For whom the Lord loveth he chasteneth, and scourgeth every son whom he receiveth. If ye endure chastening, God dealeth with you as with sons; for what son is he whom the father chasteneth not?

But if ye be without chastisement, whereof all are partakers, then are ye bastards, and not sons.

Sin 56

Hebrews, XII. 6–8

The man of least capacity is the one who shows himself incapable of self-correction.

'Alî

Humility 190

For a refutation, great king, when it has been recognized, brings him who has been refuted into knowledge of things which he did not know before.

Hermes (Discourse of Isis to Horus)

When the door of the intellect opens to thee in denial, then does denial amount to bestowal.

Supra 120

Ibn 'Aṭâ'illâh

Digestion sometimes advanced maie be
By outward cold.

Conform. 170

Thomas Norton

As many as I love, I rebuke and chasten: be zealous therefore, and repent.

Revelation, III. 19

Whom best I love I cross.

Shakespeare (*Cymbeline,* v. iv. 101)

Contradiction, sickness, scruples, spiritual aridity, and all inner and outward torments are the chisel with which God carves his statues for paradise.

St Alphonsus Liguori

Just as an unshapen stone can be fashioned into a beautiful image worthy of adoration and worship only after it has received many a stroke of the chisel, so also a distorted and inharmonious life has to pass through many a trial, suffering and tribulation, before a great change could come over it, before the life of ignorance could be transmuted into a life of immortal splendour and joy, fit to be revered and adored.

Swami Ramdas

Nay, good my fellows, do not please sharp fate
To grace it with your sorrows; bid that welcome
Which comes to punish us, and we punish it
Seeming to bear it lightly.

Death 226

Shakespeare (*Antony and Cleopatra,* IV. xii. 135)

Suffering Transmuted Into Joy

Metanoia 488

Lo! with hardship goeth ease.

Qur'ân, XCIV. 6

Some one asked, 'What is Ṣûfism?' The Shaykh said, 'To feel joy in the heart at the coming of sorrow.'

Rûmî

When sadness and suffering become intense and intolerable, know for certain that a new era is going to dawn bringing signal progress.

Swami Ramdas

Supra 118 · *Judgment* 266

That contrariness which is now in us . . . is cause of our tribulations and all our woe, and our Lord Jesus taketh them and sendeth them up to Heaven, and there are they made more sweet and delectable than heart may think or tongue may tell. And when we come thither we shall find them ready, all turned into very fair and endless worships.

Julian of Norwich

Beauty 689

All whatever makes anguish and strife in nature, that makes mere joy in God; for the whole host of heaven is set and tuned into one harmony; each angelical kingdom into a peculiar instrument, but all mutually composed together into one music, viz. into the only love-voice of God: Every string of this melody exalts and rejoices the other.

Boehme

Renun. 146

Wouldst thou know for certain whether thy sufferings are thine own or God's? Tell by these tokens. Suffering for thyself, in whatever way, the suffering hurts thee and is hard to bear. But suffering for God and God alone thy suffering hurts thee not nor does it burden thee, for God bears the load. Believe me, if there were a man willing to suffer on account of God and of God alone, then though he fell a sudden prey to the collective sufferings of all the world it would not trouble him nor bow him down, for God would be the bearer of his burden.

Eckhart

Metanoia 493

Three days before the time set for the execution, Felicitas gave birth to a child. She suffered intense pain, which she could not hide from the rude soldiers on guard.

'If you lament and cry out now,' one of them said to her, 'What will you do when you are thrown to the wild beasts?'

'I am suffering alone,' she replied, 'what I am suffering now; then Another will be with me, who will suffer for me, because I will suffer for Him.'

From the Memoirs of St Perpetua

One day Ḥasan of Baṣra and Mâlik son of Dînâr and Shakîk of Balkh came to see Râbi'a (al-'Adawîya) when she was ill. Ḥasan said, 'None is sincere in his claim (to love God) unless he patiently endure the blows of his Lord.' Râbi'a said, 'This smells of egoism.' Shaḳîk said, 'None is sincere in his claim unless he give thanks for the blows of his Lord.' Râbi'a said, 'This must be bettered.' Mâlik son of Dînâr said, 'None is sincere in his claim unless he delight in the blows of his Lord.' Râbi'a said, 'This still needs to be improved.' They said, 'Do thou speak.' She said, 'None is sincere in his claim unless he forget the blows in beholding his Lord.'

'Aṭṭâr

Holiness 914

His joy admits no sorrow nor affliction.

Dîvâni Shamsi Tabrîz, XLV

Wr. Side 74

Do not try to drive pain away by pretending that it is not real;
Pain, if you seek serenity in Oneness, will vanish of its own accord.

Seng-ts'an

Center 835

Of all this suffering and abandonment the man should make an inward joy; and he should give himself into the hands of God, and should be glad because he is able to suffer for the glory of God. And if he be true to this disposition, he shall taste such an inward joy as he never tasted before; for nothing is more joyful to the lover of God, than to feel that he belongs wholly to his Beloved.

Ruysbroeck

Man, if thou art faithful to God, and desirest nought but Him,
The harshest affliction will be for thee a Paradise.

Angelus Silesius

Judgment 266

O Lord, thou fillest me with joy in *all* that thou doest.[1]

St Teresa of Lisieux

Our Saviour's cross is the throne of delights. That Centre of Eternity, that Tree of Life in the midst of the Paradise of God!

Thomas Traherne

How long delay I, speaking thus, and embrace not the cross, that by the cross I may be made alive, and by the cross (win) the common death of all and depart out of life?

Come hither ye ministers of joy unto me, ye servants of Aegeates:[2] accomplish the desire of us both, and bind the lamb unto the wood of suffering, the man unto the maker, the soul unto the Saviour.

The Martyrdom of St Andrew

[1] Gloss on *Psalm* XCI.5 (Vulgate).
[2] Egeas, proconsul of Patrae, who persecuted St Andrew.

Sacrifice

Ye will not attain unto piety until ye spend of that which ye love. And whatsoever ye spend, Allâh is Aware thereof.

Qur'ân, III. 92

Offer unto me that which is very dear to thee—which thou holdest most covetable. Infinite are the results of such an offering!

Srimad Bhagavatam, XI. V

Death 220

He is said to have made the greatest sacrifice who has sacrificed his own mind and endeavours at the shrine of self-abnegation.

Yoga-Vasishtha

Thou wast seeking what thou shouldst offer in thy behalf; offer thyself. For what doth God ask of thee, except thyself? Since in the whole earthly creation He made nothing better than thee.

Creation 42

St Augustine

Humility 191

The sacrifices of God are a broken spirit: a broken and a contrite heart, O God, thou wilt not despise.

Psalm LI. 17

Behold a continual 'cleaving of the moon' in our hearts.[1]

Dîvâni Shamsi Tabrîz, IX

Who chooseth me must give and hazard all he hath.

Shakespeare (*Merchant of Venice*, II. vii. 9)

Death 206

Look now, here is a bargain: give one life and receive a hundred.

Dîvâni Shamsi Tabrîz, XLII

O my son, the whole affair consists in spiritual sacrifice. If you are able for this, it is well; if not, do not occupy yourself with the futilities of the Sûfis.

Abû Muḥammad Ruwaym

[1] This refers to a miracle of Muhammad at Mecca, where the moon was cloven in two, each half circumambulating the Ka'bah in opposite directions, to rejoin above the sacred House. This is analogous esoterically to the alchemical *solve et coagula* where a finite equilibrium is ruptured to be re-established on the principial plane.

Austerity not to Be Sought

Common-sense is the fundamental factor in all Sadhanas (spiritual disciplines). No rule is an eternal rule. Rules change from place to place, time to time and from one condition to another condition.... The objective of all Anushthanas (religious austerities) is rigorous mental discipline and not mere physical mortification.... Abnormal cravings are against Anushthana, not the normal requirements.

Swami Sivananda

Introd.
Conform.

If the vine-dresser be not skilled in his art, he is as likely to crop off the good branches which bear grapes as the wild shoots, and thus spoil the vineyard. So it is with those who do not understand this spiritual art; they leave the roots of vice and evil dispositions alive in the heart, and hew and lop at poor nature, and thereby destroy this noble vineyard.

Tauler

Orthod. 300
Creation
42

'Remember, Consolata, that I am good. Do not disfigure me! the world likes to think of sanctity under a mask of austerity, discipline and chains.—No, this is false! If sacrifice and penance are encountered in the life of a saint, these do not constitute his whole life. The saint, the soul who gives himself generously to me, is the happiest being in the world, for I am good, exclusively good.

'Do not forget that the Jesus whom thou sawest die on a cross at the end of his mortal career is the same person who shared the common life with all men for thirty years; the same who during his three years of preaching took part in banquets.... Do not disfigure me and do not ever forget that I am good.'

Sister Consolata

Wr. Side
474

There are some, however, who are not content with the myrrh that God gives them, but think fit to give themselves some, and create evils for themselves and sick fancies, and have indeed suffered long and much, for they take hold of all things by the wrong end. And they gain little grace from all their pain, because they are building up stones of their own laying, whether it be penances or abstinence, or prayer or meditation. According to them, God must wait their leisure, and let them do their part first, else no good will come of the work. God hath fixed it in His purpose that He will reward nothing but His own works. In the kingdom of Heaven He will crown nothing to all eternity but His works, and not thine. What He has not wrought in thee, He takes no account of.

Tauler

Action
340

No sacrifices, sufferings, and death have any place in religion, but to satisfy and fulfil that love of God which could not be satisfied without our salvation.

William Law

Love
618

By the cross of the malefactor on Christ's left hand may be understood those who have made a religious profession, and are hanging on the cross of continual exercises and outward austerities which they have bound themselves to practise; they have well deserved this cross, but it brings them no profit, because they have not died on it to self-will and other sinful failings. It is possible for them after this crucifixion to go down to eternal torment with the unjust malefactor; so that, to use a common proverb, they drag the barrow here and the wagon in the world to come.

Tauler

Holiness 914
Peace 700

Whatever sacrifice may be made in the service of the Lord, know that it is the equanimity of your soul that is the best and fittest offering. Equanimity is sweet to taste and has the supernatural power of transforming everything to ambrosia.

Equanimity expands the soul and gladdens the mind, as the sunlight fills the vault of heaven, and it is considered to be the highest devotion.

Yoga-Vasishtha

RENUNCIATION – DETACHMENT

Truth alone is the austerity of the *Kali Yuga*.

Sri Ramakrishna[1]

As can be seen in the introductory passages of the preceding chapter, suffering is inextricably linked with renunciation, both as a cause, and as a result. In the former case, however, man is the victim of the ultimate term of an inexorable cosmological process; in the latter case he is active to the suffering, which this time is strictly relative and subordinate to the end in view, namely, liberation from the tyranny of vicissitudes and disequilibriums to which he as individual is subject—hence, liberation from the domain of suffering itself, through ritual identification with the supraformal states, which are outside the realm of change.

If a great deal has been written on the subject of renunciation, it is not so much that the doctrine is difficult to grasp, as that the practice needs encouragement. 'The Lord thy God is a jealous God' (*Deuteronomy*, VI. 15), and will not suffer associates, idols, or the worship of anything considered good or desirable or real beside Himself,—He, who is infinitely desirable by virtue of being the One Reality that is. The essential to remember in renunciation is that nothing real is ever lost;[2] each withdrawal from the world is an appropriation on a higher plane of that from which one withdrew. In the final withdrawal there is no more self, no more world—only the Void, the Real. Hence the paradox: the more a person can dispense with the world, the less can it dispense with him,—for the center can dispense with the periphery, whereas a periphery without a center—manifested or unmanifested—is a sheer impossibility. Using the same pattern of analogy: 'Detachment is possible without renunciation, but renunciation has no meaning except in view of detachment' (Schuon: *Perspectives spirituelles*, p. 281).

Reality 803
Introds.
Illusion
Void
Holiness

The illusion in desire must be overcome through discernment—the urgency through renunciation. But one must be detached from one's own renunciation.

Knowl.
761

[1] Cited in *Women Saints of East and West*, London, The Ramakrishna Vedanta Centre, 1955, p. 121.

[2] Assuming it is under ritual control, for withdrawal should not imply a shrinking, nor reduction a decrease.

Indifference to Pleasure and Pain

Excessive pains and pleasures are justly to be regarded as the greatest diseases to which the soul is liable. For a man who is in great joy or in great pain, in his unseasonable eagerness to attain the one and to avoid the other, is not able to see or to hear anything rightly: but he is mad, and is at the time utterly incapable of any participation in reason.

Illusion 89

Plato (*Timaeus*, 86 C)

Someone complained to Meister Eckhart that no one could understand his sermons. He said, To understand my sermons a man requires three things. He must have conquered strife and be in contemplation of his highest good and be satisfied to do God's bidding and be a beginner with beginners and naught himself and be so master of himself as to be incapable of anger.

Eckhart

Once upon a time a lover of secret lore came to an anchorite and asked to be admitted as a pupil. Then he said to him: My son, your purpose is admirable, but do you possess equanimity or not? He replied: Indeed, I feel satisfaction at praise and pain at insult, but I am not revengeful and I bear no grudge. Then the master said to him: My son, go back to your home, for as long as you have no equanimity and can still feel the sting of insult, you have not attained to the state where you can connect your thoughts with God.

Infra 160

Isaac of Acre

> Not to be cheered by praise,
> Not to be grieved by blame,
> But to know thoroughly one's own virtues or powers
> Are the characteristics of an excellent man.

Realiz. 859
Holiness 914

Subhâshita Ratna Nidhi, Stanza 29

It was a good maxim of the old Jewish writers, 'the Holy Spirit dwells not in terrene and earthly passions.' Divinity is not so well perceived by a subtile wit, 'as by a purified sense,' as Plotinus phraseth it.

. . . Aristotle himself thought a young man unfit to meddle with the grave precepts of morality, till the heat and violent precipitancy of his youthful affections were cooled and moderated. And it is observed of Pythagoras, that he had several ways to try the capacity of his scholars, and to prove the sedateness and moral temper of their minds, before he would entrust them with the sublimer mysteries of his philosophy. The Platonists were herein so wary and solicitous, that they thought the minds of men could never be purged enough from those earthly dregs of sense and passion, in which they were so much steeped, before they could be capable of their divine metaphysics: and therefore they so much solicit 'a separation from the body' in all those that would, as Socrates speaks, 'sincerely understand divine truth;' for that was the scope of their philosophy. This was also intimated by them in their defining philosophy to be μελέτη θανάτου 'a meditation of death;' aiming herein at only a moral way of dying, by loosening the soul from the body and this sensitive life; which they thought was necessary to a right contemplation of intelligible things: and therefore, besides those ἀρεταὶ καθαρτικαὶ by which the souls of men were to be separated from sensuality and purged from fleshly filth, they devised a farther way of separation more accommodated to the condition of philosophers, which was their *mathemata,* or mathematical contemplations, whereby the souls of men might farther shake off their dependency upon sense, and learn to go as it were alone, without the crutch of any sensible or material thing to support them.

Death 215

Symb. 306

John Smith the Platonist

Nature willingly receiveth honor and reverence.
Grace faithfully attributeth all honor and glory unto God.
Nature feareth shame and contempt.
Grace rejoiceth to suffer reproach for the Name of Jesus.

The Imitation of Christ, III. liv

Inv. 1017

The human soul has in it something that is divine; but there are joined to it also the irrational parts, namely, desire and repugnance. These . . . are motive forces . . . that have to do with mortal bodies. And so, as long as the divine part of the soul is in the divine body, desire and repugnance are far away from it; but when the divine part has entered into a mortal body, they come into being as accretions on it, and it is through their presence that the soul becomes bad. . . . The soul of irrational animals consists of repugnance and desire; and that is why these animals are called 'irrational', because they are deprived of the rational part of the soul.

Hermes

What is it to me whether my neighbour is pleased with me or with another? the joy is his, not the smallest share of it is mine. If happiness springs from the joy of others, then I should have it in every event; so why am I not glad when men rejoice to honour another? . . . Some loathe me; then why shall I rejoice in being praised? Some praise me; then why shall I be cast down by blame?

Living beings are of diverse character; not even the Conquerors can content them, much less simple souls such as I. Then why think of the world? They blame a fellow-creature who gains naught, they scorn him who gains something; being thus by nature unpleasant companions, what happiness can come from them?

Śânti-deva

Who are these by whom you wish to be admired? Are not these the men whom you generally describe as mad? What do you want then? Do you want to be admired by madmen?

Suff. 126b

Epictetus

Beauty 663

Were I full of God I should care nothing whatever for the world. To respect the world shows want of self-respect. Self-respect betokens despisery of things.

Eckhart

We must learn, in our pursuit of wisdom, to listen with equanimity to the reproaches of the foolish, and to despise contempt itself.

Seneca

Man is perfect only when bestowal and denial, humiliation and honour, have become alike in his heart.

Abû ʿUthmân al-Hayrî al-Nisâbûrî

M.M. 978

Truly wise is he who is unstirred by praise or blame, by love or hatred. He is not moved by the opposites of life. Verily does he delight in the blissful Self.

Srimad Bhagavatam, XI. v

Sin 64

The Perfect Way is only difficult for those who pick and choose;
Do not like, do not dislike; all will then be clear.
Make a hairbreadth difference, and Heaven and Earth are set apart.

Seng-tsʿan

Knowl. 755

For if the quality that belongs to the celestial world is to be delivered from the malediction and the wrath of nature, it must preserve its equilibrium.

Boehme

The dervish . . . never even thinks of mankind, and when his heart has been broken away from them he is as indifferent to their reprobation as to their favour, he moves unfettered and free.

Hujwîrî

Action 346

I say not but that evermore some men shall say or think somewhat against us, the whiles we live in the travail of this life, as they did against Mary.[1] But I say, if we will give no more heed to their saying nor to their thinking, nor no more leave off our ghostly privy work for their words and their thoughts, than she did—I say then that our Lord shall answer them in spirit (if it shall be well with them that so say and so think) in such sort, that they shall within few days have shame of their words and their thoughts.

The Cloud of Unknowing, XXIII

Action 358

To work alone thou hast the right, but never to the fruits thereof. Be thou neither actuated by the fruits of action, nor be thou attached to inaction.

Infra 160

O Dhananjaya, abandoning attachment and regarding success and failure alike, be steadfast in Yoga and perform thy duties. Even-mindedness is called Yoga.

Bhagavad-Gîtâ, II. 47, 48

[1] *St Luke,* X. 40.

For where there is true love, a man is neither out of measure lifted up by prosperity, nor cast down by mishap; whether you give or take away from him, so long as he keeps his beloved, he has a spring of inward peace. Thus, even though thy outward man grieve, or weep downright, that may well be borne, if only thy inner man remain at peace, perfectly content with the will of God.

Tauler

Contem. **536**
Conform. 180

The steady-minded, undeluded knower of Brahma, being well-established in Brahma, neither rejoices on receiving the pleasant nor grieves on receiving the unpleasant.

Bhagavad-Gîtâ, v. 20

Detachment From Place and Conditions

To whom God is dearer in one thing than another, that man is a barbarian, still in the wilds, a child. He to whom God is the same in everything has come to man's estate. But he to whom creatures all mean want and exile has come into his own.

Eckhart

He is still weak for whom his native land is sweet, but he is strong for whom every country is a fatherland, and he is perfect for whom the whole world is a place of exile.[1]

Hugh of Saint-Victor

Keep thyself as a stranger and pilgrim upon the earth, who hath nothing to do with the affairs of this world.

The Imitation of Christ, I. xxiii

Pilg. 365
Death 215

In foreign countries they recognize home, and in every home they see a foreign country.

Letter to Diognetus

Suff. 126b

There is no question of time and space. Understanding depends on ripeness of mind. What does it matter if one lives in the East or in the West?

Sri Ramana Maharshi

. . . We, to whom the world is our native country, just as the sea is to the fish . . .

Dante (*De Vulgari Eloquentia*, I. vi)

Reality 790

[1] The same sentence turns up in the journal of Johannes Kelpius, q.v., in the 17th century. Perhaps these writers are paraphrasing Seneca.

Native land, and home, and all possessions,
I know you all to be but empty things;
Any thoughtless one may have you.
As for me, the devotee, I go to win the Eternal Truth.

Milarepa

He that thus knows what this world is, has great reason to be glad that he is born into it, and yet still greater reason to rejoice in being called out of it, preserved from it, and shown how to escape with the preservation of his soul.

Judgment 250

William Law

Though the spirit of man in this state be joined to a body, and made a member of this material visible world, yet itself doth belong to another country.

Metanoia 480

Benjamin Whichcote

Rise above time and space,
Pass by the world, and be to yourself your own world.

Holiness 924

Shabistarî

To be the heavenly Father's Son one has to be a stranger to the world, remote from self, heartwhole and having the mind purified.

O man, renounce thyself and so with toil-free virtue win the prize or, cleaving to thyself, with toilful virtues lose it.

Infra 143

Eckhart

Thou shouldst know that it is inner abandonment that leads men to the highest truth.

Henry Suso

For him who is sincerely devoted to the religious life, it is the same whether he refrain from worldly activities or not. . . .

For him who is freed from attachment to worldly luxuries, it is the same whether he practise asceticism or not. . . .

For him who hath attained the mastery of his mind, it is the same whether he partake of the pleasures of the world or not. . . .

Peace 700

For him whose humility and faith (with respect to his *guru*) are unshakable, it is the same whether he dwell with his *guru* or not. . . .

For him who hath attained the Sublime Wisdom, it is the same whether he be able to exercise miraculous powers or not.

Flight 941

Gampopa

Those who live in the world and try to find salvation are like soldiers that fight protected by the breastwork of a fort, while the ascetics who renounce the world in search of God are like soldiers fighting in the open field. To fight from within the fort is more convenient and safer than to fight in the open field.

Sri Ramakrishna

Live unconcerned in this World. This Divine Lesson is taught us from Heaven by the Holy Ghost upon this Ground: 1. *Cor.* 7. 29, 30, 31. *But this I say Brethren, the time is short. It remaineth, that both they that have wives, be, as though they had none; And they, that wept as though they wept not; and they, that rejoyce, as though they rejoyced not; and they, that buy, as though they bought not; & they that use this world, as not abusing it: For the fashion of this world passeth away.* The *Apostle* here divideth all this World into 4 *Heads;* 1. Relations; 2. Passions; 3. Possessions; 4. Employments, and Entertainments. *Solomon* saith in one place: *Why shouldst thou set thine heart upon that, which is not.* There is no real Difference between having a Husband, Wife, or Children, and having none; between being in Grief, or Joy, and being without Grief, or Joy; between having an Estate, and having none; between being in the height of all Employments, or Entertainments, and being out of all. This world hath nothing real. It is all a Shadow. Seeing then the various States of things on Earth have no real Difference, pass thou thorow all estates with a perfect indifference of Spirit, in a constant calm.

Peter Sterry

Rev. 959

Illusion 99
Infra 160

The man of enlightened mind who is active in the world and the illumined sage who sits in his hermitage, are both alike in their spiritual calm, and have undoubtedly reached the state of blessedness.

It is the activity or the inactivity of the mind which is the sole cause of the restlessness or tranquillity of men. Thickly-gathering desires serve to fill the mind with the vanity of their nature, which is the cause of all its woes: therefore try to weaken your worldly senses at all times. . . .

The homes of householders who have well-governed minds and have banished their sense of egoism are as good as solitary forests, cool caves or peaceful woods, O Rama-ji.

Men of pacified mind view the bright and beautiful buildings of cities in the same dispassionate light as they behold the trees of a forest. He who, in his inmost Spirit, sees the world in God, is verily the Lord of mankind!

Yoga-Vasishtha

Contem. 528

Reality 803

If a determined scholar wants to learn *Tao,* he may live in a city or hold office at the same time without any inconvenience. The work is easy and the medicine is not far away. If the secret is disclosed, it will be so simple that every one may get a good laugh.

Chang Po-tuan

Contem. 542

Such a householder may be compared to a waterfowl. It is constantly diving under water; yet, by fluttering its wings only once, it shakes off all traces of wet. . . .

One can live in the world after acquiring love of God. It is like breaking the jack-fruit after rubbing your hands with oil; the sticky juice of the fruit will not smear them.

Sri Ramakrishna

The true saint goes in and out amongst the people and eats and sleeps with them and buys and sells in the market and marries and takes part in social intercourse, and never forgets God for a single moment.

Abû Sa'îd ibn Abi 'l-Khayr

Holiness 914

When an anchorite goes into a tavern, the tavern becomes his cell, and when a haunter of taverns goes into a cell, that cell becomes his tavern.[1]

Hujwîrî

Towards the end of his life he (Abû Muhammad Ruwaym) hid himself among the rich and gained the Caliph's confidence, but such was the perfection of his spiritual rank that he was not thereby veiled from God. Hence Junayd said: 'We are devotees occupied (with the world), and Ruwaym is a man occupied (with the world) who is devoted (to God).'

Hujwîrî

Contem. 523

The happiness of solitude is not found in retreats. It may be had even in busy centres. Happiness is not to be sought in solitude or in busy centres. It is in the Self.

Sri Ramana Maharshi

O know this truth,
That neither at home nor in the forest does enlightenment dwell.
Be free from prevarication
In the self-nature of immaculate thought!

Saraha

Contem. 536

Retirement means abidance in the Self. Nothing more. It is not leaving one set of surroundings and getting entangled in another set, nor even leaving the concrete world and becoming involved in a mental world.

Sri Ramana Maharshi

A monk asked Daishu Ekai (Ta-chu Hui-hai), one of the T'ang masters, when Zen was in its heyday:

'What is great *nirvâṇa?*'

The master answered, 'Not to commit oneself to the karma of birth-and-death is great *nirvâṇa*.'

'What, then, is the karma of birth-and-death?'

Infra 160
Conform. 170

'To desire great *nirvâṇa* is the karma of birth-and-death.'

Zen *Mondo*

Visitor: I often desire to live in solitude where I can find all I want with ease, so that I may devote all my time to meditation only. Is such a desire good or bad?

Action 329
Realiz. 873
Sin 77

Maharshi: Such thoughts will bestow a *janma* (reincarnation) for their fulfilment. What does it matter where and how you are placed? The essential point is that the mind must always remain in its source. There is nothing external which is not also internal. The mind is all. If the mind is active even solitude becomes like a market place. There is no use closing your eyes. Close the mental eye and all will be right.

Sri Ramana Maharshi

[1] Sri Ramakrishna enunciates the same principle in the story of the man at a religious gathering wishing he were at a house of pleasure, contrasting with his friend in the house of pleasure wishing he were at the religious gathering.

It is permissible to take life's blessings with both hands provided thou dost know thyself prepared in the opposite event to take them just as gladly. This applies to food and friends and kindred, to anything God gives and takes away. . . . As long as God is satisfied do thou rest content. If he is pleased to want something else of thee, still rest content.

Eckhart

Conform. 180

A householder, endowed with knowledge like Janaka's,[1] can enjoy fruit both from the tree and from the ground. He can serve holy men, entertain guests, and do other things like that. I said to the Divine Mother, 'O Mother, I don't want to be a dry sâdhu.'

Sri Ramakrishna

The rational soul can therefore make a good use of even material and temporal felicity, if it does not give itself over to the creature, to the neglect of the Creator, but rather applies this felicity to the service of the Creator, of whose abounding liberality it has been bestowed.

St Augustine

Therefore, both wealth and poverty are Divine gifts: wealth is corrupted by forgetfulness, poverty by covetousness. Both conceptions are excellent, but they differ in practice. Poverty is the separation of the heart from all but God, and wealth is the preoccupation of the heart with that which does not admit of being qualified. When the heart is cleared (of all except God), poverty is not better than wealth nor is wealth better than poverty. Wealth is abundance of worldly goods and poverty is lack of them: all goods belong to God: when the seeker bids farewell to property, the antithesis disappears and both terms are transcended.

Hujwîrî

Holiness 914
Death 220
Center 835

The attainment of the onepointedness of the mind and the senses is the best of austerities. It is superior to all religious duties and all other austerities.

Śrî Śankarâchârya

Inv. 1017

Eluding the Ego

The only thing which a man must renounce if he wishes to attain the Supreme Truth is the notion of individuality—nothing else.

Swami Ramdas

Death 220

[1] Janaka (7th century B.C.?), the father of Sîtâ, was a great philosopher king, of whom we have a similar prototype in Solomon.

Sanyâsa is to renounce one's individuality. This is not the same as tonsure and ochre robes. A man may be a *grhi* (householder); yet if he does not think he is a *grhi*, he is a *sanyâsi*. On the contrary a man may wear ochre robes and wander about; yet if he thinks he is a *sanyâsi* he is not that. To think of *sanyâsa* defeats its own purpose. . . .

It consists in renunciation not of material objects but of attachment to them.

Sri Ramana Maharshi

Suff. 133

All the activity of man in the works of self-denial has no good in itself, but is only to open an entrance for the one only Good, the light of God, to operate upon us.

William Law

M.M. 973

Asceticism is for the common run of people, since it consists in making the concupiscent appetite abstain from pleasures, in renouncing the temptation to return again to that from which one is separated, in dropping the search for what one has lost, in depriving oneself of superfluous desires, in thwarting the goad of the passions, in neglecting all which does not concern the soul. But this is an imperfection as regards the path of the elect, for it presupposes an importance attached to the things of this world, an abstention from their use, an outward mortification in depriving oneself of things here, while inwardly an attachment is felt for them.

To make an issue of the world amounts to turning thyself toward thyself: it is to pass thy time struggling with thyself; it is to take account of thy feelings and to remain with thyself against thy concupiscence. . . .

In all truth, asceticism is the ardent aspiration of the heart towards Him alone; it is to place in Him the aspirations and desires of the soul; to be preoccupied uniquely with Him, without any other preoccupation, in order that He (to whom be praise!) may remove from thee the mass of these causes.

Ibn al-ʿArîf

Realiz. 887

You have wasted your life in cultivating your spiritual nature: what has become of annihilation in Unification (*al-fanâ fi'l-tawḥîd*)?

Al-Ḥallâj

Much has been discussed about the necessity of pure nourishment in order to have spiritual development. For me, true nourishment means the assimilation of pure thoughts, and devotion to the Truth or God. In vain you may take the most perfectly sâttvic meals each day; what nourishment will your soul derive from this mass of pure food if your mind remains constantly absorbed in material thoughts? . . . Everything that the senses receive is nourishment. Be careful therefore not to let yourself be encroached upon by your stomach. And try always to remain the master of what you absorb.[1]

Ananda Moyî

Some people were conversing about the future life, some saying that fish-eaters will be born into the Pure Land, others that they will not. Hônen overhearing them said, 'If it is a

[1] It is reported in a biography of Ananda Moyî that she herself is capable of both prodigious eating and prodigious fasting. Cf. also the remarkable tour de force of Abû Sa'id ibn Abi 'l-Khayr, recounted in Nicholson's *Studies in Islamic Mysticism*, pp. 71–72.

case of eating fish, cormorants would be born into the Pure Land; and if it is a case of not eating them, monkeys would be so born. But I am sure that whether a man eats fish or not, if he only calls upon the sacred name, he will be born into the Pure Land.'

Hônen

Introd.
Conform.
Inv.
1036

Now it may be asked, 'How can a man be without appetites and enjoyment so long as he is in this present state? I am hungry, and I eat; I am thirsty, I drink; I am weary, I sleep; I am cold, I warm myself; and I cannot possibly find that to be bitter nor barren of natural enjoyment which is the satisfaction of my natural desires. This I cannot alter, so long as nature is nature.' True: but this pleasure, ease, satisfaction, enjoyment, or delight, must not penetrate into the depths of thy heart, nor make up a portion of thy inner life. It must pass away with the things that caused it, and have no abiding place in thee. We must not set our affections thereon, but allow it to come and go, and not repose upon the sense of possession with content or delight in the world or the creature.

Tauler

Illusion
109

Thus, there is sensual eating and wise eating. When the body composed of the four elements suffers the pangs of hunger and accordingly you provide it with food, but without greed, that is called wise eating. On the other hand, if you gluttonously delight in purity and flavour, you are permitting the distinctions which arise from wrong thinking. Merely seeking to gratify the organ of taste without realizing when you have taken enough is called sensual eating.

Huang Po

Repletion combined with contemplation is better than hunger combined with mortification, because contemplation is the battle-field of men, whereas mortification is the playground of children.

Hujwîrî

Contem.
523

Do not therefore ascribe blame or praise to the eating (or not eating) of food, or to the drinking (or not drinking) of wine, but ascribe praise, or woe, unto those who make use properly or improperly of meat and drink.

Palladius

The people most separated from God are the ascetics by their asceticism, the devotees by their devotion, and the knowers by their knowledge.

Bâyazîd al-Bistâmî

Suff. 133
Knowl.
734

'Thou must overcome the need of remorse, the pain of penitence, the labour of confession, the love of the world, temptation of the devil, pride of the body, and annihilation of self-will which drags so many souls back that they never come to real love.'

Mechthild of Magdeburg

Always repenting of wrongs done
Will never bring my heart to rest.

Chi K'ang

Metanoia
495

In short, the Sûfîs are unanimous in recognizing the existence of mortification and discipline, but hold that it is wrong to pay regard to them. Those who deny mortification do not mean to deny its reality, but only to deny that any regard should be paid to it or that anyone should be pleased with his own actions in the place of holiness, inasmuch as mortification is the act of Man, while contemplation is a state in which one is kept by God, and a man's actions do not begin to have value until God keeps him thus. The mortification of those whom God loves is the work of God in them without choice on their part: it overwhelms and melts them away; but the mortification of ignorant men is the work of themselves in themselves by their own choice: it perturbs and distresses them, and distress is due to evil.

Action 340

Suff. 133

Hujwîrî

The Integral Effort Required

Thou shalt not plow with an ox and an ass together.

Metanoia 480

Deuteronomy, XXII. 10

Oh! how often does man give himself up in will to God, and take himself back again as quickly, and fall away from God!

Tauler

God permits not that any other thing should dwell together with Him.

St John of the Cross

Now, spiritual eating is by the mouth of desire, and desire is nothing else but will and hunger. . . . If you have many wills and many hungers, all that you eat is only the food of so many spiritual diseases, and burdens your soul with a complication of inward distempers. And under this working of so many wills it is that religious people have no more good or health and strength from the true religion than a man who has a complication of bodily distempers has from the most healthful food. For no will or hunger, be it turned which way it will or seem ever so small or trifling, is without its effect.

Sin 64

William Law

Lord, how often shall I resign myself, and wherein shall I forsake myself?

Always and at every hour; as well in small things as in great. I except nothing, but do desire that thou be found stripped of all things.

Otherwise, how canst thou be mine, and I thine, unless thou be stripped of all self-will, both within and without?

Conform. 166

The sooner thou doest this, the better it will be with thee; and the more fully and sincerely thou doest it, so much the more shalt thou please me, and so much the greater shall be thy gain.

Some there are who resign themselves, but with exceptions: for they put not their whole trust in God, therefore they study how to provide for themselves.

Faith 501

Some also at first do offer all, but afterwards, being assailed with temptation, they return again to their own ways, and therefore make no progress in the ways of virtue.

These shall not attain to the true liberty of a pure heart, nor to the favor of my sweetest friendship, unless they first make an entire resignation and a daily oblation of themselves. Without this, there neither is nor can be a fruitful union.

Suff. 132

The Imitation of Christ, III. xxxvii

The heat of charity is poor indeed where not God alone, but foreign things besides and with God, urge and excite us towards the works of virtue. Such men are fickle of heart; for in all the things which they do, nature is secretly seeking its own, often without their knowledge, for they know not themselves.

Conform. 170
Realiz. 859

Ruysbroeck

We do not wholly rely upon God, but will have earthly and fleshly things with it. This is impossible, such things offend God and move Him to anger. God must be alone, without a competitor. . . . Hence it comes that courageous persons and such as with magnanimity wholly give up themselves—that is, who wholly deny and sink themselves in God—cannot bear (without the utmost discontent) that one should deal with God so cowardly, as to hold sometimes with Him and sometimes with the world, in that they well know that this comes from the failings of such persons and that all their uneasiness is owing to their timidity or reserve towards God. They cannot well refrain from telling such persons of it, which however offends them—yea, often makes them have an aversion for those that inform them of their failings; since, according to their self-love, they would rather that we should bewail them and look on their grievance with compassion; nay, they would have us believe it to be God's workings. But these enlightened and courageous souls do not choose to act in such a manner: they must give evidence to the truth, which self-love in the others cannot endure.

Contem. 542
Knowl. 745
Sin 75

Johannes Kelpius

Who hath resisted Him, and hath had peace?

Job, IX. 4

Peace 694

Without renunciation of all worldly aims,
What gain is it to meditate the Chosen Teachings?

Milarepa

Wisdom will not enter into a malicious soul, nor dwell in a body subject to sins.

Wisdom, I. 4

Death 208

There are such who perhaps think themselves willing that God should have his due, providing that he also let them enjoy their own without any let or molestation; but they are very jealous lest he should encroach too much upon them, and are careful to maintain a *meum* and *tuum* with heaven itself, and to set bounds to God's prerogative over them, lest it should swell too much, and grow too mighty for them to maintain their own privileges under it. They would fain understand themselves to be free-born under the dominion of God himself, and therefore ought not to be compelled to yield obedience to any such laws of his as their own private seditious lusts and passions will not suffer them to give their consent unto.

Introd. *Rev.*

John Smith the Platonist

And yet, to say the truth, I had as lief have the foppery of freedom as the morality of imprisonment.

Shakespeare (*Measure for Measure,* I. ii. 142)

Flies sit at times on the sweet-meats kept exposed for sale in the shop of a confectioner; but no sooner does a sweeper pass by with a basketful of filth than the flies leave the sweet-meats and sit upon the filth-basket. But the honey-bee never sits on filthy objects, and always drinks honey from the flowers. The worldly men are like flies. At times they get a momentary taste of divine sweetness, but their natural tendency for filth soon brings them back to the dunghill of the world. The good man, on the other hand, is always absorbed in the beatific contemplation of divine beauty.

Sri Ramakrishna

Again, the worldly man is like a snake trying to swallow a mole. The snake can neither swallow the mole nor give it up.

Sri Ramakrishna

To be too fond of this world and of that which is therein, provoketh the wrath of Heaven. If thou sacrifice this fondness, thou shalt be sure of the glory and grace of thy God.

Shekel Hakodesh, 80

Woe to them that are at ease in Zion!

Amos, VI. 1

Judgment 250

Because thou art lukewarm, and neither cold nor hot, I will spue thee out of my mouth.

Revelation, III. 16

God will have all, or none; serve Him, or fall
Down before Baal, Bel, or Belial:
Either be hot, or cold: god doth despise,
Abhorre, and spew out all Neutralities.

Robert Herrick

Maria Alvarez Davila y Salazar was one of the nieces of Madre Teresa; as noble as one could be; as beautiful as one could wish. One September day, she invited her friends to accompany her on an outing: the women in litter, the gentlemen on horseback. . . .

She has them take the direction of the Convent of Saint-Joseph, she descends, the gate to the enclosure opens, Teresa of Jesus who was awaiting her appears carrying a crucifix. Maria, kneeling, kisses ardently the feet of Jesus, then without turning back to answer the farewells, she lets the Madre close the door behind her and forthwith divests herself of her finery.

Thus did Maria Davila cause among all the nobility and knights of the city a great astonishment.

From a Life of St Teresa of Avila

A wife once spoke to her husband, saying, 'My dear, I am very anxious about my brother. For the last few days he has been thinking of renouncing the world and of becoming a Sannyasin, and has begun preparations for it. He has been trying gradually to curb his desires and reduce his wants.' The husband replied, 'You need not be anxious about your brother. He will never become a Sannyasin. No one has ever renounced the world by making long preparations.' The wife asked, 'How then does one become a Sannyasin?' The husband answered, 'Do you wish to see how one renounces the world? Let me show you.' Saying this, instantly he tore his flowing dress into pieces, tied one piece round his loins, told his wife that she and all women were henceforth his mother, and left the house never to return.

Sri Ramakrishna

If it is literally true what our Lord said, that His Kingdom was not of this world, then it is a truth of the same certainty, that no one is a member of this kingdom but he that, in the literal sense of the words, renounces the spirit of this world. Christians might as well part with half the Articles of their Creed, or but half believe them, as really to refuse or but by halves enter into these self-denials.

Metanoia 484

William Law

To attain Buddhahood thus we must scatter this life's aims and objects to the wind.

Milarepa

O ye who believe! Lo! among your wives and your children there are enemies for you, therefore beware of them.

Qur'ân, LXIV. 14

If any man come to me, and hate not his father, and mother, and wife, and children, and brethren, and sisters, yea, and his own life also, he cannot be my disciple.

Death 208

St Luke, XIV. 26

Dhu 'l-Nûn: How can one attain the state of the wise?

The Bedouin Sheikh: By giving up undertakings and genealogies, by cutting short all relations.

Peace 700

Dhu 'l-Nûn

One who has only a mild spirit of renunciation says, 'Well, all will happen in the course of time; let me now simply repeat the name of God.' But a man possessed of a strong spirit of renunciation feels restless for God, as the mother feels for her own child. A man of strong renunciation seeks nothing but God. He regards the world as a deep well and feels as if he were going to be drowned in it. He looks on his relatives as venomous snakes; he wants to fly away from them. And he does go away. He never thinks, 'Let me first make some arrangement for my family and then I shall think of God.' He has great inward resolution.

Action 346 *Holy War* 407

Sri Ramakrishna

Whoso has three things is beloved of God. The first is riddance of goods; the second, of friends, and the third is riddance of self.

Eckhart

For when we have renounced the world, we have constituted God as our property and consequently we have become His property in such a way that He has become our portion and we are His peculiar heritage.

Faith 501

St Peter Damian

'Give Me all that thine is and I will give thee all that Mine is!'

Suff. 132 *Metanoia* 493

Mechthild of Magdeburg

If thou wilt be my disciple, deny thyself utterly.

The Imitation of Christ, III. lvi

If you love God, tear out your heart's love of the world.

'Alî

My son, thou canst not possess perfect liberty unless thou wholly renounce thyself.... Keep this short and complete saying: 'Forsake all and thou shalt find all.'

Void 724

The Imitation of Christ, III. XXXII

A fire can be extinguished; but the flame of desire is never extinguished. Pains affect the body for a while, then cease, and leave you at rest; but from the pain of desire you never get rest, unless you cure it by the medicine of true intelligence (νοῦς), and put desire away from you, and learn to abstain; for desire grows in strength if you follow it, but dies away if you turn from it and abstain.

Knowl. 761 *Illusion* 89

Hermes

Experience has taught us to separate the pure from the impure. Therefore, if you would ameliorate Nature, and produce a more perfect and elaborated subject, purge the body by dissolution of all that is heterogeneous, and unite the pure to the pure, the well-digested to the well-digested, and the crude to the crude, according to the natural and not the material weight.

P. State 573

Michael Sendivogius

Remember, an thou seekest aught of thine thou never shalt find God, for thou art not seeking God merely. Thou art seeking for something with God, making a candle of God, as

Conform. 170

it were, with which to find something, and then, having found it, throwing the candle away. Thus shalt thou fare: aught that thou findest with God is naught, whatsoever it be, whether profit or wages or the interior life or anything else: naught dost thou seek and naught shalt thou find.

Eckhart

When therefore you seek to state or to conceive Him, put all else aside; abstracting all, keep solely to Him; do not look for something to add, rather you have probably not yet sufficiently abstracted from Him in your conception.

M.M. 975
Realiz. 873

Plotinus

Know that no man in this life ever gave up so much that he could not find something else to let go. Few people, knowing what this means, can stand it long, and yet it is an honest requital, a just exchange. To the extent that you eliminate self from your activities, God comes into them—but not more and no less. Begin with that, and let it cost you your uttermost. In this way, and no other, is true peace to be found.

Eckhart

This pearl of eternity is the peace and joy of God within thee, but can only be found by the manifestation of the life and power of Jesus Christ in thy soul. But Christ cannot be thy power and thy life till, in obedience to His call, thou deniest thyself, takest up thy daily cross and followest Him in the regeneration. This is peremptory, it admits of no reserve or evasion, it is the one way to Christ and eternal life. But be where thou wilt, either here or at Rome or Geneva, if self is undenied, if thou livest to thine own will, to the pleasures of thy natural lust and appetites, senses, and passions, and in conformity to the vain customs and spirit of this world, thou art dead whilst thou livest, . . . a stranger to all that is holy and heavenly within thee and utterly incapable of finding the peace and joy of God in thy soul.

Supra 139

Death 208
Creation 42

William Law

Sell all that thou hast, and distribute unto the poor, and thou shalt have treasure in heaven: and come, follow me.

St Luke, XVIII. 22

For I tell thee by that Truth, which is God Himself, if thou art ever to become a man after the will of God, every thing must die in thee to which thou art cleaving, whether it be God's gifts, or the saints, or the angels, or even all that would afford thee consolation for thy spiritual wants: all must be given up. If God is to shine in on thy soul brightly, without a cloud, and accomplish His noble and glorious will in thee, thou must be free and unencumbered by all that affords thee comfort out of God.

Tauler

A man ought therefore to mount above all creatures, and perfectly to renounce himself, and to be in a sort of ecstasy of mind, and to see that thou, the Creator of all things, hast nothing amongst creatures like unto thyself.

P. State 583

The Imitation of Christ, III. xxxi

The Rupture With Habit

Death 206

He who would be what he ought to be must stop being what he is.

Eckhart

Metanoia 488

How can the habitual be severed for thee when thou hast not severed thyself from habit?

Ibn 'Aṭâ'illâh

Charity 597

If you do not first hate your body, my son, you cannot love your Self.

Hermes

Introd. *Love*
Grace 556
Metanoia 493

List ye, good people all: there is none happier than he who stands in uttermost detachment. No temporal, carnal pleasure but brings some ghostly mischief in its train, for the flesh lusts after things that run counter to the spirit and spirit lusts for things that are repugnant to the flesh.[1] He who sows the tares of love in flesh reaps death but he who sows good love-seed in the spirit reaps of the spirit eternal life. The more man flees from creatures the faster hastens to him their creator.

Eckhart

Holy War 407

Having cast off the bonds, like a fish which breaks the net in the water, like a fire that returns not to the spot already burnt up, let one walk alone like a rhinoceros.

Khaggavisâna Sutta, 28

Abandon life and the world, that you may behold the Life of the world.

Dîvâni Shamsi Tabrîz, XLII

Interiorization and Transmutation of the Faculties

What is the lesson of the *Gîtâ?* It is what you get by repeating the word ten times. As you repeat 'Gîtâ', 'Gîtâ', the word becomes reversed into 'tâgi', 'tâgi'—which implies renunciation. He alone has understood the secret of the *Gîtâ* who has renounced his attachment to 'woman and gold' and has directed his entire love to God. It isn't necessary to read the whole of the *Gîtâ*. The purpose of reading the book is served if one practises renunciation.

Sri Ramakrishna

[1] 'The Spirit of God stirs up motions and desires in the saints contrary to those of the flesh, or unrenewed part in man, and inclines them to desire and endeavour the utter destruction of it' (*Cruden's Concordance*, gloss on *Gal.* V. 17).

Abstinence is the bringing of the senses under control through the knowledge that 'all is the Brahman'. This abstinence should constantly be practised.

Tejo-bindu Upanishad, I. 17

Abstinence is detachment from the body and senses, say the Sages.

Trishikhi Brâhmana Upanishad, 28

Many of those who work on *tan* (gold medicine) are ignorant of the method of fixing *shên* (appearance). This does not depend upon swiftness or slowness. The fixing consists in condensing *shên* into the interior of the essence (*ching*). Fixing *shên,* in other words, is extinguishing all desires and returning *shên* to the mind. When this is done, one may see the whole figure of *hsing* (character, nature).

The Secret Papers in the Jade Box of Ch'ing-hua

Contem. 528
Conform. 170
Supra
Introd.
P. State
563

Shut the eyes in your head, that you may see the hidden eye.

Dîvâni Shamsi Tabrîz, XLII

Sin 77
Center 816

When young, beware of fighting; when strong, beware of sex; and when old, beware of possession.

Confucius

Jesus (Peace be upon him!) said, 'Beware of looking, for it sows desire in the heart, and it is sufficient for seduction.'

Christ in Islâm

The ascetic said, '... If you wish to see one who is more ascetic than I, climb this mountain.' Dhu 'l-Nûn went up the mountain and saw a young man seated in a cell; one of his feet, which he had cut off, was lying outside and worms were eating it. 'One day,' he said in answer to Dhu 'l-Nûn's question, 'I was sitting in this cell, when a woman passed by. My heart inclined to her and my body urged me to follow her. I put one foot outside. I heard a voice saying, "After having served and obeyed God for thirty years, art not thou ashamed to obey the Devil now?" Thereupon I cut off the foot which I had set outside, and I am waiting here to see what will happen to me. Why have you come to a sinner like me? If you wish to see a man of God, go to the top of the mountain.'

'Aṭṭâr

Suff.
120

To direct the mental current towards the basic unity of all things and to divert it from the observation of differences is 'Observance' (*niyama*); therein lies transcendent bliss. It should be practised regularly by the wise.

Tejo-bindu Upanishad, I. 18

Reality
775

The countenance of a man who longs for Paradise, and fixes his whole mind upon the one object of reaching it, will always appear as if he had a hatred and abhorrence of the world.

Hônen

Holiness 910
Suff. 126b

M.M. 978

Although enjoying full sexual development, he does not know the conjugation of male and female.

Tao Te Ching, LV

Supra 143

A lion eats meat and yet it mates only once in twelve years; but a sparrow eats grain and it indulges in sex-life day and night. Such is the difference between a Divine Incarnation and an ordinary human being.

Sri Ramakrishna

Renunciation of all manifest forms through the contemplating of the Self which is Existence and Bliss is the mode of worship of the great. It smoothly leads them towards liberation.

Tejo-bindu Upanishad, I. 19

Contem. 528

Abhyâsa consists in withdrawal within the Self every time you are disturbed by thought. It is not concentration or destruction of the mind but withdrawal into the Self.

Sri Ramana Maharshi

Death 215

I made a pillow of my mother's bones and remained in an undistracted state of tranquillity.

Milarepa

Sin 77
Supra 139

Now, as there is no one to touch harlots who have entered into a vacant house, so he who does not touch objects of sense that enter into him is an ascetic and a devotee and a 'performer of the sacrifice to the Self'.

Maitri Upanishad, VI. 10

'Perhaps your Reverence has met a certain lady?'
Mahatissa the Elder replied:
'I know not whether man or woman passed.
A certain lump of bones went by this way.'

Buddhaghosa

God delivered me from the desire for women to such a point that I cannot tell, when a woman appears to me, whether it is a woman or a wall.

Bâyazîd al-Bisṭâmî

A brother who restrains the controlling faculties should be like the Elder Chittagutta who lived in the great Kurandaka cave.... At the cave-entrance there was a great ironwood tree. But the Elder had never looked up at it. It is said that he knew that it was in blossom when each spring he saw the filaments that fell to the ground.

Buddhaghosa

Once in the season of spring she (Râbi'a al-'Adawiyya) went into her chamber and bowed her head in meditation. Her handmaid said, 'O mistress, come forth that thou mayst behold the wondrous works of God!' 'Nay,' she answered, 'do thou come within, that thou mayst behold their Maker. Contemplation of the Maker hath turned me from contemplating that which He made.'

Beauty 663

'Aṭṭâr

Jesus (Peace be upon him!) struck the ground with his hand and took up some of it and spread it out, and behold, he had gold in one of his hands and clay in the other. Then he said to his companions, 'Which of them is sweeter to your hearts?' They said, 'The gold.' He said, 'They are both alike to me.'

Christ in Islâm

Though gold and gems by the world are sought and prized,
To me they seem no more than weeds or chaff.

Fu Hsüan

It is related that Shiblî[1] cast four hundred dînârs into the Tigris. When asked what he was doing, he replied: 'Stones are better in the water.' 'But why,' they said, 'don't you give the money to the poor?' He answered: 'Glory to God! what plea can I urge before Him if I remove the veil from my own heart only to place it on the hearts of my brother Moslems? It is not religious to wish them worse than myself.'

Action 346

Hujwîrî

A husband and wife renounced the world and jointly undertook a pilgrimage to various religious shrines. Once, as they were walking on a road, the husband, being a little ahead of the wife, saw a piece of diamond on the road. Immediately he scratched the ground to hide the diamond, thinking that, if his wife saw it, she might perchance be moved by avarice and thus lose the merit of her renunciation. While he was thus busy, the wife came up and asked him what he was doing. In an apologetic tone he gave her an evasive reply. She noticed the diamond, however, and reading his thoughts, asked him, 'Why have you left the world, if you still feel the difference between the diamond and the dust?'

Sri Ramakrishna

Supra 136

As long as you do not subdue the mind, you cannot get rid of your desires; and unless you suppress your desires, you cannot control your restless mind.

Hence, knowledge of Truth, subjection of the mind and abandonment of desires are the joint causes of spiritual bliss, which is unattainable by the practice of any one of them singly.

Infra 160

Yoga-Vasishtha

A full investigation into Truth will extinguish your desires at once, and the extinction of desires will restore your mind to rest.

Knowl. 761

Yoga-Vasishtha

[1] Abû Bakr Dulaf b. Jaḥdar al-Shiblî, a famous Sufi shaykh of Bagdad.

Sin 64

God cannot be seen so long as there is the slightest taint of desire; therefore have thy small desires satisfied, and renounce the big desires by right reasoning and discrimination.

Sri Ramakrishna

Every time you attempt satisfaction of a desire the knowledge comes that it is better to desist. Repeated reminders of this kind will in due course weaken the desires.

Sri Ramana Maharshi

'Consider the last of every thing, and thou wilt depart from the dream of it.'

Niffarî

Action 329

If a man holds aloof from the desires of this world, the misfortunes of this world hold aloof from him.

Hermes

Illusion 99

When one sees this universe as ephemeral, one gains true discrimination and turns away from worldliness. The Self becomes the Saviour of self.

Srimad Bhagavatam, XI. iii

Contem. 536
Conform. 170

Keep thyself detached from all mankind; keep thyself devoid of all incoming images; emancipate thyself from everything which entails addition, attachment or encumbrance, and address thy mind at all times to a saving contemplation wherein thou bearest God fixed within thy heart as the object from which its eyes do never waver; any other discipline, fasts, vigils, prayers, or whatever it may be, subordinate to this as to its end, using thereof no more than shall answer for this purpose, so shalt thou win the goal of all perfections.

Eckhart

Spiritual Poverty

Humility 199

He who has little will receive.
He who has much will be embarrassed.

Tao Te Ching, XXII

Supra Introd.

The more we have the less we own.

Eckhart

This law is not for those whose wants are many, but for those whose wants are few.

Angelus Silesius

The bird in a forest can perch but on one bough,
And this should be the wise man's pattern.

Tso Ssŭ

Introd. *Humility*

Copia (full satisfaction of desire) is to be got by means of temperance.

Hermes

The poor man is not he whose hand is empty of provisions, but he whose nature is empty of desires.

Hujwîrî

It is only after the disappearance of all worldly interests that the universality of the transcendental Spirit is known.

Yoga-Vasishtha

Void 724

Superfluity distresses wise men more than deficiency distresses you.

Apollonius of Tyana, to the King of Babylon

He who with little is well content is rich indeed as a king; and a king, in his greatness, is poor as the pedlar, when his kingdom sufficeth him not.

Shekel Hakodesh, 175

Every thing which is more than necessary to man, is hostile to him.

Sextus the Pythagorean

It was our belief that the love of possessions is a weakness to be overcome. Its appeal is to the material part, and if allowed its way it will in time disturb the spiritual balance of the man.

Ohiyesa

Jesus said: Wretched is the body which depends upon a body, and wretched is the soul which depends upon these two.

The Gospel according to Thomas, Log. 87

A man is possessed of limited powers and is miserable; he wants to expand his powers so that he may be happy. But consider if it will be so; if with limited perceptions one is miserable, with extended perceptions the misery must increase proportionately. . . . What is the real power? Is it to increase prosperity or bring about peace? That which results in peace is the highest perfection.

Sri Ramana Maharshi

Peace 694

Desires are only the lack of something: and those who have the greatest desires are in a worse condition than those who have none or very slight ones.

Plato (*Eryxias,* 405 E)

Introd. *Illusion*

To have but few desires and to be satisfied with simple things is the sign of a superior man.

Gampopa

Holy War 396

The noblest victory of all . . . the victory over pleasure. . . .

Plato (*Laws*, 840 C)

Infra 160

God, indeed, is not in want of any thing, but the wise man is in want of God alone. He, therefore, who is in want but of few things, and those necessary, emulates him who is in want of nothing.

Sextus the Pythagorean

Nothing Real Is Lost

Allâh asketh naught of any soul save that which He hath given it.

Qur'ân, LXV. 7

Realiz. 890

Am I my own, or have I anything of my own? Am I other than Thou?

Hermes

Throw pure gold into the fire:
If it contains no alloy, what is there to burn?

Shabistarî

Suff. 133

Sin 66

We suppress nothing at all in man: Neither intelligence, nor wit, nor art; for all these come from the divine wisdom. By no means do we suppress the Word pronounced by the wisdom formed in God, but only the Beast who wants to hold sway over the divine contemplativeness, that is to say, the bestial will of egoism and individualism, which has strayed from God and which honours its own self like a false god.

Boehme

Peace 700

Inasmuch as heat comes to me without fire, I am content if He extinguish my fire.
Inasmuch as He gives light without any lamp—if your lamp is gone, why are you lamenting?

Rûmî

Realiz. 873

Renunciation is non-identification of the Self with the non-self.

Sri Ramana Maharshi

The supreme reason (why I am unattached) is that nothing really exists except the Self.

Śrî Śankarâchârya

Reality 803

Wherefore forsake all things for God, and then God will be truly given unto you in all things.

Tauler

In order to arrive at possessing everything,
Desire to possess nothing.

St John of the Cross

Void 724

Sweet lady, you have given me life and living;
For here I read for certain that my ships
Are safely come to road.

Shakespeare (*Merchant of Venice*, v. i. 286)

What thou holdest, passes away from thee, what thou losest, thou findest in thee.

St Augustine

Death 208

Cymbeline. O Imogen!
Thou hast lost by this a kingdom.
Imogen. No, my lord;
I have got two worlds by't.

Shakespeare (*Cymbeline*, v. v. 373)

M.M. 978

He is not solitary with whom is God, nor is the power of joy extinguished because his appetite is kept from things abject and vile. He rather does himself an injustice who admits to the society of his joy what is disgraceful or unworthy of his love.

Hugh of Saint-Victor

Desire is slavery; renunciation is freedom.

Hermes

The knowledge which results in renunciation (*zuhd*) consists of the realisation that what is renounced is of little value in comparison with what is received.

Al-Ghazâlî

Goodness is to be seen in its perfection only when man's virtue is fortified against desire, and he scorns all things that are alien to him. Now all earthly things which man holds in his possession to gratify his bodily desires are alien to all that part of his nature which is akin to God; and these things are rightly called 'possessions', for this reason, that they were not born with us, but we began to get possession of them at a later time. All such things then are alien to man; yes, and the body too we must regard as alien, that so we may scorn not only the objects of our greed, but also that which is the source of the vicious greed within us.

Suff. 126b

Hermes

O rejoice
Center 847 Beyond a common joy, and set it down
Holiness 902 With gold on lasting pillars. In one voyage
Pilg. 365 Did Claribel her husband find at Tunis,
And Ferdinand, her brother, found a wife
Void 724 Where he himself was lost; Prospero his dukedom
Realiz. 859 In a poor isle; and all of us ourselves,
Metanoia 480 When no man was his own.

Shakespeare (*The Tempest*, v. i. 206)

True asceticism consists in abandoning what one does not need when one has received something better.

Henry Madathanas

The world is but a single day, in which we are fasting.

Abu 'l-Faḍl Muḥammad b. al-Ḥasan al-Khuttalî

Al-Shiblî, being asked concerning abstinence, said: 'Alas for you! What value is there in that which is less than the wing of a gnat, that abstinence must needs be exercised concerning it?'[1]

Al-Kalâbâdhî

Introd. *Void* Whoso renounces things in their contingent sense possesses them as absolute and eternal.

Eckhart

Detachment and Serenity

Yoga is restraining the mind-stuff (*Chitta*) from taking various forms (*Vrittis*).... Control is by practice and non-attachment.... That effect which comes to those who have given up their thirst after objects either seen or heard, and which wills to control the objects,[2] is non-attachment. That is extreme non-attachment which gives up even the qualities, and comes from the knowledge of *Purusha*. *Peace* 705

Patanjali

He who would be serene and pure needs but one thing, detachment.

Eckhart

[1] Referring to the ḥadith (saying of Muhammad): 'This world does not weigh with God so much as a gnat's wing.'

[2] Hence the antithesis of 'quietism'. Yoga aims to control the obstacles to grace, and not—as is often contended—the 'gift of grace' itself.

The root evil is attachment which has sprung from the seed called desire—the primal cause of ignorance. The eradication of desire and attachment means liberation.

Swami Ramdas

Introd. *Death*

Detachment abideth in itself.

Eckhart

Unless the mind becomes steady there cannnot be yoga. It is the wind of worldliness that always disturbs the mind, which may be likened to a candle-flame. If that flame doesn't move at all, then one is said to have attained yoga.

Sri Ramakrishna

Contem. 532

The heart detached has no desire for anything nor has it anything to be delivered from. So it has no prayers at all; its only prayer consists in being uniform with God.

Eckhart

Conform. 180

The mind detached is of such nobility that what it sees is true, what it desires befalls and its behests must be obeyed.

Avicenna

Holiness 924

Motionless detachment makes a man superlatively Godlike. For that God is God is due to his motionless detachment, and it is from his detachment that he gets his purity and his simplicity and his immutability. . . . Prayers and good works wrought by a man in time affect no more the divine detachment than if no prayers nor virtuous works had come to pass in time; nor is God any kindlier disposed towards that wight than if his prayers and deeds had all been left undone. Further I declare, when the Son in his Godhead was pleased to be made man and was and suffered martyrdom, God's motionless detachment was no more disturbed than if he had never been made man.

Eckhart

Peace 700

Action 346

M.M. 994

VOLITION, OBEDIENCE AND CONFORMITY

Pax hominibus bonae voluntatis.

St Luke, II. 14

'So God created man in his own image, in the image of God created he him' (*Genesis*, I. 27). The end of the spiritual path is the recovery of the primordial perfection that has been lost in the Fall, and the *means* to this end is a spiritual method; but method is abused the moment it becomes mistaken for the end. Method is the raft which Buddhist and Christian texts insist has served its function once the farther shore is attained. In examining the bewildering variety, often complexity, of spiritual methods, one fact emerges: all methods are based on spiritual realities to which the soul must inwardly conform, whether or not these realities exteriorize as outward practices. Renunciation of impure desires, for example, is incumbent on all who follow a spiritual training, but all are not for that commanded to leave the world. All souls must engage in spiritual combat, but holy war on an actual battlefield is only one support among others for this combat. Gentleness, love, forgiveness and serenity are likewise indispensable conditions of soul, but the outward observance of nonviolence is not an inevitable corollary;[1] and one even finds a harmonious synthesis of these two qualities of puissance and serenity reaching its apogee in a tradition like that of the American Indian. 'As for spiritual chastity, of which physical chastity is only one support among others equally possible, it is imposed unconditionally, for without it there could be no way out from the illusory world of forms; but this spiritual chastity will assume different names depending on the path: thus in Islam it becomes "poverty", in such manner that the functions of procreation and "chastity" may even merge here on the physical plane' (Schuon: *L'Oeil du Coeur*, pp. 86–87).

Suff. 133
Orthod. 296
Holy War passim
P. State 573
Introd. *Love*

'The outward manifestations of the same virtue can differ according to the points of view and the circumstances. Thus, certain Sufis have manifested their scorn of the world by wearing poor and patched-up garments, while others have affirmed the same interior attitude by wearing sumptuous clothing. In this latter case, the Sufi's affirmation of his person is really only a submission to the impersonal truth which he incarnates; his humility is his extinction in an aspect of glory which does not belong to him' (Titus Burckhardt: *Introduction aux Doctrines ésotériques de l'Islam*, Lyon, 1955, pp. 41–42).[2] 'Blessed is the rich man that is found without blemish' (*Ecclesiasticus*, XXXI. 8).

[1] Islamic history tells how the great caliph 'Alî was on the point of killing an enemy in battle, when the man spat in his face. 'Alî turned away, refusing to kill when angered.

[2] Cf. also Schuon: *Les Stations de la Sagesse*, p. 149, note 1.

It is almost unavoidable that the neophyte in spiritual work should have an over-simplified notion of the latitude and scope of spiritual attitudes possible, since these attitudes stem from the infinity of archetypes which exist, whereas the neophyte will tend to limit the truth to the fragment he knows of it.

Apart from personal qualifications, method presupposes legitimate spiritual affiliation, with an adequate background of theoretical knowledge, and basically comprises a 'technique' combined of virtue and concentration intended to make Divine Grace operative. Virtue fashions the soul in the image of God, and concentration 'focuses' this image to receive the illumination of grace: 'Truth illuminates the will, which illuminated, vivifies the truth' (Schuon: *Perspectives spirituelles*, p. 231). So essential is concentration, the Vedanta teaches, that 'man can acquire true Divine Knowledge even without observing the rites prescribed; and one finds in fact in the *Vêda* many examples of persons who have neglected to accomplish such rites, or who have been prevented from doing so, and yet through keeping their attention perpetually concentrated and fixed on the Supreme *Brahma* (which constitutes the only really indispensable preparation), have acquired true Knowledge of It' (*Brahma-Sûtras*, III. iv. 36–38; translation and commentary by Guénon in *L'Homme et son Devenir*, p. 173; cf. also *La Métaphysique orientale*, pp. 16–17).[1]

Introd. *Holiness* *Inv.* 1017

Once it is seen that every method delineates some inward spiritual truth, to which outward practice is subordinate, and that the underlying purpose in all method is the development of spiritual concentration, to which any method is subordinate, then the relative, contingent nature of all method whatsoever can be better appreciated. In this respect a Hindu text admonishes: 'Ritual purifications, worship, sacrificial rites, mystical ecstasy, fasts and observances, silence, control of the senses, objects of meditation and meditation itself, sacred formulae, charities, . . . ritual sacrifices, long periods of penance, mortifications, pilgrimages: all these are obstacles under the garb of religion. . . . Desire for the society of saints, attempts to avoid the sinful, . . . thinking of the shape which dwells in the body, and of the Shapeless in that shape, and thinking further that in the Shapeless dwells the Principle, the Brahman, and thus (to) silence the heart. All these are but obstacles in the form of Knowledge' (*Shiva Samhitâ*, V. 6–8, 12–13; in Daniélou: *Yoga*, p. 118). Sri Ramakrishna expresses the relativity of method in a different way: 'You see, the thing is somehow or other to get into the Lake of the Nectar of Immortality. Suppose one person gets into It by propitiating the Deity with hymns and worship, and you are pushed into It. The result will be the same. Both of you will certainly become immortal' (*Gospel*, p. 217).

Renun. 143 *Void* 721

But the question of method is still left unresolved, and the whole problem seems to culminate in the words of Ibn 'Aṭâ'illâh: 'Thy seeking from Him is an imposition on Him, and thy seeking for Him is because thou art lacking Him, while thy seeking for other than Him is for thy want of awe of Him, and thy seeking from other than Him is because of thy remoteness from Him' (*Ḥikam*, no. 31). These words amount to a chastisement of intention; what is asked of us is right intention, or even perfect intention, which in truth is nothing other than perfect submission and conformity to the will of God. If one aspires to

[1] The teaching set forth here corresponds with the conception of *devekuth* in Spanish Kabbalism. This all assumes, of course, doctrinal preparation and normal attachment to some traditional form ('initiatic regularity')—the 'active' aspect of obedience, dealt with in the chapter on Orthodoxy and Ritual.

be God-like, one must conform to the likeness of the Divinity, since it is not in the nature of the Divinity to conform to one's own likeness. The key, then, to right method lies in an obedience tantamount to conformity: this is the infallible touchstone to all spiritual practice, the *sine qua non* of any discipline.

Sin 66
Pilg. 379
Illusion 85

Since intention and obedience concern volition, or the will, it is in the last analysis a reformation of the will that is involved. 'It is not this individual nature as such which constitutes the egocentric illusion; the veil which has to be pierced is uniquely the attribution, to this individual nature, of an autonomous and "a priori" character which belongs to the Essence alone' (Burckhardt, op. cit., p. 108). 'The training of the will is logically prior to the training of the intellect' (Coomaraswamy: *Gotama the Buddha*, London, 1948, p. 33). The Fall is equivalent to a turning away of the will from God; man is redeemed when his will is again concentrated, definitively, upon God. But as the author of *The Imitation of Christ* says, 'this is not the work of one day, nor children's sport; yea rather in this short word is included all perfection' (III. xxxii). Man is held to the world of phenomena, forms, or *mâyâ*, by the mind, which itself is a product of *mâyâ*. In explaining the processes of *yoga*, Alain Daniélou writes:

'The limits of the physical body can never be transgressed without knowing and thoroughly mastering the sensory impulses which govern the process of living.

'The most vital impulses delude us the most, thus protecting vital functions from dangerous interferences.... The network of the instincts binds the gross to the subtle body and keeps us imprisoned. The knots of this network are strong and complex and, without the proper technique for undoing them, we can never escape from our physical envelope but are kept always on the path of the individual and social instincts by which the continuity of physical life is assured' (op. cit., p. 3).[1]

Orthod. 280
Judgment 250
Grace 552
558

'With men this is impossible; but with God all things are possible' (*Matt.*, XIX. 26). By no power of the human will alone can an individual escape from the 'Round of Existence', but only through the initial grace of God: 'No man can come to me, except the Father which hath sent me draw him' (*John*, VI. 44); 'He hath chosen us in him before the foundation of the world' (*Eph.*, I. 4); 'and they that are with him are called, and chosen, and faithful' (*Rev.*, XVII. 14). This brings in the question of free will. From the macrocosmic and metaphysical standpoint, the Divinity alone has 'free will'. It alone being unconditioned and undetermined. 'The ineluctability of natural law is only the inverse reflection, in the mirror of Nature, of the free sovereignty of the Divine Act' (T. Burckhardt. *Études Trad.*, 1950, p. 13). 'For the scripture saith unto Pharaoh, Even for this same purpose have I raised thee up, that I might shew my power in thee, and that my name might be declared throughout all the earth. Therefore hath he mercy on whom he will have mercy, and whom he will he hardeneth' (*Romans*, IX. 17–18). From the microcosmic and 'initiatic' standpoint, man by remotion participates in the Divine

[1] 'Meditation or concentration practised haphazardly and outside of tradition will be ineffective, and even dangerous in different ways; the illusion of progress in the absence of real criteria is certainly not the least of these dangers' (Schuon: 'Des Stations de la sagesse'. *France-Asie*, juin-juillet, 1953, p. 510). 'Can the Ethiopian change his skin, or the leopard his spots? Then may ye also do good, that are accustomed to do evil' (*Jeremiah*, XIII. 23). 'The sea will ebb and flow, heaven show his face:/Young blood doth not obey an old decree:/We cannot cross the cause why we were born' (*Love's Labour's Lost*, IV. iii. 216).

Nature, which is his own immortal Essence, and at one degree or another inevitably reflects the possibility of free will;[1] consequently he is accountable for his acts: 'It must needs be that offences come; but woe to that man by whom the offence cometh!' (*Matt.*, XVIII. 7). St Augustine follows the same logic in saying that God is the author of our good acts, and that we alone are responsible for our bad acts.[2] What is pertinent to the question of practice can be resolved in the traditional formula: He who is the perfect slave of God is the master of all else.

Creation 42
Wr. Side 464
Action 340
Rev. 967

[1] 'The autonomous activity of the human will is the fundamental fact in the working of all *Karma'* (Coates and Ishizuka: *Hônen the Buddhist Saint*, vol. I, p. X).

[2] '"We" are the authors of whatever is done amiss, and therefore not really "done" at all; while of whatever is actually done, God is the author' (Coomaraswamy: *Hinduism and Buddhism*, p. 37, note 81).

Volition

Holiness 914

The measure of your holiness is proportionate to the goodness of your will. Consider then how good your will is, and the degree of your holiness will be clear to you. For every one is as holy, as he is good of heart.

Ruysbroeck

Then is God in the man, when there is nothing in him which is contrary to the will of God.

Tauler

Renun. 146

Metanoia 480

You may know with the utmost certainty that, if you have no inward peace, if religious comfort is still wanting, it is because you have more wills than one. For the multiplicity of wills is the very essence of fallen nature, and all its evil, misery and separation from God lies in it; and as soon as you return to and allow only this one will, you are returned to God and must find the blessedness of His kingdom within you.

William Law

Contem. 536

Thou seekest Paradise and desirest to come to the place where thou wilt be delivered from all pain and all anxiety. Calm thy heart, and make it pure and white: so wilt thou be, even here, this very Paradise.

Angelus Silesius

P. State 583

Introd.

Suff.

This world is an outer court of Eternity, and therefore it may well be called a Paradise, for it is such in truth. And in this Paradise, all things are lawful, save one tree and the fruits thereof. That is to say: of all things that are, nothing is forbidden and nothing is contrary to God but one thing only: that is, self-will, or to will otherwise than as the Eternal Will would have it. . . . And could a man, while on earth, be wholly quit of self-will and ownership, and stand up free and at large in God's true light, and continue therein, he would be sure of the Kingdom of Heaven.

Theologia Germanica, L, LI

Supra Introd.

It is the mind that makes one wise or ignorant, bound or emancipated. One is holy because of his mind, one is wicked because of his mind, one is a sinner because of his mind, and it is the mind that makes one virtuous. So he whose mind is always fixed on God requires no other practices, devotion, or spiritual exercises.

Sri Ramakrishna

By the will art thou lost, by the will art thou found,
By the will art thou free, captive and bound.

Angelus Silesius

The will is that which has all power: it makes heaven and it makes hell: for there is no hell but where the will of the creature is turned from God, nor any heaven but where the will of the creature worketh with God.

William Law

Judgment 266

Reduce to meekness the wild motions of the will, and make it thy care to tame the cruel beast. Thou art bound to the will: strive to unfasten the bond that cannot be broken. The will is thy Eve.

St Bonaventura

Holy War 396

Will-power should be understood to be the strength of mind which makes it capable of meeting success or failure with equanimity. . . . Success develops arrogance and the man's spiritual progress is thus arrested. Failure on the other hand is beneficial, in as much as it opens the eyes of the man to his limitations and prepares him to surrender himself. Self-surrender is synonymous with eternal happiness. Therefore one should try to gain the equipoise of mind under all circumstances. That is will-power.

Sri Ramana Maharshi

Sin 75
Renun. 136

The Will the cause of Woe

When man is punisht, he is plagued still,
Not for the fault of Nature, but of will.

Robert Herrick

. . . O Jerusalem, how often would I have gathered thy children together, as a hen gathereth her chickens under her wings, and thou wouldest not' (*Matt.*, XXIII. 37): It is not said, 'thou *couldest* not,' but 'thou *wouldest* not.'

Boehme

Grace 558
Sin 76

The soul's impurity consists in bad judgements, and purification consists in producing in it right judgements, and the pure soul is one which has right judgements.

Epictetus

Knowl. 761
Judgment 242

To make ourselves in all things conformable to the will of God means to lead an entirely angelic life: it means truly to live the life of Jesus Christ.

St Vincent de Paul

The greatest good for man is to become conformable to the will of God.

St Thomas Aquinas

The perfection of the love of God consists in the union of our will with His most holy will.

St Alphonsus Liguori

Faith 501

(The) inner voice speaks in the devotee when he has surrendered up his body, mind and soul into the keeping of the Divine who dwells within him.

Swami Ramdas

True peace consists in not separating ourselves from the will of God.

St Thomas Aquinas

Metanoia 495

The greatest treasure next to God is a good will in this world:
Even if all has been lost, by it all can be regained.

Angelus Silesius

Holiness 902

When any disturbing news is brought you, bear this in mind, that news cannot affect anything within the region of the will.

Epictetus

Knowl. 734

What need of so much news from abroad, when all that concerns either life or death is all transacting and at work within us?

William Law

You carry heaven and hell with you.

Sri Ramana Maharshi

The foundation of heaven and hell is laid in men's own souls.

John Smith the Platonist

Death 215

How do the people of the Anga country live? What are the residents of the Vanga country doing? It is in such enquiries and studies that we spend our time. We do not care to ascertain the nature of the city of Yama, the God of Death to which we are all bound to go at some time or other, sooner or later, and we do not have even the slightest desire to know about it.

Sri Chandrasekhara Bhâratî Swâmigal

Reality 773
Action 329

It is in our power to choose the better, and likewise to choose the worse. . . . The soul, when it cleaves to evil things, draws near to corporeal nature, and for this reason the man who has chosen the worse is under the dominion of Destiny.
. . . The intelligent substance in us is self-determining. The intelligent substance remains ever in the same state without change, not partaking of the nature of the things which come into being, and therefore Destiny has no hold on it.

Hermes

Creation 42
Judgment 261

The soul of man alone has received from God the faculty of voluntary movement, and in this way especially is made like to Him, and thus being liberated, as far as might be, from that hard and ruthless mistress, necessity, may justly be charged with guilt, in that it does not honour its Liberator.

Philo

Be it known that man alone of beings holds a mid-place between corruptible and incorruptible.

Dante (*De Monarchia,* III. xvi)

For the Soul is many things; and each of us is an Intellectual Cosmos, linked to this world by what is lowest in us, but by what is highest to the Divine Intellect. By all that is intellective we are permanently in that higher realm, but at the fringe of the intellectual we are fettered to the lower.

What, then, is the achieved Sage?

One whose Act is determined by that higher phase of the Soul. It does not suffice to perfect virtue to have only this spirit as cooperator in the life: the acting force of the Sage is the Divine Mind (*νοῦς*) which therefore is itself his presiding spirit, guided by a presiding spirit of its own, no other than the very Divinity.

Plotinus

Holiness 924 *Metanoia* 480 Introd. *Holiness*

Before man is life and death, good and evil, that which he shall choose shall be given him.

Ecclesiasticus, XV. 18

Do not now say, that you have this one will and one hunger and yet find not the food of life by it. For as sure as you are forced to complain, so sure is it that you have it not. . . . For God's kingdom must manifest itself with all its riches in that soul which wills nothing else; it never was nor can be lost but by the will that seeks something else.

William Law

Renun. 146 *Grace* 556

Go not after thy lusts, but turn away from thy own will.

Ecclesiasticus, XVIII. 30

Man is a holy temple of God. But all traders must first be driven out of this temple of God, that is all the fancies and imaginations which are not really of Him, and also all delight in our own will.

Tauler

Introd. *Inv. Contem.* 528

God will never give himself openly to the soul . . . except she bring her husband, her whole free will, to wit.

Eckhart

Now the free will may reach to which it pleases; both gates stand open to him. Many who are in Christ's line are also brought through imagination and lust, as Adam was, into iniquity; they are indeed called, but they persevere not in the election, for the election is set upon him who departs from sin; he is elected that dies to sin in Christ's death, and rises in Christ's resurrection, who receives God in Christ, not only in the mouth, but in divine desire in the will and new-birth, as a new fiery generation: Knowledge apprehends it not, only the earnest desire and breaking of the sinful will, that apprehends it. . . . Adam is chosen in Christ; but that many a twig withers on the tree, is not the tree's fault, for it

Infra 170

withdraws its sap from no twig. only the twig gives forth itself too eagerly with the desire: it runs on in self-will. viz. it is taken by the inflammation of the sun and the fire. before it can draw sap again in its mother. and refresh itself.

Boehme

Know. therefore. my children. that two spirits wait upon man—the spirit of truth and the spirit of deceit. And in the midst is the spirit of understanding of the mind. to which it belongeth to turn whithersoever it will.

The Testament of Judah, XX. 1. 2

The free will of man holds a place midway between the divine Spirit and the inordinate desires of the body.

St Bernard

The will stands between God and nature and must in all its workings unite either with God or nature.

William Law

Whatever you are. or whatever you feel. is all owing to the working and creating power of your own will. This is your God or your Devil. your Heaven or your hell. and you have only so much of one or the other as your will. which is the first mover. is either given up to the one or to the other.

William Law

Center 816
Introd.
Orthod.

All that exists is gained in two steps: by lifting up the foot from self-interest and setting it down on the commandments of God.

Bâyazîd al-Bisṭâmî

Intention

Give good tidings to My bondmen
Who hear advice and follow the best thereof. Such are those whom Allâh guideth. and such are men of understanding.

Qur'ân, XXXIX. 17. 18

He (the aspirant with right intention) does not ask what is allowable and pardonable. but what is commendable and praiseworthy.

William Law

That which is lawful is clear, and that which is unlawful likewise, but there are certain doubtful things between the two from which it is well to abstain.

Muhammad

'If anyone will generously renounce his own will to seek only My good pleasure, My Divine Heart will illuminate him with a vivid light to know My wishes.'

St Gertrude

He (Christ) hangs all true acquaintance with divinity upon the doing God's will, 'If any man will do his will, he shall know of the doctrine, whether it be of God' (*Jn.* VII. 17).

John Smith the Platonist

Love, and do what thou wilt.

St Augustine

Holiness 902

The all-important thing is to have a true heart, whether one's outward appearance is good or bad.

Hônen

Renun. 136

I may err but I may not be a heretic—for the first has to do with the mind and the second with the will!

Eckhart

Heterod. passim

Actions will be judged according to intentions.

Muhammad

Issuing from the north of the Old City (Patna in Central India) and going down for three *li* to the east, there is Dêvadatta's rock cave; and fifty paces from this, there is a large square black rock. Formerly, a religious mendicant, pacing backwards and forwards on it, reflected as follows:—'This body of mine suffers the bitterness of impermanency; in vain do I attain to an outlook which is not impure. I loathe this body!' Thereupon he seized a knife, meaning to kill himself; but once more he reflected: 'The World-Honoured One has set his canon against self-slaughter.' Then he further reflected: 'Although this is so, I now only desire to slay three baneful thieves,—lust, hate, and ignorance.' He then took the knife and cut his throat. At the beginning of the cut he became a Śrotâpanna; when half through, an Anâgâmin; and when quite through he became an Arhat, and attained to pari-nirvâna (and died).

The Travels of Fa-hsien

Death 226
M.M. 994

The magnetic needle always points towards the north, and hence it is that the sailing vessel does not lose her course. So long as the heart of man is directed towards God, he cannot be lost in the ocean of worldliness.

Sri Ramakrishna

Rev. 965

Faith 501 — Realiz. 870 — Suff. 133

So likewise must we simply commit ourselves to God with perfect trust in His eternal purpose: for when He pleases that it shall be accomplished to our waiting souls, then, no doubt, He will come to us, and be born in us. But when? Leave that to Him: to some He comes in their youth; to others in old age; to some in death: this leave to His Divine will, and do not take upon thyself to adopt any singular exercises, but keep the Commandments, and believe the articles of the Christian faith.

Tauler

Holy War 403

Work your work before the time, and he will give you your reward in his time.

Ecclesiasticus, LI. 38

Let not the postponement of giving despite the persistency in the appeal occasion thee despair, for He has guaranteed thee the response in the manner He chooses for thee, not the manner thou choosest for thyself, and in the time He wills, not in the time thou willest.

Ibn ʿAṭâʾillâh

God doth not promise here to man, that He
Will free him quickly from his miserie;
But in His own time, and when He thinks fit,
Then He will give a happy end to it.

Robert Herrick

Grace 552

So then it is not of him that willeth, nor of him that runneth, but of God that sheweth mercy.

Romans, IX. 16

Renun. 143

Even should one zealously strive to learn the Way,
That very striving will make one's error more.

Po Chü-i

Stumbling is the fruit of haste.

ʿAlî

Orthod. 296

Wisely and slow; they stumble that run fast.

Shakespeare (*Romeo and Juliet*, II. iii. 94)

Hasten slowly and ye shall soon arrive.

Milarepa

Symb. 306 — Contem. 542 — Heterod. 423 — Suff. 133

To go by this road, and not another, I confide to thee;
Observe but *the traces of my wheel.*
And to impart throughout an equal heat,
Too precipitantly towards earth and sky neither mount nor descend.
For in mounting too high the sky thou wilt burn,
And in descending too low the earth thou wilt destroy.

But if in the middle thy course remains,
The path is more united and the way more sure.

De Nuysement

Pilg. 365

Then, as my gift and thine own acquisition
Worthily purchas'd, take my daughter: but
If thou dost break her virgin knot before
All sanctimonious ceremonies may
With full and holy rite be minister'd,
No sweet aspersion shall the heavens let fall
To make this contract grow; but barren hate,
Sour-ey'd disdain and discord shall bestrew
The union of your bed with weeds so loathly
that you shall hate it both: therefore take heed,
As Hymen's lamps shall light you.

Shakespeare (*The Tempest*, IV. i. 13)

All Auctors writing of this *Arte*,
Saye haste is of the *Devils* parte:
The little Boke writ of the Philosophers feast,
Saith, *omnis festinatio ex parts diaboli est:*
Wherefore that Man shall soonest speede,
Which with greate Leasure wisely woll proceede.

Thomas Norton

Holy War 403

The fruits of consummation ripen through patiently following the injunctions of the *Shastras* and repeating the *mantrams* given at the Initiations without haste for success, and by perfecting oneself through long practice.

Yoga-Vasishtha

We should never pray to God that He may grant what we desire, but that His will may be accomplished in us.

St Nilus

What Thou art, thus may I be.

Taittirîya Saṁhitâ, I. 5. 7. 6

What now can be so desirable to a sober, sensible man as to have the vain, disorderly passions of his own corrupted heart removed from him, to be filled with such unity, love, and concord as flows from God, to stand united to and co-operating with the divine goodness, willing nothing but what God wills, loving nothing but what God loves, and doing all the good that he can to every creature, from a principle of love and conformity to God? Then the kingdom of God is come, and His will is done in that soul, as it is done in Heaven. Then Heaven itself is in the soul and the life and conversation of the soul is in Heaven. From such a man the curse of this world is removed; he walks upon consecrated

Charity 605

P. State 563

ground, and everything he meets, everything that happens to him, helps forward his union and communion with God. For it is the state of our will that makes the state of our life; when we receive everything from God and do everything for God, everything does us the same good and helps us to the same degree of happiness. Sickness and health, prosperity and adversity, bless and purify such a soul in the same degree; as it turns everything towards God, so everything becomes divine to it. For he that seeks God in everything is sure to find God in everything. When we thus live wholly unto God, God is wholly ours and we are then happy in all the happiness of God; for by uniting with Him in heart, and will, and spirit, we are united to all that He is and has in Himself. This is the purity and perfection of life that we pray for in the Lord's Prayer, that God's kingdom may come and His will be done in us, as it is in Heaven. And this we may be sure is not only necessary, but attainable by us, or our Saviour would not have made it a part of our daily prayer.

Holiness 914

Realiz. 870

Holiness 900

William Law

We are, in the hands of God, like blocks of marble in the hands of sculptors.

St Alphonsus Liguori

(He) who desires to attain to the unification of the human will with the Divine Will, should be as a dead body in the hands of God, acquiescing in all the vicissitudes which come to pass through His decree and all that is brought about by the might of His power.

Death 206

Junayd

He who serves God after God's will shall be rewarded according to his own will; but he who prays to God according to his own will shall not be answered in accordance with his own will, but after God's will.

Metanoia 493

Tauler

All consult Thee (O Truth) on what they will, though they hear not always what they will. He is Thy best servant, who looks not so much to hear that from Thee, which himself willeth; as rather to will that, which from Thee he heareth.

St Augustine

Have courage to look up to God and say, 'Deal with me hereafter as Thou wilt, I am as one with Thee, I am Thine. I flinch from nothing so long as Thou thinkest it good. Lead me, where Thou wilt, put on me what raiment Thou wilt. Wouldst Thou have me hold office, or eschew it, stay or fly, be poor or rich? For all this I will defend Thee before men. I will show each thing in its true nature, as it is.'

Renun. 139

Epictetus

Direct all your prayers to one thing only, that is, to conform your will perfectly to the Divine will. . . . On this rests our sanctification.

St Teresa of Avila

Suddenly, or even thou knowest, all is away and thou left barren in the boat, blowing with blundering blasts now hither and now thither, thou knowest never where nor whither.

Pilg. 366

Yet be not abashed: for he shall come, I promise thee, full soon, when he liketh, to relieve thee and doughtily deliver thee of all thy dole, far more worthily than ever he did before. Yea! and if he go again, again will he come: and each time, if thou wilt bear thee by meek suffering, will he come more worthlier and merrylier than other. And all this he doth because he will have thee made as pliant to his will ghostly as a Roan glove to thine hand bodily.

P. State 563

The Epistle of Privy Counsel, XII

Our Lady said: 'Take this gold coin. It is thine own will. Offer it in all things to my Lord and Son.'

In deep modesty and holy awe the little maid received the great coin. She looked at it and saw that on one side it bore the Descent from the Cross, on the other the whole Kingdom of Heaven with the nine choirs and above them the Throne of God. Then God's voice spoke to her, 'If thou offer Me this coin so that thou wilt never take it back, then will I loose thee from the Cross and bring thee to Me in My Kingdom.'

Suff. 132

Mechthild of Magdeburg

It would be very bitter if my life-work, the Jesuit Order, should be dissolved. The establishment of my Society has been the object of all my endeavors. . . . If it were to happen without my guilt, about fifteen minutes of meditation and recollection in prayer would banish all disquietude from my soul, even were the whole Society dissolved as salt in water!

St Ignatius Loyola

We ought to be utterly detached in our intention, having no one, nothing, in view but the Godhead as such: not happiness nor this nor that, just God and Godhead in itself. Aught beside that thou intendest is a divine impurity. Seize the actual Godhead itself. God help us so to do.

M.M. 975 994

Eckhart

Brahmaloka cannot be gained so long as there is any desire left in the person. Desirelessness alone will confer the *loka* on him. His desirelessness signifies the absence of the incentive for rebirth.

Renun. 146

Sri Ramana Maharshi

Allâh promiseth to the believers, men and women, Gardens underneath which rivers flow, wherein they will abide—blessed dwellings in Gardens of Eden. But acceptance from Allâh is greater. That is the supreme triumph.

M.M. 973 *Beauty* 663

Qur'ân, IX. 72

God revealed to Jesus (Peace be upon him!), 'When I consider the secret thoughts of a worshipper and do not find in him love of this world or of the next, I fill him with love of Me and take him under My care.'

Christ in Islâm

Suff. 120

Better is poverty at the hand of God than riches in the store-house.

Egyptian Tradition

Realiz. 859

'Thy attention to this world is nobler than thy enslavement to the next world.'

Niffarî

M.M. 978 *Judgment* 250

To escape the joys of after life, this is the world's chief joy!
To add the pain of other births, this is the world's worst sorrow! Buddha, escaped from pain of birth, shall have no joy of the 'hereafter'!

Asvaghosha

I desire not to desire, for my will is without value, since I am ignorant in any case. Therefore choose Thou for me what Thou knowest to be best and do not put my perdition in what my autonomy and free choice prefer.

Bâyazîd al-Bisṭâmî

O, these deliberate fools! when they do choose,
They have the wisdom by their wit to lose.

Shakespeare (*Merchant of Venice,* II. ix. 80)

Wish that what is expedient and not what is pleasing may happen to you.

Sextus the Pythagorean

Renun. 136

The Lord will draw us and securely lead us to Himself, in a way contrary to all our natural will, until He have divested us thereof, and consumed it and made it thoroughly subject unto the Divine will. For this is His will: that we should cease to regard our own wishes or dislikes; that it should become a light matter to us whether He give or take away, whether we have abundance or suffer want, and let all things go, if only we may receive and apprehend God Himself, that whether things please or displease us, we may leave all things to take their course and cleave to Him alone.

Tauler

Void 721

If you accept a degree, it will become for you a curtain which will stop your progress.

Bâyazîd al-Bisṭâmî

Even the love which it is given us to feel for the bodily form of Christ can keep us from receiving the Holy Ghost.[1]

Eckhart

Renun. 139 *Illusion* 109 Supra Introd.

He who fondly imagines to get more of God in thoughts, prayers, pious offices and so forth, than by the fireside or in the stall: in sooth he does but take God, as it were, and swaddle his head in a cloak and hide him under the table. For he who seeks God under settled forms lays hold of the form while missing the God concealed in it.

Eckhart

[1] 'If I go not away, the Comforter will not come unto you' (*St John,* XVI. 7).

Who serves Him in anticipation of something from Him, or in order to avert through obedience the coming of punishment, has not met the claims of His Attributes.

Ibn 'Aṭâ'illâh

O Abû Sa'id, endeavour to remove self-interest (*ṭama'*) from thy dealings with God. So long as that exists, sincerity (*ikhlâṣ*) cannot be attained. Devotions inspired by self-interest are work done for wages, but devotions inspired by sincerity are work done to serve God. . . .

Perfect love proceeds from the lover who hopes naught for himself;
What is there to desire in that which has a price?
Certainly the Giver is better for you than the gift:
How should you want the gift, when you possess the very Philosopher's Stone?

Abu 'l-Qâsim Bishr-i Yâsîn

Beauty 663

God would have Himself gratuitously worshipped, would have Himself gratuitously loved, that is chastely loved; not to be loved for the reason that He giveth anything besides Himself, but because He giveth Himself. He therefore that calleth upon God in order that he may be made rich, doth not call on God, for he calleth upon that which he wisheth to come to him.

St Augustine

Q. Mother, I have practised austerities and *japa* so much, but I have not achieved anything.

A. God is not like a fish or vegetables that you can buy Him for a price.

Śrî Sâradâ Devî

Renun. 143
Grace 552

Thou wast not, and thou wast created. What hast thou given to God? Thou wast evil, and thou wast redeemed. What hast thou given to God? What is there that thou hast not received from Him gratuitously? With reason it is called 'grace' because it is bestowed 'gratis'. What is required of thee, therefore, is this, that thou too shouldst worship Him 'gratis'; not because He gives things temporal, but because He offers thee things eternal.

St Augustine

Humility 191
Renun. 158

We must come without the will to get or the will to avoid, just as the wayfarer asks the man he meets which of two ways leads anywhere, not wanting the right hand to be the road rather than the left, for he does not wish to go one particular road, but the road which leads to his goal. We ought to approach God as we approach a guide, dealing with Him as we deal with our eyes, not beseeching them to show us one sort of things rather than another, but accepting the impressions of things as they are shown us.

Epictetus

The changing mind of man cannot be satisfied permanently with anything. What he likes at one time, he does not like at another. What he wants now, he does not want later. The only way for a man to be always happy is to submit to God's will and, leaving everything to Him, be content with the condition in which He places him. From changing

Peace 694

circumstances we cannot get real happiness. Happiness lies within.... Surrender means inner contentment and peace.

Swami Ramdas

Abû Sa'îd al-Kharrâz said: 'Gratitude means acknowledging the Benefactor, and confessing (His) lordship.'...

One of the great Ṣûfîs[1] said: 'Gratitude consists in being unconscious of gratitude through the vision of the Benefactor.'

Al-Kalâbâdhî

Renun. 139
Faith 501

We ought to be too much at one with God in will to worry very much about ways and works.

Eckhart

When a man sets his thoughts and his inward active endeavour on the virtues and on outward behaviour more than on God and on union with God: though he remains in the grace of God (for in the virtues he aims at God), yet none the less his life is unstable, for he does not feel himself to rest in God above all virtues. And therefore he possesses something that he does not know; for, Him Whom he seeks in the virtues and in the multiplicity of acts, he possesses within himself, above intention, above virtues, and above all ways and means. And therefore, if this man would overcome his fickleness, he must learn to rest above all virtues in God and in the most high Unity of God.

Realiz. 859, 887
Renun. 143

Ruysbroeck

Renun. 146

All other sacrifices that we make, whether of worldly goods, honours, or pleasures, are but small matters compared to that sacrifice and destruction of all selfishness, as well spiritual as natural, that must be made before our regeneration hath its perfect work. There is a denial of our own will and certain degrees even of self-denying virtues which yet give no disturbance to this selfishness. To be humble, mortified, devout, patient *in a certain degree,* and to be persecuted for our virtues is no hurt to this selfishness; nay, *spiritual-self* must have all these virtues to subsist upon and his life consists in seeing, knowing, and feeling the bulk, strength, and reality of them. But still in all this show and glitter of virtue there is an unpurified bottom on which they stand; there is a selfishness which can no more enter into the Kingdom of Heaven than the grossness of flesh and blood can enter into it.

Death 208

William Law

Orthod. 296

For people are by no means all called upon to follow the same route to God, as St Paul points out. Supposing then thou findest that thy best course does not lie in great bodily activity and strenuous work nor in privations, things after all of little worth unless a man is specially driven thereto by God and is strong enough to stand them without disturbing his interior life, in this case be at peace and think no more about it. Thou wilt say, perhaps, 'If it is no great matter, why have our forbears done these things and numbers of the saints?' Remember, our Lord has given them this way and he gave them too the strength for it, so

[1] Al-Ḥallâj. He had recently been executed, and Kalâbâdhî deemed it preferable to keep his citations anonymous, knowing that Sufis would readily identify the source.

they were able to pursue that way and him, what fell from him. That was the best for them, but God has not bound up man's salvation with any given way. What one way has, what possibilities, with these God has furnished all good ways without exception.... What you get in one way may be got in any other provided it is sound and good and God is the only thing in view, nor are all men able to travel the same road.... Thou shalt sometimes find it harder to keep back a single word than not to speak at all. Just as it is sometimes harder to brook some idle taunt than to bear a heavy blow, and more difficult to be alone in company than in the desert; and the giving up of some trifling thing will frequently cost more than something big, and doing some small action more than one accounted fine. So a man can quite well follow our Lord by way of his own weakness, and need not, must not, think him far away.

Realiz. 873 *Reality* 790 *Renun.* 139 *Sin* 62 *Center* 841

Eckhart

The gnostics are the laziest folk in the two worlds, because they get their harvest without ploughing.

They have made laziness their prop (and rely upon it) since God is working for them.

The vulgar do not see God's working and (therefore) never rest from toil at morn or eve.

Action 340 *Renun.* 143

Rûmi

I prayed to the Divine Mother only for love. I offered flowers at Her Lotus Feet and said with folded hands: 'O Mother, here is Thy ignorance and here is Thy knowledge: take them both and give me only pure love for Thee. Here is Thy holiness and here is Thy unholiness: take them both and give me only pure love for Thee. Here is Thy virtue and here is Thy sin: here is Thy good and here is Thy evil; take them all and give me only pure love for Thee. Here is Thy dharma and here is Thy adharma: take them both and give me only pure love for Thee.'

M.M. 978

Sri Ramakrishna

Take, O Lord, and receive all my liberty, my memory, my understanding and my whole will. Thou hast given me all that I am and all that I possess. I surrender it all to Thee that Thou mayest dispose of it according to Thy will. Give me but Thy love and Thy grace: these are sufficient for me, and I will have no more to desire.

St Ignatius Loyola

O God! if I worship Thee in fear of Hell, burn me in Hell: and if I worship Thee in hope of Paradise, exclude me from Paradise: but if I worship Thee for Thine own sake, withhold not Thine Everlasting Beauty!

Contem. 536 *Inv.* 1017

Râbi'a of Baṣra

Obedience

Suff. 132

To obey is better than sacrifice.

I. Samuel, xv. 22

Suff. 130 *Renun.* 146

The thought of suffering, humiliation, confusion and poverty is directly opposed to our natural ideas of happiness, and only by a miracle of grace can we rejoice in the midst of such evils. But such a miracle will certainly be wrought in the souls of all who give themselves up unreservedly to the accomplishment of the Divine will. The honor of God demands that all who generously devote themselves to His service shall experience contentment and happiness therein.

St Alphonsus Liguori

Peace 700

To maintain equilibrium of mind, the sovereign remedy is to submit calmly to the will and workings of God who guides and controls the destinies of the universe.

Swami Ramdas

A very powerful and efficacious medicament, and the means to purify ourselves from every imperfection, to overcome all temptations and to preserve in our heart imperturbable peace, is conformity to the will of God.

St Vincent de Paul

Resignation, or the renunciation of self-will for the will of God, is before all things needful for all men who wish to be saved.

Ruysbroeck

While we live here, we must not so much seek to enjoy God, but rather to do His will.

St Teresa of Avila

Suffer patiently, and you will receive more grace than if you now experienced the most tender and fervent devotion.

Tauler

True and perfect obedience is a virtue above all virtues. No great work can be accomplished without it; nor can there be any task, however small or insignificant, which will not be done to better purpose in obedience. . . . Obedience brings out the best of everything; it never fails or errs in any matter; and no matter what you do, if you do it in true obedience, it will not miss being good.

Grace 556

Obedience has no cares; it lacks no blessing. Being obedient, if a man purifies himself, God will come into him in course; for when he has no will of his own, then God will command for him what God would command for himself. When I give my will up to the care of my prelate, and have no will of my own, God must will for me; for if he were to neglect me, he would be neglecting himself. So it is with everything: where I do not choose for myself, God chooses for me.

Eckhart

Creation 42

I came down from heaven, not to do mine own will, but the will of him that sent me.

St John, VI. 38

Lo! I am ready with even mind to suffer all that God has decreed concerning me, whether unto life or unto death. Nothing is better, more salutary, or more pleasant to me, nor do I choose or desire anything else than that He should find me ever ready to accept the judgement of His will.

Ruysbroeck

The most dangerous kind of impatience is that in which persons imagine they would be patient under other trials than those sent by God.

St Alphonsus Liguori

Pilg. 366

In Thy good pleasure is our peace. The body by its own weight strives towards its own place. Weight makes not downward only, but to his own place. Fire tends upward, a stone downward. They are urged by their own weight, they seek their own places. Oil poured below water, is raised above the water; water poured upon oil, sinks below the oil. They are urged by their own weights to seek their own places. When out of their order, they are restless; restored to order, they are at rest. My weight, is my love; thereby am I borne, whithersoever I am borne.

St Augustine

Contem. 547

Who is rich? He who rejoices in his portion.

Talmud ('Sayings of the Fathers')

Renun. 156

When a man has his proper station in life, he does not hanker after what is beyond him.

Epictetus

Action 335

Pauline had my most intimate confidence; she cleared up all my doubts. One day I showed surprise that God does not give an equal glory in heaven to all the elect; I was afraid that not all were happy. Thereupon she had me bring Papa's large glass, which she placed beside my little thimble; then, filling them both with water, she asked me which one appeared fuller. I told her that both seemed equally full, and that it was impossible to pour in more water than they could hold. Whereupon my 'Little Mother' had me understand that in heaven the least of the elect would not envy the happiness of the highest. In this way, putting the most sublime secrets within my reach, she gave my soul the nourishment it needed.

St Teresa of Lisieux

Humility 197

Each knows his allotted grade and seeks it as a child seeks its mother's breast, and iron, the lodestone. To occupy or even to aspire to a higher grade is impossible. In the grade in which he is placed each sees the realisation of his highest hopes. He loves his own grade passionately and cannot conceive that a higher could exist. If it were not so, heaven would not be heaven but a mansion of grief and bitter disillusion. Nevertheless, those in the superior participate in the enjoyment of the lower grades.

Ibn 'Arabî

Center 847

The quality of love stilleth our will, and maketh us long only for what we have, and giveth us no other thirst.

Did we desire to be more aloft, our longings were discordant from his will who here assorteth us,

and for that, thou wilt see, there is no room within these circles, if of necessity we have our being here in love, and if thou think again what is love's nature.

Nay, 'tis the essence of this blessed being to hold ourselves within the divine will, whereby our own wills are themselves made one.[1]

Metanoia 480

Dante (*Paradiso,* III. 70)

A good man is no more solicitous whether this or that good thing be mine, or whether my perfections exceed the measure of this or that particular creature; for whatsoever good he beholds any where, he enjoys and delights in it as much as if it were his own, and whatever he beholds in himself, he looks not upon it as his property but as a common good; for all these beams come from one and the same fountain and ocean of light, in whom he loves them all with a universal love.

Introd. *Holiness* *Reality* 790

John Smith the Platonist

Seeing then God is one only God, then all whatever will live in him must be like his will and word: As a concert of music must be tuned into one harmony, though there be many strings, and manifold voices and sounds therein; so must the true human harmony be tuned with all voices into a love melody.

Beauty 679

Boehme

In my Father's house are many mansions.

St John, XIV. 2

There is one glory of the sun, and another glory of the moon, and another glory of the stars: for one star differeth from another star in glory.[2]

Realiz. 873 *Rev.* 967

I. Corinthians, XV. 41

There is not one of us but hath his known station.

Action 335

Qur'ân, XXXVII. 164

Each one hears God speak according to his own penetration of the essential Realities, for God sets each thing in its place.

Jilî

[1] 'To be absolutely perfect in one's kind is to reach that level of reference at which all kinds of perfection coincide' (Coomaraswamy: *Patron and Artist,* Wheaton College Press, 1936).

[2] This is at a far remove from the egalitarian doctrines of our day.

When the Sun of Gnosis shines forth from the heaven above, on to this most blessed road (of mystic knowledge), each is enlightened according to his capacity and finds his own place in the knowledge of the Truth.

'Aṭṭâr

Center 833

Divine things are revealed unto each created spirit in proportion to its powers.

Dionysius

'In Paradise each choir of angels performs its functions, without desiring or envying the functions of another.'

Sister Consolata

Every creature is, as it were, 'God-created' or 'finite-infinity', with the result that no creature's existence could be better than it is.... God in His infinite goodness gives being to all in the way in which each can receive it. With Him there is no jealousy; He communicates being without distinction; and, since all receive being in accord with the demands of their contingent nature, every creature rests content in its own perfection, which God has freely bestowed upon it. None desires the greater perfection of any other; each loves by preference that perfection which God has given it and strives to develop and preserve it intact.

Nicholas of Cusa

'Well, then,' said the king, 'you must make me too, O Apollonius, a member of your religious brotherhood.' 'I would do so,' said the other, 'provided only you will not be esteemed vulgar and held cheap by your subjects. For in the case of a king a philosophy that is at once moderate and indulgent makes a good temper, as is seen in your own case; but an excess of rigour and severity would seem vulgar, O king, and beneath your august station; and what is more, it might be construed by the envious as due to pride.'

Apollonius of Tyana, to an Indian king

Renun. 139
Action 335

To know one's abiding-place leads to fixity of purpose, fixity of purpose to calmness of mind, calmness of mind to serenity of life, serenity of life to careful consideration of means, careful consideration of means to the achievement of the end.

The Great Learning, I

Renun. 160

Conformity to the Divine Will contains, in a high degree, true mortification, perfect subjection, complete self-abnegation, the imitation of Jesus Christ, union with God, and, in general, all virtues, which are virtues only for this reason: because they are according to the Will of God. This conformity is the beginning and the guide of all perfection.

St Alphonsus Liguori, citing a holy Father

Supra Introd.
Beauty 676

'One is only as obedient as one is humble, and one can never be humble if one is not obedient.'

St Catherine of Siena

'Virtues exercised without obedience are less valuable to Me.'

Mechthild of Magdeburg

Introd. *Sin*

Disobedience is the mouth of hell; argumentativeness is its tongue, whetted like a sword; self-gratification is its sharp teeth; self-justification its throat; high opinion of oneself, which casts one into hell, is the belching of its all-devouring belly. But he who, through obedience, conquers the first, by one stroke cuts off all the rest and with one stride reaches heaven.

St Gregory of Sinai

Wr. Side 459

Not every one that saith unto me, Lord, Lord, shall enter into the kingdom of heaven; but he that doeth the will of my Father which is in heaven.

St Matthew, VII. 21

Suff. 133

Do not, ascribing too much importance to fasting, lose sight of obedience which is the golden road to heaven.

St Peter Damian

Humility 191

Disobedience which engenders humility and a sense of poverty is better than obedience which engenders hauteur and a sense of pride.

Ibn 'Aṭâ'illâh

Sin 62

With some people God will transform their disobedience into obedience, so that their actions will not be counted as transgressions before God. With others, their apparent disobedience is in itself obedience, since it is in conformity with the Divine Will, although the Divine Commandment contradicts what the Divine Will demands of them.

Jîlî

There are two ways, one wrong and one right. The wrong way is Man's way to God, and the right way is God's way to Man.

Abu 'l-Ḥasan al-Khurqânî

Supra Introd. *Rev.* 967

Be obedient to your superior, and your inferior will obey you.

'Alî

What Eve has damned and lost by disobedience, Mary has saved by obedience.

St Louis de Montfort

P. State 579

Nature, whose power is in her obedience to the Will of God. . . .

Michael Sendivogius

God does not will that we abound in knowledge, but that we lovingly and humbly submit ourselves in all things to His will.

Henry Suso

Orthod. 300

Verily religion with Allâh is submission (*islâm*).

Qur'ân, III. 19

Freedom and Necessity

Soul becomes free when it moves without hindrance, through Intellectual-Principle, towards The Good; what it does in that spirit is its free act; Intellectual-Principle is free in its own right.... This state of freedom belongs in the absolute degree to the Eternals in right of that eternity and to other beings in so far as without hindrance they possess or pursue The Good which, standing above them all, must manifestly be the only good they can reasonably seek....

Certainly there can be no subjection whatever in That to which reality owes its freedom, That in whose nature the conferring of freedom must clearly be vested, pre-eminently to be known as the Liberator.[1] . . . Even self-mastery is absent here, not that anything else is master over it but that self-mastery begins with Being while the Supreme is to be set in a higher order.... Where we speak of self-mastery there is a certain duality, Act against essence.

Reality 773

M.M. 994

Plotinus

What is the freedom of a godly man? Being absolutely nothing to and wanting absolutely nothing for himself but only the glory of God in all his works.

Eckhart

Where the Spirit of the Lord is, there is liberty.

II. Corinthians, III. 17

The highest peace which the soul can enjoy
Is to know itself as much as possible united with God's will.

Angelus Silesius

Peace 694

A man once asked the mystic Bayazid: Who is the true Prince?

The man who cannot choose, said Bayazid: the man for whom God's choice is the only possible choice.

Bāyazid al-Bisṭāmī

To a good man to serve the will of God, it is in the truest and best sense to serve himself, who knows himself to be nothing without or in opposition to God; *Quò minùs quid sibi arrogat homo, èo evadit nobilior, clarior, divinior.* This is the most divine life that can be, for a man to act in the world upon eternal designs, and to be so wholly devoted to the will of God, as to serve it most faithfully and entirely. This indeed bestows a kind of immortality upon these flitting and transient acts of ours, which in themselves are but the offspring of a moment.

Action 340

John Smith the Platonist

Whoso has not escaped from will, no will hath he.

Dîvâni Shamsi Tabrîz, XIII

[1] This would correspond with the Islamic Divine Name *al-Fattâḥ*.

The honour of man consists of slavery.
In having no share of Free-will.

Shabistarî

The greatest joy in Heaven is the Will of God.

Mechthild of Magdeburg

Center 847

Happiness is nothing else, as we usually describe it to ourselves, but the enjoyment of some chief good: and therefore the Deity is so boundlessly happy, because it is every way one with its own immense perfection: and every thing so much the more feelingly lives upon happiness, by how much the more it comes to partake of God, and to be made like to him. . . . And, as it is impossible to enjoy happiness without a fruition of God: so it is impossible to enjoy him without an assimilation and conformity of our natures to him in a way of true goodness and godlike perfection.

John Smith the Platonist

Reality 773 *Creation* 31

The contemplating Intellect, the first or highest, has self-disposal to the point that its operation is utterly independent; it turns wholly upon itself: its very action is itself; at rest in its good it is without need, complete, and may be said to live to its will: there the will is intellection: it is called will because it expresses the Intellectual-Principle in the willing-phase and, besides, what we know as will imitates this operation taking place within the Intellectual-Principle. Will strives towards the good which the act of Intellectual-Principle realizes. Thus that principle holds what will seeks, that good whose attainment makes will identical with Intellection.

Plotinus

Symb. 321

O mind, if you really desire the good of your master, the soul, worship always Narahari, the great Lord: for a person who is an adept in putting ornaments on a reflected image decorates always only the real figure.[1]

Śrî Śankarâchârya

Realiz. 873

It is enough that one surrenders oneself. Surrender is to give oneself up to the original cause of one's being. Do not delude yourself by imagining such source to be some God outside you. One's source is within oneself. Give yourself up to it. That means that you should seek the source and merge in it.

Sri Ramana Maharshi

Introd. *Realiz.*

What becomes of the light of the stars when the sun appears on our horizon? Certainly it perishes not, but it is ravished into and absorbed in the sun's sovereign light, with which it is happily mingled and allied; and what becomes of man's will when it is entirely delivered up to God's pleasure? It does not altogether perish, yet is it so lost and dispersed in the will of God that it appears not, and has no other will than the will of God. Consider the glorious and never sufficiently praised St Louis, embarking and setting sail for beyond

[1] I.e., the soul is but a reflection of Âtma, and therefore its own felicity is inseparable from the veneration of its Source.

seas: and see the queen, his dear wife, embarking with his majesty. Now if any one had asked of this brave princess: Madam, whither are you going? She would without doubt have replied, I go whither the king goes. . . . As for the places whither he goes, they are all indifferent to me, and of no consideration whatever, except so far as he will be in them: for I have no affection for anything but the king's presence: it is therefore the king that goes, it is he that designs the journey, but, as for me, I do not go, I only follow: I desire not the journey, but solely the presence of the king: the staying, the journeying, and all their circumstances being utterly indifferent to me.

St Francis de Sales

Sweet do the winds become unto him who desires to abide by the Moral Law: sweet do the streams flow for him: even so, may the herbs be sweet unto us: may night be sweet and dawn also sweet unto us: may the region of Earth be sweet and also Heaven, our father: may the sylvan god be sweet unto us: may the sun be sweet: and sweet may our kine become.

Rig-Veda, I. xc. 6–8

Oh, the great folly of those who resist the will of God! They must of necessity endure afflictions, because no one can prevent the accomplishment of the Divine decrees. They must suffer without deriving any benefit from their trials. They cause themselves great disquietude in this world and draw down upon themselves still greater chastisements in the life to come.

St Alphonsus Liguori

Introd. *Suff.*

Reason is according to Providence: that which is irrational is according to Necessity: and the attributes of the body are according to Destiny. This is my teaching concerning the things that are according to Providence, Necessity, and Destiny.

Hermes

All things must praise the Creator of all beings: the devils praise him in the might of wrath, and the angels and men praise him in the might of love.

Boehme

Good things and evil, life and death, poverty and riches, are from God.

Ecclesiasticus, XI. 14

Introd. *Sin*

There is no soundness in them, whom aught of Thy creation displeaseth.

St Augustine

Beauty 689

He who looks upon Nature as anything but the constant expression of God's Will, is an Atheist.

The All-Wise Doorkeeper

Supra 180 *Heterod.* 420

Those men who are devoid of mind are merely led along in the train of Destiny. They have no conception of anything incorporeal, and they do not rightly understand the

Faith 505

Introd. *Sin*

meaning of Destiny, that very power by which they are led; they complain of the bodily discipline which she imposes, and they do not recognize any other kind of happiness than that which she confers.

Action 329

Philosophers are above Destiny; for they find no joy in the happiness she gives, since they hold pleasures in subjection; and they are not harmed by the ills she inflicts, because they dwell at all times in the immaterial world.

Hermes

Judgment 239

Earth has no escape from heaven: flee she up or flee she down heaven still invades her, energising her, fructifying her whether for her weal or for her woe. God treats man the same: weening to escape him we run into his arms, for all corners are exposed to him.

Eckhart

Reality 775

It is manifest that all things, whether they offend or are offended against, whether they give delight or are given delight, publish and proclaim the unity of the Creator.

St Augustine

Let us crave from thy goodness no boon save this: be it thy will that we be kept still knowing and loving thee, and that we may never fall away from this blest way of life.

Hermes

He who has all he will, his every wish, that man has peace. None has it but the man whose will and God's are wholly one. God grant us this atonement.

Eckhart

And unto Allâh falleth prostrate whosoever is in the heavens and the earth, willingly or unwillingly, as do their shadows in the morning and the evening hours.

Qur'ân, XIII. 15

HUMILITY

The highest goodness is like water. Water is beneficent to all things but does not contend. It stays in places which others despise. Therefore it is near Tao.

Tao Te Ching, VIII

Without a deeply-grounded, perfectly objective, and quasi-'functional' knowledge of our relative nothingness before other creatures, and of 'our' absolute nothingness before God, the dichotomy between Creator and creature will remain impassable, however much we advance in 'merit'. True humility, far from being an attitude, is based on a real comprehension of the nature of things, and is the fundamental virtue associated with Purification.

Reality 803

'The nothingness of the "creature" is tangible even on the terrestrial and empirical plane, therefore in a way that is altogether immediate for anyone who refrains from choosing in appearances that which flatters his blindness: in space, as in time, we are nothing; the two "infinities" overwhelm us from all sides. If the earth is a speck of dust in the incommensurable gulf of space, and if the life of humanity is a moment between the two incalculable abysses of time, what remains of man? Man can add neither an inch to his stature nor an instant to his life; he can only be in one place at a time, so that the incommensurability of space seems to make sport of him; he can live only in the moment that destiny decrees for him, without being able to turn back or to fly forward, so that time seems to hold him by the throat without ever letting go; and for all his vainglory, man is physically but an animal scarcely able to dissimulate the pangs of his misery. And lastly, he can do nothing without God—not even sin.

Illusion 99

'There are also the psychological miseries of fallen man, his propensity to take the finite for the Infinite, the ephemeral for the Eternal, the relative for the Absolute; in a word, the fundamental inversion of his nature in relation to God' (Schuon: 'De l'humilité', *Études Traditionelles,* 1950, pp. 118, 119).

Introd. *Metanoia*

'In Hindu morality (*yama* and *niyama*), humility appears as "modesty" (*hrî*); spiritual, or esoteric, humility is "childhood" (*bâlya*), and in its highest sense, "extinction" (*nirvâna*)' (Schuon: *Perspectives spirituelles,* p. 253).

P. State 573

There is an 'alchemical' aspect to the question in what is called the 'junction of extremes', relating to the reciprocity which functions between the principial domain and its inverse analogy on the reflected plane (cf. Guénon: *Études Traditionnelles,* 1946, p. 201ff, and *passim*), where the highest and the lowest by their very opposition have a close correspondence or interdependence. This may manifest as humility or obscurity: but it can also take other appearances, such as commonness, eccentricity, or even madness. Yet the criterion to go by is that the anonymity in question will always be of suprahuman and never subhuman origin.

Introd. *Ecstasy*

The Conquest of Pride

It is useless to have lived, even for a very long time, with a spiritual preceptor if one be lacking in humility and devotion and thus be unable to develop spiritually.

Gampopa

Just as worldly people glory in their riches, certain aspirants and sâdhus are vain about their moral worth. This kind of pride constitutes a serious obstacle to spiritual realization; it has to be completely extirpated. As long as a man is vain, he remains the minuscule and ordinary incarnation of a soul and cannot know the divinity.

Swami Sivananda

Living in religion (as I can speak by experience) if one is not in a right course of prayer and other exercises between God and our soul, one's nature groweth much worse than ever it would have been, if one had lived in the world. For pride and self-love, which are rooted in the soul by sin, find means to strengthen themselves exceedingly in religion, if the soul is not in a course that may teach her and procure her true humility. For by the corrections and contradictions of the will (which cannot be avoided by any living in a religious community) I find my heart grown, as I may say, as hard as a stone; and nothing would have been able to soften it but by being put into a course of prayer, by which the soul tendeth towards God and learneth of Him the lesson of truly humbling herself.

Sin 75

Dame Gertrude More

More fearing are the gnostics in a state of spiritual expansion (*bast*) than they are in contraction (*qabd*), for few are those in the state of expansion who stay within the confines of spiritual propriety (*adab*).

Introd. *Heterod.*

Ibn 'Aṭâ'illâh

Ah! how many cedars of Lebanon, how many stars of the firmament, have we not seen fall miserable, and in the twinkling of an eye lose all their height and their brightness! Whence comes that sad and curious change? It was not for want of grace, which is wanting to no man; but it was for want of humility. They thought themselves capable of guarding their own treasures. They trusted in themselves, relied upon themselves.

Heterod. 423

St Louis de Montfort

Whoever has in his heart even so much as a rice-grain of pride, cannot enter into Paradise.

Sin 64

Muhammad

God cannot be realized if there is the slightest trace of pride.

Sri Ramakrishna

Humility is to the virtues what the chain is to the rosary: remove the chain, and all the beads escape; take away humility, and all the virtues disappear.

The Curé d'Ars

A Jnani (knower of God) and a Premika (lover of God) were once passing through a forest. On the way they saw a tiger at a distance. The Jnani said, 'There is no reason why we should flee; the Almighty God will certainly protect us.' At this the Premika said, 'No, brother, come, let us run away. Why should we trouble the Lord for what can be accomplished by our own exertions?'

Sri Ramakrishna

Abbot Nisteron the Elder, walking in the desert with a hermit, met a dragon and fled. Upon which the brother said to him: 'How now, my Father, are you afraid?' 'No, brother,' replied this holy man; 'but I am obliged to flee the sight of this dragon, because I have not virtue enough to put the demon of vanity to flight.'

Vitae Patrum, VIII. 12

The sins of those who are near (to God) are the good deeds of the righteous.

Al-Sarrâj

Renun. 143

'Set thy sin beneath thy feet, and thy virtue beneath thy sin.'

Niffari

Contrition

The Lord is nigh unto them that are of a broken heart; and saveth such as be of a contrite spirit.

Psalm XXXIV. 18

Lamentation and weeping are a mighty stock-in-trade; the Universal Mercy is the mightiest nurse.

The nurse and mother seeks a pretext (for giving relief): (she waits to see) when her child will begin to weep.

He (God) created the child, your wants, in order that it might moan and that milk might be produced for it.

He said, *'Call ye upon God!'* Refrain not thou from lamentation, in order that the milk of his lovingkindnesses may flow.

Rûmî

P. State 573

My nature is that of a kitten. It only cries, 'Mew, mew!' The rest it leaves to its mother.

Sri Ramakrishna

Sit before a representation of the sacred syllable OM; concentrate on it until your tears flow abundantly.

Swami Sivananda

Metanoia 484

God's high decree would be broken, if Lethe were passed, and such viands tasted, without some scot of penitence that may shed tears.

Dante (*Purgatorio*, XXX. 143)

By the rivers of Babylon, there we sat down, yea, we wept, when we remembered Zion.

Psalm CXXXVII. 1

Sin 66

He may make sorrow earnestly that knoweth and feeleth not only what he is, but that he is. And whoso felt never this sorrow, let him make sorrow; for he hath never yet felt perfect sorrow.

The Cloud of Unknowing, XLIV

The 'lamenter'[1] cries, for he is humbling himself, remembering his nothingness in the presence of the Great Spirit.

Black Elk

The teares of Saints more sweet by farre,
Then all the songs of sinners are.

Robert Herrick

O eye, thou makest lament for others: sit down awhile and weep for thyself!

The bough is made green and fresh by the weeping cloud, for the reason that the candle is made brighter by weeping.

Wheresoever people are lamenting, sit you there, because you have a better right to moan,

Reality 803

Inasmuch as they are concerned with parting from that which passes away, and are forgetful of the ruby of everlastingness that belongs to the mine of Reality.

Rûmî

'My heart is conquered more by your miseries than by your virtues.'

Sister Consolata

[1] See note p. 291.

One day a voice said: 'O Bâyazîd! Our Treasury is surfeited with acts of worship and devotion. If thou desirest to possess Us, bring Us something which is not found in our Treasury.' 'But my God! What can I bring you?' And the voice replied: 'Bring us distress, humility, supplication, and heartbreak.'

'Aṭṭâr

Conform. 170

Can a man be profitable unto God?

Job, xx. 2

Gold I have none, but I present my need.

Robert Herrick

What offering shall I bring Thee? For all things are from Thee. Thou givest all, and receivest nothing; for Thou hast all things, and there is nothing that Thou hast not.

Hermes

Renun. 158

No creature is noble enough to help in this work; God has not graced their nature with such excellence.

Eckhart

God is the helper of the helpless and the strength of the weak. He stood by the side of the saints in their hour of trial.

So long as the Lord of Elephants trusted in his own strength he was defeated.

The moment he forgot his own strength, and in his weakness called upon the Lord, God was at hand to help him—even before His name was half-uttered.

Draupadi in her helplessness called upon the Lord.

Dushasana was worsted in his effort to unclothe her; for the Lord became her clothing. Try as one may the power of asceticism, or physical or temporal might, a man is bound to fail.

Verily the strength of the defeated is the name of the Lord.

Gujarati Hymn

Renun. 143
Holy War 410
Inv. 1009

Let us adore him rather with thanksgiving; for words of praise are the only offering that he accepts.

Hermes

Our names we make to shine before the people,
But secretly our hearts we all have tarnished.
Each breath sees us commit sins by the thousand,
Yet not once in our life repent we one sin.
Yielding to self we sin and know no limit—
What shall we do, O God, how make repentance?
No one of us but knows his heart's sedition,
Yet we have come, thy mercy to petition.

The Mevlidi Sherif

Inv. 1009

In the courtyard there grows a strange tree,
Its green leaves ooze with a fragrant moisture.
Holding the branch I cut a flower from the tree,
Meaning to send it away to the person I love.
Its sweet smell fills my sleeves and lap.
The road is long, how shall I get it there?
Such a thing is not fine enough to send:
But it may remind him of the time that has past since he left.

Faith 509

Old Chinese Poem

Beauty 663

The things in heaven receive no benefit from the things on earth; but the things on earth receive all benefits from the things in heaven.

Hermes

Illusion 99

Lord, what is man, that thou takest knowledge of him!

Psalm CXLIV. 3

The feeling uppermost in our minds is the humbling consciousness of our utter unworthiness.

Michael Sendivogius

Remember that God has put a mark upon us, consisting in our shortcomings and diseases, to show to us that we have nothing to pride ourselves about, and that nothing comes within the reach of our full and perfect understanding; that we are far from knowing absolute truth, and that our own knowledge and power amount to very little indeed.

Paracelsus

'And forget the Lord thy God' (*Deut.* VIII. 12–14). When then wilt thou not forget God? Only when thou dost not forget thyself. For if thou rememberest thine own nothingness in all things, thou wilt also remember the transcendence of God in all things.

Realiz. 859

Philo

Judgment 239
Creation 48

Fear God out of love, and be thou perfect, and thy fear will then be whole; 'tis meet that it should be so, for He formed thee out of nothing, and from nothingness He brought thee to stand before Him continually, and fashioned thee into a beautiful form.

Yesod Hayirah, 4

Introd.
Conform.

When you preside over men, remember that divinity also presides over you.

Sextus the Pythagorean

Rev. 967

Verily, verily, I say unto you, The servant is not greater than his lord; neither he that is sent greater than he that sent him.

St John, XIII. 16

The more that holy men advance in contemplation, the more they despise what they are, and know themselves to be nothing, or next to nothing.

St Gregory the Great

This then is the sum of great knowledge for a man to know that by himself he is nothing, and that whatever he is, he is from God and is for God. . . . It is however not enough that thou acknowledge that which in thee is good to be from God, unless also on that account thou exalt not thyself above him that hath not yet, who perchance will outstrip thee when he shall have received.

St Augustine

Action 340

Only let your present and past distress make you feel and acknowledge this twofold great truth: first, that in and of yourself you are nothing but darkness, vanity, and misery; secondly, that of yourself you can no more help yourself to light and comfort than you can create an angel.

William Law

Sin 66
Grace 552
Pilg. 366

Canst thou bind the sweet influences of Pleiades, or loose the bands of Orion?

Canst thou bring forth Mazzaroth in his season? or canst thou guide Arcturus with his sons?

Knowest thou the ordinances of heaven? canst thou set the dominion thereof in the earth?[1]

Job, XXXVIII. 31–33

The past has flown away,
The coming month and year do not exist;
Ours only is the present's tiny point.

Shabistarî

Illusion 99

How poor a thing is a son of Adam! He knows not his length of days, nor understands his sicknesses: the sting of a flea will make him suffer; he smells of sweat, and dies of a cough!

'Alî

To avoid falling into this evil (of self-esteem) which threatens you, always keep your mind collected in the heart and be for ever ready to repulse these arrows of the enemy. Standing there within, like a general on the battlefield, choose a place of advantage for battle, fortify it thoroughly and never leave it, but make it your shelter from which to give battle. This place, its fortification and armament, is a profound and sincere realisation of your nothingness, of the fact that you are poor, blind, naked and rich only in weaknesses, faults and deeds that are blameworthy, foolish, vain and sinful. Having taken up this position, never let your mind wander outside your fortress.

Unseen Warfare, I. XXXIV

Contem. 536
Holy War 407
Sin 77

[1] These questions situate the vaunted achievements of modern technology in a proper perspective.

I am a fool, and full of poverty.

Shakespeare (*Love's Labour's Lost*, v. i. 381)

Life is a frost of cold felicite,
And death the thaw of all our vanitie.

Christolero's Epigrams

If thou wouldst become a pilgrim on the path
Of love
The first condition is
That thou become as humble as dust
And ashes.

Introd. *Death*

Ansârî

If the soul does not suffer great pain from the shamelessness of sin, it cannot later fully enjoy the beneficence of truth.

Diadoch of Photikos

Lord, I confesse, that Thou alone art able
To purifie this my *Augean* stable:
Be the Seas water, and the Land all Sope,
Yet if Thy Bloud not wash me, there's no hope.

Grace 552

Robert Herrick

Nothing is demanded of thee so much as dependency, and nothing so speedily brings thee bestowal as abasement and poverty.

Ibn ʿAṭâ'illâh

Introds. *Heterod.* *Realiz.*

In order that a full dissolution may be made, there is need of contrition, that calcination may afterwards produce a viscous state, when it will be fit for dissolution.

Philalethes

Realiz. 859 *Sin* 66 *Ecstasy* 641 *Conform.* 180

Meekness in itself is nought else but a true knowing and feeling of a man's self as he is. For surely, whoso might verily see and feel himself as he is, he should verily be meeked. Two things there be that be causes of this meekness, the which be these: One is the filth, the wretchedness, and the frailty of man, into the which he is fallen by sin, and the which he must always feel in some degree the whiles he liveth in this life, be he never so holy. Another is the over-abundant love and the worthiness of God in himself; in beholding of which all nature quaketh, all clerks be fools, and all saints and angels be blind. Insomuch, that were it not, through the wisdom of his Godhead, that he measured their beholding according to their ableness in nature and in grace, I cannot say what should befall them.

The Cloud of Unknowing, XIII

Reality 803 *Pilg.* 366 Infra 199

One of thine attributes is pure nothingness, which belongeth unto thee and unto the world in its entirety. If thou acknowledge thy nothingness, He will increase thee with His Being.

Shaykh Aḥmad al-ʿAlawî

The Holy Ghost flows into the soul as fast as she is poured forth in humility and so far as she has gotten the capacity. He fills all the room he can find.

Grace 556

Eckhart

Nothing is in vain or without profit to the humble soul: like the bee, it takes its honey even from bitter herbs: it stands always in a state of divine growth, and everything that falls upon it is like a dew of Heaven to it.

William Law

The best of moments for thee is a moment wherein thou witnesseth the fact of thy poverty and art therein cast back upon thy destitution.

Ibn 'Aṭâ'illâh

I am waning, that thy love may increase and wax.

Dîvâni Shamsi Tabrîz, XVIII

Metanoia 484

When I perceive that I am earth and cinders or whatever is still more worthless, it is just then that I have confidence to come before Thee, when I am humbled, cast down to the clay, reduced to such an elemental state, as seems not even to exist. And the watchful pen of Moses has recorded this my soul's condition in his memorial of me. For Abraham, he says, drew near and said, 'Now I have begun to speak to the Lord, and I am earth and ashes' (*Gen*. XVIII. 27), since it is just when he knows his own nothingness that the creature should come into the presence of his Maker.

Pilg. 366

Philo

And give good tidings to the humble,
Whose hearts fear when Allâh is mentioned.

Qur'ân, XXII. 34–35

Judgment 239

Obscurity

Men of little ability, too,
By depending upon the great, may prosper:
A drop of water is a little thing,
But when will it dry away if united to a lake?

Subhâshita Ratna Nidhi, Stanza 173

Conform. 180

If sun thou canst not be, then be the humble planet.

Tibetan Precept

Wouldst thou raise up thy head to carry the crown and diadem of pride with a name of praise, understand, that the God who alone is clothed in the garment of excellency, unto whom alone greatness belongeth, bringeth low unto dust the high, and raiseth on high those who are low.

Sin 56

Yesod Hayirah, 66

The rich man seeks gold, the poor man seeks God:
The poor man in fact finds gold, the other mud.

P. State 563

Angelus Silesius

For *Gods* Conjunctions Man maie not undoe,
But if his Grace fully consent thereto,
By helpe of this *Science*, which our *Lord* above
Hath given to such Men as he doth love;
Wherefore old *Fathers* conveniently
Called this *Science Holy Alkimy*.
Therefore noe Man shulde be too swifte,
To cast away our *Lords* blessed guift:
Consideringe how that Almighty *God*
From great Doctours hath this *Science* forbod,
And graunted it to few Men of his mercy,
Such as be faithfull trew and lowly.

Introd. *Conform.*

Thomas Norton

What Lucifer has lost by pride, Mary has gained by humility.

P. State 579

St Louis de Montfort

It is humility alone that makes the unpassable gulf between Heaven and hell. No angels in Heaven but because humility is in all their breath; no devils in hell but because the fire of pride is their whole fire of life.

Supra Introd. *Sin* 75

William Law

The end of all strife and contention is regret; but the end of humility is strength and possession.

Shekel Hakodesh, 69

(The sun dancers) also put rabbit skins on their arms and legs, for the rabbit represents humility, because he is quiet and soft and not self-asserting—a quality which we must all possess when we go to the center of the world.

Center passim

Black Elk

Bury thine existence in the earth of obscurity, for if a seed be not buried it bringeth not forth in fullness.

Pilg. 378

Ibn ʿAṭâ'illâh

To vaunt oneself is to fail of self-respect.

'Alî

Give up all externalities and show—even show of piety.

Swami Ramdas

Conform. 170

Use your light. but dim your brightness.

Tao Te Ching, LII

Learn to know all. but keep thyself unknown.

Gnostic Device

Contem. 523

Kusa mura ya:	Among the grasses.
Na mo shiranu,	An unknown flower
Shiroku saku.	Blooming white.

Zen *Haiku*

Mary was singularly hidden during her life. It is on this account that the Holy Ghost and the Church call her *Alma Mater*—'Mother secret and hidden'.[1] Her humility was so profound that she had no inclination on earth more powerful or more constant than that of hiding herself. from herself as well as from every other creature. so as to be known to God only.

St Louis de Montfort

P. State 573

We are to practise virtue. not possess it.

Eckhart

Hide the good you do. and make known the good done to you.

'Alî

Conceal your good deeds as you conceal your evil deeds.

Râbi'a of Başra

The Junction of Extremes

'Be humble, and you will remain entire.'
Be bent, and you will remain straight.
Be vacant, and you will remain full.

Metanoia 488
Void 724

[1] Antiphon to the Blessed Virgin for Advent.

Be worn, and you will remain new.
He who has little will receive.
He who has much will be embarrassed.

Introd. *Peace*

Therefore the sage keeps to One and becomes the standard for the world.

Tao Te Ching, XXII

When pride cometh, then cometh shame: but with the lowly is wisdom.

Proverbs, XI. 2

Whosoever shall exalt himself shall be abased; and he that shall humble himself shall be exalted.

St Matthew, XXIII. 12

Holy War 403

Take the lowest place and ye shall reach the highest.

Milarepa

God resisteth the proud, but giveth grace unto the humble.

James, IV. 6

Suff. 120

In pride I so easily lost Thee—
But now the more deeply I sink
The more sweetly I drink
Of Thee!

Mechthild of Magdeburg

Allâh is the Rich and ye are the poor.

Qur'ân, XLVII. 38

Is not the Tao of heaven like the drawing of a bow? It brings down the part which is high; it raises the part which is low; it lessens the part which is redundant (convex); it fills up the part which is insufficient (concave).

Tao Te Ching, LXXVII

Grace 556

Union comes by grace, the highest stooping down to inform the lowest, and therein lies our hope of future sight.

Eckhart

P. State 563

When we use the water in the sweat lodge we should think of *Wakan-Tanka* who is always flowing, giving His power and life to everything; we should even be as water which is lower than all things, yet stronger even than the rocks.

Black Elk

He who knows honour and yet keeps to humility
Will become a valley that receives all the world into it.

Tao Te Ching, XXVIII

'Wherever I gave special grace
I sought for the lowest and smallest and most hidden.
The highest mountains may not receive
The revelations of My Grace,
For the flood of My Holy Spirit
Flows by nature down into the valleys.
One finds many a wise writer of books
Who in himself, to My sight, is a fool.
And I tell thee further, it greatly honours Me
And strengthens mightily Holy Church
That unlearned lips should teach
The learned tongues of My Holy Spirit.'

Mechthild of Magdeburg

Knowl. 734

ETERNAL WISDOM: According to the natural order, we take the highest emanation of all beings from their first origin through the noblest beings down to the lowest. But the return to the origin takes place through the lowest to the highest. Therefore, if thou wouldst see Me in My uncreated Divinity, thou shoudst learn to know Me in My suffering humanity, for that is the swiftest way to eternal bliss.

Henry Suso

Symb. 306
Metanoia 488

Only seek it at the bottom of the vessel, or you will wander astray.

Michael Maier

Since the highest has so lowered itself as to become the lowest, we may expect that its blood may be the means of redeeming all its brethren.

Philalethes

Orthod. 280

The meek shall inherit the earth.

Psalm XXXVII. 11

P. State 573

Be humble and meek if thou would be exalted.

Lodan-Gawai-Roltso, Verse 66

My lowliness raises up God and the lower I humble myself the higher do I exalt God and the higher I do exalt God the more gently and sweetly he pours into me his divine gift, his divine influx. For the higher the inflowing thing the more easy and smooth is its flow. How God is raised upon my lowliness I argue thus: the more I abase and keep myself down the higher God towers above me. The deeper the trough the higher the crest. In just the same way, the more I abase and humble myself the higher God goes and the better and easier he pours into me his divine influx. So it is true that I exalt God by my lowliness.

Eckhart

If thou want the water of mercy, go, become lowly, and then drink the wine of mercy and become drunken.

Rûmî

Ecstasy 637

Blessed are the poor in spirit: for theirs is the kingdom of heaven.

St Matthew, v. 3

I have a sufficient witness to the truth of what I say—my poverty.

Plato (*Apology*, 31 C)

Tze Ch'i of Nan-poh was travelling on the Shang mountain when he saw a large tree which astonished him very much. A thousand chariot teams could have found shelter under its shade.

'What tree is this?' cried Tze Ch'i. 'Surely it must have unusually fine timber.' Then looking up, he saw that its branches were too crooked for rafters; while as to the trunk he saw that its irregular grain made it valueless for coffins. He tasted a leaf, but it took the skin off his lips; and its odour was so strong that it would make a man as it were drunk for three days together.

'Ah!' said Tze Ch'i. 'This tree is good for nothing, and that is how it has attained this size. A wise man might well follow its example.'

Holiness 910

Chuang-tse (ch. IV)

A foolish man proclaimeth his qualifications;
A wise man keepeth them secret within himself;
A straw floateth on the surface of water,
But a precious gem placed upon it sinketh.

M.M. 987

Subhâshita Ratna Nidhi, Stanza 58

Though I be free from all men, yet have I made myself servant unto all, that I might gain the more. . . . To the weak became I as weak, that I might gain the weak: I am made all things to all men, that I might by all means save some.

I. Corinthians, IX. 19, 22

Suff. 128

Who is wise? He who learns from all men.

Talmud ('Sayings of the Fathers')

Returning is the motion of Tao,
Weakness is the appliance of Tao.
All things in the universe come from existence,
And existence from non-existence.

Creation 26

Tao Te Ching, XL

Sincerity

Humility is accepting the truth from the Truth for the Truth.

Al-Kalâbâdhî

Before honour is humility.

Proverbs, XV. 33

Man is humble when the flaming fire of desires[1] has become extinct.

Al-Tirmidhî

Peace 700

One must be proud with the rich and humble with the poor. . . . A pride well placed is humility.

Abû 'Uthmân al-Hayrî al-Nisâbûrî

Introd. *Conform.*

Let your old age be childlike, and your childhood like old age; that is, so that neither may your wisdom be with pride, nor your humility without wisdom.

St Augustine

O ye who believe! When it is said unto you, Make room! in assemblies, then make room; Allâh will make way for you (hereafter). And when it is said, Come up higher! go up higher; Allâh will exalt those who believe among you, and those who have knowledge, to high ranks. Allâh is Informed of what ye do.

Qur'ân, LVIII. 11

Rev. 967 *Knowl.* 761

Though I preach the gospel, I have nothing to glory of: for necessity is laid upon me; yea, woe is unto me, if I preach not the gospel!

I. Corinthians, IX. 16

Action 335

[1] Desire includes all passions, mental, psychic and physical.

DEATH

Thou bringest forth the living from the dead, and Thou bringest forth the dead from the living.

Qur'ân, III. 27

If you love and favour your tender, delicate flesh still, do not read my book.

Boehme[1]

'Every change of state whatsoever is both a death and a birth, depending on the side it is envisaged from: death in relation to the antecedent state, birth in relation to the subsequent state. Initiation is generally described as a "second birth", which in fact it is: but this "second birth" necessarily implies death to the profane world' (Guénon: *Aperçus sur l'Initiation*, pp. 182–183), and by consequence must ultimately imply the term or fruition of an initiation which has become fully effective, fully realized. For the death in question—as the following passages will show—is something entirely other than a matter of verbal formalities. Guénon adds in the same paragraph that 'every change of state has to be considered as taking place in darkness'. This is the 'dark night' that must be traversed, the 'sundering of soul from spirit' that must be accomplished, even in this lifetime, for one who becomes what Rûmî terms 'a dead man walking'.[2]

Orthod. 280

Realiz. 870

For it is a certainty that one cannot live in reality until one has died to illusion, that 'My kingdom is not of this world', that 'flesh and blood cannot inherit the kingdom of God', and that 'friendship of the world is enmity with God'. 'There is in man something which must die, or which must be destroyed: it is the soul as desire, whose terminal point is the sensorial body' (Schuon: *L'Oeil du Coeur*, p. 209). Similarly Hermes: 'The cause of death is carnal desire' (*Lib.* I. 18), as with the Buddha: 'From lust creatures run into distress: they are tormented in the six states of existence and people the cemetery again and again' *Saddharma-pundarîka*, II. 63).[3] There is no escaping death: either one's desires are put to death, or they will deliver one into the kingdom of death.

Judgment 250

'It is only by making stepping stones of our dead selves, until we realize at last that there is literally nothing with which we can identify our Self, that we can become what we are' (Coomaraswamy: *Hinduism and Buddhism*, p. 63).[4] 'Blessed is the man on whose tomb can be written, *Hic jacet nemo*' (id., p. 30).

Renun. 158

[1] *Three Principles*, IV. 43; cit. Hobhouse, *Law*, p. 270.

[2] *Mathn*. VI. 742 (although for the Sufi it is really the world that is dead).

[3] Or, *Lotus of the True Law*, tr. H. Kern, Oxford, Clarendon Press, 1884, p. 48.

[4] 'What though I die hourly, I have each time found a better life' (Angelus Silesius: *Cherub*. I. 30).

If some of the texts in this chapter seem uncompromisingly severe, the same theme will be found treated even more inexorably if possible in certain passages of BOOK THREE, where it is a question of universal extinction. But 'the Self is not within the reach of the weak' (Alain Daniélou: *Yoga*, p. 17). 'A sharpened edge of a razor, hard to traverse, A difficult path is this—the wise declare!' (*Kaṭha Upanishad*, III. 14). 'Now to all of us it standeth on a razor's edge, either pitiful ruin . . . or life' (*Iliad*, Book X). 'A narrow path betwixt Hell's bottomless abyss,/Fine and sharp as a sword blade' (Shabistarî: *Gulshan-i-Râz;* Lederer, p. 66). 'Before reaching this station (of enlightenment), what thorny trees and swords' edges!' (Jîlî: *Al-Insân al-Kâmil*: Burckhardt, p. 38). Of the Tibetan 'Direct Path' Marco Pallis writes: 'It is not for faint hearts to think of braving its perils' (*Peaks and Lamas*, p. 144). 'Timid people are absolutely unfit for the spiritual life' (Swami Sivananda: *Yogic Exercises*, p. 97). Islamic esoterism defines a Sufi as 'one who can walk holding burning coals in his hand'. According to William Law, 'There is nothing safe in religion, but in such a course of behaviour as leaves nothing for corrupt nature to feed or live upon: which can only then be done when every degree of perfection we aim at is a degree of death to the passions of the natural man' (Hobhouse, p. 20).

Pilg. 366

An Indian holy woman was asked when one could be certain of safety on the spiritual road; she sent back, for answer, a little plate of ashes.

Infra 220

And yet it must not be forgotten that the death represented here is that 'happy death' of which St Joseph as the prototype of spiritual master is the patron, for this death is on the threshold of true life, immortality and infinite beatitude. 'It is not a dying, but a . . . transmutation' (Boehme: *Sig. Rerum*, XI. 35). Death for the Self-possessed is nothing other than the throwing off of agitation and the perfecting of serenity.

Peace 705

'Die Before Ye Die'

Die before ye die.

Muhammad

Infra 208

Die before thou diest, so as not to die when thou diest, or indeed must thou perish.

Angelus Silesius

It is well for a man to depart to the forest ere the four bearers carry him away amidst the laments of his folk. Free from commerce and hindrance, possessing naught but his body, he has no grief at the hour of death, for already he has died to the world.

Śânti-deva

Who dies not before dying, perishes when he dies.

Abraham von Frankenberg

Metanoia 488

Callimachus, die that thou mayest live.

Acts of John, 76

While living
Be a dead man.

Bunan Zenji

O reason, to gain eternal life tread everlastingly the way of death.

Dîvâni Shamsi Tabrîz, IV

Infra 220

If one dies, it results in grief for the other who lives. The way to get rid of grief is *not to live*. Kill the one who grieves. Who will remain then to suffer? The ego must die. That is the only way.

Sri Ramana Maharshi

Death will not destroy the man who does not wear upon his body the jewels of his vicious desires, as a robber does not kill a traveller who has no precious chain of gold round his neck.

Yoga-Vasishtha

Let us not grow so as to become old after being new, but let the newness itself grow.

St Augustine

Infra 232

She cannot die, for she died before her death, in order to be living when she died.

Daniel von Czepko

O Mother, what wilt Thou accomplish by killing one who is already dead?[1]

Sri Ramakrishna

A man who has attained the final degree of perfection is dead and yet not dead, but infinitely more alive in God with Whom he lives for he no longer lives by himself.

St Simeon the New Theologian

Realiz. 870

'What if anyone is dead? What if any one is ruined? Be dead yourself—be ruined yourself.' In that sense there is no pain after one's death. What is meant by this sort of death? Annihilation of the ego, though the body is alive.

Sri Ramana Maharshi

The Gnostics have a death before the general death.

Shaykh Aḥmad al-ʿAlawî

I am not a Brahmin, rajah's son or merchant; nor am I any what; I fare in the world a sage, of no-thing, homeless, self completely gone out—it is inept to ask me of my lineage.

Suttanipâta, 455–456

Void. 724

The *Ṣûfî* is he that is dead to self and living by the Truth; he has escaped from the grip of human faculties and has really attained (to God).

Hujwîrî

True perfection is not consistent with any terrene loves or worldly affections: this mundane life and spirit which acts so strongly and impetuously in this lower world, must be crucified: the soul must be wholly dissolved from this earthy body in which it is so deeply immersed, while it endeavours to enlarge its sorry tabernacle upon this material globe, and by a holy abstraction from all things that pinion it to mortality, withdraw itself and retire into a divine solitude.

John Smith the Platonist

Void 728

Die, then, to the creatures, by God's leave, and to your passions, by His command, and you will then be worthy to be the dwelling-place of the knowledge of God. The sign of your death to the creatures is that you detach yourself from them and do not look for anything from them. The sign that you have died to your passions is that you no longer seek benefit for yourself, or to ward off injury, and you are not concerned about yourself, for you have committed all things unto God. The sign that your will has been merged in the Divine Will is that you seek nothing of yourself or for yourself—God's Will is working in you. Give yourself up into the hands of God, like the ball of the polo-player, who sends it to and fro with his mallet, or like the dead body in the hands of the one who washes it, or like the child in its mother's bosom.

ʿAbd al-Qâdir al-Jîlânî

Conform. 170

[1] Sri Ramakrishna in a state of ecstasy, addressing the Divine Mother about his own approaching illness.

Him I call indeed a Brâhmana who has traversed this miry road, the impassable world, difficult to pass, and its vanity, who has gone through, and reached the other shore.

Pilg. 385

Dhammapada, XXVI. 414

The dead who have died in the Godhead are beyond our ken, like the dead are who die here to the body. That death is the soul's eternal quest. Slain in the three Persons she loses her naught and is hurled into the Godhead.

Infra 208

Eckhart

Kâhn have I bought. the price he asked, I gave.
Some cry, ''Tis great', and others jeer, ''Tis small'—
I gave in full, weighed to the utmost grain,
My love, my life, my soul, my all.

Renun. 146
Suff. 132

Mîrâ Bâî

The *Double* only *seems,* but The *One is,*
Thy-self to Self-annihilation give
That this false *Two* in that true *One* may live.

Reality 775
Metanoia 480

'Aṭṭâr

There's a special providence in the fall of a sparrow. If it be now, 'tis not to come; if it be not to come, it will be now; if it be not now, yet it will come: the readiness is all. Since no man has aught of what he leaves, what is't to leave betimes?

Shakespeare (*Hamlet,* v. ii. 232)

Only the Dead Know God

Flesh and blood cannot inherit the kingdom of God; neither doth corruption inherit incorruption.

I. Corinthians, xv. 50

The kingdom of God is for none but the thoroughly dead.

Eckhart

No creature can attain a higher grade of nature without ceasing to exist.[1]

Pilg. 378

St Thomas Aquinas

[1] This is a basic principle in alchemical transmutation.

Verily, verily, I say unto you, Except a corn of wheat fall into the ground and die, it abideth alone: but if it die, it bringeth forth much fruit.

He that loveth his life shall lose it; and he that hateth his life in this world shall keep it unto life eternal.

St John, XII. 24–25

Holy War 394

Give up thy life, if thou would'st live.

Tibetan Precept

That which thou sowest is not quickened, except it die.

I. Corinthians, XV. 36

If you are really desirous of mastering Zen, it is necessary for you once to give up your life and to plunge right into the pit of death.

Yekiwo

Pilg. 366

Before the perfect union death must precede.

Philalethes

The grave is the first stage of the journey into eternity.

Muhammad

The Portal of God is non-existence.

Chuang-tse (ch. XXIII)

Realiz. 890

The soul has a private door into divine nature at the point where for her things all come to naught.

St Augustine

Sell this present world of yours for the next world and you will gain both in entirety, but do not sell the next world for this world, for so shall you lose the two together. Act towards this world as if it were not, and towards the world to come as if it would never cease to be.

Ḥasan al-Baṣrî

Renun. 158

Reality 803

Those who cling to life die, and those who defy death live.

Uyesugi Kenshin

> To sue to live, I find I seek to die,
> And, seeking death, find life.

Shakespeare (*Measure for Measure*, III. i. 42)

Your substance will never be white, if it has not first been black.

Philalethes

Humility 191

Introd. *Renun.*

No matter what it is, if a man is afraid of losing it, he will lose it, but if he is willing to give it up, he will get it. So be ready to give up your life for the good of the people around you.

Tôyo Tenshitsu

Renun. 152a

If thou lovest thyself, thou must have no love of self; if thou wouldst save thyself, thou dost not well to be saving of self.

Śânti-deva

No one can be saved until he has been born again.

Hermes

Orthod. 280

Except a man be born again, he cannot see the kingdom of God.

St John, III. 3

Lo, in being slain I live.

Al-Ḥallâj

Introd. *Suff.*

Even God must die, if He wishes to live for thee:
How thinkest thou, without dying, to inherit His Life?

Angelus Silesius

I am crucified with Christ: nevertheless I live; yet not I, but Christ liveth in me.

Galatians, II. 20

Infra 220

If the Creator shall enter in, the creature must depart. Of this be assured.

Theologia Germanica, LIII

Realiz. 890

Whither I go, ye cannot come.

St John, VIII. 21

Thou mayest not come to heaven bodily, but ghostly. And yet it shall be so ghostly that it shall not be in a bodily manner: neither upwards nor downwards, nor on one side nor on another, behind nor before.

The Cloud of Unknowing, LIX

Inv. 1036

It is said that the highest state of prayer is when the spirit leaves the flesh and the world, and in the act of prayer, loses all matter and all form. To maintain oneself unfailingly in this state, is in reality to pray without cease.

Just as the body in dying is separated from all the goods of this life, so does the spirit which dies at the summit of prayer likewise quit all the representations that it has of the world. For without dying this death, it could never find itself and live with God.

St Maximus the Confessor

Contem. 523 *M.M.* 998

Nowhere bodily is everywhere ghostly. . . . Therefore travail fast in this nought, and in this nowhere, and leave thine outward bodily wits and all that they work in: for I tell thee truly that this work may not be conceived by them.

The Cloud of Unknowing, LXVIII, LXX

He who does not become an expert in annihilation shall not discover the beautiful face of the bride.

Abu 'l-Mawâhib ash-Shâdhilî

The soul must put itself to death.[1]

Eckhart

The seed of metals is hidden out of sight still more completely than that of animals; nevertheless, it is within the compass of our Art to extract it. The seed of animals and vegetables is something separate, and may be cut out, or otherwise separately exhibited; but metallic seed is diffused throughout the metal, and contained in all its smallest parts; neither can it be discerned from its body: its extraction is therefore a task which may well tax the ingenuity of the most experienced philosopher. . . . Let the sons of knowledge learn that the great object of our Art is the manifestation of the hidden seed of gold, which can be effected only by full and perfect volatilisation of that which is fixed, and the subsequent corruption of its particular form. To break up gold in this way is the most profound secret in the world. . . . Gold does not easily give up its nature, and will fight for its life, but our agent is strong enough to overcome and kill it, and then it also has power to restore it to life, and to change the lifeless remains into a new pure body.

Philalethes

P. State 563
Introd. *Conform.*

Any man that but man is
With nothing shall be pleas'd, till he be eas'd
With being nothing.

Shakespeare (*Richard II*, v. v. 39)

Peace 694
Void 724

No soul is rested till it is made nought as to all things that are made.

Julian of Norwich

It is the character only of a good man to be able to deny and disown himself, and to make a full surrender of himself unto God; forgetting himself, and minding nothing but the will of his Creator; triumphing in nothing more than in his own nothingness, and in the allness of the Divinity. But indeed this, his being nothing, is the only way to be all things; this, his having nothing, the truest way of possessing all things.

John Smith the Platonist

Reality 803

He that is dead is freed from sin.

Romans, VI. 7

Our own life is to be hated, and the reason is plain: it is because there is nothing lovely in it. It is a legion of evil, a monstrous birth of the serpent, the world, and the flesh; it is an apostasy from the life and power of God in the soul, a life that is death to Heaven, that is pure unmixed idolatry, that lives wholly to self and not to God; and therefore all this own life is to be absolutely hated, all this self is to be denied and mortified, if the nature, spirit,

Humility 191
Sin 66

[1] One could virtually say that every page of Eckhart proclaims death.

Renun. 146 *Metanoia* 480

tempers and inclinations of Christ are to be brought to life in us. For it is as impossible to live to both these lives at once as for a body to move two contrary ways at the same time. And therefore all these mortifications and self-denials have an absolute necessity in the nature of the thing itself.

William Law

One must be dead to see God.

Eckhart

Wr. Side 464

We have nothing for our portion but pride and blindness of spirit, hardness of heart, weakness and inconstancy of soul, concupiscence, revolted passions, and sicknesses in the body. We are naturally prouder than peacocks, more groveling than toads, more vile than unclean animals, more envious than serpents, more gluttonous than hogs, more furious than tigers, lazier than tortoises, weaker than reeds, and more capricious than weather-cocks. We have within ourselves nothing but nothingness and sin, and we deserve nothing but the anger of God and everlasting hell.

Sin 56

. . . He who is infinite Wisdom, does not give commandments without reason, and He has commanded us to hate ourselves only because we so richly deserve to be hated. Nothing is worthier of love than God and nothing is worthier of hatred than ourselves.

Renun. 139

In order to rid ourselves of self, we must die to ourselves daily. That is to say, we must renounce the operations of the powers of our soul, and of the senses of our body. We must see as if we saw not, understand as if we understood not, and make use of the things of this world as if we made no use of them at all.[1] This is what St Paul calls dying daily.[2]

St Louis de Montfort

None of you shall see his Lord before dying.

Muhammad

Seldome comes Glorie till a man be dead.

Robert Herrick

Introd. *Realiz.*

Nothing can live eternally, but that which hath lived from all eternity.

William Law

The soul and body of Christ and His divine nature were so inseparably joined together that they cannot be severed throughout all eternity. Nevertheless Christ had to die, and His soul had to be separated from His body, and once more joined to it on the third day, that His body might be glorified, and rendered as subtle as His soul and spirit.

The Glory of the World

Introd. *Metanoia*

That which is mortal cannot draw near to that which is immortal, nor that which is for a time to that which is everlasting, nor that which is corruptible to that which is incorruptible.

Hermes

[1] *I. Cor.* VII, 29–31. Cf. the gloss on this passage by Peter Sterry, *Renunciation*, p. 141.

[2] *I. Cor.* XV. 31.

The childish go after outward pleasures;
They walk into the net of widespread death.
But the wise, knowing immortality,
Seek not the stable among things which are unstable here.

Kaṭha Upanishad, IV. 2

A man must wholly die to self-hood in the death, and give himself up in the resigned will wholly into the obedience of God, as a new child of a new will; . . . without this there is only the form, viz. the history[1] that was once brought to pass, and that a man need only accept of it, and comfort himself therewith: but this will remains without, for it will be a child of an assumed grace, and not wholly die to its self-hood in the grace, and become a child of grace in the resigned will.

Metanoia 484

Boehme

The fallen soul, really dead to the kingdom of Heaven, can have no help but by a new birth of the light and Spirit of Heaven, really brought forth again in it.

William Law

No perfection can be obtained from imperfect metals, either by themselves or mixed, nor can that which is itself imperfect bring other things to perfection. . . . Geber clearly shews that the substance of our Stone cannot exist in imperfect metals; because things that are impure in themselves do not abide the fire which might purify them, while our mercury (on account of its purity) is not in the slightest degree injured by the fire. Thus we perceive that no one imperfect metal can contain the substance of our Stone. But neither is it to be found in a mixture of impure metals—for by mixing they become less pure than they were before. Moreover we said above that the substance we required was *one.*

P. State 563

Reality 775

The Golden Tract

One cannot see God and yet retain individuality.

Sri Ramana Maharshi

If by certain ones still living in this corruptible flesh, yet growing in incalculable power by a certain piercingness of contemplation, the Eternal Brightness of God is able to be seen, this is not at variance with the words of Job: 'Wisdom is hidden from the eyes of all the living' (XXVIII. 21); because he that sees Wisdom, which is God, wholly dies to this life, that henceforth he should not be held by the love thereof. He who sees God dies by the mere circumstance alone, that either by the bent of the interior, or by the carrying out of practice, he is separated with all his mind from the gratifications of this life. Hence yet further it is said of Moses: 'No man shall see me and live' (*Ex.* XXXIII. 20): as though it were plainly expressed, 'No man ever at any time sees God spiritually and lives to the world carnally'.

Realiz. 870

Ecstasy 636

St Gregory the Great

And, through this spiritual dying, his (a man's) soul is taken from him, and lifted up on high; while his body is still upon earth, his spirit and heart are already in his eternal

[1] 'That Christ once died and suffered for us, etc.' In this sense, cf. also *II. Cor.* V. 16.

Fatherland; and all his actions have a heavenly source, and seem no longer to belong to this earth. For he lives no longer according to the flesh, but according to the Spirit, not in the unfruitful works of darkness, but in the light and in the day—in works that stand the test of fire.

Action 340

The Sophic Hydrolith

Almighty God is found out by clear thought, when, the corruption of our mortality being once for all trodden underfoot, He is seen by us, taken up (into heaven) in the Brightness of His Divinity. . . . For the mind, whether of angels or men, while it gazes toward the unencompassed Light, shrinks into little by the mere fact that it is a creature.

Metanoia 484

St Gregory the Great

A grain of wheat has the air and light of this world enclosed or incorporated in it. . . . On the other hand, that great ocean of light and air, having its own offspring hidden in the heart of the grain, has a perpetual strong tendency to unite and communicate with it again. . . . But here let it be well observed that this desire on both sides cannot have its effect till the husk and gross part of the grain falls into a state of corruption and death; till this begins, the mystery of life hidden in it cannot come forth. The application here may be left to the reader.

Center 816

William Law

The substance is dissolved in a bath, and its parts reunited by putrefaction. In ashes it blossoms.

Basil Valentine

This transformation is so sacred that it must be undergone in darkness.

Realiz. 873

Black Elk

Die you must to all and everything that you have worked or done under any other spirit but that of meekness, humility, and true resignation to God. Everything else, be it what it will, has its rise from the fire of nature, it belongs to nothing else and must of all necessity be given up, lost, and taken from you again by fire, either here or hereafter.

Pilg. 366
Judgment passim

William Law

The mystery of our Art, which we conceal with so great care, is the preparation of the Mercury, which above ground is not to be found made ready to our hand. But when it is prepared, it is 'our water' in which gold is dissolved, whereby the latent life of the gold is set free, and receives the life of the dissolving Mercury, which is to gold what good earth is to the grain of wheat. When the gold has putrefied in the Mercury, there arises out of the decomposition of death a new body, of the same essence, but of a glorified substance. Here you have the whole of our Philosophy in a nutshell. There is no secret about it, except the preparation of Mercury, its mingling with the gold in the right proportions, and the regulation of the fire in accordance with its requirements. Gold by itself does not fear the fire; hence the great point is, to temper the heat to the capacity of the Mercury.

Introd. *Pilg.*
Metanoia 488
Orthod. 296

Philalethes

It is necessary, then, to reduce metallic bodies to their homogeneous water which does not wet the hands, that from this water there may be generated a new metallic species which is nobler by far than any existing metal, viz., our Celestial Ruby.

Philalethes

P. State 563

Be thou thy self in thy whole Person, the Sacrifice of a whole Burnt-Offering, ascending in a Sacred flame of heavenly love to God, the only and eternal Beauty.

Peter Sterry

Suff. 132

Up, noble Soul! Out of thyself so far thou never comest in again and enter into God so deep thou never comest out again.

Eckhart

Holiness 902

The Art of Dying

True philosophers are always occupied in the practice of dying.

Plato (*Phaedo*, 67 D)

Infra 226

Against his will he dieth that hath not learned to die. Learn to die and thou shalt learn to live, for there shall none learn to live that hath not learned to die.

The Book of the Craft of Dying

Supra 206

The lion of this world seeks a prey and provision; the lion of the Lord seeks freedom and death.

Rûmi

Thou shalt understand that it is a science most profitable, and passing all other sciences, for to learn to die. For a man to know that he shall die, that is common to all men; as much as there is no man that may ever live or he hath hope or trust thereof; but thou shalt find full few that hath this cunning to learn to die.... I shall give thee the mystery of this doctrine; the which shall profit thee greatly to the beginning of ghostly health, and to a stable fundament of all virtues.

Henry Suso

Faith 505

It is good to have a reminder of death before us, for it helps us to understand the impermanence of life on this earth, and this understanding may aid us in preparing for our own death. He who is well prepared is he who knows that he is nothing compared with *Wakan-Tanka*, who is everything; then he knows that world which is real.

Black Elk

Humility 191

It is good to prepare for death. One should constantly think of God and chant His name in solitude during the last years of one's life. If the elephant is put into the stable after its bath it is not soiled again by dirt and dust.

Inv. 1036

Sri Ramakrishna

O how wise and happy is he that now laboreth to be such an one in his life, as he will desire to be found at the hour of death!

The Imitation of Christ, I. xxiii

M.M. 998

Is not philosophy the study of death?

Plato (*Phaedo,* 81 A)

Pilg. 365

Be in the world like a traveller, or like a passer on, and reckon yourself as of the dead.

Muhammad

These all died in faith, not having received the promises, but having seen them afar off, and were persuaded of them, and embraced them, and confessed that they were strangers and pilgrims on the earth.

Renun. 139

Hebrews, XI. 13

Esteem thyself as dead upon earth, and as one to whom the whole world is crucified.

The Imitation of Christ, III. xliv

We should anyway have a wholesome loathing of life itself, always ready to part with it on short notice, to-day or to-morrow.

Hônen

As regards the community, I try to consider myself as already dead. In this way everything becomes indifferent to me and I remain in peace.

Sister Consolata

Learn now to die to the world, that thou mayest then begin to live with Christ.

The Imitation of Christ, I. xxiii

Think upon death, Christian: why think of the rest? there is nothing more profitable one can think upon than the manner in which one should die.

Angelus Silesius

Thou oughtest so to order thyself in all thy thoughts and actions, as if to-day thou wert to die.

The Imitation of Christ, I. xxiii

My noble father, Mila-Shergyal,
Hath left no trace of ever having lived;
My fond and loving mother, Nyang-Tsa-Kargyen,

Is now nought but a heap of whitened bones:
Even these are pictures of illusoriness,
Which make me seek the contemplative life.

Milarepa

Illusion 99

When thou seest in the pathway a severed head,
Which is rolling toward our field,
Ask of it, ask of it, the secrets of the heart:
For of it thou wilt learn our hidden mystery.

Dîvâni Shamsi Tabrîz, II

Supra 208

The world is impermanent. One should constantly remember death.

Sri Ramakrishna

The agony of death cometh in truth.[1]

Qur'ân, L. 19

Judgment passim

If at any time thou hast seen another man die, make account that thou must also pass the same way.

The Imitation of Christ, I. xxiii

O man, remember death at all times.

Ansârî

Knowl. 755

In all thy works remember thy last end, and thou shalt never sin.

Ecclesiasticus, VII. 40

Allâh giveth life to you, then causeth you to die, then gathereth you unto the Day of Resurrection whereof there is no doubt. But most of mankind know not.

Qur'ân, XLV. 26

Thou fool, this night thy soul shall be required of thee.

St Luke, XII. 20

Death is like an arrow that is already in flight, and your life lasts only until it reaches you.

Hermes

Action 329

Even on one's birthday morning, omens of one's death appear;
Ever be alert and watchful; waste no time, O Tingri folk. . . .
Ever transient is this world of ours; all things change and pass away;
For a distant journey even now prepare, O Tingri folk.

Phadampa Sangay

[1] The Arabic root S K R has the basic sense of intoxication or sweetness (from which is derived the word *sugar*), thus allowing an esoteric interpretation of *sakrah*, 'agony', in the sense of 'ecstasy'.

The sun exactly at noon is exactly (beginning to) go down. And a creature exactly when he is born is exactly (beginning to) die.

Hui Shih

Worldly folk who have heaped up evil *karma* during their lifetime, and who anticipate reaping, as the result, the pangs of birth, old age, illness, and death, in this world, vainly seek to evade or ameliorate the intensity and anguish thereof by means of propitiatory ceremonies and medical treatment. Neither through the power or authority of kings, nor the valour of the hero, nor the charming form of the belle, nor the wealth of the rich, nor the fleetness of the coward, nor the oratory of an able pleader, can one ward off, or retard for a moment, the Decree of Time. There are no means or methods, be they peaceful, noble, fascinating, or stern, which can buy off or stop the execution of this unalterable decree. If any there be who are truly fearful of those pangs, and sincerely seek to prevent their recurrence, and are really eager to attain a state of eternal bliss, I possess the secret rite for the attainment thereof.

Infra 220

Milarepa

Strive after the Good before thou art in danger, before pain masters thee and thy mind loses its keenness.

Kulârnava Tantra, I. 27

He who would preserve his life should call to mind the hour of death; and while he be yet in his day, let him fear the coming of his night.

Shekel Hakodesh, 117

It is astonishing that anyone, and even while seeing his like die, should forget death.

'Alî

Everyone is aware of the eternal Self. He sees so many dying but still believes himself eternal. Because it is the Truth. Unwillingly the natural Truth asserts itself. The man is deluded by the intermingling of the conscious Self with the insentient body. This delusion must end.

Metanoia 480

Sri Ramana Maharshi

Remember thy Creator in the days of thy youth, before the time of affliction come, and the years draw nigh of which thou shalt say: They please me not.

Ecclesiastes, XII. 1

Every day an angel of heaven cries:
O people there below! produce offspring to die; build to be destroyed; gather ye together to depart!

Illusion 99

'Alî

For ever it has been that mourners in their turn were mourned.

Old Chinese Poem

Putrefaction is the end
Of all that Nature doth entend.

Robert Herrick

First Clown.
A pick-axe, and a spade, a spade,
For and a shrouding sheet;
O! a pit of clay for to be made
For such a guest is meet.
(*Throws up another skull.*)

Hamlet. There's another; why may not that be the skull of a lawyer? Where be his quiddities now, his quillets, his cases, his tenures, and his tricks? Why does he suffer this rude knave now to knock him about the sconce with a dirty shovel, and will not tell him of his action of battery? Hum! This fellow might be in 's time a great buyer of land, with his statutes, his recognizances, his fines, his double vouchers, his recoveries; is this the fine of his fines, and the recovery of his recoveries, to have his fine pate full of fine dirt? will his vouchers vouch him no more of his purchases, and double ones too, than the length and breadth of a pair of indentures? The very conveyance of his lands will hardly lie in this box, and must the inheritor himself have no more, ha?

Shakespeare (*Hamlet,* v. i. 100)

The idea most vital and essential to the samurai is that of death, which he ought to have before his mind day and night, night and day, from the dawn of the first day of the year till the last minute of the last day of it.

Daidôji Yûsan

It has been proved to us by experience that if we would have pure knowledge of anything we must be quit of the body.

Plato (*Phaedo,* 66 E)

Zen has no secrets other than seriously thinking about birth-and-death.

Takeda Shingen

O ye who believe! Let not your wealth nor your children distract you from remembrance of Allâh. Those who do so, they are the losers.

Inv. 1036

And spend of that wherewith We have provided you before death cometh unto one of you and he saith: My Lord! If only thou wouldst reprieve me for a little while, then I would give alms and be among the righteous.

But Allâh reprieveth no soul when its term cometh, and Allâh is Informed of what ye do.

Qur'ân, LXIII. 8–10

And if the Wine you drink, the Lip you press,
End in the Nothing all Things end in—Yes—
Then fancy while Thou art, Thou art but what
Thou shalt be—Nothing—Thou shalt not be less.

Omar Khayyâm

Supra 208

Think often of death with attention, bringing to mind everything which must then happen. If you do this, that hour will not catch you unawares. . . . Men of this world flee from the thought and memory of death, so as not to interrupt the pleasures and enjoyments of their senses, which are incompatible with memory of death. This makes their attachment to the blessings of the world continually grow and strengthen more and more, since they meet nothing opposed to it. But when the time comes to part with life and all the pleasures and things they love, they are cast into excessive turmoil, terror and torment.

Unseen Warfare, II. ix

Because Thou lovest the Burning-ground,
I have made a Burning-ground of my heart—
That Thou, Dark One, haunter of the Burning-ground,
Mayest dance Thy eternal dance.
Nought else is within my heart, O Mother:
Day and night blazes the funeral pyre:
The ashes of the dead, strewn all about,
I have preserved against Thy coming:
Infra 232
With death-conquering Mahakala neath Thy feet
Do Thou enter in, dancing Thy rhythmic dance,
That I may behold Thee with closed eyes.

Bengali Hymn to Kâlî

The Destruction of I-Consciousness

Reality 803

The foolish man conceives the idea of 'self', the wise man sees there is no ground on which to build the idea of 'self'.[1]

Asvaghosha

Action 340

The feeling of 'I' and 'mine' is the result of ignorance.

Sri Ramakrishna

Center 841

The word 'I' does not exist for him who always sees the Self.

Shiva Samhitâ, v. 208

Renun. 143

I went to pray with the devotees, and did not feel in my place. I went among those who mortify themselves, . . . among those who fast, and still I did not feel in my place. I said, 'O my God! what then is the way to thee?' And the answer given me was, 'Abandon thyself and come.'

Bâyazîd al-Bisṭâmî

[1] Which automatically voids the reincarnationist hypothesis.

He is a Knower of the Self to whom the ideas 'me' and 'mine' have become quite meaningless.

Śrî Śankarâchârya

Introd. *Holiness*

'Between Me and thee is thy self-experience: cast it away, and I will veil thee from thyself.'

Niffarî

Peace 700

'What am I? What is the world?' This enquiry into truth is a fire. Burn up the false 'I' which identifies itself with the body in the fire of enquiry. When this is accomplished, then the 'I' remains in *Atman.*[1]

Yoga-Vasishtha

As long as the notion of first person exists, the notions of second and third persons also exist. When, through reflection upon one's reality, the notion of first person is destroyed, the notions of second and third persons also disappear, and the resulting state is alone the true State of Self, revealed as One.[2]

Sri Ramana Maharshi

Realiz. 859

This 'I' is an illusion which must be eliminated from our life and thought.

Swami Ramdas

In the samâdhi that comes at the end of reasoning and discrimination, no such thing as 'I' exists. But it is extremely difficult to attain it; 'I-consciousness' lingers so persistently. That is why a man is born again and again.

Sri Ramakrishna

Judgment 250

'I' and 'you' are the veil
Between heaven and earth.

Shabistarî

Illusion 85

In the case of ordinary people the 'I' never disappears. You may cut down the aśwattha tree, but the next day sprouts shoot up.

Sri Ramakrishna

Introd. *Conform.*

He (Abû Sa'îd) was asked, 'What is evil and what is the worst evil?' He replied, 'Evil is "thou"; and the worst evil is "thou", when thou knowest it not.'

Abû Sa'îd ibn Abi 'l-Khayr

Sin 56 80

There will be a time when this body will require four people to carry it with difficulty (to the cremation ground) but at this moment we are carrying this heavy burden ourselves by reason of the sustaining force of the sense of 'I' which we have in it. It will be clearly seen that somebody other than the body is necessary to entertain this sense of 'I' in it, to sustain it by its force and to carry its weight.

Sri Chandrasekhara Bhâratî Swâmigal

[1] This is the Maharshi's method.
[2] A profound synthesis of the whole spiritual method.

The consciousness that 'This is I' is like the shadow of night and is dispelled by the rising of the sun of true knowledge. Think neither of the entity nor the non-entity of yourself or others. Preserve the tranquillity of your mind, by ignoring the thought of both positive and negative existence, and be rid of the sense of distinction between cause and effect.

Yoga-Vasishtha

It is very difficult to separate with a knife the kernel of a coconut from the shell before the milk inside has dried up. When the milk dries up, the kernel rattles inside the shell. At that time it loosens itself from the shell. Then the fruit is called a dry coconut. The sign of a man's having realized God is that he has become like a dry coconut. He has become utterly free from the consciousness that he is the body.

Holiness 914

Sri Ramakrishna

Break the idol of clay, that you may behold the face of the Fair.

Dîvâni Shamsi Tabrîz, XLII

Is it an easy thing to obtain the Knowledge of Brahman? It is not possible unless the mind is annihilated. The guru said to the disciple, 'Give me your mind and I shall give you Knowledge.' . . . A man attains Brahmajnâna as soon as his mind is annihilated. With the annihilation of the mind dies the ego, which says 'I', 'I'.

Contem. 528

Sri Ramakrishna

The ferryman punted them dexterously out from shore. Suddenly they saw a body in the water, drifting rapidly down stream. Tripitaka stared at it in consternation. Monkey laughed. 'Don't be frightened, Master,' he said. 'That's you.' And Pigsy said, 'It's you, it's you.' Sandy clapped his hands. 'It's you, it's you,' he cried. The ferryman too joined in the chorus. 'There *you* go!' he cried. 'My best congratulations.'

Realiz. 873

Monkey

The mind exists in relation to the 'I', and vice-versa. 'I' is only an idea of the mind; it and the 'I' are identical. If the 'I' vanishes, the mind will do the same, and vice-versa. Destroy the mind by means of true knowledge (tattva-jnâna); destroy the 'I' by the consciousness that 'I am Brahman' ('aham Brahmâsmi bhâvana'), and this through constant and intense meditation. When the mind vanishes and thought ceases, name and form (nama-rûpa) disappear and the end is attained.

Swami Sivananda

Center 831

Caṇḍâlî[1] blazes at the navel,
She burns the Five Tathagatas,
She burns Locana and the others,
AHAṂ (=I) is burned and the Moon is melted.

P. State 590

Hevajratantra, I

[1] 'The Tibetan for Caṇḍâlî is gTum-mo, translatable as "mystic heat". It may be remembered that the Tibetans employed the "mystic heat" technique in order to keep the body physically warm' (note by translator).

The Cycle of Births and Deaths

For that which is born death is certain, and for the dead birth is certain. Therefore grieve not over that which is unavoidable.

Bhagavad-Gîtâ, II. 27

The living come from the dead, just as the dead come from the living.

Plato (*Phaedo*, 72 A)

Metanoia 488

That which goes out of being[1] also comes into being; and that which comes into being also goes out of being.

Hermes

Introd. *Void*

The destruction of *one* thing is the generation of another.

The Glory of the World

The living all find death unpleasant, men mourn over it. And yet, what is death, but the unbending of the bow and its return to its case; what is it, but the emptying of the corporeal envelope and the liberation of the two souls[2] imprisoned therein? After the encumbrances and vicissitudes of life the two souls leave; the body follows them into repose. This is the Great Return.—That the incorporeal has produced the corporeal, and that the body returns to incorporeity, this idea of the Eternal Round is known to many people, but only the elect draw the practical consequences from it.

Chuang-tse (ch. XXII)

Infra 226

Judgment 250

The destruction of your bodies then will be the starting-point for a rebirth, and their dissolution, a renewal of your former happiness. But your minds will be blinded, so that you will think the contrary, and will regard the punishment[3] as a boon, and the change to a better state as a degradation and an outrage. But the more righteous among you . . . look forward to the change.

Hermes

What wilt thou do with the body? This poor flesh, which thou guardest in order to feed vultures, jackals, and the like, is fitted only to be a tool for men's works. Though thou guardest it thus, pitiless Death will tear away the body and give it to the vultures; and then what wilt thou do? To a servant who will not remain, gifts of garments and the like are not given; when it has eaten, the body will depart, then why waste thy riches upon it? Pay to it its wage, then set thy thought upon thine own business; for we give not to the hireling all that he may earn. Conceive of the body as a ship that travels to and fro, and make it go at thy bidding for creatures to fulfil their end.

Śânti-deva

Renun. 143

Pilg. 365

[1] I.e. among other meanings, into non-manifestation.

[2] This would correspond with *Âtmâ* and *jîvâtmâ*.

[3] I.e. life in the body.

Illusion 85

We love and tend the body, the least agreeable and most vile of all things! For if we had to tend our neighbour's body for ten days only we could not bear it. Consider what it would be to get up in the morning and clean some one else's teeth, and then to perform some other necessary office for him. Truly it is wondrous that we should love that for which we do such mean services day by day. I stuff this bag; then I empty it; what could be more tiresome? But I am bound to serve God. That is why I stay here and put up with washing this miserable body of mine, and giving it fodder and shelter; and when I was younger, it laid other commands on me as well, and yet I bore with it. Why then, when Nature, who gave you your body, takes it away, can you not bear it?

Epictetus

As in this body the embodied soul passes through childhood, youth and old age, in the same manner it goes from one body to another; therefore the wise are never deluded regarding it. . . .

As man casts off worn-out garments and puts on others which are new, similarly the embodied soul, casting off worn-out bodies, enters into others which are new.

Bhagavad-Gîtâ, II. 13, 22

Know that the body is like a garment. Go, seek the wearer of the garment, do not lick (kiss) a garment. . . .

Flight 949

You are such that without the material body you have a spiritual body: do not, then, dread the going forth of the soul from the body.

Rûmî

Bodies are like pillow-cases. It doesn't matter whether they remain or drop off.[1]

Sri Ramakrishna

When he (the embodied self) comes to weakness—whether he come to weakness through old age or through disease—this person frees himself from these limbs just as a mango, or a fig, or a berry releases itself from its bond; and he hastens again, according to the entrance and place of origin, back to life.

Infra 225

Bṛihad-Âraṇyaka Upanishad, IV. iii. 36

Know well, O beloved Prince, that as the sun does not cease from giving his light to the other hemisphere after he has set here, so does your intelligence continue to glow even after it has run its course in this life.

Center 833

Yoga-Vasishtha

Reality 775

As fire assumes the forms of burning objects, so hath the all-pervading Lord assumed the forms of beings and things.

As the flames rise and fall but not the fire itself, so birth and death belong to the bodies but not to the Self.

Srimad Bhagavatam, XI. iii

[1] This attitude prevails in the Zen art of swordsmanship. Cf. Herrigel: *Zen in the Art of Archery,* p. 103: 'Like a petal dropping in the morning sunlight and floating serenely to earth, so must the fearless detach himself from life, silent and inwardly unmoved.'

The Self existed before the birth of this body and will remain after the death of this body. So it is with the series of bodies taken up in succession. The Self is immortal. The phenomena are changeful and appear mortal. The fear of death is of the body.

Sri Ramana Maharshi

There is no real coming and going.
For what is going but coming?

Shabistarî

Creation 26

Nobody is born or dies at any time; it is the mind that conceives its birth and death and its migration to other bodies and other worlds.

Yoga-Vasishtha

Illusion 85
Supra 220

Reabsorption of the Faculties

When a person here is deceasing, his voice goes into his mind; his mind, into his breath; his breath, into heat, the heat, into the highest divinity.[1]

Chândogya Upanishad, VI. viii. 6

The *Spirit* is the Fountain of Life, which flowes forth from God, to Feed, and Maintain the Breath of Life in the Body. When the time of Death comes, This Spirit draws back to their Head again Those streams of Life, by which It went forth into the Body.

Peter Sterry

A manifest sign of this return of things to their source has been placed in living beings by most holy Nature. Our life-breath, which we draw from above out of the air, we send up again to the place whence we received it.

Hermes

I am striving to give back the Divine in myself to the Divine in the All.

Plotinus (Last words as reported in Porphyry's *Life of Plotinus*)

Introd.

He who, at the time of death, thinking of Me alone, goes forth, leaving the body, he attains unto my Being. There is no doubt in this.

Bhagavad-Gîtâ, VIII. 5

Inv. 1017

'It shall be thus: I draw My breath and thy soul will follow on to Me as a needle to a magnet.'

Mechthild of Magdeburg

Rev. 967

[1] After death each soul must confront the Divinity. But Judgment still follows. Cf. Introd. *Judgment*.

Orthod. 300

Those who, when the law has been well preached to them, follow the law, will pass over the dominion of death, however difficult to cross.

Dhammapada, VI. 86

M.M. 998

What is soundless, touchless, formless, imperishable, tasteless, constant, odorless, without beginning, without end, higher than the great, unchanging—

By discerning That, one is liberated from the mouth of death.... If one recites this supreme secret in an assembly of Brahmans, or at a time of the ceremony for the dead, devoutly—

That makes for immortality.

Kaṭha Upanishad, III. 15, 17

Sin 77

The soule shuts up the windowes of sense, when she would have the room fill'd with the light of Reason. Reason's self must first be cast into a deep sleep[1] and die, before she can rise again in the brightnesse of the Spirit.

Peter Sterry

Indomitable faith combined with supreme serenity of mind are indispensable at the moment of death.

Gampopa

May we, the disciples of the Buddha, when we come to die suffer no mental perversion, nor come under the spell of any hallucination, nor lose the consciousness of the Truth, but, free from agony of mind and body, may we in peace of mind, like those in an ecstasy, have that holy retinue of Amida come to meet us, and, embarking safely on the ship of his Original Vow, may we have our birth into Amida Buddha's Pure Land, and sit upon the lotus of the first rank.

Introd. *Pilg.* *Center* 847

Zendô

The Happy Death

There are those to whom death is as a draught of pure water to the thirsty.

Ibn al-ʿArîf

Death is a bridge whereby the lover is joined to the Beloved.

ʿAbd al-ʿAzîz b. Sulaymân

Center 847

Courage is but being fearless of the death which is but the parting of the Soul from the body, an event which no one can dread whose delight is to be his unmingled self.

Plotinus

[1] Cp. the deep-sleep state (*suṣupta-sthâna*) described in the Vedanta.

He that hath got the mastery over his own will, feels no violence from without, finds no contests within; and like a strong man, keeping his house, he preserves all his goods in safety: and when God calls for him out of this state of mortality, he finds in himself a power to lay down his own life; neither is it so much taken from him, as quietly and freely surrendered up by him.

John Smith the Platonist

Holy War 396
P. State 563

Illusion 99

As Tullius says, in his *De Senectute,* 'natural death is, as it were, our haven after a long voyage, and our repose.' And as a good sailor, when he nears the harbour, lowers his sails, and gently and with feeble headway enters it, so should we lower the sails of our worldly occupations, and return to God with all our mind and heart, so that we may enter our haven with all gentleness and all peace. And our own nature teaches us much as to the gentleness (of natural death); because in such a death there is no pain, nor any bitterness; but as a ripe apple lightly and without violence detaches itself from the bough, so our soul, without pain, leaves the body in which it has been. Wherefore Aristotle, in his *Youth and Old Age,* says that the death which comes to old age is a death without sadness.

Dante (*Il Convito,* IV. xxviii. 1)

Introd. *Pilg.*

From you (the Druids) we learn that . . . death is the centre, not the finish, of a long life.

Lucan

Nor will I be silent as to this particular, that it appeared both to Plato and Pythagoras, that old age was not to be considered with reference to an egress from the present life, but to the beginning of a blessed life.

Phavorinus

Supra 223

Now light comes to you, O Soul, from Mind (νοῦς), and darkness from the body; and if you understand this, you ought not to grieve that you must quit the body. For it is through the body that you suffer the greatest loss, and are prevented from apprehending those things which it behoves you most to know. Nay, you ought rather to grieve that you are separated from the thought-world; for it is by the action of that world on you that you enjoy the greatest gains and are enabled to attain to those things which it behoves you to seek after and to grasp. Turn away then from the physical world, scorn and hate and fear it, and seek refuge in the thought-world, which is the root and source of your existence, and the abode appropriate to a being that stands so high in rank and authority as you; that so you may partake of everlasting life, and attain to perfect bliss.

Hermes

Sin 66

Beauty 663

Creation 42

A good name is better than precious ointment; and the day of death than the day of one's birth.

Ecclesiastes, VII. 1

This is the precious, the blessed work of Death in a Believer. It breaketh the union, it maketh a separation not between Christ, and a Believer, not between the Soul, and Body of a Believer, as they are joyned together in Christ by the band, and unity of the Eternal

Spirit, and make one Spiritual Man, or Person in Christ;[1] but between the precious, and the vile, between the Carnal and Spiritual Principle of a Saint both in Soul and Body.

Peter Sterry

There is no death more glorious than that which brings a life:
No life more precious than that which springs from death.

Angelus Silesius

The stroke of death is as a lover's pinch,
Which hurts, and is desir'd.

Shakespeare (*Antony and Cleopatra*, v. ii. 297)

We can die gladly if God will live and work in us.... We die, 'tis true, but 'tis a gentle death.

Eckhart

Knowl. 745

If you really understand Him that governs the universe and if you carry Him about within you, do you still long for paltry stones and pretty rock? What will you do, then, when you are going to leave the very sun and moon? Shall you sit crying like little children?

Renun. 158

Epictetus

Supra 220

Your fear of death is really fear of yourself: see what it is from which you are fleeing!

Rûmî

I say, since death alone delivers me,
It is of all things the best of things.

Angelus Silesius

The Prince of mankind (Muhammad) said truly that no one who has passed away from this world

Supra 215

Feels sorrow and regret for having died; nay, but he feels a hundred regrets for having missed the opportunity,

Center 847

Saying to himself, 'Why did I not make death my object—death which is the store-house of all fortunes and riches ... ?'

The grief of the dead is not on account of death; it is because they dwelt on the phenomenal forms of existence.

Rûmî

Pilg. 366

So great a good is death: could a hellhound have it,
He would straightway get himself buried alive.

Angelus Silesius

In the hope of ease to come,
Let's endure one Martyrdome.

Robert Herrick

[1] I.e., the essence (*purusha*) is reunited with the primordial substance (*prakriti*) in the 'glorified body'.

Of the lives of the martyrs one hears little: it is the virtues witnessed in their hours of suffering which are cited and praised, and which take the place of these lives, so does a beautiful death adorn an entire life.

Angelus Silesius

The Prefect: 'Do you suppose that you will ascend into heaven to receive such a recompense?'

St Justin: 'I do not suppose it, but I am fully persuaded of it.'

From the trial of St Justin before
Rusticus, the prefect of Rome

First Gaoler. Come, sir, are you ready for death?

Posthumus. Over-roasted rather; ready long ago. . . .

First Gaol. Look you, sir, you know not which way you shall go.

Post. Yes, indeed do I, fellow. . . . I tell thee . . . there are none want eyes to direct them the way I am going but such as wink and will not use them. . . .

First Gaol. Unless a man would marry a gallows and beget young gibbets, I never saw one so prone.

Shakespeare (*Cymbeline,* v. iv. 153)

Faith 509

I look forward with joy to the wild animals held in readiness for me, and I pray that they may attack me; . . . come fire and cross and grapplings with wild beasts, wrenching of bones, hacking of limbs, crushings of my whole body; only be it mine to attain unto Jesus Christ. . . . I am God's wheat, and I am ground by the teeth of wild beasts, that I may be found Christ's pure bread.

St Ignatius of Antioch

Suff. 130

Ecstasy 641

. . . O happy dagger!
This is thy sheath; there rest, and let me die.

Shakespeare (*Romeo and Juliet,* v. iii. 169)

Welcome to thee,
O sword of eternity!
Through Buddha
And through Dharma alike
Thou hast cleft thy way.

Rikyû (at the moment of committing seppuku)

M.M. 978

Chuang-tse said, 'Were I to prevail upon God to allow your body to be born again, and your bones and flesh to be renewed, so that you could return to your parents, to your wife, and to the friends of your youth,—would you be willing?'

At this, the skull opened its eyes wide and knitted its brows and said, 'How should I cast aside happiness greater than that of a king, and mingle once again in the toils and troubles of mortality?'

Chuang-tse (ch. XVIII)

When my bier moveth on the day of death,
Think not my heart is in this world.
Do not weep for me and cry 'Woe, woe!'
Thou wilt fall in the devil's snare: that is woe.
When thou seest my hearse, cry not 'Parted, parted!'
Union and meeting are mine in that hour.
If thou commit me to the grave, say not 'Farewell, farewell!'
For the grave is a curtain hiding the communion of Paradise.
After beholding descent, consider resurrection;
Why should setting be injurious to the sun and moon?
Metanoia 488
To thee it seems a setting, but 'tis a rising;
Tho' the vault seems a prison, 'tis the release of the soul.
Pilg. 378
What seed went down into the earth but it grew?
Why this doubt of thine as regards the seed of man?

Dîvâni Shamsi Tabrîz, XXIV

How then do I know but that the dead repent of having previously clung to life?

Chuang-tse (ch. II)

Say to my brethren when they see me dead,
and weep for me, lamenting me in sadness:
'Think ye I am this corpse ye are to bury?
I swear by God, this dead one is not I.
When I had formal shape, then this, my body,
Served as my garment. I wore it for a while.'

Ode by al-Ghazâlî, composed on his deathbed

Why mourn the dead? They are free from bondage. Mourning is the chain forged by the mind to bind itself to the dead.

Sri Ramana Maharshi

Umm Aiman was seen in tears and was asked if she was weeping for the death of the Prophet. She replied: 'Not for him do I weep. Know I not that he hath gone to that which is better for him than this world? But I weep for the tidings of heaven (*akhbâru 's-samâ*) that have been cut off from us.'
Rev. 959

Ibn Sa'ad

'How mournest thou not for thine own sons, when Death, the Bleeder, has pierced them with his lancet?

Since the evidence of pity is tears in the eyes, why are thine eyes without moisture and tearless?'

Metanoia 484
The Shaykh turned towards his wife and said to her, 'Old woman, verily the season of December is not like Tamûz (July).

Whether they all are dead or living, when are they absent and hid from the eye of the heart?
Center 816

Inasmuch as I see them distinct before me, for what reason should I rend my face as thou doest?

Although they are outside of Time's revolution, they are with me and playing around me.

Weeping is caused by severance or by parting; I am united with my dear ones and embracing them.

Other people see them in sleep; I see them plainly in my waking state.

I hide myself for a moment from this world, I shake the leaves of sense-perception from the tree.'

Rûmî

Holiness 934
Introd.
P. State

Tsekung hurried in and said, 'How can you sing in the presence of a corpse? Is this good manners?'

The two men looked at each other and laughed, saying, 'What should this man know about the meaning of good manners indeed?' Tsekung went back and told Confucius, asking him, 'What manner of men are these? ... They can sit near a corpse and sing, unmoved. ... '

'These men,' replied Confucius, 'play about beyond the material things; I play about within them. Consequently, our paths do not meet, and I was stupid to have sent you to mourn.[1] They consider themselves as companions of the Creator, and play about within the One Spirit of the universe. They look upon life as a huge goiter or excrescence, and upon death as the breaking of a tumor. How could such people be concerned about the coming of life and death or their sequence? They borrow their forms from the different elements, and take temporary abode in the common forms, unconscious of their internal organs and oblivious of their senses of hearing and vision. They go through life backwards and forwards as in a circle without beginning or end, strolling forgetfully beyond the dust and dirt of mortality, and playing about with the affairs of inaction. How should such men bustle about the conventionalities of this world, for the people to look at? ... (They) are strange in the eyes of man, but normal in the eyes of God. Hence the saying that the meanest thing in heaven would be the best on earth; and the best on earth, the meanest in heaven.'

Chuang-tse (ch. VI)

Creation 33
M.M.
978

Supra 223
Flight 949
Action 358
Holiness 910
Beauty 663
Metanoia 488

The truly wise mourn not either for the dead or for the living.

Bhagavad-Gîtâ, II. 11

M.M.
978

Securitas (freedom from fear) is to be got by fixing in one's mind a true opinion about physical death.

Hermes

[1] A clear distinction is made here between the esoteric and exoteric paths, showing how the attitude may differ depending on the path. But Confucius reveals the wisdom of an esoteric path, while confessing further in the same passage that 'I am one condemned by God' (in Wieger, p. 259, this reads, *'le Ciel m'a condamné à cette besogne massacrante'*) to follow the exoteric law.

The attitude of the Indian toward death, the test and background of life, is entirely consistent with his character and philosophy. Death has no terrors for him; he meets it with simplicity and perfect calm, seeking only an honorable end as his last gift to his family and descendants.

Ohiyesa

. . . My wishes be
To leave this life, not loving it, but Thee.

Robert Herrick

Our body being illusory and transitory, it is useless to give over-much attention to it.

Gampopa

Supra 215

We ought to turn our confidence towards death, and our caution towards the fear of death: what we really do is just the contrary; we fly from death, yet we pay no heed to forming judgements about death, but are reckless and indifferent.

Epictetus

Suff. 130

O Lord, he whom Thou killest doth not smell of blood,
And he whom Thou burnest doth not smell of smoke,
For he whom Thou burnest is happy in the burning,
And he whom Thou killest, rejoiceth in being killed.

Ansârî

Conform. 170
Creation 48
M.M. 978

Seek not for life on earth or in heaven. Thirst for life is delusion. Knowing life to be transitory, wake up from this dream of ignorance and strive to attain knowledge and freedom before death shall claim thee. The purpose of this mortal life is to reach the shore of immortality by conquering both life and death.

Srimad Bhagavatam, XI. xiii

Death Kills Death

The last enemy that shall be destroyed is death.

I. Corinthians, XV. 26

Introds.
Creation
Suff.

God died, that a kind of celestial exchange might be made, that men might not see death. . . . For as much as He is both God and man, wishing that we should live by that which was His, He died by that which was ours. For He had nothing himself whereby He could die, nor had we anything whereby we could live. . . . Seek for anything in God by which He may die and thou wilt not find it. But we all die, who are flesh, being men bearing on them sinful flesh. Seek for that whereby sin may live; it hath nothing. So then neither could He have death by that which was His, nor we life by that which was ours; but we have life by that which is His, and He death by what is ours. What an exchange!

St Augustine

O death, I am to thee a death.

Boehme

Judgment 266

Raising myself from death, I kill death—which kills me. I raise up again the bodies that I have created. Living in death, I destroy myself—whereof you rejoice. You cannot rejoice without me and my life.

If I carry the poison in my head, in my tail which I bite with rage lies the remedy. Whoever thinks to amuse himself at my expense, I shall kill with my gimlet eye.

Whoever bites me must bite himself first; otherwise, if I bite him, death shall bite him first, in the head; for first he must bite me—biting being the medicine of biting.

Giovanni Battista Nazari (the Mercury of the Philosophers speaking)

Metanoia passim
Sin 62
Conform. 170
Orthod. 280
Inv. 1007

The white Tiger grows angry in the western mountains and the green Dragon is irresistibly fierce in the eastern sea. Capture them with both hands and let them fight to death. They then change into a layer of the powder of purple gold.

Chang Po-tuan

Center 835

The mouth of the pit was open, and I saw a huge animal like a dragon, which let down its tail. I knew that God had sent it and that I should be saved in this way. I took hold of its tail and it dragged me out. A heavenly voice cried to me, 'This is an excellent escape of thine, O Abû Ḥamza! We have saved thee from death by means of a death.'

Abû Ḥamza al-Khurâsânî

Death, lie thou there, by a dead man interr'd.

Shakespeare (*Romeo and Juliet,* v. iii. 87)

Supra 206

Death is swallowed up in victory.
O death, where is thy sting? O grave, where is thy victory?

I. Corinthians, xv. 54–55

And call not those who are slain in the way of Allâh 'dead'. Nay, they are living, only ye perceive not.

Qur'ân, II. 154

PART II: *Combat — Action*

JUDGMENT

O hell! what have we here?
A carrion Death, within whose empty eye
There is a written scroll.

Shakespeare (*Merchant of Venice*, II. vii. 62)

The intellectual vision of the profane modern man is restricted to his physical self and material environment in such a way that corporeal death for him seems quasi-absolute. Cut off from the knowledge even of his own soul,[1] and unable with the experience of sense-data to discern its supraphysical reality, he cannot seriously envisage the possibility of a soul that survives bodily death; and he concludes that the ancient teachings on the posthumous modifications of the human soul are fables subjectively rationalized, or else he selects those elements that accord with his fancies and sentiments and recasts them into a pattern mirroring his own emotions and yearnings. In fact, he does precisely what he charges the ancients with having done. And in any case, whatever he retains of a posthumous doctrine concerns heaven and rejects hell—if one can use the term 'heaven' to describe what often amounts to little more than an 'ideal' corporeal state permanently 'etherealized', or 'astralized'.

But the body according to traditional cosmology is created hierarchically subsequent to the human soul, of which it is only a projection or exteriorization. The soul has precedence over the body in priority of creation and reality of being. And yet the soul being created—except for its immortal center—must like all creation pass through death. There is a critical moment or juncture, a 'judgment', 'discrimination', or 'separation'—which can be prefigured even in this lifetime—when the covering veils are removed and the basic intention or condition of the soul is exposed naked before its spiritual archetype. Only that which is of the spirit can return to the spirit; everything *other* will necessarily be thrown back to the degree of multiplicity, darkness or chaos that corresponds with its condition. 'Fear not them which kill the body, but are not able to kill the soul: but rather fear him which is able to destroy both soul and body in hell' (*Matt.* X. 28). The 'second death' of the *Apocalypse* may also concern the case where the soul, 'leaving the human state, will simply pass into another individual state of manifestation. This is a dreadful possibility for the

Conform. 170
Knowl. 749
Wr. Side 459

[1] 'The limitation of profane psychology lies in the fact that it deals only with psychic phenomena, without being able to apply qualitative criteria' (Burckhardt: *Du Soufisme*, p. 27). Again, quoting the same author: 'Spiritual psychology can be understood only in terms of a supraindividual element, necessarily mysterious, and inconceivable on the plane of ordinary psychology' (*Introduction aux Doctrines ésotériques de l'Islam*, p. 27).

profane person, who has every advantage to be maintained in what we have called the "prolongations" of the human state, which moreover, in all traditions, is the principal reason for funeral rites'[1] (Guénon: *Aperçus sur l'Initiation*, p. 185).

What Guénon has to say about the culmination of a macrocosmic world cycle is likewise instructive in relation to the human microcosm: 'While the positive results of the cyclical manifestation are "crystallized" to be transmuted thenceforth into the germs of the possibilities for the future cycle, which completes the process of "solidification" under its "benefic" aspect (implying essentially the "sublimation" that coincides with the final "reversion"), what cannot be thus utilized, namely, everything which constitutes the purely negative results of this same manifestation, is "precipitated" under the form of a *caput mortuum* (in the alchemical sense of this term) into the lowest "prolongations" of our state of existence, or into that part of the subtle domain which can truly be qualified as "infra-corporeal".[2] But in both cases there has equally been a passage into extra-corporeal modalities, higher for the one and lower for the other, so that it can definitely be said that corporeal manifestation itself in relation to the cycle in question really vanishes or is entirely "volatilized". Throughout all this and up to the final point one always has to consider the two terms which correspond with what the Hermetic tradition designates respectively as "coagulation" and "solution", and this from two sides at once: on the "benefic" side there is thus "crystallization" and "sublimation"; on the "malefic" side, there is "precipitation" and the final return to the indistinction of "chaos" '[3] (*Le Règne de la Quantité*, pp. 164–165).

Introd. *Wr. Side*

Schuon carries the microcosmic implications further: 'If we did not deny the nothing in ourselves, we would affirm that the nothing is, and "every house divided against itself falleth". In denying the nothing, we affirm, and in affirming, we deny the nothing. This is why it is said: "Whosoever shall seek to save his life shall lose it"; for to seek to save life is to affirm the nothing. In affirming the nothing, man becomes nothing—for one becomes what one loves; but this man becomes the nothing which exists, since that which exists cannot cease existing; and man exists. Not having created himself, neither can man annihilate himself; he can only damn himself. Damnation is identification with the nothing which exists' (*L'Oeil du Coeur*, p. 35).

Death 208
Introd. *Humility*
Introd. *Holiness*

'The fear of God is no more a matter of sentiment than is the love of God; like love, which is the tendency of our whole being towards transcendent Reality, fear is an attitude of the intelligence and the will: it consists in taking account, at each moment, of a Reality which infinitely surpasses us, against which we can do nothing, opposing which we could not live, and from whose teeth we cannot escape' (Schuon: *Perspectives spirituelles*, p. 279).

He who knows no fear in this life may equally find himself without hope in the next.

[1] A subordinate reason being to disperse the psychic residues.

[2] 'This is what the Hebrew Kabbala ... designates as the "world of rinds" (*ôlam qlippoth*); there fall the "former kings of Edom", in so far as they represent the unusable "residues" of past *Manvantaras*.'

[3] 'It must be evident that the two sides which we have called "benefic" and "malefic" correspond exactly to those of the "right" and the "left" where the "elect" and the "damned" are ranked respectively in the "Last Judgment", that is to say, specifically, in the final "discrimination" of the results of the cyclical manifestation.'

The Fear of God

The fear of the Lord is the beginning of wisdom: and the knowledge of the holy is understanding.

Proverbs, IX. 10

O mankind! Lo! the promise of Allâh is true. So let not the life of the world beguile you, and let not the beguiler beguile you with regard to Allâh.

Faith 505

Lo! Satan is an enemy to you, so treat him as an enemy. He only summoneth his faction to be companions of the Flaming Fire.

Qur'ân, XXXV. 5, 6

Terreat hic terror quos terreus alligat error.
Let this terror affright those bound in terrestrial error.

Inscription to Last Judgment at Autun, by the sculptor Giselbertus

Remember now thy Creator in the days of thy youth, while the evil days come not.

Ecclesiastes, XII. 1

A man asked Jesus (Peace be upon him!) 'Who is the best of men?' Then he took two handfuls of earth and said, 'Which of these two is the better? Men were created from earth, so the most honourable of them is the most God-fearing of them.'

Christ in Islâm

Beauty 676

If God should come and speak to us with His own voice (although He ceaseth not to speak through His Scriptures), and should say to man, thou wishest to sin, sin; do whatsoever pleaseth thee; whatever thou lovest on earth, let it be thine; . . . let no man resist thee; . . . let all these earthly things which thou hast desired abound unto thee, and live in them, not for a season, but for ever; but My face thou shalt never see. My brethren, wherefore did you groan, save because a chaste fear enduring for ever and ever hath been born in you? Why is your heart stricken? If God should say, Thou shalt never see My face, what of it? Thou wilt abound in all that earthly felicity; temporal goods will surround thee; thou wilt not lose them, nor forsake them; what more dost thou want? Chaste fear would weep indeed and groan, and would say, Nay, let all be taken away, and let me see Thy face. Chaste fear would cry out in the words of the psalm, 'O God of hosts, convert us, and shew Thy face: and we shall be saved' (*Ps.* LXXIX. 8 – Douay). Chaste fear would cry out in the words of the psalm . . . 'One thing I have asked of the Lord, this will I seek after.' What is this? 'That I may dwell in the house of the Lord all the days of my life.' What if he desire this for the sake of earthly felicity? Hear what followeth: 'that I may see the delight of the Lord,' and as His temple be protected (*Ps.* XXVI. 4 *sq.*). . . . If you ask this one thing . . .

Rev. 959

Center 847

and fear to lose this one thing only, you will not envy earthly delights, and you will look for that true happiness, and you will be in His body to Whom it is sung: 'Blessed are all they that fear the Lord: that walk in His ways' (*Ps.* CXXVII. 1).

St Augustine

Infra 250

At the very beginning (of one's religious career) it is indispensably necessary to have the most profound aversion for the interminable sequence of repeated deaths and births.

Gampopa

Faith 510

The secret of the Lord is with them that fear him; and he will shew them his covenant.

Psalm XXV. 14

The fear of God makes one secure.

'Alî

Reality 803

He who truly fears a thing flees from it, but he who truly fears God, flees unto Him.

Al-Qushayrî

Fear God, and men will fear thee, though they be stronger and mightier than thou. For when thou opposest them with the strength of thy fear of God, they in their terror will tremble in their hearts like women; while the sense of confidence, security, and tranquility will suffer thee to draw thy foot out of the snare of the fowler.

Yesod Hayirah, 3

Introd. *Conform.*

Whoever is afraid of God and has chosen fear of God as his religion, the jinn and mankind and every one who sees him are afraid of him.

Rûmî

Let us therefore fear that we may not fear.

St Augustine

As good as God is, so great He is; and as much as it belongeth to His goodness to be loved, so much it belongeth to His greatness to be dreaded. For this reverent dread is the fair courtesy that is in Heaven afore God's face. And as much as He shall then be known and loved overpassing that He is now, in so much He shall be dreaded overpassing that He is now.

Julian of Norwich

Do not the sailors of the ships fear the sea? Yet sinners fear not the Most High.

Book of Enoch, CI. 9

Infra 250 Infra 268 *Pilg.* 378

The raft on which you are borne upon this great sea (of earthly life) is made of water frozen to ice; and it is only by chance that it serves to bear you. Soon the sun will rise and shine on it, and melt the ice, and it will turn into water again, and you will be left sitting on water. But you certainly will not be able to remain in that position; you must therefore look for something to bear you up. And there is nothing that will serve that purpose, except ability to swim and to direct your course aright until you shall have reached (firm ground).

Hermes

That is one of the most lettings of grace, that a wretched man will not acknowledge his own blindness for pride of himself; or else if he knoweth it, he chargeth it not but maketh merry and game as he were over all secure.

Therefore unto all these men that are thus blinded and bounded with the love of this world and are so foul forshapen from the fairhead of man, I say and counsel that they think on their soul, . . . what mischief and what peril it is to them for to be out of grace and departed from God as they be; for there is nothing that holdeth them from the pit of hell that they should not at once fall therein, but one bare single thread of this bodily life whereby they hang. What lightlier may be lost than a single thread may be broken in two? For were the breath stopped in their body, and that may lightly fall, their soul should pass forth and anon be in hell without end.

Sin 66

Death 215

Walter Hilton

The worldly folk who scorn the scion of the Conqueror and accept him not shall be broiled in all the hells, like fire hidden under ash.

Śânti-deva

And warnings came in truth unto the house of Pharaoh

Who denied Our revelations, every one. Therefore We grasped them with the grasp of the Mighty, the Powerful.

Are your disbelievers better than those, or have ye some immunity in the scriptures?

Qur'ân, LIV. 41–43

He that layeth aside the fear of God, can never continue long in good estate, but falleth quickly into the snares of the devil.

The Imitation of Christ, I. xxiv

Holy War 407

God is marvellous in His works, and He is not mocked. I could give some instances of men who set about this matter with great levity and were heavily punished by meeting (some of them) with fatal accidents in their laboratories. For this work is no light thing, as many suppose, perhaps, because the Sages have called it child's play. Those to whom God has revealed His secrets may indeed find the experiment simple and easy. But do thou carefully beware of exposing thyself to great danger by unseasonable carelessness. Rather begin thy work with reverent fear and awe and with earnest prayer, and then thou wilt be in little danger.

Introd. *Pilg.*
Pilg. 366
P. State 573
Introd. *Death*

The Sophic Hydrolith

We may well believe that those who knew the gods and neglected them in one life may in another be deprived of the knowledge of them altogether.

Infra 261

Sallustius

Metanoia 493

Fear God that ye may not retrogress; love Him that ye may progress.

St Augustine

For thus a man begins truly to see and to be seen; to know God and to be known of him; to see God by the feeling of fear; to come under the regard of God's pity.

Richard of Saint-Victor

Action 329
Infra 250
Death 215
Inv. passim

Meditate upon, consider, and weigh deeply the serious facts contained in the biographies of previous saintly lives, the law of *karma,* the inconveniences and miseries of all *sangsâric* states of existence, the difficulties of obtaining the boon of a well-endowed human life, and the certainty of death and the uncertainty of the exact time of death; and, having weighed these in your minds, devote yourselves to the study and practice of the *mantrayânic* doctrines.

Milarepa

On Judging Others

Judge not, that ye be not judged.[1]

For with what judgment ye judge, ye shall be iudged: and with what measure ye mete, it shall be measured to you again.

St Matthew, VII. 1, 2

Action 329

For if ye forgive men their trespasses, your heavenly Father will also forgive you:

But if ye forgive not men their trespasses, neither will your Father forgive your trespasses.

St Matthew, VI. 14–15

The nature of the mind is such that it becomes what it thinks intensely upon. For example, if you think of the vices or faults of other men, you will be charged with these same vices or faults, at least for the present time. He who knows this psychological law will never allow himself to censure others or to discover faults in their conduct. He will see only

[1] Which does not advocate abandoning the dictates of intelligence! Those who habitually lean on this phrase out of conciliatory sentiments forget that judgment implies alternatives, and that apart from the discipline of a strict neutrality, which is precisely what lacks with them, their acquiescence or 'tolerance' is already a decision 'pro' for which they are responsible, and hence a *judgment*—at the risk of abusing this word to describe what is usually only a pre-*judice,* or *idée fixe,* i.e. 'misjudgment'. These apologists, who say that 'God alone can judge', are first to affirm 'the Divine in every man', easily overlooking the fact that it is precisely this Principle in man which has the capacity for judgment—otherwise as well renounce all claims to intellectuality or intelligence. Life, especially the spiritual life, is a succession of discernments and judgments, without this having to mean in any way an intrusion upon God's domain of final judgment.

the good in them. Such is the means of increasing in concentration, in yoga, and in spirituality.

Swami Sivananda

'One thing thou must do in order to arrive at this union, this purity, is to abstain from judging the will of man in anything at all that thou findest said or done, no matter by what creature, whether against thee, whether against another.'

St Catherine of Siena

Never allow yourself boldly to judge your neighbour; judge and condemn no one, especially for the particular bodily sin of which we are speaking.[1] If someone has manifestly fallen into it, rather have compassion and pity for him. Do not be indignant with him or laugh at him, but let his example be a lesson in humility to you; realizing that you too are extremely weak and as easily moved to sin as dust on the road, say to yourself: 'He fell today, but tomorrow I shall fall.' Know that, if you are quick to blame and despise others, God will mete out a painful punishment to you by letting you fall into the same sin for which you blame others.

Sin 66

Unseen Warfare, I. xix

Abu Jariya, an inhabitant of Basra, coming to Medina and being convinced of the inspired mission of Muhammad asked him, according to a Muslim historian, for some great rule of conduct.

'Speak evil of no one,' answered the Prophet.

Muhammad

Hast thou heard a word against thy neighbour? let it die within thee, trusting that it will not burst thee.

Ecclesiasticus, XIX. 10

Whoever listens to slander is himself a slanderer.

ʿAlî

The patient man hath a great and wholesome purgatory, who though he receive injuries, yet grieveth more for the malice of another, than for his own suffering; who prayeth willingly for his adversaries, and from his heart forgiveth their offences. He delayeth not to ask forgiveness of whomsoever he hath offended; he is sooner moved to compassion than to anger; he often offereth violence to himself, and laboreth to bring his body wholly into subjection to the spirit.

Suff. 126b

The Imitation of Christ, I. xxiv

He giveth his cheek to him that smiteth him: he is filled full with reproach.

Lamentations, III. 30

Humility 191

Kindle not the coals of sinners by rebuking them, lest thou be burnt with the flame of the fire of their sins.

Ecclesiasticus, VIII. 13

[1] Lust.

Action 346

For a religious devotee to try to reform others instead of reforming himself is a grievous mistake.

Gampopa

As far as possible one should not interfere in the affairs of others.

Sri Ramana Maharshi

One trafficking his advice to you is like a merchant offering heavy usury.

'Alî

Knowl. 745

When you have made trial of yourself, O such-and-such, you will be unconcerned with making trial of others.

Rûmî

Charity 597

Do not wrong or hate your neighbour, for it is not him that you wrong, you wrong yourself.

Thomas Wildcat Alford

Forgive thy neighbour if he hath hurt thee: and then shall thy sins be forgiven to thee when thou prayest.

Ecclesiasticus, XXVIII. 2

And these Thy servants who are gathered to slay me, in zeal for Thy religion and in desire to win Thy favour, forgive them, O Lord, and have mercy upon them; for verily if Thou hadst revealed to them that which Thou hast revealed to me, they would not have done what they have done; and if Thou hadst hidden from me that which Thou hast hidden from them, I should not have suffered this tribulation. Glory unto Thee in whatsoever Thou doest, and glory unto Thee in whatsoever Thou willest.

Al-Ḥallâj (before his crucifixion)

At the day of judgment . . . shall he stand to judge them, who doth now humbly submit himself to the censures of men. . . .

Then will it appear that he was wise in this world, who had learned to be a fool and despised for Christ's sake.

Metanoia 488

The Imitation of Christ, I. xxiv

Therefore if thine enemy hunger, feed him; if he thirst, give him drink: for in so doing thou shalt heap coals of fire on his head.

Romans, XII. 20 & *Proverbs*, XXV. 21–22

No man should be judged by others here in this life, for goods or for evil that they do. Nevertheless deeds may lawfully be judged, but not the men, whether they be good or evil.

The Cloud of Unknowing, XXIX

Hate roguery but not the rogue.

Swami Sivananda

If thou art an *Angel,* and hath to do with a *Devil,* use no *reviling Language,* for so the Angel himself is by the Spirit of God markt with a Character of Honour for this, that he *used no reviling Speeches to the Devil.*

Peter Sterry

Let your words be of velvet and your arguments of steel.

Swami Sivananda

Demons blaspheme God, beasts ignore Him, men love Him, while angels contemplate His light without removing their gaze. By this even canst thou know whom to call angel, man, beast, or demon.

Angelus Silesius

Contem. 547

Beware of praising anyone for qualities that he lacks: his acts betray him, and give you the lie.

'Alî

Those that do not punish bad men, wish that good men may be injured.

Pythagoras

To praise the wicked is a heinous sin.

'Alî

We in our perversity want God to be merciful in such way as not to be just. Others again, trusting all too much in their own justice, would have God just in such a way that they do not wish Him to be merciful. God shows himself to be both; He is supremely both. His mercy does not prescribe His justice, nor does His justice sweep away His mercy. He is merciful and just.

St Augustine

Wr. Side 474

Damnation is no foreign, separate, or imposed state that is brought in upon us, or adjudged to us by the will of God, but is the inborn, natural, essential state of our own disordered nature, which is absolutely impossible in the nature of the thing to be anything else but our own hell both here and hereafter, unless all sin be separated from us and righteousness be again made our natural state by a birth of itself in us. And all this, not because God will have it so by an arbitrary act of His sovereign will but because He cannot change His own nature or make anything to be happy and blessed, but only that which has its proper righteousness and is of one will and spirit with Himself.

William Law

Intra 266

Realiz. 887

O my heart, rise not as a witness against me.

Egyptian Tradition (inscribed on scarab placed on mummy's breast)

And We created not the heaven and the earth and all that is between them in vain. That is the opinion of those who disbelieve. And woe unto those who disbelieve, from the Fire!

Creation 48

Or shall We treat those who believe and do good works as those who spread corruption in the earth; or shall We treat the pious as the wicked?

A Scripture that We have revealed unto thee, full of blessing, that they may ponder its revelations, and that men of understanding may reflect.

Qur'ân, XXXVIII. 27–29

Sin 61

Be not deceived; God is not mocked: for whatsoever a man soweth, that shall he also reap.

For he that soweth to his flesh shall of the flesh reap corruption; but he that soweth to the Spirit shall of the Spirit reap life everlasting.

Galatians, VI. 7, 8

Center 838

Now in the days of Cronos there existed a law respecting the destiny of man, which has always been, and still continues to be in Heaven,—that he who has lived all his life in justice and holiness shall go, when he is dead, to the Islands of the Blessed, and dwell there in perfect happiness out of the reach of evil; but that he who has lived unjustly and impiously shall go to the house of vengeance and punishment, which is called Tartarus.

Plato (*Gorgias,* 523 A)

Then woe that day unto the deniers
Who play in talk of grave matters;
The day when they are thrust with a disdainful thrust, into the fire of hell:
This is the Fire which ye were wont to deny.
Is this magic, or do ye not see?
Endure the heat thereof, and whether ye are patient of it or impatient of it is all one for you. Ye are only being paid for what ye used to do.

Qur'ân, LII. 11–16

'Grant me grace for all men' (said Bâyazîd al-Bisṭâmî in a moment of spiritual ecstasy).

'O Bâyazîd,' said a voice, 'raise thine eyes.'

He raised his eyes and saw that the Lord, may He be exalted! was even more disposed than he himself in indulgence for His slaves. And he grew bold to ask: 'O God! be merciful even to Satan.'

Wr. Side 474

Came the reply: 'Satan is of fire; fire unto fire; make it thy study not to merit the fire.'

'Aṭṭâr

O Lord my Lord, lo, thou hast ordained in thy Law that the righteous shall inherit these things, but that the ungodly shall perish. The righteous, therefore, can endure the narrow things because they hope for the wide: those, however, who have done wickedly endure the narrow things, but yet shall not see the wide. And he said unto me:

Suff. 120

Thou art not a judge above God
Nor wise above the Most High.

Yea, rather, let the many that now are perish than that the law of God which is set before them be despised! For God did surely command them that came (into the world), when they came, what they should do to live, and what they should observe to avoid punishment. Nevertheless they were disobedient, and spake against him;

Orthod. 300

They devised for themselves vain thoughts,
they proposed to themselves wicked treacheries;
They even affirmed the Most High exists not....
In his statutes they have put no faith,
and have set at naught his commandments:
Therefore, *O Ezra,*
For the empty, empty things,
And for the full, full things! . . .
And I will not grieve over the multitude of them that perish: for
they it is who now are made like vapour,
counted as smoke,
are comparable unto the flame:
They are fired, burn hotly, are extinguished! . . .

Heterod. 420

Metanoia 493
P. State 563

Infra 261

For how long a time hath the Most High been longsuffering with the inhabitants of the world—not for their sakes, indeed, but for the sake of the times which he has ordained! . . .

Wr. Side 464

Just as the husbandman sows much seed upon the ground and plants a multitude of plants, and yet not all which were sown shall be saved in due season, nor shall all that were planted take root; so also they that are sown in the world shall not all be saved. . . .

Nay, Lord God! but
spare thy people,
compassionate thine inheritance! . . .
And he answered me and said: . . .

Thou comest far short of being able to love my creation more than I! . . . For the Most High willed not that men should come to destruction; but they—his creatures—have themselves defiled the Name of him that made them, and have proved themselves ungrateful to him who prepared life for them. . . .

Charity 602
Wr. Side 459

I considered my world, and lo! it was destroyed;
and my earth, and lo! it was in peril—
on account of the tumults of those who are (living) in it.

Introd. *Flight*

And I saw, and spared (some) with very great difficulty, and saved me a grape out of a cluster, and a plant out of a great forest. Perish, then, the multitude which has been born in vain; but let my grape be preserved, and my plant, which with much labour I have perfected!

IV Ezra, VII-IX, passim

Infra 250
Love 618

What Allâh openeth to man in the way of mercy, none can withhold; and what He withholdeth, none can release thereafter. He is the Mighty, the Wise.

Qur'ân, XXXV. 2

And John, when he saw the unchanged mind (soul) of Fortunatus,[1] said: O nature that is not changed for the better! O fountain of the soul that abideth in foulness! O essence of corruption full of darkness! O death exulting in them that are thine! O fruitless tree full of fire! O tree that bearest coals for fruit! O matter that dwellest with the madness of matter

Wr. Side 448

[1] An inveterate disbeliever.

and neighbour of unbelief! Thou hast proved who thou art, and thou art always convicted, with thy children. And thou knowest not how to praise the better things: for thou hast *them* not. Therefore, such as is thy way, such also is thy root and thy nature. Be thou destroyed from among them that trust in the Lord: from their thoughts, from their mind, from their souls, from their bodies, from their acts, their life, their conversation, from their business, their occupations, their counsel, from the resurrection unto God, from their sweet savour wherein thou wilt not share, from their faith, their prayers, from the holy bath, from the eucharist, from the food of the flesh, from drink, from clothing, from love, from care, from abstinence, from righteousness: from all these, thou most unholy Satan, enemy of God, shall Jesus Christ our God and the judge of all that are like thee and have thy character, make thee to perish.

Acts of John, 84

Illusion 85

But for those who have received no commission to govern other men, but stand in a private character without office, it is needful that they secretly judge themselves inwardly, and beware of judging all things without, for in such judgments we do commonly err, and the true position of things is generally very far otherwise from that which it appears to us, as we often come to discover afterwards. On this point remember the proverb: 'He is a wise man who can turn all things to the best.'

Tauler

No Helping Friend That Day

Not much can be done in the way of helping a man after he is dead.

Plato (*Laws,* 959 B)

Center 847

For those who respond to their Lord is the Good; as for those who do not respond to Him, verily if they had all that is in the earth, and with it the likeness thereof, they would proffer it as ransom. Such will have a woeful reckoning, and their habitation will be hell, an evil abode.

Qur'ân, XIII. 18

Supra 239
Illusion 99

Suppose that thou hadst up to this day lived always in honours and delights, what would it all avail thee if thou wert doomed to die at this instant?

All therefore is vanity, except to love God and serve Him only.

The Imitation of Christ, I. XXIV

Death 215

For what shall it profit a man, if he shall gain the whole world, and lose his own soul?

St Mark, VIII. 36

Why dost thou not provide for thyself against that great day of judgment, when no man can excuse or answer for another, but every one shall have enough to answer for himself!

The Imitation of Christ, I. xxiv

And guard yourselves against a day when no soul will in aught avail another, nor will intercession be accepted from it, nor will compensation be received from it, nor will they be helped.

Qur'ân, I. 48

Whence shall I find a kinsman, whence a friend, when the Death-god's messengers seize me? Righteousness alone can save me then.

Śânti-deva

What is it that accompanies us when we pass off from this body? Wealth and other things remain behind in the house itself. Friends and relatives come as far as the burning ground and return home from there. Only the good deeds and the evil ones done by us go with us when we go.

Conform. 166

Sri Chandrasekhara Bhâratî Swâmigal

Seeing that when we die we must go on our way alone and without kinsfolk or friends, it is useless to have devoted time (which ought to have been dedicated to the winning of Enlightenment) to their humouring and obliging, or in showering loving affection upon them.

Renun. 146

Gampopa

The Day of Judgement is decisive, and displays unto all the seal of truth. Even as now a father may not send a son, or a son his father, or a master his slave, or a friend his dearest, that in his stead he may be ill, or sleep, or eat, or be healed; so shall none then pray for another on that Day,[1] neither shall one lay a burden on another; for then every one shall bear his own righteousness or unrighteousness. . . . So shall no man then be able to have mercy on him who is condemned in the Judgement, nor overwhelm him who is victorious.

Infra 268

Holiness 902

IV Ezra, VII. 104–106, 115

O ye Compassionate Ones, ye possess the wisdom of understanding, the love of compassion, the power of (doing) divine deeds and of protecting, in incomprehensible measure. Ye Compassionate Ones, (such-and-such a person) is passing from this world to the world beyond. He is leaving this world. He is taking a great leap. No friends (hath he). Misery is great. (He is without) defenders, without protectors, without forces and kinsmen. The light of this world hath set. He goeth to another place. He entereth thick darkness. He falleth down a steep precipice. He entereth into a jungle solitude. He is pursued by *Karmic* forces. He goeth into the Vast Silence. He is borne away by the Great Ocean. He is wafted on the Wind of *Karma.* He goeth in the direction where stability existeth not. He is caught by the Great Conflict. He is obsessed by the Great Afflicting Spirit. He is awed and terrified by the

Infra 250

Action 329

[1] This does not invalidate prayers for the dead. It means rather that God's decrees are irreversible once consummated.

Messengers of the Lord of Death. Existing *Karma* putteth him into repeated existence. No strength hath he. He hath come upon a time when he hath to go alone.

O ye Compassionate Ones, defend (so-and-so) who is defenceless. Protect him who is unprotected. Be his forces and his kinsmen. Protect (him) from the great gloom of the *Bardo*.[1] Turn him from the red (or storm) wind of *Karma*. Turn him from the great awe and terror of the Lords of Death.

The Tibetan Book of the Dead (Invocation for the dead)

The Round of Existence

Conform. 166

O son of Kunti, whatever state (or being) one dwells upon in the end, at the time of leaving the body, that alone he attains, because of his constant thought of that state or being. . . .

If the embodied meets with death when Sattva[2] is predominant, then he attains the spotless regions of the knowers of the Highest.

Meeting with death in Rajas, one is born among those attached to action; and dying in Tamas, one is born in the wombs of senseless beings.[3]

Bhagavad-Gîtâ, VIII. 6; XIV. 14, 15

Introd. *Creation*

All things which have a soul change, and possess in themselves a principle of change, and in changing move according to law and to the order of destiny: natures which have undergone a lesser change move less and on the earth's surface, but those which have suffered more change and have become more criminal sink into the abyss, that is to say, into Hades and other places in the world below, of which the very names terrify men. . . . When (the soul) has communion with divine virtue and becomes divine, she is carried into another and better place, which is perfect in holiness; but when she has communion with evil, then she also changes the place of her life.

Plato (*Laws*, 904 D)

Action 329

The whole *Sangsâra*, being e'er entangled in the Web of *Karma*,
Whoever holdeth fast to it severeth Salvation's Vital Cord.
In harvesting of evil deeds the human race is busy;
And the doing so is to taste the pangs of Hell.

Milarepa

[1] The intermediate state between death and judgment.

[2] In Hindu cosmology: *Sattva* = lightness, the ascending luminous tendency or principle in things, pertaining to the supraformal domain; *Rajas* = expansion, the fiery principle, pertaining to the subtle domain; *Tamas* = heaviness, the descending obscure principle, pertaining to the corporeal domain.

[3] It is perhaps well to remind the reader here that where it is a question of rebirth into formal states of manifestation, transmigration but not reincarnation is involved, i.e. the states resemble those of our world by analogy but not identity, as a being can never pass through the same state twice.

There awaits men at death what they do not expect or think.

Heraclitus

It is not death, but a bad life, that destroys the soul.

Sextus the Pythagorean

Supra Introd.

That which nature binds, nature also dissolves: and that which the soul binds, the soul likewise dissolves. Nature, indeed, bound the body to the soul; but the soul binds herself to the body. Nature, therefore, liberates the body from the soul; but the soul liberates herself from the body. Hence there is a twofold death; the one, indeed, universally known, in which the body is liberated from the soul; but the other peculiar to philosophers, in which the soul is liberated from the body. Nor does the one entirely follow the other.[1]

Porphyry

Renun. 156

And thus I escaped from the cycle, the painful, the misery-laden.

Inscription on gold funereal tablet found near site of Sybaris

Those living souls who place their reliance on worldly objects can never taste true felicity. As is the desire or thought in the mind, so is the soul born in its next incarnation. ... The mind delights itself with the thoughts of desired objects; then, itself assimilating their natures, it takes on the same form as the one in which it delights.

Yoga-Vasishtha

Whoso doeth right it is for his soul, and whoso doeth wrong it is against it. And thy Lord is not at all a tyrant to His slaves.

Qur'ân, XLI. 46

God casts no soul away, unless it cast itself away. Every soul is its own judgment.

Boehme

Infra 266

So the soul brings forth itself and falls into deception by its own choice, and thus loses the consciousness of its freedom by subjection to the bondage of life.

Yoga-Vasishtha

The entangled creatures, attached to worldliness, talk only of worldly things in the hour of death. What will it avail such men if they outwardly repeat the name of God, take a bath in the Ganges, or visit sacred places? If they cherish within themselves attachment to the world, it must show up at the hour of death. While dying they rave nonsense. Perhaps they cry out in a delirium, 'Turmeric powder! Seasoning! Bay-leaf!' The singing parrot, when at

Death 215

[1] Explains the translator: 'The meaning of this twofold death is as follows: Though the body, by the death which is universally known, may be loosened from the soul, yet while material passions and affections reside in the soul, the soul will continually verge to another body, and as long as this inclination continues, remain connected with body. But when, from the predominance of an intellectual nature, the soul is separated from material affections, it is truly liberated from the body: though the body at the same time verges and clings to the soul, as to the immediate cause of its support.'

ease, repeats the holy names of Râdhâ and Krishna, but when it is seized by a cat it utters its own natural sound; it squawks, 'Kaa! Kaa!' It is said in the Gîtâ that whatever one thinks in the hour of death, one becomes in the after-life.

Conform. 170

Sri Ramakrishna

When our term of service is ended, when we are divested of our guardianship of the material world, and freed from the bonds of mortality, (God) will restore us, cleansed and sanctified, to the primal condition of that higher part of us which is divine. . . . But those who have lived evil and impious lives are not permitted to return to heaven. For such men is ordained a shameful transmigration into bodies of another kind, bodies unworthy to be the abode of holy mind.

Creation 42

Hermes

During this mortal life we must choose eternal love or eternal death, there is no middle choice.

Introd. *Heterod.*

St Francis de Sales

That Self is eternal; yet all men think it mortal. That Self is infinite; yet all men think it finite. Those who possess TAO are princes in this life and rulers in the hereafter. Those who do not possess TAO behold the light of day in this life and become clods of earth in the hereafter.

Chuang-tse (ch. XI)

This is a Scripture whereof there is no doubt, a guidance unto the God-fearing. . . .

Who believe in that which is revealed unto thee (Muhammad) and that which was revealed before thee, and are certain of the Hereafter.

These depend on guidance from their Lord. These are the successful.

As for the disbelievers, whether thou warn them or thou warn them not it is all one for them; they believe not.

Allâh hath sealed their hearing and their hearts, and on their eyes there is a covering. Theirs will be an awful doom.

Qur'ân, II. 2, 4–7

Created beings, although they are destroyed (in their original forms) at the periods of dissolution, yet, being affected by the good or evil acts of former existence, are never exempted from their consequences; and when Brahmâ creates the world anew, they are the progeny of his will, in the fourfold condition of gods, men, animals, and inanimate things.

Vishṇu Purâṇa

All Lust is Love degenerated, Love Corrupted. . . .

Introd. *Love*

Ixion in the Poets loved a Goddess, in the place of whom he embraced a Cloud formed into the Shape of a Divine Beauty. Thus he became the Father of the *Centaures*, half Men, half Beasts. Then he was cast into hell where he is fastened to a Wheel turning continually round, on which he is tormented day, and night. This Parable is meant of thee, O Lustful Spirit: Thou wert made for Divine Love, for the Love of the Divine Beauty. Thus hast changed, this Love into various Lusts. Thou defilest thy self with Shadows, Clouds of Darkness formed into the Empty Shapes of Beauty. Instead of the Divine, and Humane Nature in the Blessed Harmony of an Immortal Union, all thy Births, all thy Production are Horrid hateful monsters; Man, Beast, and Devil all in One Spirit, in One Person. Thy end is the Endless Circle of thy Lusts, and of the Divine wrath, as the Wheel of Eternity, a Wheel of Fire, holding thee fast tied to it, and torturing thee without any Rest, or Period.

Sin 66

Peter Sterry

Death cometh into him from every side while yet he cannot die, and before him is a harsh doom.

A similitude of those who disbelieve in their Lord: Their works are as ashes which the wind bloweth hard upon a stormy day. They have no control of aught that they have earned. That is the extreme failure.

Action 340

Qur'ân, XIV. 17, 18

'This fire does not consume, because the damned, however much torment they endure, never lose their being. I tell thee, they ask for death, but they cannot obtain it, because they cannot lose being. They do indeed lose the being of grace by their sin, but natural being, never.'

Supra Introd.

St Catherine of Siena

And when they are flung into a narrow place thereof, chained together, they pray for destruction there.

Pray not that day for one destruction, but pray for many destructions!

Qur'ân, XXV. 13, 14

Know God, O Rama, and serve those who speak of Him to you. He alone is real. Know it now or after a thousand incarnations!

Orthod. 288 *Reality* 803

Yoga-Vasishtha

A (metal) mirror that has been stained with rust by some passing incident can be made bright again by rubbing and polishing; but when a mirror has been rusted . . . so that the mirror takes on the character of a different mirror, with a nature that is fixed and ingrained,—then the polisher's skill is of no avail, and the mirror cannot be cleansed from rust except by submitting it again to the refining fire. And even so, souls stained by some passing incident can be made bright again by warnings and reproofs—so that they recall to memory the state of life in which they were before; but souls (that have become) foul and unclean in their very nature can be made bright again only by being plunged in misery, and remaining long in that condition, and undergoing it again and again.

Hermes

You will rise into the air, if you break and become dust.
If you break not, He who moulded you will break you. . . .
O friend, if you reach perfection in our assembly,
Your seat will be the throne, you will gain your desire in all things.

Introd. *Death*

But if you stay many years more in this earth,
You will pass from place to place, you will be as the dice in backgammon.

Dîvâni Shamsi Tabrîz, XLVI

A single moment's dualistic thought is sufficient to drag you back to the twelvefold chain of causation. It is ignorance which turns the wheel of causation, thereby creating an endless chain of karmic causes and results. This is the law which governs our whole lives up to the time of senility and death.

Huang Po

Attachment to this world
Brings the soldiers back,
Wandering over the battle-field.

Seami Motokiyo

There is no doubt that the man who does not worship Shiva is an animal and remains endlessly wandering in the cycle of existences.

Yoga Shikhâ Upanishad

And they say: Verily, this is nought but magic manifest!
When we are dead and have become dust and bones, shall we then, forsooth, be raised?
And our forefathers?
Say: Yea—and ye shall be brought low!

Qur'ân, XXXVII. 15–18

Heterod. 420

Reality 790

I say, that of all idiocies, that is the most stupid, most vile, and most damnable which holds that after this life there is none other; because, if we look through all the writings of the philosophers, as well as of the other wise authors, they all agree in this, that there is some part of us which is immortal. . . . And this seems to be the meaning of all law, whether of Jews, Saracens, or Tartars, or any others who live at all according to law. For that all deceived themselves, were an impossibility, horrible even to mention. . . . Again, it is confirmed by the most veracious teaching of Christ, which is the Way, the Truth, and the Light.

Dante (*Il Convito,* II. ix. 4, 6)

Beauty 676

Knowl. 761

If the soul is really immortal, what care should be taken of her, not only in respect of the portion of time which is called life, but of eternity! And the danger of neglecting her from this point of view does indeed appear to be awful. If death had only been the end of all, the wicked would have had a good bargain in dying, for they would have been happily quit not only of their body, but of their own evil together with their souls. But now, inasmuch as the soul is manifestly immortal, there is no release or salvation from evil except the attainment of the highest virtue and wisdom.

Plato (*Phaedo,* 107 C)

Death then is the dissolution of the body, and the cessation of bodily sense; and about this we have no cause to be troubled. But there is something else, which demands our anxious thought, though men in general disregard it through ignorance or unbelief.... When the soul has quitted the body, there will be held a trial and investigation of its deserts.... The everlasting existence of the soul is to its detriment in this respect, that its imperishable faculty of feeling makes it subject to everlasting punishment. Know then that we have good cause for fear and dread, and need to be on our guard, lest we should be involved in such a doom as this. Those who disbelieve will, after they have sinned, be forced to believe; they will be convinced, not by words, but by hard facts, not by mere threats, but by suffering the punishment in very deed. All things are known to God, and the punishments inflicted will vary in accordance with the character of men's offences.

Supra 239

Sin 61

Hermes

Thinketh man that he is to be left aimless?

Qur'ân, LXXV. 36

If you wish to learn in what condition the soul will be when it has quitted the body, observe in what condition it is while it is joined with the body.

Hermes

There are two patterns eternally set before men; the one blessed and divine, the other godless and wretched: but they do not see them, or perceive that in their utter folly and infatuation they are growing like the one and unlike the other, by reason of their evil deeds; and the penalty is, that they lead a life answering to the pattern which they are growing like. And if we tell them, that unless they depart from their cunning, the place of innocence will not receive them after death; and that here on earth, they will live ever in the likeness of their own evil selves, and with evil friends—when they hear this they in their superior cunning will seem to be listening to the talk of idiots.

Introd. *Creation*

Introd.

Plato (*Theaetetus,* 177 A)

For, while we lived and committed iniquity we considered not what we were destined to suffer after death!

IV Ezra, VII. 126

But in that thou moreover sayest, why do not the souls which are without God feel hell in this world? I answer: They bear it about with them in their wicked consciences, but they know it not; because the world hath put out their eyes, and its deadly cup hath cast them likewise into a sleep, a most fatal sleep. Notwithstanding which it must be owned that the wicked do frequently feel hell within them during the time of this mortal life, though they may not apprehend that it is hell, because of the earthly vanity which cleaveth unto them from without, and the sensible pleasures and amusements wherewith they are intoxicated. And moreover it is to be noted, that the outward life in every such one hath yet the light of the outward nature, which ruleth in that life; and so the pain of hell cannot, so long as that hath the rule, be revealed. But when the body dieth or breaketh away, so as the soul cannot any longer enjoy such temporal pleasure and delight, nor the light of this outward world,

Illusion 85

which is wholly thereupon extinguished as to it; then the soul stands in an eternal hunger and thirst after such vanities as it was here in love withal, but yet can reach nothing but that false will, which it had impressed in itself while in the body; and wherein it had abounded to its great loss. And now whereas it had too much of its will in this life, and yet was not contented therewith, it hath after this separation by death, as little of it; which createth in it an everlasting thirst after that which it can henceforth never obtain more, and causeth it to be in a perpetual anxious lust after vanity, according to its former impression, and in a continual rage of hunger after those sorts of wickedness and lewdness whereinto it was immersed, being in the flesh. Fain would it do more evil still, but that it hath not either wherein or wherewith to effect the same, left it.

Conform. 166
Infra 261

Boehme

The reason why . . . we live starving in the coldness and deadness of a formal, historical, hearsay-religion, is this: we are strangers to our own inward misery and wants, we know not that we lie in the jaws of death and hell; we keep all things quiet within us, partly by outward forms and modes of religion and morality, and partly by the comforts, cares and delights of this world. Hence it is that we consent to receive a Saviour . . . not because we feel an absolute want of one, but because we have been told there is one, and that it would be a rebellion against God to reject Him.

Death 208
Humility 191

William Law

Would wicked men dwell a little more at home, and descend into the bottom of their own hearts, they should soon find hell opening her mouth wide upon them, and those secret fires of inward fury and displeasure breaking out upon them, which might fully inform them of the estate of true misery, as being a short anticipation of it. But in this life wicked men for the most part elude their own misery for a time, and seek to avoid the dreadful sentence of their own consciences, by a tergiversation and flying from themselves into a converse with other things, *ut nemo in sese tentat descendere*; else they would soon find their own home too hot for them.

Realiz. 859

John Smith the Platonist

This Day We seal up their mouths, and their hands speak out to Us and their feet bear witness as to what they used to earn.[1]

Qur'ân, XXXVI. 65

Jesus said: Blessed is the lion which the man eats and the lion will become man; and cursed is the man whom the lion eats and the man will become lion.

Holy War 396

The Gospel according to Thomas, Log. 7

The race of birds was created out of innocent light-minded men, who, although their minds were directed toward heaven, imagined, in their simplicity, that the clearest demonstration of the things above was to be obtained by sight. . . . The race of wild

[1] According to a commentary, this passage dramatically suggests the metamorphosis from the central, vertical state of man to the peripheral, horizontal state of beast suffered by a being thrown back into the 'round of existence'—which moreover is corroborated by the citations from Plato and Hermes directly below.

pedestrian animals, again, came from those who had no philosophy in any of their thoughts, and never considered at all about the nature of the heavens, because they had ceased to use the courses of the head, but followed the guidance of those parts of the soul which are in the breast. In consequence of these habits of theirs they had their front-legs and their heads resting upon the earth to which they were drawn by natural affinity.... And the most foolish of them, who trail their bodies entirely upon the ground and have no longer any need of feet, he made without feet to crawl upon the earth. The fourth class were the inhabitants of the water: these were made out of the most entirely senseless and ignorant of all, whom the transformers did not think any longer worthy of pure respiration, because they possessed a soul which was made impure by all sorts of transgression; and instead of the subtle and pure medium of air, they gave them the deep and muddy sea to be their element of respiration; and hence arose the race of fishes and oysters, and other aquatic animals, which have received the most remote habitations as a punishment of their outlandish ignorance.

Knowl. 761

Plato (*Timaeus,* 91 D ff.)

If a soul, when it has entered a human body, persists in evil, it does not taste the sweets of immortal life, but is dragged back again; it reverses its course, and takes its way back to the creeping things; and that ill-fated soul, having failed to know itself, lives in servitude to uncouth and noxious bodies. To this doom are vicious souls condemned

Realiz. 859

Hermes

Even though through the days and years of life, you have piled up much merit by the practice of the *Nembutsu,* if at the time of death you come under the spell of some evil, and at the end give way to an evil heart, and lose the power of faith in and practice of the *Nembutsu,* it means that you lose that birth into the Pure Land immediately after death.... This is indeed a most terrible thing to contemplate, and one which no words can describe.

Inv. passim

Hônen

Now when the time drew near that our blessed Lord was to enter upon His last great sufferings, *viz.* the realities of that second death through which He was to pass, then it was that all the anguishing terrors of a lost soul began to open themselves in Him; then all that eternal death which Adam had brought into his soul, when it lost the light and spirit of Heaven, began to be awakened and stirring in the second Adam, who was come to stand in the last state of the fallen soul, to be encompassed with that eternal death and sensibility of hell which must have been the everlasting state of fallen man.

The beginning of our Lord's entrance into the terrible jaws of this second death may be justly dated from those affecting words, 'My soul is exceeding sorrowful, even unto death; tarry ye here with me and watch.' ... The progress of these terrors are plainly shown us in our Lord's agony in the garden, when the reality of this eternal death so broke in upon Him, so awakened and stirred itself in Him, as to force great drops of blood to sweat from His body.... This was that cup He was drinking from the sixth to the ninth hour on the Cross, nailed to the terrors of a two-fold death, when He cried out, 'My God, my God, why hast thou forsaken me?'

We are not to suppose that our Lord's agony was the terrors of a person that was going to be murdered, or the fears of that death which men could inflict upon Him; for He had told His disciples not to fear them that could only kill the body, and therefore we may be sure He had no such fears Himself. No, His agony was His entrance into the last eternal terrors of the lost soul, into the real horrors of that dreadful eternal death which man unredeemed must have died into when he left this world. We are therefore not to consider our Lord's death upon the Cross as only the death of that mortal body which was nailed to it, but we are to look upon Him with wounded hearts, as fixed and fastened in the state of that two-fold death, which was due to the fallen nature, out of which He could not come till He could say 'It is finished; Father, into thy hands I commend my spirit.' In that instant He gave up the ghost of this earthly life; and as a proof of His having overcome all the bars and chains of death and hell, He rent the rocks, opened the graves, and brought the dead to life, and triumphantly entered into that long-shut-up paradise, out of which Adam died, and in which He promised the thief he should that day be with Him.

Death 232

William Law

I have already said, and say now, and shall say (it) again: There are more who perish than shall be saved, even as the flood is greater than a drop!

IV Ezra, IX. 15, 16

Enter ye in at the strait gate: for wide is the gate, and broad is the way, that leadeth to destruction, and many there be which go in thereat:

Because strait is the gate, and narrow is the way, which leadeth unto life, and few there be that find it.

St Matthew, VII. 13, 14

There is a vast preponderance of those who wander downwards unliberated.

The Tibetan Book of the Dead

We have heard truly that the way is strait that leads to life. This is the way of penance that few find, the which therefore is called strait; for by it, and it be right, the flesh is stripped from unlawful solace of the world, and the soul is restrained from shrewd pleasure and unclean thoughts, and is only dressed to the love of God. But this is seldom found in men, for nearly none savour that which belongs to God: but they seek earthly joy and in that they are delighted, wherefore following their bodily appetite, and despising their ghostly, they forsake all the ways that are healthful to their soul, and they abhor them as strait, sharp, and unable to be borne by their lust.

Illusion 89

Richard Rolle

Death 215

Illusion 99

When the messenger (of death) comes to take thee away, let him find thee prepared. Alas! thou wilt have no opportunity for speech, for verily his terror will be before thee. Say not, 'Thou art taking me in my youth,' for thou knowest not when thy death will take place. Death comes and seizes the babe at the mother's breast as well as the aged man. Observe this, for I speak unto thee good advice, which thou shouldest ponder in thy heart.

Egyptian Tradition

'Abide not with the passing things, or they will give information concerning thee on the day of terror, and thou wilt mourn for the loss of that with which thou wast, and enter the company of those that fear.'

Niffarî

Beware lest at the time of your departure from the body you allow oblivion and fear to interpose between you and the road, that you may not wander from the road and go astray. But if you should (at any time) forget it, seek to recall it to your memory; and to that end, make use of what is told you by those who have already travelled along that road, and know it by experience; for they are the masters of right guidance; they are lights in the darkness.

Knowl. 755 *Orthod.* 288

Hermes

The body which thou hast now is called the thought-body of propensities. Since thou hast not a material body of flesh and blood, whatever may come—sounds, lights, or rays—are, all three, unable to harm thee: thou art incapable of dying. It is quite sufficient for thee to know that these apparitions are thine own thought-forms. Recognize this to be the *Bardo*.[1]

O nobly-born, if thou dost not now recognize thine own thought-forms, whatever of meditation or of devotion thou mayst have performed while in the human world—if thou hast not met with this present teaching—the lights will daunt thee, the sounds will awe thee, and the rays will terrify thee. Shouldst thou not know this all-important key to the teachings—not being able to recognize the sounds, lights, and rays—thou wilt have to wander in the *Sangsâra*.

The Tibetan Book of the Dead (Instructions for addressing a person shortly after death)

For this I say: if, looking in thy Heart,
Thou for *Self-whole* mistake thy *Shadow-part,*
That Shadow-part indeed into *The Sun*
Shall melt, but senseless of its Union:
But in that Mirror if with purgéd eyes
Thy Shadow Thou *for* Shadow recognize
Then shalt Thou back into thy Centre fall
A conscious Ray of that eternal *All.*

Heterod. **428** Supra Introd. *Center* **828** *Realiz.* 890

'Aṭṭâr

'Stay before Me in this world alone, and I will dispose thee in thy grave alone, and bring thee forth therefrom alone unto Me, and thou wilt stay before Me on the day of resurrection alone: and when thou art alone, thou wilt see only my face; and when thou seest only my face, there will be neither reckoning nor book; and when there is neither reckoning nor book, then there will be no terror; and when there is no terror, then thou wilt be one of the intercessors.'

Death 206 *M.M.* 978

Niffarî

[1] Cf. note 1, p. 250.

Those who repeat: *Lâ ilâha illâ 'llâh,*[1] will not experience solitude at the moment of death, nor on the Day of Resurrection.

Muhammad

Introd. *Holiness*

Blessed is human birth; even the dwellers in heaven desire this birth: for true wisdom and pure love may be attained only by man.

Srimad Bhagavatam, XI. xiii

There are those who pass from light to light, those who pass from darkness to darkness, those who pass from light to darkness, and those who pass from darkness to light. Of these four, be thou the first....

Contem. 547

He who would misuse the boon of human life is far more stupid than he who would employ a gold vessel inlaid with precious gems as a receptacle for filth.

Nâgârjuna

It is far better to be an animal than to be a man making no effort to attain knowledge, for ... the animals have no sins and have no need to expiate them.

Sri Chandrasekhara Bhâratî Swâmigal

Does anybody bake the sediment of sesamum oil in a vessel of lapis lazuli or use as fuel thereof loads of sandal-wood? Does the seeker of the fibre of the Arka plant plough the soil with ploughs having their edges in gold? Does anybody cut down the branches of the camphor tree to make out of them a fence to protect wild corn? The person who is born as a human being and in this plane of action and is yet unfortunate enough not to engage himself in penance does all these things.

Holiness 900

Bhartṛari

If we happen to lose a paltry sum of two rupees, it takes us three days to get over the grief caused by that loss. But we are prepared to waste the whole of this far more valuable life without seeking Knowledge. What can possibly compensate or remedy this colossal loss?

Sri Chandrasekhara Bhâratî Swâmigal

One free and well-endowed human life is more precious than myriads of non-human lives in any of the six states of existence.

Gampopa

The Lord has said that human birth is exceedingly hard to win; hard as for a turtle to pass its neck into the hole of a yoke in the ocean.

Śânti-deva

'Master,' said the turtle, 'I should not dream of accepting a reward; but there is one thing you can do for me. I have heard that the Buddha of the Western Heaven knows both the

[1] The *Shahâdah,* 'There is no divinity if not the Divinity'—central formula of Islâm, theurgic in essence, and hence really untranslatable.

past and the future. I have been perfecting myself here for about one thousand years. This is a pretty long span, and I have already been fortunate enough to learn human speech; but I still remain a turtle. I should indeed be very much obliged if you would ask the Buddha how long it will be before I achieve human form.'

Monkey

Human birth, the desire for salvation and the company of holy men are rare things on this earth. Those who are blest with all three are the most favoured of men.

Śrî Śankarâchârya

All such as stand still while they are in this state, and make no progress before death, must stand still for ever hereafter.

Tauler

Holy War 403

On Hell and Damnation

Thou shalt see the wretched people, who have lost the good of the intellect.

Dante (*Inferno*, III. 17)

Knowl. 761

Our Lady stretched out her hands, and bright rays came forth which seemed to penetrate into the earth. All at once the ground vanished, and the children found themselves standing on the brink of a sea of fire. As they peered into this dreadful place, the terrified youngsters saw huge numbers of devils and damned souls. The devils resembled hideous black animals, each filling the air with despairing shrieks. The damned souls were in their human bodies[1] and seemed to be brown in color, tumbling about constantly in the flames and screaming with terror. All were on fire within and without their bodies, and neither devils nor damned souls seemed able to control their movements. They were tossing about in the flames like fiery coals in a furnace. There was never an instant's peace or freedom from pain.

The Children of Fatima

. . . But that I am forbid
To tell the secrets of my prison-house,
I could a tale unfold whose lightest word
Would harrow up thy soul, freeze thy young blood,
Make thy two eyes, like stars, start from their spheres,
Thy knotted and combined locks to part,

[1] In damnation, unlike transmigration, the human or central state is maintained, but inverted, following the descending or 'infernal' direction of the world axis. Cf. *The Wrathful Side*, Introduction.

And each particular hair to stand an end,
Like quills upon the fretful porpentine.[1]

Shakespeare (*Hamlet*, I. v. 13)

Hear now from me of the far more dreadful Taptakumbha. It has all around it heated pans, encircled by flames of fire, filled with iron dust and boiling oil resembling flames. In these vessels are thrown by the envoys of Yama perpetrators of wicked deeds with their faces bent towards the ground; they are fried there with their bodies bursting and rendered foul with fatty excretions. With their heads, eyes and bones coming out, they are forcibly taken up by ferocious vultures and are again thrown into them. Then accompanied by hissing sounds and converted into liquid, their heads, bodies, tendons, flesh, skin and bones are mixed with oil. Then the perpetrators of iniquities are pounded in these volumes of eddying oil with a ladle by the emissaries of Yama. Thus I have described to you at length, O father, the hell Taptakumbha.

Mârkaṇḍeya Purâṇa

One of them (a monk who had died in a state of mortal sin) appeared to the surviving (monk) and said that he was damned.

'Do you suffer much?' asked the living man.

For reply, the dead man extended his hand and let fall a drop of sweat on a copper candelabrum. The candelabrum melted in less than an instant, like wax in a fiery furnace. And such a dreadful stench filled the air that the monks were obliged to abandon the monastery for three days. The monk who received this apparition left the monastery and entered the Franciscan Order. I have this fact from another Franciscan, who was in the same monastery.

Ruysbroeck

And verily the doom of the Hereafter is greater did they but know.

Qur'ân, XXXIX. 26

Metanoia 493

The most terrible pain in hell is the pain of *perdition*—namely, the loss of God. Those who have gratuitously betrayed Him in time have to bear in hell His eternal privation, and this torment surpasses all their torments. Since in scorning God they have burned for creatures with an unjust fire, they are condemned beyond that to eternal fire. Fire answers fire. . . . He who loves not God and who dies in coldness will eternally shiver to the depths of his being in a cold without end. He will be frozen for having scorned true love; he will be burned for having adored false love. Cold summons, cold answers. Fire summons, fire answers. . . . I urge you to believe me, for I know whereof I speak.

Ruysbroeck

Hell is no other, but a soundlesse pit,
Where no one beame of comfort peeps in it.

Robert Herrick

[1] The reference here is simply to purgatory.

Abandon all hope, ye who enter.

Dante (*Inferno,* III. 9)

Faith 514

Death carries off a man who is gathering flowers, and whose mind is distracted, as a flood carries off a sleeping village.

Dhammapada, IV. 47

Watch therefore, for ye know neither the day nor the hour wherein the Son of man cometh.

St Matthew, XXV. 13

Illusion 99

. . . She saith to him,
Stolen waters are sweet, and bread eaten in secret is pleasant.
But he knoweth not that the dead are there; and that her guests are in the depths of hell.

Proverbs, IX. 16–18

I have seen a place
Whose name is Eternal Hatred.
It is built in the deepest abyss
Out of stones laid there by mortal sin.
Pride was the first stone
Laid by Lucifer. . . .

Sin 75

Wr. Side passim

The whole place is upside down, the highest being in the lowest and most unworthy places. Lucifer sits in the depths bound by his sin and out of his burning mouth there pour ceaselessly all the sin, torment and infamy by which Hell and Purgatory and the earth are so piteously surrounded. . . .

How Hell burns and rages in itself, how the devils fight with souls, how souls seethe and roast, how they swim and wade in the morass and the stench among the reptiles and filth, how they lie in sulphur and pitch, neither they nor anyone else can ever tell. Seeing that I, through God's mercy and without effort, had seen this great distress, I, poor wretch, suffered so greatly from the stench and the subterranean heat that for three days I could neither sit nor walk nor use any of my five senses. I was as a being struck by lightning. But my soul suffered no harm for it had brought no infection with it which would there merit eternal death. It is possible that a pure soul among these wretched creatures might be to them an everlasting light and comfort.

Mechthild of Magdeburg

Infra 266

As to them that are ungodly of works, shrieking in anguish, they are flayed of their whole skin by the Death-god's henchmen, their bodies bathed with copper molten in the fire, their flesh cut off in gobbets by hundreds of blows from flaming swords and pikes, and they fall again and again upon beds of red-hot iron.

Śânti-deva

People who revile the *Nembutsu* will fall into hell, where their immeasurable torments shall continue for five *kalpas.*

Hônen

And ye know of those of you who broke the Sabbath, how We said unto them: Be ye apes, despised!

And We made it an example to their own and to succeeding generations, and an admonition to the God-fearing.[1]

Qur'ân, II. 65, 66

The mortal who thinks of his gains or his honours or the favour of many men will be afraid of death when it falls upon him. Whatsoever it be in which the pleasure-crazed spirit takes its delight, that thing becomes a pain a thousand times greater. Therefore the wise man will seek not for pleasure, for from desire arises terror.

Suff. 120

Śânti-deva

Verily We shall cause those who disbelieve to taste an awful doom, and verily We shall requite them the worst of what they used to do.[2]

Qur'ân, XLI. 27

The fire of Hell this strange condition hath,
To burn, not shine (as learned *Basil* saith.)

Infra 266

Robert Herrick

Theologians speak of hell. I will tell you what hell is. It is merely a state. Your state here is your eternal state. This is hell. Take an illustration. A thief who has incurred the penalty of death on being caught: picture his state of mind seeing others happy! So do we feel, and worse. And so with those in hell who see God and his friends: the height of torment, so the masters say.

Eckhart

Hell is even an enemy of the devil, for he is a strange guest therein, viz. a perjured fiend cast out of heaven: he will be lord in that wherein he was not created; the whole creation accuses him for a false perjured apostate spirit, which is departed from his order; yea even the nature in the wrath is his enemy though he be of the same property; yet he is a stranger, and will be lord, though he has lost his kingdom, and is only an inmate in the wrath of God; he that was too rich, is now become too poor; he had all when he stood in humility, and now he has nothing, and is moreover captivated in the gulf: this is his shame, that he is a king, and yet has fooled away his kingdom in pride; the royal creature remains, but the dominion is taken away; of a king he is become an executioner; what God's anger apprehends, there he is a judge, viz. an officer of God's anger, yet he must do what his Lord and Master wills.

Faith 510

Conform. 185

Boehme

[1] Traditional teaching envisages the possibility of apes 'issuing' from man, never the inverse. It is noteworthy that certain scientists adopt the same attitude; cf. Douglas Dewar and L. M. Davies, in *The Nineteenth Century and After,* April, 1943, pp. 167–173. While on this subject, a correspondence might be pointed out between the 'doctrine' of evolution and a key formula of the Black Mass: 'And the flesh was made Word'.

[2] The 'merciful' corollary to this is in the chapter, *The Primordial State.*

It is a question, what burns in hell? Doctors reply with one accord: 'self-will'.

Eckhart

Nothing burneth in hell but self-will.

Theologia Germanica, XXXIV

See here the whole truth in short. All sin, death, damnation, and hell is nothing else but this kingdom of self, or the various operations of self-love, self-esteem, and self-seeking which separate the soul from God and end in eternal death and hell.

William Law

What is there that the fire of hell shall feed upon, but thy sins?

The more thou sparest thyself now and followest the flesh, the more severe hereafter shall be thy punishment, and thou storest up greater fuel for that flame.

Suff. 120

In what things a man hath sinned, in the same shall he be the more grievously punished.

There shall the slothful be pricked forward with burning goads, and the gluttons be tormented with extreme hunger and thirst.

There shall the luxurious and lovers of pleasure be bathed in burning pitch and stinking brimstone, and the envious, like mad dogs, shall howl for very grief.

There is no sin but shall have its own proper torment.

The Imitation of Christ, I. xxiv

Now, in this present time, man is set between heaven and hell, and may turn himself towards which he will. For the more he hath of ownership, the more he hath of hell and misery; and the less of self-will, the less of hell, and the nearer he is to the Kingdom of Heaven.

Conform. 166

Theologia Germanica, LI

For a soul may not stand still alway in one state while that it is in the flesh; for it is either profiting in grace, or impairing in sin.

Walter Hilton

Man has an eternity within him, is born into this world, not for the sake of living here, not for anything this world can give him, but only to have time and place to become either an eternal partaker of a divine life with God or to have an hellish eternity among fallen angels.

Pilg. 365
Creation 48

William Law

Darkness, light, fire and air, water and earth, stand in their temporary, created distinction and strife for no other end, with no other view, but that they may obtain the one thing needful, their first condition in Heaven: and shall man that is born into time for no other end, on no other errand but that he may be an angel in eternity, think it hard to live as if there were but one thing needful for him? What are the poor politics, the earthly wisdom, the ease, sensuality, and advancements of this world for us but such fruits as must be eaten in hell? To be swelled with pride, to be fattened with sensuality, to grow great

Charity 608

through craft, and load ourselves with earthly goods is only living the life of beasts, that we may die the death of devils. On the other hand, to go starved out of this world, rich in nothing but heavenly tempers and desires, is taking from time all that we came for, and all that can go with us into eternity.

William Law

Death 220

Nothing can throw thee into the infernal abyss so much as this detested word (heed it well!) this mine and thine.

Angelus Silesius

There is of nothing so much in hell as of self-will.

Theologia Germanica, XLIX

There is no Hell but selfhood, no Paradise but selflessness.

Abu Sa'îd ibn Abi 'l-Khayr

The Fire of Heaven

Wr. Side 474

God is in the Scriptures said to be a consuming fire. But to whom? To the fallen angels and lost souls. But why and how is He so to them? It is because those creatures have lost all that they had from God but fire; and therefore God can only be found and manifested in them as a consuming fire. Now, is it not justly said that God, who is nothing but infinite love, is yet in such creatures only a consuming fire, and that though God be nothing but love, yet they are under the wrath and vengeance of God, because they have only that fire in them which is broken off from the light and love of God, and so can know or feel nothing of God but His fire in them? . . . Hell is nothing but the fire of heaven separated from its first light and majesty.

William Law

Damnation is in the essence.
A damned person could be in the highest heaven:
He would still experience hell and its torments.

Angelus Silesius

Center 835

And if a thoroughly godly-minded and God-loving man . . . were to be cast down to the depths of hell, he must needs turn it into a kingdom of heaven, and God and eternal blessedness would exist in hell.

Tauler

At the Gathering for Judgement the Faithful will say, 'O Angel, is not Hell the common road
Trodden by the believer and infidel alike? Yet we saw not any smoke or fire on our way.'
Then the Angel will reply: 'That garden which ye saw as ye passed
Was indeed Hell, but unto you it appeared a pleasaunce of greenery.
Since ye strove against the flesh and quenched the flames of lust for God's sake,
So that they became verdant with holiness and lit the path to salvation;
Since ye turned the fire of wrath to meekness, and murky ignorance to radiant knowledge;
Since ye made the fiery soul (*nafs*) an orchard where nightingales of prayer and praise were ever singing—
So hath Hell-fire become for you greenery and roses and riches without end.'

Rûmî

Metanoia 484

It is, therefore, exceeding good and beneficial to us to discover this dark, disordered fire of our soul; because when rightly known and rightly dealt with, it can as well be made the foundation of Heaven as it is of hell. For when the fire and strength of the soul is sprinkled with the blood of the Lamb, then its fire becomes a fire of light, and its strength is changed into a strength of triumphing love, and will be fitted to have a place amongst those flames of love that wait about the throne of God.

William Law

Realiz. 859 *Sin* 62 *Inv.* 1017 *Love* 614

Hell! I am to thee a conqueror; thou must serve me for the kingdom of joy: Thou shalt be my servant and minister to the kingdom of joy; thou shalt enkindle the flames of love with thy wrath, and be a cause of the spring in paradise.

Boehme

And verily We have adorned the world's firmament with lamps, and We have made them missiles for the devils.

Qur'ân, LXVII. 5

Heterod. 418

Let not any question the close and Divine Contexture of the whole Work in all the parts and conduct of it; by a firm connexion of *Causes* and *Effects*, like links in a Chain, from its *first beginning* of its *last end*; because he meeteth with an *Hell*, as well as an *Heaven* in this work of God. *Divine Love* (which transcendently excels in all Wisdom and Prudence, beyond all the highest wits of men, the richest Contrivances of Poets,) knoweth how to joynt an *Hell* into its work, with such Divine Artifice, incomprehensible to Men or Angels, that this also shall be beautiful, with delights in its place and shall give a sweetness a lustre to the whole piece.

Peter Sterry

Introd. *Sin* *M.M.* 978 *Beauty* 689

For even those things which mortals deem foul are goodly in God's sight, because they have been made subject to God's laws.

Hermes

Conform. 185

We said: O fire, be coolness and peace for Abraham.

Qur'ân, XXI. 69

Shadrach, Meshach, and Abednego, came forth of the midst of the fire.

P. State 563

And the princess, governors, and captains, and the king's counsellers, being gathered together, saw these men, upon whose bodies the fire had no power, nor was an hair of their head singed, neither were their coats changed, nor the smell of fire had passed on them.

Daniel, III. 26, 27

Renun. 139
Peace 700

For a peaceful meditation, we need not to go to the mountains and streams;
When thoughts are quieted down, fire itself is cool and refreshing.

Kwaisen, abbot of Yerin-ji, to his Zen disciples entering the 'fire-*samâdhi*' amidst flames of their besieged monastery

Judgment Day

Introd.
Realiz.

Thus shall the Day of Judgement be:
(A day) whereon is neither sun, nor moon, nor stars;
neither clouds, nor thunder, nor lightning;
neither wind, nor rain-storm, nor cloud-rack;
neither darkness, nor evening, nor morning;
neither summer, nor autumn, nor winter;
neither heat, nor frost, nor cold;
neither hail, nor rain, nor dew;
neither noon, nor night, nor dawn;
neither shining, nor brightness, nor light,

Center 832

save only the splendour of the brightness of the Most High, whereby all shall be destined to see what has been determined (for them).

IV Ezra, VII. 39–43

Heterod. 420

Woe unto the repudiators on that Day!
This is the Day of Decision, We have brought you and the men of old together.
If now ye have any wit, outwit Me.

Qur'ân, LXXVII. 37–39

Supra 242
Metanoia 480

Touching the last day, people say, God shall judge. So he shall, but not as they think. Each man is his own judge in this sense: the state he then appears in he is in eternally. . . . Those souls who all their days have spent their time in God till God has come to be their being, to them God stays their being. . . . Not so the wicked who have squandered their time on creatures; what their state is it continues to be, and this eternal lapsing from God and from his friends is called hell. Yet bear in mind that these same persons get their being from God or they would not be at all.... They are in God as 'twere a man with his life forfeit to some righteous lord whose honour he has stolen and whose friends, and plotted frequently against his life; and now his lord, who showed him only kindness in hopes of his reform, is vexed to find that he declines to mend. Holding him in the grip of justice, his lord forbears to kill. He punishes the outrage on himself.... Bound hand and foot, the man is cast into the lowest donjon among toads and reptiles and the foul water which is wont to lie in deepest donjon-keeps.... Even so it is permissible to say that man is at the court, for the donjon is the royal court as much as the hall is where the king stays with his friends; but conditions, you see, are different.... Know that grief endures eternally. I marvel that anyone who hears these words should dare to sin. Purgatory is so grievous in itself that anyone who knows the rights of it would stay no time in sin.

Supra Introd.

Eckhart

The Judgement day.

In doing justice, God shall then be known,
Who shewing mercy here, few priz'd, or none.

Robert Herrick

Supra 239

Those who have merited paradise will enter therein; the damned will go to hell. God will then say: Let those leave hell whose hearts contain faith even the weight of a mustard seed! Then they will be let out, although they have already been burned to ashes, and plunged into the river of rain-water, or into the river of life; and immediately they will be revived.[1]

Muhammad

The last judgement is a kindling of the fire both of God's love and anger, in which the matter of every substance perisheth, and each fire shall attract into itself its own, that is, the substance that is like itself: Thus God's fire of love will draw into it whatsoever is born in the love of God, or love-principle, in which also it shall burn after the manner of love, and yield itself into that substance. But the torment will draw into itself what is wrought in the anger of God in darkness, and consume the false substance; and then there will remain only the painful aching will in its own proper nature, image, and figure.

Supra 266

Supra Introd.

With what matter and form shall the human body rise?

It is sown a natural gross and elementary body, which in this lifetime is like the outward elements; yet in this gross body there is a subtle power and virtue.[2] As in the earth also there is a subtle good virtue, which is like the sun, and is one and the same with the sun; which also in the beginning of time did spring and proceed out of the divine power and virtue, from whence all the good virtue of the body is likewise derived. This good virtue of the mortal body shall come again and live for ever in a kind of transparent chrystalline material property, in spiritual flesh and blood; as shall return also the good virtue of the earth, for the earth likewise shall become chrystalline, and the divine light shine in everything that hath a being, essence, or substance. And as the gross earth shall perish and

Center 816

Supra Introd.

[1] The allusion here is to souls in purgatory.
[2] On this cf. Guénon: *Le Roi du Monde*, ch. VII: '*Luz ou le séjour d'immortalité.*'

Wr. Side 464

never return, so also the gross flesh of man shall perish and not live for ever. But all things must appear before the judgement, and in the judgement be separated by the fire; yea, both the earth, and also the ashes of the human body. For when God shall once move the spiritual world, every spirit shall attract its spiritual substance to itself. A good spirit and soul shall draw to itself its good substance, and an evil one its evil substance. But we must here understand by substance, such a material power and virtue, the essence of which is mere virtue, like a material tincture (such a thing as hath all figures, colours, and virtues in it, and is at the same time transparent), the grossness whereof is perished in all things.

Beauty 670

Boehme

Every part therefore of the bodies, perishing either in death or after it, in the grave or wheresoever, shall be restored, renewed, and from a natural and corruptible body it shall become immortal, spiritual, and incorruptible. Be it all made into powder and dust, by chance or cruelty, or dissolved into air or water, so that no part remain undispersed, yet shall it not, yet can it not be kept hidden from the omnipotency of the Creator, who will not have one hair of the head to perish. Thus shall the spiritual flesh become subject to the spirit, yet shall it be flesh still, as the carnal spirit before was subject to the flesh, and yet a spirit still.[1]

Metanoia 488

Orthod. 300

St Augustine

Take heed to yourselves, lest at any time your hearts be overcharged with surfeiting, and drunkenness, and cares of this life, and so that day come upon you unawares.

For as a snare shall it come on all them that dwell on the face of the whole earth.

Watch ye therefore, and pray always, that ye may be accounted worthy to escape all these things that shall come to pass, and to stand before the Son of man.

Illusion 99

Wr. Side 464

Inv. 1013

St Luke, XXI. 34–36

On the Day of Judgement Hell will swallow all the Damned and together with them everything in the whole World that is putrid and stinking, and the Maw of the Abyss will close upon their Heads; neither Man nor Demon nor Creature else will evermore go forth. The Issues will be sealed, and eternal Despair will do its eternal Work.

Ruysbroeck

[1] See the citation by Boehme on p. 894.

ORTHODOXY—RITUAL—METHOD

Holy scripture cries aloud for freedom from self.

Meister Eckhart[1]

All scriptures without any exception proclaim that for attaining Salvation mind should be subdued.

Sri Ramana Maharshi[2]

Freedom from self requires a method, and a method in turn means recourse to cosmic principles that transcend the limitations of the human individuality, which otherwise must find itself compromised to fighting fire with fire. For the self or ego exists by definition as a result of ignorance, and ignorance cannot overcome ignorance: 'How can Satan cast out Satan?' (*Mark*, III. 23).[3] It is true that 'if ye be led of the Spirit, ye are not under the law' (*Gal.* V. 18), but this reservation is for those who have already transcended the law; and if blind pride alone can make those under the law consider themselves above it, so will true humility have those rare ones who may in reality be above the law comport themselves in a manner exemplary to those still under it. 'All things are lawful for me, but all things are not expedient' (*I. Cor.* X. 23).

Death 220

Orthodoxy means right doctrine, and ritual could be called the correct application of it. 'The heterodoxy of a conception is really nothing else than its falsity, resulting from its disagreement with fundamental principles; more often than not this falsity is even a manifest absurdity, once the question is reduced to essentials: it could not be otherwise, since metaphysic . . . excludes everything of a hypothetical character, only admitting those things the comprehension whereof at once implies real certainty. Given these conditions, orthodoxy is the same as true knowledge, since it consists in an unbroken agreement with principles. . . . Only it should be well understood that it is less a question of having recourse to the authority of written texts than of observing the perfect coherence of traditional teaching as a whole' (Guénon: *Introduction générale à l'Étude des Doctrines hindoues*, pp. 165–166).

Knowl. 749

'The word "heresy" means choice, the having opinions of one's own, and thinking what we *like* to think' (Coomaraswamy: *Am I My Brother's Keeper?*, p. 42). But 'such as men

Introd. *Heterod.*

[1] Evans, I. 418.

[2] *Who Am I?*, p. 29.

[3] And yet there are those who would vanquish the ego while obstinately refusing submission to a legitimate traditional form, unaware of course that the ego which they wish to dominate is what lies at the base of this patent individualism. Autodetermination in spiritual matters amounts to intellectual anarchy.

themselves are, such will God himself seem to be' (John Smith the Platonist: *Select Discourses,* p. 8).[1] And the Upanishads warn that 'Whosoever shall have such a doctrine, be they gods or be they devils, they shall perish' (*Chandogya Upanishad,* VIII. viii. 4). Spiritual reality is other than what men's minds can conceive. For 'hath not God made foolish the wisdom of this world?' (*I. Cor.* I. 20). Where we are logical, the Truth is 'illogical';[2] where we are straight, the Truth is bent; we are virtuous where It is wanton, and polluted where It is virgin. What is mean with us is precious with God; where we are sober His saints are drunk, we are busy where they are idle, and heedless where they are recollected. To do as we like about our spiritual practice is to promote the perpetuation of our distraction, our suffering, our folly and spiritual blindness. The ego is a Hydra-headed monster sprouting two heads for each cut off, and it is ritual alone that can cauterize the wound and prevent new growth, and at the same time nourish the soul with the legitimate food proper to its spiritual formation. Ritual is God's way, as opposed to man's way. It is a medicine, the antidote to self-deception, a corrective, a restorative, a healing balm, a power outside of our own limitations, delegated by God through His elect on earth for the spiritual welfare of humanity and, by extension, all creation. Its usage differs following each revelation, each age and society, but its end is always the same, namely, a support capable of putting us in contact with the Divine Center, source of all Reality. Until modern times the question could hardly arise as to what ritual and what spiritual method to pursue. One normally adhered to the rite in which one was born, and one normally submitted to the method advanced by one's spiritual superiors. In the modern world, by contrast, one is more often born into profanity than into any serious rite, and if the question of a spiritual path does arise, how can the individual in his ignorance know which way to turn? An answer to this dilemma may be sought through personal prayer, the one rite which is constantly at everyone's disposal, and which serves to establish as well as maintain a rapport between the individual and the Creator.

P. State 563
Ecstasy 637
Charity 608
Center passim
Contem. 521

Ritual is not an arbitrary product of the human will. 'A rite, etymologically, is that which is accomplished in conformity with "order", and which by consequence imitates or reproduces at its level the process of manifestation itself' (Guénon: *Le Règne de la Quantité,* p. 32). 'The end of rites is always to put the human being into contact, directly or indirectly, with something that surpasses his individuality and which belongs to other (higher) states of existence' (Guénon: *Aperçus sur l'Initiation,* p. 112). 'To accomplish a rite is not only to retrace a symbol, but also to participate in a certain mode of being' (Burckhardt: *Introduction aux Doctrines ésotériques de l'Islam,* p. 99).[3] The outward sacraments precisely underscore the fact that man is composed of 'substance' as well as 'spirit', and that the body can be used as a point of departure for reverberations reaching progressively into the interior states of being. 'All things,' according to Dante, 'are arranged in a certain order, and this order constitutes the form by which the Universe

Introd. *Beauty*
Infra 275
Introd. *Rev.*
Symb. 306
Introd. *Inv.*

[1] 'I am in My servant's opinion of Me' (*ḥadîth qudsî,* cited by Ibn 'Arabî in the *Fuṣûṣu 'l-Ḥikam;* cf. Nicholson: *Studies in Islamic Mysticism,* p. 159). 'One looks on God exactly according to one's own inner feeling. Take, for instance, a devotee with an excess of tamas. He thinks that the Divine Mother eats goat' (Sri Ramakrishna, *Gospel,* p. 322,—goats being sacrificed to Kâlî).

[2] Although in appearance only; cf. Introd. *Ecstasy.*

[3] 'In the Brâhmaṇas the truth that is so potent is the truth of exactitude in the sacrifice, a ritual or ceremonial accuracy' (W. N. Brown: *Walking on the Water,* p. 6).

resembles God' (*Paradiso,* I. 103). This agrees with Tong Tshung-chu (2nd Century B.C.): 'In the Universe there is no hazard; . . . all is influence and harmony, accord and answering accord' (cf. A. Préau, *Voile d'Isis,* 1932, p. 554), and with Michael Maier: 'Like all the visible things which are in Nature, celestial bodies as well as terrestrial ones have been created in terms of number, weight and measure. There is, thus, between them, an admirable and marvelous proportion in the parts, the forces, the qualities, the quantities and their effects' (*Voile d'Isis,* 1932, p. 461); and lastly Plotinus: 'Because analogy is the law of all things, . . . things in this world could not be independent; indeed, of necessity there had to be a certain relation between them' (3rd *Ennead;* cited by A. Daniélou: *Introduction to the Study of Musical Scales,* p. 157).

Beauty 689
Reality 775

The customary charge against ritual has to do with 'abuse by the priesthood'. It is certainly true that salvation cannot be purchased simply through the automatic performance of set rites. The question of volition, intention and effort is paramount here. 'I will pray with the spirit, and I will pray with the understanding also' (*I. Cor.* XIV. 15). 'Virtues in their way delineate the truth; we must "know" with all our being, and not with the intelligence only. Spiritual sincerity—or the Truth in its wholeness—demands something of us which, apart from all question of doctrinal and ritual form, comes entirely from ourselves' (Schuon: 'Le Yoga comme principe spirituel'; in *Yoga, Science de l'Homme intégral,* edit. Jacques Masui, Paris, 1953, p. 29). 'The Pharisees had orthodoxy and regularity, but they had neither grace nor virtues. They did not have grace, because practically speaking, they put their orthodoxy and regularity in place of the living God; and they did not have virtues, because they replaced human values—the moral qualifications—with exterior observances, which thus isolated, lost their efficacity. Christ did not deny their authority—"they sit in Moses' seat"—but he condemned them in spite of it' (Schuon: *Perspectives spirituelles,* p. 107).[1]

Introds.
Conform.
Knowl.

However, 'even if disorder and confusion violate the most divine decrees and laws, there is no reason which authorizes a reversal—even in God's favour—of the order that God has instituted. For God is not divided against Himself; otherwise, how would His Kingdom stand?' (Dionysius: *Ep.* VIII. 1088 C). In the same vein is Dante's censure of Pope Nicholas the Third: 'And were it not that reverence for the Great Keys thou heldest in the glad life yet hinders me, I should use still heavier words' (*Inf.* XIX. 100). He condemned the abuses of the Papacy, not the institution. 'Know that the evil lies in the men who hold the doctrines, not in the principles on which the doctrines are based' (Hujwîrî: *Kashf,* p. 43).[2]

Rev. 965

In briefest outline, traditional regularity or orthodoxy implies attachment to an authentic living tradition with a legitimate line of representatives ('apostolic succession'). Adherence to the laws of the tradition—loosely 'exoterism'—is incumbent on every mortal seeking salvation. Within a tradition (and actually constituting its essence) is normally found an

[1] Bâyazîd al-Bisṭâmî opposes esoteric learning received from 'the Living who dieth not', to exoteric religious science, 'received from the dead who have received it from the dead' (Dermenghem: *Vies des Saints musulmans,* p. 202).

[2] In cases of this sort, infallibility of function can be compared to a traditional symbol composed of base matter, where the form is real even if the substance is 'unworthy'. In point of fact, however, known instances of abuse carry little weight when compared with the vast preponderance of spiritual conformity in all the great hieratic collectivities of the world.

M.M. 973

'esoteric' way reserved for a relative few who have the necessary intellectual or spiritual qualifications, and which envisages sanctification more especially as its goal. The word 'normally' is used advisedly, because the twentieth century has seen such an upheaval in traditional patterns that anomalies and exceptions have inevitably resulted. For the rest, it is indispensable to study carefully the works of the contemporary authorities on these questions in order to evaluate things in a proper perspective. The principle to remember in all this is that the Universal Order is not subject to change. Whatever the modifications and upheavals in sublunar existence, the dichotomy between this world and the next remains constant and quasi-absolute, and so accordingly will the basic laws of the spiritual life for bridging this gulf remain necessary and valid. We live in the *formal* realm, where *per-form*ance necessarily precedes *trans-form*ation.

Reality 773

The Chain of Transmission

From the moment you came into the world of being,
A ladder was placed before you that you might escape.

Dîvâni Shamsi Tabrîz, XII

Reality 790

I have seen the ancient Way, the Old Road that was taken by the formerly All-Awakened, and that is the path I follow.

Saṁyutta-nikâya, II. 106

And unto thee have We revealed the Scripture with the Truth, confirming whatever Scripture was before it, and safeguarding it.

Qur'ân, v. 48

A portion of Myself has become the living Soul in the world of life from time without beginning. . . .

Center 841

From Me alone comes memory, wisdom, and also their loss. I am that which is known in all the Vedas. Verily I am the Author of Vedanta and the knower of the Vedas am I.

Bhagavad-Gîtâ, XV. 7, 15

Knowl. 755 749

Think not that I am come to destroy the law, or the prophets: I am not come to destroy, but to fulfil.

St Matthew, v. 17

I do not create; I only tell of the past.

Confucius

By adhering to the Tao of the past
You will master the existence of the present.

Tao Te Ching, XIV

The one path leading up to the highest peak is the mysterious orthodox line of transmission established by Buddhas and Fathers, and to walk along this road is the essence of appreciating what they have done for us. When the monk fails to discipline himself along this road, he thereby departs from the dignity and respectability of monkhood, laying himself down in the slums of poverty and misery.

Dai-o Kokushi

Heterod. 423

The man of refinement turns his thoughts back to the past, goes back to his origin, and does not forget those through whom life has come to him.

Li Chi, Chi Yi, sect. 2

Creation 42

Trust in the Lord with all thine heart; and lean not unto thine own understanding.

Proverbs, III. 5

For no creature, howsoever rational and intellectual, is lighted of itself, but is lighted by participation of eternal Truth.

St Augustine

Rev. 961

If I bear witness of myself, my witness is not true.

St John, v. 31

Every word I have been saying is in the Scriptures themselves.

Hônen

I will lead you into no doctrine but what is strictly conformable to the letter of Scripture and the most orthodox piety.

William Law

Introd. *Heterod.*

Now if anyone says of me, Gotama the Pilgrim, knower and seer as aforesaid, that my eminent Aryan gnosis and insight have no superhuman quality, and that I teach a Law that has been beaten out by reasoning, experimentally thought out and self-expressed, if he will not recant, not repent and abandon this view, he falls into hell.

Majjhima-nikâya, I. 68

Rev. 959

I declared this imperishable Yoga to Vivasvân, and Vivasvân told it to Manu, Manu taught it to Ikshvâku.

Thus, handed down in regular succession, the royal sages knew it. This Yoga through long lapse of time has been lost in this world, O Parantapa.

That same ancient Yoga has been today declared to thee by Me, for thou art my devotee and my friend. This is the supreme secret.

Bhagavad-Gîtâ, IV. 1–3

Knower of the Unseen, He revealeth unto none His secret,

Save unto every messenger whom He hath chosen, and then He maketh a guard to go before him and a guard behind him

That He may know that they have indeed conveyed the messages of their Lord. He surroundeth all their doings, and He keepeth count of all things.

Qur'ân, LXXII. 26–28

If you think you have ascended to that high place of the heart and climbed that high and great mountain, if you feel you have seen Christ transfigured, do not believe too easily whatever you may see in him, or hear from him unless Moses and Elias are with him. For we know that every word must be confirmed by the mouth of two or three witnesses (*Deut.* XIX. 15). I suspect every truth which is not confirmed by the authority of Scripture nor do I accept Christ transfigured unless Moses and Elias bear him company.

Richard of Saint-Victor

All the mystic paths (*ṭuruq*) are utterly barred except to him who followeth in the steps of the Apostle.

Junayd

I invite all people to the marriage of the Lamb, but no-one to myself.

William Law

Unless one's *Guru* be of an unbroken (apostolic) line,
What gain is it to take Initiation?

Infra 288
280

Milarepa

In order to say a little about this dark night, I shall trust neither to experience nor to knowledge, since both may fail and deceive; but, while not omitting to make such use as I can of these two things, I shall avail myself, in all that, with the Divine favour, I have to say, or at the least, in that which is most important and dark to the understanding, of Divine Scripture; for, if we guide ourselves by this, we shall be unable to stray, since He Who speaks therein is the Holy Spirit. And if in aught I stray, whether through my imperfect understanding of that which is said in it or of matters unconnected with it, it is not my intention to depart from the sound sense and doctrine of our Holy Mother the Catholic Church; for in such a case I submit and resign myself wholly, not only to her command, but to whatever better judgement she may pronounce concerning it.

Rev.
959

St John of the Cross

All good men are united with God through means. These means are the grace of God, and the sacraments of Holy Church, and the Divine virtues, faith, hope and charity, and a virtuous life according to the commandments of God; and to these there belongs a death to sin and to the world and to every inordinate lust of nature. And through these, we remain united with Holy Church, that is, with all good men; and with these, we obey God, and are one will with Him, even as an orderly convent is united with its Superior: and without this union none can please God nor be saved.

Ruysbroeck

As the immaterial hierarchies of heavenly powers are illumined by God in just sequence, so that the Divine light penetrates from the first hierarchy to the second, from the second to the third and so to them all; so the saints, illumined by holy angels, are linked together and united by the bond of the Holy Spirit and thus become akin to them and of equal rank. Moreover, the saints—those who appear from generation to generation, from time to time, following the saints who preceded them—become linked with their predecessors through obedience to Divine commandments and, endowed with Divine grace, become filled with the same light. In such a sequence all of them together form a kind of golden chain, each saint being a separate link in this chain, joined to the first by faith, right actions and love; a chain which has its strength in God and can hardly be broken.

Conform.
180

Rev.
967

A man who does not express desire to link himself to the latest of the saints in all love and humility owing to a certain distrust of him, will never be linked with the preceding saints and will not be admitted to their succession, even though he thinks he possesses all

Heterod. 423

possible faith and love for God and for all His saints. He will be cast out of their midst, as one who refused to take humbly the place allotted to him by God before all time, and to link himself to that latest saint as God had disposed.

St Simeon the New Theologian

See, in all mortal bequests, the heritage becomes the property of the living, whilst it slips from the dying; whereas in the case of one who bequeaths unto his children the inheritance of religious knowledge, wisdom, and the Holy Law, the possession, while being transmitted to his children, is not altogether alienated from him after death.

Yesod Hayirah, 30

By the reigning sympathy and by the fact in nature that there is an agreement of like forces and an opposition of unlike, we can explain in the same way as with magic the efficacy of prayers. There is no question of a will that grants. Some influence falls from the being addressed upon the petitioner, but that being itself, sun or star, perceives nothing of it all. The prayer is answered by the mere fact that part and other part are wrought to one tone like a musical string which, being plucked at one end, vibrates at the other also. Often, too, the sounding of one string awakens what might pass for a perception in another, the result of their being in harmony and tuned to one musical scale. Now if the vibration in one lyre affects another by virtue of the sympathy existing between them, then certainly in the All, even though it is constituted in contraries, there must be one melodic system; for it contains its unisons as well, and its entire content, even to those contraries, is a kinship.

Reality 775 *Beauty* 689

Plotinus

Symb. 306 *Metanoia* 493

The impulse from below (*itharuta dil-tata*) calls forth that from above.

Zohar, I. 164

There are three sources to Ritual. Heaven and Earth are the source of its existence, our ancestors the source of its being in a class by itself, sovereigns and teachers the source of its disciplinary power. Without Heaven and Earth how could it have come to be?[1] Without our ancestors how could it have emerged? Without sovereigns and teachers how could it have disciplined men? If any of these had been lacking, there would not have been this pacifying influence among men. Thus it is that there is Ritual, the ritual serving of Heaven above, of Earth below, the reverencing of ancestors and the honouring of sovereigns and teachers.

Symb. 322

Hsun Ch'ing

The Rules which I have desired to govern my self by in this Discourse, are these, *Right Reason,* the universally acknowledged *Principles* of all sober *Philosophy* and sound *Divinity,* the *Analogy* of *Faith,* the *Letter* of the *Scriptures,* the *Spirit* of *Christ,* in the *Head* of the Church *Christ* himself, in the *Scriptures,* in the *Church,* in the *Spirits,* and *experiences* of each *Saint.* As all these answer each other in a mutual *Harmony,* like the several strings upon a well-tuned Lute: So shall I ever (as I humbly believe) take it for granted, That whatever jars upon any of these strings, is also out of tune to all the rest.

Symb. 313

[1] The translator, E. R. Hughes, adds the following note: 'This is not rodomontade. All great traditional rituals are sacramental representations of the nature of the universe.'

That what-ever is not conformable to any one of these Rules, crosseth them all. If any thing of this kind shall have faln from me, I do here humbly repent of it, and recant it. I humbly desire all those, who shall stoop to read what I have written, to believe that no Gift can be so acceptable to me, as the displacing of any *Errour* or *Falshood* in my Mind, with the gentle hand of Love, by bringing in the *Truth* which it hath *counterfeited* and *vailed.* If any Errour have seemed sweet, beautiful, or desirable to me, in any kind. I humbly undertake to have a confident assurance of this, That the Truth it self shining forth, will bring along with it far more transcendent Sweetnesses, Beauties, Agreeablenesses and Satisfactions to all my Desires, excelling it in all the impressions of a Divine Pleasantness, Power and Profit, as the true *Sun* doth a *Parelius,* or a counterfeit Image of the Sun in a Cloud.

Heterod. 418

Peter Sterry

Whether the legislator is establishing a new state or restoring an old and decayed one, in respect of Gods and temples, . . . if he be a man of sense, he will make no change in anything which the oracle of Delphi, or Dodona, or the God Ammon, or any ancient tradition has sanctioned in whatever manner, whether by apparitions or by inspired words from Heaven, in obedience to which mankind have established sacrifices in connection with mystic rites, either originating on the spot, or derived from Tyrrhenia or Cyprus or some other place, and on the strength of which traditions they have consecrated oracles and images, and altars and temples, and portioned out a sacred domain for each of them. The least part of all these ought not to be disturbed by the legislator. . . . There should be one law, which will make men in general less liable to transgress in word or deed, and less foolish, because they will not be allowed to practise religious rites contrary to law. . . . Gods and temples are not easily instituted, and to establish them rightly is the work of a mighty intellect.[1]

Rev. 959

Plato (*Laws*, 738 & 909)

These rites of the *Inipi* (Purification Lodge) are very *wakan* (sacred) and are used before any great undertaking for which we wish to make ourselves pure or for which we wish to gain strength; and in many winters past our men, and often the women, made the *Inipi* even every day, and sometimes several times in a day, and from this we received much of our power. Now that we have neglected these rites we have lost much of this power; it is not good, and I cry when I think of it. I pray often that the Great Spirit will show to our young people the importance of these rites.

Black Elk

If one only understood the meaning of these sacrifices to Heaven and Earth, and the significance of the services in ancestral worship in summer and autumn, it would be as easy to bring peace and order to a nation as to point a finger at the palm.

Confucius

[1] 'It is only when it is visualized how brief a thing human life is, and how negligible in duration, when compared with the immensity of Time and yet how important the *self* is as part of the Supreme Self, in following the way of enjoined duty, and how all engaged in a common pilgrimage have obligations to help everyone else, that a proper sense of values will be gained, and happiness or misery will be understood as temporary and fleeting, when compared with the ultimate joy (*ânanda*) that awaits every soul. The blazing of the path to this goal is the function of Dharmaśâstra' (K. V. Rangaswami Aiyangar: *Some Aspects of the Hindu View of Life according to Dharmaśâstra,* Baroda, 1952, p. 180).

Moves Walking . . . picked up a pointed stick and, after offering it to *Wakan-Tanka,* drew a line in the soft earth, from the west to the east; after offering the stick again to the heavens, he drew another line from the north to the south. Finally, the altar was completed by making two lines of tobacco on top of the two paths drawn on the ground, and then this tobacco was painted red. This altar now represented the universe and all that is in it. At its center was *Wakan-Tanka;* His presence was really there in the altar, and that is why it was made in such a careful and sacred way.

Center 831 Introd. *Symb.*

Black Elk

Symb. 313

Hebrew words have not the same force in them when translated into another tongue.

Ecclesiasticus (Prologue to the Greek translation made by the grandson of the author)

The Scholars represent ritual-and-righteousness, the farmers and fighting men food and drink. To ennoble the farmers and fighting men and bring the Scholars to ruin is equivalent to (society) abandoning ritual-and-righteousness and devoting itself to food and drink; and if ritual-and-righteousness were gone, the foundations of society would be destroyed and the upper and lower ranks be in confusion, whilst the Yin and Yang would get out of gear, the wetness and dryness of the seasons be out of order, the five cereals not be raising their heads, the myriad populace dying of starvation, with the farmers having no means of tilling the soil and the soldiers no means of fighting.

When Tzu Kung wished to abolish the sacrifice of a sheep at the announcement of the new moon, Confucius said, 'You love the sheep in this matter: I love the rite.' Tzu Kung hated the waste of a sheep, whilst Confucius was concerned about the abandonment of ritual. The fact is that if an old dyke is taken to be of no use and is let go, there is sure to be a disastrous flood; and if old rites are taken to be of no service and are let go, there is sure to be disastrous anarchy. The existence of the Scholars down the generations amounts to the old dykes of ritual-and-righteousness. To have them is of no (apparent) utility: not to have them is definitely harmful.

Action 346 *Holiness* 935

Wang Ch'ung

Initiation and Spiritual Filiation

Inv. 1031

Initiation into the Divine Name or the solemn Mantra-Diksha is one of the holiest and most significant of the sacred rituals in the spiritual life. To receive the Guru-Mantra from a realised saint and Sad-Guru is the rarest of good fortune and the most precious of the divine blessings that may be bestowed upon the aspirant. . . .

A most tremendous transformation begins to take place in the innermost core of the conscience of the initiated or the receiver of the Mantra. The initiated is himself unaware of this fact because of the veil of ignorance or Mula-Ajnana that still covers him.... But, nevertheless, this transformation starts with initiation, and like unto a seed that is sown in the earth, ultimately culminates in the grand fruit of realisation or Atma-Jnana.... The Sadhaka, after receiving initiation, must make earnest and continuous effort in the form of spiritual Sadhana if the Diksha is to become blissfully fruitful as Self-realisation. This part is the Sadhaka's sole responsibility in which task he will doubtless receive the help, guidance and grace of the Guru in the measure of his firm faith and loyalty to him....

Introd. *Inv.* *Contem.* 542

The process of initiation links you up directly with the Divine Being. Initiation or Mantra-Diksha is at one end of this golden chain and the Lord or the Highest Transcendental Atmic Experience is at the other end of it.

Swami Sivananda

Also no man coulde yet this *Science* reach,
But if *God* send a *Master* him to teach:
For it is soe *wonderfull* and soe selcouth,
That it must needes be tought from mouth to mouth:
Also he must (be he never so loath)
Receive it with a most sacred dreadfull Oath.

Infra 288

Thomas Norton

To expect a result from merely human methods is like trying to fly without wings or understand without intelligence.

Chuang-tse (ch. IV)

For let no one presume to suppose that he can make the first matter.

Michael Sendivogius

P. State 563

Without the ferment of gold no one can compose the Stone or develop the tinging virtue.

Basil Valentine

God must become man, man must become God; heaven must become one thing with the earth, the earth must be turned to heaven: If you will make heaven out of the earth, then give the earth the heaven's food, that the earth may obtain the will of heaven, that the will of the wrathful Mercury may give itself in unto the will of the heavenly Mercury.

Introd. *Creation* *Metanoia* 488

But what wilt thou do? Wilt thou introduce the poisonful Mercury (which has only a death's will in itself) into the temptation, as the false magus does? Will you send one devil to another, and make an angel of him? In deed and in truth I must needs laugh at such folly: If thou wilt keep a corrupt black devil, how dost thou think to turn the earth by the devil to heaven? Is not God the creator of all beings? Thou must eat of God's bread, if thou wilt transmute thy body out of the earthly property into the heavenly.

Wr. Side 451 Supra Introd.

Boehme

The medicine must be present before it can create anything. If the true seed or germ of the medicine is not found in the *ting* (furnace), the operation will be as futile as firing an empty kettle.

Chang Po-tuan

The noblest utterance upon *Tawḥîd* (the doctrine of Unity) is the saying of Abu Bakr aṣ-Ṣiddîq:[1] 'Glory be to Him who hath made for His creatures no means of attaining unto Knowledge of Him save through their impotence to attain (in themselves) unto that Knowledge!'

Grace 552

Junayd

Knowl. 749
Realiz. 890

'Always remember that I alone am holy and can render thee holy by transfusing My holiness into thee.'

Sister Consolata

Those who have attempted to digest common Mercury by means of artificial heat have failed as ludicrously as any one who should endeavour to incubate artificially an addled egg.

Philalethes

Visitor: 'Can anyone get any benefit by repeating sacred syllables (*mantras*) picked up casually?'
Maharshi: 'No. He must be competent and initiated in such *mantras*.'

Sri Ramana Maharshi

If your name is absent from the rolls of the Red Terrace,
In vain you learn the 'Method of Avoiding Food':
For naught you study the 'Book of Alchemic Lore'.

Po Chü-i

Where true principles lack the results are imperfect.

Henry Madathanas

Happy is the man who hath been initiated into the Greater Mysteries and leads a life of piety and religion.

Euripides

Athens hath produced many excellent, even divine inventions and applied them to the use of life, but she has given nothing better than those Mysteries by which we are drawn from an irrational and savage life and tamed, as it were, and broken to humanity. They are truly called *Initia*, for they are indeed the beginnings of a life of reason and virtue.

Cicero

Inv. 1017
Humility 191

The Effect of the Holy Sacrament.
The bread of the Lord acts in us like the philosophers' stone:
It transforms us into gold, if we are melted.

Angelus Silesius

[1] A.D. 573–634; first caliph of Islâm.

Verily, verily, I say unto you, Except ye eat the flesh of the Son of man, and drink his blood, ye have no life in you.

St John, VI. 53

By the fall of our first father we have lost our first glorious bodies, that eternal, celestial flesh and blood which had as truly the nature of paradise and Heaven in it as our present bodies have the nature, mortality and corruption of this world in them: if, therefore, we are to be redeemed there is an absolute necessity that our souls be clothed again with this first paradisaical or heavenly flesh and blood, or we can never enter into the Kingdom of God. Now this is the reason why the Scriptures speak so particularly, so frequently, and so emphatically of the powerful blood of Christ, of the great benefit it is to us, of its redeeming, quickening, life-giving virtue; it is because our first life or heavenly flesh and blood is born again in us, or derived again into us from this blood of Christ....

Creation 42

No figurative meaning of the words is here to be sought for, we must eat Christ's flesh and drink His blood in the same reality as He took upon Him the real flesh and blood of the blessed Virgin. We can have no real relation to Christ, can be no true members of His mystical body, but by being real partakers of that same kind of flesh and blood which was truly His and was His for this very end, that through Him the same might be brought forth in us. All this is strictly true of the holy sacrament, according to the plain letter of the expression; which sacrament was thus instituted that the great service of the Church might continually show us that the whole of our redemption consisted in the receiving the birth, spirit, life and nature of Jesus Christ into us, in being born of Him, and clothed with a heavenly flesh and blood from Him, just as the whole of our fall consists in our being born of Adam's sinful nature and spirit, and in having a vile, corrupt, and impure flesh and blood from him.

Infra 285

William Law

God the Father has in Christ's death and entrance into our humanity again received our self-hood into his will; and that this might be, he first tinctured the humanity with the Deity, that the humanity might be a pleasant sweet savour and offering to him in his power, for before death lay before it....

Realiz. 868

And so we all must follow him upon the path which he has made open for us; none can see God, unless God become first man in him.

Boehme

No artificial fire can infuse so high a degree of heat as that which comes from heaven.

John Mehung

The gist of the whole matter lies in the fact that the small and weak cannot aid that which is itself small and weak, and a combustible substance cannot shield another substance from combustion. That which is to protect another substance against combustion must itself be safe from danger. The latter must be stronger than the former, that is to say, it must itself be essentially incombustible. He, then, who would prepare the incombustible sulphur of the Sages, must look for our sulphur in a substance in which it is incombustible.

Basil Valentine

And this is the true key of the work—to incrudate the mature by the conjunction of an immature—being incrudated to calcine it—being calcined to dissolve it—and all this philosophically, not vulgarly.[1]

Philalethes

Pilg. 366

The words of Mercury are harsh after the songs of Apollo.

Shakespeare (*Love's Labour's Lost,* v. ii. 938)

For repairing a broken piece of bamboo ware, you must use bamboo. When you want your hen to hatch chickens, eggs and not stones must be supplied. Working on medicines of unlike kinds is merely a waste of energy. What can be better than the true lead for fulfilling the secrets of the sages?

Chang Po-tuan

If the Mercury is not properly prepared, the gold remains common gold, being joined with an improper agent; it continues unchanged, and no degree of heat will help it to put off its corporeal nature. Without our Mercury the seed (i.e., gold) cannot be sown; and if gold is not sown in its proper element, it cannot be quickened any more than the corn which the West Indians keep underground, in air-tight stone jars, can germinate.

Philalethes

To every seed his own body.

I. Corinthians, xv. 38

No one can ascend to that Heaven which is sought by you unless He Who came down from a Heaven which you seek shall not first enlighten. Ye seek an incorruptible Medicine, which shall not only transmute the body from corruption into a perfect mode but so preserve it continually; yet except in Heaven itself, never anywhere will you discover it.

Thomas Vaughan

. . . We marry
A gentler scion to the wildest stock,
And make conceive a bark of baser kind
By bud of nobler race: this is an art
Which does mend nature, change it rather, but
The art itself is nature.

Creation 42

Shakespeare (*Winter's Tale,* IV. iii. 92)

With his own self does God adorn the soul like gold adorned with a precious stone.

Eckhart

[1] As many alchemical citations appear in this book, it may be well to remind the reader that alchemy 'initiatically' is the spiritual perspective which objectifies the interior work in terms of chemical processes. Man is the oven, the heart the vessel, spirit the fire, and the elements of the soul the raw material which has to be transmuted through 'Mercurial' or spiritual influences into the final Stone or 'powder', capable in turn of transforming 'base metals' into gold—an operation which is the hallmark of the adept's realized state.

O truly sacred mysteries! O pure light! In the blaze of the torches I have a vision of heaven and of God. I become holy by initiation. The Lord reveals the mysteries; He marks the worshipper with His seal, gives light to guide his way, and commends him, when he has believed, to the Father's care, where he is guarded for ages to come. These are the revels of my mysteries! If thou wilt, be thyself also initiated, and thou shalt dance with angels around the unbegotten and imperishable and only true God, the Word of God joining with us in our hymn of praise.

Clement of Alexandria

Inv. passim

Doctrine and Method

Even as one desirous of reaching a longed-for city requireth the eyes for seeing the way and the feet for traversing the distance, so, also, whosoever desireth to reach the City of *Nirvâṇa* requireth the Eyes of Wisdom and the Feet of Method.

Prajñâ-Pâramitâ

Pilg. 365

If you know the Stone without the method of its preparation, your knowledge can be of no more use to you than if you knew the right method without being acquainted with the true Matter. Therefore our hearts are filled with gratitude to God for both kinds of knowledge.

The Glory of the World

Knowl. 734

If you do not know the whole operation from beginning to end, you know nothing at all.

Helvetius

Jesus (Peace be upon him!) said, 'He who acquires knowledge and does not act upon it is like a woman who practises immorality in secret, then becomes pregnant and her pregnancy becomes apparent and she is covered with shame. Thus shall God (Exalted is He!) cover with shame on the Day of Resurrection in the sight of witnesses him who does not act upon his knowledge.'

Christ in Islâm

Knowl. 745

To have heard and thought about the Doctrine and not practised it and acquired spiritual powers to assist thee at the moment of death is useless.

Gampopa

Judgment 250

A Christ not in us is the same thing as a Christ not ours. If we are only so far with Christ as to own and receive the history of His birth, person, and character, if this is all that we have of Him, we are as much without Him, as much left to ourselves, as little

Death 208

helped by Him as those evil spirits which cried out, 'We know thee who thou art, the holy one of God.'

William Law

To get a crop one must needs sow the grain with the husk on. . . . So rites and ceremonies are necessary for the growth and perpetuation of a religion. They are the receptacles that contain the kernel of truth, and consequently every man must perform them before he reaches the central truth.

Infra 300

Sri Ramakrishna

There are some incredulous who will not believe that this bread upon the altar may be changed, that God can do it. (How unworthy, to deny that God is capable of this.)

Eckhart

There are some who, regarding laws in their literal sense in the light of symbols of matters belonging to the intellect, are over punctilious about the latter, while treating the former with easy-going neglect. . . . Why, we shall be ignoring the sanctity of the Temple and a thousand other things, if we are going to pay heed to nothing except what is shown us by the inner meaning of things. Nay, we should look on all these outward observances as resembling the body, and their inner meaning as resembling the soul. It follows that, exactly as we have to take thought for the body, because it is the abode of the soul, so we must pay heed to the letter of the laws. If we keep and observe these, we shall gain a clearer conception of those things of which these are the symbols; and besides that we shall not incur the censure of the many and the charges they are sure to bring against us.

Symb. 306
Supra
Introd.

Philo

Just as you require food for the body, so also you require food for the soul in the shape of prayers, Japa (repetition of a consecrated word), Kirtan (religious reunions), meditation.

Swami Sivananda

For as there is body and soul in man, and in Scripture the letter and the sense, so in every sacrament there is the visible external which may be handled and the invisible within, which is believed and taught. The material external is the sacrament, and the invisible and spiritual is the sacrament's substance (*res*) or *virtus*. . . . The sacrament is the corporeal or material element set out sensibly, representing from its similitude, signifying from its institution, and containing from its sanctification, some invisible and spiritual grace.

Hugh of Saint-Victor

Spiritual perception (*al-manʿâ*) is very subtle; one can only retain it with the aid of the senses (*al-ḥiss*), and one can only make it last through spiritual conversation (*al-muḍakkara*), invocation (*aḍ-ḍikr*) and the breaking of natural habits.

Renun.
152a

Al-ʿArabî al-Ḥasanî ad-Darqâwî

Only the ignorant person disdains ritual practice (*wird*). Illumination (*wârid*) is found in the Final Abode, while ritual practice is discarded with the discarding of this abode; and the first thing to be attended to is that whose existence cannot be replaced. Ritual practice He asks of thee, while illumination thou askest of Him; and what does He ask of thee compared with what thou askest of Him?

Ibn ʿAṭâ'illâh

A man can reach God if he follows one path rightly. Then he can learn about all the other paths. It is like reaching the roof by some means or other. Then one is able to climb down by the wooden or stone stairs, by a bamboo pole, or even by a rope.

Sri Ramakrishna

Reality 790

He that thinks or holds that outward exercises of religion hurt or are too low for his degree of spirituality, shows plainly that his spirituality is only in idea; that it is something that is in him only as a speculation, or as something that is in his head and not in his heart.

The truly spiritual man is he that sees God in all things, that sees all things in God, that receives all things as from Him, that ascribes all things to Him, that loves and adores Him in and for all things, in all things absolutely resigned unto Him, doing them for Him from a principle of pure and perfect love of Him. There is no spiritual person but this ... and to such a one the outward institutions of religion are ten times more dear and valuable than to those that are less spiritual.... And to think that the spirituality of religion is hurt by the observance of outward institutions of religion is as absurd as to think that the inward spirit of charity is hurt by the observance of outward acts of charity....

Wr. Side 448

And I defy any man ... to show that the outward word or outward prayers and outward psalmody and outward teaching is consistent with a religion that is too spiritual to admit of outward institutions.

William Law

Image worship is very necessary for beginners.

Swami Sivananda

Symb. 321

The apostle (*Rom.* VI. 14; *Phil.* III. 8) doth not mean to disparage a real inward righteousness, and the strict observance of the law; but his meaning is to show how poor and worthless a thing all outward observances of the law are in comparison of a true internal conformity to Christ in the renovation of the mind and soul according to his image and likeness.

Supra Introd.

Metanoia 484

John Smith the Platonist

Observe the forms and rituals as set forth in the Scriptures, without losing sight of their inner spirit.

Srimad Bhagavatam, XI. V

The Eternal Religion, the religion of the rishis, has been in existence from time out of mind and will exist eternally. There exist in this Sanâtana Dharma all forms of worship—worship of God with form and worship of the Impersonal Deity as well. It contains all paths—the path of knowledge, the path of devotion, and so on. Other forms of religion, the modern cults, will remain for a few days and then disappear.

Reality 790

Sri Ramakrishna

The worshippers of the gods go to the gods; to the ancestors go the ancestor-worshippers; the spirit-worshippers go to the spirits; but My worshippers come unto Me. . . .

Action 338

Whatever thou doest, whatever thou eatest, whatever thou offerest as oblation, whatever thou givest and the austerities thou performest, O son of Kunti, do that as an offering to Me. . . .

Fill thy mind with Me, be thou My devotee, worship Me and bow down to Me; thus, steadfastly uniting thy heart with Me alone and regarding Me as thy Supreme Goal, thou shalt come unto Me.

Bhagavad-Gîtâ, IX. 25, 27, 34

Lo! We have shown him the way, whether he be grateful or disbelieving.

Qur'ân, LXXVI. 3

The Spiritual Master

If thou see a man of understanding, go to him early in the morning, and let thy foot wear the steps of his doors.

Holy War 405

Ecclesiasticus, VI. 36

The teacher is he who knows the Eternal Wisdom, the Veda, who is devoted to the All-Pervader Vishnu, who knows not arrogance, who knows the method of yoga, ever stands upon yoga and has become yoga itself; who is pure, who is devoted to his teachers and who has witnessed the Supreme Person, Purusha. He who possesses all these virtues is called a 'dispeller of darkness', a guru.

Beauty 676

The syllable 'gu' means darkness, the syllable 'ru' means dispeller; he is therefore called a 'guru' because he dispels darkness.

Holiness 924

The guru is the Supreme Cause, the guru is the ultimate destiny, the guru is transcendent sapience, the guru is the supreme resort, the guru is the final limit, the guru is the supreme wealth. Because he teaches 'That' (the Supreme Essence) the guru is most great.

Advaya Târaka Upanishad, 14–18

Beloved Vedas and Tantras handed down to us by tradition, as also Mantras and usages, become fruitful if communicated to us by the Guru, and not otherwise.

Supra 280

Kulârnava Tantra, XI

It is said even God cannot grant Moksha (Deliverance), but only the Guru.

P. State 579

Swami Ramdas

If you do not meet a transcendental teacher, you will have swallowed the Mahâyâna medicine in vain!

Chih Kung

He who works at the prayer (of Jesus) from hearsay or reading and has no instructor, works in vain.

St Gregory of Sinai

Can the water of the (polluted) stream clear out the dung? Can man's knowledge sweep away the ignorance of his sensual self?

How shall the sword fashion its own hilt? Go, entrust this wound to a surgeon.

Flies gather on every wound, so that no one sees the foulness of his wound.

Those flies are your thoughts and your possessions: your wound is the darkness of your states;

And if the Pîr (spiritual master) lays a plaster on your wound, at once the pain and lamentation are stilled,

So that you fancy it is healed, (whereas in reality) the ray of the plaster has shone upon the spot.

Beware! Do not turn your head away from the plaster, O you who are wounded in the back, but recognize that that (healing of the wound) proceeds from the ray: do not regard it as (proceeding) from your own constitution.

Rûmî

Only that knowledge which issues from the lips of the guru is alive; other forms are barren, powerless, and the cause of suffering.

Shiva Samhitâ, III. 11

Knowl. 734

Our Art is good and precious, nor can any one become a partaker of it, unless it be revealed to him by God, or unless he be taught by a skilled Master. It is a treasure such as the whole world cannot buy.

The Glory of the World

A man who reads about the doctrines of the Jôdo without receiving oral instruction will miss the thing really necessary to the attainment of Ôjô.[1]

Hônen

Inv. passim

The mistaken impression of the reality of the world is never to be effaced without the knowledge of its unreality derived from the *Shastras*,[2] and the living lips of a Teacher.

Yoga-Vasishtha

Illusion 85

Since there is no connection between the (mere) sound and the true meaning, . . . a wise man must resort to a good master, for without him the truth cannot be found even in millions of ages. . . . So good men who desire their own perfection always pay with their whole being full honour to their master, who is the bestower of infinite rewards. They

[1] Enlightenment.

[2] The Hindu scriptures.

abandon envy and malignancy, and pride and self-conceit, their determination set on enlightenment and the concept of weariness renounced, and thus they always honour their guru, master of the world, who bestows success in all things. . . . Thereby they gain by the grace of their guru and without any obstruction that truth supreme which is taught by all the Buddhas. It is eternal, resplendent and pure, the abode of the conquerors, the divine substance in all things and the source of all things. Just as a sun-stone shines brightly from the proximity of the sun-light which dispels the enclosing darkness, even so does the jewel of a pupil's mind, freed from the murkiness of impurity, light up from the proximity of a world-teacher who is bright with the fire of the practice of truth.

Holiness 921
Holy War 394
P. State 563
Sin 66

Anangavajra

Stand in the multitude of ancients that are wise, and join thyself from thy heart to their wisdom, that thou mayst hear every discourse of God, and the sayings of praise may not escape thee.

Ecclesiasticus, VI. 35

And with how many a prophet have there been a number of devoted men who fought (beside him). They quailed not for aught that befell them in the way of Allâh, nor did they weaken, nor were they brought low. Allâh loveth the steadfast.

Holy War 405
Pilg. 379
Faith 510

Qur'ân, III. 146

You should keep to one place, one master, one method, and one system of yoga. This is the way which leads to positive success.

Swami Sivananda

God and the *Gūru* are not really different: they are identical. He that has earned the Grace of the *Guru* shall undoubtedly be saved and never forsaken, just as the prey that has fallen into the tiger's jaws will never be allowed to escape. But the disciple, for his part, should unswervingly follow the path shown by the Master.

Sri Ramana Maharshi

If thou desire (spiritual) poverty, that depends on companionship (with a Shaykh): neither thy tongue nor thy hand avails.

Soul receives from soul the knowledge thereof, not by way of book nor from tongue.

(Even) if those mysteries (of spiritual poverty) are in the traveller's heart, knowledge of the mystery is not yet possessed by the traveller.

Creation 42

Rûmî

Even if your intelligence is above that of Yen-tzû and Min-tzû (two of the most clever pupils of Confucius), yet, merely with constrained conjecture, you will not be able to succeed unless you have a true teacher. For the work of *chin tan* (gold medicine) in case oral instructions are lacking, where and how can you fertilize the fecund womb?

Chang Po-tuan

If you wish not your head to be lost, be (lowly as) a foot: be under the protection of the Quṭb (spiritual Pole) who is possessed of discernment.

Though you be a king, deem not yourself above him: though you be honey, gather naught but his sugar-cane. . . .

As you have no strength, keep making a lamentation; since you are blind, take care, do not turn your head away from him that sees the road.

Humility 191

Rûmî

To attain tranquillity of spirit and avoid inconstancy, one must have initiation and the aid of a guide. The spiritual path is the most dreadful there is; it is strewn with innumerable pitfalls. Unless he is guided by an experienced hand, a man no matter how intelligent is sure to make some false moves. . . .

Supra 280 *Pilg.* 366

In this world, even to learn the art of stealing one needs a *guru*. How much greater is the necessity of a *guru* if one is to acquire the supreme knowledge of Brahman!

Swâmi Brahmânanda

It is very important for a person who wishes to 'lament'[1] to receive aid and advice from a *wichasha wakan* (holy man), so that everything is done correctly, for if things are not done in the right way, something very bad can happen, and even a serpent could come and wrap itself around the 'lamenter'.

Black Elk

(The) Guru simply helps you in the eradication of ignorance. Does he hand over Realisation to you? . . . The ego is a very powerful elephant and cannot be brought under control by anyone less than a lion, who is no other than the Guru in this instance; whose very look makes the elephant tremble and die.

Realiz. 873 Introd. *Conform.*

Sri Ramana Maharshi

It is necessary to seek a teacher who is not himself in error, to follow his instructions, and so learn to distinguish, in the matter of attention, defects and excesses of right and of left, encountered through diabolical suggestion. . . . If there is no such teacher in view, one must search for one, sparing no efforts.

Nicephorus the Solitary

It is impossible for us to study all the Śâstras and learn all that is laid down in them. . . . Further, as the defects inherent in us are numberless, it is impossible for us to find out all of them or to seek to get rid of them by ourselves. A Guru is therefore necessary to know about our spiritual equipments, to find out the stage in which we are at present, to decide what course of action will take us to the next higher stage and to teach and guide us aright. . . . When the Śâstras prescribe what courses of action have to be pursued by aspirants in the several stages, a Guru is necessary to know what the Śâstras prescribe and to teach us the particular course of action suited to our qualifications.

Sri Chandrasekhara Bhâratî Swâmigal

[1] 'Lamenting' or 'crying for a vision': spiritual retreat in an isolated spot, such as a mountain top.

There is no other way for overhauling the vicious worldly Samskaras (tendencies left from former actions) and the passionate nature of raw, worldly-minded persons than the personal contact with and service of the Guru.

Swami Sivananda

Supra Introd.

The fleshly soul is a dragon with hundredfold strength and cunning: the face of the Shaykh is the emerald that plucks out its eye.

Rûmî

Infra 296
Contem. 542
Conform. 170
Heterod. 423

Would you know the perfect Master? It is he who understands the regulation of the fire, and its degrees. Nothing will prove to you so formidable an impediment as ignorance of the regimen of heat and fire; for our whole Art may be looked upon as being concentrated in this one thing, seeing it is all important for the proper development of our substance that the degree of heat which is brought to bear on it should be neither too great nor too small. In regard to this point many learned men have gone grievously astray.

Thomas Norton

Judgment 250

I bow down to my most adorable Teacher who is all-knowing and has, by imparting Knowledge to me, saved me from the great ocean of births and deaths filled with Ignorance.

Śrî Śankarâchârya

Never should (a student) think ill of him (the teacher). For the teacher gives him a (new) birth in knowledge. And that is the highest birth. Mother and father engender his body only.

Apastamba Dharma Sutra, I. i. 15–18

The Guru is Brahmâ, the Guru is Vishnu, the Guru is Siva; the Guru is the supreme Godhead itself in the visible form; Obeisance to that Guru.

Invocation preceding Hindu prayer recitals

Knowl. 761

The words of the wise are as goads, and as nails fastened by the masters of assemblies, which are given from one shepherd.

Ecclesiastes, XII. 11

The learner ought to share in his teacher's thought; he should be quicker in his listening than the teacher is in his speaking.

Hermes

Renun. 160

The supreme mystery in the Veda's End (Vedânta),
Which has been declared in former time,
Should not be given to one not tranquil,
Nor again to one who is not a son or a pupil.

To one who has the highest devotion (*bhakti*) for God,
And for his spiritual teacher (*guru*) even as for God,
To him these matters which have been declared
Become manifest (if he be) a great soul (*mahâtman*)—
Yea, become manifest (if he be) a great soul!

Śvetâśvatara Upanishad, VI. 22–23

Beauty 676

To live and act in such a manner that the Guru is highly pleased with the disciple is a sure condition for the working of grace. . . . A true aspirant seeks always to please his Guru by moulding his life in accordance with his teachings.

Swami Ramdas

Conform. 170

Who can finde such a *Master* out,
As was my *Master*, him needeth not to doubt:
Which right nobil was and fully worthy laude,
He loved Justice, and he abhorred fraude;
He was full secrete when other men were lowde,
Loath to be knowne that hereof ought he Could;
When men disputed of Colours of the Rose,
He would not speake but keepe himselfe full close;
To whome I laboured long and many a day,
But he was solleyn to prove with straight assaye,
To search and know of my Disposition,
With manifold proofes to know my Condition:
And when he found unfeigned fidelity,
In my greate hope which yet nothing did see,
At last I conquered by grace divine
His love, which did to me incline.[1]

Thomas Norton

Holiness 902, 914

M.M. 987

Faith 505

He who fills duly both ears with the Veda is to be considered (like one's) father and mother; him one should never injure.

Mânava-dharma-śâstra, II. 144

Now the Brahman who obeys his Guru till the end of his body goes straight to the eternal abode of Brahma.

Mânava-dharma-śâstra, II. 244

O Lord (Krishna), thou art self-effulgent, the embodiment of Truth. Thou art the Atman, the innermost Self in all beings. Thou art the teacher of teachers.

Srimad Bhagavatam, XI. ii

[1] 'Shinran always spoke of himself as Hônen's disciple, and he had such an absolute confidence in Hônen as his master, that he once said, that if being deceived by Hônen meant his own dropping into hell, he would never regret it' (Coates and Ishizuka: *Hônen the Buddhist Saint*, Historical Introduction, p. 49).

He (the *brahmacârin*) should look upon his Guru as God. Verily is the Guru the embodiment of divinity. Accordingly the student must serve him and please him in every way.

Srimad Bhagavatam, XI. xi

Despite the identity of the master with the Self, the scriptures categorically declare that however learned or endowed with superhuman powers, one cannot realise the Self without the master's grace. What is the secret of this?

Creation 42

Though in reality the master is the very Self of all, the soul having by ignorance become differentiated and individualised, cannot find out its original Pure state of Being, without the master's grace. Hence the scriptural statement.

Sri Ramana Maharshi

Reality 773

Supra 275

Infra 300

For there is One alone Who hath no need of doctrine (*or* learning) that is to say, God, Who is over everything, for He existeth of Himself, and there is no other being who existed before Him. Now all rational beings are learners, because they are beings who have been made and created. The ranks of the celestial hosts who existed first of all, and the orders of beings who are the most exalted of all possess teachers in the Trinity, Who is exalted above everything. The orders of beings of the second group learn from the beings of the first group, and those which belong to the third group learn from those of the second group, which is above them, and in this manner each of the later groups learneth from that which is above it, even down to the lowest group of all; for those among them who are superior in respect of knowledge and excellence teach knowledge unto those who are inferior to them. Therefore those who imagine that they have no need of teachers,[1] and who will not be convinced by those who teach them things of good, are sick with the want of the knowledge which is the mother and the producer of pride. Now those who are princes and the foremost ones among these in respect of destruction are those who intentionally (*or* wilfully) fell from sojourning in heaven, and from the service thereof, and these are the devils who fly in the air because they forsook the heavenly Teacher and rebelled.

Palladius

Heterod. 430

The guide of those who have no spiritual guide is Satan.

Bâyazîd al-Bisṭâmî

Heterod. 423

Wherever thou seest one (that is) naked and destitute, know that he hath fled from the (spiritual) master.

Rûmî

Judgment 250

Do not delay lest you become worthless. Find a teacher as soon as possible, in order that you may acquire the principle of *hsüan* (mystery). If you do not learn the truth in your present life, how will you know in what womb you are going to be born in the next life?

[1] There are people ready to invent anything as an excuse for a teacher, even including an imagined contact with some dead person supposed to be in paradise, or the idea of an infallible voice that guides from within. But the doctrine as expounded in this chapter is unequivocal: there is no spiritual master outside of the orthodox channels, and no amount of self-interested rationalizing can bring one into being.

When you have decided to take up the subject, you must stick to it through to the end. The greatest of human events are life and death.

Chang Po-tuan

Holy War 405

If a person has a sincere fervour for the Lord and is eager to follow a spiritual path, he is sure to meet a true *guru* through the grace of the Lord. Consequently, the spiritual aspirant should not be troubled about the coming of a *guru.*

Swâmi Brahmânanda

Grace 558

If you go on working with the light available, you will meet your Master, as he himself will be seeking you.

Sri Ramana Maharshi

Metanoia 493

Is it right to receive investiture from the hands of more than one? Yes, it is right, provided that the second investiture is not accompanied with the intention of annulling the first.

Muḥammad ibnu 'l-Munawwar

Supra 280

Disciple.—Mahârâj, the Scriptures speak of service to the *guru* as a necessary means for spiritual realization. Up to what point is this true?

The Swâmi.—It is necessary in the preliminary stages. But after that it is your own spirit which plays the role of *guru.*

Swâmi Brahmânanda

Center 816

The grace of a Saint who is accepted as a Guru is essential to the progress of a spiritual aspirant.

Swami Ramdas

Love for the dervishes is the key which opens the door into Paradise and those who hate them are worthy of anathema.

'Aṭṭâr

The disease of worldliness has become chronic in man. It is mitigated, to a great extent, in holy company.

Sri Ramakrishna

The knowledge which has been learned from a teacher best helps one to attain his end.

Chândogya Upanishad, IV. ix. 2

The Just Proportioning of Ritual

Reality 790

The Path is one for all; the means to reach the Goal must vary with the Pilgrims.

Tibetan Precept

Introd. *Pilg.*

Meditation without sufficient preparation through having heard and pondered the Doctrine is apt to lead to the error of losing oneself in the darkness of unconsciousness.

Gampopa

Action 335

They are all too few who are fully ripe for gazing in God's magic mirror. Precious few succeed in living the contemplative life at all here upon earth. Many begin, but fail to consummate it. Because they have not rightly lived the life of Martha. As the eagle spurns its young that cannot gaze at the sun, even so fares it with the spiritual child.

He who would build high must lay firm and strong foundations.

Eckhart

Grace 556

God is preparing you to receive the nectar of Ananda. If you get it without the proper preparation to stand it, your mind and body will be shattered to pieces. So He is gradually preparing you and when He knows that you are ready to receive Him, then He comes to you in all His glory.

Swami Ramdas

Supra 285 *Conform.* 170

The greatest secret of our operation is no other than a cohobation of the nature of one thing above the other, until the most digested virtue be extracted out of the digested body of the crude one. But there are hereto requisite: Firstly, an exact measurement and preparation of the ingredients required; secondly, an exact fulfilment of all external conditions; thirdly, a proper regulation of the fire; fourthly, a good knowledge of the natural properties of the substances; and fifthly, patience, in order that the work may not be marred by overgreat haste.

Philalethes

Too swift arrives as tardy as too slow.

Shakespeare (*Romeo and Juliet,* II. vi. 15)

If there are mangoes on the top of a big tree you do not jump all at once to pluck them. It is impossible. You gradually climb up the tree by catching hold of the different branches and so reach the top of the tree. Even so you cannot jump all at once to the summit of the spiritual ladder. You will have to place your feet with caution on each rung.

Swami Sivananda

Heterod. 423

The same moderate course must be adopted in the fiery regimen of our Magistery. For it is all important that the liquid should not be dried up too quickly, and that the earth of the Sages should not be melted and dissolved too soon, otherwise your fishes would be changed into scorpions.

Basil Valentine

Use common sense all along in your spiritual discipline (sâdhanâ); I always insist on this point.

Swami Sivananda

All the words of my mouth are in righteousness; there is nothing froward or perverse in them.
They are all plain to him that understandeth, and right to them that find knowledge.

Proverbs, VIII. 8, 9

Knowl. 749

My words are very easy to know, and very easy to practise.
Yet all men in the world do not know them, nor do they practise them.
It is because they have knowledge that they do not know me.

Tao Te Ching, LXX

Knowl. 734

For my yoke is easy, and my burden is light.

St Matthew, XI. 30

Allâh would make the burden light for you, for man was created weak.

Qur'ân, IV. 28

No one should be depressed and give the false impression that the precepts of the Gospel are impossible or impracticable. God who has predestined the salvation of man has, of course, not laid commandments upon him with the intention of making him an offender because of their impracticability. No; but so that by their holiness and the necessity of them for a virtuous life they may be a blessing to us, as in this life so in eternity.

St John Chrysostom

Faith 509

The sabbath was made for man, and not man for the sabbath:
Therefore the son of man is Lord also of the sabbath.

St Mark, II. 27, 28

And be assured also that, when the Spirit of Christ is the Spirit that ruleth in you, there will be no hard sayings in the Gospel.

William Law

Suff. 130

The whole work of Philosophy,
Which to most men appears impossible,
. . . is a convenient and easy task.
If we were to shew it to the outer world
We should be derided by men, women, and children.

Abraham Lambspring

Contem. 542

Religion is no sullen Stoicism, no sour Pharisaism; it does not consist in a few melancholy passions, in some dejected looks or depressions of mind: but it consists in freedom, love, peace, life, and power; the more it comes to be digested into our lives, the more sweet and lovely we shall find it to be. Those spots and wrinkles which corrupt minds think they see in the face of religion, are indeed nowhere else but in their own deformed and misshapen apprehensions. It is no wonder when a defiled fancy comes to be the glass, if you have an unlovely reflection.

Wr. Side 448

John Smith the Platonist

In this first section of our work, nothing is to be done without hard and persevering toil; though it is quite true that afterwards the substance develops under the influence of gentle heat without any imposition of hands.

Philalethes

Our furnace is cheap, our fire is cheap, and our material is cheap—and he who has the material will also find a furnace in which to prepare it, just as he who has flour will not be at a loss for an oven in which it may be baked.

Supra 280

Basil Valentine

Let no one deem it difficult to arrive at this however hard may seem, and be, indeed, to start with, the parting from and dying to all things. Having once got into it no life is more easy, more delightful or more lovely. God is so very careful to be always with a man to guide him to himself in case of his taking the wrong way. No man ever wanted anything so much as God wants to make the soul aware of him. God is ever ready, but we are so unready. God is near to us, but we are far from him. God is in, we are out; God is at home, we are strangers.

Love 618
Center 841

Eckhart

To be capable of knowing God, and to wish and hope to know him, is the road which leads straight to the Good; and it is an easy road to travel. Everywhere God will come to meet you, everywhere he will appear to you, at places and times at which you look not for it, in your waking hours and in your sleep, when you are journeying by water and by land, in the night-time and in the day-time, when you are speaking and when you are silent; for there is nothing which is not God.

Inv. 1036
Reality 803

Hermes

Holiness 902

When we have no aim but God, nothing can part us from Him, or lead us astray. . . .

The vine-dresser loves to strip off the leaves, that thus the sun may have nothing to hinder its rays from pouring on the grapes. In like manner do all means of grace fall away from this (God-loving) man, such as images of the saints, teachings, holy exercises, set prayers, and the like. Yet let none cast these things aside before they fall away of themselves through divine grace: that is to say, when a man is drawn up above all that he can comprehend.

Tauler

'If a man is so occupied with recollecting Me that he forgets to pray to Me, I grant him a nobler gift than that which I accord to those who petition Me.'

Muhammad *(ḥadîth qudsî)*

Manasika Pooja (mental worship) is more powerful than external Pooja with flowers, etc. Arjuna thought that Bhima was not doing any kind of worship. He was proud of his external worship of Lord Siva. He offered plenty of Bael leaves. But Bhima offered to the Lord mentally the Bael leaves of all the Bael trees of the whole world. He was doing Manasika Pooja of Lord Siva. . . .

Manasika Pooja can be done by advanced students. Beginners should certainly do worship with flowers, sandal paste, incense, etc.

Swami Sivananda

The thoughtless man, even if he can recite a large portion of the law, but is not a doer of it, has no share in the priesthood, but is like a cowherd counting the cows of others.

Knowl. 745

The follower of the law, even if he can recite only a small portion of the law, but, having forsaken passion and hatred and foolishness, possesses true knowledge and serenity of mind, he, caring for nothing in this world or that to come, has indeed a share in the priesthood.

Contem. 536

Dhammapada, I. 19, 20

Religion is a matter of experience. Merely by becoming a member of a church, creed or sect, a person cannot be entitled to this experience. By reading any amount of scriptures and sacred books he cannot obtain this experience. By the observance of rites, ceremonies, or worship a man cannot come by this experience. Spiritual realisation is a question of individual effort and struggle. . . . The man of true religion when he is on the path, is mainly concerned with his own internal struggle for liberation and peace.

Introd. *Conform.*

Swami Ramdas

Giving up all Dharmas (righteous and unrighteous actions), come unto Me alone for refuge. I shall free thee from all sins; grieve not.

Inv. 1017 1009

This should never be spoken by thee to one who is devoid of austerity or without devotion, nor to one who does not render service, nor to one who speaks ill of Me.

Bhagavad-Gîtâ, XVIII. 66–67

When a flood comes from the ocean, all the land is deep under water. Before the flood, the boat could have reached the ocean only by following the winding course of the river. But after the flood, one can row straight to the ocean. One need not take a roundabout course.

Sri Ramakrishna

A ship has no need of oars when the wind swells the sails, for then the wind gives it sufficient power easily to navigate the salt sea of passions. But when the wind dies and the ship stops, it has to be set in motion by oars or by a towboat.

Introd. *Pilg.*

St Gregory of Sinai

Abandon right and a fortiori wrong; one who has reached the farther shore has no more need of rafts.

M.M. 978

Majjhima-nikâya, I. 135

If I intend to cross the sea and want a ship, that is part and parcel of the wishing to be over and having gotten to the other side I do not want the ship.

Eckhart

Let him no longer use the Law as a means of arrival when he has arrived.

St Augustine

When the fish is caught we pay no more attention to the trap.

Huang Po

Holiness 902

The law is therefore read, because we have not yet come to that Wisdom, which fills the hearts and minds of those who gaze thereon.

St Augustine

Knowl. 761

At the ultimate point, beyond which you can go no further,
You get to where there are no rules, no standards,
To where thought can accept Impartiality,
To where effect of action ceases,
Doubt is washed away, belief has no obstacle.

Seng-tsʽan

Introd. *Wr. Side*

Some say further, that we can and ought to get beyond all virtue, all custom and order, all law, precepts and seemliness, so that all these should be laid aside, thrown off and set at nought. Herein there is some truth, and some falsehood. Behold and mark: Christ was greater than His own life, and above all virtue, custom, ordinances and the like, and so also is the Evil Spirit above them, but with a difference.

Theologia Germanica, XXX

How long should a devotee perform daily devotions such as the sandhyâ? As long as his hair does not stand on end and his eyes do not shed tears at the name of God.

Sri Ramakrishna

The Necessity of Submission

Metanoia 484

The beginning of her (Wisdom) is the most true desire of discipline.

Wisdom, VI. 18

All such (Pythagorean) precepts as define what is to be done, or what is not to be done, refer to divinity as their end.

Iamblichus

Know that the main foundation of piety is this, to have right opinions and apprehensions of God, *viz*. That he is, and that he governs all things.

Epictetus

Knowl. 761

Verily, that which is Law (*dharma*) is truth. Therefore they say of a man who speaks the truth, 'He speaks the Law,' or of a man who speaks the Law, 'He speaks the truth.' Verily, both these are the same thing.

Bṛihad-Âraṇyaka Upanishad, I. iv. 14

Supra Introd.

Respect the bishop as a type of God, and the presbyters as the council of God and the college of the Apostles. Apart from these there is not even the name of a Church.

St Ignatius of Antioch

With all thy soul fear the Lord, and reverence his priests.

Ecclesiasticus, VII. 31

Judgment 239

The scribes and the Pharisees sit in Moses' seat: All therefore whatsoever they bid you observe, that observe and do; but do not ye after their works: for they say, and do not.

St Matthew, XXIII. 2, 3

Knowl. 745

They who know truth in truth, and untruth in untruth, arrive at truth, and follow true desires.

Dhammapada, I. 12

Sin 80

By the heaven full of paths!
Ye, verily, are of dissenting opinion.
He is made to turn away from it (the truth) who is himself averse.

Qur'ân, LI. 7–9

Reality 790

Men must have bounds how farre to walke; for we
Are made farre worse, by lawless liberty.

Robert Herrick

For God has not created man to be his own lord, but his servant: He will have angels under obedience, and not devils in their own fire-might.

Boehme

Conform. 180

He that refuseth instruction despiseth his own soul: but he that heareth reproof getteth understanding.

Proverbs, XV. 32

Suff. 128

I will meditate in thy precepts, and have respect unto thy ways.
I will delight myself in thy statutes: I will not forget thy word.

Psalm CXIX. 15, 16

SYMBOLISM

That which is below is as that which is above, and that which is above is as that which is below.

From the *Emerald Tablet* of Hermes Trismegistus[1]

Reality 775 803

Metaphysic presents as a first principle the oneness, unity, or non-duality of Being, and this principle has as its direct corollary the relativity of all states of existence apart from Pure Being. Relativity means interdependence, and this implies a causal sequence linking together the indefinity of created states. Symbolism is the language which renders intelligible, often with geometric formality and precision, this causal sequence. Much of the exposition of Coomaraswamy and Guénon is really an unveiling of traditional symbols, a great part of whose content, universality and metaphysical signification has been obscured or lost in the West since the Renaissance. Yet, 'it is inasmuch as he "knows immortal things by the mortal" that the man as a veritable person is distinguished from the human animal, who knows only the things as they are in themselves and is guided only by this estimative knowledge' (Coomaraswamy: 'The Christian and Oriental, or True, Philosophy of Art'; in *Why Exhibit Works of Art?*, p. 50).

Introd. *Reality*

'All that exists, whatever its modality, necessarily participates in universal principles, nor does anything exist except by participation in these principles, which are the eternal and immutable essences contained in the permanent actuality of the Divine Intellect. Consequently one can say that all things, however contingent they are in themselves, translate or represent these principles in their manner and according to their order of existence, for otherwise they would purely and simply be nothing. Thus, from one order to another, all things are linked together in correspondence contributing to the total and universal harmony, harmony itself . . . being nothing other than the reflection of principial unity in the multiplicity of the manifested world; and it is this correspondence which forms the real foundation of symbolism' (Guénon: *Autorité spirituelle et Pouvoir temporel*, p. 22).

Beauty 689

'Initiatic symbolism by its very nature defies reduction to more or less narrowly systematic formulas, such as those in which profane philosophy delights. It is the function of symbols to be the support for conceptions where the possibilities of extension are truly unlimited, while all expression is itself but a symbol; therefore one must always make allowance for that part which is inexpressible, and which in the realm of pure metaphysic is precisely that which most matters' (Guénon: *L'Ésotérisme de Dante*, pp. 69–70).

M.M. 998

[1] These words form part of an alchemical document which can be traced through the Middle Ages to Arabic sources such as Jabir, who claimed he was quoting from Apollonius of Tyana; also found in the Greco-Egyptian Leyden Papyrus (c. 300 A.D.) at Thebes. Cf. Holmyard: *Alchemy*, and Seligmann: *The Mirror of Magic*.

Introd. *Orthod.*

'Rite and symbol . . . are closely linked in their very nature. In fact, every rite necessarily includes a symbolic meaning in all its constituent elements, and inversely, every symbol produces . . . on the person who meditates upon it with the aptitudes and dispositions required, effects which are strictly comparable to those of rites properly speaking. . . . One could also say that rites are symbols "put into action", and that every ritual gesture is a symbol "acted out". . . . Basically rite and symbol are only two aspects of a single reality; and this reality in the last analysis is nothing other than the correspondence which links together all the degrees of universal Existence, in such manner that through this correspondence our human state can be put into communication with the higher states of being' (Guénon: *Aperçus sur l'Initiation,* ch. XVI). In the words of Burckhardt, 'Universal Man is himself the total symbol of God' (*De l'Homme universel,* p. 4).

Orthod. 275 *Holiness* 924

'Every thing is a symbol which serves as a direct support for spiritual realization, a *mantra*[1] for example, or a divine name, or in a secondary manner, a graphic, pictorial or sculptural symbol, such as sacred images (*pratīkas*)' (Schuon: *Perspectives spirituelles,* p. 135).

'The Byzantine, Roman and early Gothic arts are theologies: they express God, or rather, they "realize" Him on a certain plane' (id., p. 46). It will be useful to study a little further the theory of the icon to throw light on the function and meaning of symbols and images in general. The following passages are cited from *Byzantine Mosaic Decoration,* by Otto Demus (London, Kegan Paul, 1947–8): 'The relation between the prototype and its image, argued Theodore of Studium and John of Damascus, is analogous to that between God the Father and Christ His Son. The Prototype, in accordance with Neoplatonic ideas, is thought of as producing its image of necessity, as a shadow is cast by a material object, in the same way as the Father produces the Son and the whole hierarchy of the invisible and the visible world. Thus the world itself becomes an uninterrupted series of "images" which includes in descending order from Christ, the image of God, the *Proorismoi* (the Neoplatonic "ideas"), man, symbolic objects and, finally, the images of the painter, all emanating of necessity from their various prototypes and through them from the Archetype, God. This process of emanation imparts to the image something of the sanctity of the archetype: the image, although differing from its prototype *κατ' οὐσίαν* (according to its essence), is nevertheless identical with it *καθ' ὑπόστασιν* (according to its meaning), and the worship accorded to the image (*προσκύνησις τιμητική*) is passed on through the image to its prototype. . . . . A painted representation of Christ is as truly a symbolic reproduction of the Incarnation as the Holy Liturgy is a reproduction of the Passion, . . . and the artist who conceives and creates an image conforming to certain rules is exercising a function similar to that of the priest.

'Three main ideas of paramount importance for the whole subsequent history of Byzantine art emerge from this reasoning on the doctrine of images. First, the picture, if created in the "right manner", is a magical counterpart of the prototype, and has a magical identity with it;[2] second, the representation of a holy person is worthy of veneration; thirdly, every image has its place in a continuous hierarchy.

'To achieve its magical identity with the prototype, the image must possess "similarity"

[1] Sacred formula.

[2] The term 'magical' can be misleading; the Byzantine artist would consider the identity as 'spiritual' or 'real'.

(ταυτότης τῆς ὁμοιώσεως). It must depict the characteristic features of a holy person or a sacred event in accordance with authentic sources. The sources were either images of supernatural origin (ἀχειροποίητα), contemporary portraits or descriptions, or, in the case of scenic representations, the Holy Scriptures..... If this was done according to the rules the "magical identity" was established, and the beholder found himself face to face with the holy persons or the sacred events themselves through the medium of the image. He was confronted with the prototypes, he conversed with the holy persons, and himself entered the holy places, Bethlehem, Jerusalem or Golgotha' (pp. 5–7).

The author explains how the sacramental character of the icon normally demands that the sacred subjects be facing the spectator, while for the same reason the evil characters (Judas at the Last Supper, for example, or Satan at the Temptation) are represented in profile to prevent direct contact with the satanic power[1] emanating therefrom. For action, and for secondary figures, the three-quarter view is often used to combine the uses of both frontality and profile. The curved surfaces of the niches and domes allow for the dramatic sequence necessary in the large mosaic works, even while the characters remain facing the audience; and the gold background supplies the sense of depth, infinity and sanctity, at the same time isolating and centering each character as an object of veneration in itself. The material even had to be worthy of the subject matter: 'Mosaic, with its gemlike character and its profusion of gold, must have appeared, together with enamel, as the substance most worthy of becoming the vehicle of divine ideas. It is partly for this reason that mosaic played so important a part in the evolution of post-Iconoclastic painting, and indeed actually dominated it. It allowed of pure and radiant colours whose substance had gone through the purifying element of fire and which seemed most apt to represent the unearthly splendour of the divine prototypes' (p. 10).

P. State 573

'The Byzantine church itself is the "picture-space" of the icons. It is the ideal iconostasis; it is itself, as a whole, an icon giving reality to the conception of the divine world order.... (It is) an image of the Kosmos, symbolizing heaven, paradise (or the Holy Land) and the terrestrial world in an ordered hierarchy.... The building is conceived as the image of (and so as magically identical with) the places sanctified by Christ's earthly life. This affords the possibility of very detailed topographical hermeneutics, by means of which every part of the church is identified with some place in the Holy Land. The faithful who gaze at the cycle of images can make a symbolic pilgrimage to the Holy Land by simply contemplating the images in their local church.... The church is an "image" of the festival cycle as laid down in the liturgy, and the icons are arranged in accordance with the liturgical sequence of the ecclesiastical festivals' (pp. 13, 15, 16).[2]

Pilg. 387

Schuon shows the homogeneity between the artist's attitude and his work: 'In the early Church, and in the Eastern Churches up to our time, the painters of icons prepared

[1] An 'idol', precisely, is nothing other than an image in which malefic rather than benefic influences are polarized.

[2] 'The plan of the temple, which epitomizes the entire cosmos, proceeds from the spatial fixation of the celestial rhythms which rule the whole of the visible world' (Burckhardt, *Études Traditionnelles,* 1953, p. 237).

'The word *Cosmos* was, according to tradition, credited to Pythagoras, and meant originally "order", and this order is perceived as harmony, as consonance between ourselves and the Universe. This idea was developed as the correspondence between the Macrocosmos (the World) and the Microcosmos, or Man, with sometimes the Temple as link, as "proportional mean" between the two' (Matila Ghyka: *The Geometry of Art and Life,* New York, Sheed and Ward, 1946, p. 112).

themselves for the work with fasting, prayer and sacraments; they added their humble and pious inspirations to the inspiration which had set the immutable standard for the image; they scrupulously respected the symbolism (capable of an infinity of precious nuances) of forms and colours. They found creative joy, not in inventing pretentious novelties, but in recreating with love the prototypes revealed, all of which meant a spiritual and artistic perfection such as no individual genius could ever attain' (op. cit., pp. 46–47).

Action
338

'Our customary horror of all "symbolic" explanations of works of art, apart from the fact that we are no longer interested in the intangibles to which the symbols refer, arises from the fact that symbolic analysis has so often been undertaken by amateurs and "interpreted" rather fancifully than knowingly. Then, again, we have in mind the romantic vagaries of the modern symbolists, with whose *symbolisme qui cherche* our traditional *lingua franca,* that of *le symbolisme qui sait,* has very little in common. A language that can be described as a "calculus" and as "precise", demands to be studied by methods no less disciplined than those of the philologist' (Coomaraswamy: 'The Iconography of Dürer's "Knots" and Leonardo's "Concatenation" ', *The Art Quarterly,* Detroit, Vol. VII, no. 2, 1944, p. 125).[1] 'It is only when we realize that the arts and philosophies of our remote ancestors were "fully developed", and that we are dealing with the relics of an ancient *wisdom,* as valid now as it ever was, that the thought of the earliest thinkers will become intelligible to *us.* We shall only be able to understand the astounding uniformity of the folklore motives all over the world, and the devoted care that has everywhere been taken to ensure their correct transmission, if we approach these *mysteries* (for they are nothing less) in the spirit in which they have been transmitted "from the Stone Age until now",—with the confidence of little children, indeed, but not the childish self-confidence of those who hold that wisdom was born with themselves. The true folklorist must be, not so much a psychologist as a theologian and a metaphysician, if he is to "understand his material" ' (Coomaraswamy: 'On the Loathly Bride'; *Speculum,* Cambridge, Mass., Vol. XX, no. 4, 1945, pp. 403–404).

Introds.
Pilg.
Wr. Side

Introd.
Creation

'We are not, then, "reading meanings into" primitive works of art when we discuss their formal principles and final causes, treating them as symbols and supports of contemplation rather than as objects of a purely material utility, but simply *reading their meaning*' (Coomaraswamy: 'The Symbolism of the Dome', *Indian Historical Quarterly,* March, 1938, p. 32).

[1] This horror also arises from the practice, prevalent in our times, particularly in occultist circles, of making a cult out of symbols. Coomaraswamy reminds us that 'it is the elements of integrity, harmony, and lucidity in things that are called the "traces"—*vestigia*—of God in the world' (*Elements of Buddhist Iconography,* Harvard Univ. Press, 1935, p. 90, note 146). A symbol cut off from its spiritual source, and deprived of the flavour of the next world, becomes but a residue with the aroma of psychic decay or stench of idolatry.

The Divine Imprint in Manifestation

M.M. 978 — *Judgment* 250

Whatever is here, that is there.
What is there, that again is here.
He obtains death after death
Who seems to see a difference here.

Kaṭha Upanishad, IV. 10

Yonder world is in the likeness of this world, this world is the likeness of that.

Aitareya Brâhmaṇa, VIII. 2

The invisible things of Him from the creation of the world are clearly seen, being understood by the things that are made, even His eternal power and Godhead.

Romans, I. 20

We come to this learning by analogies.

Plotinus

There is on earth no (radical) diversity.
He gets death after death,
Who perceives here seeming diversity.
As a unity only is It to be looked upon—
This indemonstrable, enduring Being.

Introd.

Bṛihad-Âraṇyaka Upanishad, IV. iv. 19–20

The Eternal Father of All Things, being not less wise in the ordering, than powerful in the creation, of the world, has made the whole Universe to cohere by means of secret influences and mutual subjection and obedience, things below being analogous to things above, and *vice versa*; so that both ends of the world are nevertheless united by a real bond of natural cohesion.

The All-Wise Doorkeeper

Orthod. 285

You cannot omit the outward if you wish to know the inward. The inward is reflected in the outward world.

Ananda Moyî

Reality 775

God made this (terrestrial) world in the image of the world above; thus, all which is found above has its analogy below . . . and everything constitutes a unity.

Zohar

Were there no relation between the two worlds, no inter-connexion at all, then all upward progress would be inconceivable from one to the other. Therefore, the divine mercy gave to the World Visible a correspondence with the World of the Realm Supernal, and for this reason there is not a single thing in this world of sense that is not a symbol of something in yonder one.[1] It may well hap that some one thing in this world may symbolize several things in the World of the Realm Supernal, and equally well that some one thing in the latter may have several symbols in the World Visible. We call a thing typical or symbolic when it resembles and corresponds to its antitype under some aspect.

Infra 313

Al-Ghazâlî

Earth contains all things, in an earthly manner, which Heaven comprehends celestially.

Proclus

Our aim is not to teach how to make gold, but something far more exalted: namely, how Nature can be seen and recognized as proceeding from God, and God seen in Nature.

Georg von Welling

The divinest and the highest of the things perceived by the eyes of the body or the mind are but the symbolic language of things subordinate to Him who Himself transcendeth them all.

Dionysius

Allâh citeth symbols for men in order that they may remember.

Qur'ân, XIV. 25

Knowl. 755

From Brahmâ to a blade of grass all things are my *Gurus.*

Kaula Tantric Precept

Verily Allâh disdaineth not to coin the similitude even of a gnat.

Qur'ân, II. 26

There is no power in Nature like to that of *Similitude.* Everything draws and attracts its like to it.

Peter Sterry

All whatever is spoken, written, or taught of God, without the knowledge of the signature is dumb and void of understanding; for it proceeds only from an historical conjecture, from the mouth of another, wherein the spirit without knowledge is dumb; but if the spirit opens to him the *signature,* then he understands the speech of another. . . . For though I see one to speak, teach, preach, and write of God, and though I hear and read the same, yet this is not sufficient for me to understand him; but if his sound and spirit out of his signature and similitude enter into my own similitude, and imprint his similitude into

[1] This is one way of putting it. Metaphysically it would be more exact to say that the visible world of necessity reflects the supernal, being but an emanation from it.

Knowl. 749 mine, then I may understand him really and fundamentally, be it either spoken or written, if he has the hammer that can strike my bell.

Reality 775 By this we know, that all human properties proceed from one; that they all have but one only root and mother; otherwise one man could not understand another. . . .

Holiness 924 Man has indeed all the forms of all the three worlds lying in him; for he is a complete image of God, or of the Being of all beings.

Boehme

I determined to use (my time) for the study and investigation of those natural secrets by which God has shadowed out eternal things.

Basil Valentine

The power of a thing or an act is in the meaning and the understanding.

Black Elk

Sin 66 There is a lamentable departure of divinity from man, when nothing worthy of heaven, or celestial concerns, is heard or believed, and when every divine voice is by a *necessary* silence dumb.

Hermes

I have spoken about Mercury, Sulphur, the vessel, their treatment, etc., etc.; and, of course, all these things are to be understood with a grain of salt. You must understand that in the preceding chapters I have spoken metaphorically; if you take my words in a literal sense, you will reap no harvest except your outlay.

Philalethes

Our dull understanding can only grasp the truth by means of material representations.

Abbé Suger of Saint-Denis

In symbols there is a meaning that words cannot define.

Ibn al-Fârid

In this state we are not able to behold the truth in its own native beauty and lustre; but while we are vailed with mortality, truth must vail itself too, that it may the more freely converse with us.

John Smith the Platonist

The mode of teaching through symbols, was considered by Pythagoras as most necessary. For this form of erudition was cultivated by nearly all the Greeks, as being most ancient. But it was transcendently honored by the Egyptians, and adopted by them in the most diversified manner. Conformably to this, therefore, it will be found, that great attention was paid to it by Pythagoras, if any one clearly unfolds the significations and arcane conceptions of the Pythagoric symbols, and thus develops the great rectitude and truth they contain, and liberates them from their enigmatic form. For they are adapted according to a simple and uniform doctrine, to the great geniuses of these philosophers, and deify in a manner which surpasses human conception.

Iamblichus

Philosophers speak many things by similitude.

Geber

All names and attributes are metaphoric with us but not with Him.

Isaac ibn Latif

Beauty 663

Things here are signs; they show therefore to the wiser teachers how the supreme God is known; the instructed priest reading the sign may enter the holy place and make real the vision of the inaccessible.

Plotinus

Sacramentum means a sign, and anyone who rests content merely with the sign will never get to the interior truth. But the seven sacred rites all point us to the unique reality. Marriage, for example, is a symbol of divine and human nature, an earnest of the union of the soul with God.

Eckhart

Love 625

Unable to grasp God's essence, we seek help in words, in names, in animal forms, in figures . . . in trees and flowers, summits and sources.

Maximus of Tyre

Introd. *Center*

Ponder here and now on His qualities,
That you may behold Him Himself to-morrow.

Shabistarî

All souls do not easily recall the things of the other world; . . . they are seen through a glass dimly; and there are few who, going to the images, behold in them the realities, and these only with difficulty.

Plato (*Phaedrus,* 250)

For now we see through a glass, darkly; but then face to face: now I know in part; but then shall I know even as also I am known.

I. Corinthians, XIII. 12

Wherever you turn, by certain traces which wisdom has impressed on her works, she speaks to you, and recalls you within, gliding back into interior things by the very forms of exterior things.

St Augustine

Illusion 109

The whole world is but a glass, full of lights representing the divine wisdom.

St Bonaventura

The outward world is but a glass, or representation of the inward; and every thing and variety of things in temporal nature must have its root, or hidden cause, in something that is more inward.

William Law

P. State 583

This world is verily an outer court of the Eternal, or of Eternity, and specially whatever in Time, or any temporal things or creatures, manifesteth or remindeth us of God or Eternity; for the creatures are a guide and a path unto God and Eternity.

Theologia Germanica, L

Introd. *Orthod.* *Reality* 775 790 *Conclusion*

All our greatest philosophers and theologians unanimously assert that the visible universe is a faithful reflection of the invisible, and that from creatures we can rise to a knowledge of the Creator, 'in a mirror and in a dark manner', as it were. The fundamental reason for the use of symbolism in the study of spiritual things (is that) we know for a fact that all things stand in some sort of relation to one another; that, in virtue of this inter-relation, all the individuals constitute one universe and that in the one Absolute the multiplicity of beings is unity itself. Every image is an approximate reproduction of the exemplar; yet, apart from the Absolute image or the Exemplar itself in unity of nature, no image will so faithfully or precisely reproduce the exemplar, as to rule out the possibility of an infinity of more faithful and precise images. . . .

Faith 505 *Orthod.* 275

When we use an image and try to reach analogically what is as yet unknown, there must be no doubt at all about the image; for it is only by way of postulates and things certain that we can arrive at the unknown. . . . Following in the way of the Ancients, we are in complete agreement with them in saying that, since there is no other approach to a knowledge of things divine than that of symbols, we cannot do better than use mathematical signs on account of their indestructible certitude.

Nicholas of Cusa

Holiness 924 *Sin* 66 *Action* 346 *Pilg.* 366 *Death* 208 *Center* 828

The practice of this (spagyric) Art enables us to understand, not merely the marvels of Nature, but the nature of God Himself, in all its unspeakable glory. It shadows forth, in a wonderful manner, how man is the image of the most Holy Trinity, the essence of the Holy Trinity, and the Oneness of Substances in that Trinity, as well as the difference of Persons; the Incarnation of the Second Person of the Holy Trinity, His Nativity, Passion, Death, and Resurrection; His Exaltation and the Eternal Happiness won by Him for us men; also our purification from original sin, in the absence of which purification all good actions of men would be vain and void—and, in brief, all the articles of the Christian faith, and the reason why man must pass through much tribulation and anguish, and fall a prey to death, before he can rise again to a new life. All this we see in our Art as it were in a mirror.

The Sophic Hydrolith

Reality 803

Here take note that thou canst contemplate God in His creatures, which He has made out of nothing, whereby thou art able to discover His omnipotence. But when thou seest and considerest how admirably the creatures are fashioned and put together, and in what wonderful order they are arranged, thou art able to perceive and trace the Wisdom of God, which is ascribed to the Son.

Tauler

You must understand that all creatures are by nature endeavouring to be like God. The heavens would not revolve unless they followed on the track of God or of his likeness. If God were not in all things, nature would stop dead, not working and not wanting; for whether thou like it or no, whether thou know it or not, nature fundamentally is seeking, though obscurely, and tending towards God. No man in his extremity of thirst but would refuse the proffered draught in which there was no God. Nature's quarry is not meat or drink nor clothes nor comfort nor any things at all wherein is naught of God, but covertly she seeks and ever more hotly she pursues the trail of God therein.

Eckhart

Illusion 109
Charity 608
Holy War 403

Mankind acts for the best when it follows in the footprints of heaven, as far as its distinctive nature permits.

Dante (*De Monarchia,* I. ix)

This Self is the trace (*padanîya*) of this All, for by It one knows this All. Just as, verily, one might find by a footprint (*pada*)....

Bṛihad-Âraṇyaka Upanishad, I. iv. 7

M.M. 986

If you do not know the way, seek where his footprints are.

Rûmî

God made the universe and all the creatures contained therein, as so many glasses wherein he might reflect his own glory: he hath copied forth himself in the creation; and in this outward world we may read the lovely characters of the divine goodness, power and wisdom. In some creatures there are darker representations of God, there are the prints and footsteps of God; but in others there are clearer and fuller representations of the Divinity, the face and image of God; according to that known saying of the schoolmen, *Remotiores similitudines creaturae ad Deum dicuntur vestigium; propinquiores verò imago.* But how to find God here, and feelingly to converse with him, and being affected with the sense of the divine glory shining out upon the creation, how to pass out of the sensible world into the intellectual, is not so effectually taught by that philosophy which professed it most, as by true religion: that which knits and unites God and the soul together,[1] can best teach it how to ascend and descend upon those golden links that unite, as it were, the world to God. That divine wisdom that contrived and beautified this glorious structure, can best explain her own art, and carry up the soul back again in these reflected beams to him who is the fountain of them.

John Smith the Platonist

Creation 48
Knowl. 745
Realiz. 887
Flight 949
Knowl. 761

The heavens declare the glory of God; and the firmament sheweth his handywork.

Psalm XIX. 1

All that is in the heavens and the earth glorifieth Allâh; and He is the Mighty, the Wise.

Qur'ân, LVII. 1

[1] This expression exactly describes the sense of the words *yoga* and *religio.*

You will be led to the knowledge of the internal things which are invisible to you, by the external things which you see before you. . . . . Even so then, we can represent to ourselves in thought the Author of all that is, by contemplating and admiring the (visible) things which He has made, and ever brings into being.

Hermes

Indeed God hath copied out himself in all created being, having no other pattern to frame any thing by but his own essence; so that all created being is *umbratilis similitudo entis increati,* and is, by some stamp or other of God upon it, at least remotely allied to him.

Reality 803

John Smith the Platonist

Beauty 670

What is inclosed in the sanctuary of the arcana
Is disclosed in the testimony of phenomena.

Ibn ʿAṭâʾillâh

And God said, Let there be lights in the firmament of the heaven to divide the day from the night; and let them be for signs, and for seasons, and for days, and years.

Genesis, I. 14

He it is Who appointed the sun a splendour and the moon a light, and measured for her stages, that ye might know the number of the years, and the reckoning. Allâh created not all that save in truth. He detaileth the revelations for people who have knowledge.

Lo! in the difference of day and night and all that Allâh hath created in the heavens and the earth are portents, verily, for the God-fearing.[1]

Qur'ân, x. 5, 6

Study the beneficent sciences of medicine and astrology, and the profound art of omens.

Gampopa

All things are ultimately traceable to supernatural causes, but nevertheless are, in this present state of the world, subject to natural conditions.

Basil Valentine

Damascene says that *the various planets produce in us various temperaments, habits and dispositions.*[2] Consequently, the heavenly bodies contribute indirectly to the goodness of our understanding. Thus, even as physicians are able to judge of a man's intellect from his bodily temperament, as a proximate disposition thereto, so too can an astrologer, from the heavenly movements, as being a remote cause of this disposition. In this sense we can approve of the saying of Ptolemy: *When Mercury is in one of Saturn's regions at the time of a man's birth, and he is waxing, he bestows on him a quick intelligence of the inner nature of things.*[3]

Action 329

St Thomas Aquinas

[1] Traditional astrology used scriptural evidence such as that cited above for its point of departure.
[2] *De Fide Orth.,* II. 7.
[3] *Centiloquium,* verbum 38.

The soul of the wise man assists the work of the stars.

Ptolemy

The Sages have seen the river in which Aeneas was cleansed of his mortality—the river of Pactolus in Lydia which was changed into gold by King Midas bathing in it—the bath of Diana—the spring of Narcissus—the blood of Adonis trickling upon the snowy breast of Venus, whence was produced the anemone—the blood of Ajax, from which sprang the beautiful hyacinth flower—the blood of the Giants killed by Jupiter's thunderbolt—the tears which Althea shed when she doffed her golden robes—the magic water of Medea, out of which grass and flowers sprang forth—the Potion which Medea prepared from various herbs for the rejuvenescence of old Jason—the Medicine of Aesculapius—the magic juice, by the aid of which Jason obtained the Golden Fleece—the garden of the Hesperides, where the trees bear golden apples in rich abundance—Atalanta turned aside from the race by the three golden apples—Romulus transformed by Jupiter into a god—the transfiguration of the soul of Julius Caesar into a Comet—Juno's serpent, Pytho, born of decomposed earth after Deucalion's flood—the fire at which Medea lit her seven torches—the Moon kindled by Phaëthon's conflagration—Arcadia, in which Jupiter was wont to walk abroad—the habitation of Pluto in whose vestibule lay the three-headed Cerberus—the Pile, on which Hercules burnt those limbs which he had received from his mother, with fire, till only the fixed and incombustible elements derived from his father were left, and he became a god—and the rustic cottage whose roof was made of pure gold. Blessed, yea, thrice blessed, is the man to whom Jehovah has revealed the method of preparing that Divine Salt by which the metallic or mineral body is corrupted, destroyed, and mortified, while its soul in the meantime is revived for the glorious resurrection of the philosophical body—blessed, I say, is he to whom the knowledge of our Art is vouchsafed in answer to prayer throughout all his work for the Holy Spirit! For it should be remembered that this is the only way in which our Art of Arts is vouchsafed to man, and if you would attain it, the service of God ought to be your chief business.

Helvetius

Supra Introd.

Suff. 124

P. State passim

Pilg. 366

Death 208

Introd. *Orthod.*

Introd.

We shall show them Our signs on the horizons and within themselves until it becomes clear to them that it is the Truth.

Qur'ân, XLI. 53

The Manifold Content of Symbols

We should know that books can be understood, and ought to be explained, in four principal senses. One is called *literal,* ... the second is called *allegorical,* ... the third sense is called *moral,* ... the fourth sense is called *anagogical.* ... And in such demonstration, the literal sense should always come first, as that whose meaning includes all the rest, and without which it would be impossible and irrational to understand the others; and above all would it be impossible with the allegorical. Because in everything which has an inside and an outside, it is impossible to get at the inside, if we have not first got at the outside.

Dante (*Il Convito*, II. 2–5)

M.M. 973
Orthod. 285

The Egyptians who receive instruction begin by learning the first method of writing, called *epistolographic;* . . . then they learn the second, which is *hieratic,* used by the sacred writers or hierogrammates; finally comes the highest and last method which they learn, the *hieroglyphic* method, of which one part has a *literal* significance, by means of the first elements (syllabic or alphabetic of the words they represent), and the other a *symbolic* significance.

Clement of Alexandria

Theoretical science is divided into two parts—the *historical* and the *spiritual* meaning; and the latter into three—*tropological, allegorical, anagogical. Tropology* (morality) relates to the improvement of morals; *allegory* to another signification than that of the letter; *anagogy* by the *spiritual* conception rises to the most sublime and secret things of the celestial mysteries. The four senses may be expressed at once in the same image. Thus, for example, *Jerusalem* literally may mean, the city of the Jews, allegorically, the Christian Church; tropologically, the human soul; anagogically, the celestial city.

Cassianus Eremita

The Old Testament has a fourfold division: according to history, etiology, analogy, and allegory.

St Augustine

Whoever will be nurtured and trained up by Sophia, and learn to understand and speak the language of wisdom, must be born again of and in the Word of Wisdom, Christ Jesus, the Immortal Seed: The divine essence which God breathed into his paradisical soul must be revived, and he must become one again with that which he was in God before he was a creature, and then his Eternal Spirit may enter into that which is within the veil, and see not only the literal, but the moral, allegorical, and anagogical meaning of the wise and their dark sayings: He then will be fit to enter, not only into Solomon's porch, the outer court of natural philosophy, sense and reason, but likewise into the inward court of holy and spiritual exercises, in divine understanding and knowledge; and so he may step into the most inward and holiest place of theosophical mysteries, into which none are admitted to come, but those who have received the high and holy unction.

John Sparrow

Orthod. 280

Do not suffer yourself to be confounded by the apparent contradictions which the Sages have introduced into their writings for the purpose of keeping their secret. Select only those sayings which are agreeable to Nature; take the roses, leave the thorns.

Michael Sendivogius

Creation 42

All these figures of speech should be regarded as symbols of sublime mysteries and Divine illuminations vouchsafed to me by the Lord God. Turn thy thoughts, oh! reader, from the mere words and seek the hidden meaning that thou mayest understand.

Ibn ʿArabî

The man who reads the sacred Scriptures without insight is blind to their inner meaning.

Hônen

O ye, who have sane intellects, mark the doctrine, which conceals itself beneath the veil of the strange verses!

Dante (*Inferno*, IX. 61)

Beware that thou conceive not bodily that which is meant ghostly, although it be spoken in bodily words.

The Cloud of Unknowing, LXI

Death 208

Lesson is without-forth in the bark, meditation within-forth in the pith.

Scala Claustralium

All the revelations that ever saw any man here in bodily likeness in this life, they have ghostly meanings. And I trow that if they unto whom they were showed had been so ghostly, or could have conceived their meanings ghostly, that then they had never been showed bodily. And therefore let us pick off the rough bark, and feed us with the sweet kernel.

The Cloud of Unknowing, LVIII

Void 721

For the Biblical stories are not simply there in order to give us the life and deeds of the ancients, as Babel would have it: No, the kingdom of Christ is found figured there throughout, as well as the kingdom of hell; the visible figure indicates always the invisible, which must reveal itself in the spiritual man.

Boehme

St Augustine says, at first the scriptures will amuse and attract the child, and in the end, when he tries to understand them, they make fools of the wise, for none is so simple-minded but can find his level there nor none so wise but when he tries to fathom them will find they are beyond his depth and discover more therein. All the stories and quotations taken from them have another, esoteric, meaning. Our understanding of them is as totally unlike the thing as it is in itself and as it is in God, as though it did not exist.

Eckhart

Introd. M.M. 973

It is in no wise contrary to truth for intelligible things to be set forth in Scripture under sensible figures, since it is not said for the purpose of maintaining that intelligible things are sensible, but in order that properties of intelligible things may be understood according to similitude through sensible figures.

St Thomas Aquinas

There is a twofold meaning in every creature, a literal and a mystical, and the one is but the ground of the other.

John Smith the Platonist

Holy Scripture by the manner of its speech transcends every science, because in one and the same sentence, while it describes a fact, it reveals a mystery.

Rev. 959

St Gregory the Great

The One God hath tempered the holy Scriptures to the senses of many.

St Augustine

The *Qur'ân* has been revealed in seven interiorities.[1]

Muhammad

Every couplet (in the *Bhâgavat Purana*), nay every letter of it breathes a variety of senses.

Chaitanya

Now that which is one according to the literal sense, is however three according to the mystical and spiritual sense. For in all the books of Holy Writ in addition to the literal sense, which the words express outwardly, there is a conception of the threefold spiritual sense, namely the allegorical, whereby we learn what we should believe concerning Godhead and manhood; the moral, whereby we learn how we should live; and the anagogical, whereby we learn in what manner we must cling to God. Whence the Holy Writ teaches these things, namely the eternal begetting and incarnation of Christ, the order of living, and the union of God and the soul. The first regards faith; the second, actions (*mores*); the third, the end of both. For the first, one should sweat at the study of the doctors; for the second, the study of the preachers; for the third, the study of the contemplatives. Augustine teaches us especially the first; Gregory teaches especially the second; Dionysius however teaches the third. Anselm follows Augustine; Bernard follows Gregory; Richard follows Dionysius. For Anselm excells in reasoning; Bernard, in preaching; Richard, in contemplation. Hugh however teaches all three things.

Center 841
Realiz. 887
Orthod. 296
Orthod. 275

St Bonaventura

In other writings the words alone carry meaning: in Scripture not only the words, but the things may mean something. Wherefore just as a knowledge of the words is needed in

[1] This *ḥadîth* even, like many others, lends itself to a plurality of interpretations, since the Arabic *aḥruf* can mean both 'letters' (referring here to seven standard styles of pronunciation) and 'limits' (referring to a superposition of senses). Sacred languages, in fact, like Arabic, Hebrew, or Sanskrit, both written and spoken, are eminent vehicles of higher realities, and their symbolic content extends beyond the science of etymologies to the very letters themselves and their numerical equivalents (*gematria*),—even their shapes. Greek and Latin, while not directly revealed languages, still retain a highly rational and logical character with an intellectual basis which carries down into later European language structures. At the other extreme (barring certain monstrosities like Esperanto) is modern usage, which tends in its exaggerated forms to exalt the purely emotional content of words, 'existentially', in terms of tensions, forces and reactions bordering upon an infra-rational intuitionism, and where the 'intellectual' content is practically limited to the pedantic Latin and Greek coinages for chemical processes and pharmaceutical products.

order to know what things are signified, so a knowledge of the things is needed in order to determine *their* mystical signification.

Hugh of Saint-Victor

When the Egyptian words are spoken, the force of the things signified works in them. Therefore, my King, as far as it is in your power, (and you are all-powerful,) keep the teaching untranslated.... It is an utterance replete with workings.[1]

Hermes

Rev. 961

Solar Symbolism

Let us meditate on that excellent glory of the divine Vivifier, (Savitar);
May He enlighten our intellects.

Gâyatrî (*Rigveda,* III. lxii. 10)

Center 833

Homage to thee, O thou who art Râ when thou risest and Temu when thou settest. Thou risest, thou risest, thou shinest, thou shinest, thou who art crowned king of the gods. Thou art the lord of heaven, the lord of earth; the creator of those who dwell in the heights and of those who dwell in the depths. (Thou art) the God One who came into being in the beginning of time. Thou didst create the earth, thou didst fashion man, thou didst make the watery abyss of the sky, thou didst form Ḥâpi (the god of the Nile), thou didst create the watery abyss, and thou dost give life unto all that therein is. Thou hast knit together the mountains, thou hast made mankind and the beasts of the field to come into being, thou hast made the heavens and the earth. Worshipped be thou whom the goddess Maât embraceth at morn and at eve. Thou dost travel across the sky with heart swelling with joy.... Hail, thou Disk, lord of beams of light, thou risest and thou makest all mankind to live. Grant thou that I may behold thee at dawn each day.

Hymn to Râ when he riseth

Introd. *Waters*

O Dayspring, Splendour of the Eternal Light, and Sun of Justice, come and enlighten us, sitting in the darkness, and in the shadow of death.

Antiphon at Lauds, 21st December (Winter Solstice)

Give thanks to the Sun of the Angels, who of his grace hath to this sun of sense exalted thee.

Dante (*Paradiso,* x. 53)

[1] Cf. note p. 316.

I am a shadow, the Sun is my lord. . . .
I am the moon, and the Sun is in front of me as the guide.

Rûmî

Travelling through the creation, the virtuous man is led to the apprehension of a Master of the creation; . . . when he observes the beauty of this material sunlight, he grasps by analogy the beauty of the real sunlight.

St Gregory of Nyssa

The outward sun . . . receives its power and lustre from the inward, as a glass or resemblance of the inward. . . . The sun of the outward world . . . is a figure of the inward all-essential sun.

Boehme

And of His signs are the night and the day and the sun and the moon. Adore not the sun nor the moon; but adore Allâh who created them, if it is in truth Him whom ye worship.

Qur'ân, XLI. 37

Realiz. 890

He who, dwelling in the sun, yet is other than the sun, whom the sun does not know, whose body the sun is, who controls the sun from within—He is your Self, the Inner Controller, the Immortal.

Bṛihad-Âraṇyaka Upanishad, III. vii. 9

And this Soul of the sun, which is therefore better than the sun, whether taking the sun about in a chariot to give light to men, or acting from without, or in whatever way, ought by every man to be deemed a God.[1]

Plato (*Laws,* 899 A)

M.M. 994

There is no sensible thing in all the world more worthy to be an image of God than the sun, which with its sensible light illumines first itself, and then all celestial and elementary bodies; so God first illumines Himself with intellectual light, and then the celestial and other Intelligences.

Dante (*Il Convito,* XII. 4)

The Sun . . . is an image of the Maker who is above the heavens.

Hermes

Dost not thou know that the light of the sun is the reflexion of the Sun beyond the veil?

Rûmî

[1] 'The traditional distinction of intelligible from sensible, invisible from visible "suns" is essential to any adequate understanding of "solar mythologies" and "solar cults" ' (Coomaraswamy: *Am I My Brother's Keeper?,* p. 92).

Remember it is the heart and not the body which strives to draw near to God. By 'heart' I do not mean the flesh perceived by the senses, but that secret thing which is sometimes expressed by spirit, and sometimes by soul.

Center 816

Al-Ghazâlî

Simplicius ... who had read Empedocles, acquaints us, that he made two worlds, the one intellectual, the other sensible; and the former of these to be the exemplar and archetype of the latter. And so the writer De Placitis Philosophorum observes ... that Empedocles made *δύο ἡλίους, τον μὲν ἀρχέτυπον, τὸν δὲ φαινόμενον,* two suns, the one archetypal and intelligible, the other apparent or sensible.[1]

Ralph Cudworth

Thus the Lord Jesus in his spiritual Glories, as he is the Divine Understanding, is the *Original* form of the Sun, the *Suns Sun.* To him agrees that, which *Plutarch* delivereth to us from the ancient Philosophers, that the God of the Sun, which inhabits the Sun, excels the Sun in the sweetness, beauty, and glory of his Light, ten thousand times more, than the Sun doth this Earth, or the darkest Cloud. The Sun it self shining in these visible Heavens is the essence or essential form, framed by this pattern, sprung forth from it. The Light, the Sun-shines, and Suns which we severally take in our eyes, are so many figures, and pictures, or shadows rather of this Sun flowing from him.

Orthod. 27
Beauty 663

Peter Sterry

This great, all-bright and ever-shining sun ... is the visible image of the Divine Goodness, faintly re-echoing the activity of the Good.... And so that Good which is above all light is called a Spiritual Light because It is an Originating Beam and an Overflowing Radiance, illuminating with its fullness every Mind above the world, around it, or within it, and renewing all their spiritual powers, embracing them all by Its transcendent compass and exceeding them all by Its transcendent elevation. And It contains within Itself, in a simple form, the entire ultimate principle of light; and is the Transcendent Archetype of Light; and, while bearing the light in its womb, It exceeds it in quality and precedes it in time; and so conjoineth together all spiritual and rational beings, uniting them in one.

Reality 775

Dionysius

He is the Sun of Knowledge. One single ray of His has illumined the world with the light of knowledge.

Knowl. 761

Sri Ramakrishna

By the light of mind the human soul is illumined, as the world is illumined by the sun,—nay, in yet fuller measure.

Holiness 914

Hermes

Christ, the Divine Sun ... risen to the zenith of our hearts....

Ruysbroeck

[1] Lib. ii. cap. xx. p. 900. tom. ii. oper. Plutarchi.

Creation 26

From the first God created two worlds, the visible and the invisible, and has made a king to reign over the visible who bears within himself the characteristic features of both worlds—one in his visible half and the other in his invisible half—in his soul and his body. Two suns shine in these worlds, one visible and another intellectual. In the visible world of the senses there is the sun, and in the invisible world of the intellect there is God, Who is and is called the sun of truth. The physical world and everything in it is illumined by the physical and visible sun; but the world of the intellect and those who are in it are illumined and enlightened by the sun of truth in the intellect. Moreover, physical things are illumined by the physical sun, and things of the intellect by the sun of the intellect separately from one another, for they are not mixed with or merged into one another—neither the physical with the intellectual nor the intellectual with the physical.

St Simeon the New Theologian

Both suns shine to us.

Boehme

The sun illumines the universe, whereas the Sun of Arunachala[1] is so dazzling that the universe is obscured and an unbroken brilliance remains. But it is not realised in the present state and can be realised only if the lotus of the heart blossoms. The ordinary lotus blossoms in the light of the visible sun, whereas the subtle Heart blossoms only before the Sun of Suns. May Arunachala make my heart blossom so that His unbroken Brilliance may shine all alone!

Sri Ramana Maharshi

From the Sun who is the glory of Tabrîz seek future bliss,
For he is a sun, possessing all kinds of knowledge, on the spiritual throne.

Divâni Shamsi Tabrîz,[2] XLIV

SONG TO THE SUN

Center 816

Thou eye of the Great God
Thou eye of the God of Glory
Thou eye of the King of creation
Thou eye of the Light of the living
Pouring on us at each time

Charity 602

Pouring on us gently, generously
Glory to thee thou glorious sun
Glory to thee thou Face of the God of life.

Ortha nan Gaidheal

[1] A sacred peak in India, associated with Śiva.
[2] Shamsi Tabrîz means 'Sun of Tabrîz'.

The Worship of Images

The picture is not in the colors . . . the Principle transcends the letter.

Lankavatara Sutra, II. 118–119

Those who consider the earthen images, do not honour the clay as such, but without regard to them in this respect, honour the Immortals designated.

Divyâvadâna, ch. XXVI

The respect that is paid to the image passes over to its archetype.

St Basil

We believe that when we worship the images, the Gods are kindly and well-disposed towards us.

Plato (*Laws,* 931 A)

It is in imitation of the angelic works of art that any work of art is accomplished here. . . . A work of art indeed is accomplished in him who comprehends this.

Aitareya Brâhmaṇa, VI. 27

Beauty 670

As bodies are reflected in mirrors, so incorporeal things are reflected in bodies, and the intelligible Kosmos is reflected in the sensible Kosmos. Therefore, my King, worship the statues of the gods, seeing that these statues too have in them forms which come from the intelligible Kosmos.

Hermes

Metanoia 488

All honor that we pay the image, we refer to the Archetype, namely Him whose image it is. . . . In no wise honor we the colors or the art, but the archetype in Christ, who is in heaven.

Hermeneia of Athos, no. 445

The Pagans ascribed to God the names of various perfections found in creatures; e.g. we have the Temples of Peace, Eternity, Concord and the Pantheon; and in the middle of the Pantheon in the open air an altar was erected to the Infinite Term which is without term. All these names express what the one ineffable Name implies; and since the proper Name is Infinite it includes all the numberless names that denote particular perfections. Numerous as such names may be they are never so many or so great that they could not be added to, for each of them is to the proper and ineffable Name what the finite is to the Infinite. . . .

M.M. 994

The simple folk were led into error by this . . . for instead of regarding those sensible expressions as an image they took them for the truth itself. The result of this was that the

M.M. 973

masses became idolaters, whereas philosophers, for the most part, continued to have a correct idea of the unity of God, as can be attested to by anyone who has carefully read the ancient philosophers and Cicero's *De Deorum Natura*.

Nicholas of Cusa

Supra Introd.

Pictures and ornaments in churches are the lessons and the scriptures of the laity. Whence Gregory: It is one thing to adore a picture, and another by means of a picture historically to learn what should be adored. . . . We worship not images, nor account them to be gods, nor put any hope of salvation in them: for that were idolatry. Yet we adore them for the memory and remembrance of things done long agone.

William Durandus

It is for the advantage (*artha*) of the worshippers (*upâsaka*) (and not by any intrinsic necessity) that Brahman—whose nature is intelligence (*cin-maya*), beside whom there is no other, who is impartite and incorporeal—is aspectually conceived (*rûpa-kalpanâ*).

Râmôpaniṣad

By the visible aspect, our thought must be caught up in a spiritual surge and rise to the invisible majesty of God.

St John Damascene

Worship me in the symbols and images which remind thee of me.

Srimad Bhagavatam, XI. v

Ritual Symbols and Representations of the Universe

Center 831

In setting up the sun dance lodge, we are really making the universe in a likeness; for, you see, each of the posts around the lodge represents some particular object of creation, so that the whole circle is the entire creation, and the one tree at the center, upon which the twenty-eight poles rest, is *Wakan-Tanka,* who is the center of everything. Everything comes from Him, and sooner or later everything returns to Him. And I should also tell you why it is that we use twenty-eight poles. I have already explained why the numbers four and seven are sacred;[1] then if you add four sevens you get twenty-eight. Also the moon lives twenty-eight days, and this is our month; each of these days of the month represents something sacred to us: two of the days represent the Great Spirit; two are for Mother Earth; four are for the four winds; one is for the Spotted Eagle; one for the sun; and one for the moon; one is for the Morning star; and four for the four ages; seven are for our seven great rites; one is for the buffalo; one for the fire; one for the water; one for the rock;

[1] The four ages, the four quarters, the six directions of space plus the center, etc.

and finally one is for the two-legged people. If you add all these days up you will see that they come to twenty-eight. You should also know that the buffalo has twenty-eight ribs, and that in our war bonnets we usually use twenty-eight feathers. You see, there is a significance for everything, and these are the things that are good for men to know, and to remember.

Black Elk

And do you say 'God is invisible'? Speak not so. Who is more manifest than God? For this very purpose has he made all things, that through all things you may see him. This is God's goodness, that he manifests himself through all things.

Hermes

Creation 48

Verily, this (brick-)built Fire-altar (Agni) is this (terrestrial) world:—the waters (of the encircling ocean) are its (circle of) enclosing-stones; the men its Yagushmatîs (bricks with special formulas); the cattle its Sûdadohas (nourishment); the plants and trees its earth-fillings (between the layers of bricks), its oblations and fire-logs; Agni (the terrestrial fire) its Lokamp*rinâ* (space-filling brick)—thus this comes to make up the whole Agni, and the whole Agni comes to be the space-filler; and, verily, whosoever knows this, thus comes to be that whole (Agni) who is the space-filler.

But, indeed, that Fire-altar also is the air . . . is the sky . . . is the sun . . . is the Metres . . . is the Year . . . is the body (etc.).

Śatapatha-Brâhmaṇa, x. v. 4, 1–12

Center 841

Matohoshila[1] took up an ear of corn, and at one end of it he pushed in a stick, and at the other end of the ear he tied the plume of an eagle. . . .

'The tassel which grows upon the top of the ear of corn,' Matohoshila said, 'and which we have represented here by the eagle plume, represents the presence of the Great Spirit, for, as the pollen from the tassel spreads all over, giving life, so it is with *Wakan-Tanka,* who gives life to all things. This plume, which is alway on top of the plant, is the first to see the light of the dawn as it comes, and it sees also the night and the moon and all the stars. For all these reasons it is very *wakan* (sacred). And this stick which I have stuck into the ear of corn is the tree of life, reaching from Earth to Heaven, and the fruit, which is the ear with all its kernels, represents the people and all things of the universe. It is good to remember these things if we are to understand the rites which are to come.'

Black Elk

Introds. *Wr. Side Peace*

And why the pentangel apendes · to that prince[2] noble
I am intent you to telle, · thogh tary hit me shulde:
Hit is a signe that Salamon · set sumwhile
In bitokning of trawthe, · bi title that hit habbes,
For hit is a figure that haldes · five pointes,
And eche line umbelappes · and loukes in other,

Holiness 924

[1] 'Bear Boy', a holy man of the Lakota Sioux.
[2] Sir Gawain had the pentacle painted on his shield.

And aywhere hit is endeles; · and Englich hit callen
Overal, as I here, · the endeles knot.
Forthy hit acordes to this knight · and to his cler armes,
For ay faithful in five · and sere five sithes
Gawan was for gode knawen, · and as golde pured,
Voided of eche vilany, · with vertues ennourned
in mote;
Forthy the pentangel newe
He ber in shelde and cote,
As tulk of tale most trewe
And gentilest knight of lote.

Sir Gawayne and the Grene Knight, 623

Suff. 130

Hail, O cross, yea be glad indeed! . . . I come unto thee that hast yearned after me. I know thy mystery, for the which thou art set up: for thou art planted in the world to establish the things that are unstable: and the one part of thee stretcheth up toward heaven that thou mayest signify the heavenly word (the head of all things): and another part of thee is spread out to the right hand and the left that it may put to flight the envious and adverse power of the evil one, and gather into one the things that are scattered abroad: And another part of thee is planted in the earth, and securely set in the depth, that thou mayest join the things that are in the earth and that are under the earth unto the heavenly things. . . .

Well done, O cross, that hast bound down the mobility of the world! Well done, O shape of understanding that hast shaped the shapeless!

The Martyrdom of St Andrew

These great masters of Paris do read vast books, and turn over the leaves with great diligence, which is a very good thing; but these (spiritually enlightened men) read the true living book, wherein all things live: they turn over the pages of the heavens and the earth, and read therein the mighty and admirable wonders of God.

Tauler

WORK, ACTION AND SOCIETY

He who sees inaction in action and action in inaction, he is intelligent among men; he is a man of established wisdom and a true performer of all actions.

Bhagavad-Gîtâ, IV. 18

Outward activity (works) belongs to the temporal domain, and inner activity (contemplation) belongs to the spiritual domain. The 'contemplative' accepts in the nature of things the legitimate place of action, but often the 'active' seems unable to gauge the full validity of contemplation. And yet, 'Ye are the salt of the earth: but if the salt have lost his savour, wherewith shall it be salted? it is thenceforth good for nothing, but to be cast out, and to be trodden under foot of men' (*Matt.* V. 13). Without the existence of action, the contemplative would lose his body and life and the substantial support these offer in his spiritual work; without the existence of contemplation, the active would lose his soul. One is therefore admonished: 'Have salt in yourselves, and have peace one with another' (*Mk.* IX. 50).

From a more profound standpoint, it is God alone who acts, whence the Scholastic saying: 'To act, one must be'—in the sense that all action derives its validity from the Divine Act stemming from Pure Being, and is legitimate in the measure that it participates therein, profane activities being considered little more than agitation.

Following the Vedanta, Guénon explains that 'action cannot have the effect of liberating from action, and its consequences do not extend beyond the limits of the individuality, envisaged moreover integrally, in the extension of which it is capable.[1] Action, whatever the kind, not being opposed to ignorance which is the root of all limitation, could never banish it; knowledge alone disperses ignorance, as the light of the sun disperses darkness, and it is then that the "Self", the immutable and eternal principle of all states manifested and unmanifested, appears in its supreme reality' (*La Métaphysique orientale,* pp. 20–21).

Introd. *Holiness* *Knowl.* 761

Fanaticism has been defined as a redoubling of effort when the goal has been forgotten. It is not fortuitous that work reaches its frenzied peak precisely in the modern world where final ends have been the most forgotten: 'Modern civilization, by its divorce from any principle, can be likened to a headless corpse of which the last motions are convulsive and insignificant' (Coomaraswamy: *Am I My Brother's Keeper?,* p. 1). Analogously, it is not to be wondered at if those 'religious' bodies least noted for their intellectuality are often the ones most noted for their 'works': 'In fact, materialism and sentimentality, far from being in opposition, can scarcely exist one without the other, and together the two attain to their

Wr. Side 464

[1] Guénon explains elsewhere that action can thus lead to the salvation of the individual as such, but not beyond, to final liberation.

utmost development' (Guénon: *Orient et Occident,* p. 33). 'Spirituality has as its object, not man, but God; it is this that a certain moral absolutism seems to forget' (Schuon: *Perspectives spirituelles,* p. 250). 'To lose oneself for God is always to give oneself to mankind. To reduce all spirituality to social charity is not only to place the human above the divine, but also to believe oneself indispensable and to place an absolute value on what one is capable of giving' (id. p. 280). Furthermore, the reduction of spirituality 'to a "humble" utilitarianism—hence to a masked materialism—is an insult done to God, for it is like saying that God does not deserve our being too exclusively preoccupied with Him, while at the same time it relegates the divine gift of intelligence to the rank of superfluous things' (Schuon: *Comprendre l'Islam,* p. 164).

Introd. *Charity*

Basically, action is the exercise of energy (*prâna*), force, or power, and this may be done with art or without art. Action needs no apologists; creation could not even commence without it. The bull in the ring acts on pure impulsion; the art is with the matador, who must employ his relatively 'non-acting action' to give the bull its quietus. Action needs no vindication; what is asked is its regulation, through right doctrine.[1] 'If our daily work, whether it be a man's trade or a woman's housework, is not to constitute an obstacle on the spiritual path, then it must play the role on this path of a positive element, or more precisely, of a secondary vehicle for the realization of the Divine within us.

'An integration such as this of work with spirituality depends upon three fundamental conditions which we shall designate respectively by the terms "necessity", "sanctification" and "perfection". The first of these conditions implies that the activity to be spiritualized correspond to a necessity and not a caprice.... The second ... implies that the activity thus defined be offered effectively to God.... The third condition implies the logical perfection of the work, for it is evident that one could not offer to God an imperfect thing, nor consecrate a vile object to Him.... God is Perfection, and man, to approach God, must be perfect in action as well as in non-acting contemplation' (Schuon: 'Le sens spirituel du travail'; *Études Trad.,* 1948, pp. 238–239).[2] 'On the other hand, it must not be forgotten that the greatest possible benefit for a human collectivity is the sanctity—even if hidden—of an individual, and not any exterior work whatsoever' (Schuon: 'Des Stations de la sagesse': *France-Asie,* juin-juillet, 1953).

Introd. *Suff.*

'The Sacrifice is something to be done; "We must do what the Gods did erst" (*Śatapatha Brâhmaṇa,* VII. 2. 1. 4). It is, in fact, often spoken of simply as "Work" (*karma*). Thus just as in Latin *operare* = *sacra facere* = ἱεροποιεῖν so in India, where the emphasis on action is so strong, to do well is to do sacred things, and only to do nothing, or what being done amiss amounts to nothing, is idle and profane.... Sacrifice, thus understood, is no longer a matter of doing specifically sacred things only on particular occasions, but of sacrificing (*making sacred*) all we do and all we are; a matter of the sanctification of whatever is done naturally, by a reduction of all activities to their

[1] See Herrigel: *Zen in the Art of Archery:* 'The effortlessness of a performance for which great strength is needed is a spectacle of whose aesthetic beauty the East has an exceedingly sensitive and grateful appreciation' (p. 42).

[2] 'The organization of the Zen monastery was very significant of this point of view. To every member, except the abbot, was assigned some special work in the caretaking of the monastery, and, curiously enough, to the novices were committed the lighter duties, while to the most respected and advanced monks were given the more irksome and menial tasks' (Okakura-Kakuzo: *The Book of Tea,* p. 69).

principles. We say "naturally" advisedly, intending to imply that whatever is done naturally may be either sacred or profane according to our own degree of awareness, but that whatever is done *un*naturally is essentially and irrevocably profane' (Coomaraswamy: *Hinduism and Buddhism*, pp. 19, 25).

Work and art and sacrifice are thus acknowledged to be but different aspects of a single unifying principle, relating the temporal order with the divine; but work divorced from art and art from sacrifice become deadly solvents, and thus equally capable of dissociating these two orders. To show what this dissociation implies, Coomaraswamy as spokesman for the normal order of things, in an essay ' "A Figure of Speech, or A Figure of Thought?" ' holds up a mirror to the modern world: 'If now the orthodox doctrines reported by Plato and the East are not convincing, this is because our sentimental generation, in which the power of the intellect has been so perverted by the power of observation that we can no longer distinguish the reality from the phenomenon, the Person in the Sun from his sightly body, or the uncreated from electric light, will not be persuaded "though one rose from the dead." Yet I hope to have shown, in a way that may be ignored but cannot be refuted, that our (present-day) use of the term "aesthetic" forbids us also to speak of art as pertaining to "the higher things of life" or the immortal part of us; that the distinction of "fine" from "applied" art, and corresponding "manufacture" of art in studios and artless industry in factories, take it for granted that neither the "artist" nor the "artisan" shall be whole men; that our freedom to work or starve is not a responsible freedom but only a legal fiction that conceals an actual servitude; that our hankering after a leisure state, or state of pleasure, to be attained by a multiplication of labour-saving devices, is born of the fact that most of us are doing forced labour, working at jobs to which we could never have been "called" by any other master than the salesman; that the very few, the happy few of us whose work is a vocation and whose status is relatively secure, like nothing better than our work, and can hardly be dragged away from it; that our division of labour, Plato's "fractioning of human faculty", makes the workman a part of the machine, unable ever to make or to co-operate responsibly in the making of any whole thing; that in the last analysis the so-called "emancipation of the artist" is nothing but his final release from any obligation whatever to the God within him,[1] and his opportunity to imitate himself or any other common clay at its worst; that all wilful self-expression is auto-erotic, narcissistic and Satanic, and the more its essentially paranoiac quality developes, suicidal; that while our invention of innumerable conveniences has made our unnatural manner of living in great cities so endurable that we cannot imagine what it would be like to do without them, yet the fact remains that not even the multi-millionaire is rich enough to commission such works of art as are preserved in our museums but were originally made for men of relatively moderate means or, under the patronage of the Church, for God and all men' (*Figures of Speech or Figures of Thought,* London, Luzac, 1946, pp. 32–33).

Symb. 317

Introd. *Beauty*

'The caste system differs from the industrial "division of labor", with its "fractioning of human faculty", in that it presupposes differences in kinds of responsibility but not in degrees of responsibility; and it is just because an organisation of functions such as this, with its mutual loyalties and duties, is absolutely incompatible with our competitive industrialism, that the monarchic, feudal and caste system is always painted in such dark

Rev. 967

[1] 'At the interior of such a world, or rather, such a "setting", spiritual reality will appear as an illusion and a ludicrous luxury' (Schuon: 'L'Âme de la caste', *Études Trad.*, 1954, p. 67).

colors by the sociologist, whose thinking is determined more by his actual environment than it is a deduction from first principles' (Coomaraswamy: *Hinduism and Buddhism*, p. 27).

Perhaps the most difficult act of all, the one that often requires the most energy, and beside which all other actions are quasi-illusory, is to translate personally and effectively spiritual precepts into spiritual practice.

Knowl. 745

On 'Karma' and 'Karma Yoga'

There is ever to living creatures fear from death, and they with all their efforts seek to be born again; where there is action, there must inevitably be death.[1]

Asvaghosha

The soul is born in this world upon leaving the soul of the world (*anima mundi*) in which her existence precedes the one we all know. Thus, the Gods who consider her proceedings in all the phases of various existences and as a whole, punish her sometimes for sins committed during an anterior life.

Apuleius

Judgment 250

All things stand in the will, and everything, animate or inanimate, is the effect and produce of that will, which worketh in it and formeth it to be that which it is. And every will, wherever found, is the birth and effect of some antecedent will, for will can only proceed from will, till you come to the first working will, which is God Himself.

William Law

Conform 166
Death 223
Reality 773

A man is the creator of his own fate, and even in his foetal life he is affected by the dynamics of the works of his prior existence. . . .

This human body entombs a self which is nothing if not emphatically a worker. It is the works of this self in a prior existence which determine the nature of its organism in the next. . . . What is lotted cannot be blotted. A frightened mouse runs to its hole; a scared serpent, to a well; a terrified elephant, to its stake—but where can a man fly from his Karma?

Garuda Purana, CXIII

Be not deceived; God is not mocked: for whatsoever a man soweth, that shall he also reap.

Galatians, VI. 7

Wr. Side 459

Let the soul reflect upon what it sends ahead for the morrow.

Qur'ân, LIX. 18

Suff. 118

From the beginning of the affair discern the end thereof, so that thou mayst not be repenting on the Day of Judgement.

Rûmî

Is the reward of excellence aught save excellence?

Qur'ân, LV. 60

[1] Simply because whatever has a beginning must have an end.

Center 838 Death 206

That which hath been is now; and that which is to be hath already been; and God requireth that which is past.

Ecclesiastes, III. 15

And as Jesus passed by, he saw a man which was blind from his birth.

And his disciples asked him saying, Master, who did sin, this man, or his parents, that he was born blind?

Jesus answered, Neither hath this man sinned, nor his parents: but that the works of God should be made manifest in him.

St John, IX. 1–3

Realiz. 859

If you want to know the *Karma* causes which belong to the past, look at the *Karma* effects which belong to the present. If you want to see the *Karma* effects as they will appear in the future, look well to the causes operating here and now.

Ingwa Sûtra

Others fear what the morrow may bring.
I am afraid of what happened yesterday.

Ansârî

To regulate one's conduct in accordance with the law of cause and effect as carefully as one guardeth the pupils of one's eyes is the sign of a superior man.

Gampopa

Are ye requited aught save what ye used to earn?

Qur'ân, X. 52

Devotee: Is society right in taking the life of a murderer?

Maharshi: What is it that prompted the murderer to commit the crime? The same power awards him the punishment. Society or the State is only a tool in the hands of the power.

Sri Ramana Maharshi

Illusion 89 Judgment 250

Actions, (both enjoined and prohibited), bring about one's connection with the body; when the connection with the body has taken place pleasure and pain most surely follow; thence come attraction and repulsion, from them actions follow again, as the results of which merit and demerit appertain to an ignorant man, which again are similarly followed by the connection with the body. This transmigratory existence is thus going on continually for ever like a wheel.

Śrî Śankarâchârya

The Moving Finger writes; and, having writ,
Moves on: nor all thy Piety nor Wit
Shall lure it back to cancel half a Line,
Nor all Thy Tears wash out a Word of it.

Omar Khayyâm

If Destiny will that fortune smile upon thee, it will come to thee, though thou art far away; or if it be ills thou'rt destined to bear, thou canst not elude them, though thou be borne to the skies.

Shekel Hakodesh, 151

The threads which they (the Fates) spin are so unchangeable, that, even if they decreed to someone a kingdom which at the moment belonged to another, and even if that other slew the man of destiny, to save himself from ever being deprived by him of his throne, nevertheless the dead man would come to life again in order to fulfil the decree of the Fates.

Apollonius of Tyana

Lo! good deeds annul ill deeds. This is a reminder for the mindful.

Qur'ân, XI. 114

Yield not to weaknesses that let the hour
Of duty pass. Though thou art palsied, walk,
And rise, though thou be broken; for thy lot
Is worthlessness, if thou defer resolve
Unto the day of health.

Ibn al-Fâriḍ

Renun. 152*a*

Thy inclination to defer (spiritual) works until a leisure time comes from foolishness of soul.

Ibn 'Aṭâ'illâh

Illusion 99

A devotee who can call on God while living a householder's life is a hero indeed. God thinks: 'He who has renounced the world for My sake will surely pray to Me. He must serve Me. Is there anything very remarkable about it? People will cry shame on him if he fails to do so. But he is blessed indeed who prays to Me in the midst of his worldly duties. He is trying to find Me, overcoming a great obstacle—pushing away, as it were, a huge block of stone weighing a ton. Such a man is a real hero.'

Sri Ramakrishna

Renun. 139

External conditions such as occupation and environment do not in any way hamper the progress of the soul towards its eternal source, if the urge within is strong enough to lead us to the goal.

It does not matter whether one is a prince, a millionaire or a labourer. What is needed is intense aspiration for the Divine.

Swami Ramdas

The time of business does not with me differ from the time of prayer, and in the noise and clatter of my kitchen, while several persons are at the same time calling for different things, I possess God in as great tranquillity as if I were upon my knees at the blessed sacrament.

Brother Lawrence

Contem. 532
Inv. 1036

Many people think that one cannot arrive at spiritual perfection in leading a family life. How much truth is there to this? What wonderful opportunities for spiritual development are to be met with in all families! The love and affection between parents and children, husband and wife, brothers and sisters, parents and friends, the benediction which the poor and unfortunate bring to us, all help infinitely in establishing a spiritual life. The more your soul undergoes the trials caused by pleasures and hardships, the more it is purified and the more refined becomes your spirit of sacrifice for others. There surges up in a heart belaboured by all the changing currents of family life a profound aspiration for divine aid much more so than is generally found in the life of a hermit centered upon himself.

Ananda Moyî

Contem. 547

Curiosity, ambition, disquiet, the not adverting to, or not considering, the end for which we are in this world, are the causes why we have a thousand times more hindrance than business, more worries than work, more occupation than profit: and these are the embarrassments, that is, the silly, vain and superfluous undertakings with which we charge ourselves, that turn us from the love of God, and not the true and lawful exercises of our vocations. . . .

Conform. 180

While the plague afflicted the Milanese, S. Charles never made any difficulty in frequenting the houses and touching the persons that were infected. Yet he only frequented and touched them, so far forth as the necessity of God's work required, nor would he for the world have thrust himself into danger without true necessity, lest he should commit the sin of tempting God. So that he was never touched with any infection, God's Providence preserving him who had so pure a confidence in it, that it had no mixture either of fear or rashness. In like manner God takes care of those who go not to the court, to the bar, to war, except by the necessity of their duty; and in that case a man is neither to be so scrupulous as to abandon good and lawful affairs by not going, nor so overweening and presumptuous as to go thither or stay there without the express necessity of duty and affairs.

St Francis de Sales

Whence comes it then, that we have so many complaints, each saying that his occupation is a hindrance to him, while notwithstanding his work is of God, who hindereth no man? Whence comes this inward reproof and sense of guilt which torment and disquiet you? Dear children, know that it is not your work which gives you this disquiet. No: it is your want of order in fulfilling your work. If you performed your work in the right method, with a sole aim to God, and not to yourselves, your own likes and dislikes, and neither feared nor loved aught but God, nor sought your own gain or pleasure, but only God's glory, in your work, it would be impossible that it should grieve your conscience.

Infra 358

Tauler

Hôjô Tokimune:[1] 'I have so much of worldly affairs to look after and it is difficult to find spare moments for meditation.'

[1] Regent of Japan (1268–84), crushed Mongolian invasions, helped establish Zen Buddhism in Japan, especially among the samurai.

Bukkô Kokushi: 'Whatever worldly affairs you are engaged in, take them up as occasions for your inner reflection, and some day you will find out who this beloved Tokimune of yours is.'

Bukkô Kokushi

Infra 340
Realiz. 859

Let all your actions shew that you love and fear God, and then every labour to which you set your hand will prosper, and from beginning to end you will pursue the work successfully and joyously.

The Glory of the World

All duties, if accompanied by devotion to me, lead to the supreme good and to eternal liberation.

Srimad Bhagavatam, XI. xi

The building up of good and evil both involve attachment to form. Those who, being attached to form, do evil, have to undergo various incarnations unnecessarily; while those who, being attached to form, do good, subject themselves to toil and privation equally to no purpose. In either case it is better to achieve sudden self-realization and to grasp the fundamental Dharma.

Huang Po

M.M. 978

The ignorant man, attached to his body, is controlled by the impressions and tendencies created by his past deeds, and is bound by the law of karma. But the wise man, his desires being quenched, is not affected by deeds. He is beyond the law of karma. Since his mind rests in the Atman he is not affected by the conditions which surround him, though he may continue to live in the body and though his senses may move amongst sense objects. For he has realized the vanity of all objects, and in multiplicity sees one infinite Lord. He is like to a man who has awakened from sleep and learned that his dream was a dream.

Srimad Bhagavatam, XI. xx

Reality 775
Illusion 94

For as great as the power of the world is, it is all built upon a blind obedience; and we need only open our eyes to get quit of its power.

William Law

Illusion 99

Let fortune go and fume with fury in another place; let her find some other matter to execute her cruelty; for fortune hath no puissance against them which have devoted their lives to serve and honour the majesty of our goddess (Isis).

Apuleius

P. State 579

The one safeguard is piety. Over the pious man neither evil daemon nor destiny has dominion; for God saves the pious from every ill.

Hermes

Who repenteth and believeth and doth righteous work: as for such, Allâh will change their evil deeds to good deeds. Allâh is ever Forgiving, Merciful.

Qur'ân, XXV. 70

Faith 514

Illusion 89
Symb. 306
Knowl. 761

The majority do not resist their bodily disposition, and so the impression of the stars takes effect in them; but not always in this or that individual who, it may happen, uses his reason to resist that inclination.

St Thomas Aquinas

Peace 694

. . . O! here
Will I set up my everlasting rest,
And shake the yoke of inauspicious stars
From this world-wearied flesh.

Shakespeare (*Romeo and Juliet,* v. iii. 109)

While Devaki, Krishna's mother, was in prison, she had a vision of God Himself endowed with four hands, holding mace, discus, conch-shell, and lotus. But with all that she couldn't get out of prison. . . .

The truth is that one must reap the result of the prârabdha karma.[1] The body remains as long as the results of past actions do not completely wear away. Once a blind man bathed in the Ganges and as a result was freed from his sins. But his blindness remained all the same.

Sri Ramakrishna

Ecstasy 637
Infra 346

Yesterday *This* Day's Madness did prepare;
To-morrow's Silence, Triumph, or Despair:
Drink! for you know not whence you came, nor why:
Drink! for you know not why you go, nor where.

Omar Khayyâm

Humility 197
Inv. 1009

My birthplace? I am born of *Jigoku* (hell);
I am a nobody's dog
Carrying the tail between the legs;
I pass this world of woes,
Saying 'Namu-amida-butsu'.[2]

Saichi

The results of actions are the production, acquisition, transformation and purification of something. They produce no other results. All actions with their accessories should, therefore, be given up.[3]

Śrî Śankarâchârya

Renun. 152a

Devotee: How is (*Samâdhi*) made ever abiding?
Maharshi: By scorching the predispositions.

Sri Ramana Maharshi

[1] Actions that have begun to bear fruit (compared to an arrow already released from the bow).
[2] 'Adoration for Amida Buddha', an invocation used in the Jôdô-Shin school of Japanese Buddhism.
[3] 'By one who aspires after liberation which by its nature cannot be the result of any action' (explanatory note added by the translator, Swâmi Jagadânanda).

I will teach you dhamma: if this is, that comes to be; from the arising of this, that arises; if this is not, that does not come to be; from the ceasing of this, that ceases.

Majjhima-nikâya, II. 32

Peace 700

The more you prune a plant, the more vigorously it grows. The more you rectify your *karma*, the more it accumulates. Find the root of *karma* and cut it.

Sri Ramana Maharshi

M.M. 978

The revolving world is that great wheel (of delusion), and the human heart is its nave or axis, which by its continuous rotation, produces all this delusion within its circumference. If, by means of your manly exertion, you can put an end to the motion of your heart, you will stop the rotation of the circle of delusion at once.

Yoga-Vasishtha

Judgment 250

He who follows the way of Truth will leave the domain which undergoes judgment and become himself judge over things.

Ibn ʿArabî

Judgment 242

Concerning Right Vocation or 'Dharma'

Man attains perfection, being engaged in his own duty. Hear now how one engaged in his own duty attains perfection.

Him from Whom is the evolution of all beings, by Whom all this is pervaded, by worshipping Him with his own duty man attains perfection.

Better is one's own duty, although imperfect, than that of another well performed. He who does the duty born of his own nature incurs no sin.

O son of Kunti, one should not relinquish the duty to which he is born, though it is defective, for all undertakings are surrounded by evil as fire by smoke.

Bhagavad-Gîtâ, XVIII. 45–48

Conform. 180

Infra 340

Everything is virtuous in its nature that fulfils the purpose for which it was ordained: and the better it does this, the more virtuous it is; therefore we call him a good man who leads the contemplative or the active life for which his nature fits him: we call the horse good that runs fast and far, which he is created to do; we call the sword good that cuts hard things with ease, for which end it is made. Thus language, being ordained to express human conceptions, is good when it does this: and the more perfectly it does it the better it is.

Dante (*Il Convito*, I. v. 4)

Every one has been made for some particular work, and the desire for that work has been put into his heart.

Rûmî

Introd. *Charity*

God and nature makes nought superfluous, but all that comes into being is for some function. For no created being is a final goal in the intention of the Creator, as Creator; but rather is the proper function of that being the goal. Wherefore it comes to pass that the proper function does not come into existence for the sake of the being, but the latter for the sake of the former.

Dante (*De Monarchia*, I. iii. 22)

Let no one forget his own duty for the sake of another's, however great: let a man, after he has discerned his own duty, be always attentive to his duty.

Dhammapada, XII. 166

The courtesan Bindumatî caused the river Ganges to flow back upstream. After she had performed the feat King Asoka[1] *said to her:*

'You possess the Power of Truth! You, a thief, a cheat, corrupt, cleft in twain, vicious, a wicked old sinner who have broken the bonds of morality and live on the plunder of fools.'

Sin 62

'It is true, your Majesty; I am what you say. But even I, wicked woman that I am, possess an Act of Truth by means of which, should I so desire, I could turn the world of men and the worlds of the gods upside down.'

Said the king, 'But what is this Act of Truth? Pray enlighten me.'

Infra 338

'Your Majesty, whosoever gives me money, be he a Khattiya or a Brâhmaṇa or a Vessa or a Sudda or of any other caste soever, I treat them all exactly alike. If he be a Khattiya, I make no distinction in his favor. If he be a Sudda, I despise him not. Free alike from fawning and contempt, I serve the owner of the money. This, your Majesty, is the Act of Truth by which I caused the mighty Ganges to flow back upstream.'

Milindapañha

The proof of a Muslim's sincerity is that he payeth no heed to that which is not his business.

Muhammad

Infra 346 *Metanoia* 488

Let every man abide in the same calling wherein he was called.

Art thou called being a servant? care not for it: but if thou mayest be made free, use it rather.

For he that is called in the Lord, being a servant, is the Lord's freeman: likewise also he that is called, being free, is Christ's servant.

Ye are bought with a price; be not ye the servants of men.

Brethren, let every man, wherein he is called, therein abide with God.

I. Corinthians, VII. 20–24

[1] King of Magadha (273–232 B.C.) of Maurya dynasty. United most of what is now India; became zealous supporter of Buddhism.

Better one's own duties incomplete than those of another well performed, for he who lives by the duties of another falls from caste at once.

Mânava-dharma-śâstra, x. 97

The benefits that God contrives to give in any one way are to be found and gotten in good ways one and all, and we ought to find in one way the good things common to them, not those peculiar to that one. For man must always do one thing, he cannot do them all. He must always be one thing and in that one find all. To try and do everything, this as well as that, to give up his own method for another which for the moment he likes better, believe me, is a fertile source of instability.

Introd. *Humility*
Reality 775
Comform. 170

Eckhart

To do one's own business in a certain way may be assumed to be justice.... But when the cobbler or any other man whom nature designed to be a trader, having his heart lifted up by wealth or strength or the number of his followers, or any like advantage, attempts to force his way into the class of warriors, or a warrior into that of legislators and guardians, for which he is unfitted, and either to take the implements or the duties of the other; or when one man is trader, legislator, and warrior all in one, then I think you will agree with me in saying that this interchange and this meddling of one with another is the ruin of the State.

Wr. Side 464
Introd. *Rev.*

Plato (*Republic,* 433, 434)

Vocation leads to heaven and eternity; in case of a digression from this norm, the world is brought to ruin by confusion.

Beauty 679

Artha-śâstra, I. 3

The virtue of Heaven and Earth, the powers of the Sage, and the uses of the myriad things in Creation, are not perfect in every direction. It is Heaven's function to produce life and to spread a canopy over it. It is Earth's function to form material bodies and to support them. It is the Sage's function to teach others and to influence them for good. It is the function of created things to conform to their proper nature.

Lich-tsc

The wisdom of a scribe cometh by his time of leisure: and he that is less in action, shall receive wisdom....

But they (the craftsmen) shall strengthen the state of the world, and their prayer shall be in the work of their craft.

Supra 329

Ecclesiasticus, XXXVIII. 25, 39

The path of wisdom is for the meditative and the path of work is for the active.

Bhagavad-Gîtâ, III. 3

Tendance of heaven and of all things that are therein is nothing else than constant worship; and there is no other being, divine or mortal, that does this, but man alone. For in the reverence and adoration, the praise and worship of men, heaven and the gods of heaven

Introd. *Holiness*

Contem. 547 Metanoia 480

find pleasure.... To some men then, but to very few, men who are endowed with mind uncontaminate, has fallen the high task of raising reverent eyes to heaven. But to all who, through the intermingling of the diverse parts of their twofold being, are weighed down by the burden of the body, and have sunk to a lower grade of intelligence—to all such men is assigned the charge of tending the elements, and the things of this lower world.

Hermes

Rev. 967

Four things support the world: the learning of the wise, the justice of the great, the prayers of the good, and the valour of the brave.

Muhammad

Pursuit of one's regular duty, in one's own stage of the religious life—that, verily, is the rule!

Maitri Upanishad, IV. 3

Holiness 900

The more you are free from the common necessities of men, the more you are to imitate the higher perfections of Angels.

William Law

Whoe'er thou art, that to this work art born,
A chosen work thou hast, howe'er the world may scorn.

Contem. 547

Boehme

Art and Skill in Work

Skillfulness in action is called Yoga.

Bhagavad-Gîtâ, II. 50

The man who forged swords for the Minister of War was eighty years of age. Yet he never made the slightest slip in his work.

The Minister of War said to him, 'Is it your skill, Sir, or have you any method?'

Contem. 532

'It is *concentration,*' replied the man. 'When twenty years old, I took to forging swords. I cared for nothing else. If a thing was not a sword, I did not notice it. I availed myself of whatever energy I did not use in other directions in order to secure greater efficiency in the direction required. Still more of that which is never without use (*Tao*); so that there was nothing which did not lend its aid.'

Chuang-tse (ch. XXII)

One with such a concentrated mind rises above the tumult of the subjective as well as of the objective world. He is like the arrow-maker, who while fashioning his arrows is conscious only of his task.

Srimad Bhagavatam, XI. iii

Allâh has prescribed *Iḥsân*[1] for everything; hence, if you kill, do it well; and if you slaughter, do it well; and let each one of you sharpen his knife and let his victim die at once.

Muhammad

It is true that there is something which terrifies the eye and surprises the soul to find that Mother Nature, with her great skill and wisdom and energy, has suddenly produced a thing like a stone cave or a blessed spot. But I have often stared casually at little things of this universe—a bird, a fish, a flower, or a small plant, and even at a bird's feather, a fish's scale, a flower petal and a blade of grass—and realized how Mother Nature has also created it with all her great skill and wisdom and energy. As it is said that the lion uses the same energy to attack an elephant as to attack a wild rabbit, so does Mother Nature truly do the same thing. She uses all her energy in producing a stone cave or a blessed spot, but she also uses all her energy in producing a bird, a fish, a flower, a blade of grass, or even a feather, a scale, a petal, a leaf. Therefore, it is not alone the stone cave or the blessed spot that terrifies the eye and surprises the soul in this world.

Chin Shengt'an

Center 841

A question is raised about those angels who live with us, serving and guarding us, as to whether or not they have less joy in identity than the angels in heaven have and whether they are hindered at all in their (proper) activities by serving and guarding us. No! Not at all! Their joy is not diminished, nor their equality, because the angel's work is to do the will of God and the will of God is the angel's work. If God told an angel to go to a tree and pick off the caterpillars, the angel would be glad to do it and it would be bliss to him because it *is* God's will.

Eckhart

Conform. 185

Put your heart, mind, intellect and soul even to your smallest acts. This is the secret of success.

Swami Sivananda

I desire to have in everything a purpose (*nîyah*), even in my eating, my drinking, and my sleeping.

Al-Ghazâlî

Conform. 170
Realiz. 870

Whether therefore ye eat, or drink, or whatsoever ye do, do all to the glory of God.

I. Corinthians, x. 31

Whatever act you do is worship, when it is done with the thought of God.

Swami Ramdas

[1] Sanctifying virtue, spiritual beauty.

Ch'ing, the chief carpenter, was carving wood into a stand for hanging musical instruments. When finished, the work appeared to those who saw it as though of supernatural execution. And the prince of Lu asked him, saying, 'What mystery is there in your art?'

'No mystery, your Highness,' replied Ch'ing; 'and yet there is something.

'When I am about to make such a stand, I guard against any diminution of my vital power. I first reduce my mind to absolute quiescence. Three days in this condition, and I become oblivious of any reward to be gained. Five days, and I become oblivious of any fame to be acquired. Seven days, and I become unconscious of my four limbs and my physical frame. Then, with no thought of the Court present to my mind, my skill becomes concentrated, and all disturbing elements from without are gone. I enter some mountain forest. I search for a suitable tree. It contains the form required, which is afterwards elaborated. I see the stand in my mind's eye, and then set to work. Otherwise, there is nothing. I bring my own natural capacity into relation with that of the wood. What was suspected to be of supernatural execution in my work was due solely to this.'

Contem. 532 · Infra 358 · *Creation* 42 · Infra 340

Chuang-tse (ch. XIX)

God the Only Doer

Allâh hath created you and what ye do.

Qur'ân, XXXVII. 96

All actions are performed by the Gunas,[1] born of Prakriti. One whose understanding is deluded by egoism alone thinks: 'I am the doer.'

Bhagavad-Gîtâ, III. 27

I am the machine, and Thou, O Lord, art the Operator. I am the house and Thou art the Indweller. I am the chariot and Thou art the Driver. I move as Thou movest me; I speak as Thou makest me speak.

Sri Ramakrishna

Man is the instrument of God.

Boehme

Conform. 170

Place yourself as an instrument in the hands of God who does His own work in His own way.

Swami Ramdas

[1] Cf. note p. 88.

The Father that dwelleth in me, he doeth the works.

St John, XIV. 10

Thou workest Thine own work; men only call it theirs.

Bengali Song

For it is God which worketh in you both to will and to do.

Philippians, II. 13

Death 220

My Majesty understandeth his divine power, and behold I worked under his direction; he was my leader. I was unable to think out a plan for work without his prompting.

Queen Hatshepsut (Inscription on her standing obelisk at Karnak)

For whoso desireth power: Lo! all power is with Allâh.

Qur'ân, XXXV. 10

Say not then, in your thought concerning Him who alone is good, that anything is impossible; for to Him belongs all power.

Hermes

Sin 56
Grace 552

Without me ye can do nothing.

St John, XV. 5

For noe Man sooner maie our Worke spill,
Then he that is presuminge his purpose to fulfill:
But he that shall trewlie doe the deede
He must use providence and ever worke with dreade.

Thomas Norton

Conform. 170

Allâh enlargeth provision for whom He will of His slaves, and justly apportions it. Lo! Allâh is Aware of all things.

Qur'ân, XXIX. 62

Outward existence (*'ain*) can perform no act of itself; its acts are those of its Lord immanent in it; hence this outward existence is passive, and action cannot be attributed to it.

Ibn 'Arabî

There's a divinity that shapes our ends,
Rough hew them how we will.

Shakespeare (*Hamlet,* v. ii. 10)

Thou threwest not when thou didst throw, but Allâh threw.

Qur'ân, VIII. 17

The negligent man at morn considers what he will do, while the intelligent man considers what it is Allâh will do with him.[1]

Illusion 99

Ibn 'Aṭâ'illâh

M.M. 978

As long as there is consciousness of diversity and not of unity in the Self, a man ignorantly thinks of himself as a separate being, as the 'doer' of actions and the 'experiencer' of effects. He remains subject to birth and death, knows happiness and misery, is bound by his own deeds, good or bad.

Srimad Bhagavatam, XI. iv

It is, in fact, the indefinable power of the Lord that ordains, sustains and controls everything that happens. Why then, should we languish tormented by vexatious thought, saying 'This wise to act; but no, that way . . .', instead of meekly but happily submitting ourselves to that Power?

Sri Ramana Maharshi

Devotee: My work demands the best part of my time and energy; often I am too tired to devote myself to *Atma-chintana*.

Maharshi: The feeling 'I work' is the hindrance. Enquire, 'Who works?' Remember, 'Who am I?' The work will not bind you. It will go on automatically. Make no effort either to work or to renounce work. Your effort is the bondage. What is bound to happen will happen.

If you are destined to cease working, work cannot be had even if you hunt for it. If you are destined to work you cannot leave it; you will be forced to engage in it. So leave it to the Higher Power. You cannot renounce or hold as you choose.

Supra 329

Sri Ramana Maharshi

'Manage not thy own affairs, and I will make everything thy servant.'

Niffarî

Before you do any thing think of God, that his light may precede your energies.

Sextus the Pythagorean

Were it not for His covering Grace, no work would be worthy of acceptance.

Ibn 'Aṭâ'illâh

In vain our labours are, whatsoe're they be,
Unless God gives the *Benedicite*.

Robert Herrick

[1] 'For the "naïve" and "impenitent" man, the world is a neutral space where one chooses agreeable situations while believing it possible to avoid the disagreeable, given a certain cleverness and luck. Quite evidently, the man who is unaware that existence is an immense furnace, has no imperious reason for wishing to leave it' (Schuon: *Sentiers de Gnose*, p. 91).

What honor can our works give to God if we do not perform them according to His good pleasure? The Lord doth not desire sacrifice, said the prophet to Saul, but obedience to His will.

Conform. 180

St Alphonsus Liguori

Know that the discipline of outward acts, though it subdue nature, cannot kill it. Nature dies by ghostly acts. There are many to be found who, with the best intentions, cling on to themselves, not denying themselves. Verily I say, these persons are mistaken, for it is contrary to human reason, contrary to the habit of grace and against the nature of the Holy Ghost. As for those who see their salvation in outward practices, I do not say they will be lost, but they will get to God only through hot cleansing fires; for they follow not God who quit not themselves; keeping hold of themselves they follow their own darkness. God is no more to be found in any bodily exercise than in sin.

Renun. 143

Death 208

Eckhart

The disciples then inquired if they could engage in worldly duties, in a small way, for the benefit of others, and Jetsün said, 'If there be not the least self-interest attached to such duties, it is permissible. But such (detachment) is indeed rare; and works performed for the good of others seldom succeed if not wholly freed from self-interest. Even without seeking to benefit others, it is with difficulty that works done even in one's own interest (or selfishly) are successful. It is as if a man helplessly drowning were to try to save another man in the same predicament. One should not be over-anxious and hasty in setting out to serve others before one hath oneself realized Truth in its fullness; to be so, would be like the blind leading the blind.'

Milarepa

Those abiding in the midst of ignorance,
Self-wise, thinking themselves learned,
Hard smitten, go around deluded,
Like blind men led by one who is himself blind.

Knowl. 734

Heterod. 418

Maitri Upanishad, VII. 9

And if the blind lead the blind, both shall fall into the ditch.

St Matthew, XV. 14

The five kinds of grains are considered good plants, but if the grains are not ripe, they are worse than cockles. It is the same with regard to kindness, which must grow into maturity.

Introd. *Charity*

Mencius

Give yourself up to ever so many good works, read, preach, pray, visit the sick, build hospitals, clothe the naked, etc., yet if anything goes along with these or in the doing of them you have anything else that you will and hunger after, but that God's kingdom may come and His will be done, they are not the works of the new-born from above and so cannot be his life-giving food.

Conform. 170

William Law

If I am asked in the place of Judgment why I have *not* done something, I shall be more pleased than if I am asked why I have done something—i.e. there is egoism in every act of mine, and egoism is dualism, and dualism is worse than sin, except as regards a pious act that is done upon me and in which I have no part.

Bâyazîd al-Bisṭâmî

Man is not holier or higher for the outward works that he does. Truly God that is the Beholder of the heart rewards the will more than the deed. The deeds truly hang on the will, not the will on the deeds.

Conform. 166

Richard Rolle

Whoever is the object of divine beneficence will be solicited on all sides. If he dispenses his gifts according to the will of heaven, his happiness will be lasting; if not, it will be ephemeral.

'Alî

God does not want precious presents. Many people spend millions of rupees in opening hospitals and feeding-houses. But they do not give their hearts.

Swami Sivananda

Do not be ashamed to give a little; for to deceive is to give still less.

'Alî

Our acts in no way help to make God give or do to us anything whatever. It is this idea our Lord would have his friends get rid of, so he undermines their faith in it to make them trust in him alone who intends bestowing largesse on them not for any reason but for love, that he may be their comfort and their stay and that they, finding that they themselves are really nothing, may admire in all the great gifts of God. For the more void and passive the mind that falls on God and is upheld by him, the deeper the man is gotten into God and the more receptive is he to the precious gifts of God. Man must build on God alone.

Faith 501
Love 618
Humility 191
Conform. 170

Eckhart

Whoever puts his confidence in acts of piety is more culpable than he who sins.

Bâyazîd al-Bisṭâmî

Can the sound of the word 'service' deceive the Lord? Does He not know? Is He waiting for these people's service?

Sri Ramana Maharshi

Say: O my people! Work according to your power. Lo! I too am working. Thus ye will come to know for which of us will be the happy sequel. Lo! the wrong-doers will not be successful.

Qur'ân, VI. 135

For so long as a man is not standing in the grace of God, he is standing alone in nature. And if such a man (were it possible, which it is not) were to fulfil all the good works which have ever been done in this world, he would still, nevertheless, be living altogether idly, unprofitably, and in vain, and it would avail him nothing.

Tauler

To be right, a person must do one of two things: either he must learn to have God in his work and hold fast to him there, or he must give up his work altogether. Since, however, man cannot live without activities that are both human and various, we must learn to keep God in everything we do, and whatever the job or place, keep on with him, letting nothing stand in our way.

Eckhart

Business men boast of their skill and cunning
But in philosophy they are like little children.
Bragging to each other of successful depredations
They neglect to consider the ultimate fate of the body.
What should they know of the Master of Dark Truth
Who saw the wide world in a jade cup,
By illumined conception got clear of Heaven and Earth:
On the chariot of Mutation entered the Gate of Immutability?

Ch'ên Tzû-ang

Judgment 250
Center 816
M.M. 978
Sin 62

Rivalry in worldly increase distracteth you
Until ye visit the graves.
Nay, but ye will come to know!
Nay, but ye will come to know!
Nay, would that ye could know with the knowledge of certainty!

Qur'ân, CII. 1–4

Illusion 99
Knowl. 749

And guard yourselves against a chastisement which cannot fall exclusively on those of you who are wrong-doers.

Qur'ân, VIII. 25

The soul cannot perform living works, unless it receives from the sun, i.e., from Christ, the assistance of the light of grace; unless it secures the protection of the moon, i.e., of the Virgin Mary, the Mother of Christ; and unless it imitates the examples of the other saints. And from the coming together of these a living and perfect work is gathered together in it. Whence the order of living hangs on those three.

St Bonaventura

Symb. 317
P. State 590
579
Orthod. 275

Attain discrimination and consider the universe as ephemeral. Reflect how all beings and things are subject to birth, growth, decay, and death—how fleeting are all. Having reflected thus, leave vain things to the vain, and gain tranquillity of mind.

Srimad Bhagavatam, XI. xiii

Reality 803
Peace 700

Follow me; and let the dead bury their dead.

St Matthew, VIII. 22

Act and Essence

Thou hast learned a trade to earn a livelihood for the body: (now) set thy hand to a religious (spiritual) trade.

Illusion 99

In this world thou hast become clothed and rich: when thou comest forth from here, how wilt thou do?

Learn such a trade that hereafter the earning of God's forgiveness may come in as revenue (to thee)....

The earnings of religion are love and inward rapture—capacity to receive the Light of God, O thou obstinate one!

This vile fleshly soul desires thee to earn that which passeth away: how long wilt thou earn what is vile? Let it go! Enough!

If the vile fleshly soul desire thee to earn what is noble, there is some trick and plot behind it.

Heterod. 418

Rûmî

That Christian in whom religion rules powerfully, is not so low in his ambitions as to pursue any of the things of this world as his ultimate end: his soul is too big for earthly designs and interests; but understanding himself to come from God, he is continually returning to him again. It is not worthy of the mind of man to pursue any perfection lower than its own, or to aim at any end more ignoble than itself.... It never more slides and degenerates from itself, than when it becomes enthralled to some particular interest: as on the other side it never acts more freely or fully, than when it extends itself upon the most universal end.

Creation 42
Sin 66
Contem. 547

John Smith the Platonist

But one thing is needful: and Mary hath chosen that good part, which shall not be taken away from her.

Reality 775

St Luke, x. 42

What is thy art?
To be good.

Marcus Aurelius

Acquaint now thyself with him, and be at peace.

Job, XXII. 21

People ought not to consider so much what they are to do as what they *are;* let them but *be* good and their ways and deeds will shine brightly. If you are just, your actions will be just too. Do not think that saintliness comes from occupation; it depends rather on what one is. The kind of work we do does not make us holy but we may make it holy. However 'sacred' a calling may be, as it is a calling, it has no power to sanctify; but rather as we *are* and have the divine being within, we bless each task we do, be it eating, or sleeping, or watching, or any other. Whatever they do, who have not much of (God's) nature, they work in vain.

Eckhart

Supra 340

The bound souls never think of God. If they get any leisure they indulge in idle gossip and foolish talk, or they engage in fruitless work. If you ask one of them the reason, he answers, 'Oh, I cannot keep still; so I am making a hedge.' When time hangs heavy on their hands they perhaps start playing cards.

Sri Ramakrishna

What is our work? Our only work here is to seek God, or do such work that will enable us to realize God. 'Worldly' work is what makes you forget God. A work in which you have no thought of God at all, how can that work help you in any way to attain God? ... Doing Karma in a state of Yoga i.e. doing actions in union with God—that is Karma Yoga. ... Do not think that the world will not go on without your work. God has engaged you in work only as His agent. If we are taken away He will find somebody else for that work.[1]

Swami Ramdas

Supra 329

'Thou art much too full of business; the essential for thee is to love Me.'

Sister Consolata

Jesus (Peace be upon him!) was asked about the best work, and he said, 'Resignation to God (Exalted is He!) and love for Him.'

Christ in Islâm

Conform. 180

To outward appearance they do little who are working all the time at the virtuous life, hence the disesteem of many people, which, however, they prefer to vulgar approbation.

Eckhart

Suff. 126b

I am negligent as if being obscure;
Drifting, as if being attached to nothing.
The people in general all have something to do,
And I alone seem to be impractical and awkward.
But I value seeking sustenance from the Mother.

Tao Te Ching, xx

Humility 197

Holiness 910

Such a (disciple of philosophy) may be compared to a man who has fallen among wild beasts—he will not join in the wickedness of his fellows, but neither is he able singly to resist all their fierce natures, and therefore seeing that he would be of no use to the State or

[1] 'At the famous conversation between Bodhidharma and the Emperor Wu-ti of Liang, when Wu-ti asked what merit he had gained through the building of many temples and the creating of many priests, Bodhidharma replied with the one word, "None" ' (Reikichi Kita and Kiichi Nagaya: *How Altruism Is Cultivated in Zen*, p. 135; tr. Ruth F. Sasaki).

to his friends, and reflecting that he would have to throw away his life without doing any good either to himself or others, he holds his peace, and goes his own way. He is like one who, in the storm of dust and sleet which the driving wind hurries along, retires under the shelter of a wall; and seeing the rest of mankind full of wickedness, he is content, if only he can live his own life and be pure from evil or unrighteousness, and depart in peace and good-will, with bright hopes.

Renun. 146

Plato (*Republic*, 496 D)

In the quiet of the morning I heard a knock at my door: threw on
my clothes and opened it myself.
I asked who it was who had come so early to see me.
He said he was a peasant, coming with good intent.
He brought a present of wine and rice-soup,
Believing that I had fallen on evil days.
'You live in rags under a thatched roof
And seem to have no desire for a better lot.
The rest of mankind have all the same ambitions:
You, too, must learn to wallow in their mire.'
'Old man, I am impressed by what you say,
But my soul is not fashioned like other men's.
To drive in their rut I might perhaps learn:
To be untrue to myself could only lead to muddle.
Let us drink and enjoy together the wine you have brought:
For my course is set and cannot now be altered.'

Supra 335
Peace 694

T'ao Ch'ien

It is related that the Messiah (God bless him and grant him peace!) passed in his wandering a man asleep wrapped up in his cloak; then he wakened him and said, 'O sleeper, arise and glorify God (Exalted is He!).' Then the man said, 'What do you want from me? Verily I have abandoned the world to its people.' So he said to him, 'Sleep, then, my friend.'

Supra 340

Christ in Islâm

Someone asked the Holy Prophet—
'What dost thou say concerning the things of the world?'
The Prophet said—'What can I say about them:
Things which are acquired with hard labour,
Preserved with perpetual watchfulness,
And left with regret.'

Illusion 99

Ansârî

'Gentlemen,' said Monkey laughing, 'I will not deceive you. If I had wanted to be an Emperor, I could have had the throne in any of the ten thousand lands and nine continents under heaven. But I have got used to being a priest and leading a lazy, comfortable existence. An Emperor has to wear his hair long; at nightfall he may not doze, at the fifth drum he must be awake. Each time there is news from the frontier his heart jumps; when there are calamities and disasters he is plunged in sorrow and despair; I should never get used to it. You go back to your job as Emperor, and let me go back to mine as priest, doing my deeds and going upon my way.'

Monkey

Uneasy lies the head that wears a crown.

Shakespeare (*Henry IV, Pt. 2,* III. i. 31)

The sweetness of life lies in dispensing with formalities.

'Alî

O God! make me busy with Thee, that they may not make me busy with them. *Metanoia* 493

Râbi'a of Baṣra

'If thou obtainest not possession of Me, will not other than I obtain possession of thee?'

Niffarî

Jesus (Peace be upon him!) said, 'O company of the disciples, be pleased with what is worthless in the world along with welfare in religion, just as the people of the world are pleased with what is worthless in religion along with welfare in the world.' *P. State* 563

Christ in Islâm

Since we exercise virtues falsely, but vices actually, we shall become falsely happy and actually miserable in so far as we ourselves are concerned. This is what Democritus laughed at, what Heraclitus deplored, what Socrates desired to cure, and what God can cure.

Marsilio Ficino

Only those who take leisurely what the people of the world are busy about can be busy about what the people of the world take leisurely.

Chang Ch'ao

I created the jinn and humankind only that they might worship Me. *Contem.* 547
I seek no livelihood from them, nor do I ask that they should feed Me.
Lo! Allâh! He it is that giveth livelihood, the Lord of Steadfast Power.

Qur'ân, LI. 56–58

Let them (the spiritual aspirants) put their money into charity and place their confidence in God. *Faith* 501

Swami Sivananda

SOUL
But how can I support myself
If I must burden myself with thee?

LOVE
Ah! Unfaithful! He who has made the soul so noble
Illusion 109 That it can enjoy nothing but God
Will never allow the body to want!

Mechthild of Magdeburg

Are you not ashamed of being more cowardly and mean-spirited than runaway slaves? How do they leave their masters when they run away? What lands or servants have they to trust to? Do not they steal just a morsel to last them for the first days, and then go on their way over land or it may be sea, contriving one resource after another to keep themselves alive? And when did a runaway slave ever die of hunger? Yet you are all of a flutter and keep awake at nights for fear you should run short of necessaries. . . . How often did you boast that you could face death at any rate with a quiet mind!

'Yes, but my family will starve.'

What of that? Does their hunger lead in a different direction? Is not the way that leads below the same, and the world it leads to the same? . . . Even the richest and those who have held the highest offices must descend, nay even kings and emperors. . . . Only you will descend hungry, if it so chance, and they will burst with over-eating and over-drinking. . . .

Suff. 120

Can you not draw water, or write, or take charge of children, or be another man's doorkeeper?

But it is disgraceful, you say, to be reduced to this necessity.

First learn then what is disgraceful, and then tell us that you are a philosopher; but for the present, if another call you so, do not allow him.

Epictetus

He who becometh aware of Thee
What use hath he for life,
For children, family or earthly things?

Ansârî

'It is base to be maintained by another.'

Creation 28

Miserable man, is there any one that maintains himself? Only the Universe does that.

Epictetus

There was a Sadhu in Malabar, a tall and stout person. He was in the police service before he became a Sadhu. He used to wear only a small towel round his waist. Once when he was going for his Bhiksha, a householder, seeing his good physique, asked him why he should not work and earn his bread, instead of begging for it. The Sadhu was told that he would be given a meal if he was prepared to cut a few logs of firewood that were lying in the householder's courtyard. The Sadhu, without uttering a word, started cutting the firewood with an axe given to him and within a short time cut the whole lot and stacked them in the proper place. Then, keeping the axe there, the Sadhu simply walked away. The householder saw the Sadhu going away without taking his food. He called him back and asked him why he was going away before getting his meal. The Sadhu then replied: 'I do not take my food where I work, and I do not work where I take my food!'

Swami Ramdas

The more you advance toward God, the less He will give you worldly duties to perform.

Sri Ramakrishna

The thing to do is to make the *Nembutsu* practice the chief thing in life, and to lay aside everything that you think may interfere with it. . . . If you cannot do it and at the same time provide yourself with food and clothing, then accept the help of others and go on doing it. Or if you cannot get others to help you, then look after yourself but keep on doing it. Your wife and children and domestics are for this very purpose, of helping you to practise it, and if they prove an obstacle, you ought not to have any. Friends and property are good, if they too prove helpful, but if they prove a hindrance they should be given up. In short, there is nothing that may not help us to *Ôjô,* so long as it helps us to go on the even tenor of our way through life undisturbed.

Hônen

Inv. 1036

Renun. 139

As long as a man has . . . worldly desires, he must perform actions and consequently suffer from worry, anxiety, and restlessness. No sooner does he renounce these desires than his activities fall away and he enjoys peace of soul. . . . When love of God is awakened, work drops away of itself. If God makes some men work, let them work. It is now time for you to give up everything.

Sri Ramakrishna

Is it they (the worldly-minded) who apportion the mercy of thy Lord? It is We who apportion among them their livelihood in the life of the world, and We raise some of them above others in rank, so that some may take labour from others; and the mercy of thy Lord is better than what they amass.

Rev. 967

And were it not that mankind would have become one nation,[1] We might well have appointed for the houses of those who disbelieve in the Beneficent, roofs of silver and (silver) stairs on which to mount,

And for their houses, (silver) doors, and couches on which to recline,

And ornaments of gold. Yet all that would have been but provision for the life of the world. And the Hereafter with thy Lord is for the God-fearing.

Qur'ân, XLIII. 32–35

Destruction of the natural integrity of things, in order to produce articles of various kinds,—this is the fault of the artisan. Annihilation of TAO in order to practise charity and duty to one's neighbour,—this is the error of the sage.

Chuang-tse (ch. IX)

[1] This was attempted by the people of Babel. The whole passage cited here runs counter to the illusion of human equality and the dream of human utopias.

The Messiah (Peace be upon him!) said, 'O you who seek the world to be charitable with it, your leaving of it alone is more charitable.' And he said, 'The least thing is such that looking after it occupies one to the exclusion of glorifying God, and glorifying God is greater and more important.'

Sin 64

Christ in Islâm

Not to commit faults counts for more than to do good.

'Alî

When on one side we place all the actions of this life and on the other silence, we find that it weighs down the scales.... It is better for you to cut yourself free from the bonds of sin than to liberate slaves from their servitude.

M.M. 987 *Holiness* 900

St Isaac the Syrian

Devotee: Wars are going on in the world....

Maharshi: Can you stop the wars? He who made the world will take care of it.... Take care of yourself and the world will take care of itself.

Supra 329 340

Sri Ramana Maharshi

Placed in this world of misery, man should take no heed of the lesser or greater sights of woe which present themselves to his view. They are as the fleeting tints and hues which paint the empty vault of the sky, and soon vanish into nothing.

Illusion 99

Yoga-Vasishtha

It is better to restore one dead heart to eternal life than life to a thousand dead bodies.

Charity 597

Pir Murâd

Worldly people have no stuff in them. They are like a heap of cow-dung. Flatterers come to them and say: 'You are so charitable and wise! You are so pious!' These are not mere words but pointed bamboos thrust at them.... Arbitration and leadership? How trifling these are! Charity and doing good to others? You have had enough of these.... If you realize God, you will get everything else. First God, then charity, doing good to others, doing good to the world, and redeeming people.

Reality 775

Sri Ramakrishna

A sight of Happiness is Happiness. It transforms the Soul and makes it Heavenly, it powerfully calls us to communion with God, and weans us from the customs of this world.... I no sooner discerned this but I was (as Plato saith, *In summâ Rationis arce quies habitat*) seated in a throne of repose and perfect rest.... Whereupon you will not believe, how I was withdrawn from all endeavours of altering and mending outward things. They lay so well, methought, they could not be mended: but I must be mended to enjoy them.

Knowl. 749 *Beauty* 689

Thomas Traherne

Look within. Seek the Self! There will be an end of the world and its miseries. . . . The world is not external. Because you identify yourself wrongly with the body you see the world outside, and its pain becomes apparent to you. But they are not real. Seek the reality and get rid of this unreal feeling.

Sri Ramana Maharshi

We rather glorify God by entertaining the impressions of his glory upon us, than by communicating any kind of glory to him. Then does a good man become the tabernacle of God wherein the divine Shechinah does rest, and which the divine glory fills, when the frame of his mind and life is wholly according to that idea and pattern which he receives from the mount. We best glorify him when we grow most like to him.

John Smith the Platonist

Introd.
Peace
Beauty
670

Dhu 'l-Nûn on his travels is accosted by a *majnûna* (mad woman devotee). 'Dost thou know what generosity is?' she asks him.

'It is to give.'

'In this world, no doubt. But in religion, what is generosity?'

'It is to be instant to obey the Creator of the worlds.'

Dhu 'l-Nûn

The clear perception of the Mind Unmodified,
And the noble impulse to serve others,
Appear to be alike, but beware, and confuse them not.

Milarepa

My dear sir, who are you? What good will you do to the world? Is the world such a small thing that you think you can help it?

Sri Ramakrishna

If you cannot govern your own self, what leisure have you for governing the empire? Begone! Do not interrupt my work.

Chuang-tse (ch. XII)

Holiness
921

Reformers have come and gone; but the ancient *smritis* still stand.

Sri Ramana Maharshi

Reality
790

The dogs bark; the caravan passes.

Arabic Proverb

Contem.
528

Sambhu said to me: 'It is my desire to build a large number of hospitals and dispensaries. Thus I can do much good to the poor.' I said to him: '. . . Sambhu, let me ask you one thing. If God appears before you, will you want Him or a number of hospitals and dispensaries?'

Sri Ramakrishna

Devotee: Is it not selfishness to remain Self-realised without helping the world?

Introd. *Heterod.*

Maharshi: The Self was pointed out to you to cover the universe and also transcend it. The world cannot remain apart from the Self. If the realisation of such Self be called selfishness that selfishness must cover the world also. It is nothing contemptible.

Sri Ramana Maharshi

Beauty 679 Introd. *Knowl.*

Over-refinement of vision leads to debauchery in colour; over-refinement of hearing leads to debauchery in sound; over-refinement of charity leads to confusion in virtue; over-refinement of duty towards one's neighbour leads to perversion of principle; over-refinement of music leads to lewdness of thought; over-refinement of wisdom leads to an extension of mechanical art; and over-refinement of shrewdness leads to an extension of vice. . . .

Holiness 924

M.M. 987

Therefore, for the perfect man who is unavoidably summoned to power over his fellows, there is naught like inaction. By means of inaction he will be able to adapt himself to the natural conditions of existence. And so it is that he who respects the state as his own body is fit to support it, and he who loves the state as his own body is fit to govern it. And if I can refrain from injuring my internal economy, and from taxing my powers of sight and hearing, sitting like a corpse while my dragon-power is manifested around, in profound silence while my thunder-voice resounds, the powers of heaven responding to every phase of my will, as under the yielding influence of inaction all things are brought to maturity and thrive,—what leisure then have I to set about governing the world?[1]

Chuang-tse (ch. XI)

Count ye the slaking of a pilgrim's thirst and tendance of the Inviolable Place of Worship as (equal to the worth of) him who believeth in Allâh and the Last Day, and striveth in the way of Allâh? They are not equal in the sight of Allâh. Allâh guideth not wrongdoing folk.

Those who believe, and have left their homes and striven with their wealth and their lives in Allâh's way are of much greater worth in Allâh's sight. These are they who are triumphant.

Qur'ân, IX. 19, 20

Charity 597

Supra 340

If a householder gives in charity in a spirit of detachment, he is really doing good to himself and not to others. It is God alone that he serves—God, who dwells in all beings; and when he serves God, he is really doing good to himself and not to others. . . . Helping others, doing good to others—this is the work of God alone, who for men has created the sun and moon, father and mother, fruits, flowers, and corn. The love that you see in parents is God's love: He has given it to them to preserve His creation. The compassion that you see in the kind-hearted is God's compassion: He has given it to them to protect the helpless. Whether you are charitable or not, He will have His work done somehow or other. Nothing can stop His work.

[1] This passage shows a profound spiritual affinity with the outlook of the American Indian, without it being any question of a 'borrowing'—especially where the outward forms of the two societies in question are so evidently dissimilar. But see the reference under *I Ching* In the INDEX OF SOURCES.

What then is man's duty? What else can it be? It is just to take refuge in God and to pray to Him with a yearning heart for His vision.

Sri Ramakrishna

Contem. 547

One single glimpse of the abstraction that God is, more unifies the soul with God than all the works of holy Christendom.

Dionysius

M.M. 975

One lifetime spent in the quest for Enlightenment is more precious than all the lifetimes during an aeon spent in worldly pursuits.

Gampopa

Know that the greatest things which are done on earth are done within, in the hearts of faithful souls.

St Louis de Montfort

An hour of meditation is worth more than the good works accomplished by the two species of ponderable beings.[1]

Muhammad

Children, could we but truly stand in this holy of holies for an hour or a moment, it were a thousand times better and more profitable for us, and more pleasing and praiseworthy in the sight of the Eternal God, than forty years spent in your own self-imposed tasks.

Tauler

'Where thou hesitatest between two courses of action, always choose the one which leaves thee more alone, more in silence, more in love.'

Sister Consolata

We must know that we can have two kinds of happiness in this life, according to two different ways, one good, one best, which lead us thereto: one is the active life, and the other the contemplative. The latter (although by the active life, as has been said, we may attain to great happiness) leads us to the highest felicity and blessedness, as the Philosopher proves in the tenth of the *Ethics*.

Dante (*Il Convito*, IV. xvii. 16)

For as the soul is more worthy than the body, so the knitting of the soul to God (the life of it) by the heavenly food of charity is better than the knitting of the body to the soul (the life of it) by any earthly food in this life. This is good for to do by itself, but without the other it is never well done. This and the other is the better; but the other by itself is the best. For this by itself deserveth never salvation; but the other by itself (where the plenty of this faileth) deserveth not only salvation, but leadeth to the greatest perfection.

The Epistle of Privy Counsel, III

Introd. *Judgment*

M.M. 973

[1] Men and jinn.

Even though you may be doing something else, let that be done while you go on with the main work of life, which consists in the practice of the *Nembutsu,* and do not let it be a sort of side work to anything else.

Inv. 1017

Hônen

A spark of pure love is more precious before God, more useful for the soul and more rich in benedictions for the Church than all the other works taken together, even if, according to appearances, one does nothing.

St John of the Cross

The man who lives in silent solitude is not only not living in a state of inactivity and idleness; he is in the highest degree active, even more than the one who takes part in the life of society. He untiringly acts according to his highest rational nature; he is on guard; he ponders; he keeps his eye upon the state and progress of his moral existence. This is the true purpose of silence. And in the measure that this ministers to his own improvement it benefits others for whom undistracted submergence within themselves for the development of the moral life is impossible. . . . And he does more, and that of a higher kind, than the private benefactor, because the private, emotional charities of people in the world are always limited by the small number of benefits conferred, whereas he who confers benefits by morally attaining to convincing and tested means of perfecting the spiritual life becomes a benefactor of whole peoples. . . . The silent recluse teaches by his very silence, and by his very life he benefits, edifies and persuades to the search for God. This benefit springs from genuine silence which is illuminated and sanctified by the light of grace. But if the silent one did not have these gifts of grace which make him a light to the world, even if he should have embarked upon the way of silence with the purpose of hiding himself from the society of his kind as the result of sloth and indifference, even then he would confer a great benefit upon the community in which he lives, just as the gardener cuts off dry and barren branches and clears away the weeds so that the growth of the best and most useful may be unimpeded. And this is a great deal. It is of general benefit that the silent one by his seclusion removes the temptations which would inevitably arise from his unedifying life among people and be injurious to the morals of his neighbours.

Realiz. 859

The Russian Pilgrim

Realisation of the Self is the greatest help that can be rendered to humanity. Therefore, the saints are said to be helpful, though they remain in forests.

Sri Ramana Maharshi

In vessels which are in a state of mutiny and by sailors who are mutineers, how will the true pilot be regarded? Will he not be called by them a prater, a star-gazer, a good-for-nothing?

In deeming the best votaries of philosophy to be useless to the rest of the world . . . attribute their uselessness to the fault of those who will not use them, and not to themselves. . . . The ruler who is good for anything ought not to beg his subjects to be ruled by him; although the present governors of mankind are of a different stamp; they may be justly compared to the mutinous sailors, and the true helmsmen to those who are called by them the good-for-nothings and star-gazers.

Plato (*Republic*, 489 A)

The Way does not itself produce results in men, but those who do produce such results need the Way for the production. The foot marches on the road and so moves on: but the road which is marched on needs to be a non-marching thing, for the body needs hand and foot if it is to move and thereby depends on something which does not move. Thus it is that sometimes the thing which is useless and yet useful is needed to produce results, and the non-production of results is, as it were, what (everything) depends on.

Wang Ch'ung

Reality 773
Infra
358
Holiness
910

When the person meditating has reached mindlessness, he penetrates and traverses the entire world. Ignorant people bear a false accusation, when they charge with egoism the sadhus who meditate in caves.

Swami Sivananda

Ecstasy 636
Void
724

There are some persons, though very few such are found, who, subduing their pleasures and neglecting public affairs, conduct their life in such a way that they burn with eagerness to attain truth; but they do not have faith that it can be investigated through its human traces, in which the ambiguous minds of most natural philosophers are accustomed to trust. Therefore they give themselves to God and do not attempt anything by themselves. With open and purified eyes they wait for what may be shown by God, and this is what Socrates is said to have taught and to have done.

Marsilio Ficino

Knowl.
734
Supra
340

Some Alchemists fancy that the work from beginning to end is a mere idle entertainment; but those who make it so will reap what they have sown—viz., nothing. We know that next to the Divine Blessing, and the discovery of the proper foundation, nothing is so important as unwearied industry and perseverance in this First Operation. It is no wonder, then, that so many students of this Art are reduced to beggary; they are afraid of work, and look upon our Art as mere sport for their leisure moments. For no labour is more tedious than that which the preparatory part of our enterprise demands. Morienus earnestly entreats the King to consider this fact, and says that many Sages have complained of the tedium of our work. 'To render a chaotic mass orderly,' says the Poet, 'is matter of much time and labour'—and the noble author of the Hermetical Arcanum describes it as an Herculean task.

Philalethes

Grace 552
Orthod.
280

Introds.
Pilg.
Inv.

Whatsoever grounds there be for good works undertaken with a view to rebirth, all of them are not worth one sixteenth part of that goodwill which is the heart's release.

Itivuttaka

For the inward work is always better than the outward; and from it the outward works of virtue draw all their power and efficacy. It is as if thou hadst a noble excellent wine, of

Ecstasy 637 Realiz. 868

such virtue that a drop of it poured into a cask of water would be enough to make all the water taste like wine and turned it into good wine. This would be a great miracle: and so it is with the noble, excellent, inward work of the soul compared to the outward.

Tauler

Ecstasy 641

Do this work (of contemplation) evermore without ceasing and without discretion, and thou shalt know well how to begin and cease in all other works with a great discretion.

The Cloud of Unknowing, XLII

Peace 700

Kill thy activities and still thy faculties if thou wouldst realise this birth in thee.

Eckhart

Death 208

Stop the working of your bodily senses, and then will deity be born in you.

Hermes

Void 728

Let the spirit turn away from outward action, and fall never away from its stillness.

Śânti-deva

Sin 66

Could the soul see God as clearly as the angels do she would never have come into the body.

Eckhart

Suff. 120 Supra 329

Had I had the chance of plunging this spiritual son of mine nine times into utter despair, he would have been cleansed thoroughly of all his sins. He would thus not have been required to be born again, but would have disappeared totally, his physical body being forever dissolved; he would have attained *Nirvâṇa*. That it will not be so, and that he will still retain a small portion of his demerits, is due to Damema's[1] ill-timed pity and narrow understanding.

Marpa

Heaven and earth do nothing, yet there is nothing which they do not accomplish.

Chuang-tse (ch. XVIII)

Non-Acting Action

A man does not attain to freedom from action by non-performance of action, nor does he attain to perfection merely by giving up action.

No one can ever rest even for an instant without performing action, for all are impelled by the Gunas, born of Prakriti, to act incessantly. . . .

[1] Marpa's well-intentioned wife.

That man, who is devoted to the Self, is satisfied with Self and is content in the Self alone, for him there is nothing to do.

For him there is nothing in this world to gain by action or to lose by inaction: nor does he need to depend on any being for any object. . . .

Holiness 902

O Pârtha, there is nothing for Me to accomplish: nothing there is in the three worlds unattained or to be attained by Me, and yet I continue in action. . . .

Reality 773

One should not unsettle the understanding of the ignorant who are attached to action: the man of wisdom, by steadily performing actions, should engage (the ignorant) in all right action.

Bhagavad-Gîtâ, III (passim)

If any man acts in such a way that his deeds are able to demean him, be sure he is not acting according to the law of God's kingdom. When works are wrought according to humanity weeds and discord soon fall among them, but he whose work is wrought in the kingdom of heaven remains tranquil in every undertaking. . . . Verily I say unto you: works wrought out of the kingdom of God are dead works but works wrought in the kingdom of God are living works. The prophet says: 'God as little loves his works as he is disturbed and changed by them.' And so with the soul when she works in accordance with the law of God's kingdom. People of this sort are always the same whether they work or whether they work not, for works give nothing to them and take nothing from them.

Supra 340

Eckhart

Do not permit the events of your daily lives to bind you, but never withdraw yourselves from them. Only by acting thus can you earn the title of 'A Liberated One'.

Renun. 136 139

Huang Po

So long as egoity lasts *prayatna* (effort) is necessary. When egoity ceases to be, actions become spontaneous. The ego acts in the presence of the Self. He cannot exist without the Self.

Sri Ramana Maharshi

If you wish to work properly, you should never lose sight of two great principles: first, a profound respect for the work undertaken, and second, a complete indifference to its fruits. Thus only can you work with the proper attitude. This is called the secret of Karma-Yoga. And you can overcome all aversion for a piece of work, if you but consider it as belonging to God.

Supra 329

Swâmi Brahmânanda

If the fruits of actions do not affect the person he is free from action.

Sri Ramana Maharshi

Him the sages call wise whose undertakings are devoid of desire for results and of plans, whose actions are burned by the fire of wisdom.

Knowl. 761

Having abandoned attachment for the fruits of action, ever content and dependent on none, though engaged in action, yet he does nothing.

Bhagavad-Gîtâ, IV. 19, 20

Tao is ever inactive, and yet there is nothing that it does not do.

Tao Te Ching, XXXVII

Reality 803
Creation 31

God is ever at his work, and is himself that which he makes.

Hermes

Inv.
passim

O Saichi, what makes you work?
I work by the 'Namu-amida-butsu.'
'Namu-amida-butsu! Namu-amida-butsu!'

Saichi

Act non-action; undertake no undertaking; taste the tasteless.

Tao Te Ching, LXIII

PILGRIMAGE—DESCENT INTO HELL

Proclaim unto mankind the pilgrimage.

Qur'ân, XXII. 27

Visita *Interiora Terrae Rectificando Invenies Occultum Lapidem.*[1]

Alchemical Adage

The spiritual work is often likened to a journey, and it has already been stated in the Introduction how the structure of this book itself is ideally envisaged in terms of a progressive recovery of the Supreme Reality. The successive stages are clearly alluded to by Meister Eckhart:

'When I preach, I usually speak of disinterest and say that a man should be empty of self and all things' (this corresponds with purification, the 'initiatic death', the 'Descent into Hell', alchemical 'putrefaction'—dealt with in BOOK ONE of this work);

'and secondly, that he should be reconstructed in the simple Good that God is' (this corresponds with the Primordial State, 'true man', the accomplishment of the Lesser Mysteries, human amplitude, Hindu *bâlya*—treated of in BOOK TWO);

'and thirdly, that he should consider the great aristocracy which God has set up in the soul, such that by means of it man may wonderfully attain to God' (this refers to the Intellect, which leads to Pure Being, 'transcendent man', exaltation in the Greater Mysteries, *fanâ, Nirvâna, prâjna*—treated of in BOOK THREE);

'and fourthly, of the purity of the divine nature' (corresponding with the unconditioned, the Void, the Absolute, final deliverance, *moksha, Pari-Nirvâna, fanâ'u 'l-fanâ'*—likewise dealt with in BOOK THREE).[2]

Tao can be rendered simply as 'Way', just as in Buddhism, the *Mahâ-yâna* is the 'Great Way', and in Hinduism, the *Deva-yâna* is the 'Way of the Gods', while in Christianity Christ says, 'I am the Way.' In Judaism Moses leads the Israelites out of Egypt and shows the Way to the Holy Land, so in Islam the Prophet leads the Way into Mecca. Interiorly this is the 'Straight Way' (*ṣirâṭ al-mustaqîm*; the Sufic *ṭarîqah*=Way), the 'Strait Way' of the Gospels, and 'Red Road' of the Sioux, or 'Holy Path' (*Shôdô*) of the Japanese. All these Ways reach their confluence in the World Axis, which unites the different states of being with their uncreate Source.[3] The Way can equally be envisaged as a Pilgrimage (to

Introd.
Wr. Side

[1] 'Visit the interior of the earth: in rectifying thou wilt find the hidden stone.'

[2] Blakney, p.1. Cp. Black Elk's words (in Brown: *The Sacred Pipe*, p. 100): 'It is four steps to the end of the sacred path.'

[3] Cf. *The Way and the Mountain*, by Marco Pallis.

Jerusalem, Mecca, Lhasa, Benares, Ise), or as a Quest (for the Holy Grail, the Terrestrial Paradise, the Fountain of Immortality), or a Voyage:

'Among the emblems which were formerly those of Janus, the Papacy has retained not only the keys, but also the bark, equally attributed to Saint Peter and become the figure of the Church: its "Roman" character demanded this transmission of symbols. . . . The figure of navigation was often used in Greco-Roman antiquity: one can cite notably as examples, the expedition of the Argonauts in search of the "Golden Fleece", and the voyages of Ulysses; it also appears in Virgil and Ovid. In India likewise, this image is sometimes met with, and we shall cite a passage containing expressions which are singularly similar to those in Dante:[1] "The Yogî", says Shankarâchârya, "having crossed the sea of passions, is united with tranquillity and possesses the 'Self' in its plenitude" (*Atmâ-Bodha*). The "sea of passions" is clearly the same as the "waves of cupidity", and in both texts it is equally a question of "tranquillity": what symbolical navigation really represents is the quest of the "great peace". . . . The port towards which mankind must be directed is the "sacred isle" which rests immutable amidst the incessant agitation of waves, and which is the "Mountain of Salvation", the "Sanctuary of Peace" (Guénon: *Autorité spirituelle et Pouvoir temporel,* pp. 106–108).

Introd. *Sp. Drown.*

Peace 705

Finally, the Way can be envisaged as a journey into subterranean regions (in search of buried 'treasure', for example), which are properly speaking infernal, and then it is called a 'Descent into Hell'; the emphasis of the present chapter from its position in the book necessarily falls on this last point of view:

'Death and descent into Hell on one side, resurrection and ascent to Heaven on the other, these are as two inverse and complementary phases, of which the first is the indispensable preparation for the second, and which could likewise be recognized without difficulty in the description of the Hermetical "Great Work"; moreover the same thing is clearly affirmed in all traditional doctrines. Thus in Islam we encounter the episode of the "night journey" of Muhammad, equally including the descent into the infernal regions (*isrâ*), then the ascent to the various paradises or celestial spheres (*mirâj*); and certain parts of this "night journey" present similarities with Dante's poem[2] which are particularly striking' (Guénon: *L'Ésotérisme de Dante,* pp. 36–37). The Christian equivalent, it need hardly be said, is the Passion, Death and Resurrection of Christ.

Realiz. 859 Introd. *Judgment*

The Descent into Hell from the initiatic point of view is the exploration of the soul,—the examination, ordering and mastery of all the possibilities latent in the psychic realm, 'unleashed' as it were in the process of this work, but excluded in large measure from the consciousness of the profane person. 'One could call it a sort of "recapitulation" of antecedent states, by which the possibilities relating to the profane state will be definitively exhausted, in order that the being henceforth may develop freely the possibilities of a higher order which he carries in himself, and the realization of which belongs properly to the initiatic domain' (Guénon: *Aperçus sur l'Initiation,* p. 183).

'In alchemy,' according to Titus Burckhardt, 'the soul is considered as being substantially identical with the *materia prima* of the formal world. . . . It is therefore legitimate to "situate" symbolically the *materia prima* of the world at the bottom of the human soul by analogy with its symbolical "situation" in regard to the macrocosm, of which it is as the

[1] *De Monarchia,* III. 16.

[2] *The Divine Comedy.*

lower pole. . . . "Matter" first of all presents itself to the "artist" under a brute aspect or chaotic state, which does not mean a state of disorder, but a state "before order". . . . It is in himself that the alchemist discovers this "chaos", after having reduced his soul to its immediate substratum, in clearing from it all that is relatively "exterior". In this state, matter is both something "vile", because very common, and the thing that is "precious", since from it will be extracted the Elixir, and therefore the spiritual gold. . . . The chaos of "matter" is necessarily dark and impenetrable, since the determinations inherent in it have not yet passed from "power" to "act"; all potentiality is by definition impenetrable' ('Considérations sur l'Alchimie', *Études Trad.,* 1949, p. 117 ff.). But it is precisely when this chaos has been worked, purified and reduced to its pristine, 'virginal' condition, that the spiritual *Fiat Lux* takes place, the marriage of soul and spirit, symbolized in the union of Mercury with Sulphur, at once ordering, illuminating and transmuting what was in darkness, 'without form, and void' (*Gen.* I. 2), into a celestial mirror resplendent with the heavenly archetypes.

P. State 563

Introds. *Realiz. Rev. Inv. Center* 828

In the same article the author warns against falsely assimilating traditional ideas to the notion of a 'collective subconscious' (current in modern psychology) whose 'folkloric' residuum can engender strange confusions: 'The "passage from power to act", concerning the interior alchemy, of certain psychic possibilities hitherto latent, has no similarity with the proceedings of modern psychoanalysis: the latter occupies itself solely in bringing to the surface the lower psychic possibilities, by definition negative and atrophied, which it identifies moreover with the intimate nature of the human being. Alchemy in contrast actualizes the positive qualities of the soul; and if at the same time the psychic dross or residues mount to the surface, these will always be considered as foreign to the true nature of the soul, and will have to be dissolved in the soul's full expansion, as impurities disappear in the ocean, shadows in bright light, and refuse in the fire.'[1]

Introd. *Wr. Side*

Guénon likewise warns against being 'confused or dissolved in a sort of "cosmic consciousness" from which all transcendency is excluded. . . . Those who make this fatal error forget or are quite simply unaware of the distinction of the "Higher Waters" from the "Lower Waters". Instead of raising themselves to the Ocean above, they sink into the depths of the Ocean below; instead of concentrating all their powers to the end of directing them towards the formless world, which alone can be called "spiritual", they disperse them into the indefinitely changing and fleeting diversity of forms of the realm of subtle manifestation (which corresponds as exactly as possible with the Bergsonian conception of "reality"), and this without suspecting that what they take for a plenitude of "life" is in reality but the kingdom of death and dissolution without return" (*Le Règne de la Quantité,* pp. 234–235).

'The fact that spiritual causes can engender darkness does not at all mean that darkness always has spiritual causes,' writes Schuon.

'The criterion is in the nature of the darkness undergone and in the manner of

[1] '"Primordial images" are not of subconscious but superconscious origin' (Coomaraswamy, in PMLA, June, 1946). Moreover, psychiatry's vaunted 'discovery' of labyrinthine complexes motivated by an unrecognized and subconscious past amounts to little but an inverse fragment of the traditional laws of *Karma* concerning actions and concordant reactions, where the causal continuity resulting from past impressions (*samskâras*) precisely is to be ruptured or surmounted—not yielded to—in the process of liberation from the *samsâra*.

undergoing it. A darkness that is spiritually necessary is passive and never abolishes continuity of judgment or humility; it engenders suffering, not error; it could not tarnish the life of a saint. Man has a right to affliction, but never to sin: to lack light is a different thing from extinguishing the light one already has' (*Perspectives spirituelles,* p. 177).

'To describe the "descent into hell" summed up in the word VITRIOL, alchemy has retained archaic symbols: it speaks of a nocturnal voyage under the sea where the hero, often compared to Jonah, is swallowed by a monster. But the belly of the Leviathan becomes a womb: an egg is formed around the imprisoned man; an intense heat prevails, so violent that the hero loses all his hair: when the monster casts him out, he surfaces on the primordial sea as bald as a newly-born.

Introd. *Suff.* *Waters* 657

'He is in fact reborn, and each detail of this symbolism is charged with meaning: the sea merged with night is the obscure *materia,* the humidity of Mercury. The monster is the Ouroboros, guardian of the latent energy, analogous to the serpent of the Kundalînî in Tantric doctrine. The heat, finally, is that of passion: the victory of the hero will be in making it a heat of "auto-incubation", a renovating fervour: then the world is no longer a tomb but a womb, and the hero, fertilizing himself,[1] becomes the egg from whence he will be reborn' (Maurice Aniane: 'Notes sur l'alchimie, "yoga" cosmologique de la chrétienté médiévale' in *Yoga, Science de l'Homme intégral,* edit. Jacques Masui, Paris, 1953, pp. 254–255).

'Thus are the steps of the Work outlined: at first "mortification", the descent and dissolution in the waters, the disappearance into the womb of the Mother, the *Anima Mundi,* who devours and kills her Son,—that is to say, takes back to herself the man who has been led astray in the individual condition. It is the domination of the Woman over the Man, of the Moon over the Sun, until, in the Soul brought back to its original virginity, the luminous center, the Spirit, manifests. Then the regenerated Son is born, the solar hero: in his turn, he subjects the Moon to the Sun, the Woman to the Man, and by the consummation of the "philosophical incest", he makes his Mother his Wife and his Daughter too' (id. pp. 251–252).

Rev. 967 *P. State* 579

But in its highest sense, the Pilgrimage is a *return* from the agitation and multiplicity of this world to the peace and beatitude of the Divine Center.[2]

Metanoia 488

[1] The intervention of a spiritual influence is presupposed: the wording here simply implies that the 'performer' in this work follows the active (initiatic) way.

[2] And in this sense of a peaceful withdrawal from the world, could come under PART IV of this book.

Wayfaring

Jesus said: The world is a place of transition, full of examples: be pilgrims therein, and take warning by the traces of those that have gone before.

Christ in Islâm

Illusion 99

As travelers meet by chance on the way, so does a man meet wife, children, relatives, and friends: let him therefore be in the world and yet separate from it.

Srimad Bhagavatam, XI. xi

Renun. 139

God Most Glorious has appointed the Pilgrimage to be the monasticism of Muhammad's community.

Muhammad

Infra 387

All whom Moses calls wise are represented as sojourners. Their souls are never colonists leaving heaven for a new home. Their way is to visit earthly nature as men who travel abroad to see and learn. So when they have stayed awhile in their bodies, and beheld through them all that sense and mortality has to show, they make their way back to the place from which they set out at first. To them the heavenly region, where their citizenship lies, is their native land: the earthly region in which they became sojourners is a foreign country.

Philo

Jesus said: Become passers-by.

The Gospel according to Thomas, Log. 42

O ye who believe! What aileth you that when it is said unto you: Go forth in the way of Allâh, ye are bowed down to the ground with heaviness. Take ye pleasure in the life of the world rather than in the Hereafter? The comfort of the life of the world is but little in the Hereafter.

Qur'ân, IX. 38

Holy War 394

It is natural for man to lead the life of a pilgrim, particularly if that pilgrimage be directed towards a certain goal.

Michael Maier

You failed to go on the pilgrimage because of your ass's nature, not because you have no ass.

Dîvâni Shamsi Tabrîz, XLIII

Metanoia 488

The journey of the pilgrims is two steps and no more:
One is the passing out of selfhood,
And one towards mystical Union with the Friend.

Shabistarî

Center 833

The path of the just is as the shining light, that shineth more and more unto the perfect day.

Proverbs, IV. 18

The Descent Into Hell

Creation 26

Journey in the earth and see how He (Allâh) hath brought forth the creation.

Qur'ân, XXIX. 20

Sp. Drown. *Center* 847 *Renun.* 158

Though God be everywhere present, yet He is only present to thee in the deepest and most central part of thy soul. . . . This depth is called the centre, the *fund* or bottom of the soul. . . . Awake, then, thou that sleepest, and Christ, who from all eternity has been espoused to thy soul, shall give thee light. Begin to search and dig in thine own field for this pearl of eternity that lies hidden in it; it cannot cost thee too much, nor canst thou buy it too dear, for it is *all*; and when thou hast found it thou wilt know that all which thou hast sold or given away for it is as mere a nothing as a bubble upon the water.

William Law

Sin 66

You do not see clearly the evil in yourself, else you would hate yourself with all your soul.

Like the lion who sprang at his image in the water, you are only hurting yourself, O foolish man.

When you reach the bottom of the well of your own nature, then you will know that the vileness was from yourself.

Rûmî

Judgment 239

Tremble, since you must dwell for a trimester under the earth with the dead in order to get rid of all your imperfections.

Irenaeus Agnostus

Precious gems are profoundly buried in the earth and can only be extracted at the expense of great labour.

Ananda Moyî

Can I go forward when my heart is here?
Turn back, dull earth, and find thy centre out.

Shakespeare (*Romeo and Juliet*, II. i. 1)

Oh, thrice blessed the mortals, who, having contemplated these Mysteries, have descended to Hades: for those only will there be a future life of happiness—the others there will find nothing but suffering.

Sophocles

All flesh that is derived from the earth, must be decomposed and again reduced to earth; then the earthy salt produces a new generation by celestial resuscitation. For where there was not first earth, there can be no resurrection in our Magistery. For in earth is the balm of Nature, and the salt of the Sages.

Basil Valentine

Allâh hath caused you to grow as a growth from the earth,
And afterward He maketh you return thereto, and He will bring you forth again, a new forthbringing.

Qur'ân, LXXI. 17, 18

Death 208
P. State 563

Christ's soul must needs descend into hell, before it ascended into heaven. So must also the soul of man. But mark ye in what manner this cometh to pass. When a man truly perceiveth and considereth himself, who and what he is, and findeth himself utterly vile and wicked, and unworthy of all the comfort and kindness that he hath ever received from God, or from the creatures, he falleth into such a deep abasement and despising of himself, that he thinketh himself unworthy that the earth should bear him, and it seemeth to him reasonable that all creatures in heaven and earth should rise up against him and avenge their Creator on him, and should punish and torment him; and that he were unworthy even of that. And it seemeth to him that he shall be eternally lost and damned, and a footstool to all the devils in hell, and that this is right and just and all too little compared to his sins which he so often and in so many ways hath committed against God his Creator. . . . Whilst a man is thus in hell, none may console him, neither God nor the creature, as it is written, 'In hell there is no redemption.' . . .

Realiz. 859
Sin 66
Humility 191
Wr. Side 464

Now God hath not forsaken a man in this hell, but He is laying his hand upon him, that the man may not desire nor regard anything but the Eternal Good only, and may come to know that that is so noble and passing good, that none can search out or express its bliss, consolation and joy, peace, rest and satisfaction. And then, when the man neither careth for, nor seeketh, nor desireth, anything but the Eternal Good alone, and seeketh not himself, nor his own things, but the honour of God only, he is made a partaker of all manner of joy, bliss, peace, rest and consolation, and so the man is henceforth in the Kingdom of Heaven.

Infra 379
Conform. 170

This hell and this heaven are two good, safe ways for a man in this present time, and happy is he who truly findeth them.

For this hell shall pass away,
But Heaven shall endure for aye.

Theologia Germanica, XI

Verily he who seeks for pearls must dive to the bottom of the sea, endangering his very existence.

Gujarati Hymn

Suff. 130
Metanoia 488

Philosophers ador'd the Night, accounting it to have some great Mystery and Deity in it. The Night of Christ's Cross hath very much Mysterious and Divine in it. *They that go down into the Deep, see the Wonders of the Lord,* Psal. 107.23. 'Tis true of a deep of Woes, as well as Waters. He that is content to enter into the Cloud, and the saddest Retreat of it, meets with Wonders and Secrets of Glory.

Peter Sterry

The Lord thy God is a consuming fire.

Deuteronomy, IV. 24

Inv. 1007

The fire emanating from the man who contemplates devours him.

Hekhaloth Rabbati, III. 4

When I was within, I would have flung me into molten glass to cool me, so immeasurable there was the burning.

Dante (*Purgatorio,* XXVII. 49)

Jesus said: Whoever is near to me is near to the fire, and whoever is far from me is far from the Kingdom.

The Gospel according to Thomas, Log. 82

The (alchemical) operation begins with fire and ends with fire.

Ibn Bishrûn

Introd. *Death*

'What is Zen?'
'Boiling oil over a blazing fire.'

Zen *Mondo*

Contem. 528
Holy war 407
Introd. *Conform.*

At the beginning all sorts of bad thoughts will rise up in your mind the moment you sit down to meditate. . . . Because of this certain disciples abandon their spiritual practices. If you attempt to chase a monkey, he will try to leap on you in vengeance. And thus it is with the old impregnations (*samskâras*); bad thoughts try to avenge themselves and redouble their efforts as soon as you strive to awaken in yourself benefic and divine reflections. Your enemy is ingenious in ways to paralyse your efforts to have him expulsed from the house. Resistance is a natural law.

Swami Sivananda

While the digestion of the dead spiritual body in man goes forward, there may be seen (as in the chemical process) many variegated colours and signs, *i.e.,* all manner of sufferings, afflictions, and tribulations, including the ceaseless assaults of the Devil, the world, and the flesh.[1] But all these signs are of good omen, since they show that such a man will at length reach the desired goal. For Scripture tells us that all that are to obtain the eternal beatitude of Christ must be persecuted in this world, and we must enter into the kingdom of heaven through much tribulation and anguish.

The Sophic Hydrolith

Faith 509

It is a terrible fact that devils always get in the way of those who are striving for Buddhahood.

Hônen

Realiz. 873

Or think ye that ye will enter paradise while yet there hath not come unto you the like of (that which came to) those who passed away before you? Affliction and adversity befell them, they were shaken as with earthquake, till the messenger and those who believed along with him said: When cometh Allâh's help? Now surely Allâh's help is nigh.

Qur'ân, II. 214

Center 841

To the man about to renounce the world the devas offer many obstructions, hoping that his effort to transcend them and to attain to Brahman will come to naught.

Srimad Bhagavatam, XI. xi

I remember one time when I 'lamented',[2] and a great storm came from the place where the sun goes down, and I talked with the Thunder-beings who came with hail and thunder and lightning and much rain, and the next morning I saw that there was hail all piled up on the ground around the sacred place, yet inside it was perfectly dry. I think that they were trying to test me. And then, on one of the nights the bad spirits came and started tearing the offerings off the poles; and I heard their voices under the ground, and one of them said: 'Go and see if he is crying.' And I heard rattles, but all the time they were outside the sacred place and could not get in, for I had resolved not to be afraid, and did not stop sending my voice to *Wakan-Tanka* for aid.

Black Elk

[1] Mention must also be made of the excesses to which the unguarded imagination can lead, and which might be called *mythomania*, or the 'lunatic fringe': two examples will suffice. The first is from *Don Quixote:*

'In fine, he gave himself up so wholly to the reading of Romances, that a-Nights he would pore on 'till 'twas Day, and a-Days he would read on 'till 'twas Night; and thus by sleeping little, and reading much, the Moisture of his Brain was exhausted to that Degree, that at last he lost the Use of his Reason. A world of disorderly Notions, pick'd out of his Books, crowded into his Imagination: and now his Head was full of nothing but Inchantments, Quarrels, Battles, Challenges, Wounds, Complaints, Amours, Torments, and abundance of Stuff and Impossibilities; insomuch, that all the Fables and fantastical Tales which he read, seem'd to him now as true as the most authentick Histories' (I. i. 1; tr. Peter Motteux, revised by Ozell, Modern Library, 1930 p. 3).

The second illustration is offered in Reginald Scot's *Discoverie of Witchcraft:*

'And they have so fraid us with bull-beggars, spirits, witches, urchins, elves, hags, fairies, satyrs, pans, faunes, syrens, kit with the can sticke, tritons, centaurs, dwarfes, giantes, imps, calcars, conjurors, nymphes, changelings, incubus, Robin-goodfellow, the spoorne, the mare, the man in the oke, the hell-waine, the fier drake, the puckle, Tom Thomoe, hobgoblins, Tom Tumbler, boneless, and such other bugs, that we were afraid of our own shadowes' (cited by Washington Irving in his essay 'Stratford-on-Avon' from his *Sketch Book)*.

[2]Cf. note p. 291.

Introd. *Holy War*

In truth, to attain to interior peace, one must be willing to pass through the contrary to peace. Such is the teaching of the Sages.

Swâmi Brahmânanda

There is not one of you but shall approach it (hell). That is a fixed ordinance of thy Lord.

Qur'ân, XIX. 71

Christian, it is necessary once to be in the gulf of hell:
If living thou dost not go there, then must thou go there dead.

Death 206

Angelus Silesius

Thus the creaturely self-will dies: it enters wholly into the nothing, that it might no more live to itself, but to God.

Thus it falls out also in the philosophic work: when the artist has first seen great wonders, which the creaturely and natural will has wrought in the power of Venus, insomuch that he supposes that he is nigh thereunto: even then nature does first die in his work, and becomes a dark night unto him.

Boehme

Death 220

What we are to feel and undergo in these last purifications, when the deepest root of all selfishness, as well spiritual as natural, is to be plucked up and torn from us, or how we shall be able to stand in that trial, are both of them equally impossible to be known by us beforehand.

William Law

You also will have to cross the void. What appears to you as a chaos when all the impulses of the mind have ceased is not really the unmanifest (avyakta). This void must be traversed; it will try to swallow you up. You are then alone, with nothing more to see or hear; nothing to encourage you. You are thrown back on yourself. It is here, at this critical juncture, that presence of spirit is required. Draw strength and courage from it. The sage Uddâlaka himself experienced enormous difficulties in crossing this void.

Swami Sivananda

Yea! Jesu help thee then, for then hast thou need. For all the woe that may be without that, is not a point to that. For then is thyself a cross to thyself.

The Epistle of Privy Counsel, VIII

Amongst grosse Workes the fowlest of all
Is to clarifie our meanes Minerall.

Thomas Norton

The time is out of joint: O cursed spite,
That ever I was born to set it right!

Shakespeare (*Hamlet,* I. v. 188)

Oh, would that it had been death!
My wealth hath not availed me.
My power hath gone from me.

Qur'ân, LXIX. 27–29

Judgment
250

Faith is then as dead, and asleep, like all the other virtues; not lost, however,—for the soul truly believes all that the Church holds; but its profession of the faith is hardly more than an outward profession of the mouth. And on the other hand, temptations seem to press it down, and make it dull, so that its knowledge of God becomes to it as that of something which it hears of far away.... The soul itself is then burning in the fire, knowing not who has kindled it nor whence it comes, nor how to escape it, nor how to put it out: if it seeks relief from the fire by spiritual reading, it cannot find any.... To converse with any one is worse, for the devil then sends so offensive a spirit of bad temper that I think I could eat people up; nor can I help myself. I feel that I do something when I keep myself under control; or rather our Lord does so, when He holds back with His hand any one in this state from saying or doing something that may be hurtful to his neighbours and offensive to God.[1]

St Teresa of Avila

Now when Christ thus hides Himself, and withdraws the shining of His brightness and His heat, . . . man begins to complain because of his wretchedness: Whither has gone the ardent love, the inwardness, the gratitude, the joyful praise? And the inward consolation, the intimate joy, the sensible savour, how has he lost them? How have the fierce tempest of love, and all the other gifts which he felt before, become dead in him? And he feels like an ignorant man who has lost all his pains and his labour. And often his natural life is troubled by such a loss.

Sometimes these unhappy men are also deprived of their earthly goods, of friends, of kinsmen; and they are abandoned of all creatures, their holiness is not known or esteemed, men speak evil of their works and their whole lives, and they are despised and rejected by all their neighbours. And at times they fall into sickness and many a plague, and some into bodily temptations; or, that which is worst of all, into temptations of the spirit.

From this poverty arise a fear lest one should fall, and a kind of half-doubt. This is the utmost point at which a man can hold his ground without falling into despair.

Ruysbroeck

Because the light and wisdom of this contemplation is most bright and pure, and the soul which it assails is dark and impure, it follows that the soul suffers great pain when it receives it in itself, just as, when the eyes are dimmed by humour, and become impure and weak, they suffer pain through the assault of the bright light. And when the soul is indeed assailed by this Divine light, its pain, which results from its impurity, is immense; because, when this pure light assails the soul, in order to expel its impurity, the soul feels itself to be so impure and miserable that it believes God to be against it, and thinks that it has set itself

[1] The Plains Indians say the rattlesnake is most deadly in summer when shedding its skin, for this renders him temporarily blind. The metamorphic symbolism of the 'shedding of skins' has its correspondence with initiatic death and rebirth.

up against God. . . . Beneath the power of this oppression and weight the soul feels itself so far from being favoured that it thinks, and correctly so, that even that wherein it was wont to find some help has vanished with everything else, and that there is none who has pity upon it. . . .

A description of this suffering and pain, although in truth it transcends all description, is given by David, when he says: The lamentations of death compassed me about; the pains of hell surrounded me; I cried in my tribulation (*Psalm* XVIII. 4, 5). But what the sorrowful soul feels most in this condition is its clear perception, as it thinks, that God has abandoned it, and, in His abhorrence of it, has flung it into darkness; it is a grave and piteous grief for it to believe that God has forsaken it. It is this that David also felt so much in a like case, saying: After the manner wherein the wounded are dead in the sepulchres, being now abandoned by Thy hand, so that Thou rememberest them no more, even so have they set me in the deepest and lowest lake, in dark places and in the shadow of death, and Thy fury is confirmed upon me and all Thy waves Thou hast discharged upon me (*Psalm* LXXXVIII. 5–7).

Suff. 120

Here God greatly humbles the soul in order that He may afterwards greatly exalt it; and if He ordained not that these feelings should be quickly lulled to sleep when they arise within the soul, it would forsake the body in a very few days; but at intervals there occur times when it is not conscious of their greatest intensity. At times, however, they are so keen that the soul seems to be seeing hell and perdition opened. Of such are they that in truth go down alive into hell, for here on earth they are purged in the same manner as there, since this purgation is that which would have to be accomplished there. And thus the soul that passes through this either enters not that place at all, or tarries there but for a very short time; for one hour of purgation here is more profitable than are many there. . . .

There is added to all this (because of the solitude and abandonment caused in it by this dark night), the fact that it finds no consolation or support in any instruction nor in a spiritual master. For, although in many ways its director may show it good cause for being comforted in the blessings which are contained in these afflictions, it cannot believe him. For it is so greatly absorbed and immersed in the realization of those evils wherein it sees its own miseries so clearly, that it thinks that, as its director observes not that which it sees and feels, he is speaking in this manner because he understands it not; and so, instead of comfort, it rather receives fresh affliction, since it believes that its director's advice contains no remedy for its troubles. And, in truth, this is so; for, until the Lord shall have completely purged it after the manner that He wills, no means or remedy is of any service or profit for the relief of its affliction. . . . And if it sometimes prays it does so with such lack of strength and of sweetness that it thinks that God neither hears it nor pays heed to it: . . . When I cry and entreat, He hath shut out my prayer (*Lamentations,* III. 8). In truth this is no time for the soul to speak with God; it should rather put its mouth in the dust, as Jeremiah says, so that perchance there may come to it some present hope (*Lamentations,* III. 29), and it may endure its purgation with patience. It is God who is working here in the soul; wherefore the soul can do nothing. Hence it can neither pray nor be present at the Divine offices and pay attention to them, much less can it attend to other things and affairs which are temporal. Not only so, but it has likewise such distractions and times of such profound forgetfulness of the memory, that frequent periods pass by without its knowing what it has been doing or

Humility 191

thinking, or what it is that it is doing or is going to do, neither can it pay attention, although it desire to do so, to anything that occupies it.

St John of the Cross

When I looked for good, then evil came unto me: and when I waited for light, there came darkness.

My bowels boiled, and rested not: the days of affliction prevented me.

I went mourning without the sun: I stood up, and I cried in the congregation.

I am a brother to dragons, and a companion to owls.

My skin is black upon me, and my bones are burned with heat.

My harp also is turned to mourning, and my organ into the voice of them that weep.

Job, XXX. 26–31

I am poured out like water, and all my bones are out of joint: my heart is like wax; it is melted in the midst of my bowels.

My strength is dried up like a potsherd; and my tongue cleaveth to my jaws; and thou has brought me into the dust of death. . . .

Let my prayer come before thee: incline thine ear unto my cry;

For my soul is full of troubles: and my life draweth nigh unto the grave.

I am counted with them that go down into the pit: I am as a man that hath no strength:

Free among the dead, like the slain that lie in the grave, whom thou rememberest no more: and they are cut off from thy hand.

Thou hast laid me in the lowest pit, in darkness, in the deeps.

Thy wrath lieth hard upon me, and thou hast afflicted me with all thy waves.

Thou hast put away mine acquaintance far from me; thou hast made me an abomination unto them: I am shut up, and I cannot come forth.

Mine eye mourneth by reason of affliction: Lord, I have called daily upon thee, I have stretched out my hands unto thee.

Wilt thou shew wonders to the dead? shall the dead rise and praise thee? Infra 379

Shall thy lovingkindness be declared in the grave? or thy faithfulness in destruction?

Psalm XXII, 14–15, & LXXXVIII, 2–11

Blow, winds, and crack your cheeks! rage! blow!
You cataracts and hurricanoes, spout
Till you have drench'd our steeples, drown'd the cocks!
You sulphurous and thought-executing fires,
Vaunt-couriers to oak-cleaving thunderbolts,
Singe my white head! And thou, all-shaking thunder,
Strike flat the thick rotundity o' the world!
Crack nature's moulds, all germens spill at once
That make ingrateful man!

Shakespeare (*King Lear,* III. ii. 1)

Suff. 124

What is gross and thick must be rendered subtle and light by calcination.

This is a very slow and painful operation, because it is necessary to extract the root itself of evil; it makes the heart bleed and nature groan in torment.

Henry Madathanas

It sounds strange to say the soul must lose her God, yet I affirm that in a way it is more necessary to perfection that the soul lose God than that she lose creatures. Everything must go. The soul must subsist in absolute nothingness. It is the full intention of God that the soul shall lose her God, for as long as the soul possesses God, is aware of God, knows God, she is aloof from God. God desires to annihilate himself in the soul in order that the soul may lose herself. For that God is God he gets from creatures. When the soul became a creature she obtained a God. When she lets slip her creaturehood, God remains to himself that he is, and the soul honours God most in being quit of God and leaving him to himself.

Void 728

Eckhart

Had there been any evil in all fallen nature, whether in life, death, or hell, that had not attacked Him (Christ) with all its force, He could not have been said to have overcome it. And therefore so sure as Christ as the Son of man was to overcome the world, death, hell and Satan, so sure is it that all the evils which they could possibly bring upon Him were to be felt and suffered by Him as absolutely necessary in the nature of the thing to declare His perfection and prove His superiority over them.

Holiness 902

William Law

It is certain that, at the moment of His death, He (Christ) was forsaken and, as it were, annihilated in His soul, and was deprived of any relief and consolation, since His Father left Him in the most intense aridity, according to the lower part of His nature. Wherefore He had perforce to cry out, saying: My God! My God! Why hast Thou forsaken Me? This was the greatest desolation, with respect to sense, that He had suffered in His life. And thus He wrought herein the greatest work that He had ever wrought, whether in miracles or in mighty works, during the whole of His life, either upon earth or in Heaven, which was the reconciliation and union of mankind, through grace, with God.

St John of the Cross

I approached near unto hell, even to the gates of Proserpine, and after that I was ravished throughout all the elements, I returned to my proper place.

Apuleius

And verily ye used to wish for death before ye met it. Now ye have seen it with your eyes!

Judgment 250

Qur'ân, III. 143

—A Road indeed that never Wing before
Flew, nor Foot trod, nor Heart imagined—o'er
Waterless Deserts—Waters where no Shore—
Valleys comprising cloudhigh Mountains: these
Again their Valleys deeper than the Seas:

Whose Dust all Adders, and whose vapour Fire:
Where all once hostile Elements conspire
To set the Soul against herself, and tear
Courage to Terror—Hope into Despair,
And Madness; Terrors, Trials, to make stray
Or stop where Death to wander or delay:
Where when half dead with Famine, Toil, and Heat,
'Twas Death indeed to rest, or drink, or eat.
A road still waxing in Self-sacrifice
As it went on: still ringing with the Cries
And Groans of Those who had not yet prevail'd,
And bleaching with the Bones of those who fail'd:
Where, almost all withstood, perhaps to earn
Nothing: and, earning, never to return.—

And first the *VALE OF SEARCH:* an endless Maze,[1]
Branching into innumerable Ways
All courting Entrance: but one right: and this
Beset with Pitfall, Gulf, and Precipice,
Where Dust is Embers, Air a fiery Sleet,
Through which with blinded Eyes and bleeding Feet
The Pilgrim stumbles, with Hyaena's Howl
Around, and hissing Snake, and deadly Ghoul,
Whose Prey he falls if tempted but to droop,
Or if to wander famish'd from the Troop
For fruit that falls to ashes in the Hand,
Water that reacht recedes into the Sand.
The only word is 'Forward!' Guide in sight,
After him, swerving neither left nor right,
Thyself for thine own Victual by Day,
At night thine own Self's Caravanserai.
Till suddenly, perhaps when most subdued
And desperate, the Heart shall be renew'd
When deep in utter Darkness, by one Gleam
Of Glory from the far remote *Harîm,*
That, with a scarcely conscious Shock of Change,
Shall light the Pilgrim toward the Mountain Range
Of Knowledge. . . .

'Aṭṭâr

Wr. Side 459

Illusion 109
Orthod. 288

P. State 583

Mony clif he overclambe · in contrayes straunge,
Fer floten fro his frendes · fremedly he rides.
At eche warthe other water · ther the wighe passed

[1] This follows the traditional symbolism of labyrinths and mazes: cf. Coomaraswamy: 'The Iconography of Dürer's "Knots" and Leonardo's "Concatenation"': *The Art Quarterly,* Spring, 1944.

He fonde a foo him before. · bot ferly hit were.
And that so foule and so felle · that feght him behode.
So mony mervayl bi mount · ther the mon findes.
Hit were to tore for to telle · of the tenthe dole.
Sumwhile with wormes he werres. · and with wolves als.
Sumwhile with wodwos. · that woned in the knarres.
Bothe with bulles and beres. · and bores otherwhile.
And etaines. that him anelede · of the heghe felle:
Nade he ben dughty and drighe. · and Drighten had served.
Douteless he hade ben ded · and dreped ful ofte.
For werre wrathed him not so much. · that winter was wors. . . .

Holy War 405

Sir Gawayne and the Grene Knight, 713

My dismal scene I needs must act alone. . . .
What if this mixture do not work at all? . . .
What if it be a poison. which the friar
Subtly hath minister'd to have me dead.
Lest in this marriage he should be dishonour'd
Because he married me before to Romeo?
I fear it is: and yet. methinks. it should not.
For he hath still been tried a holy man.
I will not entertain so bad a thought.
How if. when I am laid into the tomb.
I wake before the time that Romeo
Come to redeem me? there's a fearful point!
Shall I not then be stifled in the vault.
To whose foul mouth no healthsome air breathes in.
And there die strangled ere my Romeo comes?
Or. if I live. is it not very like.
The horrible conceit of death and night.
Together with the terror of the place.
As in a vault. an ancient receptacle.
Where. for these many hundred years. the bones
Of all my buried ancestors are pack'd:
Where bloody Tybalt. yet but green in earth.
Lies festering in his shroud: where. as they say.
At some hours in the night spirits resort:
Alack. alack! is it not like that I.
So early waking. what with loathsome smells.
And shrieks like mandrakes' torn out of the earth.
That living mortals. hearing them. run mad:
O! if I wake. shall I not be distraught.
Environed with all these hideous fears.
And madly play with my forefathers' joints.
And pluck the mangled Tybalt from his shroud?

Faith 505

Orthod. 288

Supra Introd.

And, in this rage, with some great kinsman's bone,
As with a club, dash out my desperate brains?

Shakespeare (*Romeo and Juliet*, IV. iii. 19)

Everywhere there is horror, at the same time the silence itself terrifies the mind.

Virgil (*Aen.* II. 755: cited by St Jerome in describing catacombs.)

A day that will turn children grey....

Qur'ân, LXXIII. 17

I am brought to nothing.

Psalm LXXII. 22

And Cleopatra[1] going with John into her bedchamber, and seeing Lycomedes dead for her sake, had no power to speak, and ground her teeth and bit her tongue, and closed her eyes, raining down tears: and with calmness gave heed to the apostle. But John had compassion on Cleopatra when he saw that she neither raged nor was beside herself, and called upon the perfect and condescending mercy, saying: Lord Jesus Christ, thou seest the pressure of *sorrow*, thou seest the need: thou seest Cleopatra shrieking her soul out in silence, for she constraineth within her the frenzy that cannot be borne.

Charity 602

Acts of John, 24

And worse I may be yet: the worst is not,
So long as we can say, 'This is the worst.'

Shakespeare (*King Lear*, IV. i. 27)

There shall not come upon you a time but what the one that follows will be worse.[2]

Muhammad

Wr. Side 464

And this is what indeed happens, for, when the soul is most secure and least alert, it is dragged down and immersed again in another and a worse degree of affliction which is severer and darker and more grievous than that which is past.

St John of the Cross

Death cometh unto him from every side while yet he cannot die.

Qur'ân, XIV. 17

I did not die, and did not remain alive: now think for thyself, if thou hast any grain of ingenuity, what I became, deprived of both death and life.

Dante (*Inferno*, XXXIV. 25)

Then shall they begin to say to the mountains, Fall on us; and to the hills, Cover us.

St Luke, XXIII. 30

[1] Wife of Lycomedes, praetor of the Ephesians.

[2] Taken macrocosmically, this *ḥadith* shows how completely the traditional perspective is opposed to the evolutionist thesis.

And in those days shall men seek death. and shall not find it: and shall desire to die. and death shall flee from them.

Revelation, IX. 6

There he will neither die nor live.

Qur'ân, XX. 74

Return to the Womb of the Mother

Nicodemus saith unto him. How can a man be born when he is old? can he enter the second time into his mother's womb. and be born?

P. State 563

Jesus answered. Verily. verily. I say unto thee. Except a man be born of water and of the Spirit. he cannot enter into the kingdom of God.

St John, III. 4. 5

Species are not transmuted. but their subject matter rather. therefore the first work is to reduce the body into water. that is into mercury. and this is called Solution. which is the foundation of the whole art.

Roger Bacon

Knowl. 745 *Death* 208

It is not enough to play the sophister: the grain of wheat brings forth no fruit. unless it falls into the earth: all whatever will bring forth fruit must enter into its mother from whence it came first to be.

Boehme

Sp. Drown.

When the drop departed from its native home and returned.
It found a shell and became a pearl.

Dîvâni Shamsi Tabrîz, XXVII

O descendant of Bharata. the great Prakriti is My womb: in that I place the seed. from thence is the birth of all beings.

Creation 47

O son of Kunti. whatever forms are produced in all the wombs. the great Prakriti is the womb and I am the seed-giving Father.

Bhagavad-Gîtâ, XIV. 3. 4

He who wishes to enter the Kingdom of God must first enter with his body into his mother and die there.

Paracelsus

Beya mounted upon Gabricus and enclosed him in her womb in such manner that nothing more was visible of him. She embraced him with so much love that she absorbed him completely into her own nature.

Rosarium Philosophorum

I found him whom my soul loveth: I held him, and would not let him go, until I had brought him into my mother's house, and into the chamber of her that conceived me.

Song of Solomon, III. 4

The woman must reign, before she is overcome by the man.

Philalethes

Once the Little Child has become strong and robust to the point where he can withstand Water and Fire, he will put in his own belly the Mother who had begotten him.

Nicholas Flamel

P. State **563**
Center 841

The earth that's nature's mother is her tomb:
What is her burying grave that is her womb.

Shakespeare (*Romeo and Juliet*, II. iii. 9)

Death 223

Thus, O Father, Jesus has lived, though by Thy will and the will of the Holy Ghost, in three earthly abodes: in the womb of the flesh, in the womb of the baptismal water and in the dark caverns of the subterranean world.

Syrian Liturgy

Steadfastness

Son, when thou comest to the service of God, stand in justice and in fear, and prepare thy soul for temptation.

Holy War 394

Humble thy heart, and endure: incline thy ear, and receive the words of understanding: and make not haste in the time of clouds.

Conform. 170

Wait on God with patience: join thyself to God, and endure, that thy life may be increased in the latter end.

Take all that shall be brought upon thee: and in thy sorrow endure, and in thy humiliation keep patience.

Faith 509

For gold and silver are tried in the fire, but acceptable men in the furnace of humiliation.

Suff. 128

Ecclesiasticus, II. 1–5

Introd. *Death*

If any student of this Art is afraid of hard work, let him stop with his foot upon the threshold.

Philalethes

And verily We shall try you till We know those of you who really strive, and the steadfast, and until We test your record.

Qur'ân, XLVII. 31

And let us not be weary in well doing: for in due season we shall reap, if we faint not.

Galatians, VI. 9

The fifth requisite[1] in our work is patience. You must not yield to despondency, or attempt to hasten the chemical process of dissolution. . . . You need the patience of the husbandman, who, after committing the seed to the earth, does not disturb the soil every day to see whether it is growing.

Introd. *Inv.*

Philalethes

It is no hard matter to despise human comfort, when we have that which is divine.

It is much and very much, to be able to lack both human and divine comfort: and, for God's honor, to be willing cheerfully to endure desolation of heart: and to seek oneself in nothing, nor to regard one's own merit.

The Imitation of Christ, II. ix

Supra 366

When with *Haste* the Feind hath noe availe,
Then with *Despaire* your mind he will assaile:
And oft present this Sentence to your minde,
How many seeken, and how few maie finde,
Of wiser Men then ever were yee:
What suretie than to you maie be:
He woll move ye to doubt also
Whether your Teacher had it or noe:
And also how it mought so fall,
That part he tought you but not all;
Such uncertainety he woll cast out,
To set your minde with greevous doubt:
And soe your *Paines* he woll repaire
With wann hope and with much Despaire:
Against this assault is no defence,
But only the vertue of Confidence:
To whome reason shulde you leade,
That you shall have noe cause to dreade:
If you wisely call to your minde
The vertuous manners, such as you finde

[1] The other requisites are: (1) the right Mercury. (2) a well-regulated fire. (3) the right proportion of water to fire. (4) a properly-sealed vessel.

In your *Master* and your *Teacher*,
Soe shall you have noe neede to feare:
If you consider all Circumstances about.
Whether he tought you for Love or for Doubt:
Or whether Motion of him began.
For it is hard to trust such a Man:
For he that profereth hath more neede
Of you. then you of him to speede.

Thomas Norton

Orthod. 288

Heterod. 430

He said unto me. Son of man. can these bones live? And I answered. O Lord God. thou knowest.

Again he said unto me. Prophesy upon these bones. and say unto them. O ye dry bones. hear the word of the Lord.

Thus saith the Lord God unto these bones: Behold. I will cause breath to enter into you. and ye shall live.

Ezekiel, XXXVII. 3–5

Metanoia 484

And they say: When we are bones and fragments. shall we. forsooth. be raised up as a new creation?

Say: Be ye stones or iron

Or something yet more unyielding in your estimation! Then will they say: Who shall restore us? Say: He Who created you the first time. Then will they shake their heads at thee. and say: When shall that be? Say: It will perhaps be soon.

Qur'ân, XVII. 49–51

I go to make a journey
Beyond far China's shore.
And. passing. ask the pilgrims
Who trod this way before.
'Winds on the road yet more?'

Bâbâ Ṭâhir

Holy War 405

My robe is all worn out after so many years' usage.
And parts of it in shreds loosely hanging have been blown away to the clouds.

Bokuju

How hard this achievement is can be affirmed by those who have performed it.

Eobold Vogelius

Great will be the difficulties when you first try to dominate your thoughts. You must make war on them and they will do their best to defend their existence. in asserting their right to remain in the palace of your mind. which they have occupied from time immemorial: they will struggle to the end to maintain their right of seniority. Accordingly. they will assault you with violence. When you are meditating. all kinds of ugly ideas will

Holy War 396

come up and redouble in strength with each attempt to put them down. But the positive always prevails over the negative: just as darkness cannot resist the light of the sun or the leopard confront the lion, so will your obscure and negative thoughts—those invisible intruders, enemies of peace—be unable to withstand high aspirations: they will die of themselves.

Reality 803 *Sin* 82

Swami Sivananda

Think not that God will be always caressing His children, or shine upon their head, or kindle their hearts, as He does at the first. He does so only to lure us to Himself as the falconer lures the falcon with its gay hood. Our Lord works with His children so as to teach them afterwards to work themselves: as He bade Moses to make the tables of stone after the pattern of the first which He had made Himself. Thus, after a time, God allows a man to depend upon himself, and no longer enlightens, and stimulates, and rouses him. We must stir up and rouse ourselves, and be content to leave off learning, and no more enjoy feeling and fire, and must now serve the Lord with strenuous industry and at our own cost.

Beauty 670 *Suff.* 118

Tauler

The Messiah (Peace be upon Him!) said, 'Verily you will obtain what you like only by your patience with what you dislike.'

Supra 366 *Suff.* 120

Christ in Islâm

'What is My greatest joy? To see thee surmount every obstacle undaunted and continue thy act of love.'

Sister Consolata

As God's kingdom is set up, so the devil's kingdom may be pulled down, without the noise of axes and hammers. We may then attain to the greatest achievements against the gates of hell and death, when we most of all possess our souls in patience, and collect our minds into the most peaceful, composed, and united temper. The motions of true practical religion are most like that of the heavens, which though most swift, is yet most silent. . . . This fight and contest with sin and Satan is not to be known by the rattling of the chariots, or the sound of an alarm: it is indeed alone transacted upon the inner stage of men's souls and spirits: and is rather a pacifying and quieting of all those riots and tumults raised there by sin and Satan: . . . it is a captivating and subjecting all our powers and faculties to God and true goodness.

Peace 705

John Smith the Platonist

By the morning hours
And by the night when it is stillest,
Thy Lord hath not forsaken thee nor doth He hate thee,
And verily the last state will be better for thee than the first,
And verily thy Lord will give unto thee so that thou wilt be content.
Did He not find thee an orphan and protect thee?
Did He not find thee wandering and direct thee?
Did He not find thee destitute and enrich thee?

Therefor the orphan oppress not.
Therefor the beggar drive not away.
Therefor of the bounty of thy Lord be thy discourse.

Qur'ân, XCIII

Thou, O God, hast proved us: thou hast tried us, as silver is tried.
Thou broughtest us into the net: thou laidst affliction upon our loins.
Thou hast caused men to ride over our heads: we went through fire and through water: but thou broughtest us out into a wealthy place.

Psalm LXVI. 10–12

Supra 366

Then Jonah prayed unto the Lord his God out of the fish's belly.
And said, I cried by reason of mine affliction unto the Lord, and he heard me: out of the belly of hell cried I, and thou heardest my voice.
For thou hadst cast me into the deep, in the midst of the seas: and the floods compassed me about: all thy billows and thy waves passed over me.
Then I said, I am cast out of thy sight: yet I will look again toward thy holy temple.
The waters compassed me about, even to the soul: the depth closed me round about, the weeds were wrapped about my head.
I went down to the bottoms of the mountains: the earth with her bars was about me for ever: yet hast thou brought up my life from corruption, O Lord my God.
When my soul fainted within me I remembered the Lord: and my prayer came in unto thee, into thine holy temple.
They that observe lying vanities forsake their own mercy.
But I will sacrifice unto thee with the voice of thanksgiving: I will pay that that I have vowed. Salvation is of the Lord.
And the Lord spake unto the fish, and it vomited out Jonah upon the dry land.

Jonah, ch. II

Supra 378

Supra Introd.

Faith 501

There is no refuge from Allâh save toward Him.

Qur'ân, IX. 118

Reality 803
Peace 694

Those who are steadfast in the face of multiplicity, behold what light and grace are revealed to them!

Eckhart

And though after my skin worms destroy this body, yet in my flesh shall I see God.

Job, XIX. 26

Realiz. 870

To the Master it may appear easy enough: but to the beginner it must seem at first very hard and uphill work. He should not, however, despair, for in due time he will receive the reward of his diligence and aspiration: even in the dangers which the knowledge may bring upon him, he will be kept from harm by the loving hand of Providence, as I can testify from personal experience. We have with us God's Ark of the Covenant, which contains the most precious of earthly things, and is guarded by the holy Angel of the Lord.

Michael Sendivogius

Introds.
Death
Heterod.
Orthod.
280

During the course of your journey you will at all times enjoy the assistance of spiritual beings, who will see to it that you do not succumb to the perils that will beset you on your path.

Monkey

Death 206

The spiritual way ruins the body and, after having ruined it, restores it to prosperity.

Rûmî

O ye who believe! Remember Allâh's favour unto you when there came against you hosts, and We sent against them a great wind and hosts ye could not see. And Allâh is ever Seer of what ye do.

When they came upon you from above you and from below you, and when eyes grew wild and hearts reached to the throats, and ye were imagining vain thoughts concerning Allâh.

There were the believers sorely tried, and shaken with a mighty shock.

And when the hypocrites, and those in whose hearts is a disease, were saying: Allâh and His messenger promised us naught but delusion.

Qur'ân, XXXIII. 9–12

The active heat produces the colour black in what is wet; in what is dry, the colour white; and in what is white, the colour yellow.

First comes mortification, then calcination, and finally the golden tincture produced by the light of the sacred fire which illumines the purified soul.

Supra Introd.

Henry Madathanas

Faith 509

Seek therein, and be not weary: the result justifies the labour.

Basil Valentine

The real master tolerates misprints when reading history, as a good traveller tolerates bad roads when climbing a mountain, one going to watch a snow scene tolerates a flimsy bridge, one choosing to live in the country tolerates vulgar people, and one bent on looking at flowers tolerates bad wine.

Ch'en Chiju

Those who believe and obscure not their belief by wrongdoing, theirs is safety; and they are rightly guided.

Qur'ân, VI. 82

Metanoia 493

P. State 579

Wisdom inspireth life into her children, and protecteth them that seek after her, and will go before them in the way of justice.

And he that loveth her, loveth life....

If he trust to her, he shall inherit her, and his generation shall be in assurance.

For she walketh with him in temptation, and at the first she chooseth him.

She will bring upon him fear and dread and trial; and she will scourge him with the affliction of her discipline, till she try him by her laws, and trust his soul.

Then she will strengthen him, and make a straight way to him, and give him joy.

And will disclose her secrets to him, and will heap upon him treasures of knowledge and understanding of justice.

Knowl. 761

But if he go astray, she will forsake him, and deliver him into the hands of his enemy.

Heterod. 423

Ecclesiasticus, ch. IV

Let me advise you, moreover, not to neglect your fire, or move or open the vessel, or slacken the process of decoction, until you find that the quantity of the liquid begins to diminish: if this happens after thirty days, rejoice, and know that you are on the right road. Then be doubly careful, and you will, at the end of another fortnight, find that the earth has become quite dry and of a deep black. This is the death of the compound: the winds have ceased, and there is a great calm. This is that great simultaneous eclipse of the Sun and Moon, when the Sea also has disappeared. Our Chaos is then ready, from which, at the bidding of God, all the wonders of the world may successively emerge.

Sin 77 · *Contem.* 542 · *Reality* 775 · *Judgment* 268 · *P. State* 563

Philalethes

Crossing the Flood

As in a ship convey us o'er the flood.

Rig-Veda, I. 97. 8

The sely soul, at the likeness of a ship, attaineth at the last to the land of stableness, and to the haven of health.

Epistle of Discretion

Shut out worldly sights from your mind with endeavour and the utmost perseverance, and cross the perilous ocean of woe, which is the world, in the firm barque of your virtues.

Sin 77

Yoga Vasishtha

And he (Noah) said: Embark therein! In the name of God be its course and its mooring. Lo! my Lord is Forgiving, Merciful.

Inv. 1017

Qur'ân, XI. 41

The pranava (the syllable OM) is like a ferry for the human beings who have fallen into the boundless ocean of the life of this world. Many have been able to cross the ocean of deaths and rebirths (samsâra), by means of this ferry.

Waters 653

Swami Sivananda

When only my names[1] are recollected, I always protect all beings,
I, O Saviour, will ferry them across the great flood of their manifold fears.

Aryatârâbhaṭṭârikânâmâshṭottarasátakastotra, 17

[1] Amitabha, speaking through Avalokita.

Introd. *Peace*

The sail (of the boat) is love, the Holy Spirit its mast.

Tauler

Judgment 250 *Orthod.* 288

Rare indeed is this human birth. The human body is like a boat, the first and foremost use of which is to carry us across the ocean of life and death to the shore of immortality. The Guru is the skillful helmsman; divine grace is the favorable wind. If with such means as these man does not strive to cross the ocean of life and death, he is indeed spiritually dead.

Srimad Bhagavatam, XI. xiii

The world is my sea, the sailor the spirit of God,
The boat my body, the soul he who wins back his Abode.

Angelus Silesius

Thou art the Way, and Thou the Goal; Thou the Adorable One, O Lord!
Thou art the Mother tender-hearted; Thou the chastising Father;
Thou the Creator and Protector; Thou the Helmsman who dost steer
My craft across the sea of life.

Hindu Song

Now art thou in the ghostly sea, shipping over from bodilyness into ghostliness.

Supra 366 *Center* 841

Many great storms and temptations, peradventure, shall rise in this time, and thou knowest never whither to run for sorrow. All is away from thy feeling, common grace and special. Be not over-much afraid, then, although thou have matter, as thou thinkest, but have a lovely trust in our Lord, so little as thou mayest get for the time, for he is not far.

The Epistle of Privy Counsel, XII

Thus had Râm removed His slave from the shallow marsh of worldly life, and thrown him into the vast ocean of universal Life.

Swami Ramdas

Peace 700

The yogi masters the mental waves (vrittis) and reposes definitively in the 'samâdhi without qualifications' (asamprajnâta samâdhi). He encounters grave difficulties on his way, for the waves of the ocean are nothing compared with those of the mind. The true and dauntless yogi is the pilot or the captain of the human ship which dances on the terrible ocean of deaths and rebirths (samsâra). He puts an end to the mental waves by a continual concentration (dhâranâ) and meditation, thus attaining the other shore, where peace and immortality reign.

Swami Sivananda

As in a ship o'er billows, so through divers states of being; o'er manifold and grievous perils hath the Mighty Laud (*bṛhaduktha,* i.e., Agni) set his children, by these and farther shores.

Rig-Veda, X. 56. 7

And God remembered Noah, and every living thing, and all the cattle that was with him in the ark: and God made a wind to pass over the earth, and the waters asswaged.

Genesis, VIII. 1

A synonym for the Arahant is 'the Brahmin who, crossed over, gone beyond, stands on dry land.'

Holiness 902

Saṁyutta-nikâya, IV. 175

Thou, O Lord, hast rescued me from the ocean of ages, vast in its terror, and disturbed with the waves which are our births, from the mud of molestations, which is so hard to cross.

Death 223

Anangavajra

The Inner Pilgrimage

Neither by taking ship, neither by any travel on foot,
To the Hyperborean Field shall thou find the wondrous way.[1]

M.M. 986

Pindar

I believe Bethlehem, Golgotha, the Mount of Olives, and the Resurrection to be verily in the heart of him who has God.

Center 816

St Gregory of Nyssa

A pure, clean heart is the temple of God,[2] where the eternal God ever dwelleth in truth, when all that is unlike Him has been driven out.

Tauler

I wonder at those who seek His temple in this world: why do not they seek contemplation of Him in their hearts? The temple they sometimes attain and sometimes miss, but contemplation they might enjoy always. If they are bound to visit a stone, which

Contem. 536

[1] On the subject of myth and folklore, Coomaraswamy points out that when the Hero loses his immortal Bride, 'it remains for him to seek her out in that Otherworld or unknown City whence she first came, and of which the very name and place are strange to all those of whom he asks the way, for who knows "where" is overseas or underwave, or east of the sun or west of the moon, or "when" was once upon a time? The theme is infinitely varied but always the same story of . . . a separation and a reunion, enchantment and disenchantment, fall and redemption' ('On the Loathly Bride', *Speculum*, Oct., 1945, p. 401).

[2] It should not be overlooked, however, that Tauler preached this at the dedication of a church.

is looked at only once a year. surely they are more bound to visit the temple of the heart. where He may be seen three hundred and sixty times in a day and night. But the mystic's every step is a symbol of the journey to Mecca. and when he reaches the sanctuary he wins a robe of honour for every step.

Inv. passim

Muhammad b. al-Fadl

This pearl of eternity is the Church or Temple of God within thee. the consecrated place of divine worship. where alone thou canst worship God in spirit and in truth. . . . When once thou art well grounded in this inward worship. thou wilt have learnt to live unto God above time and place. For every day will be Sunday to thee. and wherever thou goest thou wilt have a priest. a church. and an altar along with thee.

Center 838

William Law

Bâyazîd. on his journey (to the Kaʿba). sought much to find some one that was the Khiẓr[1] of his time.

He espied an old man with a stature (bent) like the new moon: he saw in him the majesty and (lofty) speech of (holy) men:

His eyes sightless. and his heart (illumined) as the sun: like an elephant dreaming of Hindustân. . . .

He (the old man) said. 'Whither art thou bound. O Bâyazîd? To what place wouldst thou take the baggage of travel in a strange land?'

Bâyazîd answered. 'I start for the Kaʿba at daybreak.' 'Eh.' cried the other. 'what hast thou as provisions for the road?'

'I have two hundred silver dirhems.' said he: 'look. (they are) tied fast in the corner of my cloak.'

He said. 'Make a circuit round me seven times. and reckon this (to be) better than the circumambulation (of the Kaʿba) in the pilgrimage:

And lay those dirhems before me. O generous one. Know that thou hast made the greater pilgrimage and that thy desire has been achieved:

(That) thou hast (also) performed the lesser pilgrimage and gained the life everlasting: (that) thou hast become pure (*ṣâf*) and sped up (the Hill of) Purity (*Ṣafâ*).

By the truth of the Truth (God) whom thy soul hath seen. (I swear) that He hath chosen me above His House.

Albeit the Kaʿba is the House of His religious service. my form too. in which I was created. is the House of His inmost consciousness.

Holiness 924

Never since God made the Kaʿba hath He gone into it. and none but the Living (God) hath ever gone into this House (of mine).

When thou hast seen me. thou hast seen God: thou hast circled round the Kaʿba of Sincerity.

Orthod. 288

To serve me is to obey and glorify God: beware thou think not that God is separate from me.

[1] *Lit.* 'The green one'. A mysterious personage or 'solitary one' (*fard*) omnipresent in Sufic lore. said to have lived in the time of Abraham. sometimes identified with Elias and St George. and alluded to without name in the *Qur'ân* as the companion of Moses.

Open thine eyes well and look on me. that thou mayst behold the Light of God in man.'[1]

Rûmî

Pradakshina (the Hindu rite of going round the object of worship) is 'All is within me'. The true significance of the act of going round Arunachala[2] is said to be as effective as circuit round the world. That means that the whole world is condensed into this Hill. The circuit round the temple of Arunachala is equally good: and self-circuit (i.e. turning round and round) is as good as the last. So all are contained in the Self. Says the *Ribhu Gita*, 'I remain fixed. whereas innumerable universes. becoming concepts within my mind. rotate within me. This meditation is the highest circuit (*pradakshina*).'

Sri Ramana Maharshi

Reality 773

Why have I not performed the Pilgrimage? It is no great matter that thou shouldst tread under thy feet a thousand miles of ground in order to visit a stone house. The true man of God sits where he is. and the *Bayt al-Ma'mûr*[3] comes several times in a day and night to visit him and perform the circumambulation above his head. Look and see![4]

Abû Sa'îd ibn Abi 'l-Khayr

For a long while I used to circumambulate the Ka'ba. When I attained unto God. I saw the Ka'ba circumambulating me.

Bâyazîd al-Bisṭâmî

Center 841

I have visited in my wanderings shrines and other places of pilgrimage.
But I have not seen another shrine blissful like my own body.

Saraha

Holiness 924

The mind is the place of pilgrimage where the *devas*, the *Vedas* and all other purifying agencies become one. A bath[5] in that place of pilgrimage makes one immortal.

Śrî Śankarâchârya

If God sets the way to Mecca before any one. that person has been cast out of the Way to the Truth.

Abû Sa'îd ibn Abi 'l-Khayr

Conform. 170
Introd.
Holiness

On my first pilgrimage I saw only the temple; the second time. I saw both the temple and the Lord of the temple; and the third time I saw the Lord alone.

Bâyazîd al-Bisṭâmî

Supra
Introd.

[1] 'One can understand why a person like Râbiah Adawiyah could neglect the pilgrimage to Mecca. since she had accomplished the "pilgrimage of the heart". which means that she had attained the "peace" which "pre-exists" beneath the tumult of exteriorized existence' (Burckhardt: in *Études Trad.*, 1952. p. 311).

[2] Cf. note 1, p. 320.

[3] The celestial archetype of the Ka'ba.

[4] The biographer adds: 'All who were present looked and saw it.'

[5] 'Merging in *Brahman*, just as a man merges into water while bathing' (note by the translator. Swami Jagadânanda).

Center 816
Action 358

He who has made the pilgrimage of his own 'Self', a pilgrimage not concerned with situation, place or time, which is everywhere, in which neither heat nor cold are experienced, which procures a lasting felicity and a final deliverance from all disturbance: such an one is actionless, he knoweth all things, and he attaineth Eternal Bliss.

Śrî Śankarâchârya

Realiz. 887
Knowl. 749

In the empty heart, void of self
Can be heard the echoing cry,
'I am the Truth.'
Thus is man one with the Eternal.
Travelling, travel and traveller have become one.

Shabistarî

HOLY WAR

Think not that I am come to send peace on earth: I came not to send peace, but a sword.

St Matthew, x. 34

The traditional conception of holy war finds explicit formulation in sacred texts, e.g., the *Bhagavad-Gîtâ*: 'Nothing is higher for a Kshatriya (member of the warrior and ruling caste) than a righteous war': 'Fortunate indeed are Kshatriyas to whom comes unsought, as an open gate to heaven, such a war': 'If thou fallest in battle, thou shalt obtain heaven: if thou conquerest, thou shalt enjoy the earth. Therefore, O son of Kunti, arise and be resolved to fight' (ch. II).[1]

One can say, citing Guénon, 'that the essential reason for war, whatever the point of view and domain in which it is envisaged, is to end a disorder and re-establish order; in other words, it is the unification of a multiplicity, by use of means which belong to the world of multiplicity itself; it is in this sense, and in this sense alone, that war can be considered legitimate. On the other hand, disorder is in a sense inherent in all manifestation as such, because manifestation outside of its principle, insofar therefore as it represents non-unified multiplicity, is nothing but an indefinite series of ruptures of equilibrium. War understood in this way, and not limited in an exclusively human sense, thus represents the cosmic process of the reintegration of the manifested into principial unity; and this is why, from the point of view of manifestation itself, this reintegration appears as a destruction, as is very clearly seen when one considers certain aspects of the symbolism of Shiva in Hindu doctrine. . . .

Reality 775
Sin 62
Introd.
Sin

'The purpose of war is the establishment of peace, for even in its most ordinary sense peace is really nothing else than order, equilibrium, or harmony, these three terms being nearly synonymous and all designating under slightly different aspects the reflection of unity in multiplicity itself, when this is related to its principle. Multiplicity is then in fact not really destroyed, but "transformed". . . .

'Even in its exterior and social sense, legitimate war, directed against those who disturb order and having as its object the restoration of order, essentially constitutes a function of "justice", hence really an equilibrating function, whatever may be the secondary and transitory appearances. Yet this is but the "little holy war", which is only an image of the other, the "great holy war".[2] . . .

'The "great holy war" is the struggle of man against the enemies he carries within

[1] This teaching by extension refers to the development of one's possibilities in the domain of action.

[2] These terms relate to a saying or *ḥadîth* of the Prophet of Islam upon returning from battle: 'We have come back from the little holy war to the great holy war.'

Introd. Realiz. Knowl. 745 Holiness 914 Reality 775 Center 838

himself, that is to say, against all those elements in him which are contrary to order and unity. Moreover, it is not a question of annihilating these elements, which like everything that exists have their reason for being and their place in the total scheme of things. It is a question rather ... of "transforming" them by bringing them back to unity, of somehow reabsorbing them. Man must strive continually and above all to realize unity within himself, in everything that constitutes his nature, according to all the modalities of his human manifestation: unity of thought, unity of action, and also what is perhaps most difficult, unity between thought and action....

'For the person who has perfectly succeeded in realizing unity within himself, all opposition ceases, and consequently the state of war ceases: for there is no longer anything but absolute order, according to the total point of view which is beyond all particular points of view.... He sees unity in all things and all things in unity, in the absolute simultaneity of the "eternal present" ' (*Le Symbolisme de la Croix*, ch. VIII).

On the same subject, T. Burckhardt writes: 'If one considers that the end of a just war is true peace, one will understand the function of the "holy war" (*jihâd*) of the soul: the interior "war" is simply the abolition of another war, that which the earthly passions wage against the immortal soul or pure intellect' (*Études Trad.*, 1952, p. 311).

Metanoia 480 Death 220 Rev. 967 Renun. 146

Exhorts Coomaraswamy: 'Be your Self, at war with oneself.'[1] And elsewhere he resumes the same theme: 'Hero and Heroine are our two selves—*duo sunt in homine*—immanent Spirit ("Soul of the soul", "this self's immortal Self") and individual soul or self: Eros and Psyche. These two, cohabitant Inner and Outer Man, are at war with one another, and there can be no peace between them until the victory has been won and the soul, our self, this "I", submits. It is not without reason that the Heroine is so often described as haughty, disdainful, "Orgelleuse". Philo and Rûmî repeatedly equate this soul, our self, with the Dragon and it is this soul that we are told to "hate" if we would be disciples of the Sun of Men' ('On the Loathly Bride', *Speculum*, Oct., 1945, p. 401).

Judgment 261

If war can be holy, it can also be diabolical, a principle of separation rather than of union. Actually every war effects a 'discrimination' between good and bad elements, but in its nature the preponderant tendency of a war may be more in one direction than the other. Even two centuries ago William Law could speak of war as a 'murdering monster ... a fiery great dragon, a full figure of Satan broke loose and fighting against every redeeming virtue of the Lamb of God.... Look at all European Christendom sailing round the globe with fire and sword and every murdering art of war to seize the possessions and kill the inhabitants of both the Indies. What natural right of man, what supernatural virtue, which Christ brought down from Heaven, was not here trodden underfoot? All that you ever read or heard of heathen barbarity was here outdone by Christian conquerors' (Hobhouse, pp. 224–226).[2] And Law fully understood that profane warfare 'compels nameless numbers of unconverted sinners to fall, murdering and murdered, among flashes of fire' (id., p. 291).

Resuming the critique by Guénon for the present century: 'One of the most conspicuous results of industrial development is the continual perfecting of engines of war and the formidable increase in their powers of destruction. This alone should be enough to shatter

[1] *The Conception of Immortality in Buddhism:* a lecture given at the Academy of Music in Brooklyn, N.Y., February, 1946.

[2] For an objective and uncompromising assessment of what has been termed the 'last crusade', see John Collier: *Indians of the Americas*, New York, 1947.

the pacifist dreams of certain admirers of modern "progress"; but these dreamers and "idealists" are incorrigible and their credulity seems to know no bounds.[1] Certainly the "humanitarianism" at present so much in vogue does not deserve to be taken seriously; but it is strange that people should talk so much about an end of war at a time when the ravages it causes are greater than they have ever been before, not only because the means of destruction have been multiplied, but also because, since wars are no longer fought between comparatively small armies composed entirely of professional soldiers, all the individuals on both sides are flung against each other indiscriminately, including those who are least qualified to carry out this kind of function. Here again is a typical example of present-day confusion, and it is truly amazing, for anyone who cares to think about it, that a "mass call-up" or "general mobilization" should have come to be considered quite a natural thing and that, with very few exceptions, the minds of all should have accepted the idea of an "armed nation".... Let it be added that these generalized wars have only been made possible by the arising of another specifically modern phenomenon, that is to say by the formation of "nations", a consequence, on the one hand, of the destruction of the feudal system, and, on the other, of the simultaneous disruption of the higher unity of mediaeval Christendom' (*The Crisis of the Modern World*, ch. VII).

But from a cosmological viewpoint, even 'profane'[2] war leads in the end to stability, by ultimately bringing about the eradication of the disordered society, civilization, or world responsible for the deviation in question. According to Guénon, the end of a cycle 'is naturally "catastrophic" in the etymological sense where this word evokes the idea of a sudden and irremediable "downfall". But on the other hand, from the viewpoint where manifestation, in disappearing as such, is brought back to its principle as regards everything which is positive in its existence, this same end appears on the contrary as the "restoration" by which ... all things are no less suddenly re-established in their "primordial state". Moreover, this can be applied analogically at all degrees, whether it is a question of a being or a world: it is always the partial point of view which is "malefic", and the total point of view, or what amounts to that with respect to the other, which is "benefic", because all possible disorders are only disorders when seen in themselves and "separatively", and because these partial disorders are effaced entirely in the presence of the total order into which they are finally merged' (*Le Règne de la Quantité*, p. 271). As he writes elsewhere on several occasions: 'All the partial and transitory disequilibriums necessarily go to make up the great total equilibrium of the Universe.'

Beauty 689

[1] There is not place here to discuss the illusions inherent in pacifism, except to say that they are based on the erroneous supposition that the collectivity can conduct itself as can an individual. Cf. Schuon: 'Des Stations de la sagesse', *France-Asie*, 1953, p. 513; also his chapter on Charity in *Les Stations de la Sagesse*.

[2] 'The truth is that there is really no "profane realm" which could in any way be opposed to a "sacred realm": there is only a "profane point of view", which is really none other than the point of view of ignorance' (Guénon: *La Crise du Monde moderne*, p. 66).

Injunction to Holy War

The kingdom of heaven suffereth violence, and the violent take it by force.

St Matthew, XI. 12

Warfare is ordained for you, though it is hateful unto you: but it may be that ye hate a thing which is good for you, and it may be that ye love a thing which is bad for you. Allâh knoweth, ye know not.

Metanoia 488

Qur'ân, II. 216

When the country becomes rich and the people live in comfort, they should be forced to fight. After they have fought, they will be able to arrive at the place where the sages live.

Chang Po-tuan

Knowledge and wealth and office and rank and fortune are a mischief in the hands of the evil-natured.

Therefore the Holy War was made obligatory on the true believers for this purpose, (namely) that they might take the spear-point from the hand of the madman.

Rûmî

'If one strike thee on thy right cheek, turn to him also the other' (Matt. v, 39). . . . That these precepts pertain rather to the inward disposition of the heart than to the actions which are done in the sight of men, requiring us to cherish patience along with benevolence in the inmost heart, but in the outward action to do that which seems most likely to benefit those whose good we ought to seek, is manifest from the fact that the Lord Jesus Himself, the perfect example of patience, when He was struck in the face, answered: 'If I have spoken evil, give testimony of the evil: but if well, why strikest thou me?' (John xviii, 23). If we look only at the words, He did not in this obey His own precept, for He did not turn another part of His face to him who had struck Him, but on the contrary prevented him who had done the wrong from adding thereto. . . .

Many things must be done in correcting with a certain benevolent severity, even against their own wishes, men whose welfare rather than their wishes it is our duty to consult. . . .

Judgment 242

If the Christian teaching condemned wars of every kind, the injunction given in the Gospel to the soldiers seeking counsel as to salvation, would rather be to cast away their arms and withdraw themselves wholly from military service, whereas what was said to them was: 'Do violence to no man, neither calumniate any man; and be content with your pay' (Luke iii, 14). The command to be content with their pay in no way implies prohibition to continue in military service.

St Augustine

And he that hath no sword, let him sell his garment, and buy one.

St Luke, XXII. 36

And those who believe say: If only a sûrah were revealed! But when a decisive sûrah is revealed and war is mentioned therein, thou seest those in whose hearts is a disease looking at thee with the look of men fainting unto death. Therefor woe betide them!

Qur'ân, XLVII. 20

If, actuated by egoism, thou thinkest: 'I will not fight,' in vain is this thy resolve. Thine own nature will impel thee.

Bhagavad-Gîtâ, XVIII. 59

Action 329

Whosoever will save his life shall lose it: and whosoever will lose his life for my sake shall find it.

St Matthew, XVI. 25

Death 208

The alms of bravery are a holy war.

'Alî

(God) is never wanting to those that seek after him, and never fails those that engage in his quarrels. While we strive against sin, we may safely expect that the Divinity itself will strive with us, and derive that strength and power into us that shall at last make us more than conquerors. . . . As he who projects wickedness, shall be sure to find Satan standing at his right hand ready to assist him in it: so he that pursues after God and holiness, shall find God nearer to him that he is to himself, in the free and liberal communications of himself to him. He that goes out in God's battles, fighting under our Saviour's banner, may look upwards, and opening his eyes may see the mountains full of horses and chariots of fire round about him. God hath not so much delight in the death and destruction of men, as to see them struggling and contending for life, and himself stand by as a looker on.

John Smith the Platonist

Metanoia 493

Center 841

Contem. 547

The Lord is a man of war.

Exodus, XV. 3

For every man that Bolingbroke hath press'd
To lift shrewd steel against our golden crown,
God for his Richard hath in heavenly pay
A glorious angel: then, if angels fight,
Weak men must fall, for heaven still guards the right.

Shakespeare (*Richard II,* III. ii. 58)

Ye slew them not, but Allâh slew them.

Qur'ân, VIII. 17

Action 340

By Me alone have they already been slain; be thou merely an instrumental cause, O Savyasâchin.

Bhagavad-Gîtâ, XI. 33

Infra 396

And if Allâh had not repelled some men by others the earth would have been corrupted.

Qur'ân, II. 251

Peace is war's purpose, the scope of all military discipline, and the limit at which all just contentions aim.

St Augustine

Killing the Inward Dragon

Supra Introd.

He who has conquered himself by the Self, he is the friend of himself; but he whose self is unconquered, his self acts as his own enemy like an external foe.

Bhagavad-Gîtâ, VI. 6

Action 329

Fools of poor understanding have themselves for their greatest enemies, for they do evil deeds which bear bitter fruits.

Dhammapada, V. 66

Wr. Side 464

God has never made and formed but one enmity; but it is an irreconcilable one, which shall endure and grow even to the end. It is between Mary, His worthy Mother, and the devil—between the children and the servants of the Blessed Virgin, and the children and tools of Lucifer.

St Louis de Montfort

Death 220

There is no greater valour nor no sterner fight than that for self-effacement, self-oblivion.

Eckhart

Realiz. 859

He who conquers others is strong;
He who conquers himself is mighty.

Tao Te Ching, XXXIII

In the process of identification powerful obstacles arise.

Yoga Darshana

Conform. 166

The regenerate will, which goes out of its selfishness or self-hood again into the resignation; the same becomes also an enemy and an abominate to self-hood; as sickness is an enemy to health, and on the contrary, health an enemy to sickness: Thus the resigned will, and also the self-will are a continual enmity, and an incessant lasting war and combat.

Boehme

Be a hero on the battlefield of the Absolute (adhyâtmâ). Be a fearless soldier of the spirit. The inward battle against the mind, the senses, the subconscious impregnations (vâsanâs) and the residues of anterior states (samskâras) is more terrible than outward battle.

Swami Sivananda

Put on the whole armour of God, that ye may be able to stand against the wiles of the devil.

For we wrestle not against flesh and blood, but against principalities, against powers, against the rulers of the darkness of this world, against spiritual wickedness in high places.

Ephesians, VI. 11, 12

Wr. Side 459

We must suppose that those who secretly wrestle with us abide in another great world which, in its nature, is akin to the natural powers of our soul. For the three princes of evil, in their fight with spiritual strugglers, attack the three powers of our soul, and if a man has failed in something or does not strive at something, they overcome him in this very thing. Thus, the dragon—the prince of the abyss—rises in arms against those who keep attention on their heart, as one whose 'strength is in his loins, and his force is in the navel of his belly' (Job xl. 16). He sends the lust-loving giant of forgetfulness against them with his clouds of fiery arrows, stirs up lust in them like some turbulent sea, makes it foam and burn in them and causes their confusion by flooding them with torrents of insatiable passions. The prince of this world, who is in charge of warfare against the excitable part, attacks those who follow the path of active virtue. Using the giant of laziness, he encompasses them with all kinds of witchery of the passions and wrestles with those who always put up a courageous resistance. Thus he either vanquishes or is himself vanquished and so he gains them either crowns or shame before the faces of the angels. The prince of high places attacks those who exercise themselves in mental contemplation, by offering them fantasies; for, in company with the spiritual wickedness in high places, his task is to affect the thinking and speaking part. Using the giant of ignorance, he brings confusion into the thought striving to rise on high, darkens and frightens it, introducing into it vague fantastic images of spirits and their metamorphoses and producing phantoms of lightning and thunder, tempests and earthquakes. Thus each of the three princes, impinging upon the corresponding powers of the soul, wages war against it, conducting his attacks against the particular part allotted to him.

St Gregory of Sinai

Illusion 89

Pilg. 366

O kings, we have slain the outward enemy, but there remains within (us) a worse enemy than he.

To slay this (enemy) is not the work of reason and intelligence: the inward lion is not subdued by the hare.

This carnal self (*nafs*) is Hell, and Hell is a dragon which is not diminished by oceans.

It would drink up the Seven Seas, and still the blazing of that consumer of all creatures would not become less. . . .

It made a mouthful of and swallowed a whole world, its belly crying aloud, *'Is there any more?'*

Sin 66

God, from where place is not, sets His foot on it: then it subsides at (the command) *Be, and it is.*[1]

Inasmuch as this self of ours is a part of Hell, and all parts have the nature of the whole,

Humility 191

To God (alone) belongs this foot to kill it: who, indeed, but God should draw its bow? . . .

Deem of small account the lion who breaks the ranks (of the enemy): the lion is he who breaks himself.

Rûmî

The Sage says
That a wild beast is in the forest,
Whose skin is of blackest dye.
If any man cut off his head,
His blackness will disappear,
And give place to a snowy white.

Death 215
Metanoia 488

Abraham Lambspring

Canst thou draw out leviathan with an hook? or his tongue with a cord which thou lettest down?

Canst thou put an hook into his nose? or bore his jaw through with a thorn?

Will he make many supplications unto thee? will he speak soft words unto thee?

Will he make a covenant with thee? wilt thou take him for a servant for ever?

Wilt thou play with him as with a bird? or wilt thou bind him for thy maidens?

Shall the companions make a banquet of him? shall they part him among the merchants?

Canst thou fill his skin with barbed irons? or his head with fish spears?

Lay thine hand upon him, remember the battle, do no more.

Judgment 242

Job, XLI. 1–8

The soul must begin by warring against itself, and stirring up within itself a mighty feud; and the one part of the soul must win victory over the others, which are more in number. It is a feud of one against two, the one part struggling to mount upward, and the two dragging it down; and there is much strife and fighting between them. And it makes no small difference whether the one side or the other wins; for the one part strives towards the Good, the others make their home among evils; the one yearns for freedom, the others are content with slavery. And if the two parts are vanquished, they stay quiet in themselves, and submissive to the ruling part; but if the one part is defeated, it is carried off as a captive by the two, and the life it lives on earth is a life of penal torment. Such is the contest about the journey to the world above. You must begin, my son, by winning victory in this contest, and then, having won, mount upward.

M.M. 978
Metanoia 480
Sin 82
P. State 563
Realiz. 887

Hermes

[1] *Qur'ân,* L. 30. & XXXVI. 82.

S. Martin ascended from the military state to the clerical, for the army of the Church is higher in status than the army of the world; its warfare is a higher one, and its soldiers fight against spiritual enemies.

St Thomas Aquinas

Fasting only means the saving of bread,
Formal prayer is the business
Of old men and women,
Pilgrimage is a pleasure of the world.
Conquer the heart,
Its subjection is conquest indeed.

Ansârî

Renun. 139
143
Pilg. 387
Inv.
1017

There is no greater victory in the life of a human being than victory over the mind. He who has controlled the gusts of passion that arise within him and the violent actions that proceed therefrom is the real hero. All the disturbances in the physical plane are due to chaos and confusion existing in the mind. Therefore to conquer the mind through the awareness of the great Truth that pervades all existence is the key to real success and the consequent harmony and peace in the individual and in the world. . . . The true soldier is he who fights not the external but the internal foes.

Swami Ramdas

Supra
Introd.

He that is slow to anger is better than the mighty: and he that ruleth his spirit than he that taketh a city.

Proverbs, XVI. 32

The truly brave man, we contend, yields neither to fear nor anger, desire nor agony: he is at all times master of himself.

Ohiyesa

Valor is the conquest of one's own self.

Srimad Bhagavatam, XI. xii

Who is the man of courage and valour? It is he that subdues his concupiscence.

Ben Zoma

Certain it doubtless is, brethren, that either thou killest iniquity or art killed by iniquity. But do not seek to kill iniquity as if it were something outside thyself. Look to thyself, mark what fighteth with thee in thee, and take heed lest thy iniquity, thine enemy, defeat thee, if it have not been killed. . . . That wherewith thou delightest in the world fighteth against the mind which cleaveth to God. Let it cleave, let it adhere, let it not weaken, let it not give way: it hath great help. It conquereth that in itself which is in rebellion against itself, if it but persevere in fighting.

St Augustine

Introd.
Death
Infra
405
Faith
509

Illusion 94

How long, O man, will thy heart yet slumber? When wilt thou wake from thy sleep? The world's battles leaving, strengthen thyself, and fight with thy desire the goodly fight.

Shekel Hakodesh, 142

The real hero is he who has subjugated his mind.

Swami Sivananda

Pilg. 366

'Tis easy to break an idol, very easy: to regard the self as easy to subdue is folly, folly.

Rûmî

He who regards many things easy will find many difficulties.
Therefore the sage regards things difficult, and consequently never has difficulties.

Tao Te Ching, LXIII

The greatest victory I must now win is that over myself.

Calderón

Metanoia 480

From whence come wars and fightings among you? come they not hence, even of your lusts that war in your members?

James, IV. 1

Renun. 146

A man's enemies are the men of his own house.

Micah, VII. 6 & *St Matthew*, X. 36

The dragon is thy sensual soul: how is it dead? It is (only) frozen by grief and lack of means.

If it obtain the means of Pharaoh, by whose command the water of the river would flow,

Then it will begin to act like Pharaoh and will waylay a hundred (such as) Moses and Aaron.

That dragon, under stress of poverty, is a little worm, (but) a gnat is made a falcon by power and riches.

Renun. 152b

Keep the dragon in the snow of separation; beware, do not carry it into the sun of ʿIrâq.

So long as that dragon of thine remains frozen, (well and good); thou art a mouthful for it, when it gains release.

Mortify it and become safe from (spiritual) death; have no mercy; it is not one of them that deserve favours:

Sin 66

For (when) the heat of the sun of lust strikes upon it, that vile bat of thine flaps its wings.

Lead it manfully to the spiritual warfare and battle: God will reward thee with access (to Him). . . .

Dost thou hope, without using violence, to keep it bound in quiet and faithfulness?

Orthod. 288

How should this wish be fulfilled for any worthless one? It needs a Moses to kill the dragon.

Rûmî

No other foes have life so long as the beginningless, endless, everlasting life of my enemies the Passions. All beings may be turned by submission to kindness: but these

Passions become all the more vexatious by my submission.... Then I will not lay down my burden until these foes be smitten before my eyes....

Infra 405

Ah, when I vowed to deliver all beings within the bounds of space in its ten points from the Passions, I myself had not won deliverance from the Passions. Knowing not my own measure, I spoke like a madman. Then I will never turn back from smiting the Passions. I will grapple with them, will wrathfully make war on them all except the passion that makes for the destruction of the Passions.

Śânti-deva

Knowledge of God's adversary means being aware that thou hast an enemy, and that God will not accept from thee anything save it be as a result of warfare; and the warfare of the heart consists in making war against the enemy, and striving with him, and exhausting him.

Shaqiq of Balkh

Let them be confounded and put to shame that seek after my soul: let them be turned back and brought to confusion that devise my hurt....

Set thou a wicked man over him: and let Satan stand at his right hand.

When he shall be judged, let him be condemned: and let his prayer become sin.

Judgment 242

Let his days be few; and let another take his office.

Let his children be fatherless, and his wife a widow.

Let his children be continually vagabonds, and beg: let them seek their bread also out of their desolate places.

Let the extortioner catch all that he hath; and let the strangers spoil his labour.

Let there be none to extend mercy upon him: neither let there be any to favour his fatherless children.

Let his posterity be cut off; and in the generation following let their name be blotted out.

Psalms XXXV, 4, & CIX, 6–13

When he is drunk asleep, or in his rage,
Or in the incestous pleasure of his bed,
At gaming, swearing, or about some act
That has no relish of salvation in't;
Then trip him, that his heels may kick at heaven,
And that his soul may be as damn'd and black
As hell, whereto it goes.

Shakespeare (*Hamlet*, III. iii. 89)

And Noah said: My Lord! Leave not one of the disbelievers in the land.

If Thou shouldst leave them, they will mislead Thy slaves and will beget none save lewd ingrates.

Qur'ân, LXXI. 26–27

Paradise lies in the shadow of swords.

Muhammad

Distracting thoughts are like the enemy in the fortress. As long as they are in possession of it, they will certainly sally forth. But if you would, as and when they come out, put them to the sword, the fortress will finally be captured.

Sri Ramana Maharshi

Action 329 *Holiness* 924

It is said that a wise man rules over the stars; but this does not mean that he rules over the influences which come from the stars in the sky: but that he rules over the powers which exist in his own constitution.

Paracelsus

Contem. 528 *Judgment* 250 *Peace* 694

If thou standest resolutely here, and shrinkest not back, thou shalt see or feel great wonders. For thou shalt find Christ in thee assaulting hell, and crushing thy beasts in pieces, and that a great tumult and misery will arise in thee: also thy secret undiscovered sins will then first awake, and labour to separate thee from God, and to keep thee back. Thus shalt thou truely find and feel how death and life fight one against the other, and shalt understand by what passeth within thyself, what heaven and hell are. At all which be not moved, but stand firm and shrink not: for at length all thy creatures will grow faint, weak, and ready to die: and then thy will shall wax stronger, and be able to subdue and keep down the evil inclinations. So shall thy will and mind ascend into heaven every day, and thy creatures gradually die away. Thou wilt get a mind wholly new, and begin to be a new creature, and getting rid of the bestial deformity, recover the divine image. Thus shalt thou be delivered from thy present anguish, and return to thy original rest.

Boehme

Reality 803

He that is in thee is much greater than all that are against thee.

William Law

Suff. 126a

The devil by prayer, the flesh by chastisement,
The world, if one renounces it, is easy to conquer.

Angelus Silesius

Metanoia 480 *Center* 835 *M.M.* 978 *Realiz.* 870 *Rev.* 967

The Sages do faithfully teach us
That two strong lions, to wit, male and female,
Lurk in a dark and rugged valley.
These the Master must catch,
Though they are swift and fierce,
And of terrible and savage aspect.
He who, by wisdom and cunning,
Can snare and bind them,
And lead them into the same forest,
Of him it may be said with justice and truth
That he has merited the meed of praise before all others,
And that his wisdom transcends that of the worldly wise.[1]

Abraham Lambspring

[1] There is not space here to reproduce *The Ten Oxherding Pictures* of Chinese and Japanese origin, treating of the same theme, where the ox turns from black to white, but the reader interested can consult Suzuki: *Manual of Zen Buddhism*. Cf. also Brown: *The Sacred Pipe*, p. 85.

He, whose firm understanding obtains a command over his words, a command over his thoughts, and a command over his whole body, may justly be called a Triple-Commander. *Peace* 700

The man who exerts this triple self-command with respect to all animated creatures, wholly subduing both lust and wrath, shall by those means attain Beatitude.

Mânava-dharma-śâstra, XII. 10–11

'You gallant men pursue this way of high renown,
Why yield you? Overcome the earth, and you the stars shall crown.' *Holiness* 921

Boethius

The Race for Perfection

He who is expert at this outstrippeth all men.

Dionysius

To find the newborn King in thee all else thou mightest find must be passed by and left behind thee. *Holiness* 921 *Illusion* 109

Eckhart

Race with one another for forgiveness from your Lord.

Qur'ân, LVII. 21

Race for the prize of honour, thou must be first;
Thou wilt nothing obtain, if thou dost not gain it alone.

Angelus Silesius

Know ye not that they which run in a race run all, but one receiveth the prize? So run, that ye may obtain.

I. Corinthians, IX. 24

I press toward the mark for the prize of the high calling of God in Christ Jesus.

Philippians, III. 14

The thing to be known does not itself begin to be when we get knowledge of it: it is only for us that our knowledge makes it begin. Let us then lay hold on this beginning, and make our way thither with all speed. *Reality* 803 *Realiz.* 873

Hermes

The true runner comes to the finish and receives the prize and is crowned. And this is the way with the just.

Plato (*Republic,* 613 C)

Haste unto remembrance of Allâh.

Qur'ân, LXII. 9

This race is nothing else than the flight from creatures to unite with their creator.

Eckhart

I will run the way of thy commandments.

Psalm CXIX. 32

To him, who longs unto his CHRIST to go,
Celerity even it self is slow.

Robert Herrick

Haste thee speedily, for the time passeth.

Walter Hilton

Spit out the poison with all speed.

The Imitation of Christ, IV. x

Holiness 934

Earnest among the thoughtless, awake among the sleepers, the wise man advances like a racer, leaving behind the hack.

Dhammapada, II. 29

As with a man dangerously wounded by an arrow, there is not a moment of time to be wasted.

Gampopa

Infra 407

Faith 501

When a deer is being pursued by the hunters, it does not stop even to look around for its fellows or look back at its pursuers, but with all eagerness, hastens straight forward, and no matter how many may be following, it escapes in safety. It is with the same determination that a man fully entrusts himself to the Buddha's power, and without regard to anything else, steadfastly sets his mind upon being born into the Pure Land.

Hônen

These race for the good things, and they shall win them in the race.

Qur'ân, XXIII. 61

This is the foremost Victory, the Victory of the Dharma, which avails for this world and the other.

Asoka

Endurance

THE FOUR GREAT VOWS

However innumerable beings are, I vow to save them:
However inexhaustible the passions are, I vow to extinguish them:
However immeasurable the Dharmas are, I vow to master them:
However incomparable the Buddha-truth is, I vow to attain it.

Zen *Gatha*

Charity 608
Illusion 89
Orthod. 285
Realiz. 873

Moses said to his servant: I will not give up until I reach the meeting of the two seas.

Qur'ân, XVIII. 60

Introd. *Waters*

If you would become rich, prepare this Salt till it is rendered sweet.

The Glory of the World

Faith 509

The world is a battlefield. The garlands and the crown,
No one obtains, who does not fight with honour and renown.

Angelus Silesius

The Buddhas do but tell the Way, it is for *you* to swelter at the task.

Dhammapada, XX. 276

Pilg. 379

Conquer we shall, but we must first contend;
'Tis not the Fight that crowns us, but the end.

Robert Herrick

The life of man upon earth is a warfare.

Job, VII. 1

Ye have not yet resisted unto blood, striving against sin.

Hebrews, XII. 4

Obey not the deniers
Who would have thee compromise, that they may compromise.

Qur'ân, LXVIII. 8, 9

Renun. 146

If you are able to discriminate, it will be a certainty for you that bravery and truth are always found together, and falsehood and cowardice.

'Alî

This spiritual warfare of ours must be constant and never ceasing.

Unseen Warfare, I. XV

You must make the most strenuous efforts. Throughout this life, you can never be certain of living long enough to take another breath.

Illusion 99

Huang Po

And be not like unto her who unravelleth the thread, after she hath made it strong, to thin filaments, making your oaths a deceit between you.

Faith 510

Qur'ân, XVI. 92

It showeth weakness in one who hath caught a glimpse of Reality to fail to persevere in *sâdhanâ* till the dawning of Full Enlightenment.

Contem. 542

Gampopa

Although sesamum seed is the source of oil, and milk the source of butter, not until the seed be pressed and the milk churned do the oil and butter appear.

Although sentient beings are of the Buddha essence itself, not until they realize this can they attain *Nirvâṇa*.

Creation 42

Padma-Sambhava

Knowl. 761 745

Remind yourself that all men assert that wisdom is the greatest good, but that there are few who strenuously endeavour to obtain this greatest good.

Pythagoras

Let him who seeks, not cease seeking until he finds.

The Gospel according to Thomas, Log. 2

Death 208

It is not for any prophet to have captives until he hath made slaughter in the land. Ye desire the lure of this world and Allâh desireth (for you) the Hereafter, and Allâh is Mighty, Wise.

Qur'ân, VIII. 67

If one wants to abide in the thought-free state, a struggle is inevitable. One must fight one's way through before regaining one's original primal state. If one succeeds in the fight and reaches the goal, the enemy, namely the thoughts, will all subside in the Self and disappear entirely.

Peace 700

Sri Ramana Maharshi

Infra 410

In God's name, cheerly on, courageous friends,
To reap the harvest of perpetual peace
By this one bloody trial of sharp war.

Shakespeare (*Richard III*, v. ii. 14)

How short is the struggle! how happy the hero,
Who for eternity triumphs over devil, flesh and world!

Holiness 902

Angelus Silesius

Vigilance

When the novices forget their *dhikr*[1] for one breath, then Satan is with them, for Satan spies upon them. So when forgetfulness enters the heart, then he enters, but when the *dhikr* enters certainly he will depart.

Ibn 'Arabî

Infra 410

In the same way that the mind seeks variety in foods and other things, so does it want diversity in sâdhanâ.[2] It rebels against monotonous practices.... To cease sâdhanâ is a grave error. The practice must never be renounced under any circumstances. Bad thoughts will always be pressing to penetrate into the mental laboratory; if the aspirant suspends his sâdhanâ his mind becomes the workshop of Satan.

Swami Sivananda

It needs but a little to overthrow and destroy everything—just a slight aberration from reason. For the helmsman to wreck his vessel, he does not need the same resources, as he needs to save it: if he turn it but a little too far to the wind, he is lost; yes, and if he do it not deliberately but from mere want of attention, he is lost all the same.... Keep awake then and watch your impressions: it is no trifle you have in keeping.

Epictetus

Heterod. 423

Up soldier, to arms! Dost thou not prefer
Peace after victory, to torment after peace?

Angelus Silesius

Suff. 120

By all means, Brother, go forward, why dost thou stand still?
To stand on the way to God is to go backwards.

Angelus Silesius

Remember Lot's wife.

St Luke, XVII. 32

He who interrupts the course of his spiritual exercises and prayer is like a man who allows a bird to escape from his hand; he can hardly catch it again.

St John of the Cross

Be sober, be vigilant; because your adversary the devil, as a roaring lion, walketh about, seeking whom he may devour.

I. Peter, v. 8

What you must be particularly careful about in this operation, is to prevent the young ones of the Crow from going back to the nest when they have once left it.

Philalethes

[1] Invocation, remembrance (of God).
[2] Spiritual discipline.

Escape for thy life; look not behind thee, neither stay thou in all the plain; escape to the mountain, lest thou be consumed.

Genesis, XIX. 17

Knowl. 755 / Judgment 261 / Supra 403 / Introd. *Illusion* / *Peace* 694

An thou returnest it is not because of any truth; it is either the senses, the world or the devil. And persisting in this turning back, thou dost inevitably lapse into sin and art liable to backslide so far as to have the eternal fall. Wherefore there is no turning back, only a pressing forward and following up this possibility to its fulfilment. It never rests until fulfilled with all being. As matter never rests until fulfilled with every possible form, so intellect never rests till it is filled to the full of its capacity.

Eckhart

Faith 510

All the operations must be performed in a single vessel and without removing it from the fire.

The substance used in the preparation of the philosophers' stone must be gathered in a single place and not scattered in several places. Once the gold has lost its lustre, it is difficult to restore it.

Henry Madathanas

Pilg. 366 / *Renun.* 146

For of all paines the most grevious paine,
Is for one faile to beginn all againe.
Every man shall greate *Paine* have
When he shall first this *Arte* covet and crave,
He shall oft tymes Chaunge his desire,
With new tydings which he shall heare;
His Councell shall oftentimes him beguile,
For that season he dreadeth noe subtile wile:
And oftentymes his minde to and fro,
With new Oppinions he shall chaunge in woe:
And soe long tyme continue in Phantasie,
A greate adventure for him to come thereby:
Soe of this *Arte* be ye never so saine,
Yet he must taste of manie a bitter paine.

Thomas Norton

Orthod. 296 / *Conform.* 170

In too much water you may easily be drowned; too little water, on the other hand, soon evaporates in the heat of the sun.

Basil Valentine

When you relax your attention for a little, do not imagine that you will recover it wherever you wish.

Epictetus

And another also said, Lord, I will follow thee; but let me first go bid them farewell, which are at home at my house.

And Jesus said unto him, No man, having put his hand to the plough, and looking back, is fit for the kingdom of God.

St Luke, IX. 61, 62

There is no greater enemy to our attainment of *Ôjô* than our fellow men.

Hônen

Renun. 146

The devils, who are skilful thieves, wish to surprise us unawares, and to strip us. They watch day and night for the favorable moment. For that end they go round about us incessantly to devour us and to snatch from us in one moment, by a sin, all the graces and merits we have gained for many years. Their malice, their experience, their stratagems and their number ought to make us fear this misfortune immensely, especially when we see how many persons fuller of grace than we are, richer in virtues, better founded in experience and far higher exalted in sanctity, have been surprised, robbed and unhappily pillaged.

St Louis de Montfort

The thief Heedlessness, waiting to escape the eye of remembrance, robs men of the righteousness they have gathered, and they come to an evil lot. The Passions, a band of robbers, seek a lodging, and when they have found it they rob us and destroy our good estate of life. Then let remembrance never withdraw from the portal of the spirit; and if it depart, let it be brought back by remembering the anguish of hell.

Śânti-deva

Center 816

Knowl. 755

Judgment 261

The senses may become turbulent at any time. Reaction may set in. Beware!

Swami Sivananda

The devil sleepeth not, neither is the flesh as yet dead; therefore cease not to prepare thyself to the battle; for on thy right hand and on thy left are enemies who never rest.

The Imitation of Christ, II. ix

Satan hath engrossed them and so hath caused them to forget remembrance of Allâh.

Qur'ân, LVIII. 19

Be vigilant even when asleep.

Heinrich Khunrath

Holiness 934

Satan, that grim pseudo-alchymist, ever lies in wait to draw those whom Christ has regenerated, and made sons of God by faith through baptism, and who are warring the good warfare, and keeping faith and a good conscience, away from the right path—and in this attempt he and his faithful servants, our sinful flesh, and the wicked, seductive world, are, alas, very frequently successful.

The Sophic Hydrolith

Wr. Side 457

Heterod. 423

Do not do things off and on. Have your sadhana every day with greater and greater intensity.

Swami Ramdas

Whoso would follow God in honest works must never stand still. He must ever travel on.

Mechthild of Magdeburg

When you go abroad don't turn round at the frontier.

Pythagoras

The mind like a thief is always lying in wait.

Swami Sivananda

There is a warfare where evil spirits secretly battle with the soul by means of thoughts. Since the soul is invisible, these malicious powers attack and fight it invisibly, in accordance with its nature. And it is possible to see on both sides weapons and plans (disposition of armies and military strategy), deceptive artifices and intimidating attacks (impetuous charges aiming at intimidation), and hand to hand battles; and victories and defeats on both sides. The only thing lacking in this mental warfare we describe, as compared with physical warfare, is a definite moment of declaration of war. In physical warfare it is customary to establish a time and to conform to certain rules. But mental warfare starts suddenly, without any declaration, with an onslaught directed at the very depths of the heart.

Philotheus of Sinai

Be vigilant. The need for agitation (rajas) may try to erupt in you. Chase out the invader with blows of a whip and resume your serenity.

Swami Sivananda

The Weapon of Invocation

Inv. passim

O ye who believe! When ye meet an army, hold firm and invoke Allâh much, that ye may be successful.

Qur'ân, VIII. 45

Then said David to the Philistine, Thou comest to me with a sword, and with a spear, and with a shield: but I come to thee in the name of the Lord of hosts, the God of the armies of Israel, whom thou hast defied.

I. Samuel, XVII. 45

'The invincible weapon, always victorious, is the incessant act of love.'

Sister Consolata

We must . . . arm ourselves against our spiritual foes with spiritual weapons, such as the Word of God.

The Sophic Hydrolith

Our best, our easiest remedy is the Name of *Jesus*. It drives the Devil flying from our sides and saves us from countless evils.

The Wonders of the Holy Name

A monk should constantly call: 'Lord, Jesus Christ, Son of God, have mercy upon me!' in order that this remembering of the name of our Lord Jesus Christ should incite him to battle with the enemy. By this remembrance[1] a soul forcing itself to this practice can discover everything which is within, both good and bad. First it will see within, in the heart, what is bad, and later—what is good. This remembrance is for rousing the serpent, and this remembrance is for subduing it. This remembrance can reveal the sin living in us, and this remembrance can destroy it. This remembrance can arouse all the enemy hosts in the heart, and little by little this remembrance can conquer and uproot them. The name of our Lord Jesus Christ, descending into the depths of the heart, will subdue the serpent holding sway over the pastures of the heart, and will save our soul and bring it to life.

Knowl. 755
Pilg. 366
Sin 66
Realiz. 859

Callistus and Ignatius Xanthopoulos

Devotees by the mere repetition of His Name destroy the mighty army of ignorance.

Swami Sivananda

Though a man born in an archer's family goes to war, and loses his life, if he only repeats the sacred name and relies upon Amida's Original Vow, there is not the slightest doubt whatever that Amida will come to welcome him to his Paradise.

Hônen

Knowing that his past actions may try to overwhelm him, the devotee must be prepared to combat them. God will give him the strength: His Name will be an impenetrable armour. It will save him from all the consequences.

Action 329

Swâmi Brahmânanda

Satan was straightway driven away by the mention of the Name of Christ like a sparrow before a hawk.

St Athanasius

O God that art God above all that are called gods, . . . at whose name every idol fleeth and every evil spirit and every unclean power. . . .

Acts of John, 41

And we also have spiritual exorcisms, the name of our Lord Jesus Christ and the power of the cross. This exorcism not only drives the dragon out of his lair and casts him into the fire, but it even heals the wounds caused by him. If many used this exorcism and were not

Supra 396

[1] Cf. note 1, p. 407.

Faith 514

healed, this was due to their lack of faith and not to the ineffectualness of the exorcism. . . . The name of Jesus Christ is terrible for demons, passions of the soul and diseases. Let us adorn and protect ourselves with it.

St John Chrysostom

Renun. 160 *Wr. Side* 464 *Inv.* 1013

(The saints of Mary) shall be clouds thundering and flying through the air at the least breath of the Holy Ghost; who, detaching themselves from everything and troubling themselves about nothing, shall shower forth the rain of the Word of God and of life eternal. They shall thunder against sin; they shall storm against the world; they shall strike the devil and his crew; and they shall pierce through and through, for life or for death, with their two-edged sword of the Word of God, all those to whom they shall be sent on the part of the Most High.

St Louis de Montfort

HETERODOXY AND DEVIATION

Whatever traditions (Smṛtis) lie outside the Veda, and whatever (works) are ill-revealed (heterodox), are all without fruit after death, since they are said to rest in darkness.

Mânava-dharma-śâstra, XII. 95

The devil hath power to assume a pleasing shape.

Shakespeare (*Hamlet,* II. ii. 636)

In a work accenting spiritual orthodoxy and traditional regularity, it is well at least to indicate what constitutes the contrary, especially as the Western outlook has largely tended to dismiss Oriental traditions with a sweep of facile charges ranging from Polytheism and Pantheism, to Naturalism, Primitivism and even Demonolatry—errors real enough at their level, but inapplicable to the traditions in question. Judeo-Christian angelology could present no problem to the Hindu: what he fails to understand is the West's insistence that his pantheon is 'polytheistic', when his sacred texts tirelessly reiterate in multiple ways the supremacy of the Absolute and Unconditioned *Brahma.* This carries in Sri Ramakrishna's 'God has form and He is formless too. Further, He is beyond both form and formlessness. No one can limit Him' (*Gospel,* p. 192); and finds its Western equivalent in Eckhart's 'Eight heavens are often spoken of and nine choirs of angels; there is nothing of that where I am. You must know that expressions of that sort, which conjure up pictures in the mind, merely serve as allurements to God. In God there is nothing but God' (I. 328).[1] 'Neither Christianity nor Hinduism is polytheistic, though both are polynominal' (Coomaraswamy: 'The Indian Doctrine of Man's Last End', *Asia,* 1937, p. 380).

Introd.
Wr. Side
Reality 803
M.M. 978
994

'In general, the notion of "pantheism", read into any doctrine, arises from a confusion of the unity which is one in itself, with the merely collective totality of all

[1] This is a way of affirming the supremacy of the Absolute over the relative: the contingent domain is perfectly 'real' at its own level (cf. the note on p. 84; *Infra,* p. 428 ff.; Introd. *Realization*). But it is this transcendent perspective which enables St Teresa of Lisieux, for example, to say: 'All these images are of no use to me: I can only be nourished by the truth. This is why I have never wanted visions' (*Histoire d'une Âme,* p. 243). Cp. Sri Ramana Maharshi: 'The six subtle centres are merely mental pictures and are meant for beginners in *Yoga*' (*Self-Enquiry,* p. 19).

'(Ryôyo Shôgei) insisted that the ordinary conception of the soul's being transported to Paradise and born there was merely a figure of speech, the fact being that neither Amida, nor the sainted beings, nor the "nine ranks" are to be conceived of as existing over there at all, because the Pure Land is the ultimate and absolute reality' (Coates and Ishizuka: *Hônen,* Historical Introd., pp. 56-57).

things' (Coomaraswamy: *A New Approach to the Vedas,* p. 85). Doctrinal refutations of this error can be found, among other places, in Schuon's *De l'Unité transcendante des Religions,* pp. 54–55, and *L'Oeil du Coeur,* p. 27; but one wonders how these confusions could arise in the first place, when the *Rig-Veda,* for example, expounds: 'Heaven and Earth have not measured, nor do they measure, his omnipotence' (III. 82. 37): and the *Bhagavad-Gîtâ*: 'By My unmanifested form all this world is pervaded: all beings dwell in Me, but I do not dwell in them' (IX. 4); and the *Srimad Bhagavatam*: 'The universe does not limit thee' (XI. i): while on the Chinese side is Chuang-tse's 'What there was before the universe, was *Tao. Tao* makes things what they are, but is not itself a thing. Nothing can produce *Tao*; yet everything has *Tao* within it' (Giles, p. 291): and Tung-shan's 'He is no other than myself, And yet I am not he' (Suzuki: *Essays in Zen,* III. 219). As Sri Ramakrishna paraphrasing Śankarâchârya puts it: 'The waves belong to the Ganges: but does the Ganges belong to the waves?' (p. 600). The theme is universal, thus Egyptian: 'Unknown is His Name in Heaven . . . Who has made all things, but has not Himself been made' (G. A. Rawlinson: *The Story of Ancient Egypt,* p. 38); and Hellenistic-Jewish: 'God fills all things; He contains but is not contained' (Philo, in Lewy, p. 27); and American Indian: 'All things are the works of the Great Spirit. We should know that He is within all things . . . and even more important, we should understand that He is also above all these things and peoples' (Black Elk, *The Sacred Pipe,* p. xx): and Islamic: 'The relative stands in need of the Absolute, while the Absolute has no need of the relative' (Jâmî: *Lawâ'iḥ,* XXI); and finds its European corollary, for example, in St Augustine: 'In many places does Plotinus explain Plato thus—that that which we call the soul of this universe has the beatitude from one fount with us, namely, a light which it is not, but which made it: and from whose intellectual illustration it has all the intelligible splendour' (*De Civ. Dei,* X. ii); and Dante: 'The veracious Mirror which doth make himself reflector of all other things, and naught doth make itself reflector unto him' (*Paradiso,* XXVI. 106); in *The Epistle of Privy Counsel*: 'He is thy being and thou not his' (Ch. I); or Peter Sterry's '*There is not the lowest thing which hath not God in it;* for God fills all. Yet as the Sun-beams fall on a Dunghill, and are not polluted, but shine on the Dung-hill; so God is still himself, to himself, high and glorious in the lowest Things' (Pinto, p. 96). 'Leave me uncensured,' says Jacob Boehme, 'I do not say, that nature is God' (*Sig. Rerum,* VIII. 42); and again: 'All things have their being in God who is nevertheless not this being' (*Myst. Mag.,* XVI. 14). Two more citations will suffice, one alchemical, from the *New Chemical Light* of Michael Sendivogius: 'The Creator of the world partly acts in and through things belonging to this world, and is thereby, in a sense, included in this world. But He absolutely transcends this world by that infinite part of His activity which lies beyond the bounds of the universe, and which is too high and glorious for the body of the world' (*Hermetic Museum,* II. 139); the other from Spanish Kabbalism: 'God is all reality, but not all reality is God' (Cordovero; Scholem: *Jewish Mysticism,* pp. 252–253).

Beauty 663

The charge of naturalism so often levelled against Islam is effectively dissipated in terms of an unanswerable challenge: 'Verily, though mankind and the jinn should assemble to produce the like of this *Qur'ân,* they could not produce the like thereof though they were helpers one of another' (*Qur'ân,* XVII. 88). Hindus are charged on the one side with a doctrine of *mâyâ* that would have the whole world illusion, and they are charged on the other with a doctrine of *pantheism* that would have the whole world God—conceptions which, as Sir S. Radhakrishnan has justly remarked, are mutually exclusive, contradictory and self-annulling.[1]

Illusion 85

Veritable heterodoxy in the plan of this book is treated simply as *error*, which basically is of two kinds—'luciferian', where something other than God is associated with God, or set up in His stead; and 'satanic', where God is 'replaced' or travestied, through total inversion, by a diabolic counterfeit. The texts in the present chapter deal mainly with deviations of the first sort, while the following chapter pursues more closely the matter of full subversion.

'Heresy' comes from the Greek ἁίρεσις—choosing for oneself (in opposition to following accepted doctrine); this possibility has its origin in dualism—represented by the tree of the knowledge of good and evil (as opposed to the unitive knowledge of the tree of life—*Gen.* II. 9), which gives to the soul the illusion of separativity, engendering the tendency to individualism and rebellion against God. 'The infallible "instinct" of animals is a lesser "intellect"; the intellect of man can be called a higher "instinct". Somewhere between instinct and intellect is situated the reason, which owes its miseries to the fact that it constitutes a sort of "luciferian" duplication of the Divine Intelligence, the only intelligence there is' (Schuon: *Perspectives spirituelles*, p. 181).

M.M. 978
Introd. *Realiz.*
Introd. *Knowl.*

'When the heart is opened to the divine influx, the *ego* finds itself inundated with beauty and glory; then intervenes the tendency to individuation—the same which separates manifestation from principle—to attribute this beauty and this glory to the *ego*, hence to the manifested, which means wishing to drag these qualities into the turmoil of manifestation, whereas it is precisely their function to lead the being out of the world.... The thought of being the vilest of sinners, ... while illogical in itself, is nonetheless an attitude of strict defense against a luciferianism ever more encroaching and insidious as the individual finds himself more and more inebriated with celestial perfumes.[2] The collapse of a badly founded spiritual state (the Sufic *istidrâj*) is a dreadful possibility' (Schuon: 'De l'Humilité', *Études Trad.*, 1950, pp. 112–113).

Introd. *Sin*
Illusion 109

'Everything related to the order of metaphysic has, in itself, the power of opening up boundless horizons to anyone who has a true conception of it: this is not a hyperbole nor a figure of speech, but it must be understood quite literally, as an immediate outcome of the very universality of the principles.... The things in question are the most tremendous that exist, and compared with them everything else is mere child's play.... Everything accomplished in this domain brings into play powers that the ordinary man has no inkling of, and those who are indiscreet may find themselves at the mercy of strange reactions, at least so long as a certain level of understanding has not been reached.... Where realization has not been

[1] Unfortunately the majority of modern Western commentaries on Eastern doctrine and ritual treat these matters in a way that would justify the worst Christian abhorrence, were these doctrines and practices really what the commentaries claim! In recent years the situation has ameliorated.

[2] 'This is why in Islam it is said that the throne of Satan is situated between earth and Heaven, which can also be understood in the sense of being situated between the body and the spirit, hence in the soul.' It is characteristic of deviation to mistake the psychic plane for the spiritual.

preceded by a sufficient theoretic preparation, many confusions may arise, and there is always the possibility of losing one's way in one of those intermediate domains where there is no security against illusions' (Guénon: *East and West,* pp. 193, 198–199, 236).

Introd. *Pilg.*

A time is reached when 'one has both an infallible compass and an impenetrable suit of armour. But before getting so far, there is often a long period of striving to be gone through ... and it is then that the very greatest precautions are necessary if all confusion is to be avoided, at least under present conditions, for clearly the same dangers cannot exist in a traditional civilization, where, moreover, those who are truly gifted in an intellectual way find everything made as easy as possible for them to develop their natural abilities' (id., pp. 190–191).

P. State 563 *Conform.* 170

Satan (from the Hebrew *ṣâṭân*—adversary) is the Adversary par excellence, the principle even of Rebellion. For those to whom the reality of God is more than just a way of thinking, the reality of Satan is likewise more than just a figment of the imagination or manner of speech or abstract symbol for evil in general. Certain indications have already been given in the introductory passages to the chapters on *Sin* and *Judgment* concerning the existence of and metaphysical reasons for a seeming 'center' of negation in apparent opposition to the Divine Center or Spiritual Principle, in relation to which this negative possibility is the shadowy counterpart or 'darker side'—'the point to which all gravities from every part are drawn' (*Inferno,* XXXIV. 110). For the moment these allusions must suffice to bring into perspective the words of Guénon: 'In virtue of the law of analogy, the lowest point is as an obscure reflection or inverse image of the highest point, from which results this consequence, paradoxical in appearance only, that the most complete absence of all principle implies a kind of "counterfeit" of principle itself, which some[1] have expressed in "theological" terms by saying that "Satan is the ape of God".[2] This observation can greatly aid in the understanding of some of the darkest enigmas of the modern world' (*Le Règne de la Quantité,* pp. 9–10).

'There is in the modern world itself a secret which is better kept than any other: it is that of the formidable enterprise of suggestion which has produced and which maintains the present mentality, and which has formed, and as it were "fabricated" it in such a way that it can only deny the existence and even the possibility of any such enterprise; which assuredly is the best means—and one of truly "diabolical" cunning—for keeping this secret from ever being discovered' (id., p. 90).

Introd. *Wr. Side*

'The supreme cunning of the devil, whatever one's conception of him, is to make his existence denied' (Guénon: *L'Erreur spirite,* p. 328).

It must not be forgotten that the devil can on occasion be an accomplished theologian;

[1] Among whom Tertullian.

[2] *Daemon Est Deus Inversus* was the name given William Butler Yeats as member of the so-called 'Hermetic Order of the Golden Dawn'.

An engraving in Robert Fludd's *Utriusque Cosmi Historia* (Oppenheim, 1617) represents the World Soul as a woman linked by her right hand to God while holding in her left a chain which attaches in turn to the wrist of an ape seated at the earth's center and measuring with compasses a globe of the world.—incidentally an eloquent testimony to the substance underlying certain contemporary constructions on the 'Grand Architect of the Universe'! This is the 'angry ape' in *Measure for Measure* (II. ii. 120), the 'glassy essence', which 'Plays such fantastic tricks before high heaven/ As make the angels weep.'

one must likewise bear in mind that 'strictly speaking, there does not really exist a profane domain, but only a profane point of view' (Guénon: *Le Règne,* p. 103). In the words of Schuon: 'The so-called "neutral" character that the profane world affects to attribute to itself is purely non-existent: let one remember in this respect the words of Christ: "He that is not with me is against me"' 'Considérations générales sur les fonctions spirituelles', *Études Trad.*, 1939, p. 353).

Infra 420

Error

Reality 775

Whoever worships another divinity (than *Âtmâ* as identified with *Brahma*), thinking, 'He is one and I another,' he knows not. He is like a sacrificial animal for the gods.

Bṛihad-Âraṇyaka Upanishad, I. iv. 10

Those unto whom ye pray instead of Him own not so much as the white spot on a date-stone.

If ye pray unto them they hear not your prayer, and if they heard they could not grant it to you. On the Day of Resurrection they will disown association with you. And none can tell thee like One who is Informed!

Qur'ân, XXXV. 13–14

He that sacrificeth unto any god, save unto the Lord only, he shall be utterly destroyed.

Exodus, XXII. 20

Introd. *Sin*

Ye cannot drink the cup of the Lord, and the cup of devils: ye cannot be partakers of the Lord's table, and of the table of devils.

I. Corinthians, x. 21

Among the heretical scholars there are a few who know of something like a reflection of the mysteries (of the *Kavod*), though not of their substance.

Samuel ben Kalonymus

Introd. *Flight*

And verily in the heaven We have set mansions of the stars, and We have beautified it for beholders.

And We have guarded it from every outcast devil,

Save him who stealeth the hearing, and him doth a clear flame pursue.

Qur'ân, XV. 16–18

Knowl. 734

Philosophers are the patriarchs of heresy.

Tertullian

. . . In religion,
What damned error, but some sober brow
Will bless it and approve it with a text,
Hiding the grossness with fair ornament?

Shakespeare (*Merchant of Venice*, III. ii. 77)

Never does one triumph over error by sacrificing any right whatsoever of the truth.

St Irenaeus

Fain would they put out the light of Allâh with their mouths, but Allâh will perfect His light however much the disbelievers are averse.

Qur'ân, LXI. 8

God hates the *Duall Number*; being known
The lucklesse number of division:
And when He blest each sev'rall Day, whereon
He did His *curious operation;*
'Tis never read there (as the Fathers say)
God blest His work done on the *second day*.[1]

Robert Herrick

M.M. 978

If I myself upon a looser Creed
Have loosely strung the Jewel of Good Deed
Let this one thing for my Atonement plead:
That One for Two I never did mis-read.

Omar Khayyâm

Reality 775

I have always avoided with horror all error in matters of faith.

Eckhart

He therefore that would like as God likes, and condemn as God condemns, must have neither the eyes of the Papist nor the Protestant; he must like no truth the less because Ignatius Loyola or John Bunyan were very zealous for it, nor have the less aversion to any error, because Dr Trapp or George Fox had brought it forth.

William Law

Judgment 242

For verily, though We Ourselves should send down the angels unto them, and the dead should speak unto them, and We should gather against them all things in array, still they would not believe unless Allâh so willed. However, most of them are ignorant.

Thus have We appointed unto every prophet an adversary—devils of humankind and jinn who inspire in one another gilded words of guile. And if thy Lord willed, they would not do so; therefore abandon them and what they invent.

Qur'ân, VI. 111–112

The rejection of heretics makes the tenets of Thy Church and sound doctrine to stand out more clearly. *For there must also be heresies, that the approved may be made manifest among the weak*.[2]

St Augustine

[1] The work referred to is the creation of a 'firmament', with the corresponding division of the higher from the lower waters, hence bringing into manifestation the 'Empire of the Demiurge'.

[2] *I. Cor.*, XI. 19.

Atheism

As physicians speak of a certain disease or madness, called Hydrophobia, the symptom of those that have been bitten by a mad dog, which makes them have a monstrous antipathy to water; so all Atheists are possessed with a certain kind of madness, that may be called Pneumatophobia, that makes them have an irrational but desperate abhorrence from spirits or incorporeal substances, they being acted also, at the same time, with an Hylomania, whereby they madly doat upon matter, and devoutly worship it as the only Numen.

Wr. Side 448

Ralph Cudworth

The Lord Buddha thereupon declared unto Subhuti, 'Belief in the unity or eternity of matter is incomprehensible; and only common, worldly-minded people, for purely materialistic reasons, covet this hypothesis.'

Prajñâ-Pâramitâ

And even as they did not like to retain God in their knowledge, God gave them over to a reprobate mind, to do those things which are not convenient;

Being filled with all unrighteousness, fornication, wickedness, covetousness, maliciousness; full of envy, murder, debate, deceit, malignity; whisperers,

Backbiters, haters of God, despiteful, proud, boasters, inventors of evil things, disobedient to parents,

Without understanding, covenantbreakers, without natural affection, implacable, unmerciful:

Who knowing the judgment of God, that they which commit such things are worthy of death, not only do the same, but have pleasure in them that do them.

Wr. Side 464

Romans, I. 28–32

Verily Allâh will bring those who believe and do good works into gardens beneath which rivers flow: while those who disbelieve take their comfort in this life and eat even as the cattle eat, and the Fire is their habitation.

Qur'ân, XLVII. 12

All carnal wisdom, arts of advancement, with every pride and glory of this life, are as so many heathen idols.

Illusion 99

William Law

The demonic people know not how to follow right or how to refrain from wrong: there is neither purity, nor good conduct, nor truth in them.

They say that 'this universe is without truth, without a basis, without God, born of mutual union caused by lust.[1] What else is there?'

Creation 47

[1] *They* in our times being the evolutionists.

Holding this view, these ruined souls, of small understanding and of fierce deeds, rise as the enemies of the world for its destruction.

Filled with insatiable desires, possessed with hypocrisy, pride and arrogance, holding evil fancies through delusion, they work with unholy resolve;

Beset with immense cares, ending only in death; regarding sensual enjoyment as the highest and feeling sure that that is all there is;

Bound by a hundred ties of hope, given over to lust and anger, they strive to secure hoards of wealth by unjust means, for sensual gratification. . . .

Bewildered by many fancies, enwrapped in the net of delusion, addicted to the gratification of the senses, they fall into a foul hell.

Bhagavad-Gîtâ, XVI. 7 ff.

Superstition . . . disrobes the Deity of true majesty and perfection, and represents it as weak and infirm, clothed with such fond, feeble, and impotent passions, as men themselves are. . . . If the superstitious man thinks that God is altogether like himself, which indeed is a character most proper to such, the atheist will soon say in his heart, 'there is no God;' and will judge it, not without some appearance of reason, to be better there were none.

John Smith the Platonist

The unbeliever's argument is always shamefaced: where is a single sign that indicates the truth of that unbelief?[1]

Where in this world is to be found a single minaret in praise of the unbelievers, so that it should be a sign (of their veracity)? . . .

The unbeliever's argument is just this, that he says, 'I see no place of abode except this external world.'

He never reflects that, wherever there is anything external, that object gives information of hidden wise purposes.

Symb. 306

The usefulness of every external object is, indeed, internal: it is latent, like the beneficial quality in medicines.

Rûmî

There cannot be the smallest thing, or the smallest quality of anything in this world, but what is a quality of Heaven or hell, discovered under a temporal form.

Supra Introd.

William Law

Atheism most commonly lurks *in confinio scientiae et ignorantiae*; when the minds of men begin to draw those gross, earthly vapours of sensual and material speculations by dark and cloudy disputes, they are then most in danger of being benighted in them. . . . And certainly, were the highest happiness of mankind such a thing as might be felt by a

Knowl. 734

[1] *Contra Cartesium*
That I can think is proof
Thou art the only In-dividual.
From whose dividuality
My feignèd individuality depends.
A. K. Coomaraswamy

corporeal touch, were it of so ignoble a birth as to spring out of this earth, and to grow up out of this mire and clay,[1] we might well sit down, and bewail our unhappy fates, that we should rather be born men than brute beasts, which enjoy more of this world's happiness than we can do, without any sin or guilt.

John Smith the Platonist

Inv. passim

At the present moment many educated persons and college students have lost faith in the power of Mantra, owing to the morbid influence of the study of science. They have entirely given up Japa. It is highly deplorable indeed. When the blood is warm they become hot-headed, proud and atheistic. Their brains and minds need a thorough overhauling and drastic flushing.

Swami Sivananda

Wr. Side 448

Remember that it is useless to try to force those to believe who will not, for even the Buddha himself cannot do that.

Hônen

Jesus (Peace be upon him!) said, 'I am not incapable of raising the dead, but I am incapable of applying a remedy to the fool.'

Christ in Islâm

For verily thou canst not make the dead to hear, nor canst thou make the deaf to hear the call when they have turned to flee.

Nor canst thou guide the blind out of their error. Thou canst make none to hear save those who believe in Our revelations so that they surrender.

Qur'ân, xxx. 52–53

Introd. *Conform.*

For this people's heart is waxed gross, and their ears are dull of hearing, and their eyes they have closed; lest at any time they should see with their eyes, and hear with their ears, and should understand with their heart, and should be converted, and I should heal them.

St Matthew, xiii. 15

And when thou recitest the *Qur'ân,* We place between thee and those who believe not in the Hereafter a hidden barrier;

Judgment 242

And We place upon their hearts veils lest they should understand it, and in their ears a deafness; and when thou makest mention of thy Lord alone in the *Qur'ân,* they turn their backs in aversion.

Qur'ân, xvii. 45–46

The fool hath said in his heart, There is no God.

Psalm xiv. 1

[1] An attitude that was to be cloaked with 'respectability' two centuries later as the evolutionist hypothesis.

Deviation

There is no approach by a side-path here in the world.

Maitri Upanishad, VI. 30

He that entereth not by the door into the sheepfold, but climbeth up some other way, the same is a thief and a robber.

St John, x. 1

Any man who does not let himself be guided by a spiritual director is guilty of rebellion towards God, for without a guide he could not obtain access to the road of salvation, were he to possess by memory a thousand works of theology.

Sheikh ʿAbd-el-Hadi ben Ridouane

Orthod. 288

No success will come to him who obtains the secrets in an irregular way.

Ko Hung

If any one by means of asceticism and self-mortification shall have risen to an exalted degree of mystical experience, without having a Pîr to whose authority and example he submits himself, the Ṣûfîs do not regard him as belonging to their community.

Muḥammad ibnu 'l-Munawwar

Beware of novel affairs, for surely all innovation is error.

Muhammad

Rev. passim

For a person rightly to be adjudged a heretic he must fulfil five conditions. First, there must be an error in his reasoning. Secondly, that error must be in matters concerning the faith, either being contrary to the teaching of the Church as to the true faith, or against sound morality and therefore not leading to the attainment of eternal life. Thirdly, the error must lie in one who has professed the Catholic faith, for otherwise he would be a Jew or a Pagan, not a heretic. Fourthly, the error must be of such a nature that he who holds it must still confess some of the truth of Christ as touching either His Godhead or His Manhood; for if a man wholly denies the faith, he is an apostate. Fifthly, he must pertinaciously and obstinately hold to and follow that error.

Malleus Maleficarum

Infra 430
Conform. 170

All dalliance with what wears the mask of the authentic, all attraction towards mere semblance, tells of a mind misled by the spell of forces pulling towards unreality. The sorcery of nature is at work in this; to pursue the non-good, drawn in unreasoning impulse by its specious appearance, is to be led unknowing down paths unchosen; and what can that be but magic?

Illusion 85
Wr. Side 451

Alone in immunity from magic is he who, though drawn by the alien parts of his being, withholds his assent to their standards of worth, recognizing the good only where his authentic self sees and knows it, tranquilly possessing it, and so never charmed away.

Metanoia 480

Plotinus

Would you know whence it is that so many false spirits have appeared in the world, who have deceived themselves and others with false fire and false light, laying claim to information, illumination and openings of the divine Life, particularly to do wonders under extraordinary calls from God? It is this: they have turned to God without turning from themselves; would be alive to God before they are dead to their own nature. Now religion in the hands of self, or corrupt nature, serves only to discover vices of a worse kind than in nature left to itself. Hence are all the disorderly passions of religious men, which burn in a worse flame than passions only employed about worldly matters; pride, self-exaltation, hatred and persecution, under a cloak of religious zeal, will sanctify actions which nature, left to itself, would be ashamed to own.

Metanoia 484 Infra 428

William Law

Woe unto you, scribes and Pharisees, hypocrites! for ye compass sea and land to make one proselyte, and when he is made, ye make him twofold more the child of hell than yourselves.

Wr. Side 448

St Matthew. XXIII. 15

The men who practise severe austerities, not enjoined by the Scriptures, being possessed with hypocrisy and egoism, impelled by lust and attachment;

Torturing, senseless as they are, all the organs of the senses and Me, dwelling in the body, know them to be of demonic resolve.

Suff. 133

Bhagavad-Gîtâ, XVII. 5, 6

They who imagine truth in untruth, and see untruth in truth, never arrive at truth, but follow vain desires.

Dhammapada, I. 11

If He come & ye (Quakers) bide none other, why do we hear at all your meetings, especially when these are most godly, as you say, the voice of sobbing, of weeping, lamentation, yea anguish, sorrow, pain & ululation as for one dead? Is this the jubilant voice of the bride for her bridegroom?

Orthod. 296

Johannes Kelpius

The Docetists abstain from the Eucharist, because they allow not that It is the flesh of our Saviour, which flesh suffered for our sins, and which the Father of His goodness raised up.

Orthod. 285

St Ignatius of Antioch

You are wrong in thinking that more profit is to be had from drinking the chalice of separation than from using the one pascal lamb in peace and unity. To receive with a separatist will the chalice of the Lord of unity and peace, is to receive it in vain, for it cannot confer life to a member separated from this Church which is the body of Christ. Do not say that it is the rest of the Church which has separated from you, and that you constitute the true Church, reduced to this little part of Bohemia. Assuredly, in the unity of the Church the diversity of rites is without danger—something which no one would question. But if a person rashly and presumptuously values a particular rite above unity and peace, then be he ever so good, holy and worthy of praise, he still merits condemnation. You say that one must first obey the precepts of Christ, and only afterwards the Church, and that if the Church teaches other precepts than those of Christ, then it is not the Church but Christ whom one must obey. Now here precisely is the beginning of all presumption, when individuals judge their own private opinion concerning the divine commandments as being more in conformity with the divine Will than is the attitude of the universal Church. . . .

Therefore one must follow the Church without fail when it determines what is best in each instance, for it possesses the faith and conserves the accumulated store, even if some error may slip into its judgment. The case is the same as with a judge who, misled by a false witness, excommunicates an innocent person, without consciously committing a reprehensible act. This judge does not transgress, but on the contrary follows the rule which obliges him to judge according to witnesses and proofs. And the innocent party, in obeying the sentence and thus separating himself from the body of the Church, does not lose the grace and life which come from the sacraments, but by obeying the Church even if it is wrong, he obtains salvation, even though deprived of the sacraments.

Orthod. 275

Conform. 170

Nicholas of Cusa (to the Bohemians on double communion)

No one ever gave information to people who were not capable of understanding it without its proving a temptation to some of them.

Muhammad

M.M. 973

Do not let there be a schism in the Order, for a schism in the Order is a serious matter, Devadatta. He who splits an Order that is united sets up demerit that endures for an aeon and he is boiled in hell for an aeon. But he who unites an Order that is split sets up sublime merit and rejoices in heaven for an aeon.

Vinaya Piṭaka, II

An heretic sinneth deadly in pride, for he chooseth his rest and his delight in his own opinion and his own saying, for he weeneth it sooth; the which opinion or saying is against God and Holy Kirk. He will not leave it, but rest him therein, as in a soothfastness; and so maketh he it his god. But he beguileth himself, for God and Holy Kirk are so oned and accorded together that who so doth against that one, doth against both. And therefore he that sayeth he loveth God and keepeth His biddings, and despiseth Holy Kirk, and setteth at nought the laws and the ordinances of it made by the head and the sovereign in governance of all Christian men, he lieth. He chooseth not God, but he chooseth the love of himself, contrary to the love of God, and so he sinneth deadly. And in that, that he weeneth most for to please God, he most displeaseth Him; for he is blind and will not see. Of this blindness and of this false resting of heretics in their own feeling, speaketh the Wise Man thus, *Est via quae videtur homini recta; et novissima ejus deducunt ad mortem*. There is a

Sin 75

way which seemeth to a man rightful, and the last end of it bringeth him to endless death.[1] This way specially is called heresy, for other fleshly sinners that sin deadly and lie still therein commonly suppose aye anon amiss of themselves, and feel biting in conscience that they go not in the right way. But an heretic supposeth aye that he doth well and teacheth well, and yet no man so well; and so weeneth he that his way were the right way, and therefore feeleth he no biting of conscience, nor meekness in heart. And soothly but if God send him meekness whilst he liveth, of His mercy, at last end he goeth to hell, and yet weeneth he for to have done well and get him the bliss of heaven for his teaching.

Sin 80

Walter Hilton

Beware thou take not on thing after thy affection and liking, and leve another: for that is the condition of an heretique.

Postcript by a scribe to Julian of Norwich's *Revelations of Divine Love*

It has never been for Allâh to send a people astray after He has guided them until He has made clear unto them what they should avoid.

Qur'ân, IX. 115

Wr. Side 444
Ecstasy 637

Seeking by food to obtain Immortality
Many have been the dupe of strange drugs.
Better far to drink good wine
And clothe our bodies in robes of satin and silk.

Old Chinese Poem

Calcination by means of a heterogeneous agent can only destroy the metallic nature, in so far as it has any effect at all. Every calcination of gold, which is not succeeded by a spontaneous dissolution, without laying on of hands, is also fallacious.

Orthod. 280

Philalethes

The mind is like white linen fresh from the laundry; it takes the colour in which you dip it. If it is associated with falsehood for a long time, it will be stained with falsehood.

Action 346

Sri Ramakrishna

Lo! the hypocrites seek to beguile Allâh, but it is He Who beguileth them.

Qur'ân, IV. 142

To teach the evil-natured man knowledge and skill is to put a sword in the hand of a brigand.

Rûmî

Introd.

My teaching has a certain property which is peculiar to it; it urges on bad men to worse wickedness.

Hermes

[1] *Proverbs,* XIV. 12.

If you in any way abuse the gift of God, or use it for your own glorification, you will most certainly be called to account by the Almighty Giver, and you will think that it would have been better for you if you had never known it.

The Sophic Hydrolith

Nay, but verily man is rebellious
That he thinketh himself independent! . . .
Is he then unaware that Allâh seeth?

Introd. *Knowl.*

Qur'ân, XCVI. 6, 7, 14

Or is the Unseen theirs that they can write thereof?

Qur'ân, LXVIII. 47

Passion is of two kinds: (1) desire of pleasure and lust, and (2) desire of worldly honour and authority. He who follows pleasure and lust haunts taverns, and mankind are safe from his mischief, but he who desires honour and authority lives in cells and monasteries, and not only has lost the right way himself but also leads others into error.

Hujwîrî

If thou strivest unduly to shorten the time thou wilt produce an abortion. Many persons have, through their ignorance, or self-opinionated haste, obtained a Nihilixir instead of the hoped for Elixir.

Conform. 170

The Sophic Hydrolith

When the unclean spirit is gone out of a man, he walketh through dry places, seeking rest, and findeth none.

Then he saith, I will return into my house from whence I came out: and when he is come, he findeth it empty, swept, and garnished.

Pilg. 366 *Holy War* 407

Then goeth he, and taketh with himself seven other spirits more wicked than himself, and they enter in and dwell there: and the last state of that man is worse than the first.[1]

St Matthew, XII. 43–45

Turn thee backwards, and keep thy eyes closed: for if the Gorgon shew herself, and thou shouldst see her, there would be no returning up again.

Dante (*Inferno*, IX. 55)

Do not sit with the frigid: for you will be chilled by their breath.

Dîvâni Shamsi Tabrîz, XLVI

For it is impossible for those who were once enlightened, and have tasted of the heavenly gift, and were made partakers of the Holy Ghost,

Orthod. 280

[1] 'In Islamic esoterism it is said that one who presents himself at a certain "gate", without having reached it by a normal and legitimate way, sees it shut in his face and is obliged to turn back, but not as a mere profane person, for he can never be such again, but as a *sâher* (a sorcerer or a magician working in the domain of subtle possibilities of an inferior order)' (Guénon: *The Reign of Quantity*, p. 317: tr. Lord Northbourne).

And have tasted the good word of God, and the powers of the world to come,

If they shall fall away, to renew them again unto repentance; seeing they crucify to themselves the Son of God afresh, and put him to an open shame.

Knowl. 761

For the earth which drinketh in the rain that cometh oft upon it, and bringeth forth herbs meet for them by whom it is dressed, receiveth blessing from God:

Judgment 261

But that which beareth thorns and briars is rejected, and is nigh unto cursing; whose end is to be burned.

Hebrews, VI. 4–8

They (the disciples of Pythagoras who deviated) received the double of the wealth which they brought, and a tomb was raised to them as if they were dead by the *homacoï*; for thus all the disciples of the man were called. And if they happened to meet with them afterwards, they behaved to them as if they were other persons, but said that they were dead, whom they had modelled by education, in the expectation that they would become truly good men by the disciplines they would learn.

Iamblichus

'And Ye Shall Be as Gods'

And Satan whispered unto him and said: 'O Adam, shall I show thee the Tree of Immortality and a kingdom that fadeth not away?'[1]

Qur'ân, XX. 120

And the serpent said unto the woman, Ye shall not surely die.[2]

Genesis, III. 4

Infra 430

As this Light vainly thinketh to understand God, it imagineth itself to be God, and giveth itself out to be God, and wisheth to be accounted so, and thinketh itself to be above all things, and well worthy of all things, and that it hath a right to all things, and hath got beyond all things, such as commandments, laws, and virtue, and even beyond Christ and a Christian life, and setteth all these at nought, for it doth not set up to be Christ, but the Eternal God. And this is because Christ's life is distasteful and burdensome to nature, therefore she will have nothing to do with it; but to be God in eternity and not man, or to

[1] The true promise of God—in contrast to this false allurement—is given in XXXVIII. 49–54, cited in the chapter, *The Primordial State.*

[2] Cf. Baird T. Spalding, *Life and Teaching of the Masters of the Far East,* vol. III, Foreword: 'The church and the graveyard are often in the same field. This alone is a direct acknowledgment that Christian teachings have not been even comprehended. The Christ-man has spoken and the listening ear has heard. "If a man believes in me, he shall never die" '—a travesty on *St John,* XI. 26.

be Christ as He was after His resurrection, is all easy, and pleasant, and comfortable to nature, and so she holdeth it to be best.

Theologia Germanica, XLII

It is not good for ordinary people to say, 'I am He.' The waves belong to the water. Does the water belong to the waves?

Sri Ramakrishna

Supra
Introd.

For the man whose ego is the principal obstacle to the Realization of the unity of God, it is pure ignorance to declare that he is God.

Swami Ramdas

As regards the blind man, . . . if any one sprinkle some musk over him, he thinks it (comes) from himself and not from the kindness of his friend.

Rûmî

The master said, 'Everything that exists is God.' The pupil understood it literally, but not in the right spirit. While he was passing through the street he met an elephant. The driver shouted aloud from his high place, 'Move away! Move away!' The pupil argued in his mind, 'Why should I move away? I am God, so is the elephant God; what fear has God of himself?' Thinking thus, he did not move. At last the elephant took him up in his trunk and dashed him aside. He was hurt severely, and going back to his master, he related the whole adventure. The master said: 'All right. You are God, the elephant is God also, but God in the shape of the elephant-driver was warning you from above. Why did you not pay heed to his warnings?'

Sri Ramakrishna

Introd.
Realiz.
Conform.
180

You should not doubt the teachings of the scriptures. Flickering faith will lead to downfall. A man of weak will, who has no faith in Japa, cannot expect to have progress in the spiritual path. If he says, 'I am practising "who am I" enquiry'—this is all wild imagination. Few are fit for 'who am I' enquiry.

Swami Sivananda

Supra 423
Inv. 1017
Infra
430

Unless your mind is pure, that is, free from all desires, you cannot attain oneness with Him. Mere saying you are He, won't do. Genuine experience is necessary. Remaining always on the dual plane, subject to the Dwandwas of pleasure and pain, honour and dishonour, profit and loss, and still saying 'I am Brahman' or 'I am one with the Divine' is self-deception.

Swami Ramdas

Renun.
136

The exoteric aspect of Truth without the esoteric is hypocrisy, and the esoteric without the exoteric is heresy. So, with regard to the Law, mere formality is defective, while mere spirituality is vain.

Hujwîrî

M.M. 973
Orthod.
285

P. State 579

To acquire the power to attain the Infinite, one must first know the finite and let oneself be guided by it. As long as the soul is identified with the body, it is indispensable to let oneself be guided by what the rules enjoin or proscribe.

Ananda Moyî

To know the real Self to be one's own is the greatest attainment according to the scriptures and reasoning. To know wrongly the non-Self such as the ego etc. to be the Self is no attainment at all. One, therefore, should renounce this misconception of taking the non-Self for the Self.

Śrî Śankarâchârya

If that is your Vedanta, I spit on Vedanta.[1]

Sri Ramakrishna

Philosophers devoid of reason find
This world a mere idea of the mind;
'Tis an idea—but they fail to see
The great Idealist who looms behind.

Beauty 670

Jâmî

The False Light

Now, there are some . . . who love to be a stumbling-block among believers in the Vedas by the strategem of deceptive arguments in a circle, and false and illogical examples.

With these one should not associate. Verily, these creatures are evidently robbers, unfit for heaven. For thus has it been said:—

By the jugglery of a doctrine that denies the Soul,
By false comparisons and proofs
Disturbed, the world does not discern
What is the difference between knowledge and ignorance.

Maitri Upanishad, VII. 8

Knowl. 734

There is an exquisite subtilty, and the same is unjust.

Ecclesiasticus, XIX. 22

[1] In reply to someone who claimed: 'Virtue and vice are both unreal, for the universe is unreal: and I am the Atman. Nothing can touch me.'

He that is blessed and familiar (with spiritual mysteries) knows that (cunning) intelligence is of Iblîs,[1] while love is of Adam.

Rûmî

The Devil lurks behind the Cross.

Spanish Proverb

Since there is many a devil who hath the face of Adam, it is not well to give your hand to every hand. . . .

The vile man will steal the language of dervishes, that he may thereby chant a spell over one who is simple.

Rûmî

All that glisters is not gold.

Shakespeare (*Merchant of Venice*, II. vii. 65)

A white bone much resembles ivory; most men fail to distinguish the one from the other. So with men. The specious kind appears to have goodness, but it is not really so.

Huai Nan Tzû

There was once a goldsmith whose tongue suddenly turned up and stuck to his palate. He looked like a man in samâdhi. He became completely inert and remained so a long time. People came to worship him. After several years, his tongue suddenly returned to its natural position, and he became conscious of things as before. So he went back to his work as a goldsmith.

Sri Ramakrishna

Glass, from the nearness of gold, acquires an emerald lustre: so by the proximity of the excellent a fool attains to cleverness.

Hitopadeśa

Fools take false coins because they are like the true.
If in the world no genuine minted coin
Were current, how would forgers pass the false?
Unless there be truth, how should there be falsehood?
Falsehood receives brilliance from truth.

Rûmî

Supra Introd.

Beware of false prophets, which come to you in sheep's clothing, but inwardly they are ravening wolves.

St Matthew, VII. 15

'Behold the savage beast with the pointed tail, that passes mountains, and breaks through walls and weapons; behold him that pollutes the whole world.'

[1] Satan.

Thus began my guide to speak to me; and beckoned him to come ashore, near the end of our rocky path;

and that uncleanly image of Fraud came onward, and landed his head and bust, but drew not his tail upon the bank.

Wr. Side 459

His face was the face of a just man, so mild an aspect had it outwardly; and the rest was all a reptile's body.

Dante (*Inferno*, XVII. 1–12)

There is a heaven made by the devil
With his false cunning.

Mechthild of Magdeburg

When a man possesses the natural rest in bare vacancy,[1] whilst in all his works he has himself in mind, and he continues obstinately disobedient in his self-will, he cannot be united with God; for he lives without charity in unlikeness to God. . . . All these men are, in their own opinion, God-seeing men, and believe themselves the holiest of all men living. Yet they live contrary and unlike to God and all saints and all good men. Observe the following marks: thus you will be able to recognise them both by their words and their works. By means of the natural rest which they feel and possess in themselves in bare vacancy, they believe themselves to be free, and to be united with God without means, and to be above all the customs of Holy Church, and above the commandments of God, and above the law, and above every work of virtue which can in any way be done. . . . And they are sometimes possessed of the Fiend; and then they are so cunning that one cannot vanquish them on the grounds of reason. But through Holy Scripture and the teaching of Christ and our Faith, we may prove that they are deceived. . . . For they live contrary to God and righteousness and all saints; and they are all precursors of the Antichrist, preparing his way in every unbelief. For they would be free, without the commandments of God, and without virtues; and empty and united with God, without love and charity. And they would be God-seeing men without loving and steadfast contemplation, and the holiest of all men living without the works of holiness. And they say that they rest in Him Whom they do not love; and are uplifted into That which they neither will nor desire. . . . And they would cunningly disguise the worst, so that it should seem the best. All these are contrary to God and all His saints; but they have a likeness to the damned spirits in hell, for these too are without charity and without knowledge, and are empty of thanksgivings and praise and of all loving adherence; and this is the cause, why they remain damned in eternity. And that these folk may be like to them they lack only this, that they should fall from time into eternity, and that the justice of God be revealed in their works.

Supra 428

Knowl. 734

Judgment 261

Ruysbroeck

Beware of the midday fiend, that feigneth light as it come out of Jerusalem and is not.

Walter Hilton

[1] 'Quietism' in its most malignant form means the submergence and passive fusion of will into the subconscious and infernal, in contradistinction to active realization of the supraconscious and divine.

All the light and knowledge that may seem sometimes to rise up in unhallowed minds, is but like those fuliginous flames that arise up from our culinary fire, that are soon quenched in their own smoke; or like those foolish fires that fetch their birth from terrene exudations, that do but hop up and down, and flit to and fro upon the surface of this earth where they were first brought forth; and serve not so much to enlighten, as to delude us; nor to direct the wandering traveller into his way, but to lead him farther out of it. While we lodge any filthy vice in us, this will be perpetually twisting itself into the thread of our finest spun speculations; it will be continually climbing up into the τὸ Ἡγεμονιχὸν, the hegemonical powers of the soul, into the bed of reason, and defile it: like the wanton ivy twisting itself about the oak, it will twine about our judgments and understandings, till it hath sucked out the life and spirit of them.

John Smith the Platonist

Pilg. 365

Knowl. 734

Strive not with a man that is full of tongue, and heap not wood upon his fire.

Ecclesiasticus, VIII. 4

The devil can cite Scripture for his purpose.

Shakespeare (*Merchant of Venice*, I. iii. 99)

It is observed by some, that God never suffered the devil to assume any human shape, but with some character whereby his body might be distinguished from the true body of a man: and surely the devil cannot so exactly counterfeit an angel of light, but that by a discerning mind he may be distinguished from him; as they say a beggar can never act a prince so cunningly, but that his behaviour sometime sliding into the course way and principles of his education, will betray the meanness of his pedigree to one of a true noble extraction. A bare imitation will always fall short of the copy from whence it is taken.[1]

John Smith the Platonist

Light of knowing that is feigned by the fiend to a murk soul is aye showed atwixt two black rainy clouds. The over cloud is presumption and highing of himself; the nether cloud is down putting and lowing of his even-christian.

Walter Hilton

This False Light saith, that we should be without conscience or sense of sin. . . . We may answer and say: Satan is also without them, and is none the better for that.

Theologia Germanica, XL

Introd. *Sin*

A mere glimpse of Reality may be mistaken for complete realization.

Gampopa

[1] 'In all the more or less symbolical descriptions given of him, the Antichrist is represented as deformed. . . . These descriptions especially emphasize bodily dissymmetries, which implies essentially that these are the visible marks of the actual nature of the being to whom they are attributed; and in fact they are always signs of some inner disequilibrium' (Guénon: *Le Règne de la Quantité*, p. 266). Guénon also insists on the ignoble, repulsive and '*hideux*' character attaching to psychoanalytical interpretations and spiritualist manifestations as a 'mark' of their origin (cf. *Le Règne*, ch. XXXIV).

Supra 428

They would fain be so holy as to draw themselves up into the eternal Godhead and pass by the eternal holy humanity of our Lord Jesus Christ.

Mechthild of Magdeburg

P. State 579

The most infallible and indubitable sign by which we may distinguish a heretic, a man of bad doctrine, a reprobate, from one of the predestinate, is that the heretic and the reprobate have nothing but contempt and indifference for our Lady, endeavouring by their words and examples to diminish the worship and love of her, openly or hiddenly, and sometimes by misrepresentation.

St Louis de Montfort

Whosoever denieth the Son, the same hath not the Father.

I. John, II. 23

Supra Introd.

Behold, all that is contrary to the True Light belongeth unto the False. . . . In so far as this Light imagineth itself to be God and taketh His attributes unto itself, it is Lucifer, the Evil Spirit; but in so far as it setteth at nought the life of Christ, and other things belonging to the True Light, which have been taught and fulfilled by Christ, it is Antichrist, for it teacheth contrary to Christ.

Theologia Germanica, XL

Peace 700

Cessation of thought-processes may be mistaken for the quiescence of infinite mind, which is the true goal.[1]

Gampopa

Also, (the Waldensians) hold and teach that all oaths, whether in justice or otherwise, without exception and explanation, are forbidden by God, and illicit and sinful, interpreting thus in an excessive and unreasonable sense the words of the holy Gospel and of St James the Apostle against swearing. Nevertheless, the swearing of oaths is lawful and obligatory for the purpose of declaring the truth in justice, according not only to the same doctrine of the saints and doctors of the Church and the tradition of the same holy Catholic Church, but also to the decree of the Church published against the aforesaid error: 'If any of these should reject the religious obligation of taking an oath by a damnable superstition, and should refuse to swear, from this fact they may be considered heretics.'

It should be known, however, that these Waldensians give themselves dispensations in the matter of taking oaths; they have the right to swear an oath to avoid death for themselves or for another, and also in order not to betray their fellows, or reveal the secret of their sect. For they say that it is an inexpiable crime and a sin against the Holy Ghost to betray a 'perfect' member of their sect.

Also, from this same fount of error, the said sect and heresy declares that all judgment is forbidden by God, and consequently is sinful, and that any judge violates this prohibition of God, who in whatever case and for whatever cause sentences a man to corporal punishment, or to a penalty of blood, or to death. In this, they apply, without the necessary explanation, the words of the holy Gospel where it is written: 'Judge not, that ye be not

[1] See note p.432.

judged,' and 'Thou shalt not kill,' and other similar texts; they do not understand these or know either their meaning or their interpretation, as the holy Roman Church wisely understands them and transmits them to the faithful according to the doctrines of the fathers and doctors, and the decisions of canon law....

Judgment 242
Holy War 394

It should be known that it is exceedingly difficult to interrogate and examine the Waldensians, and to get the truth about their errors from them, because of the deception and duplicity with which they answer questions in order not to be caught.... This is the way they do it. When one of them is arrested and brought for examination, he appears undaunted, and as if he were secure and conscious of no evil in himself. When he is asked if he knows why he has been arrested, he answers very sweetly and with a smile, 'My lord, I should be glad to learn the reason from you.' Asked about the faith which he holds and believes, he answers, 'I believe everything that a good Christian ought to believe.' Questioned as to whom he considers a good Christian, he replies, 'He who believes as Holy Church teaches him to believe.' When he is asked what he means by 'Holy Church', he answers, 'My lord, that which you say and believe is the Holy Church.' If you say to him, 'I believe that the Holy Church is the Roman Church, over which the lord pope rules, and under him, the prelates,' he replies, 'I believe it,' meaning that he believes that you believe it....

When he is questioned concerning this deception and many others like it, and asked to answer explicitly and directly, he replies, 'If you will not interpret what I say simply and sanely, then I do not know how I should answer you. I am a simple and illiterate man. Do not try to ensnare me in my words.' If you say to him, 'If you are a simple man, answer simply, without dissimulation,' he says, 'Willingly.'[1]

Bernard Gui

And therefore for God's love beware in this work, and strain not thy heart in thy breast over-rudely nor out of measure; but work more with a list than with any idle strength. For the more listily thou workest, the more meek and ghostly is thy work; and the more rudely, the more bodily and beastly. And therefore beware. For surely the beastly heart that presumeth to touch the high mount of this work shall be beaten away with stones.

Conform. 170

The Cloud of Unknowing, XLVI

But the proud scorner that will take no warning is of Lucifer's regiment, who saw the mystery of God's kingdom to stand in meekness, simplicity, and deep humility, and therefore out of his pride would aspire to be above the divine love, and harmony of obedience to God's will, and so fell into the abyss of the dark world, into the outmost darkness of the first principle, which we call Hell, where he and his legions are captives; from which the Almighty God of Love deliver us.

Sin 75

John Sparrow

[1] The principal heresy of the Waldensians was contempt for ecclesiastical authority. This passage is given at length because it typifies a perversion of mind and soul which is not at all confined to this particular sect.

As regards the taking of oaths, Bernard E. Jones writes in his *Freemasons' Guide and Compendium* (London, Harrap, 1950), p. 280: 'There have been outstanding Quaker members of the Craft': and the presence of this sect (otherwise opposed to all ritual) in Irish, English and American Lodges has in fact brought the controversy of oath versus affirmation into a rather specialized milieu....

SUBVERSION—THE WRATHFUL SIDE

There is a malevolence that envies the immortal, and will not suffer recognition of that which is divine.

Hermes[1]

The world destroys Tao, and Tao destroys the world.

Chuang-tse (ch. XVI)

A recurring motif in traditional art[2] is the world tree, the *axis mundi*, encircled by a serpent,[3] whose convolutions are the type of cosmic force or animating power of the universe; they represent the cycles of existence and current of forms, and are both generous and dangerous, like the ecstatic dance of Kâlî on the inert form of Shiva. The coils of the serpent suggest a dynamic movement either upwards or downwards, giving the serpent its benefic and its malefic sense; and sometimes the vertical axis is even shown entwined with two serpents, as in the familiar form of the caduceus, which has its Indian equivalent in the Brâhmanical staff (*Brahma-danda*), and in the *kuṇḍalinî* motif of the *sushumnâ* column flanked by the two 'arteries' or *nâdîs, pingalâ* and *idâ*. The 'celestial' way (*deva-yâna, Janua Coeli*) is thus represented by the central axis or 'coronal artery', leading upwards through the 'solar gate' to liberation from the *samsaric* states or serpentine 'round' of existence, and ultimate identity with the Supreme Principle. But the same axis can also be followed in the inverse direction, which then becomes the 'infernal' way (*pitri-yâna, Janua Inferni*)[4] as 'opposed' to the celestial, and leads to lower states, and even total disintegration of the conscious being at its extreme limit, since the only 'antipode' to the possibility of the Supreme Identity is logically complete 'annihilation'.[5]

Introd.
Pilg.
Realiz. 890
Judgment 250
Realiz. 887
Introd.
Realiz.

'This direct descent of the being following the vertical axis is represented notably by the "fall of the angels"; obviously when it is a question of human beings, this can only correspond to an exceptional case' (Guénon: *Le Symbolisme de la Croix*, p. 186). But there is another import to 'fallen angels', bearing on what in Kabbalistic terminology is called 'cutting the roots of plants', and refers to

[1] *Asclepius* I. 12 b (tr. Coomaraswamy).

[2] The following lines are based on Guénon's *Symbolisme de la Croix*, ch. XXV.

[3] Prototype of the one in *Genesis*.

[4] The *sitra ahra* ('Other Side') of the Kabbala.

[5] Hence all subversive doctrines whatsoever will inescapably be nihilistic at their core. But since 'the annihilation of anything real (is) a metaphysical impossibility' (Coomaraswamy: *Hinduism and Buddhism*, p. 30), annihilation here is really tantamount to damnation, i.e., dissolution into chaos.

It is significant that initiation into a witches' coven centers on a ritual adumbrating the marriage of the postulant with Death, whereupon the soul is admitted into the company of the 'Mighty Dead'. Cf. Gerald B. Gardner: *Witchcraft Today*, London, Ryder, 1954, ch. III.

polytheism in its veritable sense, of invoking divine qualities independently of Principle,[1] which can only happen through a deviation in tradition (contrary to 'evolutionist' thought, that would have a pure 'monotheism' develop out of a more 'primitive' polytheism). An 'angel' envisaged in this respect retains little more than the inverse shadow of its true being, and 'under these conditions, he who believes that he is summoning an angel instead greatly risks seeing a demon appear before him' (Guénon: *Études Trad.,* 1946, p. 346).

In this chapter we enter a domain of heat rather than light, and more specifically, a heat whose fumes manage to penetrate the world through ruptures in the 'wall' separating the cosmos from the 'outer darkness', and manage to do so increasingly as the present cycle of humanity, the Iron Age or *Kali-Yuga,* approaches its last stages of deterioration. 'The devil is not only terrible, he is often grotesque';[2] and we find ourselves in a world animated implicitly or explicitly, secretly or openly, by spirits, elementals, materializations, etheric states, auric eggs, astral bodies, ids, ods and egos, ectoplasmic apparitions, wraiths and visions, subliminal consciousness and collective unconsciousness, doublings, dissociations, and functional disintegrations, communications, obsessions and possessions, psychasthenia, animal magnetism, hypnoidal therapeutics, vibrations, thought-forces, mind-waves and radiations, clairvoyances and audiences and levitations, telepathic dreams, premonitions, death lights, trance writings, Rochester knockings, Buddhic bodies, and sundry other emergences and extravagances of hideous nomenclature that it would be both tedious and time-consuming to elaborate upon. It suffices to know that these elements, however seemingly disparate, and notwithstanding what may in reality be concealed behind these titles, all emanate in fact from one same obscure substance and thrive by one same fiery principle,[3] at the antipodes of the true and luminous sun. 'The subconscious is a sink of psychic residues, a sort of garbage pit or compost heap, fitted only for the roots of "plants", and far removed from the light that erects them' (Coomaraswamy, in PMLA, June, 1946).

Infra 464

Conclusion

Introd. *Pilg.*

The Christian world in particular has always seemed a target for abuse, the Renaissance itself being a result of what Schuon calls the 'posthumous vengeance' of classical antiquity; and the contagion is now virtually global. Participation in evil can be more or less conscious. According to St Augustine, '*magi faciunt miracula per privatos contractus*' (*De div.,* LXXIX. 4), while John XXII in the Bull *Super illius specula* of 1326, speaks of those who 'ally themselves with death and make a pact with hell, who sacrifice to demons . . . who pose questions to demons, obtain responses from them, and have recourse to demons to satisfy their depraved desires'; and in the sixteenth century, Jean Bodin writes: 'A sorcerer is one who, through diabolic means, *knowingly* tries to accomplish something' (*De la Démonomanie des sorciers,* Paris, 1580, p. 1). On the other hand, 'it sometimes unfortunately happens that those who think they are combatting the devil—whatever their idea of him may be—are, without in the least

[1] Cf. the articles by Guénon in the *Études Traditionnelles* for 1946: 'Les "racines des plantes"', and 'Monothéisme et angélologie'.

[2] Guénon: *L'Erreur spirite,* p. 311.

[3] Thus presenting a specious façade of unity (which Analytical Psychology in its hypersensitivity to 'archetypes' would call 'the true center, or the essence of the collective unconscious').

suspecting it, quite simply transformed into his best servants!' (Guénon: *Le Règne de la Quantité,* p. 202).[1]

In the *Catholic Encyclopedia Dictionary* (N.Y., 1941), *Diabolism* is defined as: 'all kinds of intercourse or attempts to deal with the evil spirit by witchcraft, incantations, magic, spiritism, and other occult practices. The possibility of consulting and securing the help of the devil is sufficiently attested by Scripture: God forbids consultation of soothsayers (Deut., 18); to "go aside after wizards" is unlawful (Lev., 19). Witchcraft, real or alleged, has brought misfortune and cruelty into the world; many phenomena of spiritism, or spiritualism, bear unmistakable characteristics of maleficent spiritual agencies. The Church as well as the Bible warrants the belief in evil spirits, or devils, and in their power, as far as God will permit, to do harm, but forbids dealing with them, since, by reason of the perversion of their wills, they endeavor to turn men from God.'

If sorcery has remained relatively innocuous (or has at least been kept circumscribed) among the majority of Orientals and 'primitive' peoples, it is probably owing to the presence of what might be called a certain 'racial intellectuality' as a framework within which the magicians and sorcerers operate: their societies on the whole know how to identify and evaluate the powers brought into play. Everything is kept in its proper hierarchy; theurgy is put above thaumaturgy, and the superiority of the priesthood is undisputed. Above all, a quasi-generic awareness of the unicity of a Supreme Principle precludes the perspective of an irresolvable dualism.[2]

M.M. 978

It has taken all the ignorance of the passional mentality in the West to try to *replace* the kingdom of Heaven with the kingdom of this world, to put Pan in the place of the Pantocrator, and set the Adversary against God as an equally matched opponent in a contest between two for Dominion. Writing on the Inquisition, M. Verrill states: 'One is terrified at the thought of what could have happened if the Catholic Church had not continued the battle, if the Inquisition had not been established and maintained, when from all sides were springing up innumerable fantastic, extraordinary, even repulsive and horrible religions, uniting adepts by the tens of thousands. This cult of the devil, or sorcery, magic, orgies, lust, and all forms of vice and depravity would have infested the whole civilized world of that time'.[3]

[1] Papus adds an obscure twist to the preceding line of thought, by saying: 'The difference between a magus and a sorcerer is that the former knows what he is doing and what will result from it, whereas the latter is totally ignorant of these things' (*Traité élémentaire de Science occulte,* p. 126). By more objective criteria, a magician is simply one who practises magic, whereas a sorcerer's domain is limited to black magic, which operates through the inversion of symbols. This, incidentally, should not in any way be confused with the laws of inverse analogy.

[2] It is significant that Pope Alexander IV in his Bull to the Franciscan Inquisitors (1258) forbade them to judge cases of witchcraft, except where *heretical practice* could also be proved. As the Middle Ages deteriorated, this distinction became increasingly ambiguous.

[3] Cf. *Crapouillot,* Paris, no. 22, p. 8.

In the Introduction to his translation of the *Malleus Maleficarum,* the Rev. Montague Summers points out that 'the heretics were just as resolute and just as practical, that is to say, just as determined to bring about the domination of their absolutism as is any revolutionary of to-day. The aim and objects of their leaders, Tanchelin, Everwacher, the Jew Manasses, Peter Waldo, Pierre Autier, Peter of Bruys, Arnold of Brescia, and the rest, were exactly those of Lenin, Trotsky, Zinoviev, and their fellows. There were, of course, minor differences and divergences in their tenets, that is to say, some had sufficient cunning to conceal and even to deny the extremer views which others were bold enough or mad enough more openly to proclaim. But just

Today the alternative is hardly worth posing. The witch-hunting passed insidiously into the hands of the hunted, until the Inquisition was discountenanced, and the 'testing of the spirits' became a matter of derision.[1]

Various traditions[2] point to a time bordering upon the juncture of two cosmic cycles, when the powers of darkness emerge into the open and momentarily gain the ascendancy. This is the Great Profanation, the 'Flaming Night', or Reign of the Antichrist, during which normal values are reversed and falsified, and vast portions of mankind seduced into a grotesque carnival that parodies in a way both sinister and ludicrous a spiritual kingdom on earth. Guénon speaks of this phenomenon as the product of a mass of 'residues' or 'psychic corpses' 'galvanized by an "infernal" will' (*Le Règne,* p. 267), which corresponds to the *caput mortuum* already mentioned in the Introduction to the chapter on *Judgment.* The restoration of order that follows immediately upon this reversal is at the same time the beginning of a new cycle, according to traditional doctrine, since the present world as it now exists is literally replaced by 'a new heaven and a new earth'.[3]

P. State 583

'The intellectual poverty of the neo-yogic movements furnishes the irrefutable proof that there is no spirituality without orthodoxy. It is certainly not by chance that all these movements are in effect leagued against intelligence; this quality is replaced by thought which is weak and vague rather than logical, and "dynamic" rather than contemplative. All these movements are characterized by the detachment they affect in respect to pure doctrine, the incorruptibility of which they hate; for them this purity is "dogmatism"; they fail to understand that the Truth does not deny forms from without, but transcends them from within.

Introd.

Introd. *Beauty* *Realiz.* 890

below the trappings, a little way beneath the surface, their motives, their methods, their intentions, the goal to which they pressed, were all the same. Their objects may be summed up as the abolition of monarchy, the abolition of private property and of inheritance, the abolition of marriage, the abolition of order, the total abolition of all religion. It was against this that the Inquisition had to fight, and who can be surprised if, when faced with so vast a conspiracy, the methods employed by the Holy Office may not seem—if the terrible conditions are conveniently forgotten—a little drastic, a little severe? There can be no doubt that had this most excellent tribunal continued to enjoy its full prerogative and the full exercise of its salutary powers, the world at large would be in a far happier and far more orderly position to-day. Historians may point out diversities and dissimilarities between the teaching of the Waldenses, the Albigenses, the Henricians, the Poor Men of Lyons, the Cathari, the Vaudois, the Bogomiles, and the Manichees, but they were in reality branches and variants of the same dark fraternity, just as the Third International, the Anarchists, the Nihilists, and the Bolsheviks are in every sense, save the mere label, entirely identical' (pp. xvii–xviii).

[1] One of the factors responsible for the diehard longevity of traditional civilizations like those of Tibet or Bali is precisely their recognition of a 'darker side', together with their ritual sciences practiced to maintain an 'alchemical' equilibrium between various cosmological forces. Thus, for example, the Balinese Barong dance first polarizes or 'coagulates' evil, then disperses or 'dissolves' it with the 'power of points' condensed in the krisses. All this has its analogy with the carnival revels of the Middle Ages. Cf. Guénon: 'Sur la signification des fêtes "carnavalesques" ', *Études Trad.,* 1945–6. Cf. also Coomaraswamy: 'The Darker Side of Dawn' (Smithsonian Misc. Col., vol. 94, no. 1).

[2] For example, the American Indian, the Chinese *Yô kî,* the Hindu *Vishnu Purâṇa,* the Buddhist *Chakkavatti Sihanada Suttanta,* the Japanese *Daishûgwatsuzôkô,* the Hermetical teachings given in *Asclepius,* besides the more familiar references in Judaism, Christianity and Islam.

[3] All this is unavoidably too 'schematic'. A preliminary restoration (the gathering of the 'elect', seed of the new cycle) must take place even before the end of the present cycle, just as microcosmically a *ḥâl* (Islamic term for a transitory spiritual 'state') precedes the 'initiatic death' that is followed by a *maqâm* (permanent 'station'). Macrocosmically this has to do with prophecies concerning the coming of a holy Monarch who will rule the world during a period with justice, wisdom and peace.

'Orthodoxy contains and guarantees infinitely precious values which man could never draw from himself' (Schuon: *Perspectives spirituelles*, p. 151).

N.B. Some concrete examples of the subversive doctrines dealt with in the citations of the text itself are presented in the footnotes to this chapter, to give the reader a glimpse of the vast wasteland that lies in this direction.

The Origins of Subversion

I saw a star fall from heaven unto the earth: and to him was given the key of the bottomless pit.

Revelation, IX. 1

And it came to pass when the children of men had multiplied that in those days were born unto them beautiful and comely daughters. And the angels, the children of the heaven, saw and lusted after them, and said to one another: 'Come, let us choose us wives from among the children of men and beget us children.' And Semjâzâ, who was their leader, said unto them: 'I fear ye will not indeed agree to do this deed, and I alone shall have to pay the penalty of a great sin.' And they all answered him and said: 'Let us all swear an oath, and all bind ourselves by mutual imprecations not to abandon this plan but to do this thing.' Then sware they all together and bound themselves by mutual imprecations upon it. And they were in all two hundred; who descended in the days of Jared on the summit of Mount Hermon. . . .

(They) took unto themselves wives, and each chose for himself one, and they began to go in unto them and to defile themselves with them, and they taught them charms and enchantments, and the cutting of roots (of trees), and made them acquainted with plants. And they became pregnant, and they bare great giants, whose height was three thousand ells: Who consumed all the acquisitions of men. And when men could no longer sustain them, the giants turned against them and devoured mankind. And they began to sin against birds, and beasts, and reptiles, and fish, and to devour one another's flesh, and drink the blood. Then the earth laid accusation against the lawless ones.

Supra Introd.

Book of Enoch, VI, VII

There were giants in the earth in those days; and also after that, when the sons of God came in unto the daughters of men, and they bare children to them, the same became mighty men which were of old, men of renown.

And God saw that the wickedness of man was great in the earth, and that every imagination of the thoughts of his heart was only evil continually.

And it repented the Lord that he had made man on the earth, and it grieved him at his heart.

And the Lord said, I will destroy man whom I have created from the face of the earth; both man, and beast, and the creeping thing, and the fowls of the air; for it repenteth me that I have made them.

But Noah found grace in the eyes of the Lord.

Genesis, VI. 4–8

Now there was a day when the sons of God came to present themselves before the Lord, and Satan came also among them.

And the Lord said unto Satan, Whence comest thou? Then Satan answered the Lord, and said, From going to and fro in the earth, and from walking up and down in it.

Job, I. 6, 7

If history be no ancient Fable,
Free Masons came from Tower of Babel.[1]

The Freemasons; an Hudibrastic Poem

For there are certain men crept in unawares, who were before of old ordained to this condemnation, ungodly men, turning the grace of our God into lasciviousness, and denying the only Lord God. . . .

And the angels which kept not their first estate, but left their own habitation, he hath reserved in everlasting chains under darkness unto the judgment of the great day.

Even as Sodom and Gomorrha, and the cities about them in like manner, giving themselves over to fornication, and going after strange flesh, are set forth for an example, suffering the vengeance of eternal fire.

Likewise also these filthy dreamers defile the flesh, despise dominion, and speak evil of dignities. . . .

Woe unto them! for they have gone in the way of Cain. . . .

These are spots in your feasts of charity, when they feast with you, feeding themselves without fear: clouds they are without water, carried about of winds; trees whose fruit withereth, without fruit, twice dead, plucked up by the roots;

Introd.
Judgment

Raging waves of the sea, foaming out their own shame; wandering stars, to whom is reserved the blackness of darkness for ever.

Jude, passim

Abuse of Cosmological Sciences

Pan, who is the declarer of all things and the perpetual mover of all things, is rightly called goat-herd, he being the two-formed son of Hermes, smooth in his upper part, and rough and goatlike in his lower regions.

Metanoia
480

Plato (*Cratylus,* 408 C)

The lower part of the world, which we inhabit, has been subjected to deviated angels, through the law of Divine Providence to which is due the magnificent order of things.

Beauty
689

St Augustine

[1] For some members of the Fraternity this would be too modest an appraisal, tracing as they do their lineage to Adam, and even to God as Creator or 'First Architect'.

(The disbelievers) follow what the devils relate against the kingdom of Solomon. And Solomon disbelieved not, but the devils disbelieved. They teach men magical arts and what was revealed to the two angels of Babel, Hârût and Mârût. But they taught no one until they had said: Verily, we are but a temptation, so do not disbelieve. And from these two they learn that by which they cause division between man and wife, but they injure no one thereby save with the permission of Allâh. And they learn that which harmeth them and profiteth them not. And they surely know that he who traffics therein has no share in the Hereafter. And evil is the price for which they have sold their souls, did they but know.

Supra 441

Rev. 967

Qur'ân, II. 102

When there occurs some errancy in the theurgic technique, the Images which ought to be at the Autopsia are not, but others of a different kind. These, the inferiors, assume the guise of the more venerable orders, and pretend to be the very ones which they are counterfeiting, and there will be a great mass of falsehood flow forth from the perversion. . . . Do we not know that all things which are brought into view by such a mode . . . are really phantoms of what is genuine, and that they appear good to the seeing, but never are really so?

Supra Introd.

Iamblichus

It pleased Anti-Christ
To discover all the wisdom
Enoch had learned from God,
So that Anti-Christ could openly declare it
Along with his own false teaching;
For if only he could draw Enoch to himself
All the world and great honour would be his.

Heterod. 430

Mechthild of Magdeburg

When Behmen first appeared in English, many persons of this nation, of the greatest wit and abilities, became his readers; who, instead of entering into his one only design, which was their own regeneration from an earthly to an heavenly life, turned chemists and set up furnaces to regenerate metals, in search of the philosopher's stone.

William Law

But Wonder is it that *Wevers* deale with such warks,
Free Masons and *Tanners* with poore *Parish Clerks;*
Tailors and *Glasiers* woll not thereof cease,
And eke sely *Tinkers* will put them in the prease
With greate presumption; but yet some collour there was,
For all such Men as give Tincture to Glasse:
But many *Artificers* have byn over-swifte
With hasty Credence to fume away their thrifte.

Thomas Norton

Is it not lesse absurd, then strange, to see how some Men (who would have the World account them learned . . .) will not forbeare to ranke *True Magicians* with *Conjurers, Necromancers* and *Witches* (those grand *Impostors*) who *violently intrude themselves into Magick, as if Swine should enter into a faire and delicate Garden,*[1] and (being in *league* with the *Devill*) make use of his Assistance in their *workes,* to counterfeit and corrupt the admirall *wisdome* of the *Magi,* betweene whom there is as large a difference as betweene *Angels* and *Devils.*

Elias Ashmole

Trust not in *Geomantie* that superstitious Arte,
For *God* made Reason which there is set aparte.
Trust not to all *Astrologers,* I saie whie,
For that Arte is as secreat as *Alkimy.*
That other is disproved and plainely forbod,
By holy *Saincts* of the Church of *God.*
Trust not, ne love not *Negromancy,*
For it is a property of the Devill to lye.
Trust to this *Doctrine,* set herein your desires,
And now lerne the Regiment of your Fiers.

Thomas Norton

Corruption

M.M. 973

Give not that which is holy unto the dogs, neither cast ye your pearls before swine, lest they trample them under their feet, and turn again and rend you.

St Matthew, VII. 6

Heterod. 423

The written doctrines of philosophy, if poured into the dirty and defiled vessel of a false and debased mind, are altered, changed and spoilt, and turn to urine or anything fouler than that.

Epictetus

You must have noticed kites and vultures soaring very high in the sky; but their eyes are always fixed on the charnel-pits.

Sri Ramakrishna

[1] 'Paracel. de. occult Phil. cap. 11.'

If you are so anxious for trance any narcotic will bring it about. Drug-habit will be the result and not liberation. There are *vâsanas* (habits of mind) in the latent state even in trance. The *vâsanas* must be destroyed.[1]

Sri Ramana Maharshi

Death 220

The person of uncontrolled senses resists not sexual desire, strongest of worldly tempters, and thus falls into abysmal darkness as the moth falls into the flame.

Srimad Bhagavatam, XI. iii

More souls go to hell because of sins of the flesh than for any other reason.[2]

The Blessed Virgin at Fatima

Illusion 108

Trewly such Places where Lechery is used
Must for this *Arte* be utterly refused.

Thomas Norton

Know ye not that your bodies are the members of Christ? shall I then take the members of Christ, and make them the members of an harlot? God forbid.

What? know ye not that he which is joined to an harlot is one body? for two, saith he, shall be one flesh.

But he that is joined unto the Lord is one spirit.

Realiz. 887

Flee fornication. Every sin that a man doeth is without the body; but he that committeth fornication sinneth against his own body.

[1] 'There exists a drug whose use will open the gates of the World behind the Veil of Matter' (Allan Bennett, alias Brother Iehi Aour, of the Hermetic Order of the Golden Dawn; cited by John Symonds: *The Great Beast*, London, Ryder, 1951, p. 28).

'For an aspiring mystic to revert, in the present state of knowledge, to prolonged fasting and violent self-flagellation would be as senseless as it would be for an aspiring cook to behave like Charles Lamb's Chinaman, who burned down the house in order to roast a pig. Knowing as he does (or at least as he can know, if he so desires) what are the chemical conditions of transcendental experience, the aspiring mystic should turn for technical help to the specialists—in pharmacology, in biochemistry, in physiology and neurology, in psychology and psychiatry and parapsychology' (Aldous Huxley: *Heaven and Hell*, London, Chatto & Windus, 1956, p. 64).

'This society, which I shall call for distinction's sake the "Berlin Brotherhood", . . . had discovered, by repeated experiments, that spiritual forms could become visible to the material under certain conditions, the most favorable of which were somnambulism procured through the magnetic sleep. This state, they had found, could be induced sometimes by drugs, vapors, and aromal essences; sometimes by spells, as through music, intently staring into crystals, the eyes of snakes, running water, or other glittering substances; occasionally by intoxication caused by dancing, spinning around, or distracting clamors; but the best and most efficacious method of exalting the spirit into the superior world and putting the body to sleep was, as they had proved, through animal magnetism. They taught that in the realms of spiritual existence were beings who composed the fragmentary and unorganized parts of humanity, as well as beings of higher orders than humanity. . . . Thus they invoked their presence by magical rites, and sought to obtain control over them, for the purpose of wresting from them the complete understanding of and power over the secrets of nature' (*Ghost Land; or Researches into the Mysteries of Occultism*, anonymous, tr. and edit. by Emma Hardinge Britten, Boston, 1876, pp. 33–35).

[2] By this must be understood the appetitive, demiurgic, centrifugal character of the creative impulse, which as a separative force, is at the root of all remotion from Principle. In other words, Creation and the Fall both share a 'sexual' motivation, exalted from one aspect and disastrous from another. But it can also be a unifying power. Cf. the Introductions to the chapters *Illusion* and *Love*.

Creation 42

What? know ye not that your body is the temple of the Holy Ghost which is in you, which ye have of God, and ye are not your own?

Realiz. 870

For ye are bought with a price: therefore glorify God in your body, and in your spirit, which are God's.

I. Corinthians, VI. 15–20

Men must use natural love and abhor the unnatural, not intentionally destroying the seeds of human increase, or sowing them in stony places, in which they will take no root. . . . Our citizens must not fall below the nature of birds and beasts. . . . No one shall venture to touch any person of the freeborn or noble class except his wedded wife, or sow the unconsecrated and bastard seed among harlots, or in barren and unnatural lusts; at least we must abolish altogether the connection of men with men.

Plato (*Laws* VIII, 839 ff.)

But there were false prophets also among the people, even as there shall be false teachers among you, who privily shall bring in damnable heresies. . . .

And many shall follow their pernicious ways; by reason of whom the way of truth shall be evil spoken of.

And through covetousness shall they with feigned words make merchandise of you: whose judgment now of a long time lingereth not, and their damnation slumbereth not.

For if God spared not the angels that sinned, but cast them down to hell . . .

And spared not the old world, but saved Noah . . .

And turning the cities of Sodom and Gomorrha into ashes condemned them with an overthrow, making them an ensample unto those that after should live ungodly;

And delivered just Lot, vexed with the filthy conversation of the wicked . . .

The Lord knoweth how to deliver the godly out of temptations, and to reserve the unjust unto the day of judgment to be punished:

But chiefly them that walk after the flesh in the lust of uncleanness, and despise government. Presumptuous are they, selfwilled,[1] they are not afraid to speak evil of dignities.

II. Peter, ch. II

And there was Lot, when he said to his people, Lo, ye practise iniquity that has no precedent in the whole of creation!

For do ye not come in unto males, and cut off the Way, and bring abomination into your assemblies? And there was no response from his people, except to say, Bring upon us the punishment of God if thou art of the truthful!

[1] 'Do what thou wilt shall be the whole of the law' (Aleister Crowley: Symonds, op. cit., p. 61).

'"Thou shalt not" was the Mosaic law as Moses gave it out. These were the emanations of the Sephiroth, or the Tree of Life. He veiled that fact and objectified it for the people, but gave the Priests the real meaning in the Talmud' (Baird T. Spalding: *India Tour Lessons*, Volume IV on *Life and Teaching of the Masters of the Far East*, Los Angeles, DeVorss & Co., 1948, p. 96).

'Praise be to Thee, O Lord, who permittest the forbidden' (Sabbatian formula; cited by G. G. Scholem: *Major Trends in Jewish Mysticism*, New York, Schocken Books, 1941, p. 319. According to Scholem, a play on words is involved here, *Matir Asurim*, 'who deliverest the captives', being changed to *Matir Issurim*, 'who permittest the forbidden').

He said, My Lord, deliver me from the people of depravity!

And when Our messengers brought Abraham glad tidings (of a son), they said: Lo! we are about to destroy the people of that township, for its people are wrong-doers.

He said: Lo! Lot is there. They said, We best know who is there! . . .

And verily of that We have left a clear sign for people of reflection.

Qur'ân, XXIX. 28 ff.

Thou shalt not lie with mankind, as with womankind: it is abomination.

Leviticus, XVIII. 22

The ancient and venerable authors are of opinion that *Auparishtaka* is the work of a dog and is not worthy of a man. It is a low practice and opposed to the orders of the Scriptures.

Kama-Sutra

It is related concerning Jesus (Peace be upon him!) that he came on a fire which was kindled over a man in the desert. Jesus then took water to put it out, and the fire changed into a youth, and the man changed into fire. So Jesus (Peace be upon him!) wept and said, 'O Lord, restore them to their former state that I may see what their sin was.' Then that fire was removed from them, and lo! they were a man and a youth. The man said, 'O Jesus, I have been afflicted in the world by love of this youth, and desire urged me on until I sinned with him one Thursday night, after which I sinned with him another day. Then a man came upon us and said to us, 'Woe to you! Fear God!' I replied to him, 'I am not afraid, and I do not fear.' Then when I died, and the youth died, God (Great and glorious is He!) turned us into what you see. Sometimes he becomes fire and burns me, and sometimes I become fire and burn him. And this is our punishment until the Day of Resurrection.'

Judgment 261

Christ in Islâm

Knowl. 734

Professing themselves to be wise, they became fools,

And changed the glory of the uncorruptible God into an image made like to corruptible man, and to birds, and four-footed beasts, and creeping things.

Wherefore God also gave them up to uncleanness through the lusts of their own hearts, to dishonour their own bodies between themselves:

Who changed the truth of God into a lie, and worshipped and served the creature more than the Creator, who is blessed for ever. Amen.

Beauty 663

For this cause God gave them up unto vile affections: for even their women did change the natural use into that which is against nature:

And likewise also the men, leaving the natural use of the woman, burned in their lust one toward another; men with men working that which is unseemly, and receiving in themselves that recompence of their error which was meet.

Romans, I. 22–27

It is on account of such crimes that famines and earthquakes take place, and also pestilence.

Infra 464

Justinian

'Sodom would not listen to Me nor do the people nowadays listen to Me or heed My warnings. Therefore they will incur the sad experience of My wrath.'

Teresa Neumann

Be not deceived: neither fornicators, nor idolaters, nor adulterers, nor effeminate, nor abusers of themselves with mankind,

Nor thieves, nor covetous, nor drunkards, nor revilers, nor extortioners, shall inherit the kingdom of God.

I. Corinthians, VI. 9, 10

Just as, monks, the mighty ocean consorts not with a dead body; for when a dead body is found in the mighty ocean it quickly wafts it ashore, throws it up on the shore; even so, monks, whatsoever person is immoral, of a wicked nature, impure, of suspicious behaviour, of covert deeds, one who is no recluse though claiming to be such, one who is no liver of the Brahma-life though claiming to be such, one rotten within, full of lusts, a rubbish-heap of filth,—with such the Order consorts not, but gathering together quickly throws him out. Though, monks, he be seated in the midst of the Order, yet is he far away from the Order; far away is the Order from him.

Heterod. 423

Udâna, v. 5

Malevolence

And he opened the bottomless pit; and there arose a smoke out of the pit, as the smoke of a great furnace; and the sun and the air were darkened by reason of the smoke of the pit.[1]

Revelation, IX. 2

There are spirits that are created for vengeance, and in their fury they lay on grievous torments.[2]

Ecclesiasticus, XXXIX. 33

[1] 'I saw there was a great crack to go throughout the earth, and a great smoke to go as the crack went; and that after the crack there should be a great shaking: this was the earth in people's hearts, which was to be shaken before the seed of God was raised out of the earth. And it was so; for the Lord's power began to shake them, and great meetings we began to have, and a mighty power and work of God there was amongst people, to the astonishment of both people and priests' (George Fox: *Journal,* Everyman's, p. 13).

[2] 'Our aim is not to restore Hinduism, but to sweep Christianity off the surface of the earth' (Madame Blavatsky, to Mr Alfred Alexander, cited in *The Medium and Daybreak,* London, Jan. 1893, p. 23: cf. Guénon: *Le Théosophisme, Histoire d'une Pseudo-Religion,* 2nd edit., Paris, 1930, p. 6).

'Above all (we must) fight Rome and its priests, struggle everywhere against Christianity, and chase God from Heaven' (Annie Besant, closing speech at the Congress of Free Thinkers, Brussels, Sept. 1880: id., p. 7).

'Flectere si nequeo superos, Acheronta movebo' ('If I cannot prevail upon the powers above, then I shall stir up all hell'; Freud, motto chosen from Virgil's *Aeneid,* VII. 312, for his *Traumdeutung,* or *The Interpretation of Dreams*).

'We Freemasons must pursue the definite demolition of Catholicism' (*Bulletin* of the Grand Orient in 1885: Nesta H. Webster: *Secret Societies and Subversive Movements,* London, Britons Publishing Society, 1955, p. 276).

Will all great Neptune's ocean wash this blood
Clean from my hand? No, this my hand will rather
The multitudinous seas incarnadine,
Making the green one red.

Shakespeare (*Macbeth,* II. ii. 61)

Heterod. 423

And when Allâh alone is mentioned, the hearts of those who believe not in the Hereafter are repelled, and when those (whom they worship) beside Him are mentioned, behold! they are glad.

Qur'ân, XXXIX. 45

But the natural man receiveth not the things of the Spirit of God: for they are foolishness unto him: neither can he know them, because they are spiritually discerned.

I. Corinthians, II. 14

Faith 505

Therefore they say unto God, Depart from us; for we desire not the knowledge of thy ways.

What is the Almighty, that we should serve him? and what profit should we have, if we pray unto him?

Job, XXI. 14, 15

Heterod. 420

You are too senseless-obstinate, my lord,
Too ceremonious and traditional....

Shakespeare (*Richard III,* III. i. 44)

Orthod. 275

Many people of these modern days ... call every activity by the name of Dharma and begin to proclaim that there need be no distinction of castes, that all should behave in the same manner and that the good of the country depends upon such 'equality'. There are many others who repose confidence in the teachers of this school, embark on ways directly prohibited by the Śâstras and make determined efforts to create confusion of castes and bring about the ascendance of Adharma in the world.[1] The Lord has no doubt stated that

[1] 'The new revelation manifests outside and above the Churches. Its teaching is addressed to all races on earth. Everywhere the spirits proclaim the principles on which it is based. Over all regions of the globe passes the great voice which recalls man to the thought of God and the future life' (Léon Denis: *Christianisme et Spiritisme,* pp. 277–278; in Guénon: *L'Erreur spirite,* 2nd edit., Paris, 1952, pp. 365–366).

'Today, all parts of the world have been explored: humanity, which now knows itself better, aspires to a real peace. But because of the very multiplicity of ... religions, men do not always live in harmony among each other. That is why I have decided to reunite all these religions into a single one, Caodaïsm, to restore them to primordial unity' (Cao-Daï, in a spirit message given on January 13, 1927: Maurice Colinon: *Faux Prophètes et Sectes d'aujourd'hui,* Paris, Plon, 1953, p. 90).

'We cannot stop in our progress with organizations and systems either orthodox or metaphysical, for they are sectional, sectarian, and teach a doctrine that is more or less involved with the idea of separations. They are only steps in the process of man's discovery of himself. We cannot stop at any point without becoming orthodox. That prevents further progress until we break away' (Baird T. Spalding, op. cit., p. 97).

'Thus all ecclesiastical forms will be broken, all hierarchies scattered. Canons and dogmas will be thrown down with the discarded residues of theologies and interpretations. New apostles will preach the primitive Gospel in a whisper. The faithful will be gathered under new symbols which will resemble the ancient

Reality 790

He will incarnate whenever there is a decline in Dharma and a rise in Adharma. Evidently the promulgators of such teachings are helping in a way but quite unconsciously the early advent of the Lord once again in our midst.

Sri Chandrasekhara Bhâratî Swâmigal

The worship of God is an abomination to a sinner.

Ecclesiasticus, I. 32

Wisdom and goodness to the vile seem vile;
Filths savour but themselves.

Shakespeare (*King Lear,* IV. ii. 38)

Judgment 266

The beetle cannot stand the perfume of the rose.... In the same way, the man with the beetle's temperament, mentally and formally, cannot stand the truth when he hears it, for it seems to him like vanity.[1]

Ibn ʿArabî

The wicked man thinks no good of anyone; for how should he imagine that others have what he lacks himself?

ʿAlî

symbols.... A young elect will propagate the divine fire across the devastated earth, and bare-footed, heads shaven, will preach hope and charity....

'Universal persecution will have reforged Humanity, which, on a new religious footing, will gradually return to the Gospel of Love' (Georges Barbarin: *Qui sera le Maître du Monde ...?,* Paris, 1949, p. 167).

'A great ecclesiastical system, steeped in iniquity and crime, is directed by a single man; and this man claims that his role is to be the vicar of Christ on earth, and that he has a power equal to Jehovah's.... This system is spoken of in the Scriptures as *the great whore*' (Judge Rutherford: *The Truth Will Make You Free,* p. 312: Colinon, op. cit., p. 188).

'The apparitions of the Virgin Mary and other saints in the Roman Catholic religion are nothing but spiritualism and demonism in disguise' (Alexander Freytag: *Le Bien-être sur la Terre,* p. 44; id., p. 167).

[1] 'Now was I come up in spirit through the flaming sword, into the paradise of God. All things were new: and all the creation gave another smell unto me than before, beyond what words can utter. I knew nothing but pureness, and innocency, and righteousness, being renewed up into the image of God by Christ Jesus, to the state of Adam, which he was in before he fell.... I was at a stand in my mind whether I should practise physic for the good of mankind, seeing the nature and virtues of the creatures were so opened to me by the Lord. But I was immediately taken up in spirit, to see into another or more steadfast state than Adam's in innocency, even into a state in Christ Jesus that should never fall....

'I was sent to turn people from darkness to the light, that they might receive Christ Jesus: for, to as many as should receive Him in His light, I saw that He would give power to become the sons of God; which I had obtained by receiving Christ.... I was to bring people off from all the world's religions, which are vain.... And I was to bring people off from Jewish ceremonies, and from heathenish fables, and from men's inventions and windy doctrines, by which they blew the people about this way and the other way, from sect to sect; and from all their beggarly rudiments, with their schools and colleges for making ministers of Christ, who are indeed ministers of their own making but not of Christ's; and from all their images and crosses, and sprinkling of infants, with all their holy days (so called) and all their vain traditions, which they had gotten up since the apostles' days, which the Lord's power was against.... I was moved also to cry against all sorts of music....

'But the black earthly spirit of the priests wounded my life; and when I heard the bell toll to call people together to the steeple-house, it struck at my life' (George Fox: *Journal*; op. cit., pp. 17, 20–23).

Again, as the substance of the earthly Stone is nothing accounted of in the world, and rejected by the majority of mankind, so Christ, the eternal Word of the Father, and the Heavenly Triune Stone, is lightly esteemed in this world, and scarcely even looked at; nay, we may say that nothing is so profoundly and utterly despised by mankind, as the Saving Word of God.

The Sophic Hydrolith

P. State 563

Introd. *Inv.*

Your hearts were hardened and became as rocks, or worse than rocks, for hardness. For indeed there are rocks from out which rivers gush, and indeed there are rocks which split asunder so that water floweth from them. And indeed there are rocks which fall down for the fear of Allâh.

Qur'ân, II. 74

Sin 75

'If one of the souls which burn so brightly
In the love of God, were only mine,
I (the devil) would crown myself with her
And thus reward myself for all my labours.'

Mechthild of Magdeburg

Sorcery

Men of darkness are they, who make a cult of the departed and of spirits.

Bhagavad-Gîtâ, XVII. 4

Magicians, who are commonly called witches, are thus termed on account of the magnitude of their evil deeds. These are they who by the permission of God disturb the elements, who drive to distraction the minds of men, such as have lost their trust in God, and by the terrible power of their evil spells, without any actual draught or poison, kill human beings.

Malleus Maleficarum, I. 2

Faith 501

It is so general a report, and so many aver it either from their own experience or from others, that are of indubitable honesty and credit, that the silvans and fauns, commonly called incubi, have often injured women, desiring and acting carnally with them, and that certain devils whom the Gauls call *dusii* do continually practise this uncleanness, and tempt others to it, which is affirmed by such persons, and with such confidence, that it were impudence to deny it.

St Augustine

Judgment
242

Thou shalt not suffer a witch to live.

Exodus, XXII. 18

The sort of man who injures others by magic knots, or enchantments, or incantations, or any of the like practices, if he be a prophet or diviner, let him die.

Plato (*Laws* XI, 933 D)

There shall not be found among you any one that maketh his son or his daughter to pass through the fire, or that useth divination, or an observer of times, or an enchanter, or a witch,

Or a charmer, or a consulter with familiar spirits, or a wizard, or a necromancer.

For all that do these things are an abomination unto the Lord: and because of these abominations the Lord thy God doth drive them out from before thee.

Deuteronomy, XVIII. 10–12

As to that class of monstrous natures who not only believe that there are no Gods, or that they are negligent, or to be propitiated, but in contempt of mankind conjure the souls of the living and say that they can conjure the dead and promise to charm the Gods with sacrifices and prayers, and will utterly overthrow individuals and whole houses and states for the sake of money—let him who is guilty of any of these things be condemned. . . . And when he is dead let him be cast beyond the borders unburied.

Plato (*Laws*, X. 909 B)

In the Church the devil prefers to operate through the medium of witches.

Malleus Maleficarum, I. 1

The witches attend the sabbath with the pagan goddess, Diana, and untold numbers of women, riding on beasts; they traverse the open spaces in the calm of night, obedient to her orders, as to their absolute mistress.

Canon Episcopi

The night discloses to the initiated the mysteries of impudicity. Flaming torches throw light upon the utmost excesses of debauchery; the night has not darkness enough to veil it all.[1]

Clement of Alexandria

[1] 'It is too easy to give the name of Satanism to what one doesn't understand. To admit Satanism, it is first of all necessary to admit the existence of Satan. . . .

'If we adopt . . . scientific terminology, stripped of all poetry, we see that the divine Power is reduced quite simply into two fluids with contrary poles. . . . When we are told that sessions of the sabbath terminated in a general prostitution, nothing is more believable. . . . One sees therefore that what the early Christians called shameful practices were in reality but the factors necessary for the success of magical operations. . . . The kiss of the Templars on the sacral plexus was therefore only a symbolical gesture, implying that the end to be obtained by each adept of the Order was ecstasy, or more exactly, to find himself in direct rapport with the higher entities, through the intermediary of the fluid. . . . Formerly it was claimed that sorcerers invoked the dead; today the spiritualists do nothing other than this; it is only the methods that change. . . .

'Thus it is that the cult of Belphagor has come down to us, as a magical operation, the Sabbath, the Black Mass.

It has indeed lately come to Our ears, not without afflicting Us with bitter sorrow, that in some parts of Northern Germany, as well as in the provinces, townships, territories, districts, and dioceses of Mainz, Cologne, Trèves, Salzburg, and Bremen, many persons of both sexes, unmindful of their own salvation and straying from the Catholic Faith, have abandoned themselves to devils, incubi and succubi, and by their incantations, spells, conjurations, and other accursed charms and crafts, enormities and horrid offences, have slain infants yet in the mother's womb, as also the offspring of cattle, have blasted the produce of the earth, the grapes of the vine, the fruits of trees, nay, men and women, beasts of burthen, herd-beasts, as well as animals of other kinds, vineyards, orchards, meadows, pasture-land, corn, wheat, and all other cereals; these wretches furthermore afflict and torment men and women, beasts of burthen, herd-beasts, as well as animals of other kinds, with terrible and piteous pains and sore diseases, both internal and external; they hinder men from performing the sexual act and women from conceiving, whence husbands cannot know their wives nor wives receive their husbands; over and above this, they blasphemously renounce that Faith which is theirs by the Sacrament of Baptism, and at the instigation of the Enemy of Mankind they do not shrink from committing and perpetrating the foulest abominations and filthiest excesses to the deadly peril of their own souls, whereby they outrage the Divine Majesty and are a cause of scandal and danger to very many.

Pope Innocent VIII, in the Bull of 1484,
Summis desiderantes affectibus

Necromancy the Greeks call it, but necromancy or hydromancy, whatever you like to call it, is where the dead seem to speak.... They were devils that delighted in those obscene ministries, and under the names of those whom the people held divine, got place to play their impostures, and by illusive miracles to captivate all their souls.

St Augustine

Illusion
85

Now it is very easy for the Enemy to create apparitions and appearances of such a character that they shall be deemed real and actual objects.

St Athanasius

'This cult of Belphagor was without doubt a simple act of spiritualism: one invoked in the Temple the disincarnated souls.

'What would prove it to us is the presence of women among the officiants in these types of ceremonies.

'These women were probably mediums who facilitated the phenomena of materialization or the transmission of thought....

'The idea that the ceremony ended in a general prostitution is probably not exaggerated, if one assumes that they used melodies, blood, in a word everything that can overheat the atmosphere.

'But nevertheless, the result was obtained, which consisted in putting oneself for a given time in relation with the astral' (Jean Lignières: *Les Messes Noires*, Paris, Librairie 'Astra', 1947, passim).

This view of things is corroborated by a British psychiatrist of the Jungian school: 'There is little doubt that, logical thinking being predominantly a male function and intuition a female one, the persecution of witches in Europe was largely due to an over-systematized phase of man-dominated religion that led men to suspect all truly intuitive activities of women as bad and therefore as something to be eradicated at all cost—at the cost, as it has indeed turned out, of devitalizing men's own spiritual understanding, as is so evident to-day' (John Layard: *The Lady of the Hare*, London, Faber & Faber, 1944, p. 203).

Yet because such things often happen by illusion and merely in the imagination, those who suppose that all the effects of witchcraft are mere illusion and imagination are very greatly deceived.

Malleus Maleficarum, I. 1

Aaron cast down his rod before Pharaoh, and before his servants, and it became a serpent.

Then Pharaoh also called the wise men and the sorcerers: now the magicians of Egypt, they also did in like manner with their enchantments.

Exodus, VII. 10–11

Supra 442

But between this rod and that rod there is a vast difference; from this action (magic) to that action (miracle) is a great way.

This action is followed by the curse of God; that action receives in payment the mercy of God.

The infidels in contending (for equality with the prophets and saints) have the nature of an ape: the evil nature is a canker within the breast.

Introd. *Heterod.*

Rûmî

The earth hath bubbles, as the water has,
And these are of them.

Shakespeare (*Macbeth*, I. iii. 79)

The wicked, crawling wild beast makes slaves of men by his magical arts, and torments them even until now, exacting vengeance, as it seems to me, after the manner of barbarians, who are said to bind their captives to corpses until both rot together. Certain it is that wherever this wicked tyrant and serpent succeeds in making men his own from their birth, he rivets them to stocks, stones, statues and suchlike idols, by the miserable chain of daemon-worship; then he takes and buries them alive, as the saying goes, until they also, men and idols together, suffer corruption.[1]

Introd. *Sin*

Clement of Alexandria

[1] 'Whilst he spoke the professor laid his hand on my head, and continued to hold it there, at first with a seemingly slight and accidental pressure: but ere he had concluded his address, the weight of that hand appeared to me to increase to an almost unendurable extent. Like a mountain bearing down upon my shoulders, columns of fiery, cloud-like matter seemed to stream from the professor's fingers, enter my whole being, and finally crush me beneath their terrific force into a state where resistance, appeal, or even speech was impossible. A vague feeling that death was upon me filled my bewildered brain, and a sensation of an undefinable yearning to escape from a certain thraldom in which I believed myself to be held, oppressed me with agonizing force. At length it seemed as if this intense longing for liberation was gratified. I stood, and seemed to myself to stand, free of the professor's crushing hand, free of my body, free of every clog or chain but an invisible and yet quite tangible cord which connected me with the form I had worn, but which now, like a garment I had put off, lay sleeping in an easy-chair beneath me. . . . I perceived that I was treading on a beautiful crystalline form of matter, pure and transparent, and hard as a diamond, but sparkling, bright, luminous, and ethereal. There was a wonderful atmosphere, too, surrounding me on all sides. Above and about me, it was discernible as a radiant, sparkling mist, enclosing my form, piercing the walls and ceiling. . . . I saw, or seemed to see, that I was now *all force*; that I was soul loosed from the body save by the invisible cord which connected me with it; also, that I was in the realm of soul, the soul of matter; and that as my soul, and

. . . Sleep no more!
Macbeth does murder sleep.

Judgment 250

Shakespeare (*Macbeth,* II. ii. 36)

How now, you secret, black, and midnight hags! . . .
I conjure you, by that which you profess,—
Howe'er you come to know it,—answer me:
Though you untie the winds and let them fight
Against the churches; though the yesty waves
Confound and swallow navigation up;
Though bladed corn be lodg'd and trees blown down;
Though castles topple on their warders' heads;
Though palaces and pyramids do slope
Their heads to their foundations; though the treasure
Of Nature's germens tumble all together,
Even till destruction sicken; answer me
To what I ask you.[1]

Shakespeare (*Macbeth,* IV. i. 48)

. . . Come, you spirits
That tend on mortal thoughts! unsex me here,
And fill me from the crown to the toe top full
Of direst cruelty; make thick my blood,
Stop up the access and passage to remorse,
That no compunctious visitings of nature
Shake my fell purpose, nor keep peace between
The effect and it! Come to my woman's breasts,
And take my milk for gall, you murdering ministers,
Wherever in your sightless substances
You wait on nature's mischief! Come, thick night,
And pall thee in the dunnest smoke of hell,
That my keen knife see not the wound it makes,

the soul-realm in which I had now entered, was the real force which kept matter together. I could just as easily break the atoms apart and pass through them as one can put a solid body into the midst of water or air' (*Ghost Land,* pp. 22–23).

[1] 'To every séance a formulae was attached in the shape of oaths of secrecy, so tremendous that those who were sincere in their belief were never known to break them' (*Ghost Land,* p. 50).

'All this I most solemnly, sincerely promise and swear, with a firm and steadfast resolution to perform the same, without any mental reservation or secret evasion of mind whatever, binding myself under no less penalty than that of having my throat cut across, my tongue torn out by its roots, and my body buried in the rough sands of the sea, at low-water mark, where the tide ebbs and flows twice in twenty-four hours, should I ever knowingly violate this my Entered Apprentice obligation' (*Duncan's Masonic Ritual and Monitor,* New York, n.d., pp. 34–35).

Nor heaven peep through the blanket of the dark,
To cry, 'Hold, hold!'[1]

Shakespeare (*Macbeth*, I. v. 41)

O mankind! A similitude is coined, so pay ye heed to it: Lo! those on whom ye call beside Allâh will never create a fly though they combine together for the purpose.

Action 340

Qur'ân, XXII. 73

Because of the multitude of the whoredoms of the well-favoured harlot, the mistress of witchcrafts, that selleth nations through her whoredoms, and families through her witchcrafts.

Behold, I am against thee, saith the Lord of hosts; and I will discover thy skirts upon thy face, and I will shew the nations thy nakedness, and the kingdoms thy shame.

And I will cast abominable filth upon thee, and make thee vile, and will set thee as a gazingstock.[2]

Nahum, III. 4–6

When my sight descended lower on them, each seemed wondrously distorted, between the chin and the commencement of the chest:

for the face was turned towards the loins; and they had to come backward, for to look before them was denied. . . .

Reader, so God grant thee to take profit of thy reading, now think for thyself how I could keep my visage dry,

when near at hand I saw our image so contorted, that the weeping of the eyes bathed the hinder parts at their division?[3]

Certainly I wept, leaning on one of the rocks of the hard cliff, so that my Escort said to me: 'Art thou, too, like the other fools?

'Here pity lives when it is altogether dead. Who more impious than he that sorrows at God's judgment?'

Judgment 242

Dante (*Inferno*, XX. 10)

[1] 'I promise, O Beelzebub, that I will serve you all my life and give you my heart and my soul, all the faculties of my soul, all the senses of my body, all my works, all desires and yearnings, and all the affections of my heart, all my thoughts. I give you all the parts of my body, all the drops of my blood, all my nerves, all my bones, all my veins. . . . I give you my life for your service, even if I had a thousand lives I would give them all to you' (Marie de Sains: Lenormant de Chiremont: *Histoire véritable, mémorable de ce qui s'est passé sous l'exorcisme de trois filles possédées ès païs de Flandre, en la descouverte et possession de Marie de Sains*, Paris, 2 vol., 1623: cf. *Crapouillot* no. 22, p. 11).

[2] 'Our Order possesses the KEY which opens up all Masonic and Hermetic secrets, namely, the teaching of *sexual magic*, and this teaching explains, without exception, all the secrets of Nature, all the symbolism of FREEMASONRY and all systems of religion' (*Oriflamme*, organ of the *Ordo Templi Orientis*, 1904: Symonds, op. cit., p. 119).

[3] This distortion is the punishment earned by the sorcerers for having inverted the normal hierarchy of values during their lifetime.

He who has learnt to know himself ought not to set right by means of magic anything that is thought to be amiss, nor to use force to overcome necessity, but rather to let necessity go its own way according to its nature.

Hermes

Realiz. 859
Conform. 185

The Synagogue of Satan

I know the blasphemy of them which say they are Jews, and are not, but are the synagogue of Satan.[1]

Revelation, II. 9

Whereas Edom saith, We are impoverished, but we will return and build the desolate places; thus saith the Lord of Hosts, They shall build, but I will throw down.

Malachi, I. 4

Introd. *Flight*

In what temple of God he is to sit as God, it is doubtful whether it be the ruined temple of Solomon, or in the Church.[2]

St Augustine, on *II. Thes.* II. 4

Infra 464

[1] The following passage purports to be spoken by a 'Rishi' whom Spalding and his followers would have encountered in northern India: 'There are many mistakes in translation from the original texts to the context of your Bible, as well as many false prophecies. Many of these were brought about through lack of understanding of the characters and symbols with which the translators were dealing. . . . The greater majority, however, were base falsehoods, perpetrated deliberately to mystify, to mislead, and to subvert the original gospel of the House of Israel.

'The first name was Is-rael, meaning the Crystal or pure white race, the first race that ever inhabited the world, the original or root race from which all other races sprang. . . .

'The Jews are with us; we can trace their history step by step, down the ages, from the House of Judah to the tribe of Judah, and down to the present day. They are one of the standing signs of the great race that has assisted in preserving the God ideal, until all races are restored into one race, with the Christ of God, the controlling factor. . . . The culmination of this accomplishment will place the capstone upon the Great Pyramid' (Baird T. Spalding: op. cit., vol. III, pp. 169, 171, 180).

'Indeed, we may look on the Great Pyramid as the first true Masonic temple in the world, surpassing all others that have ever been built, with their secrets depicted in stone, symbolically, to be read by those who have been initiated into the secret mysteries of their religion' (Albert Churchward: *Signs and Symbols of Primordial Man*, pp. 9–10: in Basil Stewart: *History and Significance of the Great Pyramid*, London, n.d., p. 196).

[2] 'The day when the King of Israel, the King of the House of David, puts on his head the crown offered by Europe, he will become the Patriarch of the World. The number of victims sacrificed to hasten his coming will never amount to the number of those immolated over the centuries to the megalomaniac rivalry of the Goyim governments' (*The Protocols of the Elders of Zion*, 16th Session: Edit. *Revue Internationale des Sociétés Secrètes*, Paris, 1934, p. 101).

'It will suffice to show each person that the destiny of the Jews is *his* destiny' (Jean-Paul Sartre: *Réflexions sur la Question juive*, Paris, Gallimard, 1954, p. 189).

And whensoever ye depart from the Lord, ye shall walk in all evil and work the abominations of the Gentiles, going a-whoring after women of the lawless ones, while with all wickedness the spirits of wickedness work in you. For I have read in the book of Enoch, the righteous, that your prince is Satan, and that all the spirits of wickedness and pride will conspire to attend constantly on the sons of Levi, to cause them to sin before the Lord.[1]

The Testament of Dan, v. 5, 6

Supra 442 *Symb.* 313

Since by the decree of the Holy Roman Inquisition all books apertaining to the Cabala have lately been condemned, one must know that the Cabala is double; that one is true, the other false. The true and pious one is that which . . . elucidates the secret mysteries of the holy law according to the principle of anagogy (i.e. figurative interpretation). This Cabala therefore the Church has never condemned. The false and impious Cabala is a certain mendacious kind of Jewish tradition, full of innumerable vanities and falsehoods, differing but little from necromancy. This kind of superstition, therefore, improperly called Cabala, the Church within the last few years has deservedly condemned.[2]

Sixtus of Siena

[1] According to the editor in his Introduction to the *Testaments of the Twelve Patriarchs*, we have in this passage 'the most ancient authority at present known to us for the view which connects the tribe of Dan with Antichrist, and helps to explain the exclusion of this tribe from the list of the Twelve in the N.T. Apocalypse.'

[2] 'The goat shown in our frontispiece wears on his forehead the sign of the pentagram, the point facing upwards, which suffices to make it a symbol of light; with his two hands he makes the sign of occultism, and points above to the white moon of Chesed, and below to the black moon of Geburah. This sign expresses the perfect accord of mercy with justice. One of his arms is feminine, the other masculine. . . . The torch of intelligence which glows between his horns is the magic light of universal equilibrium; it is also the figure of the soul elevated above matter, while clinging to matter itself, as the flame clings to the torch. The hideous head of the animal expresses horror at sin. . . . The caduceus, which takes the place of the organ of generation, represents eternal life; the belly covered with scales is the water; the circle above it is the atmosphere; the feathers which come next signify the volatile; humanity in turn is represented by the two breasts and the androgynous arms of this sphinx of the occult sciences.

'So behold the shadows of the infernal sanctuary scattered, behold the sphinx who was the terror of the Middle Ages exposed and precipitated from his throne: *quomodo cecidisti, Lucifer?* The terrible Baphomet is nothing more (as with all the enigmatical and monstrous idols of ancient science and its dreams) than an innocent and even pious hieroglyph. . . . In the sacred gems of the Gnostic Christians who belonged to the sect of Basilides, one finds representations of the Christ under divers figures of animals in the Kabbala: sometimes it is a bull, sometimes a lion, sometimes a serpent with a lion's or bull's head; he always wears at the same time the attributes of light, like our goat, whose sign of the pentagram prevents mistaking him for the fabulous images of Satan.

'Let us say good and loudly, to combat the vestiges of Manicheanism which still show up every day among our Christians, that Satan, as a higher personality and as a power, does not exist. Satan is the personification of all errors, all perversities, and by consequence, all weaknesses. If God can be defined as he who necessarily exists, can one not define his antagonist and his enemy as he who necessarily does not exist? . . .

'As for the name of the Sabbath . . . the simplest etymology for us is that which has this word come from the Jewish Sabbath, for it is certain that the Jews, the most faithful depositaries of the secrets of the Kabbala, have almost always in magic been the grand masters of the Middle Ages.

'The Sabbath was therefore the Sunday of the Cabalists, the day of their religious feast, or rather, the night of their regular assembly. This feast, surrounded in mystery, had as a safeguard the dread of the common people and escaped persecution through terror. . . .

'Someone wrote us recently that the very respectable Father Ventura, former superior of the Theatines, examiner of bishops, etc., etc., after having read our dogma, declared that the Kabbala, in his eyes, was an

I hate him for he is a Christian.

Shakespeare (*Merchant of Venice,* I. iii. 43)

Such as these being captivated in the mystery of Babel, wonder only after their beast Mammon, upon which they ride in pride, and scorn anything but what pleases and flatters them in their admired works of covetous iniquity, gilded over with seeming holiness. But the Babylonish structure of their turba-magna-performances will fall, when it has attained the highest limit of its constellation, and no wit of man shall be able to prop it up. In the meantime the Antichrist in Babel will rage and tyrannise, and execute the sentence of wrath, or his own dismal doom, upon himself.

John Sparrow

The Powers of Darkness

And of mankind are some who say: We believe in Allâh and the Last Day, when they believe not.

They think to beguile Allâh and those who believe, and they beguile none save themselves; but they perceive not.

Heterod. 423

In their hearts is a disease, and Allâh increaseth their disease. A painful doom is theirs because they lie.

And when it is said unto them: Make not mischief in the earth, they say: We are peacemakers only.[1]

Are not they indeed the mischief-makers? But they perceive not.

Qur'ân, II. 8–12

invention of the devil. . . . My venerable masters of theology, you are more sorcerers than one thinks and than you yourselves think: and he who said: The devil is a liar like his father, might well also have some little things to say concerning the question of your paternity. . . .' (Eliphas Lévi: *Dogme et rituel de la Haute Magie,* Paris, Éditions Niclaus, reprinted 1952, pp. 282–288, passim).

'The old idea of the devil . . . is a direct violation of the fact that there is only the power of God. There is no power opposed to the ultimate good in the universal trend' (Baird T. Spalding: op. cit., vol. IV, p. 60).

[1] 'There is a great brotherhood throughout the world who have been working for peace for thousands of years. They are back of every movement for world peace and are becoming stronger and stronger all the time. There are now about 216 groups throughout the world. There is always one central unit as a first or centralizing body, and twelve units which surround that unit, giving it more force and power. They are working also for the enlightenment of the whole world. . . .

'There are many misconceptions regarding the White Brotherhood. It must first be understood that they never make themselves known as such. . . . There will come a time, however, when they will work more openly, but that will be when there are enough illumined people to know and understand just what they are doing. . . .

'If all the financial world were to get behind this movement for unity it would be the greatest influence for peace. . . . If the cooperative system should be adopted universally, there could be no booms and no more depressions. It would also be of great influence in the abolition of war. As a matter of fact, cooperation is

Rev. 961

It is not for any prophet to deceive. Whoso deceiveth will bring his deceit with him on the Day of Resurrection.

Qur'ân, III. 161

The prince of darkness is a gentleman.

Shakespeare (*King Lear,* III. iv. 147)

Then one came to me, radiant as the sun so that I might think he was an angel. He brought me a gleaming book and said, 'Take the Pax, since thou mayest not come to the Mass.' But my soul spoke with modest wisdom, 'He who himself has no peace, can give me no peace.' Then he went away and disguised himself as a poor sick man and said, 'Thou who art holy, heal me!' But I said, 'He who is sick himself can help no one!' 'But it is written, Whosoever can, shall help others.' 'It is also written: One shall help no one against the will of God.' 'But to do good,' he said, 'is not against the will of God.' But I said, 'To him in whom there is no good, no one can do good. Thou hast an eternal sickness. If thou wouldst be healed thou must go and show thyself to a Priest or Archbishop or the Pope. For I myself have no power save to sin.' Then he said angrily, 'That I will never do!' and became as black smoke and withdrew himself rudely and went away. But I was not afraid of him.

Action 340
Judgment 242

Mechthild of Magdeburg

A great sect arose which, taking for its motto the good and the happiness of man, worked in the darkness of the conspiracy to make the happiness of humanity a prey for itself.[1]

The Duke of Brunswick

coming into existence. The people who do not cooperate and align themselves with the new order will be outlawed completely . . .' (cp. *Rev.* XIII. 17).

'This life of oneness is the life of the masters, and anyone may live that life if he will drop his alliances with institutions and religions and races and nations and accept his alliance with the Universe. This is the "ark of the covenant" which enabled the Children of Israel to succeed, but when it was lost they failed to gain their liberty from opposition' (Baird T. Spalding: op. cit., vol. IV, pp. 8, 11, 14, 203).

[1] 'Situated in the initiatic current, link in a traditional chain which has included organizations of the most diverse character (as regards their modes of expression or their titles): GNOSTICS, ALCHEMISTS, CATHARI, ALBIGENSES, ROSE-CROSS, the HERMETIC, PYTHAGOREAN and PLATONIC schools, successors to the Egyptian Temple Initiations, contemporaries of the MARTINISTS, the EUDIASTS, the ANTHROPOSOPHISTS, and the ROSICRUCIAN societies, applying the same speculative methods, SUFISM is one of the INITIATIC ORDERS in which is perpetuated the TRADITION OF SYMBOLIC TEACHING in the form of MODAKARAS. . . .

'Christ, this Great Initiate, replied (*St Mark,* IV. 10, 11) to those in his group who questioned him:

' "Unto you it is given to know the mystery of the kingdom of God: but unto them that are without, all these things are done in parables." . . .

'This is why the constant practice of Symbolism must be the basis and foundation of our initiatic work, in constructing the Temple of our Ideal, Will, Perseverance and Work being the materials best suited to assure the Beauty, Solidarity and Harmony of the Edifice.

'Tradition requires us to remain faithful to a knowledge which constitutes a link and a universal language between Thinkers of all races and all latitudes, and likewise a possibility for mutual comprehension, collaboration, understanding, and assistance among all those who aspire to the same ideal of fraternity and love' (A Disciple: 'Lettre à un Fakir'; *Les Amis de l'Islam,* Mostaganem, nos. March and May, 1954).

'We must consider how we can begin to work under another form. If only the aim is achieved, it does not matter under what cover it takes place, and a cover is always necessary. For in concealment lies a great part of our strength. For this reason we must always cover ourselves with the name of another society. The lodges that

And they plotted, and Allâh plotted: and Allâh is the best of plotters.

Qur'ân, III. 54

Conform. 185

A new society of beings has formed in the midst of thickest darkness, who know one another without being seen, who understand one another without explanations, who serve one another without friendship. The object of this society is to govern the world.... (This) gigantic project (entails) a series of calamities of which the end is lost in the darkness of time, like unto those subterranean fires of which the insatiable activity devours the bowels of the earth and which escape into the air by violent and devastating explosions.[1]

The Marquis de Luchet

Infra 464

They invoke in His stead only females; they pray to none else than Satan, a rebel

Whom Allâh cursed, and he said: Surely I will take of Thy bondmen an appointed portion,

And surely I will lead them astray, and surely I will arouse desires in them, and surely I will command them and they will cut the cattle's ears, and surely I will command them and they will change Allâh's creation. Whoso chooseth Satan for a patron instead of Allâh is verily a loser and his loss is manifest.

He promiseth them and stirreth up desires in them, and Satan promiseth them only to beguile.

Qur'ân, IV. 117–120

Introd. *Beauty*

The Emperor of the dolorous realm, from mid breast stood forth out of the ice; and I in size am liker to a giant,

are under Freemasonry are in the meantime the most suitable cloak for our high purpose, because the world is already accustomed to expect nothing great from them which merits attention.... A society concealed in this manner cannot be worked against. In case of a prosecution or of treason the superiors cannot be discovered.... We shall be shrouded in impenetrable darkness from spies and emissaries of other societies (Adam Weishaupt: *Neuesten Arbeiten des Spartacus und Philo*, Munich, 1794, pp. 143, 163; in Webster, op. cit., pp. 219–220).

[1] 'Somehow they (the "Berlin Brotherhood") all seemed to me to be men without souls. They were desperate, determined seekers into realms of being with which earth had no sympathy, and which in consequence abstracted them from all human feelings or human emotions.

'Not one of them, that I can remember, ever manifested any genial qualities or seemed to delight in social exercises. They were profound, philosophic, isolated men, pursuing from mere necessity, or as a cloak to the stupendous secrets of their existence, some scientific occupation, yet in their innermost natures lost to earth and its sweet humanities; living amongst men, but partaking neither of their vices nor their virtues.

'In their companionship I felt abandoned of my kind.... If the knowledge I had purchased was indeed a reality, there were times when I deemed it was neither good nor lawful for man to possess it. I often envied the peaceful unconsciousness of the outer world, and would gladly have gone back to the simple faith of my childhood, and then have closed my eyes in eternal sleep sooner than awaken to the terrible unrest which had possessed me since I had crossed the safe boundaries of the visible, and entered upon the illimitable wastes of the invisible' (*Ghost Land*, pp. 47–48).

'One must speak, first in one way, then in another, so as not to commit oneself and to make one's real way of thinking impenetrable to one's inferiors.... One asserts that all events in the world occur from a hundred secret springs and causes, to which secret associations above all belong; one arouses the pleasure of quiet, hidden power and of insight into hidden secrets' (Adam Weishaupt: *Nachtrag von weitern Originalschriften*, Munich, 1787, I. 12, and *Einige Originalschriften des Illuminatenordens*, Munich, 1787, p. 51; Webster, op. cit., pp. 220–221).

than the giants are to his arms: mark now how great that whole must be, which corresponds to such a part.

Sin 75

If he was once as beautiful as he is ugly now, and lifted up his brows against his Maker, well may all affliction come from him.

Dante (*Inferno,* XXXIV. 28–36)

The world is governed by very different personages from what is imagined by those who are not behind the scenes.

Disraeli

There is scarce truth enough alive to make societies secure, but security enough to make fellowships accursed. Much upon this riddle runs the wisdom of the world.

Shakespeare (*Measure for Measure,* III. ii. 246)

He (Celsus) frequently calls the Christian doctrine a secret system, we must refute him on this point . . . to speak of the Christian doctrine as a secret system is altogether absurd.

Origen

Do not as Babel does, which amuses and comforts itself with the philosopher's stone, and boasts of it, but keeps only a gross mason's stone shut up in poison and death, instead of the precious philosopher's stone: What is it for Babel to have the stone, when it lies wholly shut up in Babel? It is as if a lord bestowed a country upon me, which indeed was mine, but I could not take possession of it, and remained still a poor man notwithstanding, and yet I boasted of the dominion, and so had the name, and not the power: Even thus it goes with Babel about the precious stone of the new-birth in Christ Jesus.[1]

Boehme

We steal by line and level, an't like your grace.

Shakespeare (*The Tempest,* IV. i. 241)

Knowl. 734

As God will only be conversed withal in a way of light and understanding; so the devil loves to be conversed with in a way of darkness and obscurity.

John Smith the Platonist

And (the demon) said: . . . Like as thy Christ helpeth thee in that thou doest, so also my father helpeth me in that I do; and like as for thee he prepareth vessels worthy of thine

[1] 'Let me now beg you to observe that the light of a Master Mason is darkness visible, serving only to express that gloom which rests on the prospect of futurity' (*Masonic Ritual, Ceremony of Raising to the Third Degree*: Walton Hannah: *Darkness Visible,* London, Augustine Press, 1955, p. 140). Cp. Milton:

> 'A dungeon horrible, on all sides round,
> As one great furnace flames, yet from those flames
> No light, but rather darkness visible
> Served only to discover sights of woe.'
>
> (*Paradise Lost,* Book I)

and Pope:

> 'Of darkness visible, so much be lent
> As half to show, half veil, the deep intent.'
>
> (*The Dunciad,* Book IV)

inhabiting, so also for me he seeketh out vessels whereby I may accomplish his deeds; and like as he nourisheth and provideth for his subjects, so also for me he prepareth chastisements and torments, with them that become my dwelling-places; and like as for a recompense of thy working he giveth thee eternal life, so also unto me he giveth for a reward of my works eternal destruction; and like as thou art refreshed by thy prayer and thy good works and spiritual thanksgivings, so I also am refreshed by murders and adulteries and sacrifices made with wine upon altars; and like as thou convertest men unto eternal life, so I also pervert them that obey me unto eternal destruction and torment: and thou receivest thine own and I mine.

Acts of Thomas, 76

Introd.
Heterod.

For I tell thee truly, that the devil hath his contemplatives as God hath his.[1]

The Cloud of Unknowing, XLV

Many will say to me in that day, Lord, Lord, have we not prophesied in thy name? and in thy name have cast out devils? and in thy name done many wonderful works?

And then will I profess unto them, I never knew you: depart from me, ye that work iniquity.

St Matthew, VII. 22, 23

And he opened his mouth in blasphemy against God, to blaspheme his name.[2]

Revelation, XIII. 6

[1] 'The translation of the Bible is in error where it says that man was created in the image of God. The "in" should be left out so that it reads, "Man IS the image of God." The word "in" does not appear in the original. . . . Man gets rid of the idea that he is not God by refusing to accept the negative statements. The statement, "I am God," held habitually as the secret fact within his own nature, frees him from the negative statement that he is not God. It is always better to state the Truth than the untruth. . . . God is the beginning and is the Great Servant of mankind. To receive His spirit is to become the Sons of God, and then our attitude toward the world is to bestow our great gifts upon all around us, a gracious and generous service' (Baird T. Spalding: op. cit., vol. IV, pp. 112, 117, 154).

[2] 'The word on the triangle is that Sacred and Mysterious Name you have just solemnly engaged yourself never to pronounce, unless in the presence and with the assistance of two or more Royal Arch Companions, or in the body of a lawfully-constituted Royal Arch Chapter, whilst acting as First Principal. It is a compound word, and the combination forms the word JAH-BUL-ON. It is in four languages, Chaldee, Hebrew, Syriac, and Egyptian. JAH is the Chaldee name of God, signifying "His Essence and Majesty Incomprehensible." It is also a Hebrew word, signifying "I am and shall be," thereby expressing the actual, future, and eternal existence of the Most High. BUL is a Syriac word denoting Lord or Powerful, it is in itself a compound word, being formed from the preposition Beth, in or on, and Ul, Heaven, or on High: therefore the meaning of the word is Lord in Heaven, or on High. ON is an Egyptian word, signifying Father of all, thereby expressing the Omnipotence of the Father of All, as in that well-known prayer, Our Father, which art in Heaven. The various significations of the words may be thus collected: I am and shall be: Lord in Heaven or on High:

"Father of All! In every age,
In every clime adored
By saint, by savage, and by sage,
Jehovah, Jove, or Lord."'

(*Mystical Lecture in Royal Arch Degree, Ceremony of Exaltation*; Hannah: op. cit., pp. 181–182. The author observes in a footnote that 'although Masonry claims to be founded on the Word of God, this syncretism is as blasphemous in an Old Testament setting as, say JESUS-MOLOCH-PAN would be to the Christian').

'Name that Divinity what you will, the greatest name is the word "God." Why? We can show you today that that word vibrates at the rate of one hundred and eighty-six billion beats a second, and we know people capable of intoning that word. But the beauty of it is, the moment that you realize that vibration, you *are* that vibration every time' (Baird T. Spalding: op. cit., vol. V, p. 33).

Whoever, when I am acting with pure and single heart, works against me with charms that are counter to the ṛta (*anṛta*), may he, O Indra, as he pronounces non-existence (*asat*), himself go to non-existence, like waters held in the fist.

Rig-Veda, VII. 104. 8

Holy War 394
Sin 82

Those who believe fight in the way of Allâh, and those who disbelieve fight in the way of the devil. So fight the saints of Satan. Lo! the strategy of Satan is ever weak.

Qur'ân, IV. 76

The Last Times

Creation 38

The age of hard iron came last. Straightway all evil burst forth into this age of baser vein: modesty and truth and faith fled the earth, and in their place came tricks and plots and snares, violence and cursed love of gain. Men now spread sails to the winds, though the sailor as yet scarce knew them; and keels of pine which long had stood upon high mountain-sides, now leaped insolently over unknown waves. And the ground, which had hitherto been a common possession like the sunlight and the air, the careful surveyor now marked out with long-drawn boundary-line. Not only did men demand of the bounteous fields the crops and sustenance they owed, but they delved as well into the very bowels of the earth; and the wealth which the creator had hidden away and buried deep amidst the very Stygian shades, was brought to light, wealth that pricks men on to crime. And now baneful iron had come, and gold more baneful than iron; war came, which fights with both, and brandished in its bloody hands the clashing arms. Men lived on plunder. Guest was not safe from host, nor father-in-law from son-in-law; even among brothers 'twas rare to find affection. The husband longed for the death of his wife, she of her husband; murderous stepmothers brewed deadly poisons, and sons inquired into their fathers' years before the time. Piety lay vanquished, and the maiden Astraea, last of the immortals, abandoned the blood-soaked earth.

Supra 441

And, that high heaven might be no safer than the earth, they say that the Giants essayed the very throne of heaven, piling huge mountains, one on another, clear up to the stars. Then the Almighty Father hurled his thunder-bolts, shattered Olympus, and dashed Pelion down from underlying Ossa.

Ovid

So oft as I with state of present time,
The image of the antique world compare,
When as mans age was in his freshest prime,
And the first blossome of faire vertue bare,
Such oddes I finde twixt those, and these which are,

As that, through long continuance of his course,
Me seemes the world is runne quite out of square,
From the first point of his appointed sourse,
And being once amisse growes daily wourse and wourse.

. . . .

For that which all men then did vertue call,
Is now cald vice; and that which vice was hight,
Is now hight vertue, and so vs'd of all:
Right now is wrong, and wrong that was is right,
As all things else in time are chaunged quight.
Ne wonder; for the heauens reuolution
Is wandred farre from where it first was pight,
And so doe make contrarie constitution
Of all this lower world, toward his dissolution.

Edmund Spenser

Beauty 676

When we arrive at the decadent age,[1] we find that men dug into the mountains for precious stones. They wrought metal and jade into cunning vessels and broke open oysters in search of pearls: they melted brass and iron; the whole of nature withered under the exploitation. They ripped open the pregnant and slew the young, untimely (in order to get skins and furs). The Chilin, as a result, did not visit the land. They broke down nests and despoiled the birds that had not lain, so that the phoenix no longer hovered around. They drilled wood for fire: they piled up timber to make verandahs and balustrades: they burnt forests to drive out game and drained the waters for fish. In spite of this, the furniture at the service of the people was not enough for their use, whilst the luxuries of the rulers were abundant. Thus, the world of life partially failed and things miscarried so that the larger half of creation failed of fruition. . . .

Introd. *Flight*

(They) opened up irrigation channels, for their enrichment. They laid foundations for their cities, so that they were munitioned. Captured wild beasts were domesticated: thus, there was grievous rupture of the Yin and Yang, and the succession of the four seasons failed. Thunder-bolts wrought havoc, and hailstones fell with violence. Noxious miasma and untimely hoarfrosts fell unceasingly, resulting in atrophy and the failure of nature to bear abundantly. Luxuriant grass and thick brushwood were cut down in order to get land. They cut down the jungle in order to grow ears of corn. The plants and trees that died before germination, flowering and bearing fruit, were innumerable. . . .

Rev. 967

In the course of time, the mountains and streams were divided into boundaries and frontiers: censuses of the people were taken in order to know the population of this place and that: cities were built and moats and dykes dug: barriers were erected and weapons forged, for defensive purposes: officials were created for the departments with various robes and badges and with laws: they differentiated classes and masses and distinguished the worthy from the vulgar: they organized a system of reprimands and approbations, of rewards and punishments. . . .

[1] I.e., the beginning of the Kali Yuga.

The harmonious cooperation of Heaven and Earth, the evolution of creation by the Yin and Yang depends on the spirit of man. Hence, when there is an estrangement between the classes and masses or rulers and the people, the very air of Heaven becomes noxious and disorganised: when prince and minister are not in harmony, the crops in the fields fail to ripen.

Huai Nan Tzû

The seasons alter: hoary-headed frosts
Fall in the fresh lap of the crimson rose,
And on old Hiems' thin and icy crown
An odorous chaplet of sweet summer buds
Is, as in mockery, set. The spring, the summer,
The childing autumn, angry winter, change
Their wonted liveries, and the mazed world,
By their increase, now knows not which is which.
And this same progeny of evil comes
From our debate, from our dissension:
We are their parents and original.

Illusion 85

Shakespeare (*Midsummer-Night's Dream*, II. i. 107)

This know also, that in the last days perilous times shall come.

For men shall be lovers of their own selves, covetous, boasters, proud, blasphemers, disobedient to parents, unthankful, unholy,

Without natural affection, trucebreakers, false accusers, incontinent, fierce, despisers of those that are good,

Traitors, heady, highminded, lovers of pleasures more than lovers of God;

Having a form of godliness, but denying the power thereof: . . .

Ever learning, and never able to come to the knowledge of the truth.

Heterod. 418
Knowl. 734

II. Timothy, III. 1 ff.

Property alone will confer rank, wealth will be the only source of devotion, passion will be the sole bond of union between the sexes, falsehood will be the only means of success in litigation, and women will be the objects merely of sensual gratification. Earth will be venerated only for its mineral treasures; . . . dishonesty will be the universal means of subsistence, . . . menace and presumption will be the subterfuge for learning, liberality will be devotion, simple ablution will be purification, mutual assent will be marriage, fine clothes will be dignity.

Action 346

Vishṇu Purâṇa

I have seen servants upon horses, and princes walking as servants upon the earth.

Ecclesiastes, x. 7

By the Lord, Horatio, these three years I have taken note of it: the age is grown so picked that the toe of the peasant comes so near the heel of the courtier, he galls his kibe.

Shakespeare (*Hamlet*, v. i. 148)

The pious will be deemed insane, and the impious wise: the madman will be thought a brave man, and the wicked will be esteemed as good. As to the soul, and the belief that it is immortal by nature, or may hope to attain to immortality, as I have taught you,—all this they will mock at, and will even persuade themselves that it is false. No word of reverence or piety, no utterance worthy of heaven and of the gods of heaven, will be heard or believed.

And so the gods will depart from mankind, . . . and only evil angels will remain, who will mingle with men, and drive the poor wretches by main force into all manner of reckless crime, into wars, and robberies, and frauds, and all things hostile to the nature of the soul. Then will the earth no longer stand unshaken, and the sea will bear no ships; heaven will not support the stars in their orbits, nor will the stars pursue their constant course in heaven; all voices of the gods will of necessity be silenced and dumb; the fruits of the earth will rot; the soil will turn barren, and the very air will sicken in sullen stagnation. After this manner will old age come upon the world. Religion will be no more; all things will be disordered and awry; all good will disappear.

But when all this has befallen, Asclepius, then the Master and Father, God . . . will look on that which has come to pass, and will stay the disorder by the counterworking of his will, which is the good. He will call back to the right path those who have gone astray; he will cleanse the world from evil, now washing it away with waterfloods, now burning it out with fiercest fire, or again expelling it by war and pestilence. And thus he will bring back his world to its former aspect, so that the Kosmos will once more be deemed worthy of worship and wondering reverence, and God, the maker and restorer of the mighty fabric, will be adored by the men of that day with unceasing hymns of praise and blessing. Such is the new birth of the Kosmos; it is a making again of all things good, a holy and awe-striking restoration of all nature; and it is wrought in the process of time by the eternal will of God.

P. State 583

Hermes

The time is near in which nothing will remain of Islam but its name, and of the *Qur'ân* but its mere appearance, and the mosques of Muslims will be destitute of knowledge and worship; and the learned men will be the worst people under the heavens; and contention and strife will issue from them, and it will return upon themselves.

Muhammad

There will come a time, brethren, when . . . the ten moral courses of conduct will altogether disappear, the ten immoral courses of action will flourish excessively; there will be no word for moral among such humans—far less any moral agent. Among such humans, brethren, they who lack filial and religious piety, and show no respect for the head of the clan—'tis they to whom homage and praise will be given, just as to-day homage and praise are given to the filial-minded, to the pious and to them who respect the heads of their clans. . . . The world will fall into promiscuity, like goats and sheep, fowls and swine, dogs and jackals.

Introd. *Sin*

Chakkavatti Sihanada Suttanta

For in the fatness of these pursy times
Virtue itself of vice must pardon beg.

Shakespeare (*Hamlet,* III. iv. 153)

Pilg. 366

He that keepeth hold upon religion will be as one that holdeth in his hand a coal of fire.

Muhammad

At that dreadful last epoch men will be malign, crooked, wicked, dull, conceited, fancying to have come to the limit when they have not.

Saddharma-puṇḍarîka, XII. 4

The discourses spoken by the Tathagata are profound, of profound significance, dealing with the other world and bound up with the emptiness of this world. But the time will come when they will no longer be regarded as things to be studied and mastered; on the contrary, it is those discourses that are made by poets in the poetical style, with embellished sounds, overlaid with ornament, and spoken by profane auditors, that will be considered worthy of study and the others will disappear.

Saṁyutta-nikâya, II. 267

Supra 459

Our holy things are defiled,
the name that is called upon us is profaned;
our nobles are dishonoured,
our priests burnt,
our Levites gone into captivity;
our virgins are defiled,
our wives ravished;
our righteous are seized,
our saints scattered,
our children are cast out,
our youths are enslaved,
our heroes made powerless.

IV Ezra, X. 22

Pythagoras said, that luxury entered into cities in the first place, afterwards satiety, then lascivious insolence, and after all these destruction.

Stobaeus

Our Lord, my Divine Son, is tired of being greatly offended by men through sins against holy purity. He is inclined to command a deluge of retribution. I have interceded in order that he may still exercise mercy. I therefore request prayers and penance in reparation for those sins.

The Blessed Virgin, to Sister Pierina, in Montichiari, Italy, 1947

You wish to know the secrets of La Salette? Well, here are the secrets of La Salette: Unless you do penance, you shall all perish.

Pope Pius IX

Were God to punish men for what they have earned, He would not leave a living creature on the surface of the earth.

Sin 56

Qur'ân, XXXV. 45

. . . In the course of justice none of us
Should see salvation.

Shakespeare (*Merchant of Venice*, IV. i. 199)

The earthly man is the curse of God, and is an abominate before God's holiness; he can do nothing else but seek his self-hood, for he is in the wrath of God.

Boehme

Sin 66

The smell of man offends the gods a hundred leagues away.

Payasi Suttanta

In after times none will pursue philosophy in singleness of heart.

Hermes

'Good heavens, how deeply I am often saddened at seeing the human race, which God created perfect, in His own image, and appointed to be the lords of the earth, depart so far away from me (Nature)! I allude more particularly to you, O stolid philosophaster, who presume to style yourself a practical chemist, a good philosopher, and yet are entirely destitute of all knowledge of me, of the true Matter, and of the whole Art which you profess! For, behold, you break vials, and consume coals, only to soften your brain still more with the vapours. You also digest alum, salt, orpiment, and atrament; you melt metals, build small and large furnaces, and use many vessels: nevertheless, I am sick of your folly, and you suffocate me with your sulphurous smoke.'

John A. Mehung

Creation 42

Supra 442

Now the Spirit speaketh expressly, that in the latter times some shall depart from the faith, giving heed to seducing spirits, and doctrines of devils;

Speaking lies in hypocrisy; having their conscience seared with a hot iron.

I. Timothy, IV. 1–2

After my Parinirvana, in the last kalpa of this world, there will be plenty of these goblin-heretics about, hiding themselves within the very personalities of the saints, the better to carry out their deceiving tricks. Sometimes they gain control of some great and good Master and teach under the prestige of his name. They often assert that they have received their Dharma from some notable Master, deceiving ignorant people, discouraging them and even causing them to go insane. In such deceptive ways do they spread their false and destructive heresies.[1]

Surangama Sutra

Heterod. 423 430

[1] 'I was calling my family to dinner when I stooped over to brush a mosquito off my leg, and I noticed a flash of light. I looked up and out through the screen door and I saw a blue mist behind the ash trees in the farm yard, opposite the kitchen door. Then the mist, as it descended, became the figure of the Blessed Virgin. Oh, she was so radiant, so beautiful! An artist couldn't paint her to do her justice.

'I walked out into the yard. She smiled at me as I approached her. She said, "You are not afraid, my child, you have been expecting me. My child, eat a twig. My speaking to you on Good Friday was a test to your community. . . . Only about ten percent of them are devoted and pray the Rosary very devoutly. This ten per cent does not include your parish priest, Mater Christi Home, or the Sisters. . . .

And thus I clothe my naked villany
With odd old ends stol'n forth of holy writ,
And seem a saint when most I play the devil.

Shakespeare (*Richard III*, I. iii. 336)

(The Theosophists) in pillaging at random whatever religious systems, philosophical doctrines and scientific theories have crossed their attention, have elaborated compilations wherein are to be found bits of Vedanta, morsels of Taoism, pieces of Egyptianism, samplings of Mazdeism, fragments of Christianity, scraps of Brâhmanism, shreds of Gnosticism, dregs of Hebrew Kabbala, remnants of Paracelsus, Darwin and Plato, crumbs of Swedenborg and Hegel, Schopenhauer and Spinoza,—and have propagated this on all continents in affirming that such was Buddhist Esoterism.

Supra Introd.

Augustin Chaboseau

Supra 457

That holy seat of temples shall become a sepulchre of dead bodies.

Hermes

'Sir, what do you think of Theosophy?'

MASTER: 'I have heard that man can acquire superhuman powers through it and perform miracles. I saw a man who had brought a ghost under control. The ghost used to procure various things for his master. What shall I do with superhuman powers? Can one realize God through them? If God is not realized then everything becomes false.'

Introd. *Flight*

Sri Ramakrishna

In the last time . . . every man shall speak that which pleaseth him.[1]

Apocalypse of Thomas

' "All religions must work together against the enemies of God. You must love thy neighbor (sic), you must live the Commandments. . . . My warnings at Fatima are my pleas for prayers, the Rosary, repentance for sins and making sacrifices to save souls.

' "There are many religions that claim they can see their dead come back. These are false claims, works of trained fakers. Only canonized saints appear to people. These false claims make it hard for us, as people lose faith" ' (Mrs VanHoof, and 'Our Lady of Necedah': Anne Stuart: *Our Lady of Necedah*, 1950, passim).—These false claims also make it hard for the Church. The following dispatch from Vatican City appeared in a Cairo paper: 'The *Osservatore Romano* publishes a note in which Mgr. Alfredo Ottaviani, assessor to the Holy Office, warns the faithful against miraculous manifestations—notably those of Voltago in Italy, Espis and Bouxières in France, Hamsur in Belgium, Herolsbach in Germany, and Necedah in the United States (Wisconsin).

'The prelate points out in substance that the credulity of the faithful in these manifestations can be injurious to the cause of true miracles, and he recalls how the Church has always shown a comprehensible prudence in this domain.'

[1] 'Let us explore together the secret of a deeper devotion, a more subterranean sanctuary of the soul, where the Light Within never fades, but burns, a perpetual Flame. . . . Theologies and symbols and creeds, though inevitable, are transient and become obsolescent, while the Life of God weeps (sic) on through the souls of men in continued revelation and creative newness. . . . And one knows now why Pascal wrote, in the centre of his greatest moment, the single word, "Fire". . . . Only absolutes satisfy the soul committed to holy obedience. . . . Some of us will have to enter upon a vow of renunciation and of dedication to the "Eternal Internal" which is as complete and as irrevocable as was the vow of the monk of the Middle Ages. Little groups of such utterly dedicated souls, knowing one another in Divine Fellowship, . . . ready to go the second half, obedient as a

And for this cause God shall send them strong delusion, that they should believe a lie:

That they all might be damned who believed not the truth, but had pleasure in unrighteousness.

II. Thessalonians, II. 11–12

Sin
61

Lo! thou canst not make the dead to hear, nor canst thou make the deaf to hear the call when they have turned to flee;

Nor canst thou lead the blind out of their error. Thou canst make none to hear, save those who believe Our revelations and who have surrendered.

And when the word is fulfilled concerning them, We shall bring forth a beast of the earth to speak unto them because mankind had not faith in Our revelations.

Qur'ân, XXVII. 80–82

There shall arise false Christs, and false prophets, and shall shew great signs and wonders; insomuch that, if it were possible, they shall deceive the very elect.[1]

St Matthew, XXIV. 24

shadow, sensitive as a shadow, selfless as a shadow—such bands of humble prophets can recreate the Society of Friends and the Christian church and shake the countryside for ten miles around....

'The mark of this simplified life is radiant joy. It lives in the Fellowship of the Transfigured Face.... in childlike trust listening ever to Eternity's whisper, walking with a smile into the dark.... And the Fellowship of the Horny Hands is identical with the Fellowship of the Transfigured Face, in this Mary-Martha life.... In the Fellowship cultural and educational and national and racial differences are levelled.... We find men with chilly theologies but with glowing hearts. We overleap the boundaries of church membership.... In peace and power and confidence we work upon such apparently hopeless tasks as the elimination of war from society, and set out toward world-brotherhood and interracial fraternity....

'Each one of us can live such a life of amazing power and peace and serenity.... We have all heard this holy Whisper at times.... the welling-up whispers of divine guidance.... And under the silent, watchful eye of the Holy One we all are standing, whether we know it or not. And in that Centre, in that holy Abyss where the Eternal dwells at the base of our being ... we *centre down*, as the old phrase goes, and live in that holy Silence which is dearer than life.... I should like to testify to this, as a personal experience, graciously given.... I only speak to you because it is a sacred trust, not mine but to be given to others' (Thomas R. Kelly: *A Testament of Devotion*, London, 1943, passim).

'Feeling gives us our knowledge of values.... In a Quaker meeting for worship or for business a speaker seldom remarks "I think" but generally "I feel".... The impact of Authority or Reason ... is not typical of the Quaker method' (Howard Brinton: *Friends for 300 Years*, New York, Harper & Bros., 1952, pp. 87, 121, 120). The Catholic devotional mystic and the Hindu *bhakta* are not obliged to 'think' either, but here it is the orthodoxy of their respective traditional frameworks which performs this function on their behalf.

'Be a fool, my brother in God, and seek the Black Diamond which is in thee' (A Disciple, op. cit., June, 1954).

[1] One can deceive with words, too, where it is not always easy to separate the wheat from the chaff. The following statement, for example, taken in isolation, says nothing very wrong: 'Today as always, there are men who understand the unity of sciences and the unity of creeds.

'Raising themselves above all religious fanaticism, they demonstrate that all creeds are expressions of one and the same religion' (Papus: *Traité élémentaire de Science occulte*, Paris, 7th edit., p. 470). But it is the same author who writes: 'Faithful to the eternal law, Freemasonry struggles today against the Church in the name of Science. It (the Church) wants to do away with all that is opposed to its incomplete teaching' (id., p. 468)—an idea which develops logically into the following: 'The various sects ... are moving towards the Masonic position, and when they arrive, Masonry will witness a scene which she has prophesied for ages.... Our little dogmas will have their day and cease to be, lost in the vision of a truth so great that all men are one in their littleness' (The Rev. J. Fort Newton, *The Builders*, 1949, p. 183; in Walton Hannah: *Christian by Degrees*,

Knowl. 761

These two men (Enoch and Elias) who now came out of Paradise were so wise in Divine Truth that they drove Anti-Christ powerfully away. They told him straightly who he was, from what power his signs and wonders came, how he had come from thence and what his end should be. As the misled realized what an accursed god had been given them because of their greed, and their delight in many kinds of evil, many good men and women who had been turned from the Christian way turned back again.

Mechthild of Magdeburg

P. State 579

Christ is so unknown because I am not known. . . . The world will have to drain the cup of wrath to the dregs because of the countless sins through which His Heart is offended. The star of the infernal regions will rage more violently than ever and will cause frightful destruction because he knows that his time is short and because he sees that many have already gathered around my sign.

The Blessed Virgin, to Barbara Reuss, at Marienfried, Germany

Introd. *Judgment*

For this world in which Adam lived before his Eve, such as it was before the Malediction, must return and justice reign therein: but vanity must be purged with divine fire and cast back to the world of darkness.

Boehme

Supra 442

And a command has gone forth from the presence of the Lord concerning those who dwell on the earth that their ruin is accomplished because they have learnt all the secrets of the angels, and all the violence of the Satans, and all their powers—the most secret

London, Augustine Press, 1955, p. 43). Thus, 'there is slowly rising from the ashes of orthodoxy, the actual temple not made by hands, eternal in heaven, in man. A great new race of thinkers is coming to the fore with herculean strides. Soon the tides will surge over the earth to sweep away the debris of delusion which has been strewn over the paths of those who are struggling along under the load of evolution' (Baird T. Spalding: op. cit., vol. III, p. 64).

The idea is further developed in *Ghost Land:* 'The thoughts which shone in resplendent imagery before the eyes of my associates and myself a quarter of a century ago, have gradually been leavening the lump of civilized society during that whole period of time. They have been seen in vision, felt in soul, and taught in isolated fragments by many a solitary pioneer of the new church that shall be; but chiefly has their influence been realized as the radiation of an unknown force, whose subtile potencies are making for themselves a lever of public opinion, a giant whose will is sufficient to raise up every stone in the new temple and put them all in place, a concrete and glorious whole, when the stones of thought shall have been hewn each in its separate quarry, when every stone shall be *fair and square and true,* and ready in its separate perfection to form a part of the sublime erection. . . .

'The nineteenth century is the last which could appreciate the objects of an association contemplating amongst other ideas, the reversal and obliteration of all theological myths, and the inauguration of a true spiritual kingdom, in which truth itself will be the Bible, God the high-priest, ministering spirits the acolytes, and occult science the connecting link between the past and the present, the spiritual and the natural world. The very few that in this generation are fitted for affiliation with this society will be called, as I was, without any previous knowledge of its existence; the rest of the world may and will seek it in vain' (pp. 357–358, 403).

'Thus in religion,' concludes Eliphas Lévi, 'universal and hierarchical orthodoxy, restoration of temples in all their splendour, revival of all ceremonies in their pristine pomp, hierarchical teaching of the symbol, mysteries, miracles, legends for children, light for grown men who will take good care not to scandalize the little ones in the simplicity of their belief. There in what concerns religion is our whole utopia, and it is also the desire and need of humanity' (*Dogme et Ritual,* p. 360).

ones—and all the power of those who practise sorcery, and the power of witchcraft, and the power of those who make molten images for the whole earth: And how silver is produced from the dust of the earth, and how soft metal originates in the earth.

Supra 451

Book of Enoch, LXV. 6, 7

I saw a woman sit upon a scarlet coloured beast, full of names of blasphemy, having seven heads and ten horns.

And the woman was arrayed in purple and scarlet colour, and decked with gold and precious stones and pearls, having a golden cup in her hand full of abominations and filthiness of her fornication:

And upon her forehead was a name written, MYSTERY, BABYLON THE GREAT, THE MOTHER OF HARLOTS AND ABOMINATIONS OF THE EARTH.

Revelation, XVII. 3–5

Jesus said: I have cast fire upon the world, and see, I guard it until it (the world) is afire.
Jesus said: This heaven shall pass away and the one above it shall pass away.

The Gospel according to Thomas, Log. 10, 11

We are already in the last age of the world, and verily await the consummation of the celestial movement.

Dante (*Il Convito,* II. XV. 4)

There have been, and will be again, many destructions of mankind arising out of many causes; the greatest have been brought about by the agencies of fire and water, and other lesser ones by innumerable other causes. There is a story, which even you have preserved, that once upon a time Phaëthon, the son of Helios, having yoked the steeds in his father's chariot, because he was not able to drive them in the path of his father, burnt up all that was upon the earth, and was himself destroyed by a thunderbolt. Now this has the form of a myth, but really signifies a declination of the bodies moving in the heavens around the earth, and a great conflagration of things upon the earth, which recurs after long intervals.

Plato (*Timaeus,* 22 C)

Do they not see how many a generation We destroyed before them, whom We had established in the earth more firmly than We have established you?

Creation 38

Qur'ân, VI. 6

In the most high and palmy state of Rome,
A little ere the mightiest Julius fell,
The graves stood tenantless and the sheeted dead
Did squeak and gibber in the Roman streets.

Shakespeare (*Hamlet,* I. i. 113)

A day wherein mankind will be as thickly-scattered moths
And the mountains will become as carded wool.

Illusion 99

Qur'ân, CI. 4, 5

Creation 26

I am the origin and also the dissolution of the entire universe.

Bhagavad-Gîtâ, VII. 6

Holiness 924

I am Alpha and Omega, the beginning and the end, the first and the last.

Revelation, XXII. 13

Supra Introd. Death 223

All *Philosophy* agreeth in this, that the *last end* is the *first mover*.

Peter Sterry

Beauty 689

Let not your spirit be troubled on account of the times;
For the Holy and Great One has appointed days for all things.

Book of Enoch, XCII. 2

Doctrinal Explanations of Evil

Knowl. 745

By 'men who sin knowingly'[1] Scripture means them that are weak in the *exercised* knowledge and performance of Good; and by 'them that know the Divine Will and do it not',[2] it means them that have heard the truth and yet are weak in faith to trust the Good or in action to fulfil it. And some desire not to have understanding in order that they may do good, so great is the warping or the weakness of their will. And, in a word, evil (as we have often said) is weakness, impotence, and deficiency of knowledge (or, at least, of exercised knowledge), or of faith, desire, or activity as touching the Good.

Dionysius

Nay, but those who disbelieve live in denial
And Allâh doth encompass them from behind.

Qur'ân, LXXXV. 19, 20

Reality 803

Let none therefore seek the efficient cause of an evil will; for it is not efficient but deficient, nor is there effect but defect, namely falling from that highest essence unto a lower, this is to have an evil will. The causes whereof (being not efficient but deficient) if one endeavour to seek, it is as if he should seek to see the darkness, or to hear silence.

St Augustine

Creation 38

Thus evil hath no being, nor any inherence in things that have being. Evil is nowhere *qua* evil; and it arises not through any power but through weakness. Even the devils derive their existence from the Good, and their mere existence is good. Their evil is the result of a fall

[1] *Romans*, I. 18–20.
[2] *St Luke*, XII. 47.

from their proper virtues, and is a change with regard to their individual state, a weakness of their true angelical perfections. And they desire the Good in so far as they desire existence, life, and understanding; and in so far as they do not desire the Good, they desire that which hath no being. And this is not desire, but an error of real desire.

Dionysius

Illusion 109

All things that exist, therefore, seeing that the Creator of them all is supremely good, are themselves good. But because they are not like their Creator, supremely and unchangeably good, their good may be diminished and increased. But for good to be diminished is an evil, although, however much it may be diminished, it is necessary, if the being is to continue, that some good should remain to constitute the being. For however small or of whatever kind the being may be, the good which makes it a being cannot be destroyed without destroying the being itself. . . . So long as a being is in process of corruption, there is in it some good of which it is being deprived; and if a part of the being should remain which cannot be corrupted, this will certainly be an incorruptible being, and accordingly the process of corruption will result in the manifestation of this great good. But if it do not cease to be corrupted, neither can it cease to possess good of which corruption may deprive it. But if it should be thoroughly and completely consumed by corruption, there will then be no good left, because there will be no being. Wherefore corruption can consume the good only by consuming the being. Every being, therefore, is a good; a great good, if it cannot be corrupted; a little good, if it can; but in any case, only the foolish or ignorant will deny that it is a good. And if it be wholly consumed by corruption, then the corruption itself must cease to exist, as there is no being left in which it can dwell.

St Augustine

Suff. 124
Introd.
Judgment

Supra 464

If then the rational part of a man's soul is illumined by a ray of light from God, for that man the working of the daemons is brought to naught; for no daemon and no god has power against a single ray of the light of God. But such men are few indeed; and all others are led and driven, soul and body, by the daemons, setting their hearts and affections on the workings of the daemons. This is that love which is devoid of reason, that love which goes astray and leads men astray. The daemons then govern all our earthly life, using our bodies as their instruments; and this government Hermes called 'destiny'.

Hermes (Asclepius to Ammon)

P. State 563
Illusion 89
Action 329

And do not be like those who forgot Allâh, and whom He therefore caused to forget their own souls. Such are the evil-doers.

Qur'ân, LIX. 19

Heterod. 420
Metanoia 493

And as the Devil would fain deceive all men, and draw them to himself and his works, and make them like himself, and useth much art and cunning to this end, so is it also with this false Light; and as no one may turn the Evil Spirit from his own way, so no one can turn this deceived and deceitful Light from its errors. And the cause thereof is, that both these two, the Devil and Nature, vainly think that they are not deceived, and that it standeth quite well with them. And this is the very worst and most mischievous delusion.

Theologia Germanica, XLIII

Heterod. 430

Holy War 394 Sin 82

O Prophet! Exhort the believers to fight. If there be of you twenty steadfast they shall overcome two hundred, and if there be of you a hundred (steadfast) they shall overcome a thousand of those who disbelieve, because they (the disbelievers) are a folk without intelligence.

Qur'ân, VIII. 65

Introd. Heterod.

In Divine Perfection there is no other partner.
. . . Evil appears but as the other side of Truth.

Shabistarî

God is not the author of any evil.

Sextus the Pythagorean

Introd. M.M.

God is not the author of evil; but it is the lasting on of the things made that causes evil to break out on them. And that is why God has subjected things to change; for by transmutation the things made are purged of evil. . . . God has one quality, and one alone, the quality of goodness; and he who is good is neither disdainful nor incapable.

Hermes

Sin 57

As I live, saith the Lord God, I have no pleasure in the death of the wicked; but that the wicked turn from his way and live.

Ezekiel, XXXIII. 11

What concern hath Allâh for your punishment if ye are thankful (for His mercies) and believe (in Him)? Allâh was ever Responsive, Aware.

Qur'ân, IV. 147

God did not create man for adversity.

Mechthild of Magdeburg

For God is not the author of confusion, but of peace.

I. Corinthians, XIV. 33

Judgment 266

Evil can no more be charged upon God than darkness can be charged upon the sun; because every quality is equally good, every quality of fire is as good as every quality of light, and only becomes an evil to that creature who, by his own self-motion, has separated fire from the light in his own nature.

William Law

God can't be wrathfull; but we may conclude,
Wrathfull He may be, by similitude:
God's wrathfull said to be, when He doth do
That without *wrath,* which wrath doth *force us* to.

Robert Herrick

The grudging temper does not start from heaven above, but comes into being here below, in the souls of those men who are devoid of mind.

Hermes

When there come to pass in time the things he speculated in eternity then people think that God has changed his mind, though whether he be wrathful or benignant it is we who change and he remains the same; just as the sunshine hurts weak eyes and benefits the strong ones what time the light itself remains unchanged.

Eckhart

Reality 773
Center 838

God's anger is no disturbance of mind in Him, but His judgment assigning sin the deserved punishment; and His revolving of thought is an unchanged ordering of changeable things. For God repents not of anything He does, as man does: but His knowledge of a thing ere it be done, and His thought of it when it is done, are both alike firm and fixed.

St Augustine

Metanoia 495
Creation 31

God when He's angry here with any one,
His wrath is free from perturbation;
And when we think His looks are sowre and grim,
The alteration is in us, not Him.

Robert Herrick

From eternity to eternity no spark of wrath ever was or ever will be in the holy triune God.

William Law

God is light, and in him is no darkness at all.

I. John, I. 5

Center 833

God is bliss. He is Sat-Chit-Ananda. From Him no misery ever comes. Does the sun ever give darkness? Where God is, there misery is not. He gives us only pure bliss.

Swami Ramdas

Wherever there is perturbation, there the Good cannot be, and wherever the Good is, there no perturbation at all can be; even as wherever day is, night cannot be, and wherever night is, day cannot be.

Hermes

Peace 705

God is a light incapable of receiving its contrary (darkness).

Sextus the Pythagorean

God's unchangeable goodness . . . is also the unchangeable rule of his will; neither can he any more swerve from it, than he can swerve from himself.

John Smith the Platonist

Introd.
Conform.

Charity 602

It is compassion which constitutes the very heart of the Buddha.

Hônen

Atman, which is existence, knowledge, and bliss, can have no other inherent nature or attribute. If there appear to be such, it is illusory.

Srimad Bhagavatam, XI. xvi

METANOIA

And be not conformed to this world: but be ye transformed by the renewing of your mind, that ye may prove what is that good, and acceptable, and perfect, will of God.

Romans, XII. 2

My guide . . . turned his head where he had had his feet before.

Dante (Inf., XXXIV. 78)

'There is in man something which must be converted, or which must be transmuted: it is the soul as love—the soul as will—whose center of gravitation is the *ego*' (Schuon: *L'Oeil du Coeur,* p. 209).

'*Μετάνοια,* usually rendered by "repentance", is literally "change of mind", or intellectual metamorphosis.'[1] In this chapter the term will be used in both ways, since the first idea prefigures the second, and the basic conception in both instances is that of a change, a reversal, or a turning inward. (In Arabic, the first interpretation is usually rendered by *tawbah,* which means 'repentance', but especially in the sense of a turning away from the world and a change in perspective and values, whereas the second meaning is broadly covered by the root Q L B, 'to invert', with the noun form *qalb*=inversion, transformation, heart, center, intellect, essence.) The turning is from the circumference to the center, from appearance to reality, from the sensible to the intelligible, time to eternity, the passible to the impassible, from desire to desirelessness;—and by desire is meant anything which can attach us in an inordinate manner to our terrestrial and psychic environment, anything which removes us from our 'vertical' human center to an agitated, illusory and 'horizontal' periphery, anything at all in which can be identified the elements of uneasy restlessness (not spiritual longing!), incompletion, heat, craving.

Death 220
Center passim

'*Μετάνοια* is, then, a transformation of one's whole being; from human thinking to divine understanding, . . . and the birth of a "new man" who, so far from being overwhelmed by the weight of past errors is no longer the man who committed them.'[2]

Realiz. 873

Metaphysically, *metanoia* is in rapport with the laws of inverse analogy governing the hierarchization of the states of being. Cosmologically it relates to what is called the 'junction of extremes', while microcosmically it finds expression, for example, in the Hindu teachings concerning the *kundalinî* (cf. Guénon: 'Kundalini-yoga', in *Yoga, Science de l'Homme intégral,* edit. Masui, Paris, 1953; and Arthur Avalon: *The Serpent Power).*

Introd.
Humility

[1] Coomaraswamy: 'On Being in One's Right Mind', *The Review of Religion,* Columbia University Press, November, 1942, p. 32.

[2] Id., pp. 34, 40.

'Duo Sunt in Homine'

> The mind is said to be twofold:
> The pure and also the impure;
> Impure—by connection with desire;
> Pure—by separation from desire.

Maitri Upanishad, VI. 34

God compacted (man) of . . . two substances, the one divine, the other mortal.

Hermes

Conform. 166

When a man is drawn in two opposite directions, to and from the same object, this, as we affirm, necessarily implies two distinct principles in him.

Plato (*Republic* x, 604 B)

The being of man consists of two beings, the natural and the supernatural.

Boehme

Introd. *Holiness*

There are two minds, that of all beings, and the individual mind: he that flees from his own mind flees for refuge to the mind of all in common.

Philo

Introd. *Death*

> Two men are in me: one wants what God wants;
> The other, what the world wants, the devil, and death.

Angelus Silesius

Realiz. 859

A watchful observer of his own heart and life shall often hear the voice of wisdom and the voice of folly speaking to him: he that hath his eyes opened, may see both the visions of God falling upon him, and discern the false and foolish fires of Satan that would draw away his mind from God.

John Smith the Platonist

Introd. *Rev.*

This world is God's house, wherein a gallant sumptuous feast is prepared, and all men are his guests: and . . . there are two waiters at the table which fill out the wine to them that call for it; the one a man, the other a woman: the one called Νοῦς, or mind, from whose hand all wise men drink, the other Ἀκράτεια, or intemperance, who fills the cups of the lovers of this world.

Dion Chrysostom

He it is Who created you. but one of you is a disbeliever and one of you is a believer, and Allâh is Seer of what ye do.

Qur'ân, LXIV. 2

Man has two souls. One of them comes from the first Intelligible, and partakes of the power of the Demiurgus; the other soul is put into the man by the revolution of the heavenly bodies, and into this latter soul enters subsequently the (first) soul which is able to see God.

Hermes

Action 329

Man has two spirits, a divine and an animal spirit. The former is from the breath of God; the latter from the elements of the air and the fire. He ought to live according to the life of the divine spirit and not according to that of the animal.

Paracelsus

. . . To make in himself of twain one new man, so making peace.

Ephesians, II. 15

Realiz. 887

Two birds, fast bound companions,
Clasp close the self-same tree.
Of these two, the one eats sweet fruit;
The other looks on without eating.

On the self-same tree a person, sunken,
Grieves for his impotence, deluded;
When he sees the other, the Lord (*iś*), contented,
And his greatness, he becomes freed from sorrow.

When a seer sees the brilliant
Maker, Lord, Person, the Brahma-source,
Then, being a knower, shaking off good and evil,
Stainless, he attains supreme identity (*sâmya*) (with Him).

Mundaka Upanishad, III. i. 1

M.M. 978

In India there is a most pleasant wood,
In which two birds are bound together.
One is of a snowy white; the other is red.
They bite each other, and one is slain
And devoured by the other.
Then both are changed into white doves,
And of the Dove is born a Phoenix,
Which has left behind blackness and foul death,
And has regained a more glorious life.
This power was given it by God Himself,
That it might live eternally, and never die.

Holy War 396
Center 835
P. State 563
Grace 552
Holiness 902

Inv. 1017

It gives us wealth, it preserves our life,
And with it we may work great miracles,
As also the true Philosophers do plainly inform us.

Abraham Lambspring

That which dies is not the ruling part of us, but the subject laity, and for so long as the latter will not repent and acknowledge its perversion, so long will it be held by death.

Philo

What could begin to deny self, if there were not something in man different from self?

William Law

Reality 773

There are two natures, one self-existent, and the other ever in want.

Plato (*Philebus,* 53 D)

Flight 944

Although your intellect is flying upward, the bird of your conventional notions is feeding below.

Rûmî

Suff. 128

Corporeal nature is your wife, O Soul, and intellect (*νοῦς*) is your father; and a blow given by your father's hand is better than a kiss given by your wife.

Hermes

Rev. 967

It is always right that the superior should rule, and the inferior be ruled; and Mind is superior to sensibility.

Philo

And the Spirit of the Lord will come upon thee, and thou shalt prophesy with them, and shalt be turned into another man.

I. Samuel, x. 6

Creation 42

Man, in regard to his corporeal nature, stands at the lowest point of degradation; nevertheless, in regard to his spiritual nature, he is at the summit of nobility. He takes the impress of everything to which he directs his attention, and assumes the colour of everything to which he approaches.

Jâmî

That light of the heart or attraction to God . . . has the same contrariety to all vices of the heart that light has to darkness, and must either suppress or be suppressed by them.

William Law

Renun. 146
Death 208

It is not possible, my son, to attach yourself both to things mortal and to things divine.

Hermes

No man can serve two masters: for either he will hate the one, and love the other; or else he will hold to the one, and despise the other. Ye cannot serve God and mammon.

St Matthew, vi. 24

The better (*śreyas*) is one thing, and the pleasanter (*preyas*) quite another.
Both these, of different aim, bind a person.
Of these two, well is it for him who takes the better;
He fails of his aim who chooses the pleasanter.

Kaṭha Upanishad, ii. 1

Renun. 156

The happy life is thought to be virtuous; now a virtuous life requires exertion, and does not consist in amusement. . . .

If happiness is activity in accordance with virtue, it is reasonable that it should be in accordance with the highest virtue; and this will be that of the best thing in us. . . . This activity is contemplative.

Aristotle

Beauty 676

Contem. 547

Where your treasure is, there will your heart be also.

St Matthew, vi. 21

Conform. 166

'O Râbi'a, thou hast a desire and I have a desire. I and thy desire cannot dwell together in a single heart.'

'Aṭṭâr

Allâh coineth a similitude: A man in relation to whom are several part-owners, quarrelling, and a man belonging wholly to one man. Are the two equal in similitude? Praise be to Allâh! But most of them know not.

Qur'ân, xxxix. 29

Introd. *Holy War*

Every kingdom divided against itself is brought to desolation; and every city or house divided against itself shall not stand.

St Matthew, xii. 25

Saṁsâra is just one's own thought;
With effort he should cleanse it, then.
What is one's thought, that he becomes;
This is the eternal mystery.

Maitri Upanishad, vi. 34

When the mind turns instinctively and always to God, it will be cleansed of lust, greed and wrath.

Swami Ramdas

Inv. 1036

All that we are is the result of what we have thought: it is founded on our thoughts, it is made up of our thoughts. If a man speaks or acts with an evil thought, pain follows him, as the wheel follows the foot of the ox that draws the carriage.

Action 329

P. State 563

All that we are is the result of what we have thought: it is founded on our thoughts, it is made up of our thoughts. If a man speaks or acts with a pure thought, happiness follows him, like a shadow that never leaves him.

Dhammapada, I. 1, 2

Center 835

May the outward and inward man be at one.

Plato (*Phaedrus*, 279 C)

Repentance, Renewal and Awakening

P. State 579 Introd. Conform.

As the lesser mysteries are to be delivered before the greater, thus also discipline must precede philosophy.

Pythagoras

One must be pure in heart to enter into the life of the spirit and follow the yogas.

Srimad Bhagavatam, XI. xiv

The best preparation for this study is, in my judgment, a diligent amendment of heart and life.

The Sophic Hydrolith

The Soul will not reach the true concepts of things which are separated from bodies, if it has not separated itself from the body through purification of habits and effort of speculation.

Marsilio Ficino

Beauty 676 Orthod. 285

This (Knowledge) should be imparted only to him whose mind has been pacified, who has controlled his senses and is freed from all defects, who has practised the duties (enjoined by the scriptures) and is possessed of good qualities, who is always obedient (to the teacher) and aspires after liberation only.

Śrî Śankarâchârya

Introd. Orthod.

Come, whoever is clean of all pollution and whose soul has not consciousness of sin. Come, whosoever hath lived a life of righteousness and justice. Come all ye who are pure of heart and of hand, and whose speech can be understood. Whosoever hath not clean hands, a pure soul, and an intelligible voice must not assist at the Mysteries.

Eleusinian Mysteries: Proclamation of Archon Basileus before celebrating the Greater Mysteries

For it is absurd that a man should be forbidden to enter the temples save after bathing and cleansing his body, and yet should attempt to pray and sacrifice with a heart still soiled and spotted.

Philo

Be it known to all manner of people in this wretched dwelling-place of exile abiding, that no man may be imbued with love of endless life, nor be anointed with heavenly sweetness, unless he truly be turned to God. It behoves truly he be turned to Him, and from all earthly things be altogether turned in mind, before he may be expert in the sweetness of God's love, even in little things.

Richard Rolle

For wit thou well, a bodily turning to God without the heart following, is but a figure and a likeness of virtues and no soothfastness.

Walter Hilton

Renun. 146

As a thread does not pass through an unbored pearl, so the teachings of the Queen did not influence the heart of the King. O Rama-ji, unless the pupil himself meditates and reflects and puts the teaching into practice with interest and perseverance, even the words of *Brahma* Himself fall flat on his ears. The reason is that *Atman* knows Itself and is not the object of reason or the senses.

Yoga Vasishtha, Story of Queen Chudala, II

Knowl. 745 749

True reason, such as man had in the beginning, cannot be had or acquired by any man, who has not first been purified and become passionless. Of purity we are deprived by unreasoning tendencies of the senses, and of passionlessness—by the corrupted state of the flesh.

St Gregory of Sinai

Creation 42

We must not be deceived. Without inward prayer there is no conquest over the flesh, and without this conquest over the flesh, no true inward prayer; and without this the one and the other is no conversion, no true internal life, no perfection or Christianity.

Johannes Kelpius

Only those who love the truth, who have conquered temptations, worship the Gods, are perfect masters of themselves and are accustomed to following an appropriate regimen and diet, only these can undertake alchemical operations.

Rasaratnasamuccaya, VII. 30

Renun. 136

Devotee: What does Maharshi say about *Hatha Yoga* or *Tantric* practices?

Maharshi: Maharshi does not criticise any of the existing methods. All are good for the purification of the mind. Because the purified mind alone is capable of grasping his method and sticking to its practice.

Sri Ramana Maharshi

Introd. *Conform.* *Wr. Side* 444

First purify the mind by observing right conduct, then practice concentration; this without the other will amount to nothing. There are occultists who take up concentration, while their moral worth leaves much to be desired; that is why they make no spiritual progress.

Swami Sivananda

Create in me a clean heart, O God; and renew a right spirit within me.
Cast me not away from thy presence; and take not thy holy spirit from me.

Psalm LI. 10, 11

Realiz. 868

The spark of the soul, which is sent there by God . . . is his light striking down from above, the reflection (or image) of his divine nature and ever opposed to anything ungodly: not a power of the soul, as some theologians make it, but a permanent tendency to good; aye, even in hell it is inclined to good. According to the masters, this light is of the nature of unceasing effort; it is called synderesis, that is to say, a joining to and turning from. It has two works. One is remorse for imperfection. The other work consists in ever more invoking good and bringing it direct into the soul, even though she be in hell.

Eckhart

Action 340 *Grace* 552 *Contem.* 542 Introd. *Pilg.*

Above all things beware of taking this desire of repentance to be the effect of thy own natural sense and reason, for in so doing thou losest the key of all the heavenly treasure that is in thee, thou shuttest the door against God, turnest away from Him, and thy repentance (if thou hast any) will be only a vain, unprofitable work of thy own hands, that will do thee no more good than a well that is without water. . . . When, therefore, the first spark of a desire after God arises in thy soul, cherish it with all thy care, give all thy heart into it, it is nothing less than a touch of the divine loadstone that is to draw thee out of the vanity of time into the riches of eternity. Get up, therefore, and follow it as gladly as the Wise Men of the East followed the star from Heaven that appeared to them. It will do for thee as the star did for them: it will lead thee to the birth of Jesus, not in a stable at Bethlehem in Judea, but to the birth of Jesus in the dark centre of thy own fallen soul.

William Law

Infra 488

The process is not the turning over of an oyster-shell, but the turning round of a soul passing from a day which is little better than night to the true day of being, that is, the ascent from below, which we affirm to be true philosophy.

Plato (*Republic,* VII. 521 C)

Pilg. 366

Regeneration or the renewal of our first birth and state is something entirely distinct from this first sudden conversion or call to repentance; . . . it is not a thing done in an instant, but is a certain process, a gradual release from our captivity and disorder, consisting of several stages and degrees, both of death and life, which the soul must go through before it can have thoroughly put off the old man.

William Law

Center 816

The soul is like the eye: when resting upon that on which truth and being shine, the soul perceives and understands and is radiant with intelligence: but when turned towards the twilight of becoming and perishing, then she has opinion only, and goes blinking about, and is first of one opinion and then of another, and seems to have no intelligence.

Plato (*Republic,* VI. 508 D)

Knowl. 734

That which is night to all beings, therein the self-subjugated remains awake; and in that where all beings are awake, that is night for the knower of Self.

Bhagavad-Gîtâ, II. 69

Holiness 934
Action 346

Now wilt thou be a magus? Then thou must understand how to change the night again into the day.

Boehme

By light we lose light.

Shakespeare (*Love's Labour's Lost,* V. ii. 377)

For at one and the same time the human intelligence is illuminated with regard to divine things and darkened in respect of human things.

Richard of Saint-Victor

For when the light of God shines, the human light sets.

Philo

Center 833

Dhu 'l-Nûn the Egyptian says: 'Ordinary men repent of their sins, but the elect repent of their heedlessness,' because ordinary men shall be questioned concerning their outward behaviour, but the elect shall be questioned concerning the real nature of their conduct. Heedlessness, which to ordinary men is a pleasure, is a veil to the elect.

Hujwîrî

Realiz. 859
Sin 64

Allâh is the Protecting Friend of those who believe. He bringeth them out of darkness into light. As for those who disbelieve, their patrons are false deities. They bring them out of light into darkness.

Qur'ân, II. 257

Repentance is but a kind of table-talk, till we see so much of the deformity of our inward nature as to be in some degree frightened and terrified at the sight of it. There must be some kind of an earthquake within us, something that must rend and shake us to the bottom, before we can be enough sensible either of the state of death we are in or enough desirous of that Saviour, who alone can raise us from it. . . . Sooner or later repentance must have a broken and a contrite heart; we must with our blessed Lord go over the brook Cedron, and with Him sweat great drops of sorrow before He can say for us, as He said for Himself: 'It is finished.'

William Law

Sin 66
Infra 488
Humility 191

I am not come to call the righteous, but sinners to repentance.

St Matthew, IX. 13

These things I learnt at last, after I had repented concerning Joseph. For true repentance after a godly sort destroyeth ignorance, and driveth away the darkness, and enlighteneth the eyes, and giveth knowledge to the soul, and leadeth the mind to salvation. And those things which it hath not learnt from man, it knoweth through repentance.

The Testament of Gad, v. 7–9

P. State 573

He saved us, by the washing of regeneration, and renewing of the Holy Ghost.

Titus, III. 5

Be renewed in the spirit of your mind.

Ephesians, IV. 23

Great is *Teschubah* (repentance), for it brings healing to the world. Great is *Teschubah*, for it reaches to the Throne of Glory. Great is *Teschubah*, for it brings on the Redemption.

Talmud

Supra Introd.

Awake, Mother! Awake! How long Thou hast been asleep
In the lotus of the Mulâdhâra!
Fulfil Thy secret function, Mother:
Rise to the thousand-petalled lotus within the head,
Where mighty Śiva has His dwelling;
Swiftly pierce the six lotuses
And take away my grief, O Essence of Consciousness!

Bengali Hymn to Kâli

Repentance is a great understanding.

The Shepherd of Hermas

The Inversion of Values

I turned me to the right hand, and set my mind on the other pole.

Dante (*Purgatorio*, I. 22)

Sin 66

Is not the whole of human life turned upside down; and are we not doing, as would appear, in everything the opposite of what we ought to be doing?

Plato (*Gorgias*, 481 C)

Lower thy head, proud Sicamber; burn what thou hast adored, and adore what thou hast burned!

St Remigius, to Clovis

The carnal mind is enmity against God.

Romans, VIII. 7

Render heavy what is light and light what is heavy; make earth with air and air with earth; transform fire into water and water into fire: such is the Art.

Hermetic Formula

Make the blind seeing, and the seeing blind; thus wilt thou obtain the magistery.

Avicenna

Supra 484

The body must become water, and the water body.

The Glory of the World

Invert the elements, and you will find what you seek.

The Book of Alze

Friendship of the world is enmity with God.

James, IV. 4

Ṣûfism is enmity to this world and friendship with the Lord.

Abu 'l-Ḥusayn al-Nûri

All that we are by nature is in full contrariety to this divine love, nor can it be otherwise; a death to itself is its only cure and nothing else can make it subservient to good; just as darkness cannot be altered or made better in itself or transmuted into light, it can only be subservient to the light by being lost in it and swallowed up by it.

William Law

Jesus (Peace be upon him!) said, 'The love of this world and of the next cannot stay in the heart of a believer, just as water and fire cannot stay in one vessel.'

Christ in Islâm

Oh Soul! Reverse this figure of yours.

Marsilio Ficino

Where the way is hardest, there go thou; and what the world casteth away, that take thou up. What the world doth, that do thou not; but in all things walk thou contrary to the world. So thou comest the nearest way to that which thou art seeking.

Boehme

Suff. 126b

Our life must be converted into its contrary. We must unlearn those things which we have learned; by learning them we have hitherto not known ourselves. We must learn those things we have neglected; without knowing them we cannot know ourselves. We must like what we neglect, neglect what we like, tolerate what we flee, flee what we follow. We must cry about the jest of fortune; jest about its tears.

Marsilio Ficino

Knowl. 734 755 *Realiz.* 859

If one cannot realize the topsy-turvy-ness and topsy-turvy-ness again of the technique of *hsüan* (mysterious), how can he understand that it is possible to cultivate lotus flowers in an atmosphere of fire? Bring the White Tiger back home whereby it is fed. A glittering pearl will be born which has a shape as round as the moon. Watch with care the fire and the season of the medicinal pot, and let whatever happens occur naturally and without hurry. The *tan* (medicine) will ripen at the moment when all of the *yin* is eliminated—and (by this achievement a man) can escape out of the cage (of death) with a life of ten thousand years.

Judgment 266
P. State 590
Conform. 170

Chang Po-tuan

Sublime virtue is infinitely deep and wide.
It goes reverse to all things;
And so it attains perfect peace.

Tao Te Ching, LXV

Realiz. 870

The body must become spirit, and the spirit body.

The Glory of the World

Transmute the nature, and you will find what you want. For in our Magistery we obtain first from the gross the subtle, or the spirit; then from the moist the dry, *i.e.*, earth from water. Thus we transmute the corporeal into the spiritual, and the spiritual into the corporeal, the lowest into the highest, and the highest into the lowest.

Alphidius

Fac fugiens fixum & fixum fugiens.

Pythagoras

The fixed becomes volatile, and the volatile fixed, by dissolution and coagulation.

The Glory of the World

If the cask is to hold wine, its water must first be poured out.

Eckhart

Ecstasy 636
Supra 484
Symb. 317

Let reason go. For His light
Burns reason up from head to foot. . . .
As the light of our eyes to the sun,
So is the light of reason to the Light of Lights.

Shabistarî

Peace 700

As often as the mind is outgoing, so often it should be turned within.

Sri Ramana Maharshi

Humility 199

Thou must be emptied of that wherewith thou art full, that thou mayest be filled with that whereof thou art empty.

St Augustine

These Imperfect Bodies are not reducible to *Sanity* and *Perfection*, unless the contrary be operated in them; that is, the Manifest be made Occult, and the Occult be made Manifest.

Geber

He must increase, but I must decrease.

St John, III. 30

I beseech you the executioners, crucify me thus, with the head downward and not otherwise: . . . concerning which the Lord saith in a mystery: Unless ye make the things of the right hand as those of the left, and those of the left as those of the right, and those that are above as those below, and those that are behind as those that are before, ye shall not have knowledge of the kingdom.

Symb. 306

The Martyrdom of St Peter

As the first Adam opened the gates of Hell, so the second opened the gates of Paradise.

Jacobus de Voragine

Turn to thy heart, and thy heart will find its Saviour, its God within itself.

William Law

Center 816

One does not need far-reaching vision to see into the heavens;
Do but turn from the world, and look: thus will it be accomplished.

Angelus Silesius

Restraint of the out-going mind and its absorption in the Heart is known as introversion (*antar-mukha-drshti*).

Sri Ramana Maharshi

Contem. 536

Man's best chance of finding God is where he left him.

Eckhart

The mind, in truth, is for mankind
The means of bondage and release:
For bondage, if to objects bound;
From objects free—that's called release.

Maitri Upanishad, VI. 34

Conform. 166
Renun. 152b

One cannot completely get rid of the six passions: lust, anger, greed, and the like. Therefore one should direct them to God. . . . A man cannot see God unless he gives his whole mind to Him.

Sri Ramakrishna

Sin 62
Realiz. 870

To the eye evil presents itself no less than good. The ear is importuned by one as well as the other and so with the other senses. Wherefore it behoves thee strictly to confine thyself and with all diligence to those things which are good.

Eckhart

Supra 480

Judgment 239
Beauty
663

The repentance of fear is caused by revelation of God's majesty, while the repentance of shame is caused by vision of God's beauty.

Hujwîrî

Suff.
126b

Before a man can find God, . . . all his likings and desires have to be utterly changed. . . . All things must become as bitter to thee as their enjoyment was sweet unto thee.

Tauler

He who loves that which is not expedient, will not love that which is expedient.

Sextus the Pythagorean

Death
220

Forgetfulness of self is remembrance of God.

Bâyazîd al-Bisṭâmî

Because in the school of the Spirit man learns wisdom through humility, knowledge by forgetting, how to speak by silence, how to live by dying.

Tauler

Void
724

Union with God is separation from all else, and separation from all else is union with Him.

Abu 'l-Ḥusayn al-Nûrî

The contraction (*qabḍ*) of hearts consists in the expansion (*basṭ*) of souls, and the expansion of hearts in the contraction of souls.[1]

Bâyazîd al-Bisṭâmî

According to the Sages, no body is dissolved without the coagulation of the spirit.

The Glory of the World

Repentance means, that thou shouldst be unto God a face without a back, even as thou hast formerly been unto Him a back without a face.

Ibrâhîm al-Daqqâq

Thus a man must abscond from his senses, invert his faculties and lapse into oblivion of things and of himself.

Eckhart

Center
841

Take one step out of thyself, that thou mayst arrive at God.

Abû Saʿîd ibn Abi 'l-Khayr

P. State
583

O ye who believe! Turn unto Allâh with a true turning (*tawbah*). It may be that your Lord will remit from you your evil deeds and bring you into Gardens underneath which rivers flow.

Qur'ân, LXVI. 8

[1] 'My Shaykh used to say that both *qabḍ* and *basṭ* are the result of one spiritual influence, which descends from God on Man' (Hujwîrî: *Kashf*, p. 375).

Arise, commence a new life, turn towards the Doctrine of the Buddha; trample down the hosts of the Lord of Death as an elephant doth a house of mud.

Udâna

Transcendental Intelligence is the inner state of self-realisation of Noble Wisdom. It is realised suddenly and intuitively as the 'turning about' takes place in the deepest seat of consciousness; it neither enters nor goes out—it is like the moon seen in water.

Lankavatara Sutra, VI

Realiz. 873
P. State 590

First we look at the hills in the painting,
Then we look at the painting in the hills.

Li Liweng

Beauty 670

It is a sure sign of this light (of grace) when of his own free will a person turns from mortal things to the highest good, God namely. We are in duty bound to love him for conferring on the soul such great perfection. When she has reached her limit of endeavour then will as such is free to leap over to that gnosis which is God himself. A somersault which lands the soul at the summit of her power.

Eckhart

Holiness **900**
Creation **42**

A Mutual Turning

Turn ye unto me, saith the Lord of hosts, and I will turn unto you.

Zechariah, I. 3

Resist the devil, and he will flee from you.
Draw nigh to God, and he will draw nigh to you.

James, IV. 7, 8

'O Moses! Be for Me as I desire, and I will be for thee as thou desirest.'

Ibn al-'Arif

'I (Nature) will deal with you as you deal with me.'

John A. Mehung

Remember Me, I will remember you.

Qur'ân, II. 152

Inv. passim

'Do thou think of Me, and I will ever think of thee.'

St Catherine of Siena

God is ready to receive us, if we Repent and turn unto him.

Benjamin Whichcote

Center 841

When a sentient being worships the Buddha, the Buddha sees him. When men call upon Him, He hears them. When they think upon Him, He thinks upon them.

Zendô

Those who worship Me with devotion, they are in Me and I am in them.

Bhagavad-Gîtâ, IX. 29

Peace 694

As the soul rests in God so God reposes in her.

Eckhart

Knowl. 749

The wise man follows God, and God follows the soul of the wise man.

Sextus the Pythagorean

Conform. 180

If a man obeyeth God, God will obey him.

The Abbâ of Îlîû

God loves to meet those who love to meet Him, and dislikes to meet those who dislike to meet Him.

Muhammad

Humility 191

Realiz. 859

Who is he who pleaseth God? He whom God hath pleased. God cannot displease Himself. May He please thee also, that thou mayest please Him. But He cannot please thee, except thou hast displeased thyself. But if thou displeasest thyself, remove thine eyes from thyself. For why dost thou regard thyself? For if thou truly regard thyself, thou findest in thee what will displease thee; and thou sayest to God, 'My sin is always before me' (Ps. L, 5—Douay). Let thy sin be before thee, that it may not be before God; and refuse to be before thyself, that thou mayest be before God. For as we wish that God may not turn away His face from us, so do we wish that He may turn His face from our sins.... If thou turn not away thy face from them, thou art thyself angry with thy sins, and if thou turn not away thy face from thy sins, thou dost confess them and God will pardon them.

St Augustine

Charity 608

'Show thy repentance (*tawbah*) to everything, and everything will ask pardon for thee.'

Niffarî

No Repentance in God

He (God) is not a man, that he should repent.

I. Samuel, XV. 29

God does not see in time nor is his outlook subject to renewal.

Eckhart

Creation 31

Here we repent not, but smile; not at the sin, which cometh not again to mind, but at the Worth that ordered and provided.

Dante (*Paradiso,* IX. 103)

At the sight of the Holy Ghost her (the soul's) sins are blotted out and she forgets herself and things.

Eckhart

I am not now the man I was.

Hermes

Realiz. 873

He who repents of sin is even as he who has no sin.

Muhammad

Sin 57

Lo! those who say: Our Lord is Allâh, and thereafter walk aright, there shall no fear come upon them neither shall they grieve.

Qur'ân, XLVI. 13

Him (who knows the imperishable Âtma) these two do not overcome—neither the thought 'Hence I did wrong,' nor the thought 'Hence I did right.' Verily, he overcomes them both. What he has done and what he has not done do not affect him.

Bṛihad-Âraṇyaka Upanishad, IV. iv. 22

M.M. 978

To have sinned is not sin once it is repented of. . . . God is a present God: as he finds thee so he takes thee and accepts thee—not what thou mayst have been but what thou art this instant.

Eckhart

Center 838

I remember not that I e'er estranged me from you, nor have I conscience thereof that gnaws me.

Dante (*Purgatorio,* XXXIII. 91)

Sahl b. ʿAbdallâh (al-Tustarî) and others believe that repentance consists in not forgetting your sins, but always regretting them, so that, although you have many good works to your credit, you will not be pleased with yourself on that account; since remorse for an evil action is superior to good works. . . . Junayd and others take the opposite view, that repentance consists in forgetting the sin. They argue that the penitent is a lover of God, and the lover of God is in contemplation of God, and in contemplation it is wrong to remember sin, for remembrance of sin is a veil between God and those who contemplate Him.

Contem. 523 *Renun.* 143

Hujwîrî

The meaning of repentance is, that thou shouldst repent of repentance.

Abu Muḥammad Ruwaym

He whose mind has found repose in holy meditation, has no cause for repentance.

Peace 700

Yoga-Vasishtha

BOOK TWO:
MERCY – LOVE – CONTEMPLATION

PART III: *Life – Love*

FAITH

If I forget thee, O Jerusalem, let my right hand forget her cunning.
If I do not remember thee, let my tongue cleave to the roof of my mouth; if I prefer not Jerusalem above my chief joy.

Psalm CXXXVII, 5–6

'To have the strength of one's convictions', far from being an empty phrase, expresses the recognition of a power that is altogether effective, whether for good or for evil.

The logical necessity of faith is not difficult to grasp: the spiritual domain is invisible to the temporal, and as long as the temporal is the domain that rules with us, faith is indispensable for the perseverance required if one is to break through the specious 'reality' one knows to the as yet unknown Reality. 'Through familiarity with bodies one may very easily, though very hurtfully, come to believe that all things are corporeal' (St Augustine: *Contra academicos,* XVII. 38).

'In the discussion of Faith it is too often overlooked that the greater part of our knowledge of "things", even of those by which our worldly actions are regulated, is "authoritative"; most, indeed, even of our daily activities would come to an end if we did not believe the words of those who have seen what we have not yet seen, but might see if we would do what they have done, or go where they have been; in the same way those of the Buddhist neophyte would come to an end if he did *not* "believe" in a goal not yet attained. Actually, he believes that the Buddha is telling him the truth, and acts accordingly (*Dîgha-Nikâya,* ii. 93). Only the Perfect Man is "faithless", in the sense that in his case knowledge of the Unmade has taken the place of Faith (*Dhammapada,* 97), for which there is no more need' (Coomaraswamy: *Gotama the Buddha,* p. 23).

Holiness 924

'Faith is nothing less than the consent of our whole being to the Truth, whether we have a direct intuition of it, or only an indirect idea. It is an abuse of language to reduce "faith" to the level of "belief", since the truth is precisely the inverse: belief—or theoretical knowledge—must be changed into a faith which "removes mountains". For the apostles there was practically no difference between an idea and its spiritual application; they did not separate theory from realization—which explains the use of the word "love" to designate all conformity whatsoever to the Divine Truth' (Schuon: *Perspectives spirituelles,* p. 170).

Knowl. 745

'Faith as belief and faith as gnosis has this difference, that the obscurity of faith in the case of the ordinary believer lies in the intelligence, whereas in the case of the metaphysician it lies in the will, in the participation of being: for here the seat of faith is the heart, not the mind, and the obscurity comes from our state of individuation, not from a congenital lack of intelligence. The faith of the sage—or the "gnostic"—has two veils: the body and the *ego*; it is not the intellect they veil, but ontological consciousness' (Schuon: 'Caractères de la mystique passionnelle', *Études Trad.*, 1953, p. 260).

Trust in God

Take no thought for your life, what ye shall eat, or what ye shall drink; nor yet for your body, what ye shall put on. Is not the life more than meat, and the body than raiment?

Action 346

... Consider the lilies of the field, how they grow; they toil not, neither do they spin:

And yet I say unto you, That even Solomon in all his glory was not arrayed like one of these.

Wherefore, if God so clothe the grass of the field, which to day is, and to morrow is cast into the oven, shall he not much more clothe you, O ye of little faith?

... Your heavenly Father knoweth that ye have need of all these things.

But seek ye first the kingdom of God, and his righteousness, and all these things shall be added unto you.

Reality 775
P. State 563

St Matthew, VI. 25 ff.

Those who surrender their hearts and souls to God, those who are devoted to Him and have taken refuge in Him, do not worry much about money. As they earn, so they spend. The money comes in one way and goes out the other. This is what the *Gitâ* describes as 'accepting what comes of its own accord'.

Sri Ramakrishna

If ye trusted in God as ye should, He would sustain you even as He sustains the birds, which in the morning go forth hungry, and return in the evening filled.

Muhammad

After having listened to him (William Penn) with great patience, they (the Indians) answered: 'You bid us believe in the Creator and Preserver of Heaven and Earth, though you do not believe in Him yourself, nor trust in Him. For you have now made your own the land we held in common amongst ourselves and our friends. You now take heed, night and day, how you may keep it, so that no one may take it from you. Indeed, you are anxious even beyond your span of life, and divide it among your children. This manor for this child, that manor for that child. But we have faith in God the Creator and Preserver of Heaven and Earth.... He preserveth us, and we believe and are sure that He will also preserve our children after us, and provide for them, and because we believe this, we bequeath them not a foot of land.'

P. State 583

Johannes Kelpius

Heaven doth all things gratis give.

Thomas Norton

Our hope and trust always have been, are, and will be, in Him alone.

Michael Sendivogius

The Lord is always good and full of love; He never abandons those who put their trust in Him.

Swami Ramdas

God gave (the Sages) this gift (of the Stone) that they might not be hindered in their researches by poverty, or driven to flatter the rich for the sake of gain, and thus become contemptible, and as a jest or by-word in His sight.

The Sophic Hydrolith

I perceived that God takes in hand the affairs of them that put their trust in Him and does not let their tribulation come to naught.

Dhu 'l-Nûn

Inv. 1017

Giving up all Dharmas (righteous and unrighteous actions), come unto Me alone for refuge.

Bhagavad-Gitâ, XVIII. 66

Q: How can a European best try to fulfil the conditions necessary for Japa Yoga, in the midst of all the confused life and activity in Europe?

P. State 583 *Conform.* 166 *Action* 329

Ramdas: If we depend upon God, everything becomes possible. The conditions prevailing in Europe are more or less the same as those prevailing in other parts of the world. It is not external conditions that deter us from trying to realise God, but our own mind. If the mind is averse to the practice of constantly remembering God, then the external circumstances seem to hamper us. If the mind longs to realise God, and pines for His contact, the circumstances automatically prove to be conducive and helpful. Even difficulties that face us will become helpers in our path, provided our aspiration for God is very keen. In a wonderful way, the very obstacles become our aids, and people around us will cooperate with us in our efforts to reach Him. In strange and mysterious ways, God Himself provides us with all the things necessary for our spiritual progress.

Swami Ramdas

Sin 57 *Holiness* 935

Amida[1] does not hate a man, however deeply stained with sin he may be. And though the times be ever so degenerate, let him not doubt. For even sentient beings, who will live in the period after the Law has perished, can be born into the Pure Land. How much more men of our own times!

Hônen

Reality 775 *Symb.* 306

We easily believe and hope that which we desire. The *Divine Love* hath a *complacency* in all things, as it comprehendeth them in their Divine *Root*. It hath a *good will* to all things, as they stand in the same *Divine Root* with it self. From this *Complacency*, this *Desire*, this *Divine* Root it *believeth*, it *hopeth* all things. It believeth all things to be *Divine Tabernacles*, like that in the Wilderness, which though moving through Desarts, through a *Land of Graves*, through a Land of fiery Serpents, yet answer to their *Pattern* on the top of the

[1] A name of the Buddha, invoked in Japan.

Mount; though covered with a course Tent, exposed to the fury of the Sun, and tempests in the midst of Clouds, of dust; yet are all-glorious within, composed of rich materials, bearing a Divine Figure, filled with the Divine Presence and Glory. It hopeth all things, light in the midst of darkness, a flourishing Garden of Lillies and Roses, in a ground covered, and bound up with all the darknesses and rigours of the hardest Winter; a treasure of *Honey-Combs* in the body of a *Lion*.

Pilg. 379

Peter Sterry

The likeness of those who spend their wealth in Allâh's way is as the likeness of a grain which groweth seven ears, in every ear a hundred grains. Allâh giveth increase manifold to whom He will. Allâh is All-Embracing, All-Knowing.

Qur'ân, II. 261

So will (our Lord) stir other men in spirit to give us our needful things that belong to this life, as meat and clothes and all these other, if he see that we will not leave the work of his love for business about them. And this I say in confusion of their error who say that it is not lawful for men to set them to serve God in contemplative life, except they be secure beforehand of their bodily necessaries. For they say that *God sendeth the cow, but not by the horn.* And truly they say wrong of God, as they well know. For trust steadfastly, whatsoever thou be that truly turnest thee from the world unto God, that one of these two shall God send thee, without business of thine own: and that is, either abundance of necessaries, or else strength in body and patience in spirit to bear need. What then recketh it which a man have? for all come to one in true contemplatives. And whoso is in doubt of this, either the devil of hell is in his breast and reaveth him of belief, or else he is not yet truly turned to God as he should be, be he never so clever, or show he never so holy reasons to the contrary, whatsoever he be.

Action 346

Conform. 170

Renun. 139

The Cloud of Unknowing, XXIII

Thy heavenly Father will and must provide thee well, yea, wert thou hidden in a rock.

Tauler

The trew men search and seeke all alone
In hope to finde our delectable stone,
And for that thei would that no Man shulde have losse,
They prove and seeke all at their owne Coste;
Soe their owne Purses they will not spare,
They make their Coffers thereby full bare,
With greate Patience thei doe proceede,
Trusting only in *God* to be their speede.

Infra 509

Thomas Norton

Whoso believeth in his Lord, he feareth neither loss nor oppression.

Renun. 158

Qur'ân, LXXII. 13

For His mercy is infinite, and He never forsakes those who put their trust in Him; with Him there is no respect of persons, nor does He despise the humble and contrite heart.

Michael Sendivogius

Whosoever keepeth his duty to Allâh, Allâh will appoint a way out for him,

And will provide for him in a way that he cannot foresee. And whosoever putteth his trust in Allâh, He will suffice him. Lo! Allâh bringeth His command to pass. Allâh hath set a measure for all things.

Qur'ân, LXV. 2, 3

Pilg. 366

He that believeth in me, though he were dead, yet shall he live.

St John, XI. 25

Thinketh man that We shall not assemble his bones?
Yea, verily. We are able to restore the very tips of his fingers!

Qur'ân, LXXV. 3, 4

Jesus (Peace be upon him!) said, 'Do not worry about to-morrow's food, for if to-morrow is one of your periods your provisions will come in it along with your periods; and if it is not one of your periods, do not worry about other people's periods.'

Christ in Islâm

If thou but hast the help of God, the dry wood of the forest will bear thee fruit, until thou shalt say, 'Enough'; but if He above be not with thee, then even the garden-trees will not bring forth, their fruit they will deny, until thou shalt ask, 'Where are they?'

Shekel Hakodesh, 154

Orthod. 296

Among the gifts that God grants to His slave, when he obeys Him and submits to His service and follows this way all his life . . . (is) that He will guarantee for him the means of livelihood, sending him along from state to state, without effort or anxiety.

Ibn al-ʿArîf

Judgment 239

For the Lord doth not forsake them that fear Him,
Neither in darkness, nor in bonds, nor in tribulations, nor in necessities.

The Testament of Joseph, II. 4

Renun. 146

'Abraham trusted in God' (Gen. XV. 6). To trust in God alone and join no other with Him is no easy matter, by reason of our kinship with our yoke-fellow, mortality, which works upon us to keep our trust placed in riches and repute and office and friends and health and strength and many other things. To purge away each of these, to distrust created being, which in itself is wholly unworthy of trust, to trust in God, and in Him alone, even as He alone is truly worthy of trust—this is a task for a great and celestial understanding which has ceased to be ensnared by aught of the things that surround us.

Illusion 109 *Suff.* 126b

Philo

Think what it implies to have so little confidence in that God who is able to do all things, when ye are striving, and toiling, and wearing yourselves out with anxiety, as if you meant to live for ever. All this comes from that evil principle of covetousness.... St Peter says: 'Cast all your care upon God, for He careth for you' (I. Pet. V. 7). This carefulness concerning outward things works a man three great injuries. It blinds his reason and good sense; it quenches the fire of love, and destroys all its fervour and heat; and it blocks up the ways of secret access to God. It is like a noxious vapour, or thick smoke, that rises up and chokes a man's breath.... Not that we should tempt God; for it is our duty to exercise a reasonable prudence in providing such things as are right, to the supply of our necessities and those of others, and profitable to ourselves and the community, and to see that everything be done in a discreet and seemly manner. But that which is your end when you sit and meditate in the church, should be likewise your end when you are busied in all the affairs of daily life....

Contem. 542

Now some may ask, how we can say that God forsakes none that trust Him, seeing that He often permits good men to suffer great poverty and affliction. This He does, as Bishop Albert says, for three causes: the first, that He may try them, and see whether they utterly believe and trust Him; thus God often suffers men to be brought into distress that He may teach them submission, and then succours them that they may perceive His hand and His friendship and help; in order that their love and gratitude may increase from that time forth, and they may draw closer to God and become dearer to Him. Or again, God will by these troubles shorten their purification hereafter; or again, He sends them distress for a judgment on those who might relieve them and do it not. Therefore, children, seek first the Kingdom of God, which is God Himself, and nought else.

Introd. *Suff.*

Beauty 663

Tauler

When He gives to thee He has thee witness His Benevolence, and when He denies thee He has thee witness His Constraint; and in either case He is acquainting thee with Himself and approaching thee by means of His Kindness.

Ibn 'Aṭâ'illâh

Love 618

Belief in the Unseen

From the invisible he made all things visible, himself being invisible.

2 Enoch, XLVIII. 5

Creation 48
Beauty 670

Faith is the belief of the heart in that knowledge which comes from the Unseen.

Muḥammad b. Khafîf

Faith is the peculiar power of the consciousness which makes one take his resort in what is denoted by it.

Swami Sivananda

The look which goes from God into the soul is the beginning of faith whereby I believe things not revealed to me.

Dionysius

Faith is the substance of things hoped for, the evidence of things not seen. . . .

Through faith we understand that the worlds were framed by the word of God, so that things which are seen were not made of things which do appear. . . .

He that cometh to God must believe that he is, and that he is a rewarder of them that diligently seek him.

Death 215 *Wr. Side* 464

By faith Noah, being warned of God of things not seen as yet, moved with fear, prepared an ark to the saving of his house; by the which he condemned the world, and became heir of the righteousness which is by faith.

Hebrews, XI. 1, 3, 6, 7

Jesus (Peace be upon him!) said, 'Blessed is he who abandons a present desire for a distant promise which he has not seen.'

Christ in Islâm

Center 847

Happy the man who divorceth his desire for the glory of that which cometh not nor appeareth; for no man will regret having stifled, to the glory of God, the desire that assailed him!

Shekel Hakodesh, 134

Metanoia 484

For it is to this end we hearken to that which we are to believe, before we see it, that by believing we may purify the heart, whereby we may be able to see.

St Augustine

Jesus said unto him, If thou canst believe, all things are possible to him that believeth.

And straightway the father of the child cried out, and said with tears, Lord, I believe; help thou mine unbelief.

St Mark, IX. 23, 24

Creation 42 *Void* 724 *Knowl.* 749

It is hard for us to forsake the familiar things around us, and turn back to the old home whence we came. Things seen delight us, and things unseen give rise to disbelief. Now the things that are evil are more manifest to sight; but the Good cannot be seen by things manifest; for it has no form or shape. It is impossible that an incorporeal thing should be manifested to a thing that is corporeal; because the incorporeal is like to itself, but unlike to all else.

Hermes

Nature brings forth Mineral or Metallic seed in the bowels of the earth. This is the reason why so many will not believe in its existence—because it is invisible.

Michael Sendivogius

Pilg. 378

(Changing things) you can touch and see and perceive with the senses, but the unchanging things you can only perceive with the mind.

Plato (*Phaedo,* 79 A)

The unseen God leaves those in doubt of his existence who think nothing patent but what may be known to the flesh.

Plotinus

The world of sense and colour . . . is a narrow prison.

The cause of narrowness is composition (compoundness) and number (plurality): the senses are moving towards composition.

Know that the world of Unification lies beyond sense: if you want Unity, march in that direction.

Rûmî

Reality 775

Some dull person may allege in refutation of our reasoning his inability to accomplish those chemical transformations on which it is based; but such operators would be vindicating too great an honour for their ignorance if they claimed to advance it as an argument against the truth of our Art. They must not make their own little understandings the standard or measure of the possibilities of Nature.

Philalethes

Introd. *Knowl.*

Were we then exhausted by the first creation? Yet they are in doubt about a new creation.

Qur'ân, L. 15

Of the existence of the Phoenix I had no doubt, or I could not have looked for it.

Michael Maier

(The soul) knows that God will never forsake his own life which he hath quickened in it: he will never deny those ardent desires of a blissful fruition of himself, which the lively sense of his own goodness hath excited within it: those breathings and gaspings after an eternal participation of him are but the energy of his own breath within us; if he had had any mind to destroy it, he would never have shown it such things as he hath done; he would not raise it up to such mounts of vision, to show it all the glory of that heavenly Canaan flowing with eternal and unbounded pleasures, and then precipitate it again into that deep and darkest abyss of death and non-entity. Divine goodness cannot, it will not, be so cruel to holy souls that are such ambitious suitors for his love.

John Smith the Platonist

Creation 42

Knowl. 749

Love 618

Realiz. 870

If the conviction, 'I am nothing but Existence and am ever free' were impossible to be attained, why should the *Sruti* teach us that so affectionately like a mother?

Śrî Śankarâchârya

This is the confidence that we have in him, that, if we ask any thing according to his will, he heareth us.

I. John. v. 14

Reality 790

...In hope of eternal life, which God, that cannot lie, promised before the world began.

Titus, I. 2

Charity 608

Unto Him is the return of all of you; it is a promise of Allâh in truth. Lo! He produceth creation, then He bringeth it back.

Qur'ân, x. 4

Orthod. passim

One should have faith in himself, faith in his own Guru who initiates him with the appropriate Mantra, and faith in the Ishta Devata chosen by him. It is on the tripod of this faith that the fulfilment of the aspirant's object is resting.

Swami Sivananda

To hope for anything, to trust or pray for anything but the life of God or a birth of Heaven in our souls is as useless to us as placing our hope and trust in a graven image.

William Law

Allâh hath endeared the faith to you and hath made it beautiful in your hearts, and hath made hateful unto you disbelief and lewdness and rebellion. Such are they who are the rightly guided.

Qur'ân, XLIX. 7

When thou standest before the Supreme King of kings to beg from Him thy wants, remember who thou art, and what thy acts are, whether they be straight or the reverse, and who the God is before whom thou standest, who seeth into the darkest secret places.

Judgment 239

Yesod Hayirah, 15

The good heart lies hid, the evil heart lies hid, there is an abyss in the good heart and in the bad. But these things are bare to God's sight, from whom nothing is hid.

St Augustine

Center 841

God, since His fullness is everywhere, is near us, and since His eye beholds us, since He is close beside us, let us refrain from evil-doing. It were best that our motive should be reverence, but if not, let us at least tremble to think of the power of His sovereignty.

Philo

M.M. 998

Real trust is the abandonment of trust, and that means, that God should be unto them (the Sufis) as He was when they were not yet brought into being.

Al-Kalâbâdhî

He to whom you pray is nearer to you than the neck of your camel.

Muhammad

If we believe not, yet he abideth faithful: he cannot deny himself.

II. Timothy, II. 13

Rev. 961

Patience

The device of a man who hath no device is patience.

Arabic Tradition

Pilg. 379

He that would make an end of sorrow and come to the bound of happiness must stablish firmly the root of Faith and immovably set his thought upon Enlightenment.

Śânti-deva

Though the châtak bird is about to die of a parched throat, and around it there are seven oceans, rivers, and lakes overflowing with water, still it will not touch that water. Its throat is cracking with thirst, and still it will not drink that water. It looks up, mouth agape, for the rain to fall when the star Svâti is in the ascendant. 'To the châtak bird all waters are mere dryness beside Svâti water.'

Sri Ramakrishna

Whoever knocks persistently, ends by entering.

'Alî

Acquire a firm will and the utmost patience.

Ananda Moyî

Mere performance of *tapas* (spiritual austerity) is its progress also. Steadiness is what is required.

Sri Ramana Maharshi

Introd. *Suff.*

For endurance is a mighty charm,
And patience giveth many good things.

The Testament of Joseph, II. 7

One must practise intense spiritual discipline. Can one obtain the vision of God all of a sudden, without any preparation?

Orthod. 285

A man asked of me, 'Why don't I see God?' I said to him, as the idea came to my mind:

'You want to catch a big fish. First make arrangements for it. Throw spiced bait into the water. Get a line and a rod. At the smell of the bait the fish will come from the deep water. By the movement of the water you will know that a big fish has come.'

Sri Ramakrishna

Knowl. 761

If you are fortunate you will catch beneath this wide expanse of waters a fish which will satisfy the longing of your heart. And if you fail of success, yet your mind will be stored with the precious treasures of knowledge, and you will in any case be richly rewarded for your labour.

Michael Maier

Holy War 405

When they (worldly people) need water they begin to dig a well. But as soon as they strike a stone they give up digging there and begin at another place. Perhaps they come to a bed of sand. Finding nothing but sand, they give that place up too. How can they succeed in getting water unless they continue to dig persistently where they started?

Sri Ramakrishna

Renun. 152a

Doubt is ignorance, and it never ceases for men who turn their faces away from God and who never meditate on me, the Atman.

Srimad Bhagavatam, XI. xv

Let japa and meditation be regular, and doubts will vanish of themselves in a mysterious manner.

Swami Sivananda

Many to whom this gift was imparted late in life, have, nevertheless, been refreshed and delighted by it in extreme old age.

John A. Mehung

Man, if thou desirest a noble and holy life, and unceasingly prayeth to God for it, if thou continue constant in this thy desire, it will be granted unto thee without fail, even if only in the day or hour of thy death; and if God should not give it thee then, thou shalt find it in Him in eternity: of this be assured.

St Bernard

Fidelity

Metanoia 484
Orthod. 300

There is no faith without submission, and no submission without faith.

Abû Hanîfah

Faith (*îmân*) is a confession with the tongue, a verification with the heart, and an act with the members.

Muhammad

Knowl. 745

What faith does the true Order ask of its followers? Very little: to defer to the desire of Amida, which can be expressed thus: 'Trust in me in all sincerity of heart, and you will be saved.' . . . The faith that one has in a perfect being must be perfect and absolute.

Shinran

In roving through the universe, lucky is the man who gets the seed of the creeper of faith through the grace of his *guru* and Krishna. He sows the seed like a gardener, waters it with hearing and chanting (the Holy Name). As the creeper grows it pierces through the universe, passes beyond the *Birajâ Brahma* world to the *Para-byom,* and above that to the heavenly Brindâban, where it creeps up the wishing-tree of Krishna's feet, spreads and bears fruit in the form of love. . . . If any sin against Vaishavisin is done, it uproots or tears any creeper like a wild elephant, its leaves wither. Then the gardener on earth carefully covers it, to save it from the elephant of sin. But if parasites, like love of enjoyment or . . . thirst of gain or fame, adhere to the creeper,—then these parasites flourish from the watering, while the main creeper's growth is arrested. Cut off the parasites first; then will the main branch reach the heavenly Brindâban. When the mature fruit of love drops down, the gardener tastes it, and proceeding up the creeper he reaches the wishing-tree.

Chaitanya

Orthod. 280
Inv. passim
Heterod. 423

Be attached, like the door-ring, to that same door; keep watch and be nimble and ready to spring.
Do not be the type of our breaking faith, do not recklessly make disloyalty notorious.
Since fidelity is the badge by which the dogs are known, begone and do not bring opprobrium and ill-fame upon the dogs.[1]

Rûmî

Holy War 405

Virtue's office never breaks men's troth.

Shakespeare (*Love's Labour's Lost,* v. ii. 351)

From vows made to God no man can excuse thee: such vows are a bond between thyself and God.

Eckhart

An oath to Allâh must be answered for.

Qur'ân, XXXIII. 15

There is no meaning in changing the Mantra according to our whims. If we with all faith and devotion repeat the Mantra given by our Guru, that Mantra will take us to the summit of God-realisation. What needs to be done is that we should make our heart perfectly pure

Inv. 1036

[1] The faithless have been defined as those who doubt all their former beliefs and believe all their former doubts.

so that God's love and power may be revealed in us. This can be achieved only by the Guru's grace and repetition of His Mantra.

Orthod. 288

Swami Ramdas

If a man is initiated by a human guru, he will not achieve anything if he regards his guru as a mere man. The guru should be regarded as the direct manifestation of God. Only then can the disciple have faith in the mantra given by the guru. Once a man has faith he achieves all. The śudra Ekalavya learnt archery in the forest before a clay image of Drona. He worshipped the image as the living Drona; that by itself enabled him to attain mastery in archery.[1]

Sri Ramakrishna

To be faithful to one's engagements and obligations is the sign of a superior man.

Gampopa

Seeing that unfaithfulness to the religious vows will result in one's going to the miserable states of existence, it is useless to have entered the Order if one live not a holy life.

Gampopa

Though to myself forsworn, to thee I'll faithful prove.

Shakespeare (*Love's Labour's Lost,* IV. ii. 112)

If there are men and women who hear the name of the Buddha of Endless Life and cannot believe in him, and fall into doubt, I say, they are sinners who have just come from the three wicked worlds, and are destined to return there. Sin still enchains them, making them so stupid that they cannot understand the Buddha's words.

Heterod. 420, 423

Byôdôkakkyô Sûtra

Allâh never changeth the grace He hath bestowed on any people until they first change that which is in their hearts.

Qur'ân, VIII. 53

Knowledge Through Faith

If any of you lack wisdom, let him ask of God, that giveth to all men liberally, and upbraideth not; and it shall be given him.

But let him ask in faith, nothing wavering. For he that wavereth is like a wave of the sea driven with the wind and tossed.

James, I. 5, 6

[1] One must not mistakenly conclude that faith is subjective. What it does is 'activate' the spiritual influences attaching to a certain symbol or rite.

He who desires to know what is closed to him and whose reason is not satisfied with resignation, is separated by his very desire from Union with God, from pure knowing and from true faith; he wavers between infidelity and faith.

Conform. 170

Ahmed Et-Tahâwî

Mâlunkyâputta, any one who should say, 'I will not lead the religious life under The Blessed One until The Blessed One shall elucidate to me either that the world is eternal, or that the world is not eternal, ... or that the saint neither exists nor does not exist after death;'—that person would die, Mâlunkyâputta, before the Tathâgata had ever elucidated this to him.

Introd. *Holiness*

It is as if, Mâlunkyâputta, a man had been wounded by an arrow thickly smeared with poison, and his friends and companions, his relatives and kinsfolk, were to procure for him a physician or surgeon; and the sick man were to say, 'I will not have this arrow taken out until I have learnt whether the man who wounded me belonged to the warrior caste, or to the Brahman caste, or to the agricultural caste, or to the menial caste.'...

Or again he were to say, 'I will not have this arrow taken out until I have learnt whether the arrow which wounded me was an ordinary arrow, or a claw-headed arrow, or a vekaṇḍa, or an iron arrow, or a calf-tooth arrow, or a karavîrapatta.' That man would die, Mâlunkyâputta, without ever having learnt this....

The religious life, Mâlunkyâputta, does not depend on the dogma that the world is eternal; nor does the religious life, Mâlunkyâputta, depend on the dogma that the world is not eternal. Whether the dogma obtain, Mâlunkyâputta, that the world is eternal, or that the world is not eternal, there still remain birth, old age, death, sorrow, lamentation, misery, grief, and despair, for the extinction of which in the present life I am prescribing.

Knowl. 734

Judgment 239

Accordingly, Mâlunkyâputta, bear always in mind what it is that I have not elucidated, and what it is that I have elucidated. And what, Mâlunkyâputta, have I not elucidated? I have not elucidated, Mâlunkyâputta, that the world is eternal; I have not elucidated that the world is not eternal.... I have not elucidated that the saint neither exists nor does not exist after death. And why, Mâlunkyâputta, have I not elucidated this? Because, Mâlunkyâputta, this profits not, nor has to do with the fundamentals of religion, nor tends to aversion, absence of passion, cessation, quiescence, the supernatural faculties, supreme wisdom, and Nirvana; therefore have I not elucidated it.

And what, Mâlunkyâputta, have I elucidated? Misery, Mâlunkyâputta, have I elucidated; the origin of misery have I elucidated; the cessation of misery have I elucidated; and the path leading to the cessation of misery have I elucidated. And why, Mâlunkyâputta, have I elucidated this? Because, Mâlunkyâputta, this does profit, has to do with the fundamentals of religion, and tends to aversion, absence of passion, cessation, quiescence, knowledge, supreme wisdom, and Nirvana; therefore have I elucidated it.

Peace 694

Majjhima-nikâya, Sutta 63

Then have patience. Lo! the promise of Allâh is true. And whether we let thee see a part of that which We promise them, or whether We cause thee to die, still unto Us they will be brought back.

Rev. 965

Qur'ân, XL. 77

Inv. 1003

Faith cometh by hearing, and hearing by the word of God.

Romans, x. 17

For although, unless he understands somewhat, no man can believe in God, nevertheless by the very faith whereby he believes, he is helped to the understanding of greater things.

St Augustine

Belief and wisdom are twin brothers; God accepts not the one without the other.

'Alî

Center 835 *Knowl.* 761 *Holiness* 902

God hath said: '*Kings, when they enter a city, ruin it*' (*Qur'ân*, XXVII. 34). When gnosis is established in the heart of the gnostic, the empire of doubt and scepticism and agnosticism is utterly destroyed, and the sovereignty of gnosis subdues his senses and passions so that in all his looks and acts and words he remains within the circle of its authority.

Hujwîrî

Contem. 536

Consult thy heart, and thou wilt hear the secret ordinance of God proclaimed by the heart's inward knowledge, which is real faith and divinity.

Muhammad

Salvation Through Faith

Introd. *Conform.* *Inv.* 1009

Can the Ethiopian change his skin, or the leopard his spots? then may ye also do good, that are accustomed to do evil. . . .

With men this is impossible; but with God all things are possible.

Jeremiah, XIII. 23; and *St Matthew*, XIX. 26[1]

There is nothing that so sanctifies the heart of man, that keeps us in such habitual love, prayer, and delight in God; nothing that so kills all the roots of evil in our nature, that so renews and perfects all our virtues, that fills us with so much love, goodness, and good wishes to every creature as this faith that God is always present in us with His light and Holy Spirit.

William Law

God can be realized by true faith alone.

Sri Ramakrishna

[1] This liaison is found in the marginal references of the Cambridge edition of the King James Bible.

By 'profound faith' we mean, not having the least doubt, but trusting that the Buddha who vowed never to cast off the sinner, however bad he might be, has guaranteed certain birth into the Pure Land to those who do no more than call upon His sacred name but once.[1]

Hônen

We have learned that every soul, although weighed down with sins, caught in the meshes of imperfections, held captive in exile, imprisoned in the flesh, sunk in mire, held fast in mud, fastened to the body, rendered helpless by anxieties, distressed by business considerations, grown suspicious because of fear, broken by sorrow, wandering about in the midst of errors, worried by cares, grown restless because of suspicions, and finally a stranger in the land of enemies, and according to the word of the Prophet, *defiled with the dead, . . . counted with them that go down into hell*[2]—although, I say, thus condemned and thus despairing, we have learned that nevertheless this soul can discover within herself not only the source whence she can breathe again in the hope of forgiveness, in the hope of mercy, but the source, as well, whence she may dare aspire to the nuptials of the Word, not fearing to enter into an alliance of friendship with God Himself, nor afraid to draw the sweet yoke of love with Him who is King of angels. For what is there that she may not safely dare in the presence of Him whose image, she discerns, gives her her distinction, and a likeness to whom, she knows, makes her noble. What terror, I say, has Majesty for her to whom confidence is given by virtue of her very origin? It is necessary only that she take care to preserve by an upright life the purity of the nature given her at birth.

St Bernard

Sin 66

Pilg. 366

Creation 42

Christ did make the babes and ignorant, of a very weak and mean capacity, knowing and understanding, viz. of poor fishermen, carpenters, and the like mechanics, he made apostles, and the most understanding men of all; and also of poor, disrespected, vilified people, as of women, and simple ones, he made faithful, devout, dear, godly children, who apprehended in themselves the universal without any art.

Thus likewise it goes in the philosophic work; the essentiality which lies disappeared in death, where the Mercury is wholly earthly, cold, and impotent, does now arise in power, as if the whole being and essence were become a new life, at which the artist wonders, and marvels what it is, or how it happens, and yet does also exceedingly rejoice that he sees the divine power to spring forth before his eyes in a half dead essence, and that in the curse of God.

Boehme

Humility 197 199

Introd. *Inv.*

He will not enter hell, who hath faith equal to a single grain of mustard seed in his heart.

Muhammad

You must have heard about the tremendous power of faith. It is said in the Purâna that Râma, who was God Himself—the embodiment of Absolute Brahman—had to build a

[1] Here, of course, the whole matter hinges on intention.

[2] *Baruch*, III. 11.

Introd. *Waters*

bridge to cross the sea to Ceylon. But Hanumân, trusting in Râma's name, cleared the sea in one jump and reached the other side. He had no need of a bridge.

Inv. 1017

Once a man was about to cross the sea. Bibhishana wrote Râma's name on a leaf, tied it in a corner of the man's wearing-cloth, and said to him: 'Don't be afraid. Have faith and walk on the water. But look here—the moment you lose faith you will be drowned.' The man was walking easily on the water. Suddenly he had an intense desire to see what was tied in his cloth. He opened it and found only a leaf with the name of Râma written on it. 'What is this?' he thought. 'Just the name of Râma!' As soon as doubt entered his mind he sank under the water.

Sri Ramakrishna

And when Peter was come down out of the ship, he walked on the water, to go to Jesus.

But when he saw the wind boisterous, he was afraid; and beginning to sink, he cried, saying, Lord, save me.

And immediately Jesus stretched forth his hand, and caught him, and said unto him, O thou of little faith, wherefore didst thou doubt?

St Matthew, XIV. 29–31

If you think there is uncertainty as to the efficacy of calling upon the sacred name once, then it means that there is doubt about it every time you call upon the sacred name. The Buddha's Vow was to give birth into the Pure Land to those who would call upon his name even once, and so there is efficacy in every repetition of the sacred name.

Hônen

And verily Joseph brought you of old clear proofs, yet ye ceased not to be in doubt concerning what he brought you till, when he died, ye said: Allâh will not send any messenger after him. Thus Allâh deceiveth him who is a prodigal, a doubter.

Qur'ân, XL. 34

Holy War 403

A man who plans to jump over a moat ten feet wide, must bend every effort to jump fifteen feet. And so a man who would be born into the Pure Land must concentrate his energy upon believing that he will of a certainty be so born.

Hônen

Conform. 180

What is faith? Unquestioning surrender to God's will.

Swami Ramdas

Pilg. 385

Faith is the wealth here best for man—by faith the flood is crossed.

Suttanipâta, 182, 184

May Süleyman, the lowly, find salvation,
Make faith his guide and Paradise his station.

The Mevlidi Sherif

As you call upon His sacred name, do not entertain the faintest doubt of its efficacy.

Hônen

'(The sinner) who offends Me by counting on My mercy cannot in truth say that he hopes for My mercy. But this presumption has nonetheless sweetened the milk of mercy. At the hour of death, if he recognizes his sin, if he unburdens his conscience through holy confession, he is then purified of presumption, which no longer offends Me, and mercy remains with him.

'By this mercy he can, if he wishes, gain hope. Without hope, no sinner could escape despair, and through despair he would incur with the demons eternal damnation. . . . For the sin of despair offends Me more and is more mortal to (sinners) than all the other sins which they have committed in the course of their existence.'

Judgment 261

St Catherine of Siena

And who despaireth of the mercy of his Lord save those who are astray?

Qur'ân, xv. 56

Only confidence in God is capable of giving me wings; fear by contrast freezes me and paralyzes all my being.

Sister Consolata

A man prayed to the Master to pardon his sins. He was told that it would be enough if he took care to see that his mind did not trouble him.

Metanoia 495

Talks with Sri Ramana Maharshi

Though I am worse than a dog, you have graciously undertaken to protect me. This delusion of birth and death is maintained by you. Moreover, am I the person to sift and judge? Am I the Lord here? Oh Mahesvara. It is for you to roll me through bodies (by births and deaths) or to keep me fixed at your own feet.

Conform. 170
Death 223

Tiruvachagam

'I am the cave of the penitent, and with Me is the refuge of the sinners.'

Sin 57

Niffarî

'Dost thou know what draweth Me towards thee? It is thy blind confidence in Me.'

Sister Consolata

Faith! Faith! Faith! Once a guru said to his pupil, 'Râma alone has become everything.' When a dog began to eat the pupil's bread, he said to it: 'O Râma, wait a little. I shall butter Your bread.' Such was his faith in the words of his guru.

Sri Ramakrishna

Faith in God, then, is one sure and infallible good, consolation of life, plenitude of bright hopes, dearth of ills, harvest of goods, inacquaintance with misery, acquaintance with piety, heritage of happiness, all-round betterment of the soul which is fairly stayed on Him Who is the cause of all things and can do all things yet only wills the best.

Wr. Side 474

Philo

Rev. 965 *Waters* 653

For verily I say unto you, That whosoever shall say unto this mountain, Be thou removed, and be thou cast into the sea; and shall not doubt in his heart, but shall believe that those things which he saith shall come to pass; he shall have whatsoever he saith.

St Mark, XI. 23

Holy War 410 *M.M.* 998

. . . In God's name, march:
True hope is swift, and flies with swallow's wings;
Kings it makes gods, and meaner creatures kings.

Shakespeare (*Richard III*, V. ii. 22)

Conform. 180 Supra 512

Abide not in the works of sinners: but trust in God, and stay in thy place.
For it is easy in the eyes of God on a sudden to make the poor man rich.

Ecclesiasticus, XI. 22, 23

What tremendous faith Krishnakishore had! Once, while at Vrindâvan, he felt thirsty and went to a well. Near it he saw a man standing. On being asked to draw a little water for him, the man said: 'I belong to a low caste, sir. You are a brâhmin. How can I draw water for you?' Krishnakishore said: 'Take the name of Śiva. By repeating His holy name you will make yourself pure.' The low-caste man did as he was told, and Krishnakishore, orthodox brâhmin that he was, drank that water. What tremendous faith!

Inv. 1017

Sri Ramakrishna

Contem. 542

Do not lose thy hope in God, even for the twinkling of an eye. Never interrupt thy relations with God, and thou wilt find joy the day when sinners are undone.

Dhu 'l-Nûn, citing spiritual counsel given him by a stranger

And of His portents (is this): that thou seest the earth lowly, but when We send down water thereon it thrilleth and groweth. Lo! He Who quickeneth it is verily the Quickener of the Dead. Lo! He is Able to do all things.

Death 208

Qur'ân, XLI. 39

Action 329 *M.M.* 978

Prarabdha (past karma) being responsible for all happenings, how can the obstacles to meditation be overcome?

Prarabdha concerns the mind only when extrovert, but not introvert. The earnest seeker of the Self does not worry himself about any obstacles. To think of obstacles is itself the greatest obstacle.

Sri Ramana Maharshi

The fact that men whose heavy evil *karma* would have condemned them to almost endless suffering after death should have turned to the religious life and attained *Nirvâna* showeth the virtue of the Holy *Dharma*.

Gampopa

'Above all things beware thou doubt not nor fear of any of those my things as hard and difficult to be brought to pass.'

Apuleius (Spoken by Isis) — *Orthod.* 296

Say not, 'There is a lion in the way;' say not, though religion be good, yet it is unattainable: no, but let us intend all our powers in a serious resolved pursuance of it, and depend upon the assistance of heaven, which never fails those that soberly seek for it.

John Smith the Platonist

Even the injunctions of destiny are cancelled if one takes refuge in God. Destiny strikes off with her own hands what she has written about such a person.

Śrî Sâradâ Devî

For they (the Pythagoreans) did not conceive that some things are possible to the Gods, but others impossible, as those fancy who reason sophistically; but they believed that all things are possible to the Gods. And this very assertion is the beginning of the verses, which they ascribe to Linus, and which are as follow:

All things may be the objects of our hope,
Since nothing hopeless any where is found:
All things with ease Divinity effects,
And nought can frustrate his almighty power.

Iamblichus — *Beauty* 689, *Reality* 803

PRAYER—MEDITATION—CONTEMPLATION

Man, if thou wishest to know what it is to pray sincerely:
Enter into thyself, and interrogate the Spirit of God.

Angelus Silesius (*Cherub.* I. 237)

Prayer is communication between the temporal and spiritual domains. In its most exterior usage, the word implies petition of the individual before the Creator. Meditation will then be a deepening and continuation of prayer, where individual interests are superseded by a quickening of attention upon divine qualities and 'spiritual perspectives' for their own intrinsic worth and pre-eminent reality. Contemplation could be called the act where God communes with Himself across our intellectual center, traditionally identified with the heart,[1] the juncture between the ray of the Spirit and the individual consciousness: 'According to the Divine Reality, it is not we who contemplate God, but it is God Himself who contemplates Himself in His universal Qualities for which we are the supports of manifestation; God does not become the "object" of the relative subject' (Burckhardt: *Du Soufisme*, p. 59).

Knowl. 749

Just as *samâdhi* is literally 'synthesis', as will be seen in later chapters,[2] so is the concentration that distinguishes contemplation a synthesis or con-centering of all faculties and qualities in their common Center and archetypal Source, the transcendent Intellect, omniscient Spirit, 'Eye in the World' or Solar Comprehensor.

Realiz. 890

'Man prays, and prayer fashions man. The saint has himself become prayer, the meeting place of earth and Heaven; he thus contains the universe, and the universe prays with him. He is everywhere where nature prays, and he prays with and in her: in the peaks which touch the void and eternity, in a flower which scatters its petals, or in the lost song of a bird.

Introd. *Holiness*

'Whosoever lives in prayer has not lived in vain' (Schuon: *Perspectives spirituelles*, p. 287).[3]

[1] In contradistinction to the present tendency to identify the heart with the sentiments.

[2] Notably, *Ecstasy, Peace, Void, Supreme Center.*

[3] Those interested in technical aspects of ritual and collective prayer are referred to Guénon: *Aperçus sur l'Initiation*, ch. XXIV: 'La prière et l'incantation'.

Help Through Prayer

If the yogi, being deluded, makes mistakes in life, he should burn away his sins and impurities by prayer and meditation. This yoga of prayer and meditation is the only way of atonement.

Srimad Bhagavatam, XI. xiii

Sin 57
Introd.
Orthod.

That prayer has great power which a person makes with all his might. It makes a sour heart sweet, a sad heart merry, a poor heart rich, a foolish heart wise, a timid heart brave, a sick heart well, a blind heart full of sight, a cold heart ardent. It draws down the great God into the little heart, it drives the hungry soul up into the fullness of God, it brings together two lovers, God and the soul in a wondrous place where they speak much of love.

Mechthild of Magdeburg

Metanoia
484

Take good, gracious God as he is, plat and plain as a plaster, and lay it to thy sick self as thou art.

The Epistle of Privy Counsel, II

Lord, I am like to *Misletoe,*
Which has no root, and cannot grow,
Or prosper, but by that same tree
It clings about; so I by Thee.

Robert Herrick

Orthod. 280
Introd.
Peace

And He giveth you of all ye ask of Him, and if ye would count the bounty of Allâh ye cannot reckon it. Lo! man is verily a wrong-doer, an ingrate.

Qur'ân, XIV. 34

Ask those things of God which it is worthy of God to bestow.

Sextus the Pythagorean

God's generosity is not in abrogation of His wisdom. From that it follows that He does not grant every request.

ʿAlî

Man prayeth for evil as he prayeth for good: for man was ever hasty.

Qur'ân, XVII. 11

Conform.
170

Humility 191

That I am and how that I am. as in nature and in grace. all I have is of thee. Lord. and thou it art. And all I offer it unto thee. principally to the praising of thee. for the help of all mine even Christians and of me.

The Epistle of Privy Counsel, III

Center 841

Believe and trust that as it is easy for you to breathe the air and live by it. or to eat and drink. so it is easy and even still easier for your faith to receive all spiritual gifts from the Lord. Prayer is the breathing of the soul: prayer is our spiritual food and drink.

John of Cronstadt

Prayer is a cleansing for hearts from the stain of iniquities. and an opening to the door of the Mysteries.

Prayer is the place of confidences and the wellspring of intimacies: in it widens the scope of the Mysteries. and in it shine the rays of the Luminaries.

Ibn ʿAṭâ'illâh

Faith 512

Indeed many things which we shall not be able to discover either by the experiment of work or by the investigations of reason. we shall deserve to be taught by importunate prayer. by the revelation of divine inspiration.

Richard of Saint-Victor

We are led from supplication to the object of supplication. and from the familiar intercourse we acquire a similarity to it. and from imperfection we quietly receive the Divine Perfection.

Iamblichus

Knowl. 749
Creation 31
Symb. 306
M.M. 987

Holy is God the Father of all. who is before the first beginning:
holy is God. whose purpose is accomplished by his several Powers:
holy is God. who wills to be known. and is known by them that are his own.
Holy art Thou. who by thy word hast constructed all that is:
holy art Thou. whose brightness nature has not darkened:
holy art Thou. of whom all nature is an image.
Holy art Thou. who art stronger than all domination:
holy art Thou. who art greater than all pre-eminence:
holy art Thou. who surpassest all praises.
Accept pure offerings of speech from a soul and heart uplifted to thee.
Thou of whom no words can tell. no tongue can speak. whom silence only can declare.

Hermes

On Meditation and Contemplation

Whither shall I look when I praise Thee? Upward or downward, inward or outward? For Thou art the place in which all things are contained: there is no other place beside Thee; all things are in Thee.

Hermes

Death 208
Center 816

God is neither high nor low: and who speaks otherwise, is still but badly informed on the truth.

God is neither here nor there: whosoever desires to find Him, let him chain his hands and feet, body and soul.

Angelus Silesius

M.M. 975
Infra 542

We ought not to question whence It comes; there is no whence, no coming or going in place; It either appears or does not appear. We must not run after It, but we must fit ourselves for the vision and then wait tranquilly for it as the eye waits on the rising of the sun which in its own time appears above the horizon and gives itself to our sight.

Plotinus

Conform. 170

Master your vagabond thoughts, try to raise yourself above the fluctuations of life, and you will see all your anxiety disappear concerning the place propitious for *sâdhanâ*.

Ananda Moyî

Renun. 139

What does solitude or seclusion mean?

The Self being all-extensive there can be no fixed spot for its solitary state; however to be still is to remain solitary.

Sri Ramana Maharshi

Where creature stops, there God begins. All God wants of thee is for thee to go out of thyself in respect of thy creatureliness and let God be God in thee.

Eckhart

Death 220

Real prayer can only be yours
When you have staked and gambled yourself away
And your essence is pure.

Shabistarî

Renun. 146

King. Prize you yourselves? what buys your company?
Rosaline. Your absence only.

Shakespeare (*Love's Labour's Lost,* v. ii. 225)

Metanoia 484

Not for the first time to-morrow will God grant thy supplication and thy prayer: he has granted it already in his eternity ere ever thou becamest man. Suppose thy prayer is foolish or lacking earnestness, God will deny it thee not then, he has denied it thee already in his eternity.

Center 838

Eckhart

God hears when we pray to Him. What are we without His power? In fact, without the words which He puts into our mouth, how can we pray?

Action 340

Swami Ramdas

God who's in Heav'n, will hear from thence;
If not to'th sound, yet, to the sense.

Robert Herrick

Let not thy requesting be a seeking for His giving, lest thy understanding of Him should diminish; let thy requesting be for manifesting thy servitude and establishing the rights of sovereignty.

Humility 191

How can thy subsequent seeking be a cause of His prior giving?
A predestined decree is exalted above being referred to causes.

Reality 773

Ibn ʿAṭâ'illâh

But we must beseech him to pardon us; though his children do indeed get pardon from their Father even before they ask it.

Hermes

O Lord, prayer at Thy gate
Is a mere formality:
Thou knowest what Thy slave desires.

Ansârî

For the Great Spirit is everywhere; he hears whatever is in our minds and hearts, and it is not necessary to speak to Him in a loud voice.

Center 841

Black Elk

Then thanks for all the bountiful mercies and kindnesses with which He has anticipated thy wishes, and due reflection be in the case of the petitions with which thou wouldst supplicate Him as regards the coming time! This is what is meant by the *Ordering of Prayer*; this it is which was the essential characteristic of the methods adopted by the early Pious Ones when they prayed.

Conform. 170

Yesod Hayirah, 17

When one takes God as he is divine, having the reality of God within him, God sheds light on everything. Everything will taste like God and reflect him. God will shine in him all the time. He will have the disinterest, renunciation, and spiritual vision of his beloved, ever-present Lord. He will be like one athirst with a real thirst; he cannot help drinking

P. State 563

even though he thinks of other things. Wherever he is, with whomsoever he may be, whatever his purpose or thoughts or occupation—the idea of the Drink will not depart as long as the thirst endures; and the greater the thirst the more lively, deep-seated, present, and steady the idea of the Drink will be. Or suppose one loves something with all that is in him, so that nothing else can move him or give pleasure, and he cares for that alone, looking for nothing more; then wherever he is or with whomsoever he may be, whatever he tries or does, that Something he loves will not be extinguished from his mind. He will see it everywhere, and the stronger his love grows for it the more vivid it will be. A person like this never thinks of resting because he is never tired.

Infra 532

Holiness 934

The more he regards everything as divine—more divine than it is of itself—the more God will be pleased with him. To be sure, this requires effort and love, a careful cultivation of the spiritual life, and a watchful, honest, active oversight of all one's mental attitudes toward things and people. It is not to be learned by world-flight, running away from things, turning solitary and going apart from the world. Rather, one must learn an inner solitude, wherever or with whomsoever he may be. He must learn to penetrate things and find God there, to get a strong impression of God firmly fixed in his mind.

Beauty 670
Realiz. 859
Void 728

Eckhart

Who know the world live alone.

'Alî

The greatest Saints avoided the society of men, when they could conveniently: and did rather choose to live to God in secret.

The Imitation of Christ, I. xx. 1

Mention is made of two classes of yogis: the hidden and the known. Those who have renounced the world are 'known' yogis: all recognize them. But the 'hidden' yogis live in the world. They are not known. They are like the maidservant who performs her duties in the house but whose mind is fixed on her children in the country. They are also . . . like the loose woman who performs her household duties zealously but whose mind constantly dwells on her lover.

Inv. 1036

Sri Ramakrishna

Lo! those who fear their Lord in secret, theirs will be forgiveness and a great reward.

Qur'ân, LXVII. 12

And when thou prayest, thou shalt not be as the hypocrites are: for they love to pray standing in the synagogues and in the corners of the streets, that they may be seen of men. Verily I say unto you, They have their reward.

But thou, when thou prayest, enter into thy closet, and when thou hast shut thy door, pray to thy Father which is in secret: and thy Father which seeth in secret shall reward thee openly.

Sin 77

St Matthew, VI. 5, 6

Unto the life which ye have chosen
Many interruptions come; so perform ye rites in secret.

Milarepa

Let (the student of our Art) carry on his operations with great secrecy in order that no scornful or scurrilous person may know of them: for nothing discourages the beginner so much as the mockery, taunts, and well-meant advice of foolish outsiders. Moreover, if he does not succeed, secrecy will save him from derision; if he does succeed, it will safeguard him against the persecution of greedy and cruel tyrants.

Philalethes

Peace 700

Silence is the garden of meditation.

ʿAlî

M.M. 987

Each soul must meet the morning sun, the new, sweet earth, and the Great Silence alone!

Ohiyesa

The most noble of prayers is when he who prays is transformed inwardly into that before which he kneels.

Angelus Silesius

Orthod. 285

All that has been declared (above) depends on meditation; for he who is not proficient in the knowledge of the Self reaps not the full reward of the performance of rites.

Mânava-dharma-śâstra, VI. 82

Judgment 250
Metanoia 488

If you wish to escape the pains and afflictions inherent in the wheel of deaths and rebirths (samsâra), you must devote yourself to meditation; it is the path which leads to divinity. It is the royal way to the kingdom of Brahman; it is the mysterious ladder which goes from earth to heaven, from error to truth, from darkness to light, from grief to joy, from restlessness to peace, from ignorance to knowledge, from death to immortality. Meditation leads us towards knowledge of the Self, that is to say, towards eternal peace and supreme felicity. It prepares us for the integral experience of direct intuitive knowledge.

Swami Sivananda

The soul should collect and concentrate itself in its Self.

Plato (*Phaedo*, 83 A)

Knowl. 749

God loves Himself and sings His praises, all that He can:
He kneels and He inclines, He worships Himself.

Angelus Silesius

Meditation is the chief possession of the gnostic.

Al-Muḥâsibî

Thinking is without labour and bears no fruit. Meditation labours and has its fruit. Contemplation abides untoiling and fruitful. Thinking roams about, meditation investigates, contemplation wonders. Thinking arises from the imagination, meditation from the reason, contemplation from the intelligence. Behold these three, imagination, reason, intelligence. Intelligence takes the highest place, imagination the lowest, reason lies between them. Everything which comes under the view of the lower sense, comes necessarily also under the view of the higher sense. . . . Behold then how wide is the extent of the ray of contemplation, for it embraces all things.

Richard of Saint-Victor

Infra 528
Knowl. 761

It is clear then that sense, being a composite thing, apprehends composite (or concrete) things: but intellect, being a simple and indivisible thing, apprehends simple and indivisible things (viz. universals).

Hermes

Knowl. 749

While the Soul is wholly intent upon the Glory of God in Prayer, many sweet Appearances, high Truths shew themselves clearly to the Soul, which were before Unthought of, or very difficult. They now appear, as in their Element, like Stars to him that looks stedfastly on a clear Skie in an Evening. They come forth, as out of their Bride-Chamber, ready Trim'd, prepar'd for the Soul . . . often they come thick and swarming about the Soul . . . : like Bees out of a Hive. A Man frequently learns *more* and *better at one Hour's Prayer*, than in the Study of many days.

Peter Sterry

Action 346
Knowl. 734

God allows the intelligent philosopher to bring forth hidden things through meditation on nature, and to liberate them from the darkness. . . . These hidden realities are always present, but the eyes of ordinary men do not see them—only the eyes of the intellect and of the power of imagination which perceive with a true vision.

Novum Lumen

Center 816

However, a certain number of men are not excluded from this benediction. The mystery of the resurrection dwells in their souls, and these are they who have the privilege of being counted among the number of the sons of God, for they perceive the light which reigns in the world and which the world sees not: they see it, know it, and witness it.

Robert Fludd

A magic force put to sleep by sin is latent in man. It can be awakened by the Grace of God or by the art of the Kabbala. We will find a pure and holy knowledge in us, if we manage to isolate ourselves from all outward influence and allow ourselves to be led by the inward light. At this stage of concentration the spirit distinguishes each object on which it directs its attention, being able to unite itself therein, and attain even unto God.

Jan Baptista van Helmont

Sin 77
Grace 552, 55[illegible]
Metanoia 484
Symb. 306
Realiz. 887

The intellectual working of the brain alone is not sufficient to give birth to a physician: the true physician is not he who has merely heard the truth, but he who feels the truth, who

Knowl. 745 Introd. *Holiness*

sees it before him as clearly as the light of the sun, who hears it as he would hear the noise of the cataract of the Rhine or the whistling of the storm upon the ocean, who smells it and tastes it. . . . The true philosopher sees the reality, not merely the outward appearance. He who knows the sun and the moon has a sun and a moon in him, and he can tell how they look, even if his eyes are shut. Likewise, the true physician sees in himself the whole constitution of the microcosm of man with all its parts. He sees the constitution of his patient as if the latter were a clear crystal, in which not even a single hair could escape detection. He sees him as he would the stones and pebbles at the bottom of a clear well.

Holiness 924 *Realiz.* 859

Paracelsus

On Distraction

O Krishna, the mind is restless, turbulent, strong and unyielding; I consider it as difficult to subdue as the wind.

Bhagavad-Gîtâ, VI. 34

The heart of the ordinary unenlightened man, because of his surroundings, is always liable to change, just like monkeys jumping from one branch to another. It is indeed in a state of confusion, easily moved and with difficulty controlled.

Hônen

Action 329

It is a mistake to think that the *sâdhanâ* cannot be practised for lack of time. The real cause is agitation of mind.

Swâmi Brahmânanda

'If it were just a matter of playing football with the firmament, stirring up the ocean, turning back rivers, carrying away mountains, seizing the moon, moving the Pole-star or shifting a planet, I could manage it easily enough. Even if it were a question of my head being cut off and the brain removed or my belly being ripped open and my heart cut out, or any kind of transference or transformation, I would take on the job at once,' said Monkey. 'But if it comes to sitting still and meditating, I am bound to come off badly. It's quite against my nature to sit still.'[1]

Monkey

All the gods are under the sway of the mind, but the mind never comes under the rule of any power. Even the yogis know the mind to be a terrible god, stronger than the strongest. He, therefore, who can bring the mind under subjection is indeed the god of gods.

Holy War 396

Srimad Bhagavatam, XI. xvi

[1] Here in this passage we have modern man prefigured.

Leah is the affection enflamed by divine inspiration. Rachel the reason illumined by divine revelation. . . . Each of them therefore receives her handmaid: the servant of the affections is sensation. that of reason is the imagination. But each of these handmaids is known to be necessary to her mistress. for without them the whole world would not be of use to them. For the reason would know nothing without the imagination and the affections would feel nothing without the sensibility. . . .

Introd.
Realiz.

We have seen how the two handmaids can be of service but I think we must not be silent about their vices. For Bilhah is garrulous and Zilpah drunken. For not even Rachel her mistress can repress Bilhah's loquaciousness and not even the generosity of Zilpah's mistress can satisfy her thirst. The wine Zilpah desires is the joy of pleasures. of which the more she drinks the greedier is her thirst. for the whole world will not suffice to satiate her lustful appetite. However much she drinks she is always gasping for more drink. and is rightly called Zilpah. that is a 'gasping mouth'. thirst that is never extinguished. But the imagination cries out with such importunity in the ears of the heart that Rachel can hardly ever silence her clamouring. This is the reason why. when we are saying psalms or praying. though we would shut out phantasies of thought or the images of things from the eyes of our heart. we are unable to do so. And because every day we know by experience and unwillingly suffer this clamouring. we know how great is Bilhah's garrulity. Everything we have seen or heard and especially what we ourselves have done and said is brought back to our memory and she never ceases to repeat again and again what she has already told. Moreover though the will gives no consent to hearing. she continues to tell her tale though no one is listening. She is like decrepit old men or a senile woman who will drone on without any audience and carry on a conversation as if someone were present. Hence Bilhah is not undeservedly called 'inveterate' for she copies the ways of such old people. But surely everybody knows Bilhah's garrulity and Zilpah's drunkenness unless they do not know themselves!

Illusion
109

Center
816

Realiz.
859

Richard of Saint-Victor

The mind is compared to quicksilver because its rays are scattered over divers objects. It is compared to a monkey because it leaps from object to object. It is compared to the inconstant air because it is unstable (chanchala). It is compared to an elephant in rut because of its passionate impetuosity. . . . This tendency to oscillation is innate in the mental substance.

Swami Sivananda

Romeo! humours! madman! passion! lover!
Appear thou in the likeness of a sigh:
Speak but one rime and I am satisfied:
Cry but 'Ay me!' couple but 'love' and 'dove:'
Speak to my gossip Venus one fair word.
One nickname for her purblind son and heir.
Young Adam Cupid. he that shot so trim
When King Cophetua lov'd the beggar-maid.
He heareth not. he stirreth not. he moveth not:
The ape is dead. and I must conjure him.

Void
724

I conjure thee by Rosaline's bright eyes,
By her high forehead, and her scarlet lip,
By her fine foot, straight leg, and quivering thigh,
And the demesnes that there adjacent lie,
That in thy likeness thou appear to us.[1]

Shakespeare (*Romeo and Juliet*, II. i. 8)

Judgment 250

Your mind is the cycle of births and deaths (*Samsâra*).

Sri Ramana Maharshi

Infra 536

Lead is not mercury, and mercury is not lead. Lead and mercury are procured within your body. It is essential that the body and the mind become motionless.

Hsieh Tao-kuang

Duke. Sir, your company is fairer than honest. Rest you well.
Lucio. By my troth, I'll go with thee to the lane's end. If bawdy talk offend you, we'll have very little of it. Nay, friar, I am a kind of burr; I shall stick.

Shakespeare (*Measure for Measure*, IV. iii. 189)

Renun. 146 *Metanoia* 480 *Rev.* 967

Those turbulent and unruly, uncertain and unconstant motions of passion and self-will that dwell in degenerate minds, divide them perpetually from themselves, and are always moulding several factions and tumultuous combinations within them against the dominion of reason. And the only way to unite man firmly to himself is by uniting him to God, and establishing in him a firm amity and agreement with the first and primitive being.

John Smith the Platonist

Knowl. 749

Three things prevent our hearing the eternal Word. The first is body, the second number, and the third is time. If we were rid of these three things we should be living in eternity and in the spirit, solitaries in the desert listening to the eternal Word. . . . Hearer and heard are one in the eternal Word.

Eckhart

Peace 700

Those who seek the truth by means of intellect and learning only get further and further away from it. Not till your thoughts cease all their branching here and there, not till you abandon all thoughts of seeking for something, not till your mind is motionless as wood or stone, will you be on the right road to the Gate.

Huang Po

Whatever thoughts, or concepts, or obscuring (or disturbing) passions arise are neither to be abandoned nor allowed to control one: they are to be allowed to arise without one's

[1] 'You must not be astonished at anything that comes up, no matter how shameful. Accept it all calmly, as if you were a mere spectator, uninterested, and were observing a process for which you need not feel responsible. Simply let it go on until it wearies of itself, while you listen with only half an ear. The result, in the end, is perfect stillness, which breathes without your noticing it' (Eugen Herrigel: *The Method of Zen*, London, Routledge & Kegan Paul, 1960, p. 25).

trying to direct (or shape) them. If one do no more than merely to recognize them as soon as they arise, and persist in so doing, they will come to be realized (or to dawn) in their true (or void) form through not being abandoned.

Padma-Karpo

'I allow this vehement struggle with thoughts which oppress thee, because it brings glory and salvation to souls. Offer to Me these involuntary thoughts which continuously surge up from rising to bedtime to hinder thee from loving Me, and I will transform them into graces and blessings for souls.'

Sin 62

Sister Consolata

'But I haven't dealt yet with the monster,' protested Monkey. 'The monster is in this basket,' said the Bodhisattva. 'What?' cried Pigsy and Sandy. 'A fish do all that harm?'

Illusion 85

Monkey

If once the mind can be liberated from rajas and tamas, it will guide thee. It will be thy guru.

Swami Sivananda

The good deed and the evil deed are not alike. Repel the evil deed with one which is better, then lo! he, between whom and thee there was enmity (will become) as though he was a bosom friend.

Center 835

Qur'ân, XLI. 34

We must do our business faithfully, without trouble or disquiet, recalling our mind to God mildly, and with tranquillity, as often as we find it wandering from Him.

Brother Lawrence

If at prayer we do nothing but drive away temptations and distractions, our prayer is well made.

St Francis de Sales

A desire arises in the mind. It is satisfied: immediately another comes. In the interval which separates two desires a perfect calm reigns in the mind. It is at this moment freed from all thought (sankalpa), love or hate. Complete peace (shânti) equally reigns between two mental waves (vrittis).

When the mind is concentrated on Brahman or the supreme Self, it becomes one with it, like camphor with the flame, or salt with water, or water with milk. The mind is dissolved in Brahman and becomes of the nature of Brahman. There is no longer duality. The person meditating has become Brahman, and this is the state of liberation (kaivalya).

M.M. 978

Swami Sivananda

When I said that (men's) minds were to be emptied of images, it is to be understood in this sense, that it was just as when you set a lighted taper at midday in the sunshine, the taper continues to burn, and sheds forth no less light than it did before: but its light is lost in the sunshine, because the greater light prevails over the lesser and absorbs it, so that it no longer seems to shine with a separate lustre, but is diffused and shed forth in the greater light.

Realiz. 868

Tauler

Illusion 94

We shall be filled with true forms only at that time when we (shall be) empty of dreams.

Marsilio Ficino

Realiz. 859 *Peace* 700

In (self-experience) all forms are endowed with the sameness of space.
And the mind is held steady with the nature of this same sameness.
When the mind ceases thus to be mind,
The true nature of the Innate shines forth.

Saraha

Retreat (*khalwa*) is like a smith's forge in which, by the fire of austerity,[1] the desire becomes fused, purified, delicate and gleaming like a mirror and in it appeareth the form of the Invisible. For every morning of the retreat, a veil should lift and the retreatant find himself drawing ever nearer to God, so that in forty mornings, the forty-fold veil should be lifted and the purified human nature should return from a land that is far-off to the land of proximity.

Introd. *Pilg.* *Center* 841

Al-Suhrawardî

On Concentration

The highest yoga is the control of the mind.

Srimad Bhagavatam, XI. xvi

Introd. *Holiness*

Proficiency in meditation amounts to fixity in the Real.

Sri Ramana Maharshi

Void 728

It is well to fix the thought in concentred effort: the man of wandering mind lies between the fangs of the Passions. It cannot wander if body and thought be in solitude: so it is well to forsake the world and put away vain imaginations.

Śânti-deva

[1] The same as the Hindu *tapas*.

We must learn to act without attachment. But it is rare for anyone untrained to reach the stage at which he is proof against disturbance by any act or anybody. This needs prodigiously hard work: and for God to be as present and to show as plainly to him at all times and in all company. that is for the expert and demands especially two things. One is that the man be closeted within himself where his mind is safe from images of outside things which remain external to him and. alien as they are. cannot traffic or forgather with him or find any room in him at all. Secondly. inventions of the mind itself. ideas. spontaneous notions or images of things outside or whatever comes into his head. he must give no quarter to on pain of scattering himself and being sold into multiplicity. His powers must all be trained to turn and face his inner self.

Action 358 *Holy War* 405 *P. State* 563 *Sin* 77 *Knowl.* 755 *Metanoia* 488

Thou dost object. 'But one must turn outwards to do outward works: no work is wrought except in its own mode.'—True. But to the expert soul outward modes are not merely outward things: to the interior soul all things are modes of the Deity within.

Eckhart

Introd. *Action*

Live in the world but keep the pitcher steady on your head: that is to say. keep the mind firmly on God.

Sri Ramakrishna

Renun. 139

The well-resolved mind is single and one-pointed.

Bhagavad-Gîtâ, II. 41

As a fletcher makes straight his arrow. a wise man makes straight his trembling and unsteady thought. which is difficult to guard. difficult to hold back.

Dhammapada, III. 33

Be straight. like an arrow. and escape from the bow. for without doubt every straight arrow will fly from the bow to its mark.

Rûmî

The practice of *samâdhi* is like that of shooting. An inexperienced archer will at first practise shooting at a large target. next a small one. and finally hit the bull's-eye. He will train himself to hit an object by making it smaller and smaller. first a coin. then a stick. then a few hairs. then a single hair. then a tenth. a hundredth and finally a thousandth of a hair. Thus having increased in skill. he will be able. as quick as thought. to hit a thing. no matter how small. so that even if he only heard the voice of a man or an animal sounding in the dark. he would be able at once to locate it and hit it without fail.

Ryôgon Sûtra, Vol. 1

Action 338

Can you ever see God if you do not direct your whole mind toward Him? The *Bhâgavata* speaks about Śukadeva. When he walked about he looked like a soldier with fixed bayonet. His gaze did not wander: it had only one goal and that was God. This is the meaning of yoga.

Sri Ramakrishna

Renun. 146

Realiz. 870 *Peace* 705

Even here (in this world), existence is conquered by them whose mind rests in equality, for Brahman is without imperfection and equal. Therefore they abide in Brahman.

Bhagavad-Gîtâ, v. 19

Introd. *Holiness*

What is the intention of the gnostic? The continuity of his gnosis.

Ibn al-'Arîf

Do you know what one feels in meditation? The mind becomes like a continuous flow of oil—it thinks of one object only, and that is God. It does not think of anything else.

Sri Ramakrishna

Supra 528

The *yogin* then looketh on, mentally unperturbed, at the interminable flow of thoughts as though he were tranquilly resting on the shore of a river watching the water flow past.

Padma-Karpo

M.M. 975 *Void* 724 *Metanoia* 488

There is a fine saying of one heathen philosopher to another about this, he says: 'I am aware of something in me which sparkles in my intelligence; I clearly perceive that it is somewhat but *what* I cannot grasp. Yet methinks if I could only seize it I should know all truth.' To which the other philosopher replied: 'Follow it boldly! for if thou canst seize it thou wilt possess the sum-total of all good and have eternal life!' St Augustine expresses himself in the same sense: 'I am conscious of something within me that plays before my soul and is as a light dancing in front of it; were this brought to steadiness and perfection in me it would surely be eternal life!' It hides yet it shows. It comes, but after the manner of a thief, with intent to take and to steal all things from the soul. By emerging and showing itself somewhat it purposes to decoy the soul and draw it towards itself to rob it and take itself from it. As saith the prophet: 'Lord take from them their spirit and give them instead thy spirit.'

Eckhart

Love 618

Whosoever should hold himself straightly in the 'pull' which comes from God and follow after the light as he sees it, would come to such bliss and heavenly knowledge that no heart could contain it. Then he would be as an angel, ever lovingly united to God in all things.

Mechthild of Magdeburg

By certain signs you can tell when meditation is being rightly practised. One of them is that a bird will sit on your head, thinking you are an inert thing.

Sri Ramakrishna

'Tis as when a bird is perched on your head, and your soul trembles for fear of its flitting,
So you dare not stir from your place, lest your beautiful bird should take to the air;
You dare not breathe, you suppress a cough, lest that *humâ* should fly away.

Rûmî

It must be caught in the air, before it touches the ground: otherwise it evaporates.

The Glory of the World

When Arjuna was about to shoot at the target, the eye of a fish, his eyes were fixed on the eye of the fish, and on nothing else. He didn't even notice any part of the fish except the eye. In such a state the breathing stops and one experiences kumbhaka.[1]

Sri Ramakrishna

The mind of the yogi is always fixed on God, always absorbed in the Self. You can recognize such a man by merely looking at him. His eyes are wide open, with an aimless look, like the eyes of the mother bird hatching her eggs. Her entire mind is fixed on the eggs, and there is a vacant look in her eyes.

Holiness 910

Sri Ramakrishna

When he beheld the countenance of the Ṣadr-i Jahân, you might say the bird, his spirit, flew out of his body.

Flight 944

His body fell like dry wood: his vital spirit became cold from the crown of his head to his toes.

Whatsoever they applied of incense and rose-water, he neither stirred nor spoke.

Ecstasy 636

Rûmî

Tze Ch'i of Nan-kuo sat leaning on a table. Looking up to heaven, he sighed and became silent, as though soul and body had parted.

Yen Ch'eng Tze Yu, who was standing by him, exclaimed, 'What are you thinking about that your body should become thus like dry wood, your mind like dead ashes? Surely the man now leaning on the table is not he who was here just now.'

Void 728

'My friend, ' replied Tze Ch'i, 'your question is apposite. Today I have buried myself. . . . Do you understand? . . . Ah! perhaps you only know the music of man, and not that of earth. Or even if you have heard the music of earth, you have not heard the music of heaven.'

Death 206
Beauty 679
P. State 583
Holiness 924

Chuang-tse (ch. II)

For it behoves the mind that would be led forth and let go free to withdraw itself from the influence of everything, the needs of the body, the organs of sense, specious arguments, the plausibilities of reasoning, last of all itself. For this reason Abraham glories saying 'The Lord, the God of heaven, and the God of the earth, who took me out of my father's house' (Gen. xxiv. 7): for it is not possible that he whose abode is in the body and the mortal race should attain to being with God: this is possible only for him whom God rescues out of the prison. . . . When the soul in all utterances and all actions has attained to perfect sincerity and godlikeness, the voices of the senses cease and all those abominable sounds that used to vex it. . . . All these cease when the mind goes forth from the city of the soul and finds in God the spring and aim of its own doings and intents.

Sin 77
Supra 528
Death 208
Grace 552
Renun. 158

Philo

[1] Breath retention, a process in *prânâyâma*, or breath control.

Therefore, lay aside this corpse-like body as though it were truly a corpse. Do not even murmer 'I' but enquire keenly within what it is that now shines within the heart as 'I'. Transcending the intermittent flow of diverse thoughts there arises the continuous, unbroken awareness, silent and spontaneous, as 'I—I' in the Heart. If one catches it and remains still, it will completely annihilate the sense of 'I' in the body and itself disappear as a fire of burning camphor. Sages and scriptures proclaim this to be Liberation.

Death 220
Realiz. 890

Sri Ramana Maharshi

There is a sâdhu in Hrishikesh who gets up early in the morning and stands near a great waterfall. He looks at it the whole day and says to God: 'Ah, You have done well! Well done! How amazing!' He doesn't practise any other form of japa or austerity. At night he returns to his hut.

Inv. 1017

Sri Ramakrishna

When Wang Hsichih[1] was at home, he often counted the pistils of every flower on every branch in his yard, and he would be thus occupied the whole day without saying a word, while his students stood by with towels at his side.

Chin Shengt'an

At Kugami,
In front of the Otono,
There stands a solitary pine tree,
Surely of many a generation:
How divinely dignified
It stands there!
In the morning
I pass by it:
In the evening
I stand underneath it,
And standing I gaze,
Never tired
Of this solitary pine!

Ryôkwan

The Oratory of the Heart

We may make an oratory of our heart wherein to retire from time to time to converse with Him in meekness, humility, and love. Every one is capable of such familiar

Center 816

[1] 321–379 A.D.; famous Chinese calligrapher.

conversation with God, some more, some less. He knows what we can do. Let us begin, then.

Brother Lawrence

If thou wishest to search out the deep things of God, search out the depths of thine own spirit.

Richard of Saint-Victor

Realiz. 859

MANILAL (*to the Master*): 'Well, what is the rule for concentration? Where should one concentrate?'

MASTER: 'The heart is a splendid place.'

Sri Ramakrishna

Worship me through meditation in the sanctuary of the heart.

Srimad Bhagavatam, XI. v

When the Moon of Canaan YÚSUF
In the prison of Egypt darken'd,
Nightly from her spacious Palace-
Chamber, and its rich array,
Stole ZULAIKHÁ like a fantom
To the dark and narrow dungeon
Where her buried Treasure lay.
Then to those about her wond'ring—
'Were my Palace,' she replied,
'Wider than Horizon-wide,
'It were narrower than an Ant's eye,
'Were my Treasure not inside:
'And an Ant's eye, if but there
'My Lover, Heaven's horizon were.'

Jâmî

Judgment 266

Seek for God within thine own soul.

John Smith the Platonist

'I am the companion of them that commemorate Me and the friend of them that take Me as their friend.
'If thou art with all, thou art without all when thou art without Me: and if thou art without all, thou art with all when thou art with Me.'

Muhammad

Center 841

I would rather be in hell and possess God, than be in the kingdom of heaven without God.

Tauler

Conform. 170

Hence they have born my Lord: Behold! the Stone
Is rowl'd away: and my sweet Saviour's gone! . . .
Is He, from hence, gone to the shades beneath,
To vanquish Hell, as here He conquer'd Death?
If so: I'le thither follow, without feare:
And live in Hell, if that my *Christ* stayes there.

Pilg. 366

Robert Herrick

O Lord, if Thou sendest me to Hell
I raise no protest.
And if Thou takest me to Paradise,
I go there, but not of my own choice.

If in Hell I obtain union with Thee
What care I for those who dwell in Paradise?
And were I called to Heaven without Thee
The pleasures of Paradise would then
Be worse than the fires of Hell.

Ansârî

The angels' joy is woe to me
Unless my Lord, my Love, I see.

Mechthild of Magdeburg

God is present to himself in every Creature, after the manner of a God.

Where-ever he is present, He is *entirely present* with all the Joys and Glories of eternity, ever undivided, His own Heaven to himself, in the *Depths* of *Hell* beneath as in the Heights of Heaven above, in the *dust* of the *Grave,* in a wave of the Sea, as in the most shining Cherubim or flaming Seraphim.

Peter Sterry

We should hold day with the Antipodes,
If you would walk in absence of the sun.

Shakespeare (*Merchant of Venice,* v. i. 127)

People often say to me: 'Pray for me.' And I think to myself: Why ever do ye go out? Why not stop at home and mine your own treasure? For indeed the whole truth is native in you.

Creation 42

Eckhart

Peace 694

The path to the source of your and the world's being is not without. You have to go within yourself.

Swami Ramdas

Without allowing the mind to externalize itself and go towards the objects of the senses, keep it always in its own natural abode or centre—the heart. Do not allow it to get attached to external objects. Keep it steady, free from any oscillation.

Devikalottara-Jnanachara-Vichara-Patalam

Renun. 152b
Supra
532

Dwell, O mind, within yourself:
Enter no other's home.
If you but seek there, you will find
All you are searching for.

Hindu Song

Void
724

In thy chamber thou shalt find what abroad thou shalt too often lose.

The Imitation of Christ, I. xx. 5

Supra
523

Only in the most intimate sanctuary of the soul will the mystery of the spirit be revealed.

Henry Madathanas

I'll bury thee in a triumphant grave:
A grave? O, no! a lanthorn, slaughter'd youth,
For here lies Juliet, and her beauty makes
This vault a feasting presence full of light.

Shakespeare (*Romeo and Juliet,* V. iii. 83)

Death
226
P. State
579

A spirit is within, which by deliberate skill you must separate from the body.

The Glory of the World

Action
338

There is a hidden manna altogether unknown to all but they who taste it. For this sweetness is of the heart not of the flesh, so that no carnal person can know it. 'Thou hast put gladness in my heart' (*Psalm* IV. 7). Bodily delights even as the body itself may be discerned by the bodily eye; the delights of the heart and this heart itself cannot be seen by the eye of the flesh. Why should a man know spiritual pleasure unless truly and without feigning he enters into his own heart and abides there.

Richard of Saint-Victor

Center
816
Renun.
146

A work done from within thy soul is necessary, for no door will be opened to thee by things given on loan.

A water-spring inside the house is better than an aqueduct that comes from outside.

How goodly is the Conduit which is the source of all things! It makes you independent of these other conduits.

You are quaffing drink from a hundred fountains: whenever any of those hundred yields less, your pleasure is diminished:

But when the sublime Fountain gushes from within you, no longer need you steal from the other fountains.

Rûmî

Beauty 663
Creation 28
Center
847

Whoever wishes to know the method of compounding *chin i huan tan* (gold fluid returned medicine) ought to begin by seeding and planting it in his own garden. It is needless to borrow the help of blowing or fanning (the fire): the *tan* will ripen naturally and emerge from the true womb.

Pilg. 378

Chang Po-tuan

Do thou all within. And if perchance thou seekest some high place, some holy place, make thee a temple for God within. For the temple of God is holy: which you are (I Cor. iii, 17). In a temple wouldest thou pray? Pray within thyself. Only first be thou a temple of God, because he in his temple will hear him that prayeth.

Creation 42
Knowl. 749

St Augustine

Prayer is not verbal. It is from the heart. To merge into the Heart is prayer.

Sri Ramana Maharshi

The wise man remains wholly centred in himself.

Yoga-Vasishtha

With your whole heart and soul, seek to regain Reality, nay, seek for Reality within your own heart, for Reality, in truth, is hidden within you. The heart is the dwelling-place of that which is the Essence of the universe, within the heart and soul is the very Essence of God. Like the saints, make a journey into your self; like the lovers of God, cast one glance within. As a lover now, in contemplation of the Beloved, be unveiled within and behold the Essence. Form is a veil to you and your heart is a veil. When the veil vanishes, you will become all light.

Realiz. 873
Pilg. 387
Realiz. 887

ʿAṭṭâr

Man, enter into thyself. For this Philosophers' Stone
Is not to be found in foreign lands.

Angelus Silesius

Near me is my God, who is eternal, who is the true lover, in whom are delight and satisfaction, and in whom is all wealth. All the time has this wealth immeasurable been beside me. But I left my Lord, and like a fool I courted man, who can never satisfy my desires, who, on the other hand, causes misery, fear, disease, grief, and delusion.

Srimad Bhagavatam, XI. iii (Spoken by the courtesan Pingala)

Try to enter your inner treasure-house and you will see the treasure-house of heaven. For both the one and the other are the same, and one and the same entrance reveals them both. The ladder leading to the kingdom is concealed within you, that is, in your soul. Wash yourself from sin and you will see the rungs of the ladder by which you ascend thither.

Sin 77

St Isaac of Syria

The most comfortable, best furnished, and wonderful cave, inspiring holy fear, is in your heart. It is the hridaya-guha spoken of in the Upanishads, and in which formerly dwelt Dattâtreya, Shankara, and Yajnavalkya. It is here in modern times that rishis still dwell, who have withdrawn their mind and senses from the outward world.

Swami Sivananda

Do not you believe that there is in man a deep so profound as to be hidden even to him in whom it is?

Introd. *Holiness*

St Augustine

Not knowing how near the Truth is,
People seek it far away,—what a pity!
They are like him who, in the midst of water,
Cries in thirst so imploringly.

Center 841

Hakuin

That everyone points to the chest when referring to himself by gesture, is sufficient proof that the Absolute resides as the Self in every Heart.

Sri Ramana Maharshi

Now you may ask, How can we come to perceive this direct leading of God? By a careful looking at home, and abiding within the gates of thy own soul. Therefore, let a man be at home in his own heart, and cease from his restless chase of and search after outward things.

Tauler

O you mad and miserable poeple, since you cannot grasp foreign things through anything but yourselves, how will you ever grasp outward things, when you have lost the inner ones? Why do you seek goods far away, as foreigners, when they are near, or rather within yourselves?

Introd. *Renun.*

Marsilio Ficino

Lord, I am as willing to be poor in all those things of which I have been deprived as I am ready to be rich, O Lord, if it be Thy will and to Thy glory: not my will according to nature, O Lord, but Thy will and my will according to spirit be done. For I am Thine own, O Lord, and would as well be in hell as in heaven, if it were to Thy glory. Lord, do unto me according to Thy good pleasure.

Renun. 136 *Conform.* 170

Ruysbroeck

Roaming with thee, even hell itself would be to me a heaven of bliss.

Râmâyana (Sita to Rama)

Might I but through my prison once a day
Behold this maid: all corners else o' th' earth
Let liberty make use of: space enough
Have I in such a prison.

Shakespeare (*The Tempest*, I. ii. 487)

Again and again look within thine own mind.

Padma-Sambhava

Flight 944

Seek yourself outside of the world, but in order to seek and to find yourself outside, fly outside, or rather look outside. . . . Leave behind the narrowness of this shadow and look within yourself.

Marsilio Ficino

All people have the common desire to be elevated in honor, but all people have something still more elevated in themselves without knowing it. What people usually consider as rank or honor is not true honor, for he whom Chao Meng (a powerful lord) has honored, Chao Meng can also disgrace.

Mencius

Infra 542

Learn to love solitude, and, ever alert, think of me without ceasing.

Srimad Bhagavatam, XI. viii

Look inward, and in a flash you will conquer the Apparent and the Void.

Seng-tsʻan

The Safeguarding of Prayer

That we may be constant in prayer, so must we be faithful to do all our business the whole day through with such a steadiness that nothing may make us wavering.

This prayer, which is performed in outward business, is a fruit of the inward prayer or Prayer of Silence. It is like the heat of a stove, which holds long though you put no more wood in; it is the anointing of prayer; it is the smell of the precious incense which has spread itself abroad; it is a hidden taste of the heavenly manna wherewith we are fed; it is a cooling and refreshing of the water which we have drunk; it is an impression of the Love and Presence of God in the heart itself which is continued in the performing of our business, and which serves to call a man back to his inward business when he has been scattered outwardly. And instead of suffering the mind to rove, when we rise up from inward prayer which is performed at set times, we should rather take care to preserve what we have therein received as a precious powerful water which we must be careful not to let evaporate.

Johannes Kelpius

Inv. passim

The whole life of a saint should be one great continual prayer.

Origen

When a man is always occupied with the cravings of desire and ambition, and is eagerly striving to satisfy them, all his thoughts must be mortal, and, as far as it is possible altogether to become such, he must be mortal every whit, because he has cherished his mortal part. But he who has been earnest in the love of knowledge and of true wisdom, and has exercised his intellect more than any other part of him, must have thoughts immortal and divine . . . since he is ever cherishing the divine power, and has the divinity within him in perfect order.

Plato (*Timaeus*, 90 B)

Introd.
Judgment
Knowl. 761
Metanoia
480

The Spirit of Prayer is . . . the desire of the soul turned to God.

William Law

Metanoia
484

He that thus seeks shall find: he shall live in truth, and that shall live in him; it shall be like a stream of living waters issuing out of his own soul; he shall drink of the waters of his own cistern, and be satisfied; he shall every morning find this heavenly manna lying upon the top of his own soul, and be fed with it to eternal life; he will find satisfaction within, feeling himself in conjunction with truth, though all the world should dispute against him.

John Smith the Platonist

Supra
536

In tireless prayer . . . you shall be flooded to the depths of your soul with a gracious sweetness of spirit and shall scent the inexpressible goodness of God in an interior taste and most aromatic incense, which, passing, He will minister to you; and with which you shall be satisfied when His glory shall appear. You shall be satisfied, I repeat, and with no disgust of satiety, for this immortal food is life itself.

Nicholas of Cusa

Center
847

If we carefully cultivate the life of our souls, we shall be sons and heirs of God, and shall be able to do that which now seems impossible.

Basil Valentine

The pure atmosphere which the *sâdhu* creates around himself through meditation is the river where the spirit is purified, as in a bath, of its bad thoughts and impressions.

Swami Ramdas

The whole method begins and ends with a single method: COCTION.

Here is the grand arcanum: it is a celestial spirit, progeny of the sun, moon and stars, which is rendered perfect in the saturnine object by a continual cooking, until it has reached the state of sublimation and power necessary to transform base metals into gold. This operation is accomplished by the hermetic fire. The separation of the subtle from the gross must be performed with care, by continually adding water; for the more the materials are terrestrial, the more must they be diluted and rendered mobile. Continue this method until the separated soul is reunited with the body.

Henry Madathanas

Inv.
1017
Conform.
170
Realiz.
870

Metanoia 480

If you meditate on an ideal you will acquire its nature. If you think of God day and night, you will acquire the nature of God.

Sri Ramakrishna

Prayer oneth the soul to God.

Julian of Norwich

'With this sacred pipe you will walk upon the Earth; for the Earth is your Grandmother and Mother, and She is sacred. Every step that is taken upon Her should be as a prayer. The bowl of this pipe is of red stone: it is the Earth. Carved in the stone and facing the center is this buffalo calf who represents all the four-leggeds who live upon your Mother. The stem of the pipe is of wood, and this represents all that grows upon the Earth. And these twelve feathers which hang here where the stem fits into the bowl are from *Wanbli Galeshka*, the Spotted Eagle, and they represent the eagle and all the wingeds of the air. All these peoples, and all the things of the universe, are joined to you who smoke the pipe—all send their voices to *Wakan-Tanka*, the Great Spirit. When you pray with this pipe, you pray for and with everything.'

Symb. 322

Charity 608

Black Elk (citing the White Buffalo Cow Woman
who brought the calumet to the Sioux)

We must wait on God continually, and importune Him with earnest prayer.

Michael Sendivogius

Use every means all through the day to guard your samadhi-power as you would protect a child.

Dôgen

Realiz. 873

Let me assure you that in our whole work there is nothing hidden but the regimen, of which it was truly said by the Sage that whoever knows it perfectly will be honoured by princes and potentates. I tell you plainly that if this one point were clearly set forth, our Art would become mere women's work and child's play: there would be nothing in it but a simple process of 'cooking'. Hence it has always been most carefully concealed by the Sages. But I have determined to write in a more sympathetic and kindly spirit: know then that our regimen throughout consists in coction and digestion, but that it implies a good many other processes, which those jealous Sages have made to appear different by describing them under different names.

Philalethes

These principles are like seed in the ground, they must continually be visited with heavenly influences, or else your life will be a barren field.

Thomas Traherne

This is a most difficult operation, because the metals, in which the seed is hidden, are so firmly and tightly compacted, and will not yield to violence, but only to a gentle and exquisitely subtle chemical process.

Philalethes

In order to keep our spirits active, we should do well to fix upon certain special times for the practice of the *Nembutsu*.

Hônen

Some hours of the day (should be) set apart for meditation so that the work in which we would be engaged during the other hours may be done as a spontaneous and blissful outflow of the Eternal Reality dwelling within us.

Swami Ramdas

A half-hour of silent repose after meditation is necessary: for at the time of meditation you may not obtain the desired effect: it can come a little afterwards. Remember also that turning the attention abruptly to secular things right after meditation not only does much harm in general, but more particularly hinders the growth of spirit towards realization.

Swâmi Brahmânanda

Let a gentle temperature be kept up till the husband and the wife become *one*.

The Glory of the World

Love 625

Mind is a veil that shuts you from the splendour of your immortal spirit, which is your real being. Tear up this veil by means of constant meditation and self-surrender. Retire within yourself from time to time and lose your little ego in the infinite consciousness of your supreme Self.

Swami Ramdas

Supra 528

'Appoint a day for Me and a day for thyself. Begin with My day, and My day shall transport thy day.'

Niffarî

The spirit of prayer is for all times and all occasions; it is a lamp that is to be always burning, a light to be ever shining; everything calls for it, everything is to be done in it and governed by it, because it is and means and wills nothing else but the whole totality of the soul, not doing this or that, but wholly, incessantly given up to God to be where and what and how He pleases.

William Law

Renun. 146

It is comparatively easy to meditate on 'Aham Brahma asmi', when you are seated in a fixed posture, alone in a room; but it is very, very difficult to retain this thought in a crowd and when the body is in motion. If you meditate for an hour feeling that you are Brahman and during the other twenty-three hours of the day think that you are the body, your sâdhanâ will not produce the desired results.

Swami Sivananda

Conform. 170

My prayer is my inclination, and my inclination is my prayer.

Johannes Kelpius

You must never forget the idea 'I am Brahman', the idea of the divine presence, for even a single instant. The forgetfulness of God is a veritable death, a suicide; it is the murder of the Âtman (âtmadroha). And it is the greatest sin.

Swami Sivananda

Reality 775 *Death* 208 *Heterod.* 423

'If you, my son, wish to prepare this precious Stone, you need not put yourself to any great expense. All that you want is leisure, and some place where you can be without any fear of interruption. Reduce the Matter (which is *one*) to powder, put it, together with its water, in a well-closed vessel, and expose it to continuous, gentle heat, which will then begin to operate, while the moisture favours the decomposition. . . . Kindle the fire of the Sages, and watch it well so that there may be no smoke. The rest you may leave to me (Nature).'

John A. Mehung

Do not waste your time in building mud furnaces for *tan* (medicine). The compounding of medicines should be carried out in a *yen yüeh lu* (lie-down-moon-furnace) which has a natural true fire and season, and does not require a supply of wood or charcoal and the blowing of air.[1]

Chang Po-tuan

Metanoia 488

Common mercury becomes more fusible, our quicksilver more fixed, the more it is subjected to coction.

Michael Sendivogius

By spiritual force a man can change clay or stone into gold.

Nâgârjuna

Love 614 *Death* 220 *P. State* 563

If (the love of God) but once kindle a fire within thee, my son, thou shalt then certainly feel how it consumeth all that which it toucheth; thou shalt feel it in the burning up thyself, and swiftly devouring all egoity, or that which thou callest I and Me, as standing in a separate root, and divided from the Deity, the fountain of thy being. And when this enkindling is made in thee, then the love doth so exceedingly rejoice in thy fire, as thou wouldst not for all the world be out of it; yea, wouldst rather suffer thyself to be killed, than to enter into thy something again. This fire now must grow hotter and hotter, till it shall have perfected its office with respect to thee, and therefore wilt not give over, till it come to the seventh degree. Its flame hence also will be so very great, that it will never leave thee,

[1] In *The Secret Papers in the Jade Box of Ch'ing-hua* (Davis/Yün-ts'ung: AAAS 73–13) we even have a 'furnace' diagram (p. 387, Figure 3). It shows the Sage himself seated in meditation, his robe open, with eight *kua* (broken and closed lines representing different combinations of *yin* and *yang*) forming an octagon around his navel. Other centers are marked in a way that immediately brings to mind the *kundalinî*. An inscription at the left reads: 'This *ting* (tripod) and furnace picture is as clear as is indicated here'.

though it should even cost thee thy temporal life; but it would go with thee in its sweet loving fire into death; and if thou wentest also into hell, it would break hell in pieces also for thy sake. Nothing is more certain than this; for it is stronger than death and hell.

Boehme

***Judgment* 266**

Man Created for Contemplation

Man's chief work is but to praise God. To Him it belongs to satisfy thee by His beauty, to thee to praise Him in acts of thanksgiving. If thy works be not the praise of God, thou art beginning to love thyself. . . . For it is not He that increaseth by our praises, but we. God is neither the better if thou praise Him, nor worse if thou disparage Him; but thou, by praising Him that is good, art the better; by disparaging thou art the worse, for He remaineth good, as He is.

St Augustine

Reality 773

God did not put you in this world for any need he had of you, who are wholly useless to him, but only in order to exercise in you his goodness, giving you his grace and his glory. . . . Since you have been created and put in the world with this intention, all actions contrary to this must be rejected and avoided, and those which serve for nothing to this end must be scorned, as vain and superfluous.

Consider the misfortune of the world which never thinks of this, but lives as though it believed it was only created to build houses, plant trees, amass riches and indulge in idle talk.

St Francis de Sales

Action 340

Illusion 99

He honors God in the best manner who renders his intellect as much as possible similar to God.

Sextus the Pythagorean

Fear God, and strengthen thyself to serve Him, for He formed thee to do Him honour.

Yesod Hayirah, 157

Judgment 239

Never forget the ideal of human life. . . . Since it has been given you to attain human birth, reject all the pleasures of the world. Be firmly resolved to realize God and attain infinite felicity. Do not give in, even if you have to die in this effort.

Swâmi Brahmânanda

Judgment 250

Holy War 394

For you are never your true self, till you live by your soul more than by your body, and

Center 847

you never live by your soul till you feel its incomparable excellency, and rest satisfied and delighted in the unsearchable greatness of its comprehension.

Thomas Traherne

Realiz. 887, 873; Creation 42

The entire effort of our Soul is to become God. This effort is as natural to man as that of flying is to birds. For it is inherent in all men, everywhere and always; therefore it does not follow the incidental quality of some man, but the nature of the species itself. . . . For who implanted in our Souls this (tendency toward God) but God Himself, whom we seek?

Marsilio Ficino

Holiness 900

For this is the will of God, our sanctification.

Ruysbroeck

Thou awakest us to delight in Thy praise.

St Augustine

Creation 48

And to this end did He make men, that they might contemplate heaven, and have dominion over all things under heaven.

Hermes

The creature pleasures of the soul God has no stomach for, and when she realizes this she discards the joys in which God has no share. . . . For God is spirit and he en-spirits the soul who, in her spiritual nature, belongs to an order above creatures.

Eckhart

The Human Soul . . . participates in the Divine Nature, after the manner of an eternal Intelligence: for the soul is ennobled and denuded of matter by that Sovereign Power in proportion as the Divine Light of Truth shines into it, as into an Angel; and Man is therefore called by the Philosophers the Divine Animal.

Dante (*Il Convito*, III. ii. 3)

The Bengali word *mânush* (man) implies that man possesses *hûsh*, the power to contemplate the Divine.

Ananda Moyî

Knowl. 749; Center 828, 841

The prime operation of every *Intellectual Spirit is contemplation*. The first and immediate *Object* of its *contemplation* is its *own Essence*. In this *Glass* of living and immortal Light, all other things, according to their proper *essences* in their several and *essential forms*, appear to it most clearly and delightfully, as its own *Births* and *Beauties*.

Peter Sterry

. . . Speculation, that supreme end for which the Primal Good brought into being the human race.

Dante (*De Monarchia*, I. iii. 90)

It is our end to see God through the intellect and to enjoy the seen God through the will.

Marsilio Ficino

This I am sure is so certain, that nothing is more: that there is no Happiness for Man, but in his Relation to, and conjunction with God and things Immortal: For there is no rational satisfaction.

Peace 694

Benjamin Whichcote

In the life of the Indian there was only one inevitable duty,—the duty of prayer—the daily recognition of the Unseen and Eternal. His daily devotions were more necessary to him than daily food.

Ohiyesa

You are in the world to concentrate your mind on God. . . . It is your most important duty. You forget this duty because of illusion (moha) which takes the form of family, children, money, power, position, honour, name and fame.

Illusion 85

Swami Sivananda

Having attained human birth, which is an open gateway to Brahman, one who . . . remains attached to the ties of the world is not fit to be called human. Pleasures of sense may be had in all lives: leave them, then, to the brutes! Never does the wise man yearn after them.

Srimad Bhagavatam, XI. iii

Jesus said: If you bring forth that within yourselves, that which you have will save you. If you do not have that within yourselves, that which you do not have within you will kill you.

P. State 563

The Gospel according to Thomas, Log. 70

GRACE

Allâh is gracious unto His slaves. He provideth for whom He will. And He is the Strong, the Mighty.

Qur'ân, XLII. 19

Eckhart says: 'Christ has by nature what man has by grace.' If we were God, then we could count on the sun rising. But being man, every act of life bestowed on us is a token of God's grace, since we have in our power none of the gifts of Creation, but only the uses. What could our science do with existence, if it first had to bring existence into being in order to exist?

'We can distinguish three aspects in spirituality—"virtue", "art" and "grace". In virtue and in art, man is active: with respect to grace he is passive. From the standpoint of virtue or merit, grace is gratuitous; from the standpoint of technique or art, grace is consequent to our power of concentration (although in a partial way only), in conformity with the law of causality; but in this case, even, grace still retains its "gratuity", since it never has a positive human cause, its positive cause always coming from God. Herein lies the profound meaning of the "gratuity" of grace, for it is evident that this gratuity could never signify that God is arbitrary, or that grace lacks a sufficient reason. The cause cannot come from man except in a wholly negative manner, in the sense that man can but remove the obstacles to pre-existing grace, which has for its eternal and immutable cause the Divinity. This is why *Yoga* is always associated with *Ishvara,* or God as "Being"; without the initial grace of *Ishvara, Yoga* itself would be inconceivable.[1] The Sufis analogously teach that the initiate can do nothing without the initial grace of *Allâh* (*tawfîq,* "help of God"), and that the spiritual states (*ahwâl,* from *hâl*) could never be the "productions" of any human industry. This is not to say that man cannot take measures to eliminate the obstacles to grace, for if the powers of the lower and passional soul (*nafs*) can repel a state of light, so can the powers of the higher and spiritual soul (*rûh*) repel the effects of the lower powers, and consequently all that is opposed to grace (in itself eternal). Moreover, a state of grace must be fixed in the soul by an effort at once intellectual, moral and technical, or rather, it is the soul which must be fixed in a state of grace—a fixation which the Sufis call a "station" (*maqâm*); but there are graces (*ahwâl*) which quite clearly are independent of

Introd. *Holiness*

[1] In the same way, explaining the iconography of Śiva's dance in the *Devadâruvana,* Coomaraswamy tells how Śiva, through 'seducing' the wives and daughters of the Rishis in the Deodar Forest and upsetting their ascetical regime, brings the Rishis 'to realize that it is not by asceticism, rites, or mere learning, but only by Śiva's own Grace that He can be reached' ('An Ivory Casket from Southern India': *The Art Bulletin,* Sept., 1941).

effort, and where the aspect of gratuity is directly manifest. God infinitely possesses the perfections of liberty and necessity, and these two aspects must express themselves in grace: the first is more particularly in relation with love, the second with gnosis' (Schuon: 'Le Yoga comme principe spirituel'; in *Yoga, Science de l'Homme intégral,* Paris, 1953, pp. 31–32).

Introd.
Conform.

Liberty

You may try thousands of times, but nothing can be achieved without God's grace. One cannot see God without His grace. Is it an easy thing to receive the grace of God? One must altogether renounce egotism; one cannot see God as long as one feels, 'I am the doer.' . . . God doesn't easily appear in the heart of a man who feels himself to be his own master. But God can be seen the moment His grace descends.

Action 340

Sri Ramakrishna

If we let fly an arrow, that action is not from us: we are only the bow, and the shooter of the arrow is God.

Rûmî

I can of mine own self do nothing.

St John, v. 30

No man is made suddenly sovereign in grace, but through long exercise and sly working a soul may come thereto, namely when He helpeth and teacheth a wretched soul in whom all grace lieth, for without special help and inwardly teaching of Him may no soul come thereto.

Walter Hilton

The vision of God is possible only through His Grace.

Ananda Moyî

I do believe, I shall inherit
Heaven, by Christs mercies, not my merit.

Humility 191

Robert Herrick

No creature can do more than in her lies. The soul makes headway solely by the light that God has given her, that being her own, presented her by God as a bridal gift. God comes in love with intent that the soul may arise, that in love she may energise above herself. . . . Soul does not ply the work of grace (since that is not her nature) till she is gotten yonder, where God is plying himself, where the work is as noble as the worker, his own nature, to wit.

Eckhart

But when Thou O Lord shalt command,
Then only shall I be released from self!

Mechthild of Magdeburg

He was asked, 'When shall a man be freed from his wants?' 'When God shall free him,' he replied; 'this is not affected by a man's exertion, but by the grace and help of God. First of all, He brings forth in him the desire to attain this goal. Then He opens to him the gate of repentance. Then He throws him into self-mortification, so that he continues to strive and, for a while, to pride himself upon his efforts, thinking that he is advancing or achieving something; but afterwards he falls into despair and feels no joy. Then he knows that his work is not pure, but tainted, he repents of the acts of devotion which he had thought to be his own, and perceives that they were done by God's grace and help, and that he was guilty of polytheism in attributing them to his own exertion. When this becomes manifest, a feeling of joy enters his heart. Then God opens to him the gate of certainty, so that for a time he takes anything from any one and accepts contumely and endures abasement, and knows for certain by Whom it is brought to pass, and doubt concerning this is removed from his heart. Then God opens to him the gate of love, and here too egoism shows itself for a time and he is exposed to blame, which means that in his love of God he meets fearlessly whatever may befall him and recks not of reproach; but still he thinks "I love" and finds no rest until he perceives that it is God who loves him and keeps him in the state of loving, and that this is the result of divine love and grace, not of his own endeavour. Then God opens to him the gate of unity and causes him to know that all action depends on God Almighty. Hereupon he perceives that all is He, and all is by Him, and all is His; that He has laid this self-conceit upon His creatures in order to prove them, and that He in His omnipotence ordains that they shall hold this false belief, because omnipotence is His attribute, so that when they regard His attributes they shall know that He is the Lord. What formerly was hearsay now becomes known to him intuitively as he contemplates the works of God. Then he entirely recognises that he has not the right to say "I" or "mine". At this stage he beholds his helplessness; desires fall away from him and he becomes free and calm. He wishes that which God wishes; his own wishes are gone, he is emancipated from his wants, and has gained peace and joy in both worlds.'

Metanoia 484
Pilg. 366
Suff. 126b
Knowl. 749
Reality 803
Death 220
Realiz. 873

Abû Sa'îd ibn Abi 'l-Khayr

Rarely come the Divine Intuitions except on a sudden, lest the slaves should claim them as the result of their preparations.

Ibn 'Aṭâ'illâh

Maistryefull merveylous and Archimastrye
Is the tincture of holi *Alkimy:*
A wonderfull *Science,* secrete Philosophie,
A singular grace & gifte of th'almightie:
Which never was founde by labour of Mann,
But it by Teaching, or Revelacion begann.
It was never for Mony sold ne bought,
By any Man which for it hath sought:
But given to an able Man by grace,
Wrought with greate Cost, with long laysir and space.

Rev. 959
Action 346

Thomas Norton

Holiness 924
Metanoia 488

God was made man and took upon himself by grace the nature of all things in time even as in eternity he has them all by nature.

Eckhart

O Lord, how entirely needful is thy grace for me, to begin any good work, to go on with it, and to accomplish it.

For without that grace I can do nothing, but in thee I can do all things, when thy grace doth strengthen me.

The Imitation of Christ, III. lv. 4

We thank thee, O thou Most High, with heart and soul wholly uplifted to thee;
For it is by thy grace alone that we have attained to the light, and come to know thee.

Hermes

So then it is not of him that willeth, nor of him that runneth, but of God that sheweth mercy.

Romans, IX. 16

In truth I am not worthy of it, yet this benediction depends on the grace of God.

Johann Tarnovius (replying when asked if he were a member of the Rosy Cross)

Necessity will make us all forsworn
Three thousand times within this three years' space:
For every man with his affects is born,
Not by might master'd, but by special grace.

Illusion 89

Shakespeare (*Love's Labour's Lost*, I. i. 148)

Indeed, indeed, Repentance oft before
I swore—but was I sober when I swore?
And then and then came Spring, and Rose-in-hand
My thread-bare Penitence apieces tore.

Metanoia 495

Omar Khayyâm

No single infidel would remain in the world
If he could see the shaking aside
Of those black curls,[1]
And on the earth there would not remain a faithful soul
If they were always in their place.

Illusion 85

Shabistarî

Our Art, its theory as well as its practice, is altogether a gift of God, Who gives it when and to whom He elects.

The Golden Tract

[1] 'Curls and tresses mean plurality veiling the face of Unity from its lovers' (Lederer).

Therefore, if any man desire to reach this great and unspeakable Mystery, he must remember that it is obtained not by the might of man, but by the grace of God, and that not our will or desire, but only the mercy of the Most High, can bestow it upon us. . . .

When you are in inward harmony with God's world, outward conformity will not be wanting. Yet our artist can do nothing but sow, plant, and water: God must give the increase.

The Sophic Hydrolith

Infra 558
Introd. *Inv.*

When I was preparing the substance, after discovering the true method, I was so seriously interfered with by the persons with whom I lived that I was almost on the point of giving up the whole thing in despair. At length I communicated my discovery to a friend, who faithfully executed my instructions, and brought the work to a successful issue. For which Blessed Gift may God be praised, world without end.

The Book of Alze

Contem. 528
Inv. 1007

By His hand every impossible thing is made possible; by fear of Him every unruly one is made quiet.

Rûmî

Faith 514

We can realize God only by His Grace—not by our own efforts and initiative.

Swami Ramdas

This Thought of Enlightenment[1] has arisen within me I know not how, even as a gem might be gotten by a blind man from a dunghill; it is an elixir made to destroy death in the world, an unfailing treasure to relieve the world's poverty, a supreme balm to allay the world's sickness, a tree under which may rest all creatures wearied with wandering over life's paths, a bridge open to all wayfarers for passing over hard ways, a moon of thought arising to cool the fever of the world's sin, a great sun driving away the gloom of the world's ignorance, a fresh butter created by the churning of the milk of the Good Law. For the caravan of beings who wander through life's paths hungering to taste of happiness this banquet of bliss is prepared, that will satisfy all creatures coming to it.

Śânti-deva

Introd.
Introd. *Peace*
Pilg. 385
P. State 590
Symb. 317
Center 847

But for its perfume, I should not have found the way to its taverns: and but for its resplendence,[2] the imagination would not have pictured it.

Ibn al-Fâriḍ

Ecstasy 637

Let nobody presume upon his own powers for such exaltation or uplifting of the heart or ascribe it to his own merits. For it is certain that this comes not from human deserving but is a divine gift.

Richard of Saint-Victor

[1] 'The "Thought of Enlightenment" (*Bodhi-chitta*) is to the Mahâyâna what "grace" is to Christian theology' (Explanatory note, p. 96).

[2] According to Nâbulusî the commentator, this resplendence is the human intellect, which is a flash of the Primal Intelligence.

Action 329

We who live in the world, still attached to karmas, can overcome the world by thy grace alone.

Srimad Bhagavatam, XI. ii

'If I had not stretched over thee the compassion of mercifulness, the hand of temporality would have constricted thee against gnosis.'

Niffarî

P. State 583

It is by the power of Buddha only that one can see that pure land (of Buddha) as clear as one sees the image of one's face reflected in the transparent mirror held up before one.

Amitâyur-Dhyâna-Sûtra

Necessity

Renun. 146

Many men that are reformed in faith set not their hearts for to profit in grace.

Walter Hilton

Realiz. 873 *Contem.* 523 *Center* 841 *Love* 618 *Creation* 31

It is one flash, the being-ready and the pouring-in. Nature reaching her summit, God dispenses his grace: the instant the spirit is ready God enters without hesitation or delay. . . . Thou needst not seek him here or there, he is no further off than at the door of thy heart; there he stands lingering, awaiting whoever is ready to open and let him in. Thou needst not call to him afar, he waits much more impatiently than thou for thee to open to him. He longs for thee a thousandfold more urgently than thou for him: one point the opening and the entering.

Eckhart

Now, this continual knocking of Christ at the door of the heart sets forth the case or nature of a continual, immediate divine inspiration within us: it is always with us, but there must be an opening of the heart to it: and though it is always there, yet it is only felt and found by those who are attentive to it, depend upon, and humbly wait for it.

William Law

Contem. 542

Gods Grace deserves here to be daily fed,
That, thus increast, it might be perfected.

Robert Herrick

Grace does nothing but re-form and convey into God. Grace makes the soul deiform. God, the ground of the soul and grace go together.

Eckhart

If thou knewest perfectly to annihilate thyself, and to empty thyself of all created love, then should I be constrained to flow into thee with great abundance of grace.

The Imitation of Christ, III. xlii

Humility 199

The lightning of the Son of God illuminates in an instant
The hearts which submit entirely to Him.

Angelus Silesius

When the Holy Ghost, her Spouse, has found Mary in a soul, He flies there.

St Louis de Montfort

P. State 579
Love 618

While we sojourn here our faith can continually increase as can also charity. Each it is true is in such a degree that at a given moment and given his nature he may not be in another: but while he is in one degree he is in potency to another, though never in potency to the infinite charity of Christ. But we must, by the grace of our Lord Jesus Christ work to actualize our total possibility, and to move from virtue to virtue and from degree to degree, by him who is faith and charity. Of ourselves, as ourselves, we can without him do nothing. But all that we can do, we can do in him who alone can make up for our deficiencies, that in the day of resurrection we may be found to be integral and noble members of him. Believing and loving with all our power, we can confidently beg of Him this grace of increased faith and charity, in tireless prayer approaching His throne with great trust. For He is infinitely merciful and loving and permits none to be defrauded of His holy desire.

Nicholas of Cusa

Death 223

Faith 505

Divine Grace is essential for Realisation. It leads one to God-Realisation. But such Grace is vouchsafed only to him who is a true devotee or a yogin, who has striven hard and ceaselessly on the path towards freedom.

Sri Ramana Maharshi

Holy War 405

'To tell you the truth,' said Monkey, 'although of course I am much indebted to the Master for his instruction, I have also been working very hard day and night on my own.'

Monkey

This gift is from God and not of man's deserving. But certainly no one ever receives such a great grace without tremendous labour and burning desire.

Richard of Saint-Victor

The learning of the great medicine is sometimes easy and sometimes difficult. It depends partly on us and partly on the will of heaven. If one does not perform good works and spread virtue in the *yin*[1] district, his work will be handicapped frequently by groups of devils.

Chang Po-tuan

Metanoia 484

[1] Namely, the substantial, ephemeral pole of existence, domain of the fugitive soul which has to be bridled and brought under control if the 'great medicine' is to succeed.

It is true that we can do no good thing without God's ordinary influence, except we make progress by means of a special influence from the Holy Spirit: yet, at the same time, man may do his part, inasmuch as his will has power to withstand the offers of the Holy Spirit, and to cleave to his own way. God does not justify a man without his own free will; even as our eyes cannot see except they are enlightened by the sun or any other light, yet even when we have the light we must open our eyes, or we can never see it.

Conform. 166

Tauler

The union of man and God is not a matter of grace (for grace is creature and creature has nothing to do with it), except in the ground of divinity, where the three Persons are one in nature and it (grace) is that nature itself.

Eckhart

Election and Grace

Lo! this is a Reminder, that whosoever will may choose a way unto his Lord.
Yet ye will not, unless Allâh willeth. Verily, Allâh is Knower, Wise.

Qur'ân, LXXVI. 29, 30

Supra Introd.

For Christ went before by nature, and we come after by grace. His nature is more worthy than grace, and grace is more worthy than our nature.

The Epistle of Privy Counsel, VII

Contem. 523
Peace 694
Realiz. 873

Thy son cries thus to thee by means of the things thou hast made;
but he has got from thine eternity the praises which he utters.
I have seen that which I seek;
I have found rest according to thy purpose;
by thy will I am born again.

Hermes

The soul of Man to God is as the flower to the sun; it opens at its approach, and shuts when it withdraws.

Benjamin Whichcote

How does God enter the soul? . . . First in his grace whereby a man being gratified is filled with the desire of perfecting virtue as a whole, mingled with alarm lest any creature ever filch it from him.

Beauty 676

Eckhart

For we are now so blind and unwise that we never seek God till He of His goodness shew Himself to us. And when we aught see of Him graciously, then are we stirred by the same grace to seek with great desire to see Him more blissfully.

Julian of Norwich

This treasure of God's kingdom is hidden by time and multiplicity by the soul's own activity and by her creaturehood. The more the soul departs from all this multiplicity the more God's kingdom is revealed in her. But the soul is not able for this without the help of grace. An she find it, it is grace that has aided her thereto, for grace is innate in her highest prototype. There the soul is God, using and enjoying all things God-fashion.

Eckhart

Holiness 924

Sometimes God acts as the magnet and the devotee as the needle. God attracts the devotee to Himself. Again, sometimes the devotee acts as the magnet and God as the needle. Such is the attraction of the devotee that God comes to him, unable to resist his love.

Sri Ramakrishna

Knowl. 749

This thing we tell of can never be found by seeking, yet only seekers find it.

Bâyazîd al-Bisṭâmî

For thirty years I went in search of God, and when I opened my eyes at the end of this time, I discovered that it was really He who sought for me.

Bâyazîd al-Bisṭâmî

Love 618

I am a great and glorious King in the earth. . . .
Yet at first I was of ignoble birth,
Till I was set in a high place.
To reach this lofty summit
Was given me by God and Nature.

Abraham Lambspring

Holiness 921

Art accomplishes by means of nature what nature is incapable of accomplishing without the aid of art.[1]

Henry Madathanas

Orthod. 280

However small what comes from God, if treated rightly it recurs.

Eckhart

Contem. 542

Fixing thy heart on Me, thou shalt, by My grace, overcome all obstacles; but if, through egoism, thou wilt not hear Me, thou shalt perish.

Bhagavad-Gîtâ, XVIII. 58

[1] 'Mahâyâna Buddhism . . . contends that salvation is not attainable by self-effort, but by self-effort directed towards another saving Power outside of ourselves' (Coates and Ishizuka: *Hônen the Buddhist Saint,* p. XIX).

The law was therefore given that grace might be sought; grace was given that the law might be fulfilled.

St Augustine

Wr. Side 474 *Knowl.* 749

'Many are called, but few chosen;' for they will not; they give their free will into God's anger, where they are even apprehended, and so are chosen to be 'children of wrath;' ... for God's love chooses only its likeness, and so likewise God's anger....

P. State 573 *Sin* 57

Thus also man perishes among the evil company in evil vain ways: God offers him his grace that he should repent: but evil company and the devil lead him in wicked ways, till he be even too hard captivated in the anger: and then it goes very hardly with him; he indeed was called, but he is evil: God chuses only children: Seeing he is evil, the choice passes over him; but if he again reforms and amends, the eternal choice (or election) does again receive him.

Boehme

Metanoia 493 *Faith* 510

God did not deprive thee of the operation of his love, but thou didst deprive Him of thy cooperation. God would never have rejected thee, if thou hadst not rejected his love. O all-good God, thou dost not forsake unless forsaken, thou never takest away thy gifts until we take away our hearts.

St Francis de Sales

Grace is necessary to salvation, free will equally so—but grace in order to give salvation, free will in order to receive it. Therefore we should not attribute part of the good work to grace and part to free will; it is performed in its entirety by the common and inseparable action of both; entirely by grace, entirely by free will, but springing from the first in the second.

St Bernard

THE PRIMORDIAL STATE

How dreadful is this place! this is none other but the house of God, and this is the gate of heaven.

Genesis, XXVIII. 17

Purity rightly connotes wholeness and simplicity—the elemental unity and original perfection proper to the nature of a given thing or substance, like smokeless fire, the whiteness of snow, or brilliance in gems. Man in the 'Edenic state' enjoyed the integrity of his intrinsic nature unencumbered by the obscurations of a dualistic perspective and the intrusion of blemishes extraneous to his birthright—which was, 'properly speaking, the reflection in the human domain, of principial eternity itself.'[1]

Creation 42

'The realization of the individual in his integrality is known by all traditions as the restoration of what they call the "primordial state", which is considered the state of true man, and which already escapes certain of the limitations characteristic of the ordinary state, notably, the limitation due to the temporal condition. The being who has attained this "primordial state" is still only a human individual, actually possessing no supraindividual state; and yet at this point he is liberated from time: the apparent succession of things is transmuted for him into simultaneity; he consciously possesses a faculty which is unknown to ordinary man and which one could call the "sense of eternity". This is of extreme importance, for he who cannot get away from the viewpoint of temporal succession so as to envisage all things in simultaneous mode is incapable of the least conception of a metaphysical nature. . . .

Introd. *Holiness*

'One may perhaps ask: why this designation of "primordial state"? It is because all traditions, including that of the West (for the Bible itself does not say otherwise), agree in teaching that this state was the normal one at the origin of humanity, whereas the present state is only the result of a decline, the effect of a sort of progressive materialization produced in the course of ages, over the duration of a certain cycle' (Guénon: *La Métaphysique orientale,* pp. 17–18).

Introd. *Creation*

The being who has acquired this state 'possesses it henceforth in a permanent and

[1] Guénon: *Aperçus sur l'Initiation,* p. 278. It should, perhaps, be stressed that the efforts of certain contemporary apologists to reconcile their evolutionist hypotheses with the doctrine of the Fall, by equating man's first state with the imagined existential and sub-rational spontaneity of a happy animal not yet arrived at the complexities of reason, has no basis whatsoever save in their own fancies. 'The image of God' is not that of a happy animal. The confusion here is between the instinct and the intellect, a distinction clearly set forth by Hermes, for example, where he says: 'In the irrational animals, there is instinct (φύσις) in place of intellect (νοῦς)' (*Hermetica,* I. 225): i.e. an animal's 'intellect' is precisely its instinct, in somewhat the same way that a diamond's 'intelligence' is in its brilliancy (cf. Schuon: *De l'Unité transcendante des Religions,* ch. III, sec. 6).

Holiness 902
Flight 949
Introd. *Center*

immutable manner, and nothing can make him lose it: it suffices for him to withdraw from the exterior world and enter into himself, as often as he chooses, to find again at the center of his own being the true "fountain of immortality"' (Guénon: *Aperçus sur l'Initiation*, p. 279).

It may be added that the attainment of this state posthumously corresponds to 'salvation', which is 'paradise', certainly, but not yet 'deliverance'.[1] Also, there can be no access to higher states of being until integration is achieved in this state:[2] 'The being must first of all identify the center of his own individuality (represented by the heart in traditional symbolism) with the cosmic center of the state of existence to which this individuality belongs, and which he will take as base for rising to the higher states. It is in this center that perfect equilibrium is to be found, image in the manifested world of principial immutability; through this center runs the axis which links together all the states, this axis being the "divine ray", which in its ascending course leads directly to these higher states to be attained.... This is why Dante (in the schema of the *Divine Comedy*) to be capable of ascending through the Heavens had first of all to situate himself at a point which is really the center of the terrestrial world,—a point central, moreover, to both time and space, the two conditions which essentially characterize existence in this world' (Guénon: *L'Esotérisme de Dante*, p. 62).

Center 816
Introd. *Peace*

[1] The passages in this and most of the following chapters, which refer in large part to the primordial and higher states, are placed according to doctrinal and not principial or 'existential' logic, i.e. without concern for spiritual hierarchy. For a single set of symbols, through the law of analogy, can refer simultaneously to different levels of reality, and it would be presumptuous if not impossible in all instances to essay an accurate determination of the 'level' of inspiration indicated. Cf. the Introductions to *Symbolism* and *Holiness*.

[2] Which is expressed in Christianity by the teaching that no one comes to Christ except through the Virgin: or in Islam, that no one comes to Allâh except through the Prophet.

Materia Prima and Human Perfection

If this birth really happens no creature can hinder thee, all point thee to God and this birth. We find in lightning an analogy for this. Whatever it strikes, whether tree, beast or man, it turns towards itself with the shock. A man with his back to it instantly flings round to face it: all the thousand leaves of the tree turn over to front the stroke. So with all whom this birth befalls, they are promptly turned towards this birth with everything present, be it never so earthly. Nay, even what was formerly a hindrance is now nothing but a help. Thy face is turned so full towards this birth, no matter what thou dost see and hear, thou receivest nothing save this birth in anything. All things are simply God to thee who seest only God in all things. Like one who looks long at the sun, he encounters the sun in whatever he afterwards looks at. If this is lacking, this looking for and seeing God in all and sundry, then thou lackest this birth.

Realiz. 873

Metanoia 488

Introd. *Holiness*

Eckhart

He who is detached from life, when travelling on land does not meet the rhinoceros or the tiger; when going to a battle he is not attacked by arms and weapons. The rhinoceros can find nowhere to drive his horn; the tiger can find nowhere to put his claws; the weapons can find nowhere to thrust their blades. Why is it so? Because he is beyond the region of death.

Tao Te Ching, L

The wolf also shall dwell with the lamb, and the leopard shall lie down with the kid; and the calf and the young lion and the fatling together; and a little child shall lead them.

Isaiah, IX. 6

Infra 573

The absolutely simple man, by his simplicity, bends all beings ... to the point that nothing opposes him in the six regions of space, nothing is hostile to him, fire and water do not hurt him.

Lieh-tse

Waters 653

Sword cannot pierce It, fire cannot burn It, water cannot wet It.[1] ...
It is eternal, all-pervading, unchangeable, immovable, everlasting.

Bhagavad-Gîtâ, II. 23, 24

Reality 773

Nanpo Tsek'uei said to Nü Yü, 'How is it, that in spite of your great age, you have the freshness of a child?'
Nü Yü replied, 'Through living in conformity with Tao, I have not become exhausted.'
'Could I learn this doctrine?' asked Nanpo Tsek'uei.

[1]See also St. Luke, X. 19.

'You do not have the qualifications,' said Nü Yü. 'There was Puliang I; he had the disposition required. I taught him. In three days, he had forgotten the outer world. Seven more days, and he lost the notion of objects which surrounded him. In nine more days, he had lost the notion of his own existence. Then he acquired clear penetration, and with it the science of the uninterrupted chain of momentary existence. Having acquired this knowledge, he ceased to distinguish the past from the present and the future, life from death. He understood that in reality killing does not take away life, nor does giving birth add to it, that Tao sustains the being across its endings and becomings. Hence It is justly called the Fixed Constant, since from It, the Fixed, are derived all mutations.'

Death 220 *Center* 838 *Creation* 28 *Reality* 773

Chuang-tse (ch. VI)

The Good, Asclepius, must be a thing that is devoid of all movement and all becoming, and has a motionless activity that is centred in itself; a thing that lacks nothing, and is not assailed by perturbations; a thing that is wholly filled with abundance of all that is desired. Everything that furnishes any sort of supply is called good; but the Good is the one thing which is the source of all things, and supplies all things at all times.[1]

Center 847 *Reality* 775

Hermes

His truth shall be thy shield and buckler.

Knowl. 761 Introd. *Heterod.*

Psalm XCI. 4

The Stone overcomes everything to which it is applied, and tinges foreign bodies with its own colour. . . . Know that unless you operate upon bodies until they are destroyed and their soul is extracted, with such you will never tinge any body, for nothing tinges which has not itself been tinged.

Realiz. 868 *Death* 208 *Orthod.* 280

The Book of Alze

Tinge therewith whatsoever you will, and it will multiply to you infinite treasure.

The Sophic Hydrolith

The fixed and the volatile principle must be joined in an inseparable union, which defies even the destructive force of fire.

The Glory of the World

A Yogi in Samadhi is invulnerable to all weapons; all the world cannot overpower him; and he is beyond the powers of incantations and magical diagrams.

Swatmaram Swami

For the person in concord with Unity, everything prospers; to one who has no personal interest, even the spirits are in obeisance.

Chuang-tse (ch. XII)

[1] This passage, like several others (particularly the Taoist texts) cited in the same chapter, really refers to Pure Being, at the center of all worlds; but its doctrinal sense can be restricted to our particular state, since the center of the human state is essentially identical, within the limits imposed by its particular conditions, to the center of all states, being on the One and same Axis (spirit, breath, thread, solar ray) that links all worlds together.

Your joy no man taketh from you.

St John, XVI. 22

A joyous spirit avoids and heals all mental and physical sufferings.

Swami Ramdas

When a rosebud blooms, the bees come from all sides uninvited and unasked.

Sri Ramakrishna

Quicksilver is the mother of all metals, on account of its coldness and moistness; and if it be once purified and cleansed of all foreign matter it cannot be mixed any more with grossness of any kind, neither can it be changed back into an imperfect metal. For Nature does not undo her work, and that which has once become perfectly pure can never become impure again.

A Tract of Great Price

For whosoever hath, to him shall be given, and he shall have more abundance: but whosoever hath not, from him shall be taken away even that he hath.

St Matthew, XIII. 12

Introd. *Renun.* *Illusion* 99

The fire of the Sages may be extracted from all natural things, and is called the quintessence. It is of earth, water, air, and fire. It has no cause of corruption or other contrary quality.

The Glory of the World

The invincible Star of the Metals vanquishes all things, and changes them into a nature similar to its own.

Paracelsus

Inv. 1017

Learn who are the doves of Diana, that overcome the green lion by gentleness: even the Babylonian dragon, which kills everything with its venom.

Philalethes

Infra 590

'The greatness of My glory shall overshadow thee; lions shall fear thee: bears shall protect thee; wolves shall flee before thee and I will be thy companion.'

Mechthild of Magdeburg

Know, my sons, that the Stone out of which our Art is elaborated, never touches the earth after its generation.

The Glory of the World

Infra 573

That is a certain Rule, *That every thing received, is received according to the nature and manner of the Recipient.* The Divine Understanding cloathing it self with the Images and Forms of all Objects, *deformeth* not it self, but maketh them Divine.

Peter Sterry

Symb. 321

Renun. 146
Holy War 396
Conform. 180
Holiness 921

If thou perfectly overcome thyself, thou shalt very easily bring all else under the yoke. The perfect victory is, to triumph over ourselves.

For he that keepeth himself subject, in such sort that his affections be obedient to reason, and his reason in all things obedient to me: he truly is conqueror of himself, and lord of the world.

The Imitation of Christ, III. liii. 2

The Mercury having been often sublimed, is at length fixed, and becomes capable of resisting fire: the sublimation must be repeated until at length fixation is attained.

Abraham Lambspring

Holiness 902

If the palm of one touching this cup is stained red with its wine, he will not go astray at night, the lodestar being in his hand.

Ibn al-Fâriḍ

Sin 57
62
Faith 514

When we invoke the Buddha, he gives us light. When his light shines upon us, all our sins melt away. As the fabled 'tree of the drug king' turns everything that touches it, even poisonous substances, into healing drugs, so will there be nothing of sin remaining in him upon whom the light of the Buddha falls.

Kenshin

Reality 775

It is quite true that in your common, tinkering Medicinal Art, which seeks to counteract only the separate symptoms or manifestations of disease, there is no room for an Universal Medicine. But the true physician knows that all disease (whatever shape it may assume) is simply a depression of the vital spirits, and that whatever strengthens vitality, will cut off the possibility of disease at the very source, expelling the humours which each produce their own peculiar malady, and I maintain that our Universal Medicine is a remedy of this radical kind. It gently promotes and quickens the movement of the vital spirits, and thus, by renewing the source of life, renovates and quickens the whole frame, infusing new vitality and strength into every part. For this reason adepts call it the Great Mystery of Nature, and the preventive of old age and disease. By its aid any man may live the full term of days naturally allotted to him, and need have no fear of contagion, even when the plague, or some other malignant epidemic, is striking down hundreds of his neighbours.

Helvetius

The deliverance of the vital soul (*jîva*) is delineated in the mercurial system (*rasâyana*).

Rasasiddhanta

The substance which serves for the preparation of the Arcanum must be pure, indestructible, and incombustible.

It must be pure of gross material elements, unassailable to doubt, and proof against the fire of the passions.[1]

Henry Madathanas

[1] These qualities belong to the substance. But they are imparted to the 'Artist' in the measure that he becomes identified with the substance.

The yogin who practices this yoga, surrendering himself unto me and having no other desire but me, is thwarted by nothing. His is a bliss that fades not away.[1]

Srimad Bhagavatam, XI. XX

First manage to taste divine Felicity and become immortal. Then let come what may! It will be the same to you whether you are thrown out in the street or placed on a throne. When iron is transmuted into gold by the philosophers' stone, it does not matter whether it is kept in a chest or buried in the earth.

Beauty 663

Swâmi Brahmânanda

'A soul which is truly Mine, possessed by Me, becomes like oil which refuses all fusion with a contrary liquid.'

Sister Consolata

Those we call blessed, who can purge the Queen of the Sages of her impurity, who can circulate the Catholic Virgin Earth by means of our crystalline Physico-magical Art, and who have beheld the King, with his crown on his head, and his strength of inward fire, come forth from the chamber of his crystal grave, his bodily semblance glorified with all the most beautiful hues that the world affords, like a shining carbuncle, or like a transparent, compact, and diaphanous crystal—like a salamander that has spued forth all water, and washes away the leprosy of base metals with fire.

Infra 579 *Pilg.* 378

Helvetius

He who does not know God can be consumed by fire, but he who has known God will consume fire.

Judgment 266

Bâyazîd al-Bisṭâmî

If you ask whether the substance of our Stone be dear, I tell you that the poor possess it as well as the rich.

Many have been reduced to beggary because they foolishly despised that which is highly esteemed by the Sages.

Faith 510

The Book of Alze

The substance is vile and yet most precious.

Metanoia 488

Michael Sendivogius

It is manifest to all men, the poor have more of it (*materia prima*) than the rich. The good part of it people discard, and the bad part they retain. It is visible and invisible, children play with it in the lane.

Renun. 156

Paracelsus

This kind of precious thing will be found in every house. Nevertheless, ignorant people are not able to recognize it.

Chang Po-tuan

[1] Cf. note p. 590.

Know that our Mercury is before the eyes of all men, though it is known to few. When it is prepared, its splendour is most admirable; but the sight is vouchsafed to none, save the sons of knowledge. Do not despise it, therefore, when you see it in sordid guise; for if you do, you will never accomplish our Magistery—and if you can change its countenance, the transformation will be glorious. For our water is a most pure virgin, and is loved of many, but meets all her wooers in foul garments, in order that she may be able to distinguish the worthy from the unworthy.

Knowl. 749
761

Philalethes

It is a Spiritual Essence which is neither celestial nor infernal, but an aerial, pure, and precious body, in the middle between the highest and lowest, the choicest and noblest thing under heaven. But by the ignorant and the beginner it is thought to be the vilest and meanest of things. . . . Men have it before their eyes, handle it with their hands, yet know it not, though they constantly tread it under their feet.

Waters
657

The Sophic Hydrolith

I never knew man hold vile stuff so dear.

Shakespeare (*Love's Labour's Lost,* IV. iii. 276)

Know that our Stone is lightly esteemed by the thankless multitude: but it is very precious to the Sages.

The Glory of the World

And what rule do you think I walked by? Truly a strange one, but the best in the whole world. I was guided by an implicit faith in God's goodness: and therefore led to the study of the most obvious and common things. For thus I thought within myself: God being, as we generally believe, infinite in goodness, it is most consonant and agreeable with His nature, that the best things should be most common. For nothing is more natural to infinite goodness, than to make the best things most frequent; and only things worthless scarce.

Thomas Traherne

Pythagoras, in his Fourth Table, says: How wonderful is the agreement of Sages in the midst of difference! They all say that they have prepared the Stone out of a substance which by the vulgar is looked upon as the vilest thing on earth. Indeed, if we were to tell the vulgar herd the ordinary name of our substance, they would look upon our assertion as a daring falsehood. But if they were acquainted with its virtue and efficacy, they would not despise that which is, in reality, the most precious thing in the world. God has concealed this mystery from the foolish, the ignorant, the wicked, and the scornful, in order that they may not use it for evil purposes.

Reality
790

The Glory of the World

When the inferior scholar is told of Tao,
He laughs aloud at it.
If it were not laughed at, it would not be sufficient to be Tao.

Tao Te Ching, XLI

First it is browne, roddy, and after some deale white,
And then it is called our chosen *Markasite*:
One ounce thereof is better then fifty pound
It is not to be sould in all Christian grounde;
But he that would have it he shalbe faine
To doe it make, or take himselfe the paine:
But one greate grace in that labour is saine,
Make it once well and never more againe.
Olde fathers called it thinge of vile price,
For it is nought worth by way of Marchandise:
Noe man that findeth it woll beare it awaie,
Noe more then thei would an Ounce of Claye:
Men will not beleeve that it is of high price,
No man knoweth it therefore but he be wise.
Here have I disclosed a greate secret wonder,
Which never was writ by *them* which been erth under.

Thomas Norton

Holiness 902

But you must not be led to think that anything (really) pure can be found in the physical world. If anything pure is found in this world, that thing is not *really* pure; for there can be no (true) purity in anything that is not everlasting.

Hermes

Reality 803

Hermes says that spirits cannot enter bodies unless they are purified, and then they enter only through the instrumentality of water. Arisotle says: 'I do not believe that metals can be transmuted unless they are reduced to prime matter, that is, purified of their own corruption by roasting in the fire.'

Albertus Magnus

Metals cannot be changed into other metals until they have been reduced to materia prima.

Self-will, opposed to the divine will, must cease to be so that the divine will can penetrate the heart. We must lay aside all sophistication and become like children so that the word of wisdom can resound in our being.

Henry Madathanas

Conform. 180
Infra 573

The *nü tzû* (lady) dresses in green whereas the *lang chün* (husband) puts on mourning white. If they are visible, then they cannot be used for our purpose. Things which are useful are invisible. They encounter one another in a state of obscurity whereby they make their changes indistinctly. At the moment when a flame flashes out suddenly, the *chên jên* (true man) appears to us.

Chang Po-tuan

Pilg. 378

The Agent then must be akin to the body which is to be dissolved, and, moreover, perfectly pure from all dross or alloy. Again, whereas gold is fixed and solid, the Agent must be highly volatile and spiritual; gold is thick and gross, our Agent is subtle, gold is dead, our Agent is living and life-giving: in short, our Agent should have all those qualities which gold has not, and which it is to impart to the gold. Hence we conclude that Mercury alone is the true Key of our Art; for it is in truth the dry water described by the Sages, which, though liquid, does not wet the hands, nor anything else that does not belong to the unity of its substance. Mercury is our doorkeeper, our balm, our honey, oil, urine, may-dew, mother, egg, secret furnace, oven, true fire, venomous Dragon, Theriac, ardent wine, Green Lion, Bird of Hermes, Goose of Hermogenes, two-edged sword in the hand of the Cherub that guards the Tree of Life, &c., &c.; it is our true, secret vessel, and the Garden of the Sages, in which our Sun rises and sets. It is our Royal Mineral, our triumphant vegetable Saturnia, and the magic rod of Hermes, by means of which he assumes any shape he likes. It is of this water that the Sage uses the words: 'Let Alchemists boast as much as they like, but without this water the transmutation of metals is impossible. In Nature it is not such as we use it in our Art; it is a most common thing, and yet the most precious treasure of all the world.'

Inv. 1017

Philalethes

But why do the philosophers call their gold now 'water', and now 'earth'? Do they not contradict themselves, or each other? No; our Sages, in expounding the truth, veil it under obscure and allegorical expressions, but nevertheless agree with each other so marvellously that they all seem to speak, as it were, with one mouth. They do not confound one thing with another, nor do they wish to lead the earnest enquirer astray. They express themselves in mystic phrases to hide the truth from the unworthy and impious, lest they should seem to be casting pearls before swine, and giving the holy thing to be trodden underfoot by these who think only of indulging their lustful desires. . . . The Sages, then, do well to call their gold earth or water: for they have a perfect right to term it whatever they like. . . . If any one does not perceive their meaning at first glance, he must blame his own ignorance, not their jealousy.

Reality 790

The Golden Tract

A fluidity that is not water, a limpidity that is not air; a light without fire, and a spirit without body. . . .

Ibn al-Fâriḍ

Hence the Sages call it the Phoenix and Salamander. Its generation is a resurrection rather than a birth, and for this reason it is immortal or indestructible.

Realiz. 873

Michael Sendivogius

Now there are two kinds of sulphur, the living and the combustible. Quick sulphur is the active principle of metals, and, when purged from all foreign matter, is the Matter of our Stone. But the common combustible variety is not the Matter of metals or of our Stone; rather, it is injurious to them. Common, combustible sulphur—so we are told by Avicenna and Richard the Englishman—has nothing to do with our art. However carefully prepared, it still disintegrates and destroys metals, because it has no affinity with them. . . . How then can it impart life to other things? For it has two principles of decay—its inflammability and its earthy impurity. The sulphur of the Sages, on the other hand, is living fire; it is quick, and quickens and matures lifeless substances. Common sulphur, then, cannot be the substance of the Stone.

Orthod. 280

The Golden Tract

As there are two kinds of sulphur, so there are two kinds of mercury, the common mercury and the mercury of the Sages. Common mercury is gross and crude; nor does it stand the test of fire like our mercury, but is dissipated in the form of smoke, even by gentle heat. Hence the Sages have laid down this rule: 'Our mercury is not the mercury of the vulgar herd.'

Metanoia 480
Sin 66

The Golden Tract

It is clear that our Mercury is not common mercury; for all common mercury is a male that is corporal, specific, and dead, while our Mercury is spiritual, female, living, and life-giving.

Infra 579

Philalethes

It should be noted that common gold is useless for this purpose, being unsuitable and dead.... You must, therefore, seek to obtain gold which has a pure, living spirit, and of which the sulphur is not yet weakened and sophisticated, but is pure and clear (by passing through antimony, or by the heaven and sphere of Saturn, and being purged of all its defilement): otherwise the first substance, being spiritual and ethereal, will not combine with it. For this Magistery deals only with pure bodies, and suffers no unclean thing near, on, or around it.

Suff. 124

The Sophic Hydrolith

The philosopher's gold resembles common gold neither in colour nor in substance.

The Golden Tract

Our Chalybis is the true key of our Art, without which the Torch could in no wise be kindled, and as the true magi have delivered many things concerning it, so among vulgar alchemists there is great contention as to its nature. It is the ore of gold, the purest of all spirits; a secret, internal, and yet most volatile fire, the wonder of the world, the result of heavenly virtues in the lower world—for which reason the Almighty has assigned to it a most glorious and rare heavenly conjunction, even that notable sign whose nativity is declared in the East. This star was seen by the wise men of old, and straightway they knew that a Great King was born in the world. When you see its constellation, follow it to the cradle, and there you will behold a beautiful Infant. Remove the impurities, look upon the face of the King's Son; open your treasury, give to him gold, and after his death he will bestow on you his flesh and blood, the highest Medicine in the three monarchies of the earth.

Contem. 532
Metanoia 484

Philalethes

Nature: Know that I have only one such son; he is one of seven, and the first among them; and though he is now all things, he was at first only one. In him are the four elements, yet he is not an element. He is a spirit, yet he has a body; a man, yet he performs a woman's part; a boy, yet he bears a man's weapons; a beast, and yet he has the wings of a bird. He is poison, yet he cures leprosy; life, yet he kills all things; a King, but another occupies his throne; he flees from the fire, yet fire is taken from him; he is water, but does not wet the hands; he is earth, and yet he is sown; he is air, and lives by water.[1]

Michael Sendivogius

The sleeping Father is here changed
Entirely into limpid water,
And by virtue of this water alone
The good work is accomplished.
There is now a glorified and beautiful Father,
And he brings forth a new Son.
The Son ever remains in the Father,
And the Father in the Son.
Thus in divers things
They produce untold, precious fruit.
They perish never more,
And laugh at death.

Abraham Lambspring

Pilg. 378

Death 206 232

[1] This eulogy of *materia prima* (and the primordial state identified with it) can be elucidated as follows: *Nature* is speaking of her *one son*, Mercury, here representing the principial, hence *first* or 'synthetic' nature from which the *seven* (planetary or metallic) qualities are derived, the *one* substance before its determination into *all* (outward) *things*. This is the 'quintessence' out of which proceed the *four elements*, a *spirit*, being formless, yet *body*, since still pertaining to the manifested domain. A *man*, being active in relation to the individual state, yet *woman* in virtue of the substantial, passive *part* it (qua *prakriti*) plays in relation to essence (*purusha*): a *boy*, as regards the primordial innocence (*bâlya*), but *man* in the central role of conqueror and *King* of the human state. The *winged beast* (whether lion, dragon, or griffin) symbolizes the 'fixed volatile', or union of sulphur and mercury. Its undifferentiated unity is a *poison* seen from the viewpoint of manifestation, and *kills all things* in their illusory separateness, while in reality it is *life* which *cures* the *leprosy* of egocentric illusion corrupting the soul's immortal essence. The King's *throne* is the center of the soul, *occupied by another*, the indwelling Deity. *He flees from* (i.e. is not identified with) *the fire* of the chaotic, wrathful, or obscure pole, which as inverse reflection, is *taken from him. He is water* in the sense of *mûla-prakriti*, being the undifferentiated potentiality of all forms prior to manifestation, and thus *does not wet the hands; earth* that is *sown*, i.e. seed and productive substance (*hiranyagarbha, jiva-ghana*) of formal possibilities: *air* when envisaged as Spirit (breath, *rûḥ*), and *lives* (i.e. manifests) *by water*, as in *Genesis*, I. 2: 'And the Spirit of God moved upon the face of the waters'.

Purity, Simplicity, Wholeness

Unless a man is simple, he cannot recognize God, the Simple One.

Bengali Song

Reality 775

Nil nisi parvulis.[1]

Alchemical Device

Let a Brahman reject erudition and live as a child.[2]

Bṛihad-Âraṇyaka Upanishad, III. 5

Knowl. 734

... The true metaphysical and contemplative man, who running and shooting up above his own logical or self-rational life, pierceth into the highest life: such a one, who by universal love and holy affection abstracting himself from himself, endeavours the nearest union with the divine essence that may be, *κέντρον κέντρῳ συνάψας*, as Plotinus speaks; knitting his own centre ... unto the centre of divine being.... This life is nothing else but God's own breath within him, and an *infant-Christ* (if I may use the expression) formed in his soul, who is in a sense *ἀπαύγασμα τῆς δόξης*, 'the shining forth of the Father's glory.' But yet we must not mistake; this knowledge is but here in its infancy; there is a higher knowledge, or a higher degree of this knowledge.

John Smith the Platonist

Inv. 1003

Realiz. 890

Whosoever shall not receive the kingdom of God as a little child shall in no wise enter therein.

St Luke, XVIII. 17

So long as one does not become simple like a child, one does not get divine illumination. Forget all the worldly knowledge that thou hast acquired and become as ignorant as a child, and then wilt thou get the divine wisdom.[3]

Sri Ramakrishna

O Pârtha, when a man is satisfied in the Self by Self alone and has completely cast out all desires from the mind, then he is said to be of steady wisdom....

When he completely withdraws his senses from sense-objects as the tortoise withdraws its limbs, then his wisdom becomes well-established.

Bhagavad-Gîtâ, II. 55, 58

Contem. 532

Knowl. 755

[1] 'Only for the humble ones.'

[2] That this is not the final state is made clear in the lines which follow: 'When he has rejected both the state of childhood and erudition, then he becomes a Solitary. When he has rejected both the non-solitary and the solitary, then he becomes a Brahman.'

[3] 'When the priest had done, I was moved to speak to him, and to the people very largely, shewing them the way of life and truth, and the ground of election and reprobation. The priest said he was but a child and could not dispute with me.' (*Journal of George Fox*, Everyman's, p. 42).

Ever since my Lord's Grace entered in my mind.
My mind hath never strayed seeking various distractions.

Milarepa

The true reason underlying things is invisible. unseizable. indefinable. indeterminable. Only the spirit established in the state of perfect natural simplicity can attain it in profound contemplation.

Lieh-tse

Realiz. 873

Thou art of so simple a nature. that of Thee nothing can be born other than what Thou art . . . nor can anything proceed from the supreme simplicity. other than what this. from which it proceeds. is.

St Anselm

Void 721

Perfect purity is imageless. formless. loveless.
Stripped of every quality. like the essence of God.

Angelus Silesius

Introd. *Holiness*

Seest thou not how all that is unworthy to enter paradise has been annihilated in them. while that part of them subsists which has never ceased to exist?

Ibn al-ʿArif

Center 835 *Judgment* 250 *Reality* 773 Supra 563

In the primordial state of unity and universal immobility. oppositions did not exist. All are derived from the diversification of beings. and from their contacts caused by the universal rotation. They would cease. if diversity and movement ceased. They do cease henceforth to affect the being who has reduced his distinct self and particular movement nearly to nothing. This being no longer enters into conflict with any being. because he is established in the infinite. effaced in the indefinite. He has reached and holds himself at the starting point of transformations. a neutral point where there are no conflicts. By the concentration of his nature. the nourishment of his vital spirit. and the gathering of all his powers. he is united to the principle of all geneses. His nature being entire. his vital spirit being intact. no being can injure him.

Chuang-tse (ch. XIX)

Knowl. 749

The mysteries of God. of grace. of nature. of time and eternity can no other possible way be opened in man but by this simplicity of a godly life taught in the Gospel. because only the godly life hath knowledge of God. just as the creaturely life hath only knowledge of the creature and the painful life hath knowledge of pain.

William Law

Truth by her own simplicity is known.

Robert Herrick

Infra 579

Virgin is . . . a person void of alien images. free as he was when he existed not.

Eckhart

Chastity is a lock that no one can open:
What it is within, no outsider can know.

Angelus Silesius

Those who are virgins possess even in this world the glory of the resurrection. *Realiz.* 870

St Cyprian

The chaste virgin signifies in the philosophic work the clear Deity.

Boehme

The soul assuming her real mode stands in her virgin innocence. . . . God is doing all the work and the man of the soul is absolutely idle. *Action* 340

Eckhart

The powerful energy of divine knowledge displays itself in purified souls: here we shall find the true πεδίον ἀληθέιας, as the ancient philosophy speaks, 'the land of truth.'

John Smith the Platonist

Confucius said to Lao-tse, 'To-day you are at leisure. Pray tell me about perfect Tao.'
'Purge your heart by fasting and discipline,' answered Lao-tse. 'Wash your soul as white as snow. Discard your knowledge. Tao is abstruse and difficult of discussion. . . . Attainment implies non-discussion: discussion implies non-attainment.' *Metanoia* 484 *M.M.* 987

Chuang-tse (ch. XXII)

Blessed are the pure in heart: for they shall see God. *Center* 816

St Matthew, v. 8

Who shall ascend into the hill of the Lord? or who shall stand in his holy place?
He that hath clean hands, and a pure heart: who hath not lifted up his soul unto vanity, nor sworn deceitfully.

Psalm XXIV. 3, 4

In paradise there is a perfect life without any shadow of change, also without any false evil desire, and a continual day, where the paradisical man is clear as a transparent glass, in whom the divine sun shines through and through, as gold that is thoroughly bright and pure, without any spot or foulness. *Center* 833

Boehme

There is an Art by Fire to make Glass of Ashes. The Holy Ghost is that Fire, which first reduceth this Creation, as into Ashes: then out of these Ashes raiseth a new Heaven and Earth: The same in a new Form of Gold for Glory, pouring forth Divine Beams: of Crystal for transparency, taking in the Divine Light. Then shall the glorify'd Person of him, who is the righteousness of God: then shall the Person of the Spirit, with all his Riches and Beauties, dwell unveil'd in this Natural Image thus Spiritualiz'd: as the Sun-Beams dwell in every Part and Point of clear Crystal. *Pilg.* 366 *Realiz.* 868

Peter Sterry

He who is endowed with ample virtue may be compared to an infant. . . .
He may cry all day long without growing hoarse.
It means that he is in the perfect harmony.
To know this harmony is to approach eternity:
To know eternity is to attain enlightenment.

Tao Te Ching, LV

Nothing satisfies the soul except the Father who is altogether good and absolutely simple. The more simple the soul the more like God she is. God spake never a word but one: his simple understanding. If the soul is to be simple she must withdraw from multiplicity into his one conception.

Inv. 1003

Eckhart

I cannot present myself naked before God:[1] and yet I must enter without clothing into the kingdom of heaven, since it suffers nothing foreign.

Angelus Silesius

Conform. 170

Unto the pure all things are pure.

Titus, I. 15

While we were Innocent, our Nakedness was our Purity, as a beautiful Face unveil'd, as a Jewel drawn forth from the Case.

Peter Sterry

Our purity must be like that of the crystal.

Swami Ramdas

Supra 563

However insignificant Simplicity seems, the whole world cannot make it submissive.
If princes and kings could keep to it,
All things in the world would of themselves pay homage.
Heaven and earth would unite to send down sweet dew.

Tao Te Ching, XXXVII

Know for certain that Nature is wonderfully simple: and that the characteristic mark of a childlike simplicity is stamped upon all that is true and noble in Nature. If you would imitate Nature, you should take her simplicity for your model in all the operations of Art.

Humility 197

Michael Sendivogius

When we were children we were innocent. But there was in us a seed of ignorance which grew as we grew and, over-powering us, cast away our innocent nature and led us astray. We were thereafter caught in the toils of desire and action and we move in a vicious circle of transitory pleasures and pains. It is necessary that we should hand ourselves over to the Divine and through His grace burn up this seed which is the cause of our misery and

Sin 66

[1] *II. Cor.* V. 3.

bondage, and regain our lost childhood. When once we get it back, it cannot be taken away from us. The burnt seed does not germinate. We will remain pure children for all our life.

Swami Ramdas

Supra 563

The soul shall never rest in her potential power (or nature) till she is simplified to God.

Eckhart

Peace 694

The silversmith said . . . he knew when the process of purifying was completed by seeing his own image in the silver.

St Alphonsus Liguori

Realiz. 859 887 890

Our Masters require a pure, immaculate body, that is untainted with any foreign admixture, which admixture is the leprosy of our metals.

Basil Valentine

Renun. 146

True mirth resides not in the smiling skin:
The sweetest solace is to act no sin.

Robert Herrick

This *Art* in such you only finde
As *Justice* love, with *spotles-Minde*.

Thomas Norton

The Holy Spirit does not descend until the abode of the heart and the soul has been purified and purged of all impurity.

Robert Fludd

The Philosophers warke doe not begin,
Till all things be pure without and within.

Thomas Norton

Metanoia 484

If any one complain of the difficulty of our Art, let him know that in itself it is perfectly simple, and can present no obstacle to those who love God, and are held worthy by Him of this knowledge.

The Golden Tract

Orthod. 296

A purified and enlightened *Buddhi* (Intellect) alone can entitle the sadhaka to enter the kingdom of eternity.

Swami Ramdas

Holy simplicity is the way to all wisdom:
It shows the wise they are but fools,
For when simplicity of heart
Lives in the wisdom of the senses
Much holiness comes to the human soul.

Mechthild of Magdeburg

Knowl. 734
Realiz. 870

Grown men may learn from very little children, for the hearts of little children are pure, and, therefore, the Great Spirit may show to them many things which older people miss.

Black Elk

Contem. 528 *Orthod.* 288

Abandon thought and thinking and be just as a child.
Be devoted to your master's teaching, and the Innate will become manifest.

Saraha

Zen is to have the heart and soul of a little child.

Takuan

The Pupil must regain *the child state he hath lost* ere the first sound can fall upon his ears.

Tibetan Precept

Reality 775

The object of your search should be to find a hidden thing from which, by a marvellous artifice, there is obtained a liquid by whose means gold is dissolved as gently and naturally as ice is melted in warm water. If you can find this substance, you have that out of which Nature produced gold, and though all metals and all things are derived from it, yet it takes most kindly to gold. For all other things are clogged with impurity, except gold wherein there is no uncleanness, whence in a special manner this matter is, as it were, the mother of gold.

Michael Sendivogius

Creation 42 Supra 563 *Pilg.* 378 *Heterod.* 430

It is very often said that the soul cannot be pure unless she is reduced to her original simplicity, the state God created her in; just as from copper gold cannot be made by two or three roastings, it must be reduced to its elemental nature. Things which melt on heating and solidify on cooling are purely of a watery nature. They must therefore be turned back again to water and get quite rid of their present nature. Thus heaven and science conspire to transmute it into gold. We can counterfeit silver with iron and with copper gold; the more like the more false, without riddance. It is the same with the soul. Virtues are easy to talk of, easy to feign, but to have them really is extremely rare.

Eckhart

May I be like the Sun in seeing; like Fire in brilliance; like Wind in power; like Soma in fragrance; like Lord Brihaspati in intellect; like the Asvins in beauty; like Indra-Agni in strength.[1]

Samaveda: Mantra Brahmana, II. iv. 14

[1] 'Recited by the followers of Samaveda in the daily noon-time *Sandhya* worship: after the Gayatri-repetitions are over, the worshipper looks at the Sun and recites this' (note by V. Raghavan). See *Gâyatrî* p. 317.

The Primeval Soul

No man cometh unto the Father, but by me.

St John, XIV. 6

Realiz. 890

No one can meet Allâh who has not first met the Prophet.

Muhammad

Orthod. 288

The whiteness of silver is the first degree of perfection, the yellowness of gold is the second, or highest degree.

Philalethes

Mary has two sons, a God-Man and a pure man: she is Mother of the first corporally, of the second spiritually.

Origen and St Bonaventura

'This man and that man is born in her',[1] says the Holy Ghost through the Royal Psalmist. According to the explanation of some of the Fathers, the first man that is born in Mary is the Man-God, Jesus Christ: the second is a mere man, the child of God and Mary by adoption. If Jesus Christ the Head of men is born in her, the predestinate, who are members of that Head, ought also to be born in her, by a necessary consequence.... As she is the dawn which precedes and reveals the Sun of Justice, Who is Jesus Christ, she must be seen and recognized in order that Jesus Christ may also be.... In seeing her, we see our pure nature. She is not the sun, which by the brightness of its rays blinds us because of our weakness: but she is fair and gentle as the moon, which receives the light of the sun, and tempers it to make it more suitable to our capacity. She is so charitable that she repels none of those who ask her intercession, no matter how great sinners they have been: for, as the saints say, never has it been heard since the world was the world that anyone has confidently and perseveringly had recourse to our Blessed Lady and yet has been repelled. She is so powerful that none of her petitions has ever been refused.

St Louis de Montfort

Creation 47
Symb. 317
Infra 590
Sin 57
Reality 790
Faith 501 509

I say: had Mary not borne God in ghostly fashion first, he never had been born of her in flesh. The woman said to Christ, 'Blessed is the womb which bare thee.' To which Christ replied, 'Blessed not alone the womb which bare me: blessed are they that hear the word of God and keep it.' It is more worth to God his being brought forth ghostly in the individual virgin or good soul than that he was born of Mary bodily.

Eckhart

Inv. 1003

Man must first be restored to himself, that, making in himself as it were a stepping-stone, he may rise thence and be borne up to God.

St Augustine

Realiz. 859

[1] *Psalm* LXXXVI. 5 (Douay).

M.M. 998

Our Art requires that the quicksilver should be first coagulated by means of silver into white sulphur, before the greater degree of heat is applied which, through gold, changes it into red sulphur. There must be whiteness before there is redness. Redness before whiteness spoils our whole substance.

A Tract of Great Price

Faith 512

Jesus said: Know what is in thy sight, and what is hidden from thee will be revealed to thee. For there is nothing hidden which will not be manifest.

The Gospel according to Thomas, Log. 5

Supra Introd.
Metanoia 484
Center 833

Happy the man who is busy attending to what God is saying in him. He is directly subject to the divine light-ray. The soul that stands with all her powers under the light of God is fired and inflamed with divine love. The divine light shines straight in from above, and a perpendicular sun on one's head is a thing that few can survive. Yet the highest power of the soul, her head, is held erect beneath this shaft of godly light so that there can shine in this light divine which I have oft described as being so bright, so overwhelming, so transcendent, that all lights are but darkness in comparison with this light.

Eckhart

Metanoia 488
Supra Introd.

The Dragon is a *yang* substance but comes from the place of *li* (a *yin kua*: ☲). The Tiger belongs to *yin* but is produced in the place of *k'an* (a *yang kua*: ☵). The two things are able to reverse the ordinary process in such fashion that a mother is born to her daughter. The five elements should enter wholly into the central position.

Chang Po-tuan

Introd. *Pilg.*
Center 841

O virgin mother, daughter of thy son . . . .

Dante (*Paradiso,* XXXIII. 1)

The Virgin Mary.

To work a *wonder,* God would have her shown,
At once, a Bud, and yet a *Rose full-blowne.*

Robert Herrick

Oh, most noble and most excellent heart, that hath communion with the Bride of the Emperor of Heaven! and not Bride only, but most delectable Sister and Daughter!

Dante (*Il Convito,* III. xii. 4)

Infra 590
Holiness 924

Who is she that looketh forth as the morning, fair as the moon, clear as the sun, and terrible as an army with banners?

Song of Solomon, VI. 10

King. By heaven, thy love is black as ebony.
Berowne. Is ebony like her? O wood divine!

A wife of such wood were felicity.
O! who can give an oath? where is a book?
That I may swear beauty doth beauty lack.
If that she learn not of her eye to look:
No face is fair that is not full so black.[1]

Shakespeare (*Love's Labour's Lost*, IV. iii. 247)

The sun placed in the *li* position[2] becomes female, and the moon placed in the *k'an* position becomes male. One who cannot understand this principle of topsy-turvy-ness should not talk about his knowledge of these matters, for that knowledge is only as broad as vision through a tube.

Chang Po-tuan

Metanoia 488

When a woman walks in the way of God like a man, she cannot be called a woman.

'Aṭṭâr

M.M. 978

For every woman who makes herself male will enter the Kingdom of Heaven.[3]

The Gospel according to Thomas, Log. 114

Realiz. 887

If our first fall took place when the woman received in her heart the venom of the serpent, it is not to be wondered at that our salvation was brought about when a woman conceived the flesh of the Almighty in Her womb. . . . Through a woman we were sent to destruction; through a woman salvation was restored to us.

St Augustine

Sin 62

I say with the saints, the Divine Mary is the terrestrial paradise of the New Adam.

St Louis de Montfort

Infra 583

. . . Full many a lady
I have ey'd with best regard, and many a time
The harmony of their tongues hath into bondage
Brought my too diligent ear: for several virtues
Have I lik'd several women; never any
With so full soul but some defect in her
Did quarrel with the noblest grace she ow'd,
And put it to the foil: but you, O you!
So perfect and so peerless, are created
Of every creature's best.

Shakespeare (*The Tempest*, III. i. 39)

Beauty 663

The Lord possessed me in the beginning of His ways, before He made anything, from the beginning. I was set up from eternity, and of old, before the earth was made. The depths were not as yet, and I was already conceived; neither had the fountains of waters as

Reality 790

[1] This is a clear allusion to the Black Virgins of Christian iconography.

[2] Cf. Chang Po-tuan, p. 580.

[3] Cp. Boehme, *Sig. Re*, XI. 43: 'This champion or lion is no man or woman, but he is both.'

yet sprung out: the mountains with their huge bulk had not as yet been established: before the hills I was brought forth: He had not yet made the earth, nor the rivers, nor the poles of the world. When He prepared the heavens, I was there: when with a certain law and compass He enclosed the depths: when He established the sky above, and poised the fountains of waters: when He compassed the sea with its bounds, and set a law to the waters that they should not pass their limits: when He balanced the foundations of the earth, I was with him, forming all things, and was delighted every day, playing before Him at all times, playing in the world: and my delight is to be with the children of men. . . . He that shall find me shall find life, and shall have salvation from the Lord.

Creation 33
Knowl. 761

Proverbs, VIII. 22–35 (as given in the Roman Missal for the Mass of the Immaculate Conception)

Judgment 239

Why should such fears invade me now
That writes on her? to whom do bow
The souls of all the just, whose place
Is next to God's and in his face
All creatures and delights doth see
As darling of the Trinity;
To whom the Hierarchy doth throng,
And for whom Heaven is all one song.
Joys should possess my spirit here,
But pious joys are mixed with fear:
Put off thy shoe, 'tis holy ground,
For here the flaming Bush is found,
The mystic rose, the Ivory Tower,
The morning Star and David's bower,
The rod of Moses and of Jesse,
The fountain sealed, Gideon's fleece,
A woman clothèd with the Sun,
The beauteous throne of Salomon,
The garden shut, the living spring,
The Tabernacle of the King,
The Altar breathing sacred fume,
The Heaven distilling honeycomb,
The untouched lily, full of dew,
A Mother, yet a Virgin too,
Before and after she brought forth
(Our ransom of eternal worth)
Both God and man. What voice can sing
This mystery, or Cherub's wing
Lend from his golden stock a pen
To write, how Heaven came down to men?
Here fear and wonder so advance
My soul, it must obey a trance.

Center 828
847
Beauty 679
689
Symb. 306
Center 841
M.M. 987
Ecstasy 636

A Panegyrick on the Blessed Virgin Mary

The Terrestrial Paradise

In the beginning the stars revolved in a tholiform manner.[1]

Anaxagoras

Center 831

At Meru the sun and the moon go round from left to right (*Pradakshinam*) every day, and so do all the stars.

The mountain, by its lustre, so overcomes the darkness of night, that the night can hardly be distinguished from the day.[2]

The day and the night are together equal to a year to the residents of the place.

Mahâbhârata, Vanaparvan, chs. 163, 164

It is certain that the land among these people (the Indians of Hispaniola) is as common as the sun and water; and that 'mine and thine,' the seeds of all mischief, have no place with them. They are content with so little that, in so large a country, they have rather superfluity than scarceness; so that they seem to live in a golden world, without toil, in open gardens, neither intrenched, nor shut up by walls or hedges. They deal truly with one another, without laws or books or judges.[3]

Death 220

Peter Martyr

Supposing here is a small state with few people.
Though there are various vessels I will not have them put in use.
I will make the people regard death as a grave matter and not go far away.
Though they have boats and carriages they will not travel in them;
Though they have armours and weapons they will not show them.
I will let them restore the use of knotted cords (instead of writing).
They will be satisfied with their food;
Delighted in their dress;
Comfortable in their dwellings;
Happy with their customs.

Peace 700

[1] 'Now to revolve in a tholiform manner is to revolve in a horizontal plane, like the θόλος, or "dome", of an astronomical observatory. Anaxagoras himself defined the motion more fully when he said that it was a motion, not ὑπὸ, underneath, but περὶ, around the earth' (Warren, op. cit., p. 192). This is the motion of the sky as seen from the North Pole, where the Primordial Tradition is said to have been situated.

[2] Cf. the vision Jeanne Le Royer, Sister of the Nativity (18th C.), has of the future cycle, when this present world has been totally consumed by fire: 'Renewed in its nature and ornamented with all the luminaries, the firmament will show forth a sun and stars of a substance almost spiritual and of a cool brightness which will never be eclipsed. The Earth turned into a transparent globe will have all the brilliance of the most beautiful crystal, but without the hardness' (Suzanne Jacquemin: *Les Prophéties des Derniers Temps*, Paris, 1958, p. 152).

[3] One of the first fruits of Renaissance man's new conception of the world, the conception which enabled him to breach the Pillars of Hercules, was the mass extermination of the Indians of Central America. Not the least ironical aspect of all this is that Columbus, himself a lofty idealist, was in search among other things of the Terrestrial Paradise itself.

Though the neighbouring states are within sight
And their cocks' crowing and dogs' barking within hearing;
The people (of the small state) will not go there all their lives.

Tao Te Ching, LXXX

Metanoia 488

I' the commonwealth I would by contraries
Execute all things; for no kind of traffic
Would I admit; no name of magistrate;
Letters should not be known; riches, poverty,
And use of service, none; contract, succession,
Bourn, bound of land, tilth, vineyard, none;
No use of metal, corn, or wine, or oil;

Action 346

No occupation; all men idle, all;
And women too, but innocent and pure;
No sovereignty . . . .
All things in common nature should produce
Without sweat or endeavour: treason, felony,
Sword, pike, knife, gun, or need of any engine,
Would I not have; but nature should bring forth,
Of its own kind, all foison, all abundance,
To feed my innocent people.

Shakespeare (*The Tempest*, II. i. 154)

Introd. *Holiness* *Reality* 803

You never enjoy the world aright, till the Sea itself floweth in your veins, till you are clothed with the heavens, and crowned with the stars: and perceive yourself to be the sole heir of the whole world, and more than so, because men are in it who are every one sole heirs as well as you. Till you can sing and rejoice and delight in God, as misers do in gold, and Kings in sceptres, you never enjoy the world.

Supra 563 *Charity* 597

Till your spirit filleth the whole world, and the stars are your jewels: till you are as familiar with the ways of God in all Ages as with your walk and table: till you are intimately acquainted with that shady nothing out of which the world was made: till you love men so as to desire their happiness, with a thirst equal to the zeal of your own; till you delight in God for being good to all: you never enjoy the world.

Thomas Traherne

Illusion 85

When the universe is seen correctly, it is *Atman* and all bliss, but when it is seen incorrectly, it appears as the world, full of sufferings. He whose mind has world-cognition suffers, but to him whose mind has *Atman*-cognition, the world is a garden of bliss.

Yoga-Vasishtha

A Christmas Caroll

Dark and dull night, flie hence away,
And give the honour to this Day,
That sees *December* turn'd to *May*.

. . . .

Why do's the chilling Winters morne
Smile, like a field beset with corne?
Or smell, like to a Meade new-shorne,
Thus, on the sudden?—Come and see
The cause, why things thus fragrant be.

Robert Herrick

His disciples said to Him: When will the Kingdom come? Jesus said: It will not come by expectation; they will not say: 'See, here', or: 'See, there'. But the Kingdom of the Father is spread upon the earth and men do not see it.

The Gospel according to Thomas, Log. 113

Realiz. 873
M.M. 978

The world is a mirror of infinite beauty, yet no man sees it. It is a Temple of Majesty, yet no man regards it. It is a region of Light and Peace, did not men disquiet it. It is the Paradise of God.

Thomas Traherne

Center 828

He that dwelleth in the secret place of the most High shall abide under the shadow of the Almighty.

Psalm XCI. 1

For you the world outside will now stand transformed as the very expression or manifestation of God—everywhere the Light of God will dazzle your eyes; even in the apparent diversity and activity of nature you will strangely be conscious of an all-pervading stillness and peace of the Eternal—a consciousness which is unshakably permanent. You will also feel that you are liberated from the harassing dualities of life followed by the crowning experience of an abiding state of ineffable ecstasy.

Swami Ramdas

Contem. 532
Center 835
Ecstasy passim

If . . . while remaining normally amongst the 'six dusts' of the active life, a man clothes himself with a sort of spiritual sheen, is simple, unalloyed, complete and of one piece, that man will not err to any great degree. . . . He is one who boldly and successfully displays his courageous disposition and advances without delay or hesitation. And by the very fact of so doing such a man raises up the very source and origin of his own soul and mind, and brings to a final end all those roots of existence which tie us to the cycle of life and death. For such a man there is nothing but great joy, enough to dissolve the sky and shatter the iron mountains. He is to be compared with the lotus which blossoms and becomes ever more beautiful and more deliciously scented as it gets nearer to the fire.

And if you ask, 'How can this be?' it is because the fire is itself the lotus and the lotus is itself the fire.

Hakuin

Supra 573
Knowl. 761
Faith 514
Supra 563
M.M. 978
Judgment 266

Religion, where it is in truth and in power, renews the very spirit of our minds, and doth in a manner spiritualize this outward creation to us. . . . Every particular good is a blossom of the first goodness; every created excellency is a beam from the Father of Lights: and

Reality 775

should we separate all these particularities from God, all affection spent upon them would be unchaste, and their embraces adulterous. We should love all things in God, and God in all things, because he is all in all, the beginning and original of being, the perfect idea of their goodness, and the end of their motion. . . .

Symb. 306

Thus may a good man walk up and down the world as in a garden of spices, and suck a divine sweetness out of every flower. . . . And seeing God hath never thrown the world from himself, but runs through all created essence, containing the archetypal ideas of all things in himself, and from thence deriving and imparting several prints of beauty and excellency all the world over; a soul that is truly θεοειδής, godlike, a mind that is enlightened from the same fountain, and hath its inward senses affected with the sweet relishes of divine goodness, cannot but every where behold itself in the midst of that glorious unbounded Being who is indivisibly everywhere. A good man finds every place he treads upon holy ground; to him the world is God's temple; he is ready to say with Jacob, 'How dreadful is this place! this is none other but the house of God.'[1]

Beauty 670

Holiness 914

John Smith the Platonist

Thus the Sage sees heaven reflected in Nature as in a mirror; and he pursues this Art, not for the sake of gold or silver, but for the love of the knowledge which it reveals.

Knowl. 761

Michael Sendivogius

When . . . I prayed with my heart, everything around me seemed delightful and marvellous. The trees, the grass, the birds, the earth, the air, the light seemed to be telling me that they existed for man's sake, that they witnessed to the love of God for man, that everything proved the love of God for man, that all things prayed to God and sang His praise.

Love 618
Charity 608

The Russian Pilgrim

Then darke doubts to me appeared pure,
There fownd I disclosed the *Bonds of Nature*:
The cause of Wonders were to me soe faire,
And so reasonable, that I could not dispaier.

Center 835

Thomas Norton

'Look, Master, at this realm of flowers and happy creatures—of phoenixes, cranes and deer. Is it not a better place indeed than the haunted deserts through which you and I have passed?' Tripitaka still murmured his thanks, and with a strange feeling of lightness and exhilaration they all set off up the Holy Mountain.

Pilg. 366

Monkey

Common speech often tells about the *Tao* of consenting sages. A man should search deeply into this matter. If he seeks in ordinary affairs according to the principle of topsy-turvy-ness, then the dust and the sand of the broad earth all become precious.

Introd.
Humility
Metanoia 488
Supra 563

Chang Po-tuan

[1] *Gen.* XXVIII. 17.

A new heart also will I give you, and a new spirit will I put within you: and I will take away the stony heart out of your flesh, and I will give you an heart of flesh.

Ezekiel, XXXVI. 26

Metanoia 484

No man putteth a piece of new cloth unto an old garment, for that which is put in to fill it up taketh from the garment, and the rent is made worse.

Neither do men put new wine into old bottles: else the bottles break, and the wine runneth out, and the bottles perish: but they put new wine into new bottles, and both are preserved.

St Matthew, IX. 17

Therefore if any man be in Christ, he is a new creature: old things are passed away; behold, all things are become new.

II. Corinthians, V. 17

And here, thou dear seeker, when thou seest the crimson-coloured blood of the young man (self-hood) arise out of death with the virgin's white blood (divine essence), then know that thou hast the arcanum of the whole world, and a treasure in this valley of misery, which surpasses the value of gold; take it and esteem it more excellent and sovereign than that which shall again arise from death: If thou beest born of God, then thou wilt understand what I mean.

Boehme

Realiz. 870

When nothing is grievous or hard, when all is pure joy, then verily thy child is born.

Eckhart

Man, if Paradise is not first within thee,
Then believe me for certain, thou wilt never enter in.

Angelus Silesius

Holiness 900

Paradise is yet in the world, but man is not therein, unless he be born again of God.

Boehme

All would have Christ to be their Saviour in the *next* world and to help them into Heaven when they die by His power and merits with God. But this is not willing Christ to be thy Saviour; for His salvation, if it is had, must be had in *this* world; if He saves thee it must be done in this life, by changing and altering all that is within thee, by helping thee to a new heart, as He helped the blind to see, the lame to walk, and the dumb to speak. For to have salvation from Christ is nothing else but to be made like unto Him.

William Law

M.M. 978

That man lives twice, that lives the first life well.

Robert Herrick

Death 206

Reality 790

There is but one salvation for all mankind, and that is the life of God in the soul. God has but one design or intent towards all mankind and that is to introduce or generate His own life, light, and Spirit in them, that all may be as so many images, temples and habitations of the Holy Trinity.... There is not one for the Jew, another for a Christian, and a third for the heathen. No; God is one, human nature is one, salvation is one, and the way to it is one; and that is, the desire of the soul turned to God.[1]

William Law

Holiness 924
W. Side 464

The earth is a great distilling vessel, formed by the hand of an all wise Creator, on the model of which all Sages have constructed their small distilling vessels; and if it pleased God to extinguish the central fire, or to destroy the cunning machinery, this universal frame would relapse into chaos. At the end of time, He will kindle the Central Fire into a brighter flame, will cause all the water to evaporate, will calcine the earth—and thus the earth and the water will be rendered more subtle and pure, and will form a new and more glorious earth.

Michael Sendivogius

Introd.
Pilg.

The inner creation is holy, and the outer creation, which is in the vital flux, would likewise be so had the curse not penetrated therein through the awakening of vanity. Nevertheless, if it is possible to dissociate vanity from the vital flux, this itself is holy and represents a Paradise, as will appear at the end of this world.

Boehme

Mercy and truth are met together; righteousness and peace have kissed each other.
Truth shall spring out of the earth; and righteousness shall look down from heaven.

Psalm LXXXV. 10, 11

'Mother Nature is all-powerful, and eternity is on her side. What are the inventions of man, the lofty cities which he raises on the borders of the desert, the terrible weapons that he uses to realize and defend his conquests? Nothing, but a little heaped-up dust which the great natural forces always tend to restore to its primeval form. Forsake the citadel for a few years, abandon the canon or machine gun for a few months in the prairie, and soon grass and brambles will have overgrown the stones, and rust corroded the hard steel. In how many former times have vast solitudes been peopled by powerful cities! Of them today remain no more than the ruins, and the ruins themselves finally disappear back into

[1] '"From within", every religion is the doctrine of the one Self and its terrestrial manifestation, and the means for abolishing the false self, or the means for the mysterious reintegration of our "personality" in the celestial Prototype. "From without", religions are "mythologies", or more specifically symbolisms, disposed in view of different human receptacles, and manifesting by their limitations, not a contradiction *in divinis*, but on the contrary a mercy' (Schuon: *Sentiers de Gnose*, p. 82).

the eternally virgin earth. Of what importance are the men who pass? The Spirit has only to blow on them and they will be no more! Then the sons of the Earth will repossess the Earth. And the past time will begin over again as the new time!'

Wr. Side 464

The Ghost Dancers at Wounded Knee

Of that upper earth which is under the heaven . . . the whole presents a single and continuous appearance of variety in unity. And in this fair region everything that grows—trees, and flowers, and fruits—are in a like degree fairer than any here; and there are hills, having stones in them in a like degree smoother, and more transparent, and fairer in colour than our highly-valued emeralds and sardonyxes and jaspers, and other gems, which are but minute fragments of them: for there all the stones are like our precious stones, and fairer still. The reason is, that they are pure, and not, like our precious stones, infected or corroded by the corrupt briny elements which coagulate among us, and which breed foulness and disease both in earth and stones, as well as in animals and plants. They are the jewels of the upper earth, which also shines with gold and silver and the like, and they are set in the light of day and are large and abundant and in all places, making the earth a sight to gladden the beholder's eye. And there are animals and men, some in a middle region, others dwelling about the air as we dwell about the sea; others in islands which the air flows round, near the continent; and in a word, the air is used by them as the water and the sea are by us, and the ether is to them what the air is to us. Moreover, the temperament of their seasons is such that they have no disease, and live much longer than we do, and have sight and hearing and smell, and all the other senses, in far greater perfection, in the same proportion that air is purer than water or the ether than air. Also, they have temples and sacred places in which the gods really dwell, and they hear their voices and receive their answers, and are conscious of them and hold converse with them; and they see the sun, moon, and stars as they truly are, and their other blessedness is of a piece with this.

Beauty 670
Reality 775
Flight 941
Rev. 959

Plato (*Phaedo*, 110 B ff.)

Lay up for yourselves treasures in heaven, where neither moth nor rust doth corrupt, and where thieves do not break through nor steal.

St Matthew, VI. 20

God will remit from them (the believers) the worst of what they did, and will pay them for reward the best they used to do.[1]

Qur'ân, XXXIX. 35

We will celebrate the natal day of his martyrdom in joy and gladness, both in commemoration of those who finished their contest before, and to prepare those that shall finish hereafter.

Holy War 403

A letter from the Church of Smyrna to the Church of Philomelium, reporting the martyrdom of St Polycarp

[1] The 'stern' corollary to this passage is found in *Judgment*, p. 264.

This is a reminder: that verily for the Godfearing is a beautiful place of return,
Gardens of Eden, whereof the gates are opened for them,
Wherein, reclining, they call for the plenteous fruit and drink that is therein.
And with them are chaste-eyed companions.
This is what you are promised on the Day of Reckoning.
Lo! this in truth is Our provision, which will never waste away.[1]

Holiness 902

Qur'ân, XXXVIII. 49–54

The Sphere of the Moon

Realiz. 890 Supra 579

This, verily, is the door of the heavenly world—that is, the moon. Whoever answers it, him it lets go further.

Kaushîtaki Upanishad, I. 2

'Direct thy mind to God in gratitude,' she said, 'who hath united us with the first star.'

Meseemed a cloud enveloped us, shining, dense, firm and polished, like diamond smitten by the sun.

Within itself the eternal pearl received us, as water doth receive a ray of light, though still itself uncleft.

Dante (*Paradiso*, II. 28–36)

Realiz. 870

Moses had relations with the moon while still in his body.

Zohar, I. 22 a

I have found it in some most ancient books that Thomas, the Apostle, was the author of the opinion . . . that Paradise was so high as to reach to the lunar circle.

Albertus Magnus

The moon, doubtless, is the year, and all living beings.

Śatapatha-Brâhmaṇa, VIII. iii. 3. 11

The *wiwanyag wachipi* (dance looking at the sun) is one of our greatest rites and was first held many, many winters after our people received the sacred pipe from the White Buffalo Cow Woman. It is held each year during the Moon of Fattening (June) or the Moon of Cherries Blackening (July), always at the time when the moon is full, for the growing and dying of the moon reminds us of our ignorance which comes and goes: but

[1] The satanical counterfeit to this true promise is given in *Heterodoxy*, p. 428.

when the moon is full it is as if the eternal light of the Great Spirit were upon the whole world.[1]

Black Elk

He who is adept in reducing his appetites and enjoyments day by day, resembles the crescent moon, daily increasing in brightness; he enlightens his family, as the moon sheds her lustre over the stars around her.

Yoga-Vashishtha

Renun. 158

The hidden centre of our Magnet abounds in Salt, which Salt is the menstruum in the Sphere of the Moon, and can calcine gold. This centre turns towards the Pole with an archetic appetite, in which the virtue of the Chalybs is exalted into degrees.

Philalethes

Center 816
Introd.
Realiz.

The moon shines brightly on the eve of the fifteenth day of the eighth month. This is the time at which the gold essence appears plentifully. At the moment when a single *yang* rises, start your fire without delay.

Chang Po-tuan

Contem. 542

The conjunction can take place only by means of the Moon or an imperfect body and fire; and this Moon is the sap of the water of life, which is hidden in Mercury, and is stirred up by fire; it is a spirit which enters the body, and compels it to retain its soul.

Philalethes

Orthod. 280

He whose mind is as calm as moonlight, whether at the approach of a feast or of a battle, or even at the moment of death itself, is verily a saint.

Yoga-Vasishtha

Peace 700

Such a one, being a disciple of the Buddha Śakyamuni, does obeisance to him and craves that in the moonlit shelter of the Tree of Knowledge he may seek refuge from the clouds of sorrow and death.

Buddhist Formula of Submission

Introd.
Peace

Our cup is the full moon; the wine the sun.

Ibn al-Fâriḍ

Verily ye shall see your Lord as ye see the moon on the night of its fullness, without confusion in the vision of Him.

Muhammad

[1] The moon is analogous symbolically with the center of the human state, and the full moon perfectly facing the sun represents the soul in plenitude of illumination. 'The direct light of the sun, or more particularly the ray of light which connects the moon with the sun, is thus a symbol of the Intellect, whereas the rays which the moon sends out into the night represent the intellectual intuitions which act as mediators between the moon of the Heart and the darkness of the soul' (Abû Bakr Sirâj ed-Dîn: *The Book of Certainty*, London, Rider, 1952, p. 66).

Center 816

In all our hearts thou art the full-moon shining.

The Mevlidi Sherif

Ah, Moon of my Delight who know'st no wane,
The Moon of Heav'n is rising once again.

Omar Khayyâm

At morning-tide a moon appeared in the sky,
And descended from the sky and gazed on me.
Like a falcon which snatches a bird at the time of hunting,
That moon snatched me up and coursed over the sky.
When I looked at myself, I saw myself no more,
Because in that moon my body became by grace even as soul.
When I travelled in soul, I saw nought save the moon,
Till the secret of the eternal Theophany was all revealed.
The nine spheres of heaven were all merged in that moon,
The vessel of my being was completely hidden in the sea....
Without the power imperial of Shamsu 'l-Ḥaqq of Tabrîz
One could neither behold the moon nor become the sea.

Flight 944 — Realiz. 873 — Grace 552

Dîvâni Shamsi Tabrîz, XIX

Conform. 170

In blending lead and mercury to produce *tan* (medicine), the amount of each should be such that it does no harm to the other. If you wish to inquire what the true lead is, the answer will be that the moonlight shines all day long on the western river.

Chang Po-tuan

Sin 77 — Supra 583

The sublimation of the Mercury which is here referred to, is not an artificial, but a true and natural one. It is the 'first preparation of the thin substance', by which the eclipse caused by the interposition of Earth is removed from the Moon, enabling her to receive the light of the Sun—which happens when the murky sphere of Saturn (that overshadowed the whole horizon) is removed, and Jupiter ascends the throne; then there rises upward a mist of dazzling whiteness, whence there is distilled upon the earth a pure, sweet, and fragrant dew.

Philalethes

O truly blessed night, that alone was worthy to know the time and the hour when Christ rose again from the dead. This is the night of which it is written: And the night shall be enlightened like day; and the night is my enlightening in my pleasures. The sanctification of this night, therefore, driveth away evil deeds, cleanseth offences, restoring innocence to the fallen and gladness to the mournful. It driveth out hatred, it produceth concord and curbeth tyrannies.

Paschale Praeconium

Lo! We revealed it[1] in the Night of Power.
And what can convey unto thee what the Night of Power is!
The Night of Power is better than a thousand months.
The angels and the Spirit descend therein, by permission of their Lord, from the source of all decrees.
Peace it is until the break of dawn.

Qur'ân, XCVII

Peace 705

The light of the moon shall be as the light of the sun.

Isaiah, XXX. 26

This night methinks is but the daylight sick;
It looks a little paler: 'tis a day,
Such as the day is when the sun is hid.

Shakespeare (*Merchant of Venice*, V. i. 124)

When the Reign of Jupiter comes to an end (towards the close of the fourth month) you will see the sign of the waxing moon (Crescent), and know that the whole Reign of Jupiter was devoted to the purification of the Laton. The mundifying spirit is very pure and brilliant, but the body that has to be cleansed is intensely black. While it passes from blackness to whiteness, a great variety of colours are observed; nor is it at once perfectly white; at first it is simply white—afterwards it is of a dazzling, snowy splendour. Under this Reign the whole mass presents the appearance of liquid quicksilver. This is called the sealing of the mother in the belly of the infant whom she bears. . . . Whenever you look at it you will have cause for astonishment, particularly when you see it all divided into beautiful but very minute grains of silver, like the rays of the Sun. This is the White Tincture, glorious to behold, but nothing in respect of what it may become.

Philalethes

Pilg. 366

Supra 579

M.M. 998

For thou wilt lie upon the wings of night,
Whiter than new snow on a raven's back.

Shakespeare (*Romeo and Juliet*, III. ii. 18)

Praise to Thoth, the son of Rê, the Moon beautiful in his rising, lord of bright appearings who illumines the gods.

Hymn to Thoth

Goddess of the perfection of wisdom, holy Tara who delights the heart,
Friend of the drum, perfect Queen of sacred lore who speaks kindly,
With a face like the moon, shining brilliantly, unconquered. . . .

Âryatârâbhaṭṭârikânâmâshṭottaraśatakastotra, 33, 34

Beauty 679

[1] Specifically, the *Qur'ân;* generally speaking, divine inspiration established integrally and immutably in the being, which presupposes the profound serenity of a fully receptive and acquiescent soul.—Cf. the Introduction to *Realization*.

Introd. *Holy War*

... Being an instrument of the armies on high, shining gloriously in the firmament of heaven.

Ecclesiasticus, XLIII. 9

Hear, Goddess queen, diffusing silver light,
Bull-horn'd, and wand'ring thro' the gloom of Night.
With stars surrounded, and with circuit wide
Night's torch extending, through the heav'ns you ride:
Female and male, with silv'ry rays you shine,
And now full-orb'd, now tending to decline.
Mother of ages, fruit-producing Moon,
Whose amber orb makes Night's reflected noon:
Lover of horses, splendid queen of night,
All-seeing pow'r, bedeck'd with starry light,
Lover of vigilance, the foe of strife,
In peace rejoicing, and a prudent life:
Fair lamp of Night, its ornament and friend,
Who giv'st to Nature's works their destin'd end.
Queen of the stars, all-wise Diana, hail!
Deck'd with a graceful robe and ample veil.
Come, blessed Goddess, prudent, starry, bright,
Come, moony-lamp, with chaste and spendid light,
Shine on these sacred rites with prosp'rous rays,
And pleas'd accept thy suppliants' mystic praise.

Orpheus: *Hymn to the Moon*

Supra 573

Contem. 532 *Reality* 775

O moon clad in white and vehemently shining abroad whiteness, let us learn what is the lunar radiance that we may not miss what is doubtful. For the same is the whitening snow, the brilliant eye of whiteness, the bridal procession-robe of the management of the process, the stainless chiton, the mind-constructed beauty of fair form, the whitest composition of the perfection, the coagulated milk of fulfilment, the Moon-froth of the sea of dawn, the magnesia of Lydia, the Italian stibnite, the pyrites of Achaea, that of Albania, the many-named matter of the good work, that which lulls the All to sleep, that which bears the One which is the All, that which fulfils the wondrous work.

Stephanos of Alexandria

CHARITY

I appeal to any white man, if ever he entered Logan's cabin hungry, and he gave him not to eat—if ever he came cold and naked, and he clothed him not.

Speech of a Cayuga Chief[1]

'Charity starts from the truth that the neighbour is not other than myself, since he possesses an ego; that in the sight of God he is neither more nor less "myself" than I am; that what is given to the "other" is given to "me"; that the neighbour also is made in the image of God; that he carries in himself the potentiality of the divine presence, which must be venerated in him; that the good done to the neighbour purifies us of egoistic illusion and virtually delivers us from it when it is accomplished in view of God.

'Philosophic humanitarianism, in the end atheistic, starts from the error that man and earthly well-being represent absolute values; that man is, by definition, good and that accordingly there are no men who are fundamentally evil; that there are no values incompatible with earthly well-being; that what contradicts the human individual and his comforts cannot be a good.

'True charity may sometimes be contrary to the immediate interests of men, and contrary also to earthly well-being' (Schuon: *Perspectives spirituelles*, pp. 27–28).

Charity is an impersonal benevolence towards creation as a whole, and not a personal preoccupation with the lives and affairs of others. The acts towards persons that true charity may engage in will always be characterized by an attitude of detachment and disinterestedness; there will also be an unmistakable humility or effacement,[2] in utter contrast to the condescending attitude, so prevalent today, that glorifies 'Service'.

'To act efficaciously, one must be; to be able to give the best, one must know that this best comes from God alone. Only the saints eradicate evil at its root; others do little more than transplant it. For the charity of the ordinary man is never altogether charitable: what the sinner gives with the right hand he takes back with the left, in other words he takes away indirectly what he gives directly, so that his charity is like a gift made by a thief. This is certainly not a reason for his refraining from charity; but at least let him not boast of it. The first act of charity is to rid the soul of illusions and passions and thus rid the world of a harmful being; it is to make a void, so that God can fill it, and through this plenitude give

Action 346

[1] Cf. Washington Irving: *Traits of Indian Character*.

[2] 'He that is greatest among you shall be your servant' (*Matt.* XXIII. 11): 'For though I be free from all men, yet have I made myself servant unto all, that I might gain the more' (*I Cor.*, IX. 19). This is the same as the Taoist doctrine of 'lowliness'.

Himself. The saint is effectively the void made for the passage of God' (Schuon: *Les Stations de la Sagesse*, p. 159).

The cosmological dimension and functioning of charity is well described in the following passage by M. Aniane on alchemy: 'The true role of the alchemist was double: on the one hand he aided nature, asphyxiated by the decline of humanity, to breathe the presence of God. Offering to God the prayer of the universe, he established the universe in being and renewed its existence. The texts call him *king*: secret king, he strengthened the order of time and space, the fecundity of the earth in seeds and in diamonds, as did the kings of ancient civilizations, and as the emperor of China did up to the threshold of the twentieth century. On the human plane, on the other hand, the alchemist, by "awakening" substances and gold itself to their true nature, utilized them to prepare elixirs which gave "longevity" to the body and strength to the soul: the "potable gold" was a gold *awakened* to its spiritual quality, and reflected in its order the "medicine of immortality" spoken of by St Ambrose to designate the Eucharist.

Holiness 921

'The true role of the alchemist: to celebrate analogically a mass where the species would not only be bread and wine, but the whole of nature'[1] (op. cit. Introd. *Pilgrimage*, pp. 246–247).

'If a man does not trample on a flower without reason, it is because the flower is something of God, a remote effect of the infinite Cause; he who scorns a flower indirectly scorns God. If a good man had the power to annihilate a stone, he would still not do it without a motive, for the existence of the stone—this something quasi-absolute which distinguishes it from nothingness—is a manifestation of Principle; it is therefore sacred. In every neutral contact with matter—and all the more so with his fellow creatures—man should either leave no trace, or else leave a trace that is benefic; he should enrich or pass unperceived. Even when there is necessity for destruction,—and in this case the destruction proceeds from a divine will,—man must destroy in conformity with the nature of the object, which itself will then objectify the human nothingness, in the measure that the man acts for a celestial will. The life of man being sacred, the destructions it inevitably entails are also sacred' (Schuon: *Perspectives spirituelles*, p. 227).

Holy War 394

Introd. *Suff.*

'Charity is to recognize the eternal Word in creatures' (Burckhardt: cf. *Études Traditionnelles*, 1953, p. 174).

[1] In the words of the eighteenth-century French Benedictine and Hermeticist, Dom Antoine Joseph Pernety: 'Common chemistry is the art of destroying the compounds which nature has formed, while Hermetic chemistry is the art of working with nature to perfect them' (*Études Trad.*, 1969, p. 235).

Love of the Neighbour

I have always shunned evil: I have given bread to the hungry, water to the thirsty, clothes to the naked, a ship to the stranded; to the orphan I was a father, to the widow a husband, to the roofless I gave a home.

Egyptian Tradition

By the morning hours
And by the night when it is stillest,
Thy Lord hath not forsaken thee nor doth He hate thee,
And verily the latter portion will be better for thee than the former.
And verily thy Lord will give unto thee so that thou wilt be content.
Did He not find thee an orphan and protect (thee)?
Did He not find thee wandering and direct (thee)?
Did He not find thee destitute and enrich (thee)?
Therefore the orphan oppress not,
Therefore the beggar drive not away,
Therefore of the bounty of thy Lord be thy discourse.

Grace 552

Qur'ân, Sûrah XCIII

Be not wanting in comforting them that weep, and walk with them that mourn.
Be not slow to visit the sick: for by these things thou shalt be confirmed in love.

Ecclesiasticus, VII. 38–39

I was an hungered, and ye gave me meat: I was thirsty, and ye gave me drink: I was a stranger, and ye took me in:
Naked, and ye clothed me: I was sick, and ye visited me: I was in prison, and ye came unto me.

St Matthew, XXV. 35–36

Monks, if beings knew, as I know, the ripening of sharing gifts, they would not enjoy their use without sharing them.

Itivuttaka

It is more blessed to give than to receive.

Acts, XX. 35

'Thou shalt love thy neighbour as thyself' is another full proof that God is in us of a truth and that the Holy Spirit hath as certainly an essential birth within us as the spirit of this world hath. For this precept might as well be given to a fox as to a man, if man had not

Creation 42

something quite supernatural in him. For mere nature and natural creature is nothing else but mere self, and can work nothing but to and for itself. And this not through any corruption or depravity of nature, but because it is nature's best state and it can be nothing else either in man or beast.

William Law

Heterod. 428

Unless the mind be trained to selflessness and infinite compassion, one is apt to fall into the error of seeking liberation for self alone.

Gampopa

Metanoia 484

Though I speak with the tongues of men and of angels, and have not charity, I am become as sounding brass, or a tinkling cymbal.

I. Corinthians, XIII. 1

Beauty 676

I had three manners of understanding of this light, Charity. The first is Charity unmade; the second is Charity made; the third is Charity given. Charity unmade is God; Charity made is our soul in God; Charity given is virtue. And that is a precious gift of working in which we love God, for Himself; and ourselves, in God; and that which God loveth, for God.

Julian of Norwich

Beauty 689 *Reality* 775 *Knowl.* 749

The *Divine Love* in every *Person* or *Spirit* lives not in *it self* as a *part,* but in the life or the *whole,* in the *Divine,* the *Universal Spirit,* the *Spirit* of *Love,* the *Spirit* of the *whole. I live not,* saith *St Paul, but Christ liveth in me.* Again, *If you live in the Spirit, walk in the Spirit.* Thus the Divine Love having its life in each person, in the life of the *whole,* the *Universal Spirit,* being one Spirit with that Spirit, which is the *Unity* of the *whole,* comprehendeth all things with strictest tenderest imbraces in it self, as *one self* with *it self.*

Peter Sterry

Just as a flower gives out its fragrance to whomsoever approaches or uses it, so love from within us radiates towards everybody and manifests as spontaneous service.... When we feed, clothe and attend on anybody, we feel like doing all these things to our own body, for which we do not expect any return or praise or commendation, because all bodies are our own; for, we as the all-pervading Atman or Spirit reside in all bodies.

Swami Ramdas

In everyone there is something of his fellow man. Therefore, whoever sins, injures not only himself but also that part of himself which belongs to another.

Moses ben Jacob Cordovero

Let brotherly love continue. Be not forgetful to entertain strangers: for thereby some have entertained angels unawares.

Hebrews, XIII. 1

The nobler the thing the bigger it is and the more universal. Love is as noble as it is universal. It does indeed seem hard, as our Lord commands, to love our even-christians as ourselves. The unenlightened say that we ought to love them just the same as they love themselves. Not so. We ought to love them no more than our own selves, which is not difficult.

Eckhart

Sin 66

I will cease to live as self, and will take as my self my fellow-creatures.

Śânti-deva

Introd. *Holiness*

We are all of us by birth the offspring of God—more nearly related to Him than we are to one another, for in Him we live, and move, and have our being.

William Law

Center 841

If men wish to draw near to God, they must seek Him in the hearts of men.

Abû Sa'id ibn Abi 'l-Khayr

'This is myself and this is another.'
Be free of this bond which encompasses you about,
And your own self is thereby released.

Saraha

Death 220

A man of principle is reverent in everything and as much as in anything else reverences his own self. That self of his is a branch of the tree of his parents. How could he not reverence it? To be unable to reverence oneself is an injury to one's parents, and that is an injury to one's very root. Let the root be injured and the branches along with it will die.

Rev. 967

These three (facts) are a symbol of life in the community. 'Self' includes other selves, 'son' includes other men's sons, 'wife' includes other men's wives. Let a man of principle put these three things into practice, and he will reach out to the whole of the Great Society. They constitute the Way of a great king, and it is along this line that states and families will become docile.

Li Chi, Ai Kung Wen

Holiness 921

His nature requireth that thou love all those whom He loveth, and receiveth Him in all those things wherein he giveth Himself unto thee.... Thou livest in all those whom thou lovest: and in them enjoyest all their treasures.

Thomas Traherne

And therefore whoso will speak with thee, what that he be, in what degree that he be, and thou know not what he is nor why he cometh, be soon ready with a good will for to wit what his will is.... And though thou be in prayer or devotion, that thee thinketh loath for to break off, for thee thinketh thou shouldest not leave God for man's speech, me thinketh not so in this case: for if thou be wise thou shalt not leave God but thou shalt find Him and have Him and see Him in thine even-christian as well as in prayer, but in another manner thou shalt have Him than in prayer.

Walter Hilton (Instructions for an anchoress)

M.M. 978

He who uses mankind badly, uses himself badly.

Sextus the Pythagorean

Now Scripture enjoins nothing except charity and condemns nothing except lust, and in that way informs the practices of men. . . . I mean by charity that affection of the mind which aims at the enjoyment of God for His own sake and of one's self and one's neighbour for God's sake. By lust I mean that affection of the mind which aims at the enjoyment of one's self and one's neighbour without reference to God. . . . Now in proportion as the dominion of lust is pulled down, in the same proportion that of charity is built up.

Metanoia 488

St Augustine

Charlemagne: What is the due order of the soul?

Alcuin: That it love what is higher, that is, God; that it rule what is lower, that is, the body; and that by its love it nourish and assist its fellow creatures. For the soul, cleansed and exonerated by these devotions, will fly back from this troubled and wretched life to eternal peace, and will enter into the joy of the Lord.

Introd. *Conform.* *Flight* 944

Alcuin

You cannot possibly have any ill-temper, or show any unkind behaviour to a man for whose welfare you are so much concerned, as to be his advocate with God in private. For you cannot possibly despise and ridicule that man whom your private prayers recommend to the love and favour of God.

William Law

The different organs of the human body fulfill different functions, some more noble certainly than others, but for the good of the body they must all be cared for. In the same way, try to treat with equal love all the people with whom you have relations. Make a habit of this, and soon you will perceive that all humankind is as your family. Thus the abyss between 'myself' and 'yourself' will be filled in, which is the goal of all religious worship.[1]

Ananda Moyî

Accustomed long to contemplating Love and Pity,
I have forgotten all difference between myself and others.

Milarepa

These people (the Indians of Hispaniola) love their neighbours as themselves; their discourse is ever sweet and gentle, and accompanied by a smile. I swear to your majesties, there is not in the world a better nation or a better land.

P. State 583

Christopher Columbus

Love thy Neighbour as thy *self.* Love thy *Neighbour* in thy *Jesus,* thy God. Love thy *Jesus,* thy God, in thy *Neighbour*. Let this *Neighbourhood* of Divine Love be as large as

[1] It can be seen that charity by this analogy must include chastisement of malefactors whether individual or collective, they being the corruption and disease of the body social.

the God of Love himself is. Let every other Person and Spirit, which lives and moves, and hath its being in God, within the *encompassing*, upon the *Ground* and *Root* of the *Divine Being*, be thy *Neighbour*, thy *Brother*, *another self*, as *thy self* to *thy self*; the Object to thee of an heavenly and incorruptible Love.

Peter Sterry

If ye realize the Voidness, Compassion will arise within your hearts;
If ye lose all differentiation between yourselves and others, fit to serve others ye will be;
And when in serving others ye shall win success, then shall ye meet with me;
And finding me, ye shall attain to Buddhahood.

Milarepa

Orthod. 288

He who doth choose charity for a fortress will find men of their own accord humble themselves before him; against all daily accidents and dire calamity, there's naught to shield a man like deeds of kindness.

Shekel Hakodesh, 192

P. State 563

Arise and accept an antidote to ward off old age and death; it is the knowledge that all wealth and prosperity, all pleasures and enjoyments are harmful to us unless devoted to the good of others; they tend only to sicken and enervate our frames.

Yoga-Vasishtha

The fool who's loath to spend the wealth he's got
Becomes the laughing-stock of after ages.

Old Chinese Poem

Illusion 99

Therefore, take diligent heed, lest you hide the talent committed to your care. Rather put it out at interest for the glory of God, and the good of your neighbour.

The Glory of the World

Scatter abroad what you have already amassed rather than pile up new wealth.

'Alî

Those who have riches and honours in this world are reaping the fruits of seeds which they planted in their previous life. Those who do not give alms during their present life will be extremely poor in the future. Do riches and honour come by shrewdness? Are the slow witted all poor? Poverty may come as the result of too much shrewdness.

Hakuin

Action 329

He who regards the world as he does the fortune of his own body can govern the world. He who loves the world as he does his own body can be entrusted with the world.

Tao Te Ching, XIII

Peace 698

Mercy

I am the Tathâgata, O ye gods and men! the Arhat, the perfectly enlightened one: having reached the shore myself, I carry others to the shore: being free, I make free: being comforted, I comfort: being perfectly at rest, I lead others to rest. By my perfect wisdom I know both this world and the next, such as they really are. I am all-knowing, all-seeing. Come to me, ye gods and men! hear the law. I am he who indicates the path: who shows the path, as knowing the path, being acquainted with the path. . . .

Holiness 924

I preach with ever the same voice, constantly taking enlightenment as my text. For this is equal for all; no partiality is in it, neither hatred nor affection.

I am inexorable, bear no love nor hatred towards any one, and proclaim the law to all creatures without distinction, to the one as well as the other. . . .

I re-create the whole world like a cloud shedding its water without distinction; I have the same feelings for respectable people as for the low; for moral persons as for the immoral; for the depraved as for those who observe the rules of good conduct:

For those who hold sectarian views and unsound tenets as for those whose views are sound and correct.

I preach the law to the inferior in mental culture as well as to persons of superior understanding and extraordinary faculties; inaccessible to weariness, I spread in season the rain of the law. . . .

As the rays of the sun and moon descend alike on all men, good and bad, without deficiency in one case or surplus in the other;

So the wisdom of the Tathâgata shines like the sun and moon, leading all beings without partiality.

Saddharma-puṇḍarîka, v

The Divine Mercy is perfect, in the sense that it answers every need. It is universal, in the sense that it spreads alike over those who merit it and those who do not merit it.

Love 618

Al-Ghazâlî

Wr. Side 474

He (Allâh) is the Most Merciful of those who show mercy.

Qur'ân, XII. 64

It is the part of every Child of Light, to maintain the Divine Love in his Spirit, like the Sun in the Firmament, encompassing the whole Earth, from one end to the other, shining upon all, both good and bad, upon dry and sandy Desarts, the Habitations of wild Beasts, and venemous Serpents, as well as cultivated Gardens, flourishing with wholesome Herbs, pleasant Flowers, and all sorts of fruits. Thus God himself is propounded to us for a Pattern by the Son of Good.

Peter Sterry

The sunlight is one and the same wherever it falls, but only bright surfaces like water, mirrors and polished metals can reflect it fully. So is the divine light. It falls equally and impartially on all hearts, but only the pure and clean hearts of the good and holy can fully reflect it.

Sri Ramakrishna

Introd. *Pilg.*

The sage has no self to call his own;
He makes the self of the people his self.
To the good I act with goodness;
To the bad I also act with goodness:
Thus goodness is attained.
To the faithful I act with faith;
To the faithless I also act with faith:
Thus faith is attained.

Tao Te Ching, XLIX

Introd. *Holiness*

P. State 563

Blessed are the merciful: for they shall obtain mercy.

St Matthew, v. 7

The quality of mercy is not strain'd,
It droppeth as the gentle rain from heaven
Upon the place beneath: it is twice bless'd:
It blesseth him that gives and him that takes:
'Tis mightiest in the mightiest; it becomes
The throned monarch better than his crown;
His sceptre shows the force of temporal power,
The attribute to awe and majesty,
Wherein doth sit the dread and fear of kings;
But mercy is above this sceptered sway,
It is enthroned in the hearts of kings,
It is an attribute to God himself,
And earthly power doth then show likest God's
When mercy seasons justice.

Shakespeare (*Merchant of Venice*, IV. i. 184)

Center 816

Mercy, the wise Athenians held to be
Not an Affection, but a *Deitie*.

Robert Herrick

It is of the Lord's mercies that we are not consumed, because his compassions fail not.

Lamentations, III. 22

Wr. Side 464

Let us not be justices of the peace, but angels of peace.

St Teresa of Lisieux

Judgment 242

It may be that Allâh will ordain love between you and those of them with whom ye are at enmity. Allâh is Mighty, and Allâh is Forgiving, Merciful.

Qur'ân, LX. 7

Good men are bad men's instructors,
And bad men are good men's materials.

Tao Te Ching, XXVII

Then an enemy is like a treasure found in my house, won without labour of mine; I must cherish him, for he is a helper in the way to Enlightenment.

Śânti-deva

Blessed be God . . . the Father of mercies, and the God of all comfort;
Who comforteth us in all our tribulation, that we may be able to comfort them which are in any trouble, by the comfort wherewith we ourselves are comforted of God.

II. Corinthians, I. 3, 4

Symb. 317 Introd. *Heterod.*

The Sun is he whose beneficent workings operate not only in heaven, but also upon earth, and penetrate even to the lowest depths.

Hermes

Reality 803

Tao is the source of all things, the treasure of good men, and the sustainer of bad men.

Tao Te Ching, LXII

He maketh his sun to rise on the evil and on the good, and sendeth rain on the just and on the unjust.

St Matthew, V. 45

Alike am I to all beings; hated or beloved there is none to Me.

Bhagavad-Gîtâ, IX. 29

Knowl. 761

The wisdom that is from above is first pure, then peaceable, gentle, and easy to be intreated, full of mercy and good fruits, without partiality, and without hypocrisy.

James, III. 17

Beauty 689

My mercy embraces all things.

Qur'ân, VII. 156

Hitotsu ya ni	Under one roof,
Yûjo mo netari,	Prostitutes, too, were sleeping;
Hagi to tsuki.	The hagi flowers and the moon.

Zen *Haiku*

He who realizes the Lord God, the Atman, the one existence, the Self of the universe, neither praises nor dispraises any man. Like the sun shining impartially upon all things, he looks with an equal eye upon all beings. He moves about in the world a free soul, released from all attachment.

Srimad Bhagavatam, XI. XX

Benevolence and Compassion

As the bee collects nectar and departs without injuring the flower, or its colour or scent, so let a sage dwell in his village.

Dhammapada, IV. 49

Love and pity and wish well to every soul in the world; dwell in love, and then you dwell in God; hate nothing but the evil that stirs in your own heart.

William Law

Love all things, *O Reader,* after a *divine* manner, that thou mayest be the beloved Object of all *divine things,* and *divinely beloved* by all things, that thou mayest shine with a *divine loveliness* in all *eyes,* and be received with a *divine loveliness* into *all hearts.*

Peter Sterry

Compassion is not a passion: rather a noble disposition of the soul, made ready to receive love, mercy, and other charitable passions.

Dante (*Il Convito,* II. xi. 2)

The notion of emptiness (absence of real self) engenders Compassion,
Compassion does away with the distinction between 'self and other',
The indistinction of self and other renders the service of others effective.

Death 220

Milarepa

If God alone is to be loved for Himself, then no creature is to be loved for itself; and so all self-love in every creature is absolutely condemned. And if all created beings are only to be loved in and for God, then my neighbour is to be loved as I love myself, and I am only to love myself as I love my neighbour or any other created being, that is, only in and for God. And thus the command of loving our neighbour as ourselves stands firm and yet all self-love is plucked up by the roots.

Supra 597

William Law

He who groweth (in goodness), groweth only for himself.

Qur'ân, XXXV. 18

Conform. 170

My stinginess is for God's sake, my bounty is for God's sake alone: I belong entirely to God, I do not belong to any one (else).

Rûmî

Not to care for anybody in particular is to care for mankind in general.

A Chinese Nun

This man who is within God's love and ken is none other than what God is himself. Loving thyself thou lovest all men as thyself. While thou lovest anyone less than thine own self thou dost not love thyself in truth: not till thou lovest all men as thyself, all men in one man who is both God and man. The man who loves himself and all men as himself is righteous, absolutely just.

Eckhart

M.M. 987
Action 346

The crane, the cat, and the thief walk silently and calmly, and accomplish the end that they desire; thus the holy man will always act. . . . He will not for slight purpose afflict his body, which is in the service of the Good Law; for thus it will speedily fulfil the (vain) desires of living beings. And therefore he will not cast away his life for one whose spirit of mercy is impure, but only for one whose spirit is like his own; and thus naught is lost.

Śânti-deva

Judgment 242

It is not for the Prophet, and those who believe, to pray for the forgiveness of idolaters even though they may be near of kin, after it hath become clear that they are people of hell-fire.

Qur'ân, IX. 113

Be a *Serpent* to the *evil,* but at the same time be a *Dove* to the *person.*

Peter Sterry

Realiz. 870

Have thy heart in heaven and thy hands upon the earth. Ascend in piety and descend in charity. For this is the Nature of Light and the way of the children of it.

Thomas Vaughan

Renun. 136
Inv. 1036
Contem. 532
Action 329
Introd. *Holiness*

Now, if asked, 'What is this spirit of meditation?' I reply that it is to have a sincerely benevolent and sympathetic heart at all times, whether one is talking or wagging one's elbow when writing, whether one is moving, or resting, whether one's luck is good or bad, whether one is in honour or in shame, or in gain or in loss, in right or in wrong, bundling all these things up into one verse heading and concentrating your energy with the force of an iron rock under the navel and lower part of the abdomen. . . .

If you have this 'spirit' . . . your two-edged swords will be your desk or meditation table, placed always before you. The saddle you ride on will be the cushion on which you sit in meditation. The hills, the streams, the plains will be the floor of your meditation hall. The four corners of the earth and its ten directions, the height and the depth of the universe will be to you the great 'cave' in which you are performing your meditation—they will be, in very truth, the substance of your real self.

Hakuin

He who clings to the Void
And neglects Compassion,
Does not reach the highest stage.
But he who practises only Compassion,
Does not gain release from toils of existence.
He, however, who is strong in practice of both,
Remains neither in Samsara nor in Nirvana.

Saraha

Orthod. 285
M.M. 978

Verily, we feed you for the sake of Allâh only.[1] We seek for no reward nor thanks from you.

Qur'ân, LXXVI. 9

Supra Introd.

Charity is in the heart of man, and righteousness is the path for men. Pity the man who has lost his path and does not follow it and who has lost his heart and does not know how to recover it. When people's dogs and chicks are lost, they go out and look for them, and yet the people who have lost their hearts (or original nature) do not go out and look for them. The principle of self-cultivation consists in nothing but trying to look for the lost heart....

Creation 42

Now a man is wise enough to be ashamed of a finger that is not normal, and yet he is not wise enough to be ashamed of his heart, when his heart is not normal. We say such a man has no sense of the relative importance of things.

Mencius

Judgment 250

If we really and truly understood what is our true personal dignity, we should at once spring to our feet and change our very hearts and minds. We should worship the three treasures and reverence the patriarchs. We should put filial piety first and above all else. We should preserve that proper and respectful love between husband and wife. We should be cordial in the courtesy of our behaviour towards our brothers and friends. We should never forget our duty of fidelity towards our relatives and neighbours. We should have compassion on the sick and the poor. We should work with all our might for our home and family. We should be obedient to the ordinances of our rulers. We should be merciful, honest and patient, so as to be good examples to all men. Then without appearing to instruct we should in the most natural way be able to act as guides. Then all classes of the people would associate in peace. Everyone would pass through life with smiling faces, co-operating with the Buddha and the Gods of heaven and earth.

Beauty 676
Rev. 961

Then the eight million Gods, Bonten (Brahma), Taishaku (Indra), Daikoku, Bishamon—all of them would be a protection to you. Evil demons and false gods would flee. There would be no sickness. The whole world would be at peace. The five harvest grains would mature and every home be prosperous. The children who were born would be good too.

Hakuin

P. State 583

[1] This is the inverse of humanitarian 'charity' which is for the sake of *man* only.

Where there is hate, let me bring Love—
Metanoia 488 Where there is offence, let me bring Pardon—
Where there is discord, let me bring Union—
Where there is error, let me bring Truth—
Where there is doubt, let me bring Faith—
Where there is darkness, let me bring Light—
Where there is sadness, let me bring Joy.

May it be, O lord, that I seek not so much to be consoled as to console, to be understood as to understand, to be loved as to love; because it is in giving oneself that one receives; it is in forgetting oneself that one is found; it is in pardoning that one obtains pardon; it is in dying that one is raised up to eternal life. *Death* 206

St Francis of Assisi

Cosmic Salvation

There is communion between soul and soul. The souls of the gods are in communion with those of men, and the souls of men with those of the creatures without reason. And God takes care of all; for He is higher than all. The Kosmos then is subject to God; man is subject to the Kosmos; the creatures without reason are subject to man; and God is above all, and watches over all. Introd. *Symb.*

Hermes

Every creature of God is good,[1] and nothing to be refused, if it be received with thanksgiving:
For it is sanctified by the word of God and prayer. *Inv.* 1003

I. Timothy, IV. 4, 5

It is said that no animal can enter into God our Lord:
What then are these four[2] who are round about Him?

Angelus Silesius

I feel that my mission is going to begin, my mission to have God loved as I love Him, to give my *little way* to souls. I want to spend my heaven in doing good on earth. This is not impossible, since the angels at the center of the beatific vision watch over us. No, until the end of the world, and as long as there will be souls to save, I shall be able to take no rest; but when the angel shall have said: 'Time no longer is!'[3] then shall I rest, then shall I

[1] *Genesis*, I. 31.
[2] *Revelation*, IV. 6.
[3] *Revelation*, X. 6.

delight, because the number of the elect will be complete, and all will have entered into joy and repose. My heart leaps at the thought.

St Teresa of Lisieux

May no living creatures, not even any insects,
Be bound unto *sangsâric* life; nay, not one of them;
But may I be empowered to save them all.

Milarepa

'Everything having been created for man's use, all by consequence has been made to serve and provide for the needs of rational creatures. The creature endowed with reason has not himself been created for these things, but for Me, to serve Me with all his heart and all his affection. Thou seest clearly that if man is drawn towards my Son, everything is accordingly drawn towards Him, since all the rest has been made for man.'

St Catherine of Siena

Contem. 547
Rev. 967

Every single creature has, in human nature, a stake in the eternal.

Eckhart

God's Providence, from the Fall to the restitution of all things, is doing the same thing as when He said to the dark chaos of fallen nature, 'Let there be light'; He still says and will continue saying the same thing, till there is no evil of darkness left in all that is nature and creature. God creating, God illuminating, God sanctifying, God threatening and punishing, God forgiving and redeeming, is but one and the same essential, immutable, never-ceasing working of the divine nature. . . .

That all-powerful, all-purifying blood of the Lamb . . . will never cease washing red into white, till the earth is washed into the crystal purity of that glassy sea which is before the throne of God, and all the sons of Adam clothed in such white as fits them for their several mansions in their heavenly Father's house.

William Law

Wr. Side 464

God's providence as regards Pharaoh was not terminated by his drowning.

Origen

The rectitude of Reason, was the Image of God, the Crowne, and Glory of Man. By this he grew up streight in the midst of the Creatures; uniting heightning them all by a Harmony into the same Image. So he subdued them to himselfe, himselfe with them ascended into, and rested in the Divine Image.

Knowl. 761

Thus Man was the true *Orpheus* with his Divine Hymnes in the midst of Beasts, Trees, and Stones dancing to his Musick.

Peter Sterry

Beauty 679

I tell you that, if these should hold their peace, the stones would immediately cry out.

St Luke, XIX. 40

The suffering, disease, and imperfection brought not only upon men, but also upon plants and animals, by the fall of Adam, found a remedy in that precious gift of Almighty God, which is called the Elixir, and Tincture, and has power to purge away the imperfections not only of human, but even of metallic bodies; which excels all other medicines, as the brightness of the sun shames the moon and the stars.

P. State 563

The Glory of the World

'We shall burn the sweet grass as an offering to *Wakan-Tanka*, and the fragrance of this will spread throughout heaven and earth; it will make the four-leggeds, the wingeds, the star peoples of the heavens, and all things as relatives. From You, O Grandmother Earth, who are lowly, and who support us as does a mother, this fragrance will go forth; may its power be felt throughout the universe, and may it purify the feet and hands of the two-leggeds, that they may walk forward upon the sacred earth, raising their heads to *Wakan-Tanka*!'

Black Elk (The Sioux *Inipi* Rite of Purification)

I would be a protector of the unprotected, a guide of wayfarers, a ship, a dyke, and a bridge for them who seek the further Shore; a lamp for them who need a lamp, a bed for them who need a bed, a slave for all beings who need a slave. I would be a magic gem, a lucky jar, a spell of power, a sovereign balm, a wishing-tree, a cow of plenty, for embodied beings. As the earth and other elements are for the manifold service of the countless creatures dwelling in the whole of space, so may I in various wise support the whole sphere of life lodged in space, until all be at peace. As the Blessed of old took the Thought of Enlightenment and held fast to the rule for Sons of Enlightenment in the order thereof, so do I frame the Thought of Enlightenment for the weal of the world, and so will I observe these rules in their sequence.

Grace 552

Śânti-deva

There is none but may be saved at last.

Saddharma-puṇḍarîka

Everything, by an impulse of its own nature, tends towards its perfection.

Illusion 109

Dante (*Il Convito*, I. i. 1)

One day he (Bâyazîd al-Bisṭâmî) was walking with a number of his disciples. The path was very narrow. He saw a dog coming along and turned back to let it pass. One of his disciples blamed him secretly and thought to himself, 'How can Bâyazîd, who is the king of gnostics, make way for a dog?' Bâyazîd said, 'This dog asked me with dumb eloquence, saying, "In the eternal past what fault did I commit, and what act of grace didst thou perform, that I am clad in the skin of a dog, while the robe of spiritual royalty hath been conferred on thee?" This thought came into my head and I made way for the dog.'

Action 329

'Aṭṭâr

We have to take the whole universe as the expression of the one Self. Then only our love flows to all beings and creatures in the world equally.

Reality 775

Swami Ramdas

And there is not an animal in the earth, nor a flying creature flying on two wings, but they are peoples like unto you. We have neglected nothing in the Book (of our Decrees). Then unto their Lord they will be gathered.

Qur'ân, VI. 38

Naturally magnanimous and open-minded, the red man prefers to believe that the Spirit of God is not breathed into man alone, but that the whole created universe is a sharer in the immortal perfection of its Maker.

Ohiyesa

Truly do I exist in all beings, but I am most manifest in man. The human heart is my favourite dwelling place..

Srimad Bhagavatam, XI. ii

Contem. 536

And not a drop that from our Cups we throw
For Earth to drink of, but may steal below
To quench the fire of Anguish in some Eye
There hidden—far beneath, and long ago.

Omar Khayyâm

(Pythagoras) ordered them never . . . to injure animals, but to preserve most solicitously justice towards them. . . . For through the communion of life and the same elements, and the mixture subsisting from these, they are as it were conjoined to us by a fraternal alliance.

Iamblichus

Reality 775

When the seas of Mercy begin to surge, even stones drink the Water of Life.

Rûmî

My little sisters the birds, much are ye beholden to God your Creator, and alway and in every place ye ought to praise Him for that He hath given you a double and a triple vesture; He hath given you freedom to go into every place, and also did preserve the seed of you in the ark of Noe, in order that your kind might not perish from the earth. Again, ye are beholden to him for the element of air which He hath appointed for you; moreover, ye sow not, neither do ye reap, and God feedeth you and giveth you the rivers and the fountains for your drink; He giveth you the mountains and the valleys for your refuge, and the tall trees wherein to build your nests, and forasmuch as ye can neither spin nor sew God clotheth you, you and your children: wherefore your Creator loveth you much, since He hath dealt so bounteously with you; and therefore beware, little sisters mine, of the sin of ingratitude, but ever strive to praise God.

St Francis of Assisi

When (the Apostle) was borne to the presence of God, he said: 'I know not how to utter Thy praise.' Answer came: 'O Muḥammad, if thou speakest not, I will speak: if thou deemest thyself unworthy to praise Me, I will make the universe thy deputy, that all its atoms may praise Me in thy name.'

Hujwîrî

Holiness 924

LOVE

For the Black Mole on the Cheek of my Beloved
I would give the cities of Bokhara and Samarkand.

Hâfiẓ

Love is the energizing elixir of the universe, the cause and effect of all harmonies, light's brilliance and the heat in wine and fire, it is the aroma of perfumes and the breath of the Divinity: it is the Life in all being. Love is the quickening solvent in *mâyâ*, and the coalescing agent in union. It is all that the texts have to say, and the more that remains unspoken. 'What is love? Thou shalt know when thou becomest me.'[1]

Knowl. 749

'God is Love, not because He loves, but He loves because He is Love' (Schuon: *Perspectives spirituelles*, p. 145).

Marriage presupposes love, at least where union with God is concerned, but here below love may take the form of holy chastity whose nuptials will only be celebrated *in divinis*: 'The sacred marriage, consummated in the heart, adumbrates the deepest of all mysteries. For this means both our death and beatific resurrection. The word to "marry" (*eko bhû*, become one) also means to "die", just as in Greek, *τελέω* is to be perfected, to be married, or to die. When "Each is both", no relation persists: and were it not for this beatitude (*ânanda*) there would be neither life nor gladness anywhere' (Coomaraswamy: *Hinduism and Buddhism*, p. 14).

Death 208

However, since the sexual power stands at the junction of the fiery and luminous poles of existence, there is also to be envisaged a *via amoris*, where love taken as an erotic support may not only 'be sanctified', but can even 'be sanctifying'.[2] 'In Hesychasm the absence of passions is not their death purely and simply, but their transmutation into spiritual energy; one will observe that such a conception, which finds its equivalent in Tantrism,—to name a particularly characteristic example,—clearly goes beyond the limits of mere morality and social opportunism' (Schuon: *De l'Unité transcendante des Religions*, p. 163). But whether Tantra, Hesychasm, Hermeticism, or Kabbala, an implacable chastity of intention and purity of soul are prerequisite conditions: sacred love may embrace Reality through the contemplation of an erotic symbolism or through the ritualization of terrestrial marriage, where the infant sought of God will be pre-eminently the 'spiritual child' or Hermetic Androgyne: but profane love can never clasp anything except illusion and death. 'One sees how these conceptions are situated counter to all erotic complacency, since it is a question first of all of restoring, in nature as in man, a state of

Realiz. 870

[1] Rûmi: *Mathn.*, II. Introd.

[2] These terms are taken from M. Aniane's article on alchemy in *Yoga, Science de l'Homme intégral*, p. 266.

chaste[1] submission to the divine will, which means in effect, virginity. In alchemy, the true hero, "son of the cosmos" and "saviour of the macrocosm", is he who is capable of offering a virgin soul into the embrace of transcendency' (M. Aniane, op. cit. Introd. *Pilgrimage*, p. 256).

'Love is in the depths of man as water is in the depths of the earth, and man suffers from not being able to enjoy this infinity that he carries within himself and for which he is made. *Contem.* 547

'One must dig the soil of the soul, through layers of aridity and bitterness, to find love and to live in it. *Pilg.* 366

'The profundity of love is inaccessible to man in his state of hardness, but it reveals itself outwardly in the language of art and also that of nature. In sacred art and virgin nature, the soul can taste by analogical anticipation something of the love which sleeps within it and of which it has only the nostalgia and not the experience' (Schuon: *Perspectives spirituelles*, p. 283).

[1] 'All superfluity, anything unnecessary in word or deed, is unchastity' (Eckhart, I. 314).

Our Love for God

P. State 563

All things work together for good to them that love God, to them who are the called according to his purpose.

Romans, VIII. 28

Knowl. 749
Realiz. 859

We love God with his own love; awareness of it deifies us.

Eckhart

If the seed is not subjected to heat and humidity, it becomes useless.

Knowl. 761

Cold contracts the heart and dryness hardens it, but the fire of divine love dilates it, and the water of intelligence dissolves the residue.

Henry Madathanas

In order to live a perfect act of love, I offer myself victim in a holocaust to thy merciful love, entreating thee to consume me without cease, letting my soul overflow with the waves of infinite tenderness which are within thee, that I may thus become martyr of thy love, O my God!

May this martyrdom, after having prepared me to appear before thee, then make me die, and may my soul leap without delay into the eternal embrace of thy merciful love!

Inv. 1007

I wish, O my Beloved, with each beat of my heart, to renew this offering for thee an infinite number of times, until when *the shadows retire*,[1] I can reveal my love to thee in an eternal encounter!

St Teresa of Lisieux

Suff. 132
Death 206

The pathway of love is the ordeal of fire. The shrinkers turn away from it.
Those who take the plunge into the fire attain eternal bliss.
Those who stand afar off, looking on, are scorched by the flames.
Love is a priceless thing, only to be won at the cost of death.
Those who live to die, these attain; for they have shed all thoughts of self.

Gujarati Hymn

Holiness 921

No life can express, nor tongue so much as name what this enflaming, all-conquering love of God is. It is brighter than the sun; it is sweeter than anything that is called sweet; it is stronger than all strength; it is more nutrimental than food; more cheering to the heart than wine, and more pleasant than all the joy and pleasantness of this world. Whosoever obtaineth it, is richer than any monarch on earth; and he who getteth it, is nobler than any emperor can be, and more potent and absolute than all power and authority.

Boehme

[1] *Cant.* IV. 6.

The clearest sign of the love of God is excessive pallor, associated with continuous meditation, and prolonged vigil accompanied by complete self-surrender, obediently and with great haste ere dread death come upon him (the lover); and the lover speaks of love according to the measure of the Light bestowed upon him. Hence it is said, that the sign of the love of God is the indwelling of God's Favours in the hearts of those whom God has singled out for His Love.

Holiness 910
Holy War 403

Al-Muḥâsibî

A disciple asked his teacher, 'Sir, please tell me how I can see God.' 'Come with me,' said the guru, 'and I shall show you.' He took the disciple to a lake, and both of them got into the water. Suddenly the teacher pressed the disciple's head under the water. After a few moments he released him and the disciple raised his head and stood up. The guru asked him, 'How did you feel?' The disciple said, 'Oh! I thought I should die; I was panting for breath.' The teacher said, 'When you feel like that for God, then you will know you haven't long to wait for His vision.'[1]

Sri Ramakrishna

When a man is calling upon the sacred name he ought to do it with the earnestness of the man in dire distress who is crying 'Oh save me now.'

Inv. 1036

Hônen

The Bhakta must be seized with a discontent which never ceases until the goal is reached.

Holy War 405

Swami Ramdas

I had supposed that, having passed away
    From self in concentration, I should blaze
A path to Thee; but ah! no creature may
    Draw nigh Thee, save on Thy appointed ways.
I cannot longer live, Lord, without Thee;
Thy Hand is everywhere: I may not flee.

Grace 552

Some have desired through hope to come to Thee,
    And Thou hast wrought in them their high design:
Lo! I have severed every thought from me,
    And died to selfhood, that I might be Thine.
How long, my heart's Beloved? I am spent:
I can no more endure this banishment.

Abu 'l-Ḥusayn al-Nûrî

As the hart panteth after the water brooks, so panteth my soul after thee, O God.

Psalm XLII. 1

[1] One is reminded here of the practices employed in Zen Buddhism.

Renun. 146

Thee I choose, of all the world, alone;
Wilt thou suffer me to sit in grief?
My heart is as a pen in thy hand,
Thou art the cause if I am glad or melancholy.

Conform. 185

Save what thou willest, what will have I?
Save what thou showest, what do I see? . . .

Illusion 85

When thou art hidden I am of the infidels;
When thou art manifest, I am of the faithful.

Humility 191

I have nothing, except thou hast bestowed it;
What dost thou seek from my bosom and sleeve?

Dîvâni Shamsi Tabrîz, XXX

Center 847 *Judgment* 250

The soul which has been in the presence of Christ experiences sweetness, and from this sweetness is born a chaste joy which is the embrace of divine love clasping the depths of the soul. Now all the pleasures of the earth fused into a single pleasure and poured in their entirety into a single man are nothing compared with the joy of which I speak; for here it is God who flows into our depths in all His purity, and our souls are not only filled but overflowing. Only the light of this experience can show the soul the dreadful misery of those who live without love.

Ruysbroeck

Blessed to all birds is desire of thee;
How unblest the bird that desires thee not!

Dîvâni Shamsi Tabrîz, X

Beauty 663 *Ecstasy* passim *Flight* 944 *Knowl.* 755 *M.M.* 994

Earthly love is not for the material form but for the beauty manifested upon it. . . . The soul taking that outflow from the Divine is stirred; filled with a holy ecstasy, stung by desire, it becomes Love. Before that, even Divine Mind with all Its Loveliness did not stir the soul; for until it takes the light of the Good the soul lies supine before It, cold, unquickened. But when there enters into it a glow from the Divine, the soul gathers strength, spreads true wings, and however distracted by its nearer environing, speeds its buoyant way to something greater to its memory; so long as there exists anything loftier than the near, its very nature bears it upwards, lifted by the giver of that love. Beyond Divine Mind it passes, but beyond the Good it cannot, for nothing stands above That.

Plotinus

Metanoia 488

For as bodily fire burneth and wasteth all bodily things where it cometh, right so ghostly fire, as is the love of God, burneth and wasteth all fleshly loves and likings in a man's soul.

Walter Hilton

P. State 563

My soul is a furnace: it is happy with the fire: 'tis enough for the furnace that it is the fire's house.

Rûmî

Whosoever shall be sore wounded by love
Will never become whole
Save he embrace the self-same love
Which wounded him.

Mechthild of Magdeburg

When God's nearness takes possession of a man's heart, it overwhelms all else, both the inward infiltrations of the purposes and the outward motions of the members. Thereafter that man continues, going or coming, taking or giving: there prevails in him the purpose which has ruled his mind, namely, the love of God and His nearness.

Ahmad b. 'Îsâ al-Kharrâz

Center 841 *Holiness* 902

As the tiger devours other animals, so does the 'tiger of zeal for the Lord' eat up lust, anger, and the other passions. Once this zeal grows in the heart, lust and the other passions disappear. The gopis of Vrindâvan had that state of mind because of their zeal for Krishna.

Again, this zeal for God is compared to collyrium. Râdhâ said to her friends, 'I see Krishna everywhere.' They replied, 'Friend, you have painted your eyes with the collyrium of love: that is why you see Krishna everywhere.'

Sri Ramakrishna

Ecstasy 641

'What a man loves a man is.'[1] If he loves a stone he is that stone, if he loves a man he is that man, if he loves God—nay, I durst not say more; were I to say, he is God, he might stone me. I do but teach you the scriptures.

Eckhart

Metanoia 480 *Knowl.* 749

Let no one believe that he has received the divine kiss, if he knows the truth without loving it or loves it without understanding it. But blessed is that kiss whereby not only is God recognized but also the Father is loved: for there is never full knowledge without perfect love.

St Bernard

Knowl. 745

Undisciplined love dwells in the senses for it is still entangled with earthly things. . . . Disciplined love lives in the soul and rises above the human senses and forbids the body its own will. It is modest and very still. It folds its wings and listens to an unspeakable voice and gazes into incomprehensible light and seeks eagerly the will of its Lord.

Mechthild of Magdeburg

A single atom of the love of God in a heart is worth more than a hundred thousand paradises.

Bâyazîd al-Bistâmî

Conform. 170

'Dost thou wish to do penance for thy sins? Love Me, let love be thy penance.'

Sister Consolata

[1] St Augustine.

Many do not advance in the Christian progress because they stick in penances and particular exercises, while they neglect the love of God, which is the *end*.

Suff. 133

Brother Lawrence

Love of God is the one essential thing.

Sri Ramakrishna

Many are the means described for the attainment of the highest good, such as love, performance of duty, self-control, truthfulness, sacrifices, gifts, austerity, charity, vows, observance of moral precepts. I could name more. But of all I could name, verily love is the highest: love and devotion that make one forgetful of everything else, love that unites the lover with me. What ineffable joy does one find through love of me, the blissful Self! Once that joy is realized, all earthly pleasures fade into nothingness.

Center 847

Srimad Bhagavatam, XI. viii

Center 835
Conform. 180
Beauty 689

Among those then who dwell in that world above there is no disagreement; all have one purpose; there is one mind, one feeling in them all; for the spell which binds them one to another is Love, the same in all, and by it all are wrought together into one harmonious whole.

Hermes

The soul, having conquered the multiple heavens and possessed herself of their mysterious power, is plunged into the unity of the motionless heaven, called fire or the empyrean, not because it is burning but because it is enlightening, all who are in this heaven being ablaze with the cherubic light of divine love.

Eckhart

My desire and will were rolled—even as a wheel that moveth equally—by the Love that moves the sun and the other stars.

Creation 48

Dante (*Paradiso,* XXXIII. 143)

God's Love for Us

Krishna is an Avatar of perfect love. The word Krishna is derived from *Akarshana* which means attraction. So the name itself signifies that Krishna is the Supreme Truth, manifest in human form, to attract the whole world towards Him.

Swami Ramdas

A root set in the finest soil, in the best climate, and blessed with all that sun, and air, and rain can do for it, is not in so sure a way of its growth to perfection as every man may be whose spirit aspires after all that which God is ready and infinitely desirous to give him. For the sun meets not the springing bud that stretches towards him with half the certainty as God, the source of all good, communicates Himself to the soul that longs to partake of Him.

Symb. 317

William Law

God longs as urgently for thee to go out of thyself in respect of thy creaturely nature as though his whole felicity depended on it. Why, man, what is the harm of letting God be God in thee? Go clean out of thy self for God's sake, and God will go clean out of his for thy sake. Both being gone out, what remains is simply the one.

Eckhart

Renun. 158
Metanoia **493**
Void 728
M.M. 978

In this storm of love two spirits strive together: the spirit of God and our own spirit. God, through the Holy Ghost, inclines Himself towards us; and, thereby, we are touched in love. And our spirit, by God's working and by the power of love, presses and inclines itself into God: and, thereby, God is touched. From these two contacts there arises the strife of love, at the very deeps of this meeting; and in that most inward and ardent encounter, each spirit is deeply wounded by love. These two spirits, that is, our own spirit and the Spirit of God, sparkle and shine one into the other, and each shows to the other its face. This makes each of the spirits yearn for the other in love. Each demands of the other all that it is; and each offers to the other all that it is and invites it to all that it is. This makes the lovers melt into each other. God's touch and His gifts, our loving craving and our giving back: these fulfil love. This flux and reflux causes the fountain of love to brim over: and thus the touch of God and our loving craving become one simple love. Here man is possessed by love, so that he must forget himself and God, and knows and can do nothing but love. Thereby the spirit is burned up in the fire of love, and enters so deeply into the touch of God, that it is overcome in all its cravings, and turned to nought in all its works, and empties itself; above all surrender becoming very love. And it possesses, above all virtues, the inmost part of its created being, where every creaturely work begins and ends. Such is love in itself, foundation and origin of all virtues.

Knowl. 745

Realiz. 868

Ruysbroeck

Supra 614

If thou wilt wit then what this desire is, soothly it is Jhesu. For He maketh this desire in thee and He giveth it thee, and He it is that desireth in thee and He it is that is desired. He is all, and He doth all, if thou might see Him.

Walter Hilton

Knowl. 749

It may be that we shall receive no vision or message from the Great Spirit the first time that we 'lament', yet we may try many times, for we should remember that *Wakan-Tanka* is always anxious to aid those who seek Him with a pure heart. But of course much depends on the nature of the person who cries for a vision, and upon the degree to which he has purified and prepared himself.

Black Elk

Faith 509
Grace 558

Center 816

So-called friends are in reality enemies. You will not find in this whole universe a single friend without egoism. Your true friend in need, He who is sincerely attached to you, is God, who lives within your heart.

Swami Sivananda

Or have they chosen protecting friends besides Him? But Allâh, He (alone) is the Protecting Friend. He quickeneth the dead, and He is able to do all things.

Qur'ân, XLII. 9

Creation 28
Grace 552

As Plato sometimes speaks of the divine love, it arises not out of indigency, as created love does, but out of fulness and redundancy: it is an overflowing fountain, and that love which descends upon created being is a free efflux from the almighty source of love: and it is well pleasing to him that those creatures which he hath made should partake of it.

John Smith the Platonist

Rev. 967

We are pieces of steel, and thy love is the magnet.

Dîvâni Shamsi Tabrîz, XXXII

Beauty 689

Thou (Wisdom) lovest all things that are, and hatest none of the things which thou hast made: for thou didst not appoint or make any thing hating it.

Wisdom, XI. 25

It cannot be thought that what it has once pleased God to ordain will ever be displeasing in his sight, inasmuch as he knew long before, both that it would come to pass, and that it would be pleasing to him.

Hermes

Wr. Side 474

Man, speak rightly of God: He does not hate His creature:
(This is impossible for Him), not even these ugly demons.

Angelus Silesius

This is the amiable nature of God. He is the Good, the unchangeable, overflowing fountain of good that sends forth nothing but good to all eternity. He is the Love itself, the unmixed, unmeasurable Love, doing nothing but from love, giving nothing but gifts of love to everything that He has made; requiring nothing of all his creatures but the spirit and fruits of that Love which brought them into being.

William Law

Illusion 109
Grace 556

God, despite himself, is ever hanging over some bait to lure us into him. I never give God thanks for loving me, because he cannot help it: whether he would or no it is his nature to. What I do thank him for is for not being able of his goodness to leave off loving me.

Eckhart

As a devotee cannot live without God, so also God cannot live without His devotee. Then the devotee becomes the sweetness, and God its enjoyer. The devotee becomes the lotus, and God the bee. It is the Godhead that has become these two in order to enjoy Its own Bliss. That is the significance of the episode of Râdhâ and Krishna.

Sri Ramakrishna

Creation 48

God speaks of His intense desire to meet man, for he says to David concerning those who desire Him: 'O David, it is I who desire them still more intensely.'[1]

Ibn 'Arabî

For He standeth all aloof and abideth us sorrowfully and mournfully till when we come, and hath haste to have us to Him. For we are His joy and His delight, and He is our salve and our life.

Julian of Norwich

Holy War 403

The Lord in His infinite mercy is ever anxious that we must offer our mind to Him and is very eagerly awaiting our doing so. If, therefore, we do offer our mind to Him, it is clear that He will be immensely pleased. Our mind . . . is something which is dear to Him, which He has not with Him, and which He longs to have. If we give it to Him, there can be no doubt that He will be made happy and that, as a consequence, happiness will be reflected in us also.

Sri Chandrasekhara Bhâratî Swâmigal

Contem. 528

'I thirst for thy love like a person thirsting for a spring of fresh water. . . . Write in obedience to Me, Consolata, that for an act of love I would create Paradise.'

Sister Consolata

Now you must know, God loves the soul so mightily, he who should rob God of loving the soul would rob him of his very life and being: would kill God, if one may so say; for the very love wherewith God loves the soul is what his Holy Breath is blowing in.

Eckhart

God loves me alone, and desires me so.
He dies of anguish, when I withhold myself from Him.

Angelus Silesius

Reality 803

His only desire is our sanctification.

St Catherine of Siena

The body receives nothing but its own spirit; for it has retained its soul, and what has been extracted from a body can be joined to nothing but that same body. The spirit delights in nothing so much as in its own soul, and its own body.

The Glory of the World

Creation 42
Renun. 158

[1] According to a saying of Muhammad.

My children, mark me, I pray you. Know! God loves my soul so much that his very life and being depend upon his loving me, whether he would or no. To stop God loving me would be to rob him of his Godhood; for God is love no less than he is truth; as he is good, so is he love as well. It is the absolute truth, as God lives. . . .

If anyone should ask me what God is, I should answer: God is love, and so altogether lovely that creatures all with one accord essay to love his loveliness, whether they do so knowingly or unbeknownst, in joy or sorrow.

Conform. 185

Eckhart

He that loveth not knoweth not God; for God is love. . . . He that dwelleth in love dwelleth in God, and God in him.

I. John, IV. 8, 16

M.M. 998

Love is the astrolabe of the mysteries of God.

Rûmî

God is not a mixt and compounded Being, so that His Love is one thing and Himself another: but the most pure and simple of all Beings, all Act, and pure Love in the abstract.

Creation 31

Thomas Traherne

Beauty 663

Lo, verily, not for love of all is all dear, but for love of the Self all is dear.

Bṛihad-Araṇyaka Upanishad, II. iv. 5

St *Paul* descended from the Paradise in the Third Heavens, bringeth this with him down into the World, as the Sacred Mystery, and rich ground of all Truth, from which all the Beauties and Sweetnesses of Paradise, of all the Heavens spring; That *Love is the band of Perfection.* It is *Love* then, which runneth through the whole Work of God, which frameth, informeth, uniteth it all into one *Master-piece* of Divine Love.

If God be Love, the Attributes of God are the Attributes of this Love; the Purity, Simplicity, the Soveraignty, the Wisdom, the Almightiness, the Unchangeableness, the Infiniteness, the eternity of *Divine Love.* If God be Love, his Work is the Work of Love, of a Love unmixt, unconfined, supream, infinite in Wisdom and Power, not limited in its workings by any pre-existent matter, but bringing forth freely and entirely from it self its whole work both matter and form, according to its own inclination and complacency in it self.

Reality 773

Peter Sterry

God is love, and he that has learnt to live in the Spirit of love has learnt to live and dwell in God. Love was the beginner of all the works of God, and from eternity to eternity nothing can come from God but a variety of wonders and works of love over all nature and creature.

Creation 48

William Law

Love is the nature of God, He can do nothing other:
Thus, to be God, love at each moment.

Angelus Silesius

Born at the banquet of the gods, Love has of necessity been eternally in existence, for it springs from the intention of the Soul towards its best, towards the Good: as long as Soul has been, Love has been.

Plotinus

Love, like the Spirit of God, rideth upon the wings of the wind, and is in union and communion with all the saints that are in Heaven and on earth. Love is quite pure; it has no by-ends; it seeks not its own; it has but one will, and that is to give itself into everything and overcome all evil with good. Lastly, love is the Christ of God; it comes down from Heaven; it regenerates the soul from above; it blots out all transgressions; it takes from death its sting, from the devil his power, and from the serpent his poison. It heals all the infirmities of our earthly birth; it gives eyes to the blind, ears to the deaf, and makes the dumb to speak; it cleanses the lepers and casts out devils, and puts man in Paradise before he dies.

William Law

Flight 944
Death 232
Realiz. 870

Truth seeth God, and Wisdom beholdeth God, and of these two cometh the third: that is, a holy marvelling delight in God; which is Love. Where Truth and Wisdom are verily, there is Love verily, coming of them both.[1]

Julian of Norwich

Knowl. 761

The reason for loving God is God Himself.

St Bernard

Knowl. 749

Lovable, very love, the Supreme is also self-love in that He is lovely no otherwise than from Himself and in Himself. Self-presence can hold only in the identity of associated with associating; since, in the Supreme, associated and associating are one, seeker and sought one—the sought serving as Hypostasis and substrate of the seeker—once more God's being and his seeking are identical.

Plotinus

Center 847
Creation 31

Nay, what is more, it is the 'Truth' who is Himself at once the lover and the beloved, the seeker and the sought. He is loved and sought in His character of the 'One who is all'; and He is lover and seeker when viewed as the sum of all particulars and plurality.

Jâmî

Reality 775

Lover, Beloved and Love am I in one,
Beauty and Mirror and the Eyes which see.

Abû Sa'îd ibn Abi 'l-Khayr

Knowl. 749
Center 828
816

Love dominates everything: even the Trinity
Has been subjected to it from all eternity.

Angelus Silesius

M.M. 994

[1] This doctrine has its equivalent in the *Sat-Chit-Ânanda* (Being-Knowledge-Bliss) of Hinduism.

Here the body is bold and strong for here it is at home; the world helps it, the earth is its fatherland, it is helped by all its kindred: food, drink, ease—all are opposed to spirit. The spirit is an alien here, in heaven are its kindred, its whole race; there dwell its loved ones. To succour the spirit in its distress and to impede the flesh somewhat in this strife lest it conquer the spirit, we put upon it the bridle of penitential practices to curb it, so that the spirit can control it. This is done to bring it to subjection; but to conquer and curb it a thousand times better, put thou upon it the bridle of love. With love thou overcomest it most surely, with love thou loadest it most heavily. God lies in wait for us therefore with nothing so much as with love. For love is like the fisherman's hook. To the fisherman falls no fish that is not caught on his hook. Once it takes the hook the fish is forfeit to the fisherman; in vain it twists hither and thither, the fisherman is certain of his catch. And so I say of love: he who is caught thereby has the strongest of all bonds and yet a pleasant burden. He who bears this sweet burden fares further, gets nearer therewith than by using any harshness possible to man. Moreover, he can cheerfully put up with whatever befalls, cheerfully suffer what God inflicts. Naught makes thee so much God nor God so much thine own as this sweet bond. He who has found this way will seek no other. He who hangs on this hook is so fast caught that foot and hand, mouth, eyes and heart and all that is man's is bound to be God's. . . . Who is caught in this net, who walks in this way, whatsoever he works is wrought by love, whose alone the work is: busy or idle it matters nothing. Such an one's most trivial action is more profitable, his meanest occupation is more fruitful to himself and other people and to God is better pleasing than the cumulative works of other men, who, though free from mortal sin, are yet inferior to him in love. He rests more usefully than others labour.

Suff. 120
Supra
614
Action
346

Await thou therefore this hook, so thou be happily caught, and the more surely caught so much the more surely freed.

That we may be thus caught and freed, help us O thou who art love itself.

Eckhart

Love is my bait; you must be caught by it; it will put its hook into your heart and force you to know that of all strong things nothing is so strong, so irresistible, as divine love.

It brought forth all the creation; it kindles all the life of Heaven; it is the song of all the angels of God. It has redeemed all the world; it seeks for every sinner upon earth; it embraces all the enemies of God; and from the beginning to the end of time the one work of Providence is the one work of love.

William Law

The soul inflamed by the divine splendour . . . is secretly lifted up by it as if by a hook in order to become God.

Marsilio Ficino

Fish-like in the sea behold me swimming, till he with his hook my rescue maketh.

Ḥâfiẓ

The Son of God, by a love greater than that which created the world, became man and gave His own blood and life into the fallen soul, that it might, through His life in it, be raised, quickened, and born again into its first state of inward peace and delight, glory and perfection, never to be lost any more. O inestimable truths! precious mysteries of the love of God, enough to split the hardest rock of the most obdurate heart that is but able to receive one glimpse of them! Can the world resist such love as this? Or can any man doubt whether he should open all that is within him to receive such a salvation?

Creation 42

O unhappy unbelievers, this mystery of love compels me in love to call upon you, to beseech and entreat you to look upon the Christian redemption in this amiable light! All the ideas that your own minds can form of love and goodness must sink into nothing as soon as compared with God's love and goodness in the redemption of mankind.

William Law

It is a hard and steely heart that is not softened by the divine presence and not attracted by its sweetness.

Wr. Side 448

Richard of Saint-Victor

Via Amoris

Nowadays men can not love seven night but they must have all their desires: that love may not endure by reason; for where they be soon accorded, and hasty heat, soon it cooleth. Right so fareth love nowadays, soon hot soon cold: this is no stability. But the old love was not so; men and women could love together seven years, and no lycours lusts were between them, and then was love, truth, and faithfulness: and lo, in likewise was used love in King Arthur's days.

Wr. Side 464

Le Morte d'Arthur, XVIII. XXV

Marriage is half the tradition.[1]

Muhammad

Other compacts are engraved in tables and pillars, but those with wives are inserted in children.

Pythagoras

You are permitted to refuse matrimony, in order that you may live incessantly adhering to God. If, however, as one knowing the battle, you are willing to fight, take a wife, and beget children.

Action 329

Sextus the Pythagorean

For it is better to marry than to burn.

I. Corinthians, VII. 9

[1] The other half is Patience.

The marriage ceremony is said to be the Vedic consecration for women: (also) attendance to (their) husbands, subjection to the Guru, household affairs, (and) attention to the (household sacred) fire.

Mânava-dharma-śâstra, II. 67

Brahmacharya is 'living in Brahman'. It has no connection with celibacy as commonly understood. . . . Celibacy is certainly an aid to realisation among so many other aids. . . . It is a matter of fitness of mind. Married or unmarried, a man can realise the Self, because that is here and now. If it were not so, but attainable by some efforts at some other time, and if it were new and something to be acquired, it would not be worthy of pursuit. Because what is not natural cannot be permanent either. But what I say is that the Self is here and now and alone.

Realiz. 873

Reality 803

Sri Ramana Maharshi

As there is wedlock between a man and wife so there is wedlock between God and the soul.

Eckhart

This marriage feast signifies the entrance into the highest state of union that can be between God and the soul in this life. Or in other words it is the birth-day of the Spirit of love in our souls, which, whenever we attain it, will feast our souls with such peace and joy in God as will blot out the remembrance of everything that we called peace or joy before.

William Law

Did my heart love till now? forswear it, sight!
For I ne'er saw true beauty till this night.

Shakespeare (*Romeo and Juliet,* I. v. 56)

Center 847

Better to catch one moment's glimpse of Thee
Than earthly beauties' love through life retain.

Jâmî

On my soul's lute a chord was struck by Love,
Transmuting all my being into love;
Ages would not discharge my bounden debt
Of gratitude for one short hour of love.

Jâmî

Ah! How wonderful was the yearning of the gopis for Krishna! They were seized with divine madness at the very sight of the black tamâla tree. Separation from Krishna created such a fire of anguish in Râdhâ's heart that it dried up even the tears in her eyes! Her tears would disappear in steam.

Sri Ramakrishna

But for my sighs, I should be drowned by my tears; and but for my tears, I should be burned by my sighs.

Ibn al-Fâriḍ

The soul, in hot pursuit of God, becomes absorbed in him, and she herself is reduced to naught, just as the sun will swallow up and put out the dawn.

Eckhart

Holy War 403

Râdhâ painted the picture of Sri Krishna with her own hand, but did not paint His legs lest He should run away to Mathurâ!

Sri Ramakrishna

God lays the soul in His glowing heart so that He, the great God, and she, the humble maid, embrace and are one as water with wine. . . .

Realiz. 868

My body is in long torment, my soul in high delight, for she has seen and embraced her Beloved. Through Him, alas for her! she suffers torment. As He draws her to Himself, she gives herself to Him. She cannot hold back and so He takes her to Himself. Gladly would she speak but dares not. She is engulfed in the glorious Trinity in high union. He gives her a brief respite that she may long for Him. She would fain sing His praises but cannot. She would that He might send her to Hell, if only He might be loved above all measure by all creatures. She looks at Him and says, 'Lord! Give me Thy blessing!' He looks at her and draws her to Him with a greeting the body may not know—

Contem. 536
Charity 608

Thus the body speaks to the soul
'Where hast thou been? I can bear this no more!'
And the soul replies 'Silence! Thou art a fool!
I will be with my Love
Even shouldst thou never recover!
I am His joy; He is my torment—'

Mechthild of Magdeburg

God cannot be seen with these physical eyes. In the course of spiritual discipline one gets a 'love body', endowed with 'love eyes', 'love ears', and so on. One sees God with those 'love eyes'. One hears the voice of God with those 'love ears'. One even gets a sexual organ made of love. . . . With this 'love body' the soul communes with God.

Sri Ramakrishna

Realiz. 873

Full near I came unto where dwelleth
  Laila, when I heard her call.
That voice is sweet beyond compare,
  I would that it might never cease.
She favoured me and drew me to her,
  took me in, into her precinct.
With discourse intimate addressed me,
  sat me face to face with her.

Closer drew herself towards me,
raised the cloak that hid her from me,
Made me marvel to distraction,
bewildered me with all her beauty.
She took me and amazed me,
and hid me in her inmost self,
Until I thought that she was I,
and my life she took as ransom.

Realiz. 887

Shaykh Aḥmad al-ʿAlawî

He, who with a clear eye distinguisheth the curious and close workings of the *will,* may find all its motions or affections to be the *same love* in various postures, as it rests with sweetest complacency, in the embraces of the beloved Beauty, or faints under a dispair of fruition, or an irresistable opposition in its prosecutions; as it sails on smooth Seas, with soft and prosperous gales to its haven in its eye; or wrestles with tempests of Waves and Winds, with chearful courage raising it self to surmount them. As the colours of the *Rainbow* are the same light variously reflected from the Sun, and variously falling upon the watry Cloud; so are all the motions of the Will the same *Love,* raised from the same good, beautifully shining forth, and reflecting it self variously upon the Soul in different postures of *presence* or *absence, of doubt, difficulty, impossibility* in the attainment, or *facility* and *assurance* of fruition.

Reality 775

Peter Sterry

Blissful is the dawn of Wisdom, like the virgin's wedding night;
Till experienced none can know it as it is, O Tingri folk.

Knowl. 761

Phadampa Sangay

Lo! the God-fearing are in a state secure,
Amid gardens and watersprings,
Attired in silk and gold brocade, facing one another:
Thus: and We shall wed them to houris with wide lovely eyes.

P. State 583

Qur'ân, XLIV. 51–54

One cannot explain the vision of God to others. One cannot explain conjugal happiness to a child five years old.

M.M. 998

Sri Ramakrishna

I am a full-grown Bride
I must to my lover's side!

Mechthild of Magdeburg

The Brahman is to be known by Itself alone, and to know It is as the bliss of knowing a virgin.

Knowl. 749

Daksha

There are two things in this world which delight me: women and perfumes. These two things rejoice my eyes, and render me more fervent in devotion.

Muhammad

Introd. *Beauty*

How beautiful are thy feet with shoes, O prince's daughter! the joints of thy thighs are like jewels, the work of the hands of a cunning workman.

Thy navel is like a round goblet, which wanteth not liquor: thy belly is like an heap of wheat set about with lilies.

Thy two breasts are like two young roes that are twins.

Thy neck is as a tower of ivory; thine eyes like the fishpools in Heshbon, by the gate of Bath-rabbim: thy nose is as the tower of Lebanon which looketh toward Damascus.

Thine head upon thee is like Carmel, and the hair of thine head like purple; the king is held in the galleries.

How fair and how pleasant art thou, O love, for delights! . . .

I am my beloved's, and his desire is toward me.

Song of Solomon, ch. VII

P. State 579

While the great King Solomon . . . was composing the sacred Canticle of Canticles, he had, according to the permission of those ages, a great variety of ladies and maidens attached to his service in different conditions and qualities. . . . Now under the figure of what passed in his palace, he described the various perfections of souls who in time to come were to adore, love and serve the great Pacific King Jesus Christ.

St Francis de Sales

Beauty 676

The most intense and perfect contemplation of God is through women, and the most intense union (in the sensory realm, which serves as support for this contemplation) is the conjugal act.

Ibn 'Arabî

The secret of our Art is the union of man and woman.

The Book of Alze

In that conjunction of the two sexes, or, to speak more truly, that fusion of them into one, which may be rightly named Eros, or Aphrodite, or both at once, there is a deeper meaning than man can comprehend. It is a truth to be accepted as sure and evident above all other truths, that by God, the Master of all generative power, has been devised and bestowed upon all creatures this sacrament of eternal reproduction, with all the affection, all the joy and gladness, all the yearning and the heavenly love that are inherent in its being.

Hermes

Woman, verily, O Gautama, is a sacrificial fire. In this case the sexual organ is the fuel; when one invites, the smoke; the vulva, the flame; when one inserts, the coals; the sexual pleasure, the sparks.

Creation 47

In this fire the gods offer semen. From this oblation arises a person (*puruṣa*).

Chândogya Upanishad, v. viii. 1, 2
Bṛihad Âraṇyaka Upanishad, vi. ii. 13

Realiz. 887

The goal of our Art is not reached until Sun and Moon are conjoined, and become, as it were, one body.

Morienus

Symb. 317
P. State 590

For the Sun is the Father of metals, and the Moon is their Mother: and if generation is to take place, they must be brought together as husband and wife. By itself neither can produce anything, and therefore the red and the white must be brought together.

The Glory of the World

Sun and Moon must have intercourse, like that of a man and woman: otherwise the object of our Art cannot be attained.

The Glory of the World

This procreation is the union of man and woman, and is a divine thing: for conception and generation are an immortal principle in the mortal creature.

Plato (*Symposium,* 206 C)

He who joins quicksilver to the body of magnesia, and the woman to the man, extracts the secret essence by which bodies are coloured.

P. State 563

Menabadus

Know that the secret of the work consists in male and female, *i.e.,* an active and a passive principle. In lead is found the male, in orpiment the female. The male rejoices when the female is brought to it, and the female receives from the male a tinging seed, and is coloured thereby.

Orthod. 280

Zimon

Join heaven to earth in the fire of love, and you will see in the middle of the firmament the bird of Hermes.

Philalethes

Now lay this fact to heart: the ever virgin is never fruitful. To be fruitful the soul must be wife. Spouse is the noblest title of the soul, nobler than virgin. For a man to receive God within him is good and in receiving he is virgin. But for God to be fruitful in him is still better.

Realiz. 870

Eckhart

God who sees all the rest, and brings everything to light,
Knows neither a dissolute man nor a sterile virgin.

Angelus Silesius

The Lord Jesus *hath his* Concubines, *his* Queenes, *his* Virgines; *Saints* in Remoter *Formes, Saints* in higher Formes, *Saints* unmarried *to any Forme,* who keep themselves single for the immediate imbraces of their Love.

Peter Sterry

The true sexual union is the union of the Parashakti (*kundalinî*) with Atman; anything else is but carnal connection with women.

Kulârnava Tantra, v. 111–112

Nor ever chast, except *thou* ravish mee. . . .

From the Tale of the Loathly Lady

He (the sacrificer) then offers a dish of clotted curds to Mitra and Varuṇa. Now he who performs this (Agni-ḳayana) rite comes to be with the gods: and these two, Mitra and Varuṇa are a divine pair. Now, were he to have intercourse with a human woman without having offered this (oblation), it would be a descent, as if one who is divine would become human; but when he offers this dish of clotted curds to Mitra and Varuṇa, he thereby approaches a divine mate: having offered it, he may freely have intercourse in a befitting way.

Śatapatha-Brâhmana, IX. v. 1. 54

In union with one's chosen divinity, one worships oneself, the Supreme One.

Guhyasamâjatantra, ch. 7

Knowl. 749
Contem. 523

I have often spoken of the light within the soul, which is uncreated and eke uncreaturely. It is this light I am so often hinting at in my discourses, it is the light which lays straight hold of God, bare and unveiled, as he is in himself; that is to say, it catches him in the act of self-begetting.

Eckhart

Metanoia 484

Sex life with a woman! What happiness is there in that? The realization of God gives ten million times more happiness. Gauri[1] used to say that when a man attains ecstatic love of God all the pores of the skin, even the roots of the hair, become like so many sexual organs, and in every pore the aspirant enjoys the happiness of communion with the Âtman.

Sri Ramakrishna

Center 847

With how many teeth this love doth bite!

Dante (*Paradiso,* XXVI. 51)

Ecstasy 641

If the soul were stripped of all her sheaths, God would be discovered all naked to her view and would give himself to her, withholding nothing. As long as the soul has not thrown off all her veils, however thin, she is unable to see God.

Eckhart

Sin 64
Void 721

[1] A pundit devoted to Sri Ramakrishna.

I do not wear a shirt when I sleep with the Adored One.

Rûmî

The closer the embrace, the sweeter the kiss.

Mechthild of Magdeburg

A bride, when she is to be brought forth to be married, is gloriously adorned in a great variety of precious garments, which, by enhancing her beauty, render her pleasant in the eyes of the bridegroom. But the rites of the bridal night she performs without any clothing but that which she was arrayed withal at the moment of her birth.

Basil Valentine

Orthod. 296
P. State 573

As one embraced by a darling bride knows naught of 'I' and 'thou', so self embraced by the foreknowing (solar) Self knows naught of a 'myself' within or a 'thyself' without.

Bṛihad-Âranyaka Upanishad, IV. iii. 21

Realiz. 887

He that is joined unto the Lord is one spirit.

I. Corinthians, VI. 17

Reality 790

In the embrace of this sovran one which naughts the separated self of things, being is one without distinction.

Eckhart

Either was the other's mine.

Shakespeare (*The Phoenix and the Turtle*)

She knelt down, thanked Him for His Grace, took her crown from her head and laid it against the rosy scars of His feet, and begged that she might come near Him. He took her in His Divine arms and laid His Fatherly hand on her breast and gazed into her face. Ah! how she was then kissed! And in that kiss was raised up to the highest heights above all the choirs of angels.

The smallest truth
That there I saw or heard or knew
Bore no likeness
Even to the highest wisdom
Ever known on earth.

Knowl. 734

There I saw inwardly unheard of things, so my confessor tells me. I am not expert in writing, but now I fear before God if I keep silent about these things, and before ignorant people if I speak.

Mechthild of Magdeburg

It is no wonder that this Benjamin stays all day, as it were, in the bridal chamber. . . .

Richard of Saint-Victor

Passionate love (*al-ʿishq*) is not peculiar to the human species, for it penetrates through all existing things—celestial, elemental, vegetable and mineral.

Charity 608

Avicenna

Without kâma (desire, enjoyment, love) a man has no wish for worldly profit (artha), without kâma a man does not strive after the Good (dharma), without kâma a man does not love; therefore kâma stands above the others (artha and dharma). For the sake of kâma the Ṛishis even give themselves up to asceticism, eating the leaves of trees, fruits, and roots, living on the air, and wholly bridling their senses, and others bend all their zeal to the Vedas. . . . Traders, husbandmen, herdsmen, craftsmen, as also artists, and those that carry out actions consecrated to the gods, give themselves up to their works because of kâma. Others, again, take to the sea filled with kâma; for kâma has the most varied forms: everything is steeped in kâma. No being ever was, or is, or will be, higher than the being that is filled with kâma. It is the innermost core (of the world), O king of righteousness; on it is founded dharma and artha. As butter from sour milk, so kâma comes forth from artha and dharma. . . . Better is the flower and the fruit than the wood, kâma is more excellent than artha and dharma. As honey is the sweet juice from the flower, so kâma is from these two, according to the teaching of tradition. Kâma is the womb of dharma and artha, and kâma makes up their essence. Without kâma the manifold workings of the world would not be thinkable.

Action 346

Mahâbhârata, XII, 167

The whole world is a market-place for Love,
For naught that is, from Love remains remote.
The Eternal Wisdom made all things in Love:
On Love they all depend, to Love all turn.
The earth, the heavens, the sun, the moon, the stars
The centre of their orbit find in Love.
By Love are all bewildered, stupefied,
Intoxicated by the Wine of Love.
From each, a mystic silence Love demands,
What do all seek so earnestly? 'Tis Love.
Love is the subject of their inmost thoughts,
In Love no longer 'Thou' and 'I' exist,
For self has passed away in the Beloved.
Now will I draw aside the veil from Love,
And in the temple of mine inmost soul
Behold the Friend, Incomparable Love.
He who would know the secret of both worlds
Will find the secret of them both, is Love.

Creation 48

Center 831

Ecstasy 637

Charity 597

Contem. 536

M.M. 978

ʿAṭṭâr

ECSTASY

When one attains to the state of dementation,
That is the Supreme Estate!

Maitri Upanishad, VI. 34

This chapter might with equal reason be entitled *Inebriation* or *Madness*. It has to do, both with the intensity of love, and with the 'mindlessness' of those who have transcended the purely rational faculties. 'Existence' and 'ecstasy' alike stem from a single root meaning to 'stand out', but in the sense either of a peculiarity or a norm, depending on the perspective. From the viewpoint of the Absolute, 'God's idiosyncrasy is being' (Eckhart, I. 206), since He ex-ists as the first exteriorization or determination of the Godhead. From the viewpoint of this world on the other hand, 'standing out', determined, limited—fragmented if one prefers—into the special modalities of space and time, where discursive reasoning predominates, it will be the ecstasy of intellectual synthesis, reversing as it does the process by shattering the 'veils' or successive determinations of existence that condition this world, which must then 'stand out' as 'idiosyncratic'.[1] Yet 'the madness that comes of God is superior to the sanity which is of human origin' (Plato: *Phaedrus*, 244 D; tr. Coomaraswamy): 'the wisdom of this world is foolishness with God' (*I. Cor.* III. 19). The point to remember is that 'the supralogical is superior to the logical, the logical to the illogical' (Coomaraswamy: *Figures of Speech*, p. 40).

M.M. 994
Center 847
Introd.
Orthod.

The degree to which ecstasy exteriorizes depends in some measure upon the time and place, the person and the path. There were times, says Sri Ramakrishna, when nobody could notice the depth of Râdhâ's feeling: 'People do not notice the plunge of an elephant in a big lake' (*Gospel*, p. 449). In general, the *jnânic* contemplative is by nature static and objective and retains a more implacable serenity than the *bhakta*, who sports more openly with God; but these things are relative. Ecstasy in its various forms neither proves nor disproves a divine Presence, any more than does impassibility prove or disprove an absence thereof.

Madness can take different forms. There are those who go mad from inordinate ambition or spiritual imprudence. There is the madness that seizes one in the throes of divine Inspiration, and there is also a madness that is methodical. 'Eastern hagiography knows ways of sanctification which are strange and unwonted, such as the "fools in Christ", who commit extravagant acts in order to hide their spiritual gifts from the eyes of their associates under the hideous appearance of madness, or rather, to liberate themselves from the bonds of this world in their most intimate and spiritually hampering expression,

Death 220

[1] The Arabic root J N N englobes both the idea of madness and paradise.

that of our social "self"' (Vladimir Lossky: *Essai sur la Théologie mystique de l'Eglise d'Orient*, p. 17). This is reminiscent of the *Malâmatiyah* ('People of blame') of Islam; and the phenomenon is far from rare in India.

Then, 'sometimes paradox is intentional on man's part, as in the classic example of Omar Khayyâm, whose wisdom clothed in frivolity is opposed to Pharisaism clothed in piety; if religious hypocrisy is possible, the contrary paradox must equally be so' (Schuon: *Perspectives spirituelles*, pp. 245–246).

Apart from questions of opportuneness, 'madness is after all one of the most impenetrable masks in which wisdom can be concealed, by the very fact of being the opposite extreme; this is why in Taoism the "Immortals" themselves when manifested in our world are frequently described under an aspect which is more or less extravagant, or even ridiculous' (Guénon: 'Folie apparente et sagesse cachée', *Études Traditionnelles*, 1946, p. 111).

Introd.
Humility

As for inebriation, it suffices to say here that with this Wine, the more one drinks, the more singly one sees.

Mindlessness

M.M. 978

Heavenly, formless is the Person (Purusha).
He is without and within, unborn,
Breathless, mindless (*a-manas*), pure,
Higher than the high Imperishable.

From Him is produced breath,
Mind, and all the senses,
Space, wind, light, water,
And earth, the supporter of all.

Muṇḍaka Upanishad, II. i. 2, 3

We must be transported wholly out of ourselves and given unto God.... This 'Foolish Wisdom',[1] which hath neither Reason nor Intelligence ... is the Cause of all Intelligence and Reason, and of all Wisdom and Understanding.

Dionysius

Void 724

God must be sought in estrangement, forgetfulness and non-sense; for the Godhead has in it all things *in posse* without the least likeness to anything.

Eckhart

Peace 700

Verily, while he[2] does not there think, he is verily thinking, though he does not think what is usually to be thought.

Bṛihad-Âraṇyaka Upanishad, IV. iii. 28

Contem. 532

When the intellect attains to the form of truth, it does not think, but perfectly contemplates the truth.

St Thomas Aquinas

Center 816 · *Metanoia* 484

Bewilderments of the eyes are of two kinds, and arise from two causes, either from coming out of the light or from going into the light, which is true of the mind's eye, quite as much as of the bodily eye; and he who remembers this when he sees any one whose vision is perplexed and weak, will not be too ready to laugh; he will first ask whether that soul of man has come out of the brighter life, and is unable to see because unaccustomed to the dark, or having turned from darkness to the day is dazzled by excess of light.

Plato (*Republic,* 518 A)

[1] *I. Cor.,* I. 25.
[2] The being in the Cognitive. Deep Sleep state (*prâjña, suṣupta-sthâna*).

This is the air; that is the glorious sun;
This pearl she gave me, I do feel't and see't;
And though 'tis wonder that enwraps me thus,
Yet 'tis not madness.

Shakespeare (*Twelfth-Night,* IV. iii. 1)

For when the mind of man is carried beyond itself all the limits of human reasoning are overpassed. For the whole system of human reasoning succumbs to that which the soul perceives of the divine light, when she is raised above herself and ravished in ecstasy. For what is the death of Rachel but the failure of reason?

Richard of Saint-Victor

Inebriation

Wine signifies the drink of divine Love which results from contemplating the traces of His beautiful Names. For this love begets drunkenness and the complete forgetfulness of all that exists in the world.

Symb. 306

Nâbulusî

'And thy presses shall run over with wine.' That is, thine inward ghostly wits, the which thou art wont to strain and press together by divers curious meditations and reasonable investigations about the ghostly knowing of God and of thyself, in beholding of his qualities and of thine, shall run over with wine. By the which wine in Holy Scripture is verily and mistily understood ghostly wisdom in very contemplation and high savour of the Godhead.

Contem. 528

The Epistle of Privy Counsel, IV

You know, my Friends, with what a brave Carouse
I made a Second Marriage in my house;
Divorced old barren Reason from my Bed,
And took the Daughter of the Vine to Spouse.

Supra 636

Omar Khayyâm

Behold my Mother playing with Śiva, lost in an ecstasy of joy!
Drunk with a draught of celestial wine, She reels, and yet She does not fall.

Bengali Hymn to Durgâ

Spiritual inebriation is this; that a man receives more sensible joy and sweetness than his heart can either contain or desire. Spiritual inebriation brings forth many strange gestures in men. It makes some sing and praise God because of their fulness of joy; and some weep with great tears because of their sweetness of heart. It makes one restless in all his limbs, so that he must run and jump and dance; and so excites another that he must gesticulate and clap his hands. Another cries out with a loud voice, and so shows forth the plenitude he feels within; another must be silent and melt away, because of the rapture which he feels in all his senses. At times he thinks that all the world must feel what he feels: at times he thinks that none can taste what he has attained. Often he thinks that he never could, nor ever shall, lose this well-being; at times he wonders why all men do not become God-desiring. At one time he thinks that God is for him alone, or for none other so much as for him; at another time he asks himself with amazement of what nature these delights can be, and whence they come, and what has happened to him.

Infra 641

Ruysbroeck

If they watered the earth of a tomb with such a Wine, the dead one would recover his soul and his body revive.

Realiz. 870

Ibn al-Fâriḍ

The soul says to the senses, 'Leave me! I must cool myself!' The senses answer, 'Lady! wilt thou be refreshed by the tears of Mary Magdalene? Can they suffice thee?'

SOUL

Hush! ye know not what I mean!
Hinder me not! I would drink of the unmingled wine!

M.M. 978

SENSES

Lady! In virgin chastity
The love of God is ready for thee!

SOUL

That may be so. For me
It is not the highest.

Mechthild of Magdeburg

That light of truth passes not by, but remaining fixed, inebriates the hearts of the beholders.

St Augustine

These are not drunken, as ye suppose, seeing it is but the third hour of the day.

Acts, II. 15

I am drunk with wine, and I am filled with virtue.

Introd. *Beauty*

The Book of Songs

Back in my home I drink a cup of wine
And need not fear the greed of the evening wind.

Realiz. 870

Lu Yu (*Boating in Autumn*)

Lo! the righteous verily are in delight,
On couches, gazing,
Thou wilt know in their faces the radiance of delight.
They are given to drink of a pure wine, sealed,
Whose seal is musk—For this let all those strive who strive for bliss.

Qur'ân, LXXXIII. 22–26

Holy War 403

Pour out wine till I become a wanderer from myself;
For in selfhood and existence I have felt only fatigue.

Dîvâni Shamsi Tabrîz, XXXII

Death 220

A pot of wine amidst the flowers,
Alone I drink sans company.
The moon I invite as drinking friend,
And with my shadow we are three.
The moon, I see, she does not drink,
My shadow only follows me:
I'll keep them company a while
For spring's the time for gayety.

I sing: the moon she swings her head;
I dance: my shadow swells and sways.
We sport together while awake,
While drunk, we all go our own ways.
An eternal, speechless trio then,
Till in the clouds we meet again!

Li Po (*Drinking Alone under the Moon*)

P. State 590
Metanoia 480

From drinking one cup of the pure wine,
From sweeping the dust of dung-hills from their souls,
From grasping the skirts of drunkards,
They have become Sufis.

Shabistarî

Sin 66
Orthod. 288

Wouldst thou come with me to the wine-cellar
That will cost thee much;
Even hadst thou a thousand marks
It were all spent in one hour!
If thou wouldst drink the unmingled wine
Thou must ever spend more than thou hast,
And the host will never fill thy glass to the brim!
Thou wilt become poor and naked,
Despised of all who would rather see themselves
In the dust, than squander their all in the wine-cellar.

Mechthild of Magdeburg

Renun. 146
Suff. 126b

Center 847

They shall be inebriated with the plenty of thy house; and thou shalt make them drink of the torrent of thy pleasure.

Psalm XXXV. 9

Rev. 959

And the roof of thy mouth (is) like the best wine for my beloved, that goeth down sweetly, causing the lips of those that are asleep to speak.

Song of Solomon, VII. 9

Communion with God is the true wine, the wine of ecstatic love.

Sri Ramakrishna

The taste of my wine is mild and works no poison.

Po Chü-i

And they will be honoured
In the Gardens of delight,
On couches facing one another;
A cup from a gushing spring is brought round for them,
White, delicious to the drinkers,
Wherein there is no headache nor are they made mad thereby.

Qur'ân, XXXVII. 42–47

Symb. 306

The angels, sipping pure wine from goblets,
Pour down the dregs on the world;
From the scent of these dregs man rises to heaven.

Shabistarî

M.M. 978
Conform. 170
Holiness 914

Whom Thou intoxicatest with Thy love
On him bestoweth Thou both the worlds.
But Thy mad devotee,
What use hath he for both the worlds?

Ansârî

Creation 33

I am intoxicated with Love's cup, the two worlds have passed out of my ken;
I have no business save carouse and revelry.
If once in my life I spent a moment without thee,
From that time and from that hour I repent of my life.

Dîvâni Shamsi Tabrîz, XXXI

Beauty 670

And so, carrying its gaze beyond the confines of all substances discernible by sense, (Mind) comes to a point at which it reaches out after the intelligible world, and on descrying in that world sights of surpassing loveliness, even the patterns and the originals of the things of sense which it saw here, it is seized by a sober intoxication, like those filled with Corybantic frenzy, and is inspired, possessed by a longing far other than theirs and a nobler desire.

Philo

Madness

The setting of man's own discursive faculty and the eclipsing thereof begets an *ecstasis* and a divine kind of *mania*.

Philo

Death 220
Metanoia 484

In the Kaliyuga one does not hear the voice of God, it is said, except through the mouth of a child or a madman or some such person.

Sri Ramakrishna

There is a madness which is a divine gift, and the source of the chiefest blessings granted to men. For prophecy is a madness, and the prophetess at Delphi and the priestesses at Dodona when out of their senses have conferred great benefits on Hellas, both in public and private life, but when in their senses few or none.

Plato (*Phaedrus*, 244 B)

There must be some kind of *Μανία* in all prophecy.

John Smith the Platonist

It behoves us to become ignorant of this worldly wisdom; rather must we clutch at madness.

Rûmî

Knowl. 734

I will preserve myself; and am bethought
To take the basest and most poorest shape
That ever penury, in contempt of man,
Brought near to beast; my face I'll grime with filth,
Blanket my loins, elf all my hair in knots,
And with presented nakedness outface
The winds and persecutions of the sky.

Shakespeare (*King Lear*, II. iii. 6)

Supra Introd.

Ever since the day I was banished to Hsün-yang
Half my time I have lived among the hills.
And often, when I have finished a new poem,
Alone I climb the road to the Eastern Rock.
I lean my body on the banks of white stone:
I pull down with my hands a green cassia branch.
My mad singing startles the valleys and hills:
The apes and birds all come to peep.
Fearing to become a laughing-stock to the world,
I choose a place that is unfrequented by men.

Po Chü-i (*Madly Singing in the Mountains*)

A man who has realized God shows certain characteristics. He becomes like a child or a madman or an inert thing or a ghoul.

Holiness 910

Sri Ramakrishna

It so happened to Dhu 'l-Nûn the Egyptian that a new agitation and madness was born within him.

His agitation became so great that salt from it was reaching hearts up to above the sky.

Beware, O thou of salty soil, do not put thy agitation beside the agitation of the holy lords.

The people could not endure his madness: his fire was carrying off their beards.[1]

When that fire fell on the beards of the vulgar, they bound him and put him in a prison.

There is no possibility of pulling back this rein, though the vulgar be distressed by this way.

Orthod. 296

Rûmî

The men whom he dements God uses as his ministers.

Rev. 961

Plato (*Ion*, 534 D)

(Krishnakishore) passed through a God-intoxicated state, when he would repeat only the word 'Om' and shut himself up alone in his room. His relatives thought he was actually mad, and called in a physician. Ram Kavirâj of Nâtâgore came to see him. Krishnakishore said to the physician, 'Cure me, sir, of my malady, if you please, but not of my Om.'

Inv. 1036

Sri Ramakrishna

SENSES

Ah! Lady! Comest thou there
Then are we blinded,
So fiery is the glory of the Godhead
As thou well knowest . . .
Who may abide it, even one hour?

SOUL

Fish cannot drown in the water,
Birds cannot sink in the air,
Gold cannot perish
In the refiner's fire.
This has God given to all creatures
To foster and seek their own nature,
How then can I withstand mine?
I must to God.

Renun. 158

Creation 42

Realiz. 873

Mechthild of Magdeburg

Love for God is naturally ardent and when it fills a man to overflowing, leads the soul to ecstasy. Therefore the heart of a man who experiences it cannot contain or bear it, but

[1] 'I.e. in his ecstasy he had no regard for their formal religion' (note by the translator. R. A. Nicholson).

undergoes an extraordinary change according to its own quality and the quality of the love which fills him; . . . he is as one out of his mind. A terrible death is for him a joy, and his mental contemplation of heavenly things is never broken. . . This spiritual intoxication was experienced of old by Apostles and martyrs. The first travelled far and wide over the whole world, working and suffering persecutions; the latter had their limbs cut off, shed blood like water, but, suffering the most terrible tortures, never lost courage and valiantly bore everything; being wise they were considered foolish. Yet others wandered among deserts, mountains, caves and precipices of the earth, remaining well-ordered amongst all disorder. May God grant us such disorder!

Metanoia 488
Supra 637

St Isaac of Syria

The madness of love is the greatest of heaven's blessings.

Plato (*Phaedrus,* 245 B)

This is that mystic religion which, though it has nothing in it but that same spirit, that same truth, and that same life, which always was and always must be the religion of all God's holy angels and saints in heaven, is by the wisdom of this world accounted to be madness.

Introd. *Holiness*

William Law

They say: Thou art become mad with love for thy beloved.
I reply: The savour of life is for madmen.

Al-Yâfi'î

Mad! That's the thing! Shivanath once said that one 'loses one's head' by thinking too much of God. 'What?' said I. 'Can anyone ever become unconscious by thinking of Consciousness? God is of the nature of Eternity, Purity, and Consciousness. Through His Consciousness one becomes conscious of everything; through His Intelligence the whole world appears intelligent.' Shivanath said that some Europeans had gone insane, that they had 'lost their heads', by thinking too much about God. In their case it may be true; for they think of worldly things. There is a line in a song: 'Divine fervour fills my body and robs me of consciousness.' The consciousness referred to here is the consciousness of the outer world.

Renun. 146

Sri Ramakrishna

It is related that one day when Shiblî[1] came into the bazaar, the people said, 'This is a madman.' He replied: 'You think I am mad, and I think you are sensible: may God increase my madness and your sense!

Hujwîrî

'Alî ibn 'Abdân knew a madman who wandered about in the daytime and passed the night in prayer. 'How long,' he asked him, 'hast thou been mad?'
'Ever since I *knew.*'

Al-Yâfi'î

[1] Abû Bakr Dulaf b. Jaḥdar al-Shiblî, a famous Sufi of Baghdad, d. c. A.D. 945–6.

Once more have I become mad, O Physician! Once more have I become frenzied, O Beloved!
The links of Thy chain are multiform: every single link gives a different madness.

Love 625

Rûmî

Though this be madness, yet there is method in't.

Shakespeare (*Hamlet*, II. ii. 211)

If you must be mad, why should you be mad for the things of the world? If you must be mad, be mad for God alone.

Sri Ramakrishna

Suff. 126b
Metanoia 488
Love 614
P. State 563

It is true the world will be apt enough to censure thee for a madman in walking contrary to it: And thou art not to be surprised if the children thereof laugh at thee, calling thee silly fool. For the way to the love of God is folly to the world, but is wisdom to the children of God. Hence, whenever the world perceiveth this holy fire of love in God's children, it concludeth immediately that they are turned fools, and are besides themselves. But to the children of God, that which is despised of the world is the greatest treasure.

Boehme

Flight 944
Knowl. 755
749

The mind of the philosopher alone has wings; and this is just, for he is always, according to the measure of his abilities, clinging in recollection to those things in which God abides, and in beholding which He is what He is. And he who employs aright these memories is ever being initiated into perfect mysteries and alone becomes truly perfect. But, as he forgets earthly interests and is rapt in the divine, the vulgar deem him mad, and rebuke him; they do not see that he is inspired.

Plato (*Phaedrus*, 249 D)

Suff. 130
Death 226

If once a man in intellectual vision did really glimpse the bliss and joy therein, then all his sufferings, all God intends that he should suffer, would be a trifle, a mere nothing to him; nay, I say more, it would be pure joy and pleasure.

Eckhart

When the source of immortal joy is opened within us, it flows and saturates every fibre of our being, internal and external, and makes our life at once a waveless peace and ceaseless thrill of ecstasy. Death, fear, and grief have then no significance for us.

Swami Ramdas

Center 835
Pilg. 366
Orthod. 296

How all the other passions fleet to air,
As doubtful thoughts, and rash-embrac'd despair,
And shuddering fear, and green-ey'd jealousy.
O love! be moderate; allay thy ecstasy;
In measure rain thy joy; scant this excess;
I feel too much thy blessing; make it less,
For fear I surfeit!

Shakespeare (*Merchant of Venice*, III. iii. 108)

Surely the place is holy; so that you must not wonder, if, as I proceed, I appear to be in a divine fury, for already I am getting into dithyrambics.

Plato (*Phaedrus,* 238 D)

Even if the gnostic were cast into the fire, he would not feel it, because of his absorption, and if the delights of Paradise were spread out before him, he would not turn towards them, because of the perfection of the grace that is in him and his perfect attainment, which is above all else that can be attained.

Judgment 266 *Center* 847

Al-Ghazâlî

The sum total of pleasures in the entire world is nothing, compared to the ânanda obtained through meditation.

Swami Sivananda

He who loves me is made pure; his heart melts in joy. He rises to transcendental consciousness by the rousing of his higher emotional nature. Tears of joy flow from his eyes; his hair stands on end; his heart melts in love. The bliss in that state is so intense that forgetful of himself and his surroundings he sometimes weeps profusely, or laughs, or sings, or dances; such a devotee is a purifying influence upon the whole universe.

Supra 637

Srimad Bhagavatam, XI. viii

(In the highest degree of contemplation) the Soul not only becomes happy by the gift of philosophy, but since, so to speak, it becomes God, it becomes happiness itself.

Realiz. 887

Marsilio Ficino

PART IV: *Beauty — Peace*

THE SURFACE OF THE WATERS

And the Spirit of God moved upon the face of the waters.

Genesis, I. 2

Moses means 'taken from the water'. Brahmâ and Vishnu both have the appellation *Nârâyana,* 'He who walks on the waters'; another name given Brahmâ is *Âpava*—'He who sports on the waters'. Lakshmî is called *Jaladhijâ,* 'the ocean-born'. The *Qur'ân* teaches that 'His Throne was upon the water' (XI. 7).

Creation
33

Hindu iconography pictures Vishnu asleep on a serpent couch amidst the waters, with a golden lotus issuing from his navel, and Brahmâ enthroned in the lotus. Following the same symbolism, Lakshmî, the consort or *shakti* of Vishnu, is called *Padmâ,* 'The Lotus', and *Padmâlaya,* 'She who dwells on a lotus'.[1] This recalls the epithet of the Buddha: *Mani padmê*—'Jewel in the lotus'. Agni is also 'churned from the lotus' or 'lotus-born' (*puṣkarât*).[2]

Center
816

The lotus motif is evident in the *Shrî-Yantra* designs of Hinduism, and in the Tibetan *maṇḍalas*; and the lotus in its quality of sacred flower reposing on the water and opening to the sunlight is the perfect expression of Peace and Beatitude and the unfolding or realization of the spiritual possibilities of the being.[3]

The macrocosmic corollary of the same symbolism pertains to sacred islands at the center of the sea (already alluded to in the Introduction to *Pilgrimage*). The *Montsalvat* of the Grail legends is both a 'sacred island' and a 'polar mountain', like the Hindu *Mêru*. The hyperborean *Tula* is called the 'white island', like the Hindu *shvêta dvîpa,* and the Celtic *Albion* and *Avalon*. Islamic esoterism speaks of the 'green island' (*al-jazîrah al-khaḍrah*) and the 'white mountain' (*al-jabal al-abyaḍ*), ideas which find their equivalent in the Celtic traditions.[4] There is also the Hindu 'Island of Jewels' (*maṇi-dvîpa*), with the Universal

[1] In the same order of ideas are the names of the Virgin Mary, such as Maris Stella, Ark of the Covenant, Mirror of Justice, Queen of Peace, Living Spring, Gate of Heaven, Mystical Rose, and Untouched Lily—these two flowers having essentially the same significance in the West as the lotus in the East.

[2] Cf. Coomaraswamy: *The Ṛg Veda as Land-Nâma-Bôk*, p. 4; 'Angel and Titan: An Essay in Vedic Ontology' (JAOS, Dec., 1935), passim. Guénon equates Agni at the center of the swastika with the Lamb in Christianity at the source of the four rivers (*Le Symbolisme de la Croix*, p. 177). Agni (whose root is the same as the Latin *Agnus*) has for vehicle a ram. In the *Rig Veda* is the passage: 'At the navel of the earth stands Agni'; and following the *Śatapatha Brâhmaṇa*, 'navel means the center.' Cf. *Le Roi du Monde*, ch. IX, for assimilations between the Greek *omphalos* and umbilicus, and the Sanskrit word *nâbhi* and nave(l), in relation to Axis, Pole, and other symbols attaching to the 'Center of the World'. What matters for the purposes of exposition here is the evident correlation between the ideas of surface, center, source, and summit.

[3] Cf. Schuon: *Les Stations de la Sagesse*, pp. 193–194.

[4] Cf. Guénon: *Le Roi du Monde*, passim.

Mother (*jagad-ambâ*), the *Shakti-Mâyâ*, enthroned in a jeweled palace over the recumbent forms of *Sakala Shiva* and *Nishkala Shiva*.[1]

The ocean supporting the Cosmic Lotus and the World Mountain corresponds to the 'lower waters' or 'chaos' of possibilities comprising the states of formal manifestation; its surface is the plane of reflection for the 'Celestial Ray' (*Buddhi*) whose vibration or *Fiat Lux* operates the transition from power to act, bringing order out of chaos. The surface also represents the plane of separation (Islamic *barzakh*, or Christian 'Cloud of Unknowing') between the two seas (*Gen.*, I. 7; *Qur'ân*, XVIII. 60; XXV. 53; LV. 19), the 'upper waters' being the sum of supraformal possibilities. In the midst of the waters is the firmament (*Gen.*, I. 6), as in Hinduism the *Brahmânda* or 'World Egg' which englobes the 'Golden Embryo' (*Hiranyagarbha*—archetypal seed of the possibilities that will develop during a cycle of manifestation) is represented floating on the primordial sea (*samudra*), covered by the swan *Hamsa*, vehicle of Brahmâ, equatable with the Divine Breath (*spiritus*).[2] The surface of the waters is thus the point of convergence between the finite and the Infinite.

Introds. Pilg., Realiz.

Introd. Sp. Drown.

' "Walking on the waters" symbolizes the domination of the world of forms and change' (Guénon: *Le Roi du Monde*, p. 81). The miraculous crossing of waters may be accomplished by actually walking on the water, or by levitation, or by the waters becoming shallow enough to wade through, or by the waters parting, as did the Red Sea for the Israelites, and the sea at Pamphylia for Alexander the Great, which according to Callisthenes, 'not only opened for him, but even rose and fell in homage'.[3]

Flight 941

[1] Cf. Heinrich Zimmer: *Myths and Symbols in Indian Art and Civilization*, New York, Pantheon Books, Bollingen Series VI, 1946, ch. V.

[2] Cf. Guénon: *Le Symbolisme de la Croix*, ch. XXIV; *Les États multiples de l'Être*, ch. XII; *L'Homme et son Devenir*, ch. V. The same author has indicated the symbolism underlying the Ark in the Flood (*Le Roi du Monde*, ch. XI), whose crescent shape from below combines with its inverse prototype the celestial rainbow from above to form the two halves of a sphere situated precisely at the juncture of the two seas, and enclosing (like the 'World Egg') the 'elect' or positive possibilities of formal manifestation in embryo (*Hiranyagarbha*), 'saved out' or crystallized from the preceding world cycle to be developed in the course of the future cycle.

[3] William Norman Brown: *The Indian and Christian Miracles of Walking on the Water*, Chicago, Open Court, 1928, p. 41. The question of miracles is further treated in the introductory section of the chapter, *Moving at Will—The Miracle of Flight*.

The Spirit Upon the Waters

The universe was formerly water, fluid. On it Prajâpati becoming wind, moved.

Taittirîya Saṁhitâ

Creation 26

From the Light there came forth a holy Word, which took its stand upon the watery substance.

Hermes

Introd. *Inv.*

The earth is the Lord's, and the fulness thereof; the world, and they that dwell therein.
For he hath founded it upon the seas, and established it upon the floods.

Psalm XXIV. 1, 2

The Waters (representing the principle of substance) being ripe unto conception (lit. 'in their season'), Vâyu (that is, the Wind, as physical symbol of spiration, *prâṇa*) moved over their surface. Wherefrom came into being a lovely (*vâma*) thing (that is, the world-picture), there in the Waters Mitra-Varuṇa beheld-themselves-reflected (*paryapaśyat*).

Pañcaviṁśa Brâhmaṇa, VII. 8. 1

Center 828

Surely thou art Bhâgavat who appears before me; the great Hari, whose dwelling was on the waves.... Salutation and praise to thee, O first male, the lord of creation, of preservation, of destruction! Thou art the highest object, the supreme ruler, of us thy adorers, who piously seek thee.... Let me not, O lotus-eyed, approach in vain the feet of a deity whose perfect benevolence has been extended to all.

Srimad Bhagavatam

For God's Word, who is all-accomplishing and fecund and creative, went forth, and flinging himself upon the water, which was a thing of fecund nature, made the water pregnant.

Hermes

Love 625

Thy good Spirit indeed *was borne over the waters*, not borne up by them, as if He rested upon them.

St Augustine

A holy king named Satyavrâta then reigned—a servant of the spirit which moved on the waves.

Srimad Bhagavatam

The Spirit of God, it is said, *moved over the waters*—that is to say, over that formless matter, signified by water, even as the love of the artist moves over the materials of his art. . . . According to the holy writers, the Spirit of the Lord signifies the Holy Ghost, Who is said to *move over the water*—that is to say, over what Augustine holds to mean formless matter. . . . It is the opinion . . . of Basil that the Spirit moved over the element of water, *fostering and quickening its nature and impressing vital power, as the hen broods over her chickens.*

Supra Introd.

St Thomas Aquinas

This vast egg, compounded of the elements, and resting on the waters, was the excellent natural abode of Vishnu in the form of Brahmâ, and there Vishnu, the lord of the universe, whose essence is inscrutable, assumed a perceptible form, and even he himself abided in it in the character of Brahmâ. Its womb, vast as the mountain Meru, was composed of the mountains; and the mighty oceans were the waters that filled its cavity. In that egg were the continents and seas and mountains, the planets and divisions of the universe, the gods, demons, and mankind.

M.M. 998

Vishṇu Purâṇa

They (the Egyptians) do not believe that the sun rises as a new-born babe (Horus) from the lotus, but they portray the rising of the sun in this manner to show darkly that his birth is a kindling from the waters.

Symb. 317

Plutarch

And straightway the Lord of all spoke with his own holy and creative speech, and said, 'Let the sun be'; and even as He spoke, Nature drew to herself with her own breath the fire, which is of upward-tending nature,—that fire,[1] I mean, which is unmixed and most luminous and most active and most fecund,—and raised it up aloft from the water.

Creation 31

Hermes

It is named the water of life, the purest and most blessed water, yet not the water of the clouds, or of any common spring, but a thick, permanent, salt, and (in a certain sense) dry water, which wets not the hand, a slimy water which springs out of the fatness of the earth. Likewise, it is a double mercury and Azoth which, being supported by the vapour or exudation of the greater and lesser heavenly and the earthly globe, cannot be consumed by fire. For itself is the universal and sparkling flame of the light of Nature, which has the heavenly Spirit in itself, with which it was animated at first by God.

P. State 563

The Sophic Hydrolith

The Lord sitteth upon the flood; yea, the Lord sitteth King for ever.

Holiness 921

Psalm XXIX. 10

Vishnu, with the quality of goodness and of immeasurable power, preserves created things through successive ages, until the close of the period termed a Kalpa; when the same mighty deity, invested with the quality of darkness, assumes the awful form of Rudra, and

[1] Equatable with Agni, as in the Introduction, supra.

swallows up the universe. Having thus devoured all things, and converted the world into one vast ocean, the Supreme reposes upon his mighty serpent couch amidst the deep: he awakes after a season, and again, as Brahmâ, becomes the author of creation.

Vishṇu Purâṇa

Wr. Side 464

Dominating the Waters

The wise man through earnestness, virtue, and purity, maketh himself an island which no flood can submerge.

Udâna

Holiness 902

Who sinks not in the gulf without support or stay? One who is prescient, fully synthesised, he may cross the flood so hard to pass.

Saṁyutta-nikâya, I. 53

Pilg. 385

The great rishi ... stayed the billowy river; when Viśvâmitra led Sudâs, Indra had pleasure in the Kuśikas.

Rig-Veda, III. 53. 9

Yea, the wide-spread floods Indra made into fords, easy to cross, for Sudâs.

Rig-Veda, VII. 33

The children of Israel walked upon dry land in the midst of the sea; and the waters were a wall unto them on their right hand, and on their left.

Exodus, XIV. 29

Center 835

And the priests that bare the ark of the covenant of the Lord stood firm on dry ground in the midst of Jordan, and all the Israelites passed over on dry ground, until all the people were passed clean over Jordan.

Joshua, III. 17

And Elijah took his mantle, and wrapped it together, and smote the waters, and they were divided hither and thither, so that they two went over on dry ground.

II. Kings, II. 8

Thus saith the Lord, which maketh a way in the sea, and a path in the mighty waters.

Isaiah, XLIII. 16

Thou didst walk through the sea with thine horses, through the heap of great waters.

Habakkuk, III. 15

I would not that ye should be ignorant, how that all our fathers were under the cloud, and all passed through the sea:

Orthod. 280

And were all baptized unto Moses in the cloud and in the sea.

I. Corinthians, x. 1, 2

When thou passest through the waters, I will be with thee; and through the rivers, they shall not overflow thee.

Isaiah, XLIII. 2

Now the king (Kappina), with his thousand ministers, reached the bank of the Ganges. But at this time the Ganges was full. When the king saw this, he said: 'The Ganges here is full, and swarms with savage fish. Moreover we have with us no slaves or men to make boats or rafts for us. But of this Teacher the virtues extend from the Avîci Hell beneath to the Peak of Existence above. If this Teacher be the Supremely Enlightened Buddha, may not the tips of the hoofs of these horses be wetted!'

They caused the horses to spring forward on the surface of the water. Of not a single horse was so much as the tip of the hoof wetted. On a king's highway proceeding, as it were, they went to the far shore. Farther on they reached another river. There, was needed no other Act of Truth. By the same Act of Truth, that river also, half a league in breadth, did they cross over. Then they reached the third river, the mighty river Candabhâgâ. That river also, by the same Act of Truth, did they cross over. . . .

Queen Anojâ, surrounded by a thousand chariots, reaching the bank of the Ganges and seeing no boat or raft brought for the King, by her own intuition concluded: 'The King must have crossed by making an Act of Truth. But this Teacher was reborn not for them alone. If this Teacher be the Supremely Enlightened Buddha, may our chariots not sink into the water!'

Faith 501

She caused the chariots to spring forward on the surface of the water. Of the chariots not even so much as the outer rims of the wheels were wetted. The second river also, the third river also, she crossed by the same Act of Truth.

Buddhaghosa

Their (certain Jaina sages') strength of will was so great that they could walk on water as on land.

Triṣaṣṭiśalâkâpuruṣacarita, I. 857

Said Visâkhâ the mother of Migâra: 'Most wonderful, most marvellous is the might and the power of the Tathâgata, in that though the floods are rolling on knee-deep, and though the floods are rolling on waist-deep, yet is not a single Bhikkhu wet, as to his feet or as to his robes.'

Vinaya-Piṭaka, Mahâvagga, VIII. 15

I (the Buddha) can walk on the water as if it were solid earth.

Saṁyutta-nikâya, v. 25

'Our families, from our earliest ancestors, have dwelt on the bank of this river. Now we have never heard tell that a man walked upon the water. Who then are you, and what is your magic recipe for walking upon the water without sinking?'

The miraculous man answered them. 'I am a simple and ignorant man from the south of the river. Having heard say that the Buddha was here, I was anxious to gladden myself with his wisdom and virtue. When I arrived at the southern bank, it was not the time when the river was fordable: but I asked the people who were on the bank of the river what was the depth of the water. They replied that the water would reach to my ankle, and that nothing would prevent me from crossing. I added faith to their words, and I have therefore come crossing the river. I have no extraordinary recipe.'

P. State 573

The Buddha praised him, saying, 'Well done! Well done! Truly, the man with faith in the absolute truths is able to cross the gulf of births and deaths. What is there extraordinary about it then that he should be able to cross a river several *li* wide.' Then the Buddha pronounced these stanzas: 'Faith (*śraddhâ*) can cross the gulf. . . .'

Death 223

Fa Kiu P'i Yu King

Reverend Sir, I had recourse to the Practice of Meditation, concentrated my thoughts on the Buddha, attained the Ecstasy of Joy (*piti*), obtained support on the surface of the water, and came hither as though I were treading the earth.

Jâtaka 190

If drifting in the vast ocean a man is about to be swallowed up by the Nagas, fishes, or evil beings, let his thought dwell on the power of Kwannon.[1] and the waves will not drown him.

Pilg. 385

Kwannon Sutra

I (the Buddha) crossed the flood only when I did not support myself or make any effort.

Contem. 532

Samyutta-nikâya, I. 1

The One God (Indra) stands upon the flowing streams at will.

Atharva-Veda, III. 3, 4

(Yaśoda)[2] walked on water without breaking through just as on land.

Mahâvastu

And in the fourth watch of the night Jesus went unto them, walking on the sea.

St Matthew, XIV. 25

If ye knew God as He ought to be known, ye would walk on the seas, and the mountains would move at your call.

Faith 514

Muhammad

[1] A Bodhisattva (skr. *Avalokiteśvara*) whose name is much invoked in Japan.

[2] A variant of the name Yasa, the Buddha's sixth convert.

Flight 944

Our inner journey is above the sky.
The body travels on its dusty way;
The spirit walks, like Jesus, on the sea.

Rûmî

Pilg. 378 *M.M.* 998

He is conceived in water and born in air; when he has become red in color, he walks over the water.

Michael Maier

By conquering the current called *Udâna*[1] the Yogi does not sink in water, or in swamps, he can walk on thorns, and can die at will.

Patanjali

A man after fourteen years' penance in a solitary forest obtained at last the power of walking on water. Overjoyed at this, he went to his Guru and said, 'Master, master, I have acquired the power of walking on water.' The master rebukingly replied, 'Fie, O child! is this the result of thy fourteen years' labours? Verily thou hast obtained only that which is worth a penny; for what thou hast accomplished after fourteen years' arduous labour ordinary men do by paying a penny to the boatman.'

Flight 941

Sri Ramakrishna

One day (Ḥasan of Baṣra) saw Râbi'a (al-'Adawiyya) near the riverside. Ḥasan cast his prayer-mat on to the surface of the water and said, 'O Râbi'a, come and let us pray two *rak'as* together.' Râbi'a said, 'O Ḥasan, was it necessary to offer yourself in the bazaar of this world to the people of the next? This is necessary for people of your kind, because of your weakness.' Then Râbi'a threw her prayer-mat into the air and flew up on to it and said, 'O Ḥasan, come up here that people may see us.' But that station was not for Ḥasan and he was silent. Râbi'a, wishing to gain his heart, said, 'O Ḥasan, that which you did, a fish can do just the same, and that which I did, a fly can do. The real work lies beyond both of these and it is necessary to occupy ourselves with the real work.'

'Aṭṭâr

It was the custom for Sidi 'Abderrahmân when the sea was calm to float on the waters off Algiers, seated on his prayer carpet. One day on a beach he came upon a poor shepherd who played a small flute (qashbût), and who was so absorbed in his melody that he did not even hear the Shaykh's salâm. He had promised to play for three days in a row if God would grant him the child he had long sought to have, and filled with gratitude and joy in seeing his wish accorded, he had undertaken to play for forty days. Now Sidi 'Abderrahmân did not like the flute, which certain traditions would have to be the instrument of Iblîs (the devil), who solaces his eternal anguish by wailing through its reeds.

He declared that this way of thanking God was absurd.[2] 'The Lord does not accept such homage. I am going to teach thee something which will reconcile thee to Him.' And

[1] One of the five *vâyus* or modalities of *prâna* ('breath'), having to do with expiration.

[2] The story typifies the Islamic attitude towards music, where it is classic first to reject (rationally) for its potential abuses an art form which is subsequently accepted (spiritually) for its essentially divine qualities.

he taught the shepherd the *Fâtiḥah,* together with the rites of prayer. Then he launched his rug once more and floated out to sea.

The shepherd tried to recite the formula he had just learned. But he got confused; he had forgotten a line. Heeding nothing but his zeal, he ran after the saint to have his help in recalling the words; and thus he walked on the sea. The saint, the sage, had need of a carpet to stay miraculously afloat on the waves, and here was this ignorant man walking on them barefoot. Sidi ʿAbderrahmân understood the lesson and spoke to the uneducated shepherd, this man of good will, in quoting the first ḥadîth from the *ṣahîh* of al-Bukhârî:

'Continue, O my brother, to play for Him. *Innamâ 'l-aʿmâl bi-n-nîyât.* Verily the act is in the intention.'

Conform. 170

Sufic Tradition

Supra-normal powers are encouragements which God gives as a sort of incentive for new progress and an intense sâdhanâ (ascesis).

Swami Sivananda

He who has attained the Tao can go into water without becoming wet, jump into fire without being burned, walk upon reality as if it were a void and travel on a void as if it were reality. He can be at home wherever he is and be alone in whatever surroundings. That is natural with him.

P. State 563
Metanoia 488
Renun. 139

T'u Lung

The Plane of Reflection

Have not the disbelievers seen that the heavens and the earth were of one piece?[1] Then We parted them asunder, and from the water We made every living thing.

Reality 775

Qur'ân, XXI. 30

That which is above the heavens is the masculine, and the water which is beneath the earth is the feminine.

Book of Enoch, LIV. 8, 9

But it requires profound study to become acquainted with all the secrets of our sea, and with its ebb and flow.... The whole knowledge of our Art consists in the discovery of this our sea; any Alchemist who is ignorant of it, is simply wasting his money. Our sea is derived from the mountain.

Philalethes

Supra
Introd.

[1] That such an observation should have passed as evident at the time the *Qur'ân* was revealed shows to what extent the vision of modern man has atrophied.

And We send down pure water from the sky.
That therewith We may quicken a dead land.

Qur'ân, xxv.48–49

Pilg. 385

The name Moses ... means, taken from the water, and so shall we be taken out of instability, rescued from the storm of the world-flow.

Eckhart

P. State 563

Our air, like the air of the firmament, divides the waters: and as the waters under the firmament are visible to us mortals, while we are unable to see the waters above the firmament, so in 'our work' we see the extracentral mineral waters, but are unable to see those which, though hidden within, nevertheless have a real existence. They exist but do not appear until it please the Artist.

Philalethes

Introd. *Holiness*
Flight 944
Center 832

The true earth is pure and situated in the pure heaven. . . . But we who live in these hollows are deceived into the notion that we are dwelling above on the surface of the earth; which is just as if a creature who was at the bottom of the sea were to fancy that he was on the surface of the water, and that the sea was the heaven. . . . If any man could arrive at the exterior limit, or take the wings of a bird and come to the top, then like a fish who puts his head out of the water and sees this world, he would see a world beyond; and, if the nature of man could sustain the sight, he would acknowledge that this other world was the place of the true heaven and the true light and the true earth.

Plato (*Phaedo*, 109)

BEAUTY

Ex divina pulchritudine esse omnium derivatur.

St Thomas Aquinas (*De Pulchro*)

Ars sine scientia nihil.

Jean Mignot

There is only beauty behind me.
Only beauty is before me!

Cree Song[1]

Beauty is a manifestation of the Infinite on a finite plane: its content may therefore be suprarational, but could never be irrational. Introd. *Ecstasy*

The absence of beauty is metaphysically consonant with the very structure of the modern world, which exists by a negation (in the degree possible) of Principle: and in fact this deficiency is one of the most salient characteristics of the modern industrial and utilitarian civilization in all its aspects—man, politic, ideology, religion, and form—the world over. The intuition and comprehension of beauty require the operation of specific intellectual faculties which are in a state of failure or paralysis with many people today, who can hardly conceive even the possibility of beauty beyond certain rudimentary fragments: while on the other hand they suffer a hypertrophy of counterfeit and diabolical powers which lead them to prefer chemical to natural substances, orgy to katharsis, and surrealism in everything from philosophies to Church art.[2] Sacred influences have receded in measure with the intellectual decline that helped produce the modern world; and man, having all but effectively eliminated—if not exterminated—the earthly reflections of supernatural beauty,[3] is now vitiating, physically, chemically and psychically, what

Wr. Side 474

Introd. *Action*

Introd. *Flight*

[1] Reginald and Gladys Laubin: *The Indian Tipi*, University of Oklahoma Press, 1957, p. 2.

[2] Tradition is by definition *formal*, whereas surrealism in its various guises amounts to nothing but the morbid disintegration of form. Even the syncretism of forms practised by the neo-spiritualists is in point of fact a dissolution of forms, since formal perfection means purity, whereas a mixture of elements which are mutually exclusive can only add up to chaos (cf. *Introduction*). In the present world there is at the one extreme the brutalization of form, which comes from forcing and congealing matter into the artificial and quantitative exactitudes demanded by machine technology, while at the other extreme is the revolt against mechanism in the forces of a dissolving surrealism: both extremes meet in what might be termed an 'atomic relativism'.

[3] The American Indian tradition, for example, which is as a primordial manifestation of divine beauty ennobling the face of nature.

remains of natural beauty, with an arrogance and spiritual blindness on a scale unmatched in the days of Sodom or Babel.[1]

'According to Dionysius, beauty is order: symmetry with supreme lucidity' (Eckhart, I. 366). Beauty, being archetypal order, is begotten intellectually, not calculated empirically, and the language of symbols is consequently precise, not arbitrary—a matter of truths, not tastes. In beauty is found the harmonious symmetry between being and knowing, love and knowledge. 'The first thing which strikes one in a masterpiece of traditional art is intelligence: an intelligence surprising either for its complexity or for its power of synthesis; an intelligence which envelops, penetrates and elevates.[2]

Introd. *Symb.*

'Humanly speaking, certain artists of the Renaissance are great, but with a greatness that becomes smallness before the grandeur of the sacred. In sacred art genius seems hidden; what dominates is a vast, impersonal, mysterious intelligence. The sacred work of art has a perfume of infinity, an imprint of the absolute. In it the individual talent is disciplined: it merges with the creative function of the entire tradition, which itself could never be replaced—far less surpassed—by any human resources' (Schuon: *Perspectives spirituelles*, pp. 47–48).

'It is affirmed that "beauty relates to the cognitive faculty" (St Thomas Aquinas, *Sum. Theol.*, I, 5, 4 *ad.* 1), being the cause of knowledge, for, "since knowledge is by assimilation, and similitude is with respect to form, beauty properly belongs to the nature of a formal cause" (*ib.*). Again, St Thomas endorses the definition of beauty as a cause, in *Sum. Theol.*, III, 88, 3, he says that "God is the cause of all things by his knowledge" and this again emphasizes the connection of beauty with wisdom. "It is knowledge that makes the work beautiful" (St Bonaventura, *De reductione artium ad theologiam,* 13). It is of course, by its quality of lucidity or illumination (*claritas*), which Ulrich of Strassburg explains as the "shining of the formal light upon what is formed or proportioned", that beauty is identified with intelligibility: brilliance of expression being unthinkable apart from perspicacity. Vagueness of any sort, as being a privation of due form is necessarily a defect of beauty. Hence it is that in mediaeval rhetoric so much stress is laid on the communicative nature of art, which must be always explicit.

'It is precisely this communicative character that distinguished Christian from late classical art, in which style is pursued for its own sake, and content valued only as a point of departure. . . . And whereas in the greater part of modern art one cannot fail to recognise an exhibitionism in which the artist rather exploits himself than demonstrates a truth, and modern individualism frankly justifies this self-expressionism,[3] the mediaeval artist is

Introd. *Action*

[1] A typical illustration of this mentality appears in the announcement of an industrial organization: 'New applications of the dynamics of the physical world—nuclear fission, solar energy and atomic fusion—aided by electronic automation, may bring us within our lifetimes limitless supplies of power. Consequent transformations of agriculture and industry, medicine and biology, transportation and communication, might then free all men from economic and political slavery: unite all nations in an enduring peace.' In our days it is the industrialists (or their copywriters) who have assumed the role of 'philosopher-kings'. 'Have ye a covenant on oath from Us that reacheth to the Day of Judgment, that yours shall be all that ye ordain?' (*Qur'ân,* LXVIII. 39).

[2] For a luminous account of the rapport between beauty and traditional civilization, cf. Marco Pallis: *Peaks and Lamas*.

[3] This cult of the ego is an infernal attempt to create an 'archetype' where contact with the true Archetype has been lost.

characteristically anonymous and of "unobtrusive demeanour", and it is not who speaks, but what is said that matters. . . .

'In Europe, the now despised doctrine of a necessary intelligibility reappears at a comparatively late date in a musical connection. Not only had Josquin des Prés in the fifteenth century argued that music must not only sound well but mean something,[1] but it is about this very point that the struggle between plainsong and counterpoint centred in the sixteenth century' (Coomaraswamy: *Why Exhibit Works of Art?,* ch. V).

True music like all sacred art concentrates the faculties of the soul and relates them to spiritual archetypes; the music current in the West today, like profane art in general, disperses the powers of the soul, and reinforces the bonds of illusion attaching the soul to this world. Alain Daniélou explains concisely the cosmological role of Indian music: 'The tonic is the unmoving center, the fundamental unity in relation to which all the descriptive or expressive elements are evolved. . . . This identification with the tonic, the reduction of all the world of music to its basic unmanifest unity is the essential factor which allows the utilization of music as one of the forms of Yoga, one of the ways of spiritual attainment. The absolute fixity of the tonic during a musical performance and the resulting identity of each expressive interval with a given frequency is essential for the magical effect of sounds' (*Ethnic Folkways Library* P 431).[2]

Reality 773 775 ***Inv.*** **1003**

'Music—like dance—is the art of returning earthly shadows to their celestial vibrations and divine archetypes' (Schuon: *Sentiers de Gnose,* p. 92).

Beauty and virtue are interdependent, together being the unfolding on the human level of cosmic and metaphysical realities. Virtue is the exteriorization of beauty in psychic mode, as becomes apparent when one thinks upon nobility, dignity, humility, generosity, charity, courage, fidelity, rectitude, serenity, integrity, and veracity. Man is virtuous because God is Good. 'Virtue makes of the soul a mirror of the divine beauty' (Schuon: *Perspectives spirituelles,* p. 201); and hence, as the same author says elsewhere, the soul is beautiful in proportion to its universality. Since virtues are divine in origin, they can be participated in, but not possessed, by human individuals, whose practice is limited to the removal of obstacles to virtue—e.g., a person can avoid ostentation and pride, but he cannot *qua* individual *be* humble except in a purely relative sense without risking pretension or deception.

Creation 42

'The science of virtues, which applies Divine Truth to the soul, is directly connected with spiritual realization. Its criteria are extremely subtle. It could never be abridged into a schema of moral injunctions, and its classifications are only paradigms. Its object, spiritual virtue, is what one might call a symbol lived out, the right perception of which depends upon a certain interior development, which is not necessarily the case with doctrinal comprehension' (Burckhardt: *Introduction aux Doctrines ésotériques de l'Islam,* p. 41).

'Apart from their intrinsic qualities, art forms answer to a strict utility: in order that spiritual influences may manifest without hindrance they have need of a formal setting which corresponds to them analogically; otherwise they do not shine forth, even if they still remain present. It is true that they can shine forth in spite of everything in the soul of a holy

[1] Cf. St Augustine, *Conf.* X. 33: 'When it befalls me to be more moved with the voice than the words sung, I confess to have sinned penally, and then had rather not hear music.'

[2] For a thorough analysis of the divergences between Eastern and Western musical forms, cf. Alain Daniélou: *Introduction to the Study of Musical Scales,* London, The India Society, 1943.

man, but not everyone is a saint, and a sanctuary is made to facilitate the resonances of the spirit, not impede them.[1]

'Sacred art is made to convey spiritual presences: it is made at the same time for God, for angels and for men: profane art on the contrary exists only for men, and by that very fact betrays man.

Contem. 536

'Sacred art helps man to find his own center, this center which by nature loves God' (Schuon: *Perspectives spirituelles*, pp. 36–37).

[1] 'Someone may say that angels are at ease even in a stable. But precisely, a stable is not a baroque or surrealist church.'

The Uncreate Source of Beauty

Pure aesthetic experience is theirs in whom the knowledge of ideal beauty is innate: it is known intuitively, in intellectual ecstasy without accompaniment of ideation, at the highest level of conscious being; born of one mother with the vision of God, its life is as it were a flash of blinding light of transmundane origin, impossible to analyze, and yet in the image of our very being.

Ecstasy 636
Contem. 532

Sâhitya Darpaṇa, III. 2–3

Nothing makes a thing beautiful but the presence and participation of Beauty in whatever way or manner obtained. . . . By Beauty all beautiful things become beautiful.

Plato (*Phaedo*, 100 E)

The Absolute Beauty is the Divine Majesty endued with (the attributes of) power and bounty. Every beauty and perfection manifested in the theatre of the various grades of beings is a ray of His perfect beauty reflected therein. It is from these rays that exalted souls have received their impress of beauty and their quality of perfection.

Jâmi

And my spirit is a spirit to all the spirits (of created beings); and whatsoever thou seest of beauty in the universe flows from the bounty of my nature.

Ibn al-Fârid

The Super-Essential Beautiful is called 'Beauty' because of that quality which It imparts to all things severally according to their nature, and because It is the Cause of the harmony and splendour in all things, flashing forth upon them all, like light, the beautifying communications of Its originating ray; and because it summons all things to *fare* unto Itself (from whence It hath the name of 'Fairness'),[1] and because It draws all things together in a state of mutual interpenetration. And it is called 'Beautiful' because It is All-Beautiful and more than Beautiful, and is eternally, unvaryingly, unchangeably Beautiful; incapable of birth or death or growth or decay; and not beautiful in one part and foul in another; nor yet at one time and not at another; nor yet beautiful in relation to one thing but not to another; nor yet beautiful in one place and not in another (as if It were beautiful for some but were not beautiful for others); nay, on the contrary, It is, in Itself and by Itself, uniquely and eternally beautiful, and from beforehand It contains in a transcendent manner the originating beauty of everything that is beautiful. For in the simple and supernatural nature belonging to the world of beautiful things, all beauty and all that is beautiful hath its unique and pre-existent Cause. From this Beautiful all things possess their existence, each kind being beautiful in its own manner, and the Beautiful

Reality 773

[1]A play on the Greek καλέω, to call, summon, and κάλλος, beauty.

Infra 689
Peace 694
Creation 48
Illusion 109
Reality 775

causes the harmonies and sympathies and communities of all things. And by the Beautiful all things are united together and the Beautiful is the beginning of all things, as being the Creative Cause which moves the world and holds all things in existence by their yearning for their own Beauty. And It is the Goal of all things, and their Beloved, as being their Final Cause (for it is the desire of the Beautiful that brings them all into existence), and It is their Exemplar from which they derive their definite limits; and hence the Beautiful is the same as the Good, inasmuch as all things, in all causation, desire the Beautiful and Good; nor is there anything in the world but hath a share in the Beautiful and Good. And we make bold to say that the Non-Existent also participates in the Beautiful and the Good; for then it is at once truly the Beautiful and the Good when it is praised Super-Essentially in God by the subtraction of all attributes. The One Good and Beautiful is in Its oneness the Cause of all the many beautiful and good things.

Dionysius

They call this 'loveliness-uniter' (*samyadvâma*), for all lovely things (*vâma*) come together (*samyanti*) unto it. All lovely things come together unto him who knows this.

Chândogya Upanishad, IV. xv. 2

Reality 773
Creation 28
Orthod. 288
Infra 676

He who has been instructed thus far in the things of love, and who has learned to see the beautiful in due order and succession, when he comes toward the end will suddenly perceive a nature of wondrous beauty (and this, Socrates,[1] is the final cause of all our former toils)—a nature which in the first place is everlasting, not growing and decaying, or waxing and waning; secondly, not fair in one point of view and foul in another, or at one time or in one relation or at one place fair, at another time or in another relation or at another place foul, as if fair to some and foul to others; . . . but Beauty absolute, self-sufficient, simple, and everlasting, which without diminution and without increase, or any change, is imparted to the ever-growing and perishing beauties of all other things. He who from these ascending under the influence of true love, begins to perceive that Beauty, is not far from the end. And the true order of going, or being led by another, to the things of love, is to begin from the beauties of earth and mount upwards for the sake of that other Beauty, using these as steps only, and from one going on to two, and from two to all fair forms, and from fair forms to fair practices, and from fair practices to fair notions, until from fair notions he arrives at the notion of absolute Beauty, and at last knows what the essence of Beauty is.

Plato (*Symposium*, 211)

God is beautiful, and He loves beauty.

Muhammad

Say: Who hath forbidden the adornment of Allâh which He hath brought forth for His bondmen, and the good things of His providing? . . .

Say: My Lord forbiddeth only indecencies, such of them as are apparent and such as are within, and sin and wrongful oppression, and that ye associate with Allâh that for which no

[1] Socrates is recounting the teachings transmitted to him by Diotima of Mantineia.

warrant hath been revealed, and that ye tell concerning Allâh that which ye know not.

Qur'ân, VII. 32–33

Every art is interested to adorn, and the very existence of the arts was a discovery made in behalf of ornament.

Apollonius of Tyana

Creation 48

The Object of *Love* is *loveliness* or *beauty.*

Peter Sterry

When one approaches the Wonderful one knows not whether art is Tao or Tao is art.

Hui Tsung

All that is sweet, delightful, and amiable in this world, in the serenity of the air, the fineness of seasons, the joy of light, the melody of sounds, the beauty of colours, the fragrancy of smells, the splendour of precious stones, is nothing else but Heaven breaking through the veil of this world, manifesting itself in such a degree and darting forth in such variety so much of its own nature.

William Law

Illusion 109

The Platonist Plotinus discourses on Providence[1] and from the beauty of flower and foliage proves that from the supreme God, whose beauty is unseen and ineffable, Providence reaches down even to these earthly things below: all of which things, so transitory and momentary, could not have their peculiar, richly assorted beauties, but from that intellectual and immutable Beauty forming them all.

St Augustine

Being is desirable because it is identical with Beauty, and Beauty is loved because it is Being. . . . We ourselves possess Beauty when we are true to our own being; ugliness is in going over to another order; knowing ourselves, we are beautiful; in self-ignorance, we are ugly.

Plotinus

Creation 42
Realiz. 859

The very order, disposition, beauty, change, and motion of the world and of all visible things silently proclaim that it could only have been made by God, the ineffably and invisibly great and the ineffably and invisibly beautiful.

St Augustine

Faith 505

Whenever, in the course of the daily hunt, the red hunter comes upon a scene that is strikingly beautiful or sublime—a black thundercloud with the rainbow's glowing arch above the mountain; a white waterfall in the heart of a green gorge; a vast prairie tinged with the blood-red of sunset—he pauses for an instant in the attitude of worship.

Ohiyesa

Contem. 547

[1] *Enneads,* III. ii. 13.

He is Allâh, than Whom there is no other God, the Knower of the Invisible and the Visible. He is the Beneficent, the Merciful.

He is Allâh, than Whom there is no other God, the King, the Holy, Peace, the Faithful, the Guardian, the All-Powerful, the Restorer, the Superb. Glorified be Allâh from all that they associate with Him.

Inv. 1031

He is Allâh, the Creator, the Maker, the Fashioner. His are the most Beautiful Names.[1] All that is in the heavens and the earth glorifieth Him, and He is the Mighty, the Wise.

Qur'ân, LIX. 22–24

Void 724

Men who love the body will never see the vision of the Beautiful and Good. How glorious, my son, is the beauty of that which has neither shape nor colour!

Hermes

Introd. *Reality*

That which imparts form to forms is itself formless; therefore *Tao* cannot have a name.

Chuang-tse (ch. XXII)

M.M. 994

Though the names which we give to God are many, the most high Nature of God is a Simplicity which cannot be named by any creature. But because of His incomprehensible nobility and sublimity, which we cannot rightly name nor wholly express, we give Him all these names.

Ruysbroeck

Each *Sephirah* has a specific Name, by which the angels are also named, but Thou (the unknowable Essence) hast no specific name, for Thou art the One which fills all names and gives them their true meaning.

Zohar (The 'Prayer of Elias')

Reality 773

Void 721

Contem. 532

Laotse has said: 'Thirty spokes are grouped around the hub of a wheel, and when they lose their own individuality, we have a functioning cart. We knead clay into a vessel and when the clay loses its own existence we have a usable utensil. We make a hole in the wall to make windows and doors, and when the windows and doors lose their own existence, we have a house to live in.' And so when we view a stone cave or a blessed spot and see the vertically uprising peaks, horizontally-stretching mountain passes, those that go up and form a precipice, those that go down and form a river, those that are level and form a plateau, those that are inclined and form a hillside, those that stretch across and become bridges, and those that come together and become ravines, we realize that, however incomparably manifold they are in their greatness and mystery, this mystery and grandeur arises when the parts lose their individual existence. For when they lose their own existence, there are no passes, no precipices, no rivers and no plateaux, hillsides, bridges and ravines. But it is exactly in their non-existence that the special talent in our breast and the special vision below our eyebrows wander and float at ease. And since this special talent in our breast and this special vision below our eyebrows can wander and float at ease

[1] This verse shows the correlation between the archetypes of beauty and the function of the artist in relation to God and the archetypes.

only when these things are non-existent, why, then, must we insist on going to the stone cave and to the blessed spot?

Chin Shengt'an

Pilg. 387

To comprehend the different organs of the horse is not to comprehend the horse itself. What we call the horse exists before its different organs.

Chuang-tse (ch. XXV)

Moses entered into the darkness where God was, that is into the unseen, invisible, incorporeal and archetypal essence of existing things. Thus he beheld what is hidden from the sight of the mortal nature, and, in himself and his life displayed for all to see, he has set before us, like some well-wrought picture, a piece of work beautiful and godlike, a model for those who are willing to copy it. Happy are they who imprint that image in their souls.

Philo

Since there is this Good in any good thing, It must enter from elsewhere (than the world of things): that Source must be a Good absolute and unique. Thus is revealed to us the Primarily Existent, the Good, above all that has being, Good unalloyed, all-transcending, Cause of all. The Maker, as the more consummate, must surpass the made.

Plotinus

Know how much the Lord of them is more beautiful than they: for the first author of beauty made all those things. . . .

For by the greatness of the beauty, and of the creature, the creator of them may be seen so as to be known thereby.

Wisdom, XIII. 3, 5

The real orchards and fruits are within the heart: the reflexion of their beauty is falling upon this water and earth (the external world).

Center 816

If it were not the reflexion of that delectable cypress (the heart of the saint), then God would not have called it the abode of deception.

This deception consists in that: i.e. this phantom (the external world) derives its existence from the reflexion of the heart and spirit of the holy men.

Introd. *Holiness*

All the deceived ones come to gaze on this reflexion in the opinion that this is the place of Paradise.

They are fleeing from the origins of the orchards: they are making merry over a phantom.

Illusion 85

Rûmî

Everybody is wonder-struck at the mere sight of a rich man's garden house. People become speechless at the sight of the trees, the flowers, the ponds, the drawing-room, the pictures. But alas, how few are they who seek the owner of all these!

Sri Ramakrishna

But all mistaken you and all like you
That long for that lost Eden as the true;
Fair as it was, still nothing but the Shade
And Out-court of the Majesty that made. . . .
Creation 38 For so Creation's Master-Jewel fell
From that same Eden: loving which too well,
The Work before the Artist did prefer,
And in the Garden lost the Gardener.

'Aṭṭâr

Entered into Divine Mind, herself made over to That, she (the soul) at first contemplates that Realm, but once she sees that Higher still she leaves all else aside. Thus when a man enters a house rich in beauty he might gaze about and admire the varied splendor before the master appears; but once seeing him he would ignore all else and look to him alone.

Conform. 170

Plotinus

Heav'n is most faire; but fairer He
That made that fairest Canopie.

Robert Herrick

Bliss, delight, and procreation are not what one should desire to understand. One should know the discerner of bliss, delight, and procreation.

Kaushîtaki Upanishad, III. 8

Illusion 109

Let no one deify the universe; rather let him seek after the creator of the universe.

Clement of Alexandria

Reality 775

Let each one of us leave every other kind of knowledge and seek and follow one thing only.

Plato (*Republic* x, 618 C)

Embellished for mankind is love of the joys from women and offspring, and stored-up heaps of gold and silver, and horses branded, and cattle and land. That is comfort of the life of the world. Allâh! With Him is a more excellent abode.

Qur'ân, III. 14

The Nature, then, which creates things so lovely must be itself of a far earlier beauty; we, undisciplined in discernment of the inward, knowing nothing of it, run after the outer, never understanding that it is the inner which stirs us; we are in the case of one who sees his own reflection but not realizing whence it comes goes in pursuit of it.

Sin 66

Plotinus

'I am the tree; these flowers My offshoots are.
Let not these offshoots hide from thee the tree.'

What profit rosy cheeks. forms full of grace.
And ringlets clustering round a lovely face?
When Beauty Absolute beams all around.
Why linger finite beauties to embrace?

Jâmî

Suppose a curious and fair woman. Some have seen the beauties of Heaven in such a person. It is a vain thing to say they loved too much. I dare say there are ten thousand beauties in that creature which they have not seen. They loved it not too much. but upon false causes. Nor so much upon false ones. as only upon some little ones. They love a creature for sparkling eyes and curled hair. lily breasts and ruddy cheeks: which they should love moreover for being God's Image. Queen of the Universe. beloved by Angels. redeemed by Jesus Christ. an heiress of Heaven. and temple of the Holy Ghost: a mine and fountain of all virtues. a treasury of graces. and a child of God. But these excellencies are unknown. They love her perhaps. but do not love God more: nor men as much: nor Heaven and Earth at all. And so. being defective to other things. perish by a seeming excess to that. We should be all Life and Mettle and Vigour and Love to everything: and that would poise us. . . . But God being beloved infinitely more. will be infinitely more our joy. and our heart will be more with Him. so that no man can be in danger by loving others too much. that loveth God as he ought.

Love 625
P. State 579
Infra
676

Thomas Traherne

Kings lick the earth whereof the fair are made.
For God hath mingled in the dusty earth
A draught of Beauty from His choicest cup.
'Tis *that,* fond lover—not these lips of clay—
Thou art kissing with a hundred ecstasies.
Think. then. what must it be when undefiled!

Rûmî

Center
847

He who contemplates the Lotus Feet of God looks on even the most beautiful woman as mere ash from the cremation ground.

Sri Ramakrishna

Renun.
152b

All the Sages who have written on our Art. have spoken of the work and regimen of Saturn: and their remarks have led many to choose common lead as the substance of the Stone. But you should know that *our* Saturn. or lead. is a much nobler substance than gold. It is the living earth in which the soul of gold is joined to Mercury. . . . Happy is he who can salute this planet. and call it by its right name.

Philalethes

P. State
563

What if man had eyes to see the true Beauty—the divine Beauty. I mean. pure and clear and unalloyed. not clogged with the pollutions of mortality and all the colours and vanities of human life—thither looking. and holding converse with the true Beauty simple and divine? Remember how in that communion only. beholding Beauty with the eye of the

Sin
66

Center 816
Infra 676
Realiz. 870

mind, he will be enabled to bring forth, not images of beauty, but realities (for he has hold not of an image but of a reality), and bringing forth and nourishing true virtue to become the friend of God and be immortal, if mortal man may. Would that be an ignoble life?

Plato (*Symposium*, 211 E—212 A)

The Doctrine of the Archetypes

Supra 663

Make all things according to the pattern which was shewn thee on the mount.

Exodus, xxv. 40 & *Hebrews*, viii. 5

In the Father is the exemplar of all creatures.

Eckhart

Introd. *Reality*

Things in every instance involve universals, but universals do not point to the material world. If there were no universals, things could not be described as 'things'.

Kung-sun Lung

Creation 28

(God) Himself has imparted of His own to all particular beings from that fountain of beauty—himself. For the good and beautiful things in the world could never have been what they are, save that they were made in the image of the archetype, which is truly good and beautiful.

Philo

Infra 674

(Pythagoras) adds, that the survey of all heaven, and of the stars that revolve in it, is indeed beautiful, when the order of them is considered. For they derive this beauty and order by the participation of the first and the intelligible essence. . . . And wisdom indeed, truly so called, is a certain science which is conversant with the first beautiful objects, and these divine, undecaying, and possessing an invariable sameness of subsistence; by the participation of which other things also may be called beautiful.

Iamblichus

Rev. 967
Symb. 321

There were others like Plato, a most excellent man, who maintained that there are not only as many Intelligences as there are motions of the heavens, but also as many as there are kinds of things; such as one kind for all men, another for gold, another for silver, and so on; and they say that as the Intelligences are the generators of these (motions), each of its own, so these other (Intelligences) are the generators of all other things, and the exemplars each of their own kind; and Plato calls them *Ideas*, which is as much as to say *forms*, and *universal natures*. The heathen called them gods and goddesses (although they had not so philosophical an understanding of them as Plato had), and adored their images, and built to them great temples.

Dante (*Il Convito*, ii. v. 2)

To be properly expressed a thing must proceed from within, moved by its form: it must come, not in from without but out from within.

Eckhart

Painting must be sought for beyond the shapes, but this is difficult to explain to common people.

Chang Yen-yüan

Ch'ü T'ing, the monk of the White Clouds, possessed the secret of spirit and form; he grasped the very origin of things and painted with extraordinary ease; the depth of his work was immeasurable.

Ching Hao

One should not take outward beauty for reality; he who does not understand this mystery will not obtain the truth, even though his pictures may contain likeness.

Ching Hao

Ku K'ai-chih's brush-stroke was tight and strong, connecting and continuous, moving as in a circle exceedingly swift, accomplishing the design with freedom and ease. It was like a gust of wind or a flash of lightning. The ideas existed before he took up the brush; when the picture was finished it contained them all, and thus it was filled with a divine spirit.

Chang Yen-yüan

Action 338
Introd. *Holiness*

In my poetry I am not aiming at skill, in my writing (or painting) not seeking for the strange. The boundless gift of heaven is my master.

Su Tung-p'o

People think that men alone have spirit; they do not realize that everything is inspirited. Therefore Kuo Jo-hsü[1] despised deeply the works of common men. He said that though they were called paintings, they were not painting (as art), because they transmit only the forms but not the spirit. Consequently the manner of painting which gives the resonance of the spirit and the movement of life is the foremost. And Kuo Jo-hsü said that it has been practised only by high officials and hermit scholars, which is correct.

Têng Ch'un

Charity 608

A form is made in the resigned will according to the platform or model of eternity, as it was known in the glass of God's eternal wisdom before the times of this world.

Boehme

Center 828

The painters of old painted the idea (*i*) and not merely the shape (*hsing*).

Hsieh Ho

[1] An authority on painting of the Sung period.

Art is expression informed by ideal beauty (*rasa*).

Sâhitya Darpaṇa, I. 3

That image is said to be lovely (*ramya*) which is of neither more nor less than the prescribed proportions (*mâna*).

Śukranītisâra, IV. iv. 73

There are some to whom that which captivates their heart is lovely; but for those who know, that which falls short of canonical proportion (*śâstramâna*) is not beautiful.

Śukranītisâra, IV. iv. 106

When Fabisch the sculptor showed Bernadette at Lourdes an effigy (of the Virgin) in marble destined for Massabieille, poor Bernadette remained discomfited. 'It is beautiful,' she said, to avoid hurting him; 'but it is not Her!' she added, to avoid lying. From that time she never ceased to protest against the statues that were presented to her:

'My good Mother, how they disfigure you!'

'These artists when they see You are going to be given a surprise!'

'Look at this goitre that has been made of the Holy Virgin!'

And again:

'I never said that she raised her head, but that she raised her eyes!'[1]

From a Life of St Bernadette

Paintings of modern times are brilliant and aim only at perfection, and the works of our contemporaries are so confused that they have no significance. The majority of these modern artists belong to the artisan class. . . . (They) mix their brushes and ink with dust and dirt, and their colours with mud, and in vain smear the silk. How can this be called painting?

Chang Yen-yüan

Form is not in the material: it is in the designer before ever it enters the stone; and the artificer holds it not by his equipment of eyes and hands but by his participation in his art.

Plotinus

Supra 663

The Sages have been taught of God that this natural world is only an image and material copy of a heavenly and spiritual pattern; that the very existence of this world is based upon the reality of its celestial archetype.

Michael Sendivogius

Rev. 967

You must understand then that the soul makes in the physical world nothing else than copies of the things which Mind (νοῦς) makes in the soul itself; and that Mind makes in the soul nothing else than copies of the things which the First Cause of all makes in Mind.

Hermes

Infra 679 688

The arts of music and dancing consist entirely in imitation (*monomane*).

Seami Motokiyo

[1] It is related that Bernadette when shown various representations of the Virgin chose the icon style as being the most faithful to the Original.

Forget the theatre and look at the Nô. Forget the Nô and look at the actor. Forget the actor and look at the idea. Forget the idea and you will understand the Nô.

Nô-gaku

Form is a revelation of essence.

Eckhart

Symb. 321

Paintings of the present time may possess an outward likeness, but the operations of the spirit are lacking in them. If only they strive to obtain the spirit resonance, the outward likeness will naturally be in the work.

Chang Yen-yüan

Forget all about the brush and ink. Then you shall learn the truth about landscapes.

Ching Hao

All forms of being in this corporeal world are images of pure Lights, which exist in the spiritual world.[1]

Suhrawardî

Know that all the shapes and images which you see with your bodily eyes in the world of things that come to be and cease to be are mere semblances and copies of the forms which have real existence in the thought-world,[2] those forms which are eternal and will never cease to be.

Illusion 109

Hermes

To thy care the figur'd seal's consign'd,
Which stamps the world with forms of ev'ry kind.

Orpheus: *Hymn to Apollo*

Gaze at the sky, the earth, the sea, and all the things which shine in them or above them, or creep or fly or swim beneath them. They have forms because they have rhythm: take this away, and they will no longer be. From whom then are they, save from Him, from whom rhythm is; since they have being only in so far as they are rhythmically ordered. . . . Pass, therefore, beyond the mind of the artist, so that thou mayest see the everlasting rhythm; then will wisdom shine upon thee from her inmost abode, from the very sanctuary of truth.

Infra 674

St Augustine

Every form you see has its archetype in the placeless world;
If the form perished, no matter, since its original is everlasting.
Every fair shape you have seen, every deep saying you have heard,
Be not cast down that it perished; for that is not so.

[1] Suhrawardî gives this as his own experience, gained through 'continual solitude and many ascetic practices', and then adds: 'But if any one is not satisfied with this proof, he may devote himself to ascesis, and to the service of the mystics; and perhaps he will thereby acquire a "natural disposition", and will thereupon see the Light which radiates in the world of divine Powers, and the substances of the celestial realm.'

[2] I.e., the world of archetypal Ideas. The text is an English rendering of a Latin translation from an Arabic manuscript. The editor offers τὰ νοητὰ εἴδη, τὰ ὄντως ὄντα, as a guess for the Greek original.

Creation 28 *Contem.* 536

Whereas the spring-head is undying, its branch gives water continually;
Since neither can cease, why are you lamenting?
Conceive the Soul as a fountain, and these created things as rivers:
While the fountain flows, the rivers run from it.
Put grief out of your head and keep quaffing this river-water:
Do not think of the water failing; for this water is without end.

Dîvâni Shamsi Tabrîz, XII

Beauty and Knowledge

I cannot fairly give the name of 'art' to anything irrational.

Plato (*Gorgias*, 465 A)

The works of the old masters are instruments of knowledge.

Tao-chi

Reality 775

Union in distinction makes order; order produces agreement; and proportion and agreement, in complete and finished things, make beauty.

St Francis de Sales

M.M. 994 Introd. *Reality*

The very being of God, if 'being' can be ascribed to God, is the Beautiful and the Good. . . . All things which the eye can see are mere phantoms, and unsubstantial outlines; but the things which the eye cannot see are the realities, and above all, the ideal form of the Beautiful and the Good. . . . If you are able to apprehend God, then you will apprehend the Beautiful and the Good. . . . If you seek knowledge of God, you are also seeking knowledge of the Beautiful. For there is one road alone that leads to the Beautiful, and that is piety joined with knowledge of God.

Infra 676

Hermes

The old painters worked in the same way as did the Buddha in explaining the law. He spoke by natural inspiration, without any effort, about past kalpas, their causes and effects, which manifest and dissolve in a mysterious fashion beyond human comprehension, though never contrary to truth and reason.

Ecstasy 636

Li Jih-hua

When the thought reaches the origin (or meaning) of things, the heart is inspired, and the painter's work can then penetrate into the very essence of the smallest things; it becomes inscrutable.

M.M. 998

Tao-chi

Any want of measure and symmetry in any mixture whatever must always of necessity be fatal, both to the elements and to the mixture, which is then not a mixture, but only a confused medley which brings confusion on the possessor of it. . . . Then, if we are not able to hunt the good with one idea only, with three we may catch our prey: Beauty, Symmetry, Truth are the three, and these taken together we may regard as the single cause of the mixture, and the mixture as being good by reason of the infusion of them.

Plato (*Philebus*, 64 E)

Sin is a departure from the order to the end. . . . Sin may occur in two ways in the operation of art. First, by a departure from the particular end intended by the artist, and this sin will be proper to the art: e.g., if an artist produce a bad thing, while intending to produce something good, or produce something good, while intending to produce something bad. Secondly, by a departure from the universal end of human life, and then he will be said to sin if he intend to produce a bad work and does so in effect, so that another is deceived by it. But this sin is not proper to the artist as an artist, but as a man. Consequently, for the former sin the artist is blamed as an artist, while for the latter he is blamed as a man.—On the other hand, in moral matters, where we take into consideration the order of reason to the universal end of human life, sin and evil are always due to a departure from the order of reason to the universal end of human life. Therefore man is blamed for such a sin, both as man and as a moral being.

Action 338

Sin 76

St Thomas Aquinas

We hold that all the loveliness of this world comes by communion in Ideal Form. All shapelessness whose kind admits of pattern and of form, as long as it remains outside of Reason and Idea, and has not been entirely mastered by Reason, the matter not yielding at all points and in all respects to Ideal Form, is ugly by that very isolation from the Divine Thought. But where the Ideal Form has entered, it has grouped and co-ordinated what from a diversity of parts was to become a unity: it has rallied confusion into co-operation: it has made the sum one harmonious coherence; for the Idea is a unity and what it moulds must come to unity as far as multiplicity may. And on what has thus been compacted to unity, Beauty enthrones itself, giving itself to the parts as to the sum.

This, then, is how the material thing becomes beautiful—by communicating in the thought that flows from the Divine.

Plotinus

This divine knowledge makes us amorous of divine beauty, beautiful and lovely: and this divine love and purity reciprocally exalts divine knowledge: both of them growing up together, like that Ἔρως and Ἀντέρως that Pausanias sometimes speaks of.

John Smith the Platonist

The artist (or musician) we may think of as exceedingly quick to beauty, drawn in a very rapture to it. . . . All that offends against unison or harmony repels him: he longs for measure and shapely pattern. This natural tendency must be made the starting-point: after the tone, rhythm and design in things of sense he must learn to distinguish the material forms from the Authentic Existent which is the source of the entire reasoned scheme of the

Supra 670

Infra 689 · *Knowl.* 755 · *Reality* 775

work of art; he must be led to the Beauty that manifests itself through these forms; he must be shown that what ravished him was no other than the harmony of the Intellectual World and the Beauty in that sphere—not some one shape of beauty but the All-Beauty, the Absolute Beauty; and the truths of philosophy must be implanted in him to lead him to faith in that which, all unknowing, he holds within himself. . . . His lesson must be to fall down no longer in bewildered delight before some one embodied form; he must be led, under a system of mental discipline, to beauty everywhere and be made to discern the One Principle underlying all.

Plotinus

Let our artists be those who are gifted to discern the true nature of the beautiful and graceful; then will our youth dwell in a land of health, amid fair sights and sounds, and receive the good in everything: and beauty, the effluence of fair works, shall flow into the eye and ear, like a health-giving breeze from a purer region, and insensibly draw the soul from earliest years into likeness and sympathy with the beauty of reason.

Plato (***Republic*** III, 401 C)

Virtue

Conform. 166

As Being itself under the concept of truth is the object of the intellect, so Being itself under the concept of goodness is the object of the will.

Marsilio Ficino

'Nobility' generally expresses in all things the perfection of their nature.

Dante (*Il Convito,* IV. xvi. 7)

The amiableness of virtue consisteth in this, that by it all happiness is either attained or enjoyed.

Thomas Traherne

Creation 42

Let her (the soul) strive to increase that heavenly beauty which is her birthright, and to adorn it with such shades of character and affection as it deserves.

St Bernard

Center 816 · Supra 670

Walk honestly, saith St *Paul,* Rom. 13.13. The Word is *καλῶς*, Beautifully. Write after the Copy of Divine Beauty, which dwells in the midst of thee, and shines forth from the Face of the Lord Jesus in thine Heart: Imitate, discover this in all thy Conversation.

The chief Things of Beauty, are Light and Proportion. Thy Christ in thee is both these; the Light, and the Wisdom of God. Then thou livest Beautifully, when this Light runs along thro' thy Thoughts. Affections. Actions, shining in all, and making every thing proportionable to itself.

Peter Sterry

Spiritual virtue (*al-iḥsân*) is to adore God as if thou sawest Him: and if thou seest Him not. He nevertheless sees thee.

Muhammad

Faith 505

St Dionysius being bent on lauding Mary's virtues found them so inconceivable he held his tongue.

Eckhart

P. State 579

Nothing is more becoming to woman than courtesy. And let not the miserable vulgar make another mistake in the meaning of this word, believing it to be the same as liberality; for liberality is a special and not general courtesy. Courtesy and virtue are one; and because, of old, virtue and fine manners were the custom of courts (whereas to-day the contrary is true), this word was derived from *court*: and courtesy was none other than the custom of the *court*.[1] If we should wish to-day to take such a word from the courts, especially of Italy, there would be nothing we could use except baseness.

Dante (*Il Convito,* II. xi. 3)

Wr. Side 464

An action cannot be called virtuous if it proceed not from the affection which the heart bears to the excellence and beauty of reason. . . . Is not prudence itself imprudent in an intemperate man? Fortitude, without prudence, justice and temperance, is not fortitude, but folly; and justice is unjust in the weak man who dares not do it, in the intemperate man who permits himself to be carried away with passion, and in the imprudent man who is not able to discern between the right and the wrong. Justice is not justice unless it be strong, prudent and temperate; nor is prudence prudence unless it be temperate, just and strong; nor fortitude fortitude unless it be just, prudent and temperate; nor temperance temperance unless it be prudent, strong and just. In fine, a virtue is not perfect virtue, unless it be accompanied by all the rest. . . .

Supra 674

There are certain inclinations which are esteemed virtues and are not so, but favours and advantages of nature. How many are there who are naturally sober, mild, silent, chaste and modest? Now all these seem to be virtues, and yet have no more the merit thereof than bad inclinations are blameworthy before we have given free and voluntary consent to such natural dispositions. It is no virtue to be by nature a man of little meat, yet to abstain by choice is a virtue. It is no virtue to be silent by nature, though it is a virtue to bridle one's tongue by reason. Many consider they have the virtues as long as they do not practise the contrary vices. One that has never been assaulted may truly boast that he was never a runaway, yet he has no ground to boast of his valour. . . .

Peace 700

In truth the great S. Augustine shows, in an epistle which he wrote to S. Jerome, that we may have some sort of virtue without having the rest, but that we cannot have perfect ones without having them all; whilst, as for vices, we may have some without having others, yea, it is even impossible to have them all together: so that it does not follow that he who has

Wr. Side 474

[1] This full sense of courtesy corresponds with the meaning of *adab* in Islam.

Introd. *Illusion*

lost all the virtues has by consequence all the vices, since almost every virtue has two opposite vices, which are not only contrary to the virtue but also to one another. He who has forfeited valour by rashness cannot at the same time be taxed with cowardice; nor can he who has lost liberality by prodigality, be at the same time reproached with niggardliness.

St Francis de Sales

Sin 62

In the very overcoming of temptation . . . we may draw out a hidden spiritual sweetness, as the bees suck honey from the thorn-bushes as well as from all other flowers. He who has not been tempted, knows nothing, nor lives as yet, say the wise man Solomon, and the holy teacher St Bernard. We find more than a thousand testimonies in Scripture to the great profit of temptation; for it is the special sign of the love of God towards a man for him to be tempted and yet kept from falling; for thus he must and shall of a certainty receive the crown (*James*, I. 12).

Tauler

He who would succeed in the study of this Art, should be persevering, industrious, learned, gentle, good-tempered, a close student, and neither easily discouraged nor slothful. . . . Above all, let him be honest, God-fearing, prayerful, and holy. Being thus equipped, he should study Nature, read the books of genuine Sages, who are neither imposters nor jealous churls, and study them day and night; let him not be too eager to carry out every idea practically before he has thoroughly tested it, and found it to be in harmony not only with the teaching of all the Sages, but also of Nature herself. . . . Nor let him despair though he take many false steps; for the greatest philosophers have learned most by their mistakes.

Conform. 170 *Suff.* 128

Philalethes

Of great virility and enthusiasm, good looking, courageous, learned in the scriptures, studious, sane of mind, not melancholy, keeping young, regular in food, having his senses under control, free from fear, clean, skilful, generous, helpful to all, qualified, firm, intelligent, independent, forgiveful, of good conduct and character, keeping his good deeds secret, of gentle speech, believer in the scriptures, worshipper of gods and his guru. Having no desire for other people's company, free from serious disease, such a one should be the supreme seeker qualified for all the forms of yoga. He will reach attainment within three years, without a doubt.

Metanoia 484 *Orthod.* 288 *Holiness* 914

Shiva Samhitâ, v. 23–27

Orthod. 275

Such as have borne rule in their dominions, men of great power, and endued with their wisdom, shewing forth in the prophets the dignity of prophets.

And ruling over the present people, and by the strength of wisdom instructing the people in most holy words:

Infra 679 Supra 674

Such as by their skill sought out musical tunes, and published canticles of the scriptures:

Rich men in virtue, studying beautifulness, living at peace in their houses.

All these have gained glory in their generations, and were praised in their days.

Ecclesiasticus, XLIV. 3–7

Seeing then that our souls are a region open to His invisible entrance, let us make that place as beautiful as we may, to be a lodging fit for God. Else He will pass silently into some other home, where He judges that the builder's hands have wrought something worthier.

Philo

Grace 558

What else is speculative virtue but the clarity of the intellect? What else is moral virtue but the stable ardor of appetite kindled by the clarity of the intellect? . . . What is the end of virtue? The end of moral virtue is to purify and to separate the Soul from the divisible body; that of speculative virtue, to grasp the incorporeal and universal concepts of things, whose locus is far from divisible bodies.

Marsilio Ficino

Be persuaded that the end of life, is to live conformably to divinity.

Sextus the Pythagorean

Introd. *Inv.*

Music

All songs are a part of Him, who wears a form of sound.

Vishṇu Purâṇa

Inv. 1003

Praise ye the Lord. Praise God in his sanctuary: praise him in the firmament of his power.
Praise him for his mighty acts: praise him according to his excellent greatness.
Praise him with the sound of the trumpet: praise him with the psaltery and harp.
Praise him with the timbrel and dance: praise him with stringed instruments and organs.
Praise him upon the loud cymbals: praise him upon the high sounding cymbals.
Let every thing that hath breath praise the Lord. Praise ye the Lord.

Psalm CL

Harmonies unheard create the harmonies we hear and wake the soul to the conscious ness of beauty, showing it the one essence in another kind; for the measures of our music are not arbitrary, but are determined by the Principle whose labor is to dominate matter and bring pattern into being.

Plotinus

Supra 670

Though he be gone, mine every limb beholds him
In every charm and grace and loveliness:
In music of the lute and flowing reed
Mingled in consort with melodious airs. . . .

Ibn al-Fâriḍ

Sacred music causes flight to sadness and to the evil spirits because the spirit of Jehovah sings happily in a heart filled with holy joy.

Heinrich Khunrath

Ecstasy 637

How splendid it is to drink to the sound of music!

Ibn al-Fâriḍ

Musical harmony is a most powerful conceiver. It allures the celestial influences and changes affections, intentions, gestures, notions, actions, and dispositions. . . . Fish in the lake of Alexandria are delighted with harmonious sounds; music has caused friendship between dolphins and men. The playing of the harp affects the Hyperborean swans. Melodious voices tame the Indian elephants. The elements themselves delight in music.

Cornelius Agrippa

Music . . . in remote times was not only cultivated, but venerated to such an extent that the same men were regarded as musicians, poets and sages, among whom Orpheus and Linus. . . . Because he tamed savage and unruly spirits by charming them, Orpheus gained a reputation not only for moving wild beasts, but even rocks and trees.

Charity 608

Quintilian

Allâh has not sent a Prophet except with a beautiful voice.

Muhammad

The man that hath no music in himself,
Nor is not mov'd with concord of sweet sounds,
Is fit for treasons, stratagems, and spoils;
The motions of his spirit are dull as night,
And his affections dark as Erebus:
Let no such man be trusted.

Wr. Side 448

Shakespeare (*Merchant of Venice,* v. i. 83)

Anyone who says that he finds no pleasure in sounds and melodies and music is either a liar and a hypocrite or he is not in his right senses, and is outside of the category of men and beasts. Those who prohibit music do so in order that they may keep the Divine commandment, but theologians are agreed that it is permissible to hear musical instruments if they are not used for diversion, and if the mind is not led to wickedness through hearing them.

Hujwîrî

The purpose of music, considered in relation to God, is to arouse longing for Him and passionate love towards Him and to produce states in which He reveals Himself and shows His favour, which are beyond description and are known only by experience, and, by the Ṣûfîs, these states are called 'ecstasy'. The heart's attainment of these states through hearing music is due to the mystic relationship which God has ordained between the rhythm of music and the spirit of man. The human spirit is so affected by that rhythm, that music is the cause to it of longing and joy and sorrow and 'expansion' (*inbisâṭ*) and 'contraction' (*inqibâḍ*), but he who is dull of hearing and unresponsive and hard of heart, is debarred from this joy.

Ecstasy passim

Al-Ghazâlî

Serious music preserves and restores the consonance of the parts of the Soul, as Plato and Aristotle say and as we have experienced frequently.

Marsilio Ficino

What we hear is either auspicious or inauspicious; music must not be inconsiderately executed.

Seû-mà Tshyên

Music was created for the consummation of concord in human nature, not to be the cause of voluptuousness.

Huai Nan Tzû

Music does not give rise, in the heart, to anything which is not already there: so he whose inner self is attached to anything else than God is stirred by music to sensual desire, but the one who is inwardly attached to the love of God is moved, by hearing music, to do His will. . . . The common folk listen to music according to nature, and the novices listen with desire and awe, while the listening of the saints brings them a vision of the Divine gifts and graces, and these are the gnostics to whom listening means contemplation. But finally, there is the listening of the spiritually perfect, to whom, through music, God reveals Himself unveiled.

M.M. 998

Al-Suhrawardî

The singers have hushed their notes of clear song:
The red sleeves of the dancers are motionless.
Hugging his lute, the old harper of Chao
Rocks and sways as he touches the five chords.
The loud notes swell and scatter abroad:
'Sa, sa,' like wind blowing the rain.
The soft notes dying almost to nothing:
'Ch'ieh, ch'ieh,' like the voice of ghosts talking.
Now as glad as the magpie's lucky song:
Again bitter as the gibbon's ominous cry.
His ten fingers have no fixed note:
Up and down—'kung,' chih, and yü.[1]

[1] 'Tonic, dominant and superdominant of the ancient five-note scale' (Waley).

And those who sit and listen to the tune he plays
Of soul and body lose the mastery.
P. State 583 — And those who pass that way as he plays the tune,
Suddenly stop and cannot raise their feet.

Alas, alas that the ears of common men
Should love the modern and not love the old.
Thus it is that the harp in the green window
Day by day is covered deeper with dust.

Po Chü-i (*The Harper of Chao*)

Harmony, which has motions akin to the revolutions of our souls, is not regarded by the intelligent votary of the Muses as given by them with a view to irrational pleasure, which is deemed to be the purpose of it in our day, but as meant to correct any discord which may have arisen in the courses of the soul, and to be our ally in bringing her into harmony and agreement with herself; and rhythm too was given by them for the same reason, on account of the irregular and graceless ways which prevail among mankind generally, and to help us against them.

Plato (*Timaeus*, 47 C)

Tones rise from the human heart, and music is connected with the principles of human conduct. Therefore the animals know sounds but do not know tones, and the common people know tones but do not know music. Only the superior man is able to understand music. Thus from a study of the sounds, one comes to understand the tones; from a study of the tones, one comes to understand music; and from the study of music, one comes to understand the principles of government and is thus fully prepared for being a ruler.

Confucius

Long ago the Egyptians appear to have recognized the very principle of which we are now speaking—that their young citizens must be habituated to forms and strains of virtue. These they fixed, and exhibited the patterns of them in their temples; and no painter or artist is allowed to innovate upon them, or to leave the traditional forms and invent new ones. To this day, no alteration is allowed either in these arts, or in music at all. And you will find that their works of art are painted or moulded in the same forms which they had ten thousand years ago. . . . A lawgiver may institute melodies which have a natural truth and correctness without any fear of failure. To do this, however, must be the work of God, or of a divine person; in Egypt they have a tradition that their ancient chants which have been preserved for so many ages are the composition of the Goddess Isis. . . . The love of novelty which arises out of pleasure in the new and weariness of the old, has not strength enough to corrupt the consecrated song and dance, under the plea that they have become antiquated.

Supra 676
Orthod. 275
Rev. 959

Plato (*Laws* II, 656 D)

The ancient kings were ever careful about things that affected the human heart. They tried therefore to guide the people's ideals and aspirations by means of *li*,[1] establish

[1] *Li* can mean ritual, propriety, tradition, depending on the context.

harmony in sounds by means of music, regulate conduct by means of government, and prevent immorality by means of punishments. *Li,* music, punishments and government have a common goal, which is to bring about unity in the people's hearts and carry out the principles of political order.

Music rises from the human heart. When the emotions are touched, they are expressed in sounds, and when the sounds take definite forms, we have music. Therefore the music of a peaceful and prosperous country is quiet and joyous, and the government is orderly; the music of a country in turmoil shows dissatisfaction and anger, and the government is chaotic: and the music of a destroyed country shows sorrow and remembrance of the past, and the people are distressed. Thus we see music and government are directly connected with one another.

Confucius

If the Kông (C=tonic) is disturbed, then there is disorganization: the Prince is arrogant.

If the Shâng (D) is disturbed, then there is deviation: the officials are corrupted.

If the Kyŏ (E) is disturbed, then there is anxiety: the people are unhappy.

If the Chi (G) is disturbed, then there is complaint: public services are too heavy.

If the Yù (A+) is disturbed, then there is danger: resources are lacking.

If the five degrees are all disturbed, then there is danger: ranks encroach upon each other—this is what is called impudence—and, if such is the condition, the destruction of the Kingdom may come in less than a day. . . .

Wr. Side 464

In periods of disorder, rites are altered and music is licentious. Then sad sounds are lacking in dignity, joyful sounds lack in calm. . . . When the spirit of opposition manifests itself, indecent music comes into being. . . . When the spirit of conformity manifests itself, harmonious music appears. . . . So that, under the effect of music, the five social duties are without admixture, the eyes and ears are clear, the blood and the vital spirits are balanced, habits are reformed, customs are improved, the Empire is in complete peace.

Confucius

The Pythagoreans said, that an harmonic sound is produced from the motion of the celestial bodies; and they scientifically collected this from the analogy of their intervals; since not only the ratios of the intervals of the sun and moon, and Venus and Mercury, but also of the other stars, were discovered by them.

Introd. *Orthod.*

Simplicius

The whole Pythagoric school produced by certain appropriate songs, what they called *exartysis* or adaptation, *synarmoge* or elegance of manners, and *epaphe* or contact, usefully conducting the dispositions of the soul to passions contrary to those which it before possessed. For when they went to bed they purified the reasoning power from the perturbations and noises to which it had been exposed during the day, by certain odes and peculiar songs, and by this means procured for themselves tranquil sleep, and few and good dreams. But when they rose from bed, they again liberated themselves from the torpor and heaviness of sleep, by songs of another kind. Sometimes, also, by musical sounds alone, unaccompanied with words, they healed the passions of the soul and certain

Metanoia 484

diseases, enchanting, as they say, in reality. And it is probable that from hence this name *epode,* i.e. enchantment, came to be generally used. After this manner, therefore, Pythagoras through music produced the most beneficial correction of human manners and lives.

Iamblichus

Conform. 180 *Orthod.* 285

Music expresses the harmony of the universe, while rituals express the order of the universe. Through harmony all things are influenced, and through order all things have a proper place. Music rises from heaven, while rituals are patterned on the earth. To go beyond these patterns would result in violence and disorder. In order to have the proper rituals and music, we must understand the principles of Heaven and Earth.

Confucius

P. State 583

Since the celestial spheres revolve and the planets and stars are moved, it follows that they must have musical notes and expressions with which God is glorified, delighting the souls of the angels, just as in the corporeal world our souls listen with delight to melodies and obtain relief from care and sorrow. And inasmuch as these melodies are but echoes of heavenly music, they recall to us the spacious gardens of Paradise and the pleasures enjoyed by the souls dwelling there; and then our souls long to fly up thither and rejoin their mates.

The *Rasâ'il* of the Ikhwânu 'l-Ṣafâ

Our songs are the same as His songs.

Chândogya Upanishad, I. vii. 5

The Powers which are in all things sing within me also.

Hermes

Reality 775 Infra 689

To know the science of music is nothing else than this,—to know how all things are ordered, and how God's design has assigned to each its place; for the ordered system in which each and all by the supreme Artist's skill are wrought together into a single whole yields a divinely musical harmony, sweet and true beyond all melodious sounds.

Hermes

Allâh hath created seven heavens in harmony.

Qur'ân, LXXI. 15

Creation 48 *Peace* 694

Plato saith,[1] That three sorts of Persons are led to God, The *Musician* by *Harmony,* the *Philosopher* by the *beam* of *Truth,* the *Lover* by the *light of Beauty.* All these *Conductors* to the *supream Being* meet in this *Love,* of which we speak; the *first* and *only true Beauty,* being the *first Birth,* the first *Effulgency,* the *essential Image* of the *supream Goodness,* is also the *first,* the *supream,* the *only Truth*; the *Original,* the *measure,* the *end* of *all Truth*; which by its amiable attractive Light, conducteth all Understandings in the search of Truth, and giveth them rest only in its transparent and blissful Bosom. This also is the *first,*

[1] *Phaedrus,* 248 D: cf. also Plotinus: *Enneads,* I. iii. 1.

the *only*, the *universal Harmony*, the *Musick* of all things in Heaven and on Earth: the Musick, in which all things of Earth and of Heaven, meet to make *one melodious Consort*.

Peter Sterry

In relating the things of the earth to the celestial, and those of heaven to the inferior, the Chaldeans have shown in the mutual affections between these parts of the universe (which are separated in space but not in essence) the harmony that unites them in a sort of musical accord.

Symb. 306

Philo

'Tis said, the pipe and lute that charm our ears
Derive their melody from rolling spheres. . . .
We, who are parts of Adam, heard with him
The song of angels and of seraphim.
Our memory, though dull and sad, retains
Some echo still of those unearthly strains.
Oh, music is the meat of all who love,
Music uplifts the soul to realms above.
The ashes glow, the latent fires increase:
We listen and are fed with joy and peace.

Sin 66

Rûmî

Music draws to itself the human spirits (senses), which are principally vapours of the heart, as it were, so that they almost cease to act; so entirely is the soul one thing when it listens, and the power of all (the rest of the senses) seems to fly to that sensible spirit which receives sound.

Contem. 532

Dante (*Il Convito*, II. xiv. 11)

By continuous practice of the Saman chants, in the prescribed manner and with concentration of mind, a man attains the Supreme Brahman. The songs entitled *Aparanta*, *Ullopya*, *Madraka*, *Prakari*, *Auvenaka*, *Sarobindu*, *Uttara*, the songs called *Rik*, *Gatha*, *Panika*, the music compositions associated with Daksha and Brahman,—the practice of these is indeed liberation. He who knows the inner meaning of the sound of the lute, who is expert in intervals and in modal scales and knows the rhythms, travels without effort upon the way of liberation.[1]

Realiz. 870

Yâjñavalkyà Smriti, III. iv. 112–115

And it came to pass, when the minstrel played, that the hand of the Lord came upon him.

II. Kings, III. 15

I will strike every chord in seeking spiritual transmutation, like the lute-player whose plectrum moves up and down the strings.

[1] 'This section of Yajnavalkya is important for the later philosophy of music as an aid to devotion and spiritual realisation' (Raghavan).

That, from playing the *saḥûr* tune in this fashion, the seas of Divine mercy may surge to scatter their pearls and lavish their bounty.

Rûmî

Allâh listens more intently to a man with a beautiful voice reading the *Qur'ân* than does a master of a singing-girl to her singing.

Muhammad

The music which is called Gândharva (Mârga) is that which has been from time immemorial, practised by the Gandharvas (celestial singers) and which leads surely to Mokśa (liberation), while the Gâṇa (Deśi) music is that which has been invented by composers.

Râmâmâtya

Musical innovation is full of danger to the whole State, and ought to be prohibited. So Damon tells me, and I can quite believe him:—he says that when modes of music change, the fundamental laws of the State always change with them.

Plato (*Republic* IV. 424 C)

The music of Cheng is lewd and corrupting, the music of Sung is soft and makes one effeminate, the music of Wei is repetitious and annoying, and the music of Ch'i is harsh and makes one haughty. These four kinds of music are all sensual music and undermine the people's character, and that is why they cannot be used at the sacrifices. The *Book of Songs* says, 'The harmonious sounds are *shu* and *yung* and my ancestor listened to them.' *Shu* means 'pious' and *yung* means 'peaceful'. If you have piety and peacefulness of character, you can do everything you want with a country.

Confucius

Illusion 85

Musical training is a more potent instrument than any other, because rhythm and harmony find their way into the inward places of the soul, on which they mightily fasten, imparting grace, and making the soul of him who is rightly educated graceful, or of him who is ill-educated ungraceful.

Plato (*Republic* III. 401 D)

Supra 676

Beauty addresses itself chiefly to sight; but there is a beauty for the hearing too, as in certain combinations of words and in all kinds of music, for melodies and cadences are beautiful; and minds that lift themselves above the realm of sense to a higher order are aware of beauty in the conduct of life, in actions, in character, in the pursuits of the intellect; and there is the beauty of the virtues.

Plotinus

Inv. 1003

Sound produced from ether is known as 'unstruck'. In this unstruck sound the Gods delight. The Yogis, the Great Spirits, projecting their minds by an effort of the mind into this unstruck sound, depart, attaining Liberation.

Struck sound is said to give pleasure, 'unstruck' sound gives Liberation.
This (unstruck sound) having no relation with human enjoyment does not interest ordinary men.

Sangîtă Makarandă, I. 4–6; *Nârandă Purâṇă*;
Śhivă tattvă Ratnâkară, 6, 7, 12

Consequently the most perfect, faultless harmony cannot be perceived by the ear, for it exists not in things sensible but only as an ideal conceived by the mind. . . . No man can hear it while still in the body, for it is wholly spiritual and would draw to itself the essence of the soul, as infinite light would attract all light to itself. Such infinitely perfect harmony, in consequence, would be heard only in ecstasy by the ear of the intellect, once the soul was free from the things of sense.

Nicholas of Cusa

Death 208
Ecstasy 636
Sin 77

In every strain which (the tavern-haunters) hear from the minstrel
Comes to them rapture from the unseen world.

Shabistarî

Ecstasy 637

(Pythagoras) fixed his intellect in the sublime symphonies of the world, he alone hearing and understanding, as it appears, the universal harmony and consonance of the spheres, and the stars that are moved through them, and which produce a fuller and more intense melody than any thing effected by mortal sounds. This melody also was the result of dissimilar and variously differing sounds, celerities, magnitudes, and intervals, arranged with reference to each other in a certain most musical ratio, and thus producing a most gentle, and at the same time variously beautiful motion and convolution.

Iamblichus

Introd. *Orthod.*

If any one . . . should have his terrestrial body exempt from him, and his luminous and celestial vehicle and the senses which it contains purified, either through a good allotment, or through probity of life, or through a perfection arising from sacred operations, such a one will perceive things invisible to others, and will hear things inaudible by others. With respect to divine and immaterial bodies, however, if any sound is produced by them, it is neither percussive nor destructive, but it excites the powers and energies of sublunary sounds, and perfects the sense which is co-ordinate with them.

Simplicius

The soul continues as an instrument of God's harmony, a tuned instrument of divine joy for the Spirit to strike on.

Boehme

Since the drum is often the only instrument used in our sacred rites, I should perhaps tell you here why it is especially sacred and important to us. It is because the round form of the drum represents the whole universe, and its steady strong beat is the pulse, the heart,

Symb. 322

throbbing at the center of the universe. It is as the voice of *Wakan-Tanka*, and this sound stirs us and helps us to understand the mystery and power of all things.

Black Elk

God hath men who enter Paradise through their flutes and drums.

Muhammad

Dance

When you see the type of a nation's dance, you know its character.

Confucius

Symb. 306

All the dancer's gestures are signs of things, and the dance is called rational, because it aptly signifies and displays something over and above the pleasure of the senses.

St Augustine

Action 329
Sp. Drown. 713

The Supreme Intelligence dances in the soul . . . for the purpose of removing our sins. By these means, our Father scatters the darkness of illusion (*maya*), burns the thread of causality (*karma*), stamps down evil (*mala, anava, avidya*), showers Grace, and lovingly plunges the soul in the ocean of Bliss (*ananda*). They never see rebirths, who behold this mystic dance.

Unmai Vilakkam, v. 32, 37, 39

The movement of whose body is the world, whose speech the sum of all language,
Whose jewels are the moon and stars—to that pure Śiva I bow!

Abhinaya Darpaṇa

Symb. 317

Why should not every Sûfî begin to dance, like a mote,
In the sun of eternity, that it may deliver him from decay?

Dîvâni Shamsi Tabrîz, XXIX

Love 614

God can be served in different ways. An ecstatic lover of God enjoys Him in different ways. Sometimes he says, 'O God, You are the lotus and I am the bee', and sometimes, 'You are the Ocean of Satchidânanda and I am the fish.' Sometimes, again, the lover of God says, 'I am Your dancing-girl.' He dances and sings before Him. He thinks of himself sometimes as the friend of God and sometimes as His handmaid. He looks on God sometimes as a child, as did Yaśodâ, and sometimes as husband or sweetheart, as did the gopis.

Sri Ramakrishna

He who does not dance in remembrance of the Friend has no friend.

Muhammad

Maiden! thou shalt dance merrily
Even as mine elect!

THE SOUL

I cannot dance O Lord, unless Thou lead me. *Grace* 558
If Thou wilt that I leap joyfully
Then must Thou Thyself first dance and sing!
Then will I leap for love
From love to knowledge,
From knowledge to fruition, *Knowl.* 761
From fruition to beyond all human sense.
There will I remain *Realiz.* 887
And circle evermore.

Mechthild of Magdeburg

The dancing foot, the sound of the tinkling bells,
The songs that are sung, and the various steps,
The forms assumed by our Master as He dances,
Discover these in your own heart, so shall your bonds be broken. *Contem.* 536

Tirumûlar (on Naṭarâja, representing Śiva's cosmic dance)

The Universal Harmony

Rightly is the Kosmos so named;[1] for all things in it are wrought into an ordered whole by the . . . immutable necessity that rules in it, and by the combining of the elements, and the fit disposal of all things that come into being. *Center* 835 *Reality* 773

Hermes

No teaching can be perfect without harmony. Indeed, there is nothing in which it is not found. The world itself is said to be harmoniously formed, and the very heavens revolve amidst the harmony of the spheres.

St Isidore

This science (of Astrology)[2] more than any (other) is high and noble on account of its

[1] Κόσμος = order, ornament, ordered universe (as Scott indicates).
[2] In mediaeval times astrology and astronomy formed a single cosmological science.

Symb. 306

high and noble subject (which is the movement of heaven), and high and noble by its certainty, which is without any defect, as coming from a most perfect and most regular principle. And if any conceive it to have a defect, it does not belong to it, but, as Ptolemy says, comes of our negligence and to that should be imputed.

Dante (*Il Convito,* II. xiv. 13)

Reality 775

Equality is by nature prior to inequality; . . . it is also naturally prior to diversity. Equality, it must be concluded, is eternal.

Nicholas of Cusa

Introd. *Holy War*

Opposition unites. From what draws apart results the most beautiful harmony. All things take place by strife.

Heraclitus

Fire keeps the earth from being submerged, or dissolved; air keeps the fire from being extinguished; water preserves the earth from combustion. This is what the Sages call the equilibrium of the elements, and it illustrates the aid which they render to each other.

Michael Sendivogius

It is not for the sun to overtake the moon, nor doth the night outstrip the day. They float each in an orbit.

Qur'ân, XXXVI. 40

Suff. 133

Every thing as it lieth in the whole piece, beareth its part in the Universal Consort. The Divine Musick of the whole would be changed into Confusion and Discords, All the sweet proportions of all the parts would be disordered, and become disagreeable, if any one, the least, and least considered part, were taken out of the whole. Every part is tyed to the whole, and to all the other parts, by mutual and essential *Relations.* By virtue of these Relations, All the distinct proportions, of all the parts, and of the whole, *meet* in one, on each part, filling it with, and wrapping it up in the rich Garment of the Universal Harmony, curiously wrought, with all the distinct and particular Harmonies.

Peter Sterry

Rev. 967

O Dhananjaya, there is naught else (existing) higher than I. Like pearls on a thread, all this (universe) is strung in Me.

Bhagavad-Gîtâ, VII. 7

Wind, verily, O Gautama, is that thread. By wind, verily, O Gautama, as by a thread, this world and the other world and all things are tied together.[1]

Bṛihad-Âraṇyaka Upanishad, III. vii. 2

Creation 26

See a golden *Chain,* see the *Order* of the precious *Links,* see how in a beautiful *circle* the beginning is fastned to the *end.*

Peter Sterry

[1] This luminous pneumatic thread is the same as the 'Gale of the Spirit': cf. *The Emerald Table of Hermes:* 'The Wind carried it in its womb.'

Everything is linked with everything else down to the lowest ring on the chain, and the true essence of God is above as well as below, in the heavens and on the earth, and nothing exists outside Him. And this is what the sages mean when they say: When God gave the Torah to Israel, He opened the seven heavens to them, and they saw that nothing was there in reality but His Glory; He opened the seven worlds to them and they saw that nothing was there but His Glory; He opened the seven abysses before their eyes, and they saw that nothing was there but His Glory. Meditate on these things and you will understand that God's essence is linked and connected with all worlds, and that all forms of existence are linked and connected with each other, but derived from His existence and essence.

Moses de Leon

Reality 803

Let therefore even the whole universe, that greatest and most perfect flock of the God who IS, say, 'The Lord shepherds me, and nothing shall fail me' (Ps. xxiii. 1). . . . For it cannot be that there should be any lack of a fitting portion, when God rules, whose wont it is to bestow good in fullness and perfection on all that is.

Philo

Love 618

PEACE

Where could man, scorched by fires of the sun of this world, look for felicity, were it not for the shade afforded by the tree of emancipation?

Vishṇu Purâṇa

Introd. *Center* *Holiness* 921

The equilibrium and harmony of true beauty are in direct rapport with the quality of Peace, which corresponds with the stabilization of the Beatific Vision in the Divine Center (*samâdhi*=com-posure), where the 'Real Presence' (Hebrew *Shekhînah*) of the Divinity shines in its own splendour as the 'Light of Glory'; and from which point[1] the 'Non-acting Activity' (Chinese *wei wu-wei*) of the Spirit as 'Universal Monarch' (Hindu *Chakravartî*) activates all the operations of the universe.[2]

Introd. *Holy War* *Conform.* 180 *Orthod.* 300

In Islamic *jihâd* (holy war) it is the (spiritual) sword (*sikkîn*)[3] that severs the Gordian knot of the ego and accomplishes the quietus (*sakînah*) or 'Great Peace', equivalent to the *Pax Profunda* of European Hermeticism. The word *Islâm* ('submission to the Divine Will') itself derives from a root having to do with peace, and finds its Hebrew counterpart in words like *Shlomoh* (Solomon) 'Peaceable': 'For his name shall be Solomon, and I will give peace and quietness unto Israel in his days' (*I. Chronicles,* XXII. 9), and *Salem* 'Peace'—whence Jerusalem, 'Image of Peace'.[4]

Pilg. 385 *Renun.* 136 *Orthod.* 296 *Holiness* 902

In South Indian Śaiva iconography, Śiva and Parvatî are depicted seated on the summit of Mount Kailâsa, watching the families of Brahmanical ascetics (Rishis) arduously performing their sacrificial rites in the Deodar Forest. Parvatî asks Śiva why these earnest seekers are unable to obtain release. 'He replies that it is because they are not yet at peace, but still affectible by love and wrath; they cannot cross over the sea of life to reach the farther shore so long as they can love and hate; whereas those who have freed themselves from passion and desire, even if they do not practice arduous rites, can obtain to that imperishable state of real being' (Coomaraswamy: 'An Ivory Casket from Southern India'; *The Art Bulletin,* Sept., 1941, p. 208).

Treating of the quality of serenity or peace as a spiritual dimension, a 'station of wisdom', Schuon writes: 'This virtue consists in calm, in contentment, in patience; it is the calm of that which reposes in itself, in its own quality; it is a generous relaxation,

[1] The 'Unchanging Center' (Chinese *chung-yung*—also translated 'Golden' or 'Steadfast Mean') or axis of the 'cosmic wheel', this axis being equally the trunk of the World Tree (of emancipation).

[2] The same as Aristotle's 'Motionless Mover': cf. Introduction to *Creation.*

[3] Lit. 'knife'.

[4] Cf. Guénon: *Le Roi du Monde,* ch. VI: *L'Homme et son Devenir,* ch. XXIII: *Le Symbolisme de la Croix,* ch. VII & VIII.

equilibrium, harmony; it is repose in God. This attitude undoes the knots of the soul, it removes agitation, dissipation, curiosity, restlessness, and crispation, which is the static complement of agitation. The virtue of calm derives from Divine Peace, which consists in Beatitude, in Infinite Beauty; beauty everywhere and always has at its root an aspect of calm, of existential repose, an equilibrium of possibilities; that is to say, an aspect of illimitability and happiness. The essence of the soul is beatitude; it is dissipation which renders us strangers to ourselves and casts us into poverty and ugliness, into a state of sterile dilapidation similar to a palsy, to a disordered movement which has become a state, when normally it is the static which is found at the base of the dynamic, and not inversely. Beauty contains in itself every element of happiness, from whence comes its character of peace, of plenitude, of fulfilment; now, beauty is in our very being; we live on the substance of it. It is the calm and simple, yet generous and unlimited perfection of the pond in which is reflected the serenity of the sky; it is the beauty of the water lily, of the lotus which opens to the light. It is repose in the center, blessed resignation to the Divine Will' (*France-Asie,* 1953, p. 508).

Beauty 689
Creation 42
Introd.
Waters

The Tree of Emancipation

Contem. 547

Thou madest us for Thyself, and our heart is restless, until it repose in Thee.

St Augustine

The sagely man rests in what is his proper rest; he does not rest in what is not so;—the multitude of men rest in what is not their proper rest; they do not rest in their proper rest.

Illusion 94

Chuang-tse (ch. XXXII)

You are in the world of things that come to be, and yet you seek to be at rest. But how can anything be at rest in the world of things that come to be? A boat, as long as it floats on the water, cannot be still or at rest; or if at any moment it is still, it is so only by chance, and forthwith the water begins again to shake and toss the things which float upon its surface. Then only is the boat at rest, when it is taken out of the water, and drawn up on the land, which is the place of the boat's origin, and is on a par with the boat in density and weight; then, but not till then, is the boat truly at rest. And even so, the soul, as long as it is involved in the processes of the physical world, cannot be still, nor be at rest, nor get any respite; but if it returns to its source and root, then it is still and is at rest, and reposes from the misery and debasement of its wandering in a foreign land.

Introd. *Pilg.*

Sin 66

Hermes

In returning and rest shall ye be saved.

Isaiah, XXX. 15

The end of all motion is rest.

Eckhart

He sent forth a dove from him, to see if the waters were abated from off the face of the ground;

But the dove found no rest for the sole of her foot, and she returned unto him into the ark.

Genesis, VIII. 8, 9

Reality 803

One must have confidence in the Thatness (as being the Sole Refuge) even as an exhausted crow far from land hath confidence in the mast of the ship upon which it resteth.[1]

Gampopa

[1] Once again we encounter the symbolism of the vertical axis, or World Tree, attached to the 'Ship of Life', whose deck corresponds to the 'Surface of the Waters', q.v. Cf. Coomaraswamy: 'The Symbolism of the Dome', *Indian Historical Quarterly*, Calcutta, March, 1938.

Once a bird sat on the mast of a ship. When the ship sailed through the mouth of the Ganges into the 'black waters' of the ocean, the bird failed to notice the fact. When it finally became aware of the ocean, it left the mast and flew north in search of land. But it found no limit to the water and so returned. After resting awhile it flew south. There too it found no limit to the water. Panting for breath the bird returned to the mast. Again, after resting awhile, it flew east and then west. Finding no limit to the water in any direction, at last it settled down on the mast of the ship.

Illusion 109

Sri Ramakrishna

This realization is likened to that of a crow which, although already in possession of a pond, flies off elsewhere to quench its thirst, and finding no other drinking-place returns to the one pond.

Realiz. 873
Creation 42

Padma-Sambhava

A man wandering in the sun retires to the shade of a tree and enjoys the cool atmosphere there. But after a time he is tempted to go into the hot sun. Again finding the heat unbearable, he returns to the shade. Incessantly he thus moves to and fro, from the shade into the sun and from the sun into the shade. Such a man, we say, is ignorant. A wise man would not quit the shade.

Illusion 89

Sri Ramana Maharshi

A long time ago
I went on a journey,
Right to the corner
Of the Eastern Ocean.
The road there
Was long and winding,
And stormy waves
Barred my path.
What made me
Go this way?
Hunger drove me
Into the World.
I tried hard
To fill my belly:
And even a little
Seemed a lot.
But this was clearly
A bad bargain,
So I went home
And lived in idleness.

Action 358

T'ao Ch'ien

Sleep the world away, and flee from the six dimensions:
How long wilt thou roam in thy folly and bewilderment to and fro?

Dîvâni Shamsi Tabrîz, XXXIX

I cast my hook in a single stream;
But my joy is as though I possessed a Kingdom.

Conform. 180

Chi K'ang

'Whoso sees Me, abides for Me; and whoso abides for Me, abides in the experience of Me. Whoso sees Me not, has no abode wherein he may abide.'

P. State 563

Niffarî

M.M. 978

Urged by desire I wandered in the streets
Of good and evil,
I gained nothing except feeding the fire of desire.
As long as in me remains the breath of life
Help me, for Thou alone canst hear my prayer.

Ansârî

Contem. 536

Going out were never so good, but staying at home were much better.

Theologica Germanica, IX

Reality 773

Rest is complete loss of motion. Were any creature perfectly immoveable it would be God. God is God in that he is immoveable.

Eckhart

For things find rest only in that which is the end of their being.

Philalethes

Renun. 152b

We try to acquire so many things of the world, but find no peace in them. The source of all happiness is within us. We have only to find out, know and realise it.

Swami Ramdas

Only the One is at rest in itself, receiving nothing from without.

Eckhart

Our rational nature is so great a good, that there is no good wherein we can be happy, save God.

St Augustine

Holiness 924

It (the center of the soul) is so infinite that nothing can satisfy it or give it any rest but the infinity of God.

William Law

Praised be God who has given us a mind that cannot be satisfied with the temporal.

Nicholas of Cusa

Reality 775

Sp. Drown. 713

The upward flight of the soul is always towards its perfect identity with the Great One who is the same through and in all. The river of life struggles through all obstacles and conditions to reach the vast and infinite ocean of existence.—God. It knows no rest, no freedom and no peace until it mingles with the waters of immortality and delights in the vision of infinity.

Swami Ramdas

It is not until the thoughts can find rest in nothing but God, that the man is drawn close to God Himself, and becomes His.

Tauler

Rest is unity, in which all movement is contained, for on close examination movement is seen to be rest drawn out in an orderly series.... Unity is convertible with eternity, for there cannot be more than one Eternal. God, therefore, envelops all in the sense that all is found in Him.

Nicholas of Cusa

Reality 803

Rest is more perfect than movement, and for the sake of rest the individual things are moved.

Marsilio Ficino

Religion though it hath its infancy, yet it hath no old age: while it is in its minority, it is always *in motu*; but when it comes to its maturity and full age, it will always be *in quiete*, it is then 'always the same, and its years fail not, but it shall endure for ever.'

John Smith the Platonist

All creatures are with God: the being that they have God gives them with his presence. Saith the bride in the Book of Love, 'I have run round the circle and have found no end to it, so I cast myself into the centre.'

This circle which the loving soul ran round is all the Trinity has ever wrought.... Spent with her quest she casts herself into the centre. This point is the power of the Trinity wherein unmoved it is doing all its work. Therein the soul becomes omnipotent.

Eckhart

Center 831
Action 358

From the first day, and even now,
The creature seeks nothing but the quietude of his Creator.

Angelus Silesius

There is no peace, nor ever can be, for the soul of man but in the purity and perfection of its first-created nature.

William Law

Creation 42

'I have appointed in everything a haven for the hearts that are veiled from Me: but when I appear to any heart, I become the place of its repose in everything.'

Niffarî

Illusion 85
Holiness 914

Now the city of God is called in the Hebrew Jerusalem and its name when translated is 'vision of peace'. Therefore, do not seek for the city of the Existent among the regions of the earth, since it is not wrought of wood or stone, but in a soul, in which there is no warring, whose sight is keen, which has set before it as its aim to live in contemplation and peace.

Philo

Center 832
835, 816

Every good man, in whom religion rules, is at peace and unity with himself, is as a city compacted together. Grace doth more and more reduce all the faculties of the soul into a perfect subjection and subordination to itself. The union and conjunction of the soul with

Metanoia 480
Rev. 967

God, that primitive unity, is that which is the alone original and fountain of all peace, and the centre of rest.

John Smith the Platonist

All day long the 'lamenter' sends his voice to *Wakan-Tanka* for aid, and he walks as we have described upon the sacred paths which form a cross. This form has much power in it, for whenever we return to the center, we know that it is as if we are returning to *Wakan-Tanka,* who is the center of everything; and although we may think that we are going away from Him, sooner or later we and all things must return to Him.

Judgment 268

Black Elk

Truth is the only resting-place of the soul; it is its atonement and peace with God; all is and must be disquiet, a succession of lying vanities, till the soul is again in the truth in which God at first created it. And therefore, said the Truth, 'Learn of me, for I am meek and lowly of heart; and ye shall find rest unto your souls.'

William Law

M.M. 978 Infra 705

Take no heed of time, nor of right and wrong. But passing into the realm of the Infinite, take your final rest therein.

Chuang-tse (ch. II)

Wr. Side 474 *Center* 835

Suddenly is the soul oned to God when it is truly peaced in itself: for in Him is found no wrath. And thus I saw when we are all in peace and in love, we find no contrariness, nor no manner of letting through that contrariness which is now in us.

Julian of Norwich

Lo! the God-fearing are among gardens and watersprings.
'Enter them in peace, secure.'
And We remove whatever burning venom may be in their breasts. As brethren, face to face, they rest on couches raised.
Toil cometh not unto them there, nor will they be expelled from thence.
Announce unto My slaves that verily I am the Forgiving, the Merciful.

Holiness 902

Qur'ân, xv. 45–49

The Ordering of Hearts

Cleanse your own heart, cast out from your mind, not Procrustes and Sciron,[1] but pain, fear, desire, envy, ill will, avarice, cowardice, passion uncontrolled. These things you

[1] Two robbers who were driven out of Attica by Theseus.

cannot cast out, unless you look to God alone, on Him alone set your thoughts, and consecrate yourself to His commands. If you wish for anything else, with groaning and sorrow you will follow what is stronger than you, ever seeking peace outside you, and never able to be at peace: for you seek it where it is not, and refuse to seek it where it is.

Epictetus

Supra 694

Through these rites[1] a three-fold peace was established. The first peace, which is the most important, is that which comes within the souls of men when they realize their relationship, their oneness, with the universe and all its Powers, and when they realize that at the center of the universe dwells *Wakan-Tanka,* and that this center is really everywhere, it is within each of us. This is the real Peace, and the others are but reflections of this. The second peace is that which is made between two individuals, and the third is that which is made between two nations. But above all you should understand that there can never be peace between nations until there is first known that true peace which, as I have often said, is within the souls of men.

Black Elk

Reality 775
Center 841

The ancients who wished to illustrate illustrious virtue throughout the kingdom first ordered well their own states. Wishing to order well their states, they first regulated their families. Wishing to regulate their families, they first cultivated their persons. Wishing to cultivate their persons, they first rectified their hearts. Wishing to rectify their hearts, they first sought to be sincere in their thoughts. Wishing to be sincere in their thoughts, they first extended to the utmost their knowledge. Such extension of knowledge lay in the investigation of things.

Things being investigated, knowledge became complete. Their knowledge being complete, their thoughts were sincere. Their thoughts being sincere, their hearts were then rectified. Their hearts being rectified, their persons were cultivated. Their persons being cultivated, their families were regulated. Their families being regulated, their states were rightly governed. Their states being rightly governed, the whole kingdom was made tranquil and happy.

The Great Learning, I

Realiz. 859
Knowl. 761
Rev. 967

Shall I not inform you of a better act than fasting, alms, and prayers? Making peace between one another: enmity and malice tear up heavenly rewards by the roots.

Muhammad

Blessed are the peacemakers: for they shall be called the children of God.

St Matthew, v. 9

Every one thirsts for peace, but few people understand that perfect peace cannot be obtained so long as the inner soul is not filled with the presence of God, . . . this God who is always present in all the fibres of your body and who none the less prefers to remain hidden. When through appropriate and assiduous exercises the blackness of your soul is effaced, God reveals Himself, and it is absolute peace.

Ananda Moyî

Center 841
Introd.
Suff.

[1] *Hunkapi*: 'the making of relatives': i.e.. peace with other tribes.

Introd.
Holy War

Devotee: We are pacifists. We want to bring about Peace.

Maharshi: Peace is always present. Get rid of the disturbances to Peace. This Peace is the Self.

Sri Ramana Maharshi

Serenity

Holiness
914

The quiescence of a holy man is the sign of his being a Sage.

Subhâshita Ratna Nidhi

No one has ever attained to the grandeur or glory
Of the soul which has established repose in its heart.

Angelus Silesius

There is nothing so lovely and enduring in the regions which surround us, above and below, as the lasting peace of a mind centred in God.

Yoga-Vasishtha

There is nowhere perfect rest save in a heart detached.

Eckhart

Renun. 139
Sin
77
Contem.
532
Conform. 180
Creation
42

Remember that it is not only desire of office and of wealth that makes men abject and subservient to others, but also desire of peace and leisure and travel and learning. Regard for any external thing, whatever it be, makes you subservient to another.... What, pray, is this peace of mind, which any one can hinder—I do not mean Caesar, or Caesar's friend, but a raven, a flute-player, a fever, countless other things? Nothing is so characteristic of peace of mind as that it is continuous and unhindered.... There is but one way to peace of mind (keep this thought by you at dawn and in the day-time and at night)—to give up what is beyond your control, to count nothing your own, to surrender everything to heaven and fortune, to leave everything to be managed by those to whom Zeus has given control, and to devote yourself to one object only, that which is your own beyond all hindrance.

Epictetus

No matter what path you follow, yoga is impossible unless the mind becomes quiet. The mind of a yogi is under his control; he is not under the control of his mind.

Sri Ramakrishna

Conform.
170

All is equal to the sage: he is established in quietude and tranquillity;
If a thing does not accord with him, it still accords with the will of God.

Angelus Silesius

A man said to the prophet, 'Give me a command.' He said, 'Do not get angry.' The man repeated the question several times, and he said, 'Do not get angry.'

Muhammad

Renun. 136

All God wants of man is a peaceful heart; then he performs within the soul an act too Godlike for creature to attain to or yet see. The divine Wisdom is discreetly fond and lets no creature watch.

Eckhart

Introd. *Realiz.* *Death* 208 *Void* 728

Passion is overcome only by him who has won through stillness of spirit the perfect vision. Knowing this, I must first seek for stillness; it comes through the contentment that is regardless of the world. What creature of a day should cling to other frail beings, when he can never again through thousands of births behold his beloved?

Śânti-deva

Illusion 99

Verily in the remembrance of Allâh do hearts find rest.

Qur'ân, XIII. 28

Inv. passim

Wise people, after they have listened to the laws, become serene, like a deep, smooth, and still lake.

Dhammapada, VI. 82

The Godhead gave all things up to God; it is as poor, as naked and as idle as though it were not: it has not, wills not, wants not, works not, gets not.

Eckhart

M.M. 998

In the Godhead may be no travail.

Julian of Norwich

This state of mere inherence in pure Being (Quiescence or *Mouna*) is known as the Vision of Wisdom. Such inherence means and implies the entire subsidence of the mind in the Self. Anything other than this and all psychic powers of the mind, such as thought-reading, telepathy and clairvoyance, cannot be Wisdom.

Sri Ramana Maharshi

Death 220

All Sadhanas are done with a view to still the mind. The perfectly still mind is the Universal Spirit.

Swami Ramdas

In peace is My dwelling place.

Psalm LXXVI. 2 (as rendered by Henry Suso)

The most powerful prayer, one wellnigh omnipotent, and the worthiest work of all is the outcome of a quiet mind.

Eckhart

Action 358

Renun. 143

To be a Ṣûfî is to cease from taking trouble; and there is no greater trouble for thee than thine own self, for when thou art occupied with thyself, thou remainest away from God.

Abû Sa'îd ibn Abi 'l-Khayr

'Never, above all, let anxiety gain the upper hand with thee, for if thou becomest agitated, the demon will be content and will come off victorious.'

Sister Consolata

Speaking tongues are the destruction of silent hearts.

Al-Ḥallâj

M.M. 987

He who has the good, enjoys it in silence.

The Sophic Hydrolith

Action 346

A philosopher apostrophised the soul: 'Withdraw from the restlessness of external activities!' And again: 'Flee away and hide thee from the turmoil of outward occupations and inward thoughts for they create nothing but discord!' If God is to speak his Word in the soul she must be at rest and at peace.

Eckhart

Metanoia 488

All this talk and turmoil and noise and movement and desire is outside of the veil; within the veil is silence and calm and rest.

Bâyazîd al-Bisṭâmî

The greatest number of failings in a community—experience has taught me this—come from breaking the rule of silence.

Sister Consolata (advising her community)

Fear your tongue: it is an arrow that misses the mark.
The tongue is a savage beast: leave it free, and it will wound you.

'Alî

Not that which goeth into the mouth defileth a man; but that which cometh out of the mouth, this defileth a man.

St Matthew, xv. 11

Action 329

Thy words and acts are in thy power before the word is spoken; once said, the word hath power over thee, all control from thee hath passed.

Shekel Hakodesh, 228

Whoever says what he should not say, hears what he does not want to hear.

'Alî

Renun. 152b

Abandoning without reserve all the desires born of mental fancies, and restraining completely by the mind the entire group of the senses from all directions,

With understanding held by firmness, and mind established in the Self, let him (the Yogi) (thus) by degrees attain tranquility; let him not think of anything else.

Wheresoever the restless and unsteady mind may wander away, let him withdraw it from there and bring it under the control of the Self alone.

He whose passions are quieted and mind perfectly tranquil, who has become one with Brahman, being freed from all impurities, to such a Yogi comes supreme bliss.

Bhagavad-Gitâ, VI. 24–27

Contem. 528
Realiz. 873

Children, that peace which is found in the spirit and the inner life is well worth our care, for in that peace lies the satisfaction of all our wants. In it the Kingdom of God is discovered and His righteousness is found. This peace a man should allow nothing to take from him, whatever betide, come weal or woe, honour or shame.

Tauler

Contem. 542
Center 847
Renun. 136

Satisfaction is quietness of heart under the course of destiny.

Al-Muḥâsibî

Conform. 170

All that I can offer to Jesus, is the prayer that His Holy Will be accomplished. I find myself so indifferent, so foreign to everything, that I dare compare myself to a child sleeping on the Divine Heart. From the day that I abandoned myself to Him, in asking that He occupy Himself with my whole person, I have enjoyed an enviable peace and felt a constant joy.

Sister Consolata

P. State 573

Peace is in that heart in which no wave of desire of any kind rises.

Swami Ramdas

Knowl. 755

A peaceful mind is your most precious capital.

Swami Sivananda

When a man has restrained the turbulent passions of his breast by the power of right judgment, and has spread the garment of soft compassion and sweet content over his heart and mind, let him then worship divine serenity within himself.

Yoga-Vasishtha

The mind, when it is steady in divine contemplation, expresses sattwa overcoming rajas and tamas.[1] No more is there feverish hankering after worldliness. Tranquillity comes to a heart which is no longer stirred by desires, as stillness to a fire when no more fuel is added.

Srimad Bhagavatam, XI. iii

That is, the goal of tranquillisation is to be reached not by suppressing all mind activity but by getting rid of discriminations and attachments.

Lankavatara Sutra, VIII

Renun. 160

[1] Cf. note 2, p. 250.

All things, O priests, are on fire. And what, O priests, are all these things which are on fire?

The eye, O priests, is on fire; forms are on fire; eye-consciousness is on fire; impressions received by the eye are on fire; and whatever sensation, pleasant, unpleasant, or indifferent, originates in dependence on impressions received by the eye, that also is on fire.

And with what are these on fire?

With the fire of passion, say I, with the fire of hatred, with the fire of infatuation; with birth, old age, death, sorrow, lamentation, misery, grief, and despair are they on fire.

The ear is on fire; sounds are on fire; . . . the nose is on fire; odors are on fire; . . . the tongue is on fire; tastes are on fire; . . . the body is on fire; things tangible are on fire; . . . the mind is on fire; ideas are on fire; . . . mind-consciousness is on fire; impressions received by the mind are on fire; and whatever sensation, pleasant, unpleasant, or indifferent, originates in dependence on impressions received by the mind, that also is on fire. . . .

Sin 77

Perceiving this, O priests, the learned and noble disciple conceives an aversion for the eye, conceives an aversion for forms, conceives an aversion for eye-consciousness, conceives an aversion for the impressions received by the eye, . . . conceives an aversion for the ear, conceives an aversion for sounds, . . . conceives an aversion for the nose, conceives an aversion for odors, . . . conceives an aversion for the tongue, conceives an aversion for tastes, ... conceives an aversion for the body, conceives an aversion for things tangible, ... conceives an aversion for the mind, conceives an aversion for ideas, conceives an aversion for mind-consciousness, conceives an aversion for the impressions received by the mind. . . . And in conceiving this aversion, he becomes divested of passion, and by the absence of passion he becomes free, and when he is free he becomes aware that he is free; and he knows that rebirth is exhausted, that he has lived the holy life, that he has done what it behooved him to do, and that he is no more for this world.

Mahâ-Vagga (The Buddha's Fire Sermon)

Every cup should be sweet to you which extinguishes thirst.

Sextus the Pythagorean

Action 329
Death 220

If the ego rises, all else will also rise; if it subsides, all else will also subside.[1]

Sri Ramana Maharshi

Death 223
P. State 590

Cut off and destroy in yourselves the roots of avarice, anger, ignorance, pride. When you have cut off and destroyed the 'life-death of the Six Worlds' (as it is called) that will be the cessation of both life and death. When you reach that stage your mind will come to an end and there will be real peace, the calm of non-attachment to forms and shapes of the world. Then the moon will be your pillow, so peacefully will you sleep. The sky will be your

[1] One can transcend and thus by-pass the flux of forms; one cannot 'conquer' the flux, since it is perpetual on its level of reality (cf. *Renunciation*, p. 143). Hujwîrî (*Kashf*, pp. 205–206) cites Shaykh Abû ʿAlî Siyâh of Merv: 'I saw my lower soul (*nafs*) in a form resembling my own, and some one had seized it by its hair and gave it into my hands. I bound it to a tree and was about to destroy it, when it cried out, "O Abû ʿAlî, do not trouble yourself. I am God's army: you cannot reduce me to naught." '

couch, and you will command a view of the Lotus store-world of Amida[1] and Vairocana.[2] Your body and soul will be pure. Everything will be pure. This is the first entry into Nirvana.

Hakuin

Pax Profunda

The mind of the sage being in repose becomes the mirror of the universe, the speculum of all creation.

Chuang-tse (ch. XIII)

Center 828

Having this knowledge (of the imperishable *Âtmâ*), having become calm, subdued, quiet, patiently enduring, and collected, one sees the Self in self.

Bṛihad-Âraṇyaka Upanishad, IV. iv. 23

My peace I give unto you: not as the world giveth, give I unto you.

St John, XIV. 27

The repose of the sage is not what the world calls repose. His repose is the result of his mental attitude. All creation could not disturb his equilibrium: hence his repose.

Chuang-tse (ch. XIII)

Action 346 *Holiness* 902

Nothing can repose in itself, except by returning into that from which it has gone forth. The heart is deflected from Unity into a desire for sensation in order to taste the differentiation of properties; by that are born in it the differentiation and contradiction which henceforth dominate the heart. And it cannot be relieved, except by abandoning itself such as it is in the desire for properties and by mounting once more into the most pure Calm, and by striving to silence its will, in such manner that the will beyond all sense and specific appearance is plunged into the eternal will of the Undetermined, out of which it has arisen in the beginning, i.e., out of the *Mysterium Magnum,* so that it no longer wills anything except what God wills for it.

Boehme

Sin 66 *Renun.* 152a *Conform.* 185

The soul is like a dish full of water, and the impressions like the rays of light which strike the water. Now when the water is disturbed the light seems to be disturbed too, but it is not really disturbed. So when a man has a fit of dizziness, the arts and virtues are not put to confusion, but only the spirit in which they exist: when this is at rest, they come to rest too.

Epictetus

Contem. 528 Introd. *Realiz.*

[1] Amida (Amitâbha) 'The Enlightener', one of the five Dhyânî Buddhas (Buddhas of Meditation).
[2] 'The Manifester', another of the five Dhyânî Buddhas.

A man does not seek to see himself in running water, but in still water. For only what is itself still can impart stillness into others.

Knowl. 749

Chuang-tse (ch. V)

Let the soul that is not unworthy of the vision contemplate the Great Soul; freed from deceit and every witchery and collected into calm. Calmed be the body for it in that hour and the tumult of the flesh, ay, all that is about it calm; calm be the earth, the sea, the air, and let heaven itself be still. Then let it feel how into that silent heaven the Great Soul floweth in!

Judgment 268

Plotinus

If to any the tumult of the flesh were hushed, hushed the images of earth, and waters, and air, hushed also the poles of heaven, yea the very soul be hushed to herself, and by not thinking on self surmount self, hushed all dreams and imaginary revelations, every tongue and every sign, and whatsoever exists only in transition, since if any could hear, all these say, *We made not ourselves, but He made us that abideth for ever*—If then having uttered this, they too should be hushed, having roused only our ears to Him who made them, and He alone speak, not by them, but by Himself, that we may hear His Word, not through any tongue of flesh, nor Angel's voice, nor sound of thunder, nor in the dark riddle of a similitude, but, might hear Whom in these things we love, might hear His Very Self without these, (as we two now strained ourselves,[1] and in swift thought touched on that Eternal Wisdom, which abideth over all;)—could this be continued on, and other visions of kind far unlike be withdrawn, and this one ravish, and absorb, and wrap up its beholder amid these inward joys, so that life might be for ever like that one moment of understanding which now we sighed after; were not this, *Enter into thy Master's joy?*

Sin 77
Supra 700
Beauty 663
Illusion 109
Ecstasy 636
Center 847

St Augustine

One reaches this state of mind after having the vision of God. When a boat passes by a magnetic hill, its screws and nails become loose and drop out. Lust, anger, and the other passions cannot exist after the vision of God.

Holiness 902

Sri Ramakrishna

What is the best thing of all for a man, that he may ask from the Gods?
'That he may be always at peace with himself.'

Introd.
Holy War

Contest of Homer and Hesiod

The peace of God, which passeth all understanding, shall keep your hearts and minds.

Philippians, IV. 7

If any man then has an incorporeal eye, let him go forth from the body to behold the Beautiful, let him fly up and float aloft, not seeking to see shape or colour, but rather that by which these things are made, that which is quiet and calm, stable and changeless, . . .

Center 816
Flight 944
Beauty 663

[1] St Augustine and his mother, St Monica.

that which, being one, is yet all things, which issues from itself and is contained in itself, that which is like nothing but itself.

Hermes

Reality 775, 773
M.M. 975

In tranquillity, in stillness, in the unconditioned, in inaction, we find the levels of the universe, the very constitution of TAO. . . .

Sorrow and happiness are the heresies of virtue; joy and anger lead astray from TAO; love and hate cause the loss of virtue. The heart unconscious of sorrow and happiness,—that is perfect virtue. ONE, without change,—that is perfect repose.

Chuang-tse (ch. XV)

Action 358
Renun. 136

We are taught in *Metaphysicks,* That *Being, Truth* and *Goodness,*[1] are really *one.* How sweet a rest now doth the Spirit, with its Understanding, and its Will, find to it self in every *Being,* in every *Truth,* in every *State* or *Motion* of *Being,* in every form of *Truth.* When it hath a sense of the *highest Love,* which is the same with the *highest Goodness,* designing, disposing, working all in all, even all *Conceptions* in all *Understandings,* all *Motions,* in every *Will, Humane, Angelical, Divine*? With what a joy and complacency unexpressible doth the Will, the Understanding, the whole Spirit now lie down to rest everywhere, as upon a bed of Love, as in the bosom of goodness it self?

Peter Sterry

Love 625

'Peace be with you' was the salutation of him who was the salvation of man. For it was meet that the supreme saviour should utter the supreme salutation.

Dante (*De Monarchia,* IV)

[1] This recalls the Hindu ternary *Sat-Chit-Ânanda.* Cf. note p. 623.

BOOK THREE: TRUTH–KNOWLEDGE–UNION

PART V: *Discernment – Truth*

SPIRITUAL DROWNING

Plunge in: this is the drowning!

Eckhart (I. 368)

If the spiritual work has hitherto shown itself predominantly as an effort to transcend the 'lower waters' and attain an equilibrium on the 'surface of the waters', it now becomes through inverse analogy a journey or 'immersion' into the 'higher waters' of formless possibilities[1]—supraindividual states which no longer concern the human condition as such (hence the idea of 'drowning' or 'extinction'), but to which the human being has access, at least potentially, through the centrality that is the primordial birthright of his state, and which by definition are fully realized in the plenitude of Universal Man.

Introd. *Waters*

Holiness 924

'The voyage may be accomplished, either by going upstream to the source of the waters, or by crossing these to the other bank, or else finally by descending the current to the sea,' Guénon explains in 'Le passage des eaux', *Études Trad.*, 1940. 'This use of different symbolisms, contrary in appearance only and having in reality the same spiritual signification, corresponds with the nature of metaphysic itself, which though never "systematic", is always perfectly coherent; one must simply be careful of the precise meaning with which the symbol of the river,[2] including its source, its banks and its mouth, is to be understood in each case.'

In the first case, that of 'going upstream', the river is identified with the 'World Axis': 'it is the "celestial river" which descends to earth, and is designated in the Hindu tradition by names such as *Gangâ* and *Saraswatî,* which are properly speaking the names of certain aspects of *Shakti.* In the Hebraic Kabbalah, this "River of Life" finds its correspondence in the "canals" of the Sefirothic Tree, by which the influences of the "world above" are transmitted to the "world below", and they are also in direct relationship with the *Shekinah,* which is on the whole the equivalent of *Shakti*; and there is also the case of the waters that "flow upstream", expressing a return to the celestial source, no longer represented by ascending the current, but by a reversal in the direction of this current itself. . . .'

Introd. *Wr. Side*

Introd. *Peace*

When it is a question of the four rivers of the Terrestrial Paradise that meet at the base of the 'Tree of Life', i.e. 'World Axis', 'the ascent of the current can be considered as happening in two phases: the first, on a horizontal plane, leads to the center of this world; the second, starting from there, is accomplished vertically following the axis. . . . From the

Introd. *P. State*

[1] The *Qur'ân* distinguishes these 'two seas' (*baḥrayn*) as being 'the one palatable, sweet, and the other salty, brackish' (XXV. 53).

[2] The Arabic word *bahr,* already alluded to, designates both 'sea' and 'river'.

initiatic point of view, these two successive phases have their correspondence in the domains respectively of the "Lesser Mysteries" and "Greater Mysteries".

Pilg. 385

'In the second case, where the symbolism has to do with crossing from one bank to the other, . . . the river which is thus to be crossed is more particularly the "River of Death"; the bank one leaves is the world of change, that is to say, the domain of manifested existence, . . . and the other bank is *Nirvâna,* the state of the being who is definitively freed from death.

'As for the third case, that of "descending the current", the Ocean must be considered, not as an extent of water to be crossed, but on the contrary, as the end itself to be attained, therefore as representing *Nirvâna*; the symbolism of the two banks is then different from the previous case (and this is an example of the double meaning in symbols), since it is no longer a question of passing from one to the other, but instead, of avoiding them both equally: they are respectively the "world of men" and the "world of gods", or again, the "microcosmic" and "macrocosmic" conditions'—i.e. the *Symplegades* or 'pairs of opposites' to be escaped and transcended in the process of attaining the 'metacosm'.

M.M. 978

O God, drown me in the essence of the Ocean of Divine Solitude, so that I neither see nor hear nor find nor feel except through It.

Death 220

ʿAbd as-Salâm ibn Mashîsh

The rippling tide of love which all her life had flowed secretly from God into her soul, drew it mightily back into its Source.

Mechthild of Magdeburg

The drop became a fountain, and the fountain grew into a river,
Which river became reunited to the ocean of eternity.

Sheikh Abûlfaiz Faiyazi

O thou! who art free of notion, imagination, and duality,
We are all billows in the ocean of thy being.

M.M. 978

Binavâli

All thy waves and thy billows are gone over me.

Psalm XLII. 7

Ah, noble soul, prove thy nobility! But while it is the case with thee that thou lettest not go thine own self altogether to drown in the bottomless sea of the Godhead, verily thou canst not know this divine death.

Creation 42
Renun. 146
Death 208

Eckhart

Launch out into the deep.

St Luke, v. 4

Taking the name of Kâli, dive deep down, O mind,
Into the heart's fathomless depths,
Where many a precious gem lies hid.
But never believe the bed of the ocean bare of gems
If in the first few dives you fail;
With firm resolve and self-control
Dive deep and make your way to Mother Kâli's realm.[1]

Inv. 1036
Center 816
Holy War 405

[1] I.e., *Âdyâśakti,* the Primal Energy, whose 'realm' is the 'higher waters' of Universal Possibility; *Mûla-Prakriti.*

Peace 705 — Orthod. 285 — Pilg. 366 — Introd. — Heterod.

Down in the ocean depths of heavenly Wisdom lie
The wondrous pearls of Peace, O mind;
And you yourself can gather them,
If you but have pure love and follow the scriptures' rule.
Within those ocean depths,[1] as well,
Six alligators lurk—lust, anger, and the rest—
Swimming about in search of prey.
Smear yourself with the turmeric of discrimination;
The very smell of it will shield you from their jaws.

Upon the ocean bed lie strewn
Unnumbered pearls and precious gems;
Plunge in, says Râmprasâd, and gather up handfuls there!

Râmprasâd

Full fathom five thy father lies;
  Of his bones are coral made:
Those are pearls that were his eyes:
  Nothing of him that doth fade,
But doth suffer a sea-change
Into something rich and strange.
Sea-nymphs hourly ring his knell:
(*Burden*: ding-dong.)
Hark! now I hear them,—ding-dong, bell.

Shakespeare (*The Tempest*, I. ii. 394)

Center 816

We are pearls in that Sea, therein we all abide. . . . Enter that Ocean, that your drop may become a sea which is a hundred seas of 'Omân.

Dîvâni Shamsi Tabrîz, IX, XII

Death 220 — Peace 694

If we possess God in the immersion of love—that is, if we are lost to ourselves—God is our own and we are His own: and we sink ourselves eternally and irretrievably in our own possession, which is God. . . . And this down-sinking is like a river, which without pause or turning back ever pours into the sea; since this is its proper resting-place.

Ruysbroeck

Realiz. 887

All the great rivers, that is to say, the Ganges, the Jumnâ, the Aciravatî, the Sarabhû, the Mahî—these, on reaching the great ocean lose their former names and identities and are reckoned simply as the great ocean.

Vinaya-Piṭaka, II

Void 724

What is needed is absorption in God—loving Him intensely. The 'Nectar Lake' is the Lake of Immortality. A man sinking in It does not die, but becomes immortal. Some people believe that by thinking of God too much the mind becomes deranged; but that is not true.

[1] Here the symbolism is extended to include both seas.

God is the Lake of Nectar, the Ocean of Immortality. He is called the 'Immortal' in the Vedas. Sinking in It, one does not die, but verily transcends death.

Sri Ramakrishna

Ecstasy 641
Death
232

(The self) is like a drop of water afraid of wind and earth; for by means of these twain it is made to pass away (and perish).

When it has leaped into the sea, which was its source, it is delivered from the heat of the sun and from wind and earth.

Its outward form has disappeared in the sea, but its essence is inviolate and permanent and goodly.[1]

Hark, O drop, give thyself up without repenting, that in recompense for the drop thou mayst gain the ocean.

Rûmî

Death
223
Introds. *Realiz.*
Holiness

As the drop becomes the ocean, so the soul is deified, losing her name and work, but not her essence.

Eckhart

As the flowing rivers in the ocean
Disappear, quitting name and form,
So the knower, being liberated from name and form,
Goes unto the Heavenly Person, higher than the high.

Muṇḍaka Upanishad, III. ii. 8

Void
721

I shall throw myself into the uncreate sea of the naked Godhead.

Angelus Silesius

M.M.
994

O Yeares! and Age! Farewell:
Behold I go,
Where I do know
Infinitie to dwell.

And these mine eyes shall see
All times, how they
Are lost i' th' Sea
Of vast Eternitie.

Center
838

Where never Moone shall sway
The Starres; but she,
And Night, shall be
Drown'd in one endlesse Day.

Robert Herrick

Center
833

[1] This drowning is the inverse of that drowning in the 'lower waters' (of inferior psychic possibilities), already mentioned in the Introduction to *Pilgrimage,* where precisely the death in question brings dissolution and ruin.

As these flowing rivers that tend toward the ocean, on reaching the ocean, disappear, their name and form are destroyed, and it is called simply 'the ocean'—even so of this spectator these sixteen parts that tend toward the Person, on reaching the Person, disappear, their name and form are destroyed, and it is called simply 'the Person'. That one continues without parts, immortal!

Praśna Upanishad, VI. 5

Thus let the philosopher observe, that when the three murtherers, viz. Saturn, Mars, and Mercury, are drowned in the crimson-coloured blood of the lion, they do not perish; but they are pardoned, that is, their wrath is changed into a love-desire, viz. out of Venus into Sol; for when the fiery desire enters into the watery desire, then a shining, viz. a glorious splendour, arises from and in the fire; for Venus is white, and the fire-desire is red.

Metanoia 488 *M.M.* 998

Boehme

Here, I still flow in God as a stream of time;
There, I myself shall be the sea of eternal felicity.

Angelus Silesius

Charity 608

To Tao all under heaven will come
As streams and torrents flow into a great river or sea.

Tao Te Ching, XXXII

His will is our peace; it is that sea to which all moves that it createth and that nature maketh.

Dante (*Paradiso,* III. 85)

... Divine Love; in diverse breadths and depths. with innumerable, sportful windings and turnings, flowing forth from its own full Sea of eternal Sweetness, and, through all its Chanels, hasting hither again.

Creation 26

Peter Sterry

O Thou who stealest Thy bhaktas' hearts,
Drown me deep in the Sea of Thy love!

Bengali Hymn

Love 618

The desirous soul no longer thirsts *for* God but *into* God, the pull of its desire draws it into the Infinite Sea.

Richard of Saint-Victor

All things are gathered together in one with the Divine sweetness, and the man's being is so penetrated with the Divine substance, that he loses himself therein, as a drop of water is lost in a cask of strong wine.

Realiz. 868

Tauler

'Thou shalt disappear like a drop of water in the boundless ocean.'

Sister Consolata

Drink this wine and, dying to self,
You will be freed from the spell of self.
Then will your being, as a drop,
Fall into the ocean of the Eternal.

Shabistarî

Ecstasy 637
Death 206

Dear Reader, if thou wouldst be lead to that Sea, which is as the gathering together, and confluence of all the waters of Life, of all Truths, Goodness, Joys, Beauties and Blessedness; follow the *stream* of the *Divine Love,* as it holdeth on its course, . . . bearing thee up in the bright Arms of its own Divine Power, sporting with thee all along, washing thee white as snow in its own pure floods, and bathing thy whole Spirit and Person in heavenly unexpressible sweetness.

Peter Sterry

Center 847
Creation 33

Nangtâ[1] used to instruct me about the nature of Satchidânanda Brahman. He would say that It is like an infinite ocean—water everywhere, to the right, left, above, and below. Water enveloped in water. It is the Water of the Great Cause, motionless. Waves spring up when It becomes active. Its activities are creation, preservation, and destruction. . . .

Brahman is beyond mind and speech. A salt doll entered the ocean to measure its depth; but it did not return to tell others how deep the ocean was. It melted in the ocean itself.

Sri Ramakrishna

P. State 563
Reality 773

To know all things in the cause of their existence I must soar beyond all lights, temporal and eternal, and plunge into the causeless essence which gives mind and being to my soul. Drowned in this being, aware of self and things merely as being, my soul has lost her name.

Eckhart

I am no sooner before God but I lose all other objects and do nothing else than follow the drawings and inclinations which He Himself has laid in me, to flow in God and sink down in Him—like a vessel full of water which when emptied nothing remains in it, so will I wholly empty and sink myself quite in God. This is my sole endeavor, I desire only this, and after this manner I pray.

Johannes Kelpius

M.M. 986

Though one equal to a hundred like me would not have the strength to bear the Sea, yet I cannot refrain from the drowning waters of the Sea.

May my soul and mind be a sacrifice to the Sea: this Sea has paid the blood-price of mind and soul.

Rûmî

Suff. 132

Go and plunge yourself into the calm sea of spiritual solitude, and wash your soul in the

[1] (Lit., the Naked One), a name of Totapuri, Sri Ramakrishna's sannyâsi guru.

M.M. 978

nectar of ambrosial meditation. Dive and dive deep in the depth of Unity, and fly from the salt waves of duality and the brackish water of diversity.

Yoga-Vasishtha

Pilg. 379

The Brahma-faring is lived for the plunge into Nirvana, for going beyond to Nirvana, for culmination in Nirvana.

Saṁyutta-nikâya, III. 189

Beauty 663

Thus all particular and imperfect *Beings* carry us up to the perfect and universal Being, abstracted from them all, set on his *Throne high and lifted up above them all,* from which, as their proper *Head,* they flow, by and in which, as in their proper *Root,* they subsist, being the beams of this glorious Sun, and Rivers from this full Sea.

Peter Sterry

Illusion 109

For a true and faithful servant of God shall be always pressing upward to what is before him, not suffering himself to be held back by comfort or pleasure, joy or sorrow, wealth or poverty. Through all this he shall urge onward, till he come unto the infinite ocean of the Godhead. And therein he shall be lost without his own knowledge, and dazzled by excess of light and love. There it shall be given him to know all that belongs to true perfection.

Tauler

Pilg. 385

'Say goodbye forever to all earthly creatures and take to the open sea, thanks to the unceasing act of love (invocation). Full sails away for the eternal shores!'

Sister Consolata

Even as the water of the Ganges floweth swiftly on and emptieth into the sea, so shall he who walketh in the even way of perfect understanding arrive at the cessation of death.

Udâna

Center 816

O unfathomable sink, in thy depth thou art high and in thy height profound!—How so?—That is hidden from us in thy bottomless abysm. . . .

Void 721

M.M. 998 978

The fastidious soul can rest her understanding on nothing that has name. She escapes from every name into the nameless nothingness. Escaping her own nature she falls clear of her own aught. The naught she falls into is the unknowing, which is called the dark. . . . This darkness is the incomprehensible nature of God. She sinks for evermore in the depths of this naught. She sinks and drowns; she drowns to her own aught. Her aught, surviving, sinks as naught to naught. But the naught that sinks can never comprehend the naught it sinks in. Every virtue mastered and transcended, the soul cries; 'Even so I cannot glorify and love God to the full. I die then to the virtues casting me into the naught of the Godhead to sink eternally from naught to aught.'

Eckhart

M.M. 986

My heart, when love's Sea of a sudden burst into its viewing,
Leaped headlong in, with 'Find me now who may!'

Dîvâni Shamsi Tabrîz, VII

THE VOID

Here, O Sariputra, form is emptiness and the very emptiness is form; emptiness does not differ from form, nor does form differ from emptiness; whatever is form, that is emptiness, whatever is emptiness, that is form.

Prajñâpâramitâhṛdaya[1]

Abyssus abyssum invocat.

Psalm XLI. 8

Nirvâna is the extinction of nothing except illusion. Since God alone *is*,[2] perfect extinction[3] in this sense (*fanâ' al-fanâ'*, *pari-nirvâna*) is the equivalent of Godhood. It is not that the soul becomes God, but that everything in the soul which is not 'of God' ceases henceforth to exist, virtually or finally. *Nirvâna* as 'extinction' thus corresponds with the Sanskrit term *chitta vṛitti nirodha*: cessation of the activity of the finite imagination.

Contem. 532

'According to a widespread error in the West,' writes Schuon in the chapter on *Nirvâna* in *L'Oeil du Coeur*, 'spiritual "extinction" . . . represents a "nothing", as if it were possible to realize something which is nothing. Now one of two things: either *Nirvâna* is nothingness, in which case it is unrealizable, or else it is realizable, in which case it must correspond with something real. . . . If one envisages the reabsorption of the being in God as an "annihilation", one must in all logic equally envisage the reabsorption of the terrestrial being into paradisaical existence as a passage from the "real" to the "nothing"; and inversely, if one considers "Paradise" as a "reinforcement" or "exaltation" of all that is perfect and lovable in this world below, one will likewise have to consider the state of "Supreme Extinction" as a "reinforcement" or "exaltation" of all that is positive and perfect, not just in the terrestrial world alone, but in the entire cosmos. It follows that one can regard a higher degree of reality—that of formless manifestation, or, beyond the cosmos, that of the unmanifested—either in terms of the negative aspect which it necessarily presents in relation to the lower plane where one is situated and the limitations of which it denies, or in terms of the positive aspect which it contains in itself and consequently also—and *a fortiori*—in relation to the lower plane envisaged.'

Beauty 663

Introd. *Ecstasy*

Writing on the same subject in *Time and Eternity*, Coomaraswamy underscores 'the

[1] Conze: *Buddhist Texts*, Oxford, 1954, p. 152.

[2] According to the highest formulations of all traditional doctrines—whether applied at the level of Pure Being, or transposed to the Absolute. Cf. *Reality*.

[3] Not to be confused with 'annihilation' as dissolution: cf. Introductions to *The Wrathful Side*, *Pilgrimage*, *Holiness*.

curious resistance that contemporary mentalities oppose to the concept of a static being definable only by negations of all limiting affirmations, all procedure from one experience to another. The most striking aspect of this resistance is the fact that it is almost always based on feelings: the question of the truth or falsity of a traditional doctrine is hardly ever raised, and all that seems to matter is whether one likes the doctrine or not. This is the sentimentality of those who would rather than arrive at any goal, keep on going not merely until it is reached, but "throughout all time", and who confuse their activity, which is only an unfinished procedure from potentiality to act, with a *being* in act.

Judgment 250

'... (Yet) we cannot and may not, in fact, ignore that these who speak of a static, immutable, and timeless being above the partiality of time, also speak of it as an immediately beatific experience and possession of all things that have ever been or shall ever come into being in time; not to mention the realisation of other possibilities that are not possibilities of manifestation in time; it is a more and not a less "life" that subsists in the "naught" that embraces all things, but is "none" of them. In the same way men recoil from Nirvâṇa (literally, "despiration"), although it pertains to the definition of Nirvâṇa to say that "he who finds it, finds all" (*sabbam etena labbhati, Khuddaka Pâṭha,* 8) and that it is the "supreme beatitude" (*paramaṁ sukham, Nikâyas, passim*)!' (pp. 122–123).

Center 847

Introd.

The Shattering of Forms

Know that God is beyond all senses and sensory things, beyond all shape, colour, measure and place; is wholly without form and image and, while present in all things, is above all things; therefore He is beyond all imagining.

Unseen Warfare, I. xxvi

Introd.
Heterod.

One who seeks the Dharma finds it in seeking it in nothing.

Vimalakîrti Sûtra

Infra
724

They will ask thee of the mountains. Say: My Lord will break them into scattered dust.

And leave it as an empty plain,
Wherein thou seest neither curve nor ruggedness.
On that Day they follow the summoner who deceiveth not, and voices are hushed for the Beneficent, and thou hearest but a faint murmur.

Qur'ân, xx. 105–108

Judgment
268

The stopping of becoming is Nirvana.

Saṁyutta-nikâya, II. 117

You must break the outside to let out the inside: to get at the kernel means breaking the shell. Even so to find nature herself all her likenesses have to be shattered.

Eckhart

Metanoia 488
P. State 563
M.M. 975

Being in itself is God's, not thine; if thou shouldst come to realize the truth of the matter, and to understand what is God's through stripping thyself of all that is not thine, then wouldst thou find thyself to be as the core of an onion. If thou wouldst peel it, thou peelest off the first skin, and then the second, and then the third, and so on, until there is nothing left of the onion. Even so is the slave with regard to the Being of the Truth.

Shaykh Aḥmad al-ʿAlawî

If ye pass beyond form, O friends, 'tis Paradise and rose-gardens within rose-gardens.
When thou hast broken and destroyed thine own form, thou hast learned to break the form of everything.
After that, thou wilt break every form; like Haydar (ʿAlî), thou wilt uproot the gate of Khaybar.

Rûmî

Center
847
Realiz.
890

There is, disciples, a Realm devoid of earth and water, fire and air. It is not endless space, nor infinite thought, nor nothingness, neither ideas nor non-ideas. Not this world nor

M.M. 975

that is it. I call it neither a coming nor a departing, nor a standing still, nor death, nor birth; it is without a basis, progress, or a stay; it is the ending of sorrow.

Peace 694

For that which clingeth to another thing there is a fall; but unto that which clingeth not no fall can come. Where no fall cometh, there is rest, and where rest is, there is no keen desire. Where keen desire is not, naught cometh or goeth; and where naught cometh or goeth there is no death, no birth. Where there is neither death nor birth, there neither is this world nor that, nor in between—it is the ending of sorrow.

M.M. 998

There is, disciples, an Unbecome, Unborn, Unmade, Unformed; if there were not this Unbecome, Unborn, Unmade, Unformed, there would be no way out for that which is become, born, made, and formed; but since there is an Unbecome, Unborn, Unmade, Unformed, there is escape for that which is become, born, made, and formed.

Udâna, VIII. i. 4, 3

Death 208

The first substance cannot be obtained, except by destroying the specific properties of a thing.

John A. Mehung

Illusion 109 Symb. 306 Beauty 663

Do not, I urge you, look for The Good through any of these other things; if you do, you will see not itself but its trace: you must form the idea of that which is to be grasped, cleanly standing to itself, not in any combination, the unheld in which all have hold: for no other is such, yet one such there must be.

Plotinus

M.M. 978 Sin 64

Do not cling to the notion of voidness,
But consider all things alike.
Indeed even the husk of a sesame-seed
Causes pain like that of an arrow.

Saraha

Introd. Inv.

Verily thou must sojourn and dwell in thy essence, in thy ground, and there God shall mix thee with his simple essence without the medium of any image. No image represents and signifies itself: it stands for that of which it is the image. Now seeing that thou hast no image save of what is outside thee, therefore it is impossible for thee to be beatified by any image whatsoever. . . . All things must be forsaken. God scorns to work among images.

Eckhart

Judgment 250 Death 220 Holiness 902

Those outflows whereby, if they had not been extinguished, I might have been a deva, gandharva, yakkha or a human being—those outflows are extinguished in me, cut off at the root, made like a palm-tree stump that can come to no further existence in the future. . . .

As a lotus, fair and lovely,
By the water is not soiled,
By the world am I not soiled:
Therefore, brahmin, am I Buddha.

Anguttara-nikâya, II. 37–39

For since beings have only a participation in being, we now clearly see how we arrive at God by eliminating that participation from all beings; once that is suppressed there remains only entity in its infinite simplicity, which is the essence of all beings. It is only by the most learned ignorance that the mind grasps such an entity, for nothing seems to be left once I mentally remove all that has participated being. For that very reason Denis the Great says that an understanding of God is not so much an approach towards something as towards nothing; and sacred ignorance teaches me that what seems nothing to the intellect (mind) is the incomprehensible Maximum.

Realiz. 873

Infra 724

Nicholas of Cusa

'What is the Void?' asked the Master of the Law Ch'ung-yüan. 'If you tell me that it exists, then you are surely implying that it is solid and resistant. If on the other hand you say it is something that does not exist, in that case why go to it for help?' 'One talks of the Void,' replied Shen-hui, 'for the benefit of those who have not seen their own Buddha-natures. For those who have seen their Buddha-natures the Void does not exist. It is this view about the Void that I call "going to it for help".'

Shen-hui

The One Mind alone is the Buddha, and there is no distinction between the Buddha and sentient beings, but that sentient beings are attached to forms and so seek externally for Buddhahood. By their very seeking they lose it, for that is using the Buddha to seek for the Buddha and using mind to grasp Mind.

Beauty 663

Huang Po

I have fallen in love, O mother, with the
Beautiful One, who knows no death,
knows no decay and has no form;

I have fallen in love, O mother, with the
Beautiful One, who has no middle, has
no end, has no parts and has no features;

I have fallen in love, O mother, with the
Beautiful One, who knows no birth and
knows no fear.

I have fallen in love with the Beautiful
One, who is without any family,
without any country and without any peer;
Chenna Mallikârjuna, the Beautiful, is my husband.
Fling into the fire the husbands who are subject
to death and decay.

Akka Mahâdêvî

The Plenitude of the Void

You must not call anything void, without saying what the thing in question is void of.

Hermes

Death in God is nothing but the uncreated life, that is, God himself.

Eckhart

Metanoia 480

What is he that calleth it nought? Surely it is our outer man and not our inner. Our inner man calleth it All.

The Cloud of Unknowing, LXVIII

Reality 775
M.M. 975

The virtue of love is NOTHING and ALL, or that nothing visible out of which all things proceed.... It is the holy magical root, or ghostly power from whence all the wonders of God have been wrought by the hands of his elect servants, in all their generations successively. Whosoever finds it, finds nothing and all things.

Boehme

Supra Introd.
Death 220

Covetest . . . earnestly, not for to un-be—for that were madness and despite unto God—but for to forgo the knowing and the feeling of thy being.

The Epistle of Privy Counsel, VIII

Ecstasy 636

Brahman is where reason comes to a stop. There is the instance of camphor. Nothing remains after it is burnt—not even a trace of ash.

Sri Ramakrishna

Thou shalt love the naughting,
And flee the self.

Mechthild of Magdeburg

Reality 803

Things are all made from nothing; hence their true source is nothing.

Eckhart

The cloud of forgetting of creatures, which our author (*The Cloud of Unknowing*) here mentioneth, is but the *active annihilation of creatures*, of which Father Benet Fitch[1] speaketh so much in his third book of *The Will of God*. And I generally term it a *transcending of creatures*.

Father Augustine Baker

The third grade of love is when the mind of man is rapt into the abyss of divine light, so that, utterly oblivious of all exterior things, it knows not itself and passes wholly into its

[1] William Fitch, in religion Father Benet, 1563–1611: English Capuchin, influential in the renaissance of French mysticism.

God. And so in this state is held in check and lulled to sleep the crowd of carnal desires. In this state while the mind is alienated from itself, while it is rapt unto the secret closet of the divine primacy, while it is on all sides encircled by the conflagration of divine love, and is intimately penetrated and set on fire through and through, it strips off self and puts on a certain divine condition, and being configured to the beauty gazed upon it passes into a new kind of glory.

Richard of Saint-Victor

Sin 77
Ecstasy 641
Realiz. 868

When you hear me talk about the Void (*śûnyatâ*), do not fall into the idea of vacuity.

Hui-nêng, the Sixth Patriarch

Heterod. 430

Many people are afraid to empty their minds lest they may plunge into the Void. They do not know that their own Mind is the Void. The ignorant eschew phenomena but not thought; the wise eschew thought but not phenomena.

Huang Po

Renun. 139
M.M. 978

God may be called *nothing* because he is none of all the things that we can imagine or understand; but in himself he is indeed as it were all in all, as being the cause of all other things.

Father Augustine Baker

Tao, when put in use for its hollowness, is not likely to be filled.

Tao Te Ching, IV

Dionysius declares that the soul's supreme delight is the nothingness of her prototype. And a heathen philosopher says: 'God's naught fills everywhere and his aught is nowhere.'

Eckhart

Center 847
Creation 42

This is the state of perfect union, which is termed by some a state of nothing, and by others with as much reason termed a state of totality.

Father Augustine Baker

In this secret union the loving soul flows forth and escapes from itself, and is swallowed up and as it were annihilated in the abyss of eternal love, dead to itself and living in God, knowing naught and feeling naught except the love that it savours. For it loses itself in the immense desert and darkness of the Godhead. But to lose oneself thus is to find oneself.

François Louis de Blois

St Dionysius says: 'God is nothing,' and this is also implied by St Augustine when he says: 'God is everything,' meaning: nothing is God's. So that by saying 'God is nothing' Dionysius signifies that there is no thing in his presence. It follows that the spirit must advance beyond things and thingliness, shape and shapenness, existence and existences: then will dawn in it the actuality of happiness which is the essential possession of the actual intellect.

Eckhart

Realiz. 873

Judgment 266

It was a cloud and darkness to them, but it gave light by night to these.

Exodus, XIV. 20

Sp. Drown. 713 *P. State* 563 *Holiness* 902 *Renun.* 158

For one moment sink into the ocean of God, and do not suppose that one hair of your head shall be moistened by the water of the seven seas. If the vision you behold is the Face of God, there is no doubt that from this time forward you will see clearly. When the foundations of your own existence are destroyed, have no fear in your heart that you yourself will perish.

Ḥâfiz

Knowl. 749

Absence of thoughts does not mean a blank. There must be one to know the blank. Knowledge and ignorance are of the mind. They are born of duality. But the Self is beyond knowledge and ignorance. It is light itself. There is no necessity to see the Self with another Self.

Sri Ramana Maharshi

For the Bodhisattvas, Nirvana does not mean extinction.

Lankavatara Sutra

Reality 803 *Holiness* 924

For we can unsuppose Heaven and Earth and annihilate the world in our imagination, but the place where they stood will remain behind, and we cannot unsuppose or annihilate that, do what we can. Which without us is the chamber of our infinite treasures, and within us the repository and recipient of them.

Thomas Traherne

Death 220

The soul can only be pure and white like snow if you make a void in yourself, or on the contrary if you lose yourself in the totality of creation.[1] All colours merge to produce white, which is the colour of the Formless.

Ananda Moyî

The Loved One is quite colourless, O heart;
Be not engrossed with colours, then, O heart:
All colours come from what is colourless,
And 'who can dye so well as God',[2] O heart?

Jâmî

Carry 'Nothing' in the heart. It is 'Everything'.

From *The Ten Virtues* in Japanese flower arrangement

We maintain, and it is evident truth, that the Supreme is everywhere and yet nowhere.

Plotinus

[1] 'These are the two methods of Jnâna-Yoga: that of negation (*neti, neti*), via negationis, and that of affirmation (*iti, iti*), via eminentiae, both of which lead to the realization of the identity between *âtman* and Brahman' (note by the editor). The former method relates to the viewpoint of Extinction (Part V in this book), the latter to Identity (Part VI).

[2] *Qur'ân*, II. 138.

Thirty spokes unite in one nave,
And because of the part where nothing exists we have the use of a carriage wheel.
Clay is moulded into vessels,
And because of the space where nothing exists we are able to use them as vessels.
Doors and windows are cut out in the walls of a house,
And because they are empty spaces, we are able to use them.
Therefore, on the one hand we have the benefit of existence, and on the other, we make use of non-existence.

Tao Te Ching, XI

Reality 773

M.M. 978

'Ask of Me everything, for I possess everything: do not ask of Me a single thing, for I do not approve of thy having a single thing.'

Niffarî

Renun. 158

A man may in this life reach the point at which he understands himself to be one with that which is nothing as compared with all the things that one can imagine or express in words. By common agreement, men call this Nothing 'God', and it is itself a most essential Something.

Henry Suso

Realiz. 870

Lord of himself, though not of lands,
And having nothing, yet hath all.

Sir Henry Wotton

Holiness 921

I am nothing; or if not,
Nothing to be were better.

Shakespeare (*Cymbeline,* IV. ii. 367)

Death 208

It is precisely because there is nothing within the One that all things are from it.

Plotinus

Reality 775

I do NOT teach a doctrine of extinction! Few understand this, but those who do understand are the only ones to become Buddhas. Treasure this gem!

Huang Po

Let the wicked forsake his way, and the unrighteous man his thoughts: and let him return unto the Lord, and he will have mercy upon him; and to our God, for he will abundantly pardon.

For my thoughts are not your thoughts, neither are your ways my ways, saith the Lord.

Isaiah, LV. 7, 8

Holiness 914

It is only a few more days, in this world, and each shall return to its own fountain; the blood-drop to the abysmal heart, and the water to the river, and the river to the shining sea; and the dewdrop which fell from heaven shall rise to heaven again, shaking off the dust grains which weighed it down, thawed from the earth frost which chained it here to herb

***Illusion* 99**
***Center* 816**
***Sp. Drown.* 713**
***Death* 225**

Realiz. 890 Peace 694

and sward, upward and upward ever through stars and suns, through gods, and through the parents of the gods, purer and purer through successive lives, until it enters the Nothing, which is the All, and finds its home at last.

Hypatia

Solitude

Reality 773 Center 816

That place which is everywhere at all times present (being the basis of all that evolves) and in which, in the beginning, at the end or in the middle, there is no living creature, that place (which is their own Self, the yogis) call 'Solitude'.

Tejo-bindu Upanishad, I. 23

M.M. 994

Saith the soul in the Book of Love, 'No one is God to me and I am soul to none.' By 'no one is God to me' she means merely that no entity, nothing nameable, is her God. Again the words, 'I am soul to none' mean that she is so void of self she has not got it in her to be aught to anyone. This is the state in which the soul should be: in utter destitution. The soul cries in the Book of Love, 'He is mine and I am his.' It were better she had said, 'He is not mine nor am I his,' for God who is in all is therein all his own. She can lay claim to naught: she has lost every whit whereto any wight could in anywise be aught or she withal be aught to any wight. No one is her God and she is no one's soul.

Eckhart

Sp. Drown. 713

Everything depends on this: a fathomless sinking into a fathomless nothingness.

Tauler

Death 206

For ye are dead, and your life is hid with Christ in God.

Colossians, III. 3

Brahma is the Void (*kha*), the Ancient Void of the pneuma . . . whereby I know (*veda*) what should be known.

Bṛihad-Âraṇyaka Upanishad, V. 1

Contem. 536

In dense darkness, O Mother, Thy formless beauty sparkles:
Therefore the yogis meditate in a dark mountain cave.
In the lap of boundless dark, on Mahânirvâna's waves upborne,
Peace flows serene and inexhaustible.
Taking the form of the Void, in the robe of darkness wrapped,
Who art Thou, Mother, seated alone in the shrine of samâdhi?

From the Lotus of Thy fear-scattering Feet flash Thy love's lightnings;
Thy Spirit-Face shines forth with laughter terrible and loud!

Bengali Hymn

The abyss of my heart ever invokes with great cries
The abyss of God: Say, which is the more profound?

Angelus Silesius

Center 816

Wayless and fathomless, . . . one can know of it in no other way than through itself. . . . The abyss of God calls to the abyss; that is, of all those who are united with the Spirit of God in fruitive love. This inward call is an inundation of the essential brightness, and this essential brightness, enfolding us in an abysmal love, causes us to be lost to ourselves, and to flow forth from ourselves into the wild darkness of the Godhead. And, thus united without means, and made one with the Spirit of God, we can meet God through God, and everlastingly possess with Him and in Him our eternal bliss.

Ruysbroeck

M.M. 986
Death 220
Realiz. 887
Knowl. 749

When God desires to reveal Himself to His slave by a Name or Quality, He extinguishes him, annihilating his self and his existence; then, when the creaturely light is extinguished, and the individuality is effaced, God establishes in the creaturely temple (*haykal*), but without divine localization (*hulûl*), a subtle reality which is neither detached from God nor conjoined to the creature, thus replacing what He stripped him of,—for God reveals Himself to His slaves through generosity: if He annihilated them without compensation, this would not be generosity on His part, but severity: far from Him to act thus! This subtle reality is what one calls the Holy Spirit (*ar-rûh al-quds*). And when God establishes in place of the slave a subtle reality of His Essence, His revelation is communicated to this reality, so that He never reveals Himself except to Himself, although we have called this divine subtle reality 'slave', since it takes his place. Actually, there is neither slave nor Lord, for if the slave no longer exists, the Lord ceases to be Lord: in reality, nothing remains but God alone, the Unique, the One.

Jîlî

Introd. *Renun.* Supra 724
Knowl. 749
Reality 803

At various times I have declared: I am the cause that God is God. God is gotten of the soul, his Godhead of himself; before creatures were, God was not albeit he was Godhead which he gets not from the soul. Now when God finds a naughted soul whose self and whose activity have been brought to naught by means of grace, God works his eternal work in her above grace, raising her out of her created nature. Here God naughts himself in the soul and then neither God nor soul is left. Be sure that this is God indeed.

Eckhart

God created man; and man created God. They both are the originators of forms and names only. In fact, neither God nor man was created.

Sri Ramana Marharshi

Supra Introd. *M.M.* 978

Therefore the universe is the outward visible expression of the 'Truth', and the 'Truth' is the inner unseen reality of the universe.

Jâmî

Knowl. 749 *Inv.* 1003

God can have no delight or union with any creature but because His well-beloved Son the express image of His person is found in it. This is as true of all unfallen as of all fallen creatures; the one are redeemed and the other want no redemption, only through the life of Christ dwelling in them. For as the Word, or Son of God, is the Creator of all things and by Him every thing is made that was made, so every thing that is good and holy in unfallen Angels is as much through His living and dwelling in them as every thing that is good and holy in redeemed man is through Him. And He is just as much the preserver, the strength, and glory, and life of all the thrones and principalities of Heaven as He is the righteousness, the peace, and redemption of fallen man.

Reality 790 803

This Christ of God has many names in Scripture; but they all mean only this, that He is, and alone can be, the light and life and holiness of every creature that is holy, whether in Heaven or on earth.

William Law

Contem. 532 Supra Introd.

Self-annihilation consists in this, that through the overpowering influence of the Very Being upon the inner man there remains no consciousness of aught beside Him. Annihilation of annihilation consists in this, that there remains no consciousness even of that unconsciousness. It is evident that annihilation of annihilation is involved in (the very notion of) annihilation. For if he who has attained annihilation should retain the least consciousness of his annihilation, he would not be in the state of annihilation, because the quality of annihilation and the person possessing such quality are both things distinct from the Very Being, the 'Truth' most glorious. Therefore, to be conscious of annihilation is incompatible with annihilation.

Jâmî

Now God and heaven gone, the soul is finally cut off from every influx of divinity, so his spirit is no longer given to her. Arrived at this the soul belongs to the eternal life rather than creation; her uncreated spirit lives rather than herself: the uncreated, eternally-existent which is no less than God. . . .

Humility 197

We ought to be eternally as poor as when we were not. . . . St Dionysius says, 'Be the soul never so bare the Godhead is barer': a naught from which no shoot was ever lopped nor ever shall be. . . . The Godhead is as void as though it were not.

Eckhart

Realiz. 890 *M.M.* 986

This path
No one walks along
Evening of autumnal day.

Zen *Haiku*

KNOWLEDGE

There is no means other than Knowledge for obtaining complete and final Deliverance; Knowledge alone loosens the bonds of the passions; without Knowledge, Beatitude (*Ânanda*) cannot be obtained. Action (*karma*), not being opposed to ignorance, cannot remove it; but Knowledge disperses ignorance as light disperses darkness. As soon as ignorance which is born of earthly affections is removed, the 'Self' (*Âtmâ*), by its own splendour, shines afar in an undivided state, as the sun spreads its brightness abroad when the clouds have scattered.

Śrî Śankarâchârya[1]

'There is in man something which must become conscious of itself; which must become itself, which must be purified and liberated from all that is foreign to itself; which must awaken and expand, and become all, because it is all; something which alone should be: it is the soul as knowledge, namely, the Spirit, whose "subject" is God and whose "object" is likewise God' (Schuon: *L'Oeil du Coeur,* p. 209).

Renun. 158
Infra 749

'Deliverance (in Hindu terms) is effective only insofar as it essentially implies perfect Knowledge of *Brahma*; and, inversely, this Knowledge, to be perfect, necessarily presupposes the realization of what we have ... called the "Supreme Identity". Thus, Deliverance and total and absolute Knowledge are really but one and the same thing; if it be said that Knowledge is the means of Deliverance, it must be added that in this case means and end are inseparable, for Knowledge, unlike action, carries its own fruit within itself' (Guénon: *L'Homme et son Devenir,* p. 173).

Introd. *Realiz.*

Traditional learning is basically qualitative and synthetic, concerned with essences, principles and realities behind phenomena; its fruits are integration, composition and unity. Profane academic learning—whether in the arts or sciences—is quantitative and analytical by tendency, concerned with appearances, forces and material properties; its nature is to criticize and decompose; it works by fragmentation. 'The possession of all the sciences, if unaccompanied by the knowledge of the best, will more often than not injure the possessor' (Plato: *Alcibiades II,* 144 D).

Introd. *Charity*

Thought is subservient to intellect, and insofar as it deviates from its source, 'it can only have a destructive character, similar to a corrosive acid which destroys the organic unity of beings and things. It suffices to look at the modern world, its artificial character devoid of beauty, its inhumanly abstract and quantitative structure, to understand what thought becomes when left to itself. Man, a "thinking animal", is either the divine masterpiece of nature or else its adversary; the reason for this is that "being" and "knowing" become

Infra 734

[1] *Âtmâ Bodha*: in Guénon: *L'Homme et son Devenir*, p. 174.

Introd. *Realiz.*

dissociated in the mind, which, through decadence, gives rise to all scissions' (Burckhardt: *Introduction aux Doctrines ésotériques de l'Islam,* pp. 93–94).[1]

'A strange phenomenon may be noted in the intellectual domain itself, or rather in what is left of it . . . . namely, the passion for research taken as an end in itself, quite regardless of seeing it terminate in any solution. While the rest of mankind seeks for the sake of finding and knowing, the Westerner of to-day seeks for the sake of seeking; the Gospel sentence, *Quaerite et invenietis,* is for him a dead letter, in the full force of this expression, since he calls "death" anything and everything that constitutes a definite finality, just as he gives the name "life" to what is no more than fruitless agitation. This unhealthy taste for research, real "mental restlessness" without end and without issue, shows up most particularly in modern philosophy, which for the greater part represents no more than a series of wholly artificial problems, which only exist because they are badly propounded, owing their origin and survival to nothing but carefully kept up verbal confusions; they are problems which, considering how they are formulated, are truly insoluble, but, on the other hand, no one is in the least anxious to solve them, and their sole purpose is to go on indefinitely feeding controversies and discussions which lead nowhere, and which are not meant to lead anywhere. This substitution of research for knowledge . . . is simply giving up the proper object of intelligence, and it is scarcely strange that in these conditions some people have come ultimately to abolish the very idea of truth, for the truth can only be conceived of as the end to be reached, and these people want no end to their research. It follows that there can be nothing intellectual in their efforts, even taking intelligence in its widest, not in its highest and purest sense; and if we have been able to speak of "passion for research", it is in fact because sentiment has intruded into domains where it ought never to have set foot' (Guénon: *Orient et Occident,* pp. 78–79).

Introd. *Void Action* 346

'Not only does the inferior lack the mentality of the superior, but it cannot even conceive of it exactly; few things are more painful than "psychological" interpretations which attribute to the superior man intentions which he could in no case have, and which merely reflect the smallness of their authors, as can be seen to satiety in the case of "historical criticism" or in the "science of religions"; men whose souls are fragmentary and opaque would teach us concerning the "psychology" of grandeur and of the sacred' (Schuon: *Castes et Races,* p. 15).

'Metaphysic is supra-rational, intuitive and immediate knowledge. Moreover, this pure intellectual intuition, without which there is no true metaphysic, must never be likened to the intuition spoken of by certain contemporary philosophers, which, on the contrary, is infra-rational. There is an intellectual intuition and a sensory intuition; one is above reason, but the other is below it; this latter can only grasp the world of change and becoming, namely, nature, or rather an inappreciable part of nature. The domain of intellectual intuition, by contrast, is the domain of eternal and immutable principles, it is the domain of metaphysic.

Introd. *Pilg.*

'To have a direct grasp of universal principles, the transcendent intellect itself must belong to the universal order; it is thus not an individual faculty, and to consider it as such

[1] 'According to Hindu philosophy one-sided logical abstraction ("being of being", "meaning of meaning", etc.) is a result either of rationalistic (*mânasa*) hypertrophy (*bahu-tva*) or of mental confusion (*mûdha-tva*): both antipodal to intellectual awareness (*buddha-tva*)'; Frans Vreede: *A Short Introduction to the Essentials of Living Hindu Philosophy,* Oxford University Press, 1953, p. 55.

would be contradictory, since it cannot pertain to the possibilities of the individual to transcend his own limits' (Guénon: *La Métaphysique orientale*, p. 11).

Infra 761

'This belief in the transcendent Intellect, a faculty capable, and alone capable, of direct contact with the Real, is common to all Traditional doctrines, of all ages and countries' (Marco Pallis: *Peaks and Lamas*, p. 166).

'Contemplative thought is a "vision"; it is not "in action" as is the case with passional thought.... The act of contemplation is principial, which means that its activity is in its essence, not in its operations' (Schuon: *Perspectives spirituelles*, p. 13).[1] 'When reference is made to the object of metaphysic ... this object must always be absolutely the same and can in no way be something which changes or which is subject to the influences of time and place; the contingent, the accidental, the variable, are what properly belong to the individual domain.... Where metaphysic is concerned, all that can alter with time and place are the modes of exposition, that is to say the more or less external forms which metaphysic can assume and which are capable of diverse adaptations, and also obviously, the degree of knowledge or ignorance that men, or at least the generality of them, have in respect to true metaphysic; but this always remains in essence perfectly identical with itself, for its object is essentially one, or more exactly, "without duality" as the Hindus put it; and this object, again by the very fact that it lies "beyond nature", is also beyond all change: the Arabs express this by saying that "the doctrine of Unity is unique"' (Guénon: *Introduction générale à l'Étude des Doctrines hindoues*, pp. 90–91).

'It is precisely by .... "matters of faith", and not by a difference of metaphysical basis, that one religion is distinguished from another' (Coomaraswamy: *A New Approach to the Vedas*, p. 3).

Reality 790

'If metaphysic is something sacred, this means that it should never be presented as though it were merely a profane philosophy sufficient unto itself, that is, not exceeding the limits of a play of mind. It is illogical and dangerous to talk of metaphysic without preoccupying oneself with the moral concomitants that it demands, and of which the criterion for man is his behaviour in relation to God and to his neighbour' (Schuon: *Perspectives spirituelles*, p. 234).

Beauty 676

'There are obligations which are inherent in all true knowledge, and compared with them, all outward ties seem vain and derisory; these obligations, for the very reason that they are purely inward, are the only ones that can never be shaken off' (Guénon: *Orient et Occident*, pp. 227–228).

'It is only Knowledge which leads to Knowledge. Any means whatsoever leads to Knowledge only insofar as it is a modality of Knowledge. One cannot seize Knowledge from the outside, for this outside does not exist' (Schuon: 'L'Intégration des éléments psychiques', *Études Trad.*, 1940, p. 48).

Reality 773

'Before this Divine Ray, which is transcendent Knowledge, the world and its attractions progressively withdraw, as snow melts in the sun; and no austerity could surpass in excellence the Paracletic and sanctifying miracle of pure intellectual knowledge which dissolves all the knots of ignorance' (Schuon: *De l'Unité transcendante des Religions*, p.71).

Suff. 133

[1] 'Philosophic realization does not consist in putting a theoretical view into practice, but is "simultaneous" with knowledge, i.e. with real knowledge, not with "theories about knowledge", but with "experience of knowledge" (*anubhava*), with "consciousness of reality" (*brahmabhâva*)': Frans Vreede, op. cit., p. 58.

Profane Learning

P. State 573

One loses science when losing the purity of the heart.

Nicholas Valois

Metanoia 488

You should know that these philosophers whose wisdom you so much extol have their heads where we place our feet.

Isaac of Acre

The end reached by the theologian, is the beginning of the way for the dervish.

'Abd al-Wahhâb al-Sha'rânî

Heterod. 418

Many are attracted to philosophy whose natures are imperfect and whose souls are maimed and disfigured by their meannesses. . . . And when persons who are unworthy of education approach philosophy and make an alliance with her who is a rank above them what sort of ideas and opinions are likely to be generated? Will they not be sophisms captivating to the ear, having nothing in them genuine, or worthy of or akin to true wisdom?

Plato (*Republic* VI, 495 E)

The re-integrated being, the yogi, having churned the four Wisdoms (Veda-s) and all Scriptures, enjoys their cream, their essence. The learned get only the butter milk.

Jnânasankalini Tantra, LI

With all their science, those people at Paris are not able to discern what God is in the least of creatures—not even in a fly!

Eckhart

Jejune and barren speculations may be hovering and fluttering up and down about divinity, but they cannot settle or fix themselves upon it: they unfold the plicatures of truth's garment, but they cannot behold the lovely face of it.

John Smith the Platonist

Flight 944

My brethren, the theology of this man (St Francis), founded on purity and contemplation, is a flying eagle, while our science crawls on its belly on the earth.

The Mirror of Perfection, LIII

It is much to be wished that the eyes of our self-opinionated doctors were opened, or the nebulous film, or sophistical mask, which obscures their vision, taken away, that so they might see more clearly. I am particularly alluding to the Aristotelians, and other blind theological quibblers, who spend their lives in wrangling and disputing about Divine things in a most unchristian manner, and put forth no end of manifold distinctions, divisions, and confusions, thus obscuring the Scriptural doctrine concerning the union of natures and communication of substances in Christ. If they will not believe God and His Holy Word, they might at least be enlightened by a study of our chemical Art, and of the union of two waters (viz., that of mercury and that of the Sun) which our Art so strikingly and palpably exhibits.

Infra 745

The Sophic Hydrolith

If one becomes a learned man, there is danger of his losing the disposition to practise the *Nembutsu.*

Hônen

This art is founded on the reduction of the *Corpora* into *Argentum Vivum.* It is the *Solutio Sulphuris Sapientium in Mercurio.* A knowledge not infused with life is a dead knowledge. An intelligence lacking in spirituality is only a false and borrowed light.

Heterod. 430

Henry Madathanas

Such Men presume too much upon their minde,
They weene their witts sufficient this *Arte* to finde.

Thomas Norton

For wisdom will not enter into a malicious soul, nor dwell in a body subject to sins.

Metanoia 484

Wisdom, I. 4

Philosophers are men whom too much learning and thought have made mad.

Michael Sendivogius

Of making many books there is no end; and much study is a weariness of the flesh.

Ecclesiastes, XII. 12

The knowledge of evanescent objects is not properly knowledge, but bears the same relation to reality as the mirage of the desert to water, the searcher after which obtains nothing but an increase of thirst.

Illusion 99

Azar Kaivân

In this final world age, which is the age of dissolutions, . . . the wondrous law of the one mind is being discarded everywhere, and each individual is now thinking just as he individually wishes to think. Sometimes it seems as if the real desire to know the truth is there—almost as it were by chance, but it soon turns out to be nothing but a sort of philosophical discussion or an elegant fashion of talk.

Wr. Side 464

Hakuin

Of the signs of the Hour is that knowledge shall be taken away and ignorance shall reign supreme.

Muhammad

Reality 775

Ignorant and foolish men, with a labour as vain as it is obstinate, search out the natures of things while they remain in ignorance of the One who is the Author and Maker of themselves and of all things alike. Yet they do not inquire after Him—as though without God truth might be found or happiness possessed. And, that you may be able to appreciate more clearly still how barren and indeed how pernicious such studies are, you must know that not only do they not enlighten the mind to know the truth, but they actually blind it, so that it cannot recognize the very truth. . . .

Realiz. 859
Pilg. 365
Sin 77
Judgment 261

What, then, does it profit a man to probe carefully into the nature of everything and understand it thoroughly, if he neither remembers nor knows whence he himself comes, nor whither he is going when this life is ended? For what is this mortal life but a journey? For we are passing through, and we see the things that are in this world as it were by the wayside. Does it follow, then, that we should stop and enquire into anything we see as we pass that is unusual or unfamiliar to us, and turn aside from our path for it? This is exactly what the people you are looking at are doing. Like foolish travellers, they have forgotten where they are going and have as it were sat down by the road to investigate the unfamiliar things they see. By habitually giving way to this folly they have already become such strangers to themselves that they do not remember that they are on a journey, nor do they seek their homeland. . . . No life could be more disgraceful and no end more unhappy than to have no hope of salvation when one dies, because one has been unwilling to take the path of virtue while one lived.

Hugh of Saint-Victor

Faith 512

Here a question arises: Did the disciples in this highest school of the Spirit obtain an insight into all those sciences which are learnt in the school of nature? I answer, Yes; it was given them to understand all science, whether touching the courses of the heavenly bodies, or what not, in so far as it might conduce to God's glory, or concerned the salvation of man; but those points of science which bear no fruit for the soul, they were not given to know.

Tauler

I desire no man to dislike or renounce his skill in ancient or modern languages; his knowledge of medals, pictures, paintings, history, geography or chronology; . . . but . . . all these things are to stand in their proper places. . . .

Christian redemption is quite of another nature; it has no affinity to any of these arts or sciences; it belongs not to the outward natural man, but is purely for the sake of an inward, heavenly nature.

William Law

Wr. Side 442

Jesus (Peace be upon him!) said, 'How many trees are there, yet all of them do not bear fruit; and how many fruits are there, yet all of them are not good; and how many sciences are there, yet all of them are not useful.'

Christ in Islâm

Judgment 239

He is a miserable man who knows all things, and does not know God; and he is happy who knows God, even though he know nothing else. But he who knows God and all else beside is not made more blessed thereby; for he is blessed through God alone.

St Augustine

Reality 775

Silence, O brother! put learning and culture away:
Till Thou namedst culture, I knew no culture but Thee.

Dîvâni Shamsi Tabrîz, XXXII

But thou art all my art, and dost advance
As high as learning my rude ignorance.

Shakespeare (*Sonnet* LXXVIII)

If God guides you not into the road,
It will not be disclosed by logic.

Shabistarî

Orthod. 288

The literalistic ulemas receive knowledge from transmitter to transmitter down to the Day of Judgment. The origin is far away. But the elect draw upon their knowledge from God who has placed it in their hearts.

Ibn 'Arabî

Orthod. 275
Rev. 965

Any knowledge that does not bring us this supreme bliss and freedom is not worth acquiring. We stuff our minds with knowledge of so many facts and things gained from all and sundry, or reading all kinds of books. The brain becomes a repository of learning about all the ephemeral and passing phases of life. Naturally, such a man becomes a restless being—unbalanced, confused and erratic in his behaviour and conduct. Seek, therefore, to know the true source of your life,—God. That is why you are here.

Swami Ramdas

Creation 48

Scholarship of this sort is nothing but an ear for seeking abstractions. Much learning and many arts lead merely to the delusions which are rank growing seeds of phantasy. Scholarship is nothing but this.

Hakuin

Introd.
Reality

If a man had all that sort of knowledge that ever was, he would not be at all the wiser; he would only be able to play with men, tripping them up and oversetting them with distinctions of words. He would be like a person who pulls away a stool from some one when he is about to sit down, and then laughs and makes merry at the sight of his friend overturned and laid on his back.

Plato (*Euthydemus,* 278 C)

Heterod. 418

None are so surely caught, when they are catch'd,
As wit turn'd fool: folly, in wisdom hatch'd,
Hath wisdom's warrant and the help of school
And wit's own grace to grace a learned fool.

Shakespeare (*Love's Labour's Lost,* v. ii. 69)

Sin 75

You *may*, if you squeeze hard enough, even get oil from sand,
thirsty, you *may* succeed in drinking the waters of the mirage,
perhaps, if you go far enough, you'll find a rabbit's horn,
but you'll never satisfy a fool who's set in his opinions!

Heterod. 420

Bhartṛhari

Books are comprehensible only to those who understand and things are valuable only to those who can appreciate. Give a bow of decoration to a farmer, and he will use it to chase birds away from his farm; give an emperor's robe to a southern barbarian and he will wear it to carry wood. They are simply ignorant. What else can be expected of them?

Ko Hung

Holofernes. Satis quod sufficit.

Sir Nathaniel. I praise God for you, sir: your reasons at dinner have been sharp and sententious; pleasant without scurrility, witty without affection, audacious without impudency, learned without opinion, and strange without heresy. I did converse this quondam day with a companion of the king's, who is intituled, nominated, or called, Don Adriano de Armado.

Hol. Novi hominem tanquam te: his humour is lofty, his discourse peremptory, his tongue filed, his eye ambitious, his gait majestical, and his general behaviour vain, ridiculous, and thrasonical. He is too picked, too spruce, too affected, too odd, as it were, too peregrinate, as I may call it.

Nath. A most singular and choice epithet.

(*Draws out his table-book.*)

Hol. He draweth out the thread of his verbosity finer than the staple of his argument. I abhor such fanatical phantasimes, such insociable and point-devise companions; such rackers of orthography, as to speak dout, fine, when he should say, doubt; det, when he should pronounce, debt,—d, e, b, t, not d, e, t: he clepeth a calf, cauf; half, hauf; neighbour *vocatur* nebour, neigh abbreviated ne. This is abhominable, which he would call abominable,—it insinuateth me of insanie: *anne intelligis, domine*? To make frantic, lunatic.

Nath. Laus Deo bone intelligo.

Hol. Bone? bone, for *bene*: Priscian a little scratched; 'twill serve.

Shakespeare (*Love's Labour's Lost,* v. i. 1)

He that will not seek thereby a new man born in God, and apply himself diligently thereto, let him not meddle with my writings.

I have not written anything for such a seeker, and also he shall not be able to apprehend our meaning fundamentally, though he strives never so much about it, unless he enters into the resignation in Christ; there he may apprehend the spirit of the universal, otherwise all is to no purpose; and we faithfully warn the curious critic not to amuse himself, for he will not effect anything in this way, unless he himself enters thereinto, and then it will be shewn him without much seeking; for the way is child-like.

Introd.

P. State 573

Boehme

Study is like the heaven's glorious sun,
  That will not be deep-search'd with saucy looks;
Small have continual plodders ever won,
  Save base authority from others' books.

Shakespeare (*Love's Labour's Lost*, I. i. 84)

As for the mere word-by-word explication of these dialectics, thou thyself art sufficiently expert; but to realize their true import it is necessary to renounce the Eight Worldly Ambitions, lopping off their heads, to subdue the illusion of belief in the personal ego, and, regarding *Nirvâṇa* and *Sangsâra* as inseparable, to conquer the spiritual ego by meditation in mountain solitudes. I have never valued or studied the mere sophistry of word-knowledge, set down in books in conventionalized form of questions and answers to be committed to memory (and fired off at one's opponent); these lead but to mental confusion and not to such practice as bringeth actual realization of Truth. Of such word-knowledge I am ignorant; and if ever I did know it, I have forgotten it long ago.

Death 220
M.M.
978

Milarepa (addressing a pundit)

Would you divinely know the mysteries of nature, the ground and reason of good and evil in this world, the relation and connection between the visible and invisible world, how the things of time proceed from, are influenced by, and depend upon the things and powers of eternity, there is but one only key of entrance; nothing can open the vision but seeing with the eyes of that same love which begun and carries on all that is and works in visible and invisible nature. Would you divinely know the mysteries of grace and salvation, would you go forth as a faithful witness of Gospel truths, stay till this fire of divine love has had its perfect work within you. For till your heart is an altar on which this heavenly fire never goes out, you are dead in yourself and can only be a speaker of dead words about things that never had any life within you. For without a real birth of this divine love in the essence of your soul, be as learned and polite as you will, your heart is but the dark heart of fallen Adam, and your knowledge of the kingdom of God will be only like that which murdering Cain had.

Symb.
306
Center
816
Contem. 542
Holiness
902

William Law

He who is learned is not wise;
He who is wise is not learned.

Tao Te Ching, LXXXI

For in much wisdom is much grief: and he that increaseth knowledge increaseth sorrow.

Ecclesiastes, I. 18

One man spends seventy years in learning
And fails to kindle the light.
Another, all his life learns nothing
But hears one word
And is consumed by that word.

Inv.
1007

Ansârî

Realiz. 873

Light seeking light doth light of light beguile.

Shakespeare (*Love's Labour's Lost,* I. i. 77)

O foolish people, and without understanding; which have eyes, and see not; which have ears, and hear not.

Jeremiah, v. 21

Holy words do not abide in blind hearts, but go to the Light whence they came,
While the spell of the Devil goes into crooked hearts as a crooked shoe on to a crooked foot.

Heterod. 423

Though you may learn Wisdom by rote, it becomes quit of you when you are unworthy;
And though you write it and note it, and though you brag and expound it,
It withdraws its face from you, O disputatious one: it snaps its bonds and takes flight from you.

Rûmî

Peace 694

'If I had not, of My clemency towards thee, given thee to drink of the cups of My Self-revelation unto thee, the well of every knowledge would have made thee to thirst, and the confusion of every thought would have bewildered thee.'

Niffarî

Orthod. 288

Those who do not readily drink the ambrosia of their master's instruction,
Die of thirst in the desert of multitudinous treatises.

Saraha

Action 329

When Ananda came into the presence of the Lord Buddha, he bowed down to the ground in great humility, blaming himself that he had not yet fully developed the potentialities of Enlightenment, because from the beginning of his previous lives, he had too much devoted himself to study and learning.

Surangama Sutra, I

'He who is not ashamed for the superfluity of knowledge, will never be ashamed.'

Niffarî

I would rather die of pure love than let God escape from me in dark wisdom.

Mechthild of Magdeburg

Judgment 239

Learning without religious fear is as a woman of contradictions, disobedient, and lacking in manners, one who makes her eyes look large by the use of rouge and eye-paint, and adorns herself with necklaces and ear-rings. What is the use of all her beauty and splendour, when her clothing is untidy, and her true nature is disclosed?

Yesod Hayirah, 24

You are wise, you are well-read and you have gone through spiritual practices, you have put on the garb of sanctity, you can preach and you can pose—these are all nothing if you do not experience the blissful union with the Beloved. The Saguna (Attribute) you are after is an image of your own mind and it cannot satisfy you.

Swami Ramdas

Death 220

The one thing learning has taught me is its utter powerlessness to bring me *Ôjô*.[1]

Hônen

Grace 552

Mere dry reasoning—I spit on it! I have no use for it! (*The Master spits on the ground.*)

Sri Ramakrishna

People hope to find Brahma in these books, but it is brahma (confusion) not Brahma (God) that they find there.

Sai Baba

The first step in this affair (Ṣûfism) is the breaking of ink-pots and the tearing-up of books and the forgetting of all kinds of knowledge.

Abû Sa'îd ibn Abi 'l-Khayr

It is the unlearned who are saved rather than those whose ego has not subsided despite their learning. The unlearned are saved from the unrelenting grip of self-infatuation, from the malady of a myriad whirling thoughts, from the endless pursuit of (mental) wealth; it is not from one ill alone that they are saved.

Sri Ramana Maharshi

P. State 573

Do away with learning, and grief will not be known.
Do away with sageness and eject wisdom, and the people will be more benefited a hundred times.

Tao Te Ching, XIX

Those who desire progress along the Way must first cast out the dross acquired through heterogeneous learning.

Huang Po

Know the taste of this flavour which consists in absence of knowledge.

Sahara

Cherish that which is within you, and shut off that which is without; for much knowledge is a curse.

Chuang-tse (ch. XI)

Infra 755

The Gnostic is not one who commits to memory passages from the Qur'ân, who if he forgets what he has learned, becomes ignorant. He only is the Gnostic who takes his

[1] Birth into the Pure Land.

knowledge from His Lord at all times, without having to learn it, and without studying, and this (knowledge) lasts throughout his life-time, he does not forget his knowledge, but he remembers it for ever. He has no need of a book, and he is the (true) spiritual Gnostic.

Holiness 914
902

Abû Ṭâlib al-Makkî

The false alchemists only seek to make gold, whereas the true philosophers desire nothing but knowledge; the former get only tinctures, adulterations, sterilities, while the latter inquire into the principles of things.

Heterod. 430

Becher

I am persuaded that it is only by the gift of God that this Art can be understood. If, indeed, subtlety and mental acuteness were all that is necessary for its apprehension, I have met with many strong minds, well fitted for the investigation of such subjects. But I tell you: Be simple, and not overwise, until you have found the secret. Then you will be obliged to be prudent, and you will easily be able to compose any number of books, which is doubtless more simple for him who is in the centre and beholds the thing itself, than one who is on the circumference only, and can only go by hearsay.

Grace 552
P. State 573
Center 816

Michael Sendivogius

And know, O Soul, that it will not be possible for you, when you depart from the sense-world, to take with you any knowledge of the world of composite (or concrete) things, as though such knowledge were (as in truth it is not) separated (from external things) and joined and mingled with the being of the soul itself. Grasp then the knowledge of simple things, and abandon knowledge of composite things.

Illusion 99

Hermes

The modifications of the intellect, called 'right knowledge', 'doubtful knowledge' and 'false knowledge', deviate from their existence. There is one and the same Consciousness in all of them. The differences are due to the modifications.

Realiz. 873

Śrî Śankarâchârya

One momentary glimpse of Divine Wisdom, born of meditation, is more precious than any amount of knowledge derived from merely listening to and thinking about religious teachings.

Infra 761

Gampopa

Any one in whose soul God shall put the touchstone, he will distinguish certainty from doubt.

Rev. 965

Rûmî

Intermediate between the objects of intellection and the objects of sense are the objects of opinion; and of these, some partake of the objects of intellection, but others do not.

Hermes

Knowledge has two wings, Opinion one wing: Opinion is defective and curtailed in flight.

Rûmî

As being is to becoming, so is pure intellect to opinion.

And as intellect is to opinion, so is science (gnosis) to belief, and understanding to the perception of shadows.

Plato (*Republic* VII, 534 A)

Perfected is the Word of thy Lord in truth and justice. There is naught that can change His words. He is the Hearer, the Knower.

Infra 749

If thou obeyest most of those on earth they would mislead thee far from Allâh's way. They follow naught but an opinion, and they do but guess.... And lo! a guess can never take the place of the Truth.

Qur'ân, VI. 115–116, & LIII. 28

Those who love the Truth in each thing are to be called lovers of wisdom and not lovers of opinion.

Plato (*Republic* V, 480 B)

He who pursues learning will increase every day;
He who pursues Tao will decrease every day.
He will decrease and continue to decrease,
Till he comes at non-action;
By non-action everything can be done.

Metanoia 488
Action 358

Tao Te Ching, XLVIII

The Knowledge of One's Ignorance

I am better off than he (a man reputed for wisdom) is,—for he knows nothing, and thinks that he knows; I neither know nor think that I know.... The truth is, O men of Athens, that God only is wise.

Sin 56

Plato (*Apology,* 21 D, 23 A)

The perfection of human knowledge is ignorance of Divine knowledge. You must know enough to know that you do not know.

Death 208

Hujwîrî

If any man think that he knoweth anything, he knoweth nothing yet as he ought to know.

I. Corinthians, VIII. 2

Action 340

Now when hurt toucheth a man he crieth unto Us, and afterward when We have granted him a boon from Us, he saith: Only by force of knowledge I obtained it. Nay, but it is a test. But most of them know not.

Qur'ân, XXXIX. 49

Infra 745

First, action is necessary, then knowledge, in order that thou mayest know that thou knowest naught and art no one. This is not easy to know. It is a thing that cannot be rightly learned by instruction, nor sewed on with needle nor tied on with thread. It is the gift of God.

Abû Sa'îd ibn Abi 'l-Khayr

Sin 80

Not knowing that one knows is best;
Thinking that one knows when one does not know is sickness.
Only when one becomes sick of this sickness can one be free from sickness.
The sage is never sick: because he is sick of this sickness, therefore he is not sick.

Tao Te Ching, LXXI

. . . His ignorance were wise,
Where now his knowledge must prove ignorance.

Shakespeare (*Love's Labour's Lost*, II. i. 102)

The truly learned man is he who understands that what he knows is but little in comparison with what he does not know.

'Alî

M.M. 978

Knowledge implies ignorance of what lies beyond what is known. Knowledge is always limited.

Sri Ramana Maharshi

Let me be ignorant, and in nothing good,
But graciously to know I am no better.

Shakespeare (*Measure for Measure*, II. iv. 77)

What one knows is not so much as what one does not know. There is a great variety of things.

Ko Hung

A man asked the Maharshi to say something to him. When asked what he wanted to know, he said that he knew nothing and wanted to hear something from the Maharshi.

Maharshi: You know that you know nothing. Find out that knowledge. That is liberation (*Mukti*). — *Death* 220

Sri Ramana Maharshi

When you know that you do not know anything, then you know every thing. — *Void* 724

Swami Ramdas

Gnosis is the realisation of thy ignorance when His knowledge comes. — *Rev.* 961

Junayd

The better a man will have known his own ignorance, the greater his learning will be. — *Realiz.* 859

Nicholas of Cusa

The final aim of knowledge is to hold that we know nothing, He alone being wise, who is also alone God. — Infra 749

Philo

Belief and Practice

O ye who believe! Why say ye that which ye do not?
It is most hateful in the sight of Allâh that ye say that which ye do not. — *Sin* 76

Qur'ân, LXI. 2, 3

To know and yet not to do is in fact not to know.

Wang Yang-ming

Knowledge without action is not knowledge.

Hujwîrî

It is written in the Pentateuch and the Gospel. 'Do not seek knowledge of what you do not know until you practise what you do know.' — *Faith* 512

Christ in Islâm

It is told of Ibrâhîm b. Adham that he saw a stone on which was written, 'Turn me over and read!' He obeyed, and found this inscription: 'Thou dost not practise what thou knowest; why, then, dost thou seek what thou knowest not?'

Hujwîrî

But be ye doers of the word, and not hearers only, deceiving your own selves. — *Renun.* 146

James, I. 22

That we were all, as some would seem to be,
From our faults, as faults from seeming, free!

Shakespeare (*Measure for Measure,* III. ii. 40)

Jesus (Peace be upon him!) said to the disciples, 'Verily I say unto you, the speaker of wisdom and the hearer of it are partners, and the one of them who is more worthy of it is he who verifies it by his deeds.'

Christ in Islâm

There is a big difference between a scholar and a holy man. The mind of a mere scholar is fixed on 'woman and gold', but the sâdhu's mind is on the Lotus Feet of Hari. A scholar says one thing and does another. But it is quite a different matter with a sâdhu. The words and actions of a man who has given his mind to the Lotus Feet of God are altogether different.

Holiness 914

Sri Ramakrishna

One authority says that no one is so foolish that he does not desire wisdom. Why, then, are we not all wise? Because so much is required to that end. The principal requirement is that one shall get beyond phenomenal nature and in this process he begins soon to be weary. Then he lags behind with his little wisdom, a little man. That I am a wealthy man does not mean that I am wise too, but when my nature is conformed to God's Being, so that I am wisdom itself, then only am I a wise man.

Holy War 405
Void 721
Conform. 180

Eckhart

Constant devotion to spiritual knowledge, realization of the essence of Truth, this is declared to be wisdom; what is opposed to this is ignorance.

Bhagavad-Gîtâ, XIII. 11

Philosophy is nothing else than striving through constant contemplation and saintly piety to attain to knowledge of God.

Hermes

If, after obtaining this knowledge, you give way to pride or avarice (under the pretext of economy and prudence), and thus gradually turn away from God, the secret will most certainly fade out of your mind in a manner which you do not understand. This has actually happened to many who would not be warned.

Heterod. 423

The Sophic Hydrolith

If to do were as easy as to know what were good to do, chapels had been churches, and poor men's cottages princes' palaces. It is a good divine that follows his own instructions: I can easier teach twenty what were good to be done, than be one of the twenty to follow mine own teaching. The brain may devise laws for the blood, but a hot temper leaps o'er a cold decree: such a hare is madness the youth, to skip o'er the meshes of good counsel the cripple.

Illusion 89

Shakespeare (*Merchant of Venice,* I. ii. 13)

The bane of knowledge is lack of practising it.

'Alî

Many people arrive at specific understanding, at formal, notional knowledge, but there are few who get beyond the science and the theory; yet one man whose mind is free from notions and from forms is more dear to God than the hundred thousand who have the habit of discursive reason. God cannot enter in and do his work in them owing to the restlessness of their imagination.

Eckhart

Contem. 528

All people are dead, save for those who know; and those who know are dead, save for those who practice; and those who practice are all astray, save for those who act with right intention; and those who act with right intention are all in grave danger.

Dhu 'l-Nûn

Introd. *Conform.* *Judgment* 239

No man therefore can be a man unless he be a Philosopher, nor a true Philosopher unless he be a Christian, nor a perfect Christian unless he be a Divine.

Thomas Traherne

Contem. 547

Knowledge without practice accompanying it is superior to practice without knowledge. Practice with knowledge is superior to knowledge without practice accompanying it.

Yoga-Vasishtha

Action 340

A little knowledge which you carry out in action is more profitable than much knowledge which you neglect to carry out in action. May God have mercy on him who knows and does.

Hermes

Some knowledge is needed for yoga and it may be found in books. But practical application is the thing needed, and personal example, personal touch and personal instruction.

Sri Ramana Maharshi

Orthod. 288

Then I turned my attention to the Way of the Sufis. I knew that it could not be traversed to the end without both doctrine and practice, and that the gist of the doctrine lies in overcoming the appetites of the flesh and getting rid of its evil dispositions and vile qualities, so that the heart may be cleared of all but God; and the means of clearing it is *dhikr Allah,* i.e. commemoration of God and concentration of every thought upon Him. Now, the doctrine was easier to me than the practice, so I began by learning their doctrine from the books and sayings of their Shaykhs, until I acquired as much of their Way as it is possible to acquire by learning and hearing, and saw plainly that what is most peculiar to them cannot be learned, but can only be reached by immediate experience and ecstasy and inward transformation. . . . I looked on myself as I then was. Worldly interests encompassed me on every side. . . . I realised that I stood on the edge of a precipice and would fall into Hell-fire unless I set about to mend my ways. . . . Conscious of my helplessness and having surrendered my will entirely, I took refuge with

Parts I & II *Inv.* 1009

Realiz. 859 *Humility* 191

God as a man in sore trouble who has no resource left. God answered my prayer and made it easy for me to turn my back on reputation and wealth and wife and children and friends.

Renun. 136

Al-Ghazâlî

As the Tathagata speaks so he does, as he does so he speaks. Because he speaks as he does and does as he speaks, he is therefore called Tathagata.[1]

Dîgha-Nikâya, III. 135

Wisdom without action hath its seat in the mouth; but by means of action, it becometh fixed in the heart.

Shekel Hakodesh, 43

And Moses was learned in all the wisdom of the Egyptians, and was mighty in words and in deeds.

Acts, VII. 22

Holiness 914

The Sûfî is he whose language, when he speaks, is the reality of his state.

Dhu 'l-Nûn

The two rarest things in our time are a learned man who practises what he knows and a gnostic who speaks from the reality of his state.

Abu 'l-Ḥusayn al-Nûrî

Were I indeed to define divinity, I should rather call it a *divine life*, than a *divine science*; it being something rather to be understood by a spiritual sensation, than by any verbal description. . . . Divinity indeed is a true efflux from the eternal light, which, like the sun-beams, does not only enlighten, but heat and enliven; and therefore our Saviour hath in his beatitudes connected purity of heart with the beatifical vision. And as the eye cannot behold the sun, unless it be sunlike, and hath the form and resemblance of the sun drawn in it; so neither can the soul of man behold God, unless it be Godlike, hath God formed in it, and be made partaker of the divine nature. . . .

Infra 749

To seek our divinity merely in books and writings, is to seek the living among the dead: we do but in vain seek God many times in these, where his truth too often is not so much enshrined as entombed: no; *intra te quaere Deum*, seek for God within thine own soul: he is best discerned νοερᾷ ἐπαφῇ, as Plotinus phraseth it, by an intellectual touch of him.

Contem. 536

John Smith the Platonist

The contents of this Book are not fables, but real experiments which I have seen, touched, and handled.

An Open Entrance to the Closed Palace of the King

P. State 563

For the Lord manifests himself ungrudgingly through all the universe: and you can behold God's image with your eyes, and lay hold on it with your hands.

Hermes

[1] 'He who has fully arrived', a title of the Buddha.

As our Matter, in the philosophical work, after being dissolved into its three parts or principles, must again be coagulated and reduced into its own proper salt, and into *one* essence, which is then called the salt of the Sages: so God, and His Son, must be known as *One,* by means of their essential substance, and must not be regarded as two or three Divinities, possessing more than one essence. When you have thus known God through His Son, and united them by the bond of the Holy Spirit, God is no longer invisible, or full of wrath, but you may feel His love, and, as it were, see Him with your eyes, and handle Him with your hands, in the person of Jesus Christ, His Son and express image.

The Sophic Hydrolith

Pilg. 378

Reality 775

P. State 579

That which was from the beginning, which we have heard, which we have seen with our eyes, which we have looked upon, and our hands have handled, of the Word of life. . . .

I. John, I. 1

Fear Him, love Him, and read carefully the books of His chosen Sages—and you will soon *see,* and behold with your own eyes, that I have spoken truly.

Michael Sendivogius

Center 816

I have seen the Green Catholic Lion, and the Blood of the Lion, *i.e.,* the Gold of the Sages, with my own eyes, have touched it with my hands, tasted it with my tongue, smelt it with my nose.

Heinrich Khunrath

O taste and see that the Lord is good: blessed is the man that trusteth in him.

Psalm XXXIV. 8

The Knower and the Known

God can be known only by God.

Theologia Germanica, XLII

Exalted be He Who is demonstrated only by Himself and Who exists only by His own Essence!

Ibn 'Arabî

The things of God knoweth no man, but the Spirit of God.

I. Corinthians, II. 11

Supra 743

If someone asks you what ghee is like, your answer will be, 'Ghee is like ghee.' The only analogy for Brahman is Brahman. Nothing exists besides It.

Sri Ramakrishna

Nothing but truth itself can be the exact measure of truth.

Nicholas of Cusa

Metanoia 493

He that is of God heareth God's words: ye therefore hear them not, because ye are not of God.

St John, VIII. 47

Infra 761

God is intelligence occupied with knowing itself.

Eckhart

Creation 48
Contem. 547
Center 847
841, 828

God then alone most perfectly and substantially enjoyeth Himself in the contemplation of Himself, which is the *Beatifical Vision* of the most beautiful, the most blessed *Essence* of *Essences*. This *Act* of *Contemplation* is an Intellectual and *Divine Generation*, in which the Divine Essence, with an eternity of most heightned Pleasures, eternally bringeth forth it self, within it self, into an *Image* of it self.

Peter Sterry

Reality 775

Whosoever is wise derives his wisdom from the Divine wisdom. Wherever intelligence is found it is the fruit of the Divine intelligence.

Jâmî

To be allied to wisdom is immortality.

Wisdom, VIII. 17

Beauty 676

Never did eye see the sun unless it had first become sunlike, and never can the soul have vision of the First Beauty unless itself be beautiful.

Plotinus

The proof of the sun is the sun: if thou require the proof, do not avert thy face!

Rûmî

Realiz. 890

O Light eternal who only in thyself abidest, only thyself dost understand, and self-understood, self-understanding, turnest love on and smilest at thyself!

Dante (*Paradiso*, XXXIII. 124)

He (Dhu 'l-Nûn al-Miṣrî) said, 'Real knowledge is God's illumination of the heart with the pure radiance of knowledge,' *i.e.* the sun can be seen only by the light of the sun.

'Aṭṭâr

Center 833

The Self is self-effulgent. One need give it no mental picture, any way. The thought that imagines is itself bondage, because the Self is the Effulgence transcending darkness and light; one should not think of it with the mind. Such imagination will end in bondage, whereas the Self is spontaneously shining as the Absolute. This enquiry into the Self in the form of devotional meditation, evolves into the state of absorption of the mind into the Self and leads to Liberation.

Contem. 528

Sri Ramana Maharshi

Thus saith the prophet: *Domine, in lumine tuo videbimus lumen.* Lord, we shall see Thy light by Thy light (Psa. xxxvi. 9).

Walter Hilton

God is His own brightness and is discerned through Himself alone. . . . The seekers for truth are those who envisage God through God, light through light.

Philo

. . . That ineffable light whereby the Divinity comprehends its own essence, penetrating all that immensity of being which itself is.

John Smith the Platonist

Just as one light does not depend on another in order to be revealed, so, what is one's own nature does not depend on anything else (i.e., being of the nature of Knowledge the Self does not require another knowledge in order to be known).

Infra 755

Realiz. 873

Śrî Śankarâchârya

I know God by God, and I know that which is not God by the light of God.

'Ali

It is said that one who is not a *sâdhu* is incapable of recognizing a *sâdhu,*—just as a vendor of egg-plants is incapable of appraising a diamond. Only the person who has reached through *sâdhanâ* a high spiritual state can understand a true devotee.

Swâmi Brahmânanda

St Augustine being asked the meaning of eternal life answered and said: Dost thou ask me what is eternal life? Ask eternal life, see what it says itself. None knows what heat is like the hot, nor wisdom like the wise, none knows the meaning of eternal life so well as the eternal life itself.

Eckhart

But he who is not true himself will not see the truth.

Paracelsus

He only is able to declare with spirit and power any truths or bear a faithful testimony of the reality of them who preaches nothing but what he has first seen and felt and found to be true by a living sensibility and true experience of their reality and power in his own soul. All other preaching, whether from art, hearsay, books, or education, is, at best, but playing with words and mere trifling with sacred things.

Supra 745

William Law

That which enables us to know and understand aright in the things of God, must be a living principle of holiness within us.

John Smith the Platonist

If then you do not make yourself equal to God, you cannot apprehend God; for like is known by like.

Hermes

A Buddha alone is able to understand what is in the mind of another Buddha.

Aggana Suttanta

Holiness 924

The person who knows all that God knows is a prophet.

Eckhart

All things are delivered unto me of my Father: and no man knoweth the Son, but the Father; neither knoweth any man the Father, save the Son, and he to whomsoever the Son will reveal him.

St Matthew, XI. 27

Love 618

For the Divine Nature of Christ is a magnet that draws unto itself all spirits and hearts that bear its likeness.

Tauler

Realiz. 890 *Rev.* 965

Only when thou realizest that thou art That which knows, will knowledge be truly thine; and then thy certitude will have no further need of confirmations, for the quality (*aṣ-ṣifa*) is inseparable from its subject.

Jîlî

M.M. 978

Do you not understand that he who made the inside is also he who made the outside?

The Gospel according to Thomas, Log. 89

He who knows the Truth, knows that I am speaking the truth.

Eckhart

Faith 505

Knowing demands the organ fitted to the object.

Plotinus

Introd. *Contem.*

And thus shalt thou knittingly, and in a manner that is marvellous, worship God with himself.

The Epistle of Privy Counsel, IV

O Supreme Being, O Source of beings, O Lord of beings, O God of gods, O Ruler of the universe, Thou Thyself alone knowest Thyself by Thyself.

Bhagavad-Gîtâ, X. 15

Therefore, if God's essence is to be seen at all, it must be that the intellect sees it through the divine essence itself; so that in that vision the divine essence is both the object and the medium of vision.

St Thomas Aquinas

The eternal procession is the revelation of himself to himself. The knower being that which is known.

Eckhart

Unity will not embrace plurality,
For the point of Unity has one root only.

Shabistarî

When the spirit became lost in contemplation, it said this:
'None but God has contemplated the beauty of God.'

Dîvâni Shamsi Tabriz, XXIII

Realiz. 887

They (the Sufis) are agreed that the only guide to God is God Himself.

Al-Kalâbâdhî

When he, the Spirit of truth, is come, he will guide you into all truth.

St John, XVI. 13

God is like none else, wherefore none can know him thoroughly from a likeness.

Antisthenes

M.M. 975

... That Supreme Deity who alone doth perfectly behold Himself.

Dante (*Il Convito*, II. iv. 1)

So coin not similitudes for Allâh. Lo! Allâh knoweth; ye know not.

Qur'ân, XVI. 74

To see the beauty of Lailâ requires the eyes of Majnûn.

Persian Sufic

Listen, my son; this taste cannot be told by its various parts.
For it is free from conceits, a state of perfect bliss, in which existence has its origin.

Saraha

Void 728

Brahman knows *Brahman*, and is established in Its own Self.

Yoga-Vasishtha

By no one may the Innate be explained,
In no place may it be found,
It is known of itself by merit,
And by due attendance on one's master.

Hevajratantra, I

Center 816
Orthod. 288

Introd.

As It (the Consciousness which is your Self) never ceases to exist, Its eternal immutability is self-evident and does not depend on any evidence. . . . Just as iron, water, etc., which are not of the nature of light and heat, depend for them on the sun, fire and other things other than themselves, but the sun and fire themselves, always of the nature of light and heat, do not depend for them on anything else; so, being of the nature of pure Knowledge It does not depend on an evidence to prove that It exists or that It is the Knower.

Śrî Śankarâchârya

For this science[1] there is no naturalistic evidence, its premises being as spiritual as are its inferences. . . . Indeed, there is no proof in this science except experience itself.

Shaare Tṣedek

There what we hold by faith shall be beheld, not demonstrated, but self-known in fashion of the initial truth which man believeth.

Dante (*Paradiso*, II. 43)

Realiz. 890

Recognise what God is, and what that is in you which recognises God.

Sextus the Pythagorean

God implanted in man a sight called intellect, which is capable of beholding God.

Crito

Center 816
Death 208

His divinity can in no wise be seen by human sight, but is seen by that sight with which those who see are no longer men, but beyond men.

St Augustine

Realiz. 890

Nothing at all can get into God, who is pure being, but what is also pure being.

Eckhart

Wot thou well God alone knows Himself.

Richard Rolle

God alone knows Himself.

Nicholas of Cusa

Knowledge is seeing the oneness of the Self with God.

Srimad Bhagavatam, XI. xii

He is the unseen Seer, the unheard Hearer, the unthought Thinker, the uncomprehended Comprehensor. Other than He there is no seer. Other than He there is no hearer. Other

[1] *Gematrioth*, a Kabbalistic science dealing with the esoteric values of letters, words, and numbers.

than He there is no thinker. Other than He there is no comprehensor.[1] He is your Self, the Inner Controller, the Immortal.

Bṛihad-Âraṇyaka Upanishad, III. vii. 23

Reality 803

Recollection

Not that the Good is wholly incommunicable to anything; nay, rather, while dwelling alone by Itself, and having there firmly fixed Its super-essential Ray, It lovingly reveals Itself by illuminations corresponding to each separate creature's powers, and thus draws upwards holy minds into such contemplation, participation and resemblance of Itself as they can attain—even them that holily and duly strive thereafter and do not seek with impotent presumption the Mystery beyond that heavenly revelation which is so granted as to fit their powers, nor yet through their lower propensity slip down the steep descent, but with unwavering constancy press onwards toward the ray that casts its light upon them and, through the love responsive to these gracious illuminations, speed their temperate and holy flight on the wings of a godly reverence.

Dionysius

Orthod. 296

Conform. 180

Flight 944

The soul is capable of knowing all things in her highest power.

Eckhart

Knowledge comes about insofar as the object known is within the knower.

St Thomas Aquinas

Supra 749

Knowledge is simply recollection.

Plato (*Phaedo,* 72 E)

Remembering is for those who have forgotten.

Plotinus

For what is implied in the word 'recollection', but the departure of knowledge?

Plato (*Symposium,* 208 A)

The words of wisdom are the lost things of a believer and he must claim them wherever he finds them.

Muhammad

Reality 790

Infra 761

[1] These formulae have their Islamic equivalent in the *Shahâdah: Lâ ilâha illâ 'Llâh,* 'There is no divinity if not the Divinity'—a doctrinal synthesis which can be extended to all Archetypes. Thus: there is no act if not the Act, there is no beauty if not the Beauty, there is no wisdom if not the Wisdom, etc.

As by its χρονιχοὶ πρόοδοι (as the Platonists are wont to speak) 'chronical and successive operations', (the soul) unravels and unfolds the contexture of its own indefinite intellectual powers by degrees: so by this memory and prevision it recollects and twists them all up together again into itself. And though it seems to be continually sliding from itself in those several vicissitudes and changes which it runs through in the constant variety of its own effluxes and emanations; yet is it always returning back again to its first original, by a swift remembrance of all those motions and multiplicity of operations which have begot in it the first sense of this constant efflux. As if we should see a sunbeam perpetually flowing forth from the bright body of the sun, and yet ever returning back to it again: it never loseth any part of its being, because it never forgets what itself was: and though it may number out never so vast a length of its duration, yet it never comes nearer to its old age, but carrieth a lively sense of its youth and infancy along with it, which it can at pleasure lay a fast hold on.

Creation 37

Introd. *Holiness* *Center* 841 838

But if our souls were nothing else but a complex of fluid atoms,[1] we should be continually roving and sliding from ourselves, and soon forget what we once were. The new matter that would come in to fill up that vacuity which the old had made by its departure, would never know what the old were, nor what that should be that would succeed.

John Smith the Platonist

Since all Nature is congeneric, there is no reason why we should not, by remembering but one single thing— which is what we call 'learning'—discover all the others, if we are brave and faint not in the enquiry: for it seems that to enquire and to learn are wholly a matter of remembering.

Reality 775

Plato (*Meno*, 81 C)

It is impossible to know any thing of God aright by the *Natural Image*, except you have first the *Spiritual Image*, which is God himself form'd in your Souls.

Supra 749

Peter Sterry

If any one put into your hand a flint, and asked you to bring outward and visible fire out of it for him, you would be unable to do so without the steel that belongs to it, with which you would have to elicit the spark slumbering in the stone. . . . In the same manner, the heavenly light slumbers in the human soul, and must be struck out by outward contact, namely, by the true faith, through reading and hearing, and through the Holy Spirit whom Christ restored to us.

Orthod. 280

The Sophic Hydrolith

The Comforter, which is the Holy Ghost, whom the Father will send in my name, he shall teach you all things, and bring all things to your remembrance, whatsoever I have said unto you.

Introd. *Inv.*

St John, XIV. 26

[1] In answer to Epicurus. The argument in this paragraph, which is derived from Plotinus (*Enneads*, IV. vii. 5), rejects evolutionism while anticipating it.

This sort of thing cannot be taught, my son; but God, when he so wills, recalls it to our memory.

Hermes

All the Sadhanas[1] are for keeping up a ceaseless remembrance of the great Truth which the soul has forgotten and which he is in reality.

Swami Ramdas

Inv. passim *Creation* 42

'Thy seeking of Me, that I should teach thee what thou knowest not, is like thy seeking that I should make thee ignorant of what thou knowest: wherefore, do not seek of Me, and I shall assuredly satisfy thee.'

Niffarî

Realiz. 873

Devotee: How to know the 'I'?

Maharshi: The 'I'—'I' is always there. There is no knowing it. It is not a new knowledge acquired. What is new and not here and now will be evanescent only. The 'I' is always there. There is obstruction to its knowledge and it is called ignorance. Remove the ignorance and knowledge shines forth. In fact this ignorance or even knowledge is not for *Atma*. They are only overgrowths to be cleared off. That is why *'Atma'* is said to be beyond knowledge and ignorance. It remains as it naturally is—that is all.

Sri Ramana Maharshi

M.M. 978

Real recollection consists in forgetting all but the One recollected.

Al-Kalâbâdhî

Recollecting is the non-forgetting of things with which one hath been familiar.

Padma-Karpo

By memory the soul with intellect is joined.

Orpheus

For there is not any thing which is of greater importance with respect to science, experience and wisdom, than the ability of remembering.

Iamblichus

Scientia is to be got by the mind being called back to itself and gathered together into itself.

Hermes

I will now set forth the truths which I learnt in hours of solitude, when I turned away from corporeal things, and turned to the purely spiritual things of Light, and entered into communication with the divine world, and with some of the spirits of the celestial realm. I first grasped a definite truth by mystic intuition, and then sought to demonstrate it by arguments; whereas the Peripatetics follow the reverse method, letting themselves be led by

[1] Spiritual practices.

(logical) demonstrations, without knowing beforehand the goal to which their arguments will lead them. *Our* master is Plato, especially in his works *Timaeus* and *Phaedo*; whereas Aristotle remains the loadstar of those who seek truth by the empirical method. . . . This mystic-Platonic method is a different kind of philosophy, and a shorter way than that of the Peripatetics, which loses itself in secondary questions.[1]

Supra 734

Suhrawardî

I arrived at Truth, not by systematic reasoning and accumulation of proofs but by a flash of light which God sent into my soul.

Realiz. 873

Al-Ghazâlî

Ecstasy 636

This knowledge de-ments the mind.

Eckhart

The spirit of Savouring Wisdom . . . is a ghostly touch or stirring within the unity of our spirit; and it is an inpouring and a source of all grace, all gifts and all virtues. . . . And we feel this touch in the unity of our highest powers, above reason, but not without reason.

Ruysbroeck

He who is in harmony with Nature hits the mark without effort and apprehends the truth without thinking.

Confucius

There is . . . a naked intuition of eternal truth which is always the same, which never rises nor sets, but always stands still in its vertical, and fills the whole horizon of the soul with a mild and gentle light.

Rev. 965
P. State 590

John Smith the Platonist

Beauty 676

Intuition is the outcome of the fusion of a purified heart and illumined intelligence.

Swami Ramdas

Center 816

Lighting the lamp of Knowledge in the chamber of your heart,
Behold the face of the Mother, Brahman's Embodiment.

Hindu Song

P. State 563
Realiz. 859
M.M. 998

Those who would truly know and prepare the first Matter of the Philosopher's Stone (the chief and principal mystery of this earth) must have a deep insight into the nature of things, just as those who would know the Heavenly Stone (*i.e.* the indissoluble, triune essence of the true and living God) must have a profound spiritual insight into the things of heaven.

The Sophic Hydrolith

[1] John Smith the Platonist puts it even more bluntly, where he says of Aristotle: 'He hath so defaced the sacred monuments. . . . that elder philosophy which he so corrupts' (*Disc.*, p. 114). Cf. the citation from *The Sophic Hydrolith* on pp. 734–735. In the intellectual anarchy of our times, however, the mastery of Aristotelian logic becomes the prerequisite discipline for sane thinking.

Where the mind is busied with images, time must necessarily enter into the operations of the imagination, and this has no place in the highest school of the Holy Spirit; for there neither time nor images can help us, but contact is all that is needed, the which may happen without time within the space of a moment.

Tauler

Supra 745
Creation 31

'The thing needed,' said Kung-ni, 'is abstinence of the heart.'

'And what is that?' asked Yen-hui.

'It is this,' said Kung-ni: 'to concentrate all one's intellectual energy as into a mass. Not to listen with the ears, or with the heart, but with the spirit alone. To intercept the way of the senses, to keep the mirror of the heart pure, to let the spirit occupy itself, in the inner void, with abstract objects alone. The vision of principle demands the void. To keep oneself void, that is abstinence of the heart. . . . One must stay closed, simple, in a natural purity, without artifical admixture. Thus one can manage to remain without emotion, while it is difficult to become calm again after having let oneself be moved, just as it is easier not to walk, than to efface one's tracks after having walked.

Chuang-tse (ch. IV)

Sin 77
Center 828
Void 728
Renun. 152b
P. State 573
Action 329

(The contents of memory) must be drawn together again, that they may be known; that is to say, they must as it were be collected together from their dispersion: whence the word 'cogitation' is derived. For *cogo* (collect) and *cogito* (re-collect) have the same relation to each other as *ago* and *agito, facio* and *factito*. But the mind hath appropriated to itself this word (cogitation), so that, not what is 'collected' any how, but what is 're-collected', i.e. brought together, in the mind, is properly said to be cogitated, or thought upon. . . .

Great is the power of memory, a fearful thing, O my God, a deep and boundless manifoldness; and this thing is the mind, and this am I myself. What am I then, O my God? What nature am I? A life various and manifold, and exceeding immense.

St Augustine

Holiness 924

The Self, without inside, without outside, is entirely a cognitive pleroma (*prajñâna-ghana*).[1]

Bṛihad-Âraṇyaka Upanishad, IV. V. 13

Introd. *Realiz.*

In (Christ) dwelleth all the fulness of the Godhead.

Colossians, II. 9

My Lord includeth all things in His knowledge. Will ye not then remember?

Qur'ân, VI. 80

In the eternal mirror of his works God knows all creaturely perfections both natural and ghostly.

Eckhart

Center 828

[1] This would correspond with the *Nyoraizô* or 'Treasury' of Japanese Buddhism: cf. also St Augustine (*Conf.* X. viii): '. . . the fields and spacious palaces of my memory, where are the treasures of innumerable images.'

Reality 803 *Center* 841

'His knowledge . . . remaineth in the simplicity of His presence, and comprehending the infinite spaces of that which is past and to come, considereth all things in His simple knowledge as though they were now in doing. So that, if thou wilt weigh His foreknowledge with which He discerneth all things, thou wilt more rightly esteem it to be the knowledge of a never fading instant than a foreknowledge as of a thing to come. For which cause it is not called praevidence or foresight, but rather providence, because, placed far from inferior things, it overlooketh all things, as it were, from the highest top of things.'

Boethius

Introd. *Reality*

Every god has an undivided knowledge of things divided and a timeless knowledge of things temporal; he knows the contingent without contingency, the mutable immutably, and in general all things in a higher mode than belongs to their station.

Proclus

Creation 28

The Creator of all things is brim-full with them all in one transcendent excess thereof.

Dionysius

Center 847

To desire more than this Wisdom is to be like one who seeks an elephant by following its fooprints when the elephant itself has been found.

Padma-Sambhava

A fund of omniscience exists eternally in our heart.

Tipiṭaka

The intellect is keen-eyed, possessed of foreknowledge, for God hath powdered it with His own collyrium.

Rûmî

Divine wisdom is inexhaustible; the limitation is only in the receptive faculty of the form.

Henry Madathanas

Peace 700

The mind set round with intelligence is set quite free from the intoxications (*âsrava*); that is to say, from the intoxication of sensuality (*kâma*), from the intoxication of becoming (*bhâva*), from the intoxication of delusion (*dṛishti*), from the intoxication of ignorance (*avidyâ*).

Dîgha-Nikâya

Realiz. 873

The state of true being is that from which nothing is absent; to which nothing is added and nothing still less can harm.

Thomas Vaughan

Memory (is) from the Self; . . . Understanding, from the Self; Meditation, from the Self; Thought, from the Self; Conception, from the Self; Mind, from the Self; Speech, from the Self; Name, from the Self; sacred formulae (*mantra*), from the Self; sacred works (*karma*), from the Self; indeed this whole world, from the Self.

Chândogya Upanishad, VII. xxvi. 1

The Excellence of Wisdom

He hath given me the true knowledge of the things that are: to know the disposition of the whole world, and the virtues of the elements,

The beginning, and ending, and midst of the times, the alterations of their courses, and the changes of seasons,

The revolutions of the year, and the dispositions of the stars,

The natures of living creatures, and rage of wild beasts, the force of winds, and reasonings of men, the diversities of plants, and the virtues of roots,

And all such things as are hid and not foreseen, I have learned: for wisdom, which is the worker of all things, taught me.

For in her is the spirit of understanding: holy, one, manifold, subtile, eloquent, active, undefiled, sure, sweet, loving that which is good, quick, which nothing hindereth, beneficent,

Gentle, kind, steadfast, assured, secure, having all power, overseeing all things, and containing all spirits, intelligible, pure, subtile.

For wisdom is more active than all active things: and reacheth everywhere by reason of her purity.

Wisdom, VII. 17–24

P. State 579
Supra 755
Reality 775
Creation 31
P. State 573

The Lord by wisdom hath founded the earth; by understanding hath he established the heavens.

Proverbs, III. 19

The sun and the moon are made punctual.
The stars and the trees adore.
And the sky He hath uplifted; and He hath set the measure.

Qur'ân, LV. 5–7

Beauty 689

Observe ye every thing that takes place in the heaven, how they do not change their orbits, and the luminaries which are in the heaven, how they all rise and set in order each in its season, and transgress not against their appointed order.

Book of Enoch, II. 1

Without philosophy, it is impossible to be pious;[1] but he who has learnt what things are, and how they are ordered, and by whom, and to what end, will give thanks for all things to the Maker. . . .

And he who pursues philosophy to its highest reach will learn where Reality is, and what it is: and having learnt this, he will be yet more pious.

Hermes

[1] This does not mean that a saint is necessarily a philosopher, but it does mean that the intellectual heritage of a traditional collectivity profoundly influences the spiritual formation of its members. Cf. *Orthodoxy* and *Revelation.*

Before (Pythagoras) the followers of science were called, not *philosphers,* but *wise men,* such as were those seven most ancient sages whose names are still known to fame; the first of whom was called Solon.[1] . . . Pythagoras, being asked if he called himself a wise man, denied himself that name, and said that he was not wise, but a lover of wisdom. And thence it happened afterwards that all students of wisdom were called *lovers of wisdom,* that is, philosophers; for *philo* and *sophia* in Greek are equivalent to *love* and *wisdom.* Whence we may see that these two words make up the name *philosopher,* that is to say, lover of wisdom; which we may observe is not a term of arrogance, but of humility.

Supra 743

Dante (*Il Convito,* III. xi. 2)

Piety is the knowledge of God.

Hermes

Divine wisdom is true science.[2]

Sextus the Pythagorean

Theology without alchemy is like a noble body without its right hand.

The Sophic Hydrolith

Supra 745

Amongst all things, knowledge, they say, is truly the best thing; from its not being liable ever to be stolen, from its not being purchasable, and from its being imperishable.

Hitopadeśa

Holiness 902
Grace 552
Reality 803

Knowledge is the holiest of the holies, the god of the gods, and commands the respect of crowned heads; shorn of it a man is but an animal. The fixtures and furniture of one's house may be stolen by thieves; but knowledge, the highest treasure, is above all stealing.

Garuda Purana

The ink of the scholar is more holy than the blood of the martyr.

Muhammad

The Brahman's weapon is speech; with this let the twice-born man slay his enemies.

Mânava-dharma-śâstra, XI. 33

Action 335

Wisdom is better than weapons of war.

Ecclesiastes, IX. 18

All manner of sin and blasphemy shall be forgiven unto men: but the blasphemy against the Holy Ghost shall not be forgiven unto men.

St Matthew, XII. 31

[1] The others were Chilon, Periander, Thales, Cleobulus, Bias, and Pittacus.

[2] Traditional or true sciences are qualitative and integral and have their origin in metaphysical and cosmological principles, so that there can be no question of any 'conflict between science and religion'.

Profane sciences are quantitative and divisive, and originate from the residues of traditional sciences. wherefore they cannot but be at enmity with religion. Cf. Guénon: *Le Règne de la Quantité.*

God has created nothing finer than intelligence, nothing more perfect or more beautiful; the benefits which God bestows are due to it, understanding comes from it, and the wrath of God strikes the person who scorns it.

Muhammad

Holiness 900

God rejects those who hate Him, and scorn knowledge.

Michael Sendivogius

For not to know something is far different from not wanting to know something, since not to know is a weakness, but to detest knowledge is a perversion of the will.

Hugh of Saint-Victor

Conform. 166

Nothing more ruins the world than a conceit that a little knowledge is sufficient. Which is a mere lazy dream to cover our sloth or enmity against God.

Thomas Traherne

Wisdom and true virtue being divine effluxes can never enter into any unhallowed and mortal thing.

Plotinus

Supra 749

In this present life, I reckon that we make the nearest approach to knowledge when we have the least possible intercourse or communion with the body, and are not surfeited with the bodily nature, but keep ourselves pure until the hour when God himself is pleased to release us.

Plato (*Phaedo,* 67 A)

Sin 77

Even if thou art the most sinful of the sinful, thou shalt cross over (the ocean of) sin by the bark of wisdom.

As kindled fire reduces fuel to ashes, O Arjuna, so does the wisdom fire reduce all actions (Karma) to ashes.

Nothing indeed in this world purifies like wisdom. He who is perfected by Yoga, finds it in time within himself by himself.

Bhagavad-Gîtâ, IV. 36–38

Pilg. 385

Supra 755

As fire consumes all things, so does the fire of knowledge consume all evil and ignorance.

Srimad Bhagavatam, VI. i

And he gave me a goblet filled with fire, and when I had drunk it wisdom grew in me; and God granted me understanding, and my spirit was preserved, and my mouth opened, but nothing else was added.

IV Ezra, XIV. 39, 40

Nothing is there more purifying than knowledge. Neither the practice of austerity, nor resort to places of pilgrimage, nor repetition of mantrams, nor charity, nor any other spiritual discipline, can add to the perfection already attained through knowledge.

Srimad Bhagavatam, XI. xii

With the shaft of gnosis I shall pierce through every defect.

Milindapañha, 418

Doctors debate which is the nobler, knowledge or love. Some say that love is better than knowledge. I say it is not. Our best authorities declare that knowledge is nobler than love. Love and will take God as being good. If God were not good, will would have none of him; if God were not lovely, love would scout him. But understanding would not. Knowledge is not confined either to good or to love or to wisdom or lordship. By putting names to God the soul is only dressing him up and making a figure of God; nor is this the doing of knowledge. Though God were neither good nor wise, still understanding would seize him; it strips everything off, not stopping either at wisdom or good, nor majesty nor power. It pierces to naked being and grasps God bare, ere he is clothed in thought with wisdom and goodness.

Void 721 *Love* 625

Eckhart

Wine and music rejoice the heart, but the love of wisdom is above them both.

Ecclesiasticus, XL. 20

Through the cooling influence of bhakti, one sees forms of God in the Ocean of the Absolute. These forms are meant for the bhaktas, the lovers of God. But when the Sun of Knowledge rises, the ice melts; it becomes the same water it was before. Water above and water below, everywhere nothing but water.

M.M. 994

Sri Ramakrishna

If knowledge be the mark, to know thee shall suffice.

Reality 803

Shakespeare (*Love's Labour's Lost,* IV. ii. 116)

And thus it comes to pass, that where this Love (of Wisdom) shines, all other loves grow dim and almost spent.

Dante (*Il Convito,* III. xiv. 3)

Certainly he who can love invisible things will immediately desire to know them and to see them by the intelligence. So the more Judah grows, that is the power of loving, the greater becomes Rachel's desire to bear, that is the zeal for knowledge.

Richard of Saint-Victor

For it is impossible for language, miracles, or apparitions to teach us the infallibility of God's word, or to shew us the certainty of true religion, without a clear sight into truth itself, that is into the truth of things. Which will themselves when truly seen, by the very beauty and glory of them, best discover, and prove religion.

Thomas Traherne

'I have created the soul in My image and likeness, in that I have given it memory, intelligence and volition. Intelligence is the noblest part of the soul.'

St Catherine of Siena

Creation 42

Love is inseparable from knowledge.

St Macarius of Egypt

Though a weak Christian may believe great things by an implicit faith, yet it is very desirable his faith should be turned into assurance, and that cannot be but by the riches of knowledge and understanding.

Thomas Traherne

Brethren, be not children in understanding: howbeit in malice be ye children, but in understanding be men.

I. Corinthians, XIV. 20

Gnosis . . . is the perfection of faith.

Clement of Alexandria

In tantum Deus cognoscitur, in quantum amatur.

St Bernard

The excellency of knowledge is, that wisdom giveth life to them that have it.

Ecclesiastes, VII. 12

Knowledge is the ultimate perfection of our soul, in which consists our ultimate felicity.

Dante (*Il Convito,* I. i. 1)

Creation 48

That blessed god, the Agathos Daimon, said 'soul is in body, intellect (νοῦς) is in soul, and God is in intellect.' The rarest part of matter then is air; the rarest part of air is soul; the rarest part of soul is intellect; and the rarest part of intellect is God.

Hermes

Introds. *Pilg.* *Realiz.*

If we have failed to understand, it is that we have thought of knowledge as a mass of theorems and an accumulation of propositions. But this is not a wisdom built up of theorems but one totality, not manifold detail reduced to a unity, but rather a unity working out into detail.

The true Wisdom, then, is Real Being, and Real Being is Wisdom.

Plotinus

Reality 775 *Realiz.* 859

This knowledge is more easily obtained of God than of men.

Michael Sendivogius

Supra 749

Science in God, is known to be
A Substance, not a Qualitie.

Robert Herrick

Inv. 1031

I wished, and understanding was given me: and I called upon God, and the spirit of wisdom came upon me:

And I preferred her before kingdoms and thrones, and esteemed riches nothing in comparison of her.

Renun. 152b
Center 847

Neither did I compare unto her any precious stone: for all gold in comparison of her, is as a little sand, and silver in respect to her shall be counted as clay.

I loved her above health and beauty, and chose to have her instead of light: for her light cannot be put out.

Holiness 902

Wisdom, VII. 7–10

May I reckon the wise to be the wealthy.

Plato (*Phaedrus,* 279 C)

No mention shall be made of coral, or of pearls: for the price of wisdom is above rubies.

Job, XXVIII. 18

Exaltedly pure, like the excellent nectar in the sun,
I am a shining treasure,
Wise, immortal, indestructible!

Taittirîya Upanishad, I. 10

There is no treasure like knowledge.

'Alî

Is there not one true coin for which all things ought to be exchanged?—and that is wisdom.

Plato (*Phaedo,* 69 B)

The rout and destruction of the passions, while a good, is not the ultimate good; the discovery of Wisdom is the surpassing good. When this is found, all the people will sing.

Philo

Suff. 126b

He who leaves home in search of knowledge walks in the path of God.

Muhammad

The divine mind (*sensus divinus*=intellect) is wholly of like nature with eternity. It is motionless in itself, but though stable, is yet self-moving; it is holy, and incorruptible, and everlasting, and has all attributes yet higher, if higher there be, that can be assigned to the eternal life of the supreme God, that life which stands fast in absolute reality. It is wholly filled with all things imperceptible to sense, and with all-embracing knowledge; it is, so to speak, consubstantial with God.

Reality 773
Holiness 924

Hermes

Orthod. 275

The learned ones are the heirs of the prophets—they leave knowledge as their inheritance; he who inherits it inherits a great fortune.

Muhammad

He giveth wisdom unto whom He will, and he unto whom wisdom is given, he truly hath received abundant good.

Qur'ân, II. 269

He (Bâyazîd al-Bisṭâmî) said, 'A single atom of the sweetness of gnosis in a man's heart is better than a thousand pavilions in Paradise.'

He said, 'Gnostics are a boon to Paradise, and Paradise is a bane to them.'

'Aṭṭâr

Conform. 170

When he acquires this Knowledge, the supreme purifier, a man becomes free from all merit and demerit produced by Ignorance and accumulated in many other lives. He, like the ether, does not get attached to actions in this world.

Śrî Śankarâchârya

Action 329

The best knowledge is that which enableth one to put an end to birth and death and to attain freedom from the world.

Udâna

'The sign of My gnosis is, that thou shouldst have no desire for any gnosis, nor concern thyself, after My gnosis, with the gnosis of other than Me.'

Niffarî

Center 847

This gnosis goes far beyond the knowledge of the learned, for it enters the hearts of the prophets and the saints direct from the Creative Truth Himself, nor can it be comprehended except by those who have experience of it.

Al-Ghazâlî

Supra 749

Gnosis means that the gnostic passes away from himself into God.

'Aṭṭâr

Realiz. 887

I regard the wise as my very Self.

Bhagavad-Gîtâ, VII. 18

The man of learning lives even after his death: the ignorant man is dead, while still alive.

'Alî

To seek knowledge is obligatory on every Moslem man and woman.

Muhammad

The three essentials—obedience to the Vinaya rules (Sila), meditation and intelligence—are a great summary of the whole of the Buddha's ancient Way.

Hakuin

Orthod. 300

Wisdom is the principal thing; therefore get wisdom: and with all thy getting get understanding.

Proverbs, IV. 7

And wisdom and knowledge shall be the stability of thy times, and strength of salvation.

Isaiah, XXXIII. 6

Wisdom amazes all that is sensible of her, but is herself not amazed by anything.

Apollonius of Tyana

Supra 755

There pre-exists a craving for knowledge in order that the intellect, whose being is in understanding, may be perfected by the study of truth.

Nicholas of Cusa

We beseech Thee, O Lord, mercifully pour into our minds the Holy Spirit, by whose wisdom we were created and by whose providence we are governed.

First Collect for the Saturday after Pentecost

O You who control that path where the sun comes up, look upon us with Your red and blue days, and help us in sending our voices to *Wakan-Tanka*! O You who have knowledge, give some of it to us, that our hearts may be enlightened, and that we may know all that is sacred![1]

Black Elk (Prayer from the *Inipi* Rite of Purification)

Supra 749

All wisdom is derived from God, and ever ends in Him. Any one who desires knowledge should ask it of Him, for He gives liberally, and without upbraiding.

John Cremer

The gift of knowledge is the highest of all gifts.

Sri Chandrasekhara Bhâratî Swâmigal

The Lord give thee understanding in all things.

II. Timothy, II. 7

A man's wisdom maketh his face to shine.

Ecclesiastes, VIII. 1

Dost thou hold wisdom to be anything other than truth, wherein we behold and embrace the supreme good?

St Augustine

Nothing is so peculiar to wisdom as truth.

Sextus the Pythagorean

Intellect is the satellite of Deity.

Archytas

It seems, then, that but one refuge remains for the man who is to reach the gates of

[1] One cannot fail to remark the similarity between this prayer and the Hindu *Gâyatrî* (q.v.).

salvation, and that is divine wisdom. From thence, as from a holy inviolate temple, no longer can any daemon carry him off, as he presses onward to salvation.

Clement of Alexandria

Introd.
Heterod.

The eye of Knowledge contemplates *Brahma* as It is in Itself, abounding in Bliss, pervading all things: but the eye of ignorance discovers It not, discerns It not, even as a blind man perceives not the sensible light. . . .

When the Sun of spiritual Knowledge rises in the heavens of the heart, it dispels the darkness, it pervades all, envelopes all and illumines all.

Śrî Śankarâchârya

Center 833

The miserable Passions are to be overcome by the vision of wisdom. The Passions lie not in the objects of sense, nor in the sense-organs, nor between them, nor elsewhere; where do they lie? And yet they disturb the whole world! They are but a phantom. Then cast away thy heart's terror, and labour for wisdom.

Śânti-deva

Illusion 89 85

Those (organs) attached to sensual objects cannot be curbed so much by non-indulgence as ever by knowledge.

Mânava-dharma-śâstra, II. 96

The only way to overcome (the passions) is to eradicate them. That is done by finding their source. . . . Thus, if you find the Self and abide therein there will be no trouble owing to the passions.

Sri Ramana Maharshi

Peace 700

False notions cannot be negated in any way other than (thus knowing the Self). It is these wrong notions that are the causes of delusion. These notions, bereft of their cause (ignorance), come to an absolute end, like fire bereft of fuel (when knowledge is achieved).

Śrî Śankarâchârya

A philosophy comprehensive enough to embrace the whole of knowledge is indispensable.

Gampopa

It is true knowledge (jnâna) alone that can liberate us from the subconscious impregnations (samskâras).

Swami Sivananda

Pilg. 366

That alone is entitled to the name of knowledge which makes clear to us the nature of the Self and gets us the everlasting bliss of Freedom.

Sri Chandrasekhara Bhâratî Swâmigal

The active life is a service, the contemplative life a liberty.

St Gregory the Great

Action 329, 34
M.M. 973

The highest pinnacle of knowledge is expressed in the fact that without it none can know God.

Hujwîrî

Center 847

Whilst possessed of all (objects of) desires he (the solar comprehensor) is without desire, for no desire of anything (troubles) him.

Suff. 133

Regarding this there is this verse—'By knowledge they ascend that (state) where desires have vanished: sacrificial gifts go not thither, nor the fervid practisers of rites without knowledge;'—for, indeed, he who does not know this does not attain to that world either by sacrificial gifts or by devout practices, but only to those who know does that world belong.

Śatapatha-Brâhmaṇa, x. v. 4. 15, 16

Wr. Side 474

After the Truth what is there save error?

Qur'ân, x. 32

REALITY

I AM THAT I AM.

Exodus, III. 14

God is one. And he that is one is nameless: for he does not need a name, since he is alone.

Hermes (*Fragments*, 3)

'God is an essence without duality (*advaita*), or as some maintain, without duality but not without relations (*visiṣṭâdvaita*). He is only to be apprehended as Essence (*asti*), but this Essence subsists in a twofold nature (*dvaitîbhâva*): as being and as becoming. Thus, what is called the Entirety (*kṛtsnam, pûrṇam, bhûman*) is both explicit and inexplicit (*niruktâniruktа*), sonant and silent (*śabdâśabda*), characterised and uncharacterised (*saguṇa, nirguṇa*), temporal and eternal (*kâlâkâla*), partite and impartite (*sakalâkalâ*), in a likeness and not in any likeness (*mûrtâmûrta*), shewn and unshewn (*vyaktâvyakta*), mortal and immortal (*martyâmartya*), and so forth. Whoever knows him in his proximate (*apara*) aspect, immanent, knows him also in his ultimate (*para*) aspect, transcendent; the Person seated in our heart, eating and drinking, is also the Person in the Sun. This Sun of men, and Light of lights, "whom all men see but few know with the mind",[1] is the Universal Self (*âtman*) of all things mobile or immobile. He is both inside and outside (*bahir antaś ca bhûtânâm*), but uninterruptedly (*anantaram*), and therefore a total presence, undivided in divided things. He does not come from anywhere, nor does he become anyone, but only lends himself to all possible modalities of existence. . . .

M.M. 978

Void 724
Center 816

'Whether we call him Person, or Sacerdotium, or Magna Mater, or by any other grammatically masculine, feminine or neuter names, "That" (*tat, tad ekam*) of which our powers are measures (*tanmâtrâ*) is a syzygy of conjoint principles, without composition or duality. These conjoint principles or selves, indistinguishable *ab intra*, but respectively self-sufficient and insufficient *ab extra*, become contraries only when we envisage the act of self-manifestation (*svaprakâśatvam*) implied when we descend from the silent level of the Non-duality to speak in terms of subject and object and to recognize the many separate and individual existences that the All (*sarvam*=τὸ πᾶν) or Universe (*viśvam*) presents to our physical organs of perception. And since this finite totality can be only logically and not really divided from its infinite source, "That One" can also be called an "Integral Multiplicity"[2] and "Omniform Light"[3]' (Coomaraswamy: *Hinduism and Buddhism*, pp. 10, 11).

Realiz. 890
M.M. 994

[1] *Atharva Veda*, X. 8. 14.
[2] *Ṛg Veda*, III. 54. 8.
[3] *Taittirîya Saṁhitâ*, V. 35.

'It is often asked, and quite fruitlessly, how multiplicity can come out of unity, without it being seen that the question thus posed contains no solution, for the simple reason that it is badly posed and in this form corresponds to no reality. The fact is that multiplicity does not come out of unity, any more than unity comes out of the metaphysical Zero, or than something comes out of the universal Whole, or than some possibility can be found outside of the Infinite or outside of total Possibility. Multiplicity is included in primordial unity, and it does not cease to be included there by the fact of its development in manifested mode. . . . It is unity alone, which, being the principle behind multiplicity, gives it all the reality it is capable of possessing; and even unity in its turn is not an absolute principle sufficient unto itself, but derives its own reality from the metaphysical Zero' (Guénon: *Les États multiples de l'Être*, pp. 47–48).

Burckhardt shows that 'the arguments of certain philosophers against the existence of Platonic "ideas" fall into a void once it is understood that these "ideas" or archetypes have no *existence*, following the teaching of Ibn 'Arabî,—that is to say, they are not in the nature of distinct substances' (*Introduction aux Doctrines ésotériques de l'Islam*, p. 64). What they are essentially is the informing principle that imparts its stamp to every form, the supraformal Intelligences or Powers which when conjoined with primordial matter generate the emanations we call distinct substances. If it can be shown that the 'universals' of philosophy logically resolve into mental forms that are sheer abstractions, this only demonstrates the antipodal disparity between these 'universals' and the true archetypes, of which they are at best but mental reflections, ab-stracted even of extrinsic reality.

In the kaleidoscopic Round of Existence, wave upon wave of the same form forever recurring amongst the indefinitude of other forms likewise homogeneous and equally transient, all goes to proclaim the insubstantiality of manifested forms as such and their total dependence on the immutable essences for what fugitive reality they manifest in their mutable becoming.

Illusion 85 109

'To consider the eternal ideas as nothing but simple "virtualities" in relation to the manifested beings of which they are the principial "archetypes" . . . is strictly speaking a complete reversal of the relationship between Principle and manifestation' (Guénon: 'Les Idées éternelles', *Études Trad.*, 1947, pp. 222–223).

'One might say, not that transient effects are meaningless, but that their value is not realized except to the degree that they are seen *sub specie aeternitatis*, that is *formaliter*' (Coomaraswamy: *The Transformation of Nature in Art*, p. 31).

Introd. *Void*

'There is in this world below the immense problem of separation; how to make men understand that in God they are separated from nothing?

'To the angels, our formal—or separative—world appears as a mass of debris: what in reality is united, is separated into and by form. The formal world is made of congealed essences.

Center 816

'Man escapes form—and separativity—in his own supernatural center, on the shores of the blissful and eternal essences' (Schuon: *Perspectives spirituelles*, p. 283).

The Motionless Mover

What is that which always is and has no becoming; and what is that which is always becoming and never is? That which is apprehended by intelligence and reason is always in the same state; but that which is conceived by opinion with the help of sensation and without reason, is always in a process of becoming and perishing and never really is.

Knowl. 734

Plato (*Timaeus*, 28 A)

There is nothing that stands fast, nothing fixed, nothing free from change, among the things which come into being, neither among those in heaven nor among those on earth. God alone stands unmoved, and with good reason; for he is self-contained, and self-derived, and wholly self-centred, and in him is no deficiency or imperfection. He stands fast in virtue of his own immobility, nor can he be moved by any force impinging on him from without, seeing that in him are all things, and that it is he alone that is in all things.

Introd. *Heterod.*

Hermes

Being is a term for immutability. For all things that are changed cease to be what they were, and begin to be what they were not. True being, pure being, real being has no one save Him who does not change. He has it to whom is said, 'Thou shalt change them, and they shall be changed. But Thou art always the self-same' (Ps. ci, 27 *sq.*, Douay).

St Augustine

For I am the Lord, I change not.

Malachi, III. 6

Metanoia 495

The Father of lights, with whom is no variableness, neither shadow of turning.

James, I. 17

Father of heaven and earth, Who makest time swiftly slide,
And, standing still Thyself, yet fram'st all moving laws.

Boethius

It is impossible that God should ever be willing to change; being, as is supposed, the fairest and best that is conceivable.

Plato (*Republic*, II. 381 C)

That is chiefly to be said to Be, which always exists in one and the same way; which is every way like itself; which can in no way be injured or changed; which is not subject to

P. State 563

time; which cannot at one time be other than at another. For this is what is most truly said to Be.

St Augustine

Introds. *Realiz.* *Holiness* *Creation* 42

The essences of our souls can never cease to be, because they never began to be: and nothing can live eternally, but that which hath lived from all eternity. . . . Thou (O man) begannest as time began, but as time was in eternity before it became days and years, so thou wast in God before thou wast brought into the creation: and as time is neither a part of eternity, nor broken off from it, yet come out of it; so thou art not a part of God nor broken off from Him, yet born out of Him.

William Law

Center 841

Verily, verily, I say unto you, Before Abraham was, I am.

St John, VIII. 58

Introd. *Knowl.* 749

There is only one Truth, whose existence has no need of proof, since it is itself its own proof for those who are able to perceive it. Why use complexity to seek what is simple?

Henry Madathanas

Heterod. 420

There is a sect of heretics called Sophists (*Sûfisṭâ'iyân*), who believe that nothing can be known and that knowledge itself does not exist. I say to them: 'You think that nothing can be known: is your opinion correct or not?' If they answer 'It is correct', they thereby affirm the reality of knowledge; and if they reply 'It is not correct', then to argue against an avowedly incorrect assertion is absurd.

Hujwîrî

Rev. 965

Everyone who knows that he is in doubt about something, knows a truth, and in regard to this that he knows he is certain. Therefore he is certain about a truth. Consequently everyone who doubts if there be a truth, has in himself a true thing on which he does not doubt; nor is there any true thing which is not true by truth. Consequently whoever for whatever reason can doubt, ought not to doubt that there is truth.

St Augustine

Peace 705

The Philosopher describes Truth to be that which *sistit intellectum, stayes* the *Understanding.*

Peter Sterry

M.M. 987

And thou shouldst know that all have their delight in measure as their sight sinketh more deep into the truth wherein every intellect is stilled.

Dante (*Paradiso,* XXVIII. 106)

Flight 944

Would that it were possible for you to grow wings, and soar into the air! Poised between earth and heaven, you might see the solid earth, the fluid sea and the streaming rivers, the wandering air, the penetrating fire, the courses of the stars, and the swiftness of the movement with which heaven encompasses all. What happiness were that, my son, to see all these borne along with one impulse. and to behold Him who is unmoved moving in all that moves. and Him who is hidden made manifest through his works!

Beauty 689

Hermes

The lord of all moving things is alone able to move of himself.

Plato (*Statesman*, 269 E)

Prajñâ (transcendent wisdom) remains immovable. though this does not mean the immovability or insensibility of such objects as a piece of wood or rock. It is the mind itself endowed with infinite motivities: it moves forward and backward. to the left and to the right. to every one of the ten quarters. and knows no hindrances in any direction. Prajñâ Immovable is this mind capable of infinite movements.

Takuan

There is a producer which has not been produced. a transformer which is not transformed. This non-produced has produced all beings. this non-transformed transforms all beings.... The producer is the *yin-yang*: the transformer is the cycle of the four seasons. The producer is immobile. the transformer goes and comes.

Illusion 85

Lieh-tse

The One and the Many

This universe . . . is both One and Many.

Dionysius

Ens et unum convertuntur.

Scholastic Formula

By their wordings they made him logically manifold who is but One.

Rig-Veda, x. 114. 5

Say: He Allâh, is One—
Allâh. the Absolute Plenitude:
He begetteth not nor is He begotten.
And there is none equal unto Him.

Infra 803
Creation 28
M.M. 994
998

Qur'ân, CXII

All composition originates from a simple source.

P. State 573

Marsilio Ficino

He created you from one being.

Qur'ân, XXXIX. 6

Supra 773

There must first be one from which the many arise. This one is competent to lend itself to all yet remain one, because while it penetrates all things it cannot itself be sundered; this is identity in variety.[1]

Plotinus

Creation 38

The ancients, who were our betters and nearer the gods than we are, handed down the tradition, that whatever things are said to be are composed of one and many.

Plato (*Philebus*, 16 C)

If (being) were not one, it would be bounded by something else.... So then if a multiplicity of things exist, it is necessary that they should be such as the one is.

Melissus

Introd. *Holiness*

How can all things be one, and yet everything have a distinct being of its own?

Proclus

M.M. 978

Unity (is) in multiplicity and multiplicity in Unity.

Sufic Formula

Creation 26

At one time there grew to be the one alone out of many, and at another time it separated so that there were many out of the one.

Empedocles

Creation 37

This the Sâtyakîrtas say: 'As to the Angel whom we worship, of him we say that there is one aspect in the cow, another in beasts of burden, another in the elephant, another in man (*puruṣa*), another in all existences; such is the Angel's omni-aspectuality (*sarvaṁ rûpaṁ*).' That same single aspect is the Spirit (*prâṇa*).

Jaiminîya Upaniṣad Brâhmaṇa, III. 32

Beauty 670

The Divine Light rays out immediately upon the Intelligences, and is reflected by these Intelligences upon other things.

Dante (*Il Convito*, III. xiv. 2)

Infra 790

In the Beatific Vision God manifests Himself to the elect in a general epiphany, which, nevertheless, assumes various forms according to the mental conceptions of God formed by the faithful on earth.

Ibn 'Arabî

The supreme God .... one Numen, divine force and power, which runs through the whole world, multiformly displaying itself therein. . . .

Ralph Cudworth

[1] This corresponds with the Hindu doctrine of *bhedâbheda*: 'distinction without difference', referring *in principio* to the Archetypes, but applicable by analogy to all degrees of existence.

It must be said that everything, that in any way is, is from God. For whatever is found in anything by participation must be caused in it by that to which it belongs essentially.... God is self-subsisting being itself, (which) being can be only one.... Therefore all beings other than God are not their own being, but are beings by participation. Therefore, it must be that all things which are diversified by the diverse participation of being, so as to be more or less perfect, are caused by one First Being, Who possesses being most perfectly.

Supra 773

Hence Plato said that unity must come before multitude; and Aristotle said that whatever is greatest in being and greatest in truth is the cause of every being and of every truth, just as *whatever is the greatest in heat is the cause of all heat*.[1]

St Thomas Aquinas

Beauty 663

... Thou of many forms, that art only-begotten.

Acts of Thomas, 47

Listen to the teaching of the *Shâstra* in which Vaishnavas believe: Just as Nârâyan and Krishna are one essence, so are Lakshmi and the Gopis identical and not diverse. Lakshmi in the garb of the Gopis tasted Krishna's company. In theology it is a sin to recognise a plurality of gods. The devotee meditates on one and the same God; he gives different images to the same deity.

Chaitanya

Symb. 321

(The Gods and Goddesses) have an appearance of so many several distinct deities; yet they seem to have been all really nothing else, but as Balbus in Cicero (De Natur. Deor. lib. II. cap. xxviii ...) expresses it: 'Deus pertinens per naturam cujusque rei,' God passing through, and acting in the nature of everything.

Ralph Cudworth

Introd. *Heterod.*

The Parabrahman, Devî, Shiva, and all other Deva and Devî are but one, and he who thinks them different from one another goes to Hell.

Shâktânandatarangini, III

See now the height and breadth of the eternal worth, since it hath made itself so many mirrors wherein it is reflected, remaining in itself one as before.

Dante (*Paradiso*, XXIX. 142)

Center 828

The Self is Brahmâ, the Self is Vishnu, the Self is Indra, the Self is Shiva; the Self is all this universe. Nothing exists except the Self.

Śrî Śankarâchârya

Infra 803

All mankind is in Christ one man, and the unity of Christians is one Man.

St Augustine

Holiness 924

Agni is One, only kindled in many places.
One is the Sun mightily overspreading the world.

Rev. 959

[1] *Metaph.*, Ia. 1 (993b 25).

One alone is the Dawn beaming over all this.
It is the One that has severally become all this.

Rig-Veda, VIII. 58. 2

Beauty 689

As the string in (a necklet of) gems, it is Thou in Thy Unity who penetratest all the diversity of beings and religions.

Sri Ramana Maharshi

The primal Light, that all irradiates,
By modes as many is received therein,
As are the splendours wherewith it is mated.

Dante (*Paradiso,* XXIX. 136)[1]

Indra, Mitra, Varuna, Agni, they style him.
He is also the Heavenly Bird, the winged Garutmân.
Being One, the poets many-wise name him
They call him Agni, Yama, or Mâtari-śvan.

Rig-Veda, I. 164. 46

If you asked me which form of God you should meditate upon, I should say: Fix your attention on that form which appeals to you most; but know for certain that all forms are the forms of one God alone. . . . Śiva, Kâli, and Hari are but different forms of that One. He is blessed indeed who has known all as one.

Sri Ramakrishna

Everything is enveloped in the Unity of Knowledge, symbolized by the Point.

Shaykh Aḥmad- al-'Alawî

Void 724

And after this I saw God in a Point, that is to say, in mine understanding,—by which sight I saw that He is in all things.

Julian of Norwich

Center 816, 838

From that point depend heaven and all nature.[2]

Dante (*Paradiso,* XXVIII. 41)

Beauty 663

First realize God, then think of the creation and other things. Vâlmiki[3] was given the name of Râma to repeat as his mantra, but was told at first to repeat 'marâ'. 'Ma' means God and 'râ' the world. First God and then the world. If you know one you know all. If you put fifty zeros after a one, you have a large sum; but erase the one and nothing remains. It is the one that makes the many. First one, then many. First God, then His creatures and the world.

Sri Ramakrishna

[1] Cf. also XIII. 55–60.

[2] Cp. the Sufic doctrine that the *Qur'ân* is contained in the *Fâtiḥah* (the opening chapter), which itself is contained in the *Bismillâh* ('In the Name of God': the opening words), which is contained in the letter *bâ* ب, which is contained in its diacritical point.

[3] Author of the *Râmâyana* (q.v.).

Words derive their power from the original Word.

Eckhart

Inv. 1003

All things have been derived from One.

The Emerald Table of Hermes

The One brought number into being, and number analysed the One, and the relation of number was produced by the object of numeration.

Ibn ʿArabî

No multiplicity can exist except by some participation in the One: that which is many in its parts is one in its entirety; that which is many in its accidental qualities is one in its substance; that which is many in number or faculties is one in species; that which is many in its emanating activities is one in its originating essence. There is naught in the world without some participation in the One, the Which in Its all-embracing Unity contains beforehand all things, and all things conjointly, combining even opposites under the form of oneness. And without the One there can be no Multiplicity; yet contrariwise the One can exist without the Multiplicity just as the Unit exists before all multiplied Number. And if all things be conceived as being ultimately unified with each other, then all things taken as a whole are One. . . . If you take away the One there will remain neither whole nor part nor anything else in the world; for all things are contained beforehand and embraced by the One as an Unity in Itself.

Dionysius

Center 835
Introd.
Heterod.
Beauty 689

Numbers, insofar as they are, all flow from the One:
And creatures all spring forth from God, the One.

Angelus Silesius

When man and woman become one, Thou art that One; when the units are wiped out, lo, Thou art that (Unity).

Rûmî

Love 625
Realiz. 890

One is greater than all other numbers, for it has produced the infinite variety of mathematical magnitudes; but no alteration is possible without the presence of the One which penetrates all things, and the powers of which are present in its manifestations.

Henry Madathanas

God is in all things, as their root and the source of their being. There is nothing that has not a source; but the source itself springs from nothing but itself, if it is the source of all else. God then is like the unit of number. For the unit, being the source of all numbers, and the root of them all, contains every number within itself, and is contained by none of them; it generates every number, and is generated by no other number.

Hermes

Creation 28

Supra 773

All numbers are multiples of one, all sciences converge to a common point, all wisdom comes out of one centre, and the number of wisdom is one.

Paracelsus

The zero placed first is worth nothing.
The nothing, the creature, when placed before God,
Is worth nothing: placed after Him, thence alone is its value derived.

Angelus Silesius

There are many numbers, but only One is counted.

Shabistarî

Beauty 663

The Neighbour first, and then the House.

Râbi'a of Baṣra

M.M. 994

In the 'Not Two'[1] are no separate things, yet all things are included....
The One is none other than the All, the All none other than the One.

Seng-ts'an

Infinity and Perfection do not admit of parts.

Sri Ramana Maharshi

M.M. 978

All aspects of the universe—the relative and the absolute—are but one in reality.

Hakuin

P. State 563

Things in the universe are all produced from the single *ch'i* (ethereal essence) which embodies both the will of the clear sky and the will of the clouded earth.

Hsieh Tao-kuang

All this diversity is fundamentally and basically one single unity.

Henry Suso

In the old metaphysical theology, an original and uncreated μόνας or unity is made the fountain of all particularities and numbers which have their existence from the efflux of its almighty power.

John Smith the Platonist

Void 724

The One is all things and no one of them.

Plotinus

Unum ego sum et multi in me.

Basil Valentine

Whatever happens, in any form or at any time or place, is but a variation of the One Self-existent Reality.

Yoga-Vasishtha

[1] Same as the Sanskrit *advaita* (=non-duality).

All is One.

Xenophanes

Omnia unum esse et unum omnia.[1]

Hermes

The whole is one.

Chuang-tse (ch. II)

The whole is one.

Ibn al-Fâriḍ

It is called, by *Philosophers,* one *Stone,* although it is extracted from many Bodies or Things.

Geber

In God identity is diversity.

Nicholas of Cusa

Center 835

The Unity of God is his Infiniteness.

Peter Sterry

That must be one that judgeth things to be diverse.

Aristotle

Supra 773

All is one and one is all in all.

Eckhart

All in One as One, and One in All as All, and One and all Good, is loved through the One in One, and for the sake of the One, for the love that man hath to the One.

Theologia Germanica, XLIII

Christ is all, and in all.

Colossians, III. 11

All the children of God are but ONE in Christ, which one is Christ in all.

Boehme

In God's sight all men are one man, and one man is all men.

Julian of Norwich

Holiness 924

One Reality, all comprehensive, contains within itself all realities;
The one moon reflects itself wherever there is a sheet of water,
And all the moons in the waters are embraced within the one moon;

[1] There is an inversion of this truth in the motto *e pluribus unum.*

The Dharma-body of all the Buddhas enters into my own being,
And my own being is found in union with theirs.

Yoka Daishi

Supra Introd.

Sublimity (*'uluww*) belongs to God alone. The essences (*a'yân*) of things are in themselves non-existent, deriving what existence they possess from God, who is the real substance (*'ayn*) of all that exists. Plurality consists of relations (*nisab*), which are non-existent things. There is really nothing except the Essence, and this is sublime (transcendent) for itself, not in relation to anything, but we predicate of the One Substance a relative sublimity (transcendence) in respect of the modes of being attributed to it. . . . The Substance is One, although its modes are different. None can be ignorant of this, for every man knows it of himself, and Man is the image of God. . . . The world of Nature is many forms in One Mirror; nay, One Form in diverse mirrors. Bewilderment arises from the difference of view, but those who perceive the truth of what I have stated are not bewildered.

Creation 42
Center 828

Ibn 'Arabî

Existence-Knowledge-Bliss Absolute is one, and one only. But It is associated with different limiting adjuncts on account of the different degrees of Its manifestation.

Sri Ramakrishna

Illusion 85

The colour of the water is the colour of the vessel containing it.

Junayd

It follows, therefore, that in the external world there is only One Real Being, who, by clothing Himself with different modes and attributes, appears to be endued with multiplicity and plurality to those who are confined in the narrow prison of the 'stages', and whose view is limited to visible properties and results.

Jâmî

Now there are diversities of gifts, but the same Spirit.
And there are differences of administrations, but the same Lord.
And there are diversities of operations, but it is the same God which worketh all in all.

Infra 790

I. Corinthians, XII. 4–6

There is no lord in all the world
Who lives in all His dwellings at once
Save God alone.

Mechthild of Magdeburg

In a single country of the Buddha are included all the countries of the Buddha.

Avataṃsaka Sûtra

God is in all places and in each place God is all at once.

Eckhart

Each (of the gods) contains all within itself and sees all in every other, so that everywhere there is all, and each is all, and infinite the glory!

Plotinus

God is all in all: only one, not many, one in all, and all in one.

Boehme

Multitude itself would not be contained under *being* unless it were in some way contained under one.

St Thomas Aquinas

Lord of the worlds! Thou art One, but not by number.

Zohar (The 'Prayer of Elias')

He in his unity is all things; so that we must either call all things by his name, or call him by the names of all things.

Hermes

Symb. 306

Our Arcanum . . . has no name of its own; yet there is nothing in the whole world whose name it might not with perfect propriety bear.

Philalethes

Void 724

He is the true master (*sad-guru*) who makes you perceive the Supreme Self (*paramâtman*) wherever the mind attaches itself.

Kabîr

He (God) is one as well as many. You may call the many as Maya or Lila. The multiple universe is He. One has become the many.

Swami Ramdas

Even as unity is found in each number:
So is God, the One, everywhere in things.

Angelus Silesius

I guard the original One, and rest in harmony with externals.

Chuang-tse (ch. XI)

Peace 694

He is being both to himself and to all. And in that only is he separated from all that he is being both of himself and of all. And in that is he one in all and all in him, that all things have their being in him, as he is the being of all.

The Epistle of Privy Counsel, I

M.M. 978

One in All,
All in One—
If only this is realized,
No more worry about your not being perfect!

Seng-ts'an

God, in the *holy Tongue,* they call
The Place that filleth *All in all.*

Robert Herrick

Center 841

God is present through all—not something of God here and something else there, nor all of God gathered at some one spot: there is an instantaneous presence everywhere, nothing containing, nothing left void, everything therefore fully held by Him.

Plotinus

How could localization and contact be possible (for God), seeing that He is (essentially) the existences themselves?

Jîlî

God is all in all and to each thing all things at once.

Eckhart

We who through ignorance think of ourselves as separate personalities must submit to the will and action of this infinite power, this infinite love which is Ram, the One who penetrates all.

Swami Ramdas

P. State 573

Now Nature may truly be described as being *one,* true, simple, and perfect in her own essence, and as being animated by an invisible spirit. If therefore you would know her, you, too, should be true, single-hearted, patient, constant, pious, forbearing, and, in short, a new and regenerate man.[1]

The Sophic Hydrolith

Holiness 924
Creation 48

All creatures that have flowed out from God must become united into one Man.

Eckhart

Infra 790

All saints are but one saint:
For they are one heart, spirit, mind, in one body.

Angelus Silesius

I being one become many, and being many become one.

Saṁyutta-nikâya, II. 212

Campanella teacheth us, That all *second Causes,* are *Causa prima modificata,* so many *Modifications* of the *first Cause,* so many forms and shapes in which the first Cause appears and acts. All the Works of God, are the *Divine Love,* in so many *Modes* and *Dresses.* There is *diversity* of *Manifestations,* there are *diversities* of *Operations,* which compose the whole frame and business of this Creation, which are as diverse *persons*

[1] Here we have a 'scientific' or 'alchemical' explanation for the necessity of virtues.

acting diverse parts upon this stage. But there is *one Spirit, one Lord, one God, one Love, which worketh all in all.* — *Creation* 33

Peter Sterry

It can be shown from three sources that God is one. First from His simplicity. . . . Secondly . . . from the infinity of His perfection. . . . Thirdly . . . from the unity of the world. For all things that exist are seen to be ordered to each other. — Introd. *Symb.*

St Thomas Aquinas

Twixt Truth and Errour, there's this difference known,
Errour is fruitfull, Truth is onely one. — *Wr. Side* 474

Robert Herrick

It is only evil and ignorance that have many shapes. Truth and wisdom are one and the same.

Abbot Lee Lisan

There are two substances but only one essence. They are not really two, but one and the same thing: the Sulphur is matured and well digested Mercury, the Mercury is crude and undigested Sulphur. — *M.M.* 978

Philalethes

The egg is in the hen, the hen is in the egg:
The two in One, and also the One in the two.

Angelus Silesius

One is All, by him is all, and for him is all, and in him is all. The Serpent is one: he has the two symbols. — Introd. *Wr. Side*

Chrysopeia of Cleopatra

Our water is the fire which causes both death, and, through death, a more glorious life. This water, though one, is not simple, but compounded of two things: the vessel and the fire of the Sages, and the bond which holds the two together. So when we speak of our vessel, and our fire, we mean by both expressions, our water: nor is our furnace anything diverse or distinct from our water. There is then one vessel, one furnace, one fire, and all these make up one water. The fire digests, the vessel whitens and penetrates, the furnace is the bond which comprises and encloses all, and all these three are our Mercury. — *Death* 208; *Pilg.* 378; *Contem.* 536

Philalethes

In metals, then, as in all other things, there is only one first substance, but the universal substance is modified in a vast variety of ways, according to the course of its subsequent development. Thus one thing is the mother of all things. This great fact ought always to be borne in mind in studying the works of the Sages: for nothing but mistakes and disappointment can result from a slavishly literal interpretation of their books. — *P. State* 563, 579

Michael Sendivogius

Being but one, she (wisdom) can do all things: and remaining in herself the same, she reneweth all things, and through nations conveyeth herself into holy souls, she maketh the friends of God and prophets.

Wisdom, VII. 27

Who can taste, were it but a tiny drop of the blood of Christ,
Must, in full felicity, be lost with Him in God.

Center 816

Angelus Silesius

I have drunk but a single drop, and I have understood!

Muḥammad al-Ḥarrâq

Say, how can it be that in a single drop of water,
In me, flows in its entirety the sea of God?

Sp. Drown.

Angelus Silesius

Cease to think of many things. Nature is satisfied with one thing, and he who does not know it is lost.

The Golden Tract

If one contemplates the things in mystical meditation, everything is revealed as one.

Zohar, I. 241 a

My dove, my undefiled is but one.

Song of Solomon, VI. 9

Our stone is made out of one thing, and with one thing. . . . All that is in our stone is essential to it, nor does it need any foreign ingredient. Its nature is one, and it is *one* thing.

Creation 28

Arnold of Villanova

All are really one.

Black Elk

Rev. 961

Therefore the sage keeps to One and becomes the standard for the world.

Tao Te Ching, XXII

I advise no one to approach this Art unless he knows the principle and the regimen of Nature: if he be acquainted with these, little is wanting to him except one thing, nor need he put himself to a great expense, since the stone is one, the medicine is one, the vessel one, the rule one, the disposition one.

Action 346

The Golden Tract

For blessedness lieth not in much and many, but in One and oneness.

Theologia Germanica, IX

This matter is found in one thing, out of which alone our Stone is prepared (although it is called by a thousand names), without any foreign admixture. . . . They also call it the universal Magnesia, or the seed of the world, from which all natural objects take their origin. . . . They also call it the spirit of truth that is hid in the world, and cannot be understood without the inspiration of the Holy Spirit, or the teaching of those who know it. It is found *potentially* everywhere, and in everything, but in all its perfection and fulness only in *one* thing.

The Sophic Hydrolith

Grace 552
Orthod. 288

If any one attempts to separate all things from the One, taking the term 'all things' to signify a mere plurality of things, and not a whole made up of things, he will sever the All from the One, and will thereby bring to naught the All: but that is impossible.

Hermes

M.M. 978

God spake never a word but one and that he holds so dear that he will never say another. If God stopped saying his Word, but for an instant even, heaven and earth would disappear.

Eckhart

Inv. 1003
Infra 803

Foundation of foundations and firmest pillar of all wisdom is to know that there is a First Being, that He caused all beings to be and that all beings from heaven and earth and from between them could not be saved but for the truth of His own being.

Maimonides

Perhaps the most important reason for 'lamenting' is that it helps us to realize our oneness with all things, to know that all things are our relatives; and then in behalf of all things we pray to *Wakan-Tanka* that He may give to us knowledge of Him who is the source of all things, yet greater than all things.

Black Elk

Charity 608
Introd. *Heterod.*

The Perfect Man . . . contemplates the emanation (*ṣudûr*) from himself of all that exists, and beholds the Many in his essence, even as ordinary men are conscious of their own thoughts and qualities.

Jîlî

Introd. *Holiness*

One man is identical with all and all with one. One religious practice is the same as all others, and all others the same as any one of them. This is what explains the experience of birth into the Pure Land by reliance upon Amida's power. All living beings are included in one thought. It is because of this mutual intercommunication between all things, including the Buddhas themselves, that if one but calls upon Amida's sacred name once, it has the same virtue as if he did it a million times.

Ryônin

Infra 790
Introd. *Orthod.*

Buddhas and all sentient beings are essentially one.

Chih-k'ai

The 'Eight Hundred Myriads of Gods' are nothing but different manifestations of one and the same Deity Kunitokotachi-no-Kami or the Eternally Standing Divine Being of the Earth, the Great Unity of all things in the Universe, the Primordial Being of Heaven and Earth, eternally existing from the beginning to the end of the universe.

Izawa-Nagahide

Know that knowledge to be Sâttwica,[1] by which is seen in all beings the One Immutable, inseparate in the separate.

But the knowledge which sees in all beings the distinct entities of diverse kinds as different from one another, know that knowledge to be Râjasica.

While that knowledge which is confined to one single effect, as if it were the whole, without reason, not founded on truth, and trivial, that is declared to be Tâmasica.

Bhagavad-Gîtâ, XVIII. 20–22

Persevere and be steadfast in worshipping the one true God.... He who acknowledges many gods is obliged to worship many; he undergoes grievous toils and troubles; he is distracted by cares and tormented by anxieties: and in the end he incurs destruction.

Hermes

Wâḥidiyya is that (aspect) in which the Essence appears as unifying the difference of my attributes. Here the All is both One and Many. Marvel at the plurality of what essentially is One.

Jîlî

Supra Introd.

All the inner forces and the hidden souls in man are distributed and differentiated in the bodies. It is, however, in the nature of all of them that when their knots are untied they return to their origin, which is one without any duality and which comprises the multiplicity.

Abraham Abulafia

Holiness 914

The saints behold God in a simple image and in that image they discern all things; and God himself sees himself thus, perceiving all things in himself. He need not turn, as we do, from one thing to another.

Eckhart

There is only one eternal, immutable truth. It can appear under many different aspects; but even so, it is not the truth which changes, it is we who change our manner of conceiving it.

Henry Madathanas

Center 816

O descendant of Bharata, as one sun illumines all this world, similarly He who dwells in the body illumines all bodies.

Bhagavad-Gîtâ, XIII. 33

[1] Cf. note p. 88.

Wherever you look, you will see that one unique Presence, indivisible and eternal, is manifested in all the universe, but that it is very difficult to perceive it. This is because God 'impregnates' all things.

Ananda Moyî

Every thing is of the nature of no thing.

Parmenides

Void 724

The One is a negation of negations. Every creature contains a negation: one denies that it is the other. An angel denies that it is any other creature; but God contains the denial of denials. He is that One who denies of every *other* that it is anything except himself.

Eckhart

M.M. 975

For there is only *one* substance,
In which all the rest is hidden.

Abraham Lambspring

All things are one, though they be described under various names. Let this suffice thee; seek not many utensils for thy labour. If thou knowest the substance and the method, it is enough, and thou knowest all.

Michael Maier

Inv. 1017

He that has this secret possesses all good things and great riches. One ounce of it will ensure to him both wealth and health. It is the only source of strength and recreation, and far excels the golden tincture. It is the elixir and water of life, which includes all other things.

John A. Mehung

The intricate maze of different schools of philosophy claim to clarify matters and reveal the Truth. But they, in fact, create confusions where there need be no confusion. To understand anything, there must first be the person who understands, that is the Self. Why worry about other things? Why not remain yourself and be in peace? ... Fortunate is he who does not entangle himself in the labyrinths of philosophy but goes straight to the source, the Atman from which all have sprung.

Sri Ramana Maharshi

Introd. *Knowl.*

Peace 700 *Knowl.* 734

If anyone be unacquainted with Nature's methods, he will find our Art difficult, although in reality it is as easy as to crush malt, and brew beer. In the beginning when, according to the testimony of Scripture, God made heaven and earth, there was only *one* Matter, neither wet nor dry, neither earth, nor air, nor fire, nor light, nor darkness, but one single substance, resembling vapour or mist, invisible and impalpable. It was called Hyle, or the first Matter. If a thing is once more to be made out of nothing, that 'nothing' must be united, and become *one* thing; out of this *one* thing must arise a palpable substance, out of the palpable substance *one* body, to which a living soul must be given—whence through the grace of God, it obtains its specific form.

The Glory of the World

Orthod. 296

P. State 563 *Death* 208

Beauty 663

Realise God first, and all else will come to you automatically.

Swami Ramdas

He who knows one thing, knows all things; and he who knows all things, knows one thing.

Akaranga Sutra, I. i. 4

There is only *one* thing in the whole world that enters into the composition of the Stone.

The Glory of the World

Nature uses only one substance in her work.

The Only True Way

Just as the Sun, though one, appears as many when reflected in many vessels of water, so does the one Atman, reflected in many individuals, appear to be manifold.

Srimad Bhagavatam, XI. iii

Infra 803
Center 828
Knowl. 749

Thou lurkest under all the forms of Thought,
Under the form of all Created things;
Look where I may, still nothing I discern
But Thee throughout this Universe, wherein
Thyself Thou dost reflect, and through those eyes
Of him whom MAN Thou madest, scrutinize.

Jâmî

Sanâtana Dharma

Rev. 959

All that is true, by whomsoever spoken, is from the Holy Ghost.

St Ambrose

When God is our teacher, we come to think alike.

Xenophon

They that seek the Lord understand all things.

Proverbs, XXVIII. 5

Like the bee, gathering honey from different flowers, the wise man accepts the essence of different Scriptures and sees only the good in all religions.

Srimad Bhagavatam, XI. iii

Seek ye Wisdom, even in China.

Muhammad

Knowl. 761

The sage learns no learning, but reviews what others have passed through.

Tao Te Ching, LXIII

Realiz. 873

I am debtor both to the Greeks, and to the Barbarians; both to the wise, and to the unwise.

Romans, I. 14

Lo! this, your religion, is one religion, and I am your Lord, so worship Me.

And they have broken their tradition (into fragments) among them, (yet) all are returning unto Us.

Qur'ân, XXI. 92–93

A mendicant has come to us, ever absorbed in divine moods;
Holy alike is he to Hindu and Mussalmân.

Hindu Song

The one *Sat-chit-ananda* is invoked by some as God, by some as *Allâh,* by some as *Jehovah,* by some as *Hari,* and by others as *Brahma*... Every man should follow his own religion. A Christian should follow Christianity, a Mussalmân should follow Islâm, and so on. For the Hindus the ancient path, the path of the Aryan Rishis, is the best.

Sri Ramakrishna

God, who at sundry times and in divers manners spake in time past unto the fathers by the prophets,

Hath in these last days spoken unto us by his Son, whom he hath appointed heir of all things, by whom also he made the worlds.

Hebrews, I. 1, 2

Orthod. 275

He is the Word of whom the whole human race are partakers.

St Justin

Supra 775

It is one and the same Avatara that, having plunged into the ocean of life, rises up in one place and is known as Krishna, and diving down again rises in another place and is known as Christ.

The Avataras (like Rama, Krishna, Buddha, Christ) stand in relation to the Absolute Brahma as the waves of the ocean are to the ocean.

Sri Ramakrishna

My heart has become capable of every form: it is a pasture for gazelles and a convent for Christian monks,
And a temple for idols, and the pilgrim's Ka'ba, and the tables of the Tora and the book of the Koran.

Center 816

I follow the religion of Love, whichever way his camels take. My religion and my faith is the true religion.

Ibn ʿArabî

I learned gnosis from a monk called Father Simeon.

Ibrâhîm b. Adham

His Vedânta is the same as our Tasawwuf.

Jahângîr, referring to his Hindu teacher Jadrûp

Creation 26

The Sanâtana Dharma, the Eternal Religion declared by the Rishis, will alone endure.

Sri Ramakrishna

Center 833 Holiness 924

My light is the light of sun and moon. My life is the life of heaven and earth. . . . Men may all die, but I endure for ever.

Chuang-tse (ch. XI)

Allâh is the Light of the heavens and the earth.

Qur'ân, XXIV. 35

. . . The true Light, which lighteth every man that cometh into the world.

St John, I. 9

And I saw another angel fly in the midst of heaven, having the everlasting gospel to preach unto them that dwell on the earth, and to every nation, and kindred, and tongue, and people.

Revelation, XIV. 6

Holiness 921 Center 838

For this Melchisedec, king of Salem, priest of the most high God. . . .

Without father, without mother, without descent, having neither beginning of days, nor end of life; but made like unto the Son of God; abideth a priest continually.

Now consider how great this man was, unto whom even the patriarch Abraham gave the tenth of the spoils.

Hebrews, VII. 1, 3–4

Wr. Side 441

The human race (after the Fall) was at once divided into opposite parties, some accepting the devil's sacraments and some the sacraments of Christ. . . . Hence it is clear, that from the beginning there were Christians in fact, if not in name.

Hugh of Saint-Victor

Center 841

The eternal Word or Son of God did not then first begin to be the Saviour of the world when He was born in Bethlehem of Judea; but that Word which became man in the Virgin Mary did from the beginning of the world enter as a word of life, a seed of salvation, into the first father of mankind. . . . Hence it was that so many eminent spirits, partakers of a divine life, have appeared in so many parts of the heathen world; glorious names, sons of wisdom, that shone as lights hung out by God in the midst of idolatrous darkness.

William Law

Other sheep I have, which are not of this fold.

St John, x. 16

And they say: None entereth Paradise except he be a Jew or a Christian. These are their own desires. Say: Bring your proof, if ye are truthful....

Unto God belong the East and the West, and whithersoever ye turn, there is God's Countenance. Lo! God is All-Embracing, All-Knowing.

Center 841

Qur'ân, II. 111, 115

Brother, you say there is but one way to worship and serve the Great Spirit; if there is but one religion, why do you white people differ so much about it?

Chief Red Jacket

The Christ of God was not then first crucified when the Jews brought Him to the cross; but Adam and Eve were His first real murderers; for the death which happened to them in the day that they did eat of the earthly tree was the death of the Christ of God or the divine life in their souls. For Christ had never come into the world as a second Adam to redeem it, had He not been originally the life and perfection and glory of the first Adam....

Sin 66

All the shadows and types, sacrifices and ceremonies of the Jewish religion were only so many ways of applying the benefits of Jesus Christ to that people. 'Jesus Christ, the same yesterday, to-day, and for ever,' is the same in and through all ages: He was the Saviour of Adam, the patriarchs, and the Jews, just as He is our Saviour. His body and blood, offered in their sacrifices, was their atonement, as it is ours, offered upon the cross. His flesh and blood was meat and drink or a principle of life to them, as it is to us.... God so loved man, when his fall was foreseen, that He chose him to salvation in Christ Jesus before the foundation of the world.... Therefore, when Jesus Christ came into the world, declaring the necessity of a new birth to be owned and sought by a baptism into the name of Father, Son and Holy Ghost; this was not a new kind or power of salvation, but only an open declaration of the same salvation that had been till then only typified and veiled under certain figures and shadows, as He Himself had been. And men were called, not to a new faith in Him, as then first become their inward life and light, but to a more open and plain acknowledgement of Him, who from the beginning had been the one Life and Light and only salvation of the first man and all that were to descend from him.

Introd. *Suff.*

M.M. 973

William Law

The very thing that is now called the Christian religion was not wanting amongst the ancients from the beginning of the human race, until Christ came in the flesh, after which the true religion, which already existed, began to be called 'Christian'.[1]

St Augustine

[1] 'Had (St Augustine) not retracted these brave words, the bloodstained history of Christianity might have been otherwise written!' (Coomaraswamy: *Am I My Brother's Keeper*?, p. 46).

Rev. 965

He called them gods, unto whom the word of God came, and the scripture cannot be broken.

St John, x. 35

Supra 775

And the multitude of them that believed were of one heart and of one soul: neither said any of them that ought of the things which he possessed was his own.

Acts, IV. 32

Symb. 322

Merlin made the Round Table in tokening of roundness of the world, for by the Round Table is the world signified by right, for all the world, Christian and heathen, repair unto the Round Table.

Le Morte d'Arthur, XIV. 2

We Indians know the One true God, and ... we pray to Him continually.[1]

Black Elk

Pilg. 387

Benares is to the East, Mecca to the West; but explore your own heart, for there are both Rama and Allâh.

Kabîr

Realiz. 873

It is great joy to realize that the Path to Freedom which all the Buddhas have trodden is ever-existent, ever unchanged, and ever open to those who are ready to enter upon it.

Gampopa

I proclaim the Dharma eternally.

Saddharma-pundarîka, XV

Holiness 935

The true doctrine has always existed in the world and has never perished. However, this doctrine is confided to men, some of whom break with it, while others continue it scrupulously. This is why its destiny in the world is to be sometimes brilliant, and sometimes obscure.

Chou Li

Introd. *Orthod.*

... That true inward righteousness, which judgeth not according to custom, but out of the most rightful law of God Almighty, whereby the ways of places and times were disposed, according to those times and places; itself meantime being the same always, and every where, not one thing in one place, and another in another; according to which Abraham, and Isaac, and Jacob, and Moses, and David, were righteous, and all those commended by the mouth of God; but were judged unrighteous by silly men, *judging out of man's judgment,* and measuring by their own petty habits, the moral habits of the whole human race.

St Augustine

[1] 'The Hopi Chiefs believe that at this time all the remaining "spiritual threads" of the World are looking to each other for help and encouragement, for many threads make a rope which shall lead out of this cycle and into the next' (Joseph Epes Brown, in a letter).

Hindus, Mussalmâns, and Christians are going to the same destination by different paths.

Sri Ramakrishna

We are ready to receive Jesus Christ among our gods, and to set up a shrine for him, but we cannot allow that he supersede them and hold the only place; for our gods are the ones who own this island, and the ones who came here first.

A Balinese Priest

The Truth, Râma, Krishna, Jesus, is the patrimony of all humanity.

Swami Sivananda

He alone has it (the spirit of Christ) who has changed his forms and his names from the beginning of the world and so reappeared again and again in the world.

Clementine Homilies, III. 20

Center 841

O Bhârata, whenever there is decline of virtue and predominance of vice, then I embody Myself.

For the protection of the good and for the destruction of evil-doers and for the re-establishment of Dharma[1] I am born from age to age.[2]

Bhagavad-Gîtâ, IV. 7, 8

O wonder, the Son of God has been from all eternity,
And his mother has only given birth to him today!

Angelus Silesius

P. State 579

Our fathers ... did all drink the same spiritual drink: for they drank of that spiritual Rock that followed them: and that Rock was Christ.

I. Corinthians, X. 1, 4

Out of thee shall he come forth unto me that is to be ruler in Israel; whose goings forth have been from of old, from everlasting.

Micah, V. 2

The Father is begetting his Son unceasingly.

Eckhart

It follows that He had His being, without a beginning, from all eternity, and that He will abide throughout all eternity.

The Sophic Hydrolith

[1] 'For this purpose the Son of God was manifested, that he might destroy the works of the devil' (*I. John*, III. 8).

[2] 'It may be observed here that wherever it is asserted that a given event, such as the temporal birth of Christ, is at once *unique* and *historically* true we recognize an antinomy; because, as Aristotle perceived (*Met.* VI. 2. 12, XI. 8. 3). "knowledge (ἐπιστήμη) is of that which is always or usually so, not of exceptions," whence it follows that the birth in Bethlehem can only be thought of as *historical* if it is granted, that there have also been *other* such "descents"'(Coomaraswamy: 'On the Loathly Bride', *Speculum*, Oct., 1945, note 3, pp. 402–403).

Creation 26

Before the Adam known to us God created a hundred thousand Adams.

Muhammad

Languages differ, my son, but mankind is one; and speech likewise is one. It is translated from tongue to tongue, and we find it to be the same in Egypt, Persia, and Greece.... Speech then is an image of intellect; and intellect is an image of God.

Symb. 313

Hermes

Whether one says gods or Buddhas, there is no difference between the water and waves.

Kôbô Daishi

All mystics speak the same language, for they come from the same country.

Louis-Claude de Saint-Martin

Such men there have been in all countries. Amongst the Egyptians Hermes Trismegistus holds the highest place; then come Chaldaeans, Greeks, Arabs, Italians, Gauls, Englishmen, Dutchmen, Spaniards, Germans, Poles, Hungarians, Hebrews, and many others. Though the aforementioned Sages wrote at different times, and in different languages, yet their works exhibit so marvellous an agreement, that any true philosopher may easily see that all their hearts had been gladdened by God in the discovery of this (Philosophers') Stone.

The Golden Tract

Dost thou tell me that he who has not the sacraments of God cannot be saved? I tell thee that he who has the virtue of the sacraments of God cannot perish. Which is greater, the sacrament or the virtue of the sacrament—water or faith? If thou wouldst speak truly, answer, 'faith'.[1]

Hugh of Saint-Victor

He that is not against us is for us.

St Luke, IX. 50

By whomsoever truth is said, it is said through His teaching Who is The Truth.

St Augustine

And now, O philosophy, hasten to set before me not only this one man Plato, but many others also, who declare the one only true God to be God, by His own inspiration, if so be they have laid hold of the truth.

Clement of Alexandria

Some disdain to call this Art sacred, because they say that Paynims sometimes acquire a knowledge of it, though God cannot be desirous of conferring any good thing upon them, seeing that their wilful and stubborn unbelief renders them incapable of possessing that

[1] As the same authority elsewhere points out, it is not the *lack* of the sacraments, but the *contemning* of them, which prevents salvation.

which is the cause of all good. . . . To this objection, we answer what we know to be true, that the science of this Art has never been fully revealed to anyone who has not approved himself worthy by a good and noble life, and who has not shewn himself to be deserving of this gracious gift by his love of truth, virtue, and knowledge.

Thomas Norton

Beauty 676

It is impossible that the true philosophers should ever lie.

The Golden Tract

Knowl. 745
Holiness 902

Every prophet and every saint hath a way, but it leads to God: all the ways are really one.

Rûmî

Supra 775

There are not many Gurus. There is only one Guru.

Swami Ramdas

In reality there are not two. There is only One.

Sri Ramakrishna

M.M. 978

Hence there is a single religion and a single creed for all beings endowed with understanding, and this religion is presupposed behind all the diversity of rites.

Nicholas of Cusa

P. State passim

The two are really only one; it is only the ignorant person who sees many where there is really only one.

Black Elk

We hear of two birds in the forest, yet we must understand them to be only one.

Abraham Lambspring

Metanoia 480

If one objects that this (Taoist) method is exactly that of the Zen Buddhists, we reply that under Heaven two Ways do not exist, and that the Sages are always of the same Heart.[1]

Ko Ch'ang-Kêng

If 'other' and 'others' are before your eyes,
Then a mosque is no better
Than a Christian cloister;
But when the garment of 'other' is cast off by you,
The cloister becomes a mosque.

Shabistarî

[1] It is noteworthy that Alchemy and Zen in common use a terminology which is the despair of rational logic, and which—whatever needs for secrecy are incidentally served—is aimed essentially at awakening the intellect through sudden 'vertical' intuitions. Moreover, the formulas and the *koans* once properly 'situated' prove not to be illogical.

Realiz. 890

The different paths followed by Hindus and Moslems, Śâktas and Vaishnavas, reunite in the end at the door of the divine Being.

Ananda Moyî

Amida is only another name for the great Sun Buddha (Sk. *Vairocana*, 'all-illuminating light').

Kakuban

This is the land of the gods. The people should revere them. In my essence I am the Buddha Vairocana. Let my people understand this and take refuge in the Law of the Buddhas.

The Shintô Sun Goddess to the Emperor Shômu

There is nothing but the difference of the trough and crest of a water wave between what we call Gods and what we call Buddhas.

Hakuin

Truth has many aspects. Infinite Truth has infinite expressions. Though the sages speak in divers ways, they express one and the same Truth.

Srimad Bhagavatam, XI. XV

The goal for all is the same. Yet different names are given to the goal only to suit the process preliminary to reaching the goal.

Sri Ramana Maharshi

He whom the Saivites worship as Siva (the Auspicious One), the Vedantins as the Absolute, the Bauddhas as the Buddha (the Enlightened One), the Logicians, experts in proofs, as the author of the world, those devoted to Jain doctrines as the Arhat (the Worthy), the Mimamsakas as ordained Duty,—May that Hari, the Lord of the three worlds, bestow on you the desired fruit.

Action 335

Hindu Prayer

The multiple is in the outward ceremony, while the truth alone is at the interior. The cause for the multiplicity of brotherhoods is in the multiple explanations of hieroglyphics according to times, needs and circumstances. The true community of light can only be one.

Michael Maier

Even though you may believe in the Amida Buddha, your faith is quite one-sided if you despise the many Buddhas.

Hônen

Study the teachings of the Great Sages of all sects impartially.

Gampopa

And for every nation have We appointed a ritual, . . . and your God is One God.

Qur'ân, XXII. 34

There is one Lord revealed in many scriptures.

Saraha

'My divinity is adored throughout all the world, in divers manners, in variable customs, and by many names.'

Apuleius (Spoken by Isis)

P. State 579
Inv.
passim

No nation ever existed which did not worship God and believe that He was the absolute maximum.

Nicholas of Cusa

Pythagoras was anticipated by the Indians.

Apollonius of Tyana

To the wise man every land is eligible as a place of residence; for the whole world is the country of the worthy soul.[1]

Stobaeus

Do not think you are going among infidels. Muslims attain to Salvation. The ways of Providence are infinite.[2]

Pope Pius XI, to his Apostolic Delegate to Libya

Many call Thee Father, who
Will not own me as brother too:
They speak deep words from shallow meditation.
Mankind arises from one origin;
We are alike both outward and within. . .
Christians, Jews, and heathens serve Him all,
And God has all creation in His care.

Walther von der Vogelweide

Charity
597

And unto thee have We revealed the Scripture with the truth, confirming whatever

[1] Cp. note p. 139.

[2] 'At this hour when materialistic paganism is relentlessly set upon compromising and destroying spiritual values, may the example of the faith of Abraham give courage to all of those Jews, Christians and Moslems who have learned to admire it, by inspiring them with invincible confidence in the almightiness of Him who asks but to answer those who pray to Him!' (Cardinal Tisserant, in the Preface to the June, 1951, issue of *Cahiers Sioniens*, dedicated to *Abraham, Father of the Faithful*: in *Études Trad.*, 1956, p. 4).

'We express sentiments of fraternal love towards our Muslim fellow-citizens. . . . We appreciate their deep spirit of prayer and their striking fidelity to penitential fasting. . . . We are united against tendencies towards materialism and secularism' (Closing words of the Joint Pastoral Letter of the Nigerian Bishops, the first since Independence: *Catholic Herald*, London, Oct. 21, 1960).

Scripture was before it. . . . For each We have appointed a divine law and a traced-out way. Had Allâh willed He could have made you one community.[1]

Qur'ân, v. 48

The esoteric teachings of Buddhism and the tenets of the outsiders are after all one and the same in essence. At a cursory glance the latter may seem strange and incompatible with the former; yet, tactfully utilized, there is no conflicting difference at all between them.

Iso-no-Kami-no-Yakatsugu

All such as lived according to the divine word in them, which was in all men, were Christians, such as Socrates and Heraclitus, and others among the Greeks. . . . Such as live with the word are Christians, without fear or anxiety.

St Justin

If belief in the personality of Christ is a necessary condition of salvation, we must be prepared to say that all those persons who have lived before the time of Jesus have been denied the benefit of salvation for no fault of theirs and simply because they happened to be born when Jesus was yet unborn. The same reasoning would deny salvation to those who lived even at the same time as Jesus or since that time but may not have even heard of him. Further, don't you think it very unfair on the part of God that He should suddenly wake up on a particular day and prescribe for all mankind a necessary condition of salvation? Did He forget that the people who had the misfortune to be born before that date had souls to save? If He did not forget, did He take care to prescribe for them the means necessary for them to attain salvation? If He did so prescribe, His prescription could not possibly have included a belief in the Jesus to be born. The only logical hypothesis therefore which any reasonable man can accept is that God *even when* He created the first man—if there was such a time—Himself simultaneously promulgated also the means for his salvation, for even the first man was certainly in need of salvation. We accordingly say of our Vedas that they were coeval with the first man—not in the sense that they were *created* together, for we believe that there was no first creation but that everything is beginningless, but in the sense that they were co-existing—and that they are the Revelations of God Himself. Any religion which traces its origin from a later time, any time *after* creation, and from any teacher other than God, is bound to be imperfect and short-lived.

Creation 26

Sri Chandrasekhara Bhâratî Swâmigal

Wr. Side 464

It is my personal belief, after thirty-five years' experience of it, that there is no such thing as 'Christian civilization'. I believe that Christianity and modern civilization are opposed and irreconcilable, and that the spirit of Christianity and of our ancient religion is essentially the same.

Ohiyesa

Wr. Side 459

Many cry 'Christ, Christ,' who at the judgment shall be far less near to him than he who knows not Christ.

Dante (*Paradiso*, XIX. 106)

[1] Something which man for all his efforts will never arrive at. Cf. *Introduction* and note 3, p. 20.

From the beginning of the world, there have always been God-enlightened men.

The Sophic Hydrolith

Among all peoples, in all times, God is worshiped because it is natural to do so, though not with the same rites and methods.

Marsilio Ficino

So St Augustine says, that the heathens have discoursed of certain truths, and these they have reached by virtue of the eternal laws of God which are working in all men when they speak what is true, and not by the mere light of their own nature.

Tauler

The Platonic doctrine is related to divine law . . . as the moon to the sun.

Marsilio Ficino

If it (our Art) is founded on the eternal verities of Nature, why need I trouble my head with the problem whether this or that antediluvian personage had a knowledge of it? Enough for me to know that it is now true and possible, that it has been exercised by the initiated for many centuries, and under the most distant latitudes; it may also be observed that though most of these write in an obscure, figurative, allegorical, and altogether perplexing style, and though some of them have actually mixed falsehood with truth, in order to confound the ignorant, yet they, though existing in many series of ages, differing in tongue and nation, have not diversely handled one operation, but do all exhibit a most marvellous and striking agreement in regard to the main features of their teaching—an agreement which is absolutely inexplicable, except on the supposition that our Art is something more than a mere labyrinth of perplexing words.

Philalethes

Orthod. 280

Rama, Krishna, Buddha, the great Rishis, Mahatmas and Saints point to the one goal as the highest aim of life, viz., liberation and union with God. Human life is solely intended for attaining this blessed state. The supreme Lord is seated in the hearts of all beings and creatures. . . . Verily, all the different religions are so many paths that lead mankind to the one Universal God!

Swami Ramdas

Contem. 547
Center 816

There may be many Ways (*tao*), but they all are ways of illustrating the One Mind.

Takuan

We can enjoy the divine mind through various Ideas, seek it through various traces (*vestigia*), travel toward that goal by various paths. . . . (God) so disposed the intellectual eyes and the tendencies of various Souls in different manners, in order that we may approach the different possessions of the manifold divine goods by different paths.

Marsilio Ficino

Symb. 306
Center 816

Inv. 1013

For all people will walk every one in the name of his god, and we will walk in the name of the Lord our God for ever and ever.

Micah, IV. 5

I have meditated on the different religions, endeavouring to understand them, and I have found that they stem from a single principle with numerous ramifications. Do not therefore ask a man to adopt a particular religion (rather than another), for this would separate him from the fundamental principle; it is this principle itself which must come to seek him; in it are all the heights and all the meanings elucidated; then he will understand them.

Al-Ḥallâj

The beauty of the rainbow is due to the variety of its colours.

In the same way, we regard the voices of the different believers which rise from all parts of the earth as a symphony of praises on behalf of God who can only be One.

Tierno Bokar

We do not look for gulfs when we compare religions, rather we try to find similarities and unity. This is the essential difference between the Chinese and Western view points. We firmly believe in the truism that all faiths are the paths leading towards the Ultimate Reality, just as the spokes of the wheel converge to its axis. When the people are too immersed in the dogmas and rituals of their chosen religion, it appears to them to be the only one worth following and they defend their own particular faith. However, when they have acquired enough wisdom, charity and discernment, they too are bound to perceive that the road to Heaven is nobody's monopoly and that the divine laws apply equally to all. It is the dogmas, ritual and the mode of worship that divide the faiths and not the basic essence of their beliefs.

Introd.

But I am not in favour of conversion from one faith to another, neither do I believe in the fusion of all religions into one. The Ultimate Truth is one, but it has an infinite number of aspects and what is more beautiful than that each faith should reflect only one facet of the Divine, all of them together creating a shining gem of beauty. Would the world be more beautiful if all the flowers on earth had been blended into one uniform colour or all mountains razed to make the globe monotonously flat? Each religion offers something glorious, peculiarly its own, to point out the road to the Ultimate Reality. What man or group of men would be able to prescribe a single form of religion that would satisfy all and everybody? That would be an attempt to give a finite concept of the Infinite and, of course, it would fail.

Abbot Mingzing

Void 721

Some contemplate one Name, and some another. Which of these is the best? All are eminent clues to the transcendent, immortal, unembodied Brahma: these Names are to be contemplated, lauded, and at last denied. For by them one rises higher and higher in these worlds; but where all comes to its end, there he attains to the Unity of the Person.

Maitri Upanishad, IV. 6

The tribulations and temporal sufferings of God's people have now lasted six thousand years;[1] but during this whole time, men have again and again been refreshed, comforted, and strengthened by the Spirit of God—and so it is now, and ever will be, until the great universal Sabbath and rest-day of the seventh millenium. Then this occasional spiritual refreshing will cease, and everlasting joy will reign, since God will be all in all.

P. State 583

The Sophic Hydrolith

Elias and Enoch[2] wandered from India to the sea, each followed by a great band who were all Christians and fled to them from Anti-Christ.

Wr. Side 464

Mechthild of Magdeburg

To this urn let those repair
That are either true or fair.

Shakespeare (*The Phoenix and the Turtle*)[3]

Beauty 674

And this gospel of the kingdom shall be preached in all the world for a witness unto all nations; and then shall the end come.

St Matthew, XXIV. 14

The Nature of Reality

O Voice of the great God
None is great but He.

Ortha nan Gaidheal (Rune on hearing thunder)

The thunder hymneth His praise and the angels for awe of Him.

Qur'ân, XIII. 13

All creatures are a mere naught. I say not they are small, are aught: they are absolutely naught. A thing without being is not (or is naught). Creatures have no real being, for their being consists in the presence of God. If God turned away for an instant they would all perish. I have sometimes said, and it is true, that he who has gotten the whole world plus God has gotten no more than God by himself. Having all creatures without God is no more than having one fly without God; just the same, no more nor less.

Contem. 536

Eckhart

[1] I.e., roughly the extent of the present age, or 'Kali Yuga', following another traditional chronology.

[2] It will be remembered that these names are traditionally associated with the two witnesses in *Rev.* XI.

[3] The whole of this poem conceals an eschatology which merits reflection.

Shakespeare's work in general manifests a profound grasp of Hermetic doctrine (the Lesser Mysteries), and this explains his masterful alchemy with the human soul as the basic substance of his Art: it is the *materia prima* capable of being reduced from chaos into order. No more than Dante's are his productions to be mistaken for mere literature. Cf. Paul Arnold: *Ésotérisme de Shakespeare,* Paris, 1955.

'I am He who is, thou art she who is not.'

St Catherine of Siena

In the beginning was Allâh, and beside Him there was nothing,—and He remains as He was.

Muhammad

God always is, nor has He been and is not, nor is but has not been, but as He never will not be; so He never was not.

St Augustine

Supra 773
Center 838

Being is without beginning and indestructible; it is universal, existing alone, immovable and without end; nor ever was it nor will it be, since it now *is,* all together, one, and continuous.

Parmenides

Center 841

Thou art the First and there is nothing before Thee, and Thou art the Last and there is nothing after Thee, and the Manifest and there is nothing below Thee, and the Interior and there is nothing above Thee.

Muhammad

The beginning is *Atman,* the middle is *Atman,* and the end is *Atman.* What appears other than *Atman* is mere illusion.

Yoga-Vasishtha

I am the Lord, and there is none else.

Isaiah, XLV. 5

As gold exists before it is made into ornaments, and will exist after the ornaments are melted away, and remains also as gold in the intermediate stage, when it is known by the various names of ornaments, so am I in relation to the universe—I was, I shall be, and I am.

Srimad Bhagavatam, XI. XX

Death 220

The Self is the only Reality. If the false identity vanishes the persistence of the Reality becomes apparent.

Sri Ramana Maharshi

O Friends, let us stay in that which never leaves, and we shall remain.

Marsilio Ficino

M.M. 998

When appearances and names are put away and all discrimination ceases, that which remains is the true and essential nature of things and, as nothing can be predicated as to the nature of essence, it is called the 'Suchness' of Reality. This universal, undifferentiated, inscrutable 'Suchness' is the only Reality, but it is variously characterised as Truth,

Mind-essence, Transcendental Intelligence, Noble Wisdom, etc. This Dharma of the imagelessness of the Essence-nature of Ultimate Reality is the Dharma which has been proclaimed by all the Buddhas, and when all things are understood in full agreement with it, one is in possession of Perfect Knowledge, and is on his way to the attainment of the Transcendental Intelligence of the Tathagatas.

Supra 790

Lankavatara Sutra, IV

'Relationship persists so long as subsidiary cause persists, and subsidiary cause persists so long as quest persists, and quest persists so long as thou persistest, and thou persistest so long as thou seest Me not: but when thou seest Me, thou art no more, and when thou art no more, quest is no more, and when quest is no more, subsidiary cause is no more, and when subsidiary cause is no more, relationship is no more, and when relationship is no more, limit is no more, and when limit is no more, veils are no more.'

Introd. *Illusion*

Niffarî

Our glory lies where we cease to exist.

Death 208

Sri Ramana Maharshi

He is bodiless, and yet has many bodies, or rather, is embodied in all bodies. There is nothing that He is not; for all things that exist are even He. For this reason all names are names of Him, because all things come from Him, their one Father; and for this reason He has no name, because He is the Father of all.

M.M. 994

Hermes

God is *Jehovah* cal'd; which name of His
Implies or *Essence,* or the *He* that Is.[1]

Robert Herrick

It matters not what name may carelessly be applied to mind; truly mind is one, and apart from mind there is naught else.

Padma-Sambhava

There is but one word that I know now,
And of that, my friend, I know not the name.

Saraha

Pure Consciousness cannot say 'I'.

Sri Ramana Maharshi

Allâh—may He be exalted—is exempt from all likeness as well as from every rival, contrast or opposition.

Center 835

Ibn 'Arabî

[1] These lines adumbrate the metaphysical distinction between the Absolute and Pure Being. Cf. the chapter *Mysterium Magnum.*

I went to the root of things, and found nothing but Him alone.

Mîrâ Bâî

Will contents herself with God as being good. But intellect, leaving this behind, goes in and breaks through to the root whence shoots the Son and whence the Holy Spirit blossoms forth.

M.M. 994

Eckhart

Contem. 536

Paradise is the prison of the sage, just as the world is the prison of the believer.

Yaḥyâ b. Mu'âdh al-Râzî

Beauty 663

For everything real, there is a (principial) Reality.

Muhammad

The world can be compared to ice, and the Truth (*al-haqq*) to the water from which this ice is formed. Now the name 'ice' is only lent to this coagulation, and it is the name of water which returns to it in its essential reality (*haqîqah*).

Jîlî

Satchidânanda is like an endless expanse of water. The water of the great ocean in cold regions freezes into blocks of ice. Similarly, through the cooling influence of divine love, Satchidânanda assumes forms for the sake of the bhaktas.... The heat of the sun of Knowledge melts the ice-like form of the Personal God. On attaining the Knowledge of Brahman and communing with It in nirvikalpa samâdhi, one realizes Brahman, the Infinite, without form or shape and beyond mind and words.

Symb. 317 *Knowl.* 761

Sri Ramakrishna

The formless Absolute is my Father, and God with form is my Mother.

Kabîr

The sun is the Father, and the Moon the Mother of this Stone, and the Stone unites in itself the virtues of both its parents.

Holiness 924

The Glory of the World

And I saw no difference between God and our Substance: but as it were all God: and yet mine understanding took that our Substance is in God: that is to say, that God is God, and our Substance is a creature in God.... We are enclosed in the Father, and we are enclosed in the Son, and we are enclosed in the Holy Ghost. And the Father is enclosed in us, and the Son is enclosed in us, and the Holy Ghost is enclosed in us: Almightiness, All-Wisdom, All-Goodness:[1] one God, one Lord.

Julian of Norwich

M.M. 973

God is absolute or restricted, as He pleases; and the God of religious belief is subject to limitations, for He is the God who is contained in the heart of His servant. But the absolute

[1] Here in Christian terminology is the exact equivalent of the Hindu ternary *Sat-Chit-Ânanda*. See p. 623.

God is not contained by any thing, for He is the being of all things and the being of Himself, and a thing is not said either to contain itself or not to contain itself.

Ibn 'Arabî

There are two aspects of Râdhâ-Krishna: the Absolute and the Relative. They are like the sun and its rays. The Absolute may be likened to the sun, and the Relative to the rays.

A genuine bhakta dwells sometimes on the Absolute and sometimes on the Relative. Both the Absolute and the Relative belong to one and the same Reality. It is all one—neither two nor many.

M.M. 978

Sri Ramakrishna

Nothing that exists perishes, but men are in error when they call the changes which take place 'destructions' and 'deaths'... The Father has never been made, but ever is. But the Kosmos is ever being made. For the cause of the existence of the universe is the Father; but the Father is the cause of his own existence.

Introd. *Holiness*

Hermes

All things and all ages derive their existence from the Pre-Existent. All Eternity and Time are from Him, and He who is Pre-Existent is the Beginning and the cause of all Eternity and Time and of anything that hath any kind of being. All things participate in Him nor doth He depart from anything that exists; He is before all things, and all things have their maintenance in Him; and, in short, if anything exists under any form whatever, 'tis in the Pre-Existent that it exists and is perceived and preserves its being.

Dionysius

O Arjuna, whatever is the seed of all beings, that also am I. Without Me there is no being existent, whether moving or unmoving.

Bhagavad-Gîtâ, x. 39

Discrimination is the reasoning by which one knows that God alone is real and all else is unreal. Real means eternal, and unreal means impermanent. He who has acquired discrimination knows that God is the only Substance and all else is non-existent. With the awakening of this spirit of discrimination a man wants to know God, . . . who is of the very nature of Reality.

Sri Ramakrishna

The Source, having no prior, cannot be contained by any other form of being; It is orbed around all; possessing, but not possessed, holding all, Itself nowhere held. It is omnipresent; at the same time, It is not present, not being circumscribed by anything; yet, as utterly unattached, not inhibited from presence at any point.

Introd. *Heterod. Center* 841

Plotinus

All things were made by him; and without him was not any thing made that was made.

St John, I. 3

Holiness 934 *Grace* 558 *Center* 816

Allâh! There is no god but He, the Living, the Eternal. Neither sleep overtaketh Him, nor slumber. Unto Him belongeth whatsoever is in the heavens and whatsoever is in the earth. Who is he that intercedeth with Him save by His leave? He knoweth that which is in front of them and that which is behind them, while they encompass nothing of His Knowledge save what He will. His Throne includeth the heavens and the earth, and He is never weary of preserving them. He is the Sublime, the Tremendous.

Qur'ân, II. 255 (Verse of the Throne)

A world without God is a world of nothing.

Dafydd ap Gwilym

Holiness 902 *Wr. Side* 474

All nature and all that is natural in the creature is in itself nothing else but darkness, whether it be in soul or body, in Heaven or on earth. . . . If you ask, why nature must be darkness, it is because nature is not God and therefore can have no light as it is nature. For God and light are as inseparable, as God and unity are inseparable. Everything therefore that is not God, is and can be nothing else in itself but darkness, and can do nothing but in and under and according to the nature and powers of darkness.

William Law

Knowl. 749 Supra 773

Tat. What then is real, thrice-greatest one?

Hermes. That which is not sullied by matter, my son, nor limited by boundaries, that which has no colour and no shape, that which is without integument, and is luminous, that which is apprehended by itself alone, that which is changeless and unalterable, that which is good.

Hermes

Realiz. 859

Insofar as man is a possibility of manifestation, but does not see What manifests him, he is pure absence (*'udum*): by contrast, insofar as he receives his being from the perpetual irradiation (*Tajallî*) of the Essence, he *is*.

'Abd al-Razzâq al-Qashânî

Illusion 94 Introd. *Heterod.*

The Creature truly *really is* in the proper *rank* and order of its own Being, but all that is, in the presence of the Divine Being, in comparison with it, is like a *dream*, when one awakes, less than nothing. . . . The Creature is nothing of it self, or by it self but a *momentary emanation* from God, sent forth from him, and filled with him. God is *not the Creature*, yet he is in the Creature.

Peter Sterry

There is no existence for the unreal and the real can never be non-existent. The Seers of Truth know the nature and final ends of both.

Bhagavad-Gîtâ, II. 16

M.M. 978 *Renun.* 136

Nothing is born, nothing is destroyed. Away with your dualism, your likes and dislikes. Every single thing is just the One Mind. When you have perceived this, you will have mounted the Chariot of the Buddhas.

Huang Po

Q: From all you have just said, Mind is the Buddha; but it is not clear as to what sort of mind is meant by this 'Mind which is the Buddha'.

A: How many minds have you got?

Q: But is the Buddha the ordinary mind or the Enlightened mind?

A: Where on earth do you keep your 'ordinary mind' and your 'Enlightened mind'?

Huang Po

Metanoia 480

Thou art the All and the All is in thee: and thou Art, and there is nought else that IS save thee only.

Acts of Peter, XXXIX

Realiz. 890

He is the quintessence, the essence of all essences, and yet Himself not an essence of anything.... He is also the true Catholic Magnesia, or universal seed of the world, of Whom, through Whom, and to Whom are all things in heaven and upon earth—the Alpha and Omega, the beginning and the end, says the Lord that is, and was, and is to come, the Almighty (Apoc. i).

The Sophic Hydrolith

M.M. 994

Creation 26

Think not that He is in anything, nor again that He is outside of anything; for He is limitless himself, and is the limit of all things; He is encompassed by nothing, and encompasses all things.

Hermes

Contem. 523

We must acknowledge that there is one kind of being which is always the same, uncreated and indestructible, never receiving anything into itself from without, nor itself going out to any other, but invisible and imperceptible by any sense, and of which the contemplation is granted to intelligence only.

Plato (*Timaeus*, 52 A)

Knowl. 749

He in whose heart this holy disdain of the non-Self wells up constantly and fully is a vessel of election for the direct perception of the Self, which those in this world will never know who go astray in the turmoil of an illusory universe.

Śrî Śankarâchârya

Renun. 146

How can that be real, which is not the same that it was before? Inasmuch as things change, they are illusory. But at the same time you must understand, my son, that these illusory things are dependent on Reality itself, which is above; and that being so, I say that the illusion is a thing wrought by the working of Reality.

Hermes

Illusion 85

Truth resides primarily in the Intellect, and secondarily in things according as they are related to the intellect as their source.

St.Thomas Aquinas

Everything that *IS* beareth written upon it this *Name* of God, *I AM*. All things that *be* declare a *Being*: While all things agree in this, *That they be,* they demonstrate an *universal* Being. *Being,* as it is divided and *restrained* by particular Differences (in all things particular and different one from another) by being lessened, *contracted,* and obscured, is *imperfect.* Nothing that is imperfect can subsist, exist *of it self,* or by it self: for so far as it is imperfect it *is not.*

Wr. Side 474

Peter Sterry

Sin 56

The Good is utterly alien to gods and men; but it is inseparable from God, for it is God himself. All other gods are called good merely because men have sought to honour them by giving them a title which belongs to God; but God is called the Good not by way of honouring him, but because that is his nature; for the nature of God is one and the same with the nature of the Good. God then is the Good, and the Good is God.

Hermes

Realiz. 887

All this world is nothing but the one Supreme Reality which is Niralamba, devoid of any other support but Itself. The world shines by the light of this Reality. Know that the Yogi, whose mind is turned inward, dissolves all phenomenal objects, and merges himself in that Supreme Reality and becomes one with It.

Devikalottara-Jnanachara-Vichara-Patalam

Death 220

Scrutinised through the reasoning that reality is never destroyed and unreality never born (*Bh. Gîtâ,* II. 16) you have no (real) existence. You are, therefore, O my mind, non-existent in the Self. Having both birth and death you are accepted as non-existent.

Śrî Śankarâchârya

There is only being in Self-Realisation, and nothing but being.

Sri Ramana Maharshi

M.M. 998

Not by speech, not by mind,
Not by sight can He be apprehended.
How can He be comprehended
Otherwise than by one's saying 'He is'?

Kaṭha Upanishad, VI. 12

We give thee thanks, O Lord God Almighty, which art, and wast, and art to come.

Revelation, XI. 17

The Self alone is real. The world of the senses is superimposed upon it.

Srimad-Bhagavatam, XI. xi

Creation 28

If God gave the soul his whole creation she would not be filled thereby but only with himself.

Eckhart

O! but for my love, day would turn to night.

Shakespeare (*Love's Labour's Lost,* IV. iii. 233)

That which remains after the dissolution of the mind and its created bodies, is the *Atman* alone, and that is the Supreme God, the All-Highest.

Yoga-Vasishtha

Realiz. 890

God is the Self-Existent . . . consisting of none but Himself: His acts and His Essence are beyond instrument and direction, for His Being is above 'Be' and 'He'. . . . He is the only Real Existent: all other existences exist only in the imagination and are subject to His existence.

Abu 'l-Majdûd b. Âdam Sanâ'î

M.M. 994
Illusion 85

In this evanescent panorama of life all things and objects are subject to transmutation and dissolution. The Lord alone is real with whom we are eternally united.

Swami Ramdas

Illusion 99

The finite, multiple universe has no existence apart from the infinite Self. The finite is the reading of finitude into the infinite. Seeing the finite in the infinite Self is a delusion of the mind.

Srimad Bhagavatam, XI. XX

Supra 775
Creation 42

He said, 'Even this Ṣûfism is polytheism (*shirk*).'
'Why, O Shaykh?' they asked.
He answered, 'Because Ṣûfism consists in guarding the soul from what is other than God; and there is nothing other than God.'

Abû Sa'îd ibn Abi 'l-Khayr

Renun. 146
158

Wonder of wonders! They take what *is not* as what *is,* or they see the phenomena apart from the Self.

Sri Ramana Maharshi

Although it is Total Reality, there is no perceiver of it. Wondrous is this.

Padma-Sambhava

Glory be to Him who hides Himself by the manifestations of His light, and manifests Himself by drawing a veil over His face.

Jâmî

Metanoia 488

The Deity is absolute. It transcends human words, which are of a relative nature. It is incomprehensible, and yet it permeates all things. It is everywhere. People as a rule, not knowing this truth, visit a hundred shrines day by day to worship there, and make valuable offerings month by month, and yet they are not sure to obtain any reward, though they may perchance suffer misfortunes in the world.

Shirai-Sôin

Pilg. 387

God is the Being in Whom being anything means being everything.

Parmenides

Void 724

Reality 773

God is a principle which exists by virtue of its own intrinsicality, and operates spontaneously, without self-manifestation.

Chuang-tse (gloss on ch. VI)

Thus absolute truth is indestructible. Being indestructible, it is eternal. Being eternal, it is self-existent. Being self-existent, it is infinite. Being infinite, it is vast and deep. Being vast and deep, it is transcendent and intelligent. It is because it is vast and deep that it contains all existence. It is because it is transcendent and intelligent that it embraces all existence. It is because it is infinite and eternal that it fulfills all existence. In vastness and depth, it is like the Earth. In transcendent intelligence, it is like Heaven. Infinite and eternal, it is the Infinite itself.

Tsesze

M.M. 978

Throughout all the ten regions of the universe there is no place where the Absolute is not.

Hakuin

God is ever-existent; and He makes manifest all else, but He himself is hidden, because He is ever-existent. He manifests all things, but is not manifested; He is not himself brought into being in images presented through our senses, but He presents all things to us in such images.

Symb. 306

Hermes

Death 208

I am everything that was, that is, that shall be. . . . Nor has any mortal ever been able to discover what lies under my veil.

Inscription to Isis

For thee is a proof of His Mightiness—may He be magnified!—in that He has veiled thee from Him by something not existing with Him! . . .

Illusion 85

How may it be conceived that anything should veil Him without Whom exists nothing!
O wonder! how can existence appear in non-existence?
Or how can the create endure beside Him whose attribute is the Pre-Existent?

Ibn ʿAṭâʾillâh

P. State 563

It is, indeed, nothing short of marvellous that so many seek so ordinary a thing, and yet are unable to find it.

The Glory of the World

Knowl. 749

What is there to be known apart from Him by whom all that exists is known?

Yoga Pradipa, III. 17

'O son of man, were you to come to Me with almost an earth-ful of sins, and then you met Me without joining anything with Me in the Godhead, then would I come to you with an earth-ful of forgiveness.'

Sin 57

Muhammad

O God, deliver us from preoccupation with worldly vanities, and show us the nature of things 'as they really are'.[1] Remove from our eyes the veil of ignorance, and show us things as they really are. Show not to us non-existence as existent, nor cast the veil of non-existence over the beauty of existence. Make this phenomenal world the mirror to reflect the manifestations of Thy beauty, and not a veil to separate and repel us from Thee. Cause these unreal phenomena of the universe to be for us the sources of knowledge and insight, and not the cause of ignorance and blindness.

Center 828
Illusion 109

Jâmî

[1] 'A prayer ascribed to Muḥammad' (note in text).

THE EYE OF ETERNITY—SUPREME CENTER

God's center is everywhere, His circumference nowhere.

St Bonaventura (*Itin. Mentis*, 5)

Correlations have already been made in previous chapter introductions between the concepts of centrality, origin, presence, pole, summit, interiority and simultaneity.[1]

Introds. *Contem. Ecstasy*

Samâdhi is to know the *synthesis* of all cosmic time and space—and by transposition, in *nirvikalpa samâdhi*, all existences—in an 'Eternal Present' centered in a superluminous fullness of beatitude whereof the Comprehensor is the transcendent Intellect or solar Deity (Agni) dwelling microcosmically in the 'Eye of the Heart' (Arabic *'ayn al-qalb*), also represented in Hinduism as the frontal eye of Shiva. This non-manifested 'abode' of the Supreme Principle or Universal Spirit, omniscient Self, Âtmâ, a 'point without extension' or 'moment without duration',[2] but centric and axial to all existences, where complementaries and oppositions are contained in principial equilibrium, is metacosmically the *Brahmapura* of Hinduism, equatable with the Heavenly Jerusalem or City of God, or again, the Holy Palace in the Kabbala, sanctuary of the *Shekhînah*.[3]

Introd. *P. State*

Introd. *Realiz.*

For the human being on the path of realization, the first center to be attained is that of this world, i.e. the Primordial State: 'It is in this center that "time is changed into space", since herein for our state of existence is the direct reflection of principial eternity, which excludes all succession; moreover, death cannot attain thereto, and thus it is also the "abode of immortality"; all things appear therein in perfect simultaneity in an immutable present, through the power of the "third eye", with which man has recovered the "sense of eternity" ' (Guénon: *Le Règne de la Quantité*, p. 161).

Creation 31

'God is the Eye that sees the world and which, being active where creature is passive, creates the world by its vision, which is act and not passivity; the "eye" thus becomes the metaphysical center of the world, of which it is at the same time the "sun" and the "heart". God sees not only the "exterior", but also—or rather all the more so—the "interior", and it is this latter vision which is the more real, or rather, strictly speaking, alone real, since it is the absolute or infinite Vision where God is at once the "Subject" and

[1] E.g., *Creation, Metanoia, Contemplation, The Primordial State, The Surface of the Waters, Peace, Spiritual Drowning, The Void, Reality*.

[2] Coomaraswamy: 'Gradation, Evolution, and Reincarnation'; cf. his *Time and Eternity* for a detailed exposition of the doctrine.

[3] Cf. Guénon: *L'Homme et son Devenir*, ch. III; *Le Symbolisme de la Croix*, passim.

the "Object", the Knower and the Known. The universe is only "vision" or "knowledge", in whatever mode this is realized, and its whole reality is God: the worlds are woven of visions, and the content of these visions indefinitely repeated is always the Divine, which is thus the primal Knowledge and the ultimate Reality—Knowledge and Reality being two complementary aspects of the same Divine Cause' (Schuon: *L'Oeil du Coeur*, pp. 17–18).

Knowl. 749
Illusion 94
Knowl. 761
Reality 803

The Eye of the Heart

Realiz. 890

I saw my Lord with the eye of my heart, and I said: who art Thou? He said: Thou.

Al-Ḥallâj

Death 220

When the gnostic's spiritual eye is opened, his bodily eye is shut: they see nothing but Him.

Abû Sulaymân al-Dârânî

Sin 77 *Metanoia* 480

The soul has two eyes—one looking inwards and the other outwards. It is the inner eye of the soul that looks into essence and takes being directly from God.

Eckhart

There is an eye of the soul which . . . is more precious far than ten thousand bodily eyes, for by it alone is truth seen.

Plato (*Republic* VII, 527 E)

There is a power in sight which is superior to the eyes set in the head and more far-reaching than the heavens and earth.

Eckhart

The eye of the heart, which is seventy-fold and of which these two sensible eyes are only the gleaners. . . .

Rûmî

Void 728

The light of splendour shines in the middle of the night.
Who can see it? A heart which has eyes and watches.

Angelus Ṣilesius

God gives one divine eyes; and only then can one behold Him. God gave Arjuna divine eyes so that he might see His Universal Form.

Sri Ramakrishna

Grace 556

The divine light . . . readily enters into the eye of the mind that is prepared to receive it.

Benjamin Whichcote

Holiness 924 *Reality* 803

The Yogî, whose intellect is perfect, contemplates all things as abiding in himself and thus, by the eye of Knowledge (*Jnâna-chakshus*), he perceives that everything is *Âtmâ*.

Srî Śankarâchârya

God is light, not such as these eyes see, but as the heart seeth, when thou hearest, 'He is Truth.'

St Augustine

The light of the body is the eye: if therefore thine eye be single, thy whole body shall be full of light.

St Matthew, VI. 22

What is more quiet than the single eye?

The Imitation of Christ, III. xxxi

Peace 700

They (the statesmen elect) must raise the eye of the soul to the universal light which lightens all things, and behold the absolute good; for that is the pattern according to which they are to order the State and the lives of individuals, and the remainder of their own lives also; making philosophy their chief pursuit.

Plato (*Republic* VII, 540 B)

Reality 790

Never shalt thou arrive at the unity of vision or uniformity of will, but by entering fully into the will of our Saviour Christ, and therein bringing the eye of time into the eye of eternity; and then descending by means of this united through the light of God into the light of nature.

Boehme

Conform. 180 *Flight* 949

We must shut the eyes of sense, and open that brighter eye of our understandings, that other eye of the soul, as the philosopher calls our intellectual faculty, 'which indeed all have, but few make use of it.'

John Smith the Platonist

Sin 77

If you wish to see that Face,
Seek another eye. The philosopher
With his two eyes sees double,
So is unable to see the unity of the Truth.

Shabistari

Knowl. 734 *M.M.* 978

He who would gain a golden understanding of the word of truth, should have the eyes of his soul opened, and his mind illumined by the inward light which God has kindled in our hearts from the beginning. . . . Although no man ever has, or ever can, see God with his outward bodily eyes, yet with the inward eyes of the soul He may well be seen and known.

The Sophic Hydrolith

Creation 42 *Knowl.* 745

For the outer sense alone perceives visible things and the eye of the heart alone, sees the invisible.

Richard of Saint-Victor

You must close the eyes and waken in yourself that other power of vision, the birthright of all, but which few turn to use.

Plotinus

Metanoia 484

Let us all with one accord give praise to Him, who is seated high upon the heavens, creator of all that is.
It is He that is the eye of my mind.

Hermes

Osiris, who doth meet every god within the temple of his Eye in Ȧnnu. . . .

Papyrus of Nu

'God is most great' is on my heart's lips every moment.
The heart hath gotten an eye constant in desire of thee.

Dîvâni Shamsi Tabrîz, XI

Knowl. 749

The eye by which I see God is the same as the eye by which God sees me. My eye and God's eye are one and the same.

Eckhart

Introd. *Symb.*

In these outlines, my son, I have drawn a likeness (εἰκών) of God for you, so far as that is possible; and if you gaze upon this likeness with the eyes of your heart (καρδίας ὀφθαλμοῖς), then, my son, believe me, you will find the upward path; or rather, the sight itself will guide you on your way.

Hermes

Infra 833

That sun of the intellectual world, that inner eye of the heart. . . .

Richard of Saint-Victor

I entered (into my inward self) and beheld with the eye of my soul . . . the Light Unchangeable.

St Augustine

'Open the eye of thy intelligence and look at Me.'

St Catherine of Siena

Orthod. 288

(Pythagoras) divinely healed and purified the soul, resuscitated and saved its divine part, and conducted to the intelligible its divine eye, which, as Plato says, is better worth saving than ten thousand corporeal eyes; for by looking through this alone, when it is strengthened and clarified by appropriate aids, the truth pertaining to all beings is perceived.

285

Iamblichus

. . . The eyes of your heart enlightened, that you may know what the hope is of his calling, and what are the riches of the glory of his inheritance in the saints.

Ephesians, I. 18

Infra 828

The face of our soul uncovered by opening of the ghostly eye, (we) behold as in a mirror heavenly joy.

Walter Hilton (Gloss on *II. Cor.* III. 18)

The eyes of our mind can look as easily backwards into that eternity which always hath been, as into that which ever shall be.

Infra 838

William Law

And then our Lord opened my spiritual eye and shewed me my soul in midst of my heart. I saw the Soul so large as it were an endless world, and as it were a blissful kingdom. And by the conditions that I saw therein I understood that it is a worshipful City. In the midst of that City sitteth our Lord.

Infra 832

Julian of Norwich

While thou first fastnest the Eye of thy Spirit on the Majesty of God, and then beholdest all Things, as they appear in the Light of the Divine Presence; thou indeed art in Heaven: All Things are as the Angels of God, as Divine Emanations, Divine Figures, and Divine Splendors circling thee in on every side, and God himself as a Fountain of Glories in the midst of them.

Infra 847

Peter Sterry

The Atman is self-luminous and birthless; it is existence, absolute knowledge, the eye of the eyes, One without a second.

Srimad Bhagavatam, XI. XX

Jesu, who always showest thyself unto us—for this is thy will, that we should at all times seek thee, and thyself hast given us this power, to ask and to receive, and hast not only permitted this, but hast taught us to pray: who art not seen of our bodily eyes, but art never hidden from the eyes of our soul. . . .

Faith 505

Acts of Thomas, 53

For it is with the interior eye that truth is seen.

St Augustine

For the things that are in the spiritual world can be seen by the eye of the mind alone.

Hermes

Our whole business therefore in this life is to restore to health the eye of the heart whereby God may be seen.

Contem. 547

St Augustine

He set his eye upon their hearts to shew them the greatness of his works.

Ecclesiasticus, XVII. 7

I am blind and do not see the things of this world; but when the light comes from Above, it enlightens my Heart and I can see, for the Eye of my Heart (*Chante Ishta*) sees everything; and through this vision I can help my people. The heart is a sanctuary at the Center of which there is a little space, wherein the Great Spirit (*Wakantanka*) dwells, and this is the Eye. This is the Eye of *Wakantanka* by which He sees all things, and through

Knowl. 749 *Judgment* 250 *P. State* 573

which we see Him. If the heart is not pure, *Wakantanka* cannot be seen, and if you should die in this ignorance, your soul shall not return immediately to *Wakantanka*, but it must be purified by wandering about in the world. In order to know the Centre of the Heart in which is the Mind of *Wakantanka*, you must be pure and good, and live in the manner that *Wakantanka* has taught us. The man who is thus pure contains the Universe within the Pocket of his Heart (*Chante Ognaka*).

Black Elk

Infra 832

In this abode of *Brahma* (*Brahma-pura*) there is a small lotus, a place in which is a small cavity (*dahara*) occupied by Ether (*Âkâsha*); we must seek That which is in this place, and we shall know It.

Chândogya Upanishad, VIII. i. 1

Holiness 924

If thou conceivest a small minute circle, as small as a grain of mustard seed, yet the Heart of God is wholly and perfectly therein: and if thou art born in God, then there is in thyself (in the circle of thy life) the whole Heart of God undivided.

Boehme

This *Atmâ,* which dwells in the heart, is smaller than a grain of rice, smaller than a grain of barley, smaller than a grain of mustard, smaller than a grain of millet, smaller than the germ which is in the grain of millet; this *Ätmâ,* which dwells in the heart, is also greater than the earth, greater than the atmosphere, greater than the sky, greater than all the worlds together.

Containing all works, containing all desires, containing all odors, containing all tastes, encompassing this whole world, the unspeaking, the unconcerned—this is *Âtmâ* within the heart, this is Brahma. Into It I shall enter on departing hence.

Chândogya Upanishad, III. xiv. 3, 4

Contem. 536

'Heart' is merely another name for the Supreme Spirit, because He is in all hearts.

Sri Ramana Maharshi

The Dwelling of the Tathagata is the great compassionate heart within all living beings.

Saddharma-puṇḍarîka

Introd. *Holiness*

The heart of the gnostic possesses such an amplitude that Abû Yazîd al-Bistâmî said of it: if the divine Throne with all that surrounds it were to be found a hundred million times in a corner of the heart of the gnostic, he would not feel it: and Junayd said in the same sense: if the ephemeral and the eternal are joined, there remains no further trace of the former; now, how could the heart which contains the eternal feel the existence of the ephemeral?

Ibn 'Arabî

Reality 773

O Arjuna, the Lord dwells in the heart of all beings, causing all beings to revolve, as if mounted on a wheel.

Bhagavad-Gîtâ, XVIII. 61

The nature of the universe which stilleth the centre and moveth all the rest around, hence doth begin as from its starting point. *Creation* 26

And this heaven (the *primum mobile*) hath no other *where* than the divine mind.

Dante (*Paradiso,* XXVII. 106)

Each thing a certain course and laws obeys,
Striving to turn back to his proper place; *Illusion* 109
Nor any settled order can be found,
But that which doth within itself embrace *Peace* 694
The births and ends of all things in a round.

Boethius

What a wonderful lotus it is that blooms at the heart of the wheel; who are its comprehensors?

There in the midst thunders the self-supported lion-throne, there the Great Person shines resplendent. *Reality* 803

Kabîr

He whose heart rejoices in the knowledge that he is really one with God loses his own individuality and becomes free. Be eternally satisfied with thy Beloved, and so shalt thou dwell in Him as the rose within the calyx.

'Aṭṭâr

Om mani padme hum: 'Om, the jewel in the lotus, *Hum*!' *Inv.* 1031

Tibetan Mantra

In virtue of his miraculous power, transcending human intelligence,
Residing in the centre of the smallest atom,
The Tathagata preaches the doctrine of perfect serenity. *Peace* 700

Avataṃsaka Sûtra

Soul is not in the universe, on the contrary, the universe is in Soul.

Plotinus

Soul and Godhead are one: there the soul finds that she is the kingdom of God. *Realiz.* 887

Eckhart

She is all things' place and has herself no place. *Void* 724

Eckhart

The real seekers ... do not think that the wondrous law is inside or outside or in between. *Contem.* 523

Hakuin

O valiant friend, lay hold of the skirt of Him who is exempt from 'above' and 'below'.

Rûmî

Flight 944

Once transport ourselves in spirit, outside of this universe of dimensions and localizations, and there will be no more question of trying to 'situate' the Principle.

Chuang-tse (ch. XXII)

Those who bridle their mind which travels far, moves about alone, is without a body, and hides in the chamber of the heart, will be free from the bonds of Mâra, the tempter.

Dhammapada, III. 37

The Most High is absolutely without measure, as we know,
And yet a human heart can enclose Him entirely!

Angelus Silesius

'My earth and My heaven contain Me not, but the heart of My faithful servant containeth Me.'

Muhammad

The heart is the same as Prajâpati (Lord of Creation). It is Brahma. It is all.

Bṛihad-Âraṇyaka Upanishad, V. 3

Holiness 924 *Charity* 608

Man is the mountain of mountains, the Stone of all stones, the tree of trees, the root of roots, the earth of earths. All these things he includes within himself, and God has given to him to be the preserver of all things.

The Glory of the World

Realiz. 870

He who knows That, set in the secret place (of the heart)—
He here on earth, my friend, rends asunder the knot of ignorance.

Muṇḍaka Upanishad, II. i. 10

Renun. 158

This earth is the largest thing we see around us. But larger than the earth is the ocean, and larger than the ocean is the sky. But Vishnu . . . has covered earth, sky, and the nether world with one of His feet. And that foot of Vishnu is enshrined in the sâdhu's heart. Therefore the heart of a holy man is the greatest of all.

The Gospel of Sri Ramakrishna[1]

Introd. *Heterod.*

The soul that enters into God owns neither time nor space nor anything nameable to be expressed in words. But it stands to reason, if you consider it, that the space occupied by any soul is vastly greater than heaven and earth and God's entire creation. I say more: God might make heavens and earths galore yet these, together with the multiplicity of creatures he has already made, would be of less extent than a single needle-tip compared with the standpoint of a soul atoned in God.

Eckhart

In truth, we seek for God outside of ourselves until we make the great Dis-

[1] Spoken by Ishan, a devotee, in reply to a question put by Sri Ramakrishna.

covery—which is that our heart is the sanctuary where the Lord of the universe, Vishvanâth, dwells in all His glory.

Swami Ramdas

Pilg. 387 *Contem.* 536

That God, the All-worker, the Great Self (*mahâtman*),
Ever seated in the heart of creatures,
Is framed by the heart, by the thought, by the mind—
They who know That, become immortal.

Śvetâśvatara Upanishad, IV. 17

Near and far addeth not nor subtracteth there, for where God governeth without medium the law of nature hath no relevance.

Dante (*Paradiso,* XXX. 121)

Creation 28

What is here is there, what is not here is nowhere.

Vishwasâra Tantra

Symb. 306

To know it we must be in it, beyond the mind and above our created being; in that Eternal Point where all our lines begin and end, that Point where they lose their name and all distinction, and become one with the Point itself.

Ruysbroeck

Infra 838 *Reality* 803

He, truly, indeed, is the Self (Âtmâ) within the heart, very subtile, kindled like fire, assuming all forms. This whole world is his food. On Him creatures here are woven.

Maitri Upanishad, VII. 7

Judgment 250

His throne is in heaven who teaches from within the heart.

St Augustine

Now, that golden Person who is within the sun, who looks down upon this earth from his golden place, is even He who dwells within the lotus of the heart and eats food.

Maitri Upanishad, VI. 1

Bring together in yourself all opposites of quality, heat and cold, dryness and fluidity; think that you are everywhere at once, on land, at sea, in heaven; think that you are not yet begotten, that you are in the womb, that you are young, that you are old, that you have died, that you are in the world beyond the grave; grasp in your thought all this at once, all times and places, all substances and qualities and magnitudes together; then you can apprehend God.

Hermes

P. State 579 Infra 841

This, verily, is the person (*puruṣa*) dwelling in all cities (*puriśaya*).

Bṛihad-Âraṇyaka Upanishad, II. v. 18

Infra 832

O *Wakan-Tanka,* behold the pipe! . . . You have taught us that the round bowl of the pipe is the very center of the universe and the heart of man!

Black Elk

Who is the bird of golden hue,
Who dwells in both the heart and sun,
Swan, diver-bird, surpassing bright—
Him let us worship in this fire!

Maitri Upanishad, VI. 34

In that point I say truly (is) the spirit of life, which dwells in the most secret chamber of the heart.

Dante (*Vita Nuova,* 2)

Infra 835

Logical reasoning cannot establish our duty. The shrutis are conflicting. Not a rishi whose views do not differ from those of others. The truth of religion is hidden in a cave. Follow therefore the path trodden by good men.

Vyâsa

I am seated in the hearts of all.

Bhagavad-Gîtâ, XV. 15

God . . . who dwells nowhere but in Himself, while compenetrating all things, in being neither near nor far from anything. . . .

Boehme

This center which is here, but which we know is really everywhere, is *Wakan-Tanka.*

Black Elk

Creation 42
Realiz. 887

As it becometh you to retain a glorious sense of the world, because the Earth and the Heavens and the Heaven of Heavens are the magnificent and glorious territories of God's Kingdom, so are you to remember always the unsearchable extent and illimited greatness of your own soul; the length and breadth and depth, and height of your own understanding. Because it is the House of God, a Living Temple, and a glorious Throne of the Blessed Trinity: far more magnificent and great than the Heavens; yea a person that in Union and Communion with God, is to see Eternity, to fill His Omnipresence, to possess His greatness, to admire His love; to receive His gifts, to enjoy the world, and to live in His Image. Let all your actions proceed from a sense of this greatness, let all your affections extend to this endless wideness, let all your prayers be animated by this spirit and let all your praises arise and ascend from this fountain.

Thomas Traherne

Peace 694

I found Thee not, O Lord, without, because I erred in seeking Thee without that wert within.

St Augustine

The kingdom (Law) of the Buddha is not far away, but within us; eternal truth (*Shinnyo*) can not be seen outside ourselves. Search, and you will find it. Believe, and you will soon be free from illusion. Practise, and you will at once realize the truth. Alas! for the foolish who deny this certainty! Woe to them who are too 'drunken and mad' to listen to the living sermon of nature!

Kôbô Daishi

Faith 505
Knowl. 745

And vie with one another for forgiveness from your Lord, and for a paradise as wide as are the heavens and the earth, prepared for the God-fearing.

Qur'ân, III. 133

Holy War 403

Find Buddha in your own heart, whose essential nature is the Buddha himself.

Eisai

I thought that I had arrived at the very Throne of God and I said to it: 'O Throne, they tell us that God rests upon thee.' 'O Bâyazîd,' replied the Throne, 'we are told here that He dwells in a humble heart.'

Bâyazîd al-Bisṭâmî

Humility 199

Man's heart is the central point
And heaven the circumference.

Shabistarî

The Buddhas in the numberless Buddhist kingdoms
Are nothing other than the one Buddha in the center of our soul.

Kôbô Daishi

Reality 775

The Pure Land of mystic adornment is nothing but the great Sun Buddha's Palace; and the Land of Perfect Bliss is the Land of Amida's heart. Amida is but an intellectual faculty of the great Sun Buddha, who is the substance of Amida's person. The land of mystic adornment is coextensive with the Land of Perfect Bliss, and the latter is really the former under a special aspect. All the highest possible blessednesses have their centre in that Pure Land of mystic adornment, from which the name of Amida and of his land is derived. Amida's Pure Land is really everywhere, so that the place where we meditate upon him is verily his own land. When we come to realize the truth of this, we do not need to leave this present fleeting world at all to get to the Pure Land, but we are already there. And in our present bodies and persons, just as we are, we are assimilated to Amida, and he to the great Sun Buddha, whose we are and from whom we derive our being. This, then, is the path of meditation by which, just as we are, we attain Buddhahood.

Kakuban

Infra 847
Renun. 139
Realiz. 870

In each atom a hundred suns are concealed...
The core in the centre of the heart is small,
Yet the Lord of both worlds will enter there.

Shabistarî

M.M. 978

Sp. Drown. 713

All know that the drop merges into the ocean but few know that the ocean merges into the drop.

Kabîr

If you cleave the heart of one drop of water
There will issue from it a hundred oceans.

Shabistarî

'O Arjuna, I am in the expanse of the Heart,' says Śrî Krishna. 'He who is in the Sun, is also in this man,' says a *mantra* in the Upanishads. 'The Kingdom of God is within,' says the Bible. All are thus agreed that God is within. What is to be brought down? From where? Who is to bring what, and why?

Reality 790

Realiz. 873

Realization is only the removal of obstacles to the recognition of the eternal, immanent Reality. Reality *is*. It need not be taken from place to place.

Sri Ramana Maharshi

Symb. 322

O nobly-born, these realms are not come from somewhere outside (thyself). They come from within the four divisions of thy heart, which, including its centre, make the five directions. They issue from within there, and shine upon thee. The deities, too, are not come from somewhere else: they exist from eternity within the faculties of thine own intellect. Know them to be of that nature.

The Tibetan Book of the Dead, Sixth Day

Holiness 924

The entire Universe is condensed in the body, and the entire body in the Heart. Thus the Heart is the nucleus of the whole Universe.

Sri Ramana Maharshi

Infra 841

'I am the Near, but not as one thing is near to another: and I am the Far, but not as one thing is far from another.'

Niffarî

The very small is as the very large when boundaries are forgotten;
The very large is as the very small when its outlines are not seen.

Seng-ts'an

That beyond which there is nothing greater should be called the great unit. That beyond which there is nothing smaller should be called the small unit.

Hui Shih

Thus, open out the Tao, and it envelops all space: and yet how small it is, not enough to fill the hand!

Huai Nan Tzû

P. State 579

In one soul we may be entertained and taken up with innumerable beauties. But in the Soul of Man there are innumerable infinities. One soul in the immensity of its intelligence, is greater and more excellent than the whole world. The Ocean is but the drop of a bucket to it, the Heavens but a centre, the Sun obscurity, and all Ages but as one day. It being by its understanding a Temple of Eternity, and God's omnipresence, between which and the whole world there is no proportion.

Infra 838

Thomas Traherne

There can be nothing greater in existence than the simple, absolute maximum. . . . It is above all that we can conceive, for its nature excludes degrees of 'more' and 'less'. . . . Being all that it can be, it is, for one and the same reason, as great as it can be and as small as it can be. By definition the minimum is that which cannot be less than it is; and since that is also true of the maximum, it is evident that the minimum is identified with the maximum.

Nicholas of Cusa

The limits of the soul you could not discover, though traversing every path.

Heraclitus

Nor is being such that there may ever be more than what is in one part and less in another, since the whole is inviolate.

Void 724

Parmenides

If you ask, . . . 'Where is the Place of Precious Things?', it is a place to which no directions can be given. For, if it could be pointed out, it would be a place existing in space; hence, it could not be the real Place of Precious Things. All we can say is that it is close by.

Huang Po

Infra 841

Get one (grain of *tan*, medicine) and you will be sufficient for thousands (of affairs). Do not make distinction between south, north, west and east. Fail and fail again; but build upon your previous achievements. The valuable things of life should not be toyed with.

P. State 563
Faith 509

Chang Po-tuan

As soon as one particle of dust is raised, the great earth manifests itself there in its entirety. In one lion are revealed millions of lions, and in millions of lions is revealed one lion. Thousands and thousands of them there are indeed, but know ye just one, one only.

Reality 775

Jimyo

According to what has been said, we can encompass all the vast world-systems, though numberless as grains of sand, with our One Mind. Then, why talk of 'inside' and 'outside'? Honey having the invariable characteristic of sweetness, it follows that all honey is sweet. To speak of this honey as sweet and that honey as bitter would be nonsensical! How COULD it be so? Hence we say that the Void has no inside and outside. There is only the spontaneously existing Bhûtatathatâ (*Absolute*). And, for this same reason, we say it has no centre. There is only the spontaneously existing Bhûtatathatâ.

M.M. 978
Reality 803

Huang Po

Void 724

When mind is nowhere it is everywhere. When it occupies one tenth, it is absent in the other nine tenths.

Takuan

He who knows Brahma as the real (*satya*), as knowledge (*jñâna*), as the infinite (*ananta*),
Set down in the secret place (of the heart) and in the highest heaven,
He obtains all desires,
Together with the intelligent Brahma.

Taittirîya Upanishad, II. 1

Mirror Symbolism

The Universe is the mirror of God—the mirror in which His majesty and perfection are reflected, the mirror in which He sees Himself—and the heart of man is the mirror of the Universe; if the Traveler then would know God, he must look into his own heart.

Contem. 536

ʿAzîz ibn Muhammad al-Nasafî

The Mind like a mirror is brightly illuminating and knows no obstructions,
It penetrates the vast universe to its minutest crevices;
All its contents, multitudinous in form, are reflected in the Mind,
Which, shining like a perfect gem, has no surface, nor the inside.

Supra 816

Yoka Daishi

When I saw into myself I saw God in me and everything God ever made in earth and heaven. Let me explain it better. As you know right well, anyone who faces God in the mirror of truth sees everything depicted in that mirror: all things, that is to say.

Eckhart[1]

We all, with open face beholding as in a glass the glory of the Lord, are changed into the same image from glory to glory, even as by the Spirit of the Lord.

Realiz. 887

II. Corinthians, III. 18

Realiz. 859 890

The lover is his mirror in which he is beholding his Self.

Plato (*Phaedrus*, 255 D)

God is the mirror in which thou seest thyself, as thou art His mirror in which He

[1] These words are put in the mouth of Sister Katrei, his 'Strasburg Daughter'.

contemplates His Names; now these are nothing other than Himself, in such a way that the analogy of the relationship is inverse.

Ibn ʿArabî

Man, if thou wishest to see God, there or here on earth,
Thy heart must first become a pure mirror.

Angelus Silesius

If thou canst therefore, dear heart, be such a mirror,
The true sun (God himself) in all its splendour
Will come to reflect in thee, wheresoever thou turnest.

Jesaias Rompler von Löwenhalt

Infra 833
P. State 563

Plato saith, That there are *three Kings,* round whose Thrones all things dance; *God,* the *Mind,* the *Soul*; this continual procession of the Soul from the Divine Mind, through the Angelical Mind, in the entire Image of it, with all its Divine Forms, and their Order, their Harmony, their Unity in the whole compass of their Variety, is the *mystical Dance* of the Soul round the Throne of her King, her Bridegroom, by which at once she contemplates, enjoys, springs up into his Divine Form in all its Beauties, and is filled with him. *He* in like Manner hath her ever before him, as the *Looking-Glass* of his own Beauty, lying and playing in himself, as the Image of a Flower or Tree in the Water, every way circled in by him, as she is *centered* in him.

Peter Sterry

Reality 775
P. State 579
Creation 48

St Dionysius says, fullness of joy is perfect consciousness, a balanced interchange of nature, whereby the soul beholds herself in the mirror of the Godhead. God is the mirror, unveiled to whom he will and veiled from whom he will. . . . The divine nature in the Persons is a mirror, beyond the reach of any word. In so far as the soul can project herself beyond words so far she approaches that mirror. In that mirror the union is simply one of likeness.

Eckhart

Infra 847

When he (Sulphur) is set free, he binds his gaolers, and gives their three kingdoms to his deliverer. He also gives to him a magic mirror, in which the three parts of the wisdom of the whole world may be seen and known at a glance: and this mirror clearly exhibits the creation of the world, the influences of the celestial virtues on earthly things, and the way in which Nature composes substances by the regulation of heat. With its aid, men may at once understand the motion of the Sun and Moon, and that universal movement by which Nature herself is governed—also the various degrees of heat, cold, moisture, and dryness, and the virtues of herbs and of all other things.

Michael Sendivogius

Metanoia 488
Introd.
Renun.
Holiness 924

'You are the glass which mirrors my reflection;
Your name have I inscribed with mine together.'

Introd.
Void

God said moreover: 'Well I know, my Prophet,
That gazing thus your soul will ne'er feel surfeit;

But go, and give my slaves my invitation
To come and gaze at will upon my features.'

The Mevlidi Sherif

Your eye has not strength enough
To gaze at the burning sun,
But you can see its brilliant light
By watching its reflection
Mirrored in the water.

So the reflection of Absolute Being
Can be viewed in this mirror of Not-Being,
For non-existence, being opposite Reality,
Instantly catches its reflection.

Reality 803
Humility 199

Shabistarî

All things are seen in the mirror of His intelligence, even as the shadows of the trees on the bank of a river are reflected in the limpid stream below.

Yoga-Vasishtha

Metanoia 484

Let him who desires to see God wipe his mirror and cleanse his heart.

Richard of Saint-Victor

Infra 833
P. State 590

The entire body of *Vitriol* should be recognized as nothing other than a *Mirror of Philosophical Science*. . . . It is a Mirror in which our Mercury, our Sun and Moon are seen to appear and shine.

Basil Valentine

Infra 832
Supra 816
M.M. 978

Sion is as it were a polished mirror and stands for unveiled vision with the single eye of the divine nature. Rely upon it, virtue has never seen this sight.

Eckhart

Reality 803
Knowl. 749
Realiz. 890
Void 728

For thirty years God Most High was my mirror, now I am my own mirror and that which I was I am no more, for 'I' and 'God' represents polytheism, a denial of His Unity. Since I am no more, God Most High is His own mirror. Behold, now I say that God is the mirror of myself, for with my tongue He speaks and I have passed away.

Bâyazid al-Bisṭâmî

A wise intellect is the mirror of God.

Sextus the Pythagorean

The Navel of the Universe

At the Navel of the Earth stands Agni, clothed in richest apparel.

Rig-Veda

Apollo ... is the god who sits in the centre, on the navel of the earth, and he is the interpreter of religion to all mankind.

Plato (*Republic* IV, 427 C)

Rev. 959

'On the navel of the earth I place thee!' For the navel means the centre, and the centre is safe from danger: for this reason he says, 'On the navel of the earth I place thee!'

Śatapatha-Brâhmaṇa, I. 1. 3. 23

O *Wakan-Tanka,* behold Your pipe! I hold it over the smoke of this herb. O *Wakan-Tanka,* behold also this sacred place which we have made. We know that its center is Your dwelling place. Upon this circle the generations will walk.

Black Elk

To whom then will ye liken God? or what likeness will ye compare unto him? ...
It is he that sitteth upon the circle of the earth, and the inhabitants thereof are as grasshoppers.

Isaiah, XL. 18, 22

Judgment 250

Zeus, the god of gods, who rules according to law ... collected all the gods into their most holy habitation, which, being placed in the centre of the world, beholds all created things.

Plato (*Critias,* 121 C)

The word 'navel' indicates the centre and the centre here is Avadhûtî, and here Caṇḍâlî[1] burns.

Hevajratantra, I

The world's middle seat, much famed, is thine.

Orpheus: *Hymn to the Mother of the Gods*

The pole of the world is called by the Pythagoreans the seal of Rhea.

Proclus

P. State 579

'You should prepare a necklace of otter skin, and from it there should hang a circle with a cross in the center. At the four places where the cross meets the circle there should hang

[1] Doctrine and Method combined in a 'mystic heat' that burns away ignorance. In Tibet this heat (gTum-mo) is even produced on a more exterior plane for bodily warmth.

Symb. 322 *Supra* *816*

eagle feathers which represent the four Powers of the universe and the four ages. At the center of the circle you should tie a plume taken from the breast of the eagle, for this is the place which is nearest to the heart and center of the sacred bird. This plume will be for *Wakan-Tanka*, who dwells at the depths of the heavens, and who is the center of all things.'

Black Elk (citing Kablaya, a holy man, on the Sun Dance)

Whatever moveable or immoveable thing is in this world, I am the innermost heart of it.

Yoga-Vasishtha

O centre of the compass! O inmost ground of the truth! O pivot of necessity and contingency!

Infra 838

O eye of the entire circle of existence! O point of the Koran and the *Furqân*![1]

O perfect one, and perfecter of the most perfect, who have been beautified by the majesty of God the Merciful!

Reality 773

Thou art the Pole (*Quṭb*) of the most wondrous things. The sphere of perfection in its solitude turns on thee.

Thou art transcendent; nay, thou art immanent; nay, thine is all that is known and unknown, everlasting and perishable.

Infra 841

Thine in reality is Being and not-being; nadir and zenith are thy two garments.

835 *Humility* passim

Thou art both the light and its opposite; nay, but thou art only darkness to a gnostic that is dazed.

Jîlî

The Celestial City

Flight 944 *P. State* 563

He carried me away in the spirit to a great and high mountain, and shewed me that great city, the holy Jerusalem, descending out of heaven from God.

Having the glory of God: and her light was like unto a stone most precious, even like a jasper stone, clear as crystal; . . .

And the city had no need of the sun, neither of the moon, to shine in it: for the glory of God did lighten it, and the Lamb[2] is the light thereof. . . .

And there shall be no night there; and they need no candle, neither light of the sun: for the Lord God giveth them light: and they shall reign for ever and ever.

Revelation, XXI. 10–11, 23, & XXII. 5

[1] 'Discrimination', a title of the *Qur'ân*. Cf. note 2, p. 778.

[2] Vedic *Agni*; cf. note 2, p. 649.

The sun shines not there, nor the moon and stars;
These lightnings shine not, much less this (earthly) fire!
After Him, as He shines, doth everything shine.
This whole world is illumined with His light.

Kaṭha Upanishad, v. 15
Muṇḍaka Upanishad, II. ii. 10
Śvetâśvatara Upanishad, VI. 14

That the sun does not illumine, nor the moon, nor fire; going there, they (the wise) do not return. That is My Supreme Abode.

Bhagavad-Gîtâ, xv. 6

M.M. 986

And in this Buddha-field (*Sukhâvatî*, or 'Happy Land') one has no conception at all of fire, sun, moon, planets, constellations, stars or blinding darkness, and no conception even of day and night.

Sukhâvatîvyûha Sûtra, 22

Infra 847

Here Sun and Moon lose their distinction.

Saraha

In Man's Soul is His very dwelling; and the highest light and the brightest shining of the City is the glorious love of our Lord, as to my sight.

Julian of Norwich

Supra 816
Infra 833

'The city of which we are the founders . . . exists in idea only: for I do not believe that there is such an one anywhere on earth.'
'In heaven,' I (Socrates) replied, 'there is laid up a pattern of it, methinks, which he who desires may behold, and beholding, may set his own house in order. But whether such an one exists, or ever will exist in fact, is no matter; for he (the man of understanding) will live after the manner of that city, having nothing to do with any other.'

Plato (*Republic* IX, 592 B)

Beauty 670

The Supernal Sun

On rising, the physical sun lights up the physical world and everything in it—people, animals and the rest—pouring its light equally over all; it reigns at midday and then hides again, leaving in darkness the places over which it shone. But the sun of the intellect, once it begins to shine, shines always, totally and immaterially contained in everything and at

Holiness 902

Introd. *Het.* Infra 841 *Void* 724

the same time remaining apart from its creatures, inseparably separated from them, since it is wholly in everything and at the same time is in none of the creatures exclusively (for at the same time it is elsewhere also). The whole of it is in the visible and the whole of it is in the invisible; it is totally present everywhere and yet exclusively present nowhere.

St Simeon the New Theologian

Infra 838

The Sun of the spirit is everlasting: it hath no yesterday.

Rûmî

Symb. 306 317

He has illumined appearances with the lights of His traces, and He has illumined inner selves (*as-sarâ'ir*) with the lights of His Qualities, and for that the lights of appearances set, but the lights of hearts and inner selves do not set. And thus has it been said: 'The sun of day sets at night, but the Sun of hearts never disappears.'

Ibn 'Aṭâ'illâh

'As you know the moon comes and goes, but *anpetu wi,* the sun, lives on forever; it is the source of light, and because of this it is like *Wakan-Tanka.'*

Black Elk (citing Kablaya, a holy man, on the Sun Dance)

As the sun, the eye of the whole world,
Is not sullied by the external faults of the eyes,
So the one Inner Soul of all things
Is not sullied by the evil in the world, being external to it.

Kaṭha Upanishad, v. 11

By the light of mind the human soul is illumined, as the world is illumined by the sun—nay, in yet fuller measure. For all things on which the sun shines are deprived of his light from time to time by the interposition of the earth, when night comes on; but when mind has once been interfused with the soul of man, there results from the intimate blending of mind with soul a thing that is one and indivisible, so that such men's thought is never obstructed by the darkness of error.

Holiness 902

Hermes

This divine sun shines much more brightly than all the suns in the firmament ever shone.

Tauler

He who is worshipped as Light Inaccessible, is not light that is material, the opposite of which is darkness, but light absolutely simple and infinite in which darkness is infinite light.

M.M. 978

Nicholas of Cusa

Or Phoebus' steeds are founder'd,
Or Night kept chain'd below.

Shakespeare (*The Tempest,* IV. i. 30)

In the garden of the Sages, the *Sun* sheds its genial influence both morning and evening, day and night, unceasingly.

Nicholas Flamell

Honour the King, the Eternal, in your bodies; resort unto the Lord in your hearts. For he is Understanding and knoweth the secrets of the heart, his eyes search out all men. He is the Sun by which all mankind sees. He illumines the Two Lands more than the sun.

Egyptian Tradition

Metanoia 480

The wholeness of the Sun is in the supermundane order. For there a solar world and a total light subsist, as the Oracles of the Chaldeans affirm.

Proclus

Resolution of Contraries

(The wall of) the Paradise in which Thou, God, dwellest is built of the contradictories.

Nicholas of Cusa

M.M. 978

By the friendship of contraries, and the blending of things unlike, the fire of heaven has been changed into light, which is shed on all below by the working of the Sun.

Hermes

Judgment 266
Supra 833

The true Sage brings all the contraries together and rests in the natural Balance of Heaven.

Chuang-tse (ch. II)

All contradictory truths are unified in the Truth.

Jîlî

God is (only) known by the union of contraries attributed to Him.

Sahl al-Tustarî

Kharrâz[1] . . . declared that God is not known save by His uniting all opposites in the attribution of them to Him.

Ibn ʿArabî

All this God doth, that he may eternally display the unsearchable Riches of that Variety which is in himself, that he may swallow up the Understanding of every Creature, Man or

[1] See under Abû Saʿîd al-Kharrâz.

Angel, into an admiration and adoration of the incomprehensibleness of his Wayes, his Wisdom, his Blessedness and Glory, who at once bringeth forth these Varieties (which like Morning Stars and Sons of God in the purest unmixt Light and Love, dance and sing together in his Bosom) into such fighting Contrarieties, upon the stage of the earthly and created Image here below, making that the seat of deformity, shame, woe and death, while it figureth out the highest Joys and Glories of eternal Life above, who again *gathers* up all those *jarring* and tumultuous *Contrarieties* into the first state and *supream Unity,* where the Variety is far more vast and boundless in the whole, far more full and distinct in every branch of it, where the whole is all, an eternal Melody, an eternal Beauty, an eternal Joy, unexpressibly Divine, pure and ravishing, where each branch, in its own distinct Form, is a Beauty, a Melody, a Joy, equally pure, perfect and ravishing with the whole, being crowned with the Unity and Eternity, which is the Highest Unity.

Creation 26

Beauty 689

Peter Sterry

They (the gods) make all things work together and contribute to the great whole.

Plato (*Laws* X, 905 B)

In the *Godhead* it self, the most *perfect freedom* and the most *absolute necessity* are joined together in a *Marriage,* to which the whole Heavens and Earth, with unutterable joy, sing eternall *Marriage-Songs.*

Conform. 185

Peter Sterry

'I (Nature) reconcile opposites, and calm their discord.'

John A. Mehung

Unto your Lord is your return, and He will inform you of that wherein ye used to differ.

Qur'ân, VI. 164

The repose of God is a kingdom of delight where the dreadful torment of the divine anger in the eternal nature is transformed into a divine empire of delights.

Peace 694
Suff. 130

Boehme

Such . . . Contraries joyntly to meete
In one accord is a greate Secret.

Thomas Norton

He that well understands the *Philosophers* shall finde they agree in all things, but such as are not the *Sonns* of *Art* will think they clash most fouly.

Reality 790

Elias Ashmole

When the purification has thus been performed, let water and fire become friends, which they will readily do in their earth which ascends with them.

Realiz. 870

Michael Sendivogius

This reconciliation of contrary qualities is the second great object of our calcination.

Metanoia 484

Philalethes

The bad must become like the good.

Basil Valentine

M.M. 978

Every contradiction is both false and true.

Dante (*Paradiso,* VI. 21)

In the world of Divine Unity is no room for Number,
But Number necessarily exists in the world of Five and Four.

Dîvâni Shamsi Tabrîz, XXVI

Reality 775
Heterod. 428

The knot of the heart is loosened,
All doubts are cut off,
And one's deeds (*karman*) cease
When He is seen—both the higher and the lower.

Muṇḍaka Upanishad, II. ii. 8

Action 329

It is great joy to realize that in the infinite, thought-transcending Knowledge of Reality all *sangsâric* differentiations are non-existent.

Gampopa

Ecstasy 636

I had thought that when I saw Thee here above
I would lament to Thee of many things on earth—
But now, O Lord, the sight of Thee
Has in one moment raised me above my best self.

Mechthild of Magdeburg

Infra 847

The same Truth may appear under contrary Notions, and in contrary Opinions. This is the Glory of Spiritual Things, that they can cloth themselves with all manner of Earthly Shapes. It is the Greatness and Majesty of Jesus Christ, that he passes thro' all Forms and all Conditions; and yet still is the same in the midst of them all. Is there any thing more contrary than a Cross and a Throne? And yet you may see the same Jesus in both.

Peter Sterry

Reality 790

It is by no means wonderful if this monad[1] comprehends the whole intellectual pentad, viz. essence, motion, permanency, sameness, and difference, without division, and in the most profound union; since through this union all these are after a manner one. For all things are there without separation.

Proclus

Introd. *Holiness*

Avoid entangled thoughts, that you may see the explanation in Paradise.

Dîvâni Shamsi Tabrîz, XLII

Faith 512

[1] According to Taylor, an allusion to Protogonus (=first born), the Orphic designation of the Intelligible Intellect (νοῦς νοητός): this would be equatable with the Vedantic *Buddhi*.

All doubts vanish when one gains self-control and attains tranquillity by realizing the heart of Truth. Thereupon dispute, too, is at an end.

Srimad Bhagavatam, XI. XV

The Eternal Now

Great is the grace of His promise, 'if to-day we hear His voice';[1] and this 'to-day' is extended day by day, so long as the word 'to-day' exists. Both the 'to-day' and the teaching continue until the consummation of all things; and then the true 'to-day', the unending day of God, reaches on throughout the ages.

Supra 832

Let us, then, ever listen to the voice of the divine Word. For 'to-day' is an image of the everlasting age, and the day is a symbol of light, and the light of men is the Word, through whom we gaze upon God.

Supra 833 *Inv.* 1003

Clement of Alexandria

Before heaven and earth were, *Tao* was. It has existed without change from all time. Spiritual beings drew their spirituality therefrom, while the Universe became what we can see now. To *Tao*, the zenith is not high, nor the nadir low: no point in time is long ago, nor by the lapse of ages has it grown old.

Supra 816

Chuang-tse (ch. VI)

From everlasting to everlasting, thou art God. . . .
A thousand years in thy sight are but as yesterday.

Psalm XC. 2, 4

One day is with the Lord as a thousand years, and a thousand years as one day.

II. Peter, III. 8

The *now* wherein God made the first man and the *now* wherein the last man disappears and the *now* I speak in, all are the same in God where there is but *the now*. . . . The happenings of a thousand years ago, days spent millenniums since, are in eternity no further off than is this moment I am passing now; the day to come a thousand years ahead or in as many years as you can count, is no more distant in eternity than this very instant I am in.

Eckhart

The point whereto all times are present. . . .

Dante (*Paradiso*, XVII. 17)

[1] *Hebrews*, III. 7, from *Psalm* XCV. 7.

Whatever is seen by us in time, thou, Lord God, didst not *pre*-conceive, as it is. For in the eternity in which thou dost conceive, all temporal succession coincides in one and the same Eternal Now. So there is nothing past or future where past and future coincide in the present. . . . Thou indeed, my God, who art thyself Eternity absolutely, art, and speakest (thy Word) above the now and then.

Creation 31

Nicholas of Cusa

How many of ours and our fathers' years have flowed away through Thy 'to-day', and from it received the measure and the mould of such being as they had; and still others shall flow away, and so receive the mould of their degree of being. But *Thou art still the same,*[1] and all things of to-morrow, and all beyond, and all of yesterday, and all behind it, Thou wilt do in this To-day, Thou hast done in this To-day. What is it to me, though any comprehend not this?

Introd. *Creation* *Reality* 773

St Augustine

'He at one time doth see
What are, and what have been,
And what shall after be.
Whom, since he only vieweth all,
You rightly the true Sun may call.'

Boethius

Supra 833

Before it there is no 'before' and after it there is no 'after': the beginning of the centuries is the seal of its existence.

Creation 26

Ibn al-Fâriḍ

The disciples said to Jesus: Tell us how our end will be.
Jesus said: Have you then discovered the beginning so that you inquire about the end? For where the beginning is, there shall be the end. Blessed is he who shall stand at the beginning, and he shall know the end and he shall not taste death.

Death 232

The Gospel according to Thomas, Log. 18

The seventh day (of creation) is the origin and beginning of the first.

Boehme

There is no morning or evening with the Lord.

Supra 833

Muhammad

Where every *where* and every *when* is focussed. . . .

Dante (*Paradiso,* XXIX. 12)

The Sufi is the Son of the Moment (*ibn al-waqt*). O comrade: it is not the rule of the Way to say 'To-morrow.'

Rûmî

[1] *Psalm* CII. 27.

That true day, the day not cramped by a yesterday and a to-morrow. . . .

St Augustine

Infra 847

Ah, fill the Cup:—what boots it to repeat
How Time is slipping underneath our Feet:
Unborn, TO-MORROW, and dead YESTERDAY,
Why fret about them if TO-DAY be sweet!

Omar Khayyâm

Introd. *P. State* *Flight* 949 Infra 841

Truly, a royal power this, transcending time and without place! And by the fact of its transcending time it both contains all time and is the whole of time, and the least jot of that which transcends time makes a man rich indeed, for things at the antipodes are no more distant to this power than those present here.

Eckhart

Introd. *Heterod.*

Another fullness of time. If someone had the knowledge and the power to gather up the time and all the happenings of these six thousand years and all that is to come ere the world ends to boot, all this, summed up into one present now, would be the fullness of time. This is the now of eternity, when the soul knows all things in God, as new and fresh and lovely as I find them now at present. The narrowest of the powers of my soul is more than heaven wide. To say nothing of the intellect wherein there is measureless space, wherein I am as near a place a thousand miles away as the spot I am standing on this moment. Theologians teach that the angel hosts are countless, the number of them cannot be conceived. But to one who sees distinctions apart from multiplicity and number, to him, I say, a hundred is as one. Were there a hundred Persons in the Godhead he would still perceive them as one God.

Eckhart

Beauty 689

Behold the world mingled together,
Angels with demons, Satan with the archangel.
All mingled like seed and fruit,
Infidel with faithful, and faithful with infidel.
At the point of the present are gathered
All cycles and seasons, day, month, and year.
World at beginning is world without end.

Shabistarî

All time is comprised in the present or 'now'. The past was present, the future shall be present, so that time is only a methodical arrangement of the present. The past and the future, in consequence, are the development of the present; the present comprises all present times, and present times are a regular and orderly development of it; only the present is to be found in them. The present, therefore, in which all times are included, is one: it is unity itself.

Nicholas of Cusa

That intellectual Heaven, whose Intelligences know all at once. . . .

St Augustine

Infra 841

Whatsoever, therefore, comprehendeth and possesseth the whole plenitude of unlimited life at once, to which nought of the future is wanting, and from which nought of the past hath flowed away, this may rightly be deemed eternal.

Boethius

Infra 847

Do not occupy your precious time except with the most precious of things, and the most precious of human things is the state of being occupied between the past and the future.

Ahmad b. 'Îsâ al-Kharrâz

The pure one (*ṣâfî*) is plunged in the Light of the Glorious; he is not the son of any one, free from 'times' and 'states'.

Rûmî

Death 206

Omniscience and Omnipresence

He (Agni) knows all births.

Rig-Veda, VI. XV. 13

Man is born once, I have been born many times.

Dîvâni Shamsi Tabrîz

O Arjuna, both you and I have gone through many births. I know them all, but thou knowest them not, O Parantapa.[1]

Bhagavad-Gîtâ, IV. 5

I am in heaven and in earth, in water and in air; I am in beasts and plants; I am a babe

[1] These passages treating of multiple births are applicable both 'horizontally' or 'historically', where the reference to Arjuna, for example, is one of the *transmigration* of the being through multiple states of existence, **and 'vertically' or principally, in the sense that 'there is no other transmigrant but the Lord' (Śankarâchârya:** *Brahma Sûtra Bhâṣya*, I. 1. 5)—Krishna in the citation above speaking as the universal omniscient Self: 'One as he is Person there, and many as he is in his children here' (*Śatapatha Brâhmaṇa*, X. 5. 2. 16). In neither case is there any question of 'reincarnation', the belief, incompatible metaphysically because implying a scission in the unity of the Divine Nature, that an individual being can repeat a determined state of existence. Cf. Guénon: *L'Erreur spirite*, and Coomaraswamy: 'On the One and Only Transmigrant'. Cf. also *Qur'ân*, XXIII. 100: 'Behind them is a barrier until the day when they are raised'; likewise Heraclitus: 'You cannot dip your feet twice in the same stream, for other waters are ever flowing on.'

P. State 579

in the womb, and one that is not yet conceived, and one that has been born; I am present everywhere.

Hermes

With the mind thus composed, quite purified, quite clarified, without blemish, without defilement, grown soft and workable, fixed, immovable, I directed my mind to the knowledge and recollection of former habitations... Thus I remember divers former habitations in all their modes and details.

Vinaya-Piṭaka, III

Holiness 914

Him I call indeed a Brâhmana who knows his former abodes, who sees heaven and hell, has reached the end of births, is perfect in knowledge, a sage, and whose perfections are all perfect.

Dhammapada, XXVI. 423

The Son knows all things essentially in the essence of the Father who, essentially, has the potentiality of all things that shall happen and shall not happen. He has in his Person the universal image, so that he knows all things in common with the Father and wields joint power over what has happened, what is happening and what is still to happen, as well as over what God could do an he would that never happens.

Holiness 924

Eckhart

Both what has been seen and what has not been seen, both what has been heard and what has not been heard, both what has been experienced and what has not been experienced, both the real (*sat*) and the unreal (*a-sat*) — he sees all. He sees it, himself being all.

Reality 803

Praśna Upanishad, IV. 5

To some people, God reveals Himself in the Quality of Knowledge (*al-'ilm*); for as God is revealed in the Life which penetrates everything, so does the servant taste in the unity of this life all that constitutes the nature of things. Consequently, the Essence is revealed to him in the cognitive quality, so that he will know the entire universe with the unfolding of all its worlds, from their origin until their return to principle; he knows how each thing was, how it is and how it will be; he knows that which did not exist and that which, not existing, was not non-existent; he knows how that which is not would be if it was. He has of all this a fundamental, principial and intuitive knowledge, through his own essence and by virtue of his penetration—both integral and distinctive—of the objects of knowledge; he knows in a distinctive manner in his integration, although his knowledge is realized in the Unmanifested.

Knowl. 755

Jîlî

The Father rests not, neither is present in any place, save in His Son, in Whom He rests, communicating to Him all His Essence—'at noon', which is in Eternity, where He ever begets Him.

St John of the Cross

The soul that lives in the present Now-moment is the soul in which the Father begets his only-begotten Son and in that birth the soul is born again. It is still one birth, however often the soul is reborn in God, as the Father begets his only-begotten Son.[1]

Eckhart

While yet abiding in the germ, He is repeatedly born.

Rig-Veda, VIII. 43. 9

I am the child whose father is his son and the wine whose vine is its jar. . . . I met the mothers who bore me, and I asked them in marriage, and they let me marry them.

Jîlî

P. State 579

He knoweth all that entereth the earth and all that emergeth therefrom and all that cometh down from the sky and all that ascendeth therein; and He is with you wheresoever ye may be. And Allâh is Seer of what ye do.

Qur'ân, LVII. 4

My text says: 'God sent his only-begotten Son into the world,' by which ye are to understand not the external world: it must be taken of the inner world. As surely as the Father by his simple nature begets the Son innately, so surely he begets him in the innermost recesses of the mind, which is the inner world. Here God's ground is my ground and my ground God's ground. Here I live in my own as God lives in his own. To one who even for an instant has seen into this ground, a thousand ducats of red beaten gold are worth no more than a false farthing.

Eckhart

Infra 847

My mother bore her father—lo, that is a wondrous thing—
And my father is a little child in the bosom of those who suckle it.

Badru'ddîn al-Shahîd

This being (Ens) is drawn from the eternal, but the conglomerate of the four elements is taken from time. . . .

And although the elementary conglomerate, the body (which the being has assumed or projected as an exterior degree) passes and is annihilated (for it has a beginning in time), the first being does not pass: as appears in the fact that all things are reabsorbed again into the Mother from whence they are issued and born, that is to say into the four elements.

Boehme

Introd. *Judgment* *P. State* 563

My mother bore me, yet I am older than my mother.

Michael Sendivogius

In my eternal mode of birth I have always been, am now, and shall eternally remain.

[1] It is only this once-begotten Person who can be omniprogenitive, hence omnipresent, and thus necessarily omniscient and so capable of 'remembering' his former births (cf. Coomaraswamy: 'Recollection, Indian and Platonic', p. 11).

That which I am in time shall die and come to naught, for it is of the day and passes with the day. In my birth all things were born, and I was the cause of mine own self and all things, and had I willed it had never been, nor any thing, and if I had not been then God had not been either. To understand this is not necessary.

Creation 42

Eckhart

God is His own servant and His own Master, His own child and His own Mother. It is the one Supreme Being, playing the dual parts.

Swami Ramdas

God is his own father and his own mother.

Hermes

Ponder well once for all on your own origin,
Your first mother had a father who was also her mother.
Behold the world entirely comprised in yourself.

P. State 579

Shabistarî

Among all beings—animals and non-animals—Hermes maintained that two sexes are found; and he argued, in consequence, that the Cause of All, God, comprised in Himself the masculine and feminine sexes, of which he believed Venus and Cupid were a manifestation. Valerius Romanus also shared this view and sang of an omnipotent Jupiter who was God the father and God the mother.

Creation 37

Nicholas of Cusa

He was born in the firmament, before the sky existed.

Egyptian Tradition (on Pharaoh as Divine King)

'How long has Your Reverence been following the Way?'
'Since before the era of Bhisma Râja.'[1]

Huang Po

The Father has given birth to his Son, is now giving him birth and shall go on giving him birth without stopping. This birth has been taking place in him for ever.

Reality 790

Eckhart

I have been in many shapes before I attained a congenial form. . . .
There is nothing in which I have not been.

I was with my Lord in the highest sphere
On the fall of Lucifer into the depth of hell;
I have borne a banner before Alexander. . . .

Flight 949

[1] 'This implies that he had been upon the Way since many aeons before the present world cycle began—an allusion to the eternity in which we all share by reason of our identity with the One Mind' (note by editor of text).

I am a wonder whose origin is not known.
I have been in Asia with Noah in the Ark,
I have seen the destruction of Sodom and Gomorra...
I shall be until the doom on the face of the earth....
I was originally little Gwion,
And at length I am Taliesin.

Taliesin

The soul, then, as being immortal, and having been born again many times, and having seen all things that exist, whether in this world or in the world below, has knowledge of them all; and it is no wonder that she should be able to call to remembrance all that she ever knew.

Plato (*Meno,* 81 C)

Knowl. 755

As it must be confessed that the human soul is not that which God is, so it must be presumed that among all the things which He created nothing is nearer to God than it.

St Augustine

Holiness 924

We verily created man and We know what his soul whispereth to him, and We are nearer to him than his jugular vein.

Qur'ân, L. 16

Faith 505

'I am nearer to each thing than its gnosis of itself: but its gnosis of itself does not pass beyond itself to Me, and it does not know Me, so long as its self is the object of its gnosis.'

Niffarî

Metanoia 488
Death 220

Do not desert your home to live somewhere else. The wonderful *Tao* exists not far away from your own body. It is not necessarily found in high mountains or in unknown waters.

Chang Po-tuan

Supra 816

The Prophet is closer to the believers than their selves.

Qur'ân, XXXIII. 6

God is nearer to me than I am to my own self; my life depends upon God's being near me, present in me. So is he also in a stone, a log of wood, only they do not know it. If the wood knew of God and realized his nearness like the highest angel does, then the log would be as blessed as the chief of all the angels.

Eckhart

Holiness 914

Raise the stone and there thou shalt find me. Cleave the wood and there am I.

Oxyrhynchus Papyri &
The Gospel according to Thomas, Log. 77

Since God is the universal cause of all Being, in whatever region Being can be found, there must be the Divine Presence.

St Thomas Aquinas

Reality 803

Death 220

God is nearer to you than anything else, yet because of the screen of egoism you cannot see him.

Sri Ramakrishna

It is far easier for me to lift my soul to God than my hand to my head.

Ruysbroeck

Renun. 139

If it was a matter of having to see or hear something in China or India, far beyond the seven-fold tides of the seas, one might grieve. But what we are trying to do, is to look at our own mind with our own mind—and that is something closer to us than looking at the pupils of our eyes with our eyes. And do not grieve as if it were something very deep that we are trying to look at. If it were a matter of something to be seen or listened to at the bottom of the nine-fold chasm or under the thousand-fathomed depths of the sea we might grieve—but to look at my own heart with my own heart is less than smelling my own nose with my own nostrils!

Hakuin

Realiz. 887

My soul becomes more one with God than food does with my soul.

Eckhart

For Heaven is as near to our souls as this world is to our bodies.

William Law

Realiz. 859

God is nearer to us than our own Soul: for He is the Ground in whom our Soul standeth, and He is the Mean that keepeth the Substance and the Sense-nature together so that they shall never dispart. For our Soul sitteth in God in very rest, and our Soul standeth in God in very strength, and our Soul is kindly rooted in God in endless love: and therefore if we will have knowledge of our Soul, and communing and dalliance therewith, it behoveth to seek unto our Lord God in whom it is enclosed.

Julian of Norwich

Holiness 900

We ought by rights to be seeking God in all things and finding God at all times as well as in all places, in any company and in every guise.

Eckhart

Pilg. 366

Verily if slaves and recluses experience desolation in everything, it is owing to their absence from Allâh in everything; for if they witnessed Him in everything, they would not find desolation in anything.

Ibn 'Aṭâ'illâh

Just as there can be no ice without water, so Nirvana is immediately present.

Hakuin

The Ultimate Felicity

There perfect, ripe, and whole is each desire; in it alone is every part there where it ever was, for it is not in space, nor hath it poles.

Dante (*Paradiso*, XXII. 64)

Supra 816

Where are both desires and the consummation of desires, where the desires of him who desires are possest....

Rig-Veda, IX. 113. 10, 11

The desire of our mind is to live by mind, which is continually to enter more and more into life and joy. But life and joy are infinite; and the blessed are borne into life and joy by ardent desire. They who drink of the fountain of life are satisfied in such fashion as still to thirst; and as this drink never becomes a past thing, for it is eternal, ever are they blessed both drinking and in thirst, and never shall that thirst and its satisfaction pass.[1]

Nicholas of Cusa

True happiness is a thing that never gives rise to satiety.

Hermes

From this divine eminence we see the lowness and insignificance of creatures. We feel an inkling of the perfection and stability of eternity, for there is neither time nor space, neither before nor after, but everything present in one new, fresh-springing *now* where millenniums last no longer than the twinkling of an eye. And we win participation in the manifold delights of the heavenly host. So great the joy of Mary Queen in heaven, that having but a thousandth part of it, each member of the heavenly company would taste far more than ever they have earned. There every spirit rejoices in the joy of every other, relishing it each in his degree. Every celestial habitant is, knows and loves in God, in his own self and in every other spirit whether soul or angel. And the distinctive consciousness of one God in three Persons and the Three one God gives such ineffable, amazing satisfaction that all their passionate longing is fulfilled. And just what they are full of they crave unceasingly, and what they crave is all their own in new, fresh-springing joyful ecstasy, theirs to enjoy in all security from everlasting unto everlasting.

Supra 838
P. State 579
Conform. 180
Introd. *Holiness*

Eckhart

[1] They alone can think that one would weary of a final beatitude or perfection who are unable to conceive of a felicity outside of time or space, and who therefore confuse the ecstatic instantaneity of an eternal now with the insufferable perpetuity of a ceaseless becoming. 'Could *enough* ever be applied to God, God were not God' (Eckhart, I. 247). 'We are never wearied of the daily rising and setting of the sun: we often *demand* "novelty" when our attention is distracted, but whenever we regard the realities of life, we recognize that what we really *need* is not a perpetually "novel" but a constantly "original" (*ex fonte*) experience' (Coomaraswamy: *The Ṛg Veda as Land-Nâma-Bôk*, p. 39).

And thus I saw Him, and sought Him; and I had Him, I wanted Him.

Julian of Norwich

I wish but for the thing I have.

Shakespeare (*Romeo and Juliet,* II. ii. 132)

Creation 28

You must want like a God that you may be satisfied like God.... He is infinitely Glorious, because all His wants and supplies are at the same time in his nature from Eternity.... His wants are as lively as His enjoyments: and always present with Him. For His life is perfect, and He feels them both. His wants put a lustre upon His enjoyments and make them infinite. His enjoyments being infinite crown His wants, and make them beautiful even to God Himself. His wants and enjoyments being always present are delightful to each other, stable, immutable, perfective of each other, and delightful to Him. Who being Eternal and Immutable, enjoyeth all His wants and treasures together. His wants never afflict Him, His treasures never disturb Him. His wants always delight Him; His treasures never cloy Him.

Thomas Traherne

Therein is all that souls desire and eyes find sweet.

Qur'ân, XLIII. 71

... The bread of angels whereby life is here sustained but wherefrom none cometh away sated.

Dante (*Paradiso,* II. 11)

To call an object in the world sweet and lovely
Is not yet to know the Sweetness that is God.

Angelus Silesius

Renun. 158

For the whole world before thee is as the least grain of the balance, and as a drop of the morning dew, that falleth down upon the earth.

Wisdom, XI. 23

Knowl. 743

... Your capacity
Is of that nature that to your huge store
Wise things seem foolish and rich things but poor.

Shakespeare (*Love's Labour's Lost,* V. ii. 377)

Renun. 152b

He who is satisfied with wisdom and direct vision of Truth, who has conquered the senses and is ever undisturbed, to whom a lump of earth, a stone and gold are the same, that Yogi is said to be a Yukta (a saint of established wisdom).

Bhagavad-Gîtâ, VI. 8

For let me tell you that he on whom the Most High has conferred the knowledge of this Mystery (of transmutation) esteems mere money and earthly riches as lightly as the dirt of the streets. His heart and all his desires are bent upon seeing and enjoying the heavenly reality of which all these things are but a figure.

The Sophic Hydrolith

A real sannyâsi will not enjoy any kind of bliss except the Bliss of God.

Sri Ramakrishna

With the aid of your highest faculty for wisdom (buddhi),[1] you can compare the beatitude (ânanda) resulting from meditation, with the ephemeral pleasures of the senses, and you will find the one a million times superior to the others.

Swami Sivananda

Sensual happiness is nothing but a reflection of the inherent bliss of the Self.

Sri Chandrasekhara Bhâratî Swâmigal

Compared with this loveliness, all sweetness is bitter.

Richard of Saint-Victor

It is only peculiar to God to be happy in himself alone.

John Smith the Platonist

Illusion 109

True beatitude is that which depends on nothing outside itself.

Ananda Moyî

Peace 694
Void 728

God enjoys eternalwise the contingency of things. . . . All things must needs please him for he who saw was God and what he saw was likewise God. In their eternal image which is God himself, God saw himself and saw things as a whole. God enjoyed himself, God being in himself the unique one.

Eckhart

Knowl. 749
Supra 828

Allâh hath pleasure in them and they have pleasure in Him.

Qur'ân, XCVIII. 8

Creation 42

This world Sukhavati (Pure Land), Ananda, which is the world system of the Lord Amitabha,[2] is rich and prosperous, comfortable, fertile, delightful and crowded with many Gods and men. And in this world system, Ananda, there are no hells, no animals, no ghosts, no Asuras and none of the inauspicious places of rebirth. . . .

Supra 832

Moreoever, Ananda, all the beings who have been reborn in this world-system Sukhavati, who are reborn in it, or who will be reborn in it, they will be exactly like the Paranirmitavasavartin Gods: of the same colour, strength, vigour, height and breadth, dominion, store of merit and keenness of super-knowledge; they enjoy the same dresses, ornaments, parks, palaces and pointed towers, the same kind of forms, sounds, smells, tastes and touchables, just the same kinds of enjoyments. And the beings in the

[1] Cf. note p. 837.
[2] Cf. note 1, p. 705.

world-system Sukhavati do not eat gross food, like soup or raw sugar; but whatever food they may wish for, that they perceive as eaten, and they become gratified in body and mind, without there being any further need to throw the food into the body. And if, after their bodies are gratified, they wish for certain perfumes, then the whole of that Buddha-field becomes scented with just that kind of heavenly perfume. But if someone does not wish to smell that perfume, then the perception of it does not reach him. In the same way, whatever they may wish for, comes to them, be it musical instruments, banners, flags, etc.; or cloaks of different colours, or ornaments of various kinds. If they wish for a palace of a certain colour, distinguishing marks, construction, height and width, made of various precious things, adorned with hundreds of thousands of pinnacles, while inside it various heavenly woven materials are spread out, and it is full of couches strewn with beautiful cushions,—then just such a palace appears before them. In those delightful palaces, surrounded and honoured by seven times seven thousand Apsaras, they dwell, play, enjoy and disport themselves. . . .

Holiness 902 *Heterod.* 418

And all the beings who have been born, who are born, who will be born in this Buddha-field, they all are fixed on the right method of salvation, until they have won Nirvana. And why? Because there is here no place for and no conception of the two other groups, i.e. of those who are not fixed at all, and those who are fixed on wrong ways. For this reason also that world-system is called the 'Happy Land'.

Sukhâvatîvyûha Sûtra, 15, 19, 24

Pilg. 366 Supra 832

He (Allâh) hath awarded them (the just) for all that they endured, a Garden and silk attire:

Reclining therein upon couches, they will find there neither (heat of) sun nor cold of moon.

And near unto them is the shade thereof, and the clustered fruits therein bow down.

And vessels of silver are brought round for them, and goblets of crystal,—

Crystal of silver, which they have measured as preordained.

And they are given to drink therein of a cup whose mixture is of Zanjabîl:

A spring is there, named Salsabîl.

And immortal youths go about them, whom, when thou seest, thou wouldst take for scattered pearls.

And when thou seest, there wilt thou see bliss and great sovereignty.

Their raiment will be fine green silk and gold embroidery. Bracelets of silver will they wear. Their Lord will slake their thirst with a pure drink.

Verily, this is a reward for you. Your endeavour hath found acceptance.

Qur'ân, LXXVI. 12–22

Introd. *Reality* *P. State* 583 *Peace* 694 Supra 816

That archetypal world is the true Golden Age, age of Kronos, whose very name suggests (in Greek) Exuberance (*κόρος*) and Intellect (*νοῦς*). For here is contained all that is immortal: nothing here but is Divine Mind; all is God; this is the place of every soul. Here is rest unbroken: for how can that seek change, in which all is well; what need that reach to, which holds all within itself; what increase can that desire, which stands utterly achieved? All its content, thus, is perfect, that itself may be perfect throughout, as holding nothing that is less than the divine, nothing that is less than intellective. Its knowing is not by search but by possession, its blessedness inherent, not acquired; for all belongs to it eternally and it holds the authentic Eternity imitated by Time which, circling round the Soul, makes towards the new thing and passes by the old. Soul deals with thing after thing—now Socrates; now a horse: always some one entity from among beings—but the Intellectual-Principle is all and therefore its entire content is simultaneously present in that identity: this is pure being in eternal actuality; nowhere is there any future, for every then is a now; nor is there any past, for nothing there has ever ceased to be; everything has taken its stand for ever, an identity well pleased, we might say, to be as it is.

Plotinus

Knowl. 755
Contem. 528
Supra 841
838
Introd.
Realiz.

And therefore the divinity always enjoys itself and its own infinite perfections, seeing it is that eternal and stable sun of goodness that neither rises nor sets, is neither eclipsed nor can receive any increase of light and beauty. Hence the divine love is never attended with those turbulent passions, perturbations, or wrestlings within itself, of fear, desire, grief, anger, or any such like, whereby our love is wont to explicate and unfold its affection towards its object. But as the divine love is perpetually most infinitely ardent and potent, so it is always calm and serene, unchangeable, having no such ebbings and flowings, no such diversity of stations and retrogradations as that love hath in us which ariseth from the weakness of our understandings, that do not present things to us always in the same orient lustre and beauty: neither we nor any other mundane thing (all which are in a perpetual flux) are always the same.

John Smith the Platonist

Creation 48
Supra 833
Creation 28

If we are but born into the Pure Land by embarking upon Amida's original vow, then none of our cherished desires remain unfulfilled.

Hônen

Orthod. 280
Inv. 1031

O lake, thou hast a channel to the Sea: be ashamed to seek water from the pool.

Rûmî

Sp. Drown. 713

For him who has truly known Brahman, there is nothing to learn. He who has quenched his thirst with nectar craves no other drink.

Srimad Bhagavatam, XI. xxi

Knowl. 761

All I desire I have found in Him.

Shabistarî

God only is such an almighty goodness as can attract all the powers in man's soul to itself, as being an object transcendently adequate to the largest capacities of any created being, and so unite man perfectly to himself in the true enjoyment of one uniform and simple good.

John Smith the Platonist

Creation 42

He that has once found this Art, can have nothing else in all the world to wish for, than that he may be allowed to serve his God in peace and safety.

Philalethes

Once *samâdhi* hath been tasted, hunger endeth.

Phadampa Sangay

All modes and attributes of Very Being
Are realized and present in that Being;
To see them He needs not contingent beings:
'Tis the contingent needs the Very Being.

Introd.
Renun.

Jâmî

That which is Bliss is verily the Self. Bliss and the Self are not distinct and separate but are one and identical. And *That* alone is real.

Sri Ramana Maharshi

Reality
803

Our innate nature is happiness itself and we ever have it with us. But we do not realize it. On the other hand, we begin to seek for it elsewhere in the objective world outside us, just as a person who is ignorant of a treasure buried in his own house goes about begging. The world fascinates us only because of this ignorance. There is no doubt about it.

Sri Chandrasekhara Bhâratî Swâmigal

Beauty
663

To have all that has being and is lustily to be desired and brings delight; to have it all at once and whole in the undivided soul and that in God, revealed in its perfection, in its flower, where it first burgeons forth in the ground of its existence, and all conceived where God is conceiving himself—that is happiness.

Eckhart

PART VI: *Union — Identity*

REALIZATION AND IDENTITY

ΓΝΩΘΙ ΣΕΑΥΤΟΝ[1]

Inscription to Apollo at Delphi

He who knows himself verily knows his Lord.

Muhammad

He that seeth me seeth him that sent me.

St John, XII. 45

He who sees the Word sees Me.

Saṁyutta-nikâya, III. 120

Waḥdat al-Wujûd, a Sufic formulation for the Unicity of Existence and often rendered as 'the Supreme Identity' (Vedantic *tad ekam*), refers to both the doctrine of Unity and its attainment, since 'metaphysic affirms the fundamental identity of knowing and being' (Guénon: *Introduction générale à l'Étude des Doctrines hindoues*, 4th ed., p. 144).

Reality 775

'Existence' may either be qualified, in its etymological sense of *ex-stare*, as *Being*, or unqualified signify the absolute plenitude of *universal Possibility*, in which case the corresponding realization will be that of total and unconditioned Deliverance—meaning liberation from all conditions and determinations whatsoever, including pure Being, which, though determined by nothing outside of itself, is still a modification of the Supreme Principle and hence not absolutely infinite.[2]

Introds.
Waters
Holiness
Reality 773

The possibility of full realization, even here and now, since it cannot be other than in the Here and the Now, is affirmed by all traditions, from the technical Vedantic exposition on *Moksha*, to the more cryptic allusions in the Monotheistic forms, necessarily veiled to protect against the disastrous distortions easily conceived where the collective tendency is less contemplative than passional. For never has there been and never will there be but one Supreme Identity: that of the Absolute with Itself. Only the Real, which alone *is* metaphysically, can be *realized: Âtmâ*—not *jîvâtmâ, manas, chitta, ahamkâra* or any other name by which the human ego can be designated—is *Brahma*.

Center 841
838
Heterod. 428
Reality 803
Infra 890

On the other hand, the doctrine, however inaccessible to the ordinary reaches of human comprehension, is established upon the highest truths that exist. 'To deny esoterism—and it is the thing, not the word, that matters—is to deny that there are spiritual values which not every man can understand. . . . Esoteric doctrines, far from being a luxury, answer on

[1] 'Know thyself.'

[2] Cf. Guénon: *L'Homme et son Devenir selon le Vêdânta*, and *Les États multiples de l'Être*.

Knowl. 749

the contrary to imperious needs of causality; the spiritual finality which they envisage corresponds to the profoundest aspiration of the being: it is God in us who wishes to be delivered' (Schuon: *Perspectives spirituelles*, p. 100).

Now to know one's Self, one must first know oneself. 'The Intellect being universal in essence necessarily penetrates the entire being and includes all of its constituent elements; for to exist is to know, and every aspect of our existence is a state of knowledge, or in relation to absolute Knowledge, a state of ignorance. If it is true that the reason is the central mirror of the Intellect whose organ is the subtle heart, the other faculties are no less equally planes of manifestation for the Intellect; the individual being could never be reintegrated into the Absolute without all of its faculties participating in this process in the measure necessary. Spiritual Knowledge, far from being opposed to any mode whatsoever of conformity or participation, on the contrary brings into play all that we are, hence all of the elements, psychic and even physical, which constitute our being; for nothing positive can be excluded from the process of transmutation. Nothing can be annihilated, and it is therefore necessary that the faculties or psychic energies, which form part of our reality and whose existence must have a meaning for us, be determined and canalized by the same governing Idea which determines and transforms thought. But this is possible only by placing oneself on the same ground as these psychic faculties; it is not sufficient to consider them with the reason and in a theoretical light; one has to realize the Idea in the measure possible on the plane of these very faculties, in some manner universalizing them by virtue of their respective symbolisms. Man must transpose onto a higher plane all the positive reactions provoked in him by the surrounding reality, recollecting by means of sensible things divine Realities' (Schuon: *L'Oeil du Coeur*, pp. 96–97).

Introd. *Holiness*

Beauty 663

If it should come about that this psychic terrain is sufficiently prepared, reconciled and rectified, with all concomitant elements such as doctrine, rites and grace effectively participating, there must then normally be 'a moment of discontinuity in the development of the being. This moment—absolutely unique in character—is when there occurs, through the action of the "Celestial Ray" operating on a plane of reflection, the vibration which corresponds to the cosmogonic *Fiat Lux*, and which illuminates through its irradiation the whole chaos of possibilities. Starting from this moment, order replaces chaos, light replaces darkness, act replaces potentiality, and reality replaces virtuality.[1] And when this vibration has attained its full effect, amplifying and reverberating to the limits of the being, then it is clear that the being, having now realized its total plenitude, is no longer compelled to go through any particular cycle, since it embraces them all in the perfect simultaneity of a synthetic and "non-distinctive" comprehension. It is really this that is meant by "transformation", understood as implying the "return of modified beings into unmodified Being", above and beyond all the special conditions which determine the degrees of manifested Existence' (Guénon: *Le Symbolisme de la Croix*, pp. 194–195).

Introds. *Waters* *Pilg.*

Metanoia 488 Introd. *P. State*

One can recall at this point the three basic degrees of realization, inversely contraposed, which exist apart from the innumerable sub-divisions and complex of possibilities within each state of being:—identification with the center of the state to which the individual being belongs, at the juncture between the formal and supraformal possibilities of manifestation; identification with the center of all manifested states, formal and supraformal, where they

[1] Cf. Introductions to *Pilgrimage, The Primordial State, Supreme Center.*

merge in Universal Being; identification with the Absolute, where the infinity of possibilities manifested and unmanifested is principially resolved in the Supreme Reality.[1]

This final passage out of the six-dimensioned cosmos into the unmanifested metacosm is effected through the 'seventh ray' which transpierces the Solar Gate or Sundoor, the 'narrow way' and 'eye' of that vertical axis or 'needle'[2] carrying the *sutrâtmâ* 'thread'—the 'solar ray' of which the worlds are 'woven'—the same Eye of the World that is the Center, Heart and Sun, resplendent *ab extra* but lightless, heartless, modeless, recondite and void *ab intra*, 'where God himself is naughted'. 'The liberation from good and evil that seemed impossible and is impossible for the man whom we define by what he does or thinks and who answers to the question, "Who is that?", "It's me", is possible only for him who can answer at the Sundoor to the question "Who art thou?", "Thyself". He who fettered himself must free himself, and that can only be done by verifying the assurance, "That art thou". It is as much for us to liberate him by knowing Who we are as for him to liberate himself by knowing Who he is; and that is why in the Sacrifice the sacrificer identifies himself with the victim' (Coomaraswamy: *Hinduism and Buddhism*, p. 17; cf. especially his treatment of this essential subject in 'The Symbolism of the Dome', *Indian Historical Quarterly*, Calcutta, March, 1938).

Beauty 689
Introd.
Center
M.M.
994
Introd.
Suff.
Inv.
1007

'The Christ who said, "I am the Door" is also both the "Lion of Juda" and the "Sun of men". In Byzantine churches the figure of the *Pantokrator* or Christ "in majesty" occupies the central position of the vault, which corresponds precisely with the "eye" of the dome; and this in turn ... represents, at the upper extremity of the "World Axis", the door through which is achieved the "passage out of the cosmos"' (Guénon: 'Kâla-mukha', *Études Trad.*, 1946, p. 190).[3]

'The question as to whether the Saints who have been "delivered" are "annihilated" in God or whether they remain "separate" in Him ... amounts to knowing whether the Divine Names are distinct or indistinct in God. Now every Divine Name is God, but none is any other, and above all, God is reduced to none of them. The man who "enters" into God can obviously add nothing to God, or modify anything in Him, God being immutable Plenitude. However, even the being who has realized *Parinirvâna*, that is to say, who is no longer limited by any exclusive divine Aspect but is "identified" with the Divine Essence, is still and always—or rather "eternally"—"himself"; for the divine Qualities obviously cannot but be inherent in some manner in the Essence;[4] they can thus be "extinguished" therein, but not "lost". In a word, the Saints "pre-exist" eternally in God, and their spiritual realization is only a return to themselves; analogously, every quality, every earthly pleasure, can only be a finite reflection of an infinite Perfection or Beatitude. Nothing therefore can be lost; the simple fact that we enjoy something proves that this enjoyment is found infinitely in God' (Schuon: *L'Oeil du Coeur*, pp. 68–69).

Creation
28
Infra
873
Introd.
Holiness
Center 847

'Spiritual realization is theoretically the easiest thing and practically the most difficult

[1] Cf. Guénon: *L'Homme et son Devenir*.

[2] The *foramen acus* of the Gospels: cf. *Matth.* XIX. 24.

[3] The summit of the dome is analogous microcosmically with the 'crown' of the head where the coronal artery (*sushumnâ*) joins the 'solar ray' at the aperture called the *brahma-randhra*,—all of which moreover has its ritual correspondence in the tonsuring of priests and trepanation of the dead: cf. Guénon: 'La Porte étroite' *Études Trad.*, 1938, pp. 450–451, and Coomaraswamy, op. cit.

[4] 'If it were otherwise, one would have to admit that the "Son" knew things that were unknown to the "Father", which is absurd.'

thing there is. The easiest: because it suffices to think of God; the most difficult: because human nature is the forgetfulness of God. . . .

'All spiritual paths agree in this: that there is no common measure between the means put to work and the result. "With men this is impossible: but with God all things are possible", says the Gospel. Actually, what separates man from the divine Reality is an infinitesimal barrier: God is infinitely near to man, but man is infinitely far from God. This barrier, for man, is a mountain; he stands before a mountain which he must remove with his own hands. He digs the earth, but in vain, the mountain remains: man continues to dig, however, in the name of God. And the mountain vanishes. It has never been' (Schuon: *Perspectives spirituelles*, p. 282; 'Caractères de la mystique passionnelle'. *Études Trad.*, 1953, p. 265).

Grace 552
Center 841
Inv. 1031

The same authority in another article synthesizes the essentials in a succinct formula: 'If our "being" must become "knowing", . . . our "knowing" must become "being"; if in place of "existing" it is necessary to "discern", it is likewise necessary, in place of "thinking", to "realize" ' ('The Stations of Wisdom', *Asia*, 1953, p. 172).

Introd. *Knowl.*

The great imponderable in all spiritual work is: how to be what one is, and how to cease being what one is not.

Know Thyself

I must first know myself, as the Delphian inscription says; to be curious about that which is not my concern, while I am still in ignorance of my own self, would be ridiculous.

Plato (*Phaedrus*, 229 E)

Knowl. 734

Self-knowledge is the shortest road to the knowledge of God. When ʿAlî asked Mohammad, 'What am I to do that I may not waste my time?' the Prophet answered, 'Learn to know thyself.'

ʿAzîz ibn Muhammad al-Nasafî

He who knows others is wise;
He who knows himself is enlightened.

Tao Te Ching, XXXIII

Origen says the soul's quest of God comes by self-observation. If she knew herself she would know God also.

Eckhart

Knowl. 749

To arrive at a true understanding by realizing one's true self is called (the way of) nature. To realize one's true self from understanding (of the universe) is called (the way of) culture. Who has realized his true self gains thereby understanding. Who has gained a (complete) understanding finds thereby his true self.

Tsesze

Infra 873

'Anyone who is ignorant of the nature of servantship (*ʿubûdiyyat*) is yet more ignorant of the nature of lordship (*rubûbiyyat*)', i.e., whoever does not know the way to knowledge of himself does not know the way to knowledge of God.

Hujwîrî, citing al-Tirmidhî

P. State 579

If the mind would fain ascend to the height of science, let its first and principal study be to know itself.

Richard of Saint-Victor

Our general instinct to seek and learn, our longing to possess ourselves of whatsoever is lovely in the vision, will set us enquiring into the nature of the instrument with which we search. Moreover, we shall only be obeying the ordinance of the God who bade us know ourselves.

Plotinus

Holiness 900

And is self-knowledge such an easy thing, and was he to be lightly esteemed who inscribed the text on the temple at Delphi?

Plato (*Alcibiades I*, 129 A)

Center 816

Thou believest thyself to be nothing, and yet it is in thee that the world resides.

Avicenna

Let me know myself, Lord, and I shall know Thee.

St Augustine

Death 226

Who knoweth Him, knoweth himself, and is not afraid to die.

Atharva-Veda, x. 8. 44

Whosoever would attain to the summit of his noble nature and to the vision of the sovran good, which is God himself, must have profoundest knowledge of himself and of things above himself. Thus he reaches the supreme. Beloved, learn to know thyself, it shall profit thee more than any craft of creatures.

Eckhart

As oil in sesame seeds, as butter in cream,
As water in river-beds, and as fire in the friction-sticks,
So is the Self (Âtman) apprehended in one's own self,
If one looks for Him with true austerity (*tapas*).

Introd. *Suff.*

Śvetâśvatara Upanishad, I. 15

O that we would but once learn to know ourselves!

Boehme

Withdraw into yourself and look. And if you do not find yourself beautiful yet, act as does the creator of a statue that is to be made beautiful; he cuts away here, he smooths there, he makes this line lighter, this other purer, until a lovely face has grown upon his work. So do you also: cut away all that is excessive, straighten all that is crooked, bring light to all that is overcast, labour to make all one glow of beauty and never cease chiselling your statue, until there shall shine out on you from it the godlike splendour of virtue, until you shall see the perfect goodness surely established in the stainless shrine.

Beauty 676

Plotinus

Holiness 924

Well I ween that if I knew myself as intimately as I ought, I should have perfect knowledge of all creatures.

Eckhart

Contem. 547

If anyone learn to know himself and God (i.e. our duty as men, our origin, the end of our being, and our affinity to God), he has the highest scholarship, without which it is impossible to obtain happiness, either in this world, or in the world to come.

The Sophic Hydrolith

He who knows what God is, and who knows what man is, has attained. Knowing what God is, he knows that he himself proceeded therefrom.

Chuang-tse (ch. VI)

He, knowing all, becomes the All.

Praśna Upanishad, IV. 10

The being is all that it knows.

Aristotle

In the case of God, being and knowing are identical.

Eckhart

Knowl. 745

If there were no elements of being, there would be no elements of intelligence. Verily, if there were no elements of intelligence, there would be no elements of being.

Kaushītaki Upanishad, III. 8

To think of God is to lose yourself in Him. The mind disappears and God only is.

Swami Ramdas

Supra Introd.

Only the truly intelligent understand this principle of the identity of all things. They do not view things as apprehended by themselves, subjectively; but transfer themselves into the position of the things viewed. And viewing them thus they are able to comprehend them, nay, to master them;—and he who can master them is near. So it is that to place oneself in subjective relation with externals, without consciousness of their objectivity,—this is TAO.

Chuang-tse (ch. II)

Reality 775 *Death* 220 *Center* 841

The man who is inseparate from all things enjoys divinity as God himself enjoys it.

Eckhart

In this state of absorbed contemplation there is no longer question of holding an object in view; the vision is continuous so that seeing and seen are one; object and act of vision have become identical; of all that until then filled the eye no memory remains. The first seeing is of Intellect knowing, the second that of Intellect in love. . . . The vision floods the eyes with light, but it is not a light showing some other thing, the light is itself the vision. No longer is there object seen and light to show it, no longer Intellect and object of Intellection; this is the very Radiance that brought both into being.

Plotinus

Infra 887 *Knowl.* 749 *Void* 728

'Who hath seen Me, the same hath seen the Truth.'

Muhammad

P. State 579

Thoroughly to know oneself, is above all art, for it is the highest art. If thou knowest thyself well, thou art better and more praiseworthy before God, than if thou didst not know

Knowl. 734

thyself, but didst understand the course of the heavens and of all the planets and stars, also the virtue of all herbs, and the structure and dispositions of all mankind, also the nature of all beasts, and, in such matters, hadst all the skill of all who are in heaven and on earth.

Theologia Germanica, IX

The one and only path, bhikkhus, leading to the purification of beings, to passing far beyond grief and lamentation, to the dying-out of ill and misery, to the attainment of right method, to the realization of Nirvâna, is that of the fourfold setting-up of mindfulness. . . .

Action 338

A brother—whether he departs or returns, whether he looks at or looks away from, whether he has drawn in or stretched out his limbs, whether he has donned under-robe, over-robe, or bowl, whether he is eating, drinking, chewing, reposing, or whether he is obeying the calls of nature—is aware of what he is about. In going, standing, sitting, sleeping, watching, talking, or keeping silence, he knows what he is doing. . . .

A brother when affected by a feeling of pleasure, (is) aware of it, reflecting: 'I feel a pleasurable feeling.' So, too, is he aware when affected by a painful feeling, or by a neutral feeling, or by a pleasant or painful or neutral feeling concerning material things, or by a pleasant or painful or neutral feeling concerning spiritual things. . . .

A brother, if his thought be lustful, is aware that it is so, or if his thought be free from lust, is aware that it is so; or if his thought be full of hate, or free from hate, or dull, or intelligent, or attentive, or distrait, or exalted, or not exalted, or mediocre, or ideal, or composed, or discomposed, or liberated, or bound, he is aware in each case that his thought is so, reflecting: 'My thought is lustful,' and so on. . . .

And so, too, with respect to the organ of hearing and sounds, to the organ of smell and odours, to the organ of taste and tastes, to the organ of touch and tangibles, to the sensorium and images, he is aware of the sense and of the object, of any fetter which arises on account of both, of how there comes an uprising of a fetter not arisen before, of how there comes a putting-aside of a fetter that has arisen, and of how in the future there shall arise no fetter that has been put aside. . . .

And what, bhikkhus, is right mindfulness?

Renun. 136
Holiness 914

Herein, O bikkhus, a brother, as to the body, continues so to look upon the body, that he remains ardent, self-possessed and mindful, having overcome both the hankering and the dejection common in the world. And in the same way as to feeling, thoughts and ideas, he so looks upon each, that he remains ardent, self-possessed and mindful, having overcome the hankering and the dejection that is common in the world. This is what is called right mindfulness.

Maha Satipatthana Suttanta

Orthod. 288

I did not go to the 'Maggid' of Meseritz[1] to learn Torah from him but to watch him tie his boot-laces.

A Hassidic saint

P. State 563

If somebody asks what is meant when it is said that the kneading of the elixir is done by 'assembling' together all the five roots or senses of the body, it must be answered that there

[1]Rabbi Baer the *Maggid,* or popular preacher, of Meseritz, the most important follower of Israel Baal Shem, q.v.

is the law of the five elements which prevents the increasing of desire. The eye, for instance, must not look at random, the ear must not listen at random, the tongue must not talk at random, the body must not touch at random, nor must the mind think at random. When these important laws are obeyed, then the spirit of this complex essential nature will be supplied, as it were, before one's very eyes. This essential nature is what Mencius called the 'Expansive Spirit'. From this it follows that by concentrating these in the space under the navel and preserving it, year after year, so that it becomes invincibly strong, then before you realize it, the elixir-oven will raise the elixir. . . . With this attainment comes the power to stir up the great ocean into cream, to change the hard soil into harder gold, and so it is said that 'One grain of the elixir changes iron into gold'.

Hakuin (citing the Hermit Haku-yu)

Sin 77
Contem. 542
Center 816

When thou art rid of self, then art thou self-controlled, and self-controlled art self-possessed, and self-possessed possessed of God and all that he has ever made.

Eckhart

Death 220
Introd.
Conform.

He who has exhausted all his mental constitution knows his nature. Knowing his nature, he knows heaven.

Mencius

Suff. 128

We may never come to full knowing of God till we know first clearly our own Soul. For until the time that our Soul is in its full powers we cannot be all fully holy.

Julian of Norwich

Infra 870

Man must first of all know his own soul before he can know his Lord; for his knowledge of the Lord is as the fruit of his knowledge of himself.

Ibn ʿArabî

It needeth a soul that would have knowing of ghostly things, for to have first knowing of itself. For it may not have knowing of a kind above itself but if it have knowing of itself.

Walter Hilton

P. State 579

I say, no man knows God who knows not himself first.

Eckhart

Introd.
Holiness

You ought to know yourself as you really are, so that you may understand of what nature you are and whence you have come to this world and for what purpose you were created and in what your happiness and misery consist.

Al-Ghazâlî

Contem. 547

Whosoever knows himself well knows his Maker.

ʿAlî

The enquiry 'Who am I?' is the only method of putting an end to all misery and ushering in supreme Beatitude.

Sri Ramana Maharshi

Peace 694
Infra 890

Thou seest reality in the transitory body because of ignorance. Remove this ignorance that veils thy true knowledge, and know thy Self as pure, free, divine, absolute.

Srimad Bhagavatam, XI. iv

Devotee: Why did the Self manifest as this miserable world?

Creation 48 *Center* 828

Maharshi: In order that you might seek it. Your eyes cannot see themselves. Place a mirror before them and they see themselves. Similarly with the creation.

'See yourself first and then see the whole world as the Self.'

Sri Ramana Maharshi

Introd. *Waters* *Center* 816

O nobly-born (so and so by name), the time hath now come for thee to seek the Path (in reality). Thy breathing is about to cease. Thy *guru* hath set thee face to face before with the Clear Light; and now thou art about to experience it in its Reality in the *Bardo* state, wherein all things are like the void and cloudless sky, and the naked, spotless intellect is like unto a transparent vacuum without circumference or centre. At this moment, know thou thyself; and abide in that state.

The Tibetan Book of the Dead (Instructions for addressing person at moment of death)

When one's mind is thus known in its nakedness, this Doctrine of Seeing the Mind Naked, this Self-Liberation, is seen to be exceedingly profound.

Seek, therefore, thine own Wisdom within thee.

Sp. Drown 713

It is the Vast Deep.

Padma-Sambhava

'Seek within—know thyself', these secret and sublime hints come to us wafted from the breath of Rishis through the dust of ages.

Swami Ramdas

Metanoia 480 *Contem.* 536

He is at home, but she goes outside and looks.
She sees her husband, but still asks the neighbours.
Saraha says, O fool, know yourself.
It is not a matter of meditation, or concentration or the reciting of mantras.

Saraha

The soul which is attempting to rise to the height of knowledge must make self-knowledge its first and chief concern. The high peak of knowledge is perfect self-knowledge.

Richard of Saint-Victor

You never know yourself till you know more than your body. The Image of God was not seated in the features of your face, but in the lineaments of your Soul. In the knowledge of your Powers, Inclinations, and Principles, the knowledge of yourself chiefly consisteth.

Thomas Traherne

'Thou knowest Myself in thyself, and from this knowledge thou wilt derive all that is necessary.'

St Catherine of Siena

And when he (Apollonius) had taken his seat, he (a Brahman sage) said: 'Ask whatever you like, for you find yourself among people who know everything.' Apollonius then asked him whether they knew themselves also, thinking that he, like the Greeks, would regard self-knowledge as a difficult matter. But the other, contrary to Apollonius' expectations, corrected him and said: 'We know everything, just because we begin by knowing ourselves; for no one of us would be admitted to this philosophy unless he first knew himself.' And Apollonius remembered what he had heard Phraotes[1] say, and how he who would become a philosopher must examine himself before he undertakes the task; and he therefore acquiesced in this answer, for he was convinced of its truth in his own case also.

The Life of Apollonius of Tyana

Knowl. 743

You ought, O Soul, to get sure knowledge of your own being, and of its forms and aspects. Do not think that any one of the things of which you must seek to get knowledge is outside of you; no, all things that you ought to get knowledge of are in your possession, and within you. Beware then of being led into error by seeking (elsewhere) the things which are in your possession.

Hermes

Knowl. 755

He who reflects upon himself, reflects upon his own original.

Plotinus

Creation 42

God hath stamped a copy of his own archetypal loveliness upon the soul, that man by reflecting into himself might behold there the glory of God.

John Smith the Platonist

If a man knows himself, he shall know God.

Clement of Alexandria

(The Pythagoreans) investigated, not what is *simply* good, but what is *especially* so; nor what is difficult, but what is most difficult; viz. for a man to know himself.

Iamblichus

If you know him by whom you were made, you will know yourself.

Sextus the Pythagorean

What is so much in the mind as the mind itself? But because it is in those things which it thinks of lovingly, and is lovingly habituated to sensible, that is corporeal things, it is unable to be in itself without the images of those corporeal things. Hence shameful error arises to block the way, whilst it has not the power to separate itself from the images of sensible things, so as to see itself alone. For they have marvellously cohered with it by the

Center 841

Death 220

[1] A king of India.

Sin 66 strong bond of love. And herein consists its uncleanness; since, while it strives to think of itself alone, it fancies itself to be that, without which it cannot think of itself. When therefore it is bidden to get to know itself, let it not seek itself as though it were withdrawn from itself, but let it withdraw that which it has added to itself. Infra 873

St Augustine

(Thou) hast shown me how to seek myself and know who I was, and who and in what manner I now am, that I may again become that which I was: whom I knew not, but thyself didst seek me out: of whom I was not aware, but thyself hast taken me to thee. *Creation* 42

Acts of Thomas, 15

Symb. 306 If thou know not thyself, O fairest among women, go forth, and follow after the steps of the flocks.

Canticle of Canticles, I. 7

P. State 563 If you will know yourselves, then you will be known and you will know that you are the sons of the Living Father. But if you do not know yourselves, then you are in poverty and you are poverty.

The Gospel according to Thomas, Log. 3

No one can be saved without self-knowledge.

St Bernard

Let us enter the cell of self-knowledge.

St Catherine of Siena

Judgment 239 Jesus said: Whoever knows the All but fails to know himself lacks everything.

The Gospel according to Thomas, Log. 67

Come forward now, you who are laden with vanity and gross stupidity and vast pretence, you that are wise in your own conceit and not only declare (in every case) that you perfectly know what each object is, but go so far as to venture in your audacity to add the reasons for its being what it is, as though you had either been standing by at the creation of the world, and had observed how and out of what materials its several parts were fashioned, or had acted as advisers to the Creator regarding the things He was forming—come, I say, and then, letting go all other things whatever, take knowledge of yourselves, and say clearly who you are, in body, in soul, in sense-perception, in reason and speech, in each single one, even the most minute, of the subdivisions of your being. Declare what sight is and how you see, what hearing is and how you hear, what taste, touch, smelling are, and how you act in accordance with each of them, or what are the springs and sources of these, from which is derived their very being. For pray do not, O ye senseless ones, spin your airy fables about moon or sun or the other objects in the sky and in the universe so far removed from us and so varied in their natures, until you have scrutinized and come to know yourselves. After that, we may perhaps believe you when Introd. *Holiness* Introd. *Flight* you hold forth on other subjects: but before you establish who you yourselves are, do not think that you will ever become capable of acting as judges or trustworthy witnesses in the other matters.

Philo

Men do not know themselves, and therefore they do not understand the things of their inner world. Each man has the essence of God, and all the wisdom and power of the world (germinally) in himself; he possesses one kind of knowledge as much as another, and he who does not find that which is in him cannot truly say that he does not possess it, but only that he was not capable of successfully seeking for it.

Holiness 924

Paracelsus

Children, ye shall not seek after great science. Simply enter into your own inward principle, and learn to know what you yourselves are, spiritually and naturally.

Tauler

Duke (*in disguise*). I pray you, sir, of what disposition was the duke?
Escalus. One that, above all other strifes, contended especially to know himself.[1]

Shakespeare (*Measure for Measure*, III. ii. 250)

Holy War 396

If an ordinary man, when he is about to die, could only see the five elements of consciousness as void; the four physical elements as not constituting an 'I'; the real Mind as formless and neither coming nor going; his nature as something neither commencing at his birth nor perishing at his death, but as whole and motionless in its very depths; his Mind and environmental objects as one—if he could really accomplish this, he would receive Enlightenment in a flash. He would no longer be entangled by the Triple World; he would be a World-Transcendor. He would be without even the faintest tendency towards rebirth. If he should behold the glorious sight of all the Buddhas coming to welcome him, surrounded by every kind of gorgeous manifestation, he would feel no desire to approach them. If he should behold all sorts of horrific forms surrounding him, he would experience no terror. He would just be himself, oblivious of conceptual thought and one with the Absolute. He would have attained the state of unconditioned being. This, then, is the fundamental principle.

Center 838
M.M. 978
Contem. 536
Judgment 250
Infra 890

Huang Po

The Knowledge of one's identity with the pure Self that negates the (wrong) notion of the identity of the body and the Self sets a man free even against his will[2] when it becomes as firm as the belief of the man that he is a human being.

Heterod. 428
Grace 556
Holiness 902

Śrî Śankarâchârya

[1] This is the whole theme of the play:—the closing of the houses of resort, for example, represents withdrawal from the senses, the prison represents spiritual retreat, etc. The exploration, discovery and knowledge of the self is, in fact, the central wheel on which all of Shakespeare's drama turns.

[2] Realization itself is not within reach of the volitive faculties, it being rather a matter of ripeness and maturity—volition of course being presupposed.

Illusion 94

When you wake up you will find that this whole world, above and below, is nothing other than a regarding of oneself.

Hakuin

The Spiritual Tincture

Contem. 542

Infra 870

When the soul becomes freed from everything external and is united with prayer, then prayer like a flame envelops it, as fire envelops iron and makes it all fiery. Then the soul, though still the same soul, like red hot iron, can no longer be touched by anything external.... Blessed is the man who, while still in this life, has been granted this appearance and who himself sees his image, perishable by nature, become fiery through grace.

St Elias Ekdikos

Renun. 146

Creation 42

Put a bar of iron in a fiery forge, and it will become red like the fire. Take it out, and it will lose its red colour. If you want it to remain red, you must leave it constantly in the fire. In the same way, if you want your mind to remain always filled with divine wisdom, you must leave it continually in contact with the fire of the knowledge of Brahman, by a constant and intense meditation. You must maintain the incessant undulation of a state of brahmic consciousness. You will thus be in 'the natural state' (sahaja avasthâ).

Swami Sivananda

Supra Introd.

In truth, that which puts off the human and puts on the divine is transformed into God, the same as iron in the fire takes on the appearance of the fire and is changed into it. But the essence of the soul thus deified subsists, just as the red iron does not cease to be iron. Thus the soul which before was cold is now ardent, which before was dark is now shining, which before was hard is now mollified; all is coloured with God, because the divine essence is infused in its being, all is consumed in the fire of divine love, and all is melted and passed into God, and united to him without mode, and become one single spirit with him: just as gold and bronze are melted together into an ingot of metal.

François Louis de Blois

Reality 775

The baptism of Allâh is the dyeing-vat of *Hû* ('He', the Supreme Self): therein all piebald things become of one colour.

When he (the contemplative) falls into the vat, and you say to him, 'Arise,' he says in rapture, 'I am the vat: do not blame me.'

That 'I am the vat' is the same as saying 'I am God': he has the colour of the fire, albeit he is iron.

The colour of the iron is naughted in the colour of the fire: it (the iron) boasts of its fieriness, though actually it is like one who keeps silence.

When it has become like gold of the mine in redness, then without tongue its boast is 'I am the fire.'

It has become glorified by the colour and nature of the fire: it says, 'I am the fire, I am the fire.' . . .

What fire? What iron? Close your lips: do not laugh at the beard of the assimilator's simile.

Rûmî

Peace 700

M.M. 987

If the inward penetrates the outward, and illustrates it with its sunshine, and the outward receives the sunshine of the inward, then it is tinctured, cured, and healed by the inward, and the inward illustrates it, as the sun shines through the water, or as the fire sets the iron quite through of a light glee; here now needs no other cure.

Boehme

Center 833

Just as a little drop of water mixed with a lot of wine seems entirely to lose its own identity, while it takes on the taste of wine and its color; just as iron, heated and glowing, looks very much like fire, having divested itself of its original and characteristic appearance; and just as air flooded with the light of the sun is transformed into the same splendor of light so that it appears not so much lighted up as to be light itself; so it will inevitably happen that in saints every human affection will then in some inneffable manner melt away from self and be entirely transfused into the will of God.

St Bernard

Renun. 152b

Consider the difference between iron and iron: between cold and hot iron, such is the difference between souls, between the tepid soul and that kindled by divine fire. When the iron is first cast into the fire it certainly appears to be as dark as it is cold. But after having been a time in the flame of the fire it grows warm and gradually changes its dark colour. Visibly it begins to glow, and little by little draws the likeness of fire into itself until at last it liquefies entirely and ceases altogether to be itself, changing into another kind of thing. So also, the soul absorbed in the consuming fire in the furnace of the divine love, surrounded by the glowing body of eternal desires, first kindles then grows red hot, at last liquefies completely and is altogether changed from its first state.

Richard of Saint-Victor

P. State 563

Integration

Knowl. 745 Supra 859

'The Ṣûfî is he whose thought keeps pace with his foot',[1] i.e. he is entirely present: his soul is where his body is, and his body where his soul is, and his soul where his foot is, and his foot where his soul is. This is the sign of presence without absence.

Hujwîrî

Our Conversation is in Heaven, according to the *Measure* and Degree of our *present State* and Condition. . . .

Symb. 306 *Conform.* 166

Our Conversation is in Heaven *Exemplariter & Secundum Normam. viz.* As we make Heaven the *Sampler,* and *Copy,* after which we write. And this is that which our Saviour hath taught us to pray, that *God's will may be done on Earth, as it is done in Heaven.* When we set our selves to do the Will of God here, then Heaven is come down into the World; when we endeavour to do the Will of God, desire to do it.

Benjamin Whichcote

Rev. 959

There is no need to wait until another life or for future generations. Before you is the Buddha who proves to you the possibility of penetrating to God.

Chang Po-tuan

The whole process which we employ closely resembles that followed by Nature in the bowels of the earth, except that it is much shorter.

Philalethes

I tell you of a truth, there be some standing here, which shall not taste of death, till they see the kingdom of God.

St Luke, IX. 27

Death 206

Why should it be thought a thing incredible with you, that God should raise the dead?

Acts, XXVI. 8

(The alchemists are) holy men, who in virtue of their deified spirit have tasted in this very life the first fruits of Resurrection, and have had a foretaste of the Heavenly Kingdom.

Oswald Croll

Suff. 130

The least suspicion of God-consciousness and sufferings would be all forgot. This may well happen while the soul is in the body. I say more: while yet in the body a soul may reach oblivion of its travail not to remember it again.

Eckhart

Holiness 902

Man can attain even in this life a certain and infallible felicity.

Confessio Fraternitatis

[1] Citing a Sufi named Abû Muḥammad Murta'ish.

What could man have to do with the perfection of God as the rule of his life, unless the truth and reality of the divine nature was in him? Could there be any reasonableness in this precept or any fitness to call us to be good, as God is good, unless there was that in us which is in God? Or to call us to the perfection of a heavenly Father if we were not the real children of His heavenly nature? Might it not be as well to bid the heavy stone to fly as its flying father, the eagle, doth?

Holiness 900 *Creation* 42 *Faith* 505

William Law

The man who lives in the world, hearing of the devout recluse, or going past the door of his hermitage, feels an impulse to the devout life, has recalled to his mind what man can be upon earth, that it is possible for man to get back to that primitive contemplative state in which he issued from the hands of his Creator.

P. State 583

The Russian Pilgrim

There are Christians that place and desire all their happiness in another life, and there is another sort of Christians that desire happiness in this. The one can defer their enjoyment of Wisdom till the World to come, and dispense with the increase and perfection of knowledge for a little time: the other are instant and impatient of delay, and would fain see that happiness here, which they shall enjoy hereafter. . . . For He offereth it now, now they are commanded to have their conversation in Heaven, now they may be full of joy and full of glory. *Ye are not straitened in me, but in your own bowels* (*II. Cor*. VI. 12). Those Christians that can defer their felicity may be contented with their ignorance.

M.M. 973 *Renun.* 146 *Orthod.* 296

Thomas Traherne

(The fruition of Knowledge may take place) even in this life if there be no obstruction to it (the means adopted), because it is so seen from the scriptures.

Brahma-Sutra, III. iv. 51

It is extremely foolish to think that one must wait till after one's death in expectation of obtaining all these benefits. It is also the most culpable negligence.

Hakuin

He that beholds the sun of righteousness arising upon the horizon of his soul with healing in its wings, and chasing away all that misty darkness of his own self-will and passions; such a one desires not now the star-light to know whether it be day or not, nor cares he to pry into heaven's secrets, and to search into the hidden rolls of eternity, there to see the whole plot of his salvation; for he views it transacted upon the inward stage of his own soul, and reflecting upon himself, he may behold a heaven opened from within, and a throne set up in his soul, and an almighty Saviour sitting upon it, and reigning within him: he now finds the kingdom of heaven within him, and sees that it is not a thing merely reserved for him without him, being already made partaker of the sweetness and efficacy of it. . . . It is not an airy speculation of heaven as a thing (though never so undoubtedly) to come, that can satisfy his hungry desires, but the real possession of it even in this life. Such a happiness would be less in the esteem of good men, that were only good to be enjoyed at the end of this life when all other enjoyments fail him.

Center 833 Supra 859 *Center* 838

John Smith the Platonist

The limit of the divine (in man) is to know transformation, and this is the fullness of spiritual power.

I Ching (*Hsi Tz'u* Commentary)

And it was revealed unto him by the Holy Ghost, that he should not see death, before he had seen the Lord's Christ.

St Luke, II. 26

Metallic bodies may be changed into gold through a quickening of the process which Nature uses in the heart of the earth.

The Glory of the World

Metanoia 484

Let not him who desires this knowledge for the purpose of procuring wealth and pleasure think that he will ever attain to it. Therefore, let your mind and thoughts be turned away from all things earthly, and, as it were, created anew, and consecrated to God alone. For you should observe that these three, body, soul, and spirit, must work together in harmony if you are to bring your study of this Art to a prosperous issue, for unless the mind and heart of a man be governed by the same law which develops the whole work, such an one must indubitably err in the Art.[1]

The Sophic Hydrolith

Sin 66 *Death* 208 *Pilg.* 366 *Holiness* 902

When the spirit of darkness and of foul odour is rejected, so that no stench and no shadow of darkness appear, then the body is clothed with light and the soul and spirit rejoice because darkness has fled from the body. And the soul, calling to the body that has been filled with light, says: 'Awaken from Hades! Arise from the tomb and rouse thyself from darkness! For thou hast clothed thyself with spirituality and divinity, since the voice of the resurrection has sounded and the medicine of life has entered into thee.' For the spirit is again made glad in the body, as is also the soul, and runs with joyous haste to embrace it and does embrace it. Darkness no longer has dominion over the body, since it is a subject of light and they will not suffer separation again for eternity. And the soul rejoices in her home, because after the body had been hidden in darkness, she found it filled with light. And she united with it, since it had become divine towards her, and it is now her home. For it had put on the light of divinity and darkness had departed from it. And the body and the soul and the spirit were all united in love and had become one, in which unity the mystery had been concealed. In their being united together the mystery has been accomplished, its dwelling place sealed up and a monument erected full of light and divinity.

Archelaos

P.State 563

When the spirit has been restored to the body, the sulphur to the sulphur, and the water to the earth, and all has become white, then the body retains the spirit, and there can be no further separation.

The Glory of the World

[1] In Vedantic terminology, Reality is not complete unless all of its three poles—Being, Knowledge, Bliss (*Sat-Chit-Ânanda*)—are present.

We are glad because thou hast revealed thyself to us in all thy being; we are glad because, while we are yet in the body, thou hast deigned to make us gods by the gift of thine own eternal life. . . .

We have learnt to know thee, O thou eternal constancy of that which stands unmoved, yet makes the universe revolve.

Reality 773

With such words of praise do we adore thee, who alone art good; and let us crave from thy goodness no boon save this: be it thy will that we be kept still knowing and loving thee, and that we may never fall away from this blest way of life.

Sin 56

Hermes

Realization

Modification is the means by which all beings are produced; transformation is the means by which all beings are absorbed.

Creation 26

Shi-ping-wen

From the dark I go to the varicolored. From the varicolored I go to the dark. Shaking off evil, as a horse his hairs; shaking off the body, as the moon releases itself from the mouth of Râhu; I, a perfected soul (*kṛtâtman*), pass into the uncreated Brahma-world—yea, into it I pass!

Void 728
Death 226

Chândogya Upanishad, VIII. 13

Then was Jesus led up of the Spirit into the wilderness to be tempted of the devil.

And when he had fasted forty days and forty nights, he was afterward an hungred.

Pilg. 366

And when the tempter came to him, he said, If thou be the Son of God, command that these stones be made bread.

But he answered and said, It is written, Man shall not live by bread alone, but by every word that proceedeth out of the mouth of God.

Then the devil taketh him up into the holy city, and setteth him on a pinnacle of the temple,

And saith unto him, If thou be the Son of God, cast thyself down: for it is written, He shall give his angels charge concerning thee: and in their hands they shall bear thee up, lest at any time thou dash thy foot against a stone.

Heterod. 428 430

Jesus said unto him, It is written again, Thou shalt not tempt the Lord thy God.

Again, the devil taketh him up into an exceeding high mountain, and sheweth him all the kingdoms of the world, and the glory of them;

And saith unto him, All these things will I give thee, if thou wilt fall down and worship me.

Then saith Jesus unto him, Get thee hence, Satan: for it is written, Thou shalt worship the Lord thy God, and him only shalt thou serve.

Then the devil leaveth him, and, behold, angels came and ministered unto him.[1]

St Matthew, IV. 1–11

Some traditionalists assert that his lordship the last of prophets (Muhammad)[2] was sleeping in the cave of Mount Hira, when Jebrâil (Gabriel) made his appearance in the shape of a man, and said: 'Read.' But his lordship answered: 'I am not a reader.' Then Jebrâil squeezed him so hard that he thought his death was near; but the angel again said, 'Read,' received the same answer, and again pressed his holy and prophetic lordship. After having repeated this proceeding thrice, Jebrâil exclaimed: 'Read: In the name of thy Lord Who createth, createth man from a clot. Read: And thy Lord is the Most Bounteous, Who teacheth by the pen, teacheth man that which he knew not.'[3] One of the U'lâma says: 'The first squeezing purified his august nature from all concupiscence, the second from all sinful desires, although the heart receives all good and evil impressions, such as the radiations of the faith and the inspirations of Satan. It is true that the blessed heart of his holy and prophetic lordship was free from all ignominious qualities, but his purification was necessary to divest it from human failings and to prepare it for the reception of Divine revelations.'

Rauzat-us-safa

When the great sage (Gotama the Buddha), sprung from a line of royal sages, sat down there (at the root of an Asvattha tree)[4] with his soul fully resolved to obtain the highest knowledge, the whole world rejoiced; but Mâra, the enemy of the good law, was afraid. . . . His three sons, Confusion, Gaiety, and Pride, and his three daughters, Lust, Delight, and Thirst, asked of him the reason of his despondency, and he thus made answer unto them:

Illusion 89

'This sage, wearing the armour of resolution, and having drawn the arrow of wisdom with the barb of truth, sits yonder intending to conquer my realms. . . . While, therefore, he stands within my reach and while his spiritual eyesight is not yet attained, I will assail him to break his vow as the swollen might of a river assails a dam.'

Then having seized his flower-made bow and his five infatuating arrows, . . . Mâra thus addressed the calm seer as he sat on his seat, preparing to cross to the further side of the ocean of existence:

'Up, up, O thou Kshatriya, afraid of death! follow thine own duty and abandon this law of liberation! and having conquered the lower worlds by thy arrows, proceed to gain the higher worlds of Indra. That is a glorious path to travel, which has been followed by former

[1] Christ who had by nature what saints have by grace was obviously free (like Muhammad and the Buddha in the two citations that follow) from human imperfection; but as the Word made flesh and as the Way, it was inevitable that his temporal person be submitted to the ignominies visited upon man at this time in the world cycle. His perfection manifests in the way he surmounted his trials.

[2] Islamic tradition teaches that Muhammad is the last in the line of prophets to come in this cycle; he is therefore often referred to as 'Seal of the Prophets' (*khâtimu 'n-nabîyîn*).

[3] *Qur'ân,* Chapter XCVI, the first revealed (translation Pickthall).

[4] *Ficus religiosa* or pipul tree.

leaders of men; this mendicant life is ill-suited for one born in the noble family of a royal sage to follow.' . . .[1]

Then Mâra called to mind his own army, wishing to work the overthrow of the Sakya saint; and his followers swarmed round, . . . armed with tusks and with claws, carrying headless trunks in their hands, and assuming many forms, with half-mutilated faces, and with monstrous mouths. . . .

Pilg. 366

But the great sage . . . remained untroubled . . . like a lion seated in the midst of oxen.

Then Mâra commanded his excited army of demons to terrify him; and forthwith that host resolved to break down his determination with their various powers. . . . Before these monsters standing there, so dreadful in form and disposition, the great sage remained unalarmed and untroubled, sporting with them as if they had been only rude children. . . .

Then some being of invisible shape, but of pre-eminent glory, standing in the heavens—beholding Mâra thus malevolent against the seer—addressed him in a loud voice, unruffled by enmity:

'Take not on thyself, O Mâra, this vain fatigue,—throw aside thy malevolence and retire to peace; this sage cannot be shaken by thee any more than the mighty mountain Meru by the wind. Even fire might lose its hot nature, water its fluidity, earth its steadiness, but never will he abandon his resolution, who has acquired his merit by a long course of actions through unnumbered aeons. . . . Until he attains the highest wisdom, he will never rise from his seat, just as the sun does not rise, without dispelling the darkness. . . . '

Holiness 902

Holy War 405

Having listened to his words, and having seen the unshaken firmness of the great saint, Mâra departed dispirited and broken in purpose with those very arrows by which, O world, thou art smitten in thy heart.

Thus he, the holy one, sitting there on his seat of grass at the root of the tree, . . . became the perfectly wise, the Bhagavat, the Arhat, the king of the Law, the Tathâgata, He who has attained the knowledge of all forms, the Lord of all science.

Asvaghosha

The perseverance with which I had meditated had prepared my nerves for an internal change in the whole nervous system. . . . I saw that the minuter nerves of my system were being straightened out; even the knot of the *Suṣhumnâ-Nâḍî* was loosening below the navel; and I experienced a state of supersensual calmness and clearness resembling the former states which I had experienced, but exceeding them in its depth and ecstatic intensity, and therein differing from them. Thus was a hitherto unknown and transcendent knowledge born in me. . . . What till now had been regarded as objective discrimination shone forth as the *Dharma-Kâya* (Body of Truth). I understood the *Sangsâra* and *Nirvâṇa* to be dependent and relative states; and that the Universal Cause is Mind, which is distinct from the ideas of Interestedness or Partiality. This Universal Cause, when directed along the path of Disbelief (or Selfishness), resulteth in the *Sangsâra;* while if it be directed along the path of Altruism,[2] it resulteth in *Nirvâṇa*. I was perfectly convinced that the real source of both *Sangsâra* and *Nirvâṇa* lay in the Voidness. The knowledge I now had obtained was born of my previous energetic devotions, which had served as its main cause;

M.M. 978

Charity 605

Void 728

Holy War 405

[1] Mâra then threatens with one of his arrows, which he eventually discharges, without however shaking the Buddha's resolve.

[2] I.e., universalization of the will.

and it only awaited the accident, at the crisis, of the wholesome and nourishing food, and the timely prescription contained in the scroll, to bring it forth[1].

Milarepa

Holiness 910

Blessed are they who make this passover: all things are known to them in truth and they themselves unknown to any creature.

Eckhart

Death 220

Pilg. 385
Conform. 185

In this illumination I heard the ringing of bells. My frame dissolved and my trace vanished and my name was rased out. By reason of the violence of what I experienced I became like a worn-out garment which hangs on a high tree, and the fierce blast carries it away piece by piece. I beheld naught but lightnings and thunders, and clouds raining lights, and seas surging with fire. 'The heavens and the earth crowded upon one another', and I found myself in 'darkness upon darkness'. The All-Powerful kept eradicating my faculties one after the other, and transfixing one desire after another, until I was thunderstruck by the Divine Majesty, and the supreme Beauty spouted forth from the needle's eye of the imagination; then the right Hand in the supreme aspect was extended for grasping. Things at once came into existence; obscurity departed, and after the Ark had settled on Mount Jûdî, there was heard the cry: 'O heaven and earth, come both of you, willingly or loth! They said: We come, obedient' (*Qur'ân,* XLI. 11).

Jîlî

Knowl. 755

Death 208
Conform. 170

Ecstasy passim

When, after two sleepless nights, I had passed day and night in meditating on the permutations (*Gematrioth*) or on the principles essential to a recognition of this true reality (of the holy Names contained in the Great Name of God) and to the annihilation of all extraneous thought—then I had two signs by which I knew that I was in the right receptive mood. The one sign was the intensification of natural thought on very profound objects of knowledge, a debility of the body and strengthening of the soul until I sat there, my self all soul. The second sign was that imagination grew strong within me and it seemed as though my forehead were going to burst. Then I knew that I was ready to receive the Name. I also that Sabbath night ventured at the great ineffable Name of God (the name JHWH). But immediately that I touched it, it weakened me and a voice issued from me saying: 'Thou shalt surely die and not live! Who brought thee to touch the Great Name?' And behold, immediately I fell prone and implored the Lord God saying: 'Lord of the universe! I entered into this place only for the sake of Heaven, as Thy glory knoweth. What is my sin and what my transgression? I entered only to know Thee, for has not David already commanded Solomon: Know the God of thy father and serve Him; and has not our master Moses, peace be upon him, revealed this to us in the Torah saying: Show me now Thy way, that I may know Thee, that I may there find grace in Thy sight?' And behold, I was still speaking and oil like the oil of the anointment anointed me from head to foot and very great joy seized me which for its spirituality and the sweetness of its rapture I cannot describe.

Shaare Tsedek

[1] Milarepa is referring to an ordeal he had just undergone.

When you come to a certain point of the so-called expansion of your love and vision, it undergoes a sudden and lightning change into the Universal Consciousness. It is Jiva realising that it is Brahman. Jiva is the individual soul. As soon as the ignorance which has obsessed the soul is removed, that instant it realises that it is the Supreme Spirit. In regard to this transformation, there are no intermediary stages. The stages are only, Ramdas repeats, in the process of self-purification which helps to break the barriers between us and God and grants us the knowledge that we are one with Him.

Swami Ramdas

Metanoia 488
Knowl. 749

This immediate attainment of Buddhahood by the one realization is entirely the exertion of one moment of absolute energy.

Hakuin

Center 838

If each moment wasted in the pursuit of non-Self be utilised for the pursuit of the Self, realisation of the Self will very soon ensue.

Sri Ramana Maharshi

Renun. 146 152*a*

Thou art but an atom, He, the great whole; but if for a few days
Thou meditate with care on the whole, thou becomest one with it.

Jâmî

Supra 868 870

If you would spend all your time—walking, standing, sitting or lying down—learning to halt the concept-forming activities of your own mind, you could be sure of ultimately attaining the goal.

Huang Po

Inv. 1036

You can distinctly feel the shift of mind as it leaves its seat in the brain, attempting to return to its original seat (yatâsthana); you realize that it has left its former channels to enter into new ones. . . . Its psychology is transformed. You now have a wholly new brain, a new heart, and noble sensations and feelings.

Swami Sivananda

Creation 42
Death 220

People are mines like mines of gold and silver; the more excellent of them in ignorance are the more excellent of them in Islâm when they attain knowledge.

Muhammad

Conform. 180

Humility, unostentatiousness, non-injuring, forgiveness, simplicity, service to the Guru, purity, steadfastness, self-control;

Renunciation of sense-objects as well as absence of egoism, realization of the evils of birth, death, old age, disease, pain;

Non-attachment, non-identification of self with son, wife, home and the rest; equal-mindedness in beneficial and non-beneficial happenings;

One-pointed and unwavering devotion to Me, resort to secluded places, distaste for assemblies;

Book One
Book Two

Book Three *Holiness* 914

Constant devotion to spiritual knowledge, realization of the essence of Truth, this is declared to be wisdom; what is opposed to this is ignorance.

Creation 26 *M.M.* 978

I shall declare now that which is to be known, by knowing which one attains immortality. The Supreme Brahman is beginningless; It is said to be neither Sat (existence) nor Asat (non-existence).

Center 841 *Reality* 803

With hands and feet everywhere, with eyes, heads and mouths everywhere and with ears everywhere in the universe, That alone exists enveloping all.

Introd. *Heterod.*

It shines through the functions of all the senses, and yet It is without senses; unattached, yet It sustains all; devoid of Gunas,[1] yet It is the experiencer of Gunas.

Reality 773

It exists within and without all beings; It is unmoving as well as moving, incomprehensible because of Its subtlety; It is far and also near.

775

Indivisible, yet It exists as if divided in beings; It is to be known as the Sustainer of beings; It destroys and also generates.

Center 833 816

It is the Light of lights and is said to be beyond darkness. It is knowledge, the One to be known, and the Goal of knowledge, dwelling in the hearts of all.

Bhagavad-Gîtâ, XIII. 7–17

Part VI Part V

Either the thoughts are eliminated by holding on to the root-thought 'I' or one surrenders oneself unconditionally to the Higher Power. These are the only two ways for Realisation.

Sri Ramana Maharshi

There are two ways: one is to expand your ego to infinity, and the other is to reduce it to nothing, the former by knowledge, and the latter by devotion. The Jnani says: 'I am God—the Universal Truth.' The devotee says: 'I am nothing, O God, You are everything.' In both cases, the ego-sense disappears.[2]

Swami Ramdas

Supra 868 Supra Introd.

The soul ascends from corporal things and, being caught up above herself, abides within herself . . . for the sake of the delights she finds in God. For the divine perfection invests her in him with his likeness. His fullness is poured forth without stint: angels more in number than the sands and grass and water-drops and every single angel with his own distinctive nature, not one the same as any other.

Eckhart

Rev. 965

Thus did We show Abraham the kingdom of the heavens and the earth that he might be of those possessing certainty:

Faith 505

When the night grew dark upon him he beheld a star. He said: This is my Lord. But when it set, he said: I love not things that set.

P. State 590

And when he saw the moon uprising, he said: This is my Lord. But when it set, he said: Unless my Lord guide me, I shall surely be of the folk who are astray.

[1] Cf. note p. 88.
[2] Cf. note 1, p. 726.

And when he saw the sun uprising, he said: This is my Lord. this is greater. And when it set, he said: O my people! Lo! I am free from all that ye associate (with Him).

Lo! I have turned my face toward Him who created the heavens and the earth, as one by nature upright, and I am not of the idolaters[1]. . . .

(Thus) We raise unto degrees of wisdom whom We will. Lo! thy Lord is Wise, Omniscient.

Qur'ân, VI. 75–79, 83

Infra 890

We have also a more sure word of prophecy; whereunto ye do well that ye take heed, as unto a light that shineth in a dark place, until the day dawn, and the day star arise in your hearts.

II. Peter, I. 19

Center 833

The *ḥâl* (state) is like the unveiling of that beauteous bride, while the *maqâm* (station) is the being alone with the bride.

Rûmî

Love 625

In the same manner as lovers gradually advance from that beauty which is apparent in sensible forms, to that which is divine; so the ancient priests, when they considered that there is a certain alliance and sympathy in natural things to each other, and of things manifest to occult powers, and discovered that all things subsist in all, they fabricated a sacred science from this mutual similarity. Thus they recognized things supreme in such as are subordinate, and the subordinate in the supreme.

Proclus

Beauty 663

Symb. 306
M.M. 978

We must understand that the *Order* and *Symmitry* of the *Universe* is so setled by the *Lawes* of *Creation,* that the lowest things (the *Subcelestiall* or *Elementary Region*) should be immediately subservient to the *Midle;* the *Midle* (or *Celestiall*) to those above; and these (the *Supercelestiall* or *Intelligible*) to the *Supreame Rulers* becke. With this it is further to be knowne that these *Superiours* and *Inferiours* have an *Analogicall* likenesse, and by a secret *Bond* have likewise a fast *coherence* between themselvs through insensible *Mediums,* freely combiening in *Obedience* to the same supreme *Ruler,* and also to the benefit of *Nature:* Insomuch, that if we take the said *Harmony* in the *Reverse,* we shall finde that things *Supercelestiall* may be drawne down by *Celestiall,* and *Supernaturall,* by *Naturall.* For this is the *Maxim* of old *Hermes, Quod est superius, est sicut id quod est inferius.*[2]

And upon this ground *Wisemen*[3] conceive it no way *Irrationall* that it should be possible for us to ascend by the same degrees through *each world,* to the very *Originall world* it selfe, the Maker of all things and first *Cause.*

Elias Ashmole

Rev. 967

Introd. *Orthod.*

[1] It must be remembered that Abraham lived among the Chaldeans, who were renowned astronomers, and that the sense of this allegory is purely interior.

[2] Cf. note p. 302.

[3] E.g., Cornelius Agrippa, *De Occult Philos.*, I. i.

Symb. 306 *Pilg.* 378 *P. State* 590

Oh, I swear by the afterglow of sunset,
And by the night and all that it enshroudeth,
And by the moon when she is at the full,
That ye shall journey on from plane to plane.

Qur'ân, LXXXIV. 16–19

Rev. 967 *P. State* 590 *Beauty* 689

Now, according to St Augustine, when the light of the soul eclipses creatures, it is dawn; when the angelic light eclipses the light of the soul and devours it, then it is broad day. David says, 'The righteous man mounts up and up to the perfect day.' His path is fair and smooth and pleasant and familiar. And when the psychic and angelic lights are swallowed up in the light divine, he calls that high noon. Now day is at its longest, in its prime, when the sun at its zenith pours its light into the stars and the stars pour it into the moon. These are members of the solar system. And even so the light of God embraces the angelic light and that of the soul, an orderly array, an ascending scale steadily rising in the day, all praising God in chorus.

Eckhart

Reality 790

On the narrow path of Truth,
On the Meridian line, He stands upright,
Throwing no shadow before or behind Him,
To the right hand or the left.
East and west is His Kibla cast,
Drowned in a blaze of radiant light.

Shabistarî

Infra 890

The moon and the sun and the axis of the seven heavens are swallowed
By the Canopus of the soul, when it rises from towards the southern angle.

Dîvâni Shamsi Tabrîz, XLIV

Orthod. 296 *Death* 206 Supra 859

After the attainment of God, religious duties such as the *sandhyâ* (daily worship) drop away. One day some people were sitting on the bank of the Ganges performing the *sandhyâ*. But one of them abstained from it. On being asked the reason, he said: 'I am observing *aśoucha* (temporary defilement upon birth or death of a blood relative). I cannot perform the *sandhyâ* ceremony. In my case the defilement is due to both a birth and a death. My mother, Ignorance, is dead, and my son, Self-Knowledge, has been born.'

The Gospel of Sri Ramakrishna[1]

Faith 505 *M.M.* 973

So far as the soul sinks down in faith into the unknown good, so far she is one with the unknown good and is unknown to self or any creature. She well knows that she is but knows not what she is. Not till she knows all that there is to know does she cross over to the unknown good. This crossing is obscure to many a religious.

Eckhart

The mind becomes a blank sheet; then the yogi destroys this sheet and becomes

[1] Cf. note p. 822.

identified with the supreme Purusha, the Self, the supreme Being, from whom the mind derives all its light. He thus obtains omniscience and final emancipation (kaivalya). These are things which are so much Greek to our Western psychologists; this is why they grope in darkness. They have no idea of the perfect Being or Purusha, witness of the mind's activities.

Center 833 Introd. *Judgment*

Swami Sivananda

Ânanda saw Śâriputra coming afar off, and he said to him: 'Serene and pure and radiant is your face, Brother Śâriputra! In what mood has Śâriputra been today?'

'I have been alone in Dhyâna,[1] and to me came never the thought: *I* am attaining it! *I* have got it! *I* have emerged from it!'

Death 220

Saṁyutta-nikâya, III. 235

Our Solution, then, is the reducing of our Stone to its first matter, the manifestation of its essential liquid, and the extraction of natures from their profundity, which is finished by bringing them into a mineral water; nor is this operation easy: those who have tried can bear out the truth of my words.

P. State 563

Holy War 405

Philalethes

The spirit of meditation is the combating of self-willed thinking—it is a combat against the weight of one's feelings. It is a combat against dark and deep sleepiness. It is a combat against the ideas of right and wrong, of activity and quiet, or disorder and regularity—in fact it is a combat against all the forms of the objective world of the senses—the condition which dulls the mind. By carrying on the combat with enthusiasm in the correct spirit, one may go on till there is an entirely unexpected attainment of enlightenment.

Holy War 396 *M.M.* 978

Hakuin

The truth is that a man succeeds to a great extent because of tendencies inherited from his previous births. People think he has attained the goal all of a sudden. A man drank a glass of wine in the morning. It made him completely drunk. He began to behave improperly. People were amazed to see that he could be so drunk after one glass. But another man said, 'Why, he has been drinking all night.'

Action 329 *Ecstasy* 637 Introd. *Inv.*

Sri Ramakrishna

This state is not to be got by effort, but through the dawn of knowledge, in a mysterious way. We may call it divine grace, or it may be as a result of the efforts done in a previous life. Whatever that be, when knowledge comes, we are lost in the radiance of the supreme Reality, which is universal, infinite and eternal, and thereafter we live in that state perennially.

Knowl. 761

Holiness 902

Swami Ramdas

The Master Shen-tsu asked Shen-hui: 'You say that our Original Nature has the characteristics of the Absolute. In that case it has no colour, blue, yellow or the like, that the eye can see. How then can one perceive one's Original Nature?' Shen-hui answered, 'Our Original Nature is void and still. If we have not experienced Enlightenment, erroneous

Creation 42

[1] Sanskrit for 'meditation' (called *Ch'an* in China and *Zen* in Japan).

ideas arise. But if we awaken to the erroneous nature of these ideas, both the Awakening and the wrong idea simultaneously vanish. That is what I mean by "perceiving one's Original Nature".' Shen-tsu again asked: 'Despite the light that comes from the Awakening, one is still on the plane of Birth and Destruction. Tell me by what method one can get clear of Birth and Destruction?' Shen-hui answered, 'It is only because you put into play the ideas of Birth and Destruction that Birth and Destruction arise. Rid yourself of these ideas, and there will be no substance to which you can even distantly apply these names. When the light that comes from the Awakening is quenched, we pass automatically into Non-being, and there is no question of Birth or Destruction.'

Reality 803 *M.M.* 994

Shen-hui

Moreover, Subhuti, what I have attained in Anuttara-samyak-sambodhi is the same as what all others have attained. It is something that is undifferentiated, neither to be regarded as a high state, nor is it to be regarded as a low state. It is wholly independent of any definitive or arbitrary conceptions of an individual self, other selves, living beings or an Universal Self.

Prajñâ-Pâramitâ

Sin 66

Upon learning the Art one will feel as if he had just come out of a filthy cesspool into the open sea, or left the firefly for the sunlight and moonlight—just as the hearing of peals of thunder makes one feel the insignificance of the roll of the drum, and the sight of a whale makes one feel the smallness of ordinary fish.

Ko Hung

Inv. 1031

As a result of the merit of repeating the sacred name, I have, for over ten years past, continually been gazing upon the glory of the Pure Land, and the very forms of the Buddhas and Bodhisattvas, but I have kept it secret and said nothing about it. Now however, as I draw near the end, I disclose it to you.

Hônen

Supra 868 Infra 887

I feel within me a consuming fire of heavenly love which has burned up in my soul everything that was contrary to itself and transformed me inwardly into its own nature.

William Law (spoken on his deathbed)

'Tis notorious that copper by alchemy becomes gold:
Our copper has been transmuted by this rare alchemy.

Dîvâni Shamsi Tabrîz, IV

Action 329

The Self being illumined by meditation, and then burning with the fire of Knowledge, is delivered from all accidents, and shines in its own splendour, like gold which is purified in the fire.

Śrî Śankarâchârya

Seeking the builder of the house
I have run my course in the vortex

Of countless births, never escaping the hobble (of death);
Ill is repeated birth after birth!
Householder, art seen!
Never again shalt thou build me a house.
All of thy rigging is broken,
The peak of the roof is shattered:
Its aggregations passed away,
Mind has reached the destruction of cravings.

Dhammapada, XI. 153–154 (The Buddha's Victory Song)

Beauty 663
Peace 694

From Gautama Buddha down through the whole line of patriarchs to Bodhidharma, none preached aught besides the One Mind, otherwise known as the Sole Vehicle of Liberation. Hence, though you search throughout the whole universe, you will never find another vehicle. Nowhere has this teaching leaves or branches; its one quality is eternal truth. Hence it is a teaching hard to accept.

Huang Po

Reality 790
803

There is no greater mystery than this—*viz.* ourselves being the Reality we seek to gain reality. We think that there is something hiding our Reality and that it must be destroyed before the Reality is gained. It is ridiculous. A day will dawn when you will yourself laugh at your past efforts. That which will be on the day you laugh is also here and now.

Sri Ramana Maharshi

Illusion 85
Center 838

What is worship?
To realise reality.

Ansâri

Contem. 547

Ignorance conceals the pre-existent Knowledge just as water plants cover over the surface of a pond. Clear away the plants and you have the water. You don't have to create it; it is there already. Or take another example—a cataract grows on the eye and prevents a man from seeing; remove the cataract and he sees. Ignorance is the cataract. . . . *Jnana* (Knowledge) is not something to be attained, it is eternal and self-existent. On the other hand, ignorance has a cause and an end. The root of it is the idea that the devotee is a separate being from God. Remove this and what remains is *Jnana*.

Sai Baba

Illusion 94
Sin 64
Knowl. 755
Death 220

Certain professors of education must be wrong when they say that they can put a knowledge into the soul which was not there before, like sight into blind eyes. . . . Whereas, our argument shows that the power and capacity of learning exists in the soul already.

Plato (*Republic* VII. 518 C)

Stop talking, stop thinking, and there is nothing you will not understand.
Return to the Root and you will find the Meaning;
Pursue the Light, and you will lose its source. . . .
There is no need to seek Truth; only stop having views.

Seng-ts'an

Liberation becomes artificial and therefore transitory according to the philosopher who holds that it is a change of one state into another (on the part of the Self). Again it is not reasonable that it is a union (with *Brahman*) or a separation (from Nature). As both union and separation are transitory Liberation cannot consist of the individual self going to *Brahman* or of *Brahman* coming to it. But the Self,[1] one's own real nature is never destroyed. For It is uncaused and cannot be accepted or rejected by oneself.

Śrî Śankarâchârya

Knowl. 749

'There is no relation between Me and him for whom there is a quest between himself and Me.'

Niffarî

To be satisfied in God is the highest difficulty in the whole world, and yet most easy . . . because God is. For God is not a being compounded of body and soul, or substance and accident, or power and act, but is all act, pure act, a Simple Being whose essence is to be, whose Being is to be perfect.

Thomas Traherne

Creation 42 Introd. *Void*

You were universal prior to this. But when, together with the universe, something was present with you, you became less by the addition; because the addition was not from truly subsisting Being, for to that you cannot add anything.

Porphyry

Creation 48

All the Sastras are meant only to make . . . man retrace his steps to the original source. He need not gain anything new. He must only give up his false ideas and useless accretions. Instead of doing it he tries to catch hold of something strange and mysterious because he believes that his happiness lies elsewhere. That is the mistake.

Sri Ramana Maharshi

Mind is the Buddha, while the cessation of conceptual thought is the Way. . . . Only renounce the error of intellectual or conceptual thought-processes and your nature will exhibit its pristine purity.

Huang Po

When true sanctity and purity shall ground (a man) in the knowledge of divine things, then shall the inward sciences, that arise from the bottom of his own soul, display themselves; which indeed are the only true sciences: for the soul runs not out of itself to behold temperance and justice abroad, but its own light sees them in the contemplation of its own being, and that divine essence which was before enshrined within itself.

Supra 859

Plotinus

You need not aspire for or get any new state. Get rid of your present thoughts, that is all.

Sri Ramana Maharshi

[1] That the Self Itself is Liberation is the conclusion' (note by commentator).

You will not possess intellect, till you understand that you have it.

Sextus the Pythagorean

Hermes Trismegistus said: 'O Sages! Consider the red which is perfect and that which is imperfect, and the perfect yellow and that which is imperfect, and the black which is perfect and that which is imperfect—each one of these is from one root.'

Abu 'l-Qâsim al-'Irâqî

Reality 775

Our original Buddha-Nature is, in highest truth, devoid of any atom of objectivity. It is void, omnipresent, silent, pure; it is glorious and mysterious peaceful joy—and that is all. Enter deeply into it by awaking to it yourself. That which is before you is it, in all its fullness, utterly complete. There is naught beside. Even if you go through all the stages of a Bodhisattva's progress towards Buddhahood, one by one; when at last, in a single flash, you attain to full realization, you will only be realizing the Buddha-Nature which has been with you all the time; and by all the foregoing stages you will have added to it nothing at all.[1] You will come to look upon those aeons of work and achievement as no better than unreal actions performed in a dream. That is why the Tathâgata said: 'I truly attained nothing from complete, unexcelled Enlightenment.'

Huang Po

Ecstasy passim

Realise what is present here and now. The sages did so before and still do that only. Hence they say that it looks as if newly got. Once veiled by ignorance and later revealed, Reality looks as if newly realised. But it is not new.

Sri Ramana Maharshi

And give glad tidings unto those who believe and do good works: that theirs are Gardens underneath which rivers flow: as often as they are regaled with food of the fruit thereof, they say: This is what we were nourished with aforetime, when it was given them in likeness.

Qur'ân, II. 25

Beauty 670
Symb. 306

Liberation is nothing new that is acquired.

Śrî Śankarâchârya

His disciples said to Him: When will the repose of the dead come about and when will the new world come? He said to them: What you expect has come, but you know it not.

The Gospel according to Thomas, Log. 51

P. State 583

It is of course nothing new that you gain. God is already within you, but you had forgotten Him and so temporarily lost Him. When any external matter enters your eye, you feel terrible irritation. When it is removed, your pain is relieved and you feel as if you have gained something new. In fact, you did not gain anything new. You only got back to the normal state you had lost temporarily.

Swami Ramdas

Knowl. 755
Creation 42

[1] Without the preparation of these foregoing stages, however, the ordinary mortal would be shattered by the overwhelming intensity of this final state.

This body is not a strange one, nay, but it is the very body which was dissolved and from which the sulphureity has been extracted, leaving it as ashes; it is therefore different from it in quantity but not in species. Rather is it a derived form of the first in reality.

Geber

The happiness obtained on the cessation of desire is ever the same. Whatever may have been the varieties of desire that preceded it, the bliss had on the cessation of desire is the same. If a man is affected by some disease, you may ask him: 'What is the disease you are suffering from?' When he has recovered from the disease and regained his normal health, nobody can ask him: 'What is the health you are now having?' The reason is, though diseases may be many and various, health is ever one and the same.

Sri Chandrasekhara Bhâratî Swâmigal

Of what does this true possession of God consist, when one really has him? It depends on the heart and an inner, intellectual return to God and not on steady contemplation by a given method.[1]

Eckhart

That which has a beginning must also end. . . . Realisation is not acquisition of anything new nor is it a new faculty. It is only removal of all camouflage. . . . The ultimate Truth is so simple. It is nothing more than being in the pristine state. . . . Mature minds alone can grasp the simple Truth in all its nakedness.

P. State 573
Inv. 1017

Sri Ramana Maharshi

How is it possible to find oneself when seeking others rather than oneself? . . .

By not taking the mind to be naturally a duality, and allowing it, as the primordial consciousness, to abide in its own place, beings attain deliverance.

Supra 859

Padma-Sambhava

Before a man studies Zen, to him mountains are mountains and waters are waters; after he gets an insight into the truth of Zen through the instruction of a good master, mountains to him are not mountains and waters are not waters; but after this when he really attains to the abode of rest, mountains are once more mountains and waters are waters.

Metanoia 488
Peace 694

Ch'ing-yüan

Misty rain on Mount Lu,
And waves surging in Che-chiang;
When you have not yet been there,
Many a regret surely you have;
But once there and homeward you wend,
And how matter-of-fact things look!
Misty rain on Mount Lu,
And waves surging in Che-chiang.

Su Tung-p'o

[1] I.e., Knowledge alone delivers: method and contemplation make the psychic terrain propitious for this event.

Identity

When alone the object of contemplation remains and one's own form is annihilated, this is known as 'identification'.

Yoga Darshana, III. 3

Death 220

He whose heart is steadfastly engaged in Yoga, looks everywhere with the eyes of equality, seeing the Self in all beings and all beings in the Self.

He who sees Me in all and all in Me, from him I vanish not, nor does he vanish from Me.

Bhagavad-Gîtâ, VI. 29–30

Reality 775
Metanoia 493

Would'st thou be very Christ and God? Put off, then, whatever the eternal Word did not put on.

Eckhart

Supra 873

For when we go out in love beyond and above all things, and die to all observation in ignorance and in darkness, then we are wrought and transformed through the Eternal Word, Who is the Image of the Father. In this idleness of our spirit, we receive the Incomprehensible Light, which enwraps us and penetrates us, as the air is penetrated by the light of the sun. And this Light is nothing else than a fathomless staring and seeing. What we are, that we behold; and what we behold, that we are: for our thought, our life, and our being are uplifted in simplicity, and made one with the Truth which is God. And therefore in this simple staring we are one life and one spirit with God.

Ruysbroeck

Supra 868

If then, being made of Life and Light, you learn to know that you are made of them, you will go back into Life and Light.

Hermes

Creation 42

When the soul strips off her created nature there flashes out its uncreated prototype.

Eckhart

Void 721

Unity has in all the cosmos no place of manifestation (*maẓhar*) more perfect than thyself, when thou plungest thyself into thine own essence in forgetting all relationship, and when thou seizest thyself with thyself, stripped of thy appearances, so that thou art thyself in thyself and none of the divine Qualities or created attributes (which normally pertain to thee) any longer refer to thee. It is this state of man which is the most perfect place of manifestation for Unity in all existence.

Jîlî

Center 816

When through self, by the suppressing of the mind, one sees the brilliant Self which is more subtile than the subtile, then having seen the Self through one's self, one becomes

Metanoia 480

Holiness 910
Creation 26

self-less (*nir-âtman*). Because of being selfless, he is to be regarded as incalculable (*a-saṁkhya*), without origin—the mark of liberation (*mokṣa*). This is the supreme secret doctrine (*rahasya*).

Maitri Upanishad, VI. 20

The Soul advances and is taken into unison, and in that association becomes one with the Divine Mind—but not to its own destruction; the two are one and two.

Sp. Drown.
M.M. 978

Plotinus

Reality
775

Ten is the royal number: it is born from one and nothing;
When God and creature meet, this birth takes place.

Angelus Silesius

Love 625
Infra 890

Leave me room on that Divân which leaves no room for Twain.

Jâmî

The soul is one with God and not united, so the scriptures say. For instance, if we fill a tub of water the water in the tub is united and not one: what is water is not wood and where it is wood it is not water. Now take the wood and throw it in the middle of the water, the wood is still united, nothing more, and not the same. It is different with the soul, she is one with God and not united: where God is there is the soul and where the soul is there is God.

Eckhart

Inv. 1007
Supra 870
Orthod. 275

This is that wonderful philosophical transmutation of body into spirit and of spirit into body about which an instruction has come down to us from the wise of old.

Thomas Vaughan

Conformity consists in this, that we *unite our will to God's will;* but *uniformity* means more—it requires that we *make our will one with the Divine will,* so that we desire nothing but what God desires, and will nothing but what God wills. This is the sum and substance of that perfection to which we should aspire.

St Alphonsus Liguori

Love
625

Love in union is like the honeycomb in honey.

Eckhart

What constitutes self-realisation of Noble Wisdom ... is not comparable to the perceptions attained by the sense-mind, neither is it comparable to the cognition of the discriminating and intellectual-mind. Both of these presuppose a difference between self and not-self and the knowledge so attained is characterised by individuality and generality. Self-realisation is based on identity and oneness; there is nothing to be discriminated nor predicated concerning it. But to enter into it the Bodhisattva must be free from all presuppositions and attachments to things, ideas and selfness.

Knowl.
749

Lankavatara Sutra, VII

Therefore we say that the Self-Realised Sage knows by his mind, but his mind is pure. Again we say that the vibrating mind is impure and the placid mind is pure. The pure mind is itself Brahma; therefore it follows that Brahma is nought other than the mind of the sage.

Sri Ramana Maharshi

Contem. 532
Supra
Introd.

After passing the six centres (of Yoga) the aspirant arrives at the seventh plane. Reaching it, the mind merges in Brahman. The individual soul and the Supreme Soul become one. The aspirant goes into samâdhi. His consciousness of the body disappears. He loses the knowledge of the outer world. He does not see the manifold any more. His reasoning comes to a stop.

Sri Ramakrishna

Jesus said: Blessed are the solitary and elect, for you shall find the Kingdom; because you come from it, and you shall go there again.

The Gospel according to Thomas, Log. 49

Void
728

According to the scriptures, 'No man knoweth the Father but the Son,' and hence, if ye desire to know God, ye have to be not merely like the Son, ye have to be the very Son himself.

Eckhart

Whosoever has united his will to the will of God has perfect and perpetual joy; *perfect joy,* because he has what he desires; *perpetual joy,* because no one can prevent what God wills.

St Alphonsus Liguori

Center 847
P. State
563

Dost thou hear how there comes a voice from the brooks of running water? But when they reach the sea they are quiet, and the sea is neither augmented by their in-coming nor diminished by their out-going.

Bâyazîd al-Bisṭâmî

Creation 28
Flight
949

Nirvâna is the transcendental knowledge of the sameness of all principles.

Saddharma-puṇḍarîka, v. 44

Reality
775

St Augustine says, 'The union of body and soul may be close, but closer still is the union that spirit has with spirit.'

Eckhart

Center
841

As one cannot become another one should not consider *Brahman* to be different from oneself. For if one becomes another one is sure to be destroyed.

Śrî Sankarâchârya

Illusion
85

The soul is not like God: she is identical with him.

Eckhart

Infra 890
P. State 563

Jesus said to her: I am He who is from the Same. . . . (Salome said): I am Thy disciple. (Jesus said to her): Therefore I say, if he is the Same, he will be filled with light, but if he is divided, he will be filled with darkness.

The Gospel according to Thomas, Log. 61

Reality 773

In this breaking-through I find that God and I are both the same. Then I am what I was, I neither wax nor wane, for I am the motionless cause that is moving all things.

Eckhart

Unification is the separation of the eternal from that which was originated in time.

Junayd

Knowl. 749

Therefore the cause of attainment is attainment itself.

Hujwîrî

The Sundoor

Inv. passim
Center 833

When he (the being) thus departs from this body, then he ascends upward with these very rays of the sun. With the thought of *Om*, verily, he passes up. As quickly as one could direct his mind to it, he comes to the sun. That, verily, indeed, is the world-door, an entrance for knowers, a stopping for non-knowers.

As to this there is the following verse:—

There are a hundred and one channels of the heart.
One of these passes up to the crown of the head.
Going up by it, one goes to immortality.
The others are for departing in various directions.

Judgment 250

Chândogya Upanishad, VIII. vi. 5, 6

A certain man came and knocked at a friend's door: his friend asked him, 'Who art thou, O trusty one?'

He answered, 'I'. The friend said, 'Begone, 'tis not the time: at a table like this there is no place for the raw.'

Introd.
Suff.
Pilg. 365
Contem. 542

Save the fire of absence and separation, what will cook the raw one? Who will deliver him from hypocrisy?

The wretched man went away, and for a year in travel and in separation from his friend he was burned with sparks of fire.

That burned one was cooked: then he returned and again paced to and fro beside the house of his comrade.

He knocked at the door with a hundred fears and respects, lest any disrespectful word might escape from his lips.

His friend called to him, 'Who is at the door?' He answered, ' 'Tis thou art at the door, O charmer of hearts.'

'Now,' said the friend, 'since thou art I, come in, O myself: there is not room in the house for two I's.

The double end of thread is not for the needle: inasmuch as thou art single, come into this needle.'

Supra Introd.

'Tis the thread that is connected with the needle: the eye of the needle is not suitable for the camel.

Rûmî

'Who art thou?'

'I am thyself.'

Jaiminîya Upaniṣad Brâhmaṇa, III. i. 6

In whom, when I go forth hence, shall I be going forth?

Praśna Upanishad, VI. 3

Everything is perishing except His Face:[1] unless thou art in His Face (Essence), do not seek to exist. . . .

Whosoever is uttering 'I' and 'we' at the door, he is turned back from the door and is continuing in *not*.

Death 220

Rûmî

In the naked essence man knows himself even as he is known. Which knowing, our Lord said, 'I am the door of my sheep-fold.' In these words he invites us to enter by the door of his emanation and return into the source whence we came forth.

Eckhart

Supra 859
Creation 42

He indeed neither rises nor sets, and for him that understandeth this, it is evermore high noon.

Chândogya Upanishad, III. xi. 3

Center 833
Holiness 902

This marvellous union of body, soul, and spirit, this Divine glorification and exaltation of the elect, is a consideration fraught with reverential and unspeakable awe (like the sight of the final chemical transformation); it is a sight at which the very angels will stand rapt in inexpressible wonder; and then they will see us pass into the heavens to reign with Christ, and with them, and the ministering spirits, in everlasting glory, and joy unspeakable, world without end.

The Sophic Hydrolith

Supra 870

Lord! Thou art my Beloved! My desire! My flowing stream! My Sun! And I am Thy reflection!

Mechthild of Magdeburg

Center 828

[1] *Qur'ân*, XXVIII. 88.

Inv. 1003 *Holiness* 924 *Knowl.* 749 *Action* 340

God made this name (Allâh) a mirror for man, so that when he looks in it, he knows the true meaning of 'God was and there was naught beside Him'[1], and in that moment it is revealed to him that his hearing is God's hearing, his sight God's sight, his speech God's speech, his life God's life, his knowledge God's knowledge, his will God's will, and his power God's power, and that God possesses all these attributes fundamentally; and then he knows that all the aforesaid qualities are borrowed and metaphorically applied to himself, whereas they really belong to God.

Jîlî

Introd. *Conform.* *Grace* 552

If thou wert not to attain unto Him except upon the extinction of thy wickedness and the effacement of thy claims, then wouldst thou never attain unto Him; but should He wish to unite thee unto Him, then would He cover thy qualities with His Qualities, and describe thee with His Description, and unite thee unto Him with what is from Him to thee, not with what is from thee to Him.

Ibn 'Aṭâ'illâh

This being is beyond our grasp, whereat, rejoicing greatly, let us hasten to seize it with itself: this is our highest happiness.

Eckhart

P. State 563

The one and only thing required is to free oneself from the bondage of mind and body alike, putting the Buddha's own seal upon yourself. If you do this as you sit in ecstatic meditation, the whole universe itself scattered through the infinities[2] of space turns into enlightenment. This is what I mean by the Buddha's seal.

Dôgen

Renun. 158

God summoned me before Him and said: 'With what comest thou unto Me?'
'With renunciation of the world.'
'The world for Me is only the wing of an insect. It is no great thing to renounce it.'
'I ask Thy pardon! I come with the abandonment of all self-pursuit.'
'Am I not guarantee for what I have promised?'
'I ask Thy pardon! I come with Thyself.'
'It is in this way that We receive thee.'

Bâyazîd al-Bisṭâmî

The 'I' casts off the illusion of 'I' and yet remains as 'I'. Such is the paradox of Self-Realisation.

Sri Ramana Maharshi

Symb. 306 Introd. *Holiness*

'I shall not cease to make Myself known to thee through that which is between Me and thee, until thou knowest who thou art to Me: but when thou knowest who thou art to Me, I shall make Myself known to thee through that which is between Me and everything.'

Niffarî

[1] Cf. the first citation of Muhammad on p. 804. The statements which follow are also based on a *ḥadîth*.

[2] Since infinity by definition excludes a plurality, the translation could better be rendered by a word such as 'indefinitude'.

Jesus said: If they say to you: 'From where have you originated?', say to them: 'We have come from the Light, where the Light has originated through itself'.

The Gospel according to Thomas, Log. 50

And when all things shall be subdued unto him, then shall the Son also himself be subject unto him that put all things under him, that God may be all in all.

I. Corinthians, xv. 28

For the soul to be naked she must turn away from all the images and forms spread out before her and stop at none of them. For the divine nature is no form nor semblance that she can understand. Being turned away from these towards what transcends them—divorced, that is, from images and forms—the soul receives the likeness of the formless nature of God whose real form has never been revealed to any creature. This is the secret door into the divine nature, which the soul has in the image.... Even God himself is forbidden there so far as he is subject to condition.... There the spirit loses its uses not its essence.... These are the blessed dead that are dead in God. No one can be buried and beatified in the Godhead who has not died to God.

Void 721

Sp. Drown. 713

Eckhart

This immense thought (of the 'Supreme Identity') is only befitting to him whose soul is vaster than the two worlds. As for him whose soul is only as vast as the two worlds, it befits him not. For in truth this thought is greater than the sensible world and the supra-sensible world both taken together.

Ibn 'Arabî

This breaking-through is the second death of the soul and is far more momentous than the first.

M.M. 998

Eckhart

Enter thou, for what thou art I am, and what I am thou art.

Jaiminîya Upaniṣad Brâhmaṇa, III. 14

Knowl. 749

That which is the finest essence—this whole world has that as its Self. That is Reality. That is Âtman. That art thou,[1] Śvetaketu.

Chândogya Upanishad, VI. xi. 3

P. State 563

This eternal peace is your real existence—it is not a state or truth to be attained but to be realized; because you are ever That.

Peace 705
Supra 873

Swami Ramdas

O Eternal Light of Divine Glory, since thou art in my innermost depths, since thou transcendest all things, be to me *that thou art,* a turning away from all things into the ineffable Good that thou art in thy naked self.

Eckhart

[1] In the Vedanta there are four *Mahâvâkyas,* so called: 1. 'Thou art That.' 2. 'I am Brahman.' 3. 'This Self is Brahman.' 4. '*Prajnâna* (Absolute Knowledge) is Brahman.' Cf. Sri Ramana Maharshi: *Talks,* p. 223.

He is Allâh.

Sufic Formula

I am He and He is I, except that I am what I am, and He is what He is.

Muhammad

Once He raised me up and stationed me before Him, and said to me, 'O Abû Yazîd, truly My creation desire to see thee.' I said, 'Adorn me in Thy Unity, and clothe me in Thy Selfhood, and raise me up to Thy Oneness, so that when Thy creation see me they will say, We have seen Thee: and Thou wilt be That, and I shall not be there at all.'

Bâyazîd al-Bisṭâmî

I went from God to God, until they cried from me in me, 'O Thou I!'

Bâyazîd al-Bisṭâmî

Holy War 403 *Reality* 775

Let us hasten to salvation, to the new birth. Let us, who are many, hasten to be gathered together into one love corresponding to the union of the One Being. Similarly, let us follow after unity by the practice of good works, seeking the good Monad.

Clement of Alexandria

Thou art a shadow and in love with the sun: the sun comes, the shadow is naughted speedily.

Rûmî

Creation 42 *Sin* 66

Certainly he (the first man) was created of flesh and blood, but of a flesh and blood both immutable and perpetual. . . . This matter was real, but in the same way that shadows are latent in the light, which does not alter their inability to manifest themselves in the presence of this light; but when the light is extinguished the shadows appear.[1]

Boehme

Void 728

Atoned with her creator the soul has lost her name for she herself does not exist: God has absorbed her into him just as the sunlight swallows up the dawn till it is gone.

Eckhart

Center 828 Supra 873 *Reality* 775 Introd. *Heterod.*

The Sun of my Perfection is a Glass
Wherein from *Seeing* into *Being* pass
All who, reflecting as reflected see
Themselves in Me, and Me in Them: not *Me,*
But all of Me that a contracted Eye
Is comprehensive of Infinity:
Nor yet *Themselves:* no Selves, but of The All
Fractions, from which they split and whither fall.

[1] This is 'the cool of the day' in *Genesis,* III. 8. The first Adam's propensity towards dispersion is released by the Light's withdrawal; with the second Adam the image is reversed,—the mystery of the Fall and Redemption, as adumbrated in the passages closing the present chapter. See the citation by St. Augustine on p. 270.

As Water lifted from the Deep, again
Falls back in individual Drops of Rain
Then melts into the Universal Main. *Sp. Drown.* 713
All you have been, and seen, and done, and thought,
Not *You* but *I*, have seen and been and wrought: *Action* 340
I was the Sin that from Myself rebell'd: *Center* 841
I the Remorse that tow'rd Myself compell'd:
I was the Tajidar who led the Track:
I was the little Briar that pull'd you back:
Sin and Contrition—Retribution owed,
And cancell'd—Pilgrim, Pilgrimage, and Road,
Was but Myself toward Myself: and Your *Pilg.* 387
Arrival but *Myself* at my own Door:
Who in your Fraction of Myself behold
Myself within the Mirror Myself hold
To see Myself in, and each part of Me *Creation* 48
That sees himself, though drown'd, shall ever see.
Come you lost Atoms to your Centre draw,
And *be* the Eternal Mirror that you saw:
Rays that have wander'd into Darkness wide
Return, and back into your Sun subside.

'Aṭṭâr

HOLINESS—UNIVERSAL MAN

Man is the symbol of universal Existence.

Sufic Formula[1]

In the *Introduction* it was said that tradition is characterized by a normal constant:[2] the saint or sage is none other than the culminating point, essence, pole or logical perfection of this norm.[3]

'The human being, considered in its integrality, comprises a certain sum of possibilities which constitute its corporeal or gross modality, and in addition, a multitude of other possibilities, which, extending in different directions beyond the corporeal modality, constitute its subtle modalities; however, all these possibilities taken together represent but one and the same degree of universal Existence. It follows from this that the human individuality is at once much more and much less than Westerners generally suppose it to be: much more, because they recognize in it scarcely anything except the corporeal modality, which includes but a small fraction of its possibilities; much less, however, because this individuality, far from really constituting the whole being, is but one state of this being, among an indefinity of other states, the sum of which itself is still nothing in relation to the Personality, which alone is the true being, because It alone represents its permanent and unconditioned state, and because there is nothing else which can be considered as absolutely real' (Guénon: *L'Homme et son Devenir*, p. 35).

The Personality spoken of is Universal Man (Sufic *al-Insân al-Kâmil*; *Adam Qadmôn* of the Kabbala, and Taoist *Chün-Jên*—also *Wang*, or King-Pontiff), a term which corresponds with Prototypal Existence and designates the being which has effectively realized, not only the center of its own state (*al-Insân al-Mufrad*; *Chên-Jên*), but the integrality of all states of being, whether understood as the sum of manifested states, or transposed to mean the plenitude of total possibility as explained in the preceding chapter.[4] This amplitude and exaltation characterize the great prophets and spiritual poles of mankind.

P. State passim

Man thus has capacities exceeding the angels, in that the angels (excepting the

[1] *Al-insânu ramzu 'l-wujûd* (cf. Guénon: *Aperçus sur l'Initiation*, p. 273).

[2] The Chinese 'Steadfast Mean' (Introd. *Peace*), and 'Perpetual Standard' of the alchemists.

[3] 'All believers must become mystics (i.e. spiritually centered) sooner or later, in the next life if not in this, and Sufism might be defined as the Islamic way of anticipating the next life in this' (Abû Bakr Sirâj ad-Dîn, *Islamic Quarterly*, London, April, 1956, p. 53). 'For the contemplation of truth begins in this life but is carried on perpetually in the next' (Richard of Saint-Victor: *Ben. Maj.*, I. i: tr. Kirchberger), since 'His is the praise in the Hereafter' (*Qur'ân*, XXXIV. 1).

[4] Cf. Guénon: *Le Symbolisme de la Croix*, ch. II; Burckhardt: *Introduction aux Doctrines ésotériques de l'Islam*, p. 75 ff.

Archangels) are still peripheral, i.e. identified with a single archetype, whereas man has the possibility through his potential centrality and universality of identification with the Center of all archetypes.[1] This is why only Adam can name all things.

Sanctity as such, however, may refer to a less exalted degree; broadly defined, it is 'the uninterrupted consciousness of the Divine Presence' (Burckhardt: *Introd. Doct. Islam,* p. 89), the Beatific Vision, which is what delineates the Primordial State and which presupposes that perfect conformity of will that defines the saints of even the most humble station.

'The relationship between "intellectuality" and "spirituality" is similar to the relationship between the center and the circumference, in the sense that intellectuality transcends us, whereas spirituality englobes us. Intellectuality becomes spirituality when the entire man and not only his intelligence lives in the truth' (Schuon: *Perspectives spirituelles,* p. 103).

Knowl. 745
Introd. *Beauty*

Thus according to the perspective of Zen, 'he thinks like the showers coming down from the sky; he thinks like the waves rolling on the ocean, he thinks like the stars illuminating the nightly heavens; he thinks like the green foliage shooting forth in the relaxing spring breeze. Indeed, he is the showers, the ocean, the stars, the foliage' (D. T. Suzuki, in the *Foreword* to Herrigel's *Zen in the Art of Archery,* London, 1953). Hence, if the spiritual aspirant is nurtured on the 'traces' and 'lightnings' which filter down from the Archetypes to our plane of existence, the *jîvan-mukta* by inverse analogy, in virtue of his profound and realized Essence, is him-Self the living source of Beatitude which replenishes the well-springs of life and all existence. And herein is the true functioning of that *kîmiyâ' as-sa'âdah,* 'alchemy of felicity' (cf. Guénon: *Aperçus sur l'Initiation,* p. 271) which can resuscitate the hidden processes of Nature and help all Creation breathe.

Symb. 306

Introd. *Charity*

'If man could be limited to "being", he would be holy. . . .

'*Mâyâ* in a certain sense is the possibility in Being of not being. The All-Possibility must by definition and on pain of contradiction include its own impossibility.

'It is in order not to be that Being incarnates itself in the multitude of souls; it is in order not to be that the ocean disperses itself in myriad drops of foam.

'If the soul obtains deliverance, this is because Being is' (Schuon: *Perspectives,* pp. 130, 136).

Introds. *Void* *Creation* *Sin*

Realiz. 873

'The sage really—or "effectively"—contains in himself the entire world, so that nothing exists outside of himself. His "own" soul is not nearer to him than the world; on the contrary, it will be a sort of periphery, and this precisely on account of the law of inversion. . . . What *a priori* is affirmed as "positive"—the apparent "reality" of the sensible order and the passions which belong to it—becomes "negative" in the Truth, while what appears negative from the standpoint of sensible experience—the transcendent and thus invisible Reality, with all the spiritual consequences which it entails for man—becomes "positive" in proportion as Knowledge transforms the mental and "abstract" concept into spiritual and "concrete" Life. Or again, what is "dynamic" with the ordinary man becomes "static" with the contemplative and vice versa, so that desires are reabsorbed into immutable Beatitude, while doctrinal concepts expand into Knowledge which transforms them as it were into

Metanoia 488

[1] Here is the profound metaphysic underlying the necessity of renunciation or detachment: that exclusive identification with any possibility, however exalted, precludes the possibility, ideally and essentially man's birthright, of identification with the Absolute.

Knowl. 745

realities which are "tangible", "lived", and flowing with inspiration' (Schuon: *L'Oeil du Coeur,* pp. 204–205).

Introd. *Charity*

The mastery of the microcosm has as its corollary the right ruling of the macrocosm, and this bears on the true doctrine of kingship and the consecration of kings. Universal Man by definition is King of the World, whether or not his authority is destined to be exercised from a visible throne. 'The only royal road to power is to become one's own master; the mastery of whatever else follows. This is the traditional "secret of government", Chinese and Platonic as much as it is Indian' (Coomaraswamy: *Spiritual Authority and Temporal Power,* p. 82). This doctrine recurs, more fully developed, in the chapter *Revelation.* For the present purposes, however, sovereignty is equated with the purely interior aspects of Accomplishment, Dominion, and Incorruptibility, and this irrespective of whether the mastery is over this world alone (the state of 'true man') or over all the worlds ('transcendent man': cf. Guénon, *La Grande Triade,* pp. 128–129).

Peace 698

The distinction in degrees of spiritual endowment and capacity fades before the one great distinction marking the division between God-centeredness and world-centeredness. Strictly human perfections reveal their nothingness when compared with divine realities. There is a discontinuity and inviolable mystery concealed in the reversal of values and reorientation of soul found with the great saints and sages, which is their holy secret with God and which no 'calculation' can fathom, no 'human' endeavour pierce, or 'ambition' penetrate.[1]

Introd. *Realiz.* Introd. *Renun.*

It only remains to add that what is gained spiritually in this life is gained for eternity. 'Death in *samâdhi* changes nothing essential. Of their condition thereafter little more can be said than that they are. They are certainly not annihilated, for not only is the annihilation of anything real a metaphysical impossibility, but it is explicit that "Never have I not been, or hast thou not been or ever shall not be."[2] We are told that the perfected self becomes a ray of the Sun, and a mover-at-will up and down these worlds, assuming what shape and eating what food he will; just as in John, the saved "shall go in and out, and find pasture". These expressions are consistent with the doctrine of "distinction without difference" (*bhedâbheda*) supposedly peculiar to Hindu "theism" but presupposed by the doctrine of the single essence and dual nature and by many Vedântic texts, including those of the *Brahma Sûtra,*[3] not refuted by Śaṅkara himself. The doctrine itself corresponds exactly to what is meant by Meister Eckhart's "fused but not confused".[4]

Flight 949

Reality 775 *Center* 835 *Metanoia* 480

'How that can be we can best understand by the analogy of the relation of a ray of light to its source, which is also that of the radius of a circle to its centre. If we think of such a ray or radius as having "gone in" through the centre to an undimensioned and extra-cosmic infinity, nothing whatever can be said of it; if we think of it as at the centre, it is, but in identity with the centre and indistinguishable from it; and only when it goes "out"

Realiz. 890

[1] 'To him (Hônen) the first step toward the realization of religious aspiration was a profound conviction that the so-called meditative life is quite beyond our human faculties' (Coates and Ishizuka: *Hônen the Buddhist Saint,* Kyoto, 1949, Vol. I, pp. 43–44).

[2] *Bhagavad-Gîtâ,* II. 12.

[3] *Br. Sûtra,* II. 3. 43 f.

[4] Cf. Proclus: 'Divine natures are by their summits rooted in the one, and each of them is a unity and one, through an unconfused union with the one itself' (Taylor: *Mystical Hymns of Orpheus,* p. 80); and Dionysius: 'Even so do we see, when there are many lamps in a house, how that the lights of them all are unified into one undifferentiated light, so that there shineth forth from them one indivisible brightness' (*De Div. Nom.,* II. 4).

does it have an apparent position and identity. There is then a "descent" (*avatarana*) of the Light of Lights as a light, but not as "another" light. Such a "descent" as that of Krishna or Râma differs essentially from the fatally determined incarnations of mortal natures that have forgotten Who they are; it is, indeed, *their* need that now determines the descent, and not any lack on his part who descends. Such a "descent" is of one *che solo esso a sè piace,* and is not "seriously" involved in the forms it assumes, not by any coactive necessity, but only in "sport" (*krîdâ, lîlâ*). Our immortal Self is "like the dewdrop on the lotus leaf",[1] tangent, but not adherent' (Coomaraswamy: *Hinduism and Buddhism,* pp. 30–31).

Creation 33

[1] *Chândogya Upanishad,* IV. 14. 3.

The Command to Perfection

Be ye . . . perfect, even as your Father which is in heaven is perfect.

St Matthew, v. 48

Death 206

Is he who was dead and whom We have raised to life, and set for him a light whereby he walketh among men, as him whose likeness is in darkness whence he cannot emerge?

Qur'ân, VI. 122

For in so far as a man is illumined by piety and devotion, by knowledge of God, and worship and adoration of him, . . . he surpasses other men as much as the sun outshines the other lights of heaven.

Hermes

Our concern is not merely to be sinless but to be God.

Plotinus

Metanoia 480 *Realiz.* 887

Let this mind be in you, which was also in Christ Jesus:
Who, being in the form of God, thought it not robbery to be equal with God.

Philippians, II. 5, 6

Renun. 146

A man ought to live always in perfect holiness.

Plato (*Meno*, 81 B)

Creation 42 *Beauty* 676

For you must know that to the soul in her perfection goodness would come quite natural; she would not merely practise virtues, but virtue as a whole would be her life and she would radiate it naturally. We seem to be vicious or virtuous from being now the one and now the other. This should not be: we ought to be always in a state of perfection. That is one thing to note.

Eckhart

As he which hath called you is holy, so be ye holy in all manner of conversation:
Because it is written, Be ye holy; for I am holy.[1]

I. Peter, I. 15, 16

Man is a creature who has received the order to become God.

St Basil

[1] *Leviticus*, XI. 44—a fruitless command, were sanctification possible only upon the coming of Christ.

Not life, but a good life, is to be chiefly valued.

Plato (*Crito*, 48 B)

Illusion 109

It is the height of evil not to know God.

Hermes

Wr. Side 448

Joyless in this world is he that lives sober, and he that dies not drunk will miss the path of wisdom.
Let him weep for himself—he whose life is wasted without part or lot in wine!

Ibn al-Fârid̩

Ecstasy 637

If, after having been born a human being, one give no heed to the Holy Doctrine, one resembleth a man who returneth empty-handed from a land rich in precious gems; and this is a grievous failure.

Gampopa

Contem. 547

Praise be to God for His grace in what He has given to me. I have been granted full and perfect apprehension of the Divine Essence, as I had always most earnestly desired. That one who has not attained to knowledge of the Absolute Being is not worthy to be called a man—he belongs to the type of those of whom it is said: 'They are like the beasts of the field and are even more ignorant.' But he to whom this supreme happiness has been granted has become a perfect man and the most exalted of created beings, for his own existence has become merged in that of the Absolute Being. He has become a drop in the ocean, a mote in the rays of the sun, a part of the whole. In this state, he is raised above death and the fear of punishment, above any regard for Paradise or dread of Hell. Whether woman or man, such a one is the most perfect of human beings. This is the grace of God which He gives to whom He wills.

Fât̩imah Jahânârâ Begum S̩âh̩ib

Knowl. 761
Infra 924
Sp. Drown. 713
P. State 563
Conform. 170
Grace 552

That man alone is righteous who, having naughted all created things, stands facing straight along the unswerving line into the eternal Word, where, in *the right*, he is idealised and transformed. That man is gotten where the Son is gotten and is the Son himself.

Eckhart

Even powerful beings including *Brahmâ* and *Indra* are objects of pity to that knower of the Self who has no fear about the next world nor is afraid of death.

Śrî Śankarâchârya

Of all the created things or beings of the universe, it is the two-legged men alone who, if they purify and humiliate themselves, may become one with—or may know—*Wakan-Tanka*.

Black Elk

Supra Introd.
Humility 191

Seeing all these things, what ought not we to do that we may obtain virtue and wisdom in this life? Fair is the prize, and the hope great!

Plato (*Phaedo*, 114 C)

Incorruptibility

As the ocean remains calm and unaltered though the waters flow into it, similarly a self-controlled saint remains unmoved when desires enter into him; such a saint alone attains peace, but not he who craves the objects of desire.

Bhagavad-Gîtâ, II. 70

Metanoia 484

In the loving introversion of the just man all venial sins are like to drops of water in a glowing furnace.

Ruysbroeck

Renun. 152b

Whosoever is born of God doth not commit sin; for his seed remaineth in him: and he cannot sin, because he is born of God.

I. John, III. 9

M.M. 978 *Creation* 42

The utmost a spirit can attain to in this body is to dwell in a condition beyond the necessity of virtues; where goodness as a whole comes natural to it so that not only is it possessed of virtues but virtue is part and parcel of it: it is virtuous not of necessity but of innate good nature. Arrived at this the soul has traversed and transcended all necessity for virtues: they are now intrinsic in her.

Eckhart

It is impossible that that which is divine should go astray.

Hermes

When that which is perfect is come, then that which is in part shall be done away.

I. Corinthians, XIII. 10

When the (metallic) substance attains the form of gold, it departs from the hold of the physician (artist), since it departs from its state of sickness. Once perfection has been attained, a relapse into imperfection is henceforth out of the question, even should the physician try to provoke it.

Ibn ʿArabî

Such a man behaves like a tortoise, which, once it has tucked in its limbs, never puts them out. You cannot make the tortoise put its limbs out again, though you chop it to pieces with an axe.[1]

Sri Ramakrishna

Our Arcanum, being both a spiritual and a homogenious substance, is capable of entering into a perfect atomic union with the imperfect metals, of taking up into its own nature that which is like to it, and of imparting to this Mercury its own fixity, and protecting it from the fire; so when the fire has burnt up all the impurities, that which is left is, of course, pure gold or silver, according to the quality of the Medicine—which from that time forward is (like all other gold and silver) capable of resisting the most searching ordeal.

Philalethes

Suff. 124 *P. State* 563

To the man of realization, who has fully known my being, it is indifferent whether the senses—made up of the gunas—are indrawn or turned without. What matters it to the sun whether the clouds gather together or are dispersed?

Srimad Bhagavatam, XI. XX

Turn yourself into gold and then live wherever you please.

Sri Ramakrishna

Renun. 139

For I am persuaded, that neither death, nor life, nor angels, nor principalities, nor powers, nor things present, nor things to come,

Nor height, nor depth, nor any other creature, shall be able to separate us from the love of God, which is in Christ Jesus our Lord.

Romans, VIII. 38–39

Conform. 170

. . . Tinctures when they by craft are made parfite,
So dieth (dyeth) Metalls with Colours evermore permanent,
After the qualitie of the Medycine Red or White;
That never away by eny Fire, will be brente.

George Ripley

M.M. 998

If you keep the juice of a lemon or a tamarind in a gold cup, this juice does not spoil. In a cup of copper or brass on the contrary, it goes bad at once and becomes noxious. In the same way, if sensual thoughts (vishaya vrittis) enter the pure mind of someone who practices a constant meditation, they cause no pollution nor any passional excitement. But if they come to persons of impure mind, they cause excitement when they fasten on sense objects.

Swami Sivananda

Never, my son, can a soul that has so far uplifted itself as to grasp the truly good and real slip back to the evil and unreal.

Hermes

[1] The tortoise in Oriental symbolism often represents the Primordial State.

The man, whoe'er he be, moved by these three—truth, righteousness and goodness—can no more quit these three than God can quit his Godhood.

Eckhart

Action 329

If you commit a sin, you must bear its fruit. But one who has attained perfection, realized God, cannot commit sin. An expert singer cannot sing a false note. A man with a trained voice sings the notes correctly: sâ, re, gâ, mâ, pâ, dhâ, ni.

Sri Ramakrishna

All the properties of the inner and holy body, including the outward qualities, were perfectly harmonized (in the first man): none lived in a self-desire but all reabsorbed their desire into the soul wherein the divine light was manifested and which was like the sacred sky. The light shone across all the properties and ruled equally over all the properties; all the properties reabsorbed their desire into the light, a revealed sweetness of God which interpenetrated all the properties. And through this interpenetration they were all tinctured with a delicious love, in such way that they had in themselves only this savour and desire of love.

Realiz. 868

Boehme

What time the mind is fixed on God and there abides, the senses are obedient to the mind. As one should hang a needle to a magnet and then another needle on to that, until there are four needles, say, depending from the magnet. As long as the first needle stays clinging to the magnet all the other needles will keep clinging on to that but when the leader drops the rest will go as well. So, while the mind keeps fixed on God the senses are subservient to it but if the mind should wander off from God the passions will escape and be unruly.

Rev. 967

Eckhart

Infra 921

Orthod. 275

Beauty 689

The *Throne* of the most *High* God, is a Throne of *Grace,* of *Love.* Like a *Chain* doth the whole Nature of things descend from this Throne, having its top fastned to it. What-ever the weights may be of the lowermost links of this Chain, yet *that Love* which sits upon the Throne, with a Divine delight as it lets down the Chain from it self, so draws it up again by the Order of the successive Links unto a Divine Ornament, an eternal Joy and Glory to it self. All things of *Nature* in its *Beauties,* all things of *Nature* in its *Ruine, Life* and *Death* in all forms, are a *Saints,* a *Saint* is *Christs, Christ* is *Gods.*

Peter Sterry

The soul has now no further awareness of the body and will give herself no foreign name, not man, not living being, nor anything at all; any observation of such things is beside the mark; the soul has neither time nor taste for them: This she sought and This she has found and on This she looks and not upon herself: and who she is that looks she has not leisure to know.

Death 220

Contem. 536

Once There she will barter for This nothing the universe holds—no not the heavens entire: than This there is nothing higher, nothing more blessed; above This there is no passing; all the rest however lofty lies on the downward path; she knows that This was the object of her quest, that nothing higher is. Here can be no deceit: where could she come upon truer than the truth? And the truth she affirms, that she is herself. In this happiness she knows beyond delusion that she is happy: for this is no affirmation of an excited body but of a soul become again what she was in the time of her early joy.

Plotinus

Knowl. 749 *Creation* 42

Many waters cannot quench love, neither can the floods drown it: if a man would give all the substance of his house for love, it would utterly be contemned.

Song of Solomon, VIII. 7

Center 847

He who has apprehended the beauty of the Good can apprehend nothing else: he who has seen it can see nothing else; he cannot hear speech about aught else; he cannot move his body at all; he forgets all bodily sensations and all bodily movements, and is still. But the beauty of the Good bathes his mind in light, and takes all his soul up to itself, and draws it forth from the body, and changes the whole man into eternal substance.

Hermes

P. State 563

Such at that light doth man become that to turn thence to any other sight could not by possibility be ever yielded.

For the good, which is the object of the will, is therein wholly gathered, and outside it that same thing is defective which therein is perfect.

Dante (*Paradiso,* XXXIII. 100–105)

Void 724

One who has attained the Tao is master of himself, and the universe is dissolved for him. Throw him in the company of the noisy and the dirty, and he will be like a lotus flower growing from muddy water, touched by it, yet unstained.

T'u Lung

Infra 921 914

It is related in the stories that John (the Baptist) and Jesus (Peace be upon them!) were walking in the market when a woman knocked against them. Then John said, 'I am not cognisant of that.' Jesus said, 'Praise be to God! Your body is with me, but where is your heart?' He replied, 'O cousin, if my heart found rest in something other than God for the twinkling of an eye, I should think that I had not known God.'

Christ in Islâm

Renun. 152b

It is a fair question whether our first parent, or parents (for they were two in marriage), had those natural affections ere they sinned, which we shall be free from when we are perfectly purified. If they had them, how had they that memorable bliss of paradise? Who can be directly happy that either fears or sorrows? And how could they either fear or grieve in that copious affluence of bliss, where they were out of the danger of death and sickness, having all things that a good will desired, and lacking all things that might give their happiness just cause of offence? Their love to God was unmoved, their union sincere, and thereupon exceeding delightful, having power to enjoy in full what they loved. They were in a peaceable avoidance of sin, which tranquillity kept out all external annoyance. Did they desire (do you think?) to taste the forbidden fruit, and yet feared to die? God

Center 847

forbid we should think this to be where there was no sin, for it were a sin to desire to break God's command, and to forbear it rather for fear of punishment than love of righteousness.

St Augustine

Peace 700 *Creation* 42

After the mind ceases to exist and bliss of peace has been realised, one will find it then as difficult to bring out a thought, as he now finds it difficult to keep out all thoughts. . . . The bliss of peace is too good to be disturbed. . . . The thought-free state is one's primal state and full of bliss. Is it not miserable to leave such a state for the thought-ridden and unhappy one?

Sri Ramana Maharshi

Whoso has burned in the mighty Fire of Love could never bear to cool himself with any kind of sin.

Mechthild of Magdeburg

Peace 694

You will even find it painful to rest anywhere else but there (on the summit of the holy mountain).

Richard of Saint-Victor

For in the Kingdom of Glory it is impossible to fall. No man can sin that clearly seeth the beauty of God's face: because no man can sin against his own happiness, that is, none can when he sees it clearly, willingly, and wittingly forsake it, tempter, temptation, loss, and danger being all seen.

Thomas Traherne

To the soul that has gotten and enjoyed divine perfection all that is not God has a bitter, nauseous savour.

Eckhart

Whoso is thus constrained by the deepest stirrings of a mighty love, can in no wise commit mortal sin. For when the soul is bound, it must ever love. May God bind us all in this way!

Mechthild of Magdeburg

A good man, one that is actuated by religion, lives in converse with his own reason; he lives at the height of his own being.

John Smith the Platonist

Those who have once begun the heavenward pilgrimage may not go down again to darkness and the journey beneath the earth, but they live in light always.[1]

Plato (*Phaedrus*, 256 E)

[1] This can refer both to those who have reached the Primordial State in this life, and to those who are in the way of salvation after death.

If one is once born into the Land of Bliss, he will never return to this world, but every such one will attain Buddhahood.

Hônen

When the mind, completely subdued, rests in Self alone, free from longing for all objects of desire, then he (the Sannyâsi, the Yogi) is said to be a Yukta (steadfast in Self-knowledge).

As a lamp placed in a windless spot does not flicker, the same simile is used to define a Yogi of subdued mind, practising union with the Self.

Contem. 532

In that state, when the mind is completely subdued by the practice of Yoga and has attained serenity, in that state, seeing Self by the self, he is satisfied in the Self alone.

In that state, transcending the senses, he feels that infinite bliss which is perceived by the purified understanding: knowing that and being established therein, he never falls back from his real state:

After having attained which, no other gain seems greater: being established wherein, he is not overwhelmed even by great sorrow.

Bhagavad-Gitâ, VI. 18–22

There is nothing that God lacks, so that he should desire to gain it, and should thereby become evil.

Infra 924

Hermes

How happy is he who hath reached the West when he is safe in the hand of God.

Egyptian Tradition

The soul's perfection consists in liberation from the life which is in part and admission to the life which is whole. All that is scattered in nether things is gathered together when the soul climbs up into the life where there are no opposites. The soul knows no opposition when she enters the light of intellect.

Center 835

Eckhart

Evil does not overcome him: he overcomes all evil. Evil does not burn him: he burns all evil.

P. State 563

Bṛihad-Âraṇyaka Upanishad, IV. iv. 23

Ah! God-loving soul! In thy struggles
Thou art armed with measureless might,
And with so great a power of soul
That all the peoples of the world,
All the charm of thine own body,
All the legions of the devil,
All the powers of Hell—
Cannot separate thee from God.

Mechthild of Magdeburg

Reality 773

An essential man is like eternity,
Which remains unchanged by all exteriority.

Angelus Silesius

Knowl. 734
Realiz. 887

We must distinguish, however, between our transmutative conjunction, and a sort of conjunction practised by sophists which is merely a fusing together of the two substances, and leaves each exactly what it was before. In our operation the spirit of gold infuses itself into the spirit of Mercury, and their union becomes as inseparable as that of water mixed with water.

Philalethes

Conform. 170

When He gives thee obedience, and independence in Him from it, then know that He has in sooth covered thee with His Bounties, both outward and inward.

Ibn ʿAṭâ'illâh

Infra 934

I need not say that when the soul is once got up to the top of this bright Olympus, it will then no more doubt of its own immortality, or fear any dissipation, or doubt whether any drowsy sleep shall hereafter seize upon it: no, it will then feel itself grasping fast and safely its own immortality, and view itself in the horizon of eternity.

John Smith the Platonist

Center 833
P. State 590

Thy sun shall no more go down; neither shall thy moon withdraw itself: for the Lord shall be thine everlasting light, and the days of thy mourning shall be ended.

Isaiah, LX. 20

To get established in God, one has to go beyond all the Gunas.[1] From that state there is no fall.

Swami Ramdas

. . . That Pure Land, from whose pure blessedness we shall never fall.

Hônen

P. State 583

The yogin who practices this yoga, surrendering himself unto me and having no other desire but me, is thwarted by nothing. His is a bliss that fades not away.

Srimad Bhagavatam, XI. XX

Reality 775
Center 816

A man is never born again who knows that he is the One Existence in all beings like the ether and that all beings are in him.

Śrî Śankarâchârya

Knowl. 749

'Whoso knows Me through Myself, knows Me with a gnosis that will never thereafter be denied.'

Niffarî

[1] Cf. note p. 88.

There are two kinds of sainthood. The first is merely a departure from enmity, and in this sense is general to all believers: it is not necessary that the individual should be aware of it, or realise it, for it is only to be regarded in a general sense, as in the phrase, 'The believer is the friend (*walî*) of God.' The second is a sainthood of peculiar election and choice, and this it is necessary for a man to be aware of and to realise. When a man possesses this, he is preserved from regarding himself, and therefore he does not fall into conceit; he is withdrawn from other men, that is, in the sense of taking pleasure in regarding them, and therefore they do not tempt him. He is saved from the faults inherent in human nature, although the stamp of humanity remains and persists in him: therefore he does not take delight in any of the pleasures of the soul, in such a way as to be tempted in his religion, although natural delights do persist in him. These are the special qualities of God's friendship (*wilâyah*) towards man: and if a man has these qualities, the Enemy will have no means of reaching him, to lead him astray: for God says, 'Verily, as for My servants, thou hast no authority over them' (*Qur'ân*, XV. 42).

Metanoia 484
Infra 914
Death 220

Al-Kalâbâdhi

'He that never suffereth corruption' is my name.

Papyrus of Nu

Those who obtain ultimate happiness from the divine vision never fall away from it. Because whatever at one time is, and at another time is not, is measured by time. . . . Now the vision in question . . . is not in time but in eternity. Therefore no one, having once become partaker thereof, can lose it.

Center 838

St Thomas Aquinas

The intellect of the wise man is always with divinity.

Sextus the Pythagorean

In this fountain (of life) shines a wonderful carbuncle. . . . It lights up these regions more magnificently than any worldly sun could do. It puts the night to exile and makes the day eternal, without end and without beginning; it is a sun which stays in the same place without passing through the signs of the year, or through the hours which measure the day: it is a sun which has neither noon nor midnight. The carbuncle possesses such a marvelous power that those who approach it and look at their faces in the water see all the things in the park, whichever way they turn, and know them truly, and also themselves. And once they have seen themselves therein, they will never be the victims of any illusion, so clear-sighted and wise do they return from there.

Center 833
828
Realiz. 859

John A. Mehung

Granthi (knot=bondage), snapped once, is snapped for ever.

Sri Ramana Maharshi

Inscrutability

He who has come to know God is filled with all things good; his thoughts are divine, and are not like those of the many.

Hence it is that those who have attained to the knowledge of God are not pleasing to the many, nor the many to them. They are thought mad, and are laughed at; they are hated and despised, and perhaps they may even be put to death. For evil . . . must needs dwell here on earth, where it is at home; for the home of evil is the earth, and not the whole universe, as some will blasphemously say in days to come. But the pious man will endure all things, cleaving to his knowledge of God. For to such a man all things are good, even though they be evil to others. When men devise mischief against him, he sees all this in the light of his knowledge of God; and he, and none but he, changes things evil into good.

Heterod. 420 *Beauty* 689 *Charity* 602

Hermes

When a man has reached this point we may well say, this man is God and man. All Christ has by nature he has won by grace. His body is filled with the noble nature of the soul, which she receives from God, with divine light, wherefore we may truly cry, Behold, a man divine! Pity them, my children, they are from home and no one knows them. Let those in quest of God be careful lest appearances deceive them in these people who are peculiar and hard to place; no one rightly knows them but those in whom the same light shines.[1] Namely, the light of truth. Yet it may well be that wayfarers to that same good, but who have not yet reached it, will come across these perfect of whom we have been speaking. Believe me, did I know one such, and had a convent-full of gold and precious stones, I would give the whole of it for a single fowl for him to eat. Further I declare, if all the things God ever made were mine, I would forwith give them all for the enjoyment of that man, and rightly, for they are all his. Nay, more I say: his, too, is God in the fullness of his power, and if there stood before me all who in imperfection are anhungered, I would not withhold from that man's need a single feather of the fowl, though I might feed that multitude. For, you must remember, with one in imperfection, anything he eats or drinks will drag him down and make him prone to sin. But not the virtuous man: what he eats and drinks he raises up in Christ to the Father. So look well to yourselves.

Suff. 126b *Knowl.* 749 *Orthod.* 288

. . . I warn you, you must keep a sharp look out, for they are difficult to tell; thus if they should need it, while other people fast they will be eating, while other people watch they will be sleeping, while other folks are praying they will hold their peace. In short, the things they say and do seem unaccountable, for what God makes obvious to persons on the way to their eternal happiness is foreign to those that have arrived there. These have no wants whatever: they are rich in the possession of a city of their own. I call that my own which is

Introd. *Orthod.* Introd. *Illusion*

[1] This observation has its alchemical analogy: 'Gold, which represents the true equilibrium of metallic qualities, lacks nevertheless the particular utility of this or that base metal, just as the spiritual man, who synthesizes the human virtues and who therefore is incorruptible like gold, may seem wanting in social virtues' (Burckhardt, 'Considérations sur l'Alchimie', *Études Trad.*, 1948, p. 298). Cf. *St Luke*, XVI. 8: 'The children of this world arc in their generation wiser than the children of light.'

mine eternally and no one can take from me. These people, you must know, do most valuable work. They work within, you understand, in the man of the soul. Blessed is the kingdom wherein dwells one of them; in an instant they will do more lasting good than all the outward actions ever done. See ye withhold not aught of theirs. May we recognise these people and loving God in them, with them possess the city they have won.

Eckhart

Supra 902 *Action* 346 *Center* 832

(Mary Magdalene) saw Jesus standing, and knew not that it was Jesus.

Jesus saith unto her, Woman, why weepest thou? whom seekest thou? She, supposing him to be the gardener, saith unto him, Sir, if thou have borne him hence, tell me where thou hast laid him, and I will take him away.

St John, xx. 14, 15

Praised be He Who has veiled the secret of the elect in human appearance, and Who has achieved the grandeur of sovereignty in the guise of servitude!

Ibn 'Aṭâ'illâh

Infra 924

The outward form, brethren, of him who has won the truth, stands before you, but that which binds it to rebirth is cut in twain.

Brahma-Gâla Sutta

Death 206 Supra 902

Although a jnâni living in the world may have a little blemish, yet this does not injure him. The moon undoubtedly has dark spots, but these do not obstruct its light.

Sri Ramakrishna

P. State 590

Especially well-versed is he (Mingliaotse) in the teaching of Taoism regarding 'nourishing the spirit'. Sometimes, when he is watching dancing or hearing singing which borders on the bawdy and people make ribald jokes to find out his attitude toward these things, he seems to be enjoying himself, like the romantic scholars. But when it comes to extinguishing the candle and the host asks him to stay with some girl entertainer, and when the party becomes really rowdy, he sits upright with an austere appearance, and nobody can make anything of him.

T'u Lung

His (the philosopher's) awkwardness is fearful, and gives the impression of imbecility. When he is reviled, he has nothing personal to say in answer to the civilities of his adversaries, for he knows no scandals of any one, and they do not interest him; and therefore he is laughed at for his sheepishness; and when others are being praised and glorified, in the simplicity of his heart he cannot help going into fits of laughter, so that he seems to be a downright idiot.

Plato (*Theaetetus*, 174 C)

Ecstasy 641

The God-realized man is God Himself in human form. Ignorant souls cannot recognize divinity in the man in whom the Universal Spirit is realized and attained.

Swami Ramdas

Illusion 85

Ornaments cannot be made of pure gold. Some alloy must be mixed with it. A man totally devoid of Maya will not survive more than twenty-one days. So long as the man has a body, he must have some Maya, however small it may be, to carry on the functions of the body.

Sri Ramakrishna

Supra 902 *Metanoia* 488

The soul of the lover will never forsake his beautiful one, whom he esteems above all: he has forgotten mother and brethren and companions, and he thinks nothing of the neglect and loss of his property: the rules and proprieties of life, on which he formerly prided himself, he now despises.

Plato (*Phaedrus*, 252 A)

'But Socrates rarely washed.'

Why, his body was clean and bright, nay, it was so gracious and agreeable that the handsomest and noblest were in love with him, and desired to recline by him rather than by those who were perfect in beauty. He might have never washed or bathed, if he had liked: I tell you his ablutions, if rare, were powerful.

Epictetus

Introd. *Creation* *P. State* 573 *Death* 215

The outward form is only a guide: God must become man, or else man becomes not God.

Therefore a Christian is the most simple (or plainest) man upon the earth, as Isaiah says, 'Who is so simple as my servant?' . . . He must quite die to self-hood, that the same may only hang to him from without as a garment of this world, wherein he is a stranger and pilgrim.

Boehme

Rev. 961

I am a common man, who only speak the truth.

Plato (*Ion*, 532 D)

Look not at the outward poverty of Ḥâfiẓ, for his inner self is the treasure-house of the Divine love.

Ḥâfiẓ

P. State 573 Introd. *Inv.*

One should not be deceived by the external appearance of a *jnâni*. . . . Although a *jivanmukta* (one having complete realization in this life) associated with body may, owing to his *prârabdha* (past karma), appear to lapse into ignorance or wisdom, yet he is only pure like the ether (*âkâsa*) which is always itself clear, whether covered by dense clouds or cleared of clouds by currents of air. . . . Though he remains silent like one devoid of learning, yet his suppineness is due to the implicit duality of the *vaikhari vâk* (spoken words) of the Vedas: his silence is the highest expression of the realised non-duality which is after all the true content of the Vedas. Though he instructs his disciples, yet he does not pose as a teacher in the full conviction that the teacher and disciple are mere conventions born of illusion (*mâyâ*): . . . if on the other hand he mutters words incoherently like a lunatic, it is because his experience is inexpressible like the words of lovers in embrace. If his words are many and fluent like those of an orator, they represent the recollection of his experience, since he is the unmoving non-dual One without any desire awaiting fulfilment. Although he may appear grief-stricken like any other man in bereavement, yet he evinces just the right love of and pity for the senses which he earlier controlled before he realised that they were mere instruments and manifestations of the Supreme Being. When he seems keenly interested in the wonders of the world, he is only ridiculing the ignorance born of superimposition. If he appears indulging in sexual pleasures, he must be taken to enjoy the ever-inherent Bliss of the Self, which divided Itself into the individual self and the Universal Self (and) delights in their reunion to regain Its original Nature. If he appears wrathful he means well to the offenders. All his actions should be taken to be only divine manifestations on the plane of humanity. There should not arise even the least doubt as to his being emancipated while yet alive. He lives only for the good of the world.

Illusion 85
Love 625
Wr. Side 474

Sri Ramana Maharshi

He who knows the one reality, beyond the objective world, has true knowledge. He loves me for the sake of love and does not care even for his own salvation. Such a free soul is above all rules of conduct and beyond all orders of life. Though wise, he is childlike. Though subtle, learned, and well-versed in the Scriptures, he wanders about as one who knows nothing. He causes no fear to anyone, and he is fearful of none. If vilified, he does not return the insult but remains calm. He bears enmity towards none.

Contem. 536
Action 346

Srimad Bhagavatam, XI. xi

These exalted and most noble men are just like the wood of the vine which is outwardly hard and black and dry, and good for no purpose whatever; and if we had never seen it before, we should think it of no use at all, and good for nothing but to be thrown into the fire, and burned. But in this dry wood of the vine, there lie concealed the living veins of sap, and power of yielding the noblest of all juices, and of bringing forth a greater abundance of fruit than any other sort of wood that grows. And thus it is with these beloved and lowly children, who are at all times and seasons plunged in God; they are outwardly in appearance like unto black rotten wood, seeming unto men dry and unprofitable. For there are many of these who are humble, noways remarkable for their gifts, outward or inward, nor for any extraordinary works or sayings or exercises of devotion, and who move in the narrowest sphere; but living veins from the fountain of truth lie hidden within them, forasmuch as they have asked for no earthly heritage, but God is their lot and their portion, their life and their being.

Humility 199
Conform. 180
Introd.
Renun.

Tauler

Attributes of Holiness

Knowl. 749

Beware of the discernment of the true believer, for he sees by the light of God.

Muhammad

Infra 924 *Charity* 608 *M.M.* 987

Devout men are the title and inscription of the Book of the Law. They are the demonstration of all the truths, the solution of all the mysteries. Their outward life bears us to the path of obedience; their hidden life wins us to self-denial. They began their career before the ages, and they work for eternity. They have effaced from their hearts and minds every trace of pride and hypocrisy. They have trodden the ways of God, even when they appeared powerless to move, so feeble were they. These are the friends of God: they have discovered divine secrets: they guard them with devout silence.

Hugh of Saint-Victor

Realiz. 859 *Judgment* 242 Supra 902 *Rev.* 965 *Center* 841

What are the eighteen special dharmas of a Buddha? From the night when the Tathagata knew full enlightenment, to the day when he becomes extinct in Nirvana, during all this time, the Tathagata 1. Does not stumble, 2. He is not rash or noisy in his speech, 3. He is never deprived of his mindfulness, 4. He has no perception of difference, 5. His thought is never unconcentrated, 6. His evenmindedness is not due to lack of consideration, 7. His zeal never fails, 8. His vigour never fails, 9. His mindfulness never fails, 10. His concentration never fails, 11. His wisdom never fails, 12. His deliverance never fails, 13. All the deeds of his body are preceded by cognition, and continue to conform to cognition, 14. All the deeds of his voice are preceded by cognition, and continue to conform to cognition, 15. All the deeds of his mind are preceded by cognition, and continue to conform to cognition, 16. His cognition and vision regarding the past period of time proceeds unobstructed and freely, 17. His cognition and vision regarding the future period of time proceeds unobstructed and freely, 18. His cognition and vision regarding the present period of time proceeds unobstructed and freely.

Śatasâhasrikâ prajñâramitâ IX. 1449–50

Renun. 136 *Conform.* 166

This man is known by five signs. First, he never complains. Next, he never makes excuses: when accused, he leaves the facts to vindicate him. Thirdly, there is nothing he wants in earth or heaven but what God wills himself. Fourthly, he is not moved in time. Fifthly, he is never rejoiced: he is joy itself.

Eckhart

The saints of God are known by three signs: their thought is of God, their dwelling is with God, and their business is in God. . . . If the gnostic (*'ârif*) has no bliss, he himself is in every bliss.

'Aṭṭâr

The sage keeps himself behind and he is in the front;
He forgets himself and he is preserved.

Tao Te Ching, VII

Humility 199

I am crucified with Christ: nevertheless I live: yet not I, but Christ liveth in me.

Galatians, II. 20

Death 220

This is the sign of the gnostic, that his thoughts are mostly engaged in meditation, and his words are mostly praise and glorification of God, and his deeds are mostly devotion, and his eye is mostly fixed on the subtleties of Divine action and power.

Ibrâhîm b. Adham

A man is firmly established in spiritual life when he goes into samâdhi on uttering 'Om' only once.

Sri Ramakrishna

Inv. passim

It must be realized that the outward man is able to be active and leave the inward man entirely passive and unmoved. Now in Christ too existed an outward and an inward man and also in our Lady, and what Christ and our Lady said concerning outward things was prompted by their outward man, the inner man remaining in motionless detachment. So was it when Christ said, 'My soul is sorrowful unto death.' And despite her lamentations and various things she said, Our Lady, in her inner man, stood all the while in motionless detachment. Take an illustration. The door goes to and fro upon its hinges. Now the projecting door I liken to the outward man and the hinge I liken to the inner man. As it shuts and opens the door swings to and fro while the hinge remains unmoved in the same place without undergoing any change. And likewise here.

Eckhart

P. State 579

Reality 773

We must remember that the Proficient sees things very differently from the average man; neither ordinary experiences nor pains and sorrows, whether touching himself or others, pierce to the inner hold. To allow them any such passage would be a weakness in our soul.

Plotinus

Death 226

For know that the joys of these shadows are not My joys, nor the sorrows of these phantoms My sorrows.

From the Prologue of a Wayang Purwa

Joy and grief may now and then overcast the countenance of the *jnâni*, as a cloud hides the face of the sky, but they cannot over-shadow his soul, which is bright as eternal day.

Yoga-Vasishtha

Center 833

Believe me, if my rank and station were not what they are, I should enjoy nothing so much as a solitary life, or to have joined Diogenes in his tub. For I behold this world full of vanity, greed, cruelty, venality, and iniquity; and I rejoice in the prospect of the glorious life

Contem. 523

to come. I no longer wonder, as once I did, that the true Sage, though he owns the Stone, does not care to prolong his life: for he daily sees heaven before his eyes, as you see your face in a glass. When God gives you what you desire, you will believe me, and not make yourself known to the world.

Realiz. 870

Michael Sendivogius

For he, Adeimantus, whose mind is fixed upon true being, has surely no time to look down upon the affairs of earth, or to be filled with malice and envy, contending against men; his eye is ever directed towards things fixed and immutable, which he sees neither injuring nor injured by one another, but all in order moving according to reason; these he imitates, and to these he will, as far as he can, conform himself.

Center 835 *Beauty* 689

Plato (*Republic* VI, 500 C)

The entreaty of gnostics with Allâh—may He be exalted!—is for perfection in slavery and rectitude in the obligations of sovereignty.

Introd. *Conform.* Infra 921

Ibn 'Aṭâ'illâh

Enjoying the world of sense, one is undefiled by the world of sense.
One plucks the lotus without touching the water.
So the yogin who has gone to the root of things,
Is not enslaved by the senses although he enjoys them.

Illusion 109

Saraha

The countenance of wisdom is not scowling and severe, contracted by deep thought and depression of spirit, but on the contrary cheerful and tranquil, full of joy and gladness, feelings which often prompt a man to be sportive and jocular in a perfectly refined way. Such sportiveness is in harmony with a dignified self-respect, a harmony like that of a lyre tuned to give forth a single melody by a blending of answering notes.

Heterod. 423 *Creation* 33

Philo

It is told that the most blessed Evangelist John, when he was gently stroking a partridge with his hands, suddenly saw one in the habit of a hunter coming to him. He wondered that a man of such repute and fame should demean himself to such small and humble amusements, and said: Art thou that John whose eminent and widespread fame hath enticed me also with great desire to know thee? Why then art thou taken up with such mean amusements? The blessed John said to him: What is that which thou carriest in thy hands? A bow, said he. And why, said he, dost thou not bear it about always stretched? He answered him: I must not, lest by constant bending the strength of its vigour be wrung and grow soft and perish, and when there is need that the arrows be shot with much strength at some beast, the strength being lost by excess of continual tension, a forcible blow cannot be dealt. Just so, said the blessed John, let not this little and brief relaxation of my mind offend thee, young man, for unless it doth sometimes ease and relax by some remission the force of its tension, it will grow slack through unbroken rigour and will not be able to obey the power of the spirit.

Supra 910

Cassianus Eremita

I saw Indian Brahmans living upon the earth and yet not on it, and fortified without fortifications, and possessing nothing, yet having the riches of all men.

Apollonius of Tyana

Flight 949
Void 724

The men of God are priests and prophets who have refused to accept membership in the commonwealth of the world and to become citizens therein, but have risen wholly above the sphere of sense-perception and have been translated into the world of the intelligible and dwell there registered as freemen of the commonwealth of Ideas, which are imperishable and incorporeal.

Philo

Suff. 126b
Introd. *Reality*

The day of a Saint is the Light shining from the Face of God.

Peter Sterry

Center 833
Infra 935

The wise man is always similar to himself.

Sextus the Pythagorean

Supra 902

Just as a Brahmin actor does not forget his being a Brahmin, whatever part he may be acting, so also a man should not confound himself with his body, but he should have a firm awareness of his being the Self, whatever his activity may be.

Sri Ramana Maharshi

Metanoia 480
Action 340

Our great ancestors, that is, the Companions and the Successors and the generation which followed them, ever held this path (of sanctification) to be the path of truth and right guidance. It is based on unfailing perseverance in worship, utter devotion to All-Highest God, turning away from the adornments of this world, renunciation of what most men seek after in the way of pleasure and dignity, and isolating oneself from all mankind in spiritual retreat (*khalwah*) for the sake of worship. Now these were the general practices of the Companions and the Moslems of old (*as-salaf*). Then in the second generation and afterwards, when worldliness spread and men tended to become more and more bound up with the ties of this life, those who dedicated themselves to the worship of God were distinguished from the rest by the title *aṣ-Ṣûfiyyah* (Sufis) and *al-Mutaṣawwifah* (those who aspire to be Sufis).

Ibn Khaldûn

Beauty 676
M.M. 973

O my son, one who is intimate with God has received four gifts: an honour which need be known to none, a knowledge without study, a richness without money, a joyful company without companion.

Dhu 'l-Nûn (addressed by a hermit)

(A Ṣûfî) was asked: 'Who is a Ṣûfî?' He replied: 'He who neither possesses nor is possessed.'

Al-Kalâbâdhî

Introd. *Conform.*

The man who possesses a knowledge of God, will not be very ambitious.

Sextus the Pythagorean

Judgment 242

Be ye therefore wise as serpents, and harmless as doves.

St Matthew, x. 16

Introd. *Orthod.*

'Sanctity is the forgetting of thyself in everything: in thoughts, in desires, in words.'

Sister Consolata

Beauty 663

After attaining God one forgets His external splendour, the glories of His creation. One doesn't think of God's glories after one has seen Him.

Sri Ramakrishna

The Lord Buddha . . . inquired of Subhuti, saying: 'What think you? May an Arhat (having attained to absolute quiescence of mind) thus meditate within himself, "I have obtained the condition of an Arhat"?' Subhuti replied, saying: 'No! honoured of the worlds! And why? Because there is not in reality a condition synonymous with the term Arhat. Honoured of the worlds! If an Arhat thus meditates within himself, "I hâve obtained the condition of an Arhat," there would be obvious recurrence of such arbitrary concepts as an entity, a being, a living being, and a personality.'[1]

Death 220

Prajñâ-Pâramitâ

The gods and the sages experience the Infinite continuously and eternally, without their vision being obscured at any moment. Their minds are surmised by the spectators to function; but in fact they do not. Such surmise is due to the sense of individuality in those who draw inferences. There is no *mental* function in the absence of individuality. Individuality and mind functions are co-existent. The one cannot remain without the other.

Peace 700

Sri Ramana Maharshi

Since a Tathagata, even when actually present, is incomprehensible, it is inept to say of him—of the Uttermost Person, the Supernal Person, the Attainer of the Supernal—that after dying the Tathagata is, or is not, or both is and is not, or neither is nor is not.

Saṁyutta-nikâya, III. 118

O people, it has reached me that you are afraid of your Prophet's death. Has any previous prophet lived for ever among those to whom he was sent; so that I would live for ever among you?

Muhammad

Center 838

The saint hath no fear, because fear is the expectation either of some future calamity or of the eventual loss of some object of desire, whereas the saint is the son of his time (*ibn waqtihi*): he has no future that he should fear anything: and as he hath no fear so he hath

[1] Evidently, the enlightened contemplative is not deceived regarding his true state, but the above citation is a way of affirming that the absolute Subject knows no objectification, or—from a different perspective—that the absolute Object knows no subjectification. Cf. note 2, p. 932

no hope,[1] since hope is the expectation either of gaining an object of desire or of being relieved from a misfortune, and this belongs to the future; nor does he grieve, because grief arises from the rigour of time, and how should he feel grief who is in the radiance of satisfaction (*riḍâ*) and the garden of concord (*muwâfaqat*)?

Junayd — *Center* 847

One who has realized Brahman sees Brahman everywhere and in all.

Srimad Bhagavatam, XI. xxi — *Renun.* 139

Caste distinctions drop away when one attains Self-Knowledge.

Sri Ramakrishna

The Gnostic hath not Gnosis if he know not God from every standpoint and in whatever direction he turneth. The Gnostic hath only one direction, and that is towards the Truth Itself. ***Whithersoever ye turn, there is the Face of Allâh (Qur'ân,*** II. 115).

Shaykh Ahmad al-'Alawî

'The true man in breathing, breathes through his heels, the common man in breathing breathes through his throat.'

Hakuin (citing the Hermit Haku-yu) — *Metanoia* 488

You should know, that so far as Buddha-nature is concerned, there is no difference between an enlightened man and an ignorant one. What makes the difference, is that the one realizes it, while the other is kept in ignorance of it.

Hui-nêng — *Center* 841

After reaching God one reaffirms what formerly one denied. To extract butter you must separate it from the buttermilk. Then you discover that butter and buttermilk are intrinsically related to one another. They belong to the same stuff.

Sri Ramakrishna — *Renun.* 158; *Realiz.* 873

He who sees by illumination
Discerns God first in everything.

Shabistarî

When that Sun shines upon him, the dust-bin of this world is changed for him into a rose-garden: the kernel is seen beneath the rind. No longer does the lover see any particle of himself, he sees only the Beloved: wheresoever he looks, he sees always His Face.

'Aṭṭâr — *Center* 833; *P. State* 583

St Augustine is compared to a golden cup, closed on the underside and open to the sky, even so it behoves thee to be: if thou wouldst stand with St Augustine and in the

[1] Cp. Dante's 'abandon all hope, ye that enter', for the inverse analogy, that of those condemned to damnation.

Sin 77

communion of saints, then close thy heart to everything created and be open to God as he is in himself.

Eckhart

Obj. But you will say; may we not enjoy the delights of the Creature, which is an Inferiour Image?

Ans. Yes, as a Man may have a Conversation with many Women, so that he break not the *Marriage-Union*. Thy *Fountain* must be thine own. All delights abroad must be, as *Streams* of this Fountain, not divided *Springs*. All other Images must be onely reflections of this One, & *concentred* in it.

Introd. *Contem.*

Please thy self to the full with every Content. Only let it be no Cloud to cut off: but a *Christal* to take in the Divine Glory, that this may be thine and flame in them.

Beauty 663

Peter Sterry

Holy War 405

Him I call indeed a Brâhmana who, though he has committed no offence, endures reproach, stripes, and bonds, who has endurance for his force, and strength for his army.

Renun. 136

Him I call indeed a Brâhmana who is free from anger, dutiful, virtuous, without appetites, who is subdued, and has received his last body.

Supra Introd.

Him I call indeed a Brâhmana who does not cling to sensual pleasures, like water on a lotus leaf, like a mustard seed on the point of a needle.

Realiz. 870

Him I call indeed a Brâhmana who, even here, knows the end of his own suffering, has put down his burden, and is unshackled. . . .

Conform. 170

Him I call indeed a Brâhmana who fosters no desires for this world or for the next, has no inclinations, and is unshackled.

Knowl. 734

Him I call indeed a Brâhmana who has no interests, and when he has understood (the truth), does not say How, how? and who has reached the depth of the Immortal.

M.M. 978

Him I call indeed a Brâhmana who in this world has risen above both ties, good and evil, who is free from grief, from sin, and from impurity.

P. State 590

Him I call indeed a Brâhmana who is bright like the moon, pure, serene, undisturbed, and in whom all gayety is extinct.

Dhammapada, XXVI

Infra 924

Pure, clean, void, tranquil, breathless, selfless, endless, undecaying, steadfast, eternal, unborn, independent, he abides in his own greatness.

Maitri Upanishad, VI. 28

The True King

He who exercises government by means of his virtue may be compared to the north polar star which keeps its place and all the stars turn towards it.

Confucius

Reality 773

I could be well mov'd if I were as you;
If I could pray to move, prayers would move me;
But I am constant as the northern star,
Of whose true-fix'd and resting quality
There is no fellow in the firmament.
The skies are painted with unnumber'd sparks,
They are all fire and every one doth shine,
But there's but one in all doth hold his place:
So, in the world; 'tis furnish'd well with men,
And men are flesh and blood, and apprehensive;
Yet in the number I do know but one
That unassailable holds on his rank,
Unshak'd of motion: and that I am he,
Let me a little show it. . . .

Shakespeare (*Julius Caesar,* III. i. 58)

Rev. 961

Infra 924
Supra 902

'Free, upright, and whole, is thy will, and 'twere a fault not to act according to its prompting; wherefore I do crown and mitre thee over thyself.'

Dante (*Purgatorio,* XXVII. 140)

Conform. 166

The brave and wise man, who intends to overcome his foes, must first of all strive to subdue the internal enemies of his own heart and mind, and the members of his own body. . . .
The kings of the earth in their earthly capitals are not as happy as are the lords of the cities of their own bodies, and the masters of their own minds.

Yoga-Vasishtha

Holy War 396

Many tyrants have sat on the throne, and he whom no man would think on, hath worn the crown.

Ecclesiasticus, XI. 5

Supra 910

'Twixt Kings & Tyrants there's this difference known;
Kings seek their Subjects good: Tyrants their owne.

Robert Herrick

Introd. *Rev.*

It is right that the king should govern himself before governing his subjects.

'Alî

Peace 698

Metanoia 480 Introd. *Conform.*

Whatever you honor above all things, that which you so honor will have dominion over you. But if you give yourself to the domination of God, you will thus have dominion over all things.

Sextus the Pythagorean

A dervish met a king. The king said: 'Ask a boon of me.' The dervish replied: 'I will not ask a boon from one of my slaves.' 'How is that?' said the king. The dervish said: 'I have two slaves who are thy masters: covetousness and expectation.'

Hujwîrî

Kings may rule the world, but the wise rule kings.

Shekel Hakodesh, 18

Judgment 242

He that is spiritual judgeth all things, yet he himself is judged of no man.

I. Corinthians, II. 15

The meaning of royal birth is to be the possessor of the kingdom of justice.

Mohsan Fânî

Supra 902

Not all the water in the rough rude sea
Can wash the balm from an anointed king;[1]
The breath of worldly men cannot depose
The deputy elected by the Lord.

Shakespeare (*Richard II,* III. ii. 54)

Grace 558

Men are not born Kings, but are men renown'd;
Chose first, confirm'd next, & at last are crown'd.

Robert Herrick

Sin 77 *Contem.* 542 *P. State* 563

The Stone is prepared in an empty furnace, with a threefold line of circumvallation, in a tightly closed chamber. It is subjected to continued coction, till all moisture and clouds are driven off, and the King attains to indestructible fixedness, and is no longer liable to any danger or injury, because he has become unconquerable.

Basil Valentine

Metanoia 484 *Illusion* 94 *Sin* 66

I know thee not, old man: fall to thy prayers;
How ill white hairs become a fool and jester!
I have long dream'd of such a kind of man,
So surfeit-swell'd, so old, and so profane;
But, being awak'd, I do despise my dream.
Make less thy body hence, and more thy grace;
Leave gormandising; know the grave doth gape
For thee thrice wider than for other men.

[1] Cp. the infernal counterpart to this doctrine in the citation from *Macbeth* (II. ii. 61) in *The Wrathful Side,* section *Malevolence.*

Reply not to me with a fool-born jest:
Presume not that I am the thing I was;
For God doth know, so shall the world perceive,
That I have turn'd away my former self;
So will I those that kept me company.

Shakespeare (*Henry IV, Pt.2,* v. v. 52)

Realiz. 873
Metanoia 480

Blessed be the King that cometh in the name of the Lord.

St Luke, XIX. 38

Inv. 1031

In the light of the king's countenance is life; and his favour is as a cloud of the latter rain.

Proverbs, XVI. 15

Charity 602

The virtue of a king is shown in making peace; nay, the very name of *king* confers peace.

Hermes

Peace 698

The Lord is the true God, he is the living God, and an everlasting king: at his wrath the earth shall tremble, and the nations shall not be able to abide his indignation.

Jeremiah, X. 10

Wr. Side 464

He (the musician) first uplifts his voice to laud the supreme King of the universe, and comes down thereafter to those who hold their sovereignty after His likeness. For this our kings themselves would wish, that the song should come down step by step from heaven above, and that our praise of them should be derived in due succession from the Power that has conferred on them their victories. Let the musician then address his song to that most mighty King, who is immortal, and reigns from all eternity; that primal Victor, from whom all victories come to those who follow after.

Hermes

Symb. 306
Beauty 670

Twixt Kings and Subjects ther's this mighty odds,
Subjects are taught by *Men*; Kings by the *Gods.*

Robert Herrick

M.M. 973

I am likewise a mighty king, of the Wheel that Revolveth; and a king who aboundeth in riches is in no wise happier or mightier than I. . . . Ye circle of officials of the Kingdoms of the World, if ye but served such a Kingdom as is mine, ye would be transformed into the mightiest of monarchs; and the power and wealth of all things would spring forth (for you).

Milarepa

Reality 773

And, O friend, if you reach perfection in our assembly,
Your seat will be the throne, you will gain your desire in all things.

Dîvâni Shamsi Tabrîz, XLVI

Universal Man

Death 232

I am he that liveth, and was dead; and, behold, I am alive for evermore, Amen; and have the keys of hell and of death.

Revelation, I. 18

Realiz. 887
Flight 949

His are the keys of the heavens and the earth.

Qur'ân, XLII. 12

So long as thou art regarding the holy (prophets and saints) as men, know that that view is an inheritance from Iblîs.[1]

Rûmî

P. State 563
Flight 944
Center 816

The perfect man is a spiritual being. Were the ocean itself scorched up, he would not feel hot. Were the Milky Way frozen hard, he would not feel cold. Were the mountains to be riven with thunder, and the great deep to be thrown up by storm, he would not tremble. In such case, he would mount upon the clouds of heaven, and driving the sun and the moon before him, would pass beyond the limits of this external world, where death and life have no more victory over man;—how much less what is bad for him?

Chuang-tse (ch. II)

Creation 38, 42

You must know that the Perfect Man is a copy (*nuskha*) of God, according to the saying of the Prophet, 'God created Adam in the image of the Merciful,' and in another *ḥadîth,* 'God created Adam in His own image.' . . . Further, you must know that the Essential names and the Divine attributes belong to the Perfect Man by fundamental and sovereign right in virtue of a necessity inherent in his essence.

Jîlî

Introd.
Charity

And God said, Let us make man in our image, after our likeness: and let them have dominion over the fish of the sea, and over the fowl of the air, and over the cattle, and over all the earth, and over every creeping thing that creepeth upon the earth.

Genesis, I. 26

Introd.
Realiz.

The entire man is in his being the three worlds.

Boehme

Lo! thou art of a tremendous nature.

Qur'ân, LXVIII. 4

With the man in whom all creatures end, in whom all multitudinous things have been

[1] Satan, who unlike the other angels would not prostrate himself before Adam (*Qur'ân*, VII. 11).

reduced to one in Christ: man is then one in God with Christ's humanity. Thus all creatures are one man and that man is God in Christ's Person.

Eckhart

Reality 790

It behoves the doctor, if he will be called a doctor, to study the whole process, how God has restored the *universal* in man; which is fully clear and manifest in the person of Christ, from his entrance into the humanity, even to his ascension, and sending of the Holy Ghost.

Let him follow this entire process, and then he may find the universal, provided he be born again of God.

Boehme

For them the *yin* and *yang* are always in harmony, the sun and moon shine without interruption.

Lieh-tse

Center 833

The soul never rests till she is gotten into God who is her first form and creatures never rest till they have gotten into human nature: therein do they attain to their original form, God namely. As St Dionysius hath it, 'God is the beginning and the middle and the end of all things.'

Eckhart

Illusion 109
Charity 608
Peace 694
Reality 803

The very God of peace sanctify you wholly . . . spirit and soul and body.

I. Thessalonians, v. 23

Realiz. 870

Man is in a manner all creatures.

St Gregory the Great

Our soul is of an eternal nature, made a thinking, willing, understanding creature out of that which hath willed and thought in God from all eternity; and therefore must for ever and ever be a partaker of the eternity of God.

And here you may behold the sure ground of the absolute impossibility of the annihilation of the soul. Its essences never began to be, and therefore can never cease to be; they had an eternal reality before they were in or became a distinct soul, and therefore they must have the same eternal reality in it. It was the eternal breath of God before it came into man, and therefore the eternity of God must be inseparable from it. It is no more a property of the divine omnipotence to be able to annihilate a soul than to be able to make an eternal truth become a fiction of yesterday: and to think it a lessening of the power of God to say that He cannot annihilate the soul, is as absurd as to say that it is a lessening of the light of the sun, if it cannot destroy or darken its own rays of light.

William Law

Introd.
Realiz.

I matter as much to God, as He does to me:
I help Him to maintain His being, as He does mine.

Angelus Silesius

Love 618

When Adam fell, God's Son fell: because of the rightful oneing which had been made in heaven, God's Son might not (be disparted) from Adam.

Julian of Norwich

'Thou art by nature already mine!
Nothing can come between Me and thee!
There is no angel so sublime
As to be granted for one hour
What is given thee for ever.'

Supra Introd.

Mechthild of Magdeburg

Angels and all other Creatures have their destined Ideas in the *Divine Mind.* But God himself in his own essential Image, in the Person of the Son, the *Idea of Ideas,* is the *Idea of Man.*

Peter Sterry

He it is Who hath made you regents in the earth.

Qur'ân, XXXV. 39

The angels dwell no higher than the seraphim. All above them must be human.

Mechthild of Magdeburg

I was on that day when the Names were not,
Nor any sign of existence endowed with name.
By me Names and Named were brought to view
On the day when there were not 'I' and 'We'.

Center 841

Dîvâni Shamsi Tabrîz, XVII

And whatsoever Adam called every living creature, that was the name thereof.

Genesis, II. 19

And He taught Adam all the Names.

Qur'ân, II. 31

When in the glass of Beauty I behold,
The Universe my image doth enfold.

Center 828

'Irâqî

The universe is a great man, and man is a little universe.

Sufic Saying

Man, this major world in miniature, is a unified abridgement of all that exists, and the crowning of divine works.

Creation 48

St Gregory Palamas

It is a great truth, which you should seriously consider, that there is nothing in heaven or upon the earth which does not also exist in Man, and God who is in heaven exists also in Man, and the two are but One.

Paracelsus

M.M. 978

Noble Auctors men of glorious fame,
Called our *Stone Microcosmus* by name:
For his composition is withouten doubt,
Like to this World in which we walke about:
Of Heate, of Cold, of Moyst and of Drye,
Of Hard, of Soft, of Light and of Heavy,
Of Rough, of Smooth, and of things Stable,
Medled with things fleetinge and moveable;
Of all kinds Contrary broght to one accord,
Knit by the doctrine of *God* our blessed *Lord*:
Whereby of Mettalls is made transmutation,
Not only in Colour, but transubstantiation.

Thomas Norton

Center 835
Orthod. 285
Supra 914

Man has been truly termed a 'microcosm', or little world in himself, and the structure of his body should be studied not only by those who wish to become doctors, but by those who wish to attain to a more intimate knowledge of God.

Al-Ghazâlî

Realiz. 859

Now, human nature it is that is raised above all the works of God and made a little lower than the angels (*Ps.* VIII). It contains in itself the intellectual and the sensible natures, and therefore, embracing within itself all things, has very reasonably been dubbed by the ancients the microcosm or world in miniature.

Nicholas of Cusa

Man is the microcosm in the strictest sense of the word. He is the summary of all existence. There is no creature that is not recapitulated in man. There is nothing in the universe lower than body or higher than soul.

Soul and (the spiritual) body were created together, and the soul therefore precedes the body only in dignity, not in space or in time. But the body as we know it, material and corruptible, came into existence after man's sin, and because of it. It was man, after he had transgressed, who made to himself this fragile and mortal body. This is signified by the fig leaves, which are a shade, excluding the rays of the sun, as our bodies shade our souls in the darkness of ignorance, and exclude the light of truth.

Introd. *Judgment*
Sin 66

But where, then, is that spiritual and incorruptible body which belonged to man before his sin? It is hidden in the secret recesses of our nature, and it will reappear in the future, when this mortal shall put on immortality.

Johannes Scotus Erigena

Center 816

Here is a man who, turning the emptiness of space into a sheet of paper, the waves of the ocean into an inkwell, and Mount Sumeru into a brush, writes these five characters: so-shi-sai-rai-i.[1] To such, I spread my *zagu*[2] and make my profound bow.

Hoyen of Gosozen

At the highest point of his inner self, his soul, man is more God than creature: however much he is the same as creature in his nature, in mind he is like God more than any creature. To the soul at rest in God in her potential, her essential, intellectual nature, everything comes natural as though she were created not at the will of something else but solely at her own. In this point creatures are her subjects, all submitting to her as though they were her handiwork. It was in this power the birds obeyed St Francis and listened to his preaching. And Daniel took refuge in this power, trusting himself to God alone, when he sat among the lions. Moreover, in this power it has been the custom of the saints to offer up their sufferings which, in the greatness of their love, are to them no suffering.

Conform. 166 Supra 921 *Charity* 608 *P. State* 563 *Suff.* 130

Eckhart

According to the Perfection in which God knoweth Himself, and enjoyeth Himself, so is the Perfection of this Image (of Himself). As those (*Images* of things) are, so is this clear, distinct and full. The more distinct the beam is from the first Light in its emanation, the more strong and full is the reflection. This *Divine Image* then is at once most perfectly *distinct* from its *Divine Original,* most exactly *equal* to it, and most perfectly *one with it.* As then God is, so is this essential, eternal Image of God, a compleat and distinct Person in it self, in every point with the highest and most ravishing agreeableness, answering the Divine Essence in its spring out of which it ariseth.

Center 828

If this Image were not a *Compleat Person,* Gods *knowledge* and *fruition of Himself* would be *incompleat,* without the pleasing and proportionate *returns* of an equal *Loveliness, Life* and *Love.* If this Image were not perfectly *distinct* from the bosome out of which it flourisheth, the knowledge and enjoyment of God would be *confused,* more like to the blindness, the barrenness, the cold of darkness and death, than the life and fruitfulness, the warmth of beauty, life and love, which all have their Perfection and their Joyes in the propagation of themselves into most distinct forms, and the reflection upon themselves from these forms.

Creation 48 Introds. Supra & *Realiz.* *Symb.* 306

This is the *first,* and so the most *universal Image,* the *first seat* of all *Images* of things.

Peter Sterry

Though God produced onely one creature, yet he produced all, because in it he produced the Ideas and forms of all, and that in their most perfect being, that is the Ideal, for which Reason they call this Minde, the Intelligible World.

Reality 775

Pico della Mirandola

I was Manu and the sun (*Sûrya*).

Rig-Veda, IV. xxvi. 1

[1] Chinese characters composing a Zen *mondo* ('question and answer') which reads: 'The first patriarch's motive for coming from the west.'

[2] Cloth used by Zen monks for prostration.

What you do
Still betters what is done. When you speak, sweet,
I'd have you do it ever: when you sing,
I'd have you buy and sell so; so give alms;
Pray so; and, for the ordering your affairs,
To sing them too: when you do dance, I wish you
A wave o' the sea, that you might ever do
Nothing but that; move still, still so,
And own no other function: each your doing,
So singular in each particular,
Crowns what you are doing in the present deed,
That all your acts are queens.

Shakespeare (*Winter's Tale,* IV. iii. 135)

P. State 579

Our Stone is called a little world, because it contains within itself the active and the passive, the motor and the thing moved, the fixed and the volatile, the mature and the crude—which, being homogeneous, help and perfect each other.

Philalethes

Reality 773
Knowl. 749

Man is a composite of all things spiritual.

Ezra ben Solomon

(Man is) an image which comprises everything.

Zohar, III. 139. b

The universe is composed of a part that is material and a part that is incorporeal; and inasmuch as its body is made with soul in it, the universe is a living creature.

Hermes

All things in Heaven above, and Earth beneath, meet in the Constitution of each Individual.

Peter Sterry

Conform. 166

The lights of all prophethood shone forth from his (Muhammad's) light. . . . His existence was before the nothingness (which preceded creation); his name was before the Pen. . . . All sciences are as a drop from his sea, and all wisdoms as a sip from his river, and all time is but as an hour of his enduring. In him is Reality (*al-ḥaqq*) and in him is Truth (*al-ḥaqîqah*). He is the first in Union and the last in Prophethood; the inward in Truth and the outward in Knowledge.

Al-Ḥallâj

Reality 803
Center 838
Creation 26

'My slave ceaseth not to draw nigh unto Me through devotions of freewill until I love him; and when I love him, I am the hearing wherewith he heareth and the sight wherewith he seeth and the hand wherewith he fighteth and the foot whereon he walketh.'

Muhammad

Action 340
Realiz. 890

To him who knows the Truth comes the realisation:—

'I am *Brahman*; I have no suffering and no joy; I neither long for anything, nor do I renounce anything; I am blue, I am yellow, I am white; I am in grass, leaves, trees and flowers; I am the hills, the streams, dales and peaks; I am the essence of all. When all imagination and feelings are gone, then I am the transcendental Reality. The immutable, the nameless and the formless, am I: I am the Witness-Self; I am the basis of all experience; I am the light that makes experience possible.

'I am the man who has fallen in love with a young woman and who compares her beauty to the moon; the consciousness which illumines the joy in the heart of a lover, am I. I am the taste in the dates. Gain and loss are the same to me. As the string bearing the beads remains hidden, so I am the Reality which is hidden in all beings.

'I worship the *Atman* which is the essence of living beings, the sweetness in the moon, and the splendour in the sun.'

Yoga-Vasishtha

Thus it is that man is the *hsin* (mind, heart) of heaven-and-earth, is the complement to the Five Forces. Man is the agent which produces the tastes in foods, the distinctions in sounds, and the colours in clothing.

Li Chi, Li Yun, sect. 3

Supra 900 *Flight* 949 *Center* 841

For man is a being of divine nature; he is comparable, not to the other living creatures upon earth, but to the gods in heaven. Nay, if we are to speak the truth without fear, he who is indeed a man is even above the gods of heaven, or at any rate he equals them in power. None of the gods of heaven will ever quit heaven, and pass its boundary, and come down to earth; but man ascends even to heaven, and measures it; and what is more than all beside, he mounts to heaven without quitting the earth; to so vast a distance can he put forth his power. We must not shrink then from saying that a man on earth is a mortal god, and that a god in heaven is an immortal man.

Hermes

Reality 773 *P. State* 590 *Realiz.* 890 *Beauty* 663 Supra 900 *Center* 816

'Thou art the *Quṭb* whereon the spheres of beauty revolve, and thou art the Sun by whose radiance the full-moon of perfection is replenished; thou art he for whom We set up the pattern and for whose sake We made fast the door-ring; thou art the reality symbolised by Hind and Salmâ and 'Azza and Asmâ. O thou who art endued with lofty attributes and pure qualities, Beauty doth not dumbfound thee nor Majesty cause thee to quake, nor dost thou deem Perfection unattainable: thou art the centre and these the circumference, thou art the clothed and these the splendid garments.'

Jîlî

Flight 949

I am the wind that breathes on the sea
I am the wave of the ocean
I am the murmur of the waves
I am the ox of seven combats
I am the eagle on the rocks
I am the beam of the sun

I am the fairness of plants
I am the valour of the wild boar
I am the salmon leaping
I am the stillness of the lake
I am the word of science
I am the lance-point of battle
I am the divinity who created in the head the fire.
Who throws light on the meeting in the mountains?
Who tells the ages of the moon?
Who teaches where sets the sun?
Who, if not I?

Ortha nan Gaidheal

Having taken the no-form as your form,
Going and returning are not to be elsewhere.
Having taken the no-thought as your thought,
Singing and dancing become the voice of the Law.

Hakuin

Void 724 *Ecstasy* 636

He comes from no whence and goes no whither.

Prajñâ-Pâramitâ

Introd. *Reality*

They exist without (separative) existence, and are distinct without difference.[1]

Junayd

Introd. Supra

His life (is) without death, his knowledge without ignorance, his will without aversion, his power without impotence, his hearing without deafness, his vision without blindness, his speech without aphonia, nor are any of his faculties ever annihilated by their contraries.

Ibn al-ʿArif

Center 835

He (the *Yogî* delivered in this life, *jivan-mukta*) knows that all contingent things are not different from *Atmâ* (in their principle), and that apart from *Atmâ* there is nothing, 'things differing simply (in the words of the *Veda*) in attribution, accident and name, just as earthen vessels receive different names, although they are but different forms of earth':[2] and thus he perceives that he himself is all things. . . .

M.M. 978 *Reality* 803

He is without (distinct) qualities and actionless; imperishable (*akshara*, not subject to dissolution, which exercises dominion only over the multiple), without volition (applied to a definite act or to determined circumstances); abounding in Bliss, immutable, without form; eternally free and pure. . . .

Action 358 *Void* 724

He is incorruptible, imperishable; he is the same in all things, pure, impassible, invariable. . . .

Supra 902

He is *Brahma*, after the possession of which there remains nothing to possess; after the

[1] *Kânû bilâ kawn wa bânû bilâ bawn*:—the Arabic wording of the second clause also implies separation without absence.

[2] *Chândogya Upanishad*, VI. i. 4–6.

Center 847 833

enjoyment of whose Bliss there remains no felicity to be desired; and after the attainment of the Knowledge of which there remains no knowledge to be obtained. . . .

He is *Brahma*, by which all things are illuminated, the Light of which causes the sun and all luminous bodies to shine, but which is not made manifest Itself by their light.

Realiz. 868

He himself pervades his own eternal essence (which is not different from the Supreme *Brahma*), and (simultaneously) he contemplates the whole World (manifested and unmanifested) as being (also) *Brahma*, just as fire intimately pervades a white-hot iron ball, and (at the same time) also reveals itself outwardly (by manifesting itself to the senses through its heat and its luminosity).

Śrî Śankarâchârya

Mine is the kingdom in both worlds: I saw therein none but myself, that I should hope his favour or fear him.

Center 841

Before me is no 'before', that I should follow its condition, and after me is no 'after', that I should precede its notion.

I have made all kinds of perfection mine own, and lo, I am the beauty of the majesty of the Whole: I am naught but It.

Whatsoever thou seest of minerals and plants and animals, together with Man and his qualities,

And whatsoever thou seest of elements and nature and original atoms (*haba'*)[1] whereof the substance is (ethereal as) a perfume,

And whatsoever thou seest of seas and deserts and trees and high-topped mountains,

And whatsoever thou seest of spiritual forms and of things visible whose countenance is goodly to behold,

And whatsoever thou seest of thought and imagination and intelligence and soul, and heart with its inwards,

Beauty 670
Wr. Side 459

And whatsoever thou seest of angelic aspect, or of phenomena whereof Satan is the spirit,

. . . . .

Creation 33
Reality 803

Lo, I am that whole, and that whole is my theatre: 'tis I, not it, that is displayed in its reality.

Inv. 1003

Verily, I am a Providence and Prince to mankind: the entire creation is a name, and my essence is the object named.

The sensible world is mine and the angel-world is of my weaving and fashioning; the unseen world is mine and the world of omnipotence springs from me.

And mark! In all that I have mentioned I am a slave returning from the Essence to his Lord—

Humility 191

Poor, despised, lowly, self-abasing, sin's captive, in the bonds of his trespasses.[2]

Jîlî

[1] Literally, 'fine dust'—in cosmological symbolism, *materia prima*.

[2] 'The relationship in the spiritually perfect man between the Divine Reality (*Haqiqah*) and the individuality that still subsists, is one of the most difficult of things to grasp; for the man who has reached this perfection, the

This holy god, the lord of all the gods, Amen-Râ, the lord of the throne of the two lands, the governor of Åpt; the holy soul who came into being in the beginning; the great god who liveth by Maât; the first divine matter which gave birth unto subsequent divine matter! the being through whom every god hath existence; the One One who hath made everything which hath come into existence since primeval times when the world was created; the being whose births are hidden, whose evolutions are manifold, and whose growths are unknown; the holy Form, beloved, terrible, and mighty in his risings; the lord of wealth, the power, Khepera who createth every evolution of his existence, except whom at the beginning none other existed; . . . who having made himself (to be seen, caused) all men to live; who saileth over the celestial regions and faileth not; . . . who though an old man shineth in the form of one that is young; . . . who made the heavens and the earth by his will; the greatest of the great, the mightiest of the mighty. . . .

P. State 563
Reality 775
Creation 26
Waters 651
Center 838

Papyrus of Nesi-Khonsu

O God! in Thy body I see all the gods, as well as multitudes of all kinds of beings; the Lord Brahmâ, seated on the lotus throne, all the Rishis and all the celestial serpents.

* * * *

I see Thee with diadems, maces, discus, shiningly effulgent everywhere, blazing all around like the burning fire and the sun, dazzling to the sight and immeasurable.

Thou art the Imperishable, the Supreme, the One to be known. Thou art the Supreme Refuge of this universe; Thou art the ever unchanging Guardian of the Eternal Dharma; Thou art, I know, the Ancient Being.

I see Thee without beginning, middle or end, with infinite power, with numberless arms, the sun and moon as Thine eyes, Thy mouth as the blazing fire, heating this universe with Thine own radiance.

By Thee alone the space between heaven and earth and all the quarters is pervaded. O Great Soul, seeing this, Thy wonderful and terrifying Form, the three worlds are stricken with fear.

Reality 775
790
Center 841
833
Reality 803
Judgment 239

* * * *

O boundless Form, Thou art the Primeval Deity, the Ancient Being, Thou art the Supreme Refuge of this universe; Thou art the Knower, the One to be known and the Supreme Abode. By Thee alone is this universe pervaded.

Thou art Vâyu, Yama, Agni, Varuna, the Moon; Thou art the Lord of creatures and the great Grandsire. Salutations to Thee, my salutations a thousand times, again and again my salutations to Thee!

Salutations to Thee before, salutations to Thee behind, salutations to Thee on all sides! O All, infinite in power, and immeasurable in valor, Thou pervadest all, therefore Thou art All.

Knowl. 749
P. State 590
Realiz. 890

Divine Reality in effect is no longer "veiled" by anything, whereas the individual consciousness is by definition a "veil" (*ḥijâb*) and only exists inasmuch as it "refracts" the blinding light of the divine Intellect' (Burckhardt: *Introd. Doct. ésotériques de l'Islam*, pp. 81–82). It is not the individuality as such that realizes Union, as has already been stressed in the Introduction to the preceding chapter: while neither is it excluded from intimate participation in that which essentially allows of no 'objectification'. Cf. note p. 918.

Orthod. 288

Not knowing this Thy glory and regarding Thee merely as a friend, whatever I may have said presumptuously, out of either carelessness or fondness, addressing Thee as 'O Krishna,' 'O Yâdava,' 'O Friend';

Realiz. 859

O Changless One, in whatever manner I may have been disrespectful to Thee, in jesting, in walking, in reposing, sitting, or at meals, alone, or in the presence of others; O Unfathomable One, I implore Thee to forgive all that.

* * * *

I desire to see Thee as before, with diadem, mace and discus. O Universal Form of thousand arms, do Thou manifest Thyself in that same Four-armed Form (form of Vishnu).

Bhagavad-Gîtâ, XI. 15, 17–20, 38–42, 46

On returning from this visit to Lao Tze, Confucius did not speak for three days. A disciple asked him, saying, 'Master, when you saw Lao Tze, in what direction did you admonish him?'

Introds. *Creation* *Reality*

'I saw a dragon,' replied Confucius, '—a dragon which by convergence showed a body, by radiation became colour, and riding upon the clouds of heaven, nourished the two principles of creation. My mouth was agape: I could not shut it. How then do you think I was going to admonish Lao Tze?'

Chuang-tse (ch. XIV)

Infra 935

The sun illumines earth and sky, but the saint, kindling the fire of divine wisdom, lights up the heart. He is the true friend of man. He is the Atman. He is my very Self.

Srimad Bhagavatam, XI. xix

The Eternally Awake

Metanoia 488

I sleep, but my heart waketh.

The Song of Solomon, v. 2

Death 232 *Center* 833 *Orthod.* 296

My sleep is broken; how can I slumber any more?
For now I am wide awake in the sleeplessness of yoga.
O Divine Mother, made one with Thee in yoga-sleep at last,
My slumber I have lulled asleep for evermore.
A man has come to me from a country where there is no night;
Rituals and devotions have all grown profitless for me.

Hindu Song

In the soul there exists one power which rests not day or night.

Eckhart

The disciples of Gotama are always well awake, and their thoughts day and night are always set on Buddha.

Dhammapada, XXI. 296

Men watch and rest, but thou dost watch in resting.

Theban Hymn to Ptah

Action 358

'The Enlightened and their Sons keep unfailing watch in every place. Everything is before them, I stand in their presence.'

Śânti-deva

Perpetual inspiration is as necessary to the life of goodness, holiness and happiness as perpetual respiration is necessary to animal life.

William Law

Contem. 542
Inv. 1036

Behold, he that keepeth Israel shall neither slumber nor sleep.

Psalm CXXI. 4

The gnostic praises God awake and asleep. . . . He who has tasted the sweetness of prayer sleeps no more.[1]

Abû Madyan

The Light of the World

There always are in the world a few inspired men whose acquaintance is beyond price, and who spring up quite as much in ill-ordered as in well-ordered cities. These are they whom the citizens of a well-ordered city should be ever seeking out, going forth over sea and over land to find him who is incorruptible.

Plato (*Laws*, XII, 951 B)

Renun. 139
Supra 902

And for every nation there is a messenger.

Qur'ân, X. 47

Reality 790

A supernatural person (a Buddha) is not easily found, he is not born everywhere. Wherever such a sage is born, that race prospers.

Dhammapada, XIV. 193

[1] Cp. the diabolical counterpart of this in *Macbeth*, II. ii. 36, cited in *The Wrathful Side*, section *Sorcery*.

Contem. 523
Supra 900
Wr. Side 464

Now, though many years did not pass after the age of the apostles before Satan and self got footing in the Church and set up merchandise in the house of God, yet this one heart and one spirit which then first appeared in the Jerusalem Church is that one heart and spirit of divine love to which all are called that would be true disciples of Christ. And though the practice of it is lost as to the Church in general, yet it ought not to have been lost; and therefore every Christian ought to make it his great care and prayer to have it restored in himself. And then, though born in the dregs of time or living in Babylon, he will be as truly a member of the first heavenly Church at Jerusalem as if he had lived in it in the days of the apostles.

William Law

Reality 790
P. State 563

In these wicked days, indeed, when virtue and vice are accounted alike, the ingratitude and unbelief of men keep our Art from appearing openly before the public gaze. Yet this glorious truth is even now capable of being apprehended by learned and unlearned persons of virtuous lives, and there are many persons of all nations now living who have beheld Diana unveiled.

Michael Sendivogius

Each time I come to a country where the Beloved is not respected, I find that space narrows down.

Dhu 'l-Nûn (addressed by a woman hermit)

Within a man of light there is light and he lights the whole world. When he does not shine, there is darkness.

The Gospel according to Thomas, Log. 24

How beautiful upon the mountains are the feet of him that bringeth good tidings, that publisheth peace; that bringeth good tidings of good, that publisheth salvation; that saith unto Zion, Thy God reigneth!

Isaiah, LII. 7

MOVING AT WILL—
THE MIRACLE OF FLIGHT

The sacrificer, become a bird, mounts to the celestial world.

Pañcaviṁśa Brâhmaṇa, v. 3. 5

Where domination over the waters characterizes completion of the 'Lesser Mysteries', power over air will correspondingly adumbrate attainment of the 'Greater Mysteries', with the reservation that the symbolism must in no way be defined in a rigid or exclusive manner. Also, it should be understood that certain psychic and spiritual states inwardly attach to cosmic powers which may on occasion manifest in outward phenomena; and if in fact this point were better grasped, much of the difficulty had by orientalists and ethnologists over 'borrowings' to excuse the universal recurrence of 'miracle motifs' might be eliminated. Moreover, phenomena as drawn from the passive and substantial pole of existence can only derive their sanction with respect to doctrinal criteria and orthodox canon, and apart from spiritual influence as conveyed through the mould of tradition, they have no value whatsoever: is not the devil himself 'prince of the power of the air'?[1]

Concl.

Traditional cosmology indicates that the 'materiality' of the world was less opaque in earlier periods,[2] and the dichotomy between the subtle and physical orders less pronounced, thus allowing a communication and interplay between different levels of reality which would be inconceivable in terms of the present *Kali-Yuga*. But the spiritual recovery of one's higher faculties recapitulates in a vertical sense the anterior stages of the cycle's development, meaning that the saint is essentially endowed with the capacities of primordial man, which can upon sufficient provocation exteriorize in the miracles whose beauty and symbolism recall a believing generation to higher realities.

Wr. Side 464

Needless to say, that 'sufficient provocation' is precisely what lacks in the practising atheism of contemporary society, insatiable as it is for outward powers at the expense of the soul's true functions, where these higher faculties, rejected then forgotten, atrophy, usurped by a mechanistic parody which invests the machine[3] with the powers proper to the soul, and thinks to dominate with its physical, chemical and electronic 'miracles' the forces of the universe,[4] even 'reaching for the moon' itself—a really ludicrous travesty of the

Introd. *Beauty*

[1] *Ephesians*, II. 2.

[2] Cf. Guénon: *Le Règne de la Quantité*, ch. XVII.

[3] The fact that machines work proves the existence of God: the fact that they are unlovely proves the existence of Satan.

[4] 'Will not men put forth audacious hands against the elements? They will dig up roots of plants, and investigate the properties of stones. They will dissect the lower animals,—yes, and one another also,—seeking to find out how they have come to be alive, and what manner of thing is hidden within. . . . They will cut down the woods of their native land, and sail across the sea to seek what lies beyond it. They will dig mines, and search into the

P. State
590

recovery of the primordial state![1] 'A person can only wonder at the solemn nonsense of certain pronouncements dear to the hearts of scientific "popularizers" (we should really say "scientistic"), who are pleased to assert on every occasion that modern science is pushing back without cease the limits of the known world, all of which is, in fact, exactly the contrary of the truth: never have these limits been so narrow as they are in the conceptions admitted by this so-called secular science, and never have the world and man been found so shrunken, to the point of being reduced to purely corporeal entities, deprived by hypothesis of the slightest possibility of communication with any other order of reality!' (Guénon: *Le Règne de la Quantité,* p. 116).[2]

What is meant by communication with other orders of reality? 'That the perfected possess the power of motion and manifestation at will is familiar in Christian teaching,

uttermost darkness of the depths of the earth. And all this might be borne, but they will do yet more: they will press on to the world above, seeking to discover by observation the laws of movement of the heavens. Are they then to meet with no impediment?' (Hermes: *Stobaei,* Exc. XXIII. 45). Cp. the following citations:

'For incorrigible man, persuaded by the giant, presumed in his heart to surpass by his own skill not only nature, but even the very power that works in nature, who is God' (Dante: *De Vulg. Eloq.,* I. vii).

'Why, treading as you do on earth, do you leap over the clouds? And why do you say that you are able to lay hold of what is in the upper air, when you are rooted to the ground? Why do you venture to determine the indeterminate? And why are you so busy with what you ought to leave alone. . . . you who have not yet attained to (self) knowledge?' (Philo: *On Dreams*; Lewy, p. 57).

'Our period will complete its monstrous masterpiece, by transforming man into a giant in the physical world, at the expense of his spirit, reduced to the state of a pygmy in the supernatural and eternal world' (Pope Pius XII; cited by Prof. A. Bodart at a 1962 Rotary Club lecture: *La Biologie et l'avenir de l'homme*).

'And if the Truth had followed their desires, verily the heavens and the earth and whosoever is therein had been corrupted' (*Qur'ân,* XXIII. 71).

'The pride of thine heart hath deceived thee. thou that dwellest in the clefts of the rock, whose habitation is high; that saith in his heart, Who shall bring me down to the ground?

'Though thou exult thyself as the eagle, and though thou set thy nest among the stars, thence will I bring thee down, saith the Lord' (*Obadiah,* 3, 4).

'And thy wrath is come . . . that thou shouldest . . . destroy them which destroy the earth' (*Rev.,* XI. 18).

Recounting a vision of the Last Judgment, Jeanne Le Royer, Sister of the Nativity, a French mystic whose warnings just preceded the Revolution, relates how the sun, the moon, the stars and elements, in fact 'the whole of nature seeks a reparation, regeneration, and new existence as it were to deliver it forever from the thralldom which had compelled it to serve the vanity and passions of men' (cited by Suzanne Jacquemin: *Les Prophéties des derniers Temps,* Paris, 1958, p. 152). Substantially the same message was revealed by the Blessed Virgin at La Salette.

[1] Cf. again Hermes: 'The stars which rise as exhalations from the earth do not attain to the region of heaven,—for they are not able to do that, because they rise from below,—and, as they have in them much heavy stuff, they are dragged down by their own matter, and are quickly dissipated, and being broken up, they fall down again to earth, having effected nothing except a troubling of the air above the earth' (*Stobaei,* Exc. VI, 15).

'(Nature) cannot be mastered but only humbly followed and studied, and then she showers her blessings on man' (Chushen, abbot of the Pahsienkung Taoist monastery at Sian; cited by Peter Goullart: *The Monastery of Jade Mountain,* London, Murray, 1961, p. 133).

[2] The current illusion of 'Progress' partly derives from the increasing instability of matter at this advanced moment in the cycle: the 'metabolism' or structure and balance of material properties is undergoing rapid modifications which 'conspire' to promote this illusion in the minds of those easily taken by appearances. What matter if the 'signs and wonders' spoken of in the Gospels are right now playing themselves out on a secular and technological plane, since their effect is exactly the same—namely, to deceive 'if it were possible . . . the very elect' (*St Matthew,* XXIV, 24)?

where they "shall pass in and out and find pasture";[1] and such powers are naturally proper to those who, being "joined unto the Lord, are one spirit".[2] The like is repeatedly enunciated in the Brahmanical scriptures, and often in nearly the same words. In an often recurring context the Buddha describes the four stages of contemplation (*dhyâna*) or paths of power (*ṛddhipâda*) that are the equivalent of the "Aryan Path" and are means to Omniscience, Full Awakening and Nirvâṇa. When all these stations of contemplation (*dhyâna*) have been so mastered that the practitioner can pass from one to another at will, and similarly commands the composure or synthesis (*samâdhi*) to which they lead, then in this state of unification (*eko'vadhi-bhâva*) the liberated Arhat is at once omniscient and omnipotent... The Buddha employs the old Brahmanical formula when he says that he has taught his disciples to extract from this material body another body of intellectual substance, as one might draw an arrow from its sheath, a sword from its scabbard, or a snake from its slough; it is with this intellectual body that one enjoys omniscience and is a mover-at-will as far as the Brahmaloka.

Infra 949

Introd. *Center*

'Before we ask ourselves what all this means, let us remark that supernatural no more implies unnatural than super-essential implies unessential; and that it would be unscientific to say that such attainments are impossible, unless one has made experiment in accordance with the prescribed and perfectly intelligible disciplines. To call these things "miraculous" is not to say "impossible", but only "wonderful"; and as we said before, following Plato, "Philosophy begins in wonder". Furthermore, it must be clearly understood that the Buddha, like other orthodox teachers, attaches no great importance to these powers and very strongly deprecates a cultivation of powers for their own sake and in any case forbids their public exhibition by monks who possess them. "I do, indeed," he says, "possess these three powers (*ṛddhi*) of motion-at-will, mind-reading, and teaching; but there can be no comparison of the first two of these marvels (*pratihârya*) with the much farther-reaching and far more productive marvel of my teaching."[3] It will profit us more to ask what such marvels, or those of Christ imply, than to ask whether they "really" took place on some given occasion....

'In the first place, we observe that in the Brahmanical contexts, omniscience, particularly of births, is predicated of Agni (*jâtavedas*), the "Eye in the World", and of the "all-seeing" Sun, the "Eye of the Gods", and for the very good reason that these consubstantial principles are the catalytic powers apart from which no birth could be; and further, that the power of motion at will, or what is the same thing, motion without locomotion, is predicated in the Brahmanical books of the Spirit or Universal Self (*âtman*) on the one hand, and of liberated beings, knowers of the Self and assimilated to the Self, on the other. Once we have understood that the Spirit, universal solar Self and Person, is a timeless omnipresence, it will be recognized that the Spirit, by hypothesis, is naturally possessed of all the powers that have been described.... If the "signs and wonders" are lightly dismissed, it is not because they are unreal, but because it is an evil and adulterous generation that asketh for a sign' (Coomaraswamy: *Hinduism and Buddhism,* pp. 68–71).

Center 841 Introd. *Realiz.*

Reality 773

'The justification for a miracle is in the disclosure of an evidence, while at the same time

[1] *St John,* X. 9.

[2] *I. Corinthians,* VI. 17.

[3] *Anguttara-nikâya,* I. 171. 172. This corresponds with the descent of the *Qur'ân* as the greatest miracle in Islam.

Rev. 965

the miracle appears to the certitude of the intellect as a projection of itself on the plane of facts and symbols' (Schuon: 'Nature et arguments de la foi', *Études Trad.*, 1953, p. 363).

'When we read trustworthy lives of saints, we are struck by the frequency, and we might almost say by the ease, of miracles in the Middle Ages and in antiquity. If the saints of later periods—especially those of our own times—live in a less miraculous setting, one could say that this is because the whole cosmic setting is so far removed from celestial realities and so emptied of grace as to exclude any frequency of great miracles. If a saint of our days had lived in the Middle Ages, perchance he would have walked on the waters; if, in spite of everything, he walked on the waters now—and such a supposition is contradictory—he would risk holding up the fatal evolution of events: he would make the end of the world and the completion of the divine plan impossible' (Schuon: *Perspectives spirituelles*, pp. 56–57).

Physical Flight

All bodies are not impediments to beatitude, but only the corruptible, transitory, and mortal ones; not such as God made man at first, but such as his sin procured him afterwards.

Holiness 902
Creation 42

Oh, but, they say, an earthly body is either kept on earth, or carried to earth by its natural weight, and therefore cannot be in heaven. The first men indeed were in a woody and fruitful land, which was called paradise. But because we must resolve this doubt, seeing that both Christ's body is already ascended, and that the saints at the resurrection shall do so also, let us ponder these earthly weights a little. If man's art out of a metal, that being put into the water sinks, can yet frame a vessel that shall swim, how much more credible is it for God's secret power, whose omnipotent will, as Plato says, can both keep things produced from perishing, and parts combined from dissolving (whereas the combination of corporeal and incorporeal is a stranger and harder operation than that of corporals with corporals), to take all weight from earthly things, whereby they are carried downwards, and to qualify the bodies of the blessed souls, so that, though they be terrene, yet they may be incorruptible, and fit to ascend, descend, or use what motion they will with all celerity. Or if the angels can transport bodily weights whither they please, must we think they do it with toil and feeling the burden? Why then may we not believe that the perfect spirits of the blessed can carry their bodies whither they please, and place them where they please?

Realiz. 870

Infra 949

St Augustine

I zealously persevered in my meditations. At last, I began to feel that I had obtained the power of transforming myself into any shape (desired), and of flying through the air. By day, I thus felt that I could exercise endless phenomenal powers; by night, in my dreams, I could traverse the universe in every direction unimpededly—from the summit of Mount Meru to its base—and I saw everything clearly (as I went). Likewise (in my dreams) I could multiply myself into hundreds of personalities, all endued with the same powers as myself. Each of my multiplied forms could traverse space and go to some Buddha Heaven, listen to the Teachings there, and then come back and preach the *Dharma* to many persons. I could also transform my physical body into a blazing mass of fire, or into an expanse of flowing or calm water. Seeing that I had obtained infinite phenomenal powers (even though it be but in my dreams), I was filled with happiness and encouragement at mine own success.

Thenceforth, I persevered in my devotions in a most joyous mood, until, finally, I actually could fly. Sometimes I flew over to the Min-khyüt-Dribma-Dzong (Castle lying in Shadows to the Eyebrows)[1] to meditate; and there I obtained a far greater development of

[1] Equatable esoterically with the *Âjñâ-chakra*, or sixth center in the practice of *Kuṇḍalinî*, as the editor points out. But that the meaning is to be taken literally first of all is proven in the paragraph following, where a

the Vital Warmth than ever before. Sometimes I flew back again to the Dragkar-Taso Cave.

Milarepa

By concentrating on the relation of the body to the all-pervading Ether, and, thinking of small and light objects such as the fibres of cotton-wool, the yogi is able to travel through space.

Patañjali

If from the top of Mount Sumeru a man is hurled down by an enemy, let his thought dwell on the power of Kwannon, and he will stay in the air like the sun.

Inv. 1031

Kwannon Sutra

They (certain Jaina Sages) could make the body so light that it was even lighter than air.

Triṣaṣṭiśalâkâpuruṣacarita, I. 854

Abaris (was surnamed) *a walker on air*; because being carried on the dart which was given to him by the Hyperborean Apollo, he passed over rivers and seas and inaccessible places, like one walking on the air. Certain persons likewise are of opinion, that Pythagoras did the same thing, when in the same day he discoursed with his disciples at Metapontum and Tauromenium.

Waters 653

Iamblichus

'How,' said Demetrius,[1] 'have you accomplished so long a journey in so small a fraction of the day?' And Apollonius replied: 'Imagine what you will, flying ram or wings of wax excepted, so long as you ascribe it to the intervention of a divine escort.'

Action 340

Apollonius of Tyana

But at that time, monks, the great river Ganges was rolling along full up, level to the banks. Then, monks, the Tathâgata went up to the boatman to be carried across to the other bank. He said, 'Gâutama, give me the fee for crossing.' 'My good man,' replied the Tathâgata, 'I have not the fee for crossing.' So saying, he went from that bank to the other bank on a path through the sky.

Faith 501

Lalita Vistara, XXVI

There is a country called Deccan, in which there is a monastery dedicated to Kâsyapa Buddha, made by hollowing out a great rock. It has five storeys in all. . . . There are always

discussion ensues between a farmer and his son while the Kargyütpa saint is flying overhead: 'The father said, "What is there to marvel at or to be amused about in the sight? One Nyang-Tsa-Kargyen, a very mischievous woman, had a wicked son, named Mila (Milarepa had practiced black magic in his youth before taking up the spiritual life). It is that good-for-nothing starveling. Move aside and do not allow his shadow to fall over thee, and go on leading the team." The father himself was bending his body about so as to avoid falling under my shadow. But the son said, "If a man be able to fly, I do not mind his being a good-for-nothing person: there can be nothing more wonderful than a man flying." So saying, he continued looking at me.'

[1] A Cynic philosopher, and companion of Apollonius.

Lo-han[1] in residence here. The land is uncultivated, and there are no inhabitants. Only at a great distance from the mountain are there villages, all the inhabitants of which are pagans, and know nothing of the Buddhist Faith, of Shamans, of Brahmans, or of any other of the heterodox religions. They frequently see people come flying and enter the monastery; and once when Buddhist worshippers came from the neighbouring countries to pray at this monastery, one of the villagers asked them, saying, 'Why do not you fly here? The worshippers I see here, all fly.' 'It is because our wings have not yet grown,' replied the worshippers without hesitation.

The Travels of Fa-hsien

I saw a man flying through the air, and asked him how he had attained to this degree. He answered: 'I set my feet on passion (*hawâ*) in order that I might ascend into the air (*hawâ'*).'

Dhu 'l-Nûn

Sin 62
Metanoia 488

My Shaykh used to say: 'One year a meeting of the saints of God took place in the midst of the desert, and I accompanied my spiritual director, Ḥuṣrî, to that spot. I saw some of them approaching on camels, some borne on thrones, and some flying, but Ḥuṣrî paid no heed to them. Then I saw a youth with torn shoes and a broken staff. His feet could scarcely support him, and his head was bare and his body emaciated. As soon as he appeared Ḥuṣrî sprang up and ran to meet him and led him to a lofty seat. This astonished me, and afterwards I questioned the Shaykh about the youth. He replied: 'He is one of God's saints who does not follow saintship, but saintship follows him: and he pays no attention to miracles.'

Hujwîrî

Holiness 910
Conform. 170

If thou canst walk on water
Thou art no better than a straw.
If thou canst fly in the air
Thou art no better than a fly.
Conquer thy heart
That thou mayest become somebody.

Ansârî

Waters 653
Holy War 396

Flying is no business of the spiritual yogi, who is concerned only with a knowledge of the Spirit; he is content with his spiritual knowledge and union with the Supreme, and does not meddle with the practices of the ignorant and false *Hatha-Yogis*.

Yoga-Vasishtha

Even when you see a man endowed with miraculous powers to the point of rising in the air, do not let yourself be deluded, but investigate whether he observes the divine precepts and prohibitions, whether he stays within the limits of religion and whether he accomplishes the duties this imposes upon him.

Bâyazîd al-Bisṭâmî

Heterod. 430
Orthod. 300

[1] Buddhist saints of high standing, often represented as uncouth spiritual personalities.

Dream not of lights,
Of marvels, of miracles,
For your miracles are contained
In worshipping the Truth;
All else is pride, conceit,
Illusion 85 And illusion of existence.

Shabistarî

Intellectual Flight

Center 838 841 Intellect is the swiftest of birds.

Rig-Veda, VI. 9. 5

Introd. *Realiz. Inv.* 1036 There is a channel called the Sushumnâ, leading upward, conveying the breath, piercing through the palate. Through it, by joining ($\sqrt{}$ *yuj*) the breath, the syllable *Om*, and the mind, one may go aloft.

Maitri Upanishad, VI. 21

The Comprehensor is winged.

Pañcaviṁśa Brâhmaṇa, XIV. 1. 13

Holiness 914 The truth is, that the outer form of him (the philosopher) only is in the city: his mind, disdaining the littlenesses and nothingnesses of human things, is 'flying all abroad' as Pindar says, measuring earth and heaven and the things which are under and on the earth and above the heaven, interrogating the whole nature of each and all in their entirety, but not condescending to anything which is within reach.

Plato (*Theaetetus*, 173 E)

Knowl. 734 All perceptions are mounted on lame asses: He is mounted on the wind that flies like an arrow.[1]

Rûmî

O Lord my God, thou art very great; ... who maketh the clouds his chariot: who walketh upon the wings of the wind.

Psalm CIV. 1, 3

Now it may be asked, What is this nature of the soul?—It is the consciousness (the

[1] These passages clearly show that it is the Intellect which is intended. and not merely the mind or thought or imagination.

spark or synteresis) in the soul, that is the impartible nature of the soul. So subtile is this nature of the soul that space might not exist at all for all it troubles her. For instance, if one has a friend a thousand leagues away, thither flows the soul with the best part of her powers, loving her friend there. St Augustine testifies to this. He says, 'The soul is where she loves rather than where she is giving life.' The simple nature of the soul is in no way hampered by place.

Center 841

Eckhart

The incorporeal cannot be enclosed by anything; but it can itself enclose all things; it is the quickest of all things, and the mightiest.

Center 816

Hermes

This country (*Hoa-su-cheu*) is west of *Yen-tcheou*, north of *T'ai-tcheou*, at I don't know how many myriads of stages from this country of *Ts'i*. One can neither go there by boat nor carriage; only the flight of the soul reaches there. In this country there is no leader; everything functions spontaneously. The people have neither desires nor greed, but simply natural instinct. No one there loves life, nor dreads death; each lives until his term. No enmities and no hatreds. No gains and no losses. No interests and no fears. Water does not drown them, fire does not burn them. No weapon can wound them, no hand can injure them. They rise in the air as if going up stairs, and they stretch out in the void as upon a bed. Clouds and fog do not hinder their view, the noise of thunder does not affect their hearing, no beauty or ugliness upsets their heart, no height or depth obstructs their course. The flight of the soul carries them everywhere.

P. State 583

Lieh-tse

The *Soul* in the *Will* flyeth forth upon the wings of Love, into the bosom of the beloved Object, to live there where it *loves*.

Peter Sterry

Tell me, where is the soul's abode?—Upon the pinions of the wind.

Eckhart

He whose appetites are stilled, who is not absorbed in enjoyment, who has perceived void and unconditioned freedom (Nirvâna), his path is difficult to understand, like that of birds in the air.

M.M. 986

Dhammapada, VII. 93

Love flows from God to man without effort
As a bird glides through the air
Without moving its wings—
Thus they go whithersoever they will
United in body and soul
Yet in their form separate.

Love 618

Infra 949

Mechthild of Magdeburg

The Holy Bird flew in the limitless extent of this void atmosphere, with his totality exalting God in the air of the supreme sphere.

Jîlî

The journey of the spirit is unconditioned in respect of Time and Space: our body learned from the spirit how to journey.

Rûmî

Reality 773

Unmoving, the One (*ekam*) is swifter than the mind.
The sense-powers (*deva*) reached not It, speeding on before.
Past others running, This goes standing.
In It Mâtariśvan[1] places action.

Îśâ Upanishad, 4

Knowl. 761

The sun itself, that enlightens the world, and scatters away all stench, putrefaction and corruption, is yet but darkness, and a cloud in compare with the motion of the mind and understanding.

Benjamin Whichcote

Knowl. 749

The soul is a spirit formed in the likeness of God and agreeable to him as one spirit is to another. Philosophers, again, compare the soul with fire, a thing most lofty in its nature, most theurgic in its operation, which never stops until it licks the heavens.

Eckhart

Introd. *Holiness* *Sin* 66

The soul in her totality . . . traverses the whole heaven in divers forms appearing:—when perfect and fully winged she soars upward, and orders the whole world; whereas the imperfect soul, losing her wings and drooping in her flight at last settles on the solid ground—there, finding a home, she receives an earthly frame which appears to be self-moved, but is really moved by her power; and this composition of soul and body is called a living and mortal creature. . . .

Beauty 676

The wing is the corporeal element which is most akin to the divine, and which by nature tends to soar aloft and carry that which gravitates downwards into the upper region, which is the habitation of the gods. The divine is beauty, wisdom, goodness, and the like; and by these the wing of the soul is nourished, and grows apace; but when fed upon evil and foulness and the opposite of good, wastes and falls away

Plato (Phaedrus, 246)

The most important of all the creatures are the wingeds, for they are nearest to the heavens, and are not bound to the earth as are the four-leggeds, or the little crawling people.

It may be good to mention here that it is not without reason that we humans are two-legged along with the wingeds; for you see the birds leave the earth with their wings, and we humans may also leave this world, not with wings, but in the spirit. This will help you to understand in part how it is that we regard all created beings as sacred and

[1] Epithet of Vâyu, the god of wind.

important, for everything has a *wochangi* or influence which can be given to us, through which we may gain a little more understanding if we are attentive.[1]

Black Elk

Introd. *Symb.*

In my community are people who will enter Paradise with souls like the souls of birds.

Muhammad

Transported through the practice of muni-asceticism,[2] we mount the winds; you mortals see only our bodies.

Rig-Veda, x. 136. 3

I can move through the air like a bird; I can touch with my hands the sun and moon; I have power with respect to my body even so far as unto the Brahma-world.

Saṁyutta-nikâya, v

Knowl. 745

'For I have swift and nimble wings which will ascend the lofty skies,
With which when thy quick mind is clad, it will the loathéd earth despise,
And go beyond the airy globe, and watery clouds behind thee leave,
Passing the fire which scorching heat doth from the heavens' swift course receive,
Until it reach the starry house, and get to tread bright Phoebus' ways,
Following the chilly sire's path, companion of his flashing rays,
And trace the circle of the stars which in the night to us appear,
And having stayed there long enough go on beyond the farthest sphere,
Sitting upon the highest orb partaker of the glorious light,
Where the great King his sceptre holds, and the world's reins doth guide aright,
And, firm in his swift chariot, doth everything in order set.'

Boethius

Suff. 126b

P. State 590

Beauty 670

Holiness 924

Reality 773

Self-motion is the very idea and essence of the soul.

Plato (*Phaedrus*, 245 D)

That tender one who guided the feathers of my wings to so lofty flight. . . .

Dante (*Paradiso,* xxv. 49)

God frees the simple soul and makes it wise in His love. Ah! sweet dove, thy feet are red, thy wings are smooth, thy mouth is true, thine eyes are beautiful, thy head composed, thy wanderings happy, thy flight swift.

Mechthild of Magdeburg

P. State 579

'I will bear thee through the air
To the gates of highest heaven.'

Abraham Lambspring

[1] One can understand here the importance attached to feathers and fringes in American Indian ornamentation. It may be added that they also correspond to rays of light, or showers of grace.

[2] *Muni:* a solitary contemplative.

Illusion 99
Introd. *Sin*

O heart, why art thou a captive in the earth that is passing away?
Fly forth from this enclosure, since thou art a bird of the spiritual world.
Thou art a darling bosom-friend, thou art always behind the secret veil:
Why dost thou make thy dwelling-place in this perishable abode?
Regard thine own state, go forth and journey
From the prison of the Formal world to the meadow of Ideas.
Thou art a bird of the holy world, a boon-companion in the assembly of Love.

Dîvâni Shamsi Tabrîz, XLIV

Pilg. 387
P. State 563

One who travels does so in order to open his ears and eyes and relax his spirit. He explores the Nine States and travels over the Eight Barbarian Countries, in the hope that he may gather the Divine Essence and meet great Taoists, and that he may eat of the plant of eternal life and find the marrow of rocks. Riding upon the wind and sailing upon ether, he goes coolly whithersoever the wind may carry him.

T'u Lung

God lends him wings who is not mounted on the body.

Dîvâni Shamsi Tabrîz, XIII

The spirit lifted me up between the earth and the heaven, and brought me in the visions of God to Jerusalem.

Ezekiel, VIII. 3

M.M. 998
Introd. *Holiness*

He who serves God in the 'great way' assembles all his inner power and rises upwards in his thoughts and breaks through all skies in one act and rises higher than the angels and the seraphs and the thrones, and that is the perfect worship.

Israel Baal Shem

Holiness 902
Center 841
P. State 590

I have been freed from effort and search, I have tied my sleeve to the skirt of God.

If I am flying, I behold the place to which I soar; and if I am circling, I behold the axis on which I revolve;

And if I am dragging a burden, I know whither: I am the moon, and the Sun is in front of me as the guide.

Rûmî

Infra 949
Faith 514

'A real cloud-soarer,' said the Patriarch, 'can start early in the morning from the Northern Sea, cross the Eastern Sea, the Western Sea and the Southern Sea, and land again at Ts'ang-wu. Ts'ang-wu means Ling-ling, in the Northern Sea. To do the round of all four seas in one day is true cloud-soaring.' 'It sounds very difficult,' said Monkey. 'Nothing in the world is difficult,' said the Patriarch, 'it is only our own thoughts that make things seem so.'

Monkey

The man who despises his own body shall ride up on the air like fire.

Abu 'l-Majdûd b. Âdam Sanâ'î

They that wait upon the Lord shall renew their strength; they shall mount up with wings as eagles.

Isaiah, XL. 31

The bird of my heart is a Divine bird which nests on the Throne of God. It has tired of the body's cage and become weary of life in this world. When the bird of the spirit flies from the top of this heap of dust—this world—it will find rest once more in that royal abode. When it takes its flight from this world, the lote-tree of Paradise will be its resting-place. Know that my falcon can rest only on the pinnacles of God's Throne.

Ḥâfiẓ

Peace 694

The soul is a more vigorous and puissant thing, when it is once restored to the possession of its own being, than to be bounded within the narrow sphere of mortality, or to be straightened within the narrow prison of sensual and corporeal delights; but it will break forth with the greatest vehemency, and ascend upwards towards immortality: and when it converses more intimately with religion, it can scarce look back upon its own converses, though in a lawful way, with earthly things, without being touched with a holy shamefacedness and a modest blushing.

John Smith the Platonist

Center 816

Humility 191

We ought to fly away from earth to heaven as quickly as we can; and to fly away is to become like God, as far as this is possible; and to become like him, is to become holy, just, and wise.

Plato (*Theaetetus*, 176 B)

Holiness 900

Moving at Will

Striking down what is physical, the Sunbird, the Immortal, goes where he will.

Bṛihad-Âraṇyaka Upanishad, IV. 3. 11. 12

Supra 944

I live as I please, I wander where I like;
I go in and out each day according to my pleasure;
I move where my inclination leads me;
I assume all the forms which it pleases me to assume;
I hold in my right hand the Lapis stone;
I wear in my right ear the Flower of Ankham for ornament;
I am flourishing, I am prosperous;
I am a perennial youth in the garden of immortality.

Egyptian Tradition

Holiness 924

Center 816 *Death* 225 Introd. *Realiz.*

Both he who is here in a person and who is yonder in the sun—he is one.

He who knows this, on departing from this world, proceeding on to that self which consists of food, proceeding on to that self which consists of breath, proceeding on to that self which consists of mind, proceeding on to that self which consists of understanding, proceeding on to that self which consists of bliss, goes up and down these worlds, eating what he desires, assuming what form he desires.

Taittirîya Upanishad, III. 10. 4, 5

Realiz. 890

I am the door: by me if any man enter in, he shall be saved, and shall go in and out, and find pasture.

St John, X. 9

Inv. 1003

Now that he (the seeker with resigned will) is born from within out of the speaking voice of God in God's will-spirit, he goes in the byss and abyss everywhere free, and is bound to no form; for he goes not in self-hood, but the eternal will guides him as its instrument, according as it pleases God.

Boehme

Center 828 Introd. *P. State*

This just man cannot be hindered in his introversion, for he turns inward both in fruition and in work; but he is like to a double mirror, which receives images on both sides. For in his higher part, the man receives God with all His gifts; and, in his lower part, he receives bodily images through the senses. Now he can enter into himself at will, and can practise justice without hindrance.

Ruysbroeck

Creation 33

Here have I brought thee with wit and with art; now take thy pleasure for guide; forth art thou from the steep ways, forth art from the narrow.

Dante (*Purgatorio,* XXVII. 130)

Holiness 902 Supra 944 941

The vigour of reason and intelligence is denoted by man, and the mobility of incorruptible nature by the birds. When, therefore, 'this corruptible shall have put on incorruption, and this mortal shall have put on immortality',[1] then we, being spiritual in mind and body equally, will . . . have power to be everywhere through the lightness of our incorruptible bodies. Our minds will fly by contemplation, our bodies will fly on account of incorruption.

Hugh of Saint-Victor

Center 841

The Self (Âtman), indeed, is below. The Self is above. The Self is to the west. The Self is to the east. The Self is to the south. The Self is to the north. The Self, indeed, is this whole world.

Verily, he who sees this, who thinks this, who understands this . . . he is autonomous (*sva-râj*); he has unlimited freedom in all worlds.

Chândogya Upanishad, VII. 25. 2

[1] *I. Cor.* XV. 54.

The Sages say truly
That two animals are in this forest:
One glorious, beautiful, and swift,
A great and strong deer;
The other an unicorn.[1]
They are concealed in the forest,
But happy shall that man be called
Who shall snare and capture them.

. . . .

He that knows how to tame and master them by Art,
To couple them together,
And to lead them in and out of the forest,
May justly be called a Master.
For we rightly judge
That he has attained the golden flesh,
And may triumph everywhere;
Nay, he may bear rule over great Augustus.

Abraham Lambspring

Metanoia 480
P. State 563
Holiness 921

When he has finished the banquet at *Hua-ch'ih* (Beautiful Pond) where the moon shines brightly, he (the accomplished) rides on a golden dragon to visit *Tzû-wei* (the star gods). From that time onward, when all of the *hsien* (Immortals) have been interviewed, he rides above wide mountains and wherever he likes.

Chang Po-tuan

P. State 590
Introd.
Realiz.

(The soul) hath many ways to break up her house, but her best is without a disease. This is her mystical walk, an exit only to return. When she takes the air at this door, it is without prejudice to her tenement.

Thomas Vaughan

Death 206

'Don't be angry,' said the Planet. 'As you haven't been to the Hall of Heaven before and haven't yet been given a name, the Guardians don't know who you are, and are quite right not to let you pass (through the Gate of Heaven). When you have seen the Emperor and received your appointment, they'll let you go in and out as you please.' 'That's as may be,' said Monkey, 'but at the present moment I can't get in.'

Monkey

Be content to remain, as it were, in prison for forty days and nights, even as was the good Trevisan, and employ only gentle heat. Let your delicate substance remain at the bottom, which is the womb of conception, in the sure hope that after the time appointed by the Creator for this Operation, the spirit will arise in a glorified state, and glorify its body—that it will ascend and be gently circulated from the centre to the heavens, then

Pilg. 365
Contem. 542
Pilg. 378
Conform. 170
Realiz. 870

[1] Soul and Spirit respectively, according to the alchemist, who calls the forest the Body.

descend to the centre from the heavens, and take to itself the power of things above and things below.

Philalethes

Reality 775 *M.M.* 978

You are plurality transformed into Unity,
And Unity passing into plurality:
This mystery is understood when man
Leaves the part and merges in the Whole.

Shabistarî

As a man standing on the seashore not only sees the sea but can also walk into it as often as he likes: so is it with men who have reached spiritual perfection: they can also enter the Divine Light when they wish, contemplating it and participating in it consciously in proportion to their works, their efforts and the aspirations of their desire.

Grace 556

St Simeon the New Theologian

Knowing the world to consist of consciousness, the mind of the wise man is rapt in the thought of his universality and roams free, seeing the cosmos as space in his own consciousness.

Center 816

Yoga-Vasishtha

They shall go out and in, and find their food,
And, drunk with love, in radiant darkness sleep in God.

Ruysbroeck

Grant that my soul may come forth whithersoever it pleaseth, and let it not be driven away from the presence of the great company of the gods.

Papyrus of Paqrer

With the deep, thundering roar of a lion (Milarepa) proclaimed the Truth of the realizable fact of the illusoriness of the Ego, in the full assurance of its realization, awing and subduing beings and creatures of evil and selfish disposition, and revelled in freedom in the limitless and centreless sphere of the heavens, like an unbridled lion roaming free among the mountain ranges.

Jetsün-Kahbum, or Life of Milarepa

Introd. Supra

Can earthly things seem important to him who is acquainted with the whole of eternity and the magnitude of the universe?

Cicero

Introd. *Holiness*

The WORLD is not this little Cottage of Heaven and Earth. Though this be fair, it is too small a Gift. When God made the World He made the Heavens, and the Heavens of Heavens, and the Angels, and the Celestial Powers. These also are parts of the World: So are all those infinite and eternal Treasures that are to abide for ever, after the Day of

Judgment. Neither are these, some here, and some there, but all everywhere, and at once to be enjoyed.

Center 841
838, 847

Thomas Traherne

Suppose that a being enjoyeth the possession, in various ways, of mystic power: from being one, he becometh multiform; from being multiform, he becometh one; from being visible, he becometh invisible; he passeth without hindrance to the further side of a wall or battlement, or a mountain, as if through air; he walketh on water without dividing it, as if on solid ground; he travelleth cross-legged through the sky, like the birds on the wing.

Anguttara-nikâya

It is truly beyond human nature to possess wings and fly on high at one's own will. But to receive this gift of wings, almost contrary to nature, this is surely like the possession strengthened by exercise, of a marvellous ability for contemplation, so that you may when you will, penetrate on the wing of clear sight into the difficult regions of secret knowledge, impenetrable to mere human effort. Truly we begin to be winged creatures, when having received the gift of grace divinely, we transcend the bounds of our human state, by the flight of our contemplation. For every kind of prophecy, unless it is received in ecstasy, belongs to this . . . uplifting. Is it not a thing beyond human nature to see past things which are not existing now, or to see future things which are not yet? so also to see present things which are not present to the sense, to see the secrets of another's heart, a thing not subject to any sense; to have knowledge of divine things which are above the sense sphere?

Supra
944

Richard of Saint-Victor

One evening, Dr Wm. Sanguinetti (faithful friend and personal physician of Padre Pio) tells us that he and a few others were in Padre Pio's room, when the doctor opened the following conversation:

Dr: 'Padre Pio, when God sends a saint, for instance like St Anthony to another place by bilocation, is that person aware of it?'

Padre Pio: 'Yes. One moment he is here and the next moment he is where God wants him.'

Dr: 'How is this possible?'

Padre Pio: 'By a prolongation of his personality.'

Padre Pio the Stigmatist

Even as Brahma can change his form and move at will, so amongst all beings can he change his form and move at will who is a Comprehensor thereof.

Śâṅkhâyana Araṇyaka, VII. 22

I am a happy man, indeed!
I visit the Pure Land as often as I like:
I'm there and I'm back,
I'm there and I'm back,
I'm there and I'm back,
'Namu-amida-butsu! Namu-amida-butsu!'

Center
847

Saichi

Inv.
1017

Among the gifts which Allâh (may He be glorified and exalted!) grants His slave . . . in this world . . . is that of having at his will and in all liberty the entire earth—both continents and seas: if he wishes, he can go through the air or walk on the water or cover the whole surface of the earth in less than an hour.

Ibn al-'Arîf

Center 841

Holiness 924

When my soul is in Eden with our first parents, I myself am there in a blessed manner. When I walk with Enoch, and see his translation, I am transported with him. The present age is too little to contain it. I can visit Noah in his ark, and swim upon the waters of the deluge. I can see Moses with his rod, and the children of Israel passing through the sea: I can enter into Aaron's Tabernacle, and admire the mysteries of the holy place. I can travel over the Land of Canaan, and see it overflowing with milk and honey: I can visit Solomon in his glory, and go into his temple, and view the sitting of his servants, and admire the magnificence and glory of his kingdom. No creature but one like unto the Holy Angels can see into all ages. Sure this power was not given in vain, but for some wonderful purpose: worthy of itself to enjoy and fathom. Would men consider what God hath done, they would be ravished in spirit with the glory of His doings.

Thomas Traherne

Reality 790
773
Introd.
Heterod.

Desiring the preservation of herds, Brâhmans, genii, and virtuous men—of the Vedas, of law, and of precious things—the Lord of the Universe assumes many bodily shapes: but though he pervades, like the air, a variety of beings, yet he is himself unvaried, since he has no qualities subject to change.

Srimad Bhagavatam

Holiness 914

M.M. 994

When, soul and faculties, you are as used to going up and down as a courtier is to going to and fro at court: when you recognise the various members of the heavenly company and everything God ever made and fail in nothing but know them as the good man knows the members of his household, then you will distinguish between God and Godhead.

Eckhart

REVELATION—AUTHORITY—INFALLIBILITY

No prophecy of the scripture is of any private interpretation.

II. Peter, I. 20

Revelation brings the individual, who is generically fallible, into contact with the supra-individual domain, by definition infallible.

In India, it is the *Matsya-avatâra*[1] who at the beginning of the present cycle or *Mahâyuga* reveals ('un-veils') the *Vêda*, 'by which is to be understood Science pre-eminently, following the etymological signification of this word (derived from the root *vid*, "to know"), or sacred Knowledge in its integrality: and here is a particularly clear allusion to primordial Revelation, or the "non-human" origin of Tradition. It is said that the *Vêda* lasts perpetually, being in itself anterior to all the worlds: but it is somehow hidden or enveloped during the cosmic cataclysms which separate the different cycles, and must accordingly be manifested anew.[2] The affirmation of the perpetuity of the *Vêda* is moreover in direct relationship with the cosmological theory of the primordiality of sound among the sensible qualities (as the quality proper to ether, *âkâsha*, which is the first of the elements); and this theory is not really different from that which other traditions express in speaking of creation by the word: the primordial sound is the Divine Word by which, according to the first chapter of the Hebraic *Genesis*, all things have been made. This is why it is said that the *Rishis* or sages of the first times "heard" the *Vêda*: Revelation, being a work of the Word like creation itself, is really an "audition" for the person who receives it' (Guénon: 'Quelques aspects du symbolisme du poisson', *Études Trad.*, Feb., 1936).

Introd. *Inv.*

The comprehensive nature of this 'audition' is emphasized by Guénon in another passage, alluding to the 'Night of Power' (*laylatu 'l-qadr*) wherein the *Qur'ân* is revealed: 'This "night", following the commentary of Mohyiddin ibn Arabi, is identified with the body itself of the Prophet. What is particularly noteworthy here is that "revelation" is received, not in the mind, but in the body of the being who is "delegated" to express Principle: *Et Verbum caro factum est*, says the Gospel likewise (*caro* and not *mens*), and here is another expression which is the exact equivalent, in the form proper to the Christian tradition, of what *laylatu 'l-qadr* represents in the Islamic tradition' ('Les deux nuits', *Études Trad.*, 1939, p. 161).

Introd. *Orthod.*

From a more strictly microcosmic perspective, Revelation can be considered as 'a direct

[1] The first of the manifestations of Vishnu, in the form of a fish.

[2] Also in Buddhism, the imperishable doctrine is conceived of as being preserved intact 'underseas', between cycles of manifestation, by dragon deities (cf. *Hônen*, p. 764).

communication with the higher states.... The possibility of this "Revelation" is based on the existence of faculties which are transcendent to the individual. Whatever name is given them, whether one speaks for example of "intellectual intuition" or "inspiration", it is always the same thing in the end. The first of these two terms brings to mind in a sense the "angelic" states, which in point of fact are identical with the supra-individual states of being, whereas the second recalls more particularly that action of the Holy Spirit to which Dante expressly alludes (*De Monarchia,* III. 16). One could also say that what is "inspiration" interiorly, for the person who receives it directly, becomes "Revelation" exteriorly, for the human collectivity to whom it is transmitted through his mediation, in the measure that such a transmission is possible, namely, in the measure that it can be expressed' (Guénon: *Autorité spirituelle et Pouvoir temporel,* pp. 100–101).

Introd. *Knowl. Center* 828

It is essential to understand the meaning of intellect if one is to grasp the doctrine of Revelation. 'The intellect is a receptive faculty and not a productive power: it does not "create"; it receives and transmits. It is a mirror which reflects reality in an adequate and therefore efficacious manner. With most men in the "dark age", the intellect is atrophied to the point of being reduced to a simple virtuality' (Schuon: 'Orthodoxie et intellectualité', *Études Trad.,* 1954, p. 211).

'What is Revelation for "a humanity" will analogously be intellection for an individual, and vice versa. If every man possessed intellect, not merely in a fragmentary or virtual state, but as a fully developed faculty, there would be no Revelation, since total intellection would be something natural; but as this has not been the case since the end of the Golden Age, Revelation not only is necessary but even normative as regards particular intellection, or rather, as regards its formal expression. There is no intellectuality possible outside of a revealed language, an oral or scriptural tradition, even though intellection can happen as an isolated miracle wherever the intellective faculty exists; but an extra-traditional intellection will have neither authority nor efficacity.... Revelation for the intellect functions as a principle of actualization, expression and verification (Schuon: 'De la foi', *Études Trad.,* 1953, pp. 348–349).

Introd. *P. State Holiness* 921

The Mediator as *Pontifex,* identified with the World Axis and thus effectuating contact with the higher states through his function of *bridge* or *way,* is the vehicle par excellence of Revelation.[1] 'The King is now in reality a "Highness"; his actions are no longer determined by the likes and dislikes of his sensitive part (*necessitas coactionis*), but inwardly instigated, and being thus strictly speaking "inspired", participate in the "infallibility" of whatever proceeds *ex cathedra,* "from the tripod of truth"; the burden of responsibility transferred to other shoulders no longer adds to the sum of his mortality and we can say: "O King, live for ever". When we speak of a King as "His Serene Highness" we are speaking precisely of the truly royal quality of self-possession by which a King, if he be really a King, is indeed "exalted".

Action 340

'Thus from the standpoint of Indian sociological theory and that of all traditional politics, an individual tyranny, whether that of a despot, that of an emancipated artist, or that of the self-expressive man or self-sufficient woman, effects in the long run only what is ineffectual (*akṛtâni,* "misdeeds"): all self-importance leads to the disintegration and finally the death of the body politic, collective or individual. The essence of the traditional politics

Introd. *Action*

[1] Cf. Guénon: *La Grande Triade,* ch. XVII.

amounts to this, that "Self-government" (*svarâj*) depends upon self-control (*âtmasaṁyama*), Rule on ruliness' (Coomaraswamy: *Spiritual Authority and Temporal Power in the Indian Theory of Government*, p. 85).

'But where is the notion of a real hierarchy still to be found in the modern world? Nothing and nobody is any longer in the right place; men no longer recognize any effective authority in the spiritual order or any legitimate power in the temporal; the "profane" presume to discuss what is sacred, and to contest its character and even its existence; the inferior judges the superior, ignorance sets bounds to wisdom, error prevails over truth, the human supersedes the divine, earth overtops heaven, the individual sets the measure for all things and claims to dictate to the universe laws drawn entirely from his own relative and fallible reason' (Guénon: *The Crisis of the Modern World*, p. 99; tr. Arthur Osborne).

Wr. Side 464

Infallibility relates to doctrine: 'If doctrine is infallible, this is because it is an expression of the truth, which in itself is absolutely independent of the individuals who receive it and who comprehend it. The guarantee of doctrine resides fundamentally in its "non-human" character: and one can say moreover that every truth, to whatever order it pertains, if taken from the traditional point of view, participates in this character, for it is truth only because it adheres to higher principles and is derived as a more or less immediate consequence from them, or from application to a determined domain. The truth is never made by man, as the modern "relativists" and "subjectivists" would have it, but is on the contrary imposed upon him, though not "from without" in the manner of a "physical" constraint, but in reality "from within", since man is obviously obliged to "recognize" it as truth only if he first of all "knows" it, that is to say, if it has penetrated him and if he has really assimilated it' (Guénon: *Aperçus sur l'Initiation*, p. 291).

Knowl. 749

Infallibility relates to gnosis: 'It follows from this that every man will be infallible when he expresses a truth which he really knows, namely, with which he is identified: but not inasmuch as he is a human individual will he be infallible, but inasmuch as, by reason of this identification, he represents so to speak this truth itself. Strictly, one should say in such a case, not that he expresses the truth, but rather that the truth is expressed through him' (id., p. 292).

Infallibility relates to ritual efficacy: 'This efficacy is essentially inherent in the rites themselves, insofar as they are the means of action of a spiritual influence. The rite therefore acts independently of the quality as such of the individual who performs it, and without his even having to be effectively conscious of this efficacy' (id., p. 293).

Introd. *Orthod.*

'It should also be precisely stated that doctrinal infallibility, such as it has been defined, is necessarily limited, just as the function to which it attaches, and this in several ways: first of all, it can only be applied at the interior of the traditional form to which this function relates, and it is non-existent in respect to all that pertains to any other traditional form: in other words, no one can pretend to judge a tradition in the name of another tradition.[1] . . . Secondly, if a function pertains to a certain determined order, it will entail infallibility only in what concerns this order. . . . Beyond the legitimate limits which suit each case, there is no longer infallibility, since there is then nothing to which it can be validly applied' (id., pp. 296–297).

Reality 790

[1] This would in part explain why certain contemporary Hindus, whose doctrinal grasp of their own tradition is beyond dispute, make the most lamentable errors—even absurdities—when dealing with certain aspects of Western civilization. It need hardly be added that the fault is reciprocal!

Avataric Revelation is distinguished by its cosmic proportions, being global, comprehensive, and unique, as can be seen in the complete originality of each traditional form from another.

The inspirations of the saints and sages within each form carry different dimensions and developments always homogeneous to and in terms of that form, without ever violating the doctrinal and ritual cohesion proper to that form.

All the rest is fragmentary, imperfect, or erroneous. No false prophet has ever produced anything universal and unique, but only a borrowing from, distortion, or inversion of some form already in existence.[1]

Heterod. passim

It might be said that Revelation in its entirety enables us to view the world, the cosmos, mankind, and creatures with Divine Vision, as it were, giving us patterns of knowledge and action which supersede all human opinion and conjecture whatsoever.[2]

[1] Cf. St Augustine, *De Civ. Dei*, XVIII, xviii: 'Nor can the devils create anything (whatever appearances of theirs produce these doubts) but only cast a changed shape over that which God has made, altering it only in show.'

[2] It may be well to repeat here that canonical scripture always prevails in hierarchical importance over other texts in this book, even though the passages have been collated in regard to unity and sequence of idea, rather than preponderance of authority.

Revelation

And who shall know thy thought, except thou give wisdom, and send thy Holy Spirit from above?

Knowl. 749

Wisdom, IX. 17

And it has not been (vouchsafed) to any mortal that Allâh should speak to him except by revelation or from behind a veil, or that He sendeth a messenger to reveal what He will by His leave. Lo! He is Exalted, Wise.

Symb. 306

Qur'ân, XLII. 51

I revealed my wisdom first to Brahmâ in the form of the Vedas. Brahmâ declared that wisdom unto his son Manu, from whom the seven patriarchs and sages—Bhrigu and the others—received it. From them it passed on to their sons and disciples, who, being of various temperaments and natures, understood it variously. Thus arose the several interpretations of the Vedas.

Orthod. 275

Srimad Bhagavatam, XI. viii

'Veda is called Veda (Knowledge) because by means of the Veda people learn the means which cannot be learnt either by direct perception or by reasoning.'

Sri Chandrasekhara Bhâratî Swâmigal
(citing from the Śâstras)

Since it is usual with all men of sound understandings, to call on divinity, when entering on any philosophic discussion, it is certainly much more appropriate to do this in the consideration of that philosophy which justly receives its denomination from the divine Pythagoras. For as it derives its origin from the Gods, it cannot be apprehended without their inspiring aid.

Iamblichus

We know that the revelation of these (Buddhist) mysteries was clearly not the work of man.

The Travels of Fa-hsien

All scripture is given by inspiration of God.[1]

II. Timothy, III. 16

[1] St Augustine tells us (*Civ. Dei,* XVIII. xlii) that Eleazar the high priest at Ptolemy's bidding sent six scribes from every tribe to translate the Old Testament from Hebrew into Greek. 'Their translation do we now usually call the Septuagint. The report of their divine concord therein is admirable: for Ptolemy having (to try their faith) made each one translate by himself, there was not one word of difference between them, either in sense or order, but all was one, as if only one had done it all, because indeed there was but one spirit in them all.'

'Where did you learn this from?' asked Nanpo Tsek'uei.

Infra 967 — 'I learned it from the Son of Ink,' replied Nü Yü, 'and the Son of Ink learned it from the Grandson of Learning, the Grandson of Learning from Understanding, and Understanding from Insight. Insight learned it from Practice, Practice from Song, and Song from Silence, Silence from the Void, and the Void learned it from the Seeming Beginning.'

Chuang-tse (ch. VI)

This wisdom was revealed by the inspiration of the Holy Spirit.

The Glory of the World

Inv. 1003 — God's speaking is his child-bearing.

St Augustine

Divine truth hath its humiliation and exinanition, as well as its exaltation. Divine truth becomes many times in Scripture incarnate, debasing itself to assume our rude conceptions, that so it might converse more freely with us, and infuse its own divinity into us. God having been pleased herein to manifest himself not more jealous of his own glory, than he is (as I may say) zealous of our good. *Nos non habemus aures, sicut Deus habet linguam.* If he should speak in the language of eternity, who could understand him, or interpret his meaning? or if he should have declared his truth to us only in a way of the purest abstraction that human souls are capable of, how should then the more rude and illiterate sort of men have been able to apprehend it? Truth is content, when it comes into the world, to wear our mantles, to learn our language, to conform itself as it were to our dress and fashions.

Love 618

John Smith the Platonist

By the Star when it setteth,
Your comrade erreth not, nor is deceived;
Nor doth he speak of (his own) desire.
It is naught save an inspiration that is inspired,
Which one of mighty powers hath taught him.

Orthod. 275

Qur'ân, LIII. 1–5

For the better understanding all prophetical writ, we must observe, That there is sometimes a seeming inconsistence in things spoken of, if we shall come to examine them by the strict logical rules of method: we must not therefore, in the matter of any prophetical vision, look for a constant methodical contexture of things carried on in a perpetual coherence. The prophetical spirit doth not tie itself to these rules of art, or thus knit up its dictates systematically, fitly framing one piece or member into a combination with the rest, as it were with the joints and sinews of method: for this indeed would rather argue a human and artificial contrivance than any inspiration, which, as it must beget a transportation in the mind, so it must spend itself in such abrupt kind of revelations as may argue indeed the prophet to have been inspired.[1]

Introd. *Ecstasy*

John Smith the Platonist

[1] The particular structure of the *Qur'ân,* for example, results from the translation of archetypal Idea into a temporal mode corresponding at its level with the spiritual Prototype in immediacy of concept, image and sound; cf. Schuon: *Comprendre l'Islam,* ch. II, and Burckhardt: *Introduction aux Doctrines ésotériques de l'Islam,* pp. 45–46.

Even so there came no messenger unto those before them but they said: A wizard or a madman!

Have they handed down (the saying) as a legacy one unto another? Nay, but they are froward folk.

Wr. Side 448

Qur'ân, LI. 52, 53

No State can be happy which is not designed by artists who imitate the heavenly pattern.

Beauty 670

Plato (*Republic*, 500 E)

Love without Knowledge
Is darkness to the wise soul.
Knowledge without Revelation
Is as the pain of Hell.
Revelation without death,
Cannot be endured.

Knowl. 761
734
Death
208

Mechthild of Magdeburg

The Revelation of the Scripture is from Allâh, the Mighty, the Knower,

The Forgiver of sin, the Accepter of repentance, the Stern in punishment, the Bountiful. There is no God save Him. Unto Him is the journeying.

None argue concerning the revelations of Allâh save those who disbelieve, so let not their turn of fortune in the land deceive thee.

Pilg. 365
Suff. 120

Qur'ân, XL. 2–4

Authority

No pronouncement of a prophet is ever his own; he is an interpreter prompted by Another in all his utterances, when knowing not what he does, he is filled with inspiration, as the reason withdraws and surrenders the citadel of the soul to a new visitor and tenant, the Divine Spirit which plays upon the vocal organism and raises sounds from it, which clearly express its prophetic message.

Philo

Let us set before our minds the scriptural rule that in speaking about God we should declare the Truth, not with enticing words of man's wisdom, but in demonstration of the power which the Spirit stirred up in the Sacred Writers, whereby, in a manner surpassing

Knowl. 743

speech and knowledge, we embrace those truths which, in like manner, surpass them, in that Union which exceeds our faculty, and exercise of discursive, and of intuitive reason. We must not then dare to speak, or indeed to form any conception, of the hidden super-essential Godhead, except those things that are revealed to us from the Holy Scriptures.

Dionysius

Or say they: He hath invented it? Nay, but they will not believe!
Then let them produce speech the like thereof, if they are truthful.
Or were they created out of naught? Or are they the creators?
Or did they create the heavens and the earth? Nay, but they are sure of nothing!
Or do they own the treasures of thy Lord? Or have they been given charge (thereof)?
Or have they any stairway (unto heaven) by means of which they overhear (decrees)? Then let their listener produce some warrant manifest!

Introd. *Grace*

Qur'ân, LII. 33–38

I must dare to speak the truth, when truth is my theme.

Plato (*Phaedrus,* 247 D)

Reality 775

Not on my authority, but on that of truth, it is wise for you to accept the fact that all things are one.

Heraclitus

The truth I tell is writ on many a page of the writers of the Holy Spirit.

Dante (*Paradiso,* XXIX. 40)

To those desiring to learn the great and hidden good it is authority which opens the door. And whoever enters by it and, leaving doubt behind, follows the precepts for a truly good life, and has been made receptive to teaching by them, will at length learn how pre-eminently possessed of reason those things are which he pursued before he saw their reason, and what that reason itself is, which, now that he is made steadfast and equal to his task in the cradle of authority, he now follows and comprehends, and he learns what that intelligence is in which are all things, or rather what He is who is all things, and what beyond and above all things is their prime cause.

Faith 505

Knowl. 761 Introd. *Heterod.*

St Augustine

With the ancient is wisdom.

Job, XII. 12

Orthod. 275

The ancient Non-hon-shin-ga have handed down to us in songs, *wi-gi-e,* ceremonial forms, symbols, the many things they learned of the mysteries that surround us on all sides. All these things they learned through their power of 'wa-thi-gthon', the power to search with the mind. They speak of the light of the day by which the earth and all living things that dwell thereon are influenced; of the mysteries of the darkness of night that reveal to us all the great bodies of the upper world, each of which forever travels in a circle of its own unimpeded by the others. . . .

Many of the sayings of the Non-hon-shin-ga who lived long ago have come down to us and have been treasured by the people as expressions coming from men who have been in close touch with the mysterious power whom the people had learned to worship and to reverence. Moreover, the men who uttered these sayings had long since departed for the spirit land and were regarded by their descendants as Wa-kon-da-gi, that is, sacred and mysterious persons. These sayings had been transmitted in ritual form, and during the passage of years had been jealously guarded against desecration by those persons who succeeded in memorizing them and had taken care to teach them only to such pupils as manifested a proper spirit of reverence for things sacred.[1]

Playful-calf

Beauty 670
689

We have not kept to ourselves any of the Hierarchic Utterances which have been handed down to us but have imparted them without adulteration both to yourselves and to other holy men, and will continue so to do as long as we have the power to speak and you to hear. So will we do no despite unto the tradition, unless strength fail us for the perception or the utterance of these Truths. But be these matters as God wills that we should do or speak.

Dionysius

Conform.
180

Ramdas' authority is derived from the fact that his experiences at different stages of his Sadhana bear a close resemblance with those of all saints and sages of the world who had the vision of God.

Swami Ramdas

(There is) one main characteristical distinction between the prophetical and pseudo-prophetical spirit, *viz.* That the prophetical spirit doth never alienate the mind, (seeing it seats itself as well in the rational as in the sensitive powers,) but always maintains a consistency and clearness of reason, strength, and solidity of judgment, where it comes: it doth not ravish the mind, but inform and enlighten it: but the pseudo-prophetical spirit, if indeed without any kind of dissimulation it enters into any one, because it can rise no higher than the middle region of man, which is his fancy, it there dwells as in storms and tempests, and being ἄλογόν τι in itself, is also conjoined with alienations and abruptions of mind.

John Smith the Platonist

Introd.
Ecstasy

Heterod.
430
Introd.
Pilg.

Brahman should be regarded as the Self on the evidence of the scriptures, just as religious duties are known from the same source.

Śrî Śankarâchârya

Realiz.
890

Upon Us resteth the explanation of it (the *Qur'ân*).

Qur'ân, LXXV. 19

[1] Cf. note 2, p. 22.

Orthod. 300 Infra 967

O that thou hadst hearkened to my commandments! then had thy peace been as a river, and thy righteousness as the waves of the sea.

Isaiah, XLVIII. 18

Creation 38

We are like dwarfs seated on the shoulders of giants. We see more things than the Ancients, and things farther away, but this is not due to the sharpness of our vision or the height of our build. It is because they carry us and raise us from their gigantic height.

Bernard de Chartres

Holiness 924 *Faith* 510

I, Hônen, in my original being, am really Seishi, the great Boddhisattva, who has appeared here in this Temple for the salvation of sentient beings. I am every day with all who follow me, and I shall protect them and assuredly bring them to the Land of Perfect Bliss. If I should fail to make this vow effective, I would never accept of perfect enlightenment myself.

Hônen

Holiness 921

A true king must get a commission from Heaven before he becomes king.

Tung Chung-shu

Not force but law, and that Divine, was the beginning of the Roman Empire.

Dante (*Il Convito,* IV. iv. 5)

Illusion 89

The human race when best disposed depends upon a unity in wills. But this unity cannot be unless there is one will dominating and ruling all the rest to oneness; inasmuch as the wills of mortals, because of the seductive delights of youth, have need of a directive principle,[1] as the philosopher teaches in the last *Ad Nicomachum.* Nor can that one will exist unless there be a single prince of all, whose will may be the mistress and ruler of all others. Now if all the above deductions are sound, which they are, it is necessary for the best disposition of the human race that there should be a monarch in the world, and therefore for the well-being of the world that there should be a monarchy.

Dante (*De Monarchia,* I. xv)

Imperial authority was devised for the perfecting of human life.

Dante (*Il Convito,* IV. ix. 1)

Orthod. 275

If any one doubts the *reality* of our Art, he should read the books of those ancient Sages whose good faith no one ever yet called in question, and whose right to speak on this subject cannot be challenged. If you will not believe *them,* I am not so foolish as to enter into a controversy with one who denies first principles.

Michael Sendivogius

[1] One sees how it is the tendency today, in the absence of any principle, for the 'students' (those at an age which is naturally rebellious, when passion and illusion most easily dominate the mind) of various countries to assume 'responsibility' for their governments, by way of anarchy and violence.

But if they are averse, We have not sent thee as a warder over them. Thine is only to convey (the message).

Qur'ân, XLII. 48

And if any man hear my words, and believe not, I judge him not: for I came not to judge the world, but to save the world.
He that rejecteth me, and receiveth not my words, hath one that judgeth him: the word that I have spoken, the same shall judge him in the last day.

St John, XII. 47–48

Judgment 242
Supra Introd.

And whosoever turneth away, still Allâh is the Independent, the Owner of Praise.

Qur'ân, LVII. 24

Faith 505

Infallibility

The relation of the Rituals to the rectification of the state is the relation of a balance to weight, of inked string to crookedness and straightness, of compasses and T-squares to roundness and squareness. The reason is that provided the balance is truly suspended it is impossible for it to be deceived over the weight: provided the inked string is truly applied it is impossible for it to be deceived over the crookedness and straightness (of a line).

Li Chi, Ching Chieh

Infallibility in the spiritual order and sovereignty in the temporal order are two perfectly synonymous terms.

Joseph de Maistre

Holiness 921

The Scriptures, direct experience, authority, and inference—these are the four proofs of knowledge.

Srimad Bhagavatam, XI. xii

Now I am certain that this is not an invention of my own, who am well aware that I know nothing, and therefore I can only infer that I have been filled through the ears, like a pitcher, from the waters of another.

Plato (*Phaedrus*, 235 C)

Knowl. 743

Let no man who is present at this festival find fault with my art by reason of my personal defects: but be it known that the spirit which God breathes into men of my sort is unfailing.

Hermes

Holiness 910
902

'The Order in itself can neither be spoiled nor corrupted by the fault of the subordinates or the superiors. One wishing to enter an Order should not consider the bad people that are within: he should support himself on the arm of the Order which is strong and cannot weaken, and remain faithful to it until death.'

Faith 510

St Catherine of Siena

Introd. *Orthod.*

Indeed it is hard to find anyone in this world, who gives heed to the Law itself, irrespective of the character of the man who expounds it.

Hônen

Say: I say not unto you that I possess the treasures of Allâh, nor that I have knowledge of the Unseen; and I say not unto you: Lo! I am an angel. I follow only that which is inspired in me. Say: Are the blind man and the seer equal? Will ye not then take thought?

Holiness 900

Qur'ân, VI. 50

He (the Delphian Oracle) is a god, and cannot lie; that would be against his nature.

Plato (*Apology*, 21 B)

Action 340 *Knowl.* 749

It is God that has brought you to me, Asclepius, to hear a teaching which comes from God. My discourse will be of such a nature, that by reason of its pious fervour it will be rightly deemed that there is in it more of God's working than in all that I have spoken before—or rather, that God's power has inspired me to speak.

Hermes

Inv. 1031

Blessed and praised be the name of our Lord Jesus Christ who has provided for us an image of the truth, himself namely, wherein is no possibility of error!

Eckhart

For we walk by faith not by sight: and faith will totter if the authority of the divine scriptures be shaken.

St Augustine

Renun. 146

Or believe ye in part of the Scripture and disbelieve ye in part thereof?

Qur'ân, II. 85

Reality 773 *Creation* 48

I believe in one God, sole and eternal, who moveth all the heaven, himself unmoved, with love and with desire.

And for such belief I have not only proofs physic and metaphysic, but it is given me likewise by the truth which hence doth proceed

Orthod. 275

through Moses, through the Prophets and through the Psalms, through the Gospel and through you who wrote when the glowing Spirit had made you fosterers.

Dante (*Paradiso*, XXIV. 130–138)

No one need doubt the truth or certainty of this Art. It is as true and certain, and as surely ordained by God in nature, as it is that the sun shines at noontide, and the moon shews her soft splendour at night.

The Golden Tract

Lo! those who disbelieve in the Reminder when it cometh unto them (are guilty), for lo! it is an unassailable Scripture.

Falsehood cannot come at it from before it or from behind it. A Revelation from the Wise, the Owner of Praise.

Qur'ân, XLI. 41, 42

It is a Revelation from the Lord of the Worlds.
And if he had invented any sayings in Our name,
We assuredly had taken him by the right hand
And then severed his life-artery,
And not one of you could have held Us off from him.
And lo! it is a warrant unto the God-fearing.
And lo! We assuredly know that among you there are deniers.
And lo! it is indeed an anguish for the disbelievers.
And lo! it is the Truth of Certainty.
So glorify the Name of thy Lord the Tremendous.

Qur'ân, LXIX. 43–52

Let go the things in which you are in doubt for the things in which there is no doubt.

Muhammad

The Hierarchy of Powers

These selves depend on that Self as retainers on their chieftain.

Kaushîtaki Upanishad, IV. 20

The power of the sovereign proceeds from that of Principle.

Chuang-tse (ch. XII)

Beauty 670

Every *Divine Precept* is founded in a *Divine Truth.*

Peter Sterry

There is a divinity moving you, like that contained in the stone which Euripides calls a magnet, but which is commonly known as the stone of Heraclea. This stone not only attracts iron rings, but also imparts to them a similar power of attracting other rings; and

Orthod. 275 *Holiness* 902

sometimes you may see a number of pieces of iron and rings suspended from one another so as to form quite a long chain: and all of them derive their power of suspension from the original stone. In like manner the Muse first of all inspires men herself: and from these inspired persons a chain of other persons is suspended, who take the inspiration.

Plato (*Ion*, 533 D)

Introd. *Conform.*

So, when the intellect becomes thy captain and master, the dominant senses become subject to thee.

Rûmî

Peace 698

Reason in a good man sits in the throne, and governs all the powers of his soul in a sweet harmony and agreement with itself: whereas wicked men live only ζωὴν δοξαστικὴν, being led up and down by the foolish fires of their own sensual apprehensions. In wicked men there is a democracy[1] of wild lusts and passions, which violently hurry the soul up and down with restless motions. All sin and wickedness is στάσις καὶ ὕβρις τῆς ψυχῆς, 'a sedition stirred up in the soul by the sensitive powers against reason'. It was one of the great evils that Solomon saw under the sun, 'Servants on horseback, and princes going as servants upon the ground' (Eccles. X. 7). We may find the moral of it in all wicked men, whose souls are only as servants to wait upon their senses. In all such men the whole course of nature is turned upside down, and the cardinal points of motion in this little world are changed to contrary positions: but the motions of a good man are methodical, regular, and concentrical to reason.

Wr. Side 464

John Smith the Platonist

It is fitting that the workman should use the tool, and not the tool the workman: that the rider should guide and spur the horse, and not the horse the rider: and that the sovereign should direct and govern the people, and not the people the sovereign. If in such things as these the natural order is maintained, the result is harmony and beauty: but if the relations are reversed, the result is confusion, ugliness and distortion.

Beauty passim

Hermes

Creation 38

There is no rectifying of those above by those below. Rectification must be from above downwards.

Mo Ti

The triumph of mediocre men brings down the elite.

'Alî

When the people rule over the sovereign, the times are upside down, and it cannot be but that both sovereign and people go to ruin: and even so, if the body rules over the soul, both must needs go to ruin.

Hermes

[1] One will notice the principles brought forth here distinguishing monarchy from democracy.

Once there were the heavens and the earth, there was the distinction between upper and lower, and when the first enlightened king made a permanent state, there was social organization (with their class distinctions). Two nobles cannot serve each other, neither can two commoners set each other to work. This is the mathematics of Heaven.

Hsun Ch'ing

He hath exalted some of you above others in rank.

Qur'ân, VI. 165

Conform. 180

When the upper and lower ranks in society have a family feeling for each other, this means human-heartedness.

Li Chi, Ching Chieh

The late Jagadguru has repeatedly prayed to God to vouchsafe to the people of the land the recollection of the sacred truths.... His anxiety is that faith in the Śâstras should be restored once again in the land so that the people may abandon the new ways of life, conduct and dress which they have adopted quite in violation of the rules of their respective castes and families.

It is well known that people everywhere are now suffering. It can be confidently asserted that this suffering dates from their giving up the courses of conduct observed by their ancestors. When the practice of Dharma began to decline, suffering began.

Sri Chandrasekhara Bhâratî Swamigal

Action 335

Reality 775

Those who preside over the practice of religion should be looked up to and venerated as the soul of the body.... The place of the head in the body of the commonwealth is filled by the prince, who is subject only to God and to those who exercise His office and represent Him on earth, even as in the human body the head is quickened and governed by the soul. The place of the heart is filled by the senate, from which proceeds the initiation of good works and ill. The duties of eyes, ears, and tongue are claimed by the judges and the governors of provinces. Officials and soldiers correspond to the hands. Those who always attend upon the prince are likened to the sides. Financial officers and keepers ... may be compared with the stomach and intestines.... The husbandmen correspond to the feet, which always cleave to the soil, and need the more especially the care and foresight of the head, since while they walk upon the earth doing service with their bodies, they meet the more often with stones of stumbling, and therefore deserve aid and protection all the more justly since it is they who raise, sustain, and move forward the weight of the entire body....

Then and then only will the health of the commonwealth be sound and flourishing, when the higher members shield the lower, and the lower respond faithfully and fully in like measure to the just demands of their superiors, so that each and all are as it were members one of another by a sort of reciprocity, and each regards his own interest as best served by that which he knows to be most advantageous for the others.[1]

John of Salisbury

[1] All this corresponds perfectly in mediaeval Christian terms with the principles on which the Hindu caste system is based.

I affirm that the temporal power does not receive its being from the spiritual, nor its virtue, which is its authority; nor even its efficiency absolutely. But it does receive therefrom the power of operating with greater virtue, through the light of grace which the blessing of the supreme pontiff infuses into it, in heaven and on earth.

Dante (*De Monarchia,* III. iv. 145)

Duke Ai: 'May I ask what is the art of government?'

Beauty 689

Confucius: 'The art of government simply consists in making things right, or putting things in their right places.'

Confucius

MYSTERIUM MAGNUM

The Tao that can be expressed is not the eternal Tao;
The name that can be named is not the unchanging Name.

Tao Te Ching, I

Then only will you see it, when you cannot speak of it; for the knowledge of it is deep silence, and suppression of all the senses.

Hermes (*Lib.* x. 5–6)

God is He whose Name must not even be pronounced.

Akka Chief[1]

'It is of the essence of a mystery, and above all of the Mysterium Magnum, that it cannot be communicated,[2] but only realized: all that can be communicated are its external supports or symbolic expressions; the Great Work must be done by everyone for himself. . . . The Way has been charted in detail by every Forerunner, who *is* the Way: what lies at the end of the road is not revealed, even by those who have reached it, because it cannot be told and does not appear: the Principle is not in any likeness' (Coomaraswamy: 'The Nature of Buddhist Art', in *Figures of Speech*, p. 170).

Introd.
Holy War 405
Pilg. 365
Void 721
Knowl. 749

'This supreme goal is the absolutely unconditioned state, freed from all limitation; for this very reason, it is entirely inexpressible, and all that can be said of it is only to be rendered by terms that are negative in form: negation of the limits which determine and define all existence in its relativity.[3] The acquisition of this state is what the Hindu doctrine calls "Deliverance" when considered in relation to the conditioned states, and "Union" when envisaged in relation to the supreme Principle[4].

'In this unconditioned state, moreover, all the other states of the being are recovered in principle, but transformed, being released from the special conditions which determine them insofar as particular states. What remains is all that has a positive reality, since it is here that everything has its principle; the "delivered" being is truly in possession of the plenitude of his possibilities. What disappear are uniquely the limitative conditions, whose reality is wholly negative, since they only represent a "privation" in Aristotle's sense of the word. Hence, far from being a kind of annihilation as certain Westerners believe, this final

Renun. 158
Void 724

[1] R.P. Trilles: *L'Ame du Pygmée d'Afrique*, Paris, Les Éditions du Cerf, 1945, p. 91. .

[2] 'It is inexpressible (and not the incomprehensible as is commonly believed) that the word "mystery" originally designated, for in Greek, μυστηριον derives from μυειν, which means "to be quiet", "to be silent". Stemming from the same verbal root *mu* (from whence comes the Latin *mutus*, "mute") is the word μυθος, "myth", which prior to deviating from its original sense so as to mean nothing more than a fanciful tale, signified things incapable of direct expression, which could only be suggested by a symbolical likeness, whether verbal or figurative' (Guénon: *Les États multiples de l'Être*, p. 37, note 1).

[3] In-finite = not finite, a 'negative' definition.

[4] Cf. note 1, p. 726.

state is on the contrary absolute plenitude, the supreme reality in regard to which all the rest is but illusion' (Guénon: *La Métaphysique orientale*, pp. 19–20).

Sensibilia are not 'denied' in the *via negativa*, which would be metaphysically untenable: but, to use the language of Zen, 'they are understood in relation to what they are not'.[1]

'The (Supreme) Essence (*adh-Dhât*) is God insofar as He is without "aspects", being in Himself neither the "object" nor the "subject" of any knowledge. The (Divine) Qualities (*aṣ-Ṣifât*) by contrast are the "aspects" through which God reveals (*tajalla*) Himself in a relative manner. If the Essence cannot be known by created beings, this is because relative being does not subsist in confrontation with absolute and infinite Reality. However, the Essence is knowable at each degree of reality, in the sense that It is the inner reality of all knowledge. God knows Himself by Himself in Himself without any internal distinction: and He knows Himself by Himself in the universe according to relative modes which are infinitely varied' (Burckhardt: *De l'Homme universel*, pp. 8–9).

Void 728

[1] D. T. Suzuki: *Essays in Zen Buddhism*, 3rd Series, London, Kyoto, 1934, p. 205.

Exoterism and Esoterism

A roll of a book was therein . . . written within and without.

Ezekiel, II. 9. 10

The Aristotelians assert that some of their books are esoteric and some exoteric.

Clement of Alexandria

His (Plato's) doctrines were intelligible to very few of the most devout, and quite obscure to the profane.

Apuleius

Allâh hath conferred on those who strive with their wealth and lives a rank above the sedentary.

Qur'ân, IV. 95

I have meat to eat that ye know not of.

St John, IV. 32

There are more things in heaven and earth, Horatio,
Than are dreamt of in your philosophy.

Shakespeare (*Hamlet,* I. v. 166)

Knowl. 734

A man cannot comprehend spiritual things with his ordinary intelligence.

Sri Ramakrishna

(The teacher) should keep his teaching within the bounds of the pupil's understanding. To give him what he cannot understand may scare him away, or his mind may do violence to it. Muhammad said: 'No one ever gave information to people who were not capable of understanding it without its proving a temptation to some of them': and Jesus said: 'Do not hang jewels on the necks of swine'. A man of learning is not bound to tell all he knows to everyone.[1] But each is servant to the other according to the measure of his intelligence.

Heterod. 418
Infra 987
Rev. 967

To the pupil of limited ability should be given only what is clear and suitable for him: nor should he be told that there are refinements which are being held back from him. That would abate his desire for what is clear, and confuse his heart by making him think that he is being stingily dealt with. For naturally everyone thinks that he is fit for all kinds of knowledge, and the more stupid he is the more he rejoices in the perfection of his intellect. From this it follows that the convictions of the common people should not be disturbed, because that would remove the barrier between them and acts of disobedience. With them

Orthod. 300

[1] Cf. note 2, p. 22.

Faith 512 Wr. Side 464 Action 335

one should not plunge rashly into the refinements of real knowledge, but should be content to teach them the necessary acts of worship, and faithfulness in their occupations, without raising doubts. The door of discussion should not be opened to the common people, for that might cause them to forget their proper work on which the continuance of life depends.

Al-Ghazâlî

Conform. 180

Over every possessor of knowledge is one more knowing.

Qur'ân, XII. 76

Symb. 313

And the disciples came, and said unto him, Why speakest thou unto them in parables? He answered and said unto them, Because it is given unto you to know the mysteries of the kingdom of heaven, but to them it is not given.

St Matthew, XIII. 10, 11

For all cannot be mystics or grasp the mysteries.

Shabistarî

Reality 790

He who would rightly read and understand the Old Testament should represent to himself two images: an *exoteric* Adam, the terrestrial man, and an *esoteric* Christ, and transform the two images into a single one. Then he will be able to understand the spiritual sense of all that Moses and the Prophets have said.

Boehme

Charity 602

Dhamma has been taught by me without making a distinction between esoteric and exoteric. For the Tathagata has not the closed fist of a teacher in respect of mental states.[1]

Dîgha-Nikâya, II. 100

O Lord! great is the difference between the doctrine in which earthly things are taught and that in which heavenly things are taught, and yet neither is without Christ, for neither kind can be known without the truth.

Richard of Saint-Victor

The ancient theologians covered all the sacred mysteries of divine things with poetic veils, that they might not be diffused among profane people.

Marsilio Ficino

If I have told you earthly things, and ye believe not, how shall ye believe, if I tell you of heavenly things?

St John, III. 12

Jesus said: I tell My mysteries to those who are worthy of My mysteries.

The Gospel according to Thomas, Log. 62

[1] I.e., the function of an Avatar is universal and non-limitative in scope. Thus, the parables of Christ are for each to understand according to his capacity.

I have treasured in my memory two stores of knowledge which I had from God's Apostle. One of them I have divulged, but if I divulged the other, my throat would be cut.

Abû Hurayrah

If a single atom of the Prophet manifested itself to creation, naught that is beneath the Throne would endure it.

Bâyazîd al-Bisṭâmî

Holiness 924

Jesus said to His disciples: Make a comparison to Me and tell Me whom I am like. Simon Peter said to Him: Thou art like a righteous angel. Matthew said to Him: Thou art like a wise man of understanding. Thomas said to Him: Master, my mouth will not at all be capable of saying whom Thou art like. Jesus said: I am not thy Master, because thou hast drunk, thou hast become drunk from the bubbling spring which I have measured out. And He took him, He withdrew, He spoke three words to him.

Now when Thomas came to his companions, they asked him: What did Jesus say to thee? Thomas said to them: If I tell you one of the words which He said to me, you will take up stones and throw at me: and fire will come from the stones and burn you up.

The Gospel according to Thomas, Log. 13

Ecstasy 637

Gnosis is fire and faith light, gnosis is ecstasy and faith a gift. The difference between the believer and the gnostic is that the believer sees by the light of God and the gnostic sees by means of God Himself, and the believer has a heart, but the gnostic has no heart: the heart of the believer finds rest in worship, but the gnostic finds no rest save in God.

Al-Sarrâj

Knowl. 761
749
Infra 994
Peace
694

'To glance at My face for an instant
Is worth a thousand years of devotion.'

Shabistarî

Center 847
Action 346

Via Negativa

That Self (*Atman*) is not this, it is not that (*neti, neti*).

Bṛihad-Araṇyaka Upanishad, IV. iv. 22

Void 721

From time to time I tell of the one power in the soul which alone is free. Sometimes I have called it the tabernacle of the soul; sometimes a spiritual light, anon I say it is a spark. But now I say: it is neither this nor that.

Eckhart

Flight 944
Contem. 532

In his *The Names of God,* Denis says that God who is the Maximum, is neither this nor that.

Nicholas of Cusa

Neither in 'Yea' nor 'Nay' is the Truth found.

The Hoka Priests

. . . That unity of God which we can neither conceive, as it is in itself, nor divide into this or that.

William Law

God Almighty revealed to me that I was neither that nor this.

Abû Sa'îd ibn Abi 'l-Khayr

. . . The One, who is neither I nor Thou, this nor that.

Theologia Germanica, XLIII

Do not ask if the Principle is in this or that.

Chuang-tse (ch. XXII)

Introd. *Renun.* *Void* 724

This thing or that thing is not all things: as long as I am this or that, have this or that, I am not all things nor I have not all things. Purify till thou nor art nor hast not either this or that, then thou art omnipresent, and being neither this nor that thou art all things.

Eckhart

Infra 978 *Realiz.* 890

To thy Harîm Dividuality
No entrance finds—no word of This and That.

Jâmî

Beauty 663

Eternal blessedness lieth in one thing alone, and in nought else. And if ever man or the soul is to be made blessed, that one thing alone must be in the soul. Now some might ask, 'But what is that one thing?' I answer, it is Goodness, or that which hath been made good; and yet neither this good nor that, which we can name, or perceive or show; but it is all and above all good things.

Theologia Germanica, IX

Anything we know that we are able to impart or that we can define, that is not God: for in God is neither this nor that which we can abstract nor has he limitation.

Eckhart

Conform. 170 *Illusion* 85

Grasp the skirt of his favour, for on a sudden he will flee:
But draw him not, as an arrow, for he will flee from the bow.
What delusive forms does he take, what tricks does he invent!
If he is present in form, he will flee by the way of spirit.
Seek him in the sky, he shines in water, like the moon:
When you come into the water, he will flee to the sky.

Seek him in the placeless, he will sign you to place:
When you seek him in place, he will flee to the placeless.
As the arrow speeds from the bow, like the bird of your imagination,
Know that the Absolute will certainly flee from the Imaginary.
I will flee from this and that, not for weariness, but for fear
That my gracious Beauty will flee from this and that.

Dîvâni Shamsi Tabriz, XX

Infra 978

Illusion 109

Your duty is *to be;* and not *to be this or that.*

Sri Ramana Maharshi

God is neither this nor that. As St Dionysius says, 'He who thinks that he sees God, if he sees aught sees naught of God.'[1]

Eckhart

Infra 998

. . . A motionless center . . . wherefrom is seen naught but an infinity, which is neither this nor that, neither yea nor nay.

Chuang-tse (ch. II)

Reality 773

Only by negation can we come to know these things.

Dante (*Il Convito*, XV, 3)

We cannot know what God is, but rather what He is not.

St Thomas Aquinas

There is no knowing what God is. Something we do know, namely, what God is not.

Eckhart

For when we aspire from this depth to that height, it is a part of no small knowledge if, before we can know what God is, we can yet know what He is not.

St Augustine

Knowl. 743

God is above the sphere of our esteem,
And is the best known, not defining Him.

Robert Herrick

It is impossible to say what God is in Himself, and it is more exact to speak of Him by excluding everything.

St John Damascene

Aught that a man could or would think of God, God is not at all.

Eckhart

[1] Other similar citations are found in Eckhart, vol. I, pp. 38, 112, 172, 345, 346.

Infra 978

Rase the words 'this' and 'that': duality
Denotes estrangement and repugnancy:
In all this fair and faultless universe
Naught but one Substance and one Essence see.

Reality 775

Jâmî

It is profound, it is vast.
It is neither self nor other.

Saraha

Introd. *Holiness*
Introd. *Illusion*

When you meditate or affirm, deny the three bodies and identify yourself with the Essence which dwells in you. Reject names and forms; do not confuse the physical body, or the mind, or the breath (prâna), the understanding, or the sense organs (indriyas) with the pure eternal Atman. The supreme Self is entirely distinct from all these 'vehicles' or products of illusion (Mâyâ).

Swami Sivananda

In this total abstraction does the Way of the Buddhas flourish; while from discrimination between this and that a host of demons blazes forth!

Huang Po

Center 838, 841
Introd. *Heterod.*

We may also know God to be eternal and omnipresent, not because he fills either place or time, but rather because he wanteth neither. . . . And therefore the Platonists were wont to attribute Αἰὼν or eternity to God, not so much because he had neither beginning nor end of days, but because of his immutable and uniform nature, which admits of no such variety of conceptions as all temporary things do.

John Smith the Platonist

God does not know *what* he is, because he is not any what.

Johannes Scotus Erigena

Non-Duality

Infra 994

God is neither good nor true.

Eckhart

God is He that is neither Mind nor Truth, but is the cause to which Mind and Truth, and all things, and each several thing that is, owe their existence.

Hermes

It exceedeth all things in a super-essential manner, and is revealed in Its naked truth to those alone who pass right through the opposition of fair and foul.

Dionysius

There is in reality neither truth nor error, neither yes nor no, nor any distinction whatsoever, since all—including the contraries—is One.

Chuang-tse (ch. II)

Center 835
Reality 775

He (Bâyazîd al-Bisṭâmî) was asked concerning the command to do good and shun evil. He answered, 'Be in a domain where neither of these things exists: both of them belong to the world of created beings; in the presence of Unity there is neither command nor prohibition.'

'Aṭṭâr

Confoundress! With Thy flashing sword
Thoughtlessly Thou hast put to death my virtue and my sin alike!

Hindu Song to Kâlî

If God keeps the ego in a man, then He keeps in him the sense of differentiation and also the sense of virtue and sin. But in a rare few He completely effaces the ego, and these go beyond virtue and sin, good and bad. As long as a man has not realized God, he retains the sense of differentiation and the knowledge of good and bad.

Sri Ramakrishna

Death 220
Holiness 914

There neither vice nor virtue ever entered in.

Eckhart

Now, the Self (Âtman) is the bridge, the separation for keeping these worlds apart. Over that bridge there cross neither day, nor night, nor old age, nor death, nor sorrow, nor well-doing, nor evil-doing.

Chândogya Upanishad, VIII. iv. 1

Introd.
Waters
Realiz. 890

Uncontaminated whether by virtue or by vice, self cast away, for such there's no more action needed here.

Suttanipâta, 790

Knowl. 761

The vision of God transcends virtues.

Eckhart

There a father becomes not a father; a mother, not a mother; the worlds, not the worlds; the gods, not the gods; the Vedas, not the Vedas; a thief, not a thief. There the destroyer of an embryo becomes not the destroyer of an embryo; a Câṇḍâla (the son of a Śûdra father and a Brahman mother) is not a Câṇḍâla; a Paulkasa (the son of a Śûdra father and a Kshatriya mother) is not a Paulkasa; a mendicant is not a mendicant; an ascetic is not an

Action 329

ascetic. He is not followed by good, he is not followed by evil, for then he has passed beyond all sorrows of the heart.

Bṛihad-Araṇyaka Upanishad, IV. iii. 22

One who hath here escaped attachment whether to virtue or vice, one sorrowless, to whom no dust adheres, one pure, him I call a very Brahman.

Dhammapada, XXVI. 412

It is great joy to realize that in the infinite, thought-transcending Knowledge of Reality all *sangsâric* differentiations are non-existent. . . .

It is great joy to realize that in the self-emanated, divine *Nirmâṇa-Kâya* there existeth no feeling of duality.

Gampopa

Him (who knows this) these two do not overcome—neither the thought 'Hence I did wrong,' nor the thought 'Hence I did right.' Verily, he overcomes them both. What he has done and what he has not done do not affect him.

Bṛihad-Araṇyaka Upanishad, IV. iv. 22

Holiness 924

The Perfect Man in himself stands over against all the individualisations of existence.

Jîlî

When a seer sees the brilliant
Maker, Lord, Person, the Brahma-source,
Then, being a knower, shaking off good and evil,
He reduces everything to unity in the supreme Imperishable.

Maitri Upanishad, VI. 18

Holiness 902

The man of God is beyond infidelity and religion.
To the man of God right and wrong are alike.

Dîvâni Shamsi Tabrîz, VIII

If one is aware of night, one is also aware of day. If one is aware of knowledge, one is also aware of ignorance.

But there is another state in which God reveals to His devotee that Brahman is beyond both knowledge and ignorance.

Sri Ramakrishna

Metanoia 495

What, after all, is right and what is wrong? That thought or action which takes you towards God is right, and that thought or action which takes you away from God is wrong. You can find out for yourself whether you are progressing towards God, or going away from Him. There is no thought of right and wrong after you have reached God: all thoughts cease and all duality is transcended. Your life then flows spontaneously for the good of all. You live and act in the divine consciousness. The so-called sin has no significance for the saint who has realised God. He becomes totally pure and holy. His entire life is an offering at the feet of God.

Swami Ramdas

Dhu 'l-Nûn met an old man whose face was illuminated with the secret of the gnostics, and he asked him about the way towards God.

'If thou knewest God,' replied the man, 'thou wouldst also know the way which leads to him. Abandon differences and divergences.'

Al-Yâfi'î

Knowl. 749

Objects that come into being and are capable of being made the objects of Knowledge are as unreal as those known in dream. As duality has no (real) existence Knowledge is eternal and objectless.

Srî Śankarâchârya

Illusion 94

(Pure) Existence, by itself, is beyond cause and effect.

Swami Ramdas

Where there is (discursive) knowledge there is also ignorance. The sage Vaśishtha was endowed with great knowledge and still he wept at the death of his sons. Therefore I ask you to go beyond both knowledge and ignorance. The thorn of ignorance has pierced the sole of a man's foot. He needs the thorn of knowledge to take it out. Afterwards he throws away both thorns. The jnâni says, 'This world is a "framework of illusion".' But he who is beyond both knowledge and ignorance describes it as a 'mansion of mirth'. He sees that it is God Himself who has become the universe, all living beings, and the twenty-four cosmic principles.

Sri Ramakrishna

Death 226
Sin 62
Orthod. 296
Illusion 99
Creation 33
Reality 803

'Whoso sees Me, transcends speech and silence.'

Niffarî

Speech and silence are one! There is no distinction between them. . . . Thus, the sound of the Tathâgata's voice is everlasting, nor can there be any such reality as the time before he began to preach or the time after he finished preaching. The preaching of the Tathâgata is identical with the Dharma he taught.

Huang Po

Infra 987
Center 838

O King, you say that *Atman* is infinite. Well, that which is infinite must be unconditioned by time and space. Absence of duality in *Atman* renders it incapable of being a cause. A cause exists both before and after the effect, as clay does in respect of the jar. But in *Atman* there is neither beginning nor end. Besides, a cause must modify or change itself to produce an effect. *Atman* being All and Absolute is free from the possibility of change or modification. *Atman* is indestructible and immutable. It has never fallen from its nature. As there is no duality in *Atman,* so it is neither subject nor object. Nor is there any action in it. It is eternally pure like the blue sky or space, O King, and it is your own nature.

Yoga-Vasishtha

Introd.
Knowl.
Creation 42

Infra 994

Nature and the sacred laws have schooled them (the Therapeutae) to worship the Self-existent who is better than the good, purer than the One and more primordial than the Monad.

Philo

Action 329

The three gunas[1]—sattwa, rajas, and tamas—belong to the mind and not to the Self. Rise thou above the gunas and know the Self. First, overcome rajas and tamas by developing sattwa, and then rise above sattwa by sattwa itself.

Srimad Bhagavatam, XI. vii

How can there be knowledge or ignorance in Me who am eternal and always of the nature of Pure Consciousness?

Śrî Śankarâchârya

Since I have known God, neither truth nor falsehood has entered my heart.

Abû Ḥafṣ Ḥaddâd

Center 835

The very nature of the Great Way is voidness of opposition.

Huang Po

The union (between God and the soul) may become such, that God altogether pours Himself into it, and draws it so entirely into Himself that it no longer has any distinct perception of virtue or vice, or recognises any marks by which it knows what it is itself.

Tauler

Reality 775

The mystery of Diamond hardness (is) that the dharmas or elements of existence are in reality but of one essence.

Hakuin

Death 232

Neither life nor death is any more real than empty space which a man slashes with his cold blade.

Soshun

I have put duality away, I have seen that the two worlds are one;
One I seek, One I know, One I see, One I call.

Dîvâni Shamsi Tabrîz, XXXI

Introd. *Conform.*

By counting beads, repeating prayers,
And reading the Koran,
The heathen becomes not a Mussulman.
The man to whom true infidelity (*kufr-i haqîqî*, i.e. ignoring differentiation) becomes revealed
With pretended faith becomes disgusted.

Shabistarî

[1] Cf. note p. 88.

In all things
See but One, say One, know One.

Shabistarî

Thus shalt thou go beyond both good and evil. Good actions will proceed from thee without any thought of merit, and thou shalt desist from evil actions naturally and not through a sense of evil.

Srimad Bhagavatam, XI. ii

Holiness 902

The Self is beyond knowledge and ignorance.

Sri Ramana Maharshi

He who affirms the duality (of God and the world) falls into the error of associating something with God; and he who affirms the singularity of God (in excluding from His reality all that manifests as multiple) commits the fault of enclosing Him in a (rational) unity. Beware of comparison when thou envisagest duality: and beware of abstracting the Divinity when thou envisagest Unity!

Ibn ʿArabî

As soon as the mouth is opened, evils spring forth. People either neglect the root and speak of the branches, or neglect the reality of the 'illusory' world and speak only of Enlightenment. Or else they chatter of cosmic activities leading to transformations, while neglecting the Substance from which they spring.

Huang Po

Heterod. 428
Realiz. 873

Those who do not perceive the truth think in terms of Samsara and Nirvana, but those who perceive the truth think neither of Samsara nor of Nirvana.

Cittaviśuddhiprakaraṇa

Therefore is it said: 'The perception of a phenomenon is the perception of the Universal Nature, since phenomena and Mind are one and the same.' It is only because you cling to outward forms that you come to 'see', 'hear', 'feel' and 'know' things as individual entities. True perception is beyond your powers so long as you indulge in these.

Huang Po

All are one, both the visible and the invisible.

Shabistarî

If it's already manifest, what's the use of meditation?
And if it is hidden, one is just measuring darkness.
Saraha cries: The nature of the Innate is neither existent nor non-existent.

Saraha

Supra 975

Thus those who desire the fruit of Buddhahood should renounce the notion of existence because it is deluding like a magical display, but they should also renounce the notion of

Illusion 85

Void 724

non-existence, for it is non-existent. O Wise Ones, do ye now hearken, for in so far as one renounces both extremes, the state in which one abides is neither Samsara nor Nirvana, for one has renounced these two.

Anangavajra

Realiz. 870

Whether in the body, or out of the body, I cannot tell: God knoweth.

II. Corinthians, XII. 3

It is neither inside of this world nor outside: neither beneath it nor above it: neither joined with it nor separate from it: it is devoid of quality and relation.

Rûmî

Jesus said: If those who lead you say to you: 'See, the Kingdom is in heaven,' then the birds of the heaven will precede you. If they say to you: 'It is in the sea,' then the fish will precede you. But the Kingdom is within you and it is without you.

Center 816

The Gospel according to Thomas, Log. 3

He who belongs to God and to whom God belongs is not connected with anything in the universe. The real gist of gnosis is to recognize that to God is the kingdom. When a man knows that all possessions are in the absolute control of God, what further business has he with mankind, that he should be veiled from God by them or by himself? All such veils are the result of ignorance. As soon as ignorance is annihilated, they vanish, and this life is made equal in rank to the life hereafter.

Reality 803

Hujwîrî

If they ask you: 'What is the sign of your Father in you?', say to them: 'It is a movement and a rest.'

The Gospel according to Thomas, Log. 50

The less one understands and knows about these two conditions of life—the active and the calm—the more careful should one be to value them both and remember that they are only two aspects of one uniform condition. This is what is meant when it is said that a monk who is really practising meditation does not know that he is walking when he is walking, or sitting when he is sitting.

Action 358

Hakuin

Upward, downward, the way is one and the same.

Heraclitus

I do not say it is the *Liṅga*,
I do not say it is oneness with the *Liṅga*,
I do not say it is union,
I do not say it is harmony,
I do not say it has occurred,
I do not say it has not occurred,

I do not say it is You.
I do not say it is I.
After becoming one with the *Liṅga*
in Chenna Mallikârjuna.
I say nothing whatever.

Akka Mahâdêvî

Infra 987

Hesiod is the teacher of most men: they suppose that his knowledge was very extensive. when in fact he did not know night and day. for they are one.

Heraclitus

Jesus saw children who were being suckled. He said to his disciples: These children who are being suckled are like those who enter the Kingdom. They said to Him: Shall we then. being children. enter the Kingdom? Jesus said to them: When you make the two one. and when you make the inner as the outer and the outer as the inner and the above as the below. and when you make the male and the female into a single one. so that the male will not be male and the female (not) be female. . . . then shall you enter (the Kingdom).

The Gospel according to Thomas, Log. 22

P. State 579

Although the Wisdom of *Nirvâṇa* and the Ignorance of the *Sangsâra* illusorily appear to be two things. they cannot truly be differentiated.
It is an error to conceive them otherwise than as one.

Padma-Sambhava

Devotee: What does Maharshi think of the theory of universal illusion (*Mâyâ*)?
Maharshi: What is *Mâyâ*? It is only Reality.
D.: Is not *Mâyâ* illusion?
M.: *Mâyâ* is used to signify the manifestations of the Reality. Thus *Mâyâ* is only Reality.

Sri Ramana Maharshi

Illusion 85

Eternal and temporal are not separate from one another.
For in that Being this non-existent has its being.

Shabistarî

Seek His face. I beg you. and you will enjoy forever. Do not move to touch Him. I beg you. for He is stability. Do not distract yourself over various things in order to reach Him. for He is unity itself. Halt the movement. unite the multiplicity. and immediately you will reach God. who has long since reached you wholly.

Marsilio Ficino

Supra 975
Realiz. 873

The truth is neither one nor two.

Sri Ramana Maharshi

Single nature's double name
Neither two nor one was call'd.

Shakespeare (*The Phoenix and the Turtle*)

Contem. 523

The Self is neither without nor within.

Sri Ramana Maharshi

The inward and the outward are indissolubly united and form a single great eternal Current.

Ananda Moyî

Conform. 166

When self-will vanishes in this world, contemplation is attained, and when contemplation is firmly established, there is no difference between this world and the next.

Hujwîrî

Everything is everything.[1]

Anaxagoras

The Ṣûfî is he that sees nothing except God in the two worlds.

Shiblî

Each thing hath two faces, a face of its own, and a face of its Lord: in respect of its own face it is nothingness, and in respect of the Face of God it is Being. Thus there is nothing in existence save only God and His Face, for *everything perisheth but His Face (Qur'ân,* XXVIII. 28)

Reality 803

Al-Ghazâlî

Reality 790
Creation 26

Not that the One is two, but that these two are one.

Hermes

The Trackless Trace

Introd.
Charity

A good traveller leaves no track.

Tao Te Ching, XXVII

The ultimate reality itself is not a symbol, it leaves no tracks, it cannot be communicated by letters or words, but we come to it by tracing them to the source where they come forth.

Hori Kintayû

[1]Cp. citation by Dafydd ap Gwilym, p. 808.

There is no trace of any that have penetrated the hidden depths of Its infinitude.

Dionysius

That measureless Splendour of God, which together with the incomprehensible brightness, is the cause of all gifts and of all virtues—that same Uncomprehended Light transfigures the fruitive tendency of our spirit and penetrates it in a way that is wayless.

Ruysbroeck

My place is the Placeless, my trace is the Traceless.

Dîvâni Shamsi Tabrîz, XXXI

Center 816

He whose conquest cannot be conquered again, into whose conquest no one in this world enters, by what track can you lead him, the Awakened, the Omniscient, the Trackless?

Dhammapada, XIV. 179

Holiness 902 934

Silence

One word spake the Father, which Word was His Son, and this Word He speaks ever in eternal silence, and in silence must it be heard by the soul.

St John of the Cross

Inv. 1003 *Center* 841

I am not to be perceived by means of any visible form,
Nor sought after by means of any audible sound:
Whosoever walks in the way of iniquity,
Cannot perceive the blessedness of the Lord Buddha.

Prajñâ-Pâramitâ

Supra 986 *Sin* 77

Son, when thou art quiet and silent, then art thou as God was before nature and creature; thou art that which God then was; thou art that whereof he made thy nature and creature: Then thou hearest and seest even with that wherewith God himself saw and heard in thee, before ever thine own willing or thine own seeing began.

Boehme

Realiz. 890 *Center* 838

Nothing answers to the question: what is It?

Jîlî

We worship with reverent silence the unutterable Truths and, with the unfathomable and

Infra 994

holy veneration of our mind, approach that Mystery of Godhead which exceeds all Mind and Being.

Dionysius

Blame wit and words, whose force fails to tell all that I hear Love discourse.

Dante (*Il Convito*, III. iv. 7)

Humility 191

What power have we to celebrate Thy praise?

Jâmî

Heterod. 430
Grace 552

Glorified be He and High Exalted above all that they ascribe unto Him.
. . . Vision comprehendeth Him not, but He comprehendeth all vision.

Qur'ân, VI. 100, 103

Gnosis is nearer to silence than to speech.

Abû Sulaymân al-Dârânî

Center 841

God is so present everywhere that one cannot speak:
Thus thou wilt praise Him better through silence.

Angelus Silesius

Center 847

When your heart becomes the grave of your secret, that desire of yours will be gained more quickly.

The Prophet said that any one who hides his inmost thought will soon be wedded to the object of his desire.

Death 208

When seeds are hidden in the earth, their inward secret becomes the verdure of the garden.

If gold and silver were not hidden, how would they get nourishment (grow and ripen) in the mine?

Rûmî

Void 728

None sees his labour, for he works in the dark.

Boehme

Infra 998

To have understood that nothing is gained by questioning about the Principle, but that It is to be contemplated in silence, this is what is called having obtained the Great Result.

Chuang-tse (ch. XXII)

. . . The Spirit of Truth: whom the world cannot receive, because it seeth him not, neither knoweth him.

St John, XIV. 17

The spirit of wisdom cannot be delineated with pen and ink, no more than a sound can be painted, or the wind grasped in the hollow of the hand.

John Sparrow

Training began with children, who were taught to sit still and enjoy it. They were taught to use their organs of smell, to look when there was apparently nothing to see, and to listen intently when all seemingly was quiet. A child who cannot sit still is a half-developed child.

Chief Standing Bear

The Sage does not talk, the Talented Ones talk, and the stupid ones argue.

Kung Tingan

Whoso knoweth God, his tongue flaggeth.

Sufic Saying

Knowl. 743

The wise aspire to know, the foolish to relate.

Anas b. Mâlik

Let not a prudent man unasked tell any one (the Vedas), nor (tell) one who asks improperly: though he knows, let a wise man act like a dumb man in the world.

Mânava-dharma-sâstra, II. 110

Humility 197

He who knows ten should only teach nine.

Far Eastern Saying

Whom thou seest responding to all that he is asked, and expounding all that he has witnessed, and relating all that he has learned, then take that for the proof of his ignorance.

Ibn ʿAṭâʾillâh

Knowl. 734

He who knows does not speak:
He who speaks does not know.

Tao Te Ching, LXXXI

He who speaks becomes silent before the Divine Essence.

Jîlî

In the theatre the audience remains engaged in all kinds of conversation, about home, office, and school, till the curtain goes up: but no sooner does it go up than all conversation comes to a stop, and the people watch the play with fixed attention.

Sri Ramakrishna

Contem. 532

Pythagoras said, that it was either requisite to be silent, or to say something better than silence.

Stobaeus

It is better to conceal ignorance than to put it forth into the midst.

Heraclitus

Humility 203

Verily he who speaks noble truths, and gives utterance to the Word of God, observes the vow of silence. Silence is restraint of speech.

Srimad Bhagavatam, XI. xi

The one syllable *ôm* (is) the supreme *Brâhma*; suppressions of breath (are) the highest austerity: but there is (nothing) higher than the *sâvitrî*:[1] truth is better than silence.

Mânava-dharma-śâstra, II. 83

There's something still better than silence, 'tis this—to speak the truth.

Shekel Hakodesh, 245

God is a word, an unspoken word. Augustine says: 'All scripture is vain.' We say that God is unspoken, but he is unspeakable. Grant he is somewhat: who can pronounce this word? None but the Word.

Inv. 1003

Eckhart

Reality 775 *Creation* 31

God is an Eternal Being, an Infinite Unity, the Radical Principle of all things. His Essence is Infinite Light. His Power—Omnipotence: His Will—Perfect Goodness:[2] His Wish—Absolute Reality. As we strive to think of Him, we plunge into the Abyss of Silence, of infinite Glory.

The All-Wise Doorkeeper

Infra 994

The First cannot be thought of as having definition and limit. It can be described only as transcending all things produced, transcending Being. . . . Its definition could only be 'the Indefinable', for This is a Principle not to be conveyed by any sound.

Plotinus

To things of sale a seller's praise belongs;
She passes praise: then praise too short doth blot.

Shakespeare (*Love's Labour's Lost,* IV. iii. 240)

Conform. 180

When he (who beholds the wonders of God) loses beard and moustache (abandons pride and egoism) from (contemplating) His work, he will know his (proper) station and will be silent concerning the Worker (Maker).

He will only say from his soul, 'I cannot (praise Thee duly),' because the declaration thereof is beyond reckoning and bound.

Rûmî

Realiz. 870 *Peace* 700

He (the American Indian) believes profoundly in silence—the sign of a perfect equilibrium. Silence is the absolute poise or balance of body, mind, and spirit. The man who preserves his selfhood ever calm and unshaken by the storms of existence—not a leaf, as it were, astir on the tree; not a ripple upon the surface of shining pool—his, in the mind of the unlettered sage, is the ideal attitude and conduct of life.

[1] The *Gâyatrî,* q.v.

[2] This is equatable with the *Sat-Chit-Ânanda* of Hinduism.

If you ask him: 'What is silence?' he will answer: 'It is the Great Mystery!' 'The holy silence is His voice!'

Infra 998

Ohiyesa

There is no means of communicating more than this to the people: in the river there is no room for the Sea.

I speak low according to the measure of understandings: 'tis no fault, this is the practice of the Prophet.

Rûmî

Of the heavenly things God has shown me
I can speak but a little word,
Not more than a honey-bee
Can carry away on its foot
From an overflowing jar.

Mechthild of Magdeburg

I want to write it (an exposition on the *Sefer Yetsirah*) down and I am not allowed to do it, I do not want to write it down and cannot entirely desist: so I write and I pause, and I allude to it again in later passages, and this is my procedure.

Baruch Togarmi

If the works of God were such, as that they might be easily comprehended by human reason, they could not be justly called marvellous or unspeakable.

Rev. 959

The Imitation of Christ, IV. xviii. 5

This experience of Moksha or Nirvana is indescribable. Buddha aptly says: 'Do not dip the string of thought into the Unfathomable: he who questions errs and he who answers errs.'

Swami Ramdas

The One Self, the Sole Reality, alone exists eternally. When even the Ancient Teacher, Dakshinamurthi, revealed It through speechless Eloquence, who else could have conveyed it by speech?[1]

Knowl. 749

Sri Ramana Maharshi

And you, Tat and Asclepius and Ammon, I bid you keep these divine mysteries hidden in your hearts, and cover them with the veil of silence.

Hermes

[1] 'Of that Reality our mind has no conception. Whatever can be affirmed or denied falls short of it, as Dionysius says (*Theologia Mystica* C. 5). Yet in every knowing it is at once and undividedly the implicit knower and the implicitly known. When the mind has learned to say nothing Intellect shines of itself' (Bernard Kelley: 'The Metaphysical Background of Analogy', *Aquinas Society of London*, no. 29, Blackfriars Publications, 1958, p. 23).

Holiness 910

The sage wears clothes of coarse cloth but carries jewels in his bosom.

Tao Te Ching, LXX

Peace 700

To experience but momentarily the *samâdhi* wherein all thought-processes are quiescent is more precious than to experience uninterruptedly the *samâdhi* wherein thought-processes are still present.

Gampopa

Creation 42

Do not imagine, do not think, do not analyse,
Do not meditate, do not reflect:
Keep the mind in its natural state.

Tilopa

Think the not-thought.

Dôgen

On one occasion, Euxenus (a Pythagorean teacher at Aegae) asked Apollonius why so noble a thinker as he and one who was master of a diction so fine and nervous did not write a book. He replied: 'I have not yet kept silence.'[1]

Life of Appollonius of Tyana

Jesus (Peace be upon him!) said, 'Devotion has ten parts, nine of which are found in silence and one in flight from men.'

Christ in Islâm

Knowl. 743

Be silent that the lord who gave thee language may speak.

Dîvâni Shamsi Tabrîz, XLVII

Be still, and know that I am God.

Psalm XLVI. 10

Knowl. 749 *Realiz.* 887

The vision baffles telling: we cannot detach the Supreme to state it: if we have seen something thus detached we have failed of the Supreme which is to be known only as one with ourselves.

Plotinus

The furthest from God among the devotees are those who speak the most of Him.

Bâyazîd al-Bistâmî

Creation 38

When there are no Scriptures, then Doctrine is sound.

Po Chü-i

[1] He was a young man at the time of this remark. He forthwith accomplished a total silence which lasted five years. The most difficult part according to him was to refrain from refuting error when it reached his hearing.

As a matter of fact, it is such blank scrolls as these that are the true scriptures. But I quite see that the people of China are too foolish and ignorant to believe this, so there is nothing for it but to give them copies with some writing on.

Monkey

Void 724

O Mahâmati, I say unto you: During the time that elapsed between the night of the Tathâgata's Enlightenment and the night of his entrance into Nirvâna, not one word, not one statement was given out by him.

Lankavatara Sutra, VII

It is conceived of by him by whom It is not conceived of.
He by whom It is conceived of, knows It not.
It is not understood by those who (say they) understand It.
It is understood by those who (say they) understand It not.

Kena Upanishad, II. 3

Infra 994

Why do I live among the green mountains?
I laugh and answer not, my soul is serene:
It dwells in another heaven and earth belonging to no man.
The peach trees are in flower, and the water flows on. . . .

Li Po

Silence is the language of God: it is also the language of the heart.

Swami Sivananda

Contem. 536

When you come to taste the good fruit of silence, you will no longer need lessons about it.

Unseen Warfare, I. XXV

The whole world is tormented by words
And there is no one who does without words.
But in so far as one is free from words
Does one really understand words.

Saraha

The Absolute has never been defiled, for no one as yet has been able to express it by human speech.

Sri Ramakrishna

(The Essence is) incommunicable, indivisible and ineffable, transcending all name and all understanding.

St Gregory Palamas

The Godhead

There is a reality even prior to heaven and earth:
Indeed, it has no form, much less a name:
Eyes fail to see it:
It has no voice for ears to detect:
To call it Mind or Buddha violates its nature.

Supra 987

Dai-o Kokushi

Before the One, what is there to count?

Sefer Yetsirah

(God is) above the Monad itself.

Clement of Alexandria

He is above all praise.

Ecclesiasticus, XLIII. 33

Wr. Side 474

God is the opposite of nothing with being as the intermediary.

Nicholas of Cusa

God is not onely said to be
An *Ens*, but *Supraentitie*.

Introd. *Realiz.*

Robert Herrick

Supra 978

Though the two exist because of the One, do not cling to the One.

Seng-ts'an

Jesus said: Blessed is he who was before he came into being.

The Gospel according to Thomas, Log. 19

Those who have passed into the unitive life have attained unto a Being transcending all that can be apprehended by sight or insight. . . . But the soul remains contemplating that Supreme Beauty and Holiness and contemplating itself in the beauty which it has acquired by attaining to the Divine Presence, and for such a one, things seen are blotted out, but not the seeing souls. But some pass beyond this and they are the Elect of the Elect, who are consumed by the glory of His exalted Countenance, and the greatness of the Divine Majesty overwhelms them and they are annihilated and they themselves are no more. They no longer contemplate themselves, and there remains only the One, the Real, and the meaning of His Word: 'All things perish save His Countenance' is known by experience.

Realiz. 890

Al-Ghazâlî

When Truth cometh, ecstasy itself is dispossessed.

Junayd

And who shall be filled with beholding his glory?

Ecclesiasticus, XLII. 26

Center 847

Dhu 'l-Nûn was asked: 'What is the end of the gnostic?' He answered: 'When he is as he was where he was before he was.'

Al-Kalâbâdhî

Center 838

Everything in these depths of the Holy Spirit is beyond understanding or explanation.

St Simeon the New Theologian

For God to see therein would cost him all his divine names and personal properties.

Eckhart

His greatness is unsearchable.

Psalm CXLV. 3

All other revelations are only a reflection of the sky of this supreme revelation, or a drop of its ocean; while being real, they are nonetheless annihilated under the power of this essential revelation, which is exclusively of God by virtue of His knowledge of Himself, whereas the other revelations are of God by virtue of the knowledge of other persons.

Jîlî

Knowl. 749

In truth, all possibilities resolve principially into non-existence.

Ibn 'Arabî

Creation 26

The essence of perfect Tao is profoundly mysterious; its extent is lost in obscurity.

Chuang-tse (ch. XI)

The Divine Obscurity is the primordial place
Where the suns of beauty set.
It is the Self of God Himself.

Jîlî

Realiz. 890

Lord, take me into the gloom of thy Godhead that in thy dark I may lose all my light, for nothing that can be revealed do I account as light.

Eckhart

Metanoia 484

What is the last end? It is the mystery of the darkness of the eternal Godhead which is unknown and never has been known and never shall be known. Therein God abides to himself unknown.

Eckhart

Infra 998

When one attains prema[1] one has the rope to tie God.

Sri Ramakrishna

I must mount even higher than God. into a desert.

Angelus Silesius

Void 724

God has made all things out of nothing. and that same nothing is himself.

Boehme

. . . Him whom neither being nor understanding can contain.

Dionysius

Void 721

If we fail to find God it is because we seek in semblance what has no resemblance. . . . On merging into the Godhead all definition is lost.

Eckhart

P. State 563

Supra 987

Not to know It is to know It: to know It is not to know It. But how is one to understand this. that it is by not knowing It that It is known? This is the way. says the Primordial State. The Principle cannot be heard: that which is heard is not It. The Principle cannot be seen: that which is seen is not It. The Principle cannot be uttered: that which is uttered is not It. . . . Concerning the Principle. one can neither ask nor reply what It is.

Chuang-tse (ch. XXII)

Supra 986

When we speak of the All-Transcendent Godhead as an Unity and a Trinity. It is not an Unity or a Trinity such as can be known by us or any other creature. though to express the truth of Its utter Self-Union and Its Divine Fecundity we apply the titles of 'Trinity' and 'Unity' to That Which is beyond all titles. expressing under the form of Being That Which is beyond Being. But no Unity or Trinity or Number or Oneness or Fecundity or any other thing that either is a creature or can be known to any creature. is able to utter the mystery. beyond all mind and reason. of that Transcendent Godhead which super-essentially surpasses all things. It hath no name. nor can It be grasped by the reason: It dwells in a region beyond us. where our feet cannot tread.

Dionysius

The soul is a creature receptive to everything named. but the nameless she cannot receive until she is gotten so deep into God that she is nameless herself. And then none can tell if it is she that has gotten God or God has gotten her.

Eckhart

Since the Maximum is the Absolute infinite and therefore all things without distinction. it is clear that it cannot have a proper name. . . . All affirmations. therefore. that are made of God in theology are anthropomorphic. including even those most holy Names. which enshrine the highest mysteries of divine knowledge.

Nicholas of Cusa

[1] Ultimate ecstatic love.

His being cannot be accurately described by any of the names we call him.

Hermes

Now mark! God is nameless: no one can know or say anything of him.

Eckhart

Conscious of this, the Sacred Writers celebrate It by every Name while yet they call It Nameless.

Dionysius

Reality 775

All that which we call the attributes of God are only so many human ways of our conceiving that abyssal All which can neither be spoken nor conceived by us.

William Law

Of the first principle the Egyptians said nothing, but celebrated it as a darkness beyond all intellectual conception, a thrice unknown darkness.

Damascius

This therefore is one and the best extension (of the soul) to (the highest) God, and is as much as possible irreprehensible: viz. to know firmly, that by ascribing to him the most venerable excellencies we can conceive, and the most holy and primary names and things, we ascribe nothing to him which is suitable to his dignity. It is sufficient, however, to procure our pardon (for the attempt) that we can attribute to him nothing superior.

Simplicius

God is day and night, winter and summer, war and peace, satiety and hunger: but he assumes different forms, just as when incense is mingled with incense: every one gives him the name he pleases.

Heraclitus

He is both the things that are and the things that are not: for the things that are He has made manifest, and the things that are not He contains within himself.

Such is He who is too great to be named God.

Hermes

He comprehends His own existence without this comprehension existing in any manner whatsoever.

Ibn 'Arabî

Introd. *Heterod.*

How could the Knower be known?

Bṛihad-Aranyaka Upanishad, IV. v. 14

Void 728

One and one uniting, void shines into void. Where these two abysms hang, equally spirated, de-spirated, there is the supreme being: where God gives up the ghost, darkness

Introd. *Realiz.*

reigns in the unknown known unity. This is hidden from us in his motionless deep. Creatures cannot penetrate this aught.

Eckhart

The Great Mystery

Of the heaven which is above the heavens, what earthly poet ever did or ever will sing worthily? ... There abides the very being with which true knowledge is concerned; the colourless, formless, intangible essence. . . . knowledge absolute in existence absolute.

Knowl. 761

Plato (*Phaedrus*, 247 C)

Holiness 924 *Reality* 773

It is infinite, incomprehensible, immeasurable; it exceeds our powers, and is beyond our scrutiny. The place of it, the whither and the whence, the manner and quality of its being, are unknown to us. It moves in absolute stability, and its stability moves within it.

Hermes

In its true state, mind is naked, immaculate; not made of anything, being of the Voidness; clear, vacuous, without duality, transparent; timeless, uncompounded, unimpeded, colourless; not realizable as a separate thing, but as the unity of all things, yet not composed of them; of one taste, and transcendent over differentiation.

Void 724 *Center* 835

Padma-Sambhava

Supra 978

Sameness is differentiation, differentiation is sameness.

Zen Formula

Realiz. 890 Supra 994

St Augustine says, 'The soul has a private door into divinity where for her all things amount to naught.' There she is ignorant with knowing, will-less with willing, dark with enlightenment.

Eckhart

May the Great Mystery make sunrise in your heart.

Sioux Indian

An unexpectedly glorious light will burst from your substance, and the end will arrive three days afterwards. The substance will be granulated, like atoms of gold (or motes in the Sun), and turn a deep red—a red the intensity of which makes it seem black like very pure blood in a clotted state. This is the Great Wonder of Wonders, which has not its like on earth.

Philalethes

Know, may Allâh the Exalted have mercy on thee, that the Elixir of Redness is not directly formed, but must first pass through the stage of the Elixir of Whiteness (i.e. of silver). . . . After it has become dry . . . it becomes fixed in the colour of purple, and is waxy, fusible, soluble, stable. One part of it is capable of transforming into gold a thousand parts of mercury which has been fixed by means of the Elixir of Silver. Similarly if you wish to project it upon silver it will turn it into pure gold, more precious than the gold of mines.

Abu 'l-Qâsim al-ʿIrâqî

P. State 579

I heard my noble *Master* say,
How that manie men patient and wise,
Found our *White Stone* with Exercise:
After that thei were trewlie tought,
With great labour that *Stone* they Caught;
But few (said he) or scarcely one,
In fifteene Kingdomes had our *Red Stone*.

Thomas Norton

Holy War 405

Those who find the *Mysterium magnum* will know what it is: but to the godless it is incomprehensible; because they have not the will to desire to comprehend it. They are captured by the terrestrial essence so as to render them unable to draw will in the mystery of God.

Boehme

Knowl. 749
Conform. 166
Sin 66

But as it is written,[1] Eye hath not seen, nor ear heard, neither have entered into the heart of man, the things which God hath prepared for them that love him.

I. Corinthians, II. 9

Orthod. 275

He (the Pre-existent) is not an Attribute of Being, but Being is an Attribute of Him; He is not contained in Being, but Being is contained in Him; He doth not possess Being, but Being possesses Him; He is the Eternity, the Beginning, and the Measure of Existence, being anterior to Essence and essential Existence and Eternity, because He is the Creative Beginning, Middle, and End of all things. . . . He is not This without being That: nor doth He possess this mode of being without that. On the contrary He *is* all things as being the Cause of them all, and as holding together and anticipating in Himself all the beginnings and all the fulfilments of all things; and He is above them all in that He, anterior to their existence, super-essentially transcends them all. Hence all attributes may be affirmed at once of Him, and yet He is no Thing. He possesses all shape and form, and yet is formless and shapeless, containing beforehand incomprehensibly and transcendently the beginning, middle, and end of all things, and shedding upon them a pure radiance of that one and undifferenced causality whence all their fairness comes.

Dionysius

Introd.
Heterod.
Supra 994
Center 841
Supra 975
Void 724
Beauty 663

The Essence (*Dhat*) denotes Absolute Being stripped of all modes, relations, and aspects. Not that they are outside of Absolute Being; on the contrary, they belong to it, but

[1] *Isaiah*, LXIV. 4.

Realiz. 887

they are in it neither as themselves nor as aspects of it: no, they are identical with the being of the Absolute. The Absolute is the simple essence in which no name or quality or relation is manifested. When any of these appears in it, that idea is referred to that which appears in the Essence, not to the pure Essence, inasmuch as the Essence, by the law of its nature, comprehends universals, particulars, and relations, not as they are judged to exist, but as they are judged to be naughted under the might of the transcendental oneness of the Essence.

Jîlî

Void 728

When I go back into the ground, into the depths, into the well-spring of the Godhead, no one will ask me whence I came or whither I went.

Eckhart

Verily, while he does not there see, he is verily seeing, though he does not see (what is to be seen): for there is no cessation of the seeing of a seer, because of his imperishability. It is not, however, a second thing, other than himself and separate, that he may see.

Action 358

Verily, while he does not there smell, he is verily smelling. . .
Verily, while he does not there taste, he is verily tasting. . .
Verily, while he does not there speak, he is verily speaking. . .
Verily, while he does not there hear, he is verily hearing. . .

Peace 700

Verily, while he does not there think, he is verily thinking. . .
Verily, while he does not there touch, he is verily touching. . .
Verily, while he does not there know, he is verily knowing.

Bṛihad-Araṇyaka Upanishad, IV. iii. 23–30

The bride says in the Book of Love, 'I have crossed all the mountains, aye, even my own powers, and have reached the dark power of the Father. There heard I without sound, there saw I without light, there breathed I without motion: there did I taste what savoured not, there did I touch what touched not back. Then my heart was bottomless, my soul loveless, my mind formless and my nature natureless.'

Sp. Drown. 713

Eckhart

Ultimate, unheard, unreached, unthought, unbowed, unseen, undiscriminated and unspoken, albeit listener, thinker, seer, speaker, discriminator and foreknower, of that Interior Person of all beings one should know that 'He is my Self'.

Aitareya Araṇyaka, III. ii. 4

COLOPHON

INVOCATION

In the beginning was the Word, and the Word was with God, and the Word was God.

St John, I. 1

Om *is* Brahma.

Taittirîya Upanishad, I. 8

Though this age of Kali is full of vices it possesses one great virtue that during this period, through mere chanting of the Divine Name one can obtain release from bondage and realize God. That which was attained through meditation in Satyayuga, through performance of sacrifices in Tretayuga and through personal service and worship of God in Dwaparayuga can be obtained in Kaliyuga through mere chanting of Sri Hari's Name.[1]

Srimad Bhagavatam

'Of all the names and forms of God the monogrammatic syllable Oṁ, the totality of all sounds and the music of the spheres chanted by the resonant Sun, is the best' (Coomaraswamy: *Hinduism and Buddhism,* p. 11).

Basing his practice upon this metaphysical foundation, the invoker reconstructs the heavenly Sacrifice implicit in creation by converting what has been a macrocosmic 'descent' into what is now a microcosmic 'ascent', wherein his multifarious and fragmented being is recollected, reintegrated and finally resolved through the theurgy of the Divine Name into the undifferentiated and primordial unity of the Supreme Principle, with which this Name is essentially identified.

Introds.
Suff.
Metanoia

'It is in the Divine Name that there takes place the mysterious meeting of the created and the Uncreate, the contingent and the Absolute, the finite and the Infinite' (Schuon: *De l'Unité transcendante des Religions,* p. 167).

'The Divine Name, revealed by God Himself, implies a Divine Presence which becomes operative to the extent that the Name takes possession of the mind of the person invoking. Man cannot concentrate directly on the Infinite, but by concentrating on the symbol of the Infinite he attains the Infinite Itself: for when the individual subject becomes identified with the Name to the point where all mental projection is absorbed by the form of the Name, then its Divine Essence manifests spontaneously, since this sacred form tends towards

Rev.
959
Introd.
Symb.

[1] Quoted by Swami Sivananda in *Japa Yoga,* p. 159. Hari is a name of Vishnu. A similar text occurs in the *Vishnu Dharma Uttara.*

Knowl. 749

nothing outside of itself. It has a positive affinity with its Essence alone, wherein its limits finally dissolve. Thus it is that union with the Divine Name becomes Union (*al-waṣl*) with God Himself' (Burckhardt: *Introd. aux Doctrines ésotériques de l'Islam*, p. 101).

Creation 42 Sin 66 Wr. Side passim

The cyclical opportuneness of invocation as a method relates to the place in spiritual practice for a corrective which can silence the unruly elements in the soul and burn out the negative aspects, while nourishing simultaneously all that is positive and orienting the soul towards its true felicity. Man is rightly a temple of the Holy Spirit, but fallen man houses a throng of idols—disordered reason, luciferian imagination, misplaced faculties of the soul, inordinate desires, inferior psychic entities, and demonic intruders—which have usurped his temple and turned its service more into a perpetual carnival.[1] One becomes what one thinks upon; and as the mind is incessantly on the world in this period of the Kali Yuga, a method is needed that can continually and rhythmically and inexorably turn the mind away from its habitual worldly de-mentia, *re-mind*ing[2] it in the Truth. A Name of God, divinely revealed, legitimately bestowed by the proper spiritual authority, and constantly invoked, reconsecrates the altar of the temple or heart of the aspirant and, functioning as a direct support, reanimates the Holy Spirit or Divine Breath still latent but despirated within.

Reality 775

The facility of invocation derives from the prototypal simplicity[3] of the Name itself—a monadic totality in which all forms are synthesized. What most characterizes the method is in fact its mono-tony; and yet each repetition of the Name acts as a fresh initiation into the spiritual life, being a renewal of one's aspiration, and at the same time a stablizing of one's inward center, a fixing of consciousness on Reality. Sanctity, and even moments of illumination, lie outside our grasp, but we do have the volitive power to be specifically human in the sense of maintaining a static and vertical rectitude founded upon the intellectual anticipation of eternity. Invocation is moreover as a veil through which we behold the world, if not transformed, at least at one remove from immediacy.

Grace 552

Spiritual growth is like that of a tree, which needs sunlight and moisture throughout the year, although it may only visibly increase during certain short seasons. Invocation is also compared to the process of adding salts to a solution without apparent result, until at a given moment the liquid crystallizes. 'So is the kingdom of God, as if a man should cast seed into the ground; And should sleep, and rise night and day, and the seed should spring and grow up, he knoweth not how' (*St Mark*, IV. 26, 27).

Pilg. 378

As regards technique: 'The Divine Name invoked must be correctly pronounced, otherwise the condition of formal exactitude will not be fulfilled. Furthermore, it must be "consecrated" by the adequate intention of the person, which intention will be expressed in concentration, fervour, and perseverance; the "consecration" in this case is necessarily "subjective" since the symbol here is "dynamic". Finally, he who invokes must have the right to this practice, meaning that he must receive it from a master who has in turn received it, which presupposes an initiation correctly transmitted across the centuries from the origin of the corresponding Revelation' (Schuon: *L' Oeil du Coeur*, p. 198).

Orthod. 288 275

[1] Cp. *Matt.* XXI. 12–13.

[2] The Arabic word *dhikr* (invocation) lit.='remembrance'; cf. the section *Recollection* in the chapter *Knowledge*,—also the Introduction to *Metanoia*.

[3] The law of inverse analogy is applicable here, that which is 'least' in the formal realm being greatest in the supraformal domains.

The Word that God Utters

Utterance (Vâk) brought forth all the Universe. He (Prajâpati) prounced 'Bhû' (Earth) and the Earth was born. *Creation* 31

Śatapatha-Brâhmaṇa, VI passim

From the sound of Vedas that supreme Divinity made all things.

Mânava-dharma-śâstra, I. 21

In this Universe, there is no form of knowledge which is not perceived through sound: knowledge is pierced through by sound; all this Universe is but the result of sound.

Vâkya Padîya, I. 124

For the word of God is quick, and powerful, and sharper than any two edged sword, piercing even to the dividing asunder of soul and spirit, and of the joints and marrow, and is a discerner of the thoughts and intents of the heart.

Hebrews, IV. 12

The first vibration which took place at the commencement of creation, that is, on the disturbance of equilibrium (Vaishamyavastha), was a general movement (Samanya-Spanda) in the whole mass of Prakriti. This was the Pranava-Dhvani or OM sound. OM is only the approximate representation or gross form of the subtle sound which is heard in Yoga-experience. Introds. *Rev.* *Waters* *Realiz.*

Varnamala

The Lord says, 'Stand in the gate of God's house and proclaim his word, extol his word' (*Jer.* VII. 2). The heavenly Father speaks one Word and that he speaks eternally and in this Word expends he all his might: his entire God-nature he utters in this Word and the whole of creatures. This Word lies hidden in the soul unnoticed and beyond our ken, and were it not for rumours in the ground of hearing we should never heed it: but all sounds and voices have to cease and silence, perfect stillness, reign. Introd. *Love* *M.M.* 987

Eckhart

For while all things were in quiet silence, and the night was in the midst of her course, Thy almighty word leapt down from heaven from thy royal throne. *Peace* 705

Wisdom, XVIII. 14, 15

All things have been made by the power of the divine word, which is the divine spirit or breath that emanated from the divine fountain in the beginning. This breath is the spirit or soul of the world, and is called the 'spiritus mundi'.[1]

Johannes Tritheim

[1] The philosophers' stone according to Tritheim is the *spiritus mundi* rendered visible.

Beauty 679

Sound exists in four fundamental states, *viz.* (1) Vaikhari or dense, audible sound, sound in its maximum differentiation: (2) Madhyama or an inner, subtle, more ethereal state at which it is inaudible to physical ear; (3) Pashyanti, a still higher, inner, more ethereal state; (4) Para which represents Ishwara-Sakti and is the potential (Karana) state of the sound which is Avyakta or undifferentiated. The Para sound is not, like the Vaikhari, different in different languages. It is the unchanging primal substratum of them all, the source of the universe.[1]

Swami Sivananda

Creation 26

Each creature has its being
From the One Name,
From which it comes forth,
And to which it returns,
With praises unending.

Shabistarî

Infra 1013

It is with His name and His form that this world has come into existence, and it also ends with His name.

Ananda Moyî

For as the rain cometh down, and the snow from heaven, and returneth not thither, but watereth the earth, and maketh it bring forth and bud, that it may give seed to the sower, and bread to the eater:

So shall my word be that goeth forth out of my mouth: it shall not return unto me void, but it shall accomplish that which I please, and it shall prosper in the thing whereto I sent it.

Isaiah, LV. 10, 11

Center 847

The Brahman (Supreme Being) created Brahmâ, the creator seated on the lotus; having been created, Brahmâ began to think, 'By which single syllable may I be able to enjoy all the desires, all the worlds, all the gods, all the Vedas, all the sacrifices, all the sounds, all the rewards, all the beings, stationary and moving.' He practised self-control and saw *OM*, of two syllables, of four morae, the all-pervading, omnipresent, the eternally potent Brahman, the Brahman's own symbolic syllable, of which the presiding divinity is Brahman itself; with it, he enjoyed all the desires of all the worlds, all gods, all Vedas, all sacrifices, all sounds, all rewards and all beings, both stationary and moving.

Infra 1017

Atharva-Veda (*Gopatha Brahmana*), I. 16–22

Assuredly, the nature of the ether within the space of the heart is the same as the

[1] The four states mentioned retrace inversely the successive stages of manifestation from Principle (cf. Guénon: *Man and His Becoming*).

A skillful Hindu musician, in giving full development to the potentialities of a *raga*, can convey through a kind of 'musical yoga' this fourfold withdrawal of sound from manifestation towards Principle, where the performance culminates precisely in something very like a sea of undifferentiated sound.

supreme bright power. This has been manifested in threefold wise: in fire, in the sun, and in the breath of life.

Verily, the nature of the ether within the space of the heart is the same as the syllable *Om*.

Center 816

With this syllable, indeed, that bright power is raised up from the depths, goes upwards, and is breathed forth. Verily, therein is a perpetual support for meditation upon Brahma.

Maitri Upanishad, VII. 11

In my humble simplicity, poverty and misery, God has shown me His marvels—There I saw the Creation and the ordering of the House of God which He Himself has made out of His own mouth.

Mechthild of Magdeburg

So help me heaven, the work of God who is great and wise: so help me the word of the Father which he spake when he established the whole universe in his wisdom.

Eleusinian Mysteries (Oath of Initiation)

God spoke but one word, and in virtue of that in a moment were made the sun, moon and that innumerable multitude of stars, with their differences in brightness, motion and influence. *He spoke and they were made* (*Ps.* CXLVIII. 5). A single word of God's filled the air with birds, and the sea with fishes, made spring from the earth all the plants and all the beasts we see. . . .

Creation 31

This word then, whilst most simple and most single, produces all the distinction of things: being invariable produces all fit changes, and, in fine, being permanent in his eternity gives succession, vicissitude, order, rank and season to all things.

Reality 775

Beauty 689

St Francis de Sales

God is the Word which pronounces itself. Where God exists he is saying this Word: where he does not exist he says nothing. God is spoken and unspoken.

Eckhart

M.M. 994

Now I have no doubt that it is obvious to a man of sound mind that the first thing the voice of the first speaker uttered was the equivalent of God, namely *El*, whether in the way of a question or in the way of an answer. It seems absurd and repugnant to reason that anything should have been named by man before God, since man had been made by him and for him. For as, since the transgression of the human race, every one begins his first attempt at speech with a cry of woe, it is reasonable that he who existed before that transgression should begin with joy: and since there is no joy without God, but all joy is in God, and God himself is wholly joy, it follows that the first speaker said first and before anything else 'God'.

Creation 48

Center 847

Wr. Side 474

Dante (*De Vulg. Eloquentia*, I. iv. 30)

The Name of Jesus is Origin without origin: . . . before the creation of the sun and until the burning-out thereof, the Name was pre-ordained, from everlasting to everlasting, until the end of Time and thereafter.

Center 838

St Bernardine of Siena

P. State 579

There is the Rose wherein the Word Divine made itself flesh. . . .
The name of the beauteous flower which I ever invoke, morning and evening. . . .

Dante (*Paradiso*, XXIII. 73, 88)

Orthod. 280

When the World-Honored One was about to finish his sermon, he very earnestly entrusted Ânanda with Amida's sacred name.

Zendo

Man doth not live by bread only, but by every word that proceedeth out of the mouth of the Lord.

Deuteronomy, VIII. 3

Know that it is not the growing of fruits that nourisheth men, but thy word preserveth them that believe in thee.

Wisdom, XVI. 26

Holiness 902

Father and Son expire their holy Breath, and once this sacred breath inspires a man it remains in him, for he is essential and pneumatic.

Eckhart

Holiness 900

Waters 651

Give unto the Lord the glory due unto his name: worship the Lord in the beauty of holiness.

The voice of the Lord is upon the waters: the God of glory thundereth: the Lord is upon many waters.

The voice of the Lord is powerful: the voice of the Lord is full of majesty.

The voice of the Lord breaketh the cedars: yea, the Lord breaketh the cedars of Lebanon.

He maketh them also to skip like a calf: Lebanon and Sirion like a young unicorn.

The voice of the Lord divideth the flames of fire.

The voice of the Lord shaketh the wilderness: the Lord shaketh the wilderness of Kadesh.

The voice of the Lord maketh the hinds to calve, and discovereth the forests: and in his temple doth every one speak of his glory.[1]

Psalm XXIX. 2–9

[1] Among other Biblical references, cf. *Psalms* IX. 10, XXXIV. 1, 3, LXIX. 30, LXXIV. 21, CXIX. 55; *Lamentations*, III. 55–58; *Malachi*, I. 11; *St Matthew*, XVIII. 20; *St Luke*, I. 49, X. 17; *Acts*, III. 6, IV. 12, 30; *I. John*, V. 13.

Fusion

Man becomes the food of the divinity whom he worships. *Suff.* 132

Hindu Proverb

Thus abide constantly with the name of our Lord Jesus Christ, so that the heart swallows the Lord and the Lord the heart, and the two become one. *M.M.* 978

St John Chrysostom

Offer first your life and your all; then take the name of the Lord.

Gujarati Hymn

Blessed is the person who utterly surrenders his soul for the name of YHWH to dwell therein and to establish therein its throne of glory.

Zohar

For it (the Torah) is wholly in thee and thou art wholly in it.

Abraham Abulafia

By repeating the name of Krishna or Râma a man transforms his physical body into a spiritual body. *Realiz.* 870

Sri Ramakrishna

For one body takes possession of the other; even if it be unlike to it, nevertheless, through the strength and potency added to it, it is compelled to be assimilated to the same, since like derives origin from like. *Realiz.* 868

Basil Valentine

It is in pronouncing Thy Name that I must die and live. *Death* 206

Muhammad

The Cock will swallow the Fox, and, having been drowned in the water, and quickened by the fire, will in its turn be swallowed by the Fox. *Pilg.* 378

Basil Valentine

Thou, Hari, knowest and pervadest all. Murmur, O soul, the Name of Hari, and sins will disappear. Thou art I and I am Thou, and there is no difference; such is gold and bracelet, water and wave. Infra 1009

Adi Granth

You must join the husband and wife together, that each may feed upon the other's flesh and blood, and that so they may propagate their species a thousandfold. *Love* 625

Basil Valentine

A great saint has said that he who has got God's name on his lips is a Jivan-mukta, because continuous remembrance of God eliminates the ego-sense and grants him the realisation of his immortal, changeless Self. Name is a link between the devotee and God. It brings the devotee face to face with God and enables him to attain the knowledge of his oneness with Him.

Realiz. 870

Swami Ramdas

I'm fortunate indeed!
Not dead I go,
Just as I live,
I go to the Pure Land!
'Namu-amida-butsu!'

Led by 'Namu-amida-butsu,'
While living in this world,
I go to 'Namu-amida-butsu.'

Action 340

Saichi

What does the Lord Krishna teach in holding a flute in his hands? What is the symbolic meaning of the flute? It is the symbol OM. It says: 'Empty thyself of all egoism, and I will play on the flute of thy body. Let thy will become one with My will. Take refuge in OM, thou wilt enter into My being. Listen to the moving interior music of the soul and rest in eternal Peace!'

Death 220 *Conform.* 166

Swami Sivananda

'Come unto me,' says the holy Jesus, 'all ye that labour and are heavy laden, and I will refresh you.' Here is more for you to live upon, more light for your mind, more of unction for your heart than in volumes of human instruction. Pick up the words of the holy Jesus and beg of Him to be the light and life of your soul. Love the sound of His name: for Jesus is the love, the sweetness, the compassionate goodness of the Deity itself which became man, that so men might have power to become the sons of God.

Introd. *Creation*

William Law

Saichi has his heart revealed by Amida's mirror.
How happy for the favor! 'Namu-amida-butsu!'
'Namu-amida-butsu, Namu-amida-butsu!'
'Namu-amida-butsu, Namu-amida-butsu!'

Center 828

Saichi

I have found great profit in the vocal prayers that are called ejaculatory, and especially in the words of the *Pater Noster,* which are repeated during many hours with the mouth, saying for example: '*Sanctificetur Nomen tuum*', and having the heart filled with desire for Him who is enclosed therein.

Center 816

Father Jerome Gratian

There is a prayer which may be performed at all times and in all places, which by nothing can be interrupted but by sin and unfaithfulness. . . . This incessant prayer now consists in an everlasting inclination of the heart to God, which inclination flows from Love. This love draws the presence of God into us; so that, as by the operation of divine grace the love to God is generated in us, so is also the presence of grace increased by this love, that such prayer is performed in us, without us or our cogitation. It is the same as with a person living in the air and drawing it in with his breath without thinking that by it he lives and breathes, because he does not reflect upon it. Wherefore this way is called a Mystical Way—that is, a secret and incomprehensible way. In one word, the prayer of the heart may be performed at all times, though the heart cannot think or speak at all times.

Grace 552
556

Johannes Kelpius

Now, it has elsewhere been said: 'Verily, there are two Brahmas to be meditated upon: sound and non-sound. Now, non-sound is revealed only by sound.' Now, in this case the sound-Brahma is *Om*. Ascending by it, one comes to an end in the non-sound. So one says: 'This, indeed, is the way. This is immortality. This is complete union (*sâyujvatva*) and also peacefulness (*nirvṛtatva*).'

M.M.
994
Supra
1003

Now, as a spider mounting up by means of his thread (*tantu*) obtains free space, thus, assuredly, indeed, does that meditator, mounting up by means of *Om*, obtain independence (*svâtantrya*).

Maitri Upanishad, VI. 22

The Dispeller of Sin

Of the various kinds of penances in the form of action or austerity, the constant remembrance of Krishna is the best.

Suff.
133

The singing of His name is the best means for the dissolution of various sins, as fire is the best dissolver of metals.

The most heinous sins of men disappear immediately if they remember the Lord even for a moment.

Vishṇu Purâṇa

There is power enough in the *Nembutsu*, even if pronounced but once, to destroy all the sins whose effects have persisted through eighty billions of *kalpas*.

Reality
803

Hônen

Who says: Allâh! in language truly loving
Shall see his sins, like autumn leaves, removing.

The Mevlidi Sherif

Pure or impure, in whatever condition one might be, he who remembers the Lord, is pure within and without.

Hindu Prayer (a formula recited at the opening of religious ceremonies)

Then spoke the Buddha *Nang-wa-t'ä-yä* (Infinite Brightness), the Perfect Buddha, also called *Wu-pa-mé* (Light Eternal): 'Most Compassionate *P'ʌpa Shen-rä-zig Wang-ch'yuk,*[1] by the following six letters the door of birth for the six classes of created beings may be closed: *Om mani padmé hûm.* By *Om* the gate of birth among gods (*Lh'a*) is closed; by *ma* the gate of birth among Titans (*Lh'a-ma-yin*) is closed; by *ni* the gate of birth among men is closed; by *pad* the gate of birth among brute beasts is closed; by *mé* the gate of birth among *pretas* (*Yidag*) is closed; and by *hûm* the gate of birth in hell is closed. These can empty the kingdoms of the six classes of creatures. . . . Understand them well, remember them, repeat them, impress them well upon your mind.'

Introd. *Waters*

Srong-Tsan-Gampo

As an earth-clod falling into a great lake is quickly lost, so all evil acts sink (out of sight) in the threefold Veda.

The texts (Ṛg-Veda), and the other sacrificial formulas (Yajur-Veda), and the songs of various sorts (Sâma-Veda)—this must be known as the threefold Veda: who knows this Veda, he is Veda-wise.

That primordial *brahma,* consisting of three sounds,[2] in which the triple (Veda is) contained, (is) another three-fold Veda (which is) to be kept secret: who knows it, he is Veda-wise.

Mânava-dharma-śâstra, XI. 264–266

Metanoia 493

The Lord is nigh unto all them that call upon him, to all that call upon him in truth.

Psalm CXLV. 18

Infra 1013

The person who invokes God in a society where He is not invoked, will obtain boundless pardon from God.

Muhammad

Pilg. 385

Mother! Mother! My boat is sinking, here in the ocean of this world:
Fiercely the hurricane of delusion rages on every side!
Clumsy is my helmsman, the mind; stubborn my six oarsmen, the passions;
Into a pitiless wind
I sailed my boat, and now it is sinking!
Split is the rudder of devotion; tattered is the sail of faith;
Into my boat the waters are pouring! Tell me, what shall I do?
For with my failing eyes, alas! nothing but darkness do I see.
Here in the waves I will swim,
O Mother, and cling to the raft of Thy name!

Bengali Hymn

[1] Avalokiteshvara, the Bodhisattva of Compassionate Mercy.

[2] 'Essence of Veda, *a, u, m=Om (aum)*':—note in text. 'OM is the essence of the Vedas' (Swami Sivananda: *La Pratique de la Méditation,* p. 203).

The Name itself is the best atonement for sins committed against the Name.

Swami Sivananda

If you receive wounds a thousand times, still you should by no means give up the life-giving action—that is to say, calling upon Jesus Christ who is present in our hearts.

The Russian Pilgrim

Center 816

All the sins of the body fly away if one chants the name of God and sings His glories. The birds of sin dwell in the tree of the body. Singing the name of God is like clapping your hands. As, at a clap of the hands, the birds in the tree fly away, so do our sins disappear at the chanting of God's name and glories.

Sri Ramakrishna

The Name of Jesus is as ointment poured forth:
It nourishes, and illumines, and stills the anguish of the soul.

Angelus Silesius

Center 835

Teach me thy way, O Lord: I will walk in thy truth: unite my heart to fear thy name.
I will praise thee, O Lord my God, with all my heart: and I will glorify thy name for evermore.

Psalm LXXXVI. 11, 12

Conform. 170

Ceaselessly I have performed the ritual of the Holy Name that clears all sin away.[1]

Atsumori

In the voice reciting the Nembutsu there is the light of Amida waiting to receive all sinners. Let us pray to be embraced in the light of Amitâbha, relying upon the power of the Nembutsu, which is a saving boat taking us to the golden shore where there is the consummation of happiness.

Kashiwazaki

Introd. *Sp. Drown.*

As for the Godfearing, when a glamour from the devil troubleth them, they do but invoke, and behold them seers!

Qur'ân, VII. 201

The *Nembutsu* is compared to gold, which, when burned in the fire, only has its colour improved, and receives no injury though thrown into the water. In the same way the *Nembutsu,* though said when evil passions arise, is not defiled, nor does it lose its value though said when you are in conversation with others.

Hônen

Infra 1036

Death has been snatched away from me,
And in its place the 'Namu-amida-butsu'.

Saichi

Death 232

[1] 'The teaching of Amida and his Pure Land is found throughout the Nô' (Beatrice Lane Suzuki: *Nôgaku,* p. 40).

One, however sinful, who loves God and surrenders himself to him, is his beloved and his own. The mere name of God has power to save even the most depraved. Wrongdoing is not eradicated merely by expiation if the mind continues to follow wicked desires. But when the name of God and God's love have purified the heart, then indeed are all sins completely destroyed. . . . Verily nothing is more purifying than the holy name of God.

Srimad Bhagavatam, VI. i

Sin 66

Because all the incomparable merits of all the Buddhas are embraced in a single *Nembutsu* repetition, the heart comes forth purified from the filth of all its passion, as the lotus in the muddy pond leaps forth into beauteous bloom.

Hônen

By the mere repetition of His Name, the ocean of Samsara is dried up.

Swami Sivananda

Peace 700 *Center* 847

I am happy!
The root of sinfulness is cut off:
Though still functioning, it is the same as non-existent.
How happy I am!
Born of happiness is the 'Namu-amida-butsu'.

Saichi

As clouds are blown away by the wind, the thirst for material pleasures will be driven away by the utterance of the Lord's name.

Śrî Sâradâ Devî

When a man begins to call on the sacred name, a lotus flower begins to grow in Paradise for him.

Fa-chao

'Thy act of love[1] will not be extinguished at death, but will be made eternal in Heaven!'

Sister Consolata

Orthod. 285 *Death* 206

Our meditation in this present life should be in the praise of God: for the eternal exultation of our life hereafter will be the praise of God: and none can become fit for the future life, who hath not practised himself for it now.

St Augustine

Death 215

Life is short. Time is fleeting. The body is continually decaying. There is nothing but gain and gain only, in doing Japa.

Swami Sivananda

'Remember that an act of love determines the eternal salvation of a soul. . . . Do not lose time, for each act of love represents a soul.'

Sister Consolata

[1] I.e., the invocation: 'Jesus, Mary, I love you. Save souls' (said in Italian).

The sign that a gnostic is separated from the divine presence is that he ceases to do the *dhikr*.

Dhu 'l-Nûn

Man cannot live by bread alone; but he can live repeating the Name of the Lord alone. . . . One who does not do Smarana of Hari is a Neecha (low-born). The day spent without the remembrance of His Name is a mere waste.

Swami Sivananda

Infra 1017

Consider all the time to be lost to you in which you do not think of divinity.

Sextus the Pythagorean

Despise all those things, which when liberated from the body you will not want: and exercising yourself in those things of which when liberated from the body you will be in want, invoke the Gods to become your helpers.

Pythagoras

Suff. 126b

Those who simply eat, drink and sleep and do not practise any Japa are horizontal beings only.

Swami Sivananda

Judgment 250

Save me, O God, by thy name.

Psalm LIV. 1

Faith 501

'So long as you call upon Me and hope in Me, I forgive you all that originates from you: and I will not heed, O son of man, should your sins reach the horizon of the heavens, and then you asked My pardon and I would pardon you.'

Muhammad

Sin 57

Invocation and the Last Times

The sun shall be turned into darkness, and the moon into blood, before the great and the terrible day of the Lord come.

And it shall come to pass, that whosoever shall call on the name of the Lord shall be delivered.

Joel, II. 31. 32

Wr. Side 464

And I saw heaven opened, and behold a white horse; and he that sat upon him was called Faithful and True, and in righteousness he doth judge and make war.

His eyes were as a flame of fire, and on his head were many crowns; and he had a name written, that no man knew, but he himself.

Introd. *Holy War* *M.M.* 987

Supra 1003

And he was clothed with a vesture dipped in blood: and his name is called The Word of God.

Revelation, XIX. 11–13

The hour will not surprise the one who saith: Allâh, Allâh.

Muhammad

Rev. passim

Verily in the messenger of Allâh ye have a good example for him who looketh unto Allâh and the Last Day, and invoketh Allâh much.

Qur'ân, XXXIII. 21

Pilg. 387

Quoth Our Lord, 'The hour cometh and is now, when true worshippers shall worship not only on the mountains and in the Temple but in spirit, in the place of God.' The moral of which is that we ought to pray to God not only on the hill-tops and in churches, but we ought always to be praying, at all times and everywhere. St Paul says, 'Rejoice evermore: in everything give thanks: pray without ceasing.'

Eckhart

Infra 1017

Unto you that fear my name shall the Sun of righteousness arise with healing in his wings.

Malachi, IV. 2

Supra 1009

Water suffices to put out fire, the sunrise to disperse the darkness: in the *Kali Yuga* the repetition of the Name of Hari (*Vishnu*) suffices to destroy all errors.

Vishṇu Purâṇa

Holy war 410
P. State 563

If you have but the spirit of Amida within you and call upon his name, no matter how high the castle or how deep the moat, no matter how long the spear or how sharp the sword, who can stand before you?

Tôyo Tenshitsu

Center 835
M.M. 978

We in these days can secure the wonderful treasure of Amida's Original Vow alone by walking serenely in the path that lies between the two streams of fire and water.

Hônen

'Alî once asked the Prophet: 'O Messenger of God, show me the shortest road which leads to Paradise, the one most pleasing to the Lord and the one best suited to his worshippers.'

Contem. 523

The Prophet replied: 'O 'Alî, this way is to repeat without cease the name of God in seclusion (*khalwah*): so meritorious is this practice that the world will not come to an end until no one on earth any longer performs it.'

Muhammad

Even though all other religious practices should perish, the *Nembutsu* would continue for a hundred years beyond that.

Sukhâvatîvyûha Sûtra

In this last age of the world, many will become blessed by this arcanum.

Philalethes

And I looked, and, lo, a Lamb stood on the mount Sion, and with him an hundred forty and four thousand, having his Father's name written in their foreheads.

Revelation, XIV. 1

Not only in the Kali-Yuga but in all ages, at all times, in all the worlds, by the repetition of the Name of the Lord, men have crossed the ocean of Samsara and attained to the highest eminence. In Kali-Yuga Nama-Japa is particularly suited for one's emancipation, just as meditation, performance of sacrifices and worship of the Lord were suited in the other three Yugas, for the attainment of Moksha. In this most troublesome Kali-age there is no other Sadhana for emancipation.

Swami Sivananda

Reality 790

Disciple: Mahârâj, how can one awaken the *kundalinî*?

The Swâmi: By the practice of *japa*, meditation (*dhyâna*) and other spiritual exercises. According to certain *yogins*, there are special forms of meditation or practices which awaken it, but I think the most efficacious are *japa* and *dhyâna*. The practice of *japa* is expecially suited for our present time, the iron age (*kali-yuga*). There is no spiritual practice more easy. But meditation must accompany the repetition of the *mantra*.

Swâmi Brahmânanda

In that degenerate age when all the Three Sacred Treasures have perished . . . nothing will remain but the '*tabula rasa*' *Nembutsu* with its six mystical symbols, from which has already faded away all the coloring with which it had been stained by the meditative and non-meditative disciplines. At that time, as it says in the Larger Sûtra, if a man but hears the *Nembutsu* and then repeats it, he will be sure to attain *Ôjô*. In fact even though he be a non-believer in Buddhism itself, he will certainly attain *Ôjô* by merely repeating the *Nembutsu* ten times or even once, because of the mighty power in those six mystical symbols.

Hônen

Introd. *Orthod.*

The Divine Name is a powerful boat that takes man across the whirlpools of life to the haven of his eternal and spiritual nature. His Name transforms man from the human to the Divine. The Divine Name is the one sovereign panacea for all physical, mental, and intellectual ills that have created the sense of diversity and misery in the world.

Swami Ramdas

Pilg. 385 Infra 1017

Know, therefore, that the confusion of tongues is born of the multiplicity of beliefs, and that nearly all nations are formed from particular opinions concerning the being and will of God: and herein lies the confusion, that is to say, the mystery of Babylon the Great. . . .

In preaching union, therefore, we can counsel nothing better than the desire to be again one in one, one single nation, one single tree, one single man, one single soul and one single body. We must destroy and kill in ourselves all the images of the letters, letting but a single one live, no longer desiring to know or to will anything of God but to live uniquely and simply according to what God wishes to know in and through us, so that we immerse without any other knowing the hunger and desire of our souls into the five vowels. And therein is to be found the great and holy name of God, *Jehovah* or *Jesus* (insofar as living *Word* which gives life to all things), and not by following the property of nature, the distinction of multiple wills: but it is in the unique sun of love that He is revealed.

Faith 512
Creation 48

Just as the outward sun gives light and strength to the whole world: in the same way this unique name in its power gives life and meaning to all letters. . . .

Symb. 317

When the time of the outward constellation is accomplished, then the Tower of Babel, viz. the outward man including his opinion, is overthrown and destroyed in confusion, excepting the unique soul which stands before God, naked and destitute of all.

Introd. *Judgment*

Hence there is no issue unless this soul possesses the unique spirit of the sonant letters, namely, the unformed Word, so that it is enabled in its desire to draw this to itself and put it on, in such way as to make fly apart the language that is seized and compacted, and all the images of the letters, and to introduce them into one tongue and one unique will which is God, all in all. All must return to the One which is the All, for in multiplicity there is only wrangling and disorder, but in Unity reigns eternal repose, where no enmity exists.

Reality 775
Center 835

Boehme

And to the angel of the church in Philadelphia write: These things saith he that is holy, he that is true, he that hath the key of David, he that openeth, and no man shutteth; and shutteth, and no man openeth:

Holiness 924

I know thy works: behold, I have set before thee an open door, and no man can shut it: for thou hast a little strength, and hast kept my word, and hast not denied my name.

Behold, I will make them of the synagogue of Satan, which say they are Jews, and are not, but do lie; behold, I will make them to come and worship before thy feet, and to know that I have loved thee.

Wr. Side 457

Because thou hast kept the word of my patience, I also will keep thee from the hour of temptation,[1] which shall come upon all the world, to try them that dwell upon the earth.

Behold, I come quickly: hold that fast which thou hast, that no man take thy crown.

Him that overcometh will I make a pillar in the temple of my God, and he shall go no more out: and I will write upon him the name of my God, and the name of the city of my God, which is new Jerusalem, which cometh down out of heaven from my God: and I will write upon him my new name.

Holy War 410
Holiness 902
Center 832

Revelation, III. 7–12

[1] 'It may be interesting to note, that in the writings of some Kabbalists the Great Name of God appears as the supreme object of meditation in the last hour of the martyrs. In a powerful speech of the great mystic Abraham ben Eliezer Halevi of Jerusalem (died about 1530) we find a recommendation to those who face martyrdom. He advises them to concentrate, in the hour of their last ordeal, on the Great Name of God: to imagine its radiant letters between their eyes and to fix all their attention on it. Whoever does this, will not feel the burning flames or the tortures to which he is subjected. "And although this may seem improbable to human reason, it has been experienced and transmitted by the holy martyrs"' (Scholem: *Major Trends in Jewish Mysticism*, p. 146). Other instances of invocation surmounting martyrdom are given in *The Wonders of the Holy Name* (see under *Index of Sources*), and in *Japa Yoga*, by Swami Sivananda, p. 16.

For then will I turn to the people a pure language, that they may all call upon the name of the Lord, to serve him with one consent.

Zephaniah, III. 9

In that day shall there be one Lord, and his name one.

Zechariah, XIV. 9

The Universal Elixir

My heart desires to see thee, O lord of the Acacia trees: my heart desires to see thee, O Amon. Thou art the protector of the poor, a father to the motherless, a husband to the widow. Sweet it is to speak thy name: it is like the taste of life, like the taste of bread to a child, like the breath of freedom to a prisoner. Turn thyself to us, O Eternal One, who wast here before others existed. Though thou makest me to see darkness yet cause light to shine on me that I may see thee. *Pilg.* 366

Egyptian Tradition

Chaitanya used to shed tears of joy at the very mention of Krishna's name.

Sri Ramakrishna

Anyone who dies in Benares, whether a brâhmin or a prostitute, will become Śiva. When a man sheds tears at the name of Hari, Kâli, or Râma, then he has no further need of the sandhyâ and other rites. All actions drop away of themselves. The fruit of action does not touch him. *Action* 329

Sri Ramakrishna

Why should I go to Gangâ or Gayâ, to Kâśi, Kânchi, or Prabhâs, *Pilg.*
So long as I can breathe my last with Kâli's name upon my lips? 387
What need of rituals has a man, what need of devotions any more, *Orthod.*
If he repeats the Mother's name at the three holy hours? 296
Rituals may pursue him close, but never can they overtake him.
Charity, vows, and giving of gifts do not appeal to Madan's[1] mind: *Action*
The Blissful Mother's Lotus Feet are his whole prayer and sacrifice. 346
Who could ever have conceived the power Her name possesses?
Śiva Himself, the God of Gods, sings Her praise with His five mouths!

Bengali Song

[1] Author of the song.

'There is a polish for everything that taketh away rust; and the polish for the heart is the remembrance of God.' The companions said, 'Is not repelling the infidels also like this?' Muhammad said, 'No, although one fights until one's sword be broken!'

Holy War 410

Muhammad

Orthod. 296

At the beginning of spiritual life the devotee should observe such rites as pilgrimage, putting a string of beads around his neck, and so forth. But outward ceremonies gradually drop off as he attains the goal, the vision of God. Then his only activity is the repetition of God's name, and contemplation and meditation on Him.

Sri Ramakrishna

In the sweet name, Jesus Christ, the whole process is contained.

Boehme

Contem. 547

At the touch of the philosopher's stone, the eight metals become gold. Likewise all castes, even the butcher and the untouchable, become pure by repeating Hari's name. Without Hari's name the people of the four castes are but butchers.

Tulsî Dâs

If we had sense we ought to do nothing else, in public and in private, than praise and bless God and pay Him due thanks. . . .

More than that: since most of you are walking in blindness, should there not be some one to discharge this duty and sing praises to God for all? What else can a lame old man as I am do but chant the praise of God? If, indeed, I were a nightingale I should sing as a nightingale, if a swan, as a swan: but as I am a rational creature I must praise God. This is my task, and I do it: and I will not abandon this duty, so long as it is given me; and I invite you all to join in this same song.

Epictetus

I am a man who wears no official head-dress. I am nothing but the foolish Hônen, weighed down by the ten evils, and I say that the only way for me to attain *Ojô* is by calling upon the sacred name.

Hônen

Renun. 156 *M.M.* 978 *Void* 724 *Renun.* 139

Nothing is left to Saichi.
Except a joyful heart nothing is left to him.
Neither good nor bad has he, all is taken away from him:
Nothing is left to him!
To have nothing—how completely satisfying!
Everything has been carried away by the 'Namu-amida-butsu.'
He is thoroughly at home with himself:
This is indeed the 'Namu-amida-butsu!'

Saichi

Dhikr (invocation) is a fire. . . . If it enters a dwelling, it says: It is I, not another! If it finds wood, it burns it, if it finds darkness, it changes it into light: if it finds light, it adds light to light.

Realiz. 890

Ibn 'Aṭâ'illâh

If sons and daughters of good family wish to enter upon the Samâdhi of Oneness, let them sit in a solitary place, abandon all thoughts that are disturbing, not become attached to forms and features, have the mind fixed on one Buddha, and devote themselves exclusively to reciting (*ch'eng*) his name (*ming, nâmadheya*), sitting in the proper style in the direction where the Buddha is, and facing him squarely.

Sapta-śatikâ-prajñâ-pâramitâ Sûtra

Let us offer the sacrifice of praise to God continually, that is, the fruit of our lips giving thanks to his name.[1]

Suff. 132

Hebrews, XIII. 15

There is no access to the knowledge of God except through the intermediary of His Names and His Qualities: and since every Name and every Quality is contained in the name *Allâh,* it follows that there is no access to the knowledge of God except through the way of this Name.

Knowl. 749

Jîlî

Prajâpati! thou only comprehendest all these created things, and none beside thee.
Grant us our hearts' desire when we invoke thee: may we have store of riches in possession.

Rig-Veda, x. 121

At that time Mujinni Bosatsu[2] rose from his seat, and, baring his right shoulder, turned, with his hands folded, towards the Buddha, and said this: World-honoured One, for what reason is Kwanzeon Bosatsu[3] so named?

The Buddha said to Mujinni Bosatsu: Good man, when those innumerable numbers of beings—hundred-thousands of myriads of kotis of them—who are suffering all kinds of annoyances, hearing of this Kwanzeon Bosatsu, will utter his name with singleness of mind, they will instantly hear his voice and be released.

Faith 514

Even when people fall into a great fire, if they hold the name of Kwanzeon Bosatsu, the fire will not scorch them because of the spiritual power of this Bosatsu. When they are tossed up and down in the surging waves, if they pronounce his name they will get into a shallower place.

P. State 563
Waters 653

When hundred-thousands of myriads of kotis of people go out into the great ocean in

[1] I.e., Jesus Christ. This passage, alluding to *Lev.* VII. 12, shows in what way invocation may be considered a sacrificial path. Cf. Sister Consolata, p. 1023.

[2] Akshayamati, Bodhisattva of Inexhaustible Intelligence.

[3] Avalokiteshvara; cf. note 1, p. 1010. Unfortunately space prevents printing this highly symbolic Sutra in its entirety.

order to seek such treasures as gold. silver. lapis lazuli. conch shells. cornelian. coral. amber. pearls. and other precious stones. their boats may be wrecked by black storms. and they may find themselves thrown up into the island of the Rakshasas: if among them there is even a single person who will utter the name of Kwanzeon Bosatsu all the people will be released from the disaster (which is likely to befall them at the hand) of the Rakshasas. For this reason the Bosatsu is called Kwanzeon.

Pilg. 385

Holy War 410

When. again. a man is about to suffer an injury. if he will utter the name of Kwanzeon Bosatsu. the sword or the stick that is held (by the executioner) will be at once broken to pieces and the man be released.

Wr. Side 459

When all the Yakshas and Rakshasas filling the three thousand chiliocosms come and annoy a man. they may hear him utter the name of Kwanzeon Bosatsu. and no wicked spirits will dare look at him with their evil eyes. much less inflict injuries on him.

Supra 1009

When again a man. whether guilty or innocent. finds himself bound in chains or held with manacles. he uttering the name of Kwanzeon Bosatsu will see all these broken to pieces and be released.

Center 835

When all the lands in the three thousand chiliocosms are filled with enemies. a merchant and his caravan loaded with precious treasures may travel through the dangerous passes. One of the company will say to the others: 'O good men. have no fear: only with singleness of thought utter the name of Kwanzeon Bosatsu. As this Bosatsu gives us fearlessness. utter his name and you will be delivered from your enemies.' Hearing this. all the company join in the recitation. saying. 'Kwanzeon Bosatsu be adored!' Because of this uttering the name of the Bosatsu they will be released. O Mujinni. such is the awe-inspiring spiritual power of Kwanzeon Bosatsu Makasatsu.

Illusion 89

When people are possessed of excessive lust. let them always reverentially think of Kwanzeon Bosatsu and they will be freed from it. If they are possessed of excessive anger. let them always reverentially think of Kwanzeon Bosatsu. and they will be freed from it. When they are possessed of excessive folly let them always reverentially think of Kwanzeon Bosatsu. and they will be freed from it. O Mujinni. of such magnitude is his spiritual power which is full of blessings. Therefore. let all beings always think of him.

Faith 501

If a woman desire a male child. let her worship and make offerings to Kwanzeon Bosatsu. and she will have a male child fully endowed with bliss and wisdom. If she desire a female child. she will have one graceful in features and in possession of all the characteristics (of noble womanhood). and because of her having planted the root of merit the child will be loved and respected by all beings. O Mujinni. such is the power of Kwanzeon Bosatsu.

If all beings worship and make offerings to Kwanzeon Bosatsu. they will derive benefits unfailingly from this. Therefore. let all beings hold the name of Kwanzeon Bosatsu. O Mujinni. if there is a man who holds the names of all the Bodhisattvas equal in number to sixty-two billion times as many as the sands of the Ganga. and till the end of his life makes them offerings of food and drink. clothing and bedding and medicine. what do you think? Is not the merit accumulated by such a man very great?

Mujinni said: Very great. indeed. World-honoured One!

Action 346

The Buddha said: Here is another man: if he should hold the name of Kwanzeon Bosatsu even for a while and make offerings to the Bosatsu. the merit so attained by this one is fully equal to that (of the previous one). and will not be exhausted even to the end of hundred-thousands of myriads of kotis of kalpas. Those who hold the name of Kwanzeon Bosatsu gain such immeasurable and innumerable masses of blissful merit.

Saddharma-puṇḍarika, XXV (*Kwannon Sutra*)

Whatsoever ye shall ask in my name, that will I do, that the Father may be glorified in the Son.

St John, XIV. 13

Infra 1031

In this state we perform a powerful prayer, the prayer of Jesus Christ, and through His spirit. The soul can then no more pray with cogitation and make conclusive reasonings, since she is found in a continual and working prayer. All that the soul is and what is in her prays through and in Jesus Christ: and being not intent upon her own will, nor thinking discerningly on what she prays for, she receives at once what she has need of.

Johannes Kelpius

Introd. *Contem. Conform.* 170

Let them praise the name of the Lord: for his name alone is excellent: his glory is above the earth and heaven.

Psalm CXLVIII. 13

I will bring the third part through the fire, and will refine them as silver is refined, and will try them as gold is tried: they shall call on my name, and I will hear them.

Zechariah, XIII. 9

Pilg. 366

Thy name is as ointment poured forth, therefore do the virgins love thee.

The Song of Solomon, I. 3

P. State 573

The desire of our soul is to thy name, and to the remembrance of thee.

Isaiah, XXVI. 8

The humility of the name Jesus has always interposed against the wrath of the Father.

Boehme

Shunjôbô . . . conceived the idea of having both priests and laymen add the *Namu Amida Butsu* to their given names, in order that when they would be kneeling before Emma, the king of the lower world, and were asked by him what their names were, they could reply in the language of this prayer (and so escape his wrath).

Life of Hônen

Realiz. 890

What tremendous faith Krishnakishore had! He used to say, 'By chanting "Om Krishna, Om Râma", one gets the result of a million sandhyâs'.[1] Once he said to me secretly, 'I don't like the sandhyâ and other devotions any more: but don't tell anyone.'

Sometimes I too feel that way. The Mother reveals to me that She Herself has become everything.

Sri Ramakrishna

Orthod. 296

[1] Daily ritual devotions.

At nightfall he (Dhu 'l-Nûn) entered a ruined building, where he found a jar of gold and jewels covered by a board on which was inscribed the name of God. His friends divided the gold and jewels, but Dhu 'l-Nûn said, 'Give me this board, my Beloved's name is upon it': and he did not cease kissing it all day. Through the blessing thereof he attained to such a degree that one night he dreamed and heard a voice saying to him, 'O Dhu 'l-Nûn! the others were pleased with gold and precious jewels, but thou wert pleased only with My name: therefore have I opened unto thee the gate of knowledge and wisdom.'

Beauty 663

'Aṭṭâr

So remember the name of thy Lord and devote thyself with a complete devotion.

Qur'an, LXXIII. 8

As soon as the spirit is transmuted into the body, (the Stone) receives its power. So long as the spirit is volatile, and liable to evaporate, it cannot produce any effect: when it is fixed, it immediately begins to operate. You must therefore prepare it as the baker prepares the bread. Take a little of the spirit, and add it to the body, as the baker adds leaven to the meal, till the whole substance is leavened. It is the same with our spirit, or leaven. The Substance must be continuously penetrated with the leaven, until it is wholly leavened. Thus the spirit purges and spiritualizes the body, till they are both transmuted into one. Then they transmute all things, into which they are injected, into their own nature. The two must be united by a gentle and continuous fire, affording the same degree of warmth as that with which a hen hatches her eggs. It must then be placed in a St. Mary's Bath, which is neither too warm nor too cold. The humid must be separated from the dry, and again joined to it. When united, they change mercury into pure gold and silver. Thenceforward you will be safe from the pangs of poverty. But take heed that you render thanks unto God for His gracious gift which is hidden from many. He has revealed the secret to you that you may praise His holy name, and succour your needy neighbour.

Orthod. 280

Supra 1007

P. State 563

Contem. 542

P. State 579

Charity 597

The Glory of the World

The Tincture changes everything it is mixed with into its own nature, and makes it white both within and without. By one operation and way, by one substance, and by one mixing, the whole work is accomplished, while its purity is also one, and it is perfected in two stages, each consisting of a dissolution and a coction, with the repetition of these.

It must be your first object to elicit the whiteness of the substance by means of gentle and continued coction or heat. I know that the Sages describe this simple process under a great number of misleading names. But this puzzling variety of nomenclature is only intended to veil the fact that nothing is required but simple coction. This process of coction, however, you must patiently keep up, and that with the Divine permission, until the King is crowned, and you receive your great reward.

Reality 775

Pilg. 378

Infra 1036

Orthod. 280

Holiness 921

The Book of Alze

At the beginning of the operation it is a mixture of various things, then it is placed in coction in a light fire and putrefies and changes and goes from out one nature and takes on another nature and becomes finally one nature.

Books One Two Three

Abu 'l-Qâsim al-'Irâqî

Only one vessel is required for the whole process, which should be of stone, and should be capable of resisting fire. . . . The vessel should be placed in a reverberatory alembic. This should be set over a gentle fire, the vessel being kept tightly closed, in order that it may be able to retain its companion, and permit the same to enkindle the whiteness thereof, as Lucas says.

Contem. 542

The Book of Alze

This medicine is the 'Calling upon the Name of Amida', and it is wrapped up in the six syllables Na Mu A Mi Da Butsu. It means absolute concentration on Amida's name. . . . For this medicine no capital or special wisdom is needed. All one has to do is to recite the words with your mouth. . . . Here indeed is a pivot of fundamental power.

Do I hear you say 'Too easy'? 'Such wares are intended only to deceive old men and women.' Many doubt their efficacy and ask of the wise if there is not some other way more suited for clever people. And Sakyamuni pointed straight back at the heart of man and said that within one's own heart there is to be found the true Buddha nature.

Suff. 133 *Center* 816

Hakuin

Ramdas can tell you in all truth and sincerity that there is no Sadhana easier than the repetition of the Name. It can take you to the highest spiritual eminence by giving you the all-comprehensive knowledge of God. Sri Ramakrishna had gone through various Sadhanas for twelve years. Ultimately, whenever anybody went to him for spiritual advice, he would ask them to take only the name of God. This is the essence of all Sadhanas.

Swami Ramdas

'The act of love does not constitute a cross, but to maintain it in all circumstances does indeed amount to a cross.'[1]

Sister Consolata

A man said: O Prophet of God, truly the laws of Islam are numerous. Tell me of one thing through which I can obtain reward. The Prophet replied: Let thy tongue be always moist in the remembrance of Allâh.

Muhammad

When man becomes familiar with *dhikr*, he separates himself (inwardly) from all other things. Now, at death he is separated from all that is not God. . . . What remains is the invocation alone. If this invocation is familiar to him, he finds his pleasure in it and rejoices that the obstacles which kept him from it have been removed, so that he finds himself alone with his Beloved.

Introd. *Judgment* Supra 1009

Al-Ghazâlî

Only let us pray Namu Amida Butsu and nothing else.

Seigwanji

[1] '(St) Jeanne (de Chantal) took a burin, reddened it in the fire, knelt before a crucifix and engraved the name of Jesus on her flesh, over the heart. The nuns who prepared her funeral dress discovered the scar: each letter, the height of a thumb, was well formed, except for the last one which was not finished, pain or hemorrhage having stopped the intrepid hand' (Paul Lorenz: *La Dame parfaite*, Paris, 1956, p. 83).

Of words, I am the monosyllable 'Om'. Of Yajnas (sacrifices), I am Japa.[1]

Bhagavad-Gîtâ, x. 25

Supra 1003

Let the student incline his ear to the united verdict of the Sages, who describe this work as analogous to the Creation of the World.

Philalethes

Introd. *Conform.*

Q: Is it possible that the Japa Yoga in itself can replace all other rites, so that one can do just that to the exclusion of all others?

Ramdas: The object of all Sadhanas or practices is to maintain a continuous remembrance of God. When Japa can do that, why should one do any other Sadhana or rite? Japa is an all-inclusive and all-sufficient practice by which you can be ever in tune with God and finally merge yourself in Him. It is capable of taking you by itself to the highest goal of God-realisation.

Swami Ramdas

Reality 790

All the Vedic rites, oblational and sacrificial, pass away: but this imperishable syllable *ôm* is to be known to be *Brâhma* and also *Prajâpati*.

Supra 1003

The sacrifice of invocation[2] is said to be better by tenfold than the regular sacrifice: if inaudible, it is a hundredfold better: a thousandfold, if mental.

Action 346

The four household sacrifices accompanied by the regular sacrifices, all those are not worth the sixteenth part of the sacrifice by invocation.

Charity 602

But by invocation even a Brahman (becomes), there is no doubt, perfect: whether he perform anything else or not, a Brahman is called *maitra* (well disposed to all things).

Mânava-dharma-sâstra, II. 84–87

Renun. 152a

The systematic pursuit of meritorious acts and the multiplication of supererogatory practices are habits like others: they scatter the heart. Let the disciple therefore keep to a single invocation (*dhikr*), to a single action, each according to what corresponds to him.

Al-'Arabi al-Ḥasanî ad-Darqâwî

Supra 1009

People do not know what the Name of God can do. Those who repeat it constantly alone know its power. It can purify our mind completely. No other Sadhana can do that. While the other Sadhanas can take us only to a certain stage, the Name can take us to the summit of spiritual experience.

Swami Ramdas

Faith 514

Invoke his (Amida's) Name, have sincere faith in him.
And soon the lead will change into gold.

Shinran

[1] Cf. note 1, p. 1019.

[2] *Japa-yajna*, —see reference above. Burnell uses the word 'muttering', but one knows how to weigh the credentials of an orientalist who can write in his Introduction to this translation that the metaphysical systems of India 'possess nothing of permanent value, and vanish like mists before the sun when confronted with the results of positive science'.

To him that overcometh will I give to eat of the hidden manna, and will give him a white stone, and in the stone a new name written, which no man knoweth saving he that receiveth it.

Revelation, II. 17

P. State 563

Name is not a means to an end, but an end in itself.

Swami Ramdas

Nembutsu is superior and all other practices are inferior, because all virtues are wrapt up in the one sacred name. . . . Not so is it with the other practices, which are all limited to some one aspect.

Hônen

At last we have now been brought face to face with Amida and his Original Vow, and are like men longing to cross a stream who have found a ferry. As, however, we reflect upon the passing of the days and nights, and how quickly we are drawing near to the land of shadows, we must make haste and seek deliverance with all our hearts, and, forsaking everything else, earnestly lift up our voices and invoke the sacred name, otherwise our golden opportunity will have passed and nothing be left us but remorse.

Yôkwan

Waters 653
Holy War 403
Death 215

When Amida was still in the unenlightened stage, He made a vow, that, on becoming a Buddha, He would come and welcome to His Paradise all who heard of His name and called upon it. . . . saying that if they would but turn their minds to Him and call much upon Him, He would remake them, just as if he were turning tiles and pebbles into gold.

Fa-chao

Metanoia 493

In my eyes, the meaning of Zendô's commentaries, when he speaks of the three mental states, of the five forms of prayer, and of the four-fold rule for practising the *Nembutsu*, is that they are all comprehended within the *Namu Amida Butsu*.

Hônen

Do this one thing of calling upon Amida's sacred name.

Zendô

The reading of the Lotus Sûtra is a great *Nembutsu*, and the calling upon Amida's name is a short reading of the Sûtra.

Dôken

Mantra is a combination of words that stands for the Supreme Reality. It is so set that by the utterance of it, a rhythmic sound is produced which has a marvellous effect on both the mental and the physical system. The sound of the Mantra produces mental equilibrium and physical harmony. It tunes the entire human being with the eternal music of the Divine. . . . bringing the soul in direct contact with the in-dwelling and all-pervading Reality. . . . One great advantage of Mantra Yoga over other methods is that it is a

Introd.
Orthod.
Beauty 679

discipline which is at once self-sufficient and independent. Truly, one who keeps the Mantra always on his lips can attain to the infinite power, wisdom, love and vision of God.

Swami Ramdas

'He that abides in the fear of the Lord, and cleaves to His Word, and waits faithfully on His office, will transform tin and copper into silver and gold, and will do great things with the help of God: yea, with the grace of Jehovah, he will have power to make gold out of common refuse.'

P. State 563

The Sophic Hydrolith (paraphrasing *Ecclesiast.* XI)

All the other religious practices are good, but in comparison with the *Nembutsu* they are not worth mentioning.

Zendô

And therefore it is written, that *short prayer pierceth heaven.*[1]

And why pierceth it heaven, this little short prayer of one little syllable? Surely because it is prayed with a full spirit, in the height and in the depth, in the length and in the breadth of his spirit that prayeth it. In the height it is, for it is with all the might of the spirit. In the depth it is, for in this little syllable be contained all the wits of the spirit. In the length it is, for might it ever feel as it feeleth, ever would it cry as it cryeth. In the breadth it is, for it willeth the same to all other that it willeth to itself.

Charity 605

The Cloud of Unknowing, XXXVII–XXXVIII

This Divine Name is in truth a mine of riches, it is the fount of the highest holiness and the secret of the greatest happiness that a man can hope to enjoy on this Earth.

The Wonders of the Holy Name

Take ten parts of our air: one part of living gold or living silver; put all this into your vessel: subject the air to coction, until it becomes first water, and then something which is not water. If you do not know how to do this, and how to cook air, you will go wrong, for herein is the true Matter of the Philosophers. You must take that which is, but is not seen until the operator pleases. This is the water of our dew, which is extracted from the saltpetre of the Sages, by which all things grow, exist, and are nourished, whose womb is the centre of the celestial and terrestrial sun and moon. To speak more openly, it is our Magnet, which I have already called our Chalybs, or steel. Air generates this magnet, the magnet engenders or manifests our air.

Pilg. 378
P. State 563
Contem. 542
Conform. 170
Supra Introd.
Symb. 317
Metanoia 493

Michael Sendivogius

The treasure of the six syllables[2] was given me by Oya-sama:[3]
However much one spends of it, it is never exhausted.
The treasure grows all the more as it is used:

Creation 28

[1] Perhaps from *Ecclus.* XXXV. 21: 'The prayer of him that humbleth himself, shall pierce the clouds.'
[2] *Na-mu-a-mi-da-buts(u).*
[3] A name of Amida, expressing loving-kindness.

It is the most wondrous treasure.
And I am the recipient of the good thing.
How happy I am with the favor! 'Namu-amida-butsu!'

Saichi

The Prayer of my heart gave me such consolation that I felt there was no happier person on earth than I, and I doubted if there could be greater and fuller happiness in the kingdom of Heaven. Not only did I feel this in my own soul, but the whole outside world also seemed to me full of charm and delight. Everything drew me to love and thank God people, trees, plants, animals. I saw them all as my kinsfolk. I found on all of them the magic of the Name of Jesus. Sometimes I felt as light as though I had no body and was floating happily through the air instead of walking. Sometimes when I withdrew into myself I . . . was filled with wonder at the wisdom with which the human body is made.

The Russian Pilgrim

P. State 583
Introd. *Holiness*

'O Saichi, what is your pleasure?'
'My pleasure is this world of delusion;
Because it turns into the seed of delight in the *Dharma* (*hô*).'
'Namu-amida-butsu, Namu-amida-butsu!'

Saichi

Creation 33
M.M. 978

God is the completion or the fullness of existence. Hence, the Name which denotes Him, too, is full and perfect. Therefore, the power of the Name of God is incalculable, for it is the height or the zenith of power. The Name of God can achieve anything. There is nothing impossible for it. It is the means to the realization of God Himself. Even as the name of a thing in this world generates the consciousness of that thing in the mind, the Name of God generates God-consciousness in the purified mind and becomes the direct cause of the realization of the Highest Perfection.

Swami Sivananda

Faith 514

'I prefer thy act of love to all thy prayers. . . . "Jesus, Mary, I love you, save souls" includes everything.'

Sister Consolata

The meditation 'I am Brahman' comprises sacrifice, gifts, penance, ritual, prayer, *Yoga*, and worship. . . . Since the Supreme Being abides as the Self, constant surrender of the mind by absorption in the Self is said to comprise all forms of worship. If only the mind comes under control, all else is controlled.

Sri Ramana Maharshi

Suff. 132
Contem. 532

I pray not that thou shouldest take them out of the world, but that thou shouldest keep them from the evil.
They are not of the world, even as I am not of the world.
Sanctify them through thy truth: thy word is truth.
As thou hast sent me into the world, even so have I also sent them into the world.

St John, XVII. 15–18

Renun. 139
Rev. 967

Knowl. 761 *Orthod.* 296

The *Srutis* would not have stated that the essential nature of the Self was in no way connected with *Vedic* rites and conditions required by them such as a particular class, and the rest, if they did not intend that those rites and *yajnopavîta*[1] etc., their means, should be given up. Therefore, *Vedic* actions which are incompatible with the knowledge of the identity of oneself with the supreme Self should be renounced together with their means by one who aspires after liberation: and it should be known that the Self is no other than *Brahman* as defined in the *Srutis*.[2]

Srî Śankarâchârya

Realiz. 859 *Knowl.* 734

O nobly-born, if one recognize not one's own thought-forms, however learned one may be in the Scriptures—both *Sûtras* and *Tantras*—although practising religion for a *kalpa*, one obtaineth not Buddhahood. If one recognize one's own thought-forms, by one important art and by one word, Buddhahood is obtained.

The Tibetan Book of the Dead, Fourteenth Day

Supra 1009

Lo! worship preserveth from lewdness and iniquity, but verily remembrance (invocation) of Allâh is greatest.[3]

Qur'ân, XXIX. 45

Action 346

'Put on one side, Consolata, all the virtuous practices of thy day, and on the other a day passed in one continual act of love. I would prefer this latter to any other offering.'

Sister Consolata

Lord Dhanvantari, the physician of the three worlds (who expounded the Ayurvedic Medical Science), has himself declared: 'By the medicine of the repetition of the Names Achyuta, Ananta, Govinda, all diseases are cured—this is my definite and honest declaration.' . . . The Divine Name will eradicate the disease of birth and death, and bestow on you Moksha, liberation, or immortality.

Swami Sivananda

Supra 1009

That man is pure who on the pure name calleth:
Who cries: Allâh! attains his every purpose.

The Mevlidi Sherif

He who prays without ceasing unites all good in this one thing.

St Simeon the New Theologian

[1] The sacred thread worn by the three upper castes.

[2] This teaching is applicable to *sâdhus*, who have relinquished all social ties, rank, privileges, and obligations, in pursuit of the final Reality.

[3] 'In Sumatra, as well as in some of the other islands of the Dutch East Indies, there are some teachers of mysticism who have become so advanced that they say it is no longer necessary to pray aloud: for them the secret prayer of the heart is sufficient. . . . Some mystics go so far as to say that this complete consciousness of the universal unity is in itself a universal prayer which does away with the necessity for the five daily devotional exercises of ordinary men' (R. L. Archer: 'Muhammadan Mysticism in Sumatra': *Journal Malayan Branch Royal Asiatic Society*, Singapore, Sept., 1937, pp. 91–92).

And this is the invocation of the Name; this is the possession of salvation, the receiving of kisses, the communion of the bed, the union of the Word with the soul in which every man is saved. For with such light no one can be blind, with such power no one can be weak, with such salvation none can perish.

Richard of Saint-Victor

Love 625

One may heal with holiness, one may heal with the law, one may heal with the knife, one may heal with herbs, one may heal with the Holy Word: amongst all remedies this one is the healing one that heals with the Holy Word; this one it is that will best drive away sickness from the body of the faithful: for this one is the best-healing of all remedies.

The Zendavesta, Ormazd Yast

There are many virtuous actions, but they are all particular. But the prayer of the heart is the source of all blessings. It waters the soul like gardens.

St Nilus Sorsky

Action 346

In every act of adoration there is almsgiving.

Muhammad

Charity 597

Japa is the philosopher's stone or divine elixir that makes one Godlike. Through Japa alone one can realise God in this life.

Swami Sivananda

Realiz. 870

The Grape that can with Logic absolute
The Two-and-Seventy jarring Sects confute:
  The Sovereign Alchemist that in a trice
Life's leaden metal into Gold transmute.

Omar Khayyâm

Ecstasy 637

'If a creature of good will wants to love Me and make of his life a single act of love, from rising to sleeping—doing it from his heart, of course—I for My part will commit follies for this soul. Write it down.'

Sister Consolata

'Live annihilated and enclosed in a single and continual: Jesus, Mary, I love you, save souls. Nothing else, the rest does not exist for thee.'

Sister Consolata

Void 724

The Blessed Giles of Santarem felt so much love and delight in saying the Holy Name that he was raised in the air in ecstasy.

The Wonders of the Holy Name

Flight 941

In the name of Jesus Christ of Nazareth rise up and walk.

Acts, III. 6

Signs and wonders may be done by the name of . . . Jesus.

Acts, IV. 30

Faith 509

As a pearl-oyster patiently and eagerly awaits the drops of rain when the star Swati is in ascendency, receives the drop and converts it in itself, through its own efforts and processes, into a very valuable pearl, the Sadhaka eagerly and devoutly awaits Mantra-initiation from the Guru, receives the sacred Mantra from him on the rare auspicious occasion, cherishes it and nurtures it in himself, and by his effort or process of Sadhana transforms it into a tremendous spiritual power which breaks the fort of Avidya or ignorance and opens the door to the blissful Immortal Experience.

Orthod. 280 *Grace* 556

Swami Sivananda

When a wise man is consecrated in the tradition of the way of mantras, all Buddhas become manifest to him in the mandala which is their abode.

Anangavajra

If you are not absolutely convinced that the Mind is the Buddha, and if you are attached to forms, practices and meritorious performances, your way of thinking is false and quite incompatible with the Way. . . . To practise the six pâramitâs (Perfections) and a myriad similar practices with the intention of becoming a Buddha thereby is to advance by stages, but the Ever-Existent Buddha is not a Buddha of stages. Only awake to the One Mind, and there is nothing whatsoever to be attained.

Realiz. 873

Huang Po

Peace 700

When thoughts arise, then do all things arise. When thoughts vanish, then do all things vanish.

Huang Po

It is better for men to set their gaze on the Law Blossom[1] than to study, no matter how earnestly, all the other sutras. It is better for men to set their gaze on the real Law Blossom than to build towers to contain inexhaustible treasures. It is better for men to set their gaze on the real Law Blossom than to set up a million images of the Buddha. It is better for men to set their gaze on the real Law Blossom than to study all the mysteries of the three worlds. . . . To sum up—there is nothing which surpasses the repeating of the sacred formula, providing that all attachment to things is cut off.

Hakuin

[1] Essence of the *Saddharma-puṇḍarika*, q.v., as summed up in the One Mind (*shinnyo*).

The Glory of the Name

Om!—This imperishable syllable is this whole world.
Its further explanation is:
The past, the present, the future—everything is just the word *Om*.
And whatever else that transcends threefold time—that, too, is just the word *Om*. . . .

Reality 803
Center 838

This is the Self with regard to the word *Om*, with regard to its elements. The elements (*mâtra*) are the fourths; the fourths, the elements: the letter *a*, the letter *u*, the letter *m*.

The waking state, the Common-to-all-men, is the letter *a*, the first element, from *âpti* ('obtaining') or from *âdimatvâ* ('being first').

He obtains, verily, indeed, all desires, he becomes first—he who knows this.

The dream state, the Brilliant, is the letter *u*, the second element, from *utkarṣa* ('exaltation') or from *ubhayatvâ* ('intermediateness').

He exalts, verily, indeed, the continuity of knowledge; and he becomes equal (*samâna*): no one ignorant of Brahma is born in the family of him who knows this.

The deep-sleep state, the Cognitional, is the letter *m*, the third element, from *miti* ('erecting') or from *apîti* ('immerging').

He, verily, indeed, erects (*minoti*) this whole world, and he becomes its immerging—he who knows this.

The fourth is without an element, with which there can be no dealing, the cessation of development, benign, without a second.

Introds. *Pilg.* *Realiz.*

Thus *Om* is the Self (Âtman) indeed.

He who knows this, with his self enters the Self—yea, he who knows this!

Mâṇḍûkya Upanishad, 1, 8–12

When my soul sighs,* and cries Ah and Oh:
It invokes its end and beginning.
* *a* and *ω*.[1]

Creation 26

Angelus Silesius

What a miracle! The 'Namu-amida-butsu' fills up the whole world.
And this world is given to me by Oya-sama.[2]
This is my happiness, 'Namu-amida-butsu!'

Center 816

Saichi

Blessed be the name of the Lord from this time forth and for evermore.

From the rising of the sun unto the going down of the same the Lord's name is to be praised.

Psalm CXIII. 2, 3

[1] *Alpha—omega*, like the Vedic AUM, represents the whole range of sounds. Cf. Guénon: *Le Roi du Monde*, ch. IV.

[2] Cf. note 3, p. 1026.

I will bless the name of Amon. Heaven shall hear the voice of my praise, and over the breadth of the earth shall it be heard. To the north and to the south I will declare his glory.

Let him be known to all people. Declare him to your sons and daughters, to great and to small, and to the generations that are yet unborn. Shout his name to the fishes in the stream and to the birds of the air. Declare him to the wise and to the foolish. Let him be known to all people.

Charity 608

M.M. 987

Lord of those whose tongues are silent, thou who protectest the humble. Lo, I called upon thee when I was in trouble and thou savedst me. Thou givest life and strength to the wretched and savest those who are in bonds. For thou art merciful and gracious to all who call upon thee.

Egyptian Tradition

Beauty 679

Magnify his name, and give glory to him with the voice of your lips, and with the canticles of your mouths, and with harps.

Ecclesiasticus, XXXIX. 20

M.M. 994

Judgment 239

God also hath highly exalted him, and given him a name which is above every name:

That at the name of Jesus every knee should bow, of things in heaven, and things in earth, and things under the earth.

Philippians, II. 9, 10

The unique and only subject of magic,[1] as well as the true Kabbala, is none other than *Wisdom*, the *Word*, the *Christ*. And there is no other name to invoke than that of Jesus, for there is no name on earth, nor in heaven, by which we can be saved, except for the name of Jesus, in which all things are reunited, for the Christ Jesus is all in all.

Robert Fludd

The Sages call it the living fire, because God has endowed it with His own Divine, and vitalising power. . . . No student of this Art can possibly do without it. For another Sage says: 'In this invisible fire you have the whole mystery of this Art, as the three Persons of the Holy Trinity are truly concluded in one substance.'

Supra 1017

The Glory of the World

And I will bring forth in shining light those who have loved My holy name, and I will seat each on the throne of his honour.

The Book of Noah, 108. 12

A Mantra is Divinity. It is divine power or Daivi Sakti manifesting in a sound body. The Mantra itself is Devatâ (Divinity).

Swami Sivananda

And let not the naming of God be usual in thy mouth, and meddle not with the names of saints, for thou shalt not escape free from them.[2]

Ecclesiasticus, XXIII. 10

[1] I.e., theurgy.

[2] This passage demonstrates the doctrine of efficacious power contained in divine names.

Let them praise thy great and terrible name: for it is holy.

Psalm XCIX. 3

If one loses one's being in the contemplation of the Divine Name, one can merge oneself in the ocean of heavenly beauty. God and His symbolical names are one and the same: as soon as the consciousness of the outside world disappears, the self-revealing power of the Name inevitably finds its objective expression.

Ananda Moyî

Death 220
Sp. Drown. 713

Rapture (*wajd*) is the blessed plenitude of spirit provoked by the exercise of invocation, and the blessed plenitude of soul, in communion with the spirit.

Allâh then gratifies His friend with a cup filled with wine that has no equal, and which intoxicates him with a spiritual drunkenness. His heart then seems endowed with wings, which raise him to the gardens of sanctity. At this moment the enraptured one, submerged by this indescribable magnificence, swoons away losing all consciousness.

'Abd al-Qâdir al-Jîlânî

Ecstasy 637
Flight 944

Mantra is the person's real nature. That is also the state of realization.

Sri Ramana Maharshi

Creation 42

Juliet. Bondage is hoarse, and may not speak aloud,[1]
Else would I tear the cave where Echo lies,
And make her airy tongue more hoarse than mine,
With repetition of my Romeo's name.
Romeo. It is my soul that calls upon my name.

Shakespeare (*Romeo and Juliet*, II. ii. 160)

Realiz. 887

Hearer and heard are one in the eternal Word.

Eckhart

As a mirror in which a person sees the form of himself and cannot see it without the mirror, such is the relation of God to the Perfect Man, who cannot possibly see his own form but in the mirror of the name Allâh; and he is also a mirror to God, for God laid upon Himself the necessity that His names and attributes should not be seen save in the Perfect Man.

Jîlî

Center 828
Holiness 924
Realiz. 859
Rev. 959

Take the Name as Brahman Himself.

Swami Ramdas

When I invoke the sacred name, there is neither myself nor the Buddha, but merely the invocation.

Ippen Shônin

Void 728
Reality 803

[1] Perhaps an allusion to the hermetic restrictions imposed on the playwright himself.

Charity 608
Center 847

When I have attained Buddhahood, if all beings in the ten quarters, trusting in me with the most sincere heart, should wish to be born in my Country, and should utter my Name one to ten times, and if they should not be born there, may I not attain Enlightenment.

Sukhâvatîvyûha Sûtra (Amida's Vow)

I will praise thy name continually.

Ecclesiasticus, LI. 15

Hallowed be thy name.

St Matthew, VI. 9

Supra Introd.

Nitâi[1] would employ any means to make people repeat Hari's name. Chaitanya said: 'The name of God has very great sanctity. It may not produce an immediate result, but one day it must bear fruit. It is like a seed that has been left on the cornice of a building. After many days the house crumbles, and the seed falls on the earth, germinates, and at last bears fruit.'

Sri Ramakrishna

Realiz. 873
887

The invocation of God is like a coming and going which realizes a communication ever more and more complete till there is identity between the glimmers of consciousness and the dazzling lightnings of the Infinite.

Shaykh Aḥmad al-ʿAlawî

The perfection of the Named One is eminently manifested by the fact that He reveals Himself through His Name to the person who does not know Him, so that the Name is to the Named as the outward (*aẓ-ẓâhir*) is to the inward (*al-bâṭin*): and in this respect the Name is the Named One Himself.

Jîlî

These are written, that ye might believe that Jesus is the Christ, the Son of God: and that believing ye might have life through his name.

St John, XX. 31

Believing in the name of God we are God's sons.[2]

Eckhart

Holiness 900

Paul. . . .

Unto the church of God which is at Corinth, to them that are sanctified in Christ Jesus, called to be saints, with all that in every place call upon the name of Jesus Christ our Lord, both theirs and ours:

Grace be unto you, and peace.

I. Corinthians, I. 1–3

[1] Nityânanda, a close disciple of Chaitanya, q.v.

[2] *St John*, I. 12.

St Paul bore the Name of Jesus on his forehead because he gloried in proclaiming it to all men, he bore it on his lips because he loved to invoke it, on his hands for he loved to write it in his epistles, in his heart for his heart burned with love of it.

St Thomas Aquinas

The sweet Name of Jesus is honey on the tongue:
To the ear a nuptial chant, in the heart a leap of joy.

Angelus Silesius

The noblest speech is the invocation of Allâh.

Muhammad

The name of the Lord is a strong tower: the righteous runneth into it, and is safe.

Proverbs, XVIII. 10

Renun. 152b

The Name is even superior to the Lord, because the Aguna (unqualified) and Saguna (qualified) aspects of Brahman are tasted and realised by the power of the Name.

Tulsi Dâs

M.M. 994

The Ineffable which one is accustomed to call God,
Is expressed and made known through a single Word.

Angelus Silesius

Upon thee that art the only God do I call, the more than great, the unutterable, the incomprehensible: unto whom every power of principalities is subjected: unto whom all authority boweth: before whom all pride falleth down and keepeth silence: whom devils hearing of tremble: whom all creation perceiving keepeth its bounds. Let thy name be glorified by us.

Acts of John, 79

Supra 1009
Beauty 689

Everything that God has done for the salvation of the world lies hidden in the Name of Jesus.

St Bernardine of Siena

Oh! the charm of the Name! It brings light where there is darkness, happiness where there is misery, contentment where there is dissatisfaction, bliss where there is pain, order where there is chaos, life where there is death, heaven where there is hell, God where there is Maya. He who takes refuge in that glorious Name knows no pain, no sorrow, no care, no misery. He lives in perfect Peace.

Swami Ramdas

Charity 605
Judgment 266

Whenever men gather together to invoke Allâh, they are surrounded by Angels, the Divine Favour envelops them, and Peace (*as-sakînah*) descends upon them, and Allâh remembers them in His assembly.

Muhammad

For where two or three are gathered together in my name, there am I in the midst of them.

St Matthew, XVIII. 20

Praise the Lord, call upon his name, declare his doings among the people, make mention that his name is exalted.

Isaiah, XII. 4

I do not ask of Thee, O Mother! riches, good fortune, or salvation; I seek no happiness, no knowledge. This is my only prayer to Thee: that, as the breath of life forsakes me, still may I chant Thy Holy Name.

Śrî Śankarâchârya

Methods

Realiz. 870

'When these three powers of the soul—memory, intelligence and will—are reconciled and gathered together in My Name, then all the other works, both outward and inward, which man performs, are sympathetically drawn towards Me and united in Me through the sense of love, and man thus mounts to the heights, following the crucified love.'

St Catherine of Siena

Faith 509

I used to go into samâdhi uttering the word 'Mâ'. While repeating the word I would draw the Mother of the Universe to me, as it were, like the fishermen casting their net and after a while drawing it in. When they draw in the net they find big fish inside it.

Sri Ramakrishna

Death 225
Supra 1007

Verily, as the huntsman draws in fish with his net and sacrifices them in the fire of his stomach, thus, assuredly, indeed, does one draw in these breaths[1] with *Om* and sacrifice them in the fire that is free from ill.

Maitri Upanishad, VI. 26

Metanoia 488

God can do nothing but speak the Eternal Word. If we are to be we must do and our doing is hearing the Eternal Word.

Eckhart

Sp. Drown. 713
Symb. 306

The sound Om is Brahman. . . . You hear the roar of the ocean from a distance. By following the roar you can reach the ocean. As long as there is the roar, there must also be the ocean. By following the trail of Om you attain Brahman, of which the Word is the symbol.

Sri Ramakrishna

[1] I.e., the various vital powers.

As a lamp must go out if it is not constantly fed with oil; so the inward fire of man, unless it is assiduously kept up, gradually begins to burn low, and is at length completely extinguished. Therefore it is indispensable for a Christian diligently to hear, carefully to study, and faithfully to practice the Word of God.

The Sophic Hydrolith

Contem. 542
Renun. 146
Orthod. 285

Whatsoever ye do in word or deed, do all in the name of the Lord Jesus.

Colossians, III. 17

Action 346

The *dhikr* of the heart is like the humming of bees, neither loud nor disturbing.

Ibn 'Aṭâ'illâh

Meditate on the sacred word OM, chanted within like the continuous peal of a bell.

Srimad Bhagavatam, XI. viii

Contem. 532

Whether the Nembutsu be repeated audibly or inaudibly, it has the same value as far as attaining *Ôjô* is concerned, because in either case it is a calling upon the same Buddha's sacred name.

Hônen

My rosary is my tongue, on which I repeat God's Name.

Swami Sivananda

'Forget everything, love Me continually, even if thy heart is as cold as stone. Everything depends on the incessant act of love.'

Sister Consolata

Faith 509

Abandon thyself to God until his *dhikr* triumphs over thy *dhikr.*

Abû Madyan

Contem. 528

Everything must have a nucleus around which sensations can develop. The more your spirit finds its center, the higher is the aspect of health, peace and tranquillity. And then a glimpse of the Infinite may become possible. Choose . . . a symbol or a sound as center of your thought and hold to it constantly. . . . In our times it is very difficult for a worshipper to acquire a conception of the Divine, whether through methods of yoga, whether in seeking to dissolve the individual self into the universal Self.

Ananda Moyi

Supra 1013

You are always repeating the *mantra* automatically. If you are not aware of the *ajapa* (unspoken chant) which is eternally going on, you should take to *japa.*

Sri Ramana Maharshi

Supra 1003

Keep God in remembrance till self is forgotten,
That you may be lost in the Called, without distraction of caller and call.

Dîvâni Shamsi Tabrîz, IV

Death 220
Peace 700

Reality 790 *Reality* 803

All the hundred and twenty-four thousand prophets were sent to preach one word. They bade the people say 'Allâh' and devote themselves to Him. Those who heard this word with the ear alone, let it go out by the other ear: but those who heard it with their souls imprinted it on their souls and repeated it until it penetrated their hearts and souls, and their whole being became this word. They were made independent of the pronunciation of the word, they were released from the sound and the letters. Having understood the spiritual meaning of this word, they became so absorbed in it that they were no more conscious of their own non-existence.

Abu 'l-Faḍl Ḥasan of Sarakhs

Contem. 542

Thy desire is thy prayer: and if thy desire is without ceasing, thy prayer will also be without ceasing. . . . There is interior prayer without ceasing, and this is your desire. Whatever else you do, if you do but long for that sabbath, you do not cease to pray. If you would never cease to pray, never cease to long after it. The continuance of your longing is the continuance of your prayer.

St Augustine

Holiness 900

O good Jesu Thou has bound my heart in the thought of Thy Name, and now I can not but sing it: therefore have mercy upon me, making perfect that Thou hast ordained.

Richard Rolle

Center 816

The best inscription of the Name of Jesus is that to be found in the innermost heart of hearts: the next best is in words, and finally the visible Name written or carved. If the bodily eye be constantly confronted with it, it will soon become visible to the eye of the heart, that inward spiritual eye of the soul.

St Bernardine of Siena

Contem. 532

Through the practice of *dhyana*[1] or *japa* the mind becomes one-pointed. Just as the elephant's trunk, which is otherwise restless, will become steady if it is made to hold an iron chain,—so that the elephant goes its way without reaching out for any other object,—even so the ever-restless mind, which is trained and accustomed to a Name or Form through *dhyana* or *japa*, will steadily hold on to that alone.

Sri Ramana Maharshi

P. State 573

'The purity of the act of love excludes all thought and demands virginity of spirit.'

Sister Consolata

Beauty 676 *Reality* 773 *Creation* 42

Having a firm grasp of this secret Knowledge (that the Self is *Brahman*), the Supreme Goal, and being free from defects and vanity people should always fix their minds on *Brahman* which is always the same. For no man who knows *Brahman* to be different from himself is a knower of Truth.

Śrî Śankarâchârya

[1] Cf. note p. 881.

Strike thy adversaries with the Name of Jesus; there is no weapon more powerful on earth or in heaven.

St John of the Ladder

Holy War 410

There is no secret about calling upon the sacred name except that we put our heart into the act, in the conviction that we shall be born into the Land of Perfect Bliss.

Hônen

Faith 514

It is true that the utterance of the Divine Name can absolve one from all sins and enable one to attain salvation . . . ; but that is possible only when the Name is uttered with faith and reverence and the practice is free from all taints of sin against the Name, viz. . . . ranking the Name with other virtues and practising fasting, charity, sacrifices, etc., thinking that the Name by itself is insufficient.

Swami Sivananda

Supra 1009
Suff. 133
Supra 1017

The Name of the Lord purifies both body and soul. 'I have taken the Name of God: what have I to fear? what in this world can bind me? I have become immortal in taking the Name of the Lord.' It is with this ardent faith that one must perform the spiritual exercises.

Swâmi Brahmânanda

The fact is that the *Nembutsu* practice was prescribed for all sentient beings, as set forth in Amida's Original Vow long ago, and has nothing to do with a man's being either ignorant or learned.

Hônen

Supra 1031

Call upon your Lord humbly and in secret. Lo! He loveth not aggressors.

Work not confusion in the earth after the fair ordering thereof, and call on Him in fear and hope. Lo! the mercy of Allâh is nigh unto the good.

Qur'ân, VII. 55, 56

Contem. 523
542

Walk to a solitary place to invoke *Wakanda* (the 'Great Spirit'). Do not eat or drink during three or four days. Even if you do not acquire a power, *Wakanda* will help you.

Sioux Indian

Whether or not the results of meditation are obtained is of no importance. The essential is to arrive at stability; it is the most precious thing that one can gain. In any case one must trust with confidence in the Divinity and await His grace without impatience. The same rule applies equally to *japa*: *japa* pronounced even once is a benefit, whether one is aware of it or not.

Sri Ramana Maharshi

Peace 700
Conform. 170

I trow this (ecstasy) is given to none meedfully, but freely to whom Christ will: nevertheless I trow no man receives it unless he specially love the Name of Jesu, and in so mickle honours It that he never lets It pass from his mind except in sleep. I trow that he to whom it is given to do that, may fulfil the same.

Richard Rolle

Grace 558

Holy War 405

Dissolve the thing and sublimate it, and then distil it, coagulate it, make it ascend, make it descend, soak it, dry it, and ever up to an indefinite number of operations, all of which take place at the same time and in the same vessel.

Solomon Trismosin

Praised in all times be his Holy and Venerable Name!

The Glory of the World

With every breath repeat that name, unceasing:
In Allâh's name see every task completed.[1]

The Mevlidi Sherif

Contem. 523

As an immoral woman constantly thinks of her illicit lover while living in the midst of her family, so do thou silently and ceaselessly meditate on Hari while doing thy earthly work.

Chaitanya

Even between the jaws of a tiger you must be able to cry vigorously: 'I am He (the formula *So'ham*), I am He, I am not this body!'

Swami Sivananda

Renun. 139
Action 329
Holiness 914

Now, do not say, 'There is no time or leisure to go to meditation, there is too much business and it is almost impossible to carry on one's plans for meditation when the duties of this world are so pressing.' It should be known that for a robed monk who performs the discipline of meditation in the right spirit there is no such thing as business or worldly affairs. . . . My own old teacher, Sho-ju, always used to say, 'Anyone who desires to learn how to meditate uninterruptedly, even if he has to enter a city of murders and swords, or to go into a room of weeping and wailing, or to attend the wrestling ring or the theatre hall or music hall, need not add adjustment of thought to adjustment of thought, nor does he have to make careful calculations about it—all he has to do is to bundle every one of these things into one meditation topic, and then go forward without a break and without withdrawing or retreating.'

Hakuin

[1] 'The moving (chala) japa can be done at all times, whether coming or going, standing or sitting, acting or bearing, giving or taking, sleeping or waking, during sexual acts or performance of other functions—without shyness, by uttering the name of God. All can do it. It knows no limits, or rule. It purifies the voice and gives it great strength. But those who do this japa must avoid speaking lies, nor should they condemn others, speak harsh words, talk nonsense, or too much. This japa brings success and makes the mind always glad, it leaves no room for worry, annoyance, sorrow, pain, ups and downs. He who does this japa is always protected, he makes the pilgrimage of life without effort and reaches supreme reality. All his actions are a ritual, the mind is detached and he knows no fear, being always near God. The Lord himself is the vehicle on this the safest path of yoga.

'There is no need for a rosary for this japa, but the adept keeps hidden in his clothing some small objects as reminders in case his mind becomes distracted. These reminders must not be seen by others, nor should his lips move, for this yoga must be kept secret or it loses its power' (Alain Daniélou: *Yoga, The Method of Re-integration,* London, 1949, pp. 85–86).

If you are a very active person, or travel much, you do not need a special room or special time for meditation. Do the *so'ham* japa.

Swami Sivananda

All the teachings of the Buddha Shaka are equally valuable: but in adaptability to human faculties, my religion stands supreme.

Reality 790

Hônen

The Name of *Jesus* is the shortest, the easiest and the most powerful of prayers. Everyone can say it even in the midst of his daily work. God cannot refuse to hear it.

Supra 1017

The Wonders of the Holy Name

Only repeat the name of Amida with all your heart. Whether walking or standing, sitting or lying, never cease the practice of it even for a moment. This is the very work which unfailingly issues in salvation, for it is in accordance with the Original Vow of that Buddha.

Zendô

Jesus Christ Himself saith that we should pray always and not faint or be weary, and St. Paul wills that we pray without ceasing.

But it is only the state of faith which can make prayers incessant. Thus Abraham (the father of the faithful, and the man who had the greatest faith that ever was) was therein confirmed, so he called on the name of God in all places. For as he kept himself in a continual state of prayer, so he left tokens of his prayers, worship and offerings, in all places behind him.

Faith 501

Action 346

Johannes Kelpius

They shall walk up and down in his name.

Flight 949

Zechariah, x. 12

Remember Allâh, standing, sitting and reclining.

Qur'ân, iv. 103

The *Nembutsu* may be practised whether one is walking, standing, sitting or lying, and so it may be left with everyone, according to circumstances, to do it either reclining or sitting or in any way he chooses. And as to holding the rosary or putting on the sacred scarf, this too must be decided according to circumstances. The main point is not the outward manner at all, but the fixing of one's mind on the one thing, firmly determining to have *Ôjô*, and with all seriousness to call upon the sacred name. This is the all-important thing.

Reality 803

Hônen

Taking as a bow the great weapon of the Upanishad,
One should put upon it an arrow sharpened by meditation.
Stretching it with a thought directed to the essence of That,
Penetrate that Imperishable as the mark, my friend.

Contem. 532

Realiz.
890

The mystic syllable *Om* (*praṇava*) is the bow. The arrow is the Self (*Atman*).
Brahma is said to be the mark (*lakṣya*).
By the undistracted man is It to be penetrated.
One should come to be in It, as the arrow (in the mark).

Muṇḍaka Upanishad, II. ii. 3. 4

CONCLUSION

If all the trees in the earth were pens, and the sea, with seven more seas to help it, (were ink), the words of Allâh could not be exhausted. Lo! Allâh is Mighty, Wise.

Qur'ân, XXXI. 27

If anyone will not acknowledge the force of reason, he must needs have recourse to authority.

Michael Maier (*Herm. Mus.* II. 223)

The documentation is exhaustless, but the way delineated is conclusive. Were the sources and citations expanded to fill out twelve volumes, they could only confirm and corroborate what already lies herein. No reader should regret that some favourite text has been omitted—the 'empty spaces' are also important—but rather rejoice in seeing the current of universal orthodoxy to which it attaches.

It is to be hoped that this panorama of traditional wisdom carries a sufficient testimony against the theory of 'borrowing'—a conception that can only seem plausible in a world reduced to purely relative perspectives. For those who still accept the necessity of a First Principle and the possibility of Divine Revelation, there can be but one legitimate 'borrowing'—that of a constantly renewed reference to Primordial Tradition itself.[1] Thus for example, when King Udâyana's Buddha image (Coomaraswamy: *Figures of Speech*, p. 178) is found at Khotân, legend tells us that the transference was operated by a flight through the air, which implies an intellectual omnipresence on the part of the 'borrowing' artist, in which state he is able to descry the archetype of the image already established in India with Udâyana's artist.[2]

The theory of accidental repetitions is already falling into disfavour, even with the psychologists; but here the alternative proposed is hardly felicitous, bringing in as it does by way of subversion the sinister concept of a subliminal symbolism inviting through the murky waters of the subconscious to a kind of mindless union at the bottom[3]—an inverted aspect of the picture which has already been largely dealt with in the chapter, *The Wrathful Side*. It suffices to remember that if 'nineteenth century materialism closed the mind of man to what is above him, twentieth century psychology opened it to what is below him' (Coomaraswamy, paraphrasing Guénon, in *Hinduism and Buddhism*, p. 61).

Perhaps the most recurrent charge against collations of traditional doctrine is the

[1] Traditions like their celestial Archetypes can share each other's light without losing their own brightness.

[2] Cp. *Kaṭha Upanishad*, II. 21: 'Seated, he (Âtman) fares afar; recumbent, he goes everywhere.'

[3] What is termed the 'collective subconscious' is precisely a syncretism of residues and not a synthesis of archetypes: cf. Burckhardt, 'Considérations sur l'Alchimie', *Études Trad.*, 1949, p. 119.

objection that an essential disparity between the natural (meaning non-Christian) and supernatural (meaning Christian) orders is being violated in deference to an alluring and specious resemblance in secondary or superficial details—as though the Eastern Void were something other than the Western Godhead, or as though there could be two Absolutes.

This imputation of *Naturalism* (see the section, *Man's Primordial Birthright,* in the chapter on *Creation,* and also the Introduction to *Heterodoxy*) is often associated with a tendency to relegate all non-Christian doctrines of East and West into a category with the degenerate remnants of the mystery religions, under the general heading of *Gnosticism,* in the conviction that knowledge here is set above Faith and individual effort above Grace. These things can only be pointed out in passing, as their pursuit would encroach upon the object of this anthology: to serve as an 'illustrative commentary' to the exposition of the authorities followed. Moreover, these objections have already been in part answered in the introductory sections to the chapters on *Faith, Grace,* and *Knowledge.* Most problems of this sort, however, arise over a confusion between the principial and relative planes. The uniqueness of the supraformal Truth has as its necessary corollary the plurality of revelations within the formal domain—multiple by definition.[1]

Certain perspectives are admittedly unfathomable to reason, considered as a purely human faculty without reference to a higher principle. However, reason was not the priceless gift which God bestowed upon man when He created him in His image, but on the contrary, the instrument and penalty of his fall (though *a posteriori* a faculty for his redemption). An image or true copy of God could reflect nothing less than His own perfection in Being, Knowledge and Beatitude: and it is precisely through the grace of the Divine Intellect that man is born with the supreme possibility for participation in the identity of the Knower with the Known.

The closing words from Boehme's *Signatura Rerum* may likewise serve as a signature to the present work: 'I have faithfully, with all true admonition, represented to the reader what the Lord of all beings has given me: he may behold himself in this looking glass within and without, and so he shall find what and who he is: Every reader, be he good or evil, will find his profit and benefit therein: It is a very clear gate of the great mystery of all beings: By glosses, commentaries, curiosity and self-wit, none shall be able to reach or apprehend it in its own ground: but it may very well meet and embrace the true *seeker,* and create him much profit and joy: yea be helpful to him in all natural things, provided he applies himself to it aright, and seeks in the fear of God, seeing it is now a time of seeking: for a lily blossoms upon the mountains and valleys in all the ends of the earth: "He that seeketh findeth".'

[1] In other words, the *historical* pre-eminence each of Gautama and Jesus is mutually exclusive, and cannot be made an absolute, at least outside the framework of the forms in question, without contradicting the definition of the Logos, and *a fortiori,* the Supreme Principle.

ACKNOWLEDGEMENTS

Grateful acknowledgement is extended to the following owners and holders of copyright for permission to incorporate extracts from the works indicated below. Full page listings and other details are given in the INDEX OF SOURCES.

Abbey Press, St Meinrad, Ind.: *Our Lady of Fatima's Peace Plan from Heaven.*

Advaita Ashrama, Calcutta: Śankarâchârya's *Vivekachudamani,* translated by Swami Madhavananda; *Brahma-Sûtras,* translated by Swami Vireswarananda; *The Life of Swami Vivekananda.*

George Allen & Unwin, Ltd., London: *Sufism, an Account of the Mystics of Islam,* by A. J. Arberry; *The Embossed Tea Kettle and Other Works of Hakuin Zenji,* translated by R. D. M. Shaw; *Rûmî, Poet and Mystic,* translated by Reynold A. Nicholson; *Mysticism, Christian and Buddhist,* by Daisetz Teitaro Suzuki; *A Moslem Saint of the Twentieth Century,* by Martin Lings; *The Tale of Genji,* by Lady Murasaki, translated by Arthur Waley; *Mahatma Gandhi's Ideas,* by C. F. Andrews; *The Meaning of The Glorious Qur'ân,* translated by Marmaduke Pickthall; *The Nô Plays of Japan,* translated by Arthur Waley.

American Academy of Arts and Sciences, Cambridge, Massachusetts: selections from Ko Hung, translated by Lu-Ch'iang Wu, in *Proceedings of the American Academy of Arts and Sciences,* 70, Boston, 1935; selections from Chang Po-tuan, translated by Tenney L. Davis and Chao Yün-ts'ung, *Proceedings of the American Academy of Arts and Sciences,* 73, Boston, 1939; selections from Chang Po-tuan and Hsieh Tao-kuang, translated by Tenney L. Davis and Chao Yün-ts'ung, *Proceedings of the American Academy of Arts and Sciences,* 73, Boston, 1940; selections from Ko Hung, translated by Tenney L. Davis and Ch'ên Kuo-fu, *Proceedings of the American Academy of Arts and Sciences,* 74, Boston, 1941.

American Oriental Society, New Haven, Conn.: *Spiritual Authority and Temporal Power in the Indian Theory of Government,* by Ananda K. Coomaraswamy; Coomaraswamy's 'Lîlâ', and W. Norman Brown's 'The Rigvedic Equivalent for Hell', in *Journal of the American Oriental Society,* June, 1941; Coomaraswamy's 'Recollection, Indian and Platonic', *Supplement to the Journal of the American Oriental Society,* April-June, 1944.

Amitiés Spirituelles, Paris: *Histoire et Doctrines des Rose-Croix,* by Sédir.

Anandashram, Kerala (S. India): excerpts from *Devikalottara-Jnanachara-Vichara-Patalam,* Sri Ramana Maharshi, Swami Ramdas, and Swami Sivananda, in *The Vision,* Oct.-Dec., 1954.

Muṣtafâ Mu'în al-'Arab and Fâṭimah Muḥammad Ibrâhîm (Mrs René Guénon): all material by René Guénon cited in the introductory sections and elsewhere the copyright for which belongs to the Estate of René Guénon.

Edward Arnold (Publishers) Ltd., London: *The Ordinall of Alchimy,* by Thomas Norton, being a facsimile reproduction from the *Theatrum Chemicum Britannicum* (1652); Walter von der Vogelweide's 'My Brother Man' from *I Saw the World,* translated by I. G. Colvin.

Shaikh Muhammad Ashraf, Publisher, Lahore: *Maxims of 'Alî*; *The Orations of Muhammad*, translated by Maulana Muhammad Ubaidul Akbar; *The Secret Rose Garden of Sa'd-ud-Din Mahmud Shabistari*, translated by Florence Lederer.

Asia Publishing House, Bombay: *The Dance of Shiva*, by A. K. Coomaraswamy.

Aubier, Éditions Montaigne, Paris: *Mysterium Magnum*, by Jacob Boehme, translated by N. Berdiaeff; *Oeuvres choisies*, Nicholas of Cusa, translated by Maurice de Gandillac; *Angelus Silesius, Pèlerin chérubinique*, translated by Henri Plard.

Barnes & Noble, Inc., New York: *The Muqaddimah*, by Ibn Khaldûn, translated by Franz Rosenthal.

Barrie & Rockliff, London: *Selected Mystical Writings of William Law*, edited by Stephen Hobhouse; *Northern Indian Music*, Vol. I, by Alain Daniélou.

The Chester Beatty Library, Dublin: *The Poem of the Way*, by Ibn al-Fâriḍ, translated by A. J. Arberry, published by Emery Walker Ltd., London.

Benedictine Convent of Perpetual Adoration, Clyde, Missouri: *Conformity to the Will of God*, by St Alphonsus Liguori, translated from the Italian.

Benziger Brothers, Inc., New York: *Church History*, by Rev. John Laux, copyright 1932 by Benziger Brothers.

Bharatiya Vidya Bhavan, Bombay: *Guide to Aspirants* (Swami Ramdas), Anandashram Series no. 13, 1949; *Ramdas Speaks*, Vol. I, Anandashram Series no. 14, 1955; *World is God*, by Swami Ramdas.

New Blackfriars, Cambridge, England: Dafydd ap Gwilym, in *Blackfriars*, 1942.

The Bond Wheelwright Company, Publishers, Freeport, Maine: *Muhammad's People*, by Eric Schroeder.

E. J. Brill, Ltd., Leiden: *The Gospel According to Thomas*, Coptic text established and translated by A. Guillaumont, H.-Ch. Puech, G. Quispel, W. Till and Yassah 'Abd al-Masîḥ, copyright 1959 by E. J. Brill.

Britons Publishing Company, Chulmleigh, North Devon: *Secret Societies and Subversive Movements*, by Nesta Webster.

Curtis Brown, Ltd., London: *The Wisdom of China and India*, by Lin Yutang; *The Wisdom of Confucius*, by Lin Yutang; *My Country and My People*, by Lin Yutang; *From Pagan to Christian*, by Lin Yutang.

The Buddhist Society, London: *Tao-Te-King*, translated by Ch'u Ta-Kao.

Titus Burckhardt: *Siena, The City of the Virgin*, by Titus Burckhardt, translated into English by Margaret McDonough Brown, published by Oxford University Press, London, copyright by Urs Graf-Verlag, Olten, Switzerland.

Burns & Oates, London: *The Cloud of Unknowing, And Other Treatises*, by an English Mystic of the Fourteenth Century, *With a Commentary on the Cloud by Father Augustine Baker*, edited by Dom Justin McCann; *Basic Writings of Saint Thomas Aquinas*, edited by Anton C. Pegis.

Jacques Masui and *Les Cahiers du Sud*, Marseille: *Yoga, Science de l'Homme intégral*, edited by Jacques Masui.

Cambridge University Press, London; *Râbi'a the Mystic & Her Fellow-Saints in Islâm*, by Margaret Smith; *Selected Poems from the Dîvâni Shamsi Tabrîz*, translated by Reynold A. Nicholson; *Studies in Islamic Mysticism*, by R. A. Nicholson; *Translations of Eastern Poetry and Prose*, by R. A. Nicholson; *The Doctrine of the Sûfîs*, translated by Arthur John Arberry; *The Mystical Philosophy of Muhyid Dîn-Ibnul 'Arabî*, by A. E. Affifi; *The Secret Lore of India*, by Teape; *The Travels of Fa-Hsien*, translated by H. A. Giles; *Peter Sterry, Platonist and Puritan*, by Vivian de Sola Pinto.

Cassell & Co. Ltd., London: *Gotama the Buddha*, presented by Ananda K. Coomaraswamy and I. B. Horner.

Bruno Casirer (Publishers) Ltd., Oxford: *Buddhist Texts Through the Ages,* edited by Edward Conze in collaboration with I. B. Horner, D. Snellgrove, A. Waley.

Chapman & Hall Ltd. (Associated Book Publishers Ltd.), London: *The Hitopadesa,* translated by Francis Johnson.

Mrs Laura Huxley and Chatto and Windus Ltd., London: *The Perennial Philosophy,* by Aldous Huxley.

The Chion-in Temple, Kyoto: *Hônen the Buddhist Saint,* by Shunjô, translated by Rev. Harper Havelock Coates and Rev. Ryugaku Ishizuka.

The Clarendon Press, Oxford: *The Poetical Works of Robert Herrick,* edited by F. W. Moorman, 1951; *The Apocrypha and Pseudepigrapha of the Old Testament in English,* Vol. II, edited by R. H. Charles, 1913; *The Apocryphal New Testament,* edited by M. R. James, 1924; *Epictetus,* translated by P. E. Matheson; *Hermetica,* edited by Walter Scott, 1924; *Proclus: The Elements of Theology,* translated by E. R. Dodds, 2nd edition, 1963; *Visuddhimagga, The Path of Purity,* translated by Pe Maung Tin, 1922; *The Sacred Books of the East,* edited by F. Max Müller, quotations as indicated in the INDEX OF SOURCES from Vols. I, X, XXI, XLIII, XLIX; *Centuries of Meditations,* by Thomas Traherne, edited by Bertram Dobell, 1908; *The Dialogues of Plato,* translated by Benjamin Jowett, 4th edition, 1953; all of the above by permission of the Clarendon Press, Oxford.

Clonmore & Reynolds Ltd., Dublin: *Padre Pio the Stigmatist,* by Rev. Charles Mortimer Carty.

Columbia University Press, New York: *The Philosophy of Marsilio Ficino,* by Paul Oskar Kristeller, translated by Virginia Conant, 1943; 'On Being in One's Right Mind', by A. K. Coomaraswamy, in *The Review of Religion,* 1942; a citation from the *Rig-Veda* in the same journal, 1942.

Constable & Company Ltd., London: *170 Chinese Poems,* translated by Arthur Waley; *Western Mysticism,* by Dom Cuthbert Butler.

Dacre Press, A. & C. Black Ltd., London: *The Seven Steps of the Ladder of Spiritual Love,* by Jan Van Ruysbroeck, translated by F. Sherwood Taylor.

Alain Daniélou: *Introduction to the Study of Musical Scales,* by Alain Daniélou, The India Society, London, 1943.

The John Day Company, Inc., New York: *The Winged Serpent,* by Margot Astrov, copyright © 1946 by Margot Astrov, reprinted by permission of The John Day Company, Inc., publisher; *The Importance of Living,* by Lin Yutang, copyright © 1937 by The John Day Company, Inc. and reprinted by their permission; *Monkey,* by Wu Ch'eng-en, translated by Arthur Waley, copyright © 1943 by The John Day Company, Inc. and reprinted by their permission.

J. M. Dent & Sons Ltd., London: *John of Ruysbroeck,* translated by C. A. Wynschenk Dom, edited by Evelyn Underhill; the following titles are from the Temple Classics: *The Inferno, The Purgatorio,* and *The Paradiso,* by Dante Alighieri, translated by Carlyle, Okey and Wicksteed; *Dante's Latin Works,* translated by Howell and Wicksteed; *Labyrinth of the World and Paradise of the Heart,* by Comenius, edited by Lutzom; the following titles are from the Everyman's Library: *The Little Flowers of St Francis,* translated by T. Okey, and *The Mirror of Perfection,* by Friar Leo, translated by Robert Steele; *Thomas Aquinas, Selected Writings,* edited by M. C. D'Arcy; *Le Morte d'Arthur,* by Thomas Malory; *The City of God,* by St Augustine, translated by John Healey; *Chinese Philosophy in Classical Times,* edited and translated by E. R. Hughes; *The Signature of All Things & Other Discourses,* by Jacob Boehme (Law's English edition).

DeVorss & Co., Los Angeles: *Life and Teaching of the Masters of the Far East,* by Baird T. Spaulding, Vols. III & IV.

The Divine Life Society, Rishikesh, India: *Japa Yoga,* by Swami Sivananda.

Dufour Editions Inc., Chester Springs, Pennsylvania: *The Mystics of Islam*, by Reynold A. Nicholson.

E. P. Dutton & Co., Inc., New York: from the book *The Little Flowers of St Francis* translated by T. Okey, *The Mirror of Perfection* translated by R. Steele; and, *The Life of St Francis* translated by Miss E. Gurney Salter, Everyman's Library Edition, reprinted by permission of E. P. Dutton & Co., Inc.; from the book *Selected Writings of St Thomas Aquinas*, selected and edited by the Rev. Father M. C. D'Arcy, S.J., M.A., Everyman's Library Edition, reprinted by permission of E. P. Dutton & Co., Inc.; from the book *Le Morte D'Arthur* by Sir Thomas Malory, Everyman's Library Edition, reprinted by permission of E. P. Dutton & Co., Inc.; from the book *The City of God* by St Augustine, John Healey's translation, revised and edited by R. V. G. Tasker, Everyman's Library Edition, reprinted by permission of E. P. Dutton & Co., Inc.; from *Chinese Philosophy in Classical Times* (2 vols) edited and translated by E. R. Hughes, Everyman's Library Edition, reprinted by permission of E. P. Dutton & Co., Inc.; from Dante's *Divine Comedy* translated by J. A. Carlyle, Thomas Okey and P. H. Wicksteed, Temple Classics Edition, reprinted by permission of E. P. Dutton & Co., Inc.; from the book *Mysticism* by Evelyn Underhill, reprinted by permission of E. P. Dutton & Co., Inc.

Éditions Baconnier, Marseille: *Vies des Saints Musulmans*, by Émile Dermenghem.

Éditions Gallimard, Paris: *Roman de la Rose*, by Guillaume de Lorris and Jean de Meun, adapted by André Mary, © 1949 by Éditions Gallimard; *Le Culte des Saints dans l'Islam Maghrébin*, by Émile Dermenghem, © 1954 by Éditions Gallimard; *Le Règne de la Quantité*, by René Guénon, © 1945 by Éditions Gallimard.

Éditions Payot, Paris: *Moeurs et Histoire des Peaux-Rouges*, by René Thévenin and Paul Coze; *Histoire des Doctrines ésotériques*, by J. Marquès-Rivière; *Le Yoga, Immortalité et Liberté*, by Mircea Eliade.

Éditions du Seuil, Paris: *Les Alchimistes*, by M. Caron and S. Hutin; *Tahiti*, by Jean-Marie Loursin; *La Vie de Sainte Thérèse d'Avila*, by Marcelle Auclair; *Amida*, by Henri de Lubac.

Les Éditions du Soleil Levant, Profondeville, Belgium: *St Louis de Montfort*, edited by Raymond Christoflour.

Les Éditions de La Table Ronde, Paris: *Bernadette et Lourdes*, by Michel de Saint Pierre, copyright 1953 by Les Éditions de La Table Ronde; *Ces Prêtres qui Souffrent*, by Michel de Saint Pierre, copyright 1966 by Les Éditions de La Table Ronde.

Librairie et Éditions Véga, Paris: *Introduction générale à l'Etude des Doctrines hindoues*, by René Guénon, copyright 1930 by Éditions Véga; *Orient et Occident*, by René Guénon, copyright 1930 by Éditions Véga; *Autorité spirituelle et Pouvoir temporel*, by René Guénon, copyright 1930 by Éditions Véga; *Les Etats multiples de l'Etre*, by René Guénon, copyright 1932 by Éditions Véga; *L'Eloge du Vin*, by ʿOmar Ibn al-Fâridh, translated by Émile Dermenghem, copyright 1931 by Éditions Véga.

Michel Vâlsan and Éditions Traditionnelles, Paris: all material from the *Etudes Traditionnelles* cited in the introductory sections and elsewhere the copyright for which belongs to Michel Vâlsan and Éditions Traditionnelles.

Encyclopaedia Britannica International, Ltd., London: one citation from the 11th edition and one from the 14th edition of the *Encyclopaedia Britannica*.

Faber and Faber Ltd., London: from *Writings from the Philokalia on Prayer of the Heart*, translated by E. Kadloubovsky and G. E. H. Palmer, reprinted by permission of Faber and Faber Ltd.; from *Unseen Warfare*: being the *Spiritual Combat* and *Path to Paradise* of Lorenzo Scupoli, translated by E. Kadloubovsky and G. E. H. Palmer, reprinted by permission of Faber and Faber Ltd.; from *Hugh of Saint-Victor, Selected Spiritual Writings*, translated by a Religious of C.S.M.V., reprinted by permission of Faber and Faber Ltd.; from *Plotinus: The Enneads*, translated by Stephen MacKenna, reprinted by permission of Faber and Faber Ltd.;

from *Richard of Saint-Victor, Selected Writings on Contemplation,* translated by Clare Kirchberger, reprinted by permission of Faber and Faber Ltd.

Flammarion, Éditeur, Paris: *Forgerons et Alchimistes,* by Mircea Eliade; *Essai sur la Symbolique romane,* by M. M. Davy.

Folkways Records, New York: *Islamic Liturgy,* annotations by Martin Lings, Folkways Records Album No. FR 8943, 1960.

Ganesh & Co. (Madras) Private Ltd., Madras: *The Call of the Jagadguru,* Teachings of His Holiness Śrî Jagadguru Śrî Chandrasekhara Bhârati Swâmigal of Śṛṅgeri, by R. Krishnaswami Aiyar; *The Serpent Power,* by Arthur Avalon.

Grove Press, Inc., New York: *Essays in Zen Buddhism* (First Series), by Daisetz Teitaro Suzuki, reprinted by permission of Grove Press, Inc., all rights reserved; *Manual of Zen Buddhism,* by Daisetz Teitaro Suzuki, reprinted by permission of Grove Press, Inc., all rights reserved; *The Nô Plays of Japan,* by Arthur Waley, published by Grove Press, Inc.

Harper & Row, Publishers, Incorporated, New York: from *Meister Eckhart: A Modern Translation,* by Raymond Bernard Blakney, copyright, 1941 by Harper & Row, Publishers, Incorporated; from *A Method of Prayer* by Johannes Kelpius, edited by E. Gordon Alderfer, copyright, 1951 by Harper & Row, Publishers, Incorporated; reprinted by permission of the publisher; *The Perennial Philosophy,* by Aldous Huxley; *Western Mysticism,* by Dom Cuthbert Butler; *Mysticism, Christian and Buddhist,* by Daisetz Teitaro Suzuki; *The Spirit of Saint Francis de Sales,* by Jean Pierre Camus, translated by C. F. Kelley.

The Hartford Seminary Foundation, Hartford, Connecticut: *The Forty-Two Traditions of an-Nawawi,* translated by Eric F. F. Bishop, in *The Moslem World,* 1939.

Harvard University Press, Cambridge, Massachusetts: all excerpts as indicated in the INDEX OF SOURCES: reprinted by permission of the publishers from Ananda Kentish Coomaraswamy, *The Transformation of Nature in Art,* Cambridge, Mass.: Harvard University Press, Copyright, 1934, by the President and Fellows of Harvard College, renewed 1962 by D. Luisa Runstein Coomaraswamy; reprinted by permission of the publishers from Ananda Kentish Coomaraswamy, *Elements of Buddhist Iconography,* Cambridge, Mass.: Harvard University Press, Copyright, 1935, by the President and Fellows of Harvard College, renewed 1963 by Mrs. Zlata Llamas Coomaraswamy; reprinted by permission of the publishers from Henry Osborn Taylor, *The Medieval Mind,* Cambridge, Mass.: Harvard University Press; reprinted by permission of the publishers from Henry Clark Warren, *Buddhism in Translations,* Cambridge, Mass.: Harvard University Press, Copyright 1953 by the President and Fellows of Harvard College; reprinted by permission of the publishers and The Loeb Classical Library from Clement of Alexandria, *The Exhortation to the Greeks,* translated by G. W. Butterworth, Cambridge, Mass.: Harvard University Press; reprinted by permission of the publishers and The Loeb Classical Library from Philostratus, *The Life of Apollonius of Tyana,* translated by F. C. Conybeare, Cambridge, Mass.: Harvard University Press; reprinted by permission of the publishers and The Loeb Classical Library from Apuleius, *The Golden Ass,* translated by W. Adlington and revised by S. Gaslee, Cambridge, Mass.: Harvard University Press; reprinted by permission of the publishers and The Loeb Classical Library from Boethius, *The Consolation of Philosophy,* translated by 'I.T.', and revised by H. F. Stewart, Cambridge, Mass.: Harvard University Press; reprinted by permission of the publishers and The Loeb Classical Library from Ovid, *Metamorphoses,* translated by Frank J. Miller, Cambridge, Mass.: Harvard University Press; reprinted by permission of the publishers and The Loeb Classical Library from *Philo,* translated by F. H. Colson, Cambridge, Mass.: Harvard University Press; reprinted by permission of the publishers and The Loeb Classical Library from Diogenes Laertius, *Lives of Eminent Philosophers,* translated by R. D. Hicks, Cambridge, Mass.: Harvard University Press.

W. Heffer & Sons Ltd., Cambridge, England: *Poems of a Persian Ṣûfî* (Bâbâ Ṭâhir), translated by Arthur J. Arberry.

Jean Herbert: *Aux Sources de la Joie*, by Ananda Moyî, translated by Jean Herbert, Éditions Ophyrs, Paris & Neuchâtel; *Pensées*, by Swami Ramdas, translated by Marie Honegger-Durand and Lizelle Reymond, Les Grands Maîtres spirituels dans l'Inde contemporaine, Paris & Neuchatel; *Discipline monastique*, by Swâmi Brahmânanda, translated by Odette de Saussure and Jean Herbert, same edition as above, Vol. I; *La Sagesse des Prophètes*, by Ibn 'Arabî, translated by Titus Burckhardt, Albin Michel, Paris; *Carnet de Pèlerinage*, by Swami Ramdas, translated by Jean Herbert, Albin Michel, Paris; *La Pratique de la Méditation*, by Swami Sivananda, translated by Charles Andrieu and Jean Herbert, Albin Michel, Paris; *Introduction aux Doctrines ésotériques de l'Islam*, by Titus Burckhardt, P. Derain, Lyon; *De l'Homme universel*, by 'Abd al-Karîm al-Jîlî, translated by Titus Burckhardt, P. Derain, Lyon; *Du Soufisme*, by Titus Burckhardt, P. Derain, Lyon.

Hibbert Journal (George Allen & Unwin, Ltd., London): Al-Ghazâlî's *Iḥyâ 'Ulûm al-Dîn*, I. v, translated by R. Bell in the *Hibbert Journal*, October, 1943.

Mrs Marianne Rodker and The Hogarth Press Ltd., London; *Malleus Maleficarum*, translated and edited by the Rev. Montague Summers, John Rodker, 1928.

The Hokuseido Press, Toyko: *Li Po, the Chinese Poet*, translated by Shigeyoshi Obata.

Miss I. B. Horner: *Gotama the Buddha*, presented by Ananda K. Coomaraswamy and I. B. Horner, Cassell & Co., Ltd., London, 1948.

Houghton Mifflin Company, Boston: *The Tale of Genji*, by Lady Murasaki, translated by Arthur Waley, published by Houghton Mifflin Company, Boston.

Islamic Cultural Centre, London: 'The Origins of Sufism', by Abû Bakr Sirâj ad-Dîn, in *The Islamic Quarterly*, April, 1956; a citation from Muhammad in the same review, October & December, 1954.

Johnson Publications Limited, London: *Yoga, The Method of Re-integration*, by Alain Daniélou.

The Julian Press, Inc., New York: *Reincarnation: An East-West Anthology*, compiled and edited by Joseph Head and S. L. Cranston.

P. Lethielleux, Libraire-Éditeur, Paris: *Le Dialogue de Sainte Catherine de Sienne*, translated by the R. P. J. Hurtaud, O. P., 1947, 2 vols.

Librairie Académique Perrin, Paris: *Rusbrock l'Admirable (Oeuvres choisies)*, translated by Ernest Hello.

Librairie Théatrale (L'Amicale), Paris: *La Vie Est un Songe*, by Calderón, translated by Yvette and André Camp.

Martin Lings: *The Book of Certainty*, by Abû Bakr Sirâj Ed-Dîn, Rider and Company, London, 1952.

Longmans Green & Co. Ltd., Harlow, Essex: *Introduction to the History of Sûfism*, by Arthur J. Arberry.

Lutterworth Press, London: *Cleopatra's Needles and Other Egyptian Obelisks*, by E. A. Wallis Budge.

Macmillan & Co. Ltd., London: *One Hundred Poems of Kabir*, translated by Rabindranath Tagore, by permission of The Trustees of the Tagore Estate and Macmillan & Co. Ltd.

The Macmillan Company, New York: *A Moslem Saint of the Twentieth Century*, by Martin Lings; *One Hundred Poems of Kabir*, translated by Rabindranath Tagore.

Macoy Publishing & Masonic Supply Co. Inc., Richmond, Virginia: *The Life and the Doctrines of Paracelsus*, extracted and translated by Franz Hartmann.

Mrs Ernst E. Mensel: *The Soul of the Indian*, by Charles Alexander Eastman (Ohiyesa), Houghton Mifflin Company, Boston, 1911.

Mercure de France, Paris: *Histoire des Rose-Croix,* by Paul Arnold, 1955; *Esotérisme de Shakespeare,* by Paul Arnold, 1955.

Methuen & Co. Ltd. (Associated Book Publishers Ltd.), London: *Revelations of Divine Love,* by Lady Julian of Norwich, edited by Grace Warrack; *The Fire of Love or Melody of Love,* by Richard Rolle, translated by Richard Misyn; *Mysticism,* by Evelyn Underhill.

Jean-Louis Michon: for a citation by Al-ʻArabî al-Ḥasanî ad-Darqâwî in a thesis on Aḥmad Ibn ʻAjîba prepared by Jean-Louis Michon.

The Montfort Fathers, Bay Shore, New York: *True Devotion to the Blessed Virgin Mary,* by St Louis de Montfort, translated by Frederick William Faber (1862), Revised Edition, 1946.

John Murray, London: *The Monastery of Jade Mountain,* by Peter Goullart; *Islam and the Divine Comedy,* by Miguel Asin; al-Ghazâlî's *Alchemy of Happiness,* translated by C. Field, Wisdom of the East Series; *The Mevlidi Sherif,* by Süleyman Chelebi, translated by F. Lyman MacCallum, Wisdom of the East Series; *The Persian Mystics: The Invocations of Sheikh ʻAbdullâh Ansâri of Herat,* translated by Sardar Sir Jogendra Singh, Wisdom of the East Series; *Arabian Wisdom,* Wisdom of the East Series; *The Sayings of Muhammad,* compiled and translated by Allama Sir Abdullah al-Mamun al-Suhrawardy, Wisdom of the East Series; *Nôgaku, Japanese Nô Plays,* by Beatrice Lane Suzuki, Wisdom of the East Series; *Lieh-tse,* translated by Giles, Wisdom of the East Series; *The Persian Mystics—ʻAṭṭâr,* translated by Margaret Smith, Wisdom of the East Series; *The Spirit of the Brush,* translated by Shio Sakanishi, Wisdom of the East Series; *Egyptian Religious Poetry,* translated by Margaret A. Murray, Wisdom of the East Series; *The Path of Light* (*The Bodhicharyâvatâra of Śânti-deva*) translated by L. D. Barnett, Wisdom of the East Series; *Christ in Islâm,* by James Robson, Wisdom of the East Series.

The New American Library, Inc., New York: 8 words from the *Lankavatara Sutra,* translated by D. T. Suzuki, from p. 166 of *The Teachings of the Compassionate Buddha,* edited by E. A. Burtt, Copyright, 1955, by Edwin A. Burtt, reprinted by permission of The New American Library, Inc., New York.

Oeuvre St-Canisius, Fribourg, Switzerland: *Un Appel du Christ au Monde,* by P. Lorenzo Sales.

Office Central de Lisieux, Lisieux, France: Sainte Thérèse de l'Enfant-Jésus: *Histoire d'une Âme,* 1947.

The Open Court Publishing Company, La Salle, Illinois: *The Gospel of Buddha,* by Paul Carus.

Orient Longmans Limited, New Delhi: *The Incredible Sai Baba,* by Arthur Osborne.

Oxford University Press, Indian Branch, Bombay: *The Thirteen Principal Upanishads,* translated by Robert Ernest Hume, by permission of the Oxford University Press, Indian Branch.

Oxford University Press, London: *Stories of the Holy Fathers,* by E. A. Wallis Budge; *Shâh Abdul Latîf of Bhit* by H. T. Sorley; *Shekel Hakodesh* and *Yesod Hayirah,* translated by Hermann Gollancz; *Dialogues of the Buddha,* translated by T. W. and C. A. F. Rhys Davids, from *Sacred Books of the Buddhists; Minor Anthologies of the Pâli Canon,* Part II, translated by F. L. Woodward; *Tibet's Great Yogi Milarepa,* by W. Y. Evans-Wentz, published by Oxford University Press; *The Tibetan Book of the Dead,* by W. Y. Evans-Wentz, published by Oxford University Press; *Tibetan Yoga and Secret Doctrines,* by W. Y. Evans-Wentz, published by Oxford University Press; *The Tibetan Book of the Great Liberation,* by W. Y. Evans-Wentz, published by Oxford University Press.

Marco Pallis: *Peaks and Lamas,* by Marco Pallis, Cassell & Co., Ltd., London; *The Way and the Mountain,* by Marco Pallis, Peter Owen Ltd., London.

Paulist/Newman Press, New York: *Treatise on the Love of God,* by St Francis de Sales, translated by the Rev. Henry Benedict Mackey, Copyright, 1942, The Newman Bookshop, Westminster, Maryland; *The Complete Works of Saint John of the Cross,* translated by E. Allison Peers, in three volumes, The Newman Bookshop, 1945; *The Life of St Teresa of Jesus,* translated by

David Lewis, The Newman Bookshop, 1944: *Saint Bernard on the Love of God,* translated by Rev. Terence L. Connolly, Copyright, 1943, by Abbey of Gethsemani, Trappist, Kentucky.

Penguin Books Ltd., Harmondsworth, Middlesex: *Alchemy,* by E. J. Holmyard: *The Age of Chaucer,* edited by Boris Ford.

The Philosophical Library, Inc., New York: *Hinduism and Buddhism,* by Ananda K. Coomaraswamy: *Dictionary of Last Words,* edited by Le Comte.

Laurence Pollinger Limited, London: *The Spirit of Saint Francis de Sales,* by Jean Pierre Camus, translated by C. F. Kelley, Longmans Green & Co., Ltd., Publishers, Harper & Row, Proprietors.

Présence Africaine, Paris: extract taken from *Tierno Bokar, le Sage de Bandiagara,* by A. Hampaté Bâ and M. Cardaire, published by *Présence Africaine* (Paris, 1957).

Princeton University Press, Princeton, New Jersey: *The Rhetoric of Alcuin and Charlemagne,* Wilbur Samuel Howell (Princeton University Press, 1941); *Illumination in Islamic Mysticism,* Edward Jabra Jurji (Princeton University Press, 1938); *Zen and Japanese Culture,* by D. T. Suzuki, Bollingen Series LXIV (copyright © 1959 by the Bollingen Foundation, New York).

G. P. Putnam's Sons, New York: *The Wisdom of God* (*Srimad Bhagavatam*), translated by Swami Prabhavananda, Vedanta Press, Hollywood, California (Copyright 1943, by G. P. Putnam's Sons).

Bernard Quaritch Ltd., London, and the Estate of H. A. Giles: *Chuang Tzu, Mystic, Moralist, and Social Reformer,* translated by Herbert A. Giles.

Sri Ramakrishna Math, Mylapore, Madras: *A Thousand Teachings,* Śrî Śankarâchârya, translated by Swâmi Jagadânanda.

The Ramakrishna Vedanta Centre, London: *Women Saints of East and West,* Swami Ghanananda and Sir John Stewart-Wallace (Editorial Advisers), extracts from the chapters 'Akka Mahâdevî', 'Śrî Sâradâ Devî, the Holy Mother', and 'Catherine of Siena'.

Ramakrishna Vedanta Math, 19–B, Raja Rajkrishna Street, Calcutta–6: *The Sayings of Sri Ramakrishna,* complied by Swami Abhedananda.

Swami Nikhilananda and the Ramakrishna-Vivekananda Center, New York: *The Gospel of Sri Ramakrishna,* translated by Swami Nikhilananda, Copyright, 1942, by Swami Nikhilananda: *Râja-Yoga,* by Swami Vivekananda.

Sri Ramanasramam, Tiruvannamalai, S. India: *Talks with Sri Ramana Maharshi,* 3 vols., published by T. N. Venkataraman, Tiruvannamalai: *Spiritual Instruction of Bhagavan Sri Ramana Maharshi,* published by same: *Who Am I?* (of Bhagavan Sri Ramana Maharshi), published by same: *Self-Enquiry* (Being Original Instructions of Bhagavan Sri Ramana Maharshi), published by Sri Niranjanananda Swamy, Tiruvannamalai.

Random House, Inc., and Alfred A. Knopf, Inc., New York: *One Hundred and Seventy Chinese Poems,* translated by Arthur Waley, Copyright 1919 by Alfred A. Knopf, Inc. and renewed 1947 by Arthur Waley, reprinted by permission of the publisher: *The Enneads,* by Plotinus, translated by Stephen MacKenna, 3rd ed. (Pantheon Books, a division of Random House, Inc.), all rights reserved: *Zen in the Art of Archery,* by Eugen Herrigel (Pantheon Books): *The Mirror of Magic,* by Kurt Seligmann (Pantheon Books), Copyright 1948 by Pantheon Books Inc.: *Basic Writings of Saint Thomas Aquinas* (2 vols.), edited by Anton C. Pegis, Random House, Inc., Copyright, 1945, by Random House, Inc.: *The Wisdom of China and India,* edited by Lin Yutang, Random House, Inc., Copyright, 1942, by Random House, Inc.: *The Wisdom of Confucius,* edited and translated by Lin Yutang, The Modern Library, New York, Copyright, 1938, by Random House, Inc.: *The Basic Works of Aristotle,* edited by Richard McKeon, Random House, Inc., Copyright, 1941, by Random House, Inc., by arrangement with the Oxford University Press.

Fleming H. Revell Company, Westwood, New Jersey: *The Practice of the Presence of God,* by Brother Lawrence, translated from the French.

Rider & Company (Hutchinson Publishing Group Ltd.), London: *Essays in Zen Buddhism* (First Series), by D. T. Suzuki; *Essays in Zen Buddhism* (Second Series), by D. T. Suzuki; *Manual of Zen Buddhism*, by D. T. Suzuki; *The Zen Teaching of Huang Po*, translated by John Blofeld; *Theurgy or the Hermetic Practice: A Treatise on Spiritual Alchemy*, by E. J. Langford Garstin.

Routledge & Kegan Paul Ltd., London: *Zen and Japanese Culture*, by D. T. Suzuki; *The Diamond Sutra (Chin-Kang-Ching) or Prajñâ-Pâramitâ Sûtra*, translated by William Gemmell; *Byzantine Mosaic Decoration*, by Otto Demus; *Sexual Life in Ancient India*, by Johann Jakob Meyer; *The Banquet of Dante Alighieri*, by Katharine Hillard; *Zen in the Art of Archery*, by Eugen Herrigel; *Zen in the Art of Flower Arrangement*, by Gustie Herrigel; *The Book of the Dead*, translated by E. A. Wallis Budge; *The First Philosophers of Greece*, by A. Fairbanks; *The Mystics of Islam*, by R. A. Nicholson; Ibn Khaldûn's *Muqaddimah*, translated by Franz Rosenthal; *Of Learned Ignorance*, by Nicolas Cusanus, translated by Fr. Germain Heron.

Royal Asiatic Society, London: 'Aṭṭâr's *Tadhkirat al-Awliyâ*, translated by R. A. Nicholson, in the *Journal of the Royal Asiatic Society*, April, 1906; al-Ghazâlî's *Mishkât al-Anwâr*, translated by W. H. T. Gairdner; Jâmî's *Lawâ'iḥ*, translated by E. H. Whinfield & Mîrzâ Muḥammad Ḳazvînî; *Milindapañha*, translated by E. W. Burlingame, *JRAS*, 1917.

The Scarboro Fathers, Scarboro, Ontario: *Will Fatima Save America?*, by Mons. William C. McGrath, Providence Visitor Press, Providence, R. I., 1950 (a reprint from introd. to *Fatima or World Suicide*).

Schocken Books Inc., New York: *Major Trends in Jewish Mysticism*, by Gershom G. Scholem, Copyright © 1946, 1954 by Schocken Books Inc., reprinted by permission of Schocken Books Inc.

School of Oriental and African Studies, University of London: A. K. Coomaraswamy's 'Kha and Other Words Denoting "Zero"', in the *Bulletin of the School of Oriental and African Studies*, Volume VII (1934).

Frithjof Schuon: all material by Frithjof Schuon cited in the introductory sections and elsewhere the copyright for which belongs to Frithjof Schuon.

Shanti Sadan, London: *The World Within the Mind* (*Yoga-Vasishtha*), translated by Hari Prasad Shastri, Shanti Sadan, Third Edition, 1952.

Sheed & Ward Inc., New York: *An Augustine Synthesis*, arranged by Erich Przywara, published by Sheed & Ward Inc., New York; *The Mystical Theology of St Bernard*, by Étienne Gilson, published by Sheed & Ward Inc., New York.

Sheed & Ward Ltd., London: *An Augustine Synthesis*, arranged by Erich Przywara; *The Mystical Theology of St Bernard*, by Étienne Gilson; *The Face of the Saints*, by Wilhelm Schamoni.

Sidgwick & Jackson Ltd., London: *The Wonder That Was India*, by A. L. Basham.

The Society for Promoting Christian Knowledge, London: *The Way of a Pilgrim* and *The Pilgrim Continues His Way*, translated by R. M. French; *Dionysius the Areopagite: On the Divine Names and The Mystical Theology*, translated by C. E. Rolt.

Vincent Stuart & John M. Watkins Ltd., London: *The Hermetic Museum*, reprint. 1953 from 1st English edn. 1893, 2 vols.; *Meister Eckhart*, by Franz Pfeiffer, translated by C. de B. Evans, 2 vols.; *Iamblichus' Life of Pythagoras*, translated by Thomas Taylor; *The Scale of Perfection*, by Walter Hilton, edited by Evelyn Underhill; Thomas Vaughan's *Works*, edited by A. E. Waite; *On the Prayer of Jesus*, by Bishop Ignatius Brianchaninov, translated by Father Lazarus.

D. B. Taraporevala Sons & Co. Private Ltd., Bombay: *Kâma Kalpa, or The Hindu Ritual of Love*, by P. Thomas; *Yogic Home Exercises*, by Swami Sivananda.

Thames and Hudson Ltd., London: *Major Trends in Jewish Mysticism*, by Gershom G. Scholem, published for the United Kingdom by Thames and Hudson Ltd.

The Theosophical Publishing House London Ltd., London: *The Eleusinian Mysteries and Rites*, by Dudley Wright (out of print).

Tudor Publishing Company, New York: *The Dabistan* [of Mohsan Fânî], translated by David Shea and Anthony Troyer.

Miss Grace H. Turnbull: *The Essence of Plotinus,* based on the translation by Stephen MacKenna, compiled by Grace H. Turnbull, Oxford University Press, New York.

UNESCO, Paris, and The Indian Institute of World Culture, Bangalore, India: *The Indian Heritage, an Anthology of Sanskrit Literature,* selected and translated by Dr. V. Raghavan.

University Books, Inc., New Hyde Park, New York: *The Book of the Dead,* translated by E. A. Wallis Budge.

University of California Press, Berkeley, California: *Libellus de Alchimia,* by Albertus Magnus, translated by Sister Virginia Heines.

University of Oklahoma Press, Norman, Oklahoma: *The Sacred Pipe, Black Elk's Account of the Seven Rites of the Oglala Sioux,* recorded and edited by Joseph Epes Brown, Copyright 1953 by the University of Oklahoma Press; *Civilization,* as Told to Florence Drake by Thomas Wildcat Alford, Copyright 1936 by the University of Oklahoma Press.

Walter-Verlag AG, Olten, Switzerland: *Hildegard von Bingen, Geheimnis der Liebe.*

The Viking Press, Inc., New York: *The Portable Medieval Reader,* edited by James Bruce Ross and Mary Martin McLaughlin, Copyright 1949 by The Viking Press, Inc., reprinted by permission of The Viking Press, Inc.

Some sources in spite of all efforts made still could not be contacted, to whom full appreciation is none the less hereby expressed, with credits given as completely as possible in the INDEX OF SOURCES.

Profound gratitude is owed to Frithjof Schuon, and to the late Ananda K. Coomaraswamy and the late René Guénon, whose several roles have been altogether indispensable in the formation of this work. As these three authors have been cited extensively in the introductory sections, the following partial bibliography of their writings may prove helpful to the reader:

Coomaraswamy: The bulk of his publications are disseminated throughout a mass of scholarly and academic reviews difficult to obtain, but listed below are some of the more important writings which have appeared in book form:

A New Approach to the Vedas, London, Luzac, 1933
The Transformation of Nature in Art, Cambridge, Mass., Harvard University Press, 1934
Elements of Buddhist Iconography, Cambridge, Mass., Harvard University Press, 1935
The Rg Veda as Land-Nâma-Bók, London, Luzac, 1935
Spiritual Authority and Temporal Power in the Indian Theory of Government, New Haven, Conn., American Oriental Society, 1942
Hinduism and Buddhism, New York, Philosophical Library, 1943
Why Exhibit Works of Art?, London, Luzac, 1943
Figures of Speech or Figures of Thought, London, Luzac, 1946
Am I My Brother's Keeper?, New York, John Day, 1947 (also published by Denis Dobson, London, under the title of *The Bugbear of Literacy*)
Time and Eternity, Ascona, Switzerland, Artibus Asiae, 1947

Guénon: Listed are those of his books which have been translated into English:

East and West, London, Luzac, 1941
Introduction to the Study of the Hindu Doctrines, London, Luzac, 1945
Man and His Becoming According to the Vedânta, London, Luzac, 1945
The Reign of Quantity and the Signs of the Times, London, Luzac, 1953
Symbolism of the Cross, London, Luzac, 1958
The Crisis of the Modern World, London, Luzac, 1962

Schuon: Listed are those of his books which have been published in English:

The Transcendent Unity of Religions, London, Faber & Faber, 1953. (Revised Edition in preparation, London, Perennial Books)
Spiritual Perspectives and Human Facts, London, Faber & Faber, 1954, new edition, London, Perennial Books, 1970
Language of the Self, Madras, Ganesh, 1959
Gnosis, Divine Wisdom, London, John Murray, 1959
Stations of Wisdom, London, John Murray, 1961
Understanding Islam, London, Allen & Unwin, 1963
Light on the Ancient Worlds, London, Perennial Books, 1965
In the Tracks of Buddhism, London, Allen & Unwin, 1968
Dimensions of Islam, London, Allen & Unwin, 1970
Logic and Transcendence, New York, Harper and Row, 1975

And special thanks go to Marco Pallis for his generosity in contributing a valuable Foreword.

INDEX OF SOURCES

Bibliographical references that repeat will be entered in full where they first appear under each letter of the alphabet. Variations in orthography correspond to divergences in the works cited.

INDEX OF SOURCES

Aḥmad al-ʿAlawî, Shaykh (Abu ʾl-ʿAbbâs Aḥmad ibn Muṣṭafa ʾl-ʿAlawî; 1869–1934; famous Algerian spiritual pole [*quṭb*]; founder ʿAlawîyah branch of Shâdhiliyah Order).
P. 111: cited in *Les Amis de l'Islam,* Mostaganem, Jan., 1955, p. 1.
P. 196: Martin Lings: *A Moslem Saint of the Twentieth Century,* London, Allen & Unwin, 1961, p. 137.
P. 207: ibid., p. 160.
P. 628: cited by M. Lings: *Folkways Records Album* No. FR 8943, 1960.
P. 721: Lings: *Moslem Saint,* p. 137.
P. 778: 'Le Prototype unique', in *Études Trad.,* 1938, p. 300.
P. 919: Lings: *Moslem Saint,* p. 145.
P. 1034: Frithjof Schuon: *Perspectives spirituelles et Faits humains,* Paris, Cahiers du Sud, 1953, p. 220.

Ahmad b. ʿÂsim al-Antâkî (early Sufi from Antioch, d. at Damascus in 830).
P. 80: Arberry: *Sufism,* p. 44.

Ahmad b. ʿÎsâ al-Kharrâz (d. 899; famous early Sufi teacher at Baghdad, knew Dhu ʾl-Nûn, q.v.).
P. 617: *Kitâb al-Ṣidq*; Arberry: *Sufism,* p. 56.
P. 841: Hujwîrî: *Kashf,* p. 368.

Ahmed Et-Tahâwi (ibn Muhammad; d. 933; celebrated Egyptian Islamic theologian).
P. 513: *Von Mohammed bis Ghazâlî,* tr. Joseph Hell, Jena, Eugen Diederichs, 1923, p. 41.

Aitareya Âranyaka ('Forest-book', or supplement to the *Aitareya Brâhmaṇa,* q.v.).
P. 1000: Ananda K. Coomaraswamy: *Hinduism and Buddhism,* New York, Philosophical Library, 1943, p. 31.

Aitareya Brâhmaṇa (Hindu Vedic canonical commentary).
P. 306: A. K. Coomaraswamy: 'Vedic Exemplarism', *Harvard Journal of Asiatic Studies,* I, April, 1936.
P. 321: Ananda K. Coomaraswamy: *The Transformation of Nature in Art,* Cambridge, Massachusetts, Harvard University Press, 1935, p. 8.

Akaranga Sutra (Jaina scripture).
P. 64: tr. from Prakit by Hermann Jacobi, *The Sacred Books of the East,* Vol. XXII, ed. F. Max Müller, Oxford, Clarendon Press, 1884.
P. 93: ibid.
P. 790: ibid.

Akka Mahâdêvî (12th cent. A. D. Vîraśaiva Hindu woman saint).
P. 723: *Women Saints of East and West,* ed. and pub. by The Ramakrishna Vedanta Centre, London, 1955, pp. 31–2.
P. 985: ibid., p. 40.

Albertus Magnus (Saint, called Doctor Universalis; 1206?–1280; mediaeval German Dominican scholar of vast erudition, master of St Thomas Aquinas, q.v.).
P. 569: *Libellus de Alchimia,* III; tr. by Sister Virginia Heines, Los Angeles, University of California Press, 1958, p. 10.
P. 590: *Summa Theologiae,* II. xiii. 79; in W. F. Warren: *Paradise Found,* Boston, 1885, p. 305, n. 1.

Alchemical Device
P. 573: in T. Vaughan: *Light of All Lights*; cited by Seligmann in *Mirror of Magic,* p. 137.

Alcuin (Albinus, or Flaccus; 735–804; English ecclesiastic and scholar, aided Charlemagne in revival of learning at court of Franks; abbot of Tours).
P. 600: *The Rhetoric of Alcuin and Charlemagne*; tr. by W. S. Howell, Princeton, 1941, p. 153.

Alford, Thomas Wildcat (Shawnee Indian, spokesman for those members of his tribe banished to Texas, and then to Oklahoma in 1839).
P. 244: Florence Drake: *Civilization,* University of Oklahoma Press, 1936.

'Alî (ibn Abû Ṭâlib; 600?—661; son-in-law of the Prophet and fourth Khalîfah: a fountainhead of Islamic spirituality).
P. 75: *Maxims of ʿAli,* Lahore, Ashraf Publication, n.d., p. 12.

P. 104:	ibid., p. 6	P. 128:	p. 48	P. 195:	p. 5
107:	3, 4	129:	21	199:	35
118:	7	150:	1	199:	42
125:	56	172:	71	218:	9
127:	35	184:	73	218:	5

P. 804: *De Trin.* XIV. xv, 21: *Synth.* no. 163.
P. 817: idem, VIII. 2: in Pusey, *Conf.*, p. 121.
P. 818: *Conf.*, VII. x (Pusey).
P. 819: *De ag. chr.* xxxiii, 35; *Synth.* no. 51.
P. 819: *Serm. de Script N.T.* LXXXVIII; ibid., no. 929.
P. 823: *In epist. Joannis ad Parthos*; in Coomaraswamy: 'Recollection, Indian and Platonic', Supplement to the *Journal of the American Oriental Society*, No. 3, April-June, 1944, p. 1.
P. 824: *Soliloquies*, XXXI: cited by St John of the Cross, in E. Allison Peers: *Complete Works*, Westminster, Md., Newman Bookshop, 1945, Vol. II, pp. 196–7.
P. 839: *Conf.*, I. vi (Pusey).
P. 840: *In Ps.* CXXXVIII, 16: *Synth.* no. 860.
P. 841: *Conf.*, XII. xiii (Pusey).
P. 845: *De quant. animae* xxxiv, 77: *Synth.* no. 20.
P. 860: *Solil.* II: Peers, op. cit. *supra*, I. xii.
P. 866: *De Trin.* X: *Synth.* no. 11.
P. 906: *De Civ. Dei*, XIV. x; op. cit.
P. 941: idem, XIII. xvii, xviii.
P. 960: cited by Coomaraswamy in *A New Approach to the Vedas*, London, Luzac, p. 87, n. 55.
P. 962: *De ord.* II, ix, 26; *Synth.* no. 88.
P. 966: *De Doc. Christ.* i. 41; cited by Philip H. Wicksteed in tr. of *De Monarchia*: *Dante's Latin Works*, London, Temple Classics, 1904–1940, p. 240.
P. 977: *De Trin.* VIII, ii, 3; *Synth.* no. 247.
P. 1012: *In Ps.* CXLVIII, 1; ibid., no. 750.
P. 1038: idem, XXXVII, 4; ibid., no. 622.

Avataṃsaka Sûtra (Sanskrit Buddhist text, renowned in China, brought to Japan in 735 A.D., forming basic doctrine of the Kegon Shu).
P. 782: E. Steinilber-Oberlin: *Les Sectes bouddhiques japonaises*, Paris, G. Crès, 1930, p. 53.
P. 821: ibid., p. 48.

Avicenna (Abû 'Alî al-Ḥusayn ibn 'Abdallâh ibn Sînâ; 980–1037: Persian physician, philosopher, Aristotelian, Neoplatonist).
P. 161: cited by Eckhart, in Evans: *Eckhart*, I. 342.
P. 489: In Sédir: *Histoire et Doctrines des Rose-Croix*, Bihorel-lez-Rouen, Legrand, 1932, p. 207.
P. 633: in Mircea Eliade: *Forgerons et Alchimistes*, Paris, Flammarion, 1956, p. 38.
P. 860: cited by Guénon, in *Études Trad.*, 1951, p. 56.

Azar Kaivân (d. 1673 A.D.; Parsee sage, chief of the *Abâdiân* and *Azûrhûshangiân* Sects, dwelt in Khum and Patna).
P. 735: Shea and Troyer: *Dabistan of Mohsan Fânî*, p. 57.

'Azîz ibn Muhammad al-Nasafî (7th–8th cent. Sufi master from Nasaf, Khorasan, disciple of Najm al-Dîn Kubra, the founder of the Kubrawiyah Order in central Asia, northern India).
P. 107: *Maqṣad i aqṣa*: E. H. Palmer: *Oriental Mysticism*, Cambridge, 1867, 1938, I. vi. 21.
P. 828: ibid., II. v. 41–2.
P. 859: ibid., V. i. 50–51.

B

Baal Shem, Israel (d. 1760; founder of modern Hasidism in Poland).
P. 76: cited in Schuon: *Perspectives spirituelles et Faits humains*, Paris, Cahiers du Sud, 1953, p. 263.
P. 948: Gershom G. Scholem: *Major Trends in Jewish Mysticism*, New York, Schocken Books, 1954, 3rd edn., p. 335.

Bâbâ Ṭâhir (11th cent. Persian Sufic poet).
P. 381: *Poems of a Persian Ṣûfî*, being the quatrains of Bâbâ Ṭâhir, tr. by Arthur J. Arberry, Cambridge, W. Heffer, 1937, p. 31.

Bacon, Roger (1214?–1294; English philosopher, man of science, Franciscan monk; called 'the Admirable Doctor').
P. 378: cited by E. J. Langford Garstin: *Theurgy or the Hermetic Practice, A Treatise on Spiritual Alchemy*, London, Rider, 1930, p. 61.

P. 671: Osvald Sirén: *The Chinese on the Art of Painting*, New York, Schocken Books, 1963 (original edition, Peiping, Henri Vetch, 1936), p. 28.

P. 671: *Li Tai Ming Hua Chi*, II. ii; Sirén, op. cit., p. 229.

P. 672: *Lun Hua Lu Fa* (on Hsieh Ho's, q.v., 'Six Principles of Painting'), from the *Li Tai Ming Hua Chi*; *The Spirit of the Brush*, tr. of Chinese texts, by Shio Sakanishi, London, John Murray, 1939, 1948, pp. 78, 80.

P. 673: idem; ibid., p. 78.

Ch'en chiju (1558–1639: Chinese scholar and author).

P. 384: Lin Yutang: *The Importance of Living*, p. 344.

Ch'ên Tzŭ-ang (656–698: Chinese poet).

P. 345: *One Hundred and Seventy Chinese Poems*, tr. by Arthur Waley, London, Constable, 1918–1939, p. 95.

Chih-k'ai (fl. 575 A.D.: Chinese Buddhist founder of the Tendai doctrine which later formed a school of Japanese Buddhism).

P. 787: *Hônen the Buddhist Saint*, by Shunjô; tr. by Harper Havelock Coates and Ryugaku Ishizuka, Kyoto, 1949, p. 681.

Chi K'ang (223–262 A.D.; Taoist poet).

P. 145: *Taoist Song*; Waley: *170 Chinese Poems*, p. 64.

P. 695: idem.

Chih Kung (famous 6th cent. A.D. Chinese Zen monk).

P. 289: *The Zen Teaching of Huang Po*, tr. by John Blofeld, London, Rider, 1958, p. 63.

Children of Fatima, The

P. 261: by Mary Fabyan Windeatt, as retold in *Our Lady of Fatima's Peace Plan from Heaven, The Grail*, St Meinrad, Indiana, 1950, pp. 8, 9.

Chin Shengt'an (d. 1661: Chinese literary figure and critic).

P. 339: Lin Yutang: *The Importance of Living*, p. 300.

P. 536: ibid., p. 303.

P. 667: ibid., p. 301.

Chinese Nun, A

P. 606: Lin Yutang: *The Importance of Living*, p. 297.

Ching Hao (10th cent. Chinese landscape painter).

P. 671: *Pi Fa Chi* (Notes on Brushwork); Sirén: *Chinese on Art Painting*, p. 237.

P. 671: ibid., pp. 234–5.

P. 673: *Pi Fa Chi*; Sakanishi: *Spirit of the Brush*, pp. 95–6.

Ch'ing-yüan (d. 740 A.D.; a leading Zen disciple of Hui-nêng, q.v.).

P. 886: D. T. Suzuki: *Essays in Zen Buddhism* (First Series), London, Rider, 1949, 1958, p. 24.

Chou Li (code of rules for state officials in Chou dynasty, supposedly by Chou Kung, the 'Duke of Chou', d. 1105 B.C.—Chinese author and statesman).

P. 794: P. Vulliaud: *La Kabbale juive*, Paris, 1923, I. 378, citing G. Panthier: *L'esquisse d'une histoire de la philosophie chinoise*.

Christ in Islâm (citations collected from Muslim sources).

All listings below are taken from *Christ in Islâm*, by James Robson, The Wisdom of the East Series, New York, E. P. Dutton, 1930.

P. 44:	p. 44	P. 175:	p. 78	P. 365:	p. 62	P. 736:	p. 42
75:	58	239:	60	382:	47	745:	43
106:	77	285:	43	422:	57	746:	56–7
107:	73	347:	49	447:	107–8	905:	117
108:	65	348:	70	489:	65	992:	64
153:	44	349:	67	504:	50		
155:	77	352:	67	506:	63		

Christolero's Epigrams (by T. B., 1598).

P. 196: cited by Washington Irving: *The Sketch Book of Geoffrey Crayon, Gent*, heading the chapter, 'Westminster Abbey'; London, Everyman, 1906 & 1944, p. 161.

Chrysopeia of Cleopatra ('Gold-Making': Hellenistic-Egyptian alchemical treatise of 3rd or 4th cent. A.D.).

P. 600: *Journal*; in Washington Irving: *Christopher Columbus, His Life and Voyages,* New York and London, Putnam, 1896, p. 91.

Comenius (John Amos; 1592–1670; Czech theologian and educator; influenced by Boehme, q.v.).

P. 118: *Labyrinth of the World and Paradise of the Heart,* C. XXVIII; ed. Lutzom, Temple Classics, 1905.

Confessio Fraternitatis (anonymous Rosicrucian manifesto and apology appearing in Germany in 1615, an extension of the *Fama Fraternitatis* of 1614).

P. 870: Paul Arnold: *Histoire des Rose-Croix,* Paris, Mercure de France, 1955, p. 175.

Confucius (551–479 B.C.; Chinese sage, founder of Confucianism).

P. 153: Lin Yutang: *Importance Living,* p. 18.

P. 275: Lin Yutang: *From Pagan to Christian,* London, Heinemann, 1960, p. 92.

P. 279: ibid., p. 75.

P. 682: *Yochi, Liki,* XIX; Lin Yutang: *The Wisdom of Confucius,* New York, Modern Library, 1938, p. 254.

P. 683: ibid., p. 252.

P. 683: *Yochi, Liki*; Alain Daniélou: *Introduction to the Study of Musical Scales,* London, The India Society, 1943, pp. 16–17.

P. 684: idem, XIX; Lin Yutang: *Wisdom Confucius,* p. 259.

P. 686: ibid., p. 264.

P. 688: ibid., p. 260.

P. 758: Sirén: *Chinese Art Paint.,* p. 24.

P. 921: *Analects,* Bk. II; from a translation by Charles A. Wong published in China without references to publisher or date.

P. 970: The *Liki,* XXVII; Lin Yutang: *Pagan to Christian,* p. 95.

Consolata, Sister (Pierina Betrone, 1903–1946; Italian Capuchin nun, noted for spiritual method of invocation).

All listings below are taken from P. Lorenzo Sales: *Un Appel du Christ au Monde,* Fribourg, Switzerland, Editions St Canisius, n.d.

P. 122:	p. 28	P. 382:	p. 106	P. 702:	p. 43	P. 1023:	p. 116
133:	63–4	410:	147	702:	76	1027:	137
183:	77	517:	39	703:	132–3	1028:	109
192:	45	517:	49	717:	67	1029:	101
216:	146	531:	71	718:	103	1029:	102
282:	67	567:	74	918:	128	1037:	104
347:	16	617:	87	1012:	133	1038:	122
355:	73	621:	57–8	1012:	108		

Contest of Homer and Hesiod (biographical work of a grammarian in the time of Hadrian, 76–138 A.D.).

P. 706: *Contest,* 320; in Coomaraswamy: *Spiritual Authority and Temporal Power in the Indian Theory of Government,* New Haven, Conn., American Oriental Society, 1942, p. 87.

Cordovero, Moses ben Jacob (1522–1570; Spanish Kabbalist at Safed, in Upper Galilee, authority on *Zohar,* q.v.).

P. 598: Gershom G. Scholem: *Major Trends in Jewish Mysticism,* New York, Schocken Books, 1954, 3rd. edn., p. 279.

Cremer, John (14th cent. English alchemist, said in *The Hermetic Museum* to be 'Abbot of Westminster, and Friar of the Benedictine Order').

P. 768: 'The Testament of Cremer', in *The Hermetic Museum,* London, John M. Watkins, reprint. from 1st Eng. edn. 1893, Vol. II, p. 72.

Crito (5th cent. B.C. Athenian, friend and disciple of Socrates).

P. 754: 'On Prudence', from the *Physical Eclogues* of Stobaeus, p. 198; in Thomas Taylor: *Iamblichus' Life of Pythagoras,* London, John M. Watkins, 1926 (reprint. from 1818), p. 179.

Croll, Oswald (disciple of Paracelsus, q.v.).

P. 870: *Philosophy Reformed and Improved,* London, 1657; in Mircea Eliade: *Forgerons et Alchimistes,* Paris, Flammarion, 1956, pp. 171–2.

Crowfoot, Chief (Isapwo Muksika; d. 1890; leader of the Blackfoot Confederacy).

P. 100: Le Comte: *Dictionary of Last Words,* New York, Philosophical Library, 1955.

F

G

Gampopa (d. 1152 A.D.; successor of Milarepa, q.v.).

All listings are taken from the *Rosary of Precious Gems*; in *Tibetan Yoga and Secret Doctrines*, tr. by Lâma Kazi Dawa-Samdup and ed. by W. Y. Evans-Wentz, London, Oxford University Press, 1935, 1958.

P. 56: XXVI. 7	P. 232: XVII. 1	P. 312: III. 6	P. 512: XVI. 5	P. 794: XXVIII. 10
62: V. 3	240: XV. 2	330: XVI. 4	512: XVII. 7	798: III. 5
121: VI. 8	244: XXI. 8	355: XXIV. 10	518: XXVI. 8	837: XXVIII. 3
140: XXV	249: XVII. 4	404: XXII. 4	598: X. 6	901: XIII. 1
158: XVI. 2	260: XXIV. 1	406: XIV. 6	694: XXII. 10	980: XXVIII. 3, 7
190: XVII. 9	285: XVII. 8	433: XI. 5	742: XXIV. 4	992: XXIV. 6
226: XV. 11	296: X. 4	434: XI. 3	769: XV. 6	

Garuda Purana (late Hindu scripture).

P. 329: *The Garuda Puranam*, Manmatha Nath Dutt, Calcutta, 1908, Chs. CXIII, CXV.

P. 762: ibid., ch. CXV.

Gayatri (the supreme prayer in the Vedas, which it is the duty of Brahmans to recite daily, at morning, noon, and evening).

P. 317: from W. J. Wilkins: *Hindu Mythology*, Calcutta, 1906, p. 30; and V. Raghavan: *The Indian Heritage*, London and Bangalore, Indian Institute of Culture, 1956, p. 5.

Geber (Jâbir ibn Ḥayyân; fl. 721–776; celebrated Arab scholar and alchemist, at court of Hârûn al-Rashîd at Baghdad).

P. 46: *The Investigation of Perfection*, tr. by Richard Russell (1678); in E. J. Holmyard: *Alchemy*, Harmondsworth, Middlesex, Penguin Books, 1957, p. 133.

P. 309: ibid., p. 133.

P. 491: ibid., p. 132.

P. 781: ibid., p. 132.

P. 886: *Book of Exchanges*; cited by Abu 'l-Qâsim al 'Irâqî (q.v.): *Book of Knowledge Acquired Concerning the Cultivation of Gold*, tr. by E. J. Holmyard, Paris, Paul Geuthner, 1923, pp. 50–51.

Gertrude, St (the Great; 1256?–?1311; German Benedictine nun, known for her writings, supernatural visions).

P. 171: in St Alphonsus Liguori: *Conformity to the Will of God*, tr. by Benedictine Convent of Perpetual Adoration, Clyde, Missouri, 1935, p. 28.

Al-Ghazâlî (Abû Ḥâmid; d. 1111; famous Muslim theologian, Sufi authority, from Khurâsân).

P. 159: *Iḥyâ 'Ulûm al-Dîn* ('The Revivification of Religion'), IV; Margaret Smith: *Readings from the Mystics of Islâm*, London, Luzac, 1950, no. 68.

P. 230: tr. by Martin Lings, in *Folkways Records Album* No. FR 8943, 1960.

P. 307: *Mishkât al-Anwâr* ('The Niche for Lights'); tr. by W. H. T. Gairdner, London, Royal Asiatic Society, 1924, p. 71.

P. 319: *Iḥyâ*, I. v; tr. R. Bell, *Hibbert Journal*, Oct. 1943, p. 34.

P. 339: idem, IV; A. J. Arberry: *Introduction to the History of Sûfism*, London, Longmans, Green, 1942, p. 56.

P. 602: *Divine Names*; tr. Titus Burckhardt, *Études Traditionnelles*, Paris, Chacornac, 1952, p. 310.

P. 645: *Iḥyâ*, IV; Smith, op. cit., no. 72.

P. 681: idem, II; ibid., no. 69.

P. 748: *Al-Munqidh min al-dalâl*; A. J. Arberry: *Sufism, An Account of the Mystics of Islam*, London, Allen & Unwin, 1950, p. 80.

P. 758: cited by Vaswani: 'The Sufi Spirit', in *The New Orient*, May-June, 1924, p. 11.

P. 767: *Iḥyâ*, III; Smith, op. cit., no. 70.

P. 863: *Kîmiyâ al-Sa'âda* ('The Alchemy of Felicity'); Smith, op. cit., no. 65.

P. 927: idem, Wisdom of the East edition; cited in O. Cameron Gruner: *The Canon of Medicine of Avicenna*, London, Luzac, 1930, p. iv.

P. 974: *Iḥyâ*, I. v; Bell, op. cit., p. 36.

P. 986: *Mishkât*; tr. by Martin Lings: *A Moslem Saint of the Twentieth Century*, London, Allen & Unwin, 1961, p. 123.

P. 994: idem; Smith, op. cit., p. 71.

P. 244: Nicholson: *Legacy of Islam,* p. 217; in A. J. Arberry: *Sufism,* London, Allen & Unwin, 1950, p. 60.
P. 702: Hujwîrî: *Kashf,* p. 153.
P. 802: *Dîvân*: *Muqatta'at,* L; in Schuon: *Comprendre l'Islam,* Paris, Gallimard, 1961, p. 196, n. 3.
P. 816: Guénon: *Aperçus sur l'Initiation,* Éditions Traditionnelles, Paris, 1946, p. 219.
P. 929: 'Ṭa-Sînu's-Sirâj', *Islamic Quarterly,* London, April, 1956, p. 60.

Ḥasan al-Baṣrî (d. 728; born at Medina, brought up at Baṣra; eminent early Sufi, laid foundation of the 'science of hearts', *'ilm al-qulûb*).
P. 209: Smith: *Mystics of Islâm,* no. 1.

Hasidic saint, A (Hasidism: 18th cent. revival of Jewish mysticism in Poland).
P. 862: quoted in the *Seder Hadoroth Hekhadach,* p. 35; Gershom G. Scholem: *Major Trends in Jewish Mysticism,* New York, Schocken Books, 1954, 3rd edn., p. 344.

Hatshepsut (Queen of ancient Egypt in XVIIIth dynasty; daughter of Thutmose I; built famous temple Der el-Bahri).
P. 341: E. A. Wallis Budge: *Cleopatra's Needles and Other Egyptian Obelisks,* London, Lutterworth Press, 1926, p. 15.

Hekhaloth Rabbati (Jewish apocrypha dating from early Christian era).
P. 368: Scholem: *Jewish Mysticism,* Second Lecture, note 41.

Helmont, Jan Baptista van (1577?–1644; Flemish physician, chemist, and alchemist).
P. 527: *Hortus Medicinae*; in J. Marquès-Rivière: *Histoire des Doctrines ésotériques,* Paris, Payot, 1940, p. 333.

Helvetius (Johann Friedrich Schweitzer; 1625–1709; German alchemical writer, physician to the Prince of Orange; participant in a famous transmutation).
P. 285: 'John Frederick Helvetius' Golden Calf'; in *The Hermetic Museum,* London, John M. Watkins, 1953, Vol. II, p. 288.
P. 313: ibid., pp. 277–8.
P. 566: ibid., pp. 293–4.
P. 567: ibid., p. 277.

Heraclitus (c. 500 B.C.; Greek metaphysician from Ephesus).
All listings are from A. Fairbanks' *The First Philosophers of Greece,* London, Kegan Paul, Trench, Trubner, 1898.
Pp. 74, 251, 690, 827, 962, 984, 985, 989, 997.

Hermeneia of Athos
P. 321: Fichtner: *Wandmalereien der Athosklöster,* 1931, p. 15, no. 445; in Coomaraswamy: *The Transformation of Nature in Art,* Cambridge, Mass., Harvard University Press, 1935, note 113.

Hermes (Trismegistus, 'Thrice-Greatest'; the sacerdotal wisdom of Egyptian antiquity).
All listings, except where indicated, are taken from *Hermetica,* ed. and tr. from the Greek and the Latin by Walter Scott, in four volumes, Oxford, Clarendon Press, 1924–1936.

P. 31: *Libellus* X. 2, 4a
32: *Asclepius* I. 8
38: *Lib.* I. 12
41: *Stobaei,* Excerpt XI. 40
41: *Lib.* I. 12–14
42: *De Castogatione Animae,* 9, 4
48: *Lib.* III. 1a
56: *Lib.* II. 14
57: *Lib.* VI. 2a, 2b
58: *Testimonia*; Introd., p. 111
61: *Lib.* V. 4
65: *Lib.* XII (ii). 23b
66: *De Cast. An.* 11. 5, 6a
P. 69: *De Cast. An.* 11. 6b
74: *Lib.* VII
80: *De Cast. An.* 2. 8
96: *De Cast. An.* 14. 1
97: *De Cast. An.* 13. 8
99: *Lib.* XIII. 4
101: *Stob.* Exc. XI. 4. 16
103: *Stob.* Exc. X. 3
109: *De Cast. An.* 5. 7
111: *Ascl.* III. 22a
114: *De Cast. An.* 2. 7
123: *Lib.* XII (i). 2–4
129: *Stob.* Exc. XXVII
137: *Stob.* Exc. IV A. 6–8
150: *De Cast. An.* 11. 8
152: *Lib.* IV. 6b
P. 156: *De Cast. An.* 11. 12
157: *De Cast. An.* 5. 6
158: *Lib.* V. 11
159: Honein ibn Ishâq: *Dicta philosophorum* (*Hermetica,* I. 111)
159: *Ascl.* I. 11a
168: *Stob.* Exc. XVIII. 3, 4
187: *Stob.* Exc. VIII. 7
188: *Fragments,* nos. 19, 20
188: *Ascl.,* Epilogue (*Hermetica,* I. 377)
193: *Lib.* V. 10b
193: *Ascl.,* Epilogue (*Hermetica,* I. 373)

P. 763: *Didascalicon*; *Medieval Reader*, p. 573.
P. 792: *De sacramentis*; Migne, 176; Taylor, op. cit.
P. 796: Taylor, op. cit., pp. 98–9.
P. 914: *De arrha animae*; Taylor, op. cit., p. 89.
P. 950: *Noah's Ark*, I. xv; *Selected Writings*, p. 69.

Hui-nêng (637–713; the Sixth Patriarch of Chinese Zen Buddhism in line starting from Bodhidharma, c. 500 A.D., the Indian Dhyâna Buddhist who introduced Ch'an—Zen—into China; Hui-nêng, founder of the 'Southern School', gave rise to popularity of Zen in China through relating it to Taoism).
P. 725: Osvald Sirén: *The Chinese on the Art of Painting*, New York, Schocken Books, (orig. edn., Henri Vetch, Peiping, 1936), 1963, p. 95.
P. 919: ibid., p. 100.

Hui Shih (chief minister of King Hui of Liang, 370–333 B.C.; skilled in metaphysical paradox; friend and rhetorical opponent of Chuang-tse, q.v.).
P. 218: Hughes: *Chinese Philosophy*, p. 120 (taken from Chuang-tse, ch. XXXIII).
P. 826: idem.

Hui Tsung (1082–1135; Chinese emperor, 1101–1125, next to last of Northern Sung dynasty; painter, collector, patron of arts; founded imperial academy of painting).
P. 665: *Hsüan Ho Hua P'u*; Sirén: *Chinese on Art Painting*, p. 71.

Hujwîrî ('Alî b. 'Uthmân al-Jullâbî al-Hujwîrî; d. c. 1070; celebrated Sufi anthologist, orig. of Afghanistan).
All listings are from the *Kashf al-Maḥjûb*, by Hujwîrî, tr. from the Persian by R. A. Nicholson, London, Luzac, 1911, 1936, 1959.

P. 72:	p. 32	P. 146:	p. 205	P. 492:	p. 299	P. 743:	p. 18	P. 870:	p. 39
81:	156	155:	228	496:	296	745:	12	890:	204
138:	67	157:	25	514:	289	745:	12	922:	20
142:	409	207:	35	611:	277	770:	12	943:	368–9
142:	135	427:	207	643:	156	774:	15	984:	274
143:	24	429:	14	680:	401	859:	141–2	986:	332
145:	325	487:	298						

Hymn to Râ (Egyptian tradition).
P. 317: from the Papyrus of Hu-nefer (Brit. Mus. No. 9901, sheet 1); E. A. Wallis Budge: *The Book of the Dead*, London, Routledge & Kegan Paul, 1951, pp. 12–13, 15.

Hymn to Thoth (Egyptian tradition).
P. 593: *Journal of Egyptian Archaeology*, X, p. 3; Margaret A. Murray: *Egyptian Religious Poetry*, London, John Murray, 1949, p. 99.

Hypatia (370?–415 A.D.; Neoplatonic philosopher, mathematician, of Alexandria, renowned for her modesty, eloquence, beauty, and intellect; murdered by Christian mob).
P. 728: quoted in *The Philosophy of Life*, by A. M. Baten; Head and Cranston: *Reincarnation*, New York, Julian Press, 1961, p. 86; based on *Hypatia*, by Charles Kingsley, 1853, ch. VIII.

I

Iamblichus (d. c. 333 A.D.; leading representative of Syrian Neoplatonism).
All listings, except where indicated, are from *Iamblichus' Life of Pythagoras*, tr. from the Greek by Thomas Taylor, London, John M. Watkins, 1926 (reprinted from edn. of 1818).

P. 128:	p. 13	P. 519:	p. 74–5
300:	46	522:	Garstin, op. cit., p. 72
309:	55	611:	58
428:	38	670:	28–9
443:	*Egyptian Mysteries*; in E. J. Langford Garstin: *Theurgy or the Hermetic Practice, A Treatise on Spiritual Alchemy*, London, Rider, 1930, pp. 106–8	684:	61
		687:	32–3
		757:	89
		818:	37
		865:	44
		942:	72–3
		959:	1

Ibn ʿArabî (Muḥyî al-Dîn ibn al-ʿArabî; 1165–1240; from Murcia, Spain; died in Damascus; renowned pole of Sufi metaphysic).

P. 30: *Fuṣûṣu 'l-Ḥikam* ('Bezels of Wisdom'); R. A. Nicholson: *Studies in Islamic Mysticism,* London, Cambridge University Press, 1921, p. 153.

P. 49: *Fuṣûṣ*: 'The Divine Wisdom in the Word That Is of Adam'; tr. by Titus Burckhardt: *La Sagesse des Prophètes,* Paris, Albin Michel, 1955, p. 19.

P. 94: *Fuṣûṣ,* IX; tr. Burckhardt in *Études Traditionnelles,* Paris, Chacornac, 1951, p. 27.

P. 97: ibid., p. 20.

P. 182: *Al-Futûḥât al-Makkiyya* ('The Meccan Revelations'), III. 577; in Miguel Asin: *Islam and the Divine Comedy,* London, 1926, p. 159.

P. 315: *Tarjumân al-Ashwâq* ('The Interpretation of Divine Love'), Prologue; Asin, op. cit., p. 270.

P. 335: *Futûḥât,* II. 67; Burckhardt in *Études Trad.,* 1948, pp. 294–5, n. 3.

P. 341: *Ḥikmat i ʿAliyya*; cited in Jâmî, q.v.: *Lawâ'iḥ,* XXVI, tr. by E. H. Whinfield and Mîrzâ Muḥammad Ḳazvînî, London, Royal Asiatic Society, 1906–1928, p. 37.

P. 407: cited in 'Muhammadan Mysticism in Sumatra', by R. L. Archer, *Journal Malayan Branch Royal Asiatic Society,* Singapore, 1937, p. 87.

P. 450: *Fuṣûṣ*; Burckhardt: *Sagesse des Prophètes,* p. 196.

P. 621: idem; ibid., p. 183.

P. 629: idem; ibid., p. 187.

P. 737: cited by ʿAroûsî; in Émile Dermenghem: *Vies des Saints musulmans,* Algiers, Baconnier, n.d., p. 202, n. 2.

P. 749: *Fuṣûṣ*; Burckhardt in *Études Trad.,* 1951, p. 27.

P. 776: *Futûḥât,* III. 578; Asin, op. cit., p. 159.

P. 779: *Fuṣûṣ*; Nicholson: *Studies,* p. 152.

P. 782: idem; ibid., pp. 152–3.

P. 792: Nicholson: *The Mystics of Islam,* London, George Bell, 1914, reissued London, 1963, by Routledge & Kegan Paul, p. 105.

P. 805: *Risâlat al-Aḥadîyah* ('Treatise on Unity'); in René Guénon: *L'Homme et son Devenir selon le Vêdânta,* Paris, Éditions Traditionnelles, 3rd edn., 1947, p. 81.

P. 807: *Fuṣûṣ,* 282; Nicholson: *Studies,* p. 159.

P. 820: *Fuṣûṣ*; Burckhardt: *Introduction aux Doctrines ésotériques de l'Islam,* Lyon, P. Derain, 1955, p. 110.

P. 829: idem; ibid., p. 114.

P. 835: idem; Nicholson: *Studies,* p. 152.

P. 863: idem; Burckhardt: *Sagesse des Prophètes,* p. 182.

P. 893: *Risâlat*; Guénon: *Man and His Becoming,* tr. by Richard C. Nicholson, London, Luzac, 1945, p. 160.

P. 902: *Futûḥât*: Burckhardt in *Études Trad.*. 1948, p. 294.

P. 983: *Fuṣûṣ*: 'Word of Noah'; Burckhardt: *Sagesse des Prophètes,* pp. 60–61.

P. 995: *Fuṣûṣ*; Burckhardt: *Doctrines ésotériques de l'Islam,* p. 63.

P. 997: *Risâlat*; Guénon: *Man and His Becoming,* p. 114.

Ibn al-ʿArif (d. 1141; Andalusian Sufic master, famed for science of virtues).

The citations are taken from the *Maḥâsin al-Majâlis* ('The Virtues of the Sessions'), tr. from Arabic into Spanish by Miguel Asin Palacios, and from Spanish into French by F. Cavallera, Paris, Paul Geuthner, 1933 (in translating from French into English the original Arabic text has been consulted in crucial passages to check on accuracy).

P. 144:	p. 33–4	P. 493:	p. 36	P. 534:	p. 44	P. 931:	p. 47
226:	66	504:	64	574:	60	954:	64–66

Ibn ʿAṭâ'illâh (al-Iskandarî; from Alexandria, d. 1309; eminent Sufi of Shâdhilîyah Order, famous for his spiritual maxims).

All listings, except where indicated, are from *Al-Ḥikam al-ʿAṭâ'îyah,* Cairo, 1939, unpubl. tr. by Yahyâ Abû Bakr and Whitall N. Perry.

P. 60:	no. 107	P. 89:	no. 119	P. 108:	no. 98	P. 114:	no. 52
86:	73	98:	53	113:	30	126:	92

P. 50: Smith: *Mystics of Islâm*, no. 132	P. 787: N: p. 107	P. 887: B: p. 44	P. 989: B: p. 26
122: B: p. 47	788: N: 96	892: N: 93	995: B: 56–7
182: B: 74	806: B: 51	924: N: 106	995: B: 54
184: B: 62	832: N: 86–7	930: N: 113	1000: N: 95, n. 1
729: B: 69	835: B: 38	932: N: 107–8	1019: B: 31
752: B: 33	842: B: 71	946: B: 26–7	1033: N: 106–7
784: B: 52–3	843: N: 112–13	980: N: 106	1034: B: 30
	876: N: 129, B: 77	987: B: 78	

Jimyo (T'zu-ming; 986–1040; one of the great Sung masters of Zen, whose teachings are used in the Rinzai school in Japan).

P. 827: D. T. Suzuki: *Essays in Zen Buddhism* (First Series), London, Rider, 1949, 1958, p. 35.

Jnânasankalini Tantra (Hindu Tantric text).

P. 734: Alain Daniélou: *Yoga, The Method of Re-Integration*, London, Christopher Johnson, 1949, p. 76.

John Chrysostom, St (c. 347–407; Church Doctor, most famous of Greek Fathers; b. Antioch, became patriarch of Constantinople; renowned orator, whence the name Chrysostom, 'Golden Mouthed').

P. 297: *The Pilgrim Continues His Way*, tr. from the Russian by R. M. French, London, Society for Promoting Christian Knowledge, 1941, p. 67.

P. 412: *On the Epistle to the Romans, Hom. 8*; Bishop Ignatius Brianchaninov: *On the Prayer of Jesus*, tr. by Father Lazarus, London, John M. Watkins, 1952, p. 25.

P. 1007: E. Kadloubovsky and G. E. H. Palmer: *Writings from the Philokalia on Prayer of the Heart*, London, Faber & Faber, 1951, p. 223.

John of Cronstadt (John Sergieff; 1829–1908; Russian priest with parish at Cronstadt, near St Petersburg; a genius of prayer, with great popular following).

P. 522: *My Life in Christ*; in *A Treasury of Russian Spirituality*, comp. and ed. by G. P. Fedotov, London, Sheed & Ward, 1950, p. 354.

John of the Cross, St (1542–1591; leading figure in Spanish mysticism, doctor of mystic theology).

P. 146: *Ascent of Mount Carmel*, I. v. 8; tr. by E. Allison Peers: *The Complete Works of Saint John of the Cross*, 3 vols., Westminster, Maryland, Newman Bookshop, 1945.

P. 159: idem, I. xiii. 11; ibid.

P. 277: idem, Prologue; ibid.

P. 356: in Schuon: *Perspectives spirituelles et Faits humains*, Paris, Cahiers du Sud, 1953, p. 280.

P. 373: *Dark Night of the Soul*, II. v–viii, *passim*; Peers, op. cit.

P. 374: *Ascent Carmel*, II. vii. 11; op. cit.

P. 377: *Dark Night*, II. vii. 6; op. cit.

P. 407: in Aldous Huxley: *The Perennial Philosophy*, London, Chatto & Windus, 1950, p. 334 (tr. Peers).

P. 842: *Spiritual Canticle*, I. 3; Peers, op. cit.

P. 987: *Points of Love*, no. 21; Peers, op. cit.

John Damascene, St (700?–?754; called Chrysorrhoas, 'Stream of Gold', because of his eloquence; b. Damascus, Doctor of Church, last of Greek Fathers, considered first Scholastic, and first theological encyclopaedist).

P. 322: in Schuon: *De l'Unité transcendante des Religions*, Paris, Gallimard, 1948, p. 89.

P. 977: in Marco Pallis: *Peaks and Lamas*, London, Cassell, 1939–1942, p. 378.

John of the Ladder, St (called Climacus, from his book, 'Ladder of Paradise'; c. 525–605; solitary on Mount Sinai, later abbot of the monastery).

P. 78: Kadloubovsky and Palmer: *Writings from Philokalia*, p. 28.

P. 1039: in *Yoga, Science de l'Homme intégral*, texts ed. by Jacques Masui, Paris, Cahiers du Sud, 1953, p. 191 (article on *Hesychasm*, by Antoine Bloom).

John of Salisbury (d. 1180; English ecclesiastical leader, classical scholar).

P. 969: from *Policraticus*; *The Statesman's Book of John of Salisbury*, tr. by John Dickinson, New York, Appleton-Century-Crofts, 1927.

Julian of Norwich (Lady Julian, anchoress at Norwich, 14th cent.; flower of mediaeval English contemplatives).

All listings, except where indicated, are from *Revelations of Divine Love*, by Julian of Norwich, ed. by Grace Warrack, London, Methuen, 1901–1952.

Kama-Sutra (Hindu classic on love perspectives and techniques, written by Vatsyayana—fl. 450 A.D.).
P. 447: tr. by B. N. Baṣu, Calcutta, 1945, p. 220.

Kashiwazaki (Japanese Nô play stressing Nembutsu practice).
P. 1011: Beatrice Lane Suzuki: *Nôgaku, Japanese Nô Plays*, London, John Murray, 1932, p. 40.

Kaṭha Upanishad (late Vedic scripture).
The listings below are from *The Thirteen Principal Upanishads*, tr. by Robert Ernest Hume, London, Oxford University Press, 1921–1934.
Pp. 213, 226, 306, 483, 810, 833, 834.

Al-Kâtibî (d. 1276; an Arabic source of Hermetica).
P. 69: *Hermetica*, ed. and tr. by Walter Scott, Oxford, Clarendon Press, 1936, Vol. IV, p. 273.

Kaula Tantric Precept (Indian).
P. 307: cited by Sj. Atal Bihari Ghosh; in *Tibet's Great Yogî Milarepa*, ed. by W. Y. Evans-Wentz, London, Oxford University Press, 1928, p. 214.

Kaushîtaki Upanishad (late Vedic scripture).
P. 47: as rendered by Coomaraswamy.
P. 590: Hume: *Thirteen Principal Upanishads*.
P. 668: idem.
P. 861: idem.
P. 967: as rendered by Coomaraswamy.

Kelpius, Johannes (1673–1708; German theosophical contemplative, school of Boehme, q.v.; established mystical community in Pennsylvania near Philadelphia).
All selections are from *A Method of Prayer*, by Johannes Kelpius, ed. by E. Gordon Alderfer, New York, Harper, 1951.

P. 147: p. 122, 123, 126, 127	P. 501: cited by ed., pp. 49–50	P. 717: p. 120
424: cited by ed. in Introd., p. 46	542: p. 113–14	1009: 86–88
485: p. 106–7	546: 120	1021: 108–9
		1041: 104

Kena Upanishad (late Vedic scripture).
P. 993: Hume: *Thirteen Principal Upanishads*.

Kenshin (1131–1192; a vice-bishop and head priest of the Tendai sect in Japan).
P. 566: in *Hônen*, pp. 279–80.

Khaggavisâna Sutta (Buddhist Pâli canon, 'Sutta of the Rhinoceros', delivered by Gotama Buddha as teaching typical of Pratyeka-Pachcheka-Buddhas).
P. 152: *Sutta Nipâta*, tr. by Sir M. Coomâra Swâmy , London, Trübner, 1874, p. 16.

Khunrath, Heinrich (17th cent. German alchemist and Rosicrucian).
P. 409: Device pictured in *Amphitheatrum Aeternae Sapientiae*, Hanau, 1609; Kurt Seligmann: *The Mirror of Magic*, New York, Pantheon Books, 1948, pp. 131 and 133.
P. 680: Inscription from engraving in op. cit. *supra*; ibid., p. 132.
P. 749: *Am. Aet. Sap.*, fol. 147; cited in Helvetius' 'Golden Calf', *The Hermetic Museum*; London, John M. Watkins, 1893 and 1953, Vol. II, p. 279.

Kôbô Daishi (or Kukai; 774–835 A.D.; Japanese Buddhist priest, studied in China, established Shingon sect in Japan).
P. 796: Beatrice Lane Suzuki: *Nôgaku, Japanese Nô Plays*, p. 37.
P. 825: *Shingyô Hiken*; *Hônen*, p. 444.
P. 825: E. Steinilber-Oberlin and Kuni Matsuo: *Les Sectes bouddhiques japonaises*, Paris, Éditions G. Crès, 1930, p. 107.

Ko Ch'ang-Kêng (also Po Yü-chuan; 13th cent. A.D. Chinese alchemist).
P. 797: Waley: *Notes on Chinese Alchemy*, London, 1930, p. 16; cited in Mircea Eliade: *Forgerons et Alchimistes*, Paris, Flammarion, 1956, p. 128.

Ko Hung (Taoist scholar; 3rd–4th cent. A.D.).
P. 42: 'On the Gold Medicine', tr. by Lu Ch'iang Wu; ed. by Tenney L. Davis, *American Academy of Arts and Sciences*, Boston, December, 1935, p. 238.
P. 423: ibid., pp. 250–51.
P. 738: ibid., p. 238.

P. 744: *Inner Chapters of Pao-p'u-tzŭ*; tr. by Tenney L. Davis and Ch'ên Kuo-fu, *AAAS*, Dec., 1941, p. 301.

P. 882: 'Gold Medicine'; op. cit., p. 236.

Kulârnava Tantra (Hindu Tantric text).

P. 62: Alain Daniélou: 'Le Yoga de l'Age des Conflits', in *Yoga, Science de l'Homme intégral*, ed. by Jacques Masui, Paris, Cahiers du Sud, 1953, p. 136.

P. 218: in W. Y. Evans-Wentz: *The Tibetan Book of the Dead*, London, Oxford University Press, 1927–1951, p. xxxi.

P. 288: Arthur Avalon (Sir John Woodroffe): *The Sepent Power*, Madras, Ganesh, 4th edn., 1950, p. 506.

P. 631: Mircea Eliade: *Le Yoga, Immortalité et Liberté*, Paris, Payot, 1954, p. 262.

Kung-sun Lung (scion of ruling house of Chao State in 3rd cent. B.C.; Chinese philosopher, disciple of Hui Shih, q.v.).

P. 670: *Chinese Philosophy in Classical Times*, ed. and tr. by E. R. Hughes, London & New York, Everyman's Library, 1944, p. 125.

Kung Tingan (Lin Yutang calls him 'a Chinese writer'; no other data available).

P. 989: Lin Yutang: *The Importance of Living*, London, Heinemann, 1938–1951, p. 374.

Kwaisen (d. 1582; Japanese Zen master).

P. 268: D. T. Suzuki: *Zen and Japanese Culture*, London, Routledge & Kegan Paul, 1959, p. 79.

Kwannon Sutra (in Japanese, *Kwannon-gyo*; in Chinese, *Kuan-yin Ching*; the Twenty-fifth Chapter in Kumarajiva's [343–413 A.D.] translation of the *Saddharma-pundarika*, q.v., one of the most popular sutras in Japan).

P. 655: D. T. Suzuki: *Manual of Zen Buddhism*, London, Rider, 1950, p. 36.

P. 942: ibid., p. 36.

L

Lalita Vistara (Mahâyâna Buddhist text).

P. 942: William Norman Brown: *The Indian and Christian Miracles of Walking on the Water*, Chicago, Open Court, 1928, p. 19.

Lambspring, Abraham (alchemical author described in *The Hermetic Museum* as 'A Noble Ancient Philosopher').

All listings are from *The Book of Lambspring*, as given in *The Hermetic Museum*, Vol. I, John M. Watkins, London, reprint. 1953 from 1st Eng. edn. 1893.

P. 113:	p. 291	P. 402:	p. 282	P. 566:	p. 289	P. 797:	p. 289
297:	275	482:	290	572:	304	947:	296
398:	278	559:	292	789:	274	951:	280

Lankavatara Sutra (Mahâyâna Buddhist scripture originally in Sanskrit).

P. 86: D. T. Suzuki: *Manual of Zen Buddhism*, London, Rider, 1950, p. 54.

P. 96: tr. Suzuki and Goddard: *A Buddhist Bible*, 2nd edn., ed. and publ. by Dwight Goddard, Thetford, Vermont, 1938, p. 316.

P. 321: Coomaraswamy: *The Transformation of Nature in Art*, Harvard University Press, Cambridge, 1935, note 43.

P. 493: *Buddhist Bible*, p. 309.

P. 703: ibid., p. 323.

P. 726: E. A. Burtt: *The Teachings of the Compassionate Buddha*, New York, Mentor Books, 1955, p. 166.

P. 805: *Buddhist Bible*, p. 299.

P. 888: ibid., p. 317.

P. 993: D. T. Suzuki: *Essays in Zen Buddhism* (First Series), London, Rider, 1949, 1958, p. 67.

Lao-tse (see under *Tao Te Ching*).

Law, William (1686–1761; English divine, and mystical writer, school of Boehme, q.v.; after Cambridge Platonists, most eminent of English contemplatives).

All listings, except where indicated, are from *Selected Mystical Writings of William Law*, ed. by Stephen Hobhouse, London, Rockliff, 1938–1949.

P. 26:	p. 31	P. 30:	p. 37	P. 40:	p. 143
27:	52	34:	113	44:	154

Letter to Diognetus (written about 150 by a Christian apologist, to Diognetus, a learned Greek).
P. 139: tr. by Cardinal Newman, *Grammar of Assent*, p. 466; in Laux: *Church History*, p. 46.

Li Chi ('Record of Rites'; Confucian texts, compiled at the end of the last century B.C. by Tai Shen).
P. 275: E. R. Hughes: *Chinese Philosophy in Classical Times*, London & New York, Everyman's Library, 1944, p. 279.
P. 599: ibid., p. 286.
P. 930: ibid., p. 276.
P. 965: ibid., pp. 283–4.
P. 969: ibid., p. 283.

Lieh-tse (Taoist luminary, 5th–4th cent. B.C.).
P. 337: *Tch'ung-Hu-Tchenn-King*, Ch. I D; tr. by Lionel Giles, *Taoist Teachings*, Wisdom of the East Series, London, p. 20.
P. 563: idem, II F; tr. by Léon Wieger: *Les Pères du Système taoïste*, Paris, Cathasia, 1950.
P. 574: idem, IV N; ibid.
P. 775: idem, I A; ibid.
P. 925: idem, II B; ibid.
P. 945: idem, II A; ibid.

Liguori, St Alphonsus (1696–1787; Italian prelate, founded Redemptorist Order; bishop of Sant' Agata dei Goti; Doctor of the Church; known for theological treatises and other writings).
All listings, except where indicated, are taken from *Conformity to the Will of God*, tr. from the Italian by the Benedictine Convent of Perpetual Adoration, Clyde, Missouri, 7th edn., 1935.

P. 121: p. 56
122: 42–3
124: 39
125: 50
126: 6
129: Wilhelm Schamoni: *The Face of the Saints*, London, Sheed & Ward, 1948, p. 224
P. 167: p. 16
174: 7
P. 180: p. 38
181: 47–8
183: 18–19
187: 12
343: 12
577: 50
888: 24
889: 29

Li Jih-hua (1565–1635; prominent Chinese painter, and writer on painting, poetry, philosophy, medicine).
P. 674: Osvald Sirén: *The Chinese on the Art of Painting*, New York, Schocken Books, 1963 (orig. edn., Henri Vetch, Peiping, 1936), pp. 158–9.

Li Liweng (17th cent. Chinese poet and dramatist).
P. 493: Lin Yutang: *My Country and My People*, London, Heinemann, 1936–1951, p. 235.

Li Po (d. 762 A.D.; famous Chinese poet, one of 'The Eight Immortals of the Wine Cup').
P. 639: Lin Yutang: *My Country and My People*, p. 239.
P. 993: *Li Po, the Chinese Poet*, tr. by Shigeyoshi Obata, Kanda-Tokyo, Hokuseido Press, 1935, p. 73.

Loathly Lady, The (Arthurian Romance).
P. 631: Coomaraswamy: 'On the Loathly Bride', *Speculum*, Cambridge, Mass., The Mediaeval Academy of America, Vol. XX, no. 4, Oct. 1945, p. 402.

Lodan-Gawai-Roltso (or, 'The Ocean of Delight for the Wise'; a compilation of Tibetan maxims).
P. 105: tr. by Lâma Kazi Dawa-Samdup; W. Y. Evans-Wentz: *Tibetan Yoga and Secret Doctrines*, London, Oxford University Press, 1958, p. 65.
P. 201: ibid., p. 64.

Louis de Montfort, St (Louis Marie Grignion de Montfort; 1673–1716; French ecclesiastic, founded Sisters of Wisdom, Company of Mary; promoted devotion to the Virgin Mary; canonized by Pope Pius XII in 1947).
All listings, except where indicated, are taken from *True Devotion to the Blessed Virgin Mary*, tr. by F. W. Faber, 1862, and publ. by the Montfort Fathers, Bay Shore, New York, 1946.

P. 184: p. 36
190: 64
198: 36
199: 3
212: 57–9
P. 355: *St Louis De Montfort*, ed. Raymond Christo-flour, Namur, Belgium, Éditions du Soleil

Buddhists, Vol. III, London, Oxford University Press, 1938.

Mahâ-Vagga (one of the five parts of the *Vinaya-Piṭaka*, q.v.).

P. 704: Henry Clarke Warren: *Buddhism in Translations*, Cambridge, Mass., Harvard University Press, 1906, pp. 352–3.

Mahâvastu (Buddhist Sanskrit scripture).

P. 655: William Norman Brown: *The Indian and Christian Miracles of Walking on the Water*, Chicago, Open Court, 1928, p. 25.

Maier, Michael (1568–1622; German alchemist and Rosicrucian).

P. 29: *Cantilenae intellectuales de Phoenice redivivo*; tr. by Le Masurier, Paris, 1758; in Paul Arnold: *Ésotérisme de Shakespeare*, Paris, Mercure de France, 1955, p. 127.

P. 106: 'The Secrets of Alchemy', *Hermetic Museum*, II, p. 201.

P. 201: 'Epigram upon the Practice of Basilius'; ibid, I, p. 312.

P. 365: 'Secrets of Alchemy'; op. cit., pp. 204–5.

P. 507: idem; ibid., p. 204.

P. 510: 'An Epigram Written by M. M., on Norton's Chemical Treatise'; ibid., II, p. 2.

P. 656: 'Chemical Secrets of Nature'; in Kurt Seligmann: *The Mirror of Magic*, New York, Pantheon Books, 1948, p. 145.

P. 789: 'Epigram upon Basilius'; op. cit., p. 312.

P. 798: 'Silentium post clamores'; Sédir: *Doct. Rose-Croix*, p. 294.

Maimonides (Rabbi Moses ben Maimon; 1135–1204; Jewish philosopher from Cordova, became physician to Saladin in Cairo; sought harmony between Judaism, Aristotelianism, Islamic doctrines).

P. 787: *Mishneh Torah*; cited in *Time*, January 24, 1955.

Maistre, Joseph de (Comte Joseph Marie de; 1753–1821; French philosopher, author, and statesman; in Switzerland and Italy as opponent of French Revolution).

P. 965: *Du Pape*, I. i; Monaco, Éditions du Rocher, 1957, p. 166.

Maitri Upanishad (late Vedic scripture).

All citations, except where indicated, are taken from *The Thirteen Principal Upanishads*, tr. by Robert Ernest Hume, London, Oxford University Press, 1921, 1931, 1934.

Pp. 33, 50 (tr. Coomaraswamy), 65, 154, 338, 343, 423, 430, 480, 483, 491, 802 (tr. Coomaraswamy), 823, 824, 888, 920, 944, 980, 1005, 1009, 1036.

Majjhima-nikâya (Buddhist Pâli canon, part of the *Sutta-Piṭaka*).

P. 48: *Buddhist Texts Through the Ages*, ed. by Edward Conze, Oxford, Bruno Cassirer, 1954, p. 76.

P. 276: Ananda K. Coomaraswamy: *Hinduism and Buddhism*, New York, Philosophical Library, 1943, p. 75.

P. 299: Ananda K. Coomaraswamy and I. B. Horner: *Gotama the Buddha*, London, Cassell, 1948, p. 31.

P. 335: ibid., p. 144.

P. 513: Warren: *Buddhism in Translations*, pp. 119–122.

Malleus Maleficarum (a classic on witchcraft and demonology, brought out in 1489 by Henry Kramer and James Sprenger, German theologians commissioned by a Bull of Pope Innocent VIII in 1484 to combat heresy).

P. 423: Introd. Part III; tr. by the Rev. Montague Summers, London, John Rodker, 1928, p. 198.

P. 451: Part I. Question 2; ibid., p. 14.

P. 452: Part I. Question 1; ibid., p. 5.

P. 454: idem; ibid., p. 3.

Mânava-dharma-śâstra (Vedic legislation—and cosmology—regulating all aspects of Hindu life).

All listings, except where indicated, are from *The Ordinances of Manu*, tr. from the Sanskrit by Arthur Coke Burnell, London, Trübner, 1884.

Pp. 32, 34, 63, 64, 79, 110, 126, 293, 337, 403 (tr. by Sir William Jones; in E-Wentz: *Tibetan Book of the Dead*, q.v., p. 10), 526 (Müller: *Sacred Books of the East*, I, p. 287), 626, 762, 769, 989, 990, 1003 (Alain Daniélou: *Introduction to the Study of Musical Scales*, London, The India Society, 1943, p. 97), 1010, 1024.

Mâṇḍûkya Upanishad (late Vedic scripture).

P. 1031: Hume.

Marcus Aurelius (surnamed Antoninus; 121–180 A.D.; Roman emperor, conqueror, eminent Stoic philosopher).

P. 346: *Mediations*, XI. 5, tr. by George Long, 1862; in *The Stoic and Epicurean Philosophers*, ed. by

Whitney J. Oates, New York, Random House, 1940.

Mârkaṇḍeya Purâṇa (Hindu scripture).

P. 39: ed. by M. N. Dutt; Calcutta, H. C. Dass, 1896.

P. 262: ibid.

Marpa (the master of Milarepa, q.v.; fl. 11th cent. A.D.; also known as the Translator, for the Tantric texts he obtained in India and subsequently translated into Tibetan).

P. 358: *Tibet's Great Yogi Milarepa, a Biography from the Tibetan, being the Jetsön-Kahbum,* tr. by Lâma Kazi Dawa-Samdup, ed. by W. Y. Evans-Wentz, London, Oxford University Press, 1928, pp. 130–131.

Martyr, Peter (Pietro Martire d'Anghiera; 1457–1526; Italian historian, royal chronicler to Spanish court).

P. 583: cited in Washington Irving: *Christopher Columbus, His Life and Voyages,* New York and London, Putnam, 1896, p. 94.

Martyrdom of St Andrew, The (Apocryphal New Testament).

P. 131: *Acts of Andrew*; *The Apocryphal New Testament,* tr. by Montague Rhodes James, London, Oxford University Press, 1924–1953, p. 360.

P. 324: idem; ibid., pp. 359–360.

Martyrdom of St Peter, The (Apocryphal New Testament).

P. 491: *Acts of Peter*; James: *Apoc. New Test.*, pp. 334–5.

Maximus the Confessor, St (508?–662; Christian theologian from Constantinople, abbot of Chrysopolis at Scutari, vigorous opponent of Monothelite heresy, tortured and banished for stand).

P. 210: *Capita de Caritate,* II. 61, 62.

Maximus of Tyre (12th cent.; no other data available).

P. 309: *Philosophumena, Oratio,* II; M. M. Davy: *Essai sur la symbolique romane,* Paris, Flammarion, 1955, p. 44.

Mechthild of Magdeburg (1210–1297; a foremost German contemplative in the realm of Christian gnosis).

All selections are from *The Revelations of Mechthild of Magdeburg, or, The Flowing Light of the Godhead,* tr. by Lucy Menzies, London, Longmans, Green, 1953.

P. 30:	p. 13	P. 225:	p. xxi	P. 565:	p. 52	P. 782:	p. 50
35:	6	263:	84, 86	577:	243	803:	180
59:	103	350:	49	617:	40	837:	63
73:	136–7	410:	73	617:	54	891:	9
89:	52	432:	43	627:	9–10	906:	90
91:	227	434:	244	628:	23	906:	54
127:	102	443:	181	632:	12	907:	41
128:	52–3	451:	240–41	632:	63	926:	25
145:	20	460:	53	638:	22	926:	65
150:	xxx	472:	123–4	639:	70	945:	31
175:	35	476:	123	642:	24	947:	40–41
184:	131	521:	136	689:	21	961:	12
186:	203	534:	154	713:	xxxi	991:	67
200:	108	538:	23	724:	17	1005:	63
201:	59	552:	26	740:	69		

Mehung, John A. (Jean de Meung; real name, Jean Clopinel; c. 1240–1305; French writer; author of the second part of the *Roman de la Rose*).

P. 27: 'A Demonstration of Nature'; *Hermetic Museum,* I. p. 126.

P. 86: idem; ibid., p. 125.

P. 88: idem; ibid., p. 133.

P. 283: cited by Philalethes, in *Hermetic Museum,* II, p. 269.

P. 469: 'Demonstration'; op. cit., p. 123.

P. 493: idem; ibid., p. 131.

P. 510: idem; ibid., p. 132.

P. 546: idem; ibid., pp. 131, 133.

P. 722: idem; ibid., p. 137.

P. 789: idem; ibid., p. 132.

P. 836: idem; ibid., p. 127.

P. 909: *Roman de la Rose, XVIII*; in M. Caron and S. Hutin: *Les Alchimistes*, Bourges, Éditions du Seuil, 1959, pp. 148–9, from modern French of André Mary, Paris, Gallimard, 1949.

Melissus (5th cent. B.C. Greek Eleatic philosopher and statesman, disciple of Parmenides, q.v.).

P. 776: *Fragments* 10, 17; *The First Philosophers of Greece*, by A. Fairbanks, London, Kegan Paul, Trench, Trubner, 1898.

Menabadus (alchemical author cited in *The Golden Tract*, q.v.).

P. 630: *Hermetic Museum*, I, p. 15.

Mencius (Meng-tse; 372–289 B.C.; leading Confucian sage).

P. 43: *The Works of Mencius*, Bk. VI; tr. by Charles A. Wong, printed in China without date or name of publisher.

P. 79: *The Book of Mencius*, VI. i; in Lin Yutang: *The Wisdom of Confucius*, New York, Modern Library, 1938, p. 288.

P. 119: E. R. Hughes: *Chinese Philosophy in Classical Times*, London & New York, Everyman's Library, 1944, p. 110.

P. 343: *The Book of Mencius*, VI. i; op. cit., p. 289.

P. 542: Lin Yutang: *From Pagan to Christian*, London, Heinemann, 1960, p. 89.

P. 607: *The Book of Mencius*, VI. i; op. cit., pp. 286–7.

P. 863: *The Works of Mencius*, Bk. VII; op. cit.

Mevlidi Sherif, The ('The Birth-Song of the Prophet'; famous Turkish poem by Süleyman Chelebi, d. 1421 A.D., member of Halvati order of Dervishes; sung on religious occasions).

All selections are from tr. by F. Lyman MacCallum, London, John Murray, 1943.

P. 193:	p. 38–9	P. 592:	p. 38	P. 1009:	p. 17	P. 1040:	p. 17
516:	40	830:	37	1028:	17		

Milarepa (Jetsün-Milarepa; 1052?–1135; famous Tibetan saint, one of the great gurus of the Kargyütpa School of Northern Buddhism).

All selections, except where indicated, are taken from the *Biography* ed. by Evans-Wentz, as listed above under Marpa.

P. 108:	p. 259	P. 218:	p. 259	P. 600:	p. 245		**1939–1942,**
140:	177	242:	233–4	601:	273		**p. 164**
147:	263	250:	190	605:	Marco	P. 609:	p. 257
149:	186	277:	262		Pallis:	739:	245
154:	174	343:	271		*Peaks and*	876:	208–9
172:	273	353:	299		*Lamas,*	923:	16–17
200:	273	526:	278		London,	942:	212
217:	180	574:	245		Cassell,	952:	35

Milindapañha (post-canonical Pâli Theravadin work of Northern Buddhist origin, also considered orthodox by Southern Buddhists, composed about the beginning of our era).

P. 336: tr. Burlingame: JRAS, 1917; in Brown: *Walking on Water*, p. 8.

P. 764: in Coomaraswamy: *The Symbolism of Archery*, reprint vol. X of *Ars Islamica:* MCMXLIII, p. 110.

Mîrâ Bâî (1504?–?1550; princess and Hindu saint of Northern India, famed for her devotional poetry).

P. 208: cited in Coomaraswamy: *Hinduism and Buddhism*, pp. 20–21.

P. 806: in C. F. Andrews: *Mahatma Gandhi's Ideas*, London, Allen & Unwin, 1929, p. 310.

Mirror of Perfection, The (*Speculum Perfectionis*, a compilation of documents on the life of St Francis of Assisi, q.v., by Friar Leo, completed about 1318).

P. 734: tr. by Robert Steele, London & New York, Everyman's Library, 1910, 1944, in same volume with 'Little Flowers', and 'Life', by St Bonaventura.

Mohsan Fânî (d. 1670: of Persian origin; widely travelled Oriental scholar and Sufi; at court of Shah Jahan; author of *Dabistan*, an account of Oriental religions).

P. 922: *The Dabistan of Mohsan Fânî*, tr. by David Shea and Anthony Troyer, New York, Tudor Publishing Co., 1937 (first publ. 1843), p. 82.

Monkey (Chinese classic by Wu Ch'êng-ên; 1505?–?1580 A. D.; based on the pilgrimage to India of Tripitaka—the real-life Hsüan Tsang, 7th cent. A.D. Chinese Buddhist monk—whose travels formed the subject of many legends; Monkey in this tale incarnates the human ego).

All selections are from *Monkey*, tr. by Arthur Waley. New York, John Day, 1943.

P. 222: p. 282
261: 277
349: 211
P. 384: p. 121
528: 234–5
531: 274
P. 557: p. 27
586: 282
948: 26
P. 951: p. 44
993: 287

More, Dame Gertrude (1606–1633; English Benedictine nun under the spiritual direction of Father Augustine Baker, q.v.).

P. 190: in Aldous Huxley: *The Perennial Philosophy*, London, Chatto & Windus, 1950, pp. 114–115.

Morienus (or Marianos; 7th cent. Christian scholar and alchemist of Alexandria).

P. 630: cited in *The Glory of the World*, q.v.; *Hermetic Museum*, I, p. 210.

Morte d'Arthur, Le (English prose epic, abridged, compiled and translated from the body of French Arthurian romance, by Sir Thomas Malory, fl. 1470, and first printed by Caxton in 1485).

P. 625: London & New York, Everyman's Library, 1906, 1947.
P. 794: ibid.

Moses de Leon (Moses ben Shemtob de Leon; 13th cent. Spanish Kabbalist, introduced the *Zohar*, q.v., into Spain).

P. 691: *Sefer Ha-Rimmon*; in Gershom G. Scholem: *Major Trends in Jewish Mysticism*, New York, Schocken Books, 1954, 3rd edn., p. 223.

Mo Ti (Chinese Confucian philosopher, fl. 5th–4th cent. B.C.).

P. 968: *Mo Tzu Book*, XXVIII; Hughes: *Chinese Philosophy*, pp. 47–8.

Moyî, Ananda (contemporary Hindu woman saint).

All listings, except where indicated, are from *Aux Sources de la Joie*, tr. into French by Jean Herbert (from an English transl. by H. R. Joshi called *Sat-Bani*, Calcutta, Chuckervertty, Chatterjee), Paris & Neuchatel, Editions Ophrys, 1943.

P. 34: p. 23
56: 40
121: 16
144: 28–9
306: 53–4
332: 24
366: 79
430: 14–15
P. 509: p. 60
523: 74
548: 40
552: ***Matri Darshan***, **Eng. edn. p. 116**
600: p. 45–6
P. 699: p. 66
726: 49–50
789: 48
798: 20–21
849: 105
986: 28
1004: 101
1033: ***Matri Darshan***, **pp. 46–7**
P. 1037: p. 69–70

Muhammad (Arabic *Muḥammad*; 570-632 A.D.; the Prophet of Islam).

P. 38: Titus Burckhardt: ***Introduction aux Doctrines ésotériques de l'Islam***, Lyon, P. Derain, 1955, p. 78.
P. 49: (***ḥadîth qudsî*** means: of Divine inspiration, where it is God who speaks); ibid., p. 62.
P. 58: cited in Ibn ʿAṭâ'illâh, q.v.: *Ḥikam*, no. 107.
P. 59: cited in Hujwîrî, q.v.: *Kashf al-Maḥjûb* tr. by R. A. Nicholson, London, Luzac, 1911, 1936, 1959, p. 294.
P. 81: ibid., p. 197, n. 1.
P. 94: cited by Burckhardt, from Ibn ʿArabî, in *Études Trad.*, 1951, p. 20.
P. 171: *The Sayings of Muhammad*, compiled by Sir Abdullah Suhrawardy, London, John Murray, 1941, 1945, 1949, p. 49.
P. 171: ibid., p. 49.
P. 190: in Frithjof Schuon: *Perspectives spirituelles et Faits humains*, Paris, Cahiers du Sud, 1953, p. 267.
P. 206: cited by Shaykh Aḥmad al-ʿAlawî, q.v., in Martin Lings: *A Moslem Saint of the Twentieth Century*, London, Allen & Unwin, 1961, p. 160.
P. 209: *Sayings*, p. 68.
P. 212: from Ibn ʿArabî, cited by Burckhardt: *La Sagesse des Prophètes*, Paris, Albin Michel, 1955, p. 184.
P. 216: *Sayings*, p. 118.
P. 243: recounted by M. Abdul Azîz: *The Crescent in the Land of the Rising Sun*, London, 1941.
P. 260: cited by Sheikh Tâdilî: *La Vie traditionnelle c'est la Sincérité*; tr. by A. Broudier, Paris, Éditions Traditionnelles, 1960, pp. 11–12.
P. 269: in Schuon: *L'Oeil du Coeur*, Paris, Gallimard, 1950, p. 78.
P. 298: in Kalâbâdhî, q.v.: *The Doctrine of the Ṣûfîs*, ch. XLVII; tr. by A. J. Arberry, London, Cambridge University Press, 1935.

P. 316: Burckhardt: *Du Soufisme*, Lyon, P. Derain, 1951, p. 21.
P. 336: *Sayings*, p. 49.
P. 338: Abdul Azîz: op. cit., p. 123.
P. 339: *The Forty-Two Traditions of An-Nawawî*, no. 17; tr. by Eric F. F. Bishop, *The Moslem World*, Hartford, 1939.
P. 355: Burckhardt: *Introd. Doct. Islam*, p. 106.
P. 365: F. R. D. Tholuck: *Ssufismus sive Theologia Persica pantheistica*, p. 45.
P. 377: Al-Bukhârî: *Ṣaḥîḥ*.
P. 401: Muslim.
P. 423: *Forty-Two Traditions*, no. 28.
P. 425: cited by al-Ghazâlî; *Hibbert Journal*, London, October, 1943, p. 36.
P. 467: *Sayings*, p. 106.
P. 468: Tirmidhi: *Fitan*, 73; *Islamic Quarterly*, London, Oct. & Dec. 1954, p. 232.
P. 494: In Hujwîrî: *Kashf*, p. 305.
P. 495: in Ibn al-ʿArîf, q.v.; *Maḥâsin al-Majâlis*, tr. by Miguel Asin Palacios, Paris, Paul Geuthner, 1933, p. 60, n. 54.
P. 501: A. J. Arberry: *Sufism*, London, Allen & Unwin, 1950, p. 27.
P. 509: *Sayings*, no. 206.
P. 511: in Kalâbâdhî: op. cit., ch. XXVII.
P. 514: R. A. Nicholson: *The Mystics of Islam*, London, Bell, 1914, and London, Routledge & Kegan Paul, 1963, p. 53.
P. 515: *Sayings*, no. 357.
P. 537: *ḥadîth qudsî*; cited in Rûmî, q.v.: *Mathnawî*, tr. by R. A. Nicholson, London, Luzac, 1930, Bk. III, p. 91.
P. 579: source missing, but probably Bukhârî.
P. 591: in Kalâbâdhî: op. cit., ch. XI.

P. 625: see Abû Bakr Sirâj Ed-Dîn, op. cit. *infra*, p. 62.
P. 629: Washington Irving: *Life of Mahomet*, London & New York, Everyman's Library, 1911, 1944, p. 231.
P. 655: in Hujwîrî: *Kashf*, p. 267.
P. 664: cited by Burckhardt, in *Études Trad.*, 1954, p. 160.
P. 677: Burckhardt: *Introd. Doct. Islam*, p. 85.
P. 680: in Henry George Farmer: *A History of Arabian Music*, London, Luzac, 1929, p. 25.
P. 686: ibid., p. 25.
P. 688: from Imâm Abû Hanifah (b. 702 A.D.: founder of one of the four orthodox Sunnite rites).
P. 689: Burckhardt: *Introd. Doct. Islam*, p. 104, n. 1.
P. 699: *Sayings*, no. 340.
P. 701: *Forty-Two Traditions*, no. 16.
P. 735: Bukhârî; *A Manual of Ḥadith*, ed. by Maulana Muḥammad ʿAlî, Lahore, n.d., p. 39.
P. 755: Abdul Aziz: op. cit., p. 138.
P. 762: ibid., p. 123.
P. 763: cited by Schuon, in *Études Trad.*, 1953, p. 352.
P. 766: Abdul Azîz: op. cit., p. 123.
P. 766: Bukhârî; *Manual Ḥadith*, pp. 38–9.
P. 767: in Hujwîrî: *Kashf*, p. 11.
P. 791: See *Sayings*, no. 273.
P. 796: in Abû Bakr Sirâj Ed-Dîn: *The Book of Certainty*, London, Rider, 1952, p. 41.
P. 804: in Ibn ʿAṭâ'illâh: *Ḥikam*, no. 48.
P. 804: Muslim: *Daʿwât*, 16, and Tirmidhi: *Da'wât*, 19.
P. 806: Burckhardt: *De l'Homme universel*, Lyon, P. Derain, 1953, p. 39.
P. 812: *ḥadîth qudsî*; *Forty-Two Traditions*, no. 42.
P. 822: *ḥadîth qudsî*; Nicholson: *Mystics of Islam*, p. 68.
P. 839: E. H. Palmer: *Oriental Mysticism*, Cambridge, 1867, 1938, II. i. 24.

P. 36: 'Song', from 'Summer's Last Will and Testament', in *Poets of the English Language,* vol. II, ed. by W. H. Auden and Norman Holmes Pearson, New York, Viking Press, 1950, p. 77.

Nazari, Giovanni Battista (16th cent. Italian alchemical author).

P. 233: *Della trasmutatione metallica,* Brescia, 1589; in Kurt Seligmann: *The Mirror of Magic,* New York, Pantheon Books, 1948, p. 159.

Neumann, Teresa (d. 1962; German Catholic stigmatist, known for the drama of the Passion repeatedly visited upon her person).

P. 448: William C. McGrath: *Will Fatima Save America?,* Providence, R.I., Providence Visitor Press, 1950, p. 17 (concerning an apparition of Christ in 1936).

Nicephorus the Solitary (d. c. 1340; of Mount Athos; authority in the *Philokalia,* q.v.; teacher of St Gregory Palamas, q.v.).

P. 291: E. Kadloubovsky and G. E. H. Palmer: *Writings from the Philokalia on Prayer of the Heart,* London, Faber & Faber, 1951, p. 32.

Nicholas of Cusa (1401–1464; German-born Roman Catholic prelate and philosopher in the domain of Christian gnosis; cardinal and bishop of Brixen in the Tyrol).

All listings, except where indicated, are taken from *De Docta Ignorantia,* as tr. by Fr. Germain Heron: *Of Learned Ignorance,* London, Routledge & Kegan Paul, 1954.

P. 114: II. xiii
183: II. ii
310: I. xi
322: I. xxv
425: *Lettre II aux Bohémiens sur la double communion,* tr. by Maurice de Gandillac: *Oeuvres choisies,* Paris, Aubier, 1942, pp. 357–8
543: III. xii
557: III. xii
687: II. i
690: I. vii
696: III. xii
697: II. iii
723: I. xvii
745: I. i
750: I. iii
754: I. xxvi

P. 768: Dedication to *Doc. Ig.,* p. 4
781: II. ix
797: *De pace fidei:* Gandillac, op. cit., p. 425
799: I. vii
827: I. iv
834: I. xxvi
835: *De vis. Dei,* IX; Coomaraswamy: *Time and Eternity,* Ascona (Switzerland), Artibus Asiae, 1947, p. 121
839: idem, X; ibid., p. 122
840: II. iii
844: I. xxv
847: III. xii
927: III. iii
976: I. xvi
994: II. ii
996: I. xxiv

Niffarî (Muḥammad ibn ʿAbd al-Jabbâr al-Niffarî; d. c. 965; wandering dervish probably of ʿIrâqi origin, died in Egypt; recognized as a great Sufi teacher).

All citations are taken from the *Kitâb al-Mawâqif,* 'Book of Spiritual Stayings' (marked *Ma.*), and the *Kitâb al-Mukhâṭabât,* 'Book of Spiritual Addresses' (marked *Mu.*), tr. by A. J. Arberry, London, Luzac, 1935 (with slight revision based on the original Arabic).

P. 27:	*Mu.*	LV. 10	P. 221:	*Mu.*	XXIV. 27	P. 556:	*Mu.*	I. 1	P. 805:	*Ma.*	XXXVI. 26
49:	„	XIII. 11	259:	„	XIX. 25	695:	„	XII. 6	826:	„	II. 6
60:	„	VIII. 4	259:	„	I. 22	697:	„	XII. 15	845:	„	I. 2
79:	„	XIII. 9	342:	„	VIII. 2	727:	„	XII. 14	884:	„	XLVII. 24
115:	„	XV. 9	349:	*Ma.*	VIII. 1	740:	„	I. 3	892:	*Mu.*	I. 17
156:	„	XII. 16	494:	„	LXVIII. 5	740:	„	VIII. 1	908:	„	LVI. 7
176:	„	XXXI. 1	517:	*Mu.*	XVIII. 13	757:	„	V. 3	981:	„	XIV. 7
191:	„	XV. 13	545:	„	XXVI. 6	767:	„	IV. 4			

Nilakantha (16th cent. Hindu legislator).

P. 36: Coomaraswamy: 'A Pastoral Paradise', *Bulletin Museum of Fine Arts,* Boston, August, 1930.

Nilus, St (of Sinai; d. c. 450 A.D.; solitary contemplative, one of the early Fathers whose works are recorded in the *Philokalia,* q.v.).

P. 173: *153 Texts on Prayer,* no. 31; as paraphrased by St Alphonsus Liguori: *Conformity to the Will of*

God, tr. from the Italian by the Benedictine Convent of Perpetual Adoration, Clyde, Missouri, 7th edn., 1935, p. 9.

Nilus Sorsky, St (c. 1433–1508; Russian Hesychast, led semi-eremitic life, preached invocation based on respiration).

P. 1029: Bishop Ignatius Brianchaninov: *On the Prayer of Jesus,* tr. by Father Lazarus, London, John M. Watkins, 1952, p. 60.

Noah, The Book of (Old Testament Apocrypha).

P. 1032: *The Apocrypha and Pseudepigrapha of the Old Testament in English,* ed. by R. H. Charles, Oxford, Clarendon Press, 1913, Vol. II, p. 281.

Nô-gaku (='Nô theatre').

P. 673: Saying, cited by Beatrice Lane Suzuki: *Nôgaku, Japanese Nô Plays,* London, John Murray, 1932, p. 33.

Norton, Thomas (1437?—?1514; English alchemist).

All citations are from *The Ordinall of Alchimy,* being a facsimile reproduction from the *Theatrum Chemicum Britannicum* (1652) with annotations by Elias Ashmole, q.v., London, Edward Arnold, 1928 (two passages for the sake of clarity have been rendered into contemporary English, as given in *The Hermetic Museum,* London, John M. Watkins, 1953, and are marked *H.M.*).

P. 31:	p. 78	P. 293:	p. 32	P. 445:	p. 99	P. 586:	p. 34
129:	62	341:	29	501:	3	735:	16
173:	30	370:	47	503:	17	797:	***H.M.* II.**
198:	15	381:	31-2	553:	13		**pp. 11-12**
281:	14	408:	29	569:	41-2	836:	p. 54
292:	***H.M.* II.**	443:	7	577:	3	927:	85-6
	pp. 64-5	444:	100-101	577:	48	999:	88

Novum Lumen (alchemical text in the *Museum Hermeticum*).

P. 527: *Museum Hermeticum,* p. 574, as given by Maurice Aniane: 'Notes sur l'alchimie, "yoga" cosmologique de la chrétienté médiévale', in *Yoga, Science de l'Homme intégral,* ed. by Jacques Masui, Paris, Cahiers du Sud, 1953, p. 262.

O

Ohiyesa (Charles Alexander Eastman; 1858–1939; Sioux Indian; physician and author).

All citations are from Charles Alexander Eastman: *The Soul of the Indian,* Boston & New York, Houghton & Mifflin, 1911.

P. 47:	p. 13	P. 399:	p. 115	P. 611:	p. 121	P. 991:	p. 89–90
157:	99–100	526:	46	665:	46		
232:	149	549:	45	800:	24		

Old Chinese Poem (from a series known as the Nineteen Pieces of Old Poetry, some attributed to Mei Sheng, 1st cent. B.C.).

P. 105: *One Hundred and Seventy Chinese Poems,* tr. by Arthur Waley, London, Constable, 1918–1939, p. 41.

P. 194: ibid., p. 43.

P. 218: ibid., p. 46.

P. 426: ibid., p. 46.

P. 601: ibid., p. 46.

Omar Khayyâm (d. c. 1123; Persian poet, astronomer; master of metaphysical paradox).

P. 36: *Rubâiyât,* tr. by Edward Fitzgerald, First Edition, London, 1859, verse 46.

P. 39: in R. A. Nicholson: *Studies in Islamic Mysticism,* London, Cambridge University Press, 1921, p. 77.

P. 103:	*Rubâiyât,* 1st edn., verse 5	P. 592:	„	1st edn., verse 74
219:	„ 1st edn., verse 47	611:	„	5th edn., verse 39
330:	„ 1st edn., verse 51	637:	„	5th edn., verse 59
334:	„ 5th edn., 1889, verse 74	840:	„	1st edn., verse 37
419:	from the Bodleian Quatrain; Fitzgerald, op. cit., p. 89	1029:	„	5th edn., verse 59
554:	*Rubâiyât,* 1st edn., verse 70			

Only True Way, The (*or, An Useful, Good, and Helpful Tract, Pointing out the Path of Truth*; alchemical tract, published in 1677 in *The Hermetic Museum*).
P. 790: *The Hermetic Museum*, London, John M. Watkins, 1953, Vol. I, p. 153.

Open Entrance to the Closed Palace of the King, An (by 'An Anonymous Sage and Lover of Truth'; alchemical tract attributed to Philalethes, q.v.).
P. 748: *Hermetic Museum*, II, p. 163.

Origen (Adamantius; 185?–?254 A.D.; Greek, Church Father; famed theologian and prolific writer; head of catechetical school at Alexandria).
P. 118: *De Principiis*, II. 10. 4; in Stephen Hobhouse: *Selected Mystical Writings of William Law*, London, Rockliff, 1938–1949, p. 279.
P. 462: *Contra Celsum*; Ante-Nicene Christian Library, 1869, p. 403; in Nester H. Webster: *Secret Societies and Subversive Movements*, London, Britons Publishing Society, 1955, 7th edn., p. 216.
P. 542: cited by Frithjof Schuon, in *Études Traditionnelles*, Paris, Chacornac, 1953, p. 264.
P. 579: *Speculum B. V.*, lect. III, 1, 2; in St Louis de Montfort: *True Devotion to the Blessed Virgin Mary*, tr. by F. W. Faber, 1862, and publ. by the Montfort Fathers, Bay Shore, New York, 1946, p. 103.
P. 609: *De Principiis*, III. 1. 7; in Hobhouse, op. cit., p. 287.

Orpheus (traditional founder of the Greek Mysteries; perhaps title of Thraco-Phrygian priest-kings who were incarnations of Dionysus).
P. 594: *The Mystical Hymns of Orpheus*, tr. from the Greek of Onomacritus (c. 500 B.C.) by Thomas Taylor, London, Bertram Dobell and Reeves & Turner, 1896, p. 24 ff.
P. 673: ibid., p. 84.
P. 757: *Hymn to Mnemosyne*; ibid., p. 146.
P. 831: ibid., p. 63.

Ortha nan Gaidheal ('Songs of the Gaels'; Gaelic tradition).
Pp. 320, 803, 931: all supplied to the author by D. M. Matheson, Hermitage, Berkshire.

Ovid (Publius Ovidius Naso; 43 B.C.–17 A.D.; great Roman poet).
P. 39: *Metamorphoses*, Bk. I; tr. by Frank Justus Miller, London & Cambridge, Mass., Loeb Classical Library, 1916–1956, p. 9.
P. 464: idem, Bk. I; ibid., pp. 11, 13.

Oxyrhynchus Papyri (papyri found at Oxyrhynchus in Egypt in 1896–97, including classic Greek manuscripts and important early Christian documents).
P. 845: Evelyn Underhill: *Mysticism*, London, Methuen, 1911–1960, p. 101.

P

Padma-Karpo (i.e., 'White Lotus'; 17th cent. Tibetan guru in Kargyütpa line, Tantric authority, well known throughout Bhutan).
P. 531: *Epitome of the Great Symbol*, no. 87, tr. by Lâma Kazi Dawa-Samdup; W. Y. Evans-Wentz: *Tibetan Yoga and Secret Doctrines*, London, Oxford University Press, 1935, 1958.
P. 534: idem, no. 47; ibid.
P. 757: idem, no. 20 (citing the *Prajñâ-Pâramitâ*, q.v.); ibid.

Padma-Sambhava (i.e., 'Born of the Lotus'; fl. 8th cent. A.D.; the great teacher who introduced Tantric Buddhism from India into Tibet in 747 A.D., founded Nyingma School; hero of avataric dimensions in Tibetan tradition).
All citations are from *The Seeing of Reality*, tr. by Lâma Karma Sumdhon Paul and Lâma Lobzang Mingyur Dorje; in W. Y. Evans-Wentz: *The Tibetan Book of the Great Liberation*, London, Oxford University Press, 1954.

P. 406:	p. 237	P. 760:	p. 228	P. 864:	p. 238	P. 998:	p. 211
542:	217	805:	236	886:	**228-9**		
695:	230	811:	219	985:	229		

Padre Pio (Francesco Forgione; 1887–1968; Italian Capuchin at San Giovanni Rotondo; renowned as stigmatist, healer, confessor).
P. 953: Rev. Charles Mortimer Carty: *Padre Pio the Stigmatist*, London, Burns Oates & Washbourne, 1956, p. 55.

Palladius (368?–?431; Greek Christian ecclesiastic and writer, bishop of Helenopolis).

P. 753: Coomaraswamy: *Transf. Nature in Art*, p. 193, n. 30.

Phadampa Sangay (contemporary with Milarepa, q.v.; established Shibyepa School of Tibetan Tantrism).

P. 99: Evans-Wentz: *Tibetan Bk. of Great Liberation*, p. 244.
P. 119: ibid., p. 252.
P. 217: ibid., p. 244.
P. 628: ibid., p. 249.
P. 852: ibid., p. 247.

Phavorinus (Greek Sophist and skeptic philosopher, 2nd cent. A.D.).

P. 227: Stobaeus, p. 585; in Thomas Taylor: *Iamblichus' Life of Pythagoras*, London, John M. Watkins, 1926 (reprint. from edn. 1818), p. 191.

Philalethes (Eirenaeus; 17th cent. English hermetic writer).

All the citations are taken from *The Hermetic Museum*, Vol. II, and are composed of four tracts: 'A Short Vade Mecum to the Celestial Ruby' (marked *R.*), 'The Metamorphosis of Metals' (marked *M.*), 'An Open Entrance to the Closed Palace of the King' (marked *E.*), and 'The Fount of Chemical Truth' (marked *F.*).

P. 46:	*R.*	p. 247	P. 357:	*E.*	p. 170–171	P. 571:	*E.*	p. 166–7	P. 801:	*M.*	p. 233–4
113:	*M.*	237	379:	*R.*	254	579:	*M.*	239	836:	*R.*	254
196:	*R.*	253	380:	*M.*	241	591:	*E.*	167	851:	*E.*	197–8
201:	*E.*	189	380:	*R.*	257–8	591:	*F.*	267	870:	*R.*	250
209:	*E.*	172	385:	*E.*	188	592:	*F.*	268–9	881:	*R.*	260
209:	*R.*	255	407:	*E.*	192	593:	*E.*	193	903:	*M.*	238
211:	*M.*	240, 242	426:	*R.*	256	630:	*F.*	263	908:	*F.*	267
214:	*E.*	177	507:	*M.*	229	657:	*F.*	263–4	929:	*F.*	265
215:	*R.*	250	526:	*M.*	232	658:	*E.*	168	952:	*E.*	192
282:	*M.*	245	544:	*E.*	189–90	669:	*E.*	189	998:	*E.*	195
284:	*R.*	256	545:	*M.*	241	678:	*M.*	232–3	1015:	*E.*	163
284:	*E.*	177	565:	*E.*	166	696:	*R.*	258	1024:	*E.*	167
296:	*E.*	174	568:	*R.*	262	783:	*R.*	249			
298:	*E.*	169	570:	*M.*	243	785:	*R.*	254			
308:	*E.*	183	571:	*E.*	165	785:	*F.*	263			

Philo (Judaeus; fl. late 1st cent. B.C. and early 1st cent. A.D.; famous Hellenistic-Jewish philosopher of Alexandria, 'the Jewish Plato', sought to harmonize Greek and Platonic philosophy with the Pentateuch).

P. 32: *Legum allegorium*, I. 20; in Ananda K. Coomaraswamy: *Time and Eternity*, Ascona (Switzerland), Artibus Asiae, 1947, p. 121.
P. 48: *Quis rerum divinarum heres*, 115; in Coomaraswamy: *Am I My Brother's Keeper?*, New York, John Day, 1947, p. 93.
P. 50: *On the Cherubim*; *Philo*, selections ed. by Hans Lewy (from the Loeb Classical Library tr. of F. H. Colson), Oxford, Phaidon Press, 1946, p. 69.
P. 72: *On the Migration of Abraham*; Lewy, op. cit., p. 72.
P. 168: *The Unchangeableness of God*; Lewy, pp. 29–30.
P. 194: *On the Sacrifice of Abel*; ibid., p. 87.
P. 197: *Who Is the Heir?*; ibid., p. 87.
P. 286: *Migr. Abraham*; ibid., pp. 40, 41.
P. 365: *On the Confusion of the Tongues*; ibid., p. 37.
P. 480: in Coomaraswamy: 'On Being in One's Right Mind', *The Review of Religion*, Columbia University Press, November, 1942, p. 37.
P. 482: idem.
P. 482: idem.
P. 485: *Unchangeableness God*; Lewy, p. 33.
P. 487: *Who Heir?*; ibid., p. 75.
P. 504: idem; ibid., p. 89.
P. 508: *On the Giants*; ibid., p. 31.
P. 517: *On Abraham*; ibid., p. 90.
P. 535: *Allegorical Interpretation*; ibid., p. 71.
P. 640: *On the Creation of the World*; ibid., pp. 54–5.

Plotinus (205?–270 A.D.; Roman, b. in Egypt, studied at Alexandria under Ammonius Saccas; the great exponent of Neoplatonism).

All selections, except where indicated, are from *The Enneads* as tr. by Stephen MacKenna (marked 'M'), London, Faber & Faber, 1956 (1st publ. 1917–1930), and from the compilation based on MacKenna by Grace H. Turnbull (marked 'T'): *The Essence of Plotinus*, New York, Oxford University Press, 1934.

P. 27: VI. viii. 16 (M)
63: II. iii. 18 (M)
68: I. vi. 5 (M)
113: VI. ix. 9 (Coomaraswamy)
115: V. v. 11, 12 (T)
151: VI. viii. 21 (T)
169: III. iv. 3, 6 (T)
185: VI. viii. 7, 12 (M)
186: VI. viii. 6 (M)
225: Porphyry's *Life*, in M., p. 2
226: I. vi. 6 (M)
278: IV. iv. 40, 41 (M & T)
306: VI. vii. 36 (M)
309: VI. ix. 11 (M)
424: IV. iv. 44 (T)
507: V. v. 11 (M)
523: V. v. 8 (T)
616: VI. vii. 22 (T)
623: III. v. 10 (T)

P. 623: VI. viii. 15 (M)
665: V. viii. 9, 13 (T)
667: V. v. 13 (T)
668: VI. vii. 35 (T)
668: V. viii. 2 (M)
672: V. viii. 1 (M)
675: I. vi. 2 (T)
676: I. iii. 1, 2 (M & T)
679: I. vi. 3 (T)
686: I. vi. 1 (M)
706: V. i. 2 (T)
722: V. v. 10 (M)
726: VI. viii. 16 (M)
727: V. ii. 1 (M)
750: I. vi. 9 (T)
752: V. v. 12 (M)
755: IV. iv. 7 (Coomaraswamy)
763: IV. vii. 10 (in John Smith, q.v.: *Discourses*, p. 111)

P. 765: V. viii. 4, 5 (T)
776: IV. ix. 5 (T)
780: V. ii. 1 (M)
783: V. viii. 4 (T)
784: V. v. 9 (T)
807: V. v. 9 (T)
817: I. vi. 8 (T)
821: V. v. 9 (T)
851: V. i. 4 (M)
859: IV. iii. 1 (T)
860: I. vi. 9 (M)
861: VI. vii. 35, 36 (T)
865: in Smith: *Disc.*, p. 133
884: IV. vii. 10; ibid., pp. 112–113
888: IV. iv. 2 (T)
900: I. ii. 6 (M)
905: VI. vii. 34 (T)
915: I. iv. 8 (M)
990: V. v. 6 (T)
992: VI. ix. 10 (M)

Plutarch (46?–?120 A.D.; Greek biographer).

P. 107: cited by Coomaraswamy, in *The Conception of Immortality in Buddhism*, a lecture given at the Brooklyn Academy of Music, February 28, 1946.

P. 652: *Moralia*, 355 C; in Coomaraswamy: *Figures of Speech*, p. 198.

Po Chü-i (772–846 A.D.; Chinese poet of T'ang dynasty, government official, widely celebrated for his lyrics, fame spread to Shintôist Japan).

All selections are from *One Hundred and Seventy Chinese Poems*, tr. by Arthur Waley, London, Constable, 1918–1939.

P. 172: p. 159
282: 133
P. 640: p. 159
641: 144
P. 682: p. 125–6
992: 159

Polynesian Tradition

P. 26: Jean-Marie Loursin: *Tahiti*, Paris, Éditions du Seuil, 1957, p. 42.

Porphyry (232?–?304 A.D.; Greek scholar and Neoplatonic philosopher, b. in Syria, disciple of Plotinus).

P. 251: *Auxiliaries to the Perception of Intelligibles*; Thomas Taylor: *The Mystical Hymns of Orpheus*, London, Bertram Dobell, 1896, p. 162, n. 117.

P. 884: *Auxiliaries*; E. J. Langford Garstin: *Theurgy or the Hermetic Practice, A Treatise on Spiritual Alchemy*, London, Rider, 1930, p. 78.

Prajñâ-Pâramitâ **(Sanskrit Buddhist sûtras, basic Mahâyâna canon on *Śûnyatâ* or Voidness).**

P. 96: *The Diamond Sutra* (*Vajracchedikâ*; Chinese *Chin-Kang-Ching*) *or Prajñâ-Pâramitâ Sûtra*, no. 9, tr. from the Chinese by William Gemmell; London, Kegan Paul, Trench, Trubner, 1912.

P. 285: in W. Y. Evans-Wentz: *Tibet's Great Yogî Milarepa*, London, Oxford University Press, 1928, p. 1.

P. 420: Gemmell, op. cit.

P. 882: *Diamond Sutra*; Dwight Goddard: *A Buddhist Bible*, Thetford, Vermont, 2nd edn., 1938, p. 105.

P. 918: Gemmell, op. cit.

P. 931: Harper Havelock Coates and Ryugaku Ishizuka: *Hônen the Buddhist Saint*, Kyoto, 1949, p. 89.

P. 987: Gemmell, op. cit.

Praśna Upanishad (late Vedic scripture).

P. 716: *The Thirteen Principal Upanishads*, tr. by Robert Ernest Hume, London, Oxford University Press,

Q

P. 105: *Nazhat al majâlis*, p. 69; in Émile Dermenghem: *Vies des Saints musulmans*, Algiers, Baconnier, n.d., p. 335.

Sâhitya Darpaṇa (15th cent. Indian treatise by Viśvanâtha Kavirâja on rhetoric, poetry, and dramatics).
P. 663: Coomaraswamy: *Transf. Nature in Art*, p. 49.
P. 672: ibid., p. 46.

Sahl al-Tustarî (Abû Muḥammad; 818–896; celebrated theologian and Sufi of Tustar in the Ahwâz).
P. 835: Titus Burckhardt: *Introduction aux Doctrines ésotériques de l'Islam*, Lyon, P. Derain, 1955, p. 32.

Sai Baba (1856–1918; contemporary Indian saint much venerated by both Hindus and Muslims, noted for miraculous life).
P. 741: Arthur Osborne: *The Incredible Sai Baba*, London, Rider, 1958 (Calcutta, Orient Longmans, 1957), p. 21.
P. 883: ibid., pp. 26–7.

Saichi (fl. 1800: Japanese Jôdo-Shinshû Buddhist cobbler and unlettered poet, called by Suzuki 'one of the deepest Shin followers').
All citations are from D. T. Suzuki: *Mysticism Christian and Buddhist*, London, Allen & Unwin, 1957.

P. 29:	p. 157	P. 953:	p. 162	P. 1011:	p. 186	P. 1027:	p. 199
334:	194	1008:	193	1012:	213	1027:	191
360:	164	1008:	199	1018:	210	1031:	199

Saint-Martin, Louis-Claude de (1743–1803; French mystic philosopher, Catholic, school of Boehme, q.v.).
P. 86: from *L'Homme de Désir*, p. 16, cited in his *Ecce Homo*; Lyon, Paul Derain, 1959 (reprint. from edn. 1792), p. 7.
P. 796: Evelyn Underhill: *Mysticism*, London, Methuen, 1911–1960, p. 80.

Sallustius (4th cent. A.D.; Praetorian praefect under Emperor Julian).
P. 241: *Treatise on the Gods and the World*, Ch. XVIII; Thomas Taylor: *Iamblichus on the Mysteries*, London, Stuart & Watkins, 1821–1968, pp. xx–xxi; also Joseph Head and S. L. Cranston: *Reincarnation, an East-West Anthology*, New York, Julian Press, 1961, p. 93.

Samaveda (early Vedic scripture arranged for chanting).
P. 578: V. Raghavan: *The Indian Heritage*, London & Bangalore, Indian Institute of Culture, 1956, p. 45.

Samuel ben Kalonymus (fl. 12th cent.; one of the founders of German Hasidism).
P. 418: Gershom G. Scholem: *Major Trends in Jewish Mysticism*, New York, Schocken Books, 1954, 3rd edn., p. 115.

Saṁyutta-nikâya (Buddhist Pâli canon, part of the *Sutta-Piṭaka*).
P. 275: Coomaraswamy: *Hinduism and Buddhism*, New York, Philosophical Library, 1943, pp. 45–6.
P. 387: *Buddhist Texts Through the Ages*, ed. by Edward Conze; Oxford, Bruno Cassirer, 1954, p. 44.
P. 468: ibid., p. 45.
P. 653: Coomaraswamy: op. cit., p. 83.
P. 654: ibid., p. 69.
P. 655: ibid., p. 83.
P. 718: Conze: op. cit., p. 93.
P. 721: ibid., p. 92.
P. 784: Coomaraswamy: 'The Symbolism of the Dome', *Indian Hist. Quart.*, XIV, 1938, p. 55.
P. 881: D. T. Suzuki: *Essays in Zen Buddhism* (First Series), London, Rider, 1949, 1958, p. 71.
P. 918: Conze: op. cit., no. 108.
P. 947: Coomaraswamy: *Hinduism and Buddhism*, p. 69.

Sanâtan (disciple of Chaitanya, q.v.).
P. 36: Sir Jadunath Sarkar: *Chaitanya's Life and Teachings*, Calcutta, 1932, p. 297.

Saṅgîtă Makarandă; *Nărandă Purâṇă*; *Śhivă tattvă Ratnâkară* (17th and 18th cent. Sanskrit texts on music).
P. 687: Alain Daniélou: *Northern Indian Music*. London, Christopher Johnson, 1949, vol. I, p. 37.

Śankarâchârya, Śrî (fl. 800 A.D.; famous Hindu metaphysician, leading exponent of *Advaita-Vedânta*, the doctrine of non-duality).
P. 114: *Upadeshasâhasrî*, II. xvi. 43; *A Thousand Teachings*, tr. by Swâmi Jagadânanda; Mylapore, Madras, Sri Ramakrishna Math, 1949, 2nd edn.
P. 143: ibid., II. xvii. 24.
P. 159: ibid., I. 116.

P. 186: In Chandrasekhara Swâmigal: *The Call of the Jagadguru,* ed. by R. Krishnaswami Aiyer, Madras, Ganesh, 1957, pp. 116–117.
P. 221: *Upadesh.*, II. xiv. 29.
P. 261: cited by Swâmi Brahmânanda: *Discipline monastique*, tr. by Odette de Saussure and Jean Herbert, Paris & Neuchatel, 1945, 2nd edn., vol. I, p. 72.
P. 292: *Upadesh.*, II. xvii. 89.
P. 330: ibid., II. i. 3, 4.
P. 334: ibid., II. xvii. 50.
P. 389: ibid., II. xiv. 40.
P. 390: *Âtmâ-Bodha*; in René Guénon: *Man and His Becoming According to the Vedânta,* tr. by Richard C. Nicholson; London, Luzac, 1945, ch. XXIV.
P. 430: *Upadesh.*, II. xvi. 44.
P. 484: ibid., II. xvi. 72.
P. 508: ibid., II. xviii. 3.
P. 742: ibid., II. xviii. 121.
P. 751: ibid., II. xv. 42.
P. 754: ibid., I. 93.
P. 767: ibid., II. xvi. 71.
P. 769: *Âtmâ-Bodha*; op. cit.
P. 769: *Upadesh.*, II. xix. 27.
P. 777: *Vivekachudamani,* no. 388; tr. by Swami Madhavananda, Advaita Ashrama, Mayavati, Almora, 1932, p. 170.
P. 809: *Anâtmâ-shrîvigarhanam* ('Holy Disdain for the non-Self'); tr. by René Allar, in *Études Traditionnelles,* Paris, Chacornac, 1948, p. 205.
P. 810: *Upadesh.*, II. xix. 8.
P. 816: *Âtmâ-Bodha,* op. cit.
P. 867: *Upadesh.*, II. iv. 5.
P. 882: *Âtmâ-Bodha,* op. cit.
P. 884: *Upadesh.*, II. xvi. 39–41.
P. 885: Commentary on *Brahma-Sûtras,* 4. 4. 1, 2; tr. by Swami Vireswarananda, Advaita Ashrama, Mayavati, Almora, 1948.
P. 889: *Upadesh.*, II. xv. 1.
P. 901: ibid., II. xiv. 27.
P. 908: ibid., II. xvii. 69.
P. 932: *Âtmâ-Bodha*; op. cit.
P. 963: *Upadesh.*, II. xviii. 5.
P. 981: ibid., II. ix. 7.
P. 982: ibid., II. xii. 9.
P. 1028: ibid., I. 32.
P. 1036: in Frithjof Schuon: *Language of the Self,* Madras, Ganesh, 1959, p. 71.
P. 1038: *Upadesh.*, II. xvi. 70.

Śâṅkhâyana Âraṇyaka (part of the Vedic 'forest' scripture).
P. 953: Coomaraswamy: *Hinduism and Buddhism*, p. 83, n. 269.

Śânti-deva (7th cent. follower of Nâgârjuna's, q.v., Mâdhyamika school of Mahâyâna Buddhism, and a guru in the successión of the Nâlandâ Buddhist university and monastery in Bihâr, India).
All selections are from *The Path of Light,* being the *Bodhi-charyâvatâra,* tr. from the Sanskrit by L. D. Barnett, and to which is appended the *Śikshâsamuchchaya* (marked Ś); London, John Murray, 1909, 1947.

P. 72:	p. 54, 55	P. 137:	p. 68, 84	P. 263:	p. 78	P. 555:	p. 46–7
72:	106 (Ś)	206:	85	264:	84	599:	88
100:	86	210:	90	358:	105 (Ś)	604:	69
110:	39	223:	57	401:	50–51	606:	57–8
111:	90	241:	105 (Ś)	409:	55	610:	45–6
121:	69	249:	42–3	509:	103 (Ś)	701:	82
123:	66, 75	260:	49	532:	82	769:	52
						935:	55

Santos, Lucy dos (Carmelite nun, born in Portugal in 1907, who as a child witnessed, with two cousins, the apparitions of the Blessed Virgin at Fatima, Portugal, in 1917).
P. 72: in convent chapel at Tuy, Spain, 1925; recounted in *Our Lady of Fatima's Peace Plan from Heaven, The Grail,* St Meinrad, Indiana, 1950, p. 16.

Sapta-śatikâ-prajñâ-pâramitâ Sûtra (one of the earliest Mahâyâna Buddhist 'Perfect Wisdom' sûtras).
P. 1019: D. T. Suzuki: *Essays in Zen Buddhism (Second Series),* London, Rider, 1950, 1958, p. 154.

Sâradâ Devî, Śrî (1853–1920; wife of Sri Ramakrishna, q.v.; known as the Holy Mother).
P. 177: *Women Saints of East and West,* ed. by Ramakrishna Vedanta Centre, London, 1955, p. 120.
P. 519: ibid., p. 121.
P. 1012: ibid., p. 119.

Saraha (Buddhist saint and sage, fl. India early Tantric period; reputed teacher of Nâgârjuna, q.v.; influenced Tibetan doctrines).
All selections are from 'Saraha's Treasury of Songs' (*Dohâkosha*), as given in Conze: *Buddhist Texts,* no. 188.

P. 87:	v. 54	P. 599:	v. 105	P. 753:	v. 52	P. 916:	v. 64
142:	104	607:	concl.	799:	79	978:	96
389:	48	722:	v. 75	805:	90	983:	20
532:	77	740:	56	833:	87	993:	88
578:	57	741:	51	864:	62		

Al-Sarrâj (Abû Naṣr; d. 988; b. at Ṭûs in Khorasan; wrote the *Kitâb al-Luma'*, earliest surviving systematic treatise on Sufism).
P. 191: *Kitâb al-Luma'*; in *Râbi'a the Mystic and Her Fellow-Saints in Islâm,* by Margaret Smith, London, Cambridge University Press, 1928, p. 55.
P. 975: idem; in *Readings from the Mystics of Islâm,* tr. by Margaret Smith, London, Luzac, 1950, no. 43.

Śatapatha-Brâhmaṇa (Hindu Vedic canonical commentary, attributed to Yâjñavalkya).
P. 323: tr. by Julius Eggeling, *The Sacred Books of the East,* Vol. XLIII, ed. by F. Max Müller, Oxford, Clarendon Press, 1897.
P. 590: ibid.
P. 631: ibid.
P. 770: ibid.
P. 831: ibid.
P. 1003: in Alain Daniélou: *Introduction to the Study of Musical Scales,* London, The India Society, 1943, p. 97 (apparently wrongly attributed to *Manusmṛti,* I. 21; cf. Wilkins: *Hindu Mythology,* p. 344).

Śatasâhasrikâ prajñâramitâ (Mahâyâna Buddhist sutra).
P. 914: Conze: *Buddhist Texts,* no. 140.

Scala Claustralium (treatise on contemplation by Guigo II, prior of the Grande Chartreuse, end of 12th cent.).
P. 315: cited in Dom Justin McCann's edn. of *The Cloud of Unknowing,* London, Burns Oates & Washbourne, 1924–1943, p. 74, footnote.

Scholastic Formula
P. 775: traditional.

Seami Motokiyo (1373–1455; Japanese Nô master).
P. 94: *Ebira*; in Beatrice Lane Suzuki: *Nôgaku, Japanese Nô Plays,* London, John Murray, 1932, p. 59.
P. 254: idem; ibid., p. 64.
P. 672: in Waley: *Nô Plays of Japan.*

Secret Papers in the Jade Box of Ch'ing-hua, The (Chinese alchemical text ascribed to Chang Po-tuan, q.v.).
P. 153: tr. by Tenney L. Davis and Chao Yün-ts'ung; *American Academy of Arts and Sciences,* vol. 73, no. 13, Boston, July, 1940, p. 388.

Sefer Yetsirah ('Book of Creation'; Jewish esoterism; basic Kabbalistic text, along with *Zohar,* q.v.; probably written between 3rd and 6th cent. A.D., with teachings said to go back to Abraham).
P. 994: in René Guénon: *Le Symbolisme de la Croix,* Paris, Éditions Véga, 1931, p. 45.

Seigwanji (Japanese Nô play stressing Nembutsu practice).
P. 1023: B. L. Suzuki: *Nôgaku,* pp. 40–41.

Sendivogius, Michael (d. 1646: Moravian nobleman and alchemist; associated with Alexander Seton the

Cosmopolite, a notable Scottish alchemist).

All listings are from 'The New Chemical Light', as given in *The Hermetic Museum*, Vol. II, London, John M. Watkins, 1953 (reprint. from edn. 1893).

P. 28:	p. 95–6	P. 501:	p. 147	P. 578:	p. 111	P. 765:	p. 128
32:	139	504:	128–9	586:	138	785:	148
71:	141	507:	94	588:	135–6	829:	151–2
150:	105	544:	147	672:	138	836:	144
184:	142	546:	146	690:	136	843:	122
194:	147	567:	106	735:	119	916:	111
281:	95	570:	115	742:	110	936:	82
314:	147–8	572:	124–5	749:	129	964:	104
383:	147	576:	129	763:	83	1026:	106–7

Seneca (Lucius Annaeus; 4 B.C.?–65 A.D.; famous Roman statesman and Stoic philosopher, b. Córdoba, Spain).

P. 127: *De Moribus*; cited in 'John Frederick Helvetius' Golden Calf', *The Hermetic Museum*, II, p. 275.

P. 138: *Epist.* 77; ibid., p. 271.

Seng-ts'an (c. 600 A.D.; 3rd Patriarch in Dhyana school of Chinese Buddhism).

P. 131: from the *Taisho Issaikyo* (Chinese *Tipiṭaka*, q.v.); Conze: *Buddhist Texts*, p. 295.

P. 138: idem; ibid., p. 295.

P. 300: ibid., p. 298.

P. 542: ibid., p. 296.

P. 780: ibid., p. 298.

P. 783: *Shinjin-no-Mei*, 30; Suzuki: *Manuel Zen*, p. 82.

P. 826: Conze: op. cit., p. 298.

P. 883: ibid., p. 296.

P. 994: ibid., p. 296.

Seû-mà Tshyên (Ssŭ-ma Ch'ien; 145-?87 B.C., first great historian of China).

P. 681: cited by Maurice Courant in the entry 'Chine et Corée' in the *Encyclopédie de la Musique et Dictionnaire du Conservatoire*, Paris, 1924 (Daniélou: *Musical Scales*, p. 53).

Sextus the Pythagorean (Greek; no information available beyond that given in a footnote by Thomas Taylor: 'This Sextus is probably the same that Seneca so greatly extols, and from whom he derives many of those admirable sentences with which his works abound').

All citations are from 'Select Sentences of Sextus the Pythagorean', in Thomas Taylor's tr. of *Iamblichus' Life of Pythagoras*, London, John M. Watkins, 1926 (reprint. from edn. 1818), pp. 192–198.

Pp. 93, 110, 124, 157, 158, 176, 194, 251, 342, 476, 477, 492, 494, 521, 547, 600, 625, 679, 704, 754, 762, 768, 830, 865, 885, 909, 917, 922, 1013.

Shaare Tsedek ('Gates of Justice', a book written in Palestine in 1295 by an anonymous disciple of the Spanish Kabbalist, Abraham Abulafia, q.v.).

P. 754: Gershom G. Scholem: *Major Trends in Jewish Mysticism*, New York, Schocken Books, 1954, 3rd edn., pp. 152–3.

P. 876: ibid., pp. 151–2.

Shabistarî (Sa'd al-Dîn Maḥmûd; d. 1320: one of the greatest Persian Sufi poets).

All selections are from the *Gulshan-i-Râz* ('Mystic Rose Garden'), and, except where indicated, are taken from the tr. by Florence Lederer: *The Secret Rose Garden of Sa'd Ud Din Mahmud Shabistari*, Lahore, Ashraf Publication, n.d.

P. 28:	p. 44	P. 195:	p. 45	P. 554:	p. 31	P. 817:	p. 58
36:	47	221:	63	639:	54	825:	81
73:	70	225:	46	640:	55	825:	50–51
87:	45	309:	73	687:	53	826:	v. 146 (tr.
90:	82	366:	42	717:	56		E.H.
96:	46	390:	74–5	737:	42		Whinfield,
140:	74	476:	35	753:	29		London,
158:	42	490:	58	780:	47		1880)
186:	61	523:	75	797:	79	830:	p. 50

P. 834: ibid., p. 134.
P. 952: ibid., p. 131.
P. 995: ibid., p. 132.
P. 1028: in *The Pilgrim Continues His Way,* tr. by R. M. French, London, Society for Promoting Christian Knowledge, 1930, 1941, p. 72.

Simplicius (Greek Neoplatonic philosopher of early 6th cent. A.D.; native of Cilicia; disciple of Ammonius and Damascius, summarized the works of his predecessors).
P. 683: Commentary on the second book of Aristotle's Treatise on the Heavens; Thomas Taylor: *The Mystical Hymns of Orpheus,* London, Bertram Dobell, 1896, p. 80.
P. 687: idem; in Thomas Taylor's tr. of *Iamblichus' Life of Pythagoras,* p. 33.
P. 997: *In Epictet*; Taylor: *Hymns Orpheus,* pp. xxvi–xxvii.

Sioux Indian
P. 998: spoken to William Tomkins, author of *Universal American Indian Sign Language,* San Diego, 1926.
P. 1039: private source.

Sir Gawayne and the Grene Knight (c. 1375; one of the great poems of mediaeval England).
P. 324: *The Age of Chaucer,* ed. by Boris Ford, Harmondsworth, Middlesex, Penguin Books, 1955.
P. 376: ibid.

Sivananda, Swami (Sarasvati; 1887–1963; contemporary Hindu, exponent of *Japa* and authority on meditation techniques).
The majority of citations are taken from (1) *La Pratique de la Méditation,* marked (*P*), tr. by Charles Andrieu and Jean Herbert (from *Concentration and Meditation,* 1945, which was not obtainable), Paris, Albin Michel, 1950; (2) *Japa Yoga,* marked (*J*), Rishikesh, India, The Yoga-Vedanta Forest University, 1952; other sources are acknowledged in their proper places.

P. 82: (*P*) pp. 196, 197
88: „ p. 245
90: *Yogic Home Exercises,* Bombay, D. B. Taraporevala Sons, n.d. p. 98
91: ibid., p. 41
93: (*P*) p. 25
98: „ 24
99: „ 244
127: „ 344
133: (*J*) 111
190: (*P*) 342
192: „ 203
222: „ 155
243: „ 69
244: (*J*) 164
245: (*P*) 295
281: (*J*) 107–9
286: *The Vision,* Kanhangad, Nov. 1954, p. 53
287: *Yogic Home Ex.* p. 98
290: (*P*) p. 77
292: (*J*) 61
296: *Yogic Home Ex.* p. 97
297: (*P*) p. 148
299: (*J*) lii-liii
339: *Yog. Home Ex.* p. 101
344: (*J*) p. 62
349: (*P*) 291

P. 357: (*P*) p. 125
368: „ 303
370: „ 354–5
382: „ 305–6
385: „ 246
386: „ 225
397: „ 379
400: „ 76
407: „ 266
409: *Yog. Home Ex.* p. 99
410: (*P*) p. 246
410: „ 224
411: (*J*) 141
422: „ xlix
429: „ li
486: (*P*) 19
506: (*J*) 134
508: „ 131–2
510: (*P*) 300
526: „ 122–3
529: „ 54, 65
531: „ 49
531: „ 226
541: „ 139
545: „ 259
546: „ 156
549: „ 54
620: „ 286
645: „ 61
657: (*P*) 306
703: „ 191

P. 769: (*P*) p. 294
795: „ 332
849: „ 126
868: „ 143
877: „ 394
881: „ 86
903: „ 145–6
978: „ 242
993: „ 360
1004: (*J*) 5
1008: (*P*) 142
1011: (*J*) 140
1012: „ 141
1012: „ 27
1013: „ 99
1013: „ xlix
1015: „ 142
1027: „ 88
1028: „ 138
1029: „ lviii
1030: „ 108
1032: „ 4
1037: „ 104
1039: „ 139
1040: (*P*) 250
1041: „ 244

Sixtus of Siena (fl. 1570: a converted Jew and Dominican, protected by Pius V).

P. 458: P. L. B. Drach: *De l'Harmonie entre l'Église et la Synagogue,* 1844, vol. II, p. xix.

Smith the Platonist, John (1618–1652; one of the foremost Cambridge Platonists).

All passages, except where indicated, are taken from *Select Discourses,* London, 1821.

P. 42:	p. 134	P. 211:	p. 419	P. 433:	p. 498	P. 752:	p. 5
45:	16	227:	431	462:	519	756:	89–90
57:	357, 358	256:	505	477:	426	758:	104
71:	110	287:	344	480:	497	780:	138
72:	502	298:	487	507:	109–110	817:	19
73:	501	308:	406	519:	483	849:	450–451
77:	18	311:	463–4	530:	427–8	851:	140
82:	513, 514	312:	409	537:	5	851:	444
110:	453	316:	in Hob-	543:	15	865:	410–411
112:	146		house:	573:	23–5	871:	461, 476
122:	158		*Law,*	575:	5	906:	415
124:	145		q.v.,	586:	465 ff.	908:	106–7
137:	12–13		p. 329	620:	440	949:	414
148:	382	346:	435	641:	300	960:	183–4
168:	161	353:	439	675:	24	960:	298–9
171:	12	382:	510–11	697:	475	963:	211
182:	466	395:	515, 516	698:	443	968:	416–17
185:	470	421:	47	734:	10	978:	141–2
186:	159–160	422:	54, 57	748:	3, 4, 5		
207:	380	433:	7	751:	176		

Sophic Hydrolith, The (or, *Water Stone of the Wise*; anonymous European alchemical text forming part of *The Hermetic Museum*; see under *All-Wise Doorkeeper*).

All citations are from *The Hermetic Museum,* Vol. I.

P. 67:	p. 103	P. 427:	p. 84	P. 735:	p. 100–101	P. 803:	p. 111
120:	114	451:	96	746:	117	809:	98
125:	110	484:	116	749:	99	817:	95–6
214:	110–11	502:	87	756:	95–6	849:	89
241:	79–80	555:	74–6	758:	93	860:	93–4
310:	88	564:	85	762:	119	872:	76
369:	111	568:	78	784:	75–6	891:	113
409:	107	571:	81	787:	77–8	1026:	76
411:	108	652:	78	795:	94	1037:	109
427:	75	702:	73	801:	71		

Sophocles (496?–406 B.C.; Greek tragic playwright).

P. 367: Dudley Wright: *The Eleusinian Mysteries and Rites,* London, Theosophical Publishing House, n.d., p. 108.

Soshun (1262–1336; Japanese Zen priest).

P. 982: *Hônen,* p. 70.

Spanish Proverb

P. 431: traditional.

Sparrow, John (1615?–?1665; English Hermetist).

P. 314: in the Preface to his translation of Boehme's *Signatura Rerum,* Everyman's, p. 5 (cf. reference under Boehme, and note 5, p. 21).

P. 435: ibid., pp. 6–7.

P. 459: in his Postscript to op. cit., p. 221.

P. 988: Preface, op. cit., p. 5.

Spenser, Edmund (1552?–1599; renowned Elizabethan poet).

P. 465: *Faerie Queene,* Prologue Bk. V; London, Oxford University Press, 1916–1952, from 1909 Clarendon Press text ed. by J. C. Smith, based on the London edn. of 1596.

Srimad Bhagavatam (see *Bhagavatam*).

Srong-Tsan-Gampo (d. A.D. 650; first Buddhist ruler of Tibet; considered an incarnation of Avalokiteśvara, the Bodhisattva of Compassion).
P. 1010: *Mani kabum*; William Woodville Rockhill: *The Land of the Lamas*, New York, 1891, p. 333.

Standing Bear, Chief (Luther Standing Bear; b. 1868; Lakota Sioux Indian chief).
P. 989: Chief Standing Bear: *Land of the Spotted Eagle*, Boston & New York, Houghton Mifflin, 1933, pp. 69–70.

Stephanos of Alexandria (7th cent. A.D. Greek alchemist, at court of Herakleios at Byzantium: known as a philosopher, mathematician, astronomer).
P. 594: tr. by Sherwood Taylor, in E. J. Holmyard: *Alchemy*, Harmondsworth, Middlesex, Penguin Books, 1957, pp. 28–9.

Sterry, Peter (1613–1672; one of the leading Cambridge Platonists).
All citations are from Vivian de Sola Pinto: *Peter Sterry, Platonist and Puritan*, London, Cambridge University Press, 1934.

P. 27:	p. 95	P. 253:	p. 169–170	P. 606:	p. 126	P. 785:	p. 122
31:	147	267:	140	609:	157	808:	96–7
35:	166	279:	142–3	622:	121	810:	145
39:	160, 161	307:	133	628:	136–7	819:	195
47:	169	319:	154	631:	25	829:	157
60:	52	368:	163	665:	136	836:	69–70
66:	169	474:	140	677:	177	836:	137
69:	162	503:	131–2	685:	123	837:	201–2
77:	141	527:	81	690:	152	904:	140–141
80:	184	538:	149	690:	140	917:	191
98:	94	548:	147–8	707:	140	920:	179
125:	170	565:	151	716:	121	926:	97
141:	177	575:	199	717:	122	928:	148
215:	122	576:	164	718:	145	929:	97
225:	156–7	598:	131	750:	148	945:	136
226:	17	601:	123	756:	156	967:	143
228:	197	602:	126	774:	162		
245:	128–9	605:	134	781:	146		

Stobaeus (Joannes; 5th cent. A.D. Greek anthologist, compiled extensively from Greek authors).
P. 468: *Stobaei Sententiae*, 1609, p. 247; in Taylor: *Life Pythagoras*, p. 190.
P. 799: idem, p. 231 (attributed by Taylor to Democrates or Demophilus); Taylor, op. cit., p. 190.
P. 989: idem, p. 215; ibid., pp. 189–190.

Subhâshita Ratna Nidhi (or, *Precious Treasury of Elegant Sayings*; Tibetan; attributed to the Grand Lâma of Saskya; fl. 1270 A.D.).
P. 136: W. Y. Evans-Wentz: *Tibetan Yoga and Secret Doctrines*, London, Oxford University Press, 1958, p. 60 (based on Csoma de Körös's rendering).
P. 197: ibid., p. 61.
P. 202: ibid., p. 61.
P. 700: ibid., p. 60.

Sufic Aphorism / Formula / Saying
P. 97: cited by Aldous Huxley: *The Perennial Philosophy*, London, Chatto & Windus, 1950, p. 124.
P. 776: in Guénon: *Symbolisme de la Croix*, p. 80.
P. 894: Qur'ânic.
P. 926: Burckhardt: *Introd. Doctrines Islam*, p. 77.
P. 989: *Islamic Quarterly*, London, April, 1956, p. 54.

Sufic Tradition
P. 657: J. Desparet: *Coutumes, institutions et croyances des indigènes de l'Algérie*; in Émile Dermenghem: *Le Culte des Saints dans l'Islam maghrébin*, Paris, Gallimard, 1954, p. 13.

Suger, abbé of Saint-Denis (1081?–1151; French churchman and statesman).
P. 308: *Liber de rebus in administratione sua gestis*, XXVII: P.L. 186; in M. M. Davy: *Essai sur la Symbolique romane*, Paris, Flammarion, 1955, p. 44.

T

Taittirîya Samhitâ (part of the second Veda, also known as the *Krishna*, or Black, *Yajur-Veda*, traditionally ascribed to the sage Vaiśampâyana, a teacher of Yâjñavalkyâ and pupil of Vyâsa, q.v.).
P. 48: Ananda K. Coomaraswamy: *Am I My Brother's Keeper?*, New York, John Day, 1947, p. 80.
P. 173: Coomaraswamy: *Hinduism and Buddhism*, New York, Philosophical Library, 1943, p. 17.
P. 651: J. Muir: *Original Sanskrit Texts*, London, 1868.

Taittirîya Upanishad (late Vedic scripture).
The citations are from *The Thirteen Principal Upanishads*, tr. by Robert Ernest Hume, London, Oxford University Press, 1921, 1931, 1934.
Pp. 42, 766, 828, 950.

Takeda Shingen (1521–1573; great Japanese general, student of Zen).
P. 219: D. T. Suzuki: *Zen and Japanese Culture*, London, Routledge & Kegan Paul, 1959, p. 78.

Takuan (1573–1645; abbot of Daitokuji, in Kyoto, and founder of Tôkaiji Zen temple at Tokyo).
P. 578: E. Steinilber-Oberlin and Kuni Matsuo: *Les Sectes bouddhiques japonaises*, Paris, Éditions G. Crés, 1930, p. 186.
P. 775: Suzuki: *Zen and Japanese Culture*, p. 97.
P. 801: ibid., p. 104.
P. 828: ibid., p. 108.

Taliesin (6th cent. A.D. Welsh bard).
P. 845: Joseph Head and S. L. Cranston: *Reincarnation, an East-West Anthology*, New York, Julian Press, 1961, p. 120.

Talmud (ancient body of Jewish civil and canonical law).
P. 181: cited in *Life*, International Edition, Oct. 17, 1955, p. 49.
P. 202: ibid., p. 48.
P. 488: *Yoma*, 86 a; Leo Schaya: *L' Homme et l'Absolu selon la Kabbale*, Paris, Éditions Buchet/Castel, Corrêa, 1958, p. 84.

Tao-chi (also called Shih-t'ao; Chinese monk, painter, writer, of early Ch'ing dynasty, especially known for his painting).
P. 674: *Hua Yü Lu*; Osvald Sirén: *The Chinese on the Art of Painting*, New York, Schocken Books, 1963 (orig. edn. Henri Vetch, Peiping, 1936), p. 187.
P. 674: ibid., p. 191.

T'ao Ch'ien (365–427 A.D.; Chinese Taoist poet).
P. 348: in *One Hundred and Seventy Chinese Poems*, tr. by Arthur Waley, London, Constable, 1918–1939, pp. 71–2.
P. 695: ibid., pp. 72–3.

Tao Te Ching (the sacred book of the Taoists, left to the world by Lao-tse, c. 604–531 B.C., the great Chinese sage and founder of Taoism).
All passages, except where indicated, are from the tr. by Ch'u Ta-Kao, the Buddhist Lodge, London, 1937.
Pp. 28, 29, 43 (tr. Lin Yutang), 154 (source lacking), 156, 199, 200, 202, 275, 297, 347, 360, 396, 400, 490, 563, 568, 576, 584, 601, 603, 604, 716 (tr. Arthur Waley), 725, 727, 739, 741, 743, 744, 786, 791, 859, 915, 986, 989, 992.

Tarnovius, Johann (fl. 1616; theologian and Rosicrucian, of Rostock, Germany).
P. 554: Paul Arnold: *Histoire des Rose-Croix*, Paris, Mercure de France, 1955, pp. 146–7.

Tauler (Johannes; 1300?–1361; famous Rhenish Dominican contemplative, school of Eckhart; called 'the Illuminated Doctor').
All citations, except where indicated, are from *Life and Sermons of Dr. John Tauler*, tr. by Susanna Winkworth, New York, 1858.

P. 71:	p. 423–5	P. 145:	p. 237		*Law*, q.v.,		*form.* p. 53
81:	353	146:	436		p. 287	P. 248:	p. 467
112:	336	151:	278–9	P. 172:	p. 419–420	261:	369–70
133:	282	159:	357	174:	412	266:	285
133:	240–241	166:	230	176:	321–22	298:	285, 286–7
134:	290	169:	in Hob-	180:	in Liguori,	310:	340
139:	426		house:		q.v.: *Con-*	324:	78

P. 332: p. 409–410
345: 274
355: 287
358: 399
382: 320
386: in Plard: *Angelus Silesius*, q.v., p. 364, n. 11
P. 387: in Hobhouse, op. cit., p. 260
492: p. 236
492: 384
503: 281
505: 432, 434, 437, 438
532: 387–8
537: in Liguori, op. cit., p. 32
P. 541: p. 205
558: 380
678: 468–9
696: 229
703: 440
716: 287
718: 279
728: in Underhill's Ruysbroeck, q.v., p. 257
P. 736: p. 391
752: 480
759: 385
801: 377–8
834: 286
867: 356
913: 281
982: 364–5

Tejo-bindu Upanishad (Vedic scripture; part of the *Krishna*, or Black, *Yajur-Veda*, dealing with yoga).
P. 153: Alain Daniélou: *Yoga, the Method of Re-Integration*, London, Christopher Johnson, 1949, p. 78.
P. 153: ibid., p. 78.
P. 154: ibid., p. 78.
P. 728: ibid., pp. 78–9.

Ten Virtues, The (an old Zen tradition concerning Ikebana, or Japanese flower arrangement).
P. 726: Gustie L. Herrigel: *Zen in the Art of Flower Arrangement*, London, Routledge & Kegan Paul, 1958, p. 65.

Têng Ch'un (Chinese; author of *Hua Chi*, a collection of biographies of painters of the Northern Sung period, published in 1167).
P. 671: *Hua Chi*; Sirén: *Chinese on Art Painting*, p. 89.

Teresa of Avila, St (1515–1582; famous Spanish mystical saint, co-founder with St John of the Cross, q.v., of the Order of Discalced Carmelites).
P. 149: Marcelle Auclair: *La Vie de Sainte Thérèse d'Avila*, Paris, Éditions du Seuil, 1950, p. 183.
P. 174: St Alphonsus Liguori: *Conformity to the Will of God*, tr. from the Italian by Benedictine Convent of Perpetual Adoration, Clyde, Missouri, 7th edn., 1935, p. 21.
P. 180: ibid., p. 52.
P. 371: *The Life of St Teresa of Jesus, Written by Herself*; tr. by David Lewis, Westminster, Maryland, Newman Book Shop, 1944, 5th edn., Ch. XXX, 14, 15.

Teresa of Lisieux, St (Thérèse Martin; 1873–1897; French Carmelite nun, canonized in 1925 as St Teresa, the Little Flower; celebrated for her 'Little Way').
P. 131: *Histoire d'une Âme*, Office Central de Lisieux, 1947, p. 138, and in her hand under portrait facing p. 9.
P. 181: ibid., pp. 45–6.
P. 603: *Fêtes et Saisons*, no. 63, Paris, Oct-Nov. 1951, p. 8.
P. 609: *Histoire*, p. 202.
P. 614: ibid., pp. 250–251.

Tertullian (Quintus Septimus Florens Tertullianus; 160?–?230 A.D.; Latin ecclesiastical writer, of Carthage, a leading authority and Father in the early Church).
P. 418: cited in *The Sophic Hydrolith*; *The Hermetic Museum*, London, John M. Watkins, 1953, vol. I, p. 101.

Testament of Dan, The (Old Testament Apocrypha).
P. 458: *The Apocrypha and Pseudepigrapha of the Old Testament*, ed. by R. H. Charles, Oxford, Clarendon Press, 1913, vol. II, *The Testaments of the XII Patriarchs*, p. 334.

Testament of Gad, The (same as above).
P. 488: op. cit. *supra*, p. 341.

Testament of Joseph, The (same as above).
P. 504: op. cit. *supra*.
P. 509: ibid.

Testament of Judah, The (same as above).
P. 170: op. cit. *supra*.

Theban Hymn to Ptah (Egyptian antiquity).
P. 935: Brugsch: *Rel. und Myth. der alten Aegypter*, 510; in *Hermetica*, ed. by Walter Scott, Oxford, Clarendon Press, 1926, Vol. III, p. 187, n. 1.

Yoga Pradipa (Sanskrit hatha yoga treatise).
P. 812: Daniélou: *Yoga Method Re-Integration*, p. 101.
Yoga Shikhâ Upanishad (râja yoga treatise, part of Black *Yajur-Veda*).
P. 254: Daniélou: *Yoga Method Re-Integration*, p. 96.
Yoga-Vasishtha (Hindu antiquity: teachings of the sage Vasishtha to Prince Râma, transcribed 3rd cent. B.C. by Vâlmiki, q.v. under *Râmâyana*).
All selections, except where indicated, are from *The World Within the Mind*, tr. from the Sanskrit by Hari Prasad Shastri; London, Shanti Sadan, 1937, 1946, 1952.

P. 36:	p. 57	P. 222:	p. 99	P. 591:	p. 81	P. 830:	p. 53
85:	13	224:	28	591:	64–5	832:	100–101
106:	71	225:	70	601:	37	915:	91
132:	90	251:	32, 40	700:	76	921:	43
134:	64	251:	72–3	703:	63	930:	97–8
141:	25	253:	71	718:	93	943:	51
155:	66	289:	34	747:	in Maharshi, q.v., *Talks*, p. 169	952:	67
155:	32	335:	29			981:	134 (*Queen C.*)
157:	56	352:	64				
173:	37	385:	72				
206:	75	485:	110	753:	p. 96		
221:	131 (*The Story of Queen Chudala*)	496:	76	780:	20		
		540:	11	804:	138 (*Queen C.*)		
		584:	137 (*Queen C.*)	811:	56–7		

Yoka Daishi (d. 713 A.D.; Chinese Zen master, disciple of Hui-nêng, q.v.).
P. 65: *Cheng-tao Ke*, or 'Song of Enlightenment': in D. T. Suzuki: *Manual of Zen Buddhism*, London, Rider, 1950, p. 100.
P. 782: ibid., p. 97.
P. 828: ibid., p. 96.
Yôkwan (1032-1111; abbot of Zenrinji Temple in Kyôto, a predecessor of Hônen, q.v., in the Nembutsu perspective).
P. 1025: *Hônen the Buddhist Saint*, tr. by Harper Havelock Coates and Ryugaku Ishizuka, Kyoto, 1925, 1949, pp. 40–41.

Z

Zen Formula
P. 998: Reikichi Kita and Kiichi Nagaya: *How Altruism Is Cultivated in Zen*, tr. by Ruth F. Sasaki, p. 135.
Zen *Gatha* (*Gatha*, Sanskrit term for 'verse', 'hymn'; in Buddhism designates versified portion of sûtras).
P. 405: D. T. Suzuki: *Manual of Zen Buddhism*, London, Rider, 1950, p. 14.
Zen *Haiku* (elliptic verse to provoke spiritual awakening).
P. 104: D. T. Suzuki: *Zen and japanese Culture*, London, Routledge & Kegan Paul, 1959, p. 233 (by Bashô, 1643-1694).
P. 199: by Shiki (1869-1902); ibid., p. 232.
P. 604: by Bashô; ibid., p. 229.
P. 730: by Bashô; supplied by the Ven. Professor Sohaku Ogata, Abbot of Chotokuin Temple at Shokokuji Monastery, Kyoto.
Zen *Mondo* ('question and answer').
P. 142: Daishu Ekai (disciple of Baso Dôichi=Ma-tsu Tao-i; d. 788; prominent Chinese Zen master): *Tongo Nyûmon Ron*, 'Treatise on Attaining Sudden Enlightenment': in Suzuki: *Zen and Japanese Culture*, p. 139.
P. 368: Suzuki: *Essays in Zen Buddhism* (First Series), London, Rider, 1949, 1958, p. 29.
Zendavesta, The (Zoroastrian scripture).
P. 1029: from the *Ormazd Yast*; tr. by James Darmsteter, *The Sacred Books of the East*, Vol. XXIII, ed. by F. Max Müller, Oxford, Clarendon Press, 1883.